Lecture Notes in Artificial Intelligence 10956

Subseries of Lecture Notes in Computer Science

More information about this series at http://www.springer.com/series/1244

De-Shuang Huang · M. Michael Gromiha
Kyungsook Han · Abir Hussain (Eds.)

Intelligent Computing Methodologies

14th International Conference, ICIC 2018
Wuhan, China, August 15–18, 2018
Proceedings, Part III

Editors
De-Shuang Huang
Tongji University
Shanghai
China

Kyungsook Han
Inha University
Incheon
Korea (Republic of)

M. Michael Gromiha
Indian Institute of Technology Madras
Chennai
India

Abir Hussain
Liverpool John Moores University
Liverpool
UK

ISSN 0302-9743 ISSN 1611-3349 (electronic)
Lecture Notes in Artificial Intelligence
ISBN 978-3-319-95956-6 ISBN 978-3-319-95957-3 (eBook)
https://doi.org/10.1007/978-3-319-95957-3

Library of Congress Control Number: 2018947576

LNCS Sublibrary: SL7 – Artificial Intelligence

Printed on acid-free paper

This Springer imprint is published by the registered company Springer International Publishing AG part of Springer Nature
The registered company address is: Gewerbestrasse 11, 6330 Cham, Switzerland

Preface

The International Conference on Intelligent Computing (ICIC) was started to provide an annual forum dedicated to the emerging and challenging topics in artificial intelligence, machine learning, pattern recognition, bioinformatics, and computational biology. It aims to bring together researchers and practitioners from both academia and industry to share ideas, problems, and solutions related to the multifaceted aspects of intelligent computing.

ICIC 2018, held in Wuhan, China, August 15–18, 2018, constituted the 14th International Conference on Intelligent Computing. It built upon the success of ICIC 2017, ICIC 2016, ICIC 2015, ICIC 2014, ICIC 2013, ICIC 2012, ICIC 2011, ICIC 2010, ICIC 2009, ICIC 2008, ICIC 2007, ICIC 2006, and ICIC 2005 that were held in Liverpool, UK, Lanzhou, Fuzhou, Taiyuan, Nanning, Huangshan, Zhengzhou, Changsha, China, Ulsan, Korea, Shanghai, Qingdao, Kunming, and Hefei, China, respectively.

This year, the conference concentrated mainly on the theories and methodologies as well as the emerging applications of intelligent computing. Its aim was to unify the picture of contemporary intelligent computing techniques as an integral concept that highlights the trends in advanced computational intelligence and bridges theoretical research with applications. Therefore, the theme for this conference was "Advanced Intelligent Computing Technology and Applications." Papers focused on this theme were solicited, addressing theories, methodologies, and applications in science and technology.

ICIC 2018 received 632 submissions from 19 countries and regions. All papers went through a rigorous peer-review procedure and each paper received at least three review reports. Based on the review reports, the Program Committee finally selected 275 high-quality papers for presentation at ICIC 2018, included in three volumes of proceedings published by Springer: two volumes of *Lecture Notes in Computer Science* (LNCS) and one volume of *Lecture Notes in Artificial Intelligence* (LNAI).

This volume of *Lecture Notes in Artificial Intelligence* (LNAI) includes 92 papers.

The organizers of ICIC 2018, including Tongji University, Wuhan University of Science, and Technology and Wuhan Institute of Technology, made an enormous effort to ensure the success of the conference. We hereby would like to thank the members of the Program Committee and the referees for their collective effort in reviewing and soliciting the papers. We would like to thank Alfred Hofmann, executive editor at Springer, for his frank and helpful advice and guidance throughout and for his continuous support in publishing the proceedings. In particular, we would like to thank all the authors for contributing their papers. Without the high-quality submissions from the authors, the success of the conference would not have been possible. Finally, we are especially grateful to the IEEE Computational Intelligence Society, the International Neural Network Society, and the National Science Foundation of China for their sponsorship.

May 2018

De-Shuang Huang
M. Michael Gromiha
Kyungsook Han
Abir Hussain

ICIC 2018 Organization

General Co-chairs

De-Shuang Huang, China
Huai-Yu Wu, China
Yanduo Zhang, China

Program Committee Co-chairs

Kang-Hyun Jo, Korea
Xiao-Long Zhang, China
Haihui Wang, China
Abir Hussain, UK

Organizing Committee Co-chairs

Hai-Dong Fu, China
Yuntao Wu, China
Bo Li, China

Award Committee Co-chairs

Juan Carlos Figueroa, Colombia
M. Michael Gromiha, India

Tutorial Chair

Vitoantonio Bevilacqua, Italy

Publication Co-chairs

Kyungsook Han, Korea
Phalguni Gupta, India

Workshop Co-chairs

Valeriya Gribova, Russia
Laurent Heutte, France
Xin Xu, China

Special Session Chair

Ling Wang, China

International Liaison Chair

Prashan Premaratne, Australia

Publicity Co-chairs

Hong Zhang, China
Michal Choras, Poland
Chun-Hou Zheng, China
Jair Cervantes Canales, Mexico

Sponsors and Exhibits Chair

Wenzheng Bao, Tongji University, China

Program Committee

Abir Hussain
Akhil Garg
Angelo Ciaramella
Ben Niu
Bin Liu
Bing Wang
Bingqiang Liu
Binhua Tang
Bo Li
Chunhou Zheng
Chunmei Liu
Chunyan Qiu
Dah-Jing Jwo
Daowen Qiu
Dong Wang
Dunwei Gong
Evi Syukur
Fanhuai Shi
Fei Han
Fei Luo
Fengfeng Zhou
Francesco Pappalardo
Gai-Ge Wang
Gaoxiang Ouyang
Haiying Ma
Han Zhang
Hao Lin
Hongbin Huang
Honghuang Lin
Hongjie Wu
Hongmin Cai
Hua Tang
Huiru Zheng
Jair Cervantes
Jian Huang
Jianbo Fan
Jiang Xie
Jiangning Song
Jianhua Xu
Jiansheng Wu
Jianyang Zeng
Jiawei Luo
Jing-Yan Wang
Jinwen Ma
Jin-Xing Liu
Ji-Xiang Du
José Alfredo Costa
Juan Carlos Figueroa-García
Junfeng Xia
Junhui Gao
Junqing Li
Ka-Chun Wong
Khalid Aamir
Kyungsook Han
Laurent Heutte
Le Zhang
Liang Gao
Lida Zhu
Ling Wang
Lining Xing
Lj Gong
Marzio Pennisi
Michael Gromiha Maria Siluvay
Michal Choras
Ming Li
Mohd Helmy Abd Wahab
Pei-Chann Chang

Additional Reviewers

Akila Ranjith
Xiao Yang
Cheng Liang
Alessio Ferone
José Alfredo Costa
Ambuj Srivastava
Mohamed Abdel-Basset
Angelo Ciaramella
Anthony Chefles
Antonino Staiano
Antonio Brunetti
Antonio Maratea
Antony Lam
Alfredo Pulvirenti
Areesha Anjum
Athar Ali Moinuddin
Mohd Ayyub Khan
Alfonso Zarco
Azis Ciayadi
Brendan Halloran
Bin Qian
Wenbin Song
Benjamin J. Lang
Bo Liu
Bin Liu
Bin Xin
Guanya Cai
Casey P. Shannon
Chao Dai
Chaowang Lan
Chaoyang Zhang
Zhang Chuanchao
Jair Cervantes
Bo Chen
Yueshan Cheng
Chen He
Zhen Chen
Chen Zhang
Li Cao
Claudio Loconsole
Cláudio R. M. Silva
Chunmei Liu
Yan Jiang
Claus Scholz
Yi Chen
Dhiya AL-Jumeily
Ling-Yun Dai
Dongbo Bu
Deming Lei
Deepak Ranjan Nayak
Dong Han
Xiaojun Ding
Domenico Buongiorno
Haizhou Wu
Pingjian Ding
Dongqing Wei
Yonghao Du
Yi Yao
Ekram Khan
Miao Jiajia
Ziqing Liu
Sergio Santos
Tomasz Andrysiak
Fengyi Song
Xiaomeng Fang
Farzana Bibi
Fatih Adıgüzel
Fang-Xiang Wu
Dongyi Fan
Chunmei Feng
Fengfeng Zhou
Pengmian Feng
Feng Wang
Feng Ye
Farid Garcia-Lamont
Frank Shi
Chien-Yuan Lai
Francesco Fontanella
Lei Shi
Francesca Nardone
Francesco Camastra
Francesco Pappalardo
Dongjie Fu
Fuhai Li
Hisato Fukuda
Fuyi Li
Gai-Ge Wang
Bo Gao
Fei Gao
Hongyun Gao
Jianzhao Gao
Gaoyuan Liang
Geethan Mendiz
Guanghui Li
Giacomo Donato Cascarano
Giorgio Valle
Giovanni Dimauro
Giulia Russo
Linting Guan
Ping Gong
Yanhui Gu
Gunjan Singh
Guohua Wu
Guohui Zhang
Guo-sheng Hao
Surendra M. Gupta
Sandesh Gupta
Gang Wang
Hafizul Fahri Hanafi
Haiming Tang
Fei Han
Hao Ge
Kai Zhao
Hangbin Wu
Hui Ding
Kan He
Bifang He
Xin He
Huajuan Huang
Jian Huang
Hao Lin
Ling Han
Qiu Xiao
Yefeng Li
Hongjie Wu
Hongjun Bai
Hongtao Lei
Haitao Zhang
Huakang Li
Jixia Huang
Pu Huang
Sheng-Jun Huang
Hailin Hu
Xuan Huo
Wan Hussain Wan Ishak
Haiying Wang
Il-Hwan Kim

Kamlesh Tiwari
M. Ikram Ullah Lali
Ilaria Bortone
H. M. Imran
Ingemar Bengtsson
Izharuddin Izharuddin
Jackson Gomes
Wu Zhang
Jiansheng Wu
Yu Hu
Jaya sudha
Jianbo Fan
Jiancheng Zhong
Enda Jiang
Jianfeng Pei
Jiao Zhang
Jie An
Jieyi Zhao
Jie Zhang
Jin Lu
Jing Li
Jingyu Hou
Joe Song
Jose Sergio Ruiz
Jiang Shu
Juntao Liu
Jiawen Lu
Jinzhi Lei
Kanoksak Wattanachote
Juanjuan Kang
Kunikazu Kobayashi
Takashi Komuro
Xiangzhen Kong
Kulandaisamy A.
Kunkun Peng
Vivek Kanhangad
Kang Xu
Kai Zheng
Kun Zhan
Wei Lan
Laura Yadira Domínguez Jalili
Xiangtao Chen
Leandro Pasa
Erchao Li
Guozheng Li
Liangfang Zhao
Jing Liang
Bo Li
Feng Li
Jianqiang Li
Lijun Quan
Junqing Li
Min Li
Liming Xie
Ping Li
Qingyang Li
Lisbeth Rodríguez
Shaohua Li
Shiyong Liu
Yang Li
Yixin Li
Zhe Li
Zepeng Li
Lulu Zuo
Fei Luo
Panpan Lu
Liangxu Liu
Weizhong Lu
Xiong Li
Junming Zhang
Shingo Mabu
Yasushi Mae
Malik Jahan Khan
Mansi Desai
Guoyong Mao
Marcial Guerra de Medeiros
Ma Wubin
Xiaomin Ma
Medha Pandey
Meng Ding
Muhammad Fahad
Haiying Ma
Mingzhang Yang
Wenwen Min
Mi-Xiao Hou
Mengjun Ming
Makoto Motoki
Naixia Mu
Marzio Pennisi
Yong Wang
Muhammad Asghar Nadeem
Nadir Subaşi
Nagarajan Raju
Davide Nardone
Nathan R. Cannon
Nicole Yunger Halpern
Ning Bao
Akio Nakamura
Zhichao Shi
Ruxin Zhao
Mohd Norzali Hj Mohd
Nor Surayahani Suriani
Wataru Ohyama
Kazunori Onoguchi
Aijia Ouyang
Paul Ross McWhirter
Jie Pan
Binbin Pan
Pengfei Cui
Pu-Feng Du
Iyyakutti Iyappan Ganapathi
Piyush Joshi
Prashan Premaratne
Peng Gang Sun
Puneet Gupta
Qinghua Jiang
Wangren Qiu
Qiuwei Li
Shi Qianqian
Zhi Xian Liu
Raghad AL-Shabandar
Rafał Kozik
Raffaele Montella
Woong-Hee Shin
Renjie Tan
Rodrigo A. Gutiérrez
Rozaida Ghazali
Prabakaran
Jue Ruan
Rui Wang
Ruoyao Ding
Ryuzo Okada
Kalpana Shankhwar
Liang Zhao

Zheng Tian
Zheng Sijia
Zhenyu Xuan
Fangqing Zhao
Zhipeng Cai
Xing Zhou
Xiong-Hui Zhou
Lida Zhu
Ping Zhu
Qi Zhu
Zhong-Yuan Zhang
Ziding Zhang
Junfei Zhao
Juan Zou
Quan Zou
Qian Zhu
Zunyan Xiong
Zeya Wang
Yatong Zhou
Shuyi Zhang
Zhongyi Zhou

Contents – Part III

An Improved Evolutionary Extreme Learning Machine Based on Multiobjective Particle Swarm Optimization

Jing Jiang[1], Fei Han[1(✉)], Qing-Hua Ling[1,2], and Ben-Yue Su[3]

[1] School of Computer Science and Telecommunication Engineering, Jiangsu University, Zhenjiang 212013, Jiangsu, China
hanfei@ujs.edu.cn

[2] School of Computer Science and Engineering, Jiangsu University of Science and Technology, Zhenjiang 212003, Jiangsu, China

[3] School of Computer and Information, Anqing Normal University, Anqing 246133, Anhui, China

Abstract. Extreme learning machine (ELM) proposed for training single-hidden-layer feedforward neural network has drawn much attention. Since ELM randomly initializes the hidden nodes and analytically determines the output weights, it may easily lead to redundant hidden representation and numerical instability problem. Conventional single-objective optimization methods have been applied for improving ELM, however multiple objectives associate with ELM should be considered. This paper proposes an evolutionary ELM method based on multiobjective particle swam optimization to overcome the drawbacks. The input weights and hidden biases are learnt by optimizing three objectives simultaneously including the root mean squared error, the norm of the output weights and the sparsity of hidden output matrix. The proposed method tends to select small weights which make ELM becomes robust to small input changes for improving the numerical stability. Moreover, this algorithm inhibits activated neurons and enables ELM to obtain an informative hidden representation. The proposed approach can achieve good generalization performance with sparse hidden representation. Experimental results verify that the new algorithm is highly competitive and superior to ELM and conventional evolutionary ELMs on benchmark classification problems.

Keywords: Extreme learning machine
Multiobjective particle swarm optimization · Sparsity
Generalization performance

1 Introduction

Single-hidden-layer feedforward neural network (SLFN) which is typically realized by gradient-based learning algorithm usually suffer from high computation cost and trapping into local optimal [1]. Extreme learning machine (ELM) was therefore proposed for training SLFN in [2]. In ELM, the input weights and hidden biases are randomly initialized and the output weights are analytically determined through

D.-S. Huang et al. (Eds.): ICIC 2018, LNAI 10956, pp. 1–6, 2018.
https://doi.org/10.1007/978-3-319-95957-3_1

Moore-Penrose (MP) generalized inverse. Thereby ELM has fast learning speed and good generalization performance compare to traditional gradient-based learning algorithms.

However, the performance of ELM largely sensitive to the randomly initialized parameters. Firstly, it is hard to obtain the global optimal parameters which can minimize the network error. Secondly, random initialization often lead to numerical instability problem and redundant hidden representation [3]. A variety of methods have adopted metaheuristic algorithm for improving the performance of ELM. In [4], an evolutionary ELM (E-ELM) based on differential evolution (DE) was proposed. E-ELM firstly optimized the optimal input weights and hidden biases of SLFN, and then analytically calculated its output weights through MP generalized inverse. For improving E-ELM, the method combined improved particle swarm optimization (PSO) with ELM called IPSO-ELM was proposed in [5]. IPSO-ELM tended to select the solution which lead to a smaller output weights with acceptable network error for better generalization.

In this paper, an improved evolutionary ELM based on multiobjective particle swarm optimization (MOPSO) [6] is put forward. The input weights and hidden biases constitute the decision variables, so each candidate solution of MOPSO represents a specific ELM where the output weights are calculated by MP generalized inverse. MOPSO ultimately evolves a set of candidate ELMs after iterations, and all of the ELMs are combined to establish an ensemble learning model. This work not only optimizes the network error and the norm of the output weights, but further promote the sparsity level of hidden outputs. MOPSO-ELM not only prefers smaller output weights with acceptable network error which lead to a more stable ELM, but also tends to select the input weights and hidden biases that result in sparse hidden representation. Since the proposed method combine MOPSO with ELM, it is referred to as MOPSO-ELM.

2 Preliminaries

2.1 Extreme Learning Machine

For N arbitrary distinct sample (x_i, t_i), where $x_i = [x_{i1}, x_{i2}, \cdots, x_{in}]^{\mathrm{T}} \in \mathbf{R}^n$ and $t_i = [t_{i1}, t_{i2}, \cdots, t_{im}]^{\mathrm{T}} \in \mathbf{R}^m$. A standard SLFN with $\tilde{N}$ hidden neurons and activation function $g(\cdot)$ can approximate these N samples with zero error means that

$$\sum_{j=1}^{\tilde{N}} \beta_j g_j(x_i) = \sum_{j=1}^{\tilde{N}} \beta_j g(w_j \cdot x_i + b_j) = t_i, \quad i = 1, 2, \cdots, N, \quad j = 1, 2, \cdots, \tilde{N} \tag{1}$$

where $\beta_j = [\beta_{j1}, \beta_{j2}, \cdots, \beta_{jm}]^{\mathrm{T}}$ is the output weight vector connecting all the output neurons and the j th hidden neuron, $w_i = [w_{i1}, w_{i2}, \cdots, w_{in}]^{\mathrm{T}}$ is the input weight vector connecting the j th hidden neuron and all the input neurons, and b_j is the bias of the j th hidden neuron. Equation (1) can be also written as

$$H\beta = T \tag{2}$$

where

$$H(w_1,\cdots,w_{\tilde{N}},b_1,\cdots,b_{\tilde{N}},x_1,\cdots,x_N)=\begin{bmatrix} g(w_1\cdot x_1+b_1) & \ldots & g(w_{\tilde{N}}\cdot x_1+b_{\tilde{N}}) \\ \vdots & \ddots & \vdots \\ g(w_1\cdot x_N+b_1) & \ldots & g(w_{\tilde{N}}\cdot x_N+b_{\tilde{N}}) \end{bmatrix}_{N\times\tilde{N}}$$

$$\beta=\begin{bmatrix} \beta_1^{\mathrm{T}} \\ \vdots \\ \beta_{\tilde{N}}^{\mathrm{T}} \end{bmatrix}_{\tilde{N}\times m} \quad and \quad T=\begin{bmatrix} t_1^{\mathrm{T}} \\ \vdots \\ t_N^{\mathrm{T}} \end{bmatrix}_{N\times m}$$

Therefore the SLFN can be regarded as a linear system and the output weight can be analytically determined through the generalized inverse operation [2] as

$$\beta = H^{\dagger}T \tag{3}$$

Where $H^{\dagger}=(H^TH)^{-1}H^T$ is the MP generalized inverse of the matrix H.

2.2 Particle Swarm Optimization

Particle swarm optimization (PSO) is a population-based stochastic global optimization method proposed by Eberhart and Kennedy [7]. PSO and its variants work by initializing a great many particles over the search space D. In order to find the global optimal (global best position) during the iterations, the ith particle flies with a certain velocity according to the momentum vector $V_i=(v_{i1},v_{i1},\cdots,v_{iD})$, the best position $P_{ib}=(p_{i1},p_{i1},\cdots,p_{iD})$ and the best position of the population $P_g=(p_{g1},p_{g1},\cdots,$ $p_{gD})$. The adaptive PSO [8] can be represented as

$$v_{id}(t+1)=w\cdot v_{id}(t)+c_1\cdot rand()\cdot[p_{id}(t)-x_{id}(t)]+c_2\cdot rand()\cdot[p_{gd}(t)-x_{id}(t)] \tag{4}$$

$$x_{id}(t+1)=x_{id}(t)+v_{id}(t+1),\ 1\le i\le n,\ 1\le d\le D \tag{5}$$

where c_1 and c_2 are two positive acceleration constants, $rand()$ is a random number in $(0,1)$, w is the inertia weight to keep a balance between global and local search.

3 The Proposed Evolutionary ELM

MOPSO, as an updated version of PSO proposed in [6], is designed for MOP. MOPSO essentially incorporates PSO with multiobjective optimization problems for handling multiple competing objectives. It adopts an elite achieve also called external repository that can store the nondominated solutions and guide the particles fly toward the promising directions. This paper applies MOPSO to optimize ELM for better generalization with three conflicting objectives.

Considering the root mean square error (RMSE), the norm of the output weights and the sparsity of hidden output matrix as three objectives, the modified MOPSO model is constructed as

$$\min_{w,b} F = (f(T, H\beta), \|\beta\|_2, \|H\|_1)^{\mathrm{T}} \tag{6}$$

where $w = [w_1, w_2, \cdots, w_{\tilde{N}}]^{\mathrm{T}}$ and $b = [b_1, b_2, \cdots, b_{\tilde{N}}]^{\mathrm{T}}$ are the input weights and hidden biases, $f(T, H\beta)$ is the RMSE on validation set. MOPSO-ELM tends to generate an evolutionary ELM with minimum validation error, smaller output weights and sparse hidden representation by minimizing above objectives simultaneously.

The learning procedure of MOPSO-ELM is constructed on the framework of MOPSO. For any ELM, its hidden biases and input weights are encoded as a single particle. MOPSO-ELM eventually evolves multiple ELMs simultaneously and all of these candidate are finally combined through the Bagging method. The steps of MOPSO-ELM is elaborated as follow.

1. Initialize $t = 0$ the maximum number of iterations *itermax*, the position and velocity of each P_i;
2. Compute the β_i, and evaluate P_i by three objective functions;
3. Initialize P_{ib} and determine the dominance relationship between all the particles;
4. Update the external achieve $\mathcal{REP}$;
5. Update each particle in the population;
6. Compute the β_i to the new P_i, evaluate P_i and determine the dominance relationship between all the particles;
7. $t = t + 1$ go to Step 4 until $t \geq$ *itermax*;
8. Using Bagging method to ensemble all the ELMs in $\mathcal{REP}$.

4 Experimental Results and Discussion

In this section, we evaluate the MOPSO-ELM and compare it with BP, ELM [2], PSO-ELM [9], IPSO-ELM [5] on several classification problems from UCI database. The parameters of PSO-ELM, IPSO-ELM, MOPSO-ELM are listed as follows: the population size is 200, the number of maximum iteration is 20, the initial inertial weight and the final one are set as 1.2 and 0.4, respectively. Specially, MOPSO-ELM contains a repository size of 50.

The performance with respect to BP, ELM, PSO-ELM, IPSO-ELM and MOPSO-ELM are also shown in Table 1. The training, testing and validation samples are randomly generated from the datasets in each trail. The experimental results shown are the mean values of 50 trials. From Table 1, it can be seen that the MOPSO-ELM achieves higher testing accuracy than other BP and ELM models with the same hidden neurons.

From Fig. 1, both IPSO-ELM and MOPSO-ELM are able to obtain smaller output weights than PSO-ELM and ELM. These illustrate that IPSO-ELM and MOPSO-ELM can evolve small weights which enable ELM to become robust to small changes of input patterns.

Table 1. Results of the classification problems.

Problem	Algorithm	Training times (s)	Accuracy (%)		Hidden neurons
			Training	Testing	
Diabetes	BP	0.1925	80.8281	74.9688 ± 2.9973	20
	ELM	0.0007	79.5781	76.6458 ± 2.5523	20
	PSO-ELM	1.1278	79.3906	76.5 ± 2.2171	20
	IPSO-ELM	1.1992	79.1823	76.4062 ± 2.8571	20
	MOPSO-ELM	6.4008	82.5937	77.9583 ± 3.3365	20
Satellite image	BP	Out of memory			300
	ELM	0.1309	91.1256	88.946 ± 1.114	300
	PSO-ELM	104.872	90.2354	87.994 ± 0.9173	300
	IPSO-ELM	105.644	91.664	89.176 ± 0.7799	300
	MOPSO-ELM	120.083	93.6189	89.752 ± 0.9401	300
Image segmentation	BP	229.955	99.564	95.0617 ± 1.0923	150
	ELM	0.1639	96.464	94.563 ± 1.0107	150
	PSO-ELM	17.2908	96.624	94.9086 ± 0.9156	150
	IPSO-ELM	18.6319	96.388	94.3556 ± 0.9952	150
	MOPSO-ELM	25.4958	97.7613	95.4815 ± 1.0988	150

From Fig. 2, the value of hidden activation matrix from MOPSO-ELM is notably smaller than other models including ELM, IPSO-ELM. It indicates MOPSO-ELM can generate more sparse hidden outputs than other approaches on these three benchmarks. To summarize, the results on benchmark classification problems validate the effectiveness and generalization performance of our proposal.

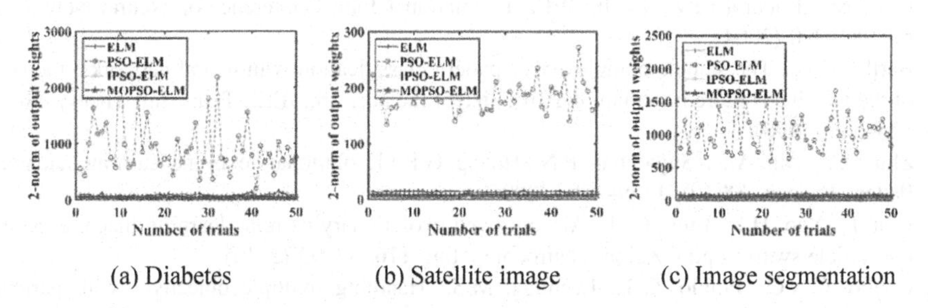

(a) Diabetes (b) Satellite image (c) Image segmentation

Fig. 1. The value of $||\beta||_2$ obtained by four ELMs

5 Conclusion

In this paper, an improved evolutionary extreme learning machine based on multiobjective particle swarm optimization was proposed. Compare to previous literatures, MOPSO-ELM which minimizes RMSE and the norm of output weights as well as promotes sparsity level of hidden neurons can achieve better generalization. Experimental

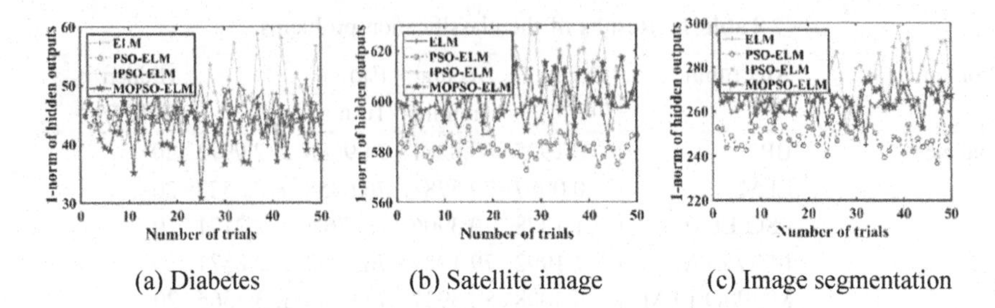

(a) Diabetes (b) Satellite image (c) Image segmentation

Fig. 2. The value of $\|H\|_1$ obtained by four ELMs

results verify the effectiveness of the proposal and the new algorithm is superior to other evolutionary ELMs on benchmark classification problems. Future research works will target a deeper study on improving MOPSO and MOPSO-ELM.

Acknowledgement. This work was supported by the National Natural Science Foundation of China [Nos. 61271385 and 61572241], the National Key R&D Program of China [No. 2017YFC0806600], the Foundation of the Peak of Six Talents of Jiangsu Province [No. 2015-DZXX-024], the Fifth "333 High Level Talented Person Cultivating Project" of Jiangsu Province [No. (2016) III-0845], and the Research Innovation Program for College Graduates of Jiangsu Province [1291170030].

References

1. Hinton, G., Salakhutdinov, R.R.: Reducing the dimensionality of data with neural networks. Science **313**, 504–507 (2006)
2. Huang, G.B., Zhu, Q.Y., Siew, C.K.: Extreme learning machine: a new learning scheme of feedforward neural networks. In: IEEE International Joint Conference on Neural Networks, pp. 985–990 (2004)
3. Bartlett, P.L.: The sample complexity of pattern classification with neural networks: the size of the weights is more important than the size of the network. IEEE Trans. Inf. Theory **44**(2), 525–536 (1998)
4. Zhu, Q.Y., Qin, A.K., Suganthan, P.N., Huang, G.B.: Evolutionary extreme learning machine. Pattern Recogn. **38**(10), 1759–1764 (2005)
5. Han, F., Yao, H.F., Ling, Q.H.: An improved evolutionary extreme learning machine based on particle swarm optimization. Neurocomputing **116**, 87–93 (2013)
6. Coello, C.A.C., Pulido, G.T., Lechuga, M.S.: Handling multiple objectives with particle swarm optimization. IEEE Trans. Evol. Comput. **8**(3), 256–279 (2004)
7. Kennedy, J., Eberhart, R.C.: Particle swarm optimization. In: IEEE International Conference on Neural Networks, pp. 1942–1948 (1995)
8. Shi, Y.H., Eberhart, R.C.: A modified particle swarm optimizer. In: IEEE Conference on Evolutionary Computation, pp. 69–73 (1998)
9. Xu, Y., Shu, Y.: Evolutionary extreme learning machine – based on particle swarm optimization. In: Wang, J., Yi, Z., Zurada, J.M., Lu, B.-L., Yin, H. (eds.) ISNN 2006. LNCS, vol. 3971, pp. 644–652. Springer, Heidelberg (2006). https://doi.org/10.1007/11759966_95

ESNMF: Evolutionary Symmetric Nonnegative Matrix Factorization for Dissecting Dynamic Microbial Networks

Yuanyuan Ma[1], Xiaohua Hu[2], Tingting He[2], Xianchao Zhu[2], Meijun Zhou[2], and Xingpeng Jiang[2(✉)]

[1] School of Information Management, Central China Normal University, Wuhan, China
[2] School of Computer, Central China Normal University, Wuhan, China
xpjiang@mail.ccnu.edu.cn

Abstract. Dynamic network is drawing more and more attention due to its potential in capturing time-dependent phenomena such as online public opinion and biological system. Microbial interaction networks that model the microbial system are often dynamic, static analysis methods are difficult to obtain reliable knowledge on evolving communities. To fulfill this gap, a dynamic clustering approach based on evolutionary symmetric nonnegative matrix factorization (ESNMF) is used to analyze the microbiome time-series data. To our knowledge, this is the first attempt to extract dynamic modules across time-series microbial interaction network. ESNMF systematically integrates temporal smoothness cost into the objective function by simultaneously refining the clustering structure in the current network and minimizing the clustering deviation in successive timestamps. We apply the proposed framework on a human microbiome datasets from infants delivered vaginally and ones born via C-section. The proposed method cannot only identify the evolving modules related to certain functions of microbial communities, but also discriminate differences in two kinds of networks obtained from infants delivered vaginally and via C-section.

Keywords: Dynamic network · Evolutionary module
Symmetric nonnegative matrix factorization · Microbiome

1 Introduction

Network is an intuitional and powerful way to depict and analyze the complex microbial system by regarding each microbe as a node and each edge as an interaction between a pair of microorganisms. Similar to social network, biological network is assumed to follow principle characteristics of complex network, e.g., scale-free [1] and small-world [2]. It has been shown that a subgraph, also called a module, exists in most biological networks. There are more dense connections among nodes within the same module, and fewer edges linking nodes between different modules. Generally speaking, modules play important roles in performing certain functions, such as potentially pathogenic microorganisms tend to gather together to cause invasive infections [3].

D.-S. Huang et al. (Eds.): ICIC 2018, LNAI 10956, pp. 7–18, 2018.
https://doi.org/10.1007/978-3-319-95957-3_2

Module extraction has become an interesting research field by which the complicated microbial module could be disentangled.

However, module extraction algorithms are usually designed for stationary scenarios where the objects to be clustered only represent a measurement at a fixed timestamp. Obviously, this assumption is too strict to be applied in the dynamic scenarios. In fact, many networks are time-dependent and their geometric structures evolve over time in the real world. For example, in early life, bacteria H. pylori is silent, but it may also become a risk factor for gastric cancer in middle-old aged population [4, 5], which is critical for exploring the pathogenesis of gastric cancer. In dynamic network, the number of modules is variable. In addition, either the emerging of new modules or dissolving of existing modules is a common phenomenon [6].

Discovering of evolving modules sets a new challenge to the traditional clustering algorithms. Although it is difficult, many pioneering work and concepts have been proposed to attack this problem. Chi et al. proposed two frameworks: preserving clustering quality (PCQ) and preserving clustering membership (PCM) for the evolutionary spectral clustering [7]. Kim and Han proposed a particle-and-density based evolutionary clustering approach to discovery the dynamic communities [6]. Lin et al. proposed Facetnet to analyze the evolutionary communities through a unified process where nonnegative matrix factorization (NMF) was used to extract the dynamic modules [8]. Ma and Dong integrated a priori information into symmetric nonnegative matrix factorization to obtain evolving modules [9]. Despite great efforts has been made, there are still some issues needed more consideration. For instance, the evolving modules obtained by the above-mentioned algorithms may not represent the true clustering structures of the dynamic networks in unsupervised task where ground-truth is not available.

In this study, we put forward a new thought for handling this problem and we apply the proposed framework on microbial network which could be referred by high-throughput sequencing technology such as 16S rRNA and metagenomics sequencing [10]. These networks obtained above are often dynamic and temporal [11–13]. Due to the fact that the numbers of local communities at different stages are unfixed, we also design an approach to select model parameters. Although SNMF has been used to some tasks such as gene modules recognition [9] and social network analysis [8], to our best knowledge, this is the first attempt to extract dynamic modules across temporal microbial interaction networks. Figure 1 demonstrates the proposed framework of this study.

The rest of this paper is organized as follows: Sect. 2 gives the detail of materials and methods. Experiments results and discussion are illustrated in Sect. 3. The conclusion is provided in Sect. 4.

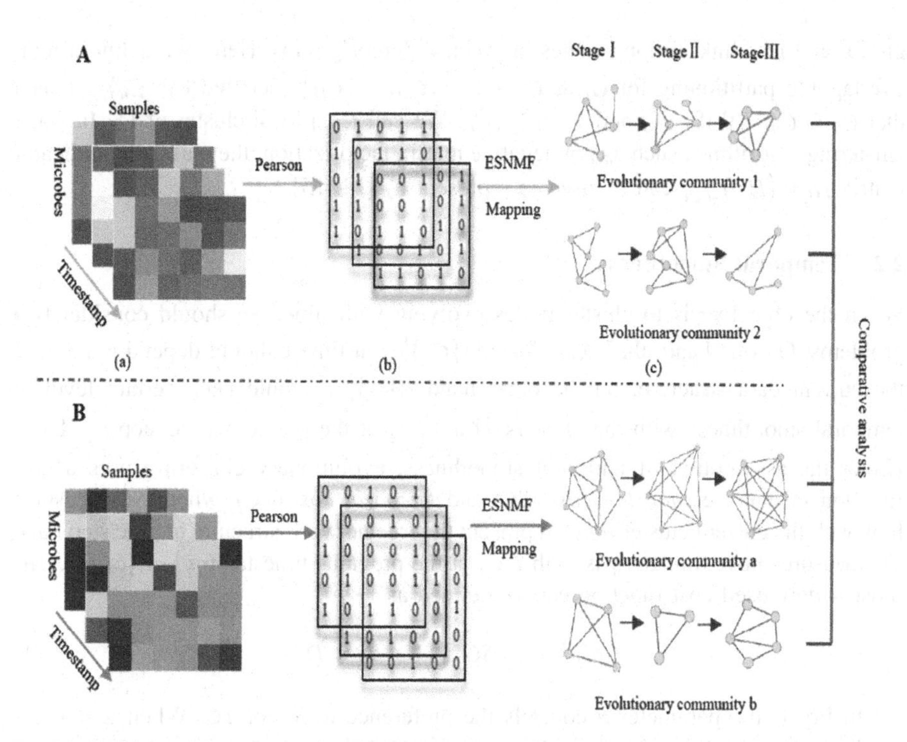

Fig. 1. An illustrative example of ESNMF on human microbiome dataset. (A) For a time-series of gut microbiome samples from infants delivered vaginally, the microbiome correlation networks of different stages can be constructed based on species abundance profiles (a). Then, at each stage the ESNMF algorithm generates a partition which is composed of some local clusters (b). We map the local clusters from different stages through the maximal mutual information and form the evolutionary communities (c). (B) The time-series of microbiome samples from infants born via C-section. The flow for discovering the evolving communities is the same with (A). After finding all dynamic communities, the comparative analysis is performed.

2 Methods and Materials

2.1 Notation

Let G_t represents the network at time t. The dynamic networks G can be defined as a sequences of networks $G_t(V_t, E_t)$, where V_t, E_t denotes the vertex set and edge set of G_t, respectively. For each network G_t, there is a corresponding adjacency matrix $W_t = (w_{ij})_{n \times n}$ in which n is the number of vertices and $w_{ij} = 1$ if the *i-th* vertex is connected to *j-th* one in the G_t, 0 otherwise. So, the dynamic network G can also be represented by a tensor $W = (w_{ijt})_{n^t \times n^t \times T}$, where n^t is the number of nodes in G_t. Noting that at any timestamp t, new nodes may emerge from G_t or existing nodes detach [6, 7].

Community finding task is to divide the nodes in a network into disjointed clusters. A good partition is that there are more dense connections among nodes within the same group, and less links among nodes between different groups. Here, we define a non-overlapping partitioning for G_t as $C_t = \{C_{1t}, C_{2t}, \ldots, C_{kt}\}$ (denoted by $(C_{it})_{i=1}^{k}$) such that $C_{it} \cap C_{jt} = \emptyset$ if $i \neq j$ and $V_t = \sum_i C_{it}$. We call C_{it} a local cluster of G_t. In some clustering algorithms, such as nonnegative matrix factorization, the clustering indicator matrix $H_t = \left(H_{ijt}\right)_{n^t \times k}$ can be used to represent C_t [14, 15].

2.2 Temporal Smoothness

When the objective is to cluster nodes evolving with time, we should consider two problems. On one hand, the local clusters $\{C_{it}\}_{i=1}^{k}$ at time t should depend mainly on the current data structure, on the other hand, $\{C_{it}\}_{i=1}^{k}$ should keep certain level of temporal smoothness with the clusters $\left\{C_{jt-1}\right\}_{j=1}^{l}$ at the previous time step $t-1$ [7]. Under the assumption of temporal smoothness, evolutionary clustering uses a cost function to balance *snapshot cost* (*SC*) and *temporal cost* (*TC*), where *SC* measures how well the current clustering C_t represent the true network structure at time step t and *TC* measures how similar C_t is with C_{t-1} at the previous timestamp $t-1$ [6, 16]. The most widely used cost function can be defined as

$$Cost = \alpha \times SC + (1 - \alpha) \times TC \tag{1}$$

In Eq. 1, the parameter α controls the preference to *SC* or *TC*. When $\alpha = 1$, (1) produces the clustering result of the current network without temporal smoothness. When $\alpha = 0$, it captures the clustering result of the previous network without *SC*. Generally, $\alpha(0 < \alpha < 1)$ produces an intermediate result between C_t and C_{t-1}.

2.3 Dynamic Network Construction

Given the abundance profiles of species in infant gut microbiome samples at different life stages, a microbial interaction network can be constructed for each stage. In this study, we use the Pearson Correlation Coefficient to quantify the interaction strength of a pair of microbes. Some insignificant links (p value > 0.05) are removed, and then the remaining links are retained to be used to construct the binary network. In our study, we retain all the connected components of each microbial network. The number of connected components in these networks are from 1 to 4. Most of the dynamic modules are from the largest connected component which has at least 3 nodes for each network.

2.4 Symmetric Nonnegative Matrix Factorization

SNMF takes a similarity matrix $A_{n \times n}$ as input and decomposes A into an low-rank matrix $H_{n \times k}$ and its transpose $H'_{n \times k}$ such that $A \approx HH'$. In this study, the adjacency matrix W_t of the network at timestamp t is taken as input of SNMF. The local clusters $\{C_{it}\}_{i=1}^{k}$ can be obtained by the Eq. 2 as below.

$$(C_t)_{ij} \leftarrow (C_t)_{ij} \frac{(W_t\, C_t)_{ij}}{(C_t\, C_t'\, C_t)_{ij}} \tag{2}$$

2.5 Evolutionary Symmetric Nonnegative Matrix Factorization

In this study, we assume that the number of microbial communities and the members of the network are changed over time and propose the evolutionary symmetric nonnegative matrix factorization stated as follows.

Following (1), the cost of evolutionary symmetric nonnegative matrix factorization (ESNMF) can be defined as

$$Cost_{ESNMF} = \alpha\, SC_{snmf} + (1-\alpha)\, TC_{snmf} \tag{3}$$

Evolutionary clustering was classified two frameworks: preserving cluster quality (PCQ) and preserving cluster membership (PCM) in [7]. Here, we select PCQ used for the downstream analysis.

Under the PCQ framework, (3) can be rewritten as

$$Cost_{ESNMF} = \alpha\, SNMF_t\,|_{C_t} + (1-\alpha)\, SNMF_{t-1}\,|_{C_t} \tag{4}$$

where $|_{C_t}$ indicates SC evaluated by the partition C_t at time t. Here,

$$SNMF_{t-1}\,|_{C_t} = \left\| W_{t-1} - C_t\, C_t' \right\|_F^2 \tag{5}$$

In order to maintain the same form as negated average association defined in [7], orthogonality was introduced into C_t, i.e. $C_t'\, C_t = I_k$ where I_k is an identity matrix. In (5), the objective is minimizing the TC. Using trace optimization, (5) is rewritten as

$$\begin{aligned} SNMF_{t-1}\,|_{C_t} &= \min_{C_t \geq 0, C_t' C_t = I_k} \left\| W_{t-1} - C_t\, C_t' \right\|_F^2 \\ &\propto \max_{C_t \geq 0, C_t' C_t = I_k} tr\left(C_t'\, W_{t-1}\, C_t \right) \end{aligned} \tag{6}$$

Relaxing the orthogonality on the rows of C_t, (4) can be reformulated as

$$\begin{aligned} Cost_{ESNMF} &= \alpha\, SNMF_t\,|_{C_t} + (1-\alpha)\, SNMF_{t-1}\,|_{C_t} \\ &\propto \max_{C_t \geq 0} tr\left\{ C_t'(\alpha\, W_t + (1-\alpha)\, W_{t-1})\, C_t \right\} \end{aligned} \tag{7}$$

Here, there are two problems needed to be considered. (a) Is the number of communities fixed at two consecutive time steps, i.e., $|C_t| = |C_{t-1}|$? (b) Is the number of nodes is invariable in each network, i.e., $|V_t| = |V_{t-1}|$? Clearly, in most cases these two assumptions are invalid. In many applications some old modules existed in the previous time $t-1$ may dissolve in the current timestamp t, or some new modules may form in the next timestamp $t+1$. Similar to (a), some key microbial species will emerge and play an important role in certain phase of disease development, but they will also become extinct after people take antibiotics. Hence, in a dynamic network, emerging or disappearing of new modules is a normal phenomenon.

When nodes to be clustered vary with time, it is key to compute $\alpha W_t + (1-\alpha) W_{t-1}$. If the old nodes existed in the previous timestamp $t-1$ disappears at the current timestamp t, we can simply remove these nodes, i.e. deleting the corresponding rows and columns of W_{t-1} and thus obtain $\overline{W}_{t-1}$. However, when new nodes appears at current timestamp t, we need to do some processing. In order to keep the same shape with W_t, we insert new rows and columns into $\overline{W}_{t-1}$ and set associated values to be zeros.

2.6 Quantify Partition for Network

In unsupervised learning task, modularity is an effective measure of the clustering results. Given the current network $G_t(V_t, E_t)$, the modularity Q_N in [17] is defined as

$$Q_N = \sum\nolimits_{i=1}^{k} \left\{ \frac{L(C_{it}, C_{it})}{L(V_t, V_t)} - \left[\frac{L(C_{it}, V_t)}{L(V_t, V_t)} \right]^2 \right\}. \quad (8)$$

The bigger modularity Q_N is, the better graph partition is. The optimal clustering can be obtained by maximizing Q_N[18].

2.7 Evolutionary Community Identification

Given the local clusters $\{C_{it-1}\}_{i=1}^{k}$ and $\{C_{jt}\}_{j=1}^{l}$ at two consecutive time steps $t-1$ and t, the mapping task is to find dynamic modules. Let $C_{it} \in \{C_{it-1}\}_{i=1}^{k}$ be the *i-th* local cluster at time $t-1$, a bipartite graph can be constructed between $\{C_{it-1}\}_{i=1}^{k}$ and $\{C_{jt}\}_{j=1}^{l}$. In this bipartite, every local module C_{it} is viewed as a vertex and the edge weight can be defined as the number of nodes that appear in both C_{it-1} and C_{jt} [9], i.e.,

$$num_{ijt} = |C_{it-1} \cap C_{jt}| \quad (9)$$

In the same manner, we can construct a bipartite graph between C_{jt} and C_{ht+1}. At last a T-partite graph is formed from C_1 to C_T.

Once a bipartite is constructed, mutual information can be used to measure the dependency degree between C_{it-1} and C_{jt}.

$$MI(C_{it-1}, C_{jt}) = \frac{num_{ijt}}{B} \log \frac{B \times num_{ijt}}{\sum_i num_{ijt} \sum_j num_{ijt}} \quad (10)$$

In Eq. 10, $B = \sum_{i=1}^{k} \sum_{j=1}^{l} num_{ijt}$, $k = |C_{t-1}|$, $l = |C_t|$ where $|S|$ denotes the number of elements in the set S. The higher of the MI is, the stronger C_{it-1} and C_{jt} are related.

In [6], Kim and Han proposed a greedy search algorithm to map local clusters from C_{t-1} to $\{C_t\}$. Notably, once a mapping between C_{it-1} and C_{jt} is done, other local clusters existed in $\{C_{it-1}\}_{i=1}^{k}$ cannot be mapped to C_{jt}. Thus, there are not more than $\min\{|C_{t-1}|, |C_t|\}$ mapping operations in the bipartite graph from C_{t-1} to C_t.

2.8 Datasets

Dataset of human gut microbiome during the first of year (HM1): This dataset is consisted of 294 gut microbiome samples from 98 full-term infants [19]. These samples are classified into two groups, one group contains 249 samples from 83 infants who were delivered vaginally, and another group contains 45 samples from 15 infants who were born by caesarean section. We divide both two groups into 3 time windows: stage I: new born, stage II: 4-months old and stage III: 12-months old. To avoid spurious correlation, we delete those OTUs presenting in less than five samples and we got 36, 55, 66 OTUs for the networks construction corresponding to three different stages. The dataset can be downloaded from [19].

3 Results and Discussion

In this section, we construct the dynamic networks and apply modularity analysis on these two groups (see Sect. 2). Due to the fact that the ground truth in microbiome clustering is unavailable, the proposed ESNMF algorithm adopts the modularity Q_N as a measurement on graph partitioning.

3.1 Parameter Selection

The proposed ESNMF algorithm contains two parameters: k_t and α_t. k_t is the cluster number in the network G_t at timestamp t. α_t is a temporal smoothness parameter that control the tradeoff between *SC* and *TC*. Noting that at each stage, the values of α_t and k_t may be different.

In our experiment, we first assign certain values to α_t, then let k_t vary in a certain intervals and record each Q_N. Lastly, the optimal combination of α_t and k_t can be obtained by searching the maximum Q_N in the two-dimension parameter space.

3.2 Analysis on HM1 Dataset

In Sect. 2, we have presented modularity as an approach to selecting the optimal parameters k_t and α_t. Figure 2 shows how the modularity of the network G_2 at stage II (4-months old) varies verse these two parameters.

As Fig. 2 shown, the maximum modularity Q_N can be achieved when $\alpha_2 = 0.9$ and $k_2 = 6$. One of the reason is that if α_2 is small, the temporal cost (*TC*) will contribute more to the whole cost and this is unexpected to be seen, on the contrary if α_2 is too large, the snapshot cost (*SC*) will play a dominant role in the objective function. When $\alpha_2 = 1$, ESNMF will degrade into the original SNMF, only capturing the clustering of the current network without temporal smoothness from the network at the previous

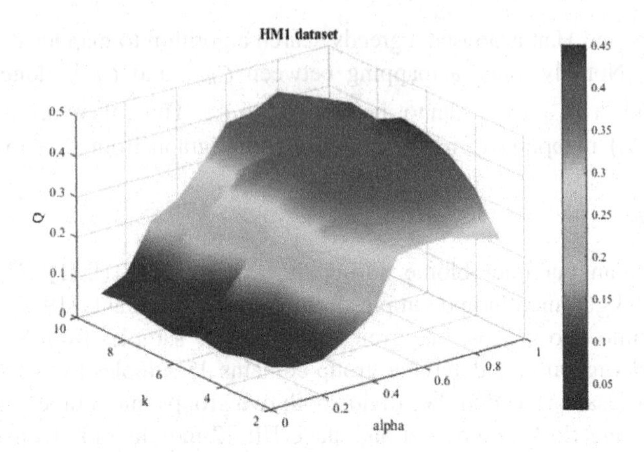

Fig. 2. The modularity of G_2 at stage II verse the two parameters: α and k

timestamp. Thus, $\alpha_2 = 0.9$ reaches a better tradeoff. In the same way, we can also obtain the optimal modularity of G_3 at stage III (12-months) and the corresponding $\alpha_3 = 0.8$ and $k_3 = 6$.

Based on the analysis above, we generate the dynamic network for certain evolutionary events. Here, we extract some mapping local communities via the maximum mutual information across the whole stages. Table 1 show two examples of dynamic communities which evolve along with growing time.

Table 1. Dynamic microbe community from infants delivered vaginally (Top 2 mapping by the optimal MI)

Dynamic Community	Local Community at Stage I (microbes)	Local Community at Stage II (microbes)	Local Community at Stage III(microbes)
$C_{4\,I} \rightarrow C_{3\,II}$	"Eubacterium" "Roseburia" "Butyrivibrio" "Faecalibacterium" "Prevotella"	"Eubacterium" "Roseburia" "Parabacteroides" "Bacteroides" "Sutterella"	NA
$C_{5\,I} \rightarrow C_{5\,II} \rightarrow C_{1\,III}$	"Rothia" "Streptococcus" "Anaerococcus" "Clostridium" "Veillonella"	"Aggregatibacter" "Fusobacterium" "Haemophilus" "Neisseria" "Rothia" "Gemella" "Granulicatella" "	"Aggregatibacter" "Fusobacterium" "Haemophilus" "Neisseria" "Rothia" "Streptococcus" "Bacteroides" "Butyrivibrio "Faecalibacterium" "Parasutterella" "Prevotella" "Proteus" "Roseburia" "Staphylococcus"

Note: Same species are colored the same.

In Table 1, $C_{i\mathrm{I}}$ is the *i-th* local community at stage I, $C_{i\mathrm{I}} \rightarrow C_{j\mathrm{II}} \rightarrow C_{k\mathrm{III}}$ denotes the evolutionary event of local communities across the whole stages. $C_{i\mathrm{I}} \rightarrow C_{j\mathrm{II}}$ indicates the dynamic community which only appears at stage I and II. As Table 1 shown, in the dynamic community $C_{4\mathrm{I}} \rightarrow C_{3\mathrm{II}}$, the microorganisms belonging to Roseburia and Eubacterium genus simultaneously appears at stage I and II.

Interestingly, in the dynamic community $C_{5I} \rightarrow C_{5II} \rightarrow C_{1\mathrm{III}}$, infants delivered vaginally grow with the microorganisms belonging to Rothia genus in the first year of life. Dominguez-Bello et al. found that infants with low levels of Rothia bacteria in their gut in their first 3 months were at higher risk for asthma [20]. That illustrates the important role of Rothia in the growth of infants. To perform certain biological function, the Rothia bacteria may interact with other beneficial bacteria such as Veillonella, Faecalibacterium and Lachnospira [20]. This is also been validated in the local clusters C_{5I} and $C_{1\mathrm{III}}$ where Rothia bacteria combine with Veillonella and Faecalibacterium, respectively. Furthermore, four genus (Aggregatibacter, Fusobacterium, Haemophilus, Neisseria) appear together at stage II and III. The Aggregatibacter, Haemophilus and Neisseria bacteria express a wide range of pathogenicity, e.g., respiratory infections, infective endocarditis, brain abscesses, meningitis [3, 21–23]. These harmful bacteria are closely linked with one another to form a main component in the evolutionary module. The dynamic module may help to understand the mechanism of certain respiratory infection diseases. Figure 3 shows the evolving community $C_{5\mathrm{I}} \rightarrow C_{5\mathrm{II}} \rightarrow C_{1\mathrm{III}}$.

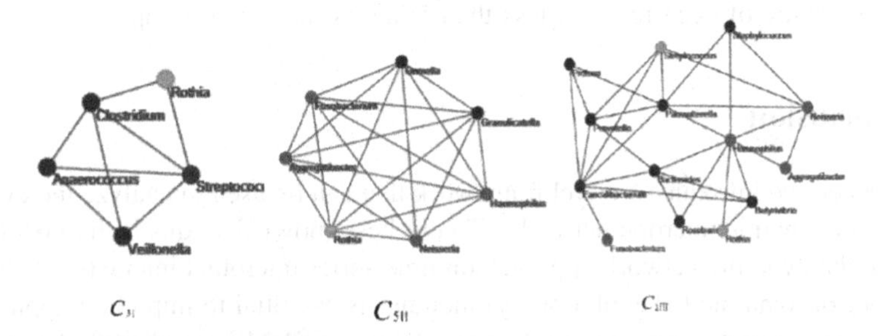

Fig. 3. The dynamic community $C_{5\mathrm{I}} \rightarrow C_{5\mathrm{II}} \rightarrow C_{1\mathrm{III}}$ identified by ESNMF. $C_{5\mathrm{I}}$ denotes the *5-th* local cluster discoveried at stage I. The green node are the ones shared by all three local clusters. The four red nodes are shared by $C_{5\mathrm{II}}$ and $C_{1\mathrm{III}}$. (Color figure online)

We also analyze the evolutionary community on the samples from 15 infants born via C-section. We expect to find that infants delivered vaginally and ones via C-section have something both in common and difference in microbial composition during these three growth phases. Table 2 shows the evolving communities in the dynamic microbe network from infants delivered by C-section.

In Table 2, we can see that the Haemophilus, Streptococcus and Veillonella bacteria appear simultaneously in the dynamic module $C_{2\mathrm{II}} \rightarrow C_{2\mathrm{III}}$, this condition is similar to $C_{5\mathrm{I}} \rightarrow C_{5\mathrm{II}} \rightarrow C_{1\mathrm{III}}$ (a dynamic module of infants delivered vaginally). A significant difference lies in lack of Streptococcus microbes in the local cluster $C_{5\mathrm{II}}$.

Table 2. Dynamic microbe community from infants delivered by C-Section (Top 2 mapping by the optimal MI)

Dynamic Community	Local Community at Stage I (microbes)	Local Community at Stage II (microbes)	Local Community at Stage III(microbes)
$C_{2II} \rightarrow C_{2III}$	NA	"Haemophilus" "Streptococcus" "Veillonella "Bacteroides" "Rothia" "Anaerococcus"	"Haemophilus" "Streptococcus" "Veillonella" "Roseburia" "Faecalibacterium" "Subdoligranulum"
$C_{5II} \rightarrow C_{4III}$	NA	"Enterobacter" "Granulicatella" "Staphylococcus"	"Enterobacter" "Granulicatella" "Blautia","Gemella" "Clostridium" "Coprobacillus" "Lachnospiraceae"

Note: Same species are colored the same.

However, in Table 2 there is no corresponding local cluster appears at stage I. The reason is that there are less microbes colonizing the gut of infants who are born via C-section during the early life (at birth). When look up the abundance profile, we only find 5 genus bacteria that have significant correlation by removing the microorganisms whose frequency of occurrence is less than 5 across the whole samples.

4 Conclusion

In this paper, we introduce a novel framework that can be used to analyze the evolving modules in dynamic microbe networks. To our best knowledge, this is the first attempt to utilize the dynamic network approach on time-series microbial interaction networks. Discovery of dynamic (or evolutionary) modules is essential to important applications such as monitoring progression of infectious disease. ESNMF can identify the evolving pattern of infant gut microbial community. Experimental studies on a microbiome dataset of human infant gut, demonstrate ESNMF can find dynamic modules that are related to specific microbial community functions, and distinguish the evolutionary modules in different networks derived from infants delivered vaginally and ones born via C-section. These modules may not be discovered by traditional static modularity analysis.

Acknowledgement. This research is supported by the National Natural Science Foundation of China (No. 61532008), the Excellent Doctoral Breeding Project of CCNU, the Self-determined Research Funds of CCNU from the Colleges' Basic Research and Operation of MOE (No. CCNU16KFY04).

References

1. Barabasi, A., Bonabeau, E.: Scale-free networks. Sci. Am. **288**(5), 60–69 (2003)
2. Watts, D.J., Strogatz, S.H.: Collective dynamics of 'small-world' networks. Nature **393** (6684), 440–442 (1998)
3. Nørskov-Lauritsen, N.: Classification, identification, and clinical significance of Haemophilus and Aggregatibacter species with host specificity for humans. Clin. Microbiol. Rev. **27**(2), 214–240 (2014)
4. Blaser, M.: Missing Microbes. Oneworld Publications, London (2014)
5. Fekete, T.: Missing microbes: how the overuse of antibiotics is fueling our modern plagues. Clin. Infect. Dis. **60**(8), 1293 (2014)
6. Kim, M.-S., Han, J.: A particle-and-density based evolutionary clustering method for dynamic networks. Proc. VLDB Endow. **2**(1), 622–633 (2009)
7. Chi, Y., Song, X., Zhou, D., Hino, K., Tseng, B.L.: On evolutionary spectral clustering. ACM Trans. Knowl. Discov. Data (TKDD) **3**(4), 17 (2009)
8. Lin, Y., Chi, Y., Zhu, S., Sundaram, H., Tseng, B.: Facetnet: a framework for analyzing communities and their evolutions in dynamic networks. In: International World Wide Web Conferences, pp. 685–694 (2008)
9. Ma, X., Dong, D.: Evolutionary nonnegative matrix factorization algorithms for community detection in dynamic networks. IEEE Trans. Knowl. Data Eng. **29**(5), 1045–1058 (2017)
10. Ma, Y., Hu, X., He, T., Jiang, X.: Multi-view clustering microbiome data by joint symmetric nonnegative matrix factorization with Laplacian regularization. In: Ma, Y., Hu, X., He, T., Jiang, X. (eds.) 2016 IEEE International Conference on Bioinformatics and Biomedicine (BIBM). IEEE (2016)
11. Gerber, G.K.: The dynamic microbiome. FEBS Lett. **588**(22), 4131–4139 (2014)
12. Kwong, W.K., Medina, L.A., Koch, H., Sing, K.W., Ejy, S., Ascher, J.S., et al.: Dynamic microbiome evolution in social bees. Sci. Adv. **3**(3), e1600513 (2017)
13. Vázquez-Baeza, Y., Gonzalez, A., Smarr, L., Mcdonald, D., Morton, J.T., Navas-Molina, J. A., et al.: Bringing the dynamic microbiome to life with animations. Cell Host Microbe **21**(1), 7 (2017)
14. Jiang, X., Hu, X., Xu, W.: Microbiome data representation by joint nonnegative matrix factorization with Laplacian regularization. IEEE/ACM Trans. Comput. Biol. Bioinform. **14**(2), 353–359 (2017)
15. Kuang, D., Ding, C., Park, H.: Symmetric nonnegative matrix factorization for graph clustering. In: Kuang, D., Ding, C., Park, H. (eds.) Proceedings of the 2012 SIAM International Conference on Data Mining. SIAM (2012)
16. Chakrabarti, D., Kumar, R., Tomkins, A.: Evolutionary clustering. In: Chakrabarti, D., Kumar, R., Tomkins, A. (eds.) Proceedings of the 12th ACM SIGKDD International Conference on Knowledge Discovery and Data Mining. ACM (2006)
17. Newman, M.E., Girvan, M.: Finding and evaluating community structure in networks. Phys. Rev. E **69**(2), 026113 (2004)
18. Feng, Z., Xu, X., Yuruk, N., Schweiger, Thomas A.J.: A novel similarity-based modularity function for graph partitioning. In: Song, I.Y., Eder, J., Nguyen, T.M. (eds.) DaWaK 2007. LNCS, vol. 4654, pp. 385–396. Springer, Heidelberg (2007). https://doi.org/10.1007/978-3-540-74553-2_36
19. Bäckhed, F., Roswall, J., Peng, Y., Feng, Q., Jia, H., Kovatcheva-Datchary, P., et al.: Dynamics and stabilization of the human gut microbiome during the first year of life. Cell Host Microbe **17**(5), 690–703 (2015)

20. Dominguez-Bello, M.G., Blaser, M.J.: Asthma: undoing millions of years of coevolution in early life? Sci. Transl. Med. **7**(307), 307fs39 (2015)
21. Page, M.I., King, E.O.: Infection due to Actinobacillus actinomycetemcomitans and Haemophilus aphrophilus. N. Engl. J. Med. **275**(4), 181–188 (1966)
22. Bieger, R.C., Brewer, N.S., Washington, J.A.: Haemophilus aphrophilus: a microbiologic and clinical review and report of 42 cases. Medicine **57**(4), 345–356 (1978)
23. Tempro, P., Slots, J.: Selective medium for the isolation of Haemophilus aphrophilus from the human periodontium and other oral sites and the low proportion of the organism in the oral flora. J. Clin. Microbiol. **23**(4), 777–782 (1986)

A Pseudo-dynamic Search Ant Colony Optimization Algorithm with Improved Negative Feedback Mechanism to Solve TSP

Jun Li[1,2(✉)], Yuan Xia[1,2], Bo Li[1,2], and Zhigao Zeng[3]

[1] College of Computer Science and Technology, Wuhan University of Science and Technology, Wuhan 430065, Hubei, China
lijun@wust.edu.cn

[2] Hubei Province Key Laboratory of Intelligent Information Processing and Real-Time Industrial System, Wuhan, Hubei 430065, China

[3] College of Computer and Communication, Hunan University of Technology, Hunan 412000, China

Abstract. Aiming at the problem that the ant colony optimization algorithm with improved negative feedback mechanism is easy to fall into the local optimum when solving the traveling salesman problem, a pseudo-dynamic search ant colony optimization algorithm with improved negative feedback mechanism is proposed. A pseudo-dynamic search state transition rule is introduced to make the ant search not limited to pheromone concentration and distance between cities, which enhance the ability to jump out of local optimum. At the same time, the weight of pheromone concentrations in the optimal paths are updated to increase the convergence speed. The simulation results of different data sets in TSPLIB standard library show that the improved algorithm not only has higher convergence accuracy, but also outperforms the local optimum obviously better than the negative feedback ant colony algorithm.

Keywords: Ant colony algorithm · Pseudo-dynamic search · Local optimum Weight · Negative feedback

1 Introduction

TSP [1] is a typical NP-hard [2] combinatorial optimization problem. Nowadays, more and more intelligent algorithms are used to solve TSP and have achieved good results. Such as: particle swarm algorithm [3], wolf group algorithm [4], ant colony optimization algorithm [5, 6], fruit fly algorithm [7]. Ant colony optimization algorithm (ACO) is the earliest intelligent algorithm that applied to solve TSP, but it is prone to easy to fall into the local optimum and low convergence accuracy when solving TSP.

By introducing an improved negative feedback mechanism, literature [8] improves the convergence accuracy of solving TSP. However, the literature [8] only considers the concentration of pheromone and the distance between nodes in the state transition rules for solving TSP, which is easy to fall into the local optimal solution.

Aiming at the problem that literature [8] is easy to fall into local optimum, a pseudo-dynamic search ant colony optimization algorithm with improved negative

D.-S. Huang et al. (Eds.): ICIC 2018, LNAI 10956, pp. 19–24, 2018.
https://doi.org/10.1007/978-3-319-95957-3_3

feedback mechanism (PACON) is proposed to optimize the literature [8]: by introducing a pseudo-dynamic search state transition rule based on the improved negative feedback mechanism, the search of ant is not limited to the concentration of pheromone and the distance between cities. Thus enhancing the ability of the algorithm to jump out of the local optimal solution and improving the convergence accuracy. In addition, an improved pheromone updating strategy is proposed to further improve the convergence speed of the algorithm by strengthening the weight of the pheromone concentration on the optimal path. In order to verify the advantages of PACON in solving TSP, we selected 10 data sets of TSPLIB standard library for comparison, including a comparison between ACO, PACON and the literature [8]. The experimental results show that PACON is more advantageous than the literature [8] in solving TSP to avoid local optimization and improve convergence accuracy.

2 PACON

2.1 Pseudo-dynamic Search of State Transition Rule Updating Strategy

In constructing the state transition formula, in addition to calculating the positive and negative feedback pheromone concentrations, we also used angle calculation rules of the pseudo-dynamic search to obtain the new formula (1).

$$S_{ij}^{k}(t) = \begin{cases} P_{ij}^{k}(t) \cdot \frac{[\theta - \psi_{ij}(t)]^{\lambda}}{\sum\limits_{s \in allowed_k} [\theta - \psi_{is}(t)]^{\lambda}}, & s \in allowed_k \\ 0, & s \notin allowed_k \end{cases} \quad (1)$$

θ represents the influence degree of random selection. λ represents the volatility of the probability of random selection. $\psi_{ij}(t)$ represents the value of angle calculation rules [9].

In order to increase the probability of the city being selected, the implementation of pseudo dynamic search is as follows: the current city a is known, the next candidate city list is $b_1, b_2, b_3, \cdots, b_i, \cdots, b_n$ (where b_i is the city number). Firstly, there are only the city of b_1, b_2, b_3 in range of the roulette selection, we can make the city b_4, b_5, b_6 also included in range of the roulette selection when the angle between the sub-path $(a, b_4), (a, b_5), (a, b_6)$ and x axis is smaller, which makes the path never explored to become the second best choice. Secondly, if the original cities b_1, b_2, b_3 are in the range of roulette selection and the probability of being selected is $b_1 > b_2 > b_3$ due to the concentration of pheromone, we can know that the angle between the sub-path $(a, b_2), (a, b_3)$ and the x axis is smaller through calculating the angle between the sub-path $(a, b_1), (a, b_2), (a, b_3)$ and the x axis, it raises the selection probability of b_2, b_3 because of smaller angle on the basis of the probability of original pheromone concentration, and achieve the pseudo-dynamic search purpose of avoiding pheromone limitations by adjusting the selected probability of the cities to affect the result of roulette selection.

2.2 Pheromone Updating Strategy

The pheromone path update in the improved ant colony algorithm [10] is as follows:

$$\tau_{ij}(t+n) \leftarrow (1-\rho)\tau_{ij}(t) \tag{2}$$

In this formula, ρ is to satisfy the evaporation rate when $0<\rho\leq 1$.

In order to enhance the speed of movement of ants to the optimal path, we add appropriate weights to the positive feedback pheromone, while keeping the update mode of the negative feedback pheromone unchanged. After evaporating, the updated pheromone concentrations are respectively shown as formula (3) and formula (4):

$$\tau_{ij}(t) \leftarrow \tau_{ij}(t) + \mu\Delta\tau_{ij}^{k} \tag{3}$$

$$\tau_{ij}(t) \leftarrow \tau_{ij}(t) + \Delta\varphi_{ij}^{k} \tag{4}$$

$\Delta\tau_{ij}^{k}$ is the pheromone mass deposited by the ants on the optimal path and $\Delta\varphi_{ij}^{k}$ is the pheromone mass deposited by the ants on the worst path.

2.3 Algorithm Description

The flow chart of the improved ant colony algorithm is shown in Fig. 1.

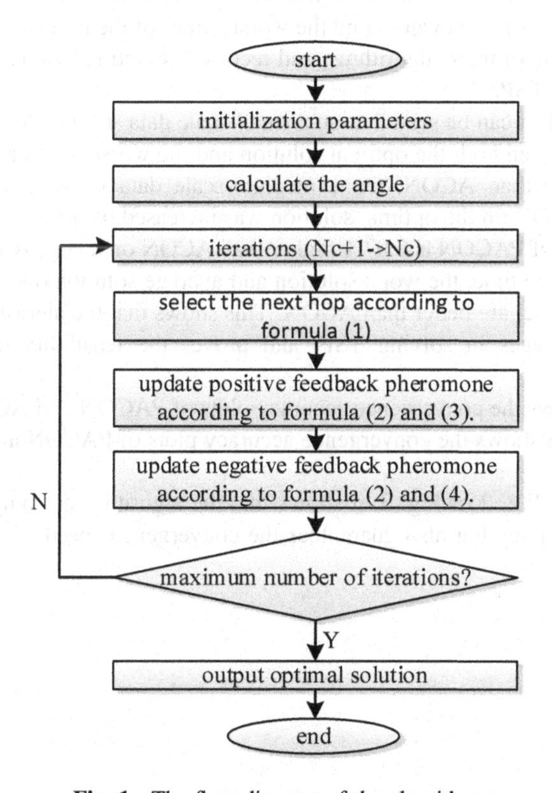

Fig. 1. The flow diagram of the algorithm

3 Simulation Experiment and Analysis

3.1 Experiment Configuration

The experiment using matlab2016a to achieve programming, the system operating environment is as follows: win7 dual-core, Intel (R) Core (i5) processor, 8 GB memory and 3.2 GHz CPU. The parameter setting of this algorithm is in accordance with literature [8], as follows: the importance factor of pheromone is $\alpha = 2$, the importance factor of heuristic function is $\beta = 8$, the maximum number of iterations is $NC_{\max} = 500$, $\rho = 0.02$, the total number of pheromones is $Q = 1$, the number of ants is $m = 35$, the range of parameters δ and θ is limited to $[5, 15]$, and the range of parameters γ and λ is limited to $[1, 5]$. After testing, we can know that the algorithm can get better results when setting $\delta = \theta = 10$ and $\gamma = \lambda = 2$.

3.2 Experiment Results

This article uses the TSPLIB standard library to verify the effectiveness of the improved algorithm for solving TSP. The data sets include bays29, gr96, bier127, kroB150, pr226, gil262, linhp318, rd400, fl417 and att532. In this paper, ACO, the improved negative feedback ant colony optimization algorithm(ACON) in the literature [8] and PACON were tested independently on each test data set 20 times to reduce the impact of random factors on the algorithm. The experiments records and compares the average values, the optimal values and the worst values of the three algorithms. Table 1 shows the test data of three algorithms, and records the optimal value, worst value and average value of TSP.

From Table 1, it can be seen that on a small-scale data set such as bays29, PACON is equal to ACON on both the optimal solution and the worst solution, but the average solution is better than ACON. On a medium-scale data set such as linhp318, the accuracy of PACON on the optimal solution was increased by 1.69% over ACON. The optimal solution of PACON is 0.07% higher than ACON on a large data set such as the att532. At the same time, the worst solution and average solution obtained by PACON in the above data set are better than ACON. This shows that the algorithm in this paper has more advantages in solving TSP, and proves the feasibility of the improved algorithm.

Figure 2 shows the convergence accuracy plots of PACON and ACON on the gr96 data set. Figure 3 shows the convergence accuracy plots of PACON and ACON on the pr226 data set.

Whether it is Fig. 2 or Fig. 3, it shows that the algorithm not only guarantees the convergence accuracy but also guarantees the convergence speed when solving TSP.

Table 1. The convergence accuracy of the algorithm

Data set	Algorithm	Algorithm convergence value		
		Optimal	Worst	Average
bays29	ACO	9168.8	9318.6	9243.2
	ACON	9094.6	9160.7	9157.4
	PACON	9094.6	9160.7	9165.6
gr96	ACO	567.3	550.1	556.4
	ACON	528.5	548.1	540.5
	PACON	522.8	546.0	533.0
bier127	ACO	125350.4	131239.8	128960.7
	ACON	120748.3	123040.9	121925.6
	PACON	120508.2	121307.7	120937.3
kroB150	ACO	28246.3	29400.7	28961.9
	ACON	27503.0	28254.1	27793.6
	PACON	27471.5	27960.7	27756.6
pr226	ACO	87431.1	92492.3	90223.4
	ACON	82901.6	86446.4	84582.7
	PACON	82519.6	85681.0	83724.4
gil262	ACO	2610.5	2693.5	2659.6
	ACON	2509.5	2592.2	2552.1
	PACON	2507.5	2555.4	2533.9
linhp318	ACO	48409.2	49634.5	49147.1
	ACON	45181.7	46861.6	46104.5
	PACON	44481.6	46568.7	45843.8
rd400	ACO	17633.5	18096.6	17791.5
	ACON	16649.2	17228.7	16833.0
	PACON	16566.2	17202.7	16818.9
fl417	ACO	13502.6	13481.2	13298.0
	ACON	12459.9	12656.5	12534.0
	PACON	12457.0	12577.3	12512.6
att532	ACO	101776.9	105446.3	103748.4
	ACON	93624.4	97079.1	95573.2
	PACON	93607.7	96986.8	95393.1

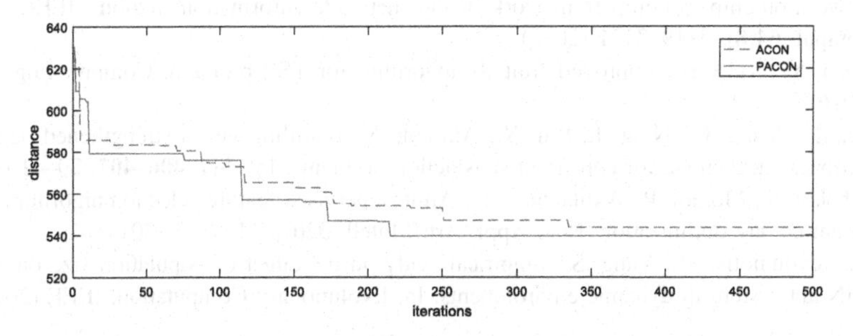

Fig. 2. Two algorithms to solve the gr96 data set

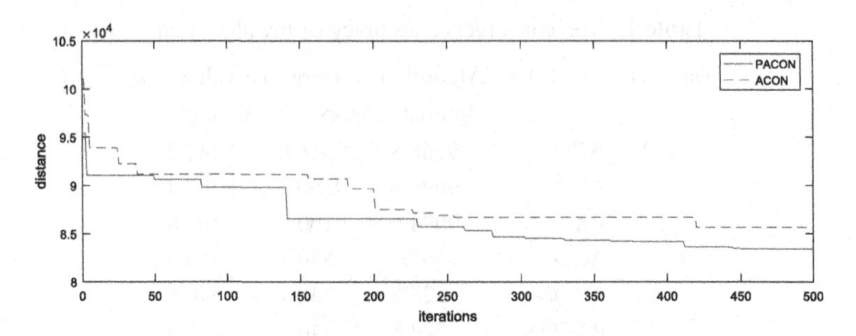

Fig. 3. Two algorithms to solve the pr226 data set

4 Conclusion

Simulation tests are conducted on 10 test data sets show that improved algorithm can effectively avoid falling into a local optimum and has a high convergence accuracy when solving TSP.

Acknowledgement. This research was financially supported by fund from Hubei Province Key Laboratory of Intelligent Information Processing and Real-time Industrial System (Wuhan University of Science and Technology) (Grant No. znxx2018QN06).

References

1. Reinelt, G.: The Traveling Salesman: Computational Solutions for TSP Applications. Springer, Heidelberg (1994). https://doi.org/10.1007/3-540-48661-5
2. Garey, M.R., Johnson, D.S.: Computers and Intractability: A Guide to the Theory of NP-Completeness. W. H Freeman, New York (1983)
3. Zhang, J.W., Si, W.J.: Improved enhanced self-tentative PSO algorithm for TSP. In: International Conference on Natural Computation, vol. 5, pp. 2638–2641. IEEE (2010)
4. Wu, H.S., Zhang, F.M., Li, H., Liang, X.L.: Discrete wolf pack algorithm for traveling salesman problem. Control Decis. **30**(10), 1861–1867 (2015)
5. Xiao, S.: Optimal travel path planning and real time forecast system based on ant colony algorithm. In: IEEE, Advanced Information Technology, Electronic and Automation Control Conference, pp. 2223–2226. IEEE (2017)
6. Hsin, H.K., Chang, E.J., Su, K.Y., Wu, A.Y.A.: Ant colony optimization-based adaptive network-on-chip routing framework using network information region. IEEE Trans. Comput. **64**(8), 2119–2131 (2015)
7. Duan, Y., Xiao, H.: Improved fruit fly algorithm for TSP problem. Comput. Eng. Appl. (2016)
8. Ye, K., Zhang, C., Ning, J., Liu, X.: Ant-colony algorithm with a strengthened negative-feedback mechanism for constraint-satisfaction problems. Inf. Sci. **406–407**, 29–41 (2017)
9. Tabakhi, S., Moradi, P., Akhlaghian, F.: An unsupervised feature selection algorithm based on ant colony optimization. Eng. Appl. Artif. Intell. **32**(6), 112–123 (2014)
10. Mavrovouniotis, M., Yang, S.: Empirical study on the effect of population size on MAX-MIN ant system in dynamic environments. In: Evolutionary Computation. IEEE (2016)

Dynamic Mutation Based Pareto Optimization for Subset Selection

Mengxi Wu[1(✉)], Chao Qian[1], and Ke Tang[2]

[1] Anhui Province Key Lab of Big Data Analysis and Application, University of Science and Technology of China, Hefei 230027, China
wmxair@mail.ustc.edu.cn, chaoqian@ustc.edu.cn
[2] Shenzhen Key Lab of Computational Intelligence, Southern University of Science and Technology, Shenzhen 518055, China
tangk3@sustc.edu.cn

Abstract. Subset selection that selects the best k variables from n variables is a fundamental problem in many areas. Pareto optimization for subset selection (called POSS) is a recently proposed approach for subset selection based on Pareto optimization and has shown good approximation performances. In the reproduction of POSS, it uses a fixed mutation rate, which may make POSS get trapped in local optimum. In this paper, we propose a new version of POSS by using a dynamic mutation rate, briefly called DM-POSS. We prove that DM-POSS can achieve the best known approximation guarantee for the application of sparse regression in polynomial time and show that DM-POSS can also empirically perform well.

Keywords: Subset selection · Pareto optimization · Sparse regression
Dynamic mutation

1 Introduction

The subset selection problem is to select a subset of size at most k from a total set of n variables for optimizing some given objectives. It arises in many applications, and is generally NP-hard [2, 8]. The greedy algorithm is widely used and plays as the simplest algorithm for subset selection [7]. Greedy algorithms iteratively select the variable with the largest marginal gain. Das and Kempe [1] proved that the greedy algorithm could achieve the best approximation guarantee on the sparse regression problem, which is a representative example of subset selection [9]. Qian et al. [10] proposed a new method based on Pareto optimization which is called POSS. It treats subset selection as a bi-objective optimization problem and optimizes the two objectives simultaneously. It was proven that POSS could achieve the best known approximation guarantee as the greedy algorithm on sparse regression. In addition, experimental results show that POSS outperforms the greedy algorithm.

In this paper, we propose a new version of POSS by using a dynamic mutation rate, briefly called DM-POSS. A dynamic mutation rate can help escape the local optimum. The (1 + 1)EA with a dynamic mutation rate was proved that it could get an asymptotically equal or better performance on many problems, such as jump functions [4], the

D.-S. Huang et al. (Eds.): ICIC 2018, LNAI 10956, pp. 25–35, 2018.
https://doi.org/10.1007/978-3-319-95957-3_4

vertex cover problem [5] in bipartite graphs, the matching problems [6] in general graphs, etc. We prove that DM-POSS can achieve the best known approximation guarantee for the application of sparse regression in polynomial time and it could have a better ability of jumping out local optima. Experimental results show its better performance.

The rest of the paper is organized as follows. Section 2 introduces the subset selection problem including the sparse regression problem and the influence maximization problems. We also introduce the POSS method in this part. In Sect. 3, we propose the DM-POSS method and analyse it theoretically. Section 4 presents the empirical results. We conclude the paper in Sect. 5.

2 Subset Selection

2.1 Subset Selection

The general subset selection problem is presented in Definition 1, which is to select the best subset with at most k variables.

Definition 1 (Subset Selection). Given all variables $V = \{X_1, \ldots, X_n\}$, an objective f and a positive integer k, the subset selection problem is to find the solution of the optimization problem:

$$\arg\min_{S \subseteq V} f(S) \text{ s.t. } |S| \leq k \tag{1}$$

Noticeably, maximizing the f is equivalent to minimizing the $-f$.

2.2 Sparse Regression

One typical subset selection problem is sparse regression [1, 10]. It aims at finding a sparse approximation solution to the linear regression problem, where the solution vector can only have a few non-zero elements. Note that all variables are normalized to have expectation 0 and variance 1.

Definition 2 (Sparse Regression). Given a predictor variable Z, all observation variables $V = \{X_1, \ldots, X_n\}$ and a positive integer k, the mean squared error of a subset $S \subseteq V$ is

$$MSE_{Z,S} = \min_{\alpha \in R^{|S|}} E\left[\left(Z - \sum_{i \in S} \alpha_i X_i\right)^2\right]$$

Sparse regression is to find a subset S, at most k variables, minimizing the mean squared error:

$$\arg\min_{S \subseteq V} MSE_{Z,S} \text{ s.t. } |S| \leq k$$

It can be replaced with the squared multiple correlation,

$$\arg\min_{S \subseteq V} R^2_{Z,S} \text{ s.t. } |S| \leq k \tag{2}$$

the corresponding objective is the squared multiple correlation

$$R^2_{Z,S} = \left(Var(Z) - MSE_{Z,S}\right)/Var(Z)$$

where $Var(Z)$ represents the variance of Z.

2.3 Influence Maximization

Influence maximization is another typical subset selection problem. It aims at finding k nodes, which are called seed nodes, to maximize the spread of influence [8].

Influence Maximization. In a social network, an active node can activate each of its neighbors once with active probability p. Influence maximization is to select at most k seed nodes as initial active nodes to maximize the number of active nodes $\sigma(S)$,

$$\arg\max_{S \subseteq V} \sigma(S) \text{ s.t. } |S| \leq k \tag{3}$$

$\sigma(S)$ is too difficult to calculate directly. Kempe et al. [8] proposed to simulate the random process of information spread to calculate it.

2.4 POSS

Qian et al. [10] proposed an algorithm for subset selection based on Pareto optimization, which is called POSS. Let a binary vector $s = \{0,1\}^n$ represent a subset S, where $s_i = 1$ means the variable ith is selected and $s_i = 0$ otherwise. POSS treats the constraint as another optimization objective. The original subset selection problem is then reformulated as a bi-objective optimization problem to minimize $f_1(s)$ and $f_2(s)$ simultaneously, where

$$f_1(s) = \begin{cases} +\infty, & s = \{0\}^n \ or \ |s| \geq 2k \\ & f(s), \ otherwise \end{cases}, \qquad f_2(s) = |s| \tag{4}$$

For sparse regression, $f_1(s) = -R^2_{Z,S}$, which means to maximize $R^2_{Z,S}$.

We need to compare the solutions with two objectives in POSS. Assume there are two solutions s and s'. When s has a smaller or equal value on both objectives, s' is worse than s. In this situation, s weakly dominates s', denoted as $s \preccurlyeq s'$. s' is strictly worse than s when s' has a smaller value on one objectives and meanwhile has a smaller or equal value on another objective. We say s dominates s', denoted as $s \preccurlyeq s'$, if s is not worse than s' and s' is not worse than s, s and s' is incomparable. It can be formulated as following:

- $s \preccurlyeq s'$ if $f_1(s) \leq f_1(s') \wedge f_2(s) \leq f_2(s')$
- $s \prec s'$ if $s \preccurlyeq s' \wedge f_1(s) < f_1(s') \vee f_2(s) < f_2(s')$
- s and s' is incomparable, if i $s \not\preccurlyeq s' \wedge s' \not\preccurlyeq s$.

POSS then solves the bi-objective problem by a simple multi-objective evolutionary algorithm. It starts from an initial empty set $s_0 = \{0\}^n$. There is an archive P, which only has s_0 at the initial time. In each iteration, a parent s is selected from P uniformly at random. A new solution s' is generated by flipping each bit of s with probability $1/n$. After evaluating the two objectives of s', if there is no solution in P dominating s', s' is added into P and the solutions weakly dominated by s' from P are removed. After T iterations, the best solution satisfying the size constraint in P is selected as the final solution.

Algorithm 1. POSS

Input: the set $V = \{X_1, \ldots, X_n\}$, a given objective f, an positive integer parameter k, the number of iteration T
Output: a subset of V with at most k variables
1: Initialize $s = \{0\}^n$, $P = \{s\}$, $t = 0$;
2: **while** $t < T$ **do**
3: Select s from P uniformly at random;
4: Generate s' by flipping each bit of s with probability $1/n$;
5: Evaluate $f_1(s)$ and $f_2(s)$;
6: **if** $\nexists z \in P,\ z \preccurlyeq s'$ **then**
7: $Q = \{z \in P | s' \preccurlyeq z\}$;
8: $P = (P \backslash Q) \cup \{s'\}$;
9: **end if**
10: $t = t + 1$;
11: **end while**
12: **return** $\arg\min_{s \in P, |s| \le k} f(s)$.

3 DM-POSS

3.1 DM-POSS

In this section, we propose a new version of POSS, which is called DM-POSS, by using a dynamic mutation rate. A dynamic mutation rate could have a better ability of jumping out local optima. In DM-POSS, a new solution s' is generated by the heavy-tailed mutation operator $fmut_\beta$ [3]. $fmut_\beta$ creates an offspring by flipping each bit of the parent s independently with probability α/n, where $\alpha \in \{1, 2, \ldots, n/2\}$ is a random positive integer that follows a power-law distribution $D^\beta_{n/2}$ as shown in Definition 3.

Definition 3 (Discrete Power-law Distribution). Let $\beta > 1$ be a constant. For a random variable X with the distribution $D^\beta_{n/2}$ on $\{1, 2, \ldots, n/2\}$, it holds that $\Pr[X = \alpha] = \left(C^\beta_{n/2}\right)^{-1} \alpha^{-\beta}$, where the normalization constant is $C^\beta_{n/2} = \sum_{i=1}^{n/2} i^{-\beta}$.

The heavy-tailed mutation operator has a dynamic mutation rate. It was proved that such a mutation operator can bring a better ability of jumping out of local optima in jump function and some combinatorial optimization. DM-POSS is described in Algorithm 2, which main procedure is the same as POSS except the mutation operator, i.e., line 4 in Algorithm 1.

Algorithm 2. DM-POSS

Input: the set $V = \{X_1, \ldots, X_n\}$, a given objective f, an positive integer parameter k, a mutation ratio β, the number of iteration T,
Output: a subset of V with at most k variables
1: Initialize $s = \{0\}^n$, $P = \{s\}$, $t = 0$;
2: **while** $t < T$ **do**
3: Select s from P uniformly at random;
4: Choose $\alpha \in \{1,2,\ldots,n/2\}$ according to $D^{\beta}_{n/2}$;
5: Generate s' by flipping each bit of s with probability α/n;
6: Evaluate $f_1(s)$ and $f_2(s)$;
7: **if** $\nexists z \in P,\ z \preccurlyeq s'$ **then**
8: $Q = \{z \in P | s' \preccurlyeq z\}$;
9: $P = (P \backslash Q) \cup \{s'\}$;
10: **end if**
11: $t = t + 1$;
12: **end while**
13: **return** $\arg\min_{s \in P, |s| \le k} f(s)$.

3.2 Theoretical Analysis

In this section, we theoretically investigate the performance of DM-POSS on a typical subset selection application, sparse regression. We prove in Theorem 1 that DM-POSS can also achieve the approximation guarantee on sparse regression, which was previously obtained by the greedy algorithm [7] and POSS [10].

Theorem 1. For sparse regression, DM-POSS with $E[T] \le 2ek^2n\left(1 + \frac{1}{2^{\beta-1}-1}\right)$ finds a set S of variables with $|S| \le k$ and $R^2_{Z,S} \ge (1 - e^{-\gamma_{\emptyset,k}}) \cdot OPT$.

Note that $E[T]$ denotes the average number of iterations T in DM-POSS, which is the expected running time of DM-POSS. OPT represents the optimal function value of Eq. 2. $\gamma_{\emptyset,k}$ is defined by the submodularity ratio presented in Definition 4 [10].

Definition 4 (Submodularity Ratio). Let f be a non-negative set function. The submodularity ratio of f with respect to a set U and a parameter $k \ge 1$ is

$$\gamma_{\emptyset,k}(f) = \min_{L \subseteq U, S: |S| \le k, S \cap L = \emptyset} \frac{\sum_{x \in S} f(L \cup \{x\}) - f(L)}{f(L \cup S) - f(L)} \tag{5}$$

So that $\gamma_{\emptyset,k}(f) = \min_{S:|S|\leq k} \frac{\sum_{x\in S} f(x)-f(\emptyset)}{f(S)-f(\emptyset)}$.

The proof of Theorem 1 is inspired from proof of Theorem 1 in [10].

Proof of Theorem 1. If there exists $s \in P$, that $|s| \leq j$ and $R^2_{Z,s} \geq (1-(1-\frac{\gamma_{\emptyset,k}}{k})^j) \cdot OPT$. We can flip the specific bit of s to get a new solution s' that $s' \leq j+1$ and $R^2_{Z,s'} \geq \left(1-(1-\frac{\gamma_{\emptyset,k}}{k})^j\right)$. Let $Jmax = \max\left\{j \in [0,k] | \exists s \in P, |s| \leq j \wedge R^2_{Z,s} \geq (1-(1-\frac{\gamma_{\emptyset,k}}{k})^j) \cdot OPT\right\}$. DM-POSS starts from $P = \{s_0\}$, where $s_0 = \{0\}^n$, so that $Jmax$ = 0 at initial time. $Jmax$ can increase one by one from 0 to k by flipping the specific 0 bit of the specific solution s in P.

In DM-POSS, s is selected with probability at least $1/Pmax$. The heavy-tailed mutation operator provides the probability at least $\frac{\alpha}{n} \cdot \left(1-\frac{\alpha}{n}\right)^{n-1}$ for the specific 0 bit to flip. As $\Pr[X=\alpha] = \left(C^\beta_{n/2}\right)^{-1} \alpha^{-\beta}$, the average probability should be

$$\sum_{\alpha=1}^{n/2} \left(C^\beta_{n/2}\right)^{-1} \alpha^{-\beta} \cdot \frac{\alpha}{n} \cdot \left(1-\frac{\alpha}{n}\right)^{n-1} \tag{6}$$

The first term is $\left(C^\beta_{n/2}\right)^{-1} \frac{1}{n} \cdot \left(1-\frac{1}{n}\right)^{n-1}$. We can prove that $\frac{1}{1^\beta}$

$$\begin{aligned}
C^\beta_{n/2} &= \frac{1}{1^\beta} + \left(\frac{1}{2^\beta}+\frac{1}{3^\beta}\right) + \left(\frac{1}{4^\beta}+\ldots+\frac{1}{7^\beta}\right) + \left(\frac{1}{8^\beta}+\ldots+\frac{1}{16^\beta}\right) + \ldots \\
&< \frac{1}{1^\beta} + \left(\frac{1}{2^\beta}+\frac{1}{2^\beta}\right) + \left(\frac{1}{4^\beta}+\ldots+\frac{1}{4^\beta}\right) + \left(\frac{1}{8^\beta}+\ldots+\frac{1}{8^\beta}\right) + \ldots \\
&= \frac{1}{1^\beta} + \left(\frac{1}{2^{\beta-1}}\right)^1 + \left(\frac{1}{2^{\beta-1}}\right)^2 + \left(\frac{1}{2^{\beta-1}}\right)^3 + \ldots
\end{aligned}$$

The right of the equation is a geometric progression, its sum is $\frac{\left(\frac{1}{2^{\beta-1}}\right)^{n'}}{1-\frac{1}{2^{\beta-1}}}$. n' can never be larger than $n/2$, so the upper bound is $\frac{1}{1-\frac{1}{2^{\beta-1}}} = 1 + \frac{1}{2^{\beta-1}-1}$.

The lower bound is

$$\left(C^\beta_{n/2}\right)^{-1} \frac{1}{n} \cdot \left(1-\frac{1}{n}\right)^{n-1} \geq \frac{1}{en} \cdot \frac{2^{\beta-1}-1}{2^{\beta-1}} \tag{7}$$

When the $Jmax$ has reached k, we can get the desired solution, which means we need $ekn\left(1+\frac{1}{2^{\beta-1}-1}\right) \cdot Pmax$ expected iterations. Note that due to the definition of f_1 and the fact that the solutions in P are incomparable, we have $Pmax \leq 2k$. It implies there exists one solution s in P satisfying that $|s| \leq k$ and $R^2_{Z,s} \geq \left(1-(1-\frac{\gamma_{\emptyset,k}}{k})^j\right) \cdot OPT \geq (1-e^{-\gamma_{\emptyset,k}}) \cdot OPT$ after $2ek^2n(1+\frac{1}{2^{\beta-1}-1})$ times of iterations.

4 Experiment

To empirically investigate the performance of DM-POSS, we conduct experiments on 10 datasets[1] of sparse regression and 4 datasets[2] of influence maximization. The previous best methods, the greedy algorithm and POSS, are selected for comparison.

We set the size constraint $k = 8$. For POSS, $T = 2^2n$ as suggested in [10]. For DM-POSS, T is also set to $2ek^2n$ for fairness. We change the parameter β from 1.5 to 3.0 with interval 0.5. POSS and DM-POSS are repeated for 10 times for average. All of the experiments are coded in Java and run on an identical configuration: an experiments are run on the server with 2.60 GHz Octo-Co Intel Xeon E5-2650 and 64G memory.

4.1 Experiment on Sparse Regression

The information about the datasets is summarized in Table 1.

The results of all experiments are shown in Table 2. In each data set, the symbol "•" denote that DM-POSS is better than the greedy algorithm and POSS and "○" denote that DM-POSS is worse than any of them. We can observe that DM-POSS with different values of β performs not worse than POSS or the greedy algorithm.

The value of the R^2 over the running time of POSS and DM-POSS on 5 datasets is shown in Fig. 1. We only plot the DM-POSS with the best performance on each dataset. The black line represents the Greedy algorithm and the green dotted line represents POSS. The red dotted line represents DM-POSS with the best results. Notice that the x-axis denotes the current number of iterations (divided by n). It shows DM-POSS usually performs worse than POSS at the beginning but finally achieves a better performance. We also conclude that the best choice of β depends on the dataset.

We realize that DM-POSS with specific values of β can find better optimal solutions. In most cases, it performs as well as POSS. We suppose the choice of β depends on the dataset. Although it is difficult to find a universal optimal β for different datasets, we find in experiments that DM-POSS can perform at least as well as POSS within only a few search of β. We should do more research on how to find the β which helps improve the quality of results.

4.2 Experiments on Influence Maximization

The detailed information about the datasets is shown in Table 3 indicating the number of nodes n and edges m on 4 real social networks.

---

[1] The datasets are from https://www.csie.ntu.edu.tw/~cjlin/libsvmtools/datasets/.

[2] The datasets are from https://snap.stanford.edu/data/.

Table 1. Datasets of sparse regression

Dataset	#Inst	#Feat	Dataset	#Inst	#Feat
Aloi	108,000	128	Clean1	476	166
Convtype	581,012	54	Mushrooms	8,124	112
Protein	21,516	356	Sensorless	58,509	48
Seismic	98,528	50	Triazing	208	60
W5a	49,749	300	Year	515,345	90

Table 2. Results of sparse regression

Dataset	Greedy	POSS	DM-POSS			
			$\beta = 1.5$	$\beta = 2$	$\beta = 2.5$	$\beta = 3$
Aloi	.073	.074 ± .002	.075 ± .009•	.075 ± .001•	.075 ± .002•	.075 ± .009•
Clean1	.386	.386 ± .006	.390 ± .006•	.387 ± .007•	.386 ± .008•	.389 ± .005•
Mushroom	.991	.991 ± .000	.991 ± .000	.991 ± .000	.991 ± .000	.992 ± .000•
Sensorless	.222	.223 ± .000	.223 ± .000•	.223 ± .000•	.223 ± .000•	.223 ± .001•
Triazing	.316	.327 ± .000	.328 ± .000•	.328 ± .000•	327 ± .000	.327 ± .000
Convtype	.249	.249 ± .000	.249 ± .000	.249 ± .000	.249 ± .000	.249 ± .000
Protein	.140	.140 ± .000	.140 ± .000	.140 ± .000	.140 ± .000	.140 ± .000
Seismic	.431	.431 ± .000	.431 ± .000	.431 ± .000	.431 ± .000	.431 ± .000
W5a	0.311	.312 ± .000	.312 ± .000	.312 ± .000	.312 ± .000	.312 ± .000
Year	0.187	.190 ± .000	.190 ± .000	.190 ± .000	.190 ± .000	.190 ± .000

We set the active probability $p = 1/d_{in}$, where d_{in} is the in-degree of the inactive node. For estimating $\sigma(S)$, we simulate the process of influence spread 30 times independently and use the average as an estimation. But for the final output solutions of the algorithms, we average over 10,000 times for accurate estimation. All algorithms are repeated for 10 times for average.

Figure 2 shows the relationship between performance and running time of POSS and DM-POSS. Notice that the x-axis denotes the current number of iterations (divided by kn). The black line represents Greedy algorithm and the pink line represents POSS. The dotted lines in different colors represent different values of β in DM-POSS.

We can summarize that DM-POSS can improve the quality of optimum solutions greatly in most cases. For NetHEPT and p2p04, DM-POSS always performs better than POSS. More research should be carried out to find the best value of β.

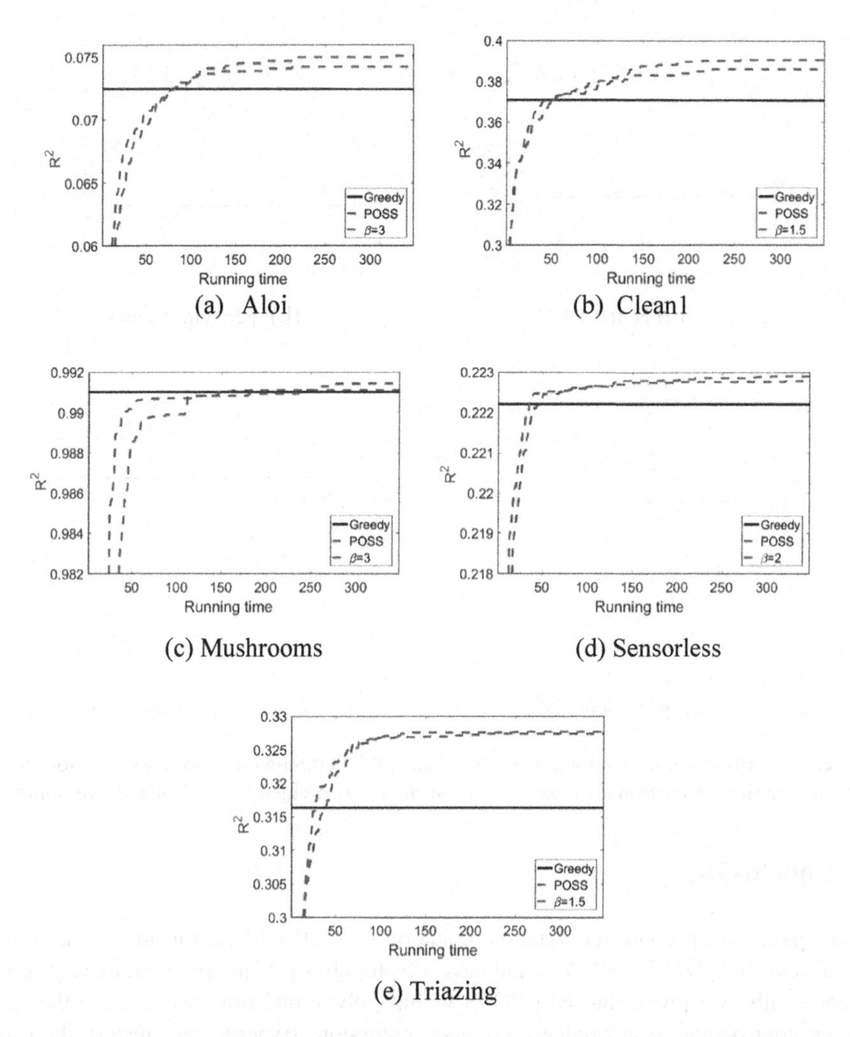

Fig. 1. Performance v.s. running time of POSS and DM-POSS with specific β on sparse regression. Note that running time is measured by the number of iterations divided by n. (Color figure online)

Table 3. Datasets of influence maximization

Dataset	#Nodes	#Edges
NetHEPT	9,877	51,971
p2p_Gnutella04	10,879	39,994
p2p_Gnutella05	8,846	31,839
p2p_Gnutella06	8,717	31,525

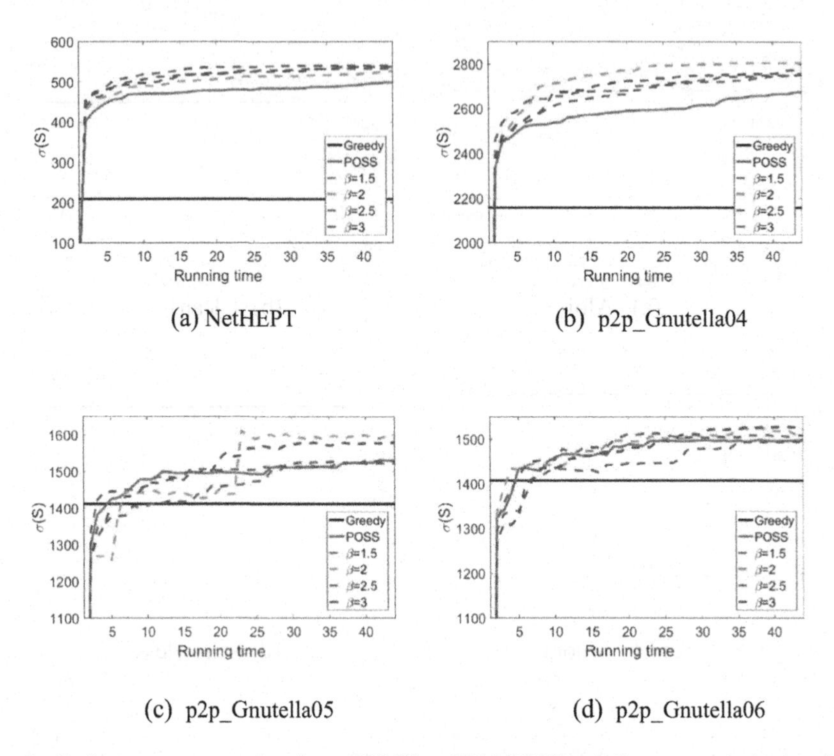

Fig. 2. Performance v.s. running time of POSS and DM-POSS on influence maximization. Note that running time is measured by the number of iterations divided by n. (Color figure online)

5 Conclusion

In this paper, we propose the dynamic mutation based Pareto optimization method for subset selection (DM-POSS). It could have a better ability of jumping out local optima. Theoretically, we prove that DM-POSS within polynomial time can achieve the best known approximation guarantee on sparse regression. Experimental results show its better performance, which however, requires a proper β value for the heavy-tailed mutation operator. We will study on the selection of β in the future work.

References

1. Das, A., Kempe, D.: Submodular meets spectral: Greedy algorithms for subset selection, sparse approximation and dictionary selection. In: 28th International Conference on Machine Learning, Bellevue, WA, pp. 1057–1064 (2011)
2. Davis, G., Mallat, S., Avellaneda, M.: Adaptive Greedy approximations. Constr. Approx. **13** (1), 57–98 (1997)
3. Doerr, B., Le, H.P., Makhmara, R., Nguyen, T.D.: Fast genetic algorithms. In: 19th ACM Genetic and Evolutionary Computation Conference, Berlin, Germany, pp. 777–784 (2017)

4. Droste, S., Jansen, T., Wegener, I.: On the analysis of the (1+1) evolutionary algorithm. Theoret. Comput. Sci. **276**(1–2), 51–58 (2002)
5. Fischer, S., Wegener, I.: The one-dimensional Ising model: mutation versus recombination. Theoret. Comput. Sci. **344**(2–3), 208–225 (2005)
6. Giel, O., Wegener, I.: Evolutionary algorithms and the maximum matching problem. In: 20th Annual Symposium on Theoretical Aspects of Computer Science, London, UK, pp. 415–426 (2003)
7. Gilbert, A.C., Muthukrishnan, S., Strauss, M.J.: Approximation of functions over redundant dictionaries using coherence. In: 14th Annual ACM-SIAM symposium on Discrete Algorithms, Baltimore, MA, pp. 243–252 (2003)
8. Kempe, D., Kleinberg, J., Tardos, E.: Maximizing the spread of influence through a social network. In: 9th ACM SIGKDD International Conference on Knowledge Discovery and Data Mining, Washington, D.C., pp. 137–146 (2003)
9. Miller, A.: Subset Selection in Regression. CRC Press, Boca Raton (2002)
10. Qian, C., Yu, Y., Zhou, Z.H.: Subset selection by Pareto optimization. In: Advances in Neural Information Processing Systems 28, Montreal, Canada, pp. 1774–1782 (2015)

Energy-Aware Fault-Tolerant Scheduling Under Reliability and Time Constraints in Heterogeneous Systems

Tian Guo[1,2], Jing Liu[1,2](✉), Wei Hu[1,2], and Mengxue Wei[1,2]

[1] College of Computer Science and Technology, Wuhan University of Science and Technology, Wuhan, China
Idealer@126.com

[2] Hubei Province Key Laboratory of Intelligent Information Processing and Real-time Industrial System, Wuhan 430065, China

Abstract. As heterogeneous systems have been deployed widely in various fields, the reliability become the major concern. Thereby, fault tolerance receives a great deal of attention in both industry and academia, especially for safety critical systems. Such systems require that tasks need to be carried out correctly in a given deadline even when an error occurs. Therefore, it is imperative to support fault-tolerance capability for systems. Scheduling is an efficient approach to achieving fault tolerance by allocating multiple copies of tasks on processors. Existing fault-tolerant scheduling algorithms realize fault tolerance without energy limit. To address this issue, this paper proposes an energy-aware fault-tolerant scheduling algorithm DRB-FTSA-E. The algorithm adopts the active replication strategy and uses a high utilization of energy consumption to complete a set of tasks with given reliability and time constraints. It finds out all schemes that meet time and system reliability constraints, and chooses the scheme with the maximum utilization of energy consumption as the final scheduling scheme. Experimental simulation results show that the proposed algorithm can effectively achieve the maximum utilization of energy consumption while meeting the reliability and time constraints.

Keywords: Fault-tolerant · Scheduling algorithm · Reliability
Time constraint · Energy consumption

1 Introduction

Since 1970s, real-time systems have been widely used in many aspects of human life [11]. Typical applications include automotive electronics systems, engineering process control systems, etc. With the rapid progress of computer technology, real-time systems tend to be isomerization [2]. Heterogeneous systems have been widely used in various security-critical real-time systems, such as aircraft and spacecraft control systems and critical patient life support systems. These systems usually require strict time deadline and high reliability. Thus, it is necessary to provide fault-tolerance capability for systems.

D.-S. Huang et al. (Eds.): ICIC 2018, LNAI 10956, pp. 36–46, 2018.
https://doi.org/10.1007/978-3-319-95957-3_5

Failures can be divided into transient failures and permanent failures. Transient failures are more frequent than permanent failures [10]. In this paper, we aim at handling transient failures. At present, the research of fault-tolerant scheduling is based on task replication technology: one copy or multiple copies. The former way is adopted because we consider strict time constraint applications in this paper, and multiple copies for each tasks may lead to miss deadline. Besides fault tolerance and time constraint, energy consumption is also an important factor when designing reliable systems. The greater number of replicas improves the system reliability but also increases energy consumption. How to reduce energy consumption under the condition of ensuring reliability and time constraints is a challenge for us.

There are many related studies on fault-tolerant scheduling, such as [7, 8, 13, 20, 21]. Zhao et al. [21] used the DAG task graph, and considered the precedence between tasks, but the communication time between tasks was not taken into consideration. Studies [1, 2, 5, 22] made a fault-tolerance analysis on priority of tasks and could handle multiple failures. Their purpose is to reduce the schedule length (i.e., makespan) as much as possible on the basis of meeting the requirement of reliability without deadline constraint. The algorithm proposed by [19] combined reliability, time constraint, task precedence constraints and makespan together into the scheduling problem. It improved system reliability as much as possible when a given deadline was met, but energy consumption was not considered. There was a lot of energy consumption while achieving high reliability. Therefore, we consider energy consumption with achieving the goal of the maximum utilization of energy consumption while meeting the reliability and time constraints based on study [19].

This paper studies a task scheduling problem which considers several aspects: reliability, time constraint and task precedence, aiming at maximizing the utilization of energy consumption. An energy-aware fault-tolerant scheduling algorithm DRB-FTSA-E is proposed to solve the task scheduling problem under reliability and time constraints. Experimental results show that our algorithm can effectively achieve the goal of the maximum utilization of energy consumption while meeting the given constraints.

Contributions of this paper are summarized as follows:

- We propose a scheduling algorithm DRB-FTSA-E to solve the task scheduling problem under reliability and time constraints. The algorithm uses the active replication strategy to tolerant faults and an utilization of energy consumption to decide the final choice.
- Experimental simulation results show that the proposed algorithm can achieve a high utilization of energy consumption under the given system reliability and time constraint.

The remainder of this paper is organized as follows. Section 2 presents models and defines the problem studied in this paper. Section 3 proposes an energy-aware fault-tolerant scheduling algorithm. Section 4 gives the simulation results and evaluates the performance of the proposed algorithm. Section 5 summarizes this paper.

2 Models and Problem Definition

In this section, we describe the models and problem considered in this paper.

2.1 System Model

We consider a heterogeneous system with multiple processors which have different capabilities. Supposing there are M processors in a system: $p_1, p_2, p_3, \ldots, p_M$ and P represents the set of processors, we have $P = \{p_1, p_2, p_3, \ldots, p_M\}$. Every two processors can communicate with each other. B_{ij} represents the communication bandwidth from P_i to P_j, and $B_{ij} = B_{ji}$. In other words, bandwidth between a pair of processors is symmetrical. Moreover, bandwidth between different processors is different. The bandwidth is the amount of data transmitted within per unit time between processors. Communication contention is not considered here.

2.2 Task Model

We use a directed acyclic graph (DAG) $G = <V, E>$ to represent task model, where V is the set of tasks and contains N tasks $v_1, v_2, v_3, \ldots, v_N$. E is the set of edges corresponding to precedence relations between tasks, $E \subseteq V \times V$. For example, $e_{12} = (v_1, v_2)$ represents the communication relationship from tasks v_1 to v_2, and indicates that v_2 cannot be executed if v_1 is not executed completely. W is an execution time matrix of $|V| \times |P|$, where $|V|$ and $|P|$ are the number elements of sets V and P, respectively. w_{ik} represents the time overhead of task v_1 on processor p_k.

C is a communication time matrix of $|V| \times |V|$. c_{ij} represents the communication time from task v_i to task v_j. If two tasks are mapped to the same processor, the communication time is zero. $data(v_i, v_j)$ represents the amount of communication data from task v_i to v_j. If task v_i is mapped to processor p_m and v_j is mapped to process p_k, the communication time from v_i to v_j is computed by $c_{ij} = data(v_i, v_j)/B_{mk}$. In order to provide transient fault-tolerance, some tasks have a backup copy. For task $v \in V$, we use notations v to represent the primary task, and v^b to represent its backup copy. All the tasks have a common deadline, denoted by D. It is assumed that tasks are performed in a non-preemptive manner. If there exists an edge e_{ij}, it means that v_i is the predecessor of v_j and v_j is the successor of v_i. $pred(v_j)$ denotes the set of all predecessors of v_j. The task can only be executed until the data of all its predecessors arrive. $succ(v_i)$ denotes the set of all the successors of task v_i. A task without any predecessor is described as an entry task, defined as v_{entry}, and a task without any successor is an exit task, defined as v_{exit}. In this paper, we assume there is only one entry task, but there can be multiple exit tasks.

Denote $ft(v_i)$ to be the finish time of task v_i, and $Exit(G)$ to be the set of all exit tasks in a DAG G. G is completed only when all exit tasks are successfully executed. So its *makespan* is obtained by

$$makespan = \max(ft(v_{exit}), v_{exit} \in Exit(\mathrm{G})) \quad (1)$$

2.3 Fault Model

Based on the common exponential distribution assumption in the reliability research [5, 6], for every processor p_j, the arrival of failures follows a Poisson distribution with λ_j which is a positive real number and equal to the expected number of occurrence of failures in unit time t [19]. Different processors have different λ values: $\Lambda = \{\lambda_1, \lambda_2, \ldots, \lambda_M\}$. So the failure distribution in unit time t can be defined as

$$f(k, \lambda_j) = \frac{\lambda_j^k e^{-\lambda_j t}}{k!}, \tag{2}$$

where k is the number of actual failures in unit time t. The definition of reliability of task v_i is the possibility of successful execution of task v_i. That is

$$R_i = f(0, \lambda_j) = e^{-\lambda_j w_{ij}} \tag{3}$$

If the task v_i is fault-tolerant and there is a replica task, the new reliability of v_i is

$$R_i' = 1 - (1 - R_i(f))^2 \tag{4}$$

So, it can be concluded that the reliability of a system with N tasks is

$$R = \prod_{i=1}^{N} R_i. \tag{5}$$

2.4 Power Model

We consider a power model adopted in some other power management researches in view of reliability [15]. Consequently, the power consumption of a processor can be approximated by

$$P_{core} = P_s + P_d = P_s + P_{ind} + C_{ef} f^3, \tag{6}$$

Assume that the processor is in a fixed frequency when tasks are executed. The energy consumption is proportional to the total effective execution time denoted by *totExecutionTime* on all processors. Define $P_{core} = 1$ unit of energy, and the total energy consumption Q_{total} for executing all the tasks are

$$Q_{total} = totExecutionTime. \tag{7}$$

2.5 Problem Definition

The problem addressed in this article is defined as follows: Given a DAG $G = <V, E>$, a system model containing M processors $P = \{p_1, p_2, p_3, \ldots, p_M\}$, a common deadline D and a system reliability target R_{target}, the goal is to maximize the utilization of energy consumption when reliability target and time constraint are met.

3 Algorithm

In this section, we propose DRB-FTSA-E to solve our problem. Firstly we introduce some basic notations. Secondly, we present the proposed algorithm in detail.

3.1 Basic Notation

Here we give some notations and their relationships.

$est(v_i, p_k)$, the earliest execution start time of task v_i on processor p_k. If v is an entry task, $est(v_i, p_k) = 0$. Otherwise,

$$est(v_i, p_k) = \max\{drt(v_i, p_k), avail(p_k)\}, \tag{8}$$

Where $drt(v_i, p_k)$ is the time when v_i receives all data from its predecessors and $avail(p_k)$ is the instant that processor p_k becomes idle.

$$drt(v_i, p_k) = \max_{v_x \in pred(v_i)}\{\max\{(aft(v_x, pro(v_x)) + c_{xi}), \\ (baft(v_x^b, pro(v_x^b)) + c)\}\}, \tag{9}$$

where $pro(v_x)$ is the executed processor of v_i, $aft(v_x, pro(v_x))$ is the actual finish time of primary task v_x on processor $pro(v_x)$, $baft(v_x^b, pro(v_x^b))$ is the actual finish time of backup task v_x^b on processor $pro(v_x^b)$ and $pred(v_i)$ is the set of all predecessors of v_i.

$eft(v_i, p_k)$ is the earliest execution finish time of task v_i on processor p_k and can be obtained by

$$eft(v_i, p_k) = est(v_i, p_k) + w_{ik} \tag{10}$$

$et(p_k)$, the total execution time of all tasks on processor p_k, can be calculated by

$$et(p_k) = \sum_{v_i \in V_{pk}} W(v_i, pro(v_i)), pro(v_i) = p_k, \tag{11}$$

where V_{p_k} is the set of tasks executed on p_k, including backup copy. Therefore, the total effective execution time on all processors, denoted by *totExecutionTime*, can be obtained by

$$totExecutionTime = \sum_{k=1}^{M} et(p_k). \tag{12}$$

$ft(v_i)$, a task is mapped to a processor which makes the task get the minimum earliest finish time among all processors. So the finish time of task v_i that are not fault-tolerant can be obtained by

$$ft(v_i) = aft(v_i, pro(v_i)). \tag{13}$$

When a task fails, its finish time is the finish time of its backup and can be computed by

$$ft(v_i) = baft\left(v_i^b, pro\left(v_i^b\right)\right). \tag{14}$$

$ranku(v_i)$, a upward rank value, which determine the executing priority of a task v_i. If v is an exit task, its upward rank is computed by $ranku(v_{exit}) = \overline{w_{exit}}$. Otherwise,

$$ranku(v_i) = \overline{w}_i + \max_{v_j \in succ(v_i)}\left(\overline{c}_{ij} + ranku\left(v_j\right)\right), \tag{15}$$

where $\overline{w_i}$ is the average execution time of task v_i on all processors and $\overline{c_{ij}}$ is the average communication time from task v_i to v_j, They can be calculated as follows:

$$\overline{w_i} = \frac{\sum_{j=1}^{M} w_{ij}}{M}, \tag{16}$$

$$\overline{c}_{ij} = \frac{data\left(v_i, v_j\right)}{\overline{B}}, \tag{17}$$

where $\overline{B}$ is the average transfer rate among processors. It can be computed by

$$\overline{B} = \frac{\sum_{m=1}^{M} \sum_{k=1}^{M} B_{mk}}{M^2}. \tag{18}$$

We determine the priority of a task by the upward rank. If $ranku(v_1) > ranku(v_2)$, then priority of v_1 is higher than that of v_2. When the values of *ranku* of any two tasks is equal, the task with smaller index executes earlier.

$$\varepsilon_k = \left|\frac{\Delta R}{\Delta Q}\right| = \left|\frac{R_{schem(k)} - R_{schem(k-1)}}{Q_{schem(k)} - Q_{schem(k-1)}}\right|, \tag{19}$$

where $R_{schem(k)}$ is calculated by Eqs. (3) and (4), $Q_{schem(k)}$ is calculated by Eq. (7), and ε_k is utilization of energy consumption.

3.2 Deadline-Reliability-Based Fault-Tolerant Scheduling Algorithm for Energy (DRB-FTSA-E)

In this subsection, we introduce the main idea of DRB-FTSA-E. First, DRB-FTSA-E uses the algorithm *FindRlist* from [19] to find out the maximum number (denoted by K) of tasks that need to be fault-tolerant. Then, it generates $K+1$ fault-tolerant scheduling schemes when the task number k for fault tolerance varies from 0 to K, and calculates the makespan, system reliability and energy consumption of each scheme. Here, the first scheme means that the algorithm doesn't tolerate faults for any task. The second scheme means that it supports fault-tolerance for one task which has the highest priority in these k tasks, and so on. Finally, it analyzes the $K+1$ schemes to find a scheme which achieves the maximum utilization of energy consumption and meets the reliability target and time constraint.

Algorithm 1. DRB-FTSA-E

Input: A DAG $G = <V, E>$, a common deadline D, $P = \{p_1, p_2, ..., p_M\}$, target reliability R_{target}, execution time matrix W, communication time matrix C;

Output: A scheduling scheme of achieving the maximum utilization of energy consumption;

1: calculate $ranku(v_i)$ for $\forall v_i \in V$ by (15) and store all the tasks in a list $priorityList$ in the descending order of their $ranku$ values.

2: $rlist \leftarrow \varnothing, makespan \leftarrow 0$;

3: **for** $r = 0$ to $N-1$ **do**

4: $v \leftarrow priorityList[r]$; // the task with maximal $ranku(v_i)$ in $priorityList$

5: $makespan \leftarrow$ MapTasksToProcessors($v, makespan$);

6: **end for**

7: $rlist \leftarrow$ FindRlist($priorityList, D, makespan$);

8: max_fault_num $\leftarrow$ the number of tasks in $rlist$;

9: **for** $k = 0$ to max_fault_num **do**

10: $tempRlist \leftarrow \varnothing$;

11: **for** $mf = 0$ to $k-1$ **do**

12: $tempRlist[mf] \leftarrow rlist[mf]$;

13: **end for**

14: **for** $r = 0$ to $N-1$ **do**

15: $v \leftarrow priorityList[r]$;

16: $makespan \leftarrow$ MapTasksToProcessors($v, makespan$);

17: **if** $v \in tempRlist$ **then**

18: $makespan \leftarrow$ MapTasksToProcessors($v^b, makespan$);

19: **end if**

20: **end for**

21: calculate R by (5) and Q_{total} by (7);

22: the corresponding result $(k, makespan, R, Q_{total})$ is obtained;

23: **end for**

24: select the scheduling scheme which has the maximum ε_k by (19) from the scheduling schemes that make $R >= R_{target}$ as the final scheduling scheme.

Algorithm 2. MapTasksToProcessors(v_i, *makespan*)

Input: v_i, *makespan*
Output: *makespan*
1: **for** $j = 1$ to M **do**
2: calculate $eft(v_i, p_j)$ by (10);
3: **end for**
4: $pro(v_i) \leftarrow$ the processor with minimum eft of v_i;
5: $aft(v_i, pro(v_i)) \leftarrow eft(v_i, pro(v_i))$
6: **if** $makespan < aft(v_i, pro(v_i))$ **then**
7: $makespan \leftarrow aft(v_i, pro(v_i))$
8: **end if**
9: return *makespan*;

The pseudocode of DRB-FTSA-E is outlined in Algorithm 1. The **input** is a DAG $G = <V, E>$, a common deadline D, $P = \{p_1, p_2, \ldots, p_M\}$, target reliability $R_{t\arg et}$, execution time matrix W, and communication time matrix C; the **output** is a scheduling scheme of achieving the maximum utilization of energy consumption. Firstly, the *ranku* values of all tasks are calculated by Eq. (15), and tasks are stored in a list *priorityList* in the descending order of their *ranku* values (lines 1). Next, *rlist* and *makespan* are initialized (line 2). *rlist* is used to store the tasks that can be backed up. Then, DRB-FTSA-E maps all tasks to appropriate processors in pre-allocated way and calculates the *makespan* by Algorithm 2 (lines 3–6). Lines 7–8 use the algorithm *FindRlist* from [19] to find out which tasks need fault-tolerance and get the maximum number K of tasks which can be fault-tolerant. DRB-FTSA-E uses a **for** loop to obtain $K + 1$ scheduling schemes when the task number K for fault tolerance varies from 0 to K, and calculates *makespan*, system reliability R and energy consumption Q_{total} of each scheme. (lines 9–23). Finally, DRB-FTSA-E picks out the scheduling scheme which has the maximum utilization of energy consumption from all the scheduling schemes that satisfy time constraint and $R \geq R_{target}$ as the final scheduling scheme (line 24).

4 Experiments

4.1 Experimental Setup

To evaluate the performance of our proposed algorithm, we conduct extensive experiments which are carried out through simulation programmed by Java programming language running on Microsoft Windows 7 with an Intel i3 processor, 2.40 GHz CPU and 4 GB RAM. For the sake of comparison, we use the following algorithms HEFT [17], FTSA [1] and DB-FTSA [19] as baselines. HEFT is a famous scheduling algorithm to reduce makespan. FTSA is a fault-tolerant scheduling algorithm where all tasks are fault-tolerant once. DB-FTSA is a fault-tolerant scheduling algorithm with time constraints and tolerates faults with the maximum number while time constraint is satisfied. We evaluate the algorithms in terms of makespan, system reliability, energy consumption and utilization of energy consumption ε_k. In order to gain more reliability with the largest utilization of energy consumption, we choose the scheme that provides the largest system reliability increasing per unit energy consumption. That is, we should select the scheme with maximum ε_k.

4.2 Experience Results and Analysis

In this subsection, we present our simulation results to evaluate the performance of our proposed algorithm. We carried out more than 100 sets of experiments on three sets P_1, P_2 and P_3 of different processors. It is assumed that our reliability target is $R_{target} = 90\%$. Here we pick out several representative data to show the difference

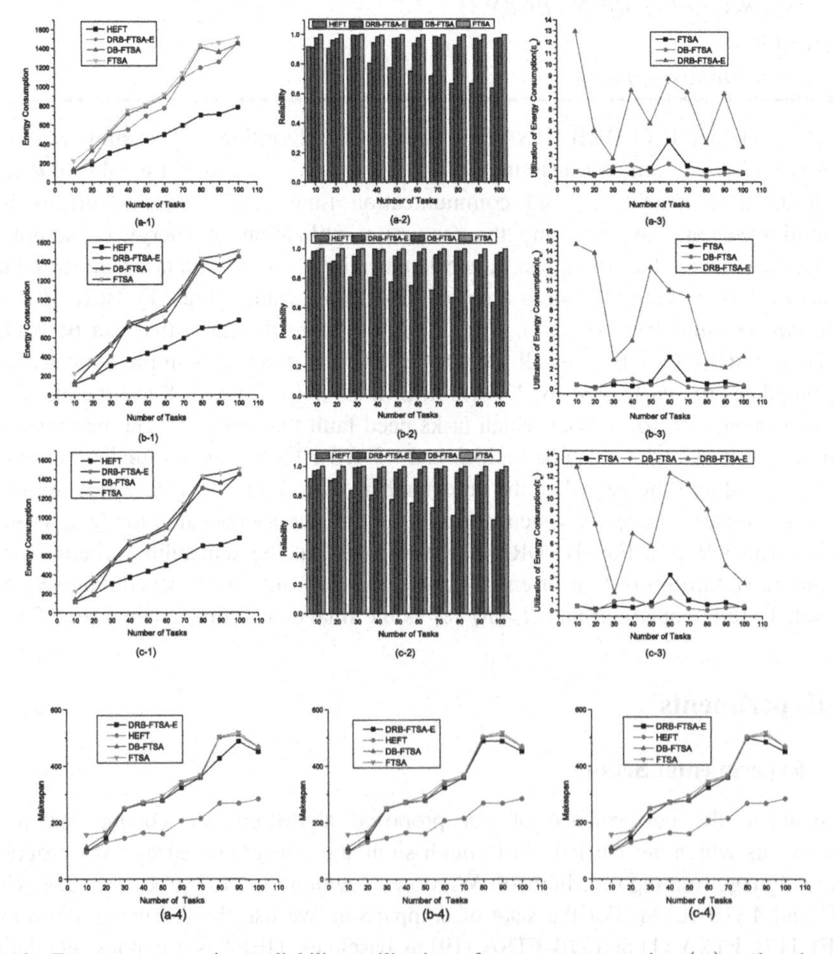

Fig. 1. Energy consumption, reliability, utilization of energy consumption (ε_k) and makespan with the number of tasks increasing between multiple algorithms on three sets of processors. Results in the first, second and third rows are based on P_1, P_2, and P_3.

between several comparison algorithms in some respects including energy consumption, reliability and utilization of energy consumption (ε_k) in Fig. 1.

Figure 1(a-1)–(c-1) show that three sets of energy consumption from FTSA, DB-FTSA and DRB-FTSA-E, and they are roughly the same. In most cases, the energy consumption obtained by DRB-FTSA-E is lower than that of obtained by FTSA and DB-FTSA. Of course the energy consumption of HEFT is always lowest. Figure 1(a-2)–(c-2) show that DRB-FTSA-E can obtain higher system reliability while reliability target is obtained, and which is far more than the reliability target. Figure 1(a-4)–(c-4) shows the makespan in different algorithms. Figure 1(a-3)–(c-3) show that the utilization of energy consumption obtained by FTSA, DB-FTSA and DRB-FTSA-E, where the result from DRB-FTSA-E is always higher than that of the other two algorithms, while the energy consumption and system reliability are roughly the same. That is to say, our algorithm DRB-FTSA-E can get the maximum utilization of energy consumption.

5 Conclusion

In this paper, we propose an algorithm DRB-FTSA-E to achieve the goal of the maximum utilization of energy consumption while meeting the time constraint and reliability target. An active replication scheme in the algorithm is adopted to achieve transient fault-tolerance, which probably leads to an increase in the schedule length. With a time constraint, not all tasks can be backed up. DRB-FTSA-E supports fault-tolerance for parts of tasks and chooses the scheme which achieves the maximum energy utilization from serval scheduling solutions. The experimental results indicate that DRB-FTSA-E efficiently improves the utilization of energy consumption and is suitable for fault-tolerant scheduling.

Acknowledgment. The authors would like to express their sincere gratitude to the editors and the referees. This work was supported by the National Natural Science Foundation of China (Grant Nos. 61602350, 61602349), the Open Foundation of Hubei Province Key Laboratory of Intelligent Information Processing and Real-time Industrial System (2016znss26C).

References

1. Benoit, A., Hakem, M., Robert, Y.: Fault tolerant scheduling of precedence task graphs on heterogeneous platforms. In: IEEE International Symposium on Parallel and Distributed Processing, pp. 1–8 (2008)
2. Broberg, J., Ståhl, P.: Dynamic fault tolerance and task scheduling in distributed systems (2016)
3. Cui, X.T., Wu, K.J., Wei, T.Q., Sha, H.M.: Worst-case finish time analysis for dag-based applications in the presence of transient faults. J. Comput. Sci. Technol. **31**(2), 267–283 (2016)
4. Deng, F., Tian, Y., Zhu, R., Chen, Z.: Fault-tolerant approach for modular multilevel converters under submodule faults. IEEE Trans. Ind. Electron. **63**(11), 7253–7263 (2016)

5. Girault, A., Kalla, H., Sighireanu, M., Sorel, Y.: An algorithm for automatically obtaining distributed and fault-tolerant static schedules. In: 2003 Proceedings of the International Conference on Dependable Systems and Networks, pp. 159–168 (2006)
6. Guo, H., Wang, Z.G., Zhou, J.L.: Load balancing based process scheduling with fault-tolerance in heterogeneous distributed system. Chin. J. Comput. **28**(11), 1807–1816 (2005)
7. Guo, Y., Zhu, D., Aydin, H.: Generalized standby-sparing techniques for energy-efficient fault tolerance in multiprocessor real-time systems. In: IEEE International Conference on Embedded and Real-Time Computing Systems and Applications, pp. 62–71 (2013)
8. Guo, Y., Zhu, D., Aydin, H., Yang, L.T., Member, S., Antonio, S.: Energy-efficient scheduling of primary/backup tasks in multiprocessor real-time systems (extended version) (2013)
9. Haque, M.A., Aydin, H., Zhu, D.: On reliability management of energy-aware real-time systems through task replication. IEEE Trans. Parallel Distrib. Syst. **28**(3), 813–825 (2017)
10. Iyer, R.K.: Measurement and modeling of computer reliability as affected by system activity. ACM Trans. Comput. Syst. **4**(3), 214–237 (1986)
11. Levitin, G., Xing, L., Dai, Y.: Optimizing dynamic performance of multistate systems with heterogeneous 1-out-of-n warm standby components. IEEE Trans. Syst. Man Cybern. Syst. **PP**(99), 1–10 (2016)
12. Liu, J., Wang, S., Zhou, A., Kumar, S., Yang, F., Buyya, R.: Using proactive fault-tolerance approach to enhance cloud service reliability. IEEE Trans. Cloud Comput. **PP**(99), 1 (2016)
13. Luo, W., Yang, F., Pang, L., Qin, X.: Fault-tolerant scheduling based on periodic tasks for heterogeneous systems. In: Yang, L.T., Jin, H., Ma, J., Ungerer, T. (eds.) ATC 2006. LNCS, vol. 4158, pp. 571–580. Springer, Heidelberg (2006). https://doi.org/10.1007/11839569_56
14. Song, Y.D., Yuan, X.: Low-cost adaptive fault-tolerant approach for semi-active suspension control of high speed trains. IEEE Trans. Ind. Electron. **PP**(99), 1 (2016)
15. Sridharan, R., Mahapatra, R.: Reliability aware power management for dual-processor real-time embedded systems. In: Design Automation Conference, pp. 819–824 (2010)
16. Tabbaa, N., Entezari-Maleki, R., Movaghar, A.: A fault tolerant scheduling algorithm for dag applications in cluster environments. Commun. Comput. Inf. Sci. **188**, 189–199 (2011)
17. Topcuouglu, H., Hariri, S., Wu, M.Y.: Performance-effective and low-complexity task scheduling for heterogeneous computing. IEEE Trans. Parallel Distrib. Syst. **13**(3), 260–274 (2002)
18. Treaster, M.: A survey of fault-tolerance and fault-recovery techniques in parallel systems. ACM Computing Research Repository (CoRR 501002, 1–11) (2005)
19. Wei, M., Liu, J., Li, T., Xu, X., Hu, W., Zhao, D.: Fault-tolerant scheduling of real-time tasks on heterogeneous systems. In: 2017 12th IEEE Conference on Industrial Electronics and Applications (ICIEA), pp. 1006–1011. IEEE (2017)
20. Xie, G.Q., Ren-Fa, L.I., Liu, L., Yang, F.: Dag reliability model and fault-tolerant algorithm for heterogeneous distributed systems. Chin. J. Comput. **36**(10), 2019–2032 (2013)
21. Zhao, B., Aydin, H., Zhu, D.: Shared recovery for energy efficiency and reliability enhancements in real-time applications with precedence constraints. ACM Trans. Des. Autom. Electron. Syst. **18**(2), 1–21 (2013)
22. Zhao, L., Ren, Y., Yang, X., Sakurai, K.: Fault-tolerant scheduling with dynamic number of replicas in heterogeneous systems. In: IEEE International Conference on High Performance Computing and Communications, pp. 434–441 (2011)
23. Zhu, D., Aydin, H.: Reliability-aware energy management for periodic real-time tasks. In: IEEE Real Time and Embedded Technology and Applications Symposium, pp. 225–235 (2007)

A Novel Many-Objective Optimization Algorithm Based on the Hybrid Angle-Encouragement Decomposition

Yuchao Su, Jia Wang(✉), Lijia Ma, Xiaozhou Wang, Qiuzhen Lin, and Jianyong Chen

College of Computer Science and Software Engineering, Shenzhen University, Shenzhen, People's Republic of China
Jia.wang@szu.edu.cn

Abstract. For many-objective optimization problems (MaOPs), the problem of balancing the convergence and the diversity during the search process is often encountered but very challenging due to its vast range of searching objective space. To solve the above problem, we propose a novel many-objective evolutionary algorithm based on the hybrid angle-encouragement decomposition (MOEA/AD-EBI). The proposed MOEA/AD-EBI combines two types of decomposition approaches, i.e., the angle-based decomposition and the encouragement-based boundary intersection decomposition. By coordinating the above two decomposition approaches, MOEA/AD-EBI is expected to effectively achieve a good balance between the convergence and the diversity when solving various kinds of MaOPs. Extensive experiments on some well-known benchmark problems validate the superiority of MOEA/AD-EBI over some state-of-the-art many-objective evolutionary algorithms.

Keywords: MaOPs · MOEA · Hybrid angle-encouragement decomposition

1 Introduction

Multi-objective evolutionary algorithms (MOEAs) have become the popular and effective approaches for tackling multi-objective optimization problems (MOPs). However, due to the curse of dimensionality in many-objective optimization problems (MaOPs), their performance will deteriorate significantly with the increase of optimization objectives [1]. To solve the above challenges in MOEAs, this paper presents a hybrid angle-encouragement decomposition method to better solve MaOPs, which is called MOEA/AD-EBI. The proposed hybrid approach includes two types of decomposition, i.e., an angle-based decomposition approach and an encouragement-based boundary intersection (EBI) decomposition approach. The angle-based one tries to select appropriate solutions from the feasible region of each subproblem, which helps to well maintain the diversity. The encouragement-based one is used when the subproblem is not associated to any near solution, which aims to ensure the convergence without considering the diversity. These two decomposition approaches are hybridized to better solve MaOPs. When compared to several competitive MaOEAs, the

D.-S. Huang et al. (Eds.): ICIC 2018, LNAI 10956, pp. 47–53, 2018.
https://doi.org/10.1007/978-3-319-95957-3_6

experiments have validated the superior performance of MOEA/AD-EBI when solving the WFG [2] and DTLZ [3] test problems with 4, 6, 8, and 10 objectives.

The rest of this paper is organized as follows. The details of MOEA/AD-EBI are given in Sect. 2, while the experimental results are provided in Sect. 3. Finally, the conclusions and future work are presented in Sect. 4.

2 The Proposed MOEA/AD-EBI

2.1 Angle-Based Decomposition Approach

It is difficult to balance the convergence and the diversity when solving MaOPs. If the convergence is emphasized, the algorithm may easily fall into local optimal. However, when the diversity is preferred, the convergence speed may be decreased. In our approach, the angle-based decomposition approach is used to guarantee the diversity first, which selects the solutions from the feasible region of each subproblem in objective space. The angle-based decomposition model is defined by

$$\begin{gathered}\arg\min_{x\in\Omega^k}(g(x|w^k, z^*)) \\ \text{where } \Omega^k = \{x|\arg\min_{w\in W}(\text{angle}(F(x), w, z^*)) = w^k\} \\ \text{angle}(F(x), w, z^*) = \arccos\left|(\textstyle\sum_{i=1}^{m}(f_i(x) - z_i^*)\cdot w_i^k)/(\sqrt{\sum_{i=1}^{m}(f_i(x) - z_i^*)^2}\cdot\sqrt{\sum_{i=1}^{m}(w_i^k)^2})\right|\end{gathered} \tag{1}$$

where x is a decision vector, N is the population size, $W = \{w^1, w^2, \ldots, w^N\}$ are the weight vectors and $k = 1,\ldots,N$, $z^* = (z_1^*, z_2^*, \ldots, z_m^*)$ is an ideal point in objective space and $i = 1,\ldots,m$, m indicates the number of objectives. $F(x)$ is a set of optimization objectives, each of which is represented by $f_i(x)$ ($m > 3$). Apparently, Eq. (1) defines the feasible region Ω^k for kth subproblem. Then, the pseudo code of running this angle-based decomposition approach is provided in Algorithm 1 with the input $\boldsymbol{Q}$ (a solution set formed by the current population and their offspring), where $\boldsymbol{R}$ is used to preserve the weight vectors (subproblems) that are not associated to any solution and $\boldsymbol{S}$ is adopted to store the solutions associated to the weight vectors (subproblems). In line 1 of Algorithm 1, $\boldsymbol{R}$ and $\boldsymbol{S}$ are all set as an empty set. For each subproblem in line 2, if its feasible region has the solutions, the one with the best aggregated value using Eq. (1) will be added into $\boldsymbol{S}$ in line 4. Otherwise, the weight vectors will be added into $\boldsymbol{R}$ in line 6. At last, these two sets ($\boldsymbol{R}$ and $\boldsymbol{S}$) are returned in line 9.

Algorithm 1. AD(Q)	
1	$\boldsymbol{R}=\varnothing, \boldsymbol{S}=\varnothing$;
2	**for** i=1 to N
3	**if** $\Omega^i \neq \varnothing$ // these sets are obtained by Eq. (1)
4	$\boldsymbol{S}=\boldsymbol{S} \cup \arg\min(g(x \mid w^i, z^*))$; // $x \in \Omega^i$
5	**else**
6	$\boldsymbol{R}=\boldsymbol{R} \cup w^i$;
7	**end if**
8	**end for**
9	**return** [$\boldsymbol{R}$, $\boldsymbol{S}$];

2.2 The EBI Decomposition Approach

When some subproblems which are not associated with solutions using the angle-based decomposition, the EBI decomposition is further used to select the solutions from the current population and their offspring without considering the diversity. Using this approach, the solutions in boundaries of population are preferred to extend the entire true PF. This EBI approach is defined by

$$\begin{gathered}\arg\min_{x\in\Omega}(g^{ebi}(x|w,z^*)) = \arg\min_{x\in\Omega}(d_1^x - \theta d_2^x) \\ \text{where } d_1^x = \frac{\|(F(x)-z^*)w\|}{\|w\|} \quad d_2^x = \left\|F(x) - (z^* + d_1^x \frac{w}{\|w\|})\right\| \quad \theta \geq 0\end{gathered} \tag{2}$$

where d_1^x is the distance of the idea point z^* and the foot point from $F(x)$ to the weight vector w, d_2^x is the distance of $F(x)$ and the weight vector w. However, different from the original PBI approach, the EBI approach in Eq. (2) changes the sign of θ and redefines Ω, which consists of the current population and their offspring. The EBI approach is employed for the subproblems that are not associated with any solution in their feasible regions. Under this case, the EBI approach aims to accelerate the convergence speed or to find the solutions for exploring the boundaries of current population. To further clarify the running of EBI, its pseudo code is given in Algorithm 2, with the input $\boldsymbol{Q}$ (a solution set) and $\boldsymbol{R}$ (a set of weight vectors that are not associated to solutions returned by Algorithm 1). Algorithm 2 is only a supplemental method for Algorithm 1, as it is difficult for the angle-based decomposition approach to always work well in the whole evolutionary process (*i.e.*, some subproblems are often not associated in the early evolutionary stage due to the crowded population). In this case, the EBI approach can be used for solution association, so as to speed up the convergence and to explore the boundaries of current population. In line 1 of Algorithm 2, the set $\boldsymbol{S}$ as an empty set. For each subproblem in line 2, select the ith subproblem from $\boldsymbol{R}$ in line 3 and then use Eq. (2) to select a solution from $\boldsymbol{Q}$ to associate with this subproblem, which is added into $\boldsymbol{S}$ and removed from $\boldsymbol{Q}$ in line 4. At last, return the set $\boldsymbol{S}$ with the associated solutions for $\boldsymbol{R}$ in line 6.

Algorithm 2. EBI(Q, R)

1	$S=\varnothing$;
2	**for** i=1 to $\|R\|$
3	Select the ith subproblem from R;
4	Use Eq. (2) to select a solution from Q to associate with the subproblem, which is added into S and removed from Q;
5	**end for**
6	**return** S;

2.3 The Proposed MOEA/AD-EBI Algorithm

Since the angle-based and EBI decomposition approaches have been introduced as above, we further give the pseudo code of MOEA/AD-EBI in Algorithm 3, where *FEs* and *maxFEs* respectively indicate the counter for current function evaluations and the number of maximal function evaluations, T is the number of neighbors for subproblems, G and $maxG$ are respectively the current generation and the maximal generation. In line 1 of Algorithm 3, the weight vectors are initialized, the evolutionary population P is randomly generated, and all the individuals are evaluated to update the idea point z^*. In line 2, if *FEs* is smaller than *maxFEs*, the following evolutionary process is executed. In line 3, the T value is decreased with the increasing number of generations, which is set to the maximal value of $|P| \times (1 - G/maxG)$ and 3 (please note that the constraint 3 is to ensure the minimal number of neighbors). The offspring set Q is initialized as an empty set. In lines 4–8, all the individuals are evolved using the simulated binary crossover (SBX) [4] and polynomial mutation (PM) [4]. It's notable that the mating pool in MOEA/AD-EBI is only composed of the T neighboring solutions. The new offspring are preserved in Q and evaluated to update the idea point z^*. In line 9, the parent population P and the offspring population Q are combined into Q. Then, the angle-based decomposition approach (**AD**) is adopted in line 10 to select solutions for each subproblem and R is the set of weight vectors that have no solution in their feasible regions. The running of **AD** (Algorithm 1) has been introduced in Sect. 2.1. When $|P| < N$ in line 11, R is not empty as the angle-based decomposition approach only selects one solution for each subproblem. In line 12, the EBI function (Algorithm 2) is further used to select appropriate solutions from the un-associated subproblems to ensure convergence. At last, the final population P is returned as the approximation set.

Algorithm 3. MOEA/AD-EBI

1	Initialize weight vectors and population ***P***, evaluate each individual and update z^*;
2	**while** $FEs<maxFEs$
3	$T=\max(\lvert \boldsymbol{P}\rvert \times (1-G/maxG), 3)$ and $\boldsymbol{Q} = \varnothing$;
4	**for** $i=1$ to N
5	Select one solution from the T neighbor solutions;
6	Generate one offspring by the solution associated to the i-th subproblem and its neighbors using SBX [4] and PM [4];
7	Evaluate each offspring, add it into ***Q***, and update z^*;
8	**end for**
9	$\boldsymbol{Q}=\boldsymbol{P}\cup \boldsymbol{Q}$;
10	$[\boldsymbol{P}, \boldsymbol{R}]$=AD(***Q***); // **Algorithm 1**
11	**if** $\lvert \boldsymbol{P}\rvert<N$
12	$\boldsymbol{P}=\boldsymbol{P}\cup$ EBI(***Q***, ***R***); // **Algorithm 2**
13	**end if**
14	**end while**
15	**return** ***P***;

3 Experimental Studies

In this section, the performance of MOEA/AD-EBI is presented and compared with other seven MaOEAs on solving the WFG [2] and DTLZ [3] benchmark problems.

All the comparison results based on the HV metric are summarized in Table 1 for all the *DTLZ* and *WFG* test problems, in which the indexes "*better/similar/worse (WFG)*", "*better/similar/worse(DTLZ)*" and "*better/similar/worse*" indicate that MOEA/AD-EBI respectively performs better than, similarly to, and worse than that of the compared algorithms using Wilcoxon's rank sum test with $\alpha = 0.05$ on the *WFG*, *DTLZ1-DTLZ4* and all test problems, respectively. The indexes "*best/all*" in the last row indicates the proportion that the corresponding algorithm achieves the best results among all the compared algorithms when solving all the 52 cases (*i.e.*, 9 *WFG* and 4 *DTLZ* test problems with 4, 6, 8, and 10 objectives). The summary results in Table 1 show that NSGA-III, MOEA/DD and SRA obtain the very promising performance when solving *DTLZ1-DTLZ4*. However, they perform poorly on most of *WFG* test

Table 1. All the comparison summary on *DTLZ1-DTLZ4* and *WFG1-WFG9*

	NSGA-III	MOEA/DD	Two_Arch2	SRA	VaEA	R&D	CSS	AD-EBI
better/similar/worse(WFG)	30/5/1	34/1/1	27/4/5	31/4/1	32/2/2	36/0/0	33/1/2	
better/similar/worse(DTLZ)	6/9/1	2/13/1	16/0/0	9/8/1	16/0/0	16/0/0	16/0/0	
better/similar/worse	36/14/2	36/14/2	43/4/5	38/12/2	48/2/2	52/0/0	49/1/2	
best/all	12/52	15/52	6/52	12/52	1/52	0/52	1/52	43/52

problems. Obviously, MOEA/AD-EBI outperforms all the compared algorithms on most of test problems and can well solve the *WFG* and *DTLZ1-DTLZ4* test problems simultaneously.

4 Conclusions and Future Work

In this paper, we have presented the hybrid angle-encouragement decomposition approach, which is embedded into a general framework of MOEAs. The proposed algorithm includes the angle-based decomposition and the EBI decomposition approaches. The former one defines the feasible region for each subproblem and only selects the solutions in this area for association, which helps to maintain the diversity first and also consider the convergence when several solutions are under the feasible region. When the subproblems are not associated to any solution using the angle-based decomposition method, the latter one (EBI) is executed to only consider the convergence and to extend the boundaries of current population, as there are no solutions under the feasible region. When compared to several competitive MaOEAs (i.e., NSGA-III [6], MOEA/DD [7], Two_Arch2 [8], SRA [5], VaEA [9], MaOEAR&D [10], and MaOEA-CSS) [11], the experiments confirm the superiority of MOEA/AD-EBI when solving different types of MaOPs.

Our future work will further study the performance of our proposed hybrid decomposition approach in other decomposition-based MOEAs. Furthermore, the extension of MOEA/AD-EBI to solve some real-world applications will also be studied.

References

1. Hughes, E.: Evolutionary many-objective optimization: many once or one many? In: Proceedings of the 2005 IEEE Congress on Evolutionary Computation, Edinburgh, UK, pp. 222–227 (2005)
2. Huband, S., Barone, L., While, L., Hingston, P.: A scalable multi-objective test problem toolkit. In: Coello Coello, C.A., Hernández Aguirre, A., Zitzler, E. (eds.) EMO 2005. LNCS, vol. 3410, pp. 280–295. Springer, Heidelberg (2005). https://doi.org/10.1007/978-3-540-31880-4_20
3. Deb, K., Thiele, L., Laumanns, M., Zitzler, E.: Scalable test problems for evolutionary multi-objective optimization. In: Abraham, A., Jain, L., Goldberg, R. (eds.) Evolutionary Multiobjective Optimization. AI&KP, pp. 105–145. Springer, London (2005). https://doi.org/10.1007/1-84628-137-7_6
4. Deb, K., Agrawal, R.B.: Simulated binary crossover for continuous search space. Complex Syst. **9**, 115–148 (1995)
5. Li, B., Tang, K., Li, J., Yao, X.: Stochastic ranking algorithm for many-objective optimization based on multiple indicators. IEEE Trans. Evol. Comput. **20**(6), 924–938 (2016)
6. Deb, K., Jain, H.: An evolutionary many-objective optimization algorithm using reference-point based non-dominated sorting approach, part I: solving problems with box constraints. IEEE Trans. Evol. Comput. **18**(4), 577–601 (2014)

7. Li, K., Deb, K., Zhang, Q.F., Kwong, S.: An evolutionary many-objective optimization algorithm based on dominance and decomposition. IEEE Trans. Evol. Comput. **19**(5), 694–716 (2015)
8. Wang, H., Jiao, L., Yao, X.: Two_Arch2: an improved two-archive algorithm for many-objective optimization. IEEE Trans. Evol. Comput. **19**(4), 524–541 (2015)
9. Xiang, Y., Zhou, Y.R., Li, M.Q.: A vector angle-based evolutionary algorithm for unconstrained many-objective optimization. IEEE Trans. Evol. Comput. **21**(1), 131–152 (2017)
10. He, Z., Yen, G.: Many-objective evolutionary algorithm: object space reduction and diversity improvement. IEEE Trans. Evol. Comput. **20**(1), 145–160 (2016)
11. Tian, Y., Zhang, X., Jin, Y.: A knee point driven evolutionary algorithm for many-objective optimization. IEEE Trans. Evol. Comput. **19**(6), 761–776 (2015)

Solving Bi-criteria Maximum Diversity Problem with Multi-objective Multi-level Algorithm

Li-Yuan Xue[1], Rong-Qiang Zeng[2,3(✉)], Hai-Yun Xu[3], and Yi Wen[3]

[1] EHF Key Laboratory of Science, School of Electronic Engineering, University of Electronic Science and Technology of China, Chengdu 611731, Sichuan, People's Republic of China
xuely2012@gmail.com

[2] School of Mathematics, Southwest Jiaotong University, Chengdu 611731, Sichuan, People's Republic of China
zrq@swjtu.edu.cn

[3] Chengdu Library and Information Center, Chinese Academy of Sciences, Chengdu 610041, Sichuan, People's Republic of China
{xuhy, wenyi}@clas.ac.cn

Abstract. The multi-level paradigm is a simple and useful approach to tackle a number of combinatorial optimization problems. In this paper, we investigate a multi-objective multi-level algorithm to solve the bi-criteria maximum diversity problem. The computational results indicate that the proposed algorithm is very competitive in comparison with the original multi-objective optimization algorithms.

Keywords: Bi-objective optimization · Hypervolume contribution Indicator · Multi-level approach · Local search · Maximum diversity problem

1 Introduction

The maximum diversity problem is to identify a subset M of m elements from a set N with n elements, which have many practical applications such as machine scheduling [1], facility location [10], traffic management [12], product design [15], and so on. In our work, we concentrate on solving bi-criteria maximum diversity problem, which is formalized as follows [13, 16]:

$$Maximize \quad f_k(x) = \frac{1}{2}\sum_{i=1}^{n}\sum_{j=1}^{m} d_{ij}^{k} x_i x_j \tag{1}$$

$$Subject\ to \quad \sum_{i=1}^{n} x_i = m \tag{2}$$

where $D^k = (d_{ij}^k)$ is an $n \times n$ matrix of constants and x is an n-vector of binary (zero-one) variables, i.e., $x_i \in \{0, 1\}$ $(i = 1, \ldots, n)$, $k = 1, 2$.

D.-S. Huang et al. (Eds.): ICIC 2018, LNAI 10956, pp. 54–62, 2018.
https://doi.org/10.1007/978-3-319-95957-3_7

Due to its extensive range of applications, a large number of heuristic and meta-heuristic algorithms have been proposed to deal with this problem in the literature, such as scatter search [2], variable neighborhood search [8], hybrid evolutionary algorithm [11] tabu search [16], etc. In this paper, we integrate the multi-level approach into the hypervolume-based optimization to solve the bi-criteria maximum diversity problem. The computational results show that the proposed algorithm is very promising.

The remaining part of this paper is organized as follows. In the next section, we present the basic notations and definitions of bi-objective optimization. Then, we investigate the details of the proposed multi-objective multi-level algorithm, which is used to solve the bi-criteria maximum diversity problem in Sect. 3. In Sect. 4, the experimental results are reported, and the concluding remarks are given in the last section.

2 Bi-objective Optimization

Without loss of generality, let X be the search space of the optimization problem under consideration and $Z = \Re^2$ be the corresponding objective space with a maximizing vector function $Z = f(X)$. Then, the dominance relations between two solutions x_1 and x_2 are presented below [9]:

Definition 1 *(Pareto Dominance). A decision vector x_1 is said to dominate another decision vector x_2 (written as $x_1 \succ x_2$), if $f_i(x_1) \geq f_i(x_2)$ for all $i \in \{1,2\}$ and $f_j(x_1) > f_j(x_2)$ for at least one $j \in \{1,2\}$.*

Definition 2 *(Pareto Optimal Solution). $x \in X$ is said to be Pareto optimal if and only if there does not exist another solution $x' \in X$ such that $x' \succ x$.*

Definition 3 *(Non-Dominated Solution). $x \in S$ $(S \subset X)$ is said to be non-dominated if and only if there does not exist another solution $x' \in S$ such that $x' \succ x$.*

Definition 4 *(Pareto Optimal Set). S is said to be a Pareto optimal set if and only if S is composed of all the Pareto optimal solutions.*

Definition 5 *(Non-Dominated Set). S is said to be a non-dominated set if and only if any two solutions $x_1 \in S$ and $x_2 \in S$ such that $x_1 \not\succ x_2$ and $x_2 \not\succ x_1$.*

In fact, we aim to find the Pareto optimal set, which keeps the best compromise between two objectives. However, it is very difficult or even impossible to obtain the Pareto optimal set in a reasonable time for the NP-hard problems. Therefore, we try to obtain a non-dominated set which is as close to the Pareto optimal set as possible.

3 Multi-objective Multi-level Algorithm

The multi-level approach is integrated into the hypervolume-based optimization method to tackle the bi-criteria maximum diversity problem. The general scheme of Multi-Objective Multi-Level Algorithm (MOMLA) is presented in Algorithm 1, in which the main steps are detailed in the following subsections.

Algorithm 1. Multi-Objective Multi-Level Algorithm

Input: N (Population size)
Output: A: (Pareto approximation set)
Step 1 - Initialization: $P \leftarrow N$ randomly generated solutions
Step 2: $A \leftarrow \Phi$
Step 3 - Fitness Assignment: Assign a fitness value for each solution $x \in P$
Step 4:
while Running time is not reached **do**
  repeat
    Hypervolume-Based Local Search: $x \in P$
    1) $x^* \leftarrow$ one randomly chosen unexplored neighbors of x
    2) $P \leftarrow P \bigcup x^*$
    3) compute x^* fitness: $HC(x^*, P)$
    4) update all $z \in P$ fitness values
    5) $\omega \leftarrow$ worst solution in P
    6) $P \leftarrow P \backslash \{w\}$
    7) update all $z \in P$ fitness values
    8) **if** $w \neq x^*$, Progress $\leftarrow$ True
  until all neighbors are explored or Progress = True
    9) $A \leftarrow$ Non-dominated solutions of $A \bigcup P$
  Multi-Level Approach:
  while $D_i > D * r$ **do**
    repeat
      1) $D_i \leftarrow D$
      2) $Num \leftarrow 0$
      3) **for** $i \leftarrow 1, \ldots, D_i$ **do**
        a) $Num \leftarrow Num + 1$ if and only if the ith variable of $x_k = 0$ (or 1)
        b) $D_i \leftarrow D - Num$
        c) Coarsening level (G)
        d) Hypervolume-Based Local Search: $x'_k \in P$, $k = 1, \cdots, N$
        end for
  end while
end while
Step 5: Return A

In MOMLA, all the solutions in an initial population are randomly generated. Specifically, we randomly select the variables of each solution according to the cardinality of the subset M, and the selected variables are assigned a value 1, while the remain variables are assigned a value 0. Then, each solution in the population is optimized by the hypervolume-based local search procedure. Afterwards, we employ the multi-level approach to further improve the entire population to obtain a high-quality Pareto approximation set at last.

3.1 Hypervolume-Based Local Search

After the initialization, we use the Hypervolume Contribution (*HC*) indicator defined in [4] to realize the fitness assignment for each solution in the population. Based on the dominance relation and two objective function values, the *HC* indicator calculates the hypervolume contribution for each solution by dividing the whole population into the non-dominated set and the dominated set.

In the hypervolume-based local search procedure, an unexplored neighbor solution x^* of x is generated by implementing the two-flip move based neighborhood, and the fitness value of x^* is calculated by the *HC* indicator. With respect to the fitness values, the worst solution ω is chosen and deleted from the population P. Then, the hypervolume-based local search procedure will repeat until the termination criterion is satisfied so as to obtain a set of efficient solutions.

3.2 Multi-level Approach

The multi-level approach is known to be highly effective to deal with the graph partitioning problems, especially handling the large instances [5, 6]. In our work, we aim to further optimize the quality of the entire population by the multi-level approach.

In the multi-level approach, we simplify the original graph with the size D by counting the exact number *Num* of the variables with same value for all the solutions in the population and coarsening the current graph down to a smaller one. After deleting all the variables with same value, we obtain a sub-graph with the size $D - Num$ from the original one.

Afterwards, each solution x_i is represented as x_i', which is applied to hypervolume-based local search for further improvements. The whole procedure will repeat until the size D_i of the current graph is no bigger than $D * r$, where r is a pre-defined number used to control the levels in Algorithm 1. In our case, we set this number r to 0.3.

4 Experimental Results

In this section, we provide the computational results of three multi-objective optimization algorithms on 18 groups of the benchmark instances of bi-criteria maximum diversity problem. All the algorithms are programmed in C++ and compiled using Dev-C++ 5.0 compiler on a PC running Windows 7 with Core 2.50 GHz CPU and 4 GB RAM.

4.1 Parameters Settings

In order to carry out the experiments on the bi-criteria maximum diversity problem, we respectively use two single-objective benchmark instances of max-cut problem with the same size provided in [7][1] and two single-objective benchmark instances of

[1] More information about the benchmark instances of max-cut problem can be found on this website: http://www.stanford.edu/~yyye/yyye/Gset/.

unconstrained binary quadratic programming problem with the same size provided in [14][2], to generate one bi-criteria maximum diversity problem instance.

The instances generated by the benchmark instances of max-cut problem with the size from 800 to 2000 are presented in Table 1 below.

Table 1. Single-objective benchmark instances of max-cut problem used for generating bi-criteria maximum diversity problem instances.

	Size	Instance 1	Instance 2	Subset *m*
bo_mdp_800_01	800	g1.rud	g2.rud	0.3n
bo_mdp_800_02	800	g1.rud	g2.rud	0.5n
bo_mdp_800_03	800	g1.rud	g2.rud	0.7n
bo_mdp_1000_01	1000	g53.rud	g54.rud	0.3n
bo_mdp_1000_02	1000	g53.rud	g54.rud	0.5n
bo_mdp_1000_03	1000	g53.rud	g54.rud	0.7n
bo_mdp_2000_01	2000	g26.rud	g27.rud	0.3n
bo_mdp_2000_02	2000	g26.rud	g27.rud	0.5n
bo_mdp_2000_03	2000	g26.rud	g27.rud	0.7n

The instances generated by the benchmark instances of unconstrained binary quadratic programming problem with the size from 3000 to 5000 are presented in Table 2 below.

Table 2. Single-objective benchmark instances of unconstrained binary quadratic programming problem used for generating bi-criteria maximum diversity problem instances.

	Size	Instance 1	Instance 2	Subset *m*
bo_mdp_3000_01	3000	P3000.1	P3000.2	0.3n
bo_mdp_3000_02	3000	P3000.1	P3000.2	0.5n
bo_mdp_3000_03	3000	P3000.1	P3000.2	0.7n
bo_mdp_4000_01	4000	P4000.1	P4000.2	0.3n
bo_mdp_4000_02	4000	P4000.1	P4000.2	0.5n
bo_mdp_4000_03	4000	P4000.1	P4000.2	0.7n
bo_mdp_5000_01	5000	P5000.1	P5000.2	0.3n
bo_mdp_5000_02	5000	P5000.1	P5000.2	0.5n
bo_mdp_5000_03	5000	P5000.1	P5000.2	0.7n

[2] More information about the benchmark instances of unconstrained binary quadratic programming problem can be found on this website: http://www.soften.ktu.lt/-gintaras/ubqop_its.html.

In these two tables, the fifth column refers to the cardinality of the subset M, where the numbers 0.3, 0.5, 0.7 are the ratios of the cardinality of the subset M to the cardinality of the size of the corresponding instance.

Besides, the algorithms need to set a few parameters, we mainly discuss two important ones: the running time and the population size, more details about the parameter settings for multi-objective optimization algorithms can be found in [3, 17]. The exact information about the parameter settings is presented in Tables 3 and 4.

Table 3. Parameter settings used for bi-criteria maximum diversity problem instances: instance size (*S*), vertices (*V*), edges(*E*), population size (*P*) and running time (*T*).

	Size (*S*)	Vertices (*V*)	Edges (*E*)	Population (*P*)	Time (*T*)
bo_mdp_800_01	800	800	19176	20	40″
bo_mdp_800_02	800	800	19176	20	40″
bo_mdp_800_03	800	800	19176	20	40″
bo_mdp_1000_01	1000	1000	5914	25	50″
bo_mdp_1000_02	1000	1000	5914	25	50″
bo_mdp_1000_03	1000	1000	5914	25	50″
bo_mdp_2000_01	2000	2000	19990	50	100″
bo_mdp_2000_02	2000	2000	19990	50	100″
bo_mdp_2000_03	2000	2000	19990	50	100″

Table 4. Parameter settings used for bi-criteria maximum diversity problem instances: instance size (*S*), population size (*P*) and running time (*T*).

	Size (*S*)	Population (*P*)	Time (*T*)
bo_mdp_3000_01	800	30	300″
bo_mdp_3000_02	800	30	300″
bo_mdp_3000_03	800	30	300″
bo_mdp_4000_01	1000	40	400″
bo_mdp_4000_02	1000	40	400″
bo_mdp_4000_03	1000	40	400″
bo_mdp_5000_01	2000	50	500″
bo_mdp_5000_02	2000	50	500″
bo_mdp_5000_03	2000	50	500″

4.2 Performance Assessment Protocol

In this paper, we evaluate the effectiveness of three different multi-objective optimization algorithms with the performance assessment package provided by Zitzler et al.[3]. The quality assessment protocol works as follows: First, we create a set of 20

[3] More information about the performance assessment package can be found on this website: http://www.tik.ee.ethz.ch/pisa/assessment.html.

runs with different initial populations for each strategy and each benchmark instance of bi-criteria maximum diversity problem. Then, we generate the reference set RS^* based on the 60 different sets $A_0, \ldots, A_{59}$ of non-dominated solutions.

According to two objective function values, we define a reference point $z = [r_1, r_2]$, where r_1 and r_2 represent the worst values for each objective function in the reference set RS^*. Afterwards, we assign a fitness value to each non-dominated set A_i by calculating the hypervolume difference between A_i and RS^*. In fact, the hypervolume difference between these two sets should be as close as possible to zero [18].

4.3 Computational Results

In this subsection, we provide the computational results on 18 groups of bi-criteria maximum diversity problem instances, which are generated by three multi-objective optimization algorithms. The information about these three algorithms are described in Table 5.

Table 5. The abbreviation for three multi-objective optimization algorithms.

Abbreviation	Algorithm description
IBMOLS	Indicator-Based Multi-Objective Local Search [3]
HBMOLS	Hypervolume-Based Multi-Objective Local Search [4]
MOMLA	Multi-Objective Multi-Level Algorithm

The computational results are summarized in Tables 6 and 7. In these two tables, there is a value both **in bold** and **in grey box** at each line, which is the best result obtained on the considered instance. The values both ***in italic*** and **bold** at each line refer to the corresponding algorithms which are **not** statistically outperformed by the algorithm obtaining the best result (with a confidence level greater than 95%).

From Tables 6 and 7, we can obviously observe that the computational results obtained by the MOMLA algorithm statistically outperforms the other two algorithms on almost all the instances. In addition, the most significant result is achieved on the instance bo_mdp_1000_03, where the average hypervolume difference value obtained by MOMLA is much smaller than the values obtained by the other two algorithms.

Actually, the multi-level approach has the distinct contribution to the performance of MOMLA. Moreover, this approach allow the algorithm to have the ability to jump out of the local optima by extracting the fixed variables to coarsen the original graph down to a smaller sub-graph. Therefore, MOMLA has a better performance than the other two algorithms.

Table 6. The computational results on the benchmark instances of bi-criteria maximum diversity problem with the size from 800 to 2000.

Instance	Algorithm		
	IBMOLS	HBMOLS	MOMLA
bo_mdp_800_01	0.160735	**0.113603**	**0.113439**
bo_mdp_800_02	0.182491	0.154866	**0.114377**
bo_mdp_800_03	0.196334	0.123552	**0.100146**
bo_mdp_1000_01	0.077453	0.050448	**0.046278**
bo_mdp_1000_02	0.073729	0.074411	**0.062152**
bo_mdp_1000_03	0.048081	0.043592	**0.031946**
bo_mdp_2000_01	**0.115277**	**0.110358**	**0.101176**
bo_mdp_2000_02	0.793274	0.728867	**0.674233**
bo_mdp_2000_03	0.68417	0.653463	**0.620013**

Table 7. The computational results on the benchmark instances of bi-criteria maximum diversity problem with the size from 3000 to 5000.

Instance	Algorithm		
	IBMOLS	HBMOLS	MOMLA
bo_mdp_3000_01	0.120914	**0.106331**	**0.101593**
bo_mdp_3000_02	0.169448	**0.149739**	**0.141877**
bo_mdp_3000_03	0.119639	0.116802	**0.091568**
bo_mdp_4000_01	0.489317	0.465037	**0.431062**
bo_mdp_4000_02	0.77545	0.745198	**0.684170**
bo_mdp_4000_03	0.584992	0.596386	**0.568310**
bo_mdp_5000_01	0.145733	**0.138421**	**0.133499**
bo_mdp_5000_02	0.144389	0.132989	**0.128909**
bo_mdp_5000_03	0.757027	0.767232	**0.717718**

5 Conclusions

In this paper, we have investigated a new effective multi-objective multi-level algorithm to solve the bi-criteria maximum diversity problem. The proposed algorithm consists of two main components: the hypervolume-based optimization procedure and the multi-level approach, which is dedicated to generating high-quality Pareto approximation set. To achieve this goal, we have carried out the experiments on 18 groups of the benchmark instances of maximum diversity problem. The experimental results indicate that the MOMLA algorithm are very competitive.

Acknowledgments. The work in this paper was supported by the Fundamental Research Funds for the Central Universities (Grant No. A0920502051722-53) and supported by the West Light Foundation of Chinese Academy of Science (Grant No: Y4C0011001).

References

1. Alidaee, B., Kochenberger, G.A., Ahmadian, A.: 0-1 quadratic programming approach for the optimal solution of two scheduling problems. Int. J. Syst. Sci. **25**, 401–408 (1994)
2. Aringhieri, R., Cordone, R.: Comparing local search metaheuristics for the maximum diversity problem. J. Oper. Res. Soc. **62**(2), 266–280 (2011)
3. Basseur, M., Liefooghe, A., Le, K., Burke, E.: The efficiency of indicator-based local search for multi-objective combinatorial optimisation problems. J. Heuristics **18**(2), 263–296 (2012)
4. Basseur, M., Zeng, R.-Q., Hao, J.-K.: Hypervolume-based multi-objective local search. Neural Comput. Appl. **21**(8), 1917–1929 (2012)
5. Benlic, U., Hao, J.-K.: An effective multilevel tabu search approach for balanced graph partitioning. Comput. Oper. Res. **38**, 1066–1075 (2011)
6. Benlic, U., Hao, J.-K.: A multilevel memetic approach for improving graph k-partitions. IEEE Trans. Evol. Comput. **15**(2), 624–642 (2011)
7. Benlic, U., Hao, J.-K.: Breakout local search for the max-cut problem. Eng. Appl. Artif. Intell. **26**, 1162–1173 (2013)
8. Brimberg, J., Mladenovic, N., Urosevic, D., Ngai, E.: Variable neighborhood search for the heaviest k-subgraph. Comput. Oper. Res. **36**(11), 2885–2891 (2009)
9. Coello, C.A., Lamont, G.B., Van Veldhuizen, D.A.: Evolutionary Algorithms for Solving Multi-Objective Problems. Genetic and Evolutionary Computation. Springer, New York (2007). https://doi.org/10.1007/978-0-387-36797-2
10. Freitas, A., Guimaraes, F., Silva, R.C.P.: Memetic self-adaptive evolution strategies applied to the maximum diversity problem. Optim. Lett. **8**(2), 705–714 (2014)
11. Gallego, M., Duarte, A., Laguna, M., Marti, R.: Hybrid heuristics for the maximum diversity problem. Comput. Optim. Appl. **200**(1), 36–44 (2009)
12. Gallo, G., Hammer, P., Simeone, B.: Quadratic knapsack problems. Math. Program. **12**, 132–149 (1980)
13. Liefooghe, A., Verel, S., Paquete, L., Hao, J.-K.: Experiments on local search for bi-objective unconstrained binary quadratic programming. In: Proceedings of the 8th International Conference on Evolutionary Multi-criterion Optimization (EMO 2015), Guimarães, Portugal, pp. 171–186 (2015)
14. Lü, Z., Glover, F., Hao, J.-K.: A hybrid metaheuristic approach to solving the UBQP problem. Eur. J. Oper. Res. **207**, 1254–1262 (2010)
15. Palubeckis, G.: Iterated tabu search for the maximum diversity problem. Optim. Lett. **189**(1), 371–383 (2007)
16. Wang, Y., Hao, J., Glover, F., Lü, Z.: A tabu search based memetic algorithm for the maximum diversity problem. Eng. Appl. Artif. Intell. **27**, 103–114 (2014)
17. Wu, Q., Wang, Y., Lü, Z.: A tabu search based hybrid evolutionary algorithm for the max-cut problem. Appl. Soft Comput. **34**, 827–837 (2015)
18. Zitzler, E., Thiele, L.: Multiobjective evolutionary algorithms: a comparative case study and the strength pareto approach. Evol. Comput. **3**, 257–271 (1999)

A Modified Teaching-Learning Optimization Algorithm for Economic Load Dispatch Problem

Ge Yu(✉) and Jinhai Liu

College of Information Science and Engineering, Northeastern University, Shenyang 110004, China
yugeneu@stumail.neu.edu.cn

Abstract. For the original Teaching-learning algorithm, it is weak in global search and prone to local search when solving complex optimization problems of high dimension. A modified algorithm based on space reverse-solution is proposed in this paper. Improvement of teacher phrase is based on the chaotic mapping and that of student phrase is based on the multi learning strategy. Then Self-learning phrase is added. The modified algorithm is applied to the complex high-dimensional benchmark functions for simulation experiments. Finally, the modified algorithm is applied to two typical power load distribution problems including 13 units and 40 units. The validity of the algorithm is verified from the aspects of convergence speed, convergence accuracy and stability.

Keywords: Improved teaching-learning optimization · Economic load dispatch Self-learning method · Reverse-solution · Chaotic mapping

1 Introduction

Teaching-learning algorithm (TLA) was proposed by Indian scholar RAO [1] in 2011. Once presented, TLA algorithm shows its strong competitiveness and it is widely used. It is a novel intelligent optimization algorithm through the simulation of the teacher's teaching process and student's learning process.

In recent year, economic development has been accompanied by the depletion of energy. How to efficiently use energy has become a common problem faced by mankind. How to optimize the economic dispatch of power load is important to solve the problem of high power generation cost and power consumption in power system. Power plants can save huge amounts of funds through excellent allocation methods. The ED problem [2] refers to the optimization of the generating cost of generating units in a certain period of time, and at the same time, it is necessary to meet various constraints. In order to complete such a high dimensional complex nonlinear mixed integer optimization problem, it is impossible to obtain the optimal solution in theory. With the limitations of the traditional linear programming, quadratic programming cannot optimize the ED problem [3] to get better results, and artificial intelligence technology emerges prominently.

D.-S. Huang et al. (Eds.): ICIC 2018, LNAI 10956, pp. 63–69, 2018.
https://doi.org/10.1007/978-3-319-95957-3_8

2 Modified Teaching-Learning Optimization Algorithm

In the original algorithm, the population consists of the students and the subjects included in the requirements. After initialization, the best qualified individual is selected as the teacher. The Initial population is generated randomly

$$x_i = low + rand(up - low) \tag{1}$$

Where *up* is the maximum and *low* is the minimum of the range of the value of the control variables. *rand* is a random number generated between 0 and 1. The reverse learning technology was first proposed by *Tizhoosh* [4] in 2005, It can enlarge the search space and accelerate the convergence.

$$x_i^o = up(i) + low(i) - x_i \tag{2}$$

This is definition of the global inverse solution. x_i is a selected individual. Each half of the better solution is obtained from the contemporary solution space and the reverse solution space, and the next iteration is repeated and repeated.

2.1 Modification of Random Number *Rand* in Teacher Phrase Based on Chaotic Map

Chaotic mapping [2] is a state of motion with randomness obtained from a particular equation in a nonlinear system. In this paper, Logistic mapping is used. Chaotic mapping can improve the algorithm, which is easy to fall into local optimum due to its simple variation mechanism and average diversity. After initializing the population, the chaotic mapping is initialized, can be expressed as follows:

$$\begin{aligned} & cx(1) = rand; \\ & while\,(cx(1) == 0.25||cx(1) == 0.5||cx(1) == 0.75) \\ & \qquad cx(1) = rand; \\ & end \end{aligned} \tag{3}$$

Where *cx* is a D-dimensional vector, in student phrase, *lf* instead of step length *rand*, can be calculated as:

$$\begin{aligned} & cx(v+1) = 4 * cx(v) * (1 - cx(v)) \\ & lf = round(1 + cx(v+1)); \\ & v = v + 1 \end{aligned} \tag{4}$$

The variation process of the student phrase is such as the follow formula:

$$\begin{aligned}
&\textit{if fitness}(b1) < \textit{fitness}(i) \\
&\qquad x_{new}(i,:) = x(i,:) + lf * (x(b1,:) - x(i,:)); \\
&\quad \textit{else} \\
&\qquad x_{new}(i,:) = x(i,:) + lf * (x(i,:) - x(b1,:)); \\
&\textit{end}
\end{aligned} \tag{5}$$

In order to increase the communication between individuals, the teacher phrase is added to the passive collection variation, can be expressed as:

$$x_{newi} = x_i + r_1 * (X_{teacher} - T_F X_{mean}) + r_2(b_1 - x_i) \tag{6}$$

Where r_2 is the average distribution from 0 to 1, b_1 is a random individual of a population, $r_2(R - x_i)$ is a passive set.

2.2 Modification of Student Phrase Based on Random Multi Learning Strategy

In student phrase, judge which one of two different variants to choose by a probability P, add a learning method by comparing two different ones other than themselves, learn from this difference, and strengthen the global search to find the optimal value realization as follows, Where b_1, b_2 are two individuals different from x.

$$\begin{aligned}
&\textit{if rand} < 0.5 \\
&\quad x_{newi} = x_i + rand * (x_i - b_1)\; f(x) < f(b_1) \\
&\quad x_{newi} = x_i + rand * (b_1 - x_i)\; f(b_1) < f(x) \\
&\textit{else} \\
&\quad x_{newi} = x_i + rand * (b_2 - b_1)\; f(b_2) < f(b_1) \\
&\quad x_{newi} = x_i + rand * (b_1 - b_2)\; f(b_1) < f(b_2)
\end{aligned} \tag{7}$$

2.3 The Self-learning Phase

In order to enhance students' ability of learning and innovation and increase population diversity, the self-learning phrase is added. And each individual are self-studied in some subjects (dimensions), and three probabilities are put forward, P_1 is the probability of choosing the learning object, the P_2 is the probability of self-learning adjustment method, and the P_3 is the probability of innovative learning. P_1, P_2 and P_3 will be assigned specific values in the following experiments [5]. The specific implementation is as follows:

$$\begin{aligned}
&\textit{if rand} < P_1 \quad x(i,j) = x(b1,j) \quad b1 = 1,2,\ldots,M \quad j = 1,2,\ldots,D \\
&\textit{else if rand} < P_2 \quad x_{new}(i,j) = x_{new}(i,j) + rand * Step(j) \\
&\textit{else if rand} < P_3 \quad x_{new}(i,j) = low(j) + rand * (up(j) - low(j)) \\
&\textit{else } x_{new}(i,j) = x(b1,j) \\
&\textit{end}
\end{aligned} \tag{8}$$

Where *Step* is Self-adjustment step, It is defined as:

$$Step = \min Step + (\max Step - \min Step)(1 - gen/Ngen) \tag{9}$$

Where $\min Step = (up - low)/3000$, *gen* is the current iteration, *Ngen* is the maximum allowed iteration algebra.

3 Simulation Experiment Applied to the Benchmark Function

The number of population is M = 30, the problem dimension is D = 40, which runs 30 times independently, and the maximum allowable iteration is 2000 generations. Average running time (T), the optimal value of the best solution (B), the average value of the optimal solution (M), the worst value (W). The standard deviation (SD). The 30 average convergence curve is shown in Fig. 1, and the convergent data is shown in

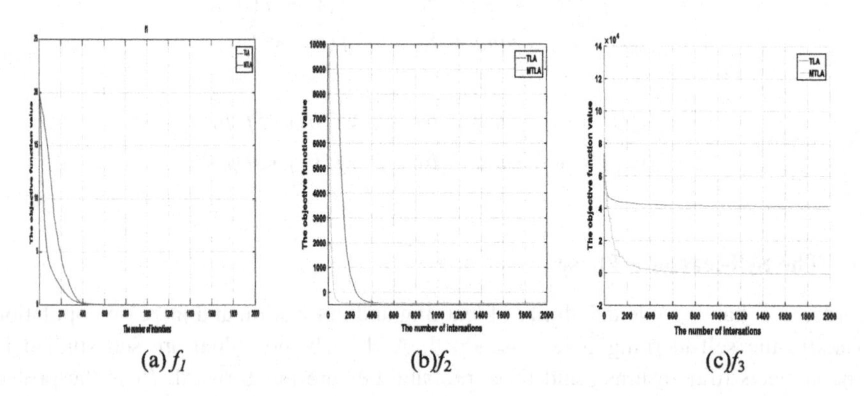

(a) f_1 (b) f_2 (c) f_3

Fig. 1. Benchmark function convergence curve

Table 1. The complex constrained benchmark functions are selected as f_1 to f_3:

f_1: *Schwefel's* $f_1 = \sum_{i=1}^{D} |xi| + \prod_{i=1}^{D} |xi|$. where $-10 \le x_i \le 10$, The optimal value 0 is reached at $x_i = 0$.

f_2: *Sphere Shift Function*, where $-100 \le x_i \le 100$, The optimal value −450 is reached at $x_i = 0$.

f_3: *Rosenbrock Shift Function* where $-100 \leq xi \leq 100$, The optimal value -390 is reached at $xi = 0$.

The results of the functions of each function are shown in Table 1.

Table 1. Simulation results of constrained benchmark functions

Function	*Algorithm*	T(s)	B	M	W	SD
f_1	TLA	4.4835	8.8330e−15	1.2530e−13	7.3859e−13	1.6738e−13
	MTLA	12.2135	2.7064e−279	8.7921e−270	2.0743e−268	0
f_2	TLA	263.8273	−2.9735e+02	−2.7950e+02	−2.5472e+02	9.2145e−03
	MTLA	266.5104	−4.4999e+02	−4.4999e+02	−4.4999e+02	1.38132e−11
f_3	TLA	5251.6195	2.2902e+09	1.8514e+10	3.5376e+10	8.3812e+09
	MTLA	195.1729	−3.9939e+02	−399.3890	−6.3408e+02	68.244

The average convergence graphs of the three test functions of each algorithm are as follows:

As we can see from the Fig. 1, MTLA can quickly find the global optimal value. MTLA has strong search ability in solving large scale and special complex benchmark functions. In Table 1, MTLA for these three special complex test functions is far superior to the original TLA in terms of convergence accuracy and stability. MTLA algorithm has been in a downward state and has strong local search capability. It combines the global search capability of reverse technology to maintain population diversity and avoid "precocious".

4 Application of Power Load Distribution

Our goal is to optimize the determination of the best power generation unit, which makes the total cost of generating electricity optimal on the premise of satisfying the load demand and generating set limit. The modeling process can be divided into two parts: the objective function and the constraint condition [6].

4.1 Simulation Experiment Applied to ED

The modified TLA algorithm is applied to optimize the ED problem with the valve point effect with 13 and 40 generating units, and the total electricity demand of 13 units is 1800 MW. The population size is 500. The max *Ngen* allowed is 500, run 30 times independently. The total electricity demand of 40 units is 10500 MW. the population size is 200. The maximum allowed is 200, run 30 times independently. The 30 average convergence curve is shown in Fig. 2, and the convergent data is shown in Table 2.

The results of the convergence table and curve convergence chart can be obtained: MTLA is superior to the original algorithm in terms of search efficiency, convergence stability, or the search for the global optimal solution, the search fastness, or the

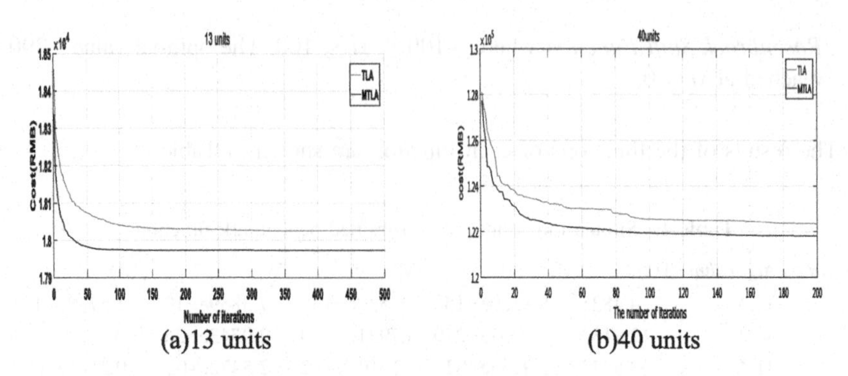

(a)13 units (b)40 units

Fig. 2. Average convergence curve

Table 2. The results of ED problem

	Algorithm	T(s)	B(RMB)	M(RMB)	W(RMB)	SD
13 unit $Pd = 1800$	TLA	105.5335	17988	18395	18179	69.0678
	MTLA	128.4333	17972	17980	18046	13.8750
40 unit $Pd = 10500$	TLA	51.0858	12248	12263	12309	795.8352
	MTLA	116.6114	12169	12180	12184	25.3462

performance stability. MTLA has a good global search capability for complex problems, mainly due to the self-learning of some subjects (dimensions) of each individual in the MTLA algorithm.

5 Conclusions

In the existing space reverse technology for improving the premise, the chaos mapping and self-learning phase are added, which increases the convergence speed and improves the convergence precision. At the same time, it optimizes the power load distribution problems, reduces costs and saves energy. But at the same time, there are still some shortcomings in the algorithm, such as weak search ability in the late stage, easy falling into the local optimum. We hope to continue research and improve the algorithm.

Acknowledgement. This work was supported by National Key R&D Program of China (2017YFF0108800) and the National Natural Science Foundation of China (61473069, 61627809).

References

1. Rao, R.V., Savsani, V.J., Vakharia, D.P.: Teaching–learning-based optimization: a novel method for constrained mechanical design optimization problems. Comput. Aided Des. **43**(3), 303–315 (2011)
2. Yu, K., Chen, X., Wang, X., Wang, Z.: Parameters identification of photovoltaic models using self-adaptive teaching-learning-based optimization. Energy Convers. Manag. **145**, 233–246 (2017)
3. Wu, Z., Fu, W., Xue, R., et al.: A novel global path planning method for mobile robots based on teaching-learning-based optimization. Information **7**(3), 39 (2016). https://doi.org/10.3390/info7030039
4. Hoehn, L., Mouron, C., et al.: Hierarchies of chaotic maps on continua. Ergodic Theory Dyn. Syst. **34**(6), 1897–1913 (2014)
5. Chen, D., Zou, F., Li, Z., et al.: An improved teaching–learning-based optimization algorithm for solving global optimization problem. Inf. Sci. **297**(C), 171–190 (2015)
6. Rao, R.V., Rai, D.P.: Optimization of fused deposition modeling process using teaching-learning-based optimization algorithm. Eng. Sci. Technol. Int. J. **19**(1), 587–603 (2016)

A Novel Multi-population Particle Swarm Optimization with Learning Patterns Evolved by Genetic Algorithm

Chunxiuzi Liu, Fengyang Sun, Qingbei Guo(✉), Lin Wang(✉), and Bo Yang

Shandong Provincial Key Laboratory of Network Based Intelligent Computing, University of Jinan, Jinan 250022, China
ise_guoqb@ujn.edu.cn, wangplanet@gmail.com

Abstract. In recent years, particle swarm optimization (PSO) and genetic algorithm (GA) have been applied to solve various real-world problems. However, the original PSO is based on single population whose learning patterns (inertia weights, learning factors) has no potentials in evolution. All particles in the population interact and search according to a fixed pattern, which leads to the reduction of population diversity in the later iterations and premature convergence on complex and multi-modal problems. Therefore, a novel multi-population PSO with learning patterns evolved by GA is proposed to improve diversity and exploration capabilities of populations. Meanwhile, the local search of PSO particles which start in the same position also evolved by GA independently maintains exploitation ability inside each sub population. Experimental results show that the accuracy is comparable and our method improves the convergence speed.

Keywords: Co-evolutionary computation · Genetic algorithm
Particle swarm optimization

1 Introduction

Many optimization problems can be solved by abstraction and modeling [1]. How to solve these complicated problems effectively emerges a challenging research topic. Among existing techniques, particle swarm optimization (PSO) algorithm was proposed by Kennedy and Eberhart [2, 3] which is a simulation of the migration and aggregation of the birds' foraging process. It has been widely used as an optimization tool in various applications such as materials science [4], communication, medicine, energy, and finance [5–9].

The particle swarm algorithm imitates the clustering behavior of insects, herds, birds, and schools of fishes. The individuals in the population chase food with inner cooperation. In nature, due to various regions and landscapes the same species are different from interaction manners and foraging habits. However, the single-population mode of PSO leads to the simplification of the population searching patterns. On the other hand, various populations of the same species in nature facilitate gene diversity

D.-S. Huang et al. (Eds.): ICIC 2018, LNAI 10956, pp. 70–80, 2018.
https://doi.org/10.1007/978-3-319-95957-3_9

through genetic operations such as crossover and mutation, which inspired Genetic Algorithm. Nevertheless, genetic operations only at the individual level also limits the evolving pattern of the entire population of GA.

To address this issue, a new multi-population PSO with learning patterns evolved by GA is proposed, which improves the diversity of population and exploration ability, as well as introduced to optimize the different solutions. Meanwhile, the local search of PSO particles which start in the same position also evolved by GA independently maintains exploitation ability inside each sub population.

The rest of the paper is organized as follows. Section 2 reviews optimization algorithms and evolutionary computation methods in various fields. After that, the method of PSO-GA co-evolution are described in Sect. 3. Section 4 provides details about our experiments. The study is concluded in Sect. 5.

2 Related Works

Kennedy and Eberhart [2, 3] proposed the particle swarm optimization (PSO) algorithm. And then the researchers attempt to integrate the mechanisms of different optimization algorithms into it [10]. Eslami et al. [11] proposed a novel technique for optimizing power system stabilizers by combining improved PSO with chaos. A multi-layer particle swarm optimization (MLPSO) improved the performance of traditional particle swarm optimization by increasing the two layers of swarms to multiple layers [12].

As a global random search method, the genetic algorithm has been successfully applied in many fields because of its strong robustness. It has been proved that genetic algorithm is powerful for optimization problems [13] and able to find high parallelism and global optimum solution [14]. There are many existing ways to improve the performance, such as adjusting the population size [15], improving the genetic operator [16] and optimizing the control parameters [17, 18]. It also has been used in Cement Hydration [19, 20], moving objects [21] and CT imaging [22, 23]. An adaptive genetic algorithm based mobile robot path planning method is proposed in [24, 25]. It is generally believed that the improvement made on a single species which makes it difficult to achieve a balance between population diversity and selection process, and that is why we come up with the multiple population evolutions [26] as an excellent solution.

The idea of co-evolution was put forward in recent years for the insufficiency of genetic algorithms. It was first proposed by Ehrlich and Raven [27]. Although there are many existing improvement methods [28–30] also with neural networks [31, 32], it is difficult to obtain satisfied results if individual evolution is controlled by individual fitness. Cao et al. [33] introduced co-evolutionary ideas in multi-population genetic evolutionary algorithms, improved the global convergence of genetic algorithms through cooperative competition between different populations. In the process, it is prone to the phenomenon of coordinated diminishing and degradation. Wang et al. [34] added a perturbation adjustment item to the right of the multi-population co-evolutionary model for this synergistic diminish and degeneracy phenomenon, without affecting evolution.

3 Methodology

In this section, particle swarm optimization is described in detail. Then, genetic algorithm is introduced briefly. Finally, the model of multi-population PSO learning patterns evolved by GA is illustrated in the last subsection.

3.1 Particle Swarm Optimization

The algorithm of particle swarm optimization is a global optimization method based on swarm intelligence theory. In each iteration, the particle updates its position and speed by tracking two best values. One is the optimum solution found by the particle itself called *pbest*. The other is the optimum solution found by the entire population which is *gbest*. In each iteration step, after the vector of velocity and position are updated, the particle updates its new vector of the ith dimension as follows:

$$v_{k+1}^{i} = \omega v_{k}^{i} + \varphi_1 r(gbest_{k}^{i} - x_{k}^{i}) + \varphi_2 r(pbest_{k}^{i} - x_{k}^{i}) \quad (1)$$

$$x_{k+1}^{i} = x_{k}^{i} + v_{k+1}^{i} \quad (2)$$

where ω is inertial weight and k is the iteration number. φ_1 and φ_2 are learning factors. x is the particle position and v is the particle velocity vector. r is a random positive number range from [0,1]. The *pbest* is the best current position of the particle and the *gbest* is the best new position found in the entire population.

3.2 Genetic Algorithm

Genetic Algorithm is a computational model of natural selection and genetic mechanism of Darwin's biological evolution. It is a method for searching for optimal solutions by resembling natural evolutionary processes. The basic process of traditional GA optimization includes:

- Encoding;
- Selection, also referred to as copying;
- Crossover;
- Mutation.

After a selection, crossover and mutation operation, the average fitness value of the population will gradually decrease. It has the characteristics of independence of problem model, global optimality, implicit parallelism and strong robustness to solve nonlinear problems.

3.3 Multi-population PSO Learning Patterns Evolved by GA

However, original genetic algorithms have the disadvantages of slow convergence rate and obtaining partial optimal solutions only [35]. In solving complex problems, its premature convergence which results in locally optimal solutions frequently is a major issue associated with the PSO [36].

The model proposed in this study is to perform the optimization process and co-evolution process successively in each iteration of the algorithm. The optimized process is to update the starting position by finding the local best or global best with the particle swarm algorithm using the jump step, and then evolve the PSO learning patterns through the genetic algorithms. Overall the two independent evolutions do not influence each other. The complete algorithm of the PSO learning patterns evolved by GA is shown in Algorithm 1.

Algorithm 1: Algorithm of PSO-GA Coevolution.

Input: Population size, jump step, parameter α
Output: The best solution
1 Initialize population and PSO parameters;
2 Define fitness F;
3 **While** *not termination condition* **do**
4 Evaluate the fitness for each particle according to its position vector x;
5 Update the best position x for each particle;
6 Update the best position $gbest$ for the whole population or *lbest* for the neighbor;
7 **for** n=1 **to** *initial starting positions size* **do**
8 **for** i=1 **to** *d* **do**
9 $v_{k+1}^i = \omega v_k^i + \varphi_1 r(gbest_k^i - x_k^i) + \varphi_2 r(pbest_k^i - x_k^i)$;
10 $v^i = \min(VMAX^i, \max(-VMAX^i, v^i))$;
11 $x_{k+1}^i = x_k^i + v_{k+1}^i$;
12 **end**
13 **end**
14 Record the best fitness value;
15 Competition selection;
16 **if** $p_c \rangle rand$ **then**
17 uniform crossover of PSO learning patterns and starting positions;
18 **end**
19 **if** $p_c \rangle rand$ **then**
20 uniform mutation of PSO learning patterns and starting positions;
21 **end**
22 Select the current best for new generation(elitism);
23 **end**
24 **Return** The best solution;

Figure 1 shows the diagram of the skeleton of our method. The population is divided into two major categories. One is the PSO with learning patterns population and the other one is starting positions population.

Figure 2 presents the trace of particles in sub-population of PSO, the GA evolution of PSO learning patterns and PSO starting positions. The two populations use genetic algorithm to co-evolve and then generate the next generation, select the most potential individuals in the respective populations to perform genetic operations.

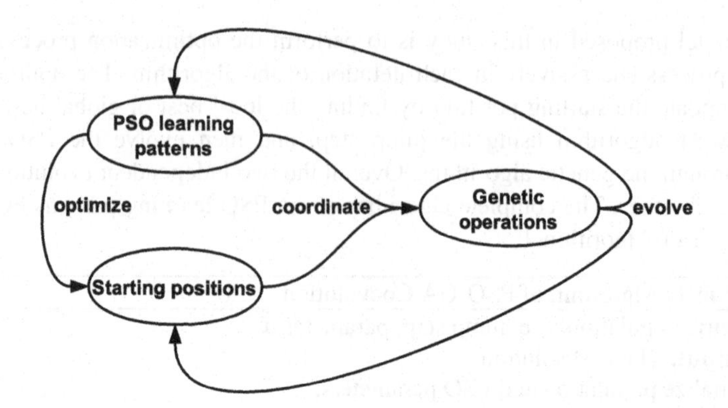

Fig. 1. The diagram of the skeleton of our method.

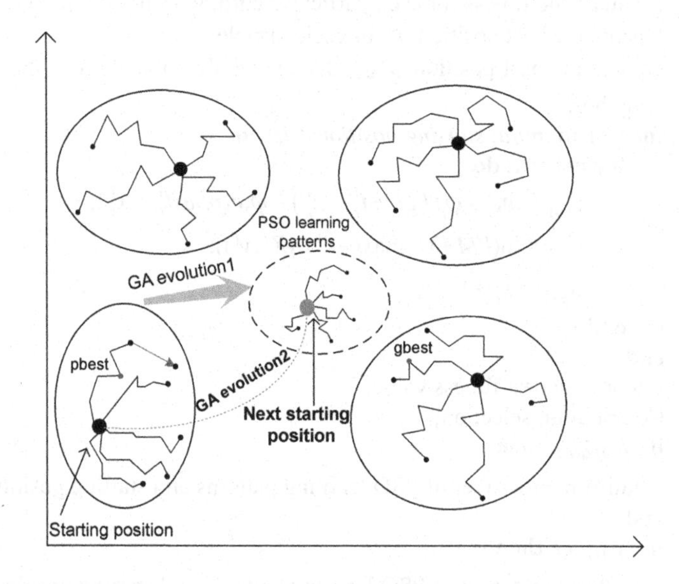

Fig. 2. The trace of particles in sub-population of PSO, the GA evolution of PSO learning patterns and PSO starting positions.

4 Experiments

This study attempts to test the combination of LPSO with GA and the combination of GPSO with GA against diverse benchmark functions to improve the performance of the particle swarm optimization and genetic algorithm in solving complex problems. LPSO is the local version of PSO with ring neighborhood and inertia weight. GPSO is PSO with inertia weight. The total 9 benchmark functions (2 unimodal and 6 multi-modal) were chosen [37]. All the functions were tested on 10 and 30 dimensions. The algorithms are listed below

- LPSO with GA.
- GPSO with GA.
- Genetic Algorithm (GA).
- Weighted fully informed particle swarm with URing topology (FIPS (URing)).

In the following methods, genetic algorithm is a widely used evolutionary algorithm. The FIPS is a fully informed PSO algorithm. The URing topologies are used with the weighted FIPS based on the goodness.

All experiments were conducted in MATLAB R2016a using the same machine with a Core (TM) 2 2.71 GHz CPU, 4G memory, and Windows 10 operating system.

Table 1 presents the formulas and properties of these functions. Tables 2 and 3 depict the optimization results for all the methods in relation to 10-D and 30-D problems. The best result is marked in bold. Table 3 shows that the multi-modal functions F7 and F9 is the best solutions in all methods which improves that our model is more efficiently to solve the complex problems.

Table 1. Benchmark functions used.

Function name	Formula	Search range	Best value
Quartic function with noise	$F_1(x) = \sum_{i=1}^{D} ix_i^4 + random[0,1)$	$[-1.28,1.28]^D$	0
Step	$F_2(x) \sum_{i=1}^{D} (\lfloor x_i + 0.5 \rfloor)^2$	$[-100,100]^D$	0
Rosenbrock	$F_3(x) = \sum_{i=1}^{D-1} (100(x_i^2 - x_{i+1})^2 + (x_i - 1)^2)$	$[-2,2]^D$	0
Griewanks	$F_4(x) = \sum_{i=1}^{D} \frac{x_i^2}{4000} - \prod_{i=1}^{D} \cos(\frac{x_i}{\sqrt{i}}) + 1$	$[-600,600]^D$	0
Noncontinuous rastrigin	$F_5(x) = \sum_{i=1}^{D} (y_i^2 - 10\cos(2\pi y_i) + 10)$ $y_i = \begin{cases} x_i & \lvert x_i \rvert \langle \frac{1}{2} \\ \frac{round(2x_i)}{2} & \lvert x_i \rvert \geq \frac{1}{2} \end{cases}$	$[-5,5]^D$	0
Shifted rosenbrock	$F_6(x) = \sum_{i=1}^{D-1} (100 - z_i^2 - z_{i+1}^2) + (z_i - 1)^2) + 390$ $z = x - o + 1$	$[-100,100]^D$	390
Shifted rotated weierstrass	$F_7(x) = \sum_{i=1}^{D} \left(\sum_{k=0}^{kmax} [a^k \cos(2\pi b^k (z_i + 0.5))] \right)$ $- \sum_{i=1}^{kmax} [a^k \cos(2\pi b^k * 0.5)] + 90$ $a = 0.5, b = 3, kmax = 20$ $z = (x - o) * M$	$[-0.5,0.5]^D$	90
Shifted expanded griewanks plus rosenbrocks	$F_8(x) = F_4(F_3(z_1, z_2)) + F_4(F_3(z_2, z_3)) + \ldots$ $+ F_4(F_3(z_{D-1}, z_D)) + F_4(F_3(z_D, z_1)) - 130$ $z = x - o + 1$	$[-3,1]^D$	−130
Shifted rotated expanded Scaffers	$F_9(x) = F(z_1, z_2) + F(z_2, z_3) + \ldots + F(z_{D-1}, z_D)$ $+ F(z_D, z_1) - 300$ $z = (x - o) * M$	$[-100,100]^D$	−300

Table 2. Results for 10-D problems on F1 to F9.

	F_1	F_2	F_3	F_4	F_5	F_6	F_7	F_8	F_9
LPSO+GA									
Best	4.90E−03	0.00E+00	3.15E−01	8.90E−03	3.60E+00	3.94E+02	9.22E+01	−1.29E+02	−2.97E+02
Worst	2.08E−02	0.00E+00	1.84E+00	2.01E−01	1.21E+01	1.07E+03	9.62E+01	−1.28E+02	−2.96E+02
Mean	1.00E−02	0.00E+00	9.25E−01	6.95E−02	8.17E+00	4.79E+02	9.43E+01	−1.29E+02	−2.97E+02
Std	3.90E−03	0.00E+00	3.37E−01	5.17E−02	2.21E+00	1.45E+02	8.02E−01	2.80E−01	2.21E−01
GPSO+GA									
Best	3.40E−03	0.00E+00	5.21E−01	9.79E−04	4.00E+00	3.91E+02	9.27E+01	−1.29E+02	−2.97E+02
Worst	1.74E−02	0.00E+00	1.69E+00	2.49E−01	1.50E+01	3.75E+03	9.59E+01	−1.28E+02	−2.96E+02
Mean	9.10E−03	0.00E+00	1.03E+00	7.88E−02	8.90E+00	6.28E+02	9.44E+01	−1.29E+02	−2.97E+02
Std	3.40E−03	0.00E+00	2.76E−01	5.62E−02	2.50E+00	6.21E+02	8.26E−01	2.78E−01	2.65E−01
GA									
Best	4.25E−02	0.00E+00	5.70E−01	1.77E−10	6.00E+00	3.90E+02	9.42E+02	−1.29E+02	−2.96E+02
Worst	3.68E−01	8.00E+00	4.39E+00	6.90E−02	2.10E+01	5.57E+02	1.00E+02	−1.25E+02	−2.95E+02
Mean	1.35E−01	2.33E+00	1.67E+00	1.32E−02	9.97E+00	4.06E+02	9.77E+01	−1.28E+02	−2.95E+02
Std	8.01E−02	2.11E+00	9.81E−01	1.79E−02	3.61E+00	4.04E+01	1.60E+00	1.14E+00	5.19E−02
FIPS(URing)									
Best	2.95E−04	0.00E+00	7.20E−01	1.76E−04	1.21E+00	3.90E+02	9.72E+01	−1.29E+02	−2.97E+02
Worst	9.92E−04	0.00E+00	1.60E+00	3.88E−02	6.05E+00	3.98E+02	9.95E+01	−1.28E+02	−2.96E+02
Mean	5.91E−04	0.00E+00	1.15E+00	1.65E−02	3.54E+00	3.94E+02	9.87E+01	−1.29E+02	−2.96E+02
Std	2.22E−04	0.00E+00	2.07E−01	1.00E−02	9.16E−01	1.70E+00	5.51E−01	2.70E−01	2.30E−01

Table 3. Results for 30−D problems on F1 to F9.

	F_1	F_2	F_3	F_4	F_5	F_6	F_7	F_8	F_9
LPSO+GA									
Best	1.07E−01	61.00E+00	8.57E+00	1.71E+00	6.77E+01	4.46E+05	1.11E+02	−1.19E+02	−2.87E+02
Worst	5.90E−01	1.78E+03	1.71E+01	9.90E+00	1.06E+02	1.92E+08	1.18E+02	−1.11E+02	−2.87E+02
Mean	2.46E−01	5.18E+02	1.30E+01	3.75E+00	8.65E+01	2.31E+07	1.16E+02	−1.15E+02	−2.87E+02
Std	1.07E−01	4.19E+02	2.18E+00	1.91E+00	1.17E+01	3.61E+07	1.71E+00	2.15E+00	2.49E−01
GPSO+GA									
Best	3.40E−03	0.00E+00	5.21E−01	9.79E−04	4.00E+00	3.91E+02	9.27E+01	−1.29E+02	−2.97E+02
Worst	1.74E−02	0.00E+00	1.69E+00	2.49E−01	1.50E+01	3.75E+03	9.59E+01	−1.28E+02	−2.96E+02
Mean	9.10E−03	0.00E+00	1.03E+00	7.88E−02	8.90E+00	6.28E+02	9.44E+01	−1.29E+02	−2.97E+02
Std	3.40E−03	0.00E+00	2.76E−01	5.62E−02	2.50E+00	6.21E+02	8.26E−01	2.78E−01	2.65E−01
GA									
Best	4.25E−02	0.00E+00	5.70E−01	1.77E−10	6.00E+00	3.90E+02	9.42E+02	−1.29E+02	−2.96E+02
Worst	3.68E−01	8.00E+00	4.39E+00	6.90E−02	2.10E+01	5.57E+02	1.00E+02	−1.25E+02	−2.95E+02
Mean	1.35E−01	2.33E+00	1.67E+00	1.32E−02	9.97E+00	4.06E+02	9.77E+01	−1.28E+02	−2.95E+02
Std	8.01E−02	2.11E+00	9.81E−01	1.79E−02	3.61E+00	4.04E+01	1.60E+00	1.14E+00	5.19E−02
FIPS(URing)									
Best	2.95E−04	0.00E+00	7.20E−01	1.76E−04	1.21E+00	3.90E+02	9.72E+01	−1.29E+02	−2.97E+02
Worst	9.92E−04	0.00E+00	1.60E+00	3.88E−02	6.05E+00	3.98E+02	9.95E+01	−1.28E+02	−2.96E+02
Mean	5.91E−04	0.00E+00	1.15E+00	1.65E−02	3.54E+00	3.94E+02	9.87E+01	−1.29E+02	−2.96E+02
Std	2.22E−04	0.00E+00	2.07E−01	1.00E−02	9.16E−01	1.70E+00	5.51E−01	2.70E−01	2.30E−01

Figures 3 and 4 show the convergence curve for 10-D and 30-D benchmark functions. The convergence speed of our method has more faster convergence rate than other methods especially in function F7 and F9 and has the advantage of improving the results.

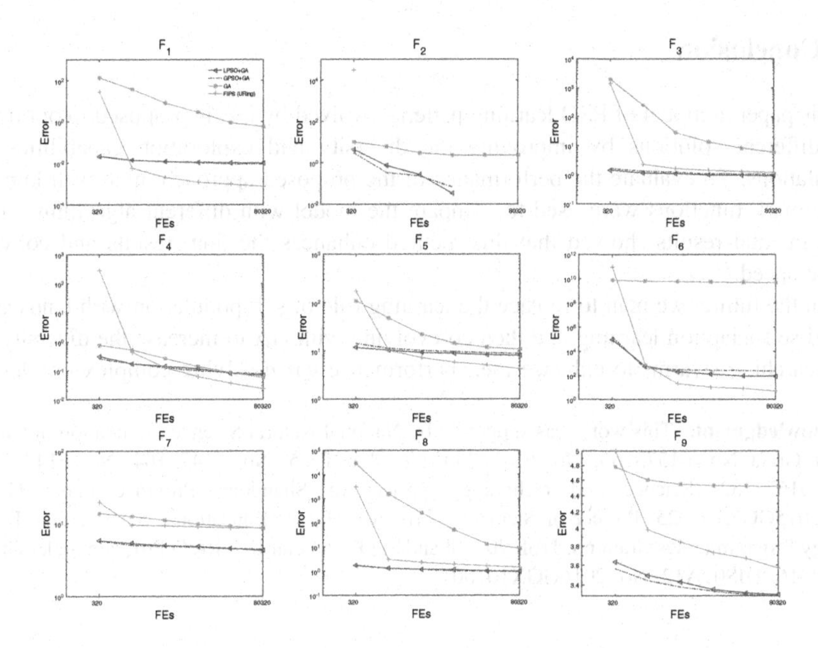

Fig. 3. The mean convergence characteristics of 10-D benchmark functions.

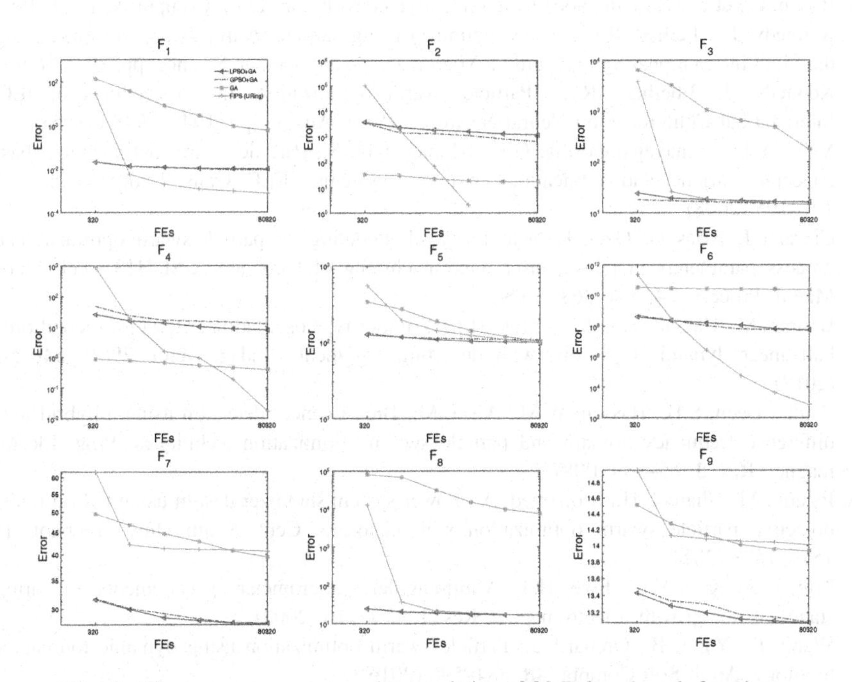

Fig. 4. The mean convergence characteristics of 30-D benchmark functions.

5 Conclusion

In this paper, a model of PSO learning patterns evolved by GA is proposed to optimize the different solutions by improving the diversity and exploration capabilities of populations. To evaluate the performance of the proposed approach, nine well-known benchmark functions were used to compare the model with different algorithms. The experimental results showed that this method enhances the final results and convergence speed.

In the future, we plan to replace the learning rule of sub-population with landscape based self-adaption learning and then co-evolving with GA to increase the diversity of the searching swarms to enhance their performance when solving complex problems.

Acknowledgment. This work was supported by National Natural Science Foundation of China under Grant No. 61573166, No. 61572230, No. 81671785, No. 61472164, No. 61472163, No. 61672262. Science and technology project of Shandong Province under Grant No. 2015GGX101025. Project of Shandong Province Higher Educational Science and Technology Program under Grant no. J16LN07. Shandong Provincial Key R&D Program under Grant No. 2016ZDJS01A12, No. 2016GGX101001.

References

1. Rosen, C., et al.: New methods for competitive coevolution. Evol. Comput. **5**, 1–29 (1997)
2. Kennedy, J., Eberhart, R.C.: A new optimizer using particle swarm theory. In: Proceedings the Sixth International Symposium on Micromachine and Human Science, pp. 39–43 (1995)
3. Kennedy, J., Eberhart, R.C.: Particle swarm optimization. In: Proceedings of IEEE International Conference on Neural Networks, Perth, Austria, pp. 1942–1948 (1995)
4. Valle, Y.D., Venayagamoorthy, G.K., Mohagheghi, S.: Particle swarm optimization: basic concepts, variants and applications in power systems. IEEE Trans. Evol. Comput. **12**, 171–195 (2008)
5. Ciurana, J., Arias, G., Ozel, T.: Neural network modeling and particle swarm optimization of process parameters in pulsed laser micromachining of hardened AISI H13 steel. Mater. Manuf. Process. **24**, 358–368 (2009)
6. Bohner, M., Hassan, T.S., Li, T.: Fite-Hille-Wintner-type oscillation criteria for second-order half-linear dynamic equations with deviating arguments. Indag. Math. **29**(2), 548–560 (2018)
7. Zainud-Deen, S.H., Hassen, W.M., Ali, E.M.: Breast cancer detection using a hybrid finite difference frequency domain and particle swarm optimization techniques. Prog. Electromagnet. Res. **3**, 35–46 (2008)
8. Eslami, M., Shareef, H., Mohamed, A.: Power system stabilizer design using hybrid multi–objective particle swarm optimization with chaos. J. Cent. South Univ. Technol. **18**, 1579–1588 (2011)
9. Lim, K.S., Koo, V.C., Ewe, H.T.: Multi-angular scatterometer measurements for various stages of rice growth. Electromagnet. Res. **1**, 159–176 (2008)
10. Wang, L., Yang, B., Orchard, J.: Particle swarm optimization using dynamic tournament topology. Appl. Soft Comput. **48**, 584–596 (2016)

11. Eslami, M., Shareef, H., Mohamed, A.: Power system stabilizer design using hybrid multi-objective particle swarm optimization with chaos. J. Cent. South Univ. Technol. **18**, 1579–1588 (2011)
12. Wang, L., Yang, B., Chen, Y.H.: Improving particle swarm optimization using multi-layer searching strategy. Inf. Sci. **274**, 70–94 (2014)
13. Alexandre, H.F.D., Vasconcelos, J.A.: Multi-objective genetic algorithms applied to solve optimization problems. IEEE Trans. Magn. **38**, 1133–1136 (2002)
14. Cmara, M., Ortega, J., Toro, F.D.: A single front genetic algorithm for parallel multi-objective optimization in dynamic environments. Neurocomputing **72**, 3570–3579 (2009)
15. Arabas, J., Michalewicz, Z., Mulawka, J.: Genetic algorithms with varying population size. In: Proceedings of the 1st IEEE International Conference on Evolutionary Computation, pp. 73–78 (1994)
16. Goldberg, D.E.: Genetic Algorithms in Search, Optimization and Machine Learning. Addison–Wesley Publishing Company, Reading (1989)
17. Srinvas, M., Patnaik, L.M.: Adaptive probabilities of crossover and mutation in genetic algorithms. IEEE Trans. Syst. Man Cybern. **24**, 656–666 (1994)
18. Eiben, A.E., Hinterding, R., Michalewicz, Z.: Parameter control in evolutionary algorithms. IEEE Trans. Evol. Comput. **3**, 124–141 (1999)
19. Wang, L., Yang, B., Wang, S., Liang, Z.F.: Building image feature kinetics for cement hydration using gene expression programming with similarity weight tournament selection. IEEE Trans. Evol. Comput. **19**, 679–693 (2015)
20. Zhou, J., Chen, L., Chen, C.L.P., Zhang, Y., Li, H.X.: Fuzzy clustering with the entropy of attribute weights. Neurocomputing **198**, 125–134 (2016)
21. Yu, Z.Q., Liu, Y., Yu, X.H., Ken, Q.P.: Scalable distributed processing of K nearest neighbor queries over moving objects. IEEE Trans. Knowl. Data Eng. **27**, 1383–1396 (2015)
22. Chen, Y., et al.: Discriminative feature representation: an effective postprocessing solution to low dose CT imaging. Phys. Med. Biol. **62**, 2103–2131 (2017)
23. Liu, J., et al.: 3D feature constrained reconstruction for low dose CT imaging. IEEE Trans. Circuits Syst. Video Technol. (2016)
24. Liu, C.L., Liu, H.W., Yang, J.Y.: A path planning method based on adaptive genetic algorithm for mobile robot. J. Inf. Comput. Sci. **8**, 808–814 (2011)
25. Han, S.Y., Chen, Y.H., Tang, G.Y.: Fault diagnosis and fault-tolerant tracking control for discrete-time systems with faults and delays in actuator and measurement. J. Frankl. Inst. **354**, 4719–4738 (2017)
26. Potts, J.C., Giddens, T.P., Yadav, S.B.: The development and evaluation of an improved genetic algorithms based on migration and artificial selection. IEEE Trans. Syst. Man Cybern. **24**, 73–86 (1994)
27. Ehrlich, P.R., Raven, P.H.: Butterflies and plants: a study in coevolution. Evolution **18**, 586–608 (1964)
28. Rosen, C., et al.: New methods for competitive coevolution. Evol. Comput. **5**, 1–29 (1997)
29. Wang, L., Orchard, J.: Investigating the evolution of a neuroplasticity network for learning. IEEE Trans. Syst. Man Cybern. Syst. 1–13 (2017)
30. Potts, J.C., Giddens, T.P., Yadav, S.B.: The development and evaluation of an improved genetic algorithms based on migration and artificial selection. IEEE Trans. Syst. Man Cybern. **24**, 73–86 (1994)
31. Chen, C.L.P., Liu, Z.: Broad learning system: an effective and efficient incremental learning system without the need for deep architecture. IEEE Trans. Neural Netw. Learn. Syst. **29**, 10–24 (2017)

32. Wang, L., Yang, B., Chen, Y.H., Zhang, X.Q., Orchard, J.: Improving neural-network classifiers using nearest neighbor partitioning. IEEE Trans. Neural Netw. Learn. Syst. **28**, 2255–2267 (2017)
33. Cao, X.B., Luo, W.J., Wang, X.F.: A co–evolution pattern based on ecological population competition model. J. Softw. **12**, 556–562 (2001)
34. Wang, B., Gao, Y., Xie, J.: A genetic algorithms based on global coordination and local evolution. Comput. Eng. **31**, 29–31 (2005)
35. Tsai, C.C., Huang, H.C., Chan, C.K.: Parallel elite genetic algorithm and its application to global path planning for autonomous robot navigation. IEEE Trans. Ind. Electron. **58**, 4813–4821 (2011)
36. Zhao, S.Z., Liang, J.J., Suganthan, P.N., Tasgetiren, M.F.: Dynamic multi–swarm particle swarm optimizer with local search for large scale global optimization. In: Proceedings of 2008 IEEE Congress on Evolutionary Computation, Hong Kong, pp. 3845–3852 (2008)
37. Suganthan, P.N., et al.: Problem definitions and evaluation criteria for the cec 2005 special session on real–parameter optimization. Technical report, Nanyang Technological University, Singapore and KanGAL Report Number 2005005 (2005)

Classification of Hyperspectral Data Using a Multi-Channel Convolutional Neural Network

Chen Chen[1], Jing-Jing Zhang[1], Chun-Hou Zheng[2(✉)], Qing Yan[1], and Li-Na Xun[1]

[1] College of Electrical Engineering and Automation, Anhui University, Hefei 230601, Anhui, China
[2] College of Computer Science and Technology, Anhui University, Hefei 230601, Anhui, China
zhengch99@126.com

Abstract. In recent years, deep learning is widely used for hyperspectral image (HSI) classification, among them, convolutional neural network (CNN) is most popular. In this paper, we propose a method for hyperspectral data classification by multi-channel convolutional neural network (MC-CNN). In this framework, one dimensional CNN (1D-CNN) is mainly used to extract the spectral feature of hyperspectral images, two dimension CNN (2D-CNN) is mainly used to extract the spatial feature of hyperspectral images, three-dimensional CNN (3D-CNN) is mainly used to extract part of the spatial and spectral information. And then these features are merged and pull into the full connection layer. At last, using neural network classifiers like logistic regression, we can eventually get class labels for each pixel. For comparison and validation, we compare the proposed MC-CNN algorithm with the other three deep learning algorithms. Experimental results show that our MC-CNN-based algorithm outperforms these state-of-the-art algorithms. Showcasing the MC-CNN framework has huge potential for accurate hyperspectral data classification.

Keywords: Deep learning · Hyperspectral image classification Convolutional neural network · Full connection layer · Logistic regression

1 Introduction

With the development of spectral sensors and remote sensing technology, we can obtain both spectral and spatial information of hyperspectral data at the same time. Therefore hyperspectral image has been widely employed in a range of successful applications. Classification of each pixel in hyperspectral image plays a crucial role in those applications. However, there are still some disadvantages in the classification process: First, hyperspectral data has a very high dimension. Second, limited number of

C. Chen, J.-J. Zhang—These authors contributed equally to the paper as first authors.
Q. Yan, L.-N. Xu—These authors contributed equally to the paper as second authors.

D.-S. Huang et al. (Eds.): ICIC 2018, LNAI 10956, pp. 81–92, 2018.
https://doi.org/10.1007/978-3-319-95957-3_10

training samples. Third, large spatial variability of spectral signature [1]. In order to solve these shortcomings, we need to extract richer and more abstract features.

In the past 20 years, a large number of different classification methods have been proposed to deal with hyperspectral data classification, typical classifiers include these based on K-Nearest Neighbor (KNN) [2], Support Vector Machine (SVM) [3] and Spectral Angle Mapping (SAM) [4]. Traditional classifiers like K-Nearest Neighbor and Spectral Angle Mapping can be attributed to single-layer classifiers, SVM with kernels can be attributed to have two layers [5], due to fewer layers, causing the extracted features to not represent the raw data well and has a large variability. From the physical sense, hyperspectral data includes not only some independent and disordered spectral feature vectors, these vectors have a certain relationship in the spatial position. The closer the pixels in the spatial, the greater the possibility of belonging to the same class. To improve classification accuracy, an idea is to design classifiers using both spatial and spectral features. Machine learning methods with multiple layers of processing extract more invariant and abstract information of data, thus are believed to have the ability of yielding higher classification effect than these traditional, shallower classifiers.

With the development of deep learning techniques, this question has been satisfactorily resolved. Deep learning framework has a multi-layer structure, so this classifier can potentially lead to progressively more complex and abstract features. Deep learning methods have been shown to yield promising performance in many field including regression or classification tasks that involve language [6], speech [7] and image [8]. The commonly used deep learning framework includes Stacked Autoencoders (SAEs) [9, 10], Deep Belief Networks (DBNs) [11] and Convolutional neural networks (CNNs) [12–14]. Of course these deep learning methods also have some shortcomings, such as SAEs and DBNs, need to pull the raw data into the column vector, this will destroy the original data structure. Two dimension CNN has extracted more complete spatial information, but spectral information extraction is not enough. One dimension CNN has extracted more complete spectral information, but ignore the information on the spatial. Although three dimension CNN can automatically extract spatial information and spectral information, but it is challenging to achieve an optimal balance between convolution kernel depth and number of hyperspectral image bands, therefore we can not get more complete spectral and spatial information. Shortcomings of the above methods may lead to undesirable classification results, so we need to extract more adequate features to solve the problem of the information is not enough.

In this paper, we present a novel approach, introducing MC-CNN into hyperspectral data classification. Here, we mainly used three different convolution channels for feature extraction, they are 1D-CNN, 2D-CNN and 3D-CNN [15] respectively. That is to say, we used three different algorithms to convolution the same data cube, then the data of each convolution channel is pulled into the column vector. Finally, the combined column vector is used as the input of the full connection layer, and then used the logical regression classifier to get the final classification results. For validation and comparison, we test the proposed framework along with three other deep learning-based hyperspectral classification algorithms-namely, 2D-CNN [16], DC-CNN [17], and 3D-CNN [18], respectively. The next section will detail the experimental algorithms. The proposed approach on two benchmark hyperspectral image datasets in a number of experimental settings. Experimental results demonstrate that our proposed MC-CNN-based framework outperforms these state-of-the-art framework and sets a new record.

The remainder of this paper is organized as follow. The second part mainly describes the MC-CNN based in hyperspectral image classification framework. The third part describes the specific experimental step and experimental results. The last part of this paper mainly analyzes the insights and reflections of the experimental, then summarize the whole paper, and describe some of the follow-up work.

2 Multi-Channel Convolutional Neural Network

2.1 One Dimensional CNN

1D-CNN is composed of convolution layers and pool layers, but the size of the convolution kernel and the pool layer are one-dimensional. The loss of network is minimized using stochastic gradient descent with back propagation. 1D-CNN framework as shown in Fig. 1.

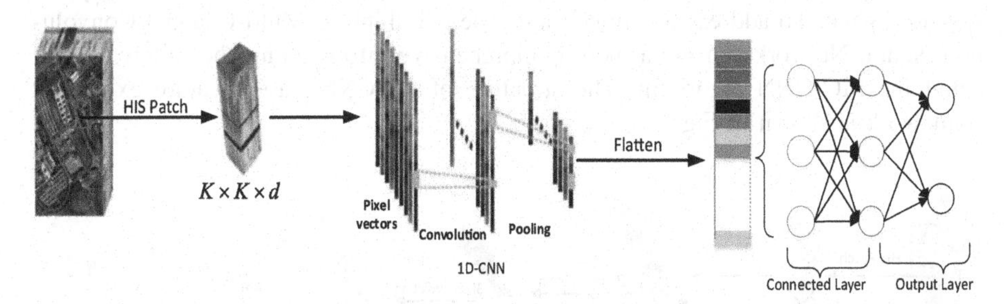

Fig. 1. Illustration of the one-dimension convolutional neural network (1D-CNN) based hyperspectral imagery classification framework.

2.2 Two Dimensional CNN

Since 1D-CNN convolution kernels are one-dimensional, and the convolution object is a pixel vector, thus only spectral features can be extracted. However 2D-CNN used 2D kernels for the 2D convolution operation, so we can extract the enough spatial features, but some spectral features are ignored. 2D-CNN is also composed of convolution layers and pooling layers. Convolution layer is mainly used to extract more abstract features, the pooling layer is mainly used to reduce the spatial resolution. 2D-CNN is using 2-dimensional convolution kernel to calculate feature map and the original data does not need to be rearranged into pixel vectors. The structure of 2D-CNN based feature extraction framework is shown in Fig. 2.

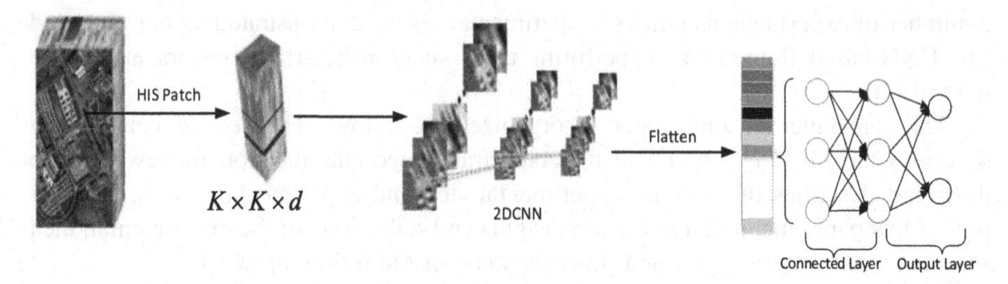

Fig. 2. Illustration of the two-dimension convolutional neural network (2D-CNN) based hyperspectral imagery classification framework.

2.3 Three Dimensional CNN

Similar to 2D-CNN, 3D-CNN is also consists of convolution layer and pooling layer. 3D-CNN uses 3D kernels for the 3D convolution operation, can extract some spectral features and some spatial features simultaneously. However, the spectral depth of 3D kernel is more difficult to determine. In other words, it is difficult to get the best match between the depth of the convolution kernel and the spectral dimension of hyperspectral image. To address this important issue, we think of Multi-Channel Convolution Neural Network can get a more comprehensive information. The following will detail the MC-CNN algorithm. The structure of 3D-CNN based feature extraction framework is shown in Fig. 3.

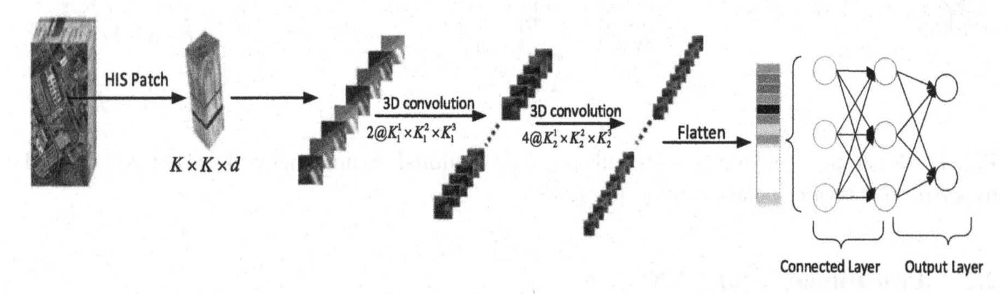

Fig. 3. Illustration of the three-dimension convolutional neural network (3D-CNN) based hyperspectral imagery classification framework.

2.4 Multi-Channel CNN

In this paper, we mainly use Multi-Channel CNN based in hyperspectral image feature extraction. and then combine the information obtained from each channel. We simultaneously use 1D-CNN, 2D-CNN, and 3D-CNN to extract information from feature cube, then pull the data cube generated by the convolution into the column vector, and these column vectors are concatenated together, the final classification step is carried out. The structure of MC-CNN based feature extraction framework is shown in Fig. 4.

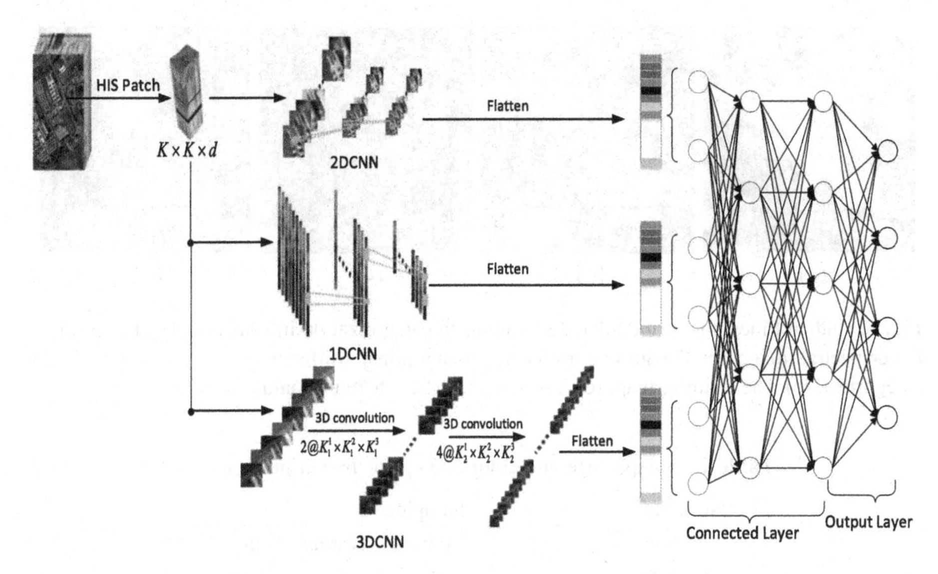

Fig. 4. Illustration of the Multi-dimension convolutional neural network (MC-CNN) based hyperspectral imagery classification framework.

3 Experimental Steps and Results

3.1 Experiment Datasets

In order to verify the effectiveness of the proposed algorithm, we compare it with three other deep learning based hyperspectral data classification algorithm. Using two different public available hyperspectral data in this paper.

3.1.1 Indian Pines Scene

Indian Pines scene was gathered by AVIRS sensor over the Indian Pines test site in North-western Indiana and consists of 145 × 145 pixels with a moderate spatial resolution of twenty meters. In this paper, we keep all the 220 bands in our experiments. The ground truth available is designated into sixteen classes and is not all mutually exclusive.

In this paper, we randomly chose 5% of the available samples as training samples for each class from the original hyperspectral data, the rest is used as a test sample. Indian Pines image and related ground truth categorization information are shown in Fig. 5. The number of training, test and total samples for each class are show in Table 1.

3.1.2 Pavia University Scene

PaviaU acquired by the ROSIS sensor during a flight campaign over Pavia University. The number of spectral bands is 103 after the 12 noisiest ones are removed, and is 610 × 340 pixels in size. The ground truth available is designated into nine classes and

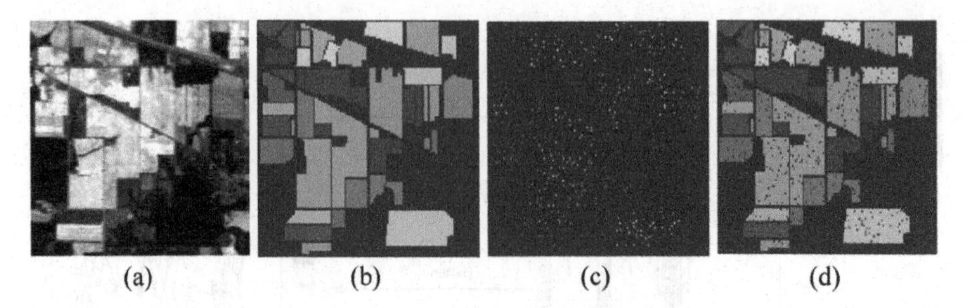

Fig. 5. Indian Pines image and related ground truth categorization information. (a) The original hyperspectral image. (b) The ground truth categorization map (different colors represent different categories). (c) The training map. (d) The test map. (Color figure online)

Table 1. Sample size and color coding for Indian pines scene.

No	Class	Samples		
	Name	Training	Testing	Total
1	Alfalfa	3	43	46
2	Corn-no till	72	1356	1428
3	Corn-min till	42	788	830
4	Corn	12	225	237
5	Grass-pasture	25	458	483
6	Grass-trees	37	693	730
7	Grass-pasture-mowed	2	26	28
8	Hay-windrowed	24	454	478
9	Oat	1	19	20
10	Soybean-no till	49	923	972
11	Soybean-min till	123	2332	2455
12	Soybean-clean	30	563	593
13	Wheat	11	194	205
14	Woods	64	1201	1265
15	Grass-trees-drives	20	366	386
16	Stone-steel-towers	5	88	93
Total		520	9729	10249

is not all mutually exclusive. We randomly choose 2 label samples as training for each class, and the rest was used for testing, as display in Table 2 and Fig. 6.

3.2 Experiment Setup

In this section, for validation and comparison, we compared MC-CNN with recently published deep learning hyperspectral image classification methods: 2D-CNN, 3D-CNN, and DC-CNN. Kappa statistic (K), average accuracy (AA) and overall accuracy (OA) were adopted to assess the classification performance of each method. In order to

Table 2. Sample size and color coding for Pavia University scene.

No	Class	Samples		
	Name	Training	Testing	Total
1	Asphalt	2	6629	6631
2	Meadows	2	18647	18649
3	Gravel	2	2097	2099
4	Trees	2	3062	3064
5	Painted metal sheets	2	1343	1345
6	Bare soil	2	5027	5029
7	Bitumen	2	1328	1330
8	Self-blocking bricks	2	3680	3682
9	Shadows	2	945	947
Total		18	42758	42776

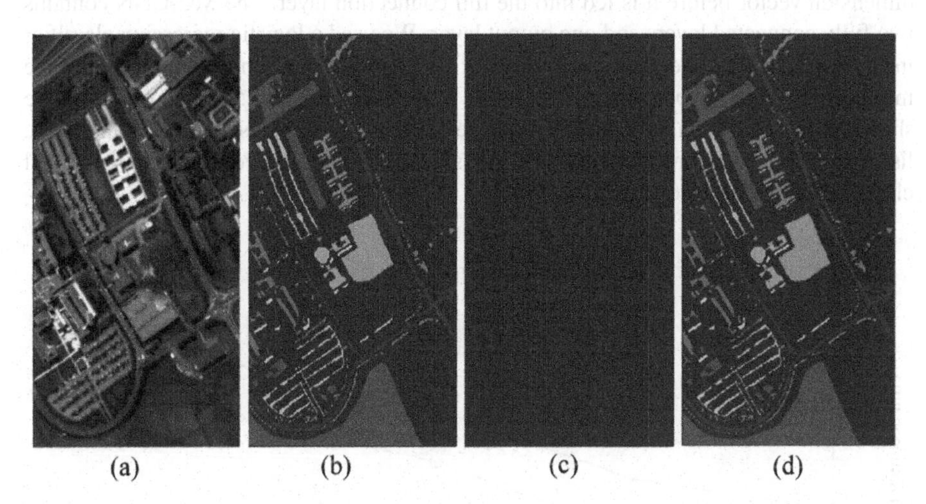

(a) (b) (c) (d)

Fig. 6. PaviaU image and related ground truth categorization information. (a) The original hyperspectral image. (b) The ground truth categorization map (different colors represent different categories). (c) The training map. (d) The test map. (Color figure online)

make the experimental results more convincing, we made the experiment 10 times for every method, take the average as the final experimental result. Detailed experimental step and parameters will be described later in detail.

3.2.1 Experimental Results for Indian Pines Scene

We extracted the composite features from Indian Pines dataset. The framework of MC-CNN is shown in Fig. 4. We extract data cube with size $11 \times 11 \times 220$. The structure parameters of each convolution channel are set as follows:

(1) Channel one: This 1D-CNN contains two pairs of convolutional and max pooling layers. The data cube is input into the convolutional layer followed by a max

pooling layer, the first convolutional layer using 300 convolutional kernels and the second convolutional layer using 200 convolutional kernels. The neuron in the pooling layer combines a small 2×1 strip of the convolutional layer.

(2) Channel two: This 2D-CNN contains two pairs of convolutional and max pooling layers. The data cube is input into the convolutional layer followed by a max pooling operational, the first convolutional layer using 300 convolutional kernels and the second convolutional layer using 200 convolutional kernels. The size of the convolutional layer is 3×3 and the max pooling layer is 2×2.

(3) Channel three: This 3D-CNN contains two pairs of convolutional and max pooling layers. The data cube is input into the convolutional layer followed by a max pooling operational, the first convolutional layer using 2 convolutional kernels and the second convolutional layer using 4 convolutional kernels. The size of the convolutional layer is $3 \times 3 \times 7$ and the max pooling layer is $5 \times 5 \times 5$.

Finally, we need to flatten the three-channel generated data cube into a 1-dimension vector before it is fed into the full connection layer. The MC-CNN contains two fully connected layers and one output layer. We used a logistic regression classifier in the output layer, and then used the back propagation algorithm to train the whole model. We varied the percentage of training samples from 1% to 5%, and the result are shown in Fig. 7. When we randomly choose 5% training samples, AA, OA, and K are listed in Table 3. In order to make the classification results more intuitive, the visual classification results are shown in Fig. 8.

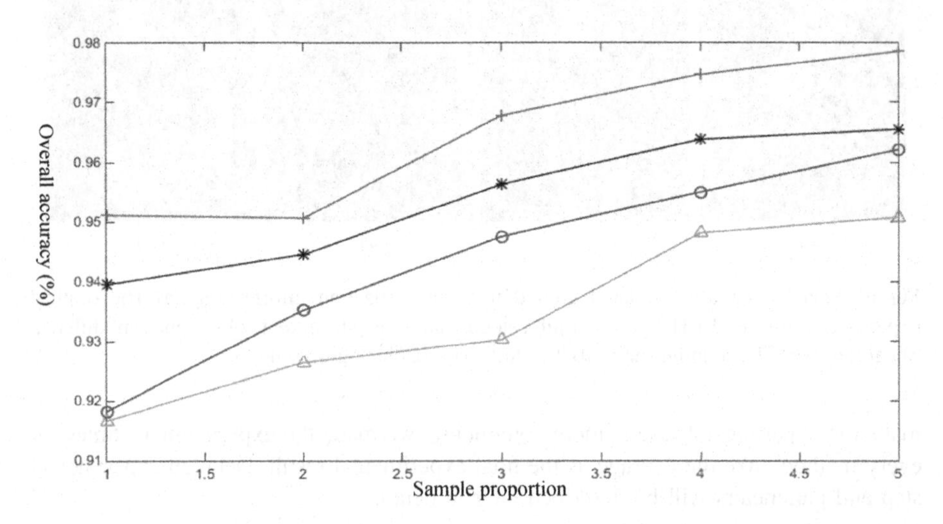

Fig. 7. Influence of sample proportion. (a) Green line represents 2D-CNN. (b) Blue line represents 3D-CNN. (c) Black line represent DC-CNN. (d) Red line represent MC-CNN. (Color figure online)

3.2.2 Experimental Results for Pavia University Scene

On this hyperspectral data, we extracted features in the same way as Indian pines scene. The framework of MC-CNN is shown in Fig. 4. We extracted data cube with size

Table 3. Sample size and color coding for Indian pines scene.

Class	Methods			
	2D-CNN	DC-CNN	3D-CNN	MC-CNN
1	0.9302	0.6744	1.0000	0.9070
2	0.9270	0.9639	0.9558	0.9683
3	0.9084	0.9810	0.9454	0.9721
4	0.8622	0.9778	0.9244	0.9689
5	0.9716	0.9432	0.9825	1.0000
6	0.9942	0.9885	0.9827	0.9942
7	0.8077	1.0000	0.9615	0.8462
8	0.9802	1.0000	0.9758	0.9912
9	0.8947	1.0000	0.9474	1.0000
10	0.9404	0.9025	0.9523	0.9567
11	0.9601	0.9704	0.9640	0.9889
12	0.8988	0.9218	0.8952	0.9627
13	0.9897	0.9845	1.0000	1.0000
14	0.9933	0.9950	0.9917	0.9825
15	0.9508	0.9863	0.9727	0.9754
16	0.9318	0.9773	0.8977	0.9659
K	0.9437	0.9606	0.9568	**0.9756**
OA	0.9507	0.9655	0.9621	**0.9786**
AA	0.9336	0.9542	0.9593	**0.9675**

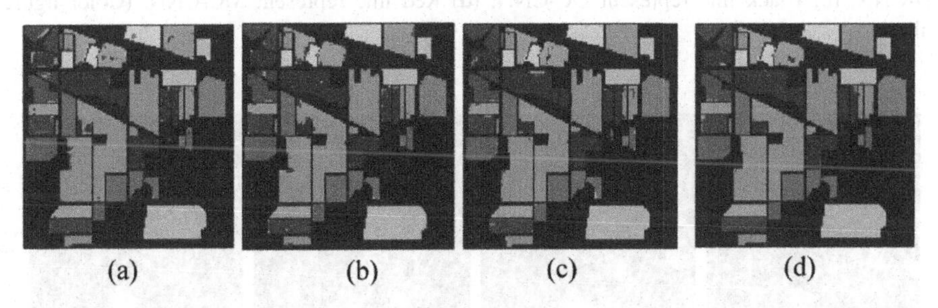

Fig. 8. The visual classification results. (a) 2D-CNN, OA = 0.9507. (b) 3D-CNN, OA = 0.9621. (c) DC-CNN, OA = 0.9655. (d) MC-CNN, OA = 0.9786.

$7 \times 7 \times 103$. In order to evaluate the performance of this paper is presented MC-CNN deep learning framework, we used a small number of labeled pixel as training samples, so only randomly select 2 label pixels from each category as training, the rest is used as a test.

In this part of the experiment, same as the previous experiment on Indian Pines dataset, there are three feature extraction channels, they are 1D-CNN, 2D-CNN and 3D-CNN respectively. All parameters involved in the compared methods are tuned in the same way as for the previous Indian Pines experiment. In other words, each channel has the same structure and parameters as Indian Pines Scene experiment.

Same as Table 3, Pavia University image test set are shown in Table 4. We varied the number of training samples from 2 to 10, and the result are shown in Fig. 9. For the same reason, we examine the classification effect from a visual perspective. The visual classification results are shown in Fig. 10.

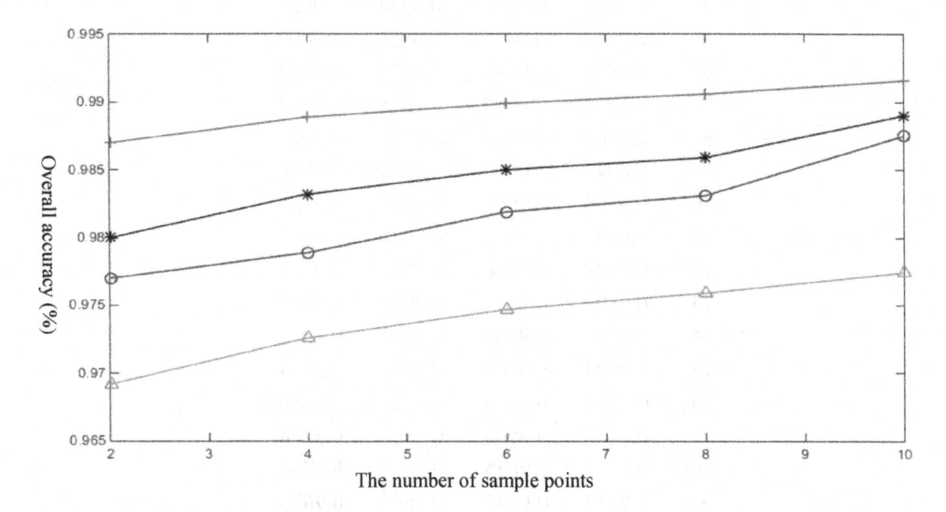

Fig. 9. Influence of sample points. (a) Green line represents 2D-CNN. (b) Blue line represents 3D-CNN. (c) Black line represent DC-CNN. (d) Red line represent MC-CNN. (Color figure online)

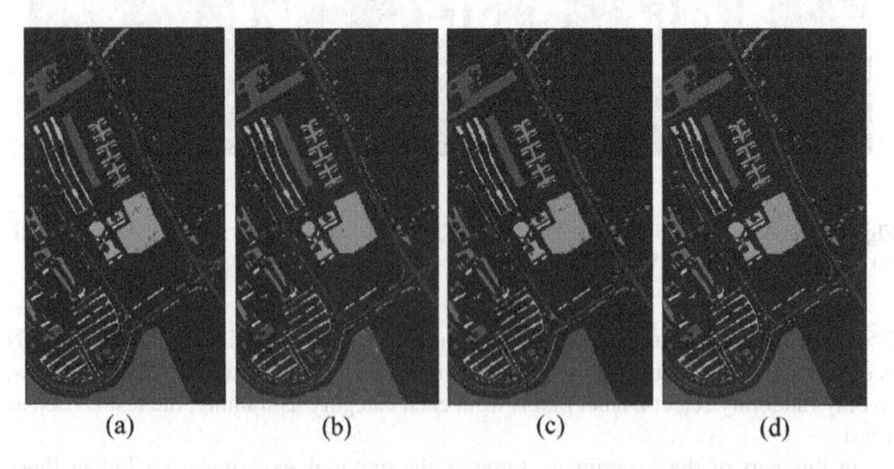

Fig. 10. The visual classification results. (a) 2D-CNN, OA = 0.9692. (b) 3D-CNN, OA = 0.9770. (c) DC-CNN, OA = 0.9800. (d) MC-CNN, OA = 0.9870.

Table 4. Sample size and color coding for Pavia University

Class	Methods			
	2D-CNN	DC-CNN	3D-CNN	MC-CNN
1	0.9689	0.9751	0.9964	0.9860
2	0.9932	0.9944	0.9927	0.9988
3	0.8584	0.9208	0.9871	0.9375
4	0.9889	0.9817	0.9909	0.9915
5	0.9993	0.9963	0.9985	0.9940
6	0.9491	0.9837	0.9946	0.9928
7	0.8336	0.9149	0.8983	0.9360
8	0.9530	0.9576	0.8438	0.9592
9	1.0000	0.9926	0.9704	0.9926
K	0.9592	0.9735	0.9696	**0.9827**
OA	0.9692	0.9800	0.9770	**0.9870**
AA	0.9494	0.9686	0.9636	**0.9765**

According to Table 4, MC-CMM acquired the best result of 0.9827 Kappa coefficient, which is higher than the second best achieved by DC-CNN. In Fig. 10, we will find MC-CNN caused the minimum number of pixels to be misclassified. Through Fig. 9, we can get a full view that the MC-CNN framework shows more competitive under a small training set.

4 Conclusion

In this paper, In order to improve classification performance, a novel MC-CNN hyperspectral image classification model has been proposed that takes full advantage of both spatial and spectral features. We proposed MC-CNN framework promote the hyperspectral image classification accuracy by utilizing multi-channel in deep feature learning. This combined approach solves the drawbacks of each feature extraction channel, thus forming complementary. Therefore, for hyperspectral image classification, our proposed MC-CNN algorithm has been proven to provide statistically higher accuracy than 2D-CNN, 3D-CNN and DC-CNN.

In the future research, we plan to investigate MC-CNN based hyperspectral classification that can exploit unlabeled samples. Because unlabeled samples are much easier to access than labeled samples in hyperspectral image, and marking hyperspectral images requires a lot of manpower and material resources. An integration of semi supervised and unsupervised classification algorithms based on MC-CNN is desirable to better solve this problem.

Acknowledgments. This work is supported by Anhui Provincial Natural Science Foundation (grant number 1608085MF 136), the National Science Foundation for China (Nos. 61602002 & 61572372).

References

1. Camps-Valls, G., Bruzzone, L.: Kernel-based methods for hyperspectral image classification. IEEE Trans. Geosci. Remote Sens. **43**, 1351–1362 (2005)
2. Samaniego, L., Bardossy, A., Schulz, K.: Supervised classification of remotely sensed imagery using a modified k-NN technique. IEEE Trans. Geosci. Remote Sens. **46**, 2112–2125 (2008)
3. Guo, B., Gunn, S.R., Damper, R.I., Nelson, J.B.: Customizing kernel functions for SVM-based hyperspectral image classification. IEEE Trans. Image Process. **17**, 622–629 (2008). A Publication of the IEEE Signal Processing Society
4. Shafri, H.Z.M., Suhaili, A., Mansor, S.: The performance of maximum likelihood, spectral angle mapper, neural network and decision tree classifiers in hyperspectral image analysis. J. Comput. Sci. **3**, 419–423 (2007)
5. Bengio, Y., Courville, A., Vincent, P.: Representation learning: a review and new perspectives. IEEE Trans. Pattern Anal. Mach. Intell. **35**, 1798 (2013)
6. Yu, D., Deng, L., Wang, S.: Learning in the deep-structured conditional random fields. In: NIPS Workshop on Deep Learning for Speech Recognition and Related Applications, pp. 1848–1852 (2009)
7. Mohamed, A.R., Sainath, T.N., Dahl, G., Ramabhadran, B., Hinton, G.E., Picheny, M.A.: Deep belief networks using discriminative features for phone recognition. In: IEEE International Conference on Acoustics, Speech and Signal Processing, pp. 5060–5063 (2011)
8. Krizhevsky, A., Sutskever, I., Hinton, G.E.: ImageNet classification with deep convolutional neural networks. Commun. ACM **60**, 2012 (2013)
9. Schölkopf, B., Platt, J., Hofmann, T.: Greedy layer-wise training of deep networks. In: International Conference on Neural Information Processing Systems, pp. 153–160 (2007)
10. Chen, Y., Lin, Z., Zhao, X., Wang, G., Gu, Y.: Deep learning-based classification of hyperspectral data. IEEE J. Sel. Top. Appl. Earth Obs. Remote Sens. **7**, 2094–2107 (2014)
11. Hinton, G.E., Osindero, S., Teh, Y.W.: A fast learning algorithm for deep belief nets. Neural Comput. **18**, 1527–1554 (2014)
12. Geng, Y., Liang, R.Z., Li, W., Wang, J., Liang, G., Xu, C., Wang, J.Y.: Learning convolutional neural network to maximize Pos@Top performance measure (2016)
13. Geng, Y., et al.: A novel image tag completion method based on convolutional neural transformation. In: Lintas, A., Rovetta, S., Verschure, P.F.M.J., Villa, A.E.P. (eds.) ICANN 2017. LNCS, vol. 10614, pp. 539–546. Springer, Cham (2017). https://doi.org/10.1007/978-3-319-68612-7_61
14. Zhang, G., et al.: Learning convolutional ranking-score function by query preference regularization. In: Yin, H., et al. (eds.) IDEAL 2017. LNCS, vol. 10585, pp. 1–8. Springer, Cham (2017). https://doi.org/10.1007/978-3-319-68935-7_1
15. Ji, S., Yang, M., Yu, K.: 3D convolutional neural networks for human action recognition. IEEE Trans. Pattern Anal. Mach. Intell. **35**, 221–231 (2013)
16. Yu, S., Jia, S., Xu, C.: Convolutional neural networks for hyperspectral image classification. Neurocomputing **219**, 88–98 (2016)
17. Zhang, H., Li, Y., Zhang, Y., Shen, Q.: Spectral-spatial classification of hyperspectral imagery using a dual-channel convolutional neural network. Remote Sens. Lett. **8**, 438–447 (2017)
18. Li, Y., Zhang, H., Shen, Q.: Spectral-spatial classification of hyperspectral imagery with 3D convolutional neural network. Remote Sens. **9**, 67 (2017)

Improved Sub-gradient Algorithm for Solving the Lower Bound of No-Wait Flow-Shop Scheduling with Sequence-Dependent Setup Times and Release Dates

Nai-Kang Yu[1], Rong Hu[1(✉)], Bin Qian[1], Zi-Qi Zhang[1], and Ling Wang[2]

[1] School of Information Engineering and Automation, Kunming University of Science and Technology, Kunming 650500, China
ronghu@vip.163.com

[2] Department of Automation, Tsinghua University, Beijing 10084, China

Abstract. In this paper, three different lower bound models of the no-wait flow-shop scheduling problem (NFSSP) with sequence dependent setup times (SDSTs) and release dates (RDs) are proposed to minimize the makespan time criterion, which is a typical NP-hard combinatorial optimization problem. First, a mixed integer programming model for flow shop scheduling problems is proposed. Secondly, some variables and constraints are relaxed based on the model proposed in this paper. Thirdly, in order to improve the lower bound of this model, a Lagrangian relaxation (LR) model for the flow shop scheduling problem is designed and an improved subgradient algorithm is proposed to solve this problem. The algorithm simulation experiment shows the effectiveness of the algorithm proposed by the article about flow shop scheduling problem and can calculate a tight lower bound.

Keywords: Lower bound · No-wait flow-shop scheduling
Sequence dependent setup times · Release dates · Lagrangian relaxation
Subgradient

1 Introduction

Flow-shop scheduling problems (FSSPs) have caused extensive research and discussion in the field of operations research and computer science This paper focuses on no-wait flow-shop scheduling problem (NFSSP) [1], a general description of which is: each job is processed continuously on the machine in a sequence and the job cannot wait once it starts being processed. Normally, setup times (SDSTs) is considered to be included in the processing time of the part and the release times (RDs) is also set to zero. However, in the metallurgical, chemical and pharmaceutical industries, setup time needs to be considered separately and the time of arrival is not zero. Cho [2] has proved that no-wait flow-shop scheduling problem is NP-hard problem when the number of machines exceeds two. Therefore, scheduling problems with sequence-dependent setup times (SDSTs) and release dates (RDs) [3, 4] have been extensively studied by

D.-S. Huang et al. (Eds.): ICIC 2018, LNAI 10956, pp. 93–101, 2018.
https://doi.org/10.1007/978-3-319-95957-3_11

scholars. For the makespan time criterion, the NFSSP with SDSTs and RDs can be classified as $F_m|no - wait, ST_{sd}, r_j|C_{\max}$.

The no-wait flow shop scheduling problem can be modeled as the Asymmetric Traveling Salesman Problem (ATSP) [5], as a branch of mixed integer programming, the solving method of its large-scale problem has always been the focus of research. The reason why it is very difficult to solve large-scale combinatorial optimization problems by using mathematical programming model, the main existing factors are that the number of variables increases with the geometric scale of the problem scale, and the nonlinear constraints are also very difficult to deal with. So some intelligent algorithms have been proposed to solve the above problems. For instance, Qian et al. [6] developed a hybrid DE approach for FSSPs with makespan criterion. However, the solution obtained by the intelligent algorithm cannot evaluate its solution, so a mathematical model that can be accurately described is proposed to evaluate the algorithm effectively. Szwarc and Gupta [7] proposed a model to the problem of flow shop with setup time in polynomial time. Tang et al. [8] proposed a steel-making process integer programming model and applied LR to solve it. Luh and Hoitomt [9] presented LR algorithms for three manufacturing scheduling problems. Held et al. [10] proposed the subgradient method to solve the Lagrangian relaxation model. The models in the above literature all achieved good performance. Now the mathematic programming model of NFSSPs with SDSTs and RDs has not been further studied.

This paper develops a mixed integer programming model to minimize the makespan time criterion of the NFSSPs with SDSTs and RDs. In the proposed lower bound model, a certain lower bound is obtained by relaxing some constraints. Based on this, a method of improving the subgradient is proposed to improve the algorithm and obtain a relatively tight lower bound.

This article is organized as follows. In Sect. 2, This section mainly introduces the NFSSP with SDSTs and RDs. In Sect. 3, the lower bound model is described in detail. A method of improving subgradient is proposed to obtain a relatively compact lower bound. In Sect. 4, experimental results and comparisons are introduced and analyzed. Finally, in Sect. 5, we end this paper and draw some conclusions and future work.

2 Formulation of NFSSP with SDSTs and RDs

The general description of The NFSSP with SDSTs and RDs is as follows: There are n jobs and m machines, each of n jobs will be processed on machines $1, 2, \ldots, m$, and the processing time of each job is deterministic. At any time, it must be satisfied that the jobs cannot be preemptively processed and each machine only can process one job at a time. Jobs must be machined continuously on each machine, which is what no-wait conditions should mean, In a flowshop with SDSTs, In a flowshop with SDSTs, Setup time must be performed at the completion of one job and at the beginning of another job on each machine. In a flow-shop with RDs, If the machine is ready to process the job but the job has not been released yet, the machine must be left idle.

2.1 Definition of NFSSP

The NFSSP with SDSTs and RDs problem, it consists of a set $J = \{j|j_1, j_2, \ldots, j_n\}$ of n jobs and a set $M = \{k|k_1, k_2, \ldots, k_m\}$ of m machines Also, let $\pi = (j_1, j_2, \ldots, j_n)$ denote a certain permutation of jobs to be processed. Let $p_{j_i,k}$ denote the processing time of job j_i on machine k, s_{ij}^k represent the setup time of machine k between the already processed job i and job j to be processed. Finally, let r_i represent the release time of the job i, indicating that the job i cannot be processed before its release time.

Figure 1 shows a simple scheduling example. The example represents 2 machines and 3 jobs. For convenience, it is assumed all the release time is 0, that is, $r_i = 0$, for $i = 1, 2, 3$. In this instance, Fig. 1 shows a NFSSP with SDSTs Gantt chart with n = 3 and m = 2, using different color squares to represent the processing time and setting time, respectively.

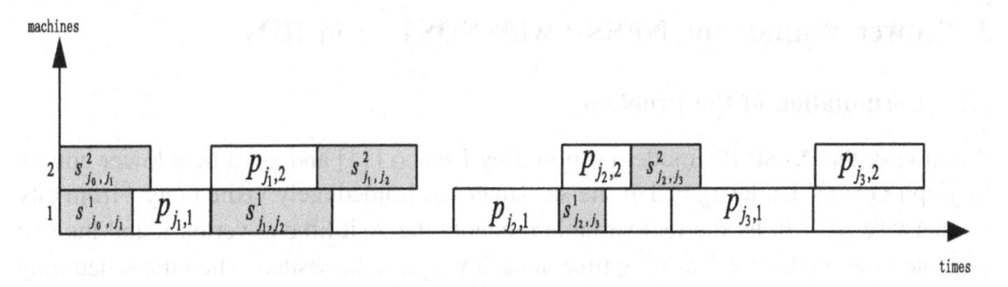

Fig. 1. An example of NFSSP with SDSTs when n = 3 and m = 2

2.2 A Graph Model

It is well-known that the no-wait flow shop scheduling problem can be reduced to an asymmetric traveling salesman problem (ATSP) with $n + 1$ cities. Similarly, it can be seen that all feasible sequences for minimizing makespan can be reduced to ATSP models with some time constraints. Let G = (V,E) denote a complete directed graph without loops, where $V = J \cup \{0\}$ is the cities set consisting of virtual city 0 and the variable c_{ij} is introduced to denote the cost between two cities i,j, where $(i,j) \in E$. In addition, according to the time constraints of the scheduling model, for each vertex i, let r_i denote the release time, indicating that the next vertex is ready to be visited. That is, when the salesman to visit a vertex can not be less than r_i, In this model, n jobs processing sequence $\pi = (j_1, j_2, \ldots, j_n)$ is actually corresponds to a Hamilton tour $H = (0, j_1, j_2, \ldots, j_n, 0)$ in the graph theory model. So the sequence model for n jobs is equivalent to finding the smallest cost of $n + 1$ cities Hamilton tour in graph G. The Hamilton Loop of the flowshop is shown in Fig. 2.

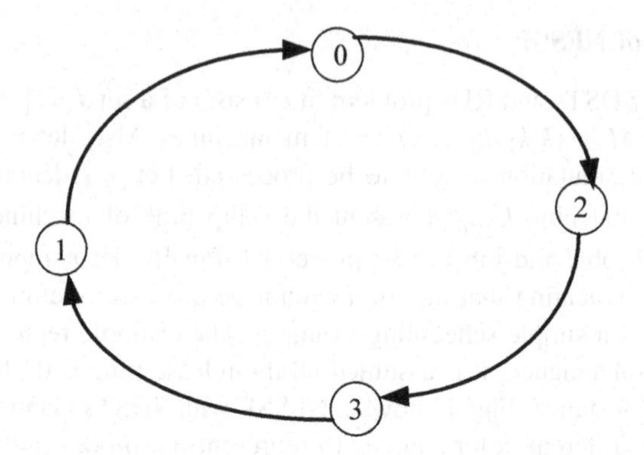

Fig. 2. The Hamiltonian loop for the scheduling model

3 Lower Bounds for NFSSP with SDSTs and RDs

3.1 Formulation of the Problem

We modify the ATSP-RT model proposed by Bianco [11] and get a new lower bound, In graph G = (V,E), let $x_{ij} = 1$ if the salesman has immediately visited city j from city i, otherwise $x_{ij} = 0$; let the real variable t_i denote the visit time at vertex i, and the real variable s denote the total waiting time at each vertex to be visited. Then the scheduling problem can be formulated by the following mixed integer program (MIP)

$$\text{MIP}: \quad LB\_C_{\max} = \min \sum\nolimits_{i=0}^{n} \sum\nolimits_{j=0}^{n} c_{ij}x_{ij} + s \tag{1}$$

s.t.

$$r_i + \sum\nolimits_{j=0}^{n} c_{ij}x_{ij} - \sum\nolimits_{i=0}^{n} \sum\nolimits_{j=0}^{n} c_{ij}x_{ij} - s \leq 0, \ \forall i = 0, 1, \cdots, n \tag{2}$$

$$\sum\nolimits_{j=0}^{n} x_{ij} = 1, \ \forall i = 0, 1, \cdots, n \tag{3}$$

$$\sum\nolimits_{i=0}^{n} x_{ij} = 1, \ \forall j = 0, 1, \cdots, n \tag{4}$$

$$u_i - u_j + ((n+1) - 1)x_{ij} \leq (n+1) - 2, \forall i, j = 0, \cdots, n \tag{5}$$

$$u_i - u_j + ((n+1) - 1)x_{ij} + ((n+1) - 3)x_{ji} \leq (n+1) - 2, \ \forall i, j = 0, \cdots, n \text{ and } i \neq j \tag{6}$$

$$s_{\min} \leq s \leq s_{\max} \tag{7}$$

$$r_0 = 0 \tag{8}$$

$$x_{ij} \in \{0, 1\}, \forall i, j = 0, 1, \cdots, n \text{ and } i \neq j \tag{9}$$

$$u_i \in \Re, \forall i = 1, \cdots, n \tag{10}$$

$$s_{\min} = \max\{LB - \sum\nolimits_{i=0}^{n} \max_{j=1,\cdots,n \text{ and } j \neq i}(c_{ij}), 0\} \tag{11}$$

$$s_{\max} = UB - \sum\nolimits_{i=0}^{n} \min_{j=1,\cdots,n \text{ and } j \neq i}(c_{ij}) \tag{12}$$

Constraints (2) propose a lower bound on the visiting time at each vertex and ensure that the time spent on the tour will not be less than the time taken from vertex i to successive vertex; Constraints (3), (4) are typical assignment constraints (AP) Assignment constraints corresponding to the scheduling problem, that is, each job can only be processed by one machine at a time, and each machine can only process one job at a time. However, under the constraints of constraints (3) and (4), it is possible to have subtours, Subtour elimination of constraints is a problem worth noting in the process of problem modeling Dantzig et al. [12] proposed a relationship between arcs and nodes to eliminate subtours, but these constraints are exponential in number, Miller et al. [13] proposed an improved subtour elimination method, that is, 'MTZ' subtour elimination; Constraints (5) and (6) are designed to prevent the occurrence of subtour in the ATSP model based on 'MTZ' subtour elimination. Constraint (7) shows that the total wait time at all vertices is between the maximum value $s_{\max}$ and minimum value $s_{\min}$, and the calculation used in this paper is given an upper bound and lower bound, calculated in the manner given in constraint (11), (12). When the variable type of x_{ij} are changed to continuous, the whole mixed integer programming model relaxes into a linear programming model, so the first lower bound is obtained. Based on the value of the first lower bound, increase the value of $s_{\min}$ to calculate the second lower bound.

3.2 Improved Lagrangian Relaxation Model

Lagrange relaxation algorithm is widely used in many optimization problems, not only scheduling problems, but also network optimization, knapsack problem and so on. In lagrangian relaxation algorithm, mostly using sub-gradient method to optimize. However, this method has some problems such as slow convergence. The step-size and direction of the algorithm in the iteration process will affect the performance of the algorithm. In the subgradient algorithm iteration process, there will be the so-called 'zigzagging phenomenon', When the iterative algorithm proceeds to the later stage, the subgradient at the λ_k and subgradient at λ_{k-1} become an obtuse angle which makes the value of λ_{k+1} very close to the value of λ_{k-1}, In other words, the two iterations have not significantly improved the results. That is, there will be the last value and the current value is very close, making the iterative route showing jagged, resulting in a very slow convergence. So this paper improves the algorithm from the direction and the step-size.

In order to obtain tighter lower bounds, a Lagrangian relaxation model is proposed and update the Lagrange multiplier vector λ, $\lambda = (\lambda_1, \lambda_2, \ldots, \lambda_m) \in R_{+}^{m}$, by using a subgradient. In Lagrange relaxation model, some constraints are called 'complicated constrains' which means these constraints are difficult to solve. For non-smooth objective function, using subgradient algorithm to optimize has obvious advantages.

The 'complicated constraint' (2) in the model proposed in the previous section is added to the objective function by the multiplier λ, and a new subtour elimination constraint is used instead of the constraints (5) and (6). The Lagrange mode

$$\begin{aligned} LM: d = \max_{\lambda \in R_+^m} z(\lambda) = \max(\min \sum\nolimits_{i=0}^{n} \sum\nolimits_{j=0}^{n} c_{ij}x_{ij} + s + \lambda^T (r_i + \sum\nolimits_{j=0}^{n} c_{ij}x_{ij} \\ - \sum\nolimits_{i=0}^{n} \sum\nolimits_{j=0}^{n} c_{ij}x_{ij} - s)) \end{aligned} \tag{13}$$

s.t.

$$\sum\nolimits_{j=0}^{n} x_{ij} = 1, \forall i = 0, 1, \cdots, n \tag{14}$$

$$\sum\nolimits_{i=0}^{n} x_{ij} = 1, \forall j = 0, 1, \cdots, n \tag{15}$$

$$\sum\nolimits_{i \in S} \sum\nolimits_{j \in \{0,1,\cdots,n\} - S} x_{ij} + x_{ji} \geq 2, \forall \varnothing \neq S \subset \{0, 1, \cdots, n\} \tag{16}$$

$$s_{\min} \leq s \leq s_{\max} \tag{17}$$

$$x_{ij} \geq 0, \ \forall i, j = 0, \ 1, \cdots, n \text{ and } i \neq j \tag{18}$$

$$\lambda_k \geq 0, k \in V \tag{19}$$

The value d in (13) is a new lower bound of the problem presented in this paper. The procedure of the subgradient algorithm is given as follows:

Step1: Let Lagrange multiplier $\lambda = 0.01$, loop = 1.
Step2: Calculate the subgradient s^{loop}, If $s^{loop} = 0$ or s^{loop} less than the setting accuracy, the algorithm is terminated and the current solution is the optimal solution, otherwise go to step 3.
Step3: $\lambda^{loop+1} = \max\{\lambda^{loop} + \xi^{loop} s^{loop}, 0\}$, Where ξ^t denotes the step-size.
Step4: loop = loop + 1, go to step 2.

From the algorithm flow, it can be concluded that the subgradient algorithm is easy to implement, and each iteration can proceed in the direction of the subgradient. The difficulty is how to select the appropriate step-size ξ. This paper adopts the following way to update the step:

$$\xi^{loop} = \varphi \frac{ub - d^{loop}}{\left\| \xi^{loop} \right\|}, 0 \prec \varphi \prec 2 \tag{20}$$

Taking the above subgradient calculation method, the algorithm has better convergence. Where *ub* is a feasible upper bound that can be corrected in the iteration, *d* represents a lower bound of the original problem, φ is a parameter used to adjust the step size when the d^{loop} does not change in several steps.

4 Test and Comparisons

4.1 Experimental Setup

In order to test the performance of the proposed lower bound model, this paper takes 6 groups of randomized experiments to do computational experiments. The combination of the number of jobs n and the number of machines m is as follows: $\{30, 50\} \times \{5, 10, 20\}$, The release time is uniformly distributed between the interval $[0, 150n\alpha]$, where α is the control parameter, the values of α are set to 0, 0.2, 0.4, 0.6, 0.8, 1 and 1.5, respectively. The processing time and setting time of the workpiece are the integers uniformly distributed in the interval $[1, 100]$, $[0, 100]$. Our algorithm is coded in Python3.0 and tested on a PC with Intel Core 3.60 GHz and 16.0-GB memory, and the mixed linear programming problems MIP are solved by the solver GUROBI.

4.2 Test Results and Comparisons

In order to compare the performance of the three kinds of lower bounds proposed in this paper, the computer experiments were carried out using the experimental design proposed in the previous section. The results are shown in Table 1, LB indicates the lower bound, *Lb*01 is the lower bound model proposed in Sect. 3.1, *Lb*02 is an improved model based on *Lb*01. *Lb*03 is Lagrange relaxation model, using the sub-gradient algorithm proposed in this paper to improve the conditions for the termination of iteration is to achieve the maximum number of iterations. As can be seen from the table, the lower bound given by the LR model is generally better than the lower bound given by the MIP model.

Table 1. Comparisons of *LB*01, *LB*02 and *LB*03 under different *r*

r	0	0	0	0.2	0.2	0.2	0.4	0.4	0.4	0.6	0.6	0.6
Lower bound	LB01	LB02	LB03	LB01	LB02	LB03	LB01	LB02	LB03	LB01	LB02	LB03
30,5	3481.47	3487.06	3594.49	3481.47	3487.06	3643.44	3481.47	3487.06	3640.09	3481.47	3487.06	3640.09
30,10	4573.70	4581.79	4740.03	4573.70	4581.79	4693.45	4573.70	4581.79	4645.19	4573.70	4581.79	4547.72
30,20	5723.62	5753.45	5721.69	5723.62	5753.45	5594.63	5723.62	5753.45	5449.64	5723.62	5753.45	5899.34
50,5	5582.40	5594.65	6002.61	5582.40	5594.65	6003.12	5582.40	5594.65	6003.08	5582.40	5594.65	6003.03
50,10	7073.18	7081.20	7146.26	7073.18	7081.20	7047.27	7073.18	7081.20	7007.34	7073.18	7081.20	6902.77
50,20	8830	8830.01	9391.40	8830	8830.01	9321.76	8830	8830.01	9199.36	8830	8830.01	8647.83
r	0.8	0.8	0.8	1.0	1.0	1.0	1.5	1.5	1.5			
Lower bound	LB01	LB02	LB03	LB01	LB02	LB03	LB01	LB02	LB03			
30,5	3481.47	3487.06	3638.65	3481.47	3487.06	3640.10	3481.47	3487.06	3639.67			
30,10	4573.70	4581.79	4562.93	4573.70	4581.79	4496.44	4573.70	4581.79	4400.08			
30,20	5723.62	5753.45	5732.29	5723.62	5753.45	5626.59	5723.62	5753.45	5534.60			
50,5	5582.40	5594.65	6001.94	5582.40	5594.65	5999.86	5582.40	5594.65	6003.91			
50,10	7073.18	7081.20	6908.87	7073.18	7081.20	8102.19	7073.18	7081.20	7994.28			
50,20	8830	8830.01	8470.06	8830	8830.01	8327.25	8830	8830.01	8658.85			

It can be seen from Table 1 that *Lb*01 is the value of the lower bound of the proposed integer programming model. When the model selects the appropriate variables and writes the complete constraints, a good value can be obtained. On the basis of *Lb*01, improve the method proposed in the previous section to improve the model, we can see that the *Lb*02 will be larger than the *Lb*01, indicating that the improvement of model 1 is effective. *Lb*03 is based on the Lagrangian relaxation model. It can also be seen from the table that a considerable part of the value is better than the *Lb*02.

5 Conclusions and Future Research

In this paper, we propose three different lower bound models for n the NFSSP with SDSTs and RDs problem, including mixed integer programming model and Lagrangian relaxation model. The significance lies in that it is very difficult to find the optimal solution to a series of combinatorial optimization problems such as scheduling problem. However, it is an effective way to evaluate the quality of solution by the difference of upper bound and lower bound. The accurate mathematical model can make up for the blindness of the search mechanism of intelligent algorithm. In this paper, the validity of the lower bound model proposed in this paper can be obtained through simulation experiments. The future research direction will start with improving the lower bound model and calculating the lower bound model of other combinatorial optimization problems.

Acknowledgements. This research is partially supported by the National Science Foundation of China (51665025), National Natural Science Fund for Distinguished Young Scholars of China (61525304), and the Applied Basic Research Foundation of Yunnan Province (2015FB136).

References

1. Graham, R.L., Lawler, E.L., Lenstra, J.K., et al.: Optimization and approximation in deterministic sequencing and scheduling: a survey. Ann. Discret. Math. **5**(1), 287–326 (1979)
2. Sahni, S., Cho, Y.: Complexity of scheduling shops with no wait in process. Math. Oper. Res. **4**(4), 448–457 (1979)
3. Urlings, T., Ruiz, R., Stützle, T.: Shifting representation search for hybrid flexible flowline problems. Eur. J. Oper. Res. **207**(2), 1086–1095 (2010)
4. Oguz, C., Salman, F.S., Yalçin, Z.B.: Order acceptance and scheduling decisions in make-to-order systems. Int. J. Prod. Econ. **125**(1), 200–211 (2010)
5. Wismer, D.A.: Solution of the flowshop-scheduling problem with no intermediate queues. Oper. Res. **20**(3), 689–697 (1972)
6. Qian, B., Wang, L., Hu, R., et al.: A hybrid differential evolution method for permutation flow-shop scheduling. Int. J. Adv. Manuf. Technol. **38**(7–8), 757–777 (2008)
7. Szwarc, W., Gupta, J.N.D.: A flow-shop problem with sequence-dependent additive setup times. Naval Res. Logist. **34**(5), 619–627 (2015)
8. Tang, L., Luh, P.B., Liu, J., et al.: Steel-making process scheduling using Lagrangian relaxation. Int. J. Prod. Res. **40**(1), 55–70 (2002)

9. Luh, P.B., Hoitomt, D.J.: Scheduling of manufacturing systems using the lagrangian relaxation technique. IEEE Trans. Autom. Control **38**(6), 1066–1079 (1993)
10. Held, M., Wolfe, P., Crowder, H.P.: Validation of subgradient optimization. Math. Program. **6**(1), 62–88 (1974)
11. Bianco, L., Ricciardelli, S., Rinaldi, G., et al.: Scheduling tasks with sequence-dependent processing times. Naval Res. Logist. **35**(2), 177–184 (1988)
12. Chvátal, V., Cook, W., Dantzig, G.B., et al.: Solution of a large-scale traveling-salesman problem. J. Oper. Res. Soc. Am. **2**(4), 393–410 (1954)
13. Miller, C.: Integer programming formulations and traveling salesman problems. J. ACM **7**, 326–329 (1960)

Cells Counting with Convolutional Neural Network

Run-xu Tan[1], Jun Zhang[1(✉)], Peng Chen[2], Bing Wang[3], and Yi Xia[1]

[1] College of Electrical Engineering and Automation, Anhui University, Hefei 230601, Anhui, China
wwwzhangjun@163.com

[2] Institute of Health Sciences, Anhui University, Hefei 230601, Anhui, China

[3] School of Electrical and Information Engineering, Anhui University of Technology, Ma Anshan 243032, China

Abstract. In this paper, we focus on the problem of cells objects counting. We propose a novel deep learning framework for small object counting named Unite CNN (U-CNN). The U-CNN is used as a regression model to learn the characteristics of input patches. The result of our model output is the density map. Density map can get the exact count of cells, and we can see the location of cell distribution. The regression network predicts a count of the objects that exit inside this frame. Unite CNN learns a multiscale non-linear regression model which uses a pyramid of image patches extracted at multiple scales to perform the final density prediction. We use three different cell counting benchmarks (MAE, MSE, GAME). Our method is tested on the cell pictures under microscope and shown to outperform the state of the art methods.

Keywords: Deep learning · Cell counting · Unite CNN

1 Introduction

Cell counting has many uses in medicine. Some routine tests such as the Clinical test of hospital, inflammation of the body, need to count the number of white blood cells in the blood, and another cell related research will be used. The classical approach to counting involves fine-tuning edge detectors to segment objects from the background [1] and counting each one. A large challenge here is dealing with overlapping objects which require methods such as the watershed transformation [2]. These approaches have many hyperparameters specifically for each task and are complicated to build. We designed a new network model, and trained on the Linux system to get the model for counting.

Using the principle and method of deep learning, we first designed a separate network structure, which has five convolutional layers and four average pooling layers. The cell data into two parts, one is the training set and the other is the test set. The training set includes 40 cell pictures, and the test set is 10 cell pictures. Although the regression model of our individual network is fast in counting, the accuracy of counting needs to be improved. In order to solve this problem, we proposed a Unite CNN model by amplifying training data set and adjusting the network parameters to optimize the

D.-S. Huang et al. (Eds.): ICIC 2018, LNAI 10956, pp. 102–111, 2018.
https://doi.org/10.1007/978-3-319-95957-3_12

network structure; through the optimization of the network, we get the best network model, and then use the method of cross validation [3] to verify the accuracy and stability of the system.

1.1 Related Work

Deep learning is a new method of image recognition in recent ten years. It originates from artificial neural network, it is a deep network structure. Deep learning has great advantages in the field of image recognition. For example, Hinton and his students at the ImageNet competition in 2012 [4], They used a convolutional neural network in deep learning to do a picture classification experiment. In this experiment, they need to divide one million pictures into one thousand classes. The error rate of the experimental classification is 15%, which is nearly 11 percentage points higher than that of second. It fully proved the superiority of deep learning. Compared with the traditional identification methods, the advantages of deep learning are that it can obtain the sample features from large amounts of data automatically, reduce the manual design steps. The final output model is obtained by continuous iteration and extraction of the characteristics of the sample.

Various methods have been proposed to traditional image counting. These method can be divided into the following categories: Detection-based method [5], Regression-based method [6] and Density estimation-based method. Here we focus on reviewing papers based on regression model counting, because our method is also part of the group. These methods define the mapping from the coordinate map of the input image to the object count. *Lempitsky et al.* introduces a counting method [7], which learns from the linear mapping of the image local feature to the density map of the object. By successful learning, we can provide object counting by simply integrating multiple regions in the estimated target density map.

We design network models by deep learning, which considers the problem of cell counting as the task of object density estimation. This method is quite different from traditional methods. We also see some good models and algorithms in the crowd counting, for example *Zhang et al.* [8] proposed a method to predict the density map with the structure of CNN, and they used two different loss function for training a switchable learning process. They extracted features using different size of filters and combined them as final count estimates. We used a simpler way to train our model, and we got better results.

2 Proposed Method

2.1 Cells Counting Model

The data we used was a red blood cell image observed under a microscope. In the experiment, we used 50 pictures of the original cells to train the network, and the number of cells in each picture is about 200. At the same time before training, we need to annotate the cells in the raw images. Annotated images provide the coordinates of the cells in the network training process, and they can be used as ground truth to ensure the accuracy of network learning.

There are five parts of our network structure. The first part and the second part respectively have two convolutional layers, the size of the convolution filter is 3 * 3, the number of convolution kernels in the first part is 64, and the number of convolution kernels in the second part is 128. The third part to the fifth part consists of three convolutional layers, the size of the convolution filter is also 3 * 3, the number of convolution filters in the third part is 256, and the number of convolution filters in the fourth and fifth parts is 512. For the four parts, A pooling layer is added behind each part. In order to ensure that the count will not cause loss because the max-pooling layer, we add the average pooling layers to the network structure. All the previous layers are followed by rectified linear units. Our Cells Counting network model is displayed in Fig. 1.

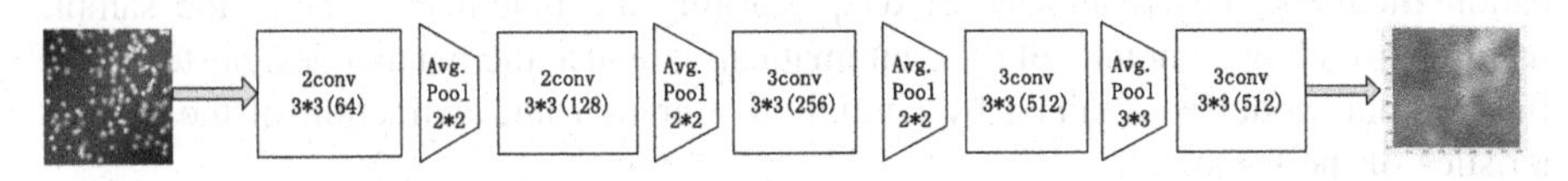

Fig. 1. Our novel cells counting model. The images of the input cells output the corresponding density map through the network, and the cell count is realized by the density map.

We propose an improved method that other people also have done, *Zhang et al.* introduced the new Crowd CNN count structure [13]. We conducted a comparison of the two methods. First of all, in our model, a convolutional layer and the nonlinear activation structure of alternating layers, can extract the deep features better than the single layer structure convolutional. Second, assuming that all data has C channels, then the individual 7 * 7 convolutional layer will contain 7 * 7 * c = 49c2 parameters, while the three 3 * 3 convolutional layers have the combination of 3 * (3 * 3 * c) = 27c2 parameters. Intuitively, it is better to choose a convolutional layer with a small filter instead of a convolutional layer with a large filter. The former can express more powerful features of input data, and less parameters are used. The only downside is that in the reverse propagation, the middle convolutional layer may cause more memory to be occupied. So the size of the convolution filters we used is 3 * 3. We do not need any extra regressor, our cells counting model is learned in an end-to-end manner to predict the object density maps directly. Finally, our experimental results will be shown in the Sect. 4.

2.2 Unite CNN

In the use of a single cell counting model, the input features need to be geometrically corrected, using annotated cell coordinates as ground truth. Through the iterative learning characteristics of the cell images, and loss rate is reducing constantly. The results of this part are shown in the fourth part. We used two methods to evaluate the results of the experiment, it called absolute error (MAE) and mean squared error (MSE), which are defined as follows:

$$MAE = \frac{1}{N}\sum_{1}^{N}|z_i - \hat{z}_i|, MSE = \sqrt{\frac{1}{N}\sum_{1}^{N}(z_i - \hat{z}_i)^2} \quad (1)$$

Where N is the number of test images, z_i is the actual number of cells in the ith image, and $\hat{z}_i$ is the estimated number of cells in the ith image. Roughly speaking, MAE indicates the accuracy of the estimates, and MSE indicates the robustness of the estimates.

Technically, the perspective distortion of image display leads to the features extracted from the same object, but there is great value difference in the depth of the different scenes. Therefore, the model with a single regression function will produce erroneous results. In order to solve this problem, we propose Unite CNN. We set the stride of the fourth avg-pooling layer to 1. The resolution of the output picture of the network is 1/8 of the resolution of the input picture. We use the hole technique to deal with the mismatch of the receiving field caused by the removal of the stride in the fourth avg-pooling layers. The convolution filter with holes can have any large receiving domain, regardless of the size of its kernel. Using holes, we double the receptive field of convolutional layers after the fourth avg-pool layer, thereby enabling them to operate with their originally trained receptive field. As illustrated in Fig. 2. Our model uses two ways, two head and three head (UCNN-h2, UCNN-h3). Each head uses Cells Counting model, then the outputs of the different heads are concatenated and passed to the three full convolutional layers, with 512 neurons each one, which are followed by a ReLU and a dropout layer. Our architecture uses a fully connected layer FC8 as the end, which has 324 neurons. The result of the final output is the density map of the cells. To train UCNN model we use the loss function defined in Eq. (2):

$$l(\Theta) = \frac{1}{2N}\sum_{n=1}^{N}\left\|R(P_n|\Theta) - D_{gt}^{(P_n)}\right\|_2^2 \quad (2)$$

Where N represents the number of training images being divided, and $D_{gt}^{(P_n)}$ represents the ground-truth density for associated training patch P_n. Recall that Θ encodes the network parameters.

$$\mathrm{GAME(L)} = \frac{1}{\mathrm{N}}\sum\nolimits_{\mathrm{n}=1}^{\mathrm{N}}\left(\sum\nolimits_{\mathrm{l}=1}^{4^{\mathrm{l}}}\left|\mathrm{D}_{\mathrm{I_n}}^{\mathrm{l}} - \mathrm{D}_{\mathrm{I_n^{gt}}}^{\mathrm{l}}\right|\right) \quad (3)$$

Where N is the total number of images, $D_{I_n}^{l}$ corresponds to the estimated object density map count for the image n and region l, and $D_{I_n^{gt}}^{l}$ is the corresponding ground truth density map. For a specific level L, the GAME(L) subdivides the image using a grid of 4^L non-overlapping regions, and the error is computed as the sum of the mean absolute errors in each of these subregions. This metric provides a spatial measurement of the error. Note that a GAME(0) is equivalent to the mean absolute error (MAE).

For visual significance estimation, our network design uses Hydra CNN [9] and Deep network [10] for reference. They put forward a different network structure, using

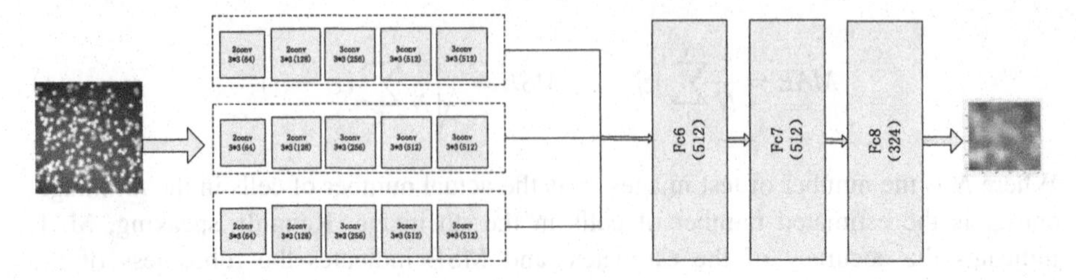

Fig. 2. Unite CNN, our network is designed to enhance the robustness of the scale change by training the patch of the multiscale image in pyramid.

a multiple input strategy. It combines the features of different views of the whole input image to return a visual saliency map. In our UCNN model, the output from layer is upsampled to the size of the input image using bilinear interpolation to obtain the final cells density map. In order to achieve the final count, we can get the total count in the image by predicting the density map. Finally, our experimental results will be shown in the Sect. 4 (Fig. 3).

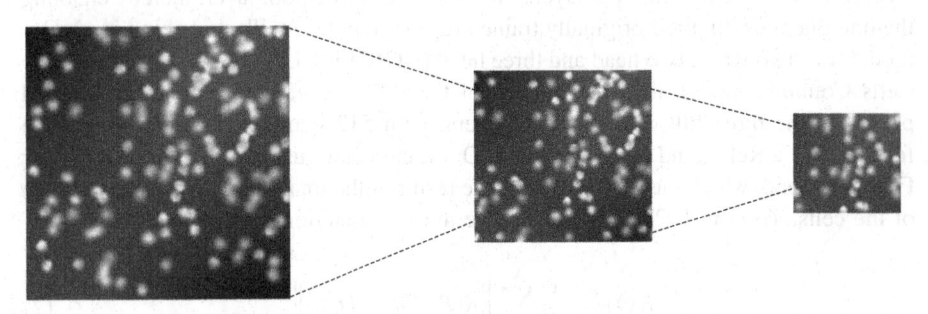

Fig. 3. Our experimental data will be processed pyramid before input, forming different sizes of patch, so that we can expand data volume and increase the stability of the experimental results.

3 Experiments

We evaluate our UCNN model on red cells dataset under microscope. We first need to process each cell and locate the location of each cell in the original picture to get dotted map. It has achieved competitive and superior performance in our cells dataset. In the end, we also demonstrate the generalizability of such a simple model in the transfer learning setting. Implementation of the proposed network and its training are based on the Caffe framework [15] developed by source code.

In a manner similar to recent works [11], we evaluate the performance of our approach using 5-fold cross validation. We randomly divide the dataset into five splits with each split containing 10 images for test, and 40 images for training. In each fold of the cross validation, we consider five splits (40 images) for training the network and the remaining split (10 images) for validating its performance. For training our models, we scale the images in order to make the largest size equal to 800 pixels. To report the results the MAE and the MSE are used. We randomly extract 1200 image patches of 150×150 pixels with their corresponding ground truth. We also augment the training data by flipping each sample. To do the test, we densely scan the image with a stride of 10 pixels. We generate the ground truth object density maps with the code provided in [12], which places a Gaussian Kernel(with a covariance matrix of $\sum = 15 \cdot 1_{2x2}$) in the center of each annotated object (Fig. 4).

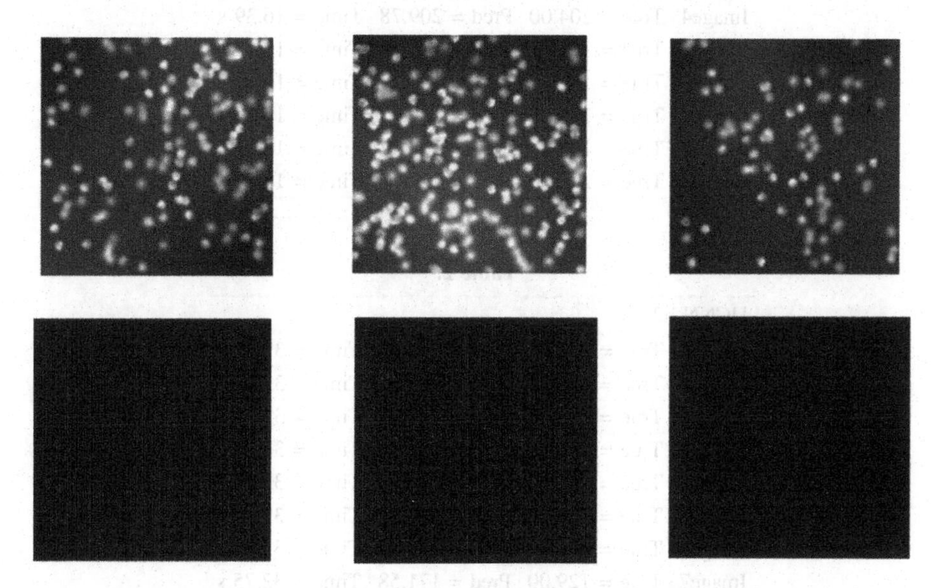

Fig. 4. The picture shows the original map and the dotted map, and the second columns are annotated as the ground truth in the training.

3.1 Cells Counting Results

We show the true results of ten pictures on the test and experimental results in Tables 1, 2 and 3, and the time for our three network framework of the experiment. We can see obviously when the network framework becomes complicated, in the same 50000 iterations of the case, it will increase the training time.

The Tables 1, 2 and 3 picture of the true count and prediction count were compared. Figure 5. shows the three methods we proposed the predictive value of fitting with the actual value. We got the first and third network (Cells Counting model and UCNN-h3) framework of the error was very small and the predicted values and the real value was close, but the network framework second (UCNN-h2) appeared a great error, the experiment effect is not good. This estimation error could possibly be a consequence of the insufficient number of training images with such large cells in the dataset.

3.2 Experiments Analysis

In this section, we have an overall analysis of all the data of the experiment. The criteria used include MAE, MSE, and GAME0-3.

Table 1.

Cells counting model			
Image0	True = 122.00	Pred = 128.15	Time = 16.44 s
Image1	True = 206.00	Pred = 207.09	Time = 16.39 s
Image2	True = 300.00	Pred = 292.70	Time = 16.37 s
Image3	True = 196.00	Pred = 197.26	Time = 16.38 s
Image4	True = 204.00	Pred = 209.78	Time = 16.39 s
Image5	True = 168.00	Pred = 171.28	Time = 16.41 s
Image6	True = 168.00	Pred = 170.23	Time = 16.60 s
Image7	True = 129.00	Pred = 140.22	Time = 16.37 s
Image8	True = 197.00	Pred = 194.42	Time = 16.39 s
Image9	True = 223.00	Pred = 220.18	Time = 16.37 s

Table 2.

UCNN-h2			
Image0	True = 122.00	Pred = 163.32	Time = 32.75 s
Image1	True = 206.00	Pred = 208.61	Time = 32.77 s
Image2	True = 300.00	Pred = 252.62	Time = 32.77 s
Image3	True = 196.00	Pred = 201.66	Time = 32.84 s
Image4	True = 204.00	Pred = 210.52	Time = 32.77 s
Image5	True = 168.00	Pred = 188.46	Time = 32.87 s
Image6	True = 168.00	Pred = 187.90	Time = 32.79 s
Image7	True = 129.00	Pred = 171.58	Time = 32.75 s
Image8	True = 197.00	Pred = 199.89	Time = 32.76 s
Image9	True = 223.00	Pred = 214.87	Time = 32.82 s

Table 3.

UCNN-h3			
Image0	True = 122.00	Pred = 125.64	Time = 46.83 s
Image1	True = 206.00	Pred = 206.96	Time = 46.86 s
Image2	True = 300.00	Pred = 292.30	Time = 46.72 s
Image3	True = 196.00	Pred = 198.10	Time = 46.95 s
Image4	True = 204.00	Pred = 207.45	Time = 46.75 s
Image5	True = 168.00	Pred = 171.44	Time = 47.09 s
Image6	True = 168.00	Pred = 168.64	Time = 46.75 s
Image7	True = 129.00	Pred = 140.07	Time = 47.76 s
Image8	True = 197.00	Pred = 194.69	Time = 46.75 s
Image9	True = 223.00	Pred = 220.70	Time = 46.77 s

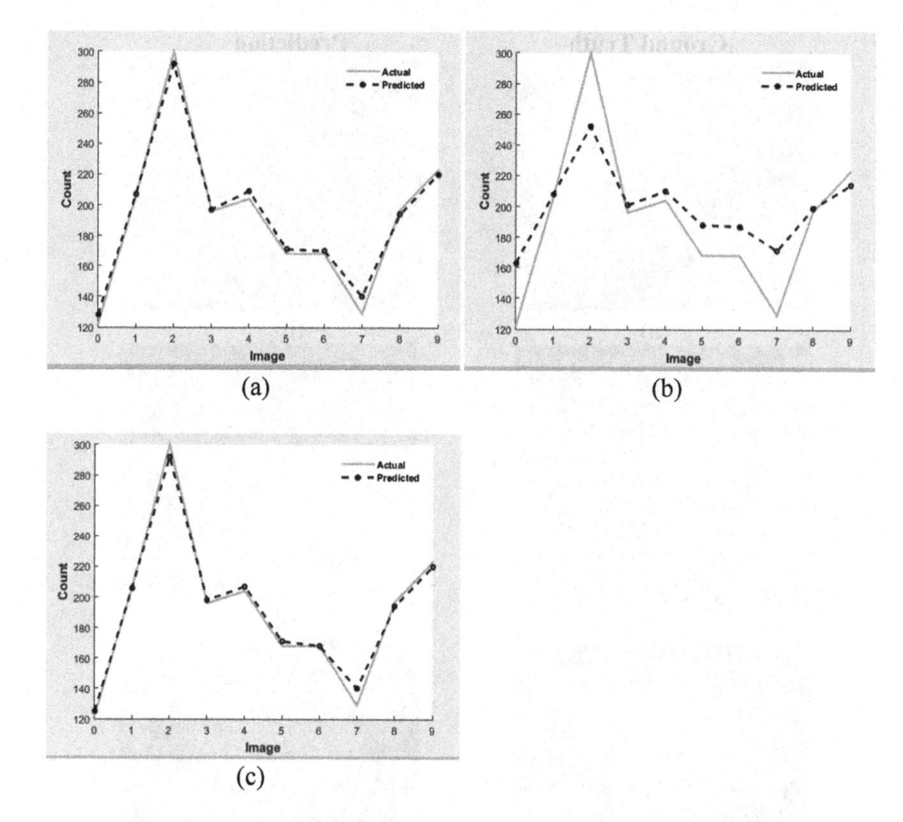

Fig. 5. Actual count vs. prediction count for cells dataset. (a) (b) (c) show the results of cells counting model, UCNN-h2, and UCNN-h3 respectively.

Table 4 shows the experimental results of the three methods we proposed at the same evaluation standard. Under this data, the best result is our UCNN-h3, which effectively reduces the value of MAE. Analyzing the results, We can see that when the angle of view is constant, the cell count error will be larger when the cell density is big. With regard to the cell count and the results provided, our network model has a certain advantage in counting the red cell dataset under the microscope.

Table 4.

Method	MAE	MSE	GAME0	GAME1	GAME2	GAME3
Cell Counting model	7.31	8.74	4.37	24.36	34.31	39.58
UCNN-h2	47.66	54.68	19.78	36.01	45.65	55.24
UCNN-h3	6.72	8.05	3.76	23.57	33.30	38.45

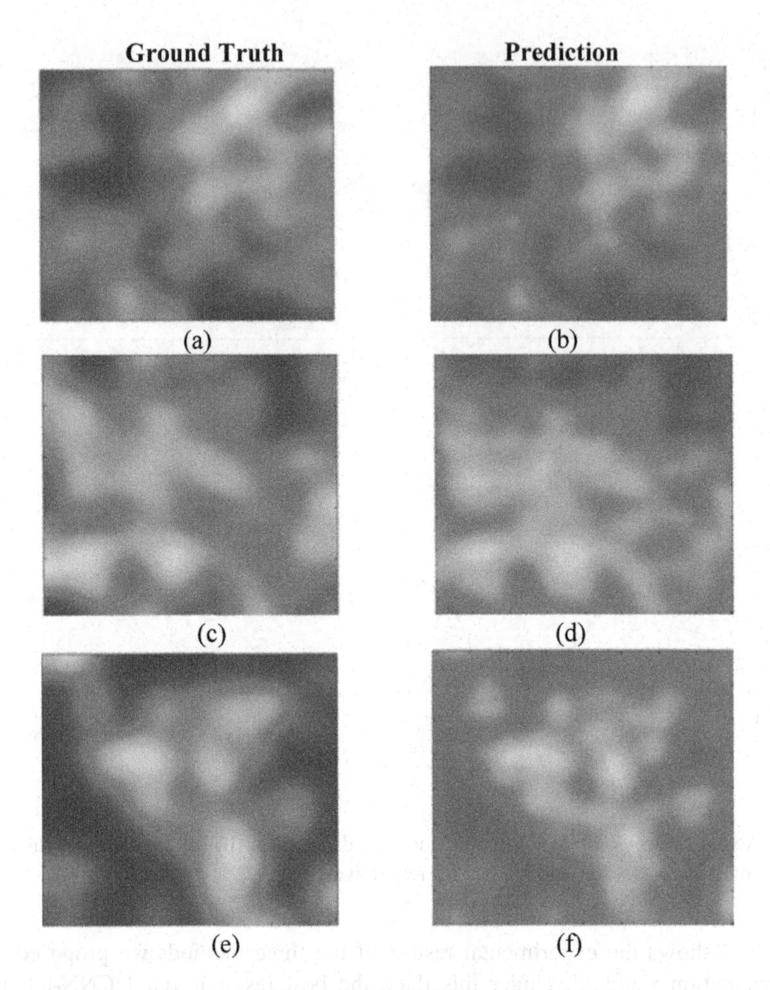

Fig. 6. Qualitative results of the UCNN-h3 in the cells dataset. (a) (c) (e) represent the density of the cells' true marked. (b) (d) (f) represent the predicted density map of the experiment.

Figure 6 shows some quantitative results. These results are obtained by our UCNN-h3 and are the best results of our experiments. The first column is our ground truth density map, and the second column is the predicted density map.

4 Conclusion

In this paper, we propose a network framework based on deep learning that can be used to count high density cell images. We use the Vgg network [14] and make a multi - column combination. Cell Counting model and UCNN-h3 have little error in cell image counting. At the same time, we also show that there are many problems to be solved in the high-density image counting, and the expansion of data volume can also improve the experimental results effectively. At present, our method has achieved good results in our own dataset.

Acknowledgement. This work is supported by Anhui Provincial Natural Science Foundation (grant number 1608085MF136).

References

1. Sezgin, M., Sankur, B.: Survey over image thresholding techniques and quantitative performance evaluation. J. Electron. Imaging **13**(1), 146–166 (2004)
2. Beucher, S.: Watershed, hierarchical segmentation and waterfall algorithm. In: Serra, J., Soille, P. (eds.) Mathematical Morphology and Its Applications to Image Processing, pp. 69–76. Springer, Dordrecht (1994). https://doi.org/10.1007/978-94-011-1040-2_10
3. Glorot, X., Bengio, Y.: Understanding the difficulty of training deep feedforward neural networks. J. Mach. Learn. Res. **9**, 249–256 (2010)
4. Krizhevsky, A., Sutskever, I., Hinton, G.E.: ImageNet classification with deep convolutional neural networks. In: International Conference on Neural Information Processing Systems, vol. 60, pp. 1097–1105. Curran Associates Inc. (2012)
5. Lempitsky, V.S., Zisserman, A.: Learning to count objects in images. In: International Conference on Neural Information Processing Systems, vol. 43, pp. 1324–1332. Curran Associates Inc. (2010)
6. Pham, V.Q., Kozakaya, T., Yamaguchi, O., Okada, R.: COUNT forest: co-voting uncertain number of targets using random forest for crowd density estimation. In: IEEE International Conference on Computer Vision, pp. 3253–3261. IEEE (2015)
7. Lempitsky, V.S., Zisserman, A.: Learning to count objects in images. In: International Conference on Neural Information Processing Systems, vol. 43, pp. 1324–1332. Curran Associates Inc. (2010)
8. Zhang, C., Li, H., Wang, X., Yang, X.: Cross-scene crowd counting via deep convolutional neural networks. In: Computer Vision and Pattern Recognition, pp. 833–841. IEEE (2015)
9. Oñoro-Rubio, D., López-Sastre, R.J.: Towards perspective-free object counting with deep learning. In: Leibe, B., Matas, J., Sebe, N., Welling, M. (eds.) ECCV 2016. LNCS, vol. 9911, pp. 615–629. Springer, Cham (2016). https://doi.org/10.1007/978-3-319-46478-7_38
10. Boominathan, L., Kruthiventi, S.S.S., Babu, R.V.: CrowdNet: a deep convolutional network for dense crowd counting, pp. 640–644 (2016)
11. Idrees, H., Saleemi, I., Seibert, C., Shah, M.: Multi-source multi-scale counting in extremely dense crowd images. In: IEEE Conference on Computer Vision and Pattern Recognition, vol. 9, pp. 2547–2554. IEEE Computer Society (2013)
12. Guerrero-Gómez-Olmedo, R., Torre-Jiménez, B., López-Sastre, R., Maldonado-Bascón, S., Oñoro-Rubio, D.: Extremely overlapping vehicle counting. In: Paredes, R., Cardoso, J.S., Pardo, X.M. (eds.) IbPRIA 2015. LNCS, vol. 9117, pp. 423–431. Springer, Cham (2015). https://doi.org/10.1007/978-3-319-19390-8_48
13. Zhang, Y., Zhou, D., Chen, S., Gao, S., Ma, Y.: Single-image crowd counting via multi-column convolutional neural network. In: Computer Vision and Pattern Recognition, pp. 589–597. IEEE (2016)
14. Simonyan, K., Zisserman, A.: Very deep convolutional networks for large-scale image recognition. Comput. Sci. (2014)
15. Jia, Y., Shelhamer, E., Donahue, J., et al.: Caffe: convolutional architecture for fast feature embedding, pp. 675–678 (2014)

Improved Adaptive Incremental Error-Minimization-Based Extreme Learning Machine with Localized Generalization Error Model

Wen-wen Han(✉), Peng Zheng, Zhong-Qiu Zhao, and Wei-dong Tian

College of Computer and Information,
Hefei University of Technology, Hefei, China
shirley.annie0714@sina.com

Abstract. Extreme learning machine (ELM) is a new type of learning algorithms for single-hidden layer feed-forward neural networks (SLFNs). The AIE-ELM aims to adaptively choose the number of hidden layer nodes for different data sets. It is an incremental extreme learning machine which achieves adaptive growth of hidden nodes and can incrementally update output weights by minimizing the training error. In order to enhance the generalization ability of AIE-ELM algorithm, this paper extends the AIE-ELM by introducing the localized generalization error model (referred to as AIEL-ELM), which takes the output sensitivity with input perturbations into account. Experimental results on several benchmark data sets verify that our proposed method can obtain the optimal number of hidden layer nodes and achieve a significant improvement of classification/regression performance and generalization ability compared with previous works.

Keywords: Extreme learning machine · Error-minimization
Local generalization error · Hidden layer nodes

1 Introduction

Recently, extreme learning machine proposed by Huang [5] has attracted the attention of many scholars. ELM is a kind of learning algorithm for feed-forward neural network, which overcomes some common problems of traditional feedforward neural network, e.g. converging slowly and falling into local optimums easily [4, 6]. For ELM, input weights and thresholds are randomly initialized while output weights can be obtained by matrix inverse. Therefore, we only need to determine the number of hidden layer nodes. However, this is not an easy task. When the number of hidden layer nodes is small, ELM may not be able to effectively learn from the training samples. And when the number of hidden layer nodes is large, it will increase the computation complexity and even result in overfitting. As a result, in order to improve the classification/regression performance and generalization ability, it is important to reduce the influence of hidden layer nodes on prediction or choose an appropriate number of hidden layer nodes. Meanwhile, the generalization ability of a trained ELM

D.-S. Huang et al. (Eds.): ICIC 2018, LNAI 10956, pp. 112–122, 2018.
https://doi.org/10.1007/978-3-319-95957-3_13

correlates with the prediction of out-of-sample data points which indicates that the data processing is crucial as well.

Zhang [16] proposed a novel incremental extreme learning machine with adaptive growth of hidden nodes and incremental updating of output weights using error-minimization-based method (AIE-ELM). It selects the optimal network structure by minimizing the training error during the incremental phase. However, when two classifiers are trained simultaneously and a small training error is arrived, the one with the large r training mean square error will produce nondeterministic outputs which tend to be deviated more from the target outputs. As a result, a small change in the inputs may have a great effect on the classification/regression results. Thus, it may not be enough to select classifiers or regression models with minimal errors.

In this paper, inspired by the local generalization error model (LGEM) introduced by Yeung [14], we proposed an improved AIE-ELM base d on this model to obtain better initial parameters and boost identification performance. LGEM optimizes on the number of hidden nodes and reduces output sensitivity to input fluctuations to earn better generalization capability. Experimental results on several dataset from the University of California at Irvine (UCI) Machine Learning Repository prove that the proposed algorithm outperforms the original AIE-ELM on both classification and regression tasks. The rest of this paper is organized as follows. In Sect. 2, the related works are reviewed. The original AIE-ELM is presented in Sect. 3. And in Sect. 4, the proposed improved AIE-ELM based on LGEM (AIEL-ELM) is described in details. Finally, experimental results are presented in Sect. 5 and some conclusion and directions for future work are highlighted in Sect. 6.

2 Related Work

In the past few years, large quantities of scholars have done lots of studies on the hidden layer structure optimization of ELM and achieved many significant achievements. To choose the optimal number of hidden layer nodes, two aspects of methods have mainly been introduced, namely pruned method and growing method. Rong et al. [11] proposed a fast pruned extreme learning machine (PELM) and applied it to classification successfully. Similarly, Michel et al. [9] proposed an optimally P-ELM (OP-ELM). And an ensemble of independent OP-ELM is integrated into OP-ELM-ER-NCL [10] with a selection criterion of negative correlation learning. These methods all try to obtain the optimal number of hidden nodes with pruning method. However, for pruning method, it is difficult to determine the initial sizes of neural networks. Meanwhile, in order to get better classification results, usually a larger size of initial neural network is required. So the initial stage will to some extent determine the overall performance, which does not sound reasonable enough. Huang and Chen [3] proposed an incremental extreme learning machine (I-ELM) based on enhanced random search. The main innovation of I-ELM is that it allows the hidden nodes to be randomly added one by one, and the output weights of the existing network are frozen when a new hidden node is added. To get a better convergence speed of I-ELM, the convex incremental extreme learning machine (CI-ELM) was proposed by Huang and Chen [2]. In CI-ELM, a novel incremental method was adopted, which originated from the concept of the Barron's convex optimization. Different from the original incremental method, CI-ELM can properly adjust the output weights of the current network when adding a new hidden node. And EFI-ELM [17], which is the error feedback incremental extreme learning machine proposed by Zou et al., solves the overfitting problem of the Meta-ELM [8]. However, both I-ELM and its extensions choose the optimal network architecture by adding hidden node one by one until the number of hidden nodes reaches the maximum, which is an obviously inefficient strategy. Then, Feng et al. [1] proposed an error minimized extreme learning machine (EM-ELM) based on the incremental learning in the growth of hidden nodes to solve this problem. EM-ELM achieves to add random hidden nodes in a single or group way. These methods all try to select the optimal network architecture with growing method. But for growing method, a unified maximum number of nodes weaken the generalization ability of the algorithm. To this end, we choose to improve the classification performance and generalization capabilities of ELM from both the network structure and the data. For the network structure, we learn from the AIE-ELM algorithm [16] to add hidden nodes adaptively rather than by setting a fixed value. For the data, we employ LGEM, which considers output sensitivity to input perturbations, to reduce the impact of the data on generalization ability.

3 AIE-ELM

The AIE-ELM algorithm is an improved error minimization based extreme learning machine. Instead of adding hidden nodes to the network in a single or group way, it takes the mechanism of generating hidden nodes randomly through adaptive methods. Moreover, the newly generated hidden nodes which are more likely to gain better performance will replace the existing ones.

The AIE-ELM can be divided into two phases.

Phase I is "Initialization Phase", which is to initialize a network with one hidden node. Firstly, a hidden node is generated and the node parameters $(a_1, b_1) \in R^d \times R$ are assigned randomly, where a_1 is the input weight and b_1 is the hidden layer bias. Then, for the input x with N training samples, calculate the hidden node output $G_1(x) = [G(a_1, b_1, x_1), \cdots, G(a_1, b_1, x_N),]^T$. Next, calculate the optimal output weight $\widehat{\beta}_1 = [G_1(x)]^{\dagger} T$ where $G_1(x)^{\dagger}$ is the Moore-Penrose generalized inverse of $G_1(x)$ and T is the target output of the ELM. Finally, the approximated function $f_1(x) = \widehat{\beta}_1 \cdot G_1(x)$ and its corresponding error $e_1 = \|f_1 - T\|$ can be obtained.

Phase II is "Recursively Growing Phase". In this phase, a temporary network φ_n is built, which selects the hidden-node number from a pool of $1, 2 \cdots \mathrm{L}_{n-1}$ and owns a residual error less than or equal to e_{n-1}. And then generate a network $\tilde{\varphi}_n^1$ with one hidden node and compare it with the network f_{n-1} according to the residual error $\tilde{e}_n^1$ and e_{n-1} to decide which one is retained. If the retained network is $\tilde{\varphi}_n^1$, then this network is just the one we are looking for in the growing phase. Otherwise, continue to generate a network $\tilde{\varphi}_n^2$ which contains two hidden nodes and compare it with f_{n-1}. This procedure will be repeated until one of these obtained temporary networks outperforms f_{n-1} or all of these networks with different number of hidden nodes (ranging from 1 to L_{n-1}) have been constructed. The best one obtained at this step is denoted as φ_n eventually.

4 The Improved ELM

Due to the input fluctuations, the AIE-ELM shows some weaknesses in learning from the unseen samples near the training samples in the entire input space. To solve this problem, the LGEM [14] is introduced. It provides the generalization error bound which may provide additional optimization guidelines for the AIEELM. In this section, we introduce the LGEM briefly at first and then provide the details of the proposed AIEL-ELM which is the improved AIE-ELM based on LGEM.

4.1 LGEM

The LGEM proposed by Yeung [14] takes the stochastic sensitivity measure (STSM) to find the generalization error bound of out-of-sample data points, which are in the Q-neighborhood of the samples, during the training process. Assuming that each unseen sample has the same probability of occurrence without any prior knowledge, the Q-neighborhood of the training sample x_b can be described as:

$$S_Q(x_b) = \{x | x = x_b + \Delta x_i, 0 < |\Delta x_i| \leq Q, \forall i = 1, 2, \cdots n\} \tag{1}$$

where n is the number of input features, Q denotes a predefined real number and Δx_i is a random variable owning zero mean uniform distributions which can be viewed as the input perturbation. In $S_Q(x_b)$, the samples other than x_b are considered to be unseen samples. S_Q is defined to be the union of all $S_Q(x_b)$, and name it as the Q-union. For $0 \leq Q_1 \leq \cdots \leq Q_k \leq \infty$, the following relationship should be satisfied: $D \subseteq S_{Q_1} \subseteq \cdots \subseteq S_{Q_k} \subseteq U$ where D is the training set and U is the input space.

The LGEM removes the generalization error of unknown samples far from the training samples from the input space T. Therefore, the localized generalization error can be defined as:

$$R_{SM}(Q) = \int_{S_Q} (f(x) - F(x))^2 p(x) dx \tag{2}$$

where $f(x)$ denotes the actual output of the classifier for the input x within S_Q, $F(x)$ represents the target output and $p(\cdot)$ is the real unknown probability density function.

According to the Hoeffding's inequality [13], the bound of RSM with a confidence level η can be described as below.

$$R_{SM}(Q) \leq \left(\sqrt{R_{emp}} + \sqrt{E_{S_Q}\left((\Delta y)^2 \right) + A} \right)^2 + \varepsilon = R^*_{SM}(Q) \tag{3}$$

where $\Delta y = f(x) - f(x_b), R_{emp} = \frac{1}{N}\sum_{b=1}^{N}(f(x_b) - F(x_b))^2$ and $\varepsilon = B\sqrt{\frac{\ln\eta}{-2N}}$. A is the range of the target outputs, B denotes the possible maximum value of the mean square error (MSE), and N is the number of the training samples. R_{emp} is the training error, and the term $E_{S_Q}\left((\Delta y)^2\right)$ denotes the ST-SM for the classifier, which can be an indicator to quantify the output changes based on the input changes.

4.2 The AIEL-ELM

According to [14], two methods can be utilized to compare two classifiers. One is to compare the Q-values of both by fixing the values of the $R^*_{SM}(Q)$. The other is to keep the Q-values and compare the value of the $R^*_{SM}(Q)$ of both. For example, given two classifiers f_1,f_2 and a predefined constant a, if $Q_1 < Q_2$ while $R^*_{SM}(Q_1) = R^*_{SM}(Q_2) = a$, then the generation ability of classifier f_2 is better than that of f_1. In the same way, with the same Q-values, the smaller the R_{SM} value is, the stronger the generalization ability is.

According to Eq. (3), if we fix $R^*_{SM}(Q) = a$, we can obtain the following equality:

$$\sqrt{E_{S_Q}\left((\Delta y)^2\right)} = \sqrt{a - \varepsilon} - \sqrt{R_{emp}} - A \tag{4}$$

Because of the large number of samples, ε is usually a tiny constant, and $a \geq \varepsilon$. The ST-SM for the classifier is always positive no matter what activation function is used. And with sigmoid activation function, the R^*_{SM} for ELM is as follows:

$$\sqrt{\frac{Q^2}{3}\sum\nolimits_{j=1}^{M}\beta_j^2\sum\nolimits_{k=1}^{n}w_{kj}^2} - \sqrt{a - \varepsilon} - \sqrt{R_{emp}} - A = 0 \tag{5}$$

Therefore, set Q^* as the real solution of Eq. (5) and k is the number of hidden nodes. We can calculate the Q-value of the network with the equation as follows:

$$h(k, Q^*) = \begin{cases} 0, & R_{emp} \geq a \\ Q^*, & else \end{cases} \tag{6}$$

And in our improved AIE-ELM, Eq. (5) is adopted to calculate the Q-values in each iteration during the growing phase.

Therefore, the process of the improved AIE-ELM based on LGEM (AIEL-ELM) is given as follows.

Algorithm AIEL-ELM. Given a set of training data $\{x_i, t_i\}_{i=1}^{N}$, the maximum number of hidden nodes L_{max}, the expected learning accuracy ε and the constant a.

Initialization Phase
Let n = 1.

1. Initialize a network with a hidden node and the parameter of the node is $(a_1, b_1) \in R^d \times R$.
2. Calculate the output of hidden layer $G_1(x)$ and optimal output weight $\hat{\beta}_1 = [G_1(x)]^{\dagger} T$.
3. Obtain the output of the network $f_1(x)$ and its corresponding error e_1.
4. Calculate the Q-value and determine whether the Q-value satisfies $R^*_{SM}(Q_1) = \text{a}$. If it meets the equality, the current network is the one we are looking for; else go to the growing phase.

Growing Phase
Let n = 2.

1. Build a temporary network.
   For k = 1: L_{n-1}
   (1) Randomly generate φ_n with k hidden nodes and denote it as $(\tilde{a}^k_n, \tilde{b}^k_n) \in R^d \times R$.
   (2) Calculate the output of the current network $\tilde{\varphi}^k_n = \sum_{i-1}^{k} \tilde{\beta}^i_n G(\tilde{a}^i_n, \tilde{b}^i_n, x)$, and its corresponding error $\tilde{e}^k_n = \|\tilde{\varphi}^k_n - T\|$ where $\tilde{\beta}^i_n$ is calculated based on the selected data set.
   (3) Compare f_{n-1} with $\tilde{\varphi}^k_n(x)$, only if $\tilde{e}^k_n > e_{n-1}$, then stop; otherwise, consider the Q-value. If $R_{emp} = \frac{1}{\mathrm{N}} \sum_{\mathrm{b}=1}^{\mathrm{N}} \left(\tilde{\varphi}^k_n(\mathrm{x_b}) - \mathrm{F}(\mathrm{x_b})\right)^2 < a$ where F(·) is the target output of the training data, calculate the Q-value for the current network; otherwise, set the Q-value as zero.
   Denote by φ_n the final obtained network in step (1) of Growing phase and by $G_{nb}(x)$ the hidden-layer output of $\varphi_n(x)$.
2. Incremental step
   (1) Randomly generate a hidden node $(a_n, b_n) \in R^d \times R$, add it to φ_n, the hidden layer output is defined as $G_n(x) = \left[G_{nb}(x)^T, G(a_n, b_n, x)\right]^T$.
   (2) Compute the output weight $\tilde{\beta}_n$ according to the corresponding data set.
   (3) Obtain the output of the incremental network $f_n(x) = \tilde{\beta}_n \cdot G_n(x)$ and its corresponding training error $e_n = \|f_n - T\|$.
3. Iteration step
   Set n = n + 1 and return to the beginning of Growing phase while $L_n < L_{max}$ and $e_n > \varepsilon$.

Finally, we need to calculate the $\mathrm{argmax}_k\, h(k, Q^*)$ for each φ_n generated in the growing phase, and choose the network with max value of $\mathrm{argmax}_k\, h(k, Q^*)$ as the best one. By combing AIE-ELM and LGEM, out-of-sample data points are properly handled and the generalization ability of the AIE-ELM algorithm is improved, which will be validated in Sect. 5.

5 Experiments and Analysis

In this section, we validate the performance of the proposed AIEL-ELM algorithm with eleven benchmark classification data sets chosen from UCI Machine Learning Repository. The specifications of the data sets are provided in Table 1.

Table 1. The basic information of the benchmark data sets.

Data sets	Attributes	Samples	Classes
Classification			
Sonar target	60	208	2
Pima diabetes	8	768	2
Ionosphere	34	351	2
Waveform	21	5000	3
Breast Cancer Wisconsin	32	569	2
Optical digit	64	5620	10
Regression			
Abalone	8	4177	-
Servo	4	167	-
Auto-MPG	8	398	-
delta ailerons	5	7129	-
delta elevators	6	9517	-

In our experiments, each data set is divided into two subsets, including training subset and testing subset. To reduce the effects of large values, all attributes have been scaled to [0, 1]. Our experiments consist of three parts with different purposes. In experiment 1, we verify our proposed method by comparing it with AIE-ELM on the regression tasks, in terms of average testing root mean square errors (RMSE) and the estimated optimal number of hidden nodes. And in experiment 2, we compare the classification performance of our algorithm with ASLGEM-ELM algorithm [12]. It aims to prove that the AIEL-ELM which combines the minimum error and local generalization error performs better. Finally, in experiment 3, we compare AIEL-ELM with other ELM algorithms on regression problems to prove the effectiveness of our proposed method. All experiments are conducted with MATLAB R 2014a as well as a 2.80 GHz CPU and 8 GB RAM PC.

The experiments for each data set were all repeated for 10 times with different randomly selected training and testing samples.

The parameters adopted in our experiments are set as follows. According to Eq. (3), the following parameters, including the difference between the range of the target output (A), the possible maximum value of the MSE (B) and the confidence level of the R^*_{SM} bound (η) are determined once the training data set is given. The constant a, which is the threshold of the R^*_{SM}, is set as 0.25. It indicates that the obtained temporary network whose training error is greater than 0.25 will be considered to have poor generalization capability. And these temporary networks will be excluded by setting the Q-value to 0.

5.1 Performance Verification of Local Generalization Error

In this subsection, we compare the proposed AIEL-ELM with AIE-ELM on regression problems. Table 2 shows the network structures of both algorithms and corresponding performance on five benchmark regression data sets. The stopping RMSE are presented in the table and the maximum of hidden nodes is set as 200. And the #Nodes denotes the average number of hidden layer nodes for the ten trials. From the table, most of the number of hidden nodes in the optimal network obtained by our model is less than that determined by the AIEELM. Meanwhile, the RMSEs of our method are lower on all the five data sets. Due to the small stop RMSE, the # Nodes of AIEL-ELM is almost the same as the original AIE-ELM on data sets with large quantities of samples, such as the delta elevators data set. It can be obviously observed that the proposed AIEL-ELM owns better generalization capability and achieves a more compact network structure.

Table 2. Performance verification of local generalization error

Dataset	Stop RSME	Algorithms	Training time(s)	Testing RSME	#Nodes
Abalone	0.08	AIE-ELM	0.0141	0.0786 ± 0.0023	10.5
		AIEL-ELM	0.0236	0.0765 ± 0.0009	9.3
Servo	0.11	AIE-ELM	0.0345	0.1349 ± 0.0103	24.6
		AIEL-ELM	0.0415	0.1194 ± 0.0154	12.4
Auto-MPG	0.08	AIE-ELM	0.0126	0.0804 ± 0.0032	20.6
		AIEL-ELM	0.0324	0.0801 ± 0.0044	18.2
Delta ailerons	0.05	AIE-ELM	0.0173	0.0456 ± 0.0026	13.5
		AIEL-ELM	0.0312	0.0431 ± 0.0011	9.5
Delta elevators	0.06	AIE-ELM	0.0289	0.0564 ± 0.0024	5.8
		AIEL-ELM	0.0411	0.0561 ± 0.0008	**5.2**

5.2 Performance Verification of Minimum Error

In this subsection, we compare the AIEL-ELM and ASLGEM-ELM to verify the improvements by combining minimum error and local generalization error on classification problems. Table 3 shows experimental results of both algorithms, including training time, testing accuracy and the estimated optimal number of hidden layer nodes. The maximum number of hidden nodes L_{max} in the AIEL- ELM is adaptive, while the one in ASLGEM-ELM is fixed. As a result, the training time of them is quite different and the training time of ASLGEM-ELM is much longer than that of AIEL-ELM on most of the data sets. From the table, it can be observed that our algorithm achieves better identification rates than ASLGEM-ELM with the estimate d optimal number of hidden nodes. In other words, the improvement of algorithm is not dependent on the local generalization error constraint, but on the joint constraint of local generalization error and minimum error.

Table 3. Performance verification of minimum error

Data sets	AIEL-ELM			ASLGEM-ELM		
	Training time(s)	Testing accuracy	#Nodes	Training time(s)	Testing accuracy	#Nodes
Sonar target	2.1684	**0.8474 ± 0.0193**	53.5	8.2969	0.8448 ± 0.0113	45.2
Pima diabetes	9.7616	**0.8427 ± 0.0028**	38.0	81.2188	0.8385 ± 0.0031	35.8
Ionosphere	3.6398	**0.9051 ± 0.0018**	35.4	9.6563	0.9034 ± 0.0021	36.3
Waveform	49.6694	**0.8560 ± 0.0050**	100.5	250.8125	0.8516 ± 0.0062	69.5
Breast Cancer Wisconsin	13.1037	**0.9706 ± 0.0008**	9.2	66.9063	0.9680 ± 0.0012	10.4
Optical digit	61.7706	**0.9783 ± 0.0011**	264.5	290.8750	0.9765 ± 0.0009	270.4

5.3 Comparison with Other ELMs

In this subsection, we compare our algorithm with other existing ELM algorithms, including I-ELM [3], CS-ELM [7] and OI-ELM [15]. In our experiments, the maximum number of hidden nodes L_{max} of I-ELM, CS-ELM and OI-ELM are set to be 100 on the same data set in order to obtain compact network structures while the number of hidden layer nodes is adaptively optimized based on different data sets in our proposed algorithm. Table 4 shows the experimental results of these four algorithms on five regression data sets. From the table, it can be found that the testing RMSEs of our

Table 4. Comparison with other ELMs

Dataset	Algorithms	Training time(s)	Testing RSME	#Nodes
Abalone	I-ELM	0.4820	0.1022 ± 0.0030	40.6
	CS-ELM	0.0375	0.0781 ± 0.0014	12.4
	OI-ELM	0.0942	0.0794 ± 0.0012	11.5
	AIEL-ELM	0.5146	**0.0765 ± 0.0009**	9.3
Servo	I-ELM	0.1398	0.1354 ± 0.0135	60.2
	CS-ELM	0.0078	0.1214 ± 0.0177	16.8
	OI-ELM	0.0555	0.1348 ± 0.0172	15.5
	AIEL-ELM	0.2365	**0.1194 ± 0.0154**	12.4
Auto-MPG	I-ELM	0.1422	0.1004 ± 0.0053	32.4
	CS-ELM	0.0125	0.0806 ± 0.0056	13.5
	OI-ELM	0.0106	0.0925 ± 0.0038	21.6
	AIEL-ELM	0.1248	**0.0801 ± 0.0044**	18.2
Delta ailerons	I-ELM	0.0235	0.0686 ± 0.0012	34.5
	CS-ELM	0.0342	0.0480 ± 0.0007	21.5
	OI-ELM	0.0218	0.0462 ± 0.0009	13.7
	AIEL-ELM	0.0312	**0.0431 ± 0.0011**	9.5
Delta elevators	I-ELM	0.0492	0.0764 ± 0.0014	8.5
	CS-ELM	0.0523	0.0670 ± 0.0005	6.8
	OI-ELM	0.0482	0.0643 ± 0.0012	7.1
	AIEL-ELM	0.0411	**0.0561 ± 0.0008**	5.2

AIEL-ELM are lower than those of other algorithms. All these results demonstrate that the proposed AIEL-ELM algorithm has better performance and generalization ability than others which is able to adaptively generate more appropriate network structure with fewer hidden layer nodes and optimal network parameters.

6 Conclusion

In this paper, we introduce the LGEM into the AIE-ELM algorithm, and propose the AIEL-ELM algorithm which achieves better classification/regression and generalization performance by automatic ally determining the optimal number of nodes in the hidden layer. Experimental results demonstrate that our proposed algorithm obtains the best performance with the combination of minimum error and local generalization error. However, in AIEL-ELM, we only consider the input perturbation and the weights perturbation will be taken into account in future works.

References

1. Feng, G., Huang, G.B., Lin, Q., Gay, R.: Error minimized extreme learning machine with growth of hidden nodes and incremental learning. IEEE Trans. Neural Netw. **20**(8), 1352–1357 (2009)
2. Huang, G.B., Chen, L.: Convex incremental extreme learning machine. Neurocomputing **70** (16), 3056–3062 (2007)
3. Huang, G.B., Chen, L.: Enhanced random search based incremental extreme learning machine. Neurocomputing **71**(16–18), 3460–3468 (2008)
4. Huang, G.B., Chen, L., Siew, C.K.: Universal approximation using incremental constructive feedforward networks with random hidden nodes. IEEE Trans. Neural Netw. **17**(4), 879 (2006)
5. Huang, G.B., Zhu, Q.Y., Siew, C.K.: Extreme learning machine: a new learning scheme of feedforward neural networks. In: Proceedings of IEEE International Joint Conference on Neural Networks, vol. 2, pp. 985–990 (2005)
6. Huang, G.B., Zhu, Q.Y., Siew, C.K.: Extreme learning machine: theory and applications. Neurocomputing **70**(1), 489–501 (2006)
7. Lan, Y., Soh, Y.C., Huang, G.B.: Constructive hidden nodes selection of extreme learning machine for regression. Neurocomputing **73**(16–18), 3191–3199 (2010)
8. Liao, S., Feng, C.: Meta-ELM: ELM with ELM hidden nodes. Neurocomputing **128**(5), 81–87 (2014)
9. Miche, Y., Sorjamaa, A., Bas, P., Simula, O., Jutten, C., Lendasse, A.: OP-ELM: optimally pruned extreme learning machine. IEEE Trans. Neural Netw. **21**(1), 158–162 (2010)
10. Mozaffari, A., Azad, N.L.: Optimally pruned extreme learning machine with ensemble of regularization techniques and negative correlation penalty applied to automotive engine coldstart hydrocarbon emission identification. Neurocomputing **131**(7), 143–156 (2014)
11. Rong, H.J., Ong, Y.S., Tan, A.H., Zhu, Z.: A fast pruned-extreme learning machine for classification problem. neurocomputing. Neurocomputing **72**(1–3), 359–366 (2008)
12. Wang, X.Z., Shao, Q.Y., Miao, Q., Zhai, J.H.: Architecture selection for networks trained with extreme learning machine using localized generalization error model. Neurocomputing **102**(2), 3–9 (2013)

13. Hoeffding, W.: Probability inequalities for sums of bounded random variables. Am. Stat. Assoc. **58**(301), 13–30 (1963)
14. Yeung, D.S., Ng, W.W.Y., Wang, D., Tsang, E.C.C., Wang, X.Z.: Localized generalization error model and its application to architecture selection for radial basis function neural network. IEEE Trans. Neural Netw. **18**(5), 1294–1305 (2007)
15. Ying, L.: Orthogonal incremental extreme learning machine for regression and multiclass classification. Neural Comput. Appl. **27**(1), 111–120 (2016)
16. Zhang, R., Lan, Y., Huang, G.-B., Soh, Y.C.: Extreme learning machine with adaptive growth of hidden nodes and incremental updating of output weights. In: Kamel, M., Karray, F., Gueaieb, W., Khamis, A. (eds.) AIS 2011. LNCS (LNAI), vol. 6752, pp. 253–262. Springer, Heidelberg (2011). https://doi.org/10.1007/978-3-642-21538-4_25
17. Zou, W., Yao, F., Zhang, B., Guan, Z.: Improved meta-elm with error feedback incremental elm as hidden nodes. Neural Comput. Appl. **8**, 1–8 (2017)

Fully Complex-Valued Wirtinger Conjugate Neural Networks with Generalized Armijo Search

Bingjie Zhang[1], Junze Wang[2], Shujun Wu[1], Jian Wang[1], and Huaqing Zhang[1](✉)

[1] College of Science, China University of Petroleum, Qingdao 266580, China
bingjie_zhang_1993@163.com, {wushujun, wangjiannl, zhhq}@upc.edu.cn
[2] College of Computer and Communication Engineering, China University of Petroleum, Qingdao 266580, China
junzewang@outlook.com

Abstract. Conjugate gradient (CG) method has been verified to be one effective strategy for training neural networks in terms of its low memory requirements and fast convergence. In this paper, an efficient CG method is proposed to train fully complex neural networks based on Wirtinger calculus. We adopt two ways to enhance the training performance. One is to construct a sufficient descent direction during training by designing a fine tuning conjugate coefficient. Another technique is to pursue the optimal learning rate instead of a fixed constance in each iteration which is determined by employing a generalized Armijo search. To verify the effectiveness and the convergent behavior of the proposed algorithm, the illustrated simulation has been performed on the complex benchmark noncircular signal.

Keywords: Complex network · Conjugate gradient · Armijo search Wirtinger

1 Introduction

Currently, the complex-valued neural networks (CVNNs) have been applied successfully in computational intelligence, pattern recognition and signal processing [1–3]. Differ from the real-valued neural networks, CVNNs employ the complex-valued parameters and variables to deal with information in complex domain. Particularly, CVNNs can reduce the number of parameters and operation, thus, they perform better in dealing with classification problems [4, 5]. There are two different kinds of CVNNs, the fully complex-valued neural networks (FCVNNs) [6–8] and the split complex-valued neural networks (SCVNNs) [9], based on the fully complex activation function and the split complex activation function, respectively. Compared with the split complex activation functions, the fully complex activation functions have significant advantages (e.g. analytic property and boundedness almost everywhere in complex domain) and have been used in multi-layer perceptions [6], radial basis function networks [7] and extreme learning machines [8].

D.-S. Huang et al. (Eds.): ICIC 2018, LNAI 10956, pp. 123–133, 2018.
https://doi.org/10.1007/978-3-319-95957-3_14

In the real-valued neural networks, the BP algorithm based gradient descent method (BPG) is the most widely used learning strategy [10–12]. As an extension from the real field to the complex domain, the complex BPG has been applied to train the SCVNNs and FCVNNs, which are called the split complex BPG (SCBPG) [9] and the fully complex BPG (FCBPG) [13], respectively. These two algorithms are both based on the gradient descent method in most cases [14–17]. However, this method converges slowly due to its consecutive steps in the updating direction of negative gradient of the objective function, especially when the iteration point is close to the optimum point. We note that the convergent rate is essentially determined by the updating direction and the learning rate. In order to accelerate the convergence, some modifications have been proposed in [18, 19]. Unfortunately, these changes still do not solve the problems very well, especially when the training procedure meets steep valleys. One important reason of this phenomenon is that the gradient method has the simple first-order convergence rate.

In order to significantly accelerate the convergent rates of BPG, the CG method and Newton method have been considered, which owe the quadratic convergence [20, 21]. Although the Newton method has the fastest convergence among these three methods, it has to calculate the Hessian matrix and its inverse, which will consume a large amount of computational memory. As a compromise algorithm, the CG method not only reaches the superlinear convergence, but also can be easily calculated without second derivative [22, 23].

With fast convergence and low computational memory, the CG method has attracted more and more attention in training real-valued neural networks. In the early stage, the linear CG method and the nonlinear CG method were introduced to solve the linear system which coefficient matrix is positive definite [24] and the massive nonlinear optimization problems [25], respectively. According to the different conjugate direction parameters, the CG methods can be categorized with three common different kinds of methods including the Hestenes-Stiefel (HS) [24, 26], Fletcher-Reeves (FR) [25] and Polak-Ribière-Polyak (PRP) [27–29] CG methods. To accelerate the convergent rate of CG method, [30] proposed a modified conjugate direction combined with Armijo search method. As for the FCVNNs, there are a few literatures that apply the conjugate gradient method. In this paper, we attempt to build a fully complex-valued neural network model based on a modified CG method.

In fact, the learning rate is also a crucial factor that affects convergence speed besides the descent direction. It is a common technique to set the learning rate as a constant in training CVNNs. In [14–17], the learning rates in the weight updating formulas for the complex-valued neural networks are all set to be small positive constants. However, the BP algorithm with fixed learning rate is prone to leading to poor convergence. As an improvement, the exact line search methods contribute to obtain the optimal learning step for each iteration in the process of training real-valued neural networks [31]. However, for BP neural networks, these methods are more time-consuming due to the amount of calculation of enormous functions and corresponding gradients. Especially when the iteration point is far away from the optimum point, these methods are not effective and reasonable. To reduce the calculation burden, the inexact line search methods are proposed, including Wolf search method [32] and Armijo search method [33]. For unconstrained optimization problems, a three terms CG method with generalized Armijo step size rule is proposed in [30]. As an inexact line search method, this generalized Armijo step size rule performs well in searching the suitable learning rate. It not only can guarantee that the objective function has rapid descent property, but also can make the eventual formed iterative sequence convergent. This search method has been proposed for training the real-valued BP neural networks in [34]. However, extending it to complex domain will be confronted with great difficulties. In this paper, we try to combine the generalized Armijo search method with a modified CG method to train a fully complex-valued neural network.

In this paper, a modified CG method with generalized Armijo rule for fully complex BP neural networks based on Wirtinger calculus is proposed, which is inspired by [30, 34]. Compared with the traditional FCBPG, it enormously accelerates the convergent rate and guarantees the sufficient descent property of error function. By Wirtinger calculus, it overcomes the conflict between the boundedness and analyticity of the complex activation function. The main contributions are as follows:

(A) To efficiently speed up the convergence, a modified fully complex-valued CG method with generalized Armijo search (FCVCGGA) is proposed to train FCVNNs.
Instead of the traditional conjugate direction coefficient, a special conjugate parameter based on this modified CG method is constructed. It not only can accelerate the convergent rates, but also can obtain the sufficient descent property of the objective function. Besides, the learning rate is not set to be a constant, but obtained by the generalized Armijo search, which also can speed up the training process. The illustrated simulations in Sect. 3 verify the high efficiency results.

(B) The framework of Wirtinger calculus is applied to the derivation. It resolves the conflict between the boundedness and differentiability of the activation function in the complex domain and greatly simplifies the complexity in describing and analyzing the algorithms.

The remaining sections of this paper are organized as follows: In Sect. 2, some notations and useful rules are given and the structure of fully CVNNs and the proposed algorithm, FCVCGGA, is introduced. Section 3 gives the experiment to support the convergence properties. Section 4 summarizes the paper.

2 Algorithm Description

2.1 Preliminaries and Fully Complex Network Structure

For simplicity, the following notations is given. $\overline{(z)}$ and $|z|$ stand for the complex conjugate and the module of a complex number z, separately. The norm of a complex vector $\mathbf{z}$ is expressed as $\|\mathbf{z}\|$, and the Frobenius norm of a complex matrix $\mathbf{Z}$ is written as $||\mathbf{Z}||$. In addition, with the Schwarz symmetry principle, we make reference to $f(\bar{z}) = \overline{f(z)}$ proposed in [6, 35, 36]. The rules of Wirtinger calculus are introduced in [15, 37–39]:

Without loss of generality, a three-layer common fully complex-valued neural network is considered with l input nodes, m hidden notes and one output node in this paper. Write $\mathbf{v}_i = (v_{i1}, v_{i2}, \cdots, v_{il})^T \in \mathbb{C}^l$ as the weights between the input and hidden layers for $i = 1, 2, \cdots, m$ and $\mathbf{V} = (\mathbf{v}_1, \mathbf{v}_2, \cdots, \mathbf{v}_m)^T \in \mathbb{C}^{m \times l}$ as the weight matrix. Denote the weight vector connecting the hidden and output layers as $\mathbf{u} = (u_1, u_2, \cdots, u_m)^T \in \mathbb{C}^m$. For brevity, we denote all weights as $\mathbf{w} = (\mathbf{u}^T, \mathbf{v}_1^T, \cdots, \mathbf{v}_m^T)^T \in \mathbb{C}^{m(l+1)}$.

Suppose that the activation functions for the hidden and output layers are set to be $f, g : \mathbb{C} \to \mathbb{C}$, separately, and write $\mathbf{F}(\mathbf{d}) = (f(d_1), f(d_2), \cdots, f(d_m))^T$ as a vector-valued function of $\mathbf{d} = (d_1, d_2, \cdots, d_m)^T \in \mathbb{C}^m$. Then, for any complex-valued training vector $\mathbf{z} \in \mathbb{C}^l$, the actual output can be described as follows:

$$y = g(\mathbf{u} \cdot \mathbf{F}(\mathbf{V}\mathbf{z})). \tag{1}$$

Assume that the K training samples are given as a set $\{\mathbf{z}^k, o^k\}_{k=1}^K \subset \mathbb{C}^l \times \mathbb{C}$, where $\mathbf{z}^k$ is the complex-valued input and o^k is the corresponding ideal output. Then, the error function that needs to be minimized is expressed by

$$E(\mathbf{w}) = \sum_{k=1}^K (o^k - y^k)(\bar{o}^k - \bar{y}^k), \tag{2}$$

where

$$y^k = g(\mathbf{u} \cdot \mathbf{F}(\mathbf{V}\mathbf{z}^k)). \tag{3}$$

According to the Schwarz symmetry principle, we notice that $f(\bar{z}) = \overline{f(z)}$, $g(\bar{z}) = \overline{g(z)}$. Thus, the conjugate of the actual output for the $k-$ th sample, $\overline{y^k}$, is expressed as

$$\overline{y^k} = g\left(\bar{\mathbf{u}} \cdot \mathbf{F}\left(\bar{\mathbf{V}}\overline{\mathbf{z}^k}\right)\right). \tag{4}$$

By the Wirtinger calculus [14, 40], there are two different forms about the gradients of $E(\mathbf{w})$ with respect to $\mathbf{w}$ and $\bar{\mathbf{w}}$. By taking partial derivative with respect to $\mathbf{w}$ and treating $\bar{\mathbf{w}}$ as a constant vector in E, we define the gradient form as $\nabla_{\mathbf{w}}E$. While taking partial derivative with respect to $\bar{\mathbf{w}}$ and treating $\mathbf{w}$ as a constant vector, we define the gradient as $\nabla_{\bar{\mathbf{w}}}E$. Thus, the direction of fastest change of $E(\mathbf{w})$ with respect to $\mathbf{w}$ can be defined as $\nabla_{\bar{\mathbf{w}}}E$. We note that $\nabla_{\bar{\mathbf{w}}}y^k = 0$ since the output y^k is not the function of $\bar{\mathbf{w}}$. Then using the chain rule of the Wirtinger calculus, we can get

$$\frac{\partial E(\mathbf{w})}{\partial \bar{\mathbf{u}}} = \sum_{k=1}^K (y^k - o^k) g'\left(\bar{\mathbf{u}} \cdot \mathbf{F}\left(\bar{\mathbf{V}}\overline{\mathbf{z}^k}\right)\right) \mathbf{F}\left(\bar{\mathbf{V}}\overline{\mathbf{z}^k}\right), \tag{5}$$

$$\frac{\partial E(\mathbf{w})}{\partial \bar{\mathbf{v}}_i} = \sum_{k=1}^K (y^k - o^k) g'\left(\bar{\mathbf{u}} \cdot \mathbf{F}\left(\bar{\mathbf{V}}\overline{\mathbf{z}^k}\right)\right) \overline{u_i} f'\left(\overline{\mathbf{v}_i} \cdot \overline{\mathbf{z}^k}\right) \overline{\mathbf{z}^k}, \tag{6}$$

where $i = 1, 2, \ldots, m$.

Write

$$\nabla_{\bar{\mathbf{w}}}E(\mathbf{w}) = \left(\left(\frac{\partial E(\mathbf{w})}{\partial \bar{\mathbf{u}}}\right)^T, \left(\frac{\partial E(\mathbf{w})}{\partial \overline{\mathbf{v}_1}}\right)^T, \ldots, \left(\frac{\partial E(\mathbf{w})}{\partial \overline{\mathbf{v}_m}}\right)^T\right)^T. \quad (7)$$

For comparison, we simply introduce the other two related algorithms for fully complex-valued BP neural networks based on Wirtinger calculus: the traditional gradient descent method (FCBPG) and the common HS CG method (FCVHSCG) obtained by simple complex-valued transformation. Then, our improved CG method with generalized Armijo search (FCVCGGA) is proposed.

For any random initial weight vector $\mathbf{w}^0$, the FCVCGGA algorithm is introduced as follows in detail.

2.2 The Modified Fully Complex-Valued CG Method with Generalized Armijo Search (FCVCGGA)

To efficiently accelerate the convergent rates of the above two algorithms, the FCVCGGA algorithm is proposed in this paper. It combines the CG method with the generalized Armijo search to greatly improve the convergent behaviors. The weights update rule of FCVCGGA is as follows:

$$\mathbf{w}^{n+1} = \mathbf{w}^n + \eta_n \overline{\mathbf{d}^n},\ n \in \mathbb{N}, \quad (8)$$

$$\overline{\mathbf{d}^n} = \begin{cases} -\nabla_{\bar{\mathbf{w}}}E(\mathbf{w}^n), & n = 0, \\ -\nabla_{\bar{\mathbf{w}}}E(\mathbf{w}^n) + \alpha_n \overline{\mathbf{d}^{n-1}}, & n \geq 1, \end{cases} \quad (9)$$

$$\begin{cases} (\nabla_{\bar{\mathbf{w}}}E(\mathbf{w}^n))^T \nabla_{\mathbf{w}}E(\mathbf{w}^n) > \left|\alpha_n (\nabla_{\mathbf{w}}E(\mathbf{w}^n))^T \overline{\mathbf{d}^{n-1}} + \alpha_n (\nabla_{\bar{\mathbf{w}}}E(\mathbf{w}^n))^T \mathbf{d}^{n-1}\right|, \\ \left|(\nabla_{\mathbf{w}}E(\mathbf{w}^n))^T \overline{\mathbf{d}^n} + (\nabla_{\bar{\mathbf{w}}}E(\mathbf{w}^n))^T \mathbf{d}^n\right| \geq 2(1+\delta_n)|\alpha_n| \|\nabla_{\bar{\mathbf{w}}}E(\mathbf{w}^n)\| \|\mathbf{d}^{n-1}\|. \end{cases} \quad (10)$$

where $\alpha_n (n \geq 1)$ is the constructed conjugate direction coefficient in the n-th training iteration. It guarantees the sufficient descent direction of error function. Actually (10) gives a range of α_n:

$$\alpha_n \in [-\alpha_n(\delta_n), \alpha_n(\delta_n)], \quad (11)$$

$$\alpha_n(\delta_n) = \frac{1}{1+\delta_n - |\cos\theta_n|} \frac{\|\nabla_{\bar{\mathbf{w}}}E(\mathbf{w}^n)\|}{\|\mathbf{d}^{n-1}\|}, \quad (12)$$

where θ_n is the complex angle between $\nabla_{\bar{\mathbf{w}}}E(\mathbf{w}^n)$ and $\mathbf{d}^{n-1}$ in the n-th training iteration, and is also the complex conjugate angle between $\nabla_{\mathbf{w}}E(\mathbf{w}^n)$ and $\overline{\mathbf{d}^{n-1}}$. δ_n is the value in the n-th training iteration, which satisfies $\delta_n \geq \delta > 0$ (δ is a positive constant).

The adaptive learning step of each iteration is automatically determined in terms of the following generalized Armijo search technique.

Let $\mu_1, \mu_2 \in (0,1)$ $(\mu_1 \leq \mu_2)$, γ_1 and γ_2 are positive constants. By the generalized Armijo search, the learning rate η_n in (8) satisfies that

$$E(\mathbf{w}^n + \eta_n \overline{\mathbf{d}^n}) \leq E(\mathbf{w}^n) + \mu_1 \eta_n ((\nabla_{\mathbf{w}} E(\mathbf{w}^n))^T \overline{\mathbf{d}^n} + (\nabla_{\bar{\mathbf{w}}} E(\mathbf{w}^n))^T \mathbf{d}^n), \tag{13}$$

and

$$\eta_n \geq \gamma_1 \quad \text{or} \quad \eta_n \geq \gamma_2 \eta_n^* > 0, \tag{14}$$

where η_n^* satisfies

$$E(\mathbf{w}^n + \eta_n^* \overline{\mathbf{d}^n}) > E(\mathbf{w}^n) + \mu_2 \eta_n^* ((\nabla_{\mathbf{w}} E(\mathbf{w}^n))^T \overline{\mathbf{d}^n} + (\nabla_{\bar{\mathbf{w}}} E(\mathbf{w}^n))^T \mathbf{d}^n). \tag{15}$$

The proposed algorithm, FCVCGGA, is distinguished from the other two common algorithms, FCBPG and FCVHSCG, on both the learning rate and descent direction. The details are listed as follows (Table 1).

Table 1. Comparison of three methods

Methods	Descent direction	Learning rate
FCBPG	$\overline{\mathbf{d}^n} = -\nabla_{\bar{\mathbf{w}}} E(\mathbf{w}^n), \quad n \in \mathbb{R}.$	Constant value [14]
FCVHSCG	$\overline{\mathbf{d}^n} = \begin{cases} -\nabla_{\bar{\mathbf{w}}} E(\mathbf{w}^n), & \text{n} = 0, \\ -\nabla_{\bar{\mathbf{w}}} E(\mathbf{w}^n) + \alpha_n \overline{\mathbf{d}^{n-1}}, & n \geq 1, \end{cases}$ where $\alpha_n = \frac{\|(\nabla_{\mathbf{w}} E(\mathbf{w}^n))^T (\nabla_{\bar{\mathbf{w}}} E(\mathbf{w}^n) - \nabla_{\bar{\mathbf{w}}} E(\mathbf{w}^{n-1}))\|}{\|(\mathbf{d}^{n-1})^T (\nabla_{\bar{\mathbf{w}}} E(\mathbf{w}^n) - \nabla_{\bar{\mathbf{w}}} E(\mathbf{w}^{n-1}))\|}$	Constant value
FCVCGGA	$\overline{\mathbf{d}^n} = \begin{cases} -\nabla_{\bar{\mathbf{w}}} E(\mathbf{w}^n), & n = 0, \\ -\nabla_{\bar{\mathbf{w}}} E(\mathbf{w}^n) + \alpha_n \overline{\mathbf{d}^{n-1}}, & n \geq 1, \end{cases}$ where α_n satisfies (10)	Generalized armijo search

We note that a constant learning rate is one common choice for FCBPG and FCVHSCG. Although it requires no additional computational burden to find the optimal learning rate, it is prone to poor convergent performance. As an improvement, some exact line search methods, such as Fibonacci search method, golden section line search and so on, have been proposed in solving complex optimization problems [41, 42]. These methods can automatically reach an optimal learning rate in each training epoch, however, they are more time-consuming in many real applications and meaningless for FCVNNs. As a trade-off, a generalized Armijo search method is presented in this paper to look for the suitable learning rate under a considerable less computational cost. Thus, in this paper, we just compare two common algorithms, FCBPG and FCVHSCG, with the proposed algorithm, FCVCGGA.

3 Experiments

In this subsection, we consider the complex benchmark noncircular signal [43] to compare the performance of the FCBPG, FCVHSCG and FCVCGGA algorithms. The signal is described as

$$\begin{aligned} z(t) = & 1.79z(t-1) - 1.85z(z-2) + 1.27z(t-3) \\ & - 0.41z(t-4) + 0.2z(t-5) + 2n(t) + 0.5\overline{n(t)} \\ & + n(t-1) + 0.9\overline{n(t-1)}. \end{aligned} \tag{16}$$

where $n(t)$ is the complex-valued doubly white circular Gaussian noise. According to the above formula, 100 inputs are selected as the training data set.

The three neural networks based on FCBPG, FCVHSCG and FCVCGGA have the identical structure with $5-6-1$ (input, hidden and output nodes). Fully complex-valued activation function, $tanh(\cdot)$, is applied for all neurons of hidden and output layers. We stochastically choose the initial weights (both the real part and imaginary part) in the interval $[-1, 1]$ with uniform distribution. The positive constant δ of FCVCGGA is set to be 0.001 with 10-fold cross-validation. The stop criteria are all set to be: the maximum iterations, $1,000$, or the error is less than 0.001.

Before each contrastive experiment, we make the preliminary numerical experiments with 10-fold cross-validation to search the suitable learning rates for FCBPG and FCVHSCG algorithms. Actually, during training the neural networks, it is a common strategy to apply the cross-validation to search the optimal learning rate. With different learning rates, some error curves are drawn in Fig. 1 for FCBPG and FCVHSCG algorithms. Table 2 shows the results of training error comparison for the three algorithms. To show the effects clearly, the logarithmic coordinate axis is used.

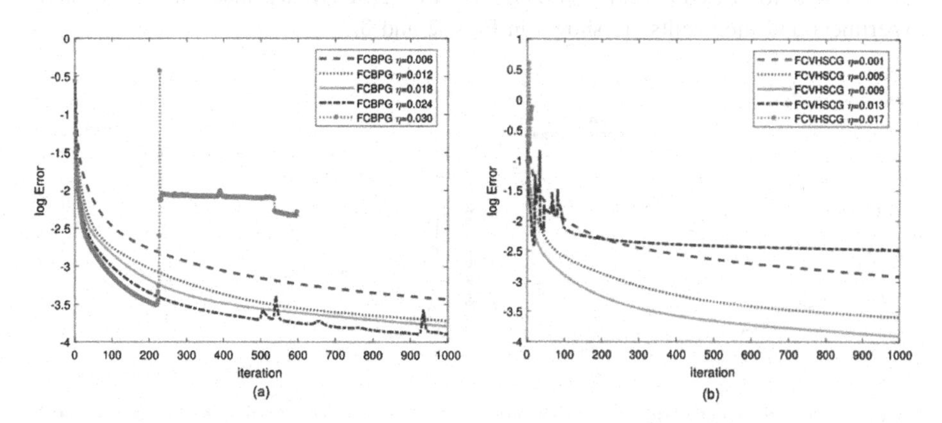

Fig. 1. Error curves for FCBPG and FCVHSCG with different learning rates: (a) FCBPG, (b) FCVHSCG.

Table 2. Error comparison for three training algorithms with different learning rates.

Algorithm	Learning rate	Error
FCBPG	0.006	3.2318×10^{-2}
	0.012	2.4225×10^{-2}
	0.018	2.2640×10^{-2}
	0.024	2.0294×10^{-2}
	0.030	*NaN*
FCVHSCG	0.001	5.4204×10^{-2}
	0.005	2.7507×10^{-2}
	0.009	2.0144×10^{-2}
	0.013	8.3795×10^{-2}
	0.017	*NaN*
FCVCGGA		1.4487×10^{-2}

As shown in Fig. 1, the curves of error functions are steadily declining when the learning rates are small and with the increase of learning rates, the oscillation risk is increasing. From Table 2, it is easy to see that the training errors of FCBPG and FCVHSCG algorithms are affected greatly by the different learning rates. For FCBPG, the minimum error is 2.0294×10^{-2} when the learning rate is 0.024 in Table 2, however, form Fig. 1(a), the corresponding training error curve is oscillating. Thus, the optimal learning rate of FCBPG is set to be 0.018 here. For FCVHSCG, the optimal learning rate is similarly chosen as 0.009. Moreover, the important thing is that among those three algorithms, the FCVCGGA algorithm performs best by comparing the magnitude of error function.

Then, by the identical structures and the selected suitable learning rates ($\eta = 0.0.018$ for FCBPG and $\eta = 0.009$ for FCVHSCG), we make the contrastive experiment and the results are shown in Figs. 2 and 3.

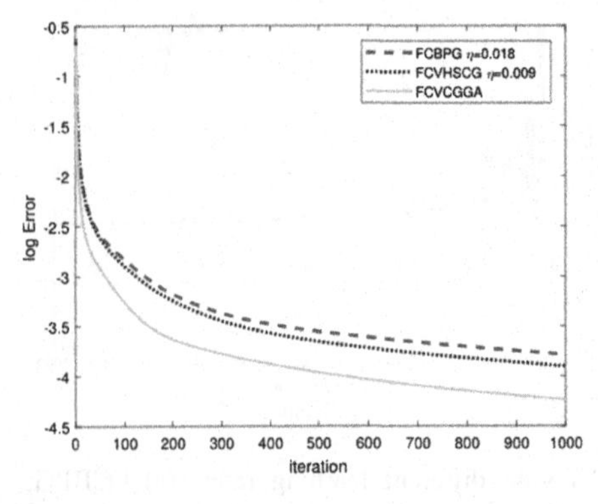

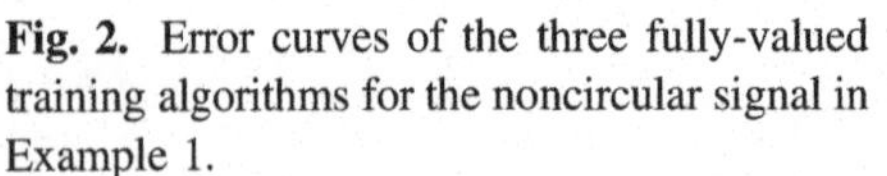

Fig. 2. Error curves of the three fully-valued training algorithms for the noncircular signal in Example 1.

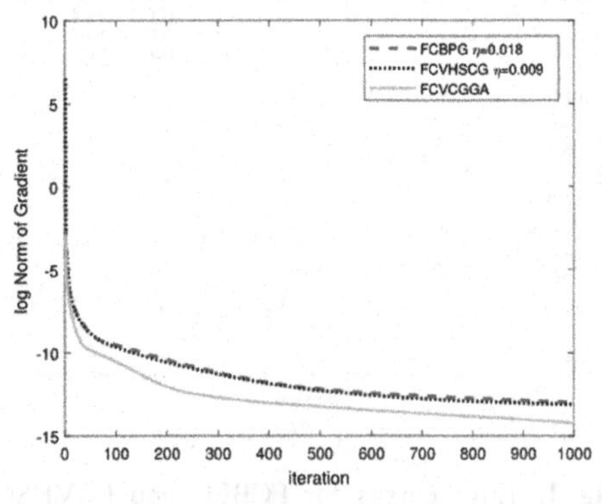

Fig. 3. Norm of gradient curves of the three fully-valued training algorithms for the noncircular signal in Example 1.

From Figs. 2 and 3, we can observe that the error function of the proposed algorithm, FCVCGGA, has the monotone decreasing property and the norm of its gradient with respect to weights approaches to zero, respectively. These simulations effectively support the theoretical results. In addition, the algorithm, FCVCGGA, evidently performs better than the other two algorithms, FCBPG and FCVHSCG.

4 Conclusion

In this paper, the FCVHSCG and FCVCGGA algorithms based on Wirti-nger calculus are proposed in training FCVNNs. By Wirtinger calculus, the conflict between the boundedness and analyticity of the complex activation function is overcome. Then, some simulations are given to verify the convergence properties. However, we only give a complex version of the typical HS CG method to train FCVNNs in this paper. In the next work, the PRP, FR and other CG methods can also be transformed from real domain to complex domain and used to train FCVNNs. Besides, their convergence analyses are also important to be considered in the practical applications.

Acknowledgments. This work was supported in part by the National Natural Science Foundation of China (No. 61305075), the Natural Science Foundation of Shandong Province (No. ZR2015A-L014, ZR201709220208) and the Fundamental Research Funds for the Central Universities (No. 15CX08011A, 18CX02036A).

References

1. Chen, S., Hong, X., Khalaf, E., Alsaadi, F.E., Harris, C.J.: Complex-valued B-spline neural network and its application to iterative frequency-domain decision feedback equalization for Hammerstein communication systems. In: 2016 International Joint Conference on Neural Networks, pp. 4097–4104 (2016)
2. Liu, Y.S., Huang, H., Huang, T.W., Qian, X.S.: An improved maximum spread algorithm with application to complex-valued RBF neural networks. Neurocomputing **216**, 261–267 (2016)
3. Fink, O., Zio, E., Weidmann, U.: Predicting component reliability and level of degradation with complex-valued neural networks. Reliab. Eng. Syst. Saf. **121**, 198–206 (2014)
4. Aizenberg, I.: Complex-Valued Neural Networks with Multivalued Neurons. Springer, Heidelberg (2011). https://doi.org/10.1007/978-3-642-20353-4
5. Nitta, T.: Solving the XOR problem and the detection of symmetry using a single complex-valued neuron. Neural Netw. **16**, 1101–1105 (2003)
6. Kim, T., Adali, T.: Approximation by fully complex multilayer perceptrons. Neural Comput. **15**, 1641–1666 (2003)
7. Savitha, R., Suresh, S., Sundararajan, N.: Metacognitive learning in a fully complex-valued radial basis function neural network. Neural Comput. **24**, 1297–1328 (2012)
8. Li, M., Huang, G., Saratchandran, P., Sundararajan, N.: Fully complex extreme learning machine. Neurocomputing **68**, 306–314 (2005)
9. Nitta, T.: An extension of the back-propagation algorithm to complex numbers. Neural Netw. **10**, 1391–1415 (1997)

10. Zhao, Z.Z., Xu, Q.S., Jia, M.P.: Improved shuffled frog leaping algorithm-based BP neural network and its application in bearing early fault diagnosis. Neural Comput. Appl. **27**, 375–385 (2016)
11. Xie, L.: The heat load prediction model based on BP neural network-markov model. Procedia Comput. Sci. **107**, 296–300 (2017)
12. Li, Z.K., Zhao, X.H.: BP artificial neural network based wave front correction for sensor-less free space optics communication. Opt. Commun. **385**, 219–228 (2017)
13. Li, H., Adali, T.: Complex-valued adaptive signal processing using nonlinear functions. EURASIP J. Adv. Sig. Process. **2008**, 1–9 (2008)
14. Zhang, H.S., Liu, X.D., Xu, D.P., Zhang, Y.: Convergence analysis of fully complex backpropagation algorithm based on Wirtinger calculus. Cogn. Neurodyn. **8**(3), 261–266 (2014)
15. Xu, D.P., Zhang, H.S., Mandic, D.P.: Convergence analysis of an augmented algorithm for fully complex-valued neural networks. Neural Netw. **69**, 44–50 (2015)
16. Zhang, H.S., Xu, D.P., Zhang, Y.: Boundedness and convergence of split-complex back-propagation algorithm with momentum and penalty. Neural Process Lett. **39**(3), 297–307 (2014)
17. Zhang, H.S., Zhang, C., Wu, W.: Convergence of batch split-complex backpropagation algorithm for complex-valued neural networks. Discrete Dyn. Nat. Soc. 1–16 (2009)
18. Papalexopoulos, A.D., Hao, S.Y., Peng, T.M.: An implementation of a neural-network-based load forecasting-model for the EMS. IEEE Trans. Power Syst. **9**, 1956–1962 (1994)
19. Lu, C.N., Wu, H.T., Vemuri, S.: Neural network based short-term load forecasting. Trans. Power Syst. **8**, 336–342 (1993)
20. Saini, L.M., Soni, M.K.: Artificial neural network-based peak load forecasting using conjugate gradient methods. IEEE Trans. Power Syst. **17**, 907–912 (2002)
21. Goodband, J.H., Haas, O.C.L., Mills, J.A.: A comparison of neural network approaches for on-line prediction in IGRT. Med. Phys. **35**, 1113–1122 (2008)
22. Hagan, M.T., Demuth, H.B., Beale, M.H.: Neural Network Design. PWS Publisher, Boston (1996)
23. Nocedal, J., Wright, S.J.: Numerical Optimization. Springer, New York (2006). https://doi.org/10.1007/978-0-387-40065-5
24. Hestenes, M.R., Stiefel, E.L.: Method of Conjugate Gradients for Solving Linear Systems. National Bureau of Standards, Washington (1952)
25. Fletcher, R., Reeves, C.M.: Function minimization by conjugate gradients. Comput. J. **7**, 149–154 (1964)
26. Liu, H., Li, X.Y.: A modified HS conjugate gradient method. In: 2011 International Conference on Multimedia Technology, pp. 5699–5702 (2011)
27. Polak, E., Ribiere, G.: Note sur la convergence de directions conjugates. Rev. Francaise d'Informatique et de Rech. Operationnelle **16**, 35–43 (1969)
28. Polyak, B.T.: The conjugate gradient method in extremal problems. USSR Comput. Math. Math. Phys. **9**, 94–112 (1969)
29. Wan, Z., Hu, C., Yang, Z.: A spectral PRP conjugate gradient methods for nonconvex optimization problem based on modified line search. Discrete Cont. Dyn. Syst. Ser. B **16**, 1157–1169 (2017)
30. Sun, Q.Y., Liu, X.H.: Global convergence results of a new three terms conjugate gradient method with generalized armijo step size rule. Math. Numer. Sinica **26**, 25–36 (2004)
31. Magoulas, G.D., Vrahatis, M.N., Androulakis, G.S.: Effective backpropagation training with variable stepsize. Neural Netw. **10**, 69–82 (1997)
32. Wang, A.P., Chen, Z.: Global convergence of a modified HS conjugate gradient method under wolfe-type line search. J. Anhui Univ. **2**, 150–156 (2015)

33. Dong, X.L., Yang, X.M., Huang, Y.Y.: Global convergence of a new conjugate gradient method with Armijo search. J. Henan Normal Univ. **43**(6), 25–29 (2015)
34. Wang, J., Zhang, B.J., Sun, Z.Q., Hao, W.X., Sun, Q.Y.: A novel conjugate gradient method with generalized Armijo search for efficient training of feedforward neural networks. Neurocomputing **275**, 308–316 (2018)
35. Needham, T.: Visual complex analysis. Am. Math. Mon. **105**, 195–196 (1998)
36. Novey, M.P.: Complex ICA using nonlinear functions. IEEE Trans. Sig. Process. **56**(9), 4536–4544 (2008)
37. Wirtinger, W.: Zur formalen theorie der funktionen von mehr komplexen ver▤nderlichen. Mathematische Annalen **97**, 357–375 (1927)
38. Kreutz-Delgado, K.: The complex gradient operator and the CR-calculus. Mathematics 1–74 (2009)
39. Mandic, D.P., Goh, S.L.: Complex Valued Nonlinear Adaptive Filters: Noncircularity. Widely Linear and Neural Models. Wiley, Hoboken (2009)
40. Brandwood, D.H.: A complex gradient operator and its application in adaptive array theory. In: IEE Proceedings H - Microwaves, Optics and Antennas, vol. 130, pp. 11–16 (1983)
41. Orozco-Henao, C., Bretas, A.S., Chouhy-Leborgne, R., Herrera-Orozco, A.R., Marín-Quintero, J.: Active distribution network fault location methodology: a minimum fault reactance and Fibonacci search approach. Electr. Power Energy Syst. **84**, 232–241 (2017)
42. Vieira, D.A.G., Lisboa, A.C.: Line search methods with guaranteed asymptotical convergence to an improving local optimum of multimodal functions. Eur. J. Oper. Res. **235**, 38–46 (2014)
43. Xia, Y., Jelfs, B., Hulle, M.M.V., Principe, J.C., Mandic, D.P.: An augmented echo state network for nonlinear adaptive filtering of complex noncircular signals. IEEE Trans. Neural Netw. **22**, 74–83 (2011)

Cross-Domain Attribute Representation Based on Convolutional Neural Network

Guohui Zhang[1], Gaoyuan Liang[2(✉)], Fang Su[3], Fanxin Qu[4], and Jing-Yan Wang[5]

[1] Huawei Technologies Co., Ltd, Shanghai, China
guohuizhang354@outlook.com
[2] Jiangsu University of Technology, Changzhou 213001, Jiangsu, China
gaoyuanliang@outlook.com
[3] Shaanxi University of Science and Technology, Xi'an, China
[4] Northwestern Polytechnical University, Xi'an, China
[5] Provincial Key Laboratory for Computer Information Processing Technology, Soochow University, Suzhou 215006, China
jimjywang@gmail.com

Abstract. In the problem of domain transfer learning, we learn a model for the prediction in a target domain from the data of both some source domains and the target domain, where the target domain is in lack of labels while the source domain has sufficient labels. Besides the instances of the data, recently the attributes of data shared across domains are also explored and proven to be very helpful to leverage the information of different domains. In this paper, we propose a novel learning framework for domain-transfer learning based on both instances and attributes. We proposed to embed the attributes of different domains by a shared convolutional neural network (CNN), learn a domain-independent CNN model to represent the information shared by different domains by matching across domains, and a domain-specific CNN model to represent the information of each domain. The concatenation of the three CNN model outputs is used to predict the class label. An iterative algorithm based on gradient descent method is developed to learn the parameters of the model. The experiments over benchmark datasets show the advantage of the proposed model.

Keywords: Convolutional neural network · Domain-Transfer learning Attribute embedding

1 Introduction

In the machine learning problems, domain transfer learning has recently attracted much attention [23, 26]. Transfer learning refers to the learning problem of a predictive model for a target domain, by leveraging the data from both the target domain and one or

The study was supported by Provincial Key Laboratory for Computer Information Processing Technology, Soochow University, China (Grant No. KJS1324).

D.-S. Huang et al. (Eds.): ICIC 2018, LNAI 10956, pp. 134–142, 2018.
https://doi.org/10.1007/978-3-319-95957-3_15

more auxiliary domains. One shortage of traditional transfer learning methods is that the attributes of the data are not used by the classification model. But the attributes of the data actually has the nature of stability across the domains. Thus using the attributes of the data is critical for the transfer learning [19, 21]. Peng et al. [19] proposed to represent the attribute vectors of each data point by using an attribute dictionary. Each data point is reconstructed by the elements in the dictionary, and the reconstruction coefficients are used as the new representation of the attributes. The attribute vector of a data point is mapped to the new representation vector by a linear transformation matrix so that the new representation vector is linked to the attribute vector. To leverage the auxiliary and the target domains, the same attribute representation method is applied to both auxiliary and target domains. The learning process is regularized by the class-intra similarity in the auxiliary domains, and by the neighborhood in the target domain. Su et al. [21] proposed a low-rank attribute embedding method for the problem of person re-identification of multiple cameras. The proposed method tries to solve the problem of multiple cameras based person re-identification as a multi-task learning problem. The proposed method uses both the low-level features with mid-level attributes as the input of the identification model. The embedding of attributes maps the attributes to a continuous space to explore the correlative relationship between each pair of attributes and also recovers the missing attributes. Both these two methods of attribute representation are based on the linear transformation. However, a simple linear function may be insufficient to represent the attributes effectively.

In this paper, we propose a novel attribute embedding method for attributes for the problem of domain transfer learning. The embedding of attributes is based on convolutional neural network (CNN) model [12, 13, 25]. The convolutional output of the input data is further mapped to the attribute vector. In this way, the attribute embedding vector not only represents the attributes of a data point but also contains the pattern of the input data constructed by the CNN model, which has been proven to be a powerful representation model. To construct the classification model for each domain, we also learn a domain independent convolutional representation and a domain-specific convolutional representation. The domain-independent convolutional representation maps the data of different domains to a shared data space to capture the patterns shared over all the domain. The domain-specific convolutional representation is used to represent the patterns specifically contained by each domain. The classification model of each domain is based on the three types of convolutional representations, i.e., attribute embedding, domain-independent and domain-specific representations. To learn the parameters of the models, we propose to minimize the mapping errors of the attributes, the classification errors across different domains, the mismatching of different domains in the domain-independent representation space, and the dissimilarity between the neighboring data points in the target domain. The joint minimization problem is solved by an alternate optimization strategy and the gradient descent algorithm.

2 Method

In the problem setting of cross-domain learning, we assume we have T domains. The first T − 1 domain are the auxiliary domains, while the T-th domain is the target domain. The problem is to learn an effective model for the classification of the target domain. The input data sets of the T domains are denoted as $X_t|_{t=1}^{T}$, where $X_t = \{(X_t^i, a_i^t, y_i^t)\}$ is the data set of the t-th domain. $X_i^t = [x_{i1}^t, \cdots, x_{i|X_i^t|}^t] \in R^{d \times |X_i^t|}$ is the input matrix of the i-th data point of the t-th domain, and each column of the matrix is a feature vector of a instance, and $a_i^t \in \{1,0\}^{|a|}$ is its binary attribute vector, and $y_i^t \in \{1,0\}^{|y|}$ is its class label vector.

We propose to embed the attributes of each input data point to a vector and use the convolutional representation of the input data as the embedding vector. Given the input matrix of a data point, X, we represent it by a CNN model composed of a convolutional layer, a activation layer, and max-pooling layer, denoted as $f_a(X)$. Since this convolutional representation of X is used as its attribute vector embedding, we propose to map it to the attribute vector a by a linear mapping function,

$$\mathrm{f_a(X)} \leftarrow \Theta^T a \tag{1}$$

where $\Theta \in R^{|a| \times m}$ is the mapping matrix. To reduce the mapping errors, we proposed to minimize the Frobenius norm distance between the convolutional representations and the mapping results for all the data points of all domains,

$$\min_{\Theta, f_a} \sum_{t=1}^{T} \left(\sum_{i=1}^{n_t} \|f_a(X_i^t - \Theta a_i^t)\|_F^2 \right) \tag{2}$$

To predict the class labels for the data points in multiple domains, we proposed the data of each domain to represent the data into a domain-independent convolutional representation and a domain-specific convolutional representation. The domain-independent convolutional representation function is shared across all the domains. It tries to extract features relevant to the class labels, but independent of the specific domains. The domain-independent convolutional recreation function is also based on a CNN model, denoted $\mathrm{f_0(X)}$. Since the $\mathrm{f_0(X)}$ outputs are domain-independent, we hope the representations of data points from different domains can be similar to each other. To this end, we impose that the distribution of the base representations of different domains is of the same. We use the mean vector of the representations of each domain as the presentation of the distribution of the domain. For the t-th domain, the mean vector is given as $\frac{1}{\mathrm{n}_t} \sum_{i=1}^{n_t} f_0(X_i^t)$. To reduce the mismatch among the domains, we proposed to minimize the Frobenius norm distances between the mean vectors of each pair of domains,

$$\min_{f_0} \sum_{t,t'=1,t<t'}^{T} \left\| \frac{1}{n_t} \sum_{i=1}^{n_t} f_0(X_i^t) - \frac{1}{n_{t'}} \sum_{i=1}^{n_{t'}} f_0(X_i^{t'}) \right\|_F^2 \tag{3}$$

To predict the class labels for the data points of different domains, we also consider the representation of the data points according to the domains. This is the domain-specific representation. The representation is also based on CNN models, and the CNN model of the t-th domain of a data point X is denoted as $f_t(X)$.

To estimate the class label from a data point of the t-th domain, we concatenate both the domain-independent and domain-specific convolutional representations of the input data, $f_0(X)$ and $f_t(X)$, and also the attribute embedding of the data, $f_a(X)$. They are concatenated to a longer vector,

$$f(X) = \begin{bmatrix} f_0(X) \\ f_t(X) \\ f_a(X) \end{bmatrix} \in R^{m_0 + m_t + m_a} \tag{4}$$

and the longer vector is transformed to a |y|-dimensional vector of scores of classification by a matrix $U = \begin{bmatrix} U_0 \\ U_t \\ U_a \end{bmatrix} \in R^{(m_0 + m_t + m_a) \times |y|}$ in a classification function, $h_t(X) = U^T f(X) = U_0^T f_0(X) + U_t^T f_t(X) + U_a^T f_a(X)$, $t = 1, \cdots, T$, where U_0, U_t, and U_a are the transformation matrices for the domain-independent representation, domain-specific representation, and the attribute embedding.

To learn the parameters of the model, we propose the following minimization,

$$\begin{aligned} \min_{U_t, W_t|_{t=0}^T, U_a, W_a, \Theta} \Bigg\{ & o(U_t, W_t|_{t=0}^T, U_a, W_a, \Theta) = \sum_{t=1}^{T-1} \left(\sum_{i=1}^{n_t} \|y_i^t - h_t(X_i^t)\|_F^2 \right) \\ & + \sum_{i=1}^{l_T} \|y_i^T - h_t(X_i^T)\|_F^2 + C_1 \sum_{t=1}^{T} \left(\sum_{i=1}^{n_t} \|f_a(X_i^t) - \Theta a_i^t\|_F^2 \right) \\ & + C_2 \sum_{t,t'=1,t<t'}^{T} \left\| \frac{1}{n_t} \sum_{i=1}^{n_t} f_0(X_i^t) - \frac{1}{n_{t'}} f_0(X_i^{t'}) \right\|_F^2 \\ & + C_3 \sum_{i,i'=1}^{n_T} M_{ii'} \|f(x_i^T) - f(X_{i'}^T)\|_F^2 \Bigg\} \end{aligned} \tag{5}$$

We explain the objective terms of the objective function o as follows,

- The classification function is used to predict the class labels, thus we propose to reduce the prediction errors measured by the Frobenius norm distance between the class label vectors and the outputs of ht(X) for the data pints with available label vectors. The first two terms are the classification error terms.

- The third term is attribute mapping error term of (2).
- The fourth term is the cross-domain matching term of the domain-independent CNN model of (3).
- For the unlabeled data points in the target domain, we also regularize them by imposing their representations to be constant with the labeled data points in the neighborhood, so that the supervision information can also be propagated to them. To this end, we hope for any neighboring two data points in the target domain, their overall representation vectors are close to each other. To this end, in the last term, we minimize the Frobenius norm distance between the representations of neighboring data points in the target domain, where $M_{ii'} = 1$ if X_i and $X_{i'}$ are a neighbor to each other and 0 otherwise.

  $C_k, k = 1, \cdots, 3$ are the tradeoff weights of different regularization terms. To solve this problem, we proposed to use the sub-gradient descent algorithm. The filters of the CNN models, the mapping matrix, and the transformation matrices are updated according to the direction of gradient function of the objective,

$$\Phi \leftarrow \Phi - \tau \nabla_{\Phi} o, \tag{6}$$

  where $\Phi = \{f_a, f_0, f_1, \cdots, f_T, \Theta, U_0, U_a, U_1, \cdots, U_T\}$, τ is the descent step, and $\nabla_{\Phi} o$ is the sub-gradient of o regarding Φ. In an iterative, the parameters of Φ are updated alternately until a maximum iteration number is reached or converge is achieved.

3 Experiment

In this section, we evaluate the proposed method over several domain-transfer problems.

3.1 Data Sets and Experimental Setting

In the experiments, we use three datasets as follows. CUHK03 data set was developed for the problem of person re-identification problems [15]. It contains 13,164 images of 1,360 persons. For each image, we annotate it by 108 attributes, including gender (male/female), wearing long hair, etc. The images are captured by six different cameras. The problem of person re-identification is to train a classifier over the images of some cameras, and then use the classifier to identify an image captured from other cameras. We treat each camera as a domain, and we use each domain as a target domain in turn. Bankrupt prediction data contains the stock price wave data of 3 years of 374 companies of three different countries, China, USA, and UK. We collected this data for the problem of prediction of company bankrupt. Each company is also labeled by a list of business type attributes. Each company is treated as a data points, presented by a set of short-term frames, and a list of binary attributes of business types. Moreover, each country is treated as domain. The prediction problem of this data set is to predict if a given company will be in bankrupt within the future 3 years. Spam email data set is for the spam email detection competition of the ECML/PKDD Discovery Challenge 2006 [3]. It contains texts of emails of 15 email users, and for each user, there are 400 emails.

Among the 400 emails of each user, half of them are spam emails, while the remaining half are non-spam emails. Each email text is composed of a set of words. Moreover, we also apply a topic classifier and a sentiment classifier to each email text to extract attributes of the text and use the extracted attributes as additional information. Each user is treated as a domain, and we also use each user as a target domain in turn. In our experiments, given a data set of several domains, we treat each domain as a target domain in turn, while treating the other domains as the auxiliary domains to help train the model. The data points in a target domain are further split into a training set and a test set with equal sizes randomly. Meanwhile, for the training set of the target domain, we further split it into equal-sized subsets. One subset is used as a labeled set, and the other set is used as an unlabeled set. We train the model over the data points of the auxiliary domains and the training set of the target domain and then test it over the test set of the target domain. The classification rate over the test set is used as the performance measure. The average classification rate over different target domains is reported and compared.

3.2 Results

In the experiments, we first compare the proposed domain transfer convolutional attribute embedding (DTCAE) algorithm to some state-of-the-art domain-transfer attribute representation methods, and then study the properties of the proposed algorithm experimentally.

Attribute embedding for domain-transfer learning problem is a new topic and there are only two existing methods. In the experiment, we compare the proposed algorithm against the two existing methods, which are the Joint Semantic and Latent Attribute Modelling (JSLAM) method proposed by Peng et al. [19], and the Multi-Task Learning with Low-Rank Attribute Embedding (MTL-LORAE) method proposed by Su et al. [21]. The comparison results over the three benchmark data sets are shown in Fig. 1. According to the reported average accuracies over the benchmark datasets, our algorithm DTCAE achieves the best performance over all the three datasets. For example, over the CUHK03 data set, the DTCAE is the only compared method which has an average accuracy higher than 0.800. Meanwhile, over the spam email dataset, only DTCAE obtains an average accuracy higher than 0.900.

Since the proposed algorithm DTCAE is an iterative algorithm. The variables are updated alternately. We are also interested in the convergence of the algorithm. Thus we plot the average classification rates with a different number of iterations. The curves over the three benchmark data sets are plotted in Fig. 2. According to the curves of Fig. 2, when more iterations are used to update the variables of the model, the average classification rates increase stably. This is not surprising because a larger number of iterations reaches a smaller objective function. This verifies the effectiveness of the proposed model and its corresponding objective function. Moreover, we also observe that when the iteration number is larger than 100, the change of the performance is very small. This means that the algorithm converges and no more iteration is needed to improve the performance.

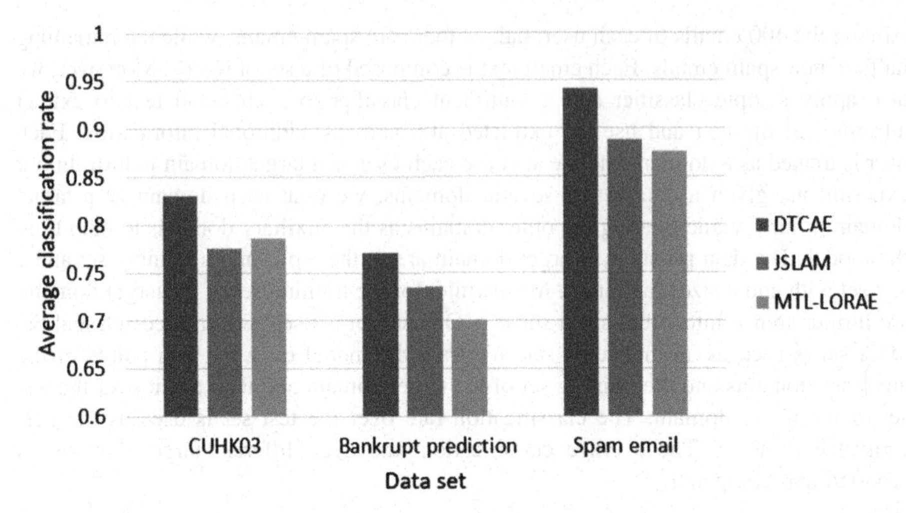

Fig. 1. Comparison results over the benchmark data sets.

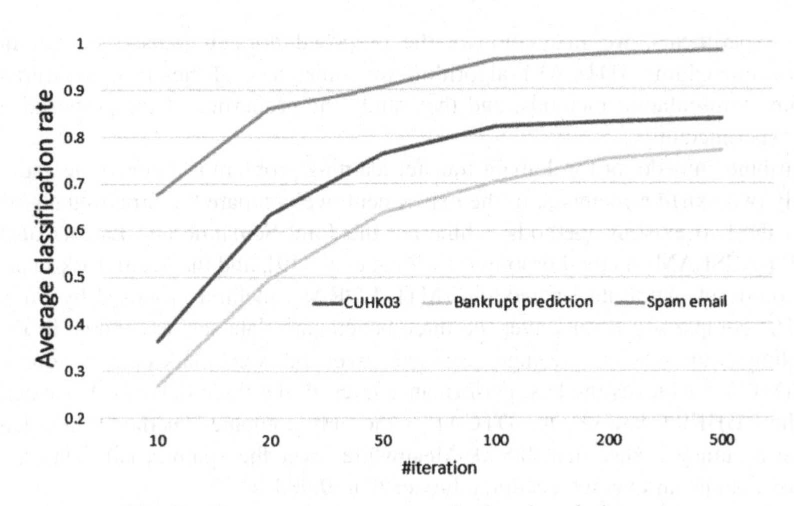

Fig. 2. Convergence curves over the benchmark data sets.

4 Conclusion

In this paper, we propose a novel model for the problem of cross-domain learning problem with attribute data. We use a CNN model to map the input data to its attributes. Moreover, a domain-independent and domain-specific CNN model are also used to represent the data input itself. The attribute embedding, the domain-independent, and domain-specific representations are concatenated as the new representation of the data points, and we further a linear layer to map the new representation to the class labels. Moreover, we also impose the domain-independent representations of data points of different domains to be in a common distribution, and the neighboring data points of target domain to be similar to each other. We model the learning problem as a minimization problem and solve it by an iterative algorithm. The experiments on three benchmark data sets show its advantages. In the future, we plan to apply the proposed method other applications, such as biometrics [20, 27], network analysis [24], human pose estimation [9–11], mobile computing [8, 16], mathematics [1, 2, 17, 18], etc. The similar approach can also be adopted in other related fields such as systems [6, 7, 14], and system security [4, 5, 22].

References

1. Bai, C., Bellier, O., Guo, L., Ni, X.: Splitting of operations, manin products, and rota–baxter operators. Int. Math. Res. Not. **2013**(3), 485–524 (2013)
2. Bai, C., Guo, L., Ni, X.: Generalizations of the classical Yang-Baxter equation and O-operators. J. Math. Phys. **52**(6), 063515 (2011)
3. Bickel, S.: ECML-PKDD discovery challenge 2006 overview. In: ECML-PKDD Discovery Challenge Workshop, pp. 1–9 (2006)
4. Chen, Y., Khandaker, M., Wang, Z.: Pinpointing vulnerabilities. In: Proceedings of the 12th ACM Asia Conference on Computer and Communications Security, pp. 334–345. ACM, Abu Dhabi (2017)
5. Chen, Y., Khandaker, M., Wang, Z.: Secure in-cache execution. In: Dacier, M., Bailey, M., Polychronakis, M., Antonakakis, M. (eds.) RAID 2017. LNCS, vol. 10453, pp. 381–402. Springer, Cham (2017). https://doi.org/10.1007/978-3-319-66332-6_17
6. Chen, Y., Wang, Z., Whalley, D., Lu, L.: Remix: on-demand live randomization. In: Proceedings of the Sixth ACM Conference on Data and Application Security and Privacy, pp. 50–61. ACM, New Orleans (2016)
7. Chen, Y., Zhang, Y., Wang, Z., Xia, L., Bao, C., Wei, T.: Adaptive android kernel live patching. In: Proceedings of the 26th USENIX Security Symposium (USENIX Security 2017). USENIX Association, Vancouver, BC, August 2017
8. Cui, P., Liu, H., He, J., Altintas, O., Vuyyuru, R., Rajan, D., Camp, J.: Leveraging diverse propagation and context for multi-modal vehicular applications. In: 2013 IEEE 5th International Symposium on Wire-less Vehicular Communications (WiVeC), pp. 1–5. IEEE (2013)
9. Ding, M., Fan, G.: Articulated Gaussian kernel correlation for human pose estimation. In: Proceedings of the IEEE Conference on Computer Vision and Pattern Recognition Workshops, pp. 57–64 (2015)
10. Ding, M., Fan, G.: Generalized sum of gaussians for real-time human pose tracking from a single depth sensor. In: 2015 IEEE Winter Conference on Applications of Computer Vision (WACV), pp. 47–54. IEEE (2015)
11. Ding, M., Fan, G.: Articulated and generalized Gaussian kernel correlation for human pose estimation. IEEE Trans. Image Process. **25**(2), 776–789 (2016)
12. Geng, Y., Liang, R.Z., Li, W., Wang, J., Liang, G., Xu, C., Wang, J.Y.: Learning convolutional neural network to maximize pos@ top performance measure. In: ESANN 2017 – Proceedings, pp. 589–594 (2016)
13. Geng, Y., et al.: A novel image tag completion method based on convolutional neural transformation. In: Lintas, A., Rovetta, S., Verschure, P.F.M.J., Villa, A.E.P. (eds.) ICANN 2017. LNCS, vol. 10614, pp. 539–546. Springer, Cham (2017). https://doi.org/10.1007/978-3-319-68612-7_61

14. Jin, Y., Wang, T., Zhang, H., Zhang, Y., Zhao, J., Tong, R.: Localized quasi(bi) harmonic field and its applications. J. Adv. Mech. Des. Syst. Manuf. **11**(4), JAMDSM0047 (2017)
15. Li, W., Zhao, R., Xiao, T., Wang, X.: DeepReID: Deep filter pairing neural network for person re-identification. In: Proceedings of the IEEE Conference on Computer Vision and Pattern Recognition, pp. 152–159 (2014)
16. Liu, H., He, J., Cui, P., Camp, J., Rajan, D.: Astra: application of sequential training to rate adaptation. In: 9th Annual IEEE Communications Society Conference on Sensor, Mesh and Ad Hoc Communications and Networks (SECON), 2012, pp. 443–451. IEEE (2012)
17. Ni, X., Bai, C.: Prealternative algebras and prealternative bialgebras. Pac. J. Math. **248**(2), 355–391 (2010)
18. Ni, X., Bai, C.: Pseudo-hessian lie algebras and l-dendriform bialgebras. J. Algebra **400**, 273–289 (2014)
19. Peng, P., Tian, Y., Xiang, T., Wang, Y., Pontil, M., Huang, T.: Joint semantic and latent attribute modelling for cross-class transfer learning. IEEE Trans. Pattern Anal. Mach. Intell. **40** (2017)
20. Shao, H., Chen, S., Zhao, J.Y., Cui, W.C., Yu, T.S.: Face recognition based on subset selection via metric learning on manifold. Front. Inf. Technol. Electron. Eng. **16**(12), 1046–1058 (2015)
21. Su, C., Yang, F., Zhang, S., Tian, Q., Davis, L., Gao, W.: Multi-task learning with low rank attribute embedding for multi-camera person re-identification. IEEE Trans. Pattern Anal. Mach. Intell. **40**(5), 1167–1181 (2017)
22. Wang, X., Chen, Y., Wang, Z., Qi, Y., Zhou, Y.: SecPod: a framework for virtualization-based security systems. In: Proceedings of the 2015 USENIX Annual Technical Conference, pp. 347–360 (2015)
23. Yang, L., Zhang, J.: Automatic transfer learning for short text mining. Eurasip J. Wirel. Commun. Netw. **2017**(1), 42 (2017)
24. Yu, T., Yan, J., Zhao, J., Li, B.: Joint cuts and matching of partitions in one graph. arXiv preprint arXiv:1711.09584 (2017)
25. Zhang, G., et al.: Learning convolutional ranking-score function by query preference regularization. In: Yin, H., Gao, Y., Chen, S., Wen, Y., Cai, G., Gu, T., Du, J., Tallón-Ballesteros, A.J., Zhang, M. (eds.) IDEAL 2017. LNCS, vol. 10585, pp. 1–8. Springer, Cham (2017). https://doi.org/10.1007/978-3-319-68935-7_1
26. Zhang, L., Yang, J., Zhang, D.: Domain class consistency based transfer learning for image classification across domains. Inf. Sci. **418–419**, 242–257 (2017)
27. Zhou, L., Lin, Y., Feng, B., Zhao, J., Tang, J.: Phylogeny analysis from gene-order data with massive duplications. BMC Genom. **18**(7), 13 (2017)

Convolutional Neural Network for Short Term Fog Forecasting Based on Meteorological Elements

Ting-ting Han[1], Kai-chao Miao[2], Ye-qing Yao[2], Cheng-xiao Liu[2], Jian-ping Zhou[2], Hui Lu[2], Peng Chen[3], Xia Yi[3], Bing Wang[4], and Jun Zhang[1(✉)]

[1] School of Electronic Engineering and Automation, Anhui University, Hefei 230601, Anhui, China
wwwzhangjun@163.com

[2] Anhui Meteorological Bureau, Hefei 230031, Anhui, China

[3] Institute of Health Sciences, Anhui University, Hefei 230601, Anhui, China

[4] School of Electrical and Information Engineering, Anhui University of Technology, Ma Anshan 243032, China

Abstract. Fog is the main weather phenomenon that causes low visibility, which makes traffic and outdoor work extremely dangerous. It is urgent to improve the accuracy of fog forecast. In this paper, ground observation meteorological elements time series data is converted into 2D image format, then we train a simple convolution neural network to predict the existing of short time fog. Different experiments is arranged to validate the performance of the proposed method, which obtained the best prediction recall 71.43% and 71.47% for next four and two hours respectively. Contrasting traditional numerical prediction and model prediction method, the application of convolutional neural network method to fog prediction is our first attempt.

Keywords: Fog forecast · Meteorological elements time series
Convolution neural network

1 Introduction

The short-term fog forecast has drawn more and more attention with the increasing popularization of modern means of transport, the visibility dependence of expressways, high-speed railways, airports, waterways and the like has become increasingly prominent [1]. Higher accuracy is required in fog forecasting. Over the past few years, research at home and abroad are devoted to numerical prediction methods [2]. However, most experiments show that one-dimensional model fail to predict fog events well. In the paper [3], some scholars have developed three-dimensional model that apply to different regions. It did not consider the impact of large-scale environment on the fogging process. The newly proposed method also should not simply extract related information of data by manually designed features, which has great limitations on the prediction of fog, such as recent work on fog prediction. The accuracy of fog forecasting has a lot of room to improvement. In this paper, convolution neural network is

D.-S. Huang et al. (Eds.): ICIC 2018, LNAI 10956, pp. 143–148, 2018.
https://doi.org/10.1007/978-3-319-95957-3_16

used for short-term fog forecast based on meteorological elements with extracting the various internal structure characteristics of the time series data of meteorological elements automatically. In recent years, researches also have try to use CNN to classify time series to predict price of stock and do some research on human activity identification [5] and the effectiveness of the application has been demonstrated [4]. However, there are no application for meteorological elements time series classification by using convolution neural network. Our goal is to use the state-of-the-art convolution neural networks to learn the internal structural features of the time series of meteorological elements to classify whether there is fog.

In this study, meteorological data are cut into time series of the same length as data set firstly. We have designed a four-layer convolution neural network to classify the data to achieve a 1–4 h forecast of short-term fog and obtain the satisfying result.

2 Methodology

2.1 Convolution Neural Network Framework

In this section, we introduce the proposed framework of convolutional neural network for multivariate meteorological elements time series classification. In the first part, convolution and pooling operation are used to extract feature from the input data, a trainable fully connected is followed, which categorize the features learned from the previous section. For better understanding, a neural network architecture for 6-variate time series input data is described in Fig. 1.

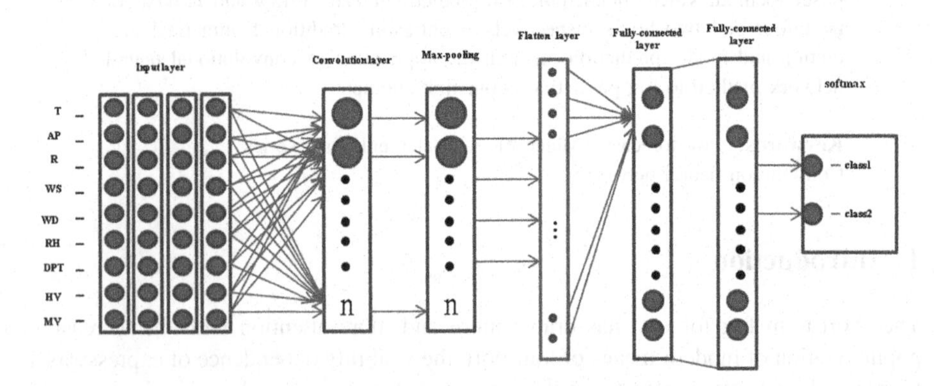

Fig. 1. The proposed CNN architecture for meteorological elements time series classification

The input layer is a N * k multivariate meteorological element time series, N denotes the variable number of input time series, which is six. Variable are temperature, Air pressure, rainfall, wind speed, wind direction, Relative humidity respectively (devoted by T, AP, R, WS, WD, RH}), K denotes the length of the time series.

Convolutional operations on multivariate time series are implemented with convolution kernels. Here are mainly based on experiments and related knowledge. we set convolution kernel is in the form of W * H, the number of filter is n, convolution stride is S, The output of the convolution layer is calculated according to the following equation:

$$Y_j^l(x,y) = f(\sum_{u=0}^{W-1}\sum_{v=0}^{H-1} X_i^{(l-1)}(x,y)W_{ji}^{(l)}(u,v) + \theta_j^{(l)}) \tag{1}$$

With Y and X represent the output and input feature map. W_{ji} represents the weight of the first j convolution kernel corresponding to the first i input of preceding layer, $\theta_j^{(l)}$ represents the first j feature map bias, f is nonlinear activation function. W and H represent the high and wide of convolution kernel respectively.

Pooled layer also known as the sub-sampling layer, we often place the pool behind the convolution. Currently there are several popular methods, Max-pooling is used in this paper, which takes the maximum of the feature point in the neighborhood.

The fully connected layer acts as a classifier throughout the convolution operation, whose purpose is to translate the feature maps learned by the convolutional layers and pool layers into the class they belong to. The formula is as follows:

$$Y_j = f(\sum\nolimits_{i-1}^{I} W_{ji}X_i + \theta_j) \tag{2}$$

$f()$ denotes the activation function. In this article, Relu is used as the activation function for convolution layer and full-connected layers.

Soft-max function is used in the last layer. Generally, soft-max is used for multi-classification. In this paper, we only need to make a binary classification. The meteorological data will be divided into two categories by convolution neural network model. In the process of training, the training accuracy can be calculated. We choose cross entropy as the loss function, The formula is as follows:

$$Loss = -\sum\nolimits_{i}^{K} y_i \ln f_i(x) \tag{3}$$

y represents the real value. $f(x)$ denotes value obtain by soft-max. i is the first i node output from soft-max. The structure of each layer is described on the Tables 1 and 2 respectively when the input time series length is 4 and 6.

Table 1. The structure of CNN for 4 steps.

Layers	Filter size	Filter number	Stride	Activation
convolution	2	200	1	ReLu
Max-pooling	1, 2	–	2	–
Full-Connected	–	100	–	ReLu
Full-Connected	–	200	–	ReLu
Full-Connected	–	–	–	Soft-max

Table 2. The structure of CNN for 6 steps.

Layers	Filter size	Filter number	Stride	Activation
convolution	3	100	1	ReLu
Max-pooling	2	–	2	–
Full-Connected	–	200	–	ReLu
Full-Connected	–	100	–	ReLu
Full-Connected	–	–	–	Soft-max

We evaluate the classification accuracy of the convolution neural network model on the test data, recall is used to evaluate the precision of each class. The formula of accuracy and recall are as follow:

$$Recall = TP/(TP + FN) \tag{4}$$

$$Accuracy = (TP + TN)/(TP + TN + FP + FN) \tag{5}$$

where TP(TN) represents the number of positive samples correctly classified, FN(FP) is the number of positive (negative) samples misclassified.

3 Experiment

3.1 Experimental Data

The selected range of our data is the meteorological elements observation data of returned every other hour from 81 National meteorological observation station in November 2016 in Anhui Province. Meteorological elements returned hourly are pressure, temperature, wind speed, wind direction, dew point temperature, rainfall, visibility per minute, hourly visibility. The way of sample is that we pick the meteorological element data for the first few hours as a time series for each hour. The label is determined by the hourly visibility of the current time point. We use the visibility of 1000 m as the boundary. When the visibility is greater than 1000 m, these samples are considered as no fog, other samples are divided into foggy. Because we want to compare the effect of sequence length on accuracy, we make the data as time series samples of three different time steps of length, which is 2, 4 ,6 respectively, In order to achieve 1–4 h predication, we also made four kinds of samples for each of the above three time series with different lengths. The number of positive and negative samples with different sequence lengths for the next 4 h prediction is the same.

The label of each type of sample is also determined by the visibility value of the predicted time point. For example, the current time is 0:00, we use a visibility value of 2:00 to determine a label, so this sample is to predict whether there are fog two hours later. Positive and negative samples of the meteorological data are too uneven, selection of training set and test set is optional. In the positive sample and the negative sample, we choose 1000 as the training set, making the ratio of the positive and negative samples be artificially defined as 1:1. The remaining part of the total sample is taken as a test set. Similarly, Since the positive and negative samples of the data are unbalanced, the positive and negative samples of the test set are also unbalanced. The above description is presented in Table 3.

Table 3. Sample number of training set and test set.

	Overall sample	Train	Test samples	series length
Positive (1–4 h)	3734	1000	2734	2.4.6
Negative (1–4 h)	54347	1000	53347	2.4.6

3.2 Experimental Result

The effect of sequence length on classification accuracy is shown in Fig. 2, We compared the positive recalls of the future one-four hour prediction when the sequence lengths were 2, 4, 6, respectively. When predicting the next 1–2 h of fog conditions, choosing a length of 2 for the sequence can obtain the highest positive recall rate with 71.43% and 72.46% respectively. In addition, in 3–4 h prediction of future, the sequence length should choose 3, which may be attributed to the increase of uncertainties in meteorological data when the time series increases.

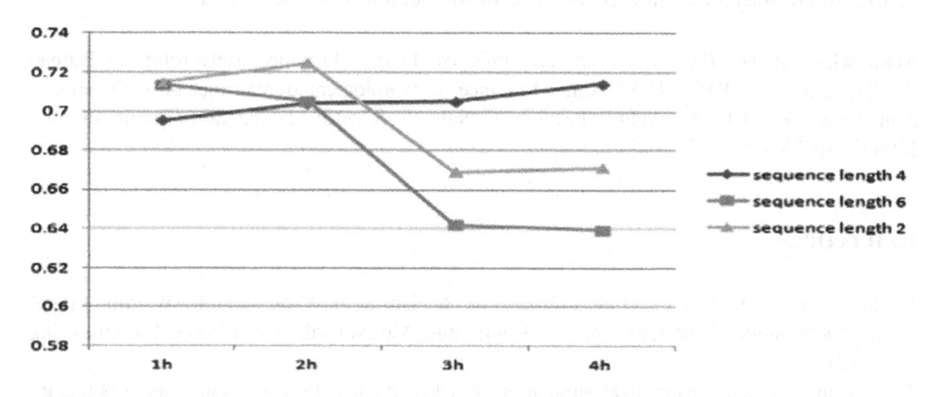

Fig. 2. Positive recall trend as time increasing of different series.

we also compared the accuracy of hourly forecasts with the increase of the predicted time in predicting the next 4 h on different time series. We can see that the forecasting accuracy is decreasing as the prediction time increases, this experimental result is consistent with our hypothesis. Overall, the size of the sequence length has a significant effect on the classification accuracy. In Fig. 2, As the length of the sequence increases, the recall rate of positive samples is on the whole declining. That is, when the sequence length is 6, the recall rate in each hour is less than 2 in the time series.

From the above analysis, we can obtain the best experimental result of whether there is fog in the next four hours by classifying the time series of meteorological elements. When we predict the next 1–2 h, we choose a time series length of 2.

Our best positive sample has a recall rate of 71.43% and 71.47%. When forecast the next 3–4 h. We choose the length of the time series as 4, with a recall of 70.51% and 70.51%. The above is shown in Table 4.

Table 4. Positive recall alignment of different time series for next four hour

Method	Recall (positive) 1	Recall (positive) 2	Recall (positive) 3	Recall (positive) 4
Ours	71.43%	71.47%	70.51%	70.51%

In this article, we design a 4-layer convolutional neural network framework to classify meteorological element time series into 2 categories to achieve one to four hour short-term fog forecasting. The experimental results show the best result of positive samples classification recall of CNN for future 4 h prediction. We can know that CNN can extract the internal structure of meteorological elements time series well, which made a foreshadowing for later research. In addition, as the number of positive and negative samples of meteorological data is extremely unbalanced, which was the problem mainly for neural network. In future work, we try to balance the sample using simple re-sampling method. Moreover, we can increase the amount of training data to verdict, there may be a greater increase in the accuracy of the forecast.

Acknowledgments. This work was supported by Jiangsu Province Meteorological Bureau BeiJiGe grant Nos. BJG201707, Anhui Province Meteorological Bureau meteorologist special grant Nos. KY201704, Anhui Provincial Natural Science Foundation (grant number 1608085MF136).

References

1. Hu, S., Wang, X., Dai, C., et al.: Influence mechanism of mass fog on highway traffic safety. In: International Conference on Transportation, Mechanical, and Electrical Engineering (2012)
2. Musson-Genon, L.: Numerical simulation of a fog event with a one-dimensional boundary layer model. Mon. Weather Rev. **115**(2), 592 (2009)
3. Yang, J.B., Nguyen, M.N., San, P.P., et al.: Deep convolutional neural networks on multichannel time series for human activity recognition. In: International Conference on Artificial Intelligence, pp. 3995–4001. AAAI Press (2015)
4. Keogh, E., Kasetty, S.: On the need for time series data mining benchmarks: a survey and empirical demonstration. Data Min. Knowl. Disc. **7**(4), 349–371 (2003)
5. Geetha, A., Nasira, G.M.: Data mining for meteorological applications: decision trees for modeling rainfall prediction. In: IEEE International Conference on Computational Intelligence and Computing Research, pp. 1–4 (2015)

RBF Neural Network Adaptive Sliding Mode Control of Rotary Stewart Platform

Tan Van Nguyen[1] and Cheolkeun Ha[2(✉)]

[1] Department of Mechanical and Aerospace Engineering, Ulsan University, Ulsan, Korea
nvtan@hueic.edu.vn

[2] Department of Mechanical Engineering, University of Ulsan, Ulsan, Korea
cheolkeun@gmail.com

Abstract. The stewart platform is widely applied in the industry. However, Rotary Stewart Platform (RSP) has very little research for this type. Moreover, Inverse Kinematic (IK) solution in papers published previously is complex and unclear. Therefore, in this paper, first, we design and build the mathematical model and check it in Simmechnics. Second, the robust control of the RSP proposed in this paper precisely tracks a command under the platform uncertainties. The inverse kinematic solution of the platform, derived in this paper, supports for control design of the platform. Radial Basis Function (RBF) neural network adaptive sliding mode controller is used to achieve the satisfactory tracking performance and the system stability. Stability of the system is guaranteed through Lyapunov theory. The simulation is conducted to illustrate the effectiveness of the proposed control for the RSP.

Keywords: Rotary Stewart Platform · Inverse kinematic
Adaptive sliding mode · Simmechanics

1 Introduction

Stewart Platform was originally designed in 1965 as a flight simulator, and it is still commonly used for that purpose [1]. This kind of the Stewart platform is activated by linear actuators. Hence, the actual use of this type is generally in the case of low speed and high load conditions such as vehicle motion and flight simulator, and motion bed for machine tools [7]. Since then, a wide variety of applications have benefited from this design. Many researches such as kinematics [2], dynamics [3], work space estimation [4], path planning [5] and force sensing applications [6] have been carried out since the original advent of Stewart platform.

The Stewart Platform has been used for a variety of applications due to their advantages over serial manipulators like higher end effector accuracy, rigidity, and load-to-weight ratio. In the last decades, most of the researches has been particularly aimed at Gough Stewart platform with linear actuators, but the rotary type has been less studied.

The 6-DOF parallel manipulator with rotary actuators was initially introduced by Hunt [8] in 1983. Thereafter, several species of this kind of parallel manipulator have

D.-S. Huang et al. (Eds.): ICIC 2018, LNAI 10956, pp. 149–162, 2018.
https://doi.org/10.1007/978-3-319-95957-3_17

been suggested such as the prototypes constructed by Sarkissian [9], where they have a few differences in linkage and joint configurations. Recently, commercial designs have been introduced by Servos & Simulation Inc. [6]. It is considered to be a famous stewart platform, due to its inclusion of several closed loop structures, is highly complicated mechanism to make kinematic and dynamic analysis. In addition, the rotary type of the Stewart platform has more complex dynamic system than the typical one.

The control response of the RSP is faster than the linear Stewart platform. So it is suitable in application of flight simulators and in the field of robot climbing stairs carrying patients and the elderly (Fig. 1).

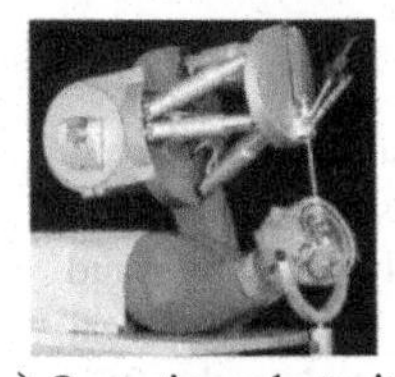

a) Operating robot with phantom

b) Motion generator for flight simulator

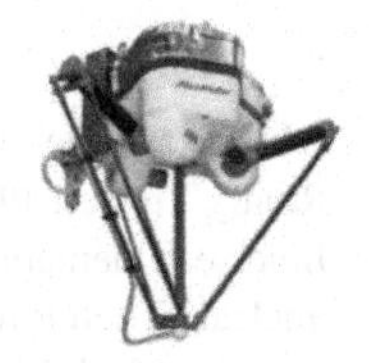

c) The Delta robot in the FlexPicker IRB 340

Fig. 1. Examples of the using of Hexapod system. Source: a) Wapler, M., Urban, V., Weisener, T., Stallkamp, J., Durr, M., and Hiller, A., "A Stewart platform for precision surgery", Transactions of the Institute of Measurement and Control 25, 4 (2003) pp. 329–334, b) ABB Ltd

However, RSP is not much attracted by researchers because its mechanism is relatively complex. Recently, some researches in RSP have been conducted for inverse kinematic [16, 17]. An intelligent neural network controller is recommended in application to nonlinear systems for precise control purposes of RSP based on its nonlinear approximation and learning abilities. An adaptive controller of a stewart platform with unknown dynamics was introduced in [10, 12]. A simple structure of neural network is radial basis function neural network (RBF). RBF neural network was used in order to reduce the chattering of the servo system [13].

In this paper, first, we analyze the characteristics of classical robots, in order to find the structure of robot with the degree of freedom. Second, the inverse kinetic of the Stewart platform is calculated to find the relationship between the center position of the motion platform and the rotation angle of the legs. Third, a robust adaptive sliding mode control is introduced. The RBF neural network is integrated with the sliding mode control to stabilize the feedback control system and improve the tracking performance against the bounded unknown external disturbances. The robust adaptive sliding mode control makes sure that the trajectory tracking error converges to zero over the external disturbances.

2 The Rotary Stewart Platform

2.1 Definition of Stewart Platform Geometry

This section introduces the geometric definitions to describe the RSP. Figure 2a shows a simplified description of the RSP consisting of a base plate and a motion platform with their respective right-handed coordinates (CSs). The base plate and the motion platform are connected by six legs. Each leg consists of a link and a crank. They are connected by spherical joints. The links are connected to the motion platform by universal joints and the cranks are connected to base platform by revolute joints, respectively. For convenience in computing, some notations are defined as follows: the body-fixed frame {B} and the motion frame {P} are attached to the stationary base plate in the bottom and the motion platform in the upper part, respectively. The frame origins are the mass center of the plate. The generalized position of the frame {P} relative to the frame {B} consists of two vectors; ${}^{B}\chi_{P} = \left[{}^{B}X_p, {}^{B}Y_p, {}^{B}Z_p\right]^T$ as a position vector of the origin of the frame {P} relative to the frame {B} and ${}^{B}\Omega_P = [\varphi \quad \theta \quad \psi]^T$ as an Euler angles representing an orientation of the frame {P} relative to the frame {B}. The Euler angles are defined by the rotation as follows: φ about X_P, θ about Y_P and ψ about Z_P. The rotation matrix describing an orientation of the frame {P} relative to the frame {B} can be written as.

$$
{}^{B}R_p = \begin{bmatrix} C_\psi C_\theta & C_\psi S_\theta S_\varphi - S_\psi C_\varphi & S_\psi S_\varphi + C_\varphi C_\psi S_\theta \\ S_\psi C_\theta & C_\varphi C_\psi + S_\varphi S_\theta S_\psi & C_\varphi S_\theta S_\psi - C_\psi S_\varphi \\ -S_\theta & C_\theta S_\varphi & C_\theta C_\varphi \end{bmatrix}
$$

Where $C_\varphi = \cos(\varphi); C_\theta = \cos(\theta); C_\psi = \cos(\psi); \quad S_\varphi = \sin(\varphi); S_\theta = \sin(\theta); S_\psi = \sin(\psi)$.

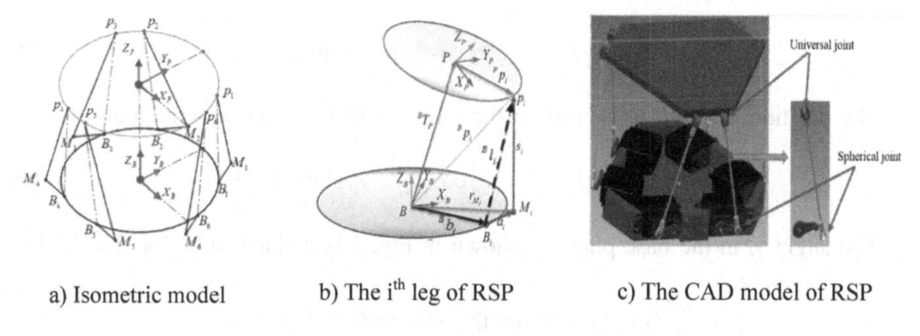

a) Isometric model b) The ith leg of RSP c) The CAD model of RSP

Fig. 2. Modeling of Rotary Stewart Platform

The generalized center position of the motion platform can be defined as

$$q = \left[{}^{B}\chi_{p}\ {}^{B}\Omega_{p}\right]^{T} \tag{1}$$

The ith leg vector ${}^{B}l_i$ with respect to the base frame {B} can be described as shown in Fig. 3b

$${}^{B}l_i = {}^{B}T_P + {}^{B}R_P\,{}^{P}p_i - {}^{B}b_i = \begin{bmatrix} l_{ix} & l_{iy} & l_{iz} \end{bmatrix}^T \quad \text{for}\, i = 1, 2, \ldots, 6 \tag{2}$$

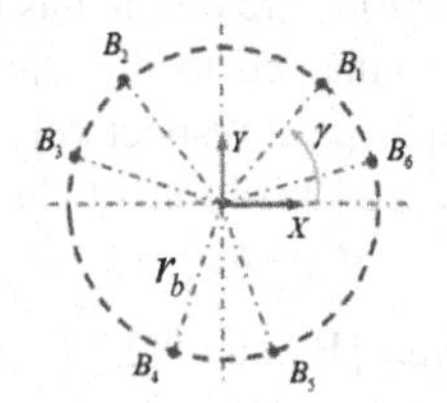

a) Positions in the base platform

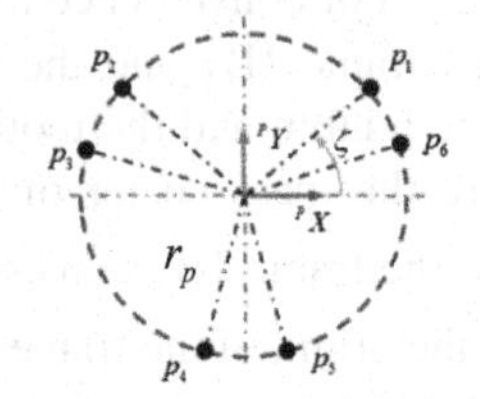

b) Positions in the top platform

Fig. 3. Representation of the coordinates

The length of leg is defined as the Euclidean norm of the leg vector Fig. 2b

$$l_i = \left\| {}^{B}l_i \right\|_2 \tag{3}$$

The point $B_i(i = 1, \ldots, 6)$ is attached to the base plate and located on the circle with radius r_b. The points $p_i(i = 1, \ldots, 6)$ is attached to the motion platform and located on the circle with radius r_p as shown in Fig. 3.

The position of point B_i located on the base platform can be described as

$${}^{B}b_i = (x_{B_i}\, y_{B_i}\, z_{B_i})^T = \begin{bmatrix} r_b\cos(\gamma_i) & r_b\sin(\gamma_i) & 0 \end{bmatrix}^T \tag{4}$$

The position of point p_i located on the motion platform can be written as

$${}^{P}p_i = (x_{p_i}\, y_{p_i}\, z_{p_i})^T = \begin{bmatrix} r_p\cos(\gamma_i) & r_p\sin(\gamma_i) & 0 \end{bmatrix}^T \tag{5}$$

The angle γ_i in the base platform shown in Fig. 3 is defined to be for *(j = 1...3)*

$$\begin{aligned} \gamma_i &= \frac{2\pi}{3}(j-1) + \gamma \quad \text{for the odd legs}(i = 1, 3, 5) \\ \gamma_i &= \frac{2\pi}{3}(j-1) - \gamma + \pi \quad \text{for the even legs}(i = 2, 4, 6) \end{aligned} \tag{6}$$

The angle ζ_I in the motion platform shown in Fig. 3 is described to be

$$\begin{aligned} \zeta_i &= \frac{2\pi}{3}(j-1)+\zeta \quad \text{for the odd legs}(i=1,3,5) \\ \zeta_i &= \frac{2\pi}{3}(j-1)-\zeta+\pi \quad \text{for the even legs } (i=2,4,6) \end{aligned} \tag{7}$$

From Eqs. (6) and (7), we get the value table of angles as below

i	B_1,p_1	B_2,p_2	B_3,p_3	B_4,p_4	B_5,p_5	B_6,p_6
γ_i	47°	133°	167°	253°	287°	373°
ζ_i	39°	141°	159°	261°	279°	381°

2.2 Inverse Kinematic

An important problem in the RSP control is to solve the inverse kinematics (IK) of the RSP. From the IK the desired input to the RSP can be obtained which is to find the lengths of all legs from a given desired position and orientation of the motion platform. In order to calculate the solution of the IK problem, it is needed that the frame {Bi} is attached to each servo motor with its origin at Bi and the axis of rotation ZBi pointing backward the origin of the base CS. The center of the spherical joint in Fig. 2c, attached to the end of servo crank, rotates the angle αi. ${}^{B}s_i$ is the vector from the origin Mi to the pi, and the ${}^{B}a_i$ is the vector from the origin Bi to Mi in the base frame {B}. Also δ is the length of each crank.

The ${}^{B}r_{M_i}$ vector from the origin in the base frame {B} to M_i point is calculated by

$${}^{B}r_{M_i} = {}^{B}b_i + {}^{B}R_{B_i}\,{}^{B_i}a_i \tag{8}$$

where α_i is the rotational angle of the ith servo crank.

The rotation matrix ${}^{B}R_{B_i}$ expressed by the rotational orientation of frame $\{B_i\}$ relative to the frame {B} can be described as

$${}^{B}R_{B_i} = R_Z(\lambda_i)R_Y\left(\frac{\pi}{2}\right) \tag{9}$$

where λ_i angle is defined by rotation of the axis of Z_{Bi} in Fig. 4b, and the ${}^{B_i}a_i$ vector from the origin B_i to M_i is described in the frame $\{B_i\}$ as

$${}^{B_i}a_i = \delta[-\sin\alpha_i \quad \pm\cos\alpha_i \quad 0]^T \tag{10}$$

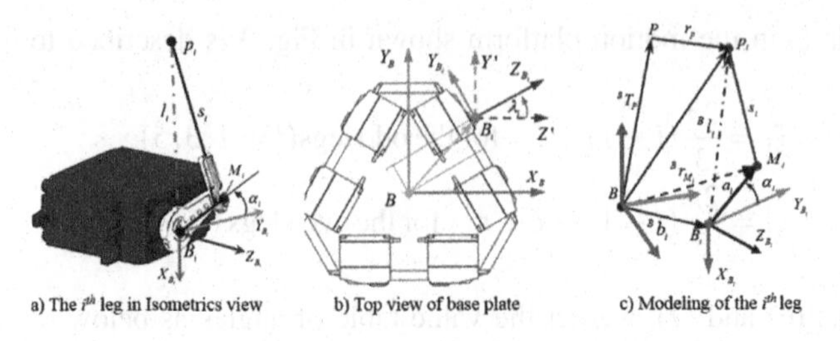

a) The ith leg in Isometrics view b) Top view of base plate c) Modeling of the ith leg

Fig. 4. Geometric description and coordinate transform of the ith leg

Also the ${}^{B}a_i$ vector is expressed by the transform of the frame $\{B_i\}$ relative to the frame {B} described as

$$ {}^{B}a_i = {}^{B}R_{B_i}\, {}^{B_i}a_i = \delta[\mp \cos\alpha_i \sin\lambda_i \quad \pm \cos\alpha_i \cos\lambda_i \quad \sin\alpha_i]^T \tag{11} $$

where sign ($\pm$) in Eqs. (10) and (11) is applied to the odd legs in case of the above row and to the even legs in case of the under row.

The relationship among ${}^{B}l_i$, ${}^{B}s_i$ and ${}^{B}a_i$ can be presented in Fig. 4 as

$$ {}^{B}s_i = {}^{B}l_i - {}^{B}a_i \quad \text{for } i = 1, \ldots, 6 \tag{12} $$

Moreover, the relationship among ${}^{B}l_i$, ${}^{B}s_i$ and ${}^{B}a_i$ can be presented as

$$ {}^{B}l_i^T\, {}^{B}l_i - {}^{B}s_i^T\, {}^{B}s_i + {}^{B}a_i^T\, {}^{B}a_i = 2\, {}^{B}l_i^T\, {}^{B}a_i \tag{13} $$

Equation (13) can be written as

$$ {}^{B}l_i^T\, {}^{B}l_i - {}^{B}s_i^T\, {}^{B}s_i + {}^{B}a_i^T\, {}^{B}a_i = 2\delta\left[\mp l_{ix} \sin\lambda_i \pm l_{iy} \cos\lambda_i\right] \cos\alpha_i + 2.\delta l_{iz} \sin\alpha_i \tag{14} $$

Let $\begin{cases} m_i = 2\delta l_{iz} \\ n_i = 2\delta\left[\mp l_{ix} \sin\lambda_i \pm l_{iy} \cos\lambda_i\right] \\ \Delta_i = {}^{B}l_i^T\, {}^{B}l_i - {}^{B}s_i^T\, {}^{B}s_i + {}^{B}a_i^T\, {}^{B}a_i \end{cases}$

Equation (13) together with Eq. (14) leads to

$$ \Delta_i = \sqrt{m_i^2 + n_i^2}\, \sin(\alpha_i + \varepsilon) \quad \textit{with} \quad \varepsilon = \arctan(n_i/m_i) $$

$$ \alpha_i + \varepsilon = \arcsin\left(\frac{\pm\Delta_i}{\sqrt{m_i^2 + n_i^2}}\right) \tag{15} $$

From Eq. (15), the rotation angle of the ith crank, α_i is expressed as

$$ \alpha_i = \arcsin\left(\frac{\pm\Delta_i}{\sqrt{m_i^2 + n_i^2}}\right) - \arctan\left(\frac{n_i}{m_i}\right) \tag{16} $$

3 RBF Neural Network and Sliding Mode Control

Using the computed torque method, the RSP system can be formulated as follows [10]

$$M(q)\ddot{q} + V_m(q,\dot{q})\dot{q} + G(q) = \tau - F(\dot{q}) - \tau_d \tag{17}$$

where $M(q)$ is the inertia matrixes with the nxn positive definite and $V_m(q,\dot{q})$ coriolis and centripetal Also $G(q)$ and $F(\dot{q})$ are an nx1gravity vector, an *nx1* friction force vector, respectively. In addition, τ, τ_d is an *nx1* control input vector, and an *nx1* unknown disturbance vector, respectively.

The tracking error is defined as

$$e(t) = q_d(t) - q(t) \tag{18}$$

The sliding variable is defined as:

$$s = \dot{e} + \Lambda e \tag{19}$$

where $\Lambda > 0$, $\Lambda = \Lambda^T = diag(\lambda_1, \lambda_2, \ldots, \lambda_6)$, $e = q_d - q$ and $\dot{e} = \dot{q}_d - \dot{q}$

Equation (19) can be rewritten as

$$\dot{s} = \ddot{q}_d - \ddot{q} + \Lambda\dot{e} \tag{20}$$

The Lyapunov function is selected as

$$L = \frac{1}{2}s^T M s \tag{21}$$

The time derivative of L becomes

$$\dot{L} = s^T M\dot{s} + \frac{1}{2}s^T \dot{M} s \tag{22}$$

From Eq. (22), we can be written as

$$M\dot{s} = f(x) - V_m(q,\dot{q})s - \tau + \tau_d \tag{23}$$

It is noted that $f(x) = M(\ddot{q}_d + \Lambda\dot{e}) + V_m(q,\dot{q})(\dot{q}_d + \Lambda e) + G(q) + F(\dot{q})$ can be approximated by the RBF network. The network input is selected based on the expression of *f*(*x*)

$$x = \begin{bmatrix} e^T & \dot{e}^T & q_d^T & \dot{q}_d^T & \ddot{q}_d^T \end{bmatrix}$$

The controller is designed as

$$\tau = \hat{f}(x) + K_v s \tag{24}$$

where K_v is a symmetric positive definite constant matrix, $\hat{f}(x)$ is the output of RBF network. Here $\hat{f}(x)$ is the approximation of $f(x)$.

From Eqs. (21) and (22), the following is obtained as

$$\begin{aligned} M\dot{s} &= f(x) - V_m(q,\dot{q})s - \hat{f}(x) - K_v s + \tau_d \\ &= -(V_m(q,\dot{q}) + K_v)s + \zeta_0 \end{aligned} \tag{25}$$

where $\tilde{f}(x) = f(x) - \hat{f}(x)$ and $\zeta_0 = \tilde{f}(x) + \tau_d$

Substituting Eq. (25) into Eq. (21), it is obtained as follows.

$$\begin{aligned} \dot{L} &= s^T(-(V_m(q,\dot{q}) + K_v)s + \zeta_0) + \frac{1}{2}s^T\dot{M}s \\ &= -s^T K_v s + s^T \zeta_0 \end{aligned} \tag{26}$$

This means that stability of the control system depends on ζ_0 with K_v given.

3.1 Function Approximation Using RBF Neural Network

RBF network is introduced to approximate the uncertain *f*(*x*). With the RBF neural network, the ideal expression of the uncertain function [14, 15]

$$f(x) = W^{*T}\varphi(\mathbf{x}) + \varepsilon \tag{27}$$

Here Kernels $\phi_j(\mathbf{x})$ is selected as

$$\phi_j(\mathbf{x}) = \exp\left[-\|\mathbf{x} - \boldsymbol{\mu}_j\|^2/2\sigma_j^2\right], j = 1, \ldots, n \tag{28}$$

where X is the input state of the network, $\varphi(\mathbf{x}) = [\phi_1 \quad \phi_2 \quad \ldots \quad \phi_n]^T$

The output vector y can be expressed as

$$y = w^{*T}\varphi(\mathbf{x}) \tag{29}$$

where ε, W^* are the approximation error and the desired weight vector of RBF network, respectively.

$$W^* = \left(w_1^*, w_2^*, \ldots, w_m^*\right)^T$$

3.2 Design of Sliding Mode Controller

The approximated uncertainty $\hat{f}(x)$ is expressed by the RBF neural network as [14].

$$\hat{f}(x) = \hat{W}^T \varphi(x) \tag{30}$$

A proposed controller is designed as

$$\tau = \hat{f}(x) + K_v s - v \tag{31}$$

where v is the robust input required to overcome the network approximation error and the disturbance τ_d

From Eq. (25), We are obtained as

$$\zeta_0 = \tilde{f}(x) + \tau_d = \tilde{W}^T \varphi(x) + \varepsilon + \tau_d \tag{32}$$

where the weight error is expressed as $\tilde{W} = \hat{W} - W^*$ and the uncertainty error is obtained as $\tilde{f}(x) = \tilde{W}^T \varphi(x) + \varepsilon$

Substituting Eqs. (31), (33) into Eq. (26) yields

$$M\dot{s} \ = -(V_m(q,\dot{q}) + K_v)s + \xi \tag{33}$$

where the term is defined as $\xi = \tilde{W}^T \varphi(x) + \varepsilon + \tau_d + v$

The robust feedback term $\boldsymbol{v}$ in Eq. (33) is designed to be

$$v = -(\varepsilon_N + b_d) sign(s) \quad \text{with } \varepsilon \leq \varepsilon_N \text{ and } \tau_d \leq b_d$$

Now let's select Lyapunov function as

$$L = \frac{1}{2} s^T M s + \frac{1}{2} tr\left(\tilde{W}^T F_w^{-1} \tilde{W}\right) \tag{34}$$

From substituting Eq. (33) into Eq. (34), the following can be obtained as

$$\begin{aligned} \dot{L} = & -s^T K_v s + \frac{1}{2} s^T \left[\dot{M} - 2V_m(q,\dot{q})\right] s \\ & - tr\left[\tilde{W}^T \left(F_w^{-1} \dot{\tilde{W}} + \varphi(x)\right)\right] + s^T(\varepsilon + \tau_d + v) \end{aligned} \tag{35}$$

Because the RSP has the characteristics of $\dot{M} - 2V_m(q,\dot{q}) = 0$, by choice of $\dot{\tilde{W}} = -F_w \varphi s^T$ Eq. (35) can be rewritten as

$$\dot{L} = -s^T K_v s + s^T(\varepsilon + \tau_d + v) \tag{36}$$

$$\begin{aligned} s^T(\varepsilon + \tau_d + v) &= s^T(\varepsilon + \tau_d) - s^T(\varepsilon_N + b_d) sign(s) \\ &= s^T(\varepsilon + \tau_d) - \|s\|(\varepsilon_N + b_d) \leq 0 \end{aligned}$$

Therefore the closed-loop system stability is obtained from the result of $\dot{L} \leq 0$.

4 Simulation and Results Rotary Stewart Platform

4.1 Simulation Rotary Stewart Platform in Simulink

We built schematic block of RSP in Simulink as shown in block diagram in Fig. 5.

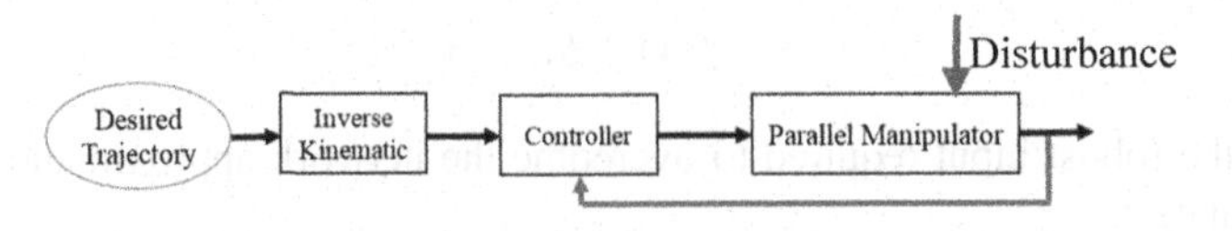

Fig. 5. Schematic block of RSP in Simulink

Block diagram of the system is built in Simulink, and it consists of inverse kinematic, controller and plant as shown in Fig. 6.

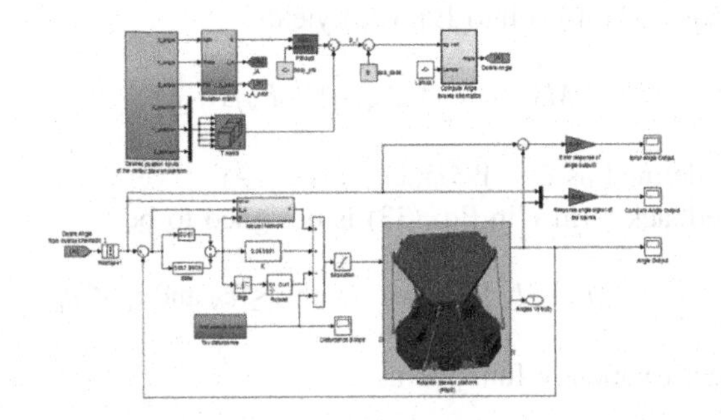

Fig. 6. Block diagram of the system in Simulink

Assuming that, an unknown disturbance vector τ_d is expressed as

$$\tau_d = [0.15\sin(10,t); -0.25\sin(10,t); 0.15\text{square}(5pi,t); -0.15\text{sawtooth}(10pi,t); 0.25\text{random}(10,t); -0.25\sin(10,t)]^T$$

4.2 Simulation and Results

Check the Inverse Kinematics

First, we check the function inverse kinematics via forward kinematic as shown in Fig. 7.

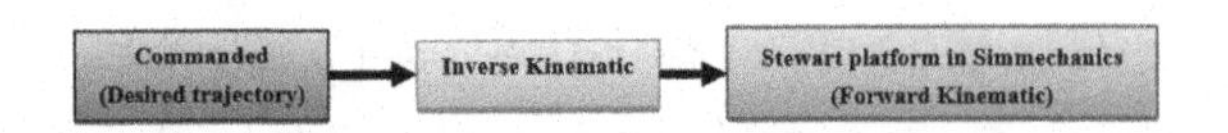

Fig. 7. Block diagram for checking inverse kinematics

Forward kinematic is used by the model of RSP in Simmechanics with desired trajectory of the center of motion platform given below.

$$x(t) = 10\cos(\pi t); y(\mathrm{t}) = 10\sin(\pi t); \mathrm{z}(t) = 115.94 + 10\cos(2\pi t)[mm]$$
$$\varphi(t) = 5\sin(1.5\pi t)[\mathrm{deg}]; \theta(t) = 5\cos(1.5\pi t)[\mathrm{deg}]; \quad \psi(t) = -5\sin(1.5\pi t)[\mathrm{deg}]$$

The equation of the inverse kinematics is correct when the signal from forward kinematics similar to commended signal input. The results are shown in Fig. 8.

The results of Fig. 8 shows that the inverse kinematic derived in this paper is true.

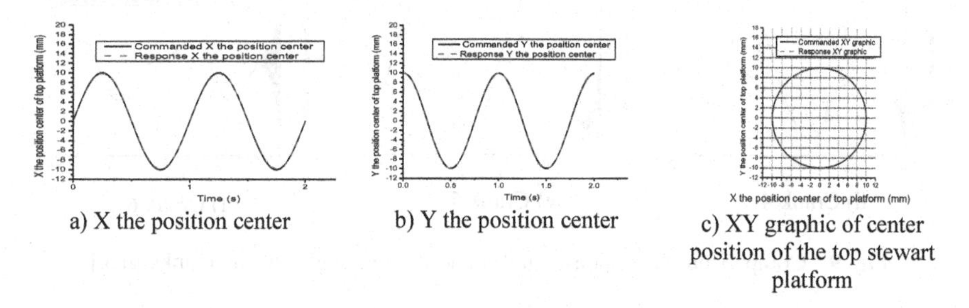

a) X the position center b) Y the position center c) XY graphic of center position of the top stewart platform

Fig. 8. The results of position center of the top motion platform

Control Six Cranks Using RBF Neural Network Adaptive Sliding Mode Controller

From the results the inverse kinematic shown in Sect. 4.2 shows that if the six angles of six cranks are controlled, then we can control the center position of the top motion platform. Therefore, in this paper, we can apply RBF neural network adaptive sliding mode control to control of the six angles of six cranks of the RSP.

- Comparison angle response results

Block diagram of the system is built in Simulink as shown Fig. 6. The simulation results using RBF neural network adaptive sliding mode controller for the six cranks are shown in Fig. 9 where the input is a desired trajectory of the center of motion platform as shown in Sect. 4.2.

- Responses of angle errors

From the system simulation in Simulink, we get responses of angle errors as shown in Fig. 10 with parameters $v = \{\varepsilon = 0.1, b_d = 0.2\}$, $\Lambda = 1000 \cdot diag([6 \quad 6 \quad 6 \quad 6 \quad 6 \quad 6])$, and K_v= 0.014516.

From Figs. 7 and 8 show that the feedback signal well follows the desired trajectory, although the external disturbances are quite large. System stability and response signal convergence to zero. Figure 11 shows the responses of angle accuracy of the robust element in two cases $\varepsilon = 0.1, b_d = 0.2$ and $\varepsilon = 0.001, b_d = 0.1$. We can see that, if ε, b_d is smaller, then the control is higher accuracy.

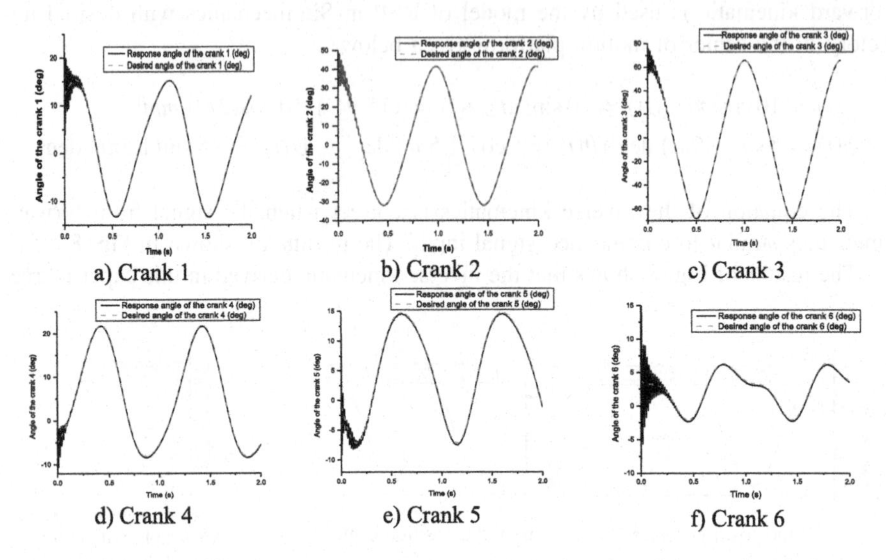

a) Crank 1 b) Crank 2 c) Crank 3

d) Crank 4 e) Crank 5 f) Crank 6

Fig. 9. Comparison the response angles and desired angles of the cranks [deg]

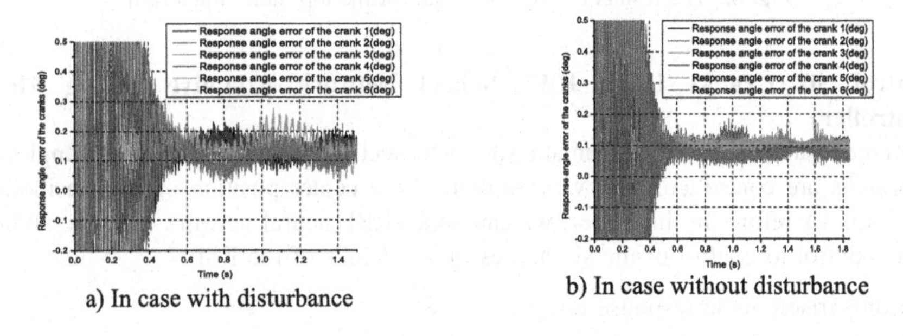

a) In case with disturbance b) In case without disturbance

Fig. 10. Response angle error of the cranks

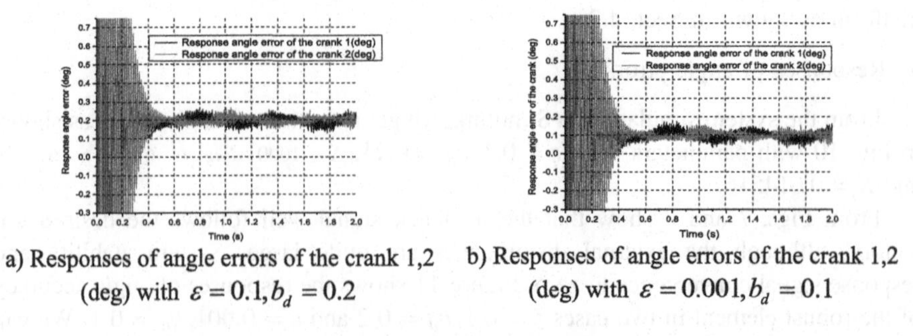

a) Responses of angle errors of the crank 1,2 (deg) with $\varepsilon = 0.1, b_d = 0.2$

b) Responses of angle errors of the crank 1,2 (deg) with $\varepsilon = 0.001, b_d = 0.1$

Fig. 11. Responses of angle errors of the cranks 1, 2

Control input signal of the system is shown in Fig. 12.

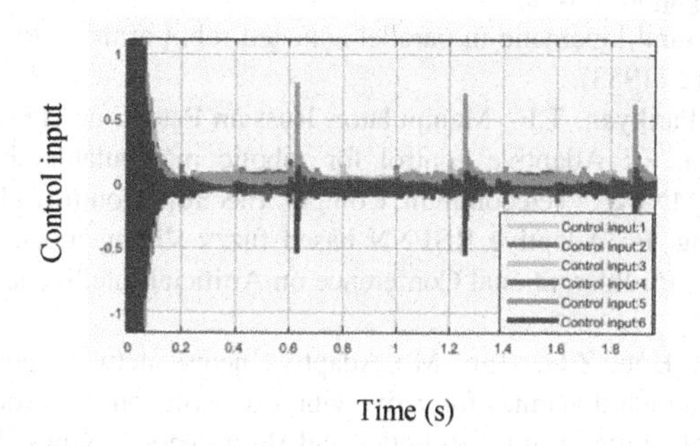

Fig. 12. Control input signal

5 Conclusions

In this paper, the IK solution of the RSP is verified through Simmechanics model. This means that control signals of the crank motors are converted into the desired trajectory of the motion center through the IK solution. As a result, the trajectory of the center position of the motion platform in Simmechanics coincides well with the desired trajectory. The adaptive RBF neural network sliding mode controller is performed to achieve the satisfactory tracking and the system stability. Based on IK solution, the input of the RSP is described. As the results in the simulation, it is shown that the adaptive RBF neural network sliding mode controller can provide robustness for the RSP against the disturbance. Furthermore, the Lyapunov stability was applied for the robust sliding mode controller design in order to select the parameters for the controller which keeps the system stable.

References

1. Stewart, D.: A platform with six degrees of freedom. Proc. Inst. Mech. Eng. **180**, 371–386 (1965)
2. Merlet, J.P.: Solving the forward kinematics of a Gough-type parallel manipulator with internal analysis. Int. J. Robot. Res. **23**, 221–235 (2004)
3. Bingul, Z., Karahan, O.: Dynamic modeling and simulation of Stewart platform. In: Küçük, S. (ed.) Serial and Parallel Robot Manipulators - Kinematics, Dynamics, Control and Optimization. Mechatronics Engineering, Kocaeli University, Turkey, March 2012. ISBN 978-953-51-0437-7
4. Merlet, J.: Designing a parallel manipulator for a specific workspace. Int. J. Robot. Res. **16** (4), 545–556 (1997)
5. Dasgupta, B., Mruthyunjaya, T.: Singularity-free path planning for the Stewart platform manipulator. Mech. Mach. Theory **33**(6), 711–725 (1998)
6. Kang, C.: Closed-form force sensing of a 6-axis force transducer based on the Stewart platform. Sens. Actuators A-Phys. **90**(1–2), 31–37 (2001)

7. Chung, N.C.: 3-axis and 5-axis machining with stewart platform. A Thesis Submitted for the Degree of Doctor of Philosophy, Department of Mechanical Engineering, National University of Singapore (2012)
8. Hunt, K.H.: Structural kinematic in parallel actuated robot arms. J. Mech. Transm. Autom. Des. **105**, 705–712 (1983)
9. Sarkissian, Y.L., Parikyan, T.F.: Manipulator. Russian Patent, no. 1585144 (1990)
10. Jing, M., Wenhui, Z.: Adaptive control for robotic manipulators base on RBF neural network. TELKOMNIKA (Telecommun. Comput. Electron. Control) **11**(3), 521–528 (2013)
11. Fei, L., Shaosheng, F.: Adaptive RBFNN based fuzzy sliding mode control for two link robot manipulator. In: International Conference on Artificial Intelligence and Computational Intelligence (2009)
12. Ma, J., Yang, T., Hou, Z.G., Tan, M.: Adaptive neural network controller of a stewart platform with unknown dynamics for active vibration isolation. In: Proceedings of the 2008 IEEE International Conference on Robotics and Biomimetics, Bangkok, Thailand (2009)
13. Yang, J., Cui, Y., Chen, M.: Sliding mode control based on RBF neural network for parallel machine tool. Open Autom. Control Syst. J. **6**, 575–582 (2014)
14. Wang, H., Kong, H., Yu, M.: RBF-neural-network-based sliding mode controller of automotive steer-by-wire systems. In: International Conference on Natural Computation (ICNC) (2015)
15. Zhang, H., Du, M., Bu, W.: Sliding mode controller with RBF neural network for manipulator trajectory tracking. IAENG Int. J. Appl. Math. **45**(4), 334–342 (2015)
16. Szufnarowski, F.: Stewart platform with fixed rotary actuators: a low cost design study (2013). https://techfak.uni-bielefeld.de/~fszufnar/publications/Szufnarowski2013.pdf
17. Johnson, J., Pillai, R., Vignesh, S.M., Srinivas, A.L.: Design and kinematic analysis of a 6-DOF parallel manipulator for pharmaceutical application. J. Chem. Pharm. Sci. **9**, 3283–3288 (2016)

Smart Robotic Wheelchair for Bus Boarding Using CNN Combined with Hough Transforms

Sarwar Ali(✉), Shamim Al Mamun, Hisato Fukuda, Antony Lam, Yoshinori Kobayashi, and Yoshonori Kuno

Graduate School of Science and Engineering, Saitama University, Saitama, Japan
{sarwar_ali, shamim, fukuda, antonylam, yoshinori, kuno}@cv.ics.saitama-u.ac.jp

Abstract. In recent times, several smart robotic wheelchair research studies have been conducted for the sake of providing a safe and comfortable ride for the user with real-time autonomous operations like object recognition. Further reliability support is essential for such wheelchairs to perform in real-time, common actions like boarding buses or trains. In this paper, we propose a smart wheelchair that can detect buses and precisely recognize bus doors and whether they are opened or closed for automated boarding. We use a modified simple CNN algorithm (i.e. modified Tiny-YOLO) as a base network on the CPU for fast detection of buses and bus doors. After that, we feed the detected information of our Hough line transform based method for accurate localization information of open bus doors. This information is indispensable for our bus-boarding robotic wheelchair to board buses. To evaluate the performance of our proposed method, we also compare the accuracy of our modified Tiny-YOLO and our proposed combined detection method with the original ground truth.

Keywords: Bus door detection · CNN · Hough transformation

1 Introduction

In today's world, the elderly and disabled are increasing day by day. According to [1], 8.5% of the world's population are aged over 65 and not all of them live a healthy life. Moreover, about 15% of the people in the world are disabled [2]. So, there are growing demands for wheelchairs that can support them. Powered wheelchairs have been developed for people that lack muscle control (e.g., spinal cord injury) and have difficulties in operating unpowered wheelchairs. However, clinical studies show that many people have struggled in operating powered wheelchairs [3]. To provide a better quality of life for the user, researchers have come up with "Smart Robotic Wheelchairs" that have user-friendly interfaces and autonomous functions that can navigate and sense information from their environment and respond in useful ways [4].

However, to have autonomous functions, the wheelchair must be able to navigate safely with avoiding obstacles and maneuvering through doorways or any narrow space along with reducing the workloads of caregivers. Most recent smart wheelchairs are used in indoor environments but the smart wheelchairs should provide autonomous functions in outdoor areas such as roads and other places to move around. The method

D.-S. Huang et al. (Eds.): ICIC 2018, LNAI 10956, pp. 163–172, 2018.
https://doi.org/10.1007/978-3-319-95957-3_18

in [5] investigated the primary concerns in developing outdoor safe navigation for smart wheelchairs by detecting any given terrain's smoothness and recognize categories of the terrain for a specified given path. Additionally, an intelligent wheelchair system was developed in [6, 7] for outdoor navigation with real-time detection and avoidance of obstacles and traffic without collisions. But in addition to moving about in an outdoor environment, wheelchair users sometimes also need to travel long distances and so might need an easy mode of transportation like riding buses or trains. If we consider this type of situation, there are still some difficult barriers for a robotic smart wheelchairs to overcome for stable autonomous operation.

For autonomous operations in such environments where the wheelchair would have to climb up or down to certain heights, LiDAR processing is essential for accurate measurement of heights and distances. In bus boarding, the wheelchair system needs precise detection of the bus door's distance and the height of the bus doorstep for boarding on the bus. In our previous work [8], a single laser bidirectional sensing based approach to step detection and step height measurement was used by our smart wheelchair to detect bus doorsteps. Our wheelchair provides for autonomous movement in outdoor terrains and can climb onto steps like stairways or at bus doorways without any need for additional support from lifts/ramps. Moreover, our ongoing research is based on boarding buses by improving further accurate measurement of the height of bus doorsteps.

Nevertheless, using only LiDAR processing has some drawbacks. Before measurement of doorstep heights and distances, our wheelchair should be able to perform real-time operations like detecting bus doors and door positions accurately and precisely. Performing such real-time detection using only LiDAR would be difficult. Therefore, smart wheelchairs require a sensing system that can make use of cameras as well. This is because they could provide fast, precise, and accurate detection in conjunction with LiDAR processing for further reliability support. In the camera processing part the smart wheelchair at first needs to detect the presence of buses at bus stops. After detection of the bus, it must identify the bus doors and which door is open. If the wheelchair recognizes the open door accurately and precisely with approximate estimates of the bus door width, then further detection of appropriate and exact locations of that door with respect to itself can be performed by LiDAR processing for accurate boarding of the bus.

Our main goal in this paper is to focus on the vision part, involving camera processing for the wheelchair. In recent times, Convolutional Neural Networks (CNN) have been used in many applications of computer vision (CV) and have outperformed many algorithms in visual perception or object recognition [9, 10], but most of these networks need specialized, high-powered, costly hardware (e.g. NVIDIA GPUs) to achieve high performance [11]. There are various kinds of CNN algorithms for object detection, such as R-CNN, Faster R-CNN, Mask R-CNN, SSD, YOLO, and from their speed vs. accuracy trade-offs we found that YOLO is best for fast detection [12, 13], compared to others. Standard YOLO still requires GPUs but there is a tiny version for the CPU called, Tiny-YOLO that can be implemented in low-powered machines like those of smart wheelchairs with a tolerable limited speed in processing. We note that although YOLO has faster detection speeds, the bounding boxes of detected objects sometimes do not give accurate and precise values of the object locations according to ground truth in real-time applications.

In this paper, we propose a bus boarding wheelchair system that can get onto a bus using CNN based image recognition for reliable and precise localization of bus doors. This is an extension of our ongoing work on a bus boarding wheelchair system in terms of the camera processing, vision component. The visual processing work presented here is then used in a smart six wheeled Bus-boarding Mobility Robot (BMR) wheelchair we are developing in collaboration with Toyota Motor Corporation and the University of Tokyo [14] with autonomous functionalities that allow for riding buses. Specifically, we use deep learning for image recognition to identify buses, open doors, and closed doors. Before boarding a given bus, the system needs to conduct in real-time, proper detection operations. Therefore, for real-time processing, we modified the Tiny-YOLO to run at a fair amount of speed. Once our system can detect a given bus with an open door using our fast and modified Tiny-YOLO, we then apply a Hough line transform algorithm to get accurate and precise localization of the door line. In the end, after determining the bus door's position, we feed the information to our BMR wheelchair for LiDAR processing for fine-tuned estimates of bus doorway dimensions to complete the bus boarding process. Moreover, we also compare the Tiny-YOLO detection approach and our proposed combined detection method with the original ground truth to show that our method detects better localization of the bus door.

The subsequent sections describe our wheelchair system, our proposed methodology, and shows the precision over the YOLO bounding boxes on detected class objects. In the last two sections, we discuss our results and conclusions.

2 Our Wheelchair System

We are collaborating with Toyota Motor Corporation and the University of Tokyo to develop a new autonomous wheelchair that can overcome the steps ahead of its path like bus doorsteps or escalators. Figure 1 shows our proposed smart wheelchair, which is called the Bus-boarding Mobility Robot (BMR).

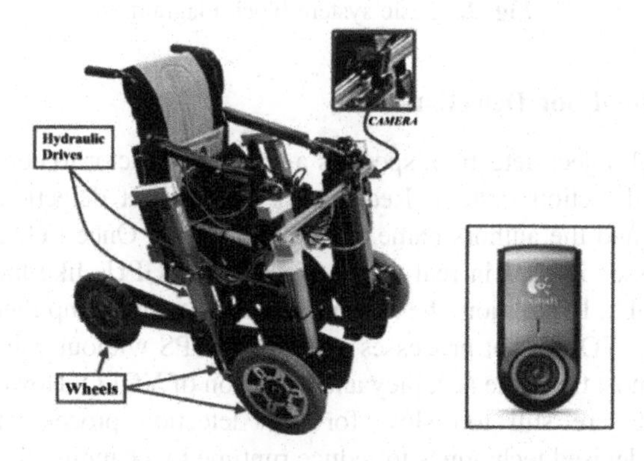

Fig. 1. Bus boarding smart wheelchair system.

This configuration makes it so the wheelchair can step onto a specific height with the front wheels and simultaneously balance itself with the rear wheel. Once the front wheel has a good grip on the upper portion of the step, it moves for-ward, while the middle wheel is lifted up and the wheelchair is balanced with the help of the rear wheel. Moreover, the wheelchair is equipped a wide vision camera (Logitech c905) for capturing frames. The camera is mounted at 75 cm above the ground plane along with our bidirectional sensing system [8].

3 Our Proposed Methodology

Our proposed system in Fig. 2 illustrates, the modified YOLO version we use for primary detection of our predefined class of objects, namely, "bus", "open-door" and "close-door" of the bus. As the localization is not perfectly performed by this system, we introduce our refinement method to get the precise localization of the bus door. Subsequent sections briefly describe our methodology for getting precise bounding box information using our system.

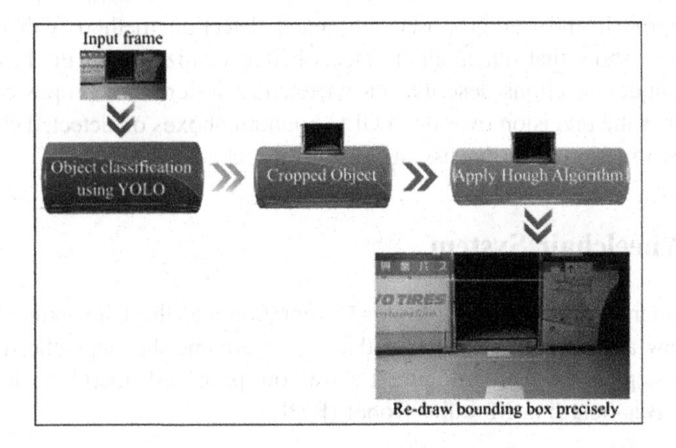

Fig. 2. Basic system block diagram.

3.1 Bus and Bus-Door Detection

In real-time visual object detection, speed is a significant factor to consider along with accuracy of the detection system. Recently, a new object detection algorithm was proposed in [13] and the authors name it You Only Look Once (YOLO). The YOLO model can processes images in real-time at 45 FPS on GPUs like the GeForce GTX Titan X by NVIDIA. In addition, the system comes up with a simplified architecture of the network, Tiny-YOLO that processes at nearly 5 FPS without using any GPU and still maintains almost the same accuracy and precision of YOLO. However, despite that fact, such speeds are still too slow for our detection process. However, some researchers have devised techniques to reduce runtime by changing the network's filter sizes and layers [15, 16]. Therefore, we first doubled the numbers of filters in the first convolutional layer to extract enough local information and visual features and replaced some 3 × 3 filters with 1 × 1 filters for reducing the filter size in the following 2 layers. This boosted the network's speed but reduced the accuracy and precision. To increase the accuracy and precision we downsampled the image to make sure the last layer remains the same as Tiny-YOLO. Our modified network is implemented in Python using Tensorflow with a 3.6 GHz Intel Core i7 CPU.

For our research purposes, we trained the network with 3 classes: "bus" and bus door with "closed door" and "opened door" states. For training, we collected images from different sources like Google, the VOC 2007 and VOC 2012 datasets, and also from our camera and arranged them such that around 600 images would be annotated for each class to train the system. Moreover, we changed the batch size to 128, subdivision to 4 to achieve an improved learning rate. To generate better visual features, the training was done on an NVIDIA GeForce GTX Titan X for 40000 epochs. During training, the image was divided into S×S splits/grids and the output of the last layer gives a feature vector, which represents region predictions. These predictions were encoded in the last layer as an $S \times S \times (B \times 5 + C)$ tensor. Where, B is the collection of predicted bounding boxes in each of the grid cells and their locations are represented by 5 location parameters x, y, w, h, and class label confidence c. C represents number of classes. In our framework, S = 7, B = 2, and C = 3. Each bounding box (BBox) requires 6 parameter values: x, y, w, h, class (C), and confidence (c) as shown in Eq. (1).

$$BBox = (C, \quad x, y, w, h, c) \tag{1}$$

where (x, y) and (w, h) represent the minimum coordinate values of the bounding box and width and height of the bounding box, respectively. From these values, the maximum coordinate is $(x + w, y + h)$. Finally, the confidence score for each box represents the intersection over the union (IoU) between the predicted box and ground truth box (manually obtained) with the probability of the detected object.

The class scores for detecting our three class labels (Bus, Close-door, Open-door) were approximately 70%. Figure 3, shows the detected class labels on frames for different types of buses and their orientations. The state of the door of bus is vital information for our boarding wheelchair.

3.2 Precise Bounding Box Method

Although, YOLO is fast, it is not capable of providing accurate shape information beyond a rectangular bounding box for localizing the object in the frame. This is because different camera angles can distort the shape of even doors, which have straight line features. Fortunately, the Hough line transform is effective for recognizing basic line shape in an image [17]. Therefore, we use the Hough line transform for refined estimate of the door's shape.

Whenever the CNN network detects any open door of the bus, the system crops the image within the bounding information (BBox) to our next layer. Now, the image contains mostly bus door pixels where we apply the Hough line transform algorithm to get straight lines. To avoid redundant lines, a mask is applied as shown in Fig. 4 so that

Fig. 3. Detection of our classes for different buses.

we can focus on only the lines near the door's edge. After applying the mask, the layer processes the image and uses the Hough transform to find the best three lines that fit with the Open-door shape. Moreover, we redraw the bounding box with the Hough detected lines to detect the bus door precisely before boarding onto the bus. For precise boarding, we use our sensor framework for measuring the door width and height of the steps [8].

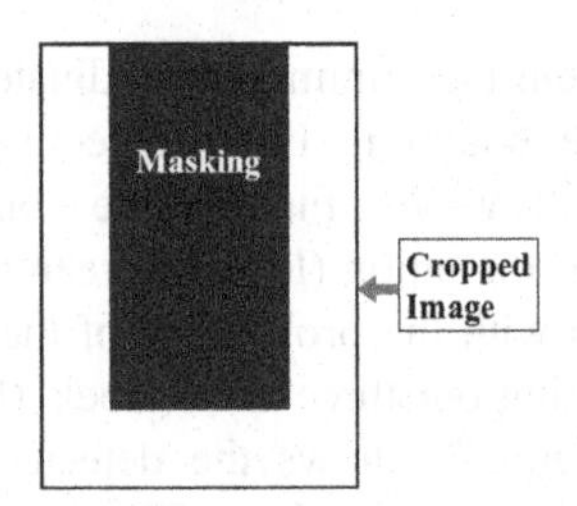

Fig. 4. Masking in cropped image.

4 Experiment and Result Discussion

Performing bus boarding experiments are always time consuming and difficult to manage so we set up a mock bus structure in our laboratory to confirm the effectiveness of our proposed method for fast and precise detection of buses and bus doors (in Fig. 5). Our modified object detection system runs at 10FPS, which is fast enough to run in real-time. From Fig. 5, we can also see the detected class as "open door" and "close door" in experiment's setup. But, the detected bounding boxes for the classes are displaced from the actual locations of the objects. Considering these inaccuracies, we applied the Hough line method to correct the localization of the object. Figure 6 demonstrates the effectiveness of our method for improved bounding boxes.

Fig. 5. Experimental setup for detection of bus with our BMR wheelchair.

Fig. 6. Demonstration of ground truth, YOLO, and our Hough line approach. (Color figure online)

The red, green, and blue boxes represent the ground truth, modified Tiny-YOLO, and our proposed method respectively. Experimental evaluation has been conducted, for comparing the performance of our proposed method and modified Tiny-YOLO with ground truth shown in Fig. 7 and 8. In our experiments, we conducted 22 trials of videos with our mock-up to get a dataset of bounding box values for our tests.

For comparing the effectiveness, we calculated the intersection over union (IoU) between bounding box values from the ground truth with the modified Tiny-YOLO, and our proposed method using Eq. 2.

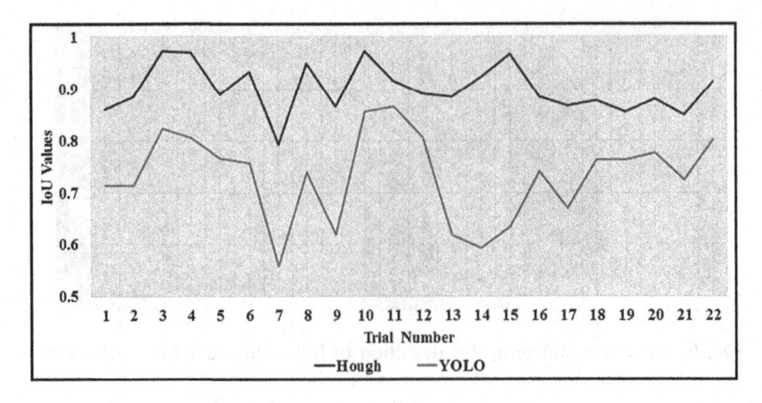

Fig. 7. IoU score comparison over 22 trials.

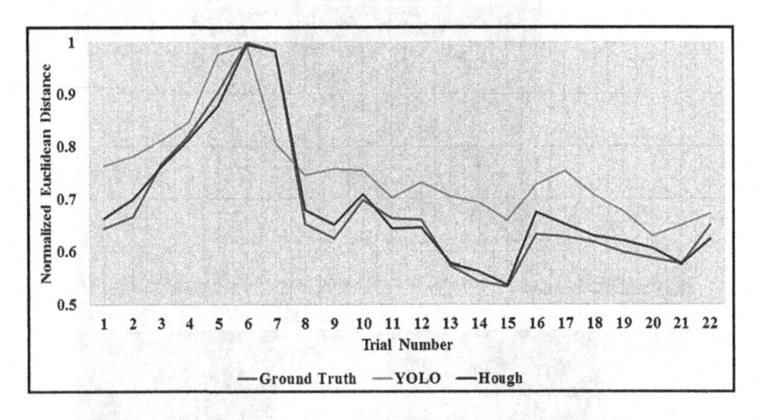

Fig. 8. Normalized Euclidean distance comparison over 22 trials.

$$IoU = \frac{Area\ of\ Overlap}{Area\ of\ Union} \tag{2}$$

Figure 7 illustrates the IoU scores of all 22 trials for our modified Tiny-YOLO and our proposed Hough Line method with respect to the ground truth. Moreover, Table 1 shows the average value of IoU of our image recognition system and proposed method. The average value of IoU of our proposed method indicates an increase in bounding box accuracy by nearly 20% over the default image recognition system with respect to the ground truth.

In addition, we measured the Euclidean distance between the minimum and maximum values of the coordinates for the bounding boxes from the ground truth, modified Tiny-YOLO, and our proposed method. From Table 1, we can see from the average Euclidean distances, our method is very close to the ground truth values. This is also true for the standard deviation and variance of the Euclidean distances, which are also very similar to the ground truth. Moreover, Fig. 8 shows a Normalized

Table 1. Different types of comparison between three types of bounding box values. (The units for the Euclidean distance are in pixel values)

Compared terms	Ground truth	Modified Tiny-YOLO	Our proposed method
Average of IoU value	1.00	0.73	0.90
Average of Euclidean distance	2067.29	2277.04	2088.80
Standard deviation of Euclidean distance	400.45	283.47	379.90
Variance of Euclidean distance	160364.24	80354.40	144326.80

Euclidean distance comparison and we can also see that our method is very closed to the ground truth. So our proposed method works better than the CNN for localizing accurate door positions.

5 Conclusion and Future Work

The main goal of this paper is to propose a bus boarding smart wheelchair that can localize the detected bus door precisely using a vision based system. Visual detection is typically costly in terms of time complexity. Therefore, we are aimed to reduce the computation cost of the CNN based detection method for running a real time system. We also showed that a purely CNN based detection method based on bounding boxes has some inaccuracies in terms of object localization. Our method of localizing a class object (bus door) significantly improves over this. Moreover, we achieved a 90% of IoU in comparison with the ground truth, which was a significant improvement over the modified YOLO. Additionally, we successfully boarded our wheelchair onto the bus using our bidirectional sensing system using a single LiDAR. Our proposed method supports the BMR for precisely localizing the bus door so that BMR can board with less computation cost.

In the future, we are planning to speed up our computation process for detection by using a simpler version of a CNN that can be run on small hardware like a low-cost CPU in real environment. Moreover, we will do more experiments with our bus boarding so that it can safely board onto buses using only a vision based system and do measurement of the door width and height.

Acknowledgement. This work was partly supported by the Saitama Prefecture Leading-edge Industry Design Project and JSPS KAKENHI Grant Number JP26240038 and in collaboration with Dr. Tomoyuki Takahata and Professor Iaso Shimoyama, at the University of Tokyo and Toyota Motor Corporation.

References

1. National Institute of Health, 28 March 2016. https://www.nih.gov/news-events/news-releases/worlds-older-population-grows-dramatically. Accessed 1 Nov 2017
2. Overview of Disability, 20 September 2017. http://www.worldbank.org/en/topic/disability. Accessed 15 Nov 2017
3. Fehr, L., Langbein, W.E., Skaar, S.B.: Adequacy of power wheelchair control interfaces for persons with severe disabilities: a clinical survey. J. Rehabil. Res. Dev. **37**(3), 353–360 (2000)
4. Bourke, T.: Development of a Robotic Wheelchair, November 2001. http://www.tbrk.org/papers/uowhonours01.pdf. Accessed 20 Nov 2017
5. Mamun, S.A., Suzuki, R., Lam, A., Kobayashi, Y., Kuno, Y.: Terrain recognition for smart wheelchair. In: Huang, D.S., Han, K., Hussain, A. (eds.) ICIC 2016. LNCS (LNAI), vol. 9773, pp. 461–470. Springer, Cham (2016). https://doi.org/10.1007/978-3-319-42297-8_43
6. Sharma, V., Simpson, R.C., Lopresti, E.F., Schmeler, M.: Clinical evaluation of semiautonomous smart wheelchair architecture (drive-safe system) with visually impaired individuals. J. Rehabil. Res. Dev. **49**(1), 35–50 (2012)
7. Burhanpurkar, M., Labbé, M., Gong, X., Guan, C., Michaud, F., Kelly, J.: Cheap or robust? The practical realization of self-driving wheelchair technology. In: Proceedings of the IEEE International Conference on Rehabilitation Robotics, July 2017
8. Mamun, S.A., Lam, A., Kobayashi, Y., Kuno, Y.: Single laser bidirectional sensing for robotic wheelchair step detection and measurement. In: Huang, D.S., Hussain, A., Han, K., Gromiha, M. (eds.) ICIC 2017. LNCS (LNAI), vol. 10363, pp. 37–47. Springer, Cham (2017). https://doi.org/10.1007/978-3-319-63315-2_4
9. Krizhevsky, A., Sutskever, I., Hinton, G.E.: Imagenet classification with deep convolutional neural networks. In: NIPS, vol. 1, p. 4 (2012)
10. Hinton, G., Li, D., Yu, D.: Deep neural networks for acoustic modeling in speech recognition: the shared views of four research groups. IEEE Sig. Process. Mag. **29**(6), 82–97 (2012)
11. Abuzaid, F.: Optimizing CPU Performance for Convolutional Neural Networks (2015). http://cs231n.stanford.edu/reports/2015/pdfs/fabuzaid_final_report.pdf. Accessed 22 Nov 2017
12. Zero to Hero: Guide to Object Detection using Deep Learning: Faster R-CNN, YOLO, SSD. http://cv-tricks.com/object-detection/faster-r-cnn-yolo-ssd. Accessed 2 Jan 2018
13. Redmon, J., Divvala, S., Girshick, R., Farhadi, A.: You only look once: unified, real-time object detection, In: CVPR, June 2016
14. Ishikawa, M., et al.: Travel Device, Patent Number: WO2016/006248 A1 (2016)
15. Ning, G.: YOLO CPU Running Time Reduction: Basic Knowledge and Strategies, 7 March 2016. http://guanghan.info/blog/en/my-works/yolo-cpu-running-time-reduction-basic-knowledge-and-strategies. Accessed 10 Jan 2018
16. Iandola, F.N., Han, S., Moskewicz, M.W., Ashraf, K., Dally, W.J., Keutzer, K.: Squeezenet: Alexnet-level accuracy with 50x fewer parameters and <0.5 mb model size. In: ICLR, 4 November 2016
17. Zhao, X., Liu, P., Zhang, M.: A novel line detection algorithm in images based on improved Hough transform and wavelet lifting transform. In: IEEE International Conference on ICITIS, 17–19 December 2010

A Question Answering System Based on Deep Learning

Lu Liu and Jing Luo(✉)

College of Computer Science and Technology, Wuhan University of Science and Technology, Wuhan 430065, China
luojing@wust.edu.cn

Abstract. The goal of the question answering system is to automatically answer questions posed by humans being in the natural language text. Recently, the restricted domain question answering system has become a research hotspot. In this paper, we examine how to efficiently use deep learning method to improve the performance of the question answering system, and design a legal question answering system. Firstly, initial results are obtained by using the vector space model, and then the results are rendered through similarity between the answers. Finally, in order to optimize the system, the deep convolutional neural network is adopted to obtain a one-dimensional sentence vector, which is used to replace the keyword vector for the answer candidate. Experimental results show that the proposed method outperforms the traditional keyword vector method.

Keywords: Question answering system · Retrieval model
Convolutional neural network

1 Introduction

With the development of the Internet, modern society has grown up to be an information-based society. The Internet is an important source of information, where the amount of data is vast and constantly growing, resulting in finding the information that we want relies on search engines is more and more difficult in a huge amount of data information. The traditional search engines that we usually use to find information always return all the relevant webpages based on the keywords entered by the users, which contain many unusual data, making users waste a lot of time to browse the unrelated pages. In order to satisfy users' need for getting information quickly and accurately, question answering system was proposed.

The question answering system allows people to interact with computers more simply and obtain information they want more directly.

Compared to the traditional search engines, the question answering system has two advantages. On the one hand, people usually use natural language to ask questions when they use the question answering system. Compared with keyword queries, using natural language not only accords with the users' habits, but it also can more clearly express the significance of the users' questions. On the other hand, the question

D.-S. Huang et al. (Eds.): ICIC 2018, LNAI 10956, pp. 173–181, 2018.
https://doi.org/10.1007/978-3-319-95957-3_19

answering system always returns a concise and accurate answer to the user, resulting in that users do not need to be screened as before.

This paper briefly describes our question answering system. This system uses the traditional keyword vector method and the simple deep learning method to model the sentences numerically, and then match the questions according to cosine similarity, and finally return the answer.

2 Related Work

Question answering system is a way of information retrieval and has a long history of development, which supports the use of natural language to answer questions asked by users in natural language. In the 1960s, along with the development of artificial intelligence, the question answering system which is one of its branches has also begun to flourish. In 1950, Turing, the father of the famous computer science, published a milestone paper that put forward the concept of "Turing Test" for the first time. In 1966, Eliza's birth was thought to be the real beginning of the question answering system. After entering the 1990s, the question answering system entered the open field and a new era was based on texts. Due to the development of the Internet, information has grown exponentially, resulting in a large number of electronic documents, providing objective conditions for the research of the question answering system during this period. Guo and Fan [1] proposed a question answering system based on natural language understanding. Yuan and Wang [2] proposed a method for the answer extraction in the Chinese question answering system. Mao et al. [3] proposed an answer extraction way for domain question answering system. Methods, Zhang and Zhao [4] designed a Chinese question answering system. Komiya et al. [5] proposed a question answering system based on a question and answer website corpus, and Shen et al. [6] implemented a question answering system based on an online social network.

At present, the application of deep learning in the field of NLP is mainly focused on the representation of words, sentences and discourses and related applications. For example, Mikolov et al. [12] used a neural network model to obtain a new type of vector representation called Word Embedding or word vector. This vector is a low-dimensional, dense, continuous vector representation that also contains semantic and grammatical information of words. Deep neural network model is also widely used in sentence modeling based on the expression of word vectors [13, 14]. Sentence expressions are applied to a large number of natural language processing tasks and have achieved outstanding results in some tasks.

In order to advance the construction of applications of AI to legal informatics problems, firstly we started with assistant lawyers and designed a question answering system based on legal question answer pairs, which can return a reference reply based on the user's case description.

3 Our System

Our system consists of four steps, i.e. data preprocessing, keyword selection, modeling of sentence, matching of sentence (retrieval) and answer candidate list. The system architecture is illustrated in Fig. 1.

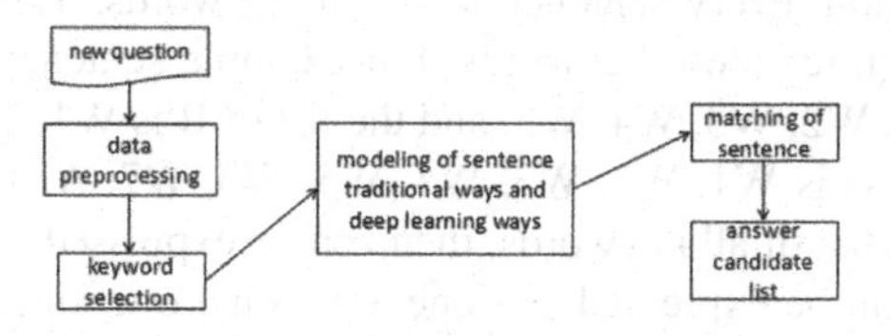

Fig. 1. System architecture

3.1 Data Preprocessing

In data preprocessing, our main work is to segment the Chinese words. The difficulties in Chinese word segmentation mainly lie in the identification of Chinese ambiguity, the identification of entity names, the discovery of new words and hot words, and so on. The word segmentation technique used in this article is the Ansj word segmentation tool. Ambiguity is based on the "best practice rules + statistics" approach. Although there is still some ambiguity that cannot be identified, it is fully capable of meeting engineering applications.

3.2 Keyword Selection

Keywords refer to the important words in a Chinese sentence, which are often more able to express the meaning of the sentence. The method that is easiest to understand and most basic is TF-IDF for extracting keywords. Term frequency (TF) refers to the frequency which the word appears in the file, which is a normalization of the number of words to prevent it from biasing towards long files. For a word W_i in a particular file (d_j), its importance is expressed as formula (1).

$$tfi,j = \frac{ni,j}{\sum_k nk,j} \quad (1)$$

In this formula, the numerator is the number of occurrences of the word Wi in the file (d_j), and the denominator is the sum of the occurrences of all the words in the file (d_j). The inverse document frequency (IDF) of the word Wi can be obtained by dividing the total number of documents by the number of documents containing the W_i, then obtaining the quotient logarithm. As showed in formula (2).

$$idfi = \log \frac{|D|}{|\{j : wi \in dj\}|} \quad (2)$$

In this formula, |D| is the total number of files in the corpus, $|\{j : W_i \in d_j\}|$ is the total number of files contained the word W_i. Finally, the TF-IDF value of the word is (1) * (2)(tf * idf).

3.3 Modeling of Sentence

After keyword selection, every sentence has their keywords. The traditional sentence modeling method requires these keywords. For example sentences A and B, the keywords list of A isW1, W2, W3, W4, W5, and the list of B is W1, W2, W3, W6, W7, so the list of all keywords is W1, W2, W3, W4, W5, W6, W7. As for sentence A, if the keyword is not in the list of all keywords, then zero is expressed, and vice versa is one. So the vector of A can be expressed as <one, one, one, one, one, zero, zero> and the vector of B is <one, one, one, zero, zero, one, one>.

Then we will talk about the deep learning method, which is based on Convolutional Neural Network (CNN) [7]. First of all, using the Word2Vec trains word vector model and all vectors of keywords are stored in this model, then the word vector matrix is built, which is based on word vectors. The row of the matrix that represents the vector of a word, and the column of the matrix represents the order of words in a sentence. After this, a convolution window will be used to convolute the word vector matrix and get several Feature Maps. The column of the Feature Map is one and the size of the convolution window is h * k, where h represents the number of longitudinal words, and K represents the dimension of the word vector. Finally, the maximum values of these feature maps will be picked out, which will be used to form a one-dimensional vector that can model the sentence. The process of the deep learning method is shown in Fig. 2.

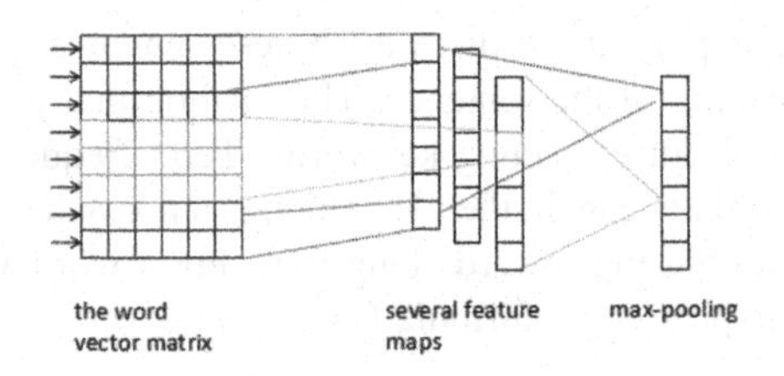

Fig. 2. Deep learning method

In our system, through Word2vec training, each word is represented as a 100-dimensional vector, then we establish the word vector matrix and the index between the word and the vector. The word vector matrix and the sentence index list are the input data in the CNN model. According to the index list, the CNN model seeks the word vector matrix and converts the index list into the word vector matrix of the sentence. Next, using the convolution window (h * k) mentioned above to convolute the word matrix, which the value of h are two, three and four and the value of k is one hundred. The number of each convolution window is five, so we will get fifteen Feature Maps and finally obtain a one-dimensional vector, which the length of is fifteen.

3.4 Matching of Sentence (Retrieval)

In a question answering system, the process of retrieval is actually the process of matching sentences. In the retrieval process, the vector space model (VSM) is usually used to calculate the similarity between sentences. The model is one of the most commonly used models in text modeling and the most commonly used to measure similarity between sentences is the cosine similarity in text processing. For example, the sentences A and B mentioned above, their similarity can be calculated by the formula (3)

$$\cos(\theta) = \frac{\sum_{i=1}^{n} (X_i * Y_i)}{\sqrt{\sum_{i=1}^{n} (X_i)^2 * \sum_{i=2}^{n} (Y_i)^2}} \tag{3}$$

In addition to cosine similarity, the similarity between sentences also can be calculated by word similarity based on the Synonymy Thesaurus. The word encoding format in Synonymy Thesaurus is shown in Table 1.

Table 1. Word encoding format

Coded bits	1	2	3	4	5	6	7	8
Symbol example	D	a	1	5	B	0	2	=\#\@
Symbol nature	Big class	Middle class	Small class		Word group	Atomic word group		
Level	1	2	3		4	5		

In the Table 1, "=" represents equality, "#" represents inequality, and "@" represents self-enclosed, independent, meaning that there are no synonyms and related words in the Synonymy Thesaurus.

In Chinese, a word usually has several meanings. The calculation of the similarity between two words needs to consider all the meanings of the words. So we need to calculate the similarity between the two meanings, which one meaning is from a word and the other meaning is from other word. Then take the largest value as the similarity between words. For example, the words A and B, the similarity is Sim(A, B) and the calculation process is as follows:

1. Two words are not in the same tree, Sim(A, B) = 0.1
2. Two words are on the second floor in the same tree. The word similarity is calculated by formula (4).

$$\mathrm{Sim}(A, B) = a \times \cos(n \times \frac{\pi}{180})(\frac{n - k + 1}{n}) \tag{4}$$

3. Two words are on the third floor in the same tree. The word similarity is calculated by formula (5).

$$\mathrm{Sim}(A,B) = b \times \cos(n \times \frac{\pi}{180})(\frac{n-k+1}{n}) \tag{5}$$

4. Two words are on the fourth floor in the same tree. The word similarity is calculated by formula (6).

$$\mathrm{Sim}(A,B) = c \times \cos(n \times \frac{\pi}{180})(\frac{n-k+1}{n}) \tag{6}$$

5. Two words are on the fifth floor in the same tree. The word similarity is calculated by formula (7).

$$\mathrm{Sim}(A,B) = d \times \cos(n \times \frac{\pi}{180})(\frac{n-k+1}{n}) \tag{7}$$

When codes of two meanings are in the same row, their similarity is calculated by the serial number. If the serial number is "=", their similarity is one; if the serial number is "#", their similarity is point five; if the serial number is "@", their similarity is not considered, because "@" means the meaning has no synonym in the Synonymy Thesaurus. For example, the codes of proud (骄傲) are "Da13A01=" and "Ee34D01=" and the codes of careful (仔细) are "Ee26A01=" and "Ee28A01=". The similarity of two meanings is calculated by the thought above and the results are shown in Table 2.

Table 2. The results of similarity

	Da13A01=	Ee34D01=
Ee26A01=	0.1	0.48392
Ee28A01=	0.1	0.51008

From Table 2, we can know the similarity of proud (骄傲) and careful (仔细) is 0.51008, which is the maximum of the four digits.

Now, the similarity of sentences can be obtained based on the similarity of words. E.g., the words contained in the sentence A are A_1, A_2...A_m and the words contained in the sentence B are B_1, B_2...B_n, then we can calculate the similarity of word A_i(1 i $\leq$ m) and word B_j(1 $\leq$ j $\leq$ n) and obtain a matrix, which the size of is m * n. the maximum value of the i-th row of the matrix is Sim(A_i, B), finally the average of the maximum values is the similarity of sentences A and B.

3.5 Answer Candidate List

Through the above ideas, when the system enters a question, system will get sentence similarities between the entered question and questions in the data set, but the difference is that the traditional method uses the keyword vector and the deep learning method uses the vector learnt by the Convolutional neural network. Then sort the question answer pairs according to the similarities and return the top five. Finally, using the sentence vectors learned by the deep learning method to calculate the similarity of answers and sort the answer again, which is the answer candidate list.

4 Experimental Results

Our experiment has mainly done a dual system. One system mainly uses the traditional method. The word segmentation tool segments the Chinese sentence and the keywords in the sentence will be picked out. Then the word vector will be established by the keywords and the system calculates the sentence similarity based on the word vector and the word similarity. Finally, the system sorts the question answer pairs according the similarity and returns the top five. The other system uses deep learning method. Compared with the traditional system, this system has no word segmentation and picking out keywords. It directly uses the deep learning method to learn and return the sentence vector, and then the sentence vector is used to calculate the similarity and the system sorts the question answer pairs by the similarity. In the end, the system returns the top five.

In our experiments, the data comes from the legal question and answer documents, containing 400000 question answer pairs and 1000 pairs of data are taken as test data. We use Recall, Precision and F1-score as three metrics to measure the systems. Their calculation methods are as shown in formulas (8), (9), and (10), and the values of three metrics are showed in Table 3.

$$R = \frac{\text{Number of correct information extracted}}{\text{Number of information in the sample}} \tag{8}$$

$$P = \frac{\text{Number of correct information found}}{\text{Total number}} \tag{9}$$

$$F = \frac{2PR}{P+R} \tag{10}$$

Table 3. The values of three indicators

Method	R (Recall)	P (Precision)	F
Traditional	72.7(%)	78.8(%)	75.6(%)
Deep learning	75.4(%)	80.3(%)	77.7(%)

From the Table 3, we can find that the question answering system has a relatively high accuracy when they use deep learning method, but the recall rate is low whatever the method is. The reason for this phenomenon is that the data is a closed domain question answer pairs based on the law. For a query question, there may be many similar questions in the data set, so the accuracy rate can be relatively high. But the answers to the same question, the words and narratives are different, resulting the recall rate is low.

5 Conclusions

This paper mainly describes a feasibility study of a question answering system in the legal field and implements through basic traditional methods and simple deep learning methods, which is the biggest change comparing with the traditional method. But only using these methods is not enough for our study. In the future, in order to achieve a better system, we could consider feature fusion and implement a question answering system based on deep learning method completely. In addition, we can also achieve a question answering system based on the Knowledge Graph and the answer generation, because the answer returned by the question answering system is usually not completed.

References

1. Guo, Q.L., Fan, X.Z.: Question answer system based on natural language understanding. Comput. Eng. **30**(3), 419–422 (2004)
2. Yuan, C., Wang, C.: Parsing model for answer extraction in Chinese question answering system. In: International Conference on Natural Language Processing and Knowledge Engineering, pp. 238–243. IEEE (2005)
3. Mao, C.L., Li, L.N., Yu, Z.T.: Research on answer extraction method for domain question answering system (QA). In: International Conference on Computational Intelligence and Security, pp. 79–83. IEEE Computer Society (2009)
4. Zhang, K., Zhao, J.: A Chinese question answering system with question classification and answer clustering. In: Seventh International Conference on Fuzzy Systems and Knowledge Discovery, pp. 2692–2696. IEEE (2010)
5. Komiya, K., Abe, Y., Morita, H.: Question answering system using Q&A site corpus query expansion and answer candidate evaluation. Springerplus **2**(1), 396 (2013)
6. Shen H., Liu, G., Vithlani, N.: Social Q&A: an online social network based question and answer system. In: International Conference on Computer Communication and Networks, pp. 1–8. IEEE (2015)
7. Kim, Y.: Convolutional neural networks for sentence classification. arXiv preprint arXiv: 1408.5882 (2014)
8. Oquab, M., Bottou, L., Laptev, I.: Learning and transferring mid-level image representations using convolutional neural networks. In: IEEE Conference on Computer Vision and Pattern Recognition, pp. 1717–1724. IEEE Computer Society (2014)
9. Kadam, A.D., Joshi, S.D., Shinde, S.V.: Question answering search engine short review and road-map to future QA search engine. In: International Conference on Electrical, Electronics, Signals, Communication and Optimization, pp. 1–8. IEEE (2015)
10. Mollá, D., Vicedo, J.L.: Question answering in restricted domains: an overview. Comput. Linguist. **33**(1), 41–61 (2007)
11. Do, P.K., Nguyen, H.T., Tran, C.X., et al.: Legal question answering using ranking SVM and deep convolutional neural network. arXiv preprint arXiv:1703.05320 (2017)

12. Mikolov, T., Chen, K., Corrado, G., et al.: Efficient estimation of word representations in vector space. In: Computer Science (2013)
13. Socher, R., Lin, C.C., Manning, C., et al.: Parsing natural scenes and natural language with recursive neural networks. In: Proceedings of International Conference on Machine Learningpp. 129–136. Omnipress, Haifa (2011)
14. Kalchbrenner, N., Grefenstette, E., Blunsom, P.: A convolutional neural network for modelling sentences. In: Proceedings of ACL, pp. 655–665. Association for Computational Linguistics, Baltimore (2014)

A Deep Clustering Algorithm Based on Self-organizing Map Neural Network

Yanling Tao[1], Ying Li[2], and Xianghong Lin[2(✉)]

[1] College of Social Development and Public Administration, Northwest Normal University, Lanzhou 730070, China
[2] College of Computer Science and Engineering, Northwest Normal University, Lanzhou 730070, China
linxh@nwnu.edu.cn

Abstract. Clustering is one of the most basic unsupervised learning problems in the field of machine learning and its main goal is to separate data into clusters with similar data points. Because of various redundant and complex structures for the raw data, the general algorithm usually is difficult to separate different clusters from the data and the effect is not obvious. Deep learning is a technology that automatically learns nonlinear and more conducive clustering features from complex data structures. This paper presents a deep clustering algorithm based on self-organizing map neural network. This method combines the feature learning ability of stacked auto-encoder from the raw data and feature clustering with unsupervised learning of self-organizing map neural network. It is aim to achieve the greatest separability for the data space. Through the experimental analysis and comparison, the proposed algorithm has better recognition rate, and improves the clustering performance on low and high dimension data.

Keywords: Clustering algorithm · Deep neural networks
Stacked auto-encoders · Self-organizing map neural network

1 Introduction

In our daily life, from information retrieval, translation to the recommendation of social networks and then to fingerprints, face recognition, more and more machine learning technologies have made the machine show the intelligence of human beings. Among these technologies, such as image recognition, segmentation, autonomous vehicles, speech recognition, question answering matching, precision push and other applications, deep learning technology is more and more adopted [1–3]. The key factor behind these success is that deep learning which is data-driven, and automatically learn representative and hierarchical abstract features from enough training data, rather than manually generated features.

Now, deep learning has become the leading method for learning supervised labelled data [4, 5]. The boom of deep learning is just rely on unsupervised learning, which uses unsupervised learning to pre-training specific deep networks, and then combines supervised learning to fine-tune the network. But most of the deep neural networks that

D.-S. Huang et al. (Eds.): ICIC 2018, LNAI 10956, pp. 182–192, 2018.
https://doi.org/10.1007/978-3-319-95957-3_20

can be applied to practice are purely supervised learning in the later period. Human and animal discrimination of unknown objects is by observing its structure and establishing connections with known objects in the brain, rather than telling them the names of each object. For ordinary data, therefore, unsupervised learning is hopeful to train a more general model for learning and experimenting with other tasks. In many applications, data labels may be unavailable or unreliable.

Clustering is one of the most fundamental unsupervised machine learning problems. Its main goal is to separate data into clusters of similar data points [6]. Besides having its own applications, it is beneficial for multiple other fundamental tasks. For instance, it can serve for automatic data labeling for supervised learning and as a pre-processing step for data visualization and analysis. However, the performance of clustering algorithms is dependent on the type of the input data, such that different problems and datasets could require different similarity measures and different separation techniques [7]. As a result, dimensionality reduction and representation learning have been extensively used alongside clustering, in order to map the input data into a feature space where separation is easier with respect to the problem's context. In 2006, Torre and Kanade [8] combined with dimension reduction and clustering, first clustered data with K-means, then projected data to lower dimension of variance maximization among groups. Dilokthanakul et al. [9] proposed a variant of the variational auto-encoder model with a Gaussian mixture as a prior distribution, with the goal of performing unsupervised clustering through deep generative models. For the high dimension and complex data, it is particularly important from data to learn the non-linear and effective features that are more friendly for clustering.

In this paper, we present a deep clustering algorithm based on self-organizing map neural network, which combines self-organizing map network in clustering algorithm and stacked auto-encoder in deep neural network. In the algorithm, the raw input data are transformed into abstract features by multiple hidden layers of stacked auto-encoder and then the self-organizing map neural network is used to cluster the unlabeled features to discover the intrinsic links of the data. The experimental results on different standard datasets show the effectiveness of the proposed method.

2 Related Works

2.1 Auto-Encoder

Auto-Encoder (AE) is a kind neural network structure of unsupervised learning which is mainly used for feature learning and dimensionality reduction [10, 11]. It is a special three layer neural network consisting of the input layer, the hidden layer and the reconstruction layer, in which the target value is similar to the input. An auto-encoder has two parts: (1) the encoder: the input $\boldsymbol{x}$ is converted to the feature $\boldsymbol{h}$ by the mapping function $f(\bullet)$; (2) the decoder: the feature $\boldsymbol{h}$ is mapping to the reconstruction data $\boldsymbol{r}$ by another mapping function $g(\bullet)$. The network structure of an auto-encoder is shown in Fig. 1. In the training process of auto-encoder, the back propagation algorithm is used to adjust the weights and biases of the network to reduce the difference between reconstruction vectors and input vectors, so that the reconstruction value is as similar as the input value, that is, $\boldsymbol{x} \approx \boldsymbol{r}$. The target vectors can be any value for the general neural network, but the target value is equal to the input value for the auto-encoder.

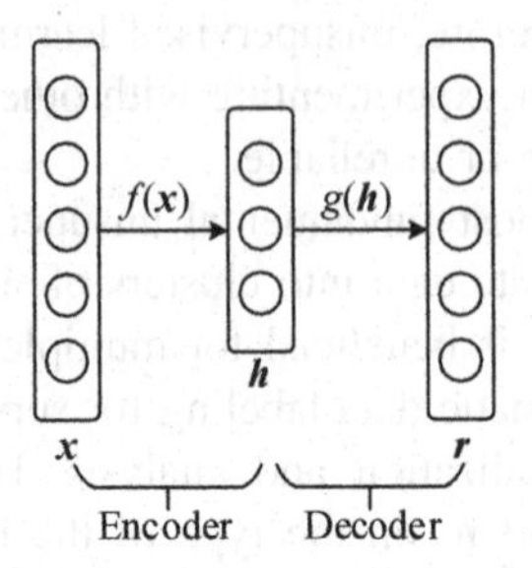

Fig. 1. The structure of an auto-encoder network

2.2 Stacked Auto-Encoder

Stacked Auto-Encoder (SAE) is a deep learning model commonly, which is made up of multiple auto-encoders in series [12]. The training methods of SAEs include two types: bottom-up training and greedy layer-wise training. In this paper, we choose the layer-wise method to train the stacked auto-encoder. The training method is as follows: firstly the raw input vectors is sent to the bottom auto-encoder to learning hidden representations until the convergence (the reconstruction error is minimal); then the hidden representations of this layer as input is sent to the next layer to learn higher level representations; once again, the same process is repeated until all the layers are trained. Finally, the hidden representations of the top layer auto-encoder is the output of the stacked auto-encoder, which can be further entered into other networks, such as SVM for classification. A stacked auto-encoder with a depth of l has an input layer, an output layer and multiple hidden layers. Its network structure is shown in Fig. 2.

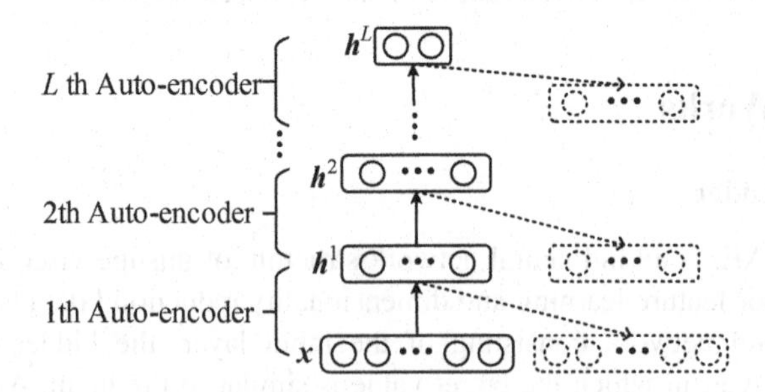

Fig. 2. The network structure of a stacked auto-encoder

An auto-encoder maps the input $\boldsymbol{x}$ into the hidden feature $\boldsymbol{h}$, and then reconstructs the $\boldsymbol{h}$ from the $\boldsymbol{x}$ through the decoding function $g(\bullet)$, which is defined as follows:

$$\boldsymbol{h} = \sigma(\boldsymbol{W}_E\boldsymbol{x} + \boldsymbol{b}_E) \tag{1}$$

$$\boldsymbol{r} = \sigma(\boldsymbol{W}_D\boldsymbol{h} + \boldsymbol{b}_D) \tag{2}$$

where $f(\bullet)$ and $g(\bullet)$ are the encode function and the decode function respectively, and $\theta = \{\boldsymbol{W}_E, \boldsymbol{W}_D, \boldsymbol{b}_E, \boldsymbol{b}_D\}$ is the parameter of the model. Feature extraction is performed by minimizing training errors between the raw data and the reconstructed data, as follows

$$Loss(\boldsymbol{x}, \boldsymbol{r}) = \|\boldsymbol{x} - \boldsymbol{r}\|_2^2 \tag{3}$$

The raw input data is coded as the feature, and then the feature is reconstructed with input data, so as to form a network that reduces the training error through a loss function. The feature $\boldsymbol{h}$ of this layer is used as the input training next layer auto-encoder, a stacked auto-encoder is formed through the greed layer-wise training.

2.3 Self-organizing Map Neural Network

The Self-Organizing Map (SOM) neural network is a special type of network models [13, 14]. It maps the input data from n-dimensional space to low dimensional (usually 1 or 2 dimensional) and discrete map with unsupervised learning, while maintaining the original topological relationship. The theory of SOM neural network comes from observing the operation of brain, and whether the human sensory information can be mapped to brain space or the relationship between other stimuli depending on the spatial relationship between a two-dimensional mapping of neuron. A SOM neural network usually consists of two layers of neurons: input layer and output layer (competitive layer) and the input layer is fully connected to the two dimensional output layer neurons by the weights and its network structure is shown in Fig. 3.

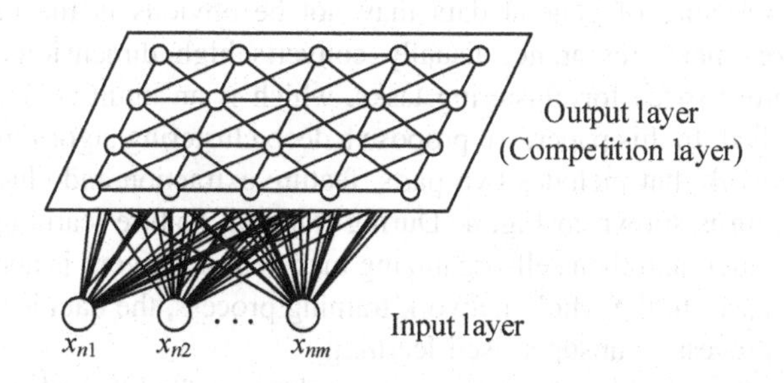

Fig. 3. The structure of a self-organizing map neural network

Similar to artificial neural networks, the SOM neural network performs two modes: training and mapping. The training uses an input sample to construct a representation (a competition process), and the mapping automatically classifies the new input vectors. The SOM is different from other artificial neural networks which uses competitive learning with unsupervised training instead of error correction learning (such as gradient descent based backpropagation). In the networks, neurons adjust the weights mainly based on transverse feedback connections. On the other hand, unsupervised learning does not need to know the target value. Nodes in the network represent similar groups by clustering. According to the output distribution generated by training process, the number and composition of clusters can be intuitively determined.

During the training process, the input data is sent into the network through the neurons of the input layer. The input data $\boldsymbol{X} = \{\boldsymbol{x}_1, \boldsymbol{x}_2, \cdots, \boldsymbol{x}_N\}$, where N represents the number of the input data. An input vector is associated with a two-dimensional output layer nodes. Assuming that the competition layer has T neurons, there is a weight vector $\boldsymbol{W} = \{\boldsymbol{w}_1, \boldsymbol{w}_2, \cdots, \boldsymbol{w}_T\}$. The connection weight vector of the i-th neuron of the output layer $\boldsymbol{w}_i = \{w_{ij}\}(i = 1, \cdots, T)$, where w_{ij} indicates that the j-th neuron of the input layer corresponds to the weight of the i-th neuron of output layer. In the training of the t time, for the input data $\boldsymbol{x}_p(p = 1, \cdots, N)$, if the output layer neuron c is the winning neuron, then

$$c = \arg\min_i \|\boldsymbol{x}_p - \boldsymbol{w}_i\|, \quad i = 1, 2, \cdots, T \tag{4}$$

where $\|\bullet\|$ is the distance metric function and T is the number of neurons in output layer. Then, the update rule is used to modify the weight vector of the winning neuron as follows:

$$\boldsymbol{w}_i(t+1) = \begin{cases} \boldsymbol{w}_i(t) + a(t)\left[\boldsymbol{x}_p - \boldsymbol{w}_i(t)\right], & i = c \\ \boldsymbol{w}_i(t), & i \neq c \end{cases} \tag{5}$$

where $a(t)$ is a variable learning rate, which decreases with the increase of time.

3 A Deep Clustering Algorithm Based on SOM Network

The clustering structure of general data may not be obvious in the original feature space. Therefore, many researches usually converts high dimensional data to low dimensional feature space for clustering tasks, which is an intuitive solution and has been widely studied. In this paper, we propose a deep clustering algorithm based on the SOM neural network that includes two parts, feature extraction and clustering, and its algorithm structure is shown in Fig. 4. During training, feature learning is done by a stacked auto-encoder, and then self-organizing map neural network is applied to cluster data in feature space. In the whole network training process, the data is unlabeled, that is, the learning process is unsupervised learning.

In this algorithm, feature extraction stage relies on the stacked auto-encoder to reduce data dimension by greed layer-wise training, gradient descent method is used to optimize the training loss, and the mean square error is used to reduce the loss. The features learned by the last layer of the auto-encoder, as the latest features, are sent to the SOM neural network to cluster. In the clustering process, the network selected as the neuron with the smallest distance between weights as winning neuron through calculate between input and output layer weights of all the distance, adjust the weights

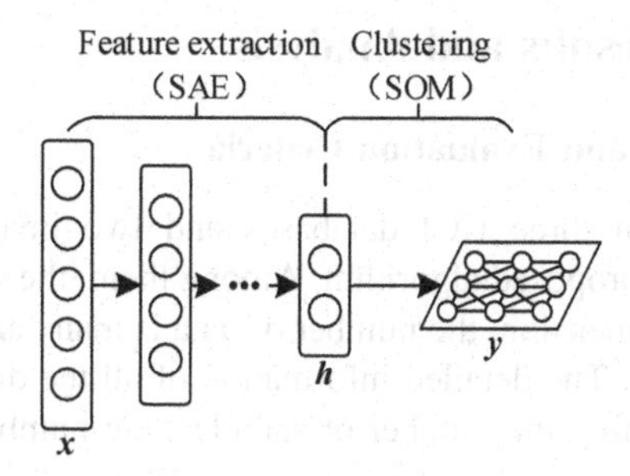

Fig. 4. The structure of a deep clustering algorithm based on the SOM network

of the winning neuron and decrease the learning rate, until the stopping condition is met (learning rate is zero or the maximum number of iterations). A detailed description of the deep clustering process based on the SOM neural network is described as Algorithm 1.

Algorithm 1. The deep clustering process based on SOM neural network

Input: $\boldsymbol{X} = \{\boldsymbol{x}_1, \boldsymbol{x}_2, \ldots, \boldsymbol{x}_N\}$, batch size b, layer number of SAE L
Output: Divide $\boldsymbol{H}$ into k group $\{C_1, \ldots, C_k\}$, where $\bigcup_{i=1}^{k} C_i = \boldsymbol{H}$

1. Initialize weights $\boldsymbol{W}^l$ and biases $\boldsymbol{b}^l$ of stacked auto-encoder of per layer
2. **for** $l \leftarrow 1$ **to** L **do**
3. **for** $i \leftarrow 1$ **to** N **do**
4. Calculate $\boldsymbol{h}^l_i$ by encoding $\boldsymbol{x}_i$
5. Calculate reconstruction $\boldsymbol{r}^l_i$ by decoding $\boldsymbol{h}^l_i$
6. **if** ($l = 1$)
7. Calculate loss between $\boldsymbol{x}_i$ and $\boldsymbol{r}^l_i$ $L(\boldsymbol{x}) = \|\boldsymbol{x} - \boldsymbol{r}^l\|^2$
8. **else**
9. Calculate loss between $\boldsymbol{h}^l_i$ and $\boldsymbol{r}^l_i$ $L(\boldsymbol{h}^l) = \|\boldsymbol{h}^l - \boldsymbol{r}^l\|^2$
10. **end if**
11. Update parameter $\boldsymbol{\theta} = \{\boldsymbol{W}^l, \boldsymbol{b}^l\}$ using gradient descent
12. **end for**
13. **end for**
14. Initialize the weights of output layer neurons $w_i (i = 1, \cdots, T)$
15. Initialize the learning rate η
16. **repeat**
17. Calculate the distance between $\boldsymbol{h}_p$ and $\boldsymbol{w}_i$, and determine the winning neuron c, $c = \arg\min_i \|\boldsymbol{h}_p - \boldsymbol{w}_i\|, (p = 1, \cdots, N)$
18. Adjust the weight of the winning neuron $\boldsymbol{w}_i$, $\boldsymbol{w}_i(t+1) = \boldsymbol{w}_i(t) + \eta[\boldsymbol{h}_p - \boldsymbol{w}_i(t)]$
19. Reduce the learning rate η
20. **until** the cessation condition is satisfied

4 Experimental Results and Analysis

4.1 Data Description and Evaluation Criteria

We select the dataset in three UCI databases and two image datasets (COIL-20, MNIST) to evaluate the proposed algorithm. Among them, the selected data dimension contains low and high dimension, the number of data is many and few, and the number of data is also much less. The detailed information of all the data in the experiment is listed in Table 1, containing the number of samples, the number of attributes and the number of classes.

Table 1. Detailed introduction of experimental data

Dataset	Number of samples	Number of attributes	Number of classes
Iris	150	4	3
Wine	178	13	3
Isolet	7797	617	26
COIL-20	1440	1024	20
MNIST	70000	784	10

We use the standard unsupervised evaluation criterion to evaluate and compare the clustering algorithms [15]. For all algorithms, we set the clusters number as the real labels number of data, and evaluate their performance with unsupervised clustering accuracy:

$$ACC = \max_{m} \frac{\sum_{i=1}^{n} \mathbf{1}\{l_i = m(c_i)\}}{N} \tag{6}$$

where N is the total number of data samples, l_i the real labels of the i-th data, C_i is the prediction value of the i-th data generated by the clustering algorithm, and m is the mapping range of all possible pairs between the clusters and labels. Intuitively, this metric requires the clustering allocation of unsupervised algorithms and the allocation of real labels, and then finds the best match between them. The best match between the two is to be effectively calculated by the Hungarian solution [16].

4.2 Experimental Parameter Setting

For all the algorithm involved in the experiments, we repeat the 10 random experiments to the average value as the final result of the experiment. In the experimental data set, the SAE network structure is set up 100-50-10 for the first two data sets (Iris and Wine), and the SAE network structure is set up D-500-300-30 for the latter three data, where D is the dimension of data space and the neurons between all layers are all connected. For stacked auto-encoders, we set that the weights is initialized to the Gaussian distribution with zero mean and standard deviation of 0.01, the number of iterations is 10000, the mini-batch size is 100, and the learning rate is 0.001. For SOM networks, the learning rate is attenuated from 0.01 to 0 per 10 times, and the maximum number of iterations is 100. In the experiment, we analyze the influence of the error curve of the auto-encoder in the training process, the data characteristics and the parameters of the model, and whether the data is encoded on the accuracy of the algorithm, and compared the differences between the different algorithms.

4.3 Experimental Analysis and Comparison

Figure 5 shows the reconstruction error curve of each layer AE on the MNIST dataset during the training process of a stacked auto-encoder. It can be seen from the graph that the reconstruction error tends to decrease with increase of epochs, and it begins to descend rapidly, then slow down until reaches a steady state. Moreover, with the increase of the number of encoder layers, the overall error trend is smaller, and the training set is smaller than the test set. Therefore, with the increase of the iterations and the layers, the features extracted from the SAE are more abstract and less redundant.

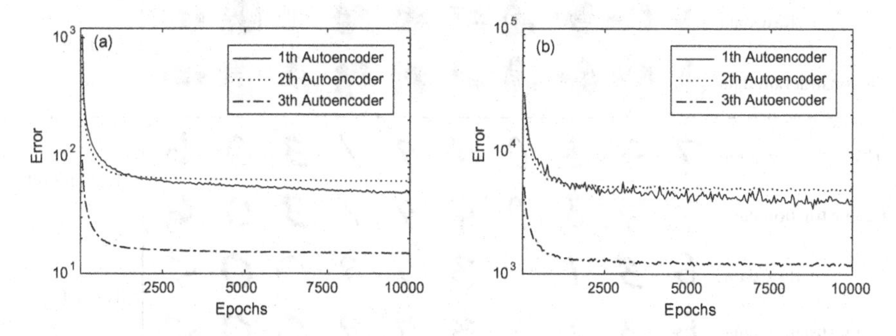

Fig. 5. The change curve of the reconstruction error of each layer of a SAE on the MNIST dataset. (a) The change curve of the reconstruction error on the training set of a SAE (b) the change curve of the reconstruction error on the test set of a SAE.

Figure 6 shows the change curve of the clustering accuracy that the MNIST dataset is coded as different number of feature. In order to test the influence of the features number on the clustering accuracy, we set the number of neurons of the last layer and change the other layers in the SAE. The network structure is set to 784-500-300-H, and H represents the number features (the variables) that are encoded. From the graph we can see that the number of the clustering accuracy decreases at first, then decreases when the feature number increases gradually, so the suitable feature number for the SAE is 30.

Figure 7 the contrast of the raw and reconstructed data after the SAE dimension reduction on two data sets, where reconstructed data is obtained by the reconstruction of 30 feature number after the SAE dimensionality reduction. It can be seen that there is a certain difference between the original and restructured pictures, but the reconstruction of the picture can see clearly the information displayed. For the data between

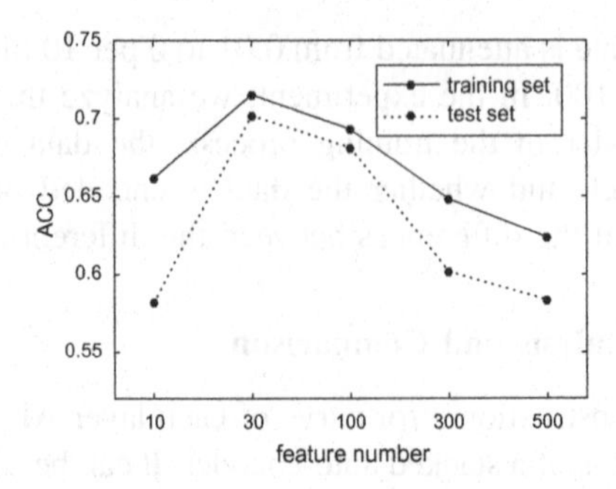

Fig. 6. Comparison of clustering accuracy of different number of features

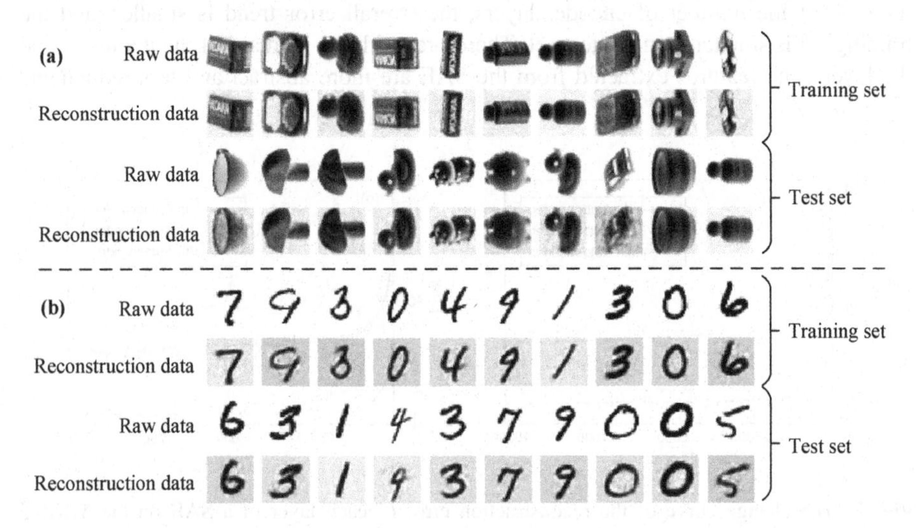

Fig. 7. A partial contrast diagram of reconstruction data and the original data. (a) The contrast diagram of MNIST dataset before and after reconstruction. (b) The contrast diagram of COIL-20 dataset before and after reconstruction.

the training set and the test set, the differences rebuilt from the naked eye are not obvious. As a result, it has some effect to reduce the dimension of data through SAE.

Figure 8 shows the part clustering effect of the deep clustering algorithm base on SOM neural network for MNIST dataset of the training set, where each row corresponds to one cluster and the image is randomly selected from the top 10 of each cluster to sort from left to right. It is observed from the graph that there is a gap between the clusters assignment and the real labels in MNIST dataset, where "1" is divided into two clusters, mainly concentrated in the different positions, and "4" and "9" can not be correctly classified in the algorithm.

Table 2 shows the result of the clustering accuracy comparison of 5 different datasets. The *D* in the network structure represents the number of properties of the raw data, and the clusters number is related to the real labels of data sets. It can be seen from the table that no matter the clustering performance of any data set after SAE coding is

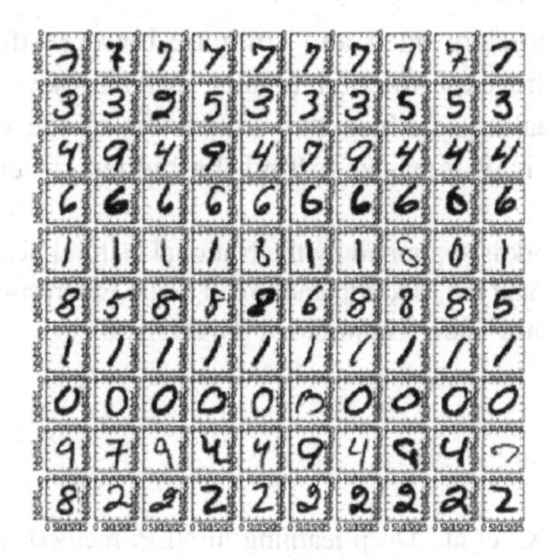

Fig. 8. Partial effect graph of MNIST dataset clustering

Table 2. Comparison of clustering accuracy of different clustering algorithms

Dateset	SOM		Proposed method		Network structure
	Training set	Test set	Training set	Test set	
Iris	88.75%	89.47%	94.49%	93.93%	*D*-100-50-10-3
Wine	91.52%	90.59%	96.24%	95.56%	*D*-100-50-10-3
Isolet	58.82%	57.68%	66.42%	64.7%	*D*-500-100-30-26
COIL-20	68.83%	67.32%	71.72%	70.85%	*D*-500-100-30-20
MNIST	52.89%	52.92%	71.54%	70.61%	*D*-500-100-30-10

better than the performance of clustering directly, the accuracy rate is greatly improved. This shows that using SAE to extract the potential features of the data, and then clustering in the feature space of the dataset is beneficial. The number of clusters set by the clustering experimental data in the table is the number of their data sets. It can be seen from the table that no matter the clustering performance of any data set after SAE coding is better than that of original data, the accuracy rate is greatly improved. This shows that using SAE to extract the potential features of the data, and then clustering in the feature space of the dataset is beneficial.

5 Conclusion

A lot of application data in life do not contain labels, which requires unsupervised learning to learn from the data of complex structures. Clustering is an important part of unsupervised learning, which aims to discover the underlying structure within the cluster of unlabeled data. As the basic model of deep learning, the deep neural network shows excellent performance in various learning tasks. We proposes a deep clustering algorithm based on self-organizing map neural network, which can learn effective features from complex data and perform unsupervised clustering on the low dimensional features. We used to validate the algorithm of 5 different data sets, analysis of influence of related parameters on it and compare between different algorithm clustering accuracy. From the experimental results, we can see that the algorithm can effectively extract the features, reduce the dimension of the data, and it has higher clustering accuracy, that is, it is easier to find the hidden structure of the data.

Acknowledgment. The work is supported by the National Natural Science Foundation of China under Grant No. 61762080, and the Medium and Small Scale Enterprises Technology Innovation Foundation of Gansu Province under Grant No. 17CX2JA038.

References

1. Lin, Y., Hang, L., Li, X., et al.: Deep learning in NLP: methods and applications. J. Univ. Electron. Sci. Technol. China **46**(6), 913–919 (2017)
2. Gheisari, M., Wang, G., Bhuiyan, M.Z.A.: A survey on deep learning in big data. In: IEEE International Conference on Computational Science and Engineering, pp. 173–180. IEEE, Guangzhou, China (2017)
3. Shen, D., Wu, G., Suk, H.I.: Deep learning in medical image analysis. Ann. Rev. Biomed. Eng. **19**(1), 221–248 (2017)
4. Schmidhuber, J.: Deep learning in neural networks: an overview. Neural Netw. **61**, 85–117 (2014)
5. LeCun, Y., Bengio, Y., Hinton, G.: Deep learning. Nature **521**(7553), 436–444 (2015)
6. Jain, A.K.: Data clustering: a review. ACM Comput. Surv. **31**(3), 264–323 (2000)
7. Xu II, R.: D.W.: Survey of clustering algorithms. IEEE Trans. Neural Netw. **16**(3), 645–678 (2005)
8. Torre, F.D.L., Kanade, T.: Discriminative cluster analysis. In: Caruana, R., Niculescu-Mizil, A. (eds.) Proceedings of the 23rd International Conference on Machine Learning, pp. 241–248. ACM (2006)
9. Dilokthanakul, N., Mediano, P.A.M., Garnelo, M., et al.: Deep unsupervised clustering with gaussian mixture variational autoencoders. arXiv preprint arXiv:1611.02648 (2016)
10. Rumelhart, D.E., Hinton, G.E., Williams, R.J.: Learning internal representations by error propagation. Nature **323**(6088), 533–536 (1986)
11. Badino, L., Canevari, C., Fadiga, L., et al.: An auto-encoder based approach to unsupervised learning of subword units. In: IEEE International Conference on Acoustics, Speech and Signal Processing, pp. 7634–7638. IEEE, Florence, Italy (2014)
12. Bengio, Y.: Learning deep architectures for AI. Found. Trends Mach. Learn. **2**, 1–127 (2009)
13. Kohonen, T.: Automatic formation of topological maps of patterns in a self-organizing system. In: Oja, E., Simula, O. (eds.) Proceedings of 2SCIA, Scandinavian Conference on Image Analysis, pp. 214–220. Helsinki, Finland (1981)
14. Kohonen, T.: Self-organized formation of topologically correct feature maps. Biol. Cybern. **43**(1), 59–69 (1982)
15. Yang, Y., Xu, D., Nie, F., et al.: Image clustering using local discriminant models and global integration. IEEE Tran. Image Process. **19**(10), 2761–2773 (2010)
16. Kuhn, H.W.: The Hungarian method for the assignment problem. Nav. Res. Logistics **2**(1–2), 83–97 (1955)

Short-Term Load Forecasting Based on RBM and NARX Neural Network

Xiaoyu Zhang[1], Rui Wang[1(✉)], Tao Zhang[1], Ling Wang[2], Yajie Liu[1], and Yabing Zha[1]

[1] College of System Engineering, National University of Defense Technology, Changsha 410073, People's Republic of China
ruiwangnudt@gmail.com

[2] Department of Automation, Tsinghua University, Beijing 100084, People's Republic of China

Abstract. In recent years, DBN applied to load forecasting as a hot issue has aroused the concern of many scholars at home and abroad. A new method based on RBM and NARX neural network for short-term load forecasting is brought forward in this paper. In order to test the performance of this model, the historical load data of a town in the UK is used. The obtained results are compared with DBN and NARX neural network based on the same dataset. Experimental results show that the proposed method significantly improves the predication accuracy.

Keywords: Short-term load forecasting · Deep belief network
Restricted Boltzmann machine · NARX neural network

1 Introduction

Short term load forecasting plays a vital role in power system operation and analysis, such as unit commitment and economic dispatch. According to the load forecasting, the system can optimize operation time of generating units, i.e., the starting and stopping time, their output [1]. A precise load forecasting is helpful in minimizing the total consumption of the generating units. Therefore, improving the accuracy of short term load forecasting is of great importance in the modern power system.

According to the forecasting models, scholars usually categorize approaches for load forecasting into two groups: statistical and artificial intelligence models. Statistical models include regression analysis [2], Kalman filtering [3], autoregressive integrated moving average (ARIMA) [4], Box–Jenkins models [5], state space model [6], exponential smoothing [7] and so on. Artificial intelligence models include data mining approaches [8], artificial neural networks (ANNs) [9], and support vector machines (SVM) [10] etc. Compared with statistical models, artificial intelligence models, as data-driven techniques, are usually more popular and adaptive.

Amongst all model used for load forecasting, deep belief network (DBN) [11] has shown promising performance. The deep belief network has a deep architecture that can represent multiple features of input patterns hierarchically with the pre-trained

D.-S. Huang et al. (Eds.): ICIC 2018, LNAI 10956, pp. 193–203, 2018.
https://doi.org/10.1007/978-3-319-95957-3_21

restricted Boltzmann machine (RBM). It has been applied in a variety of fields such as classification tasks [12], dimensionality reduction [13] and image processing [14].

The DBN has been widely studied, but existing studies focus mainly on the training process of DBN. Research on the network structure of DBN is rarely reported in the literature. After the pre-training stage of DBN, the obtained parameters can respectively be expanded for feedforward neural networks with more hidden layers. Generally, neural network (NN) models have two basic structures: feedforward and feedback. Compared with the feedforward NN, recurrent neural network (RNN) with a feedback structure has been shown to excel at time series forecast. Therefore, in this paper, we propose a new method for short-term load prediction based on RBM and NARX.

The rest of the paper is organized as follows. Section 2 describes background methodologies used in this paper including the deep belief network, NARX neural network; Sect. 3 introduces the proposed RBM-NARX network; In Sect. 4 presents experimental results; Sect. 5 concludes this paper and identifies future studies.

2 Methodology

2.1 Deep Belief Network

The network structure of a typical DBN is a deep neural network as shown in Fig. 1 which is composed of several layers of RBM and a layer of neural network [15]. Regarding DBN, the training process consists of a layer-wise pre-training process and a fine-tuning process. The former is used to provide good initial values for all parameters, while the latter is used to search the optimum based on the given initial states of the network.

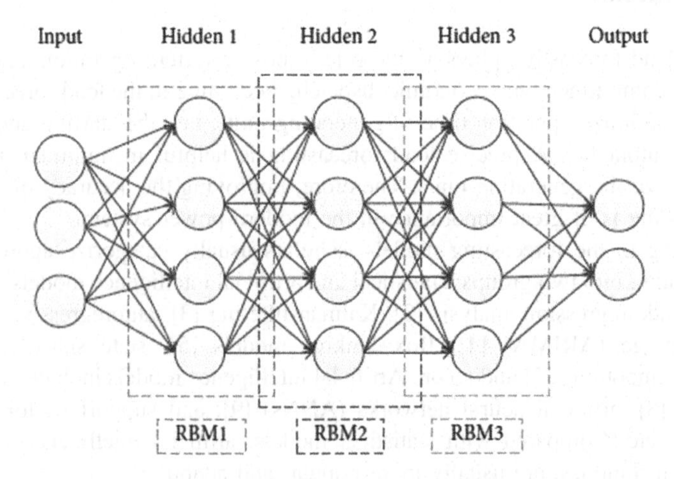

Fig. 1. Illustration of a typical deep belief network structure.

(1) **Pre-training**. The network parameters of DBN in each hidden layer can be initialized during the pre-training process, achieving a better local optimum or even the global optimal region. This process is obtained through an unsupervised greedy optimization algorithm by using the restricted Boltzmann machine (RBM).

RBM is a stochastic neural network that can learn a distribution over its set of inputs [16]. The network generally consists of two different layers of units: visible nodes and hidden units, with weighted connection between them. Connections between nodes are bidirectional and symmetric, as shown in Fig. 2.

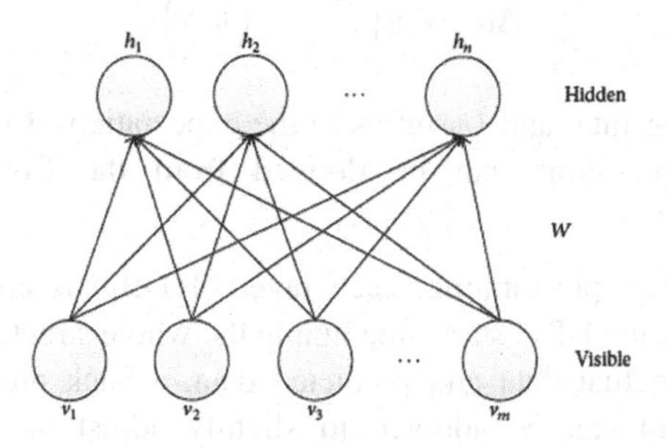

Fig. 2. Illustration of a typical restricted Boltzmann machine structure.

The RBM is an energy model. The energy function of visible layer and hidden layer is defined as:

$$E(v,h) = -\sum_{ij} w_{ij} v_i h_j - \sum_i a_i v_i - \sum_j b_j h_j \tag{1}$$

where v_i and h_j represent the states of visible node i and hidden node j, respectively. w_{ij} is the connection weight between the visible layer and hidden layer. a_i and b_j are the bias between them.

For binary state nodes v_i and $h_j \in \{0,1\}$, the state of h_j is set to 1 with probabilities:

$$p_{h_j} = p(h_j = 1|v) = \sigma\left(a_i + \sum_i w_{ij} v_i\right) \tag{2}$$

where $\sigma(x)$ is the logistic sigmoid function $1/(1+\exp(-x))$. The state of visible units v_i is set as 1 with probability:

$$p_{v_i} = p(v_i = 1|h) = \sigma\left(b_j + \sum_j w_{ij} h_j\right) \tag{3}$$

The training process of the RBM is described as follows. Firstly, a training sample is assigned to the visible nodes, and the $\{v_i\}$ is obtained. Then, the hidden nodes state that $\{h_j\}$ are sampled according to probabilities. This process is repeated once more to update the visible, hidden nodes and the one-step "reconstructed" states v_i' and h_j'. The related parameters are updated as follows:

$$\Delta w_{ij} = \eta\left(\langle v_i h_j \rangle - \left\langle v_i' h_j' \right\rangle\right) \tag{4}$$

$$\Delta a_i = \eta\left(\langle v_i \rangle - \left\langle v_i' \right\rangle\right) \tag{5}$$

$$\Delta b_j = \eta\left(\langle h_j \rangle - \left\langle h_j' \right\rangle\right) \tag{6}$$

where η is the learning rate, and $\langle \cdot \rangle$ refers to the expectation of the training data. The above-mentioned expressions can be derived from the Contrastive Divergence (CD) algorithm.

(2) **Fine tuning**. After pre-training, each layer of DBN is configured with initial parameters. Then the DBN starts fine tuning the whole structure. Based on the loss function of the actual data and predicted data, a back propagation or gradient descent algorithm can be adopted to slightly adjust the network parameters throughout the whole network, resulting in the optimal states of the parameters. The loss function is defined as follows:

$$L(y, y') = \|y - y'\|_2^2 \tag{7}$$

where y denotes the actual data and y' denotes the predicted data.

More generally, the DBN is a special BP neural network that the parameters of hidden layers are initialized by RBM, instead of being randomly assigned.

2.2 NARX Neural Network

NARX neural network is a class of recurrent neural network (RNN). Unlike BP neural network, the feedback in this recurrent model is from the output to the input. This makes the NARX neural network sensitive to historical data, and thus, enables the network to have a dynamic memory function. The schematic diagram is shown in Fig. 3.

According to the NARX structure, the dynamic change of this model can be mathematically expressed as follows:

$$y(n+1) = f\left[y(n), \cdots, y(n-d_y+1); u(n), u(n-1), \cdots, u(n-d_u+1)\right] \tag{8}$$

where $u(n)$ and $y(n)$ denote the input and output of the model at discrete time step n, respectively, while $d_u \geq 1$ and $d_y \geq 1$, $d_u \leq d_y$, are the input-memory and output-memory orders, respectively. $f(\cdot)$ is a nonlinear mapping function which can be approximated by a standard MLP model. Without loss of generality, the typical three layer structure is used in this study.

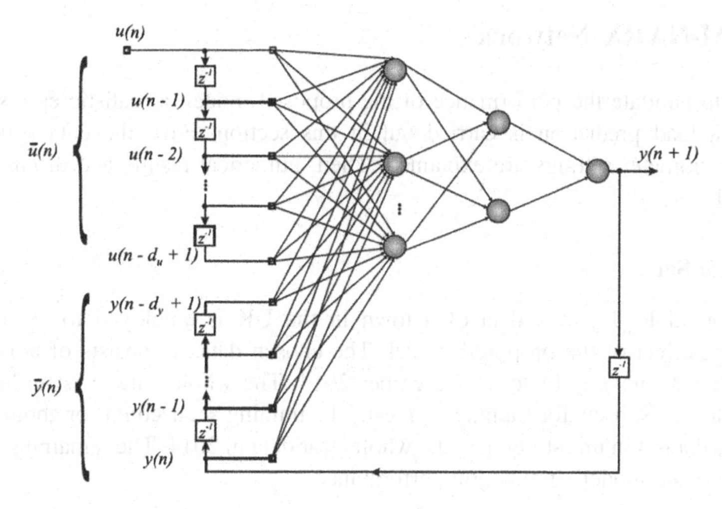

Fig. 3. Illustration of a typical NARX neural network structure.

3 RBM-NARX Network

Compared with generalized neural network which only learns a static input-output mapping relationship, the NARX neural network with feedback structure behaves like a dynamic system which is more suitable to model temporal sequences. Therefore, in this study we develop a new method based RBM and NARX neural network for short-term load forecasting. The new method is denoted as RBM-NARX network.

3.1 RBM-NARX Optimization

NARX neural network inherits the feature of generalized neural network to some extent, which has many defects. For example, a slow convergence rate, being easy to fall into local optima. The deficiency is partly due to the randomly initialized parameters. Therefore, we propose to apply RBM to initialize the weights and thresholds of NARX networks, by which we expect to improve the training speed and generalization ability of NARX neural networks. The main steps are discussed below.

3.2 RBM-NARX Algorithm

The basic steps of RBM-NARX algorithm is as follows:

Step 1 determines the primary structure of NARX neural network;
Step 2 applies RBMs to initialize the parameter of NARX neural network;
Step 3 trains the NARX neural network by gradient descent algorithm;
Step 4 forecasts load output based on the trained network.

The significance of the new model is about the initialization of connection weights and threshold using RBM. This is expected to be helpful in improving the training speed and convergence.

4 RBM-NARX Network

In order to validate the performance of the proposed model, a realistic case study of short-term load prediction is carried out in this section. First, the data source and relevant parameter settings are explained. Then, numerical results and discussion are presented.

4.1 Data Set

The historical load power data of a town in the UK is employed to examine the forecasting effect of the proposed model. The chosen dataset consists of hourly load data from 1 January 2014 to 31 December 2014. The whole dataset is further partitioned into two subsets for training and test. The training set account for about 80% of the whole dataset in this study, i.e., the whole year data in 2014. The remaining data are used to test the model's prediction performance.

4.2 Model Implementation

(1) **Parameter settings**. As indicated above, the input of the forecast model is the historical load power data. To construct the RBM-NARX model, the original time series data should be transformed into a suitable form to train the model. In this study, the state space reconstruction technique with the delay embedding theorem [17] is employed to process the original data. The dynamic system in discrete time can be depicted as:

$$X(t+1) = F(X(t)) \tag{9}$$

where $X(t)$ is the system state at time step t and F is a nonlinear vector valued function. By this theorem, the one-dimensional chaotic data is supposed to compress the information of higher dimension. Hence, the time series data $X(t)$ can be reconstructed as follows:

$$X(t) = [X(t), X(t-\tau), \cdots, X(t-(m-1)\tau)] \tag{10}$$

where τ is time delay and m is the embedding dimension. Therefore, constructing the delay embedding comes down to finding the optimal values of parameters τ and m. For a given dataset, these two parameters can be determined by the mutual information function and false nearest neighbour method [18]. In this study, τ is determined as 6 and m is 10, which can be accomplished by the utility functions mutual and false nearest in TISEAN toolbox [19]. After determining the value of τ and m, the reconstructed time series is generated which is then used to train the RBM-NARX network model. Due to the dimension of reconstructed delay vectors is 10, the number of input nodes for the RBM-NARX model is also 10. In order to accelerate the model training process, the original data are often normalized into [0, 1].

In addition, the trial and error method is employed to investigate the number of hidden layer nodes in this paper. The test results of the hidden nodes numbers for the

NARX model are shown in Fig. 4. From the figure, when the number of hidden nodes is 24, the model has the best performance according to the MAPE index. Therefore, the optimal structure of NARX network is 10-24-1.

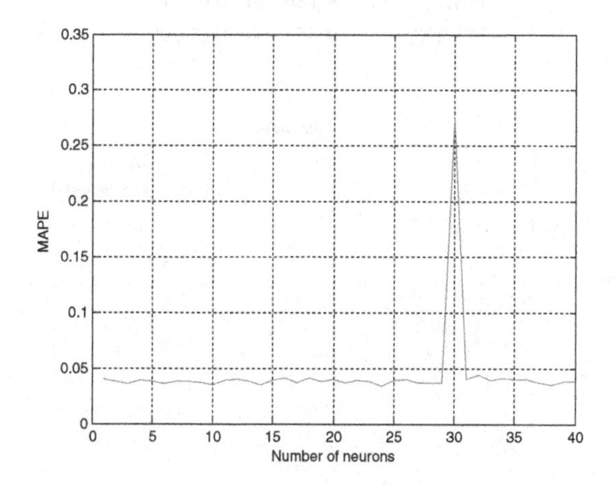

Fig. 4. The trial and error results of NARX model.

(2) **Model evaluation**. Two metrics are calculated to evaluate the effectiveness of the proposed model including MAPE and MSE which are defined as follows:

$$MSE = \frac{1}{N}\sum_{i=1}^{N}(X(i) - X'(i))^2 \tag{11}$$

$$MAPE = \frac{1}{N}\sum_{i=1}^{N}\left|\frac{X(i) - X'(i)}{X(i)}\right| \tag{12}$$

where N is the forecast horizon, $X(i)$ represents the actual value and $X'(i)$ is the predicted value at time instance i.

4.3 Experimental Results

In this section, we apply the proposed RBM-NARX model to short-term load forecasting. Meanwhile, the DBN model and NARX neural network are employed to short-term load forecasting with the same dataset as well. The obtained results by each prediction method are shown in Table 1. Furthermore, for RBM-NARX network model, the forecasting results for test data are shown in Fig. 5. In order to have a clearer understanding of this, the results for the last week of the test data are also shown in Fig. 6. The forecasting results of by the DBN model and NARX model for the last week of test data are shown in Figs. 7 and 8, respectively.

Table 1. Forecast results by different models.

Model	MSE	MAPE
RBM-NARX	7.69e−04	0.0341
DBN	9.18e−04	0.0381
NARX	9.55e−04	0.0399

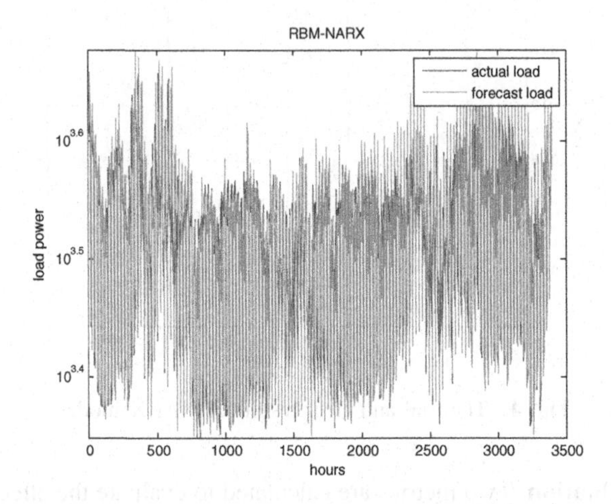

Fig. 5. The forecast results by RBM-NARX model.

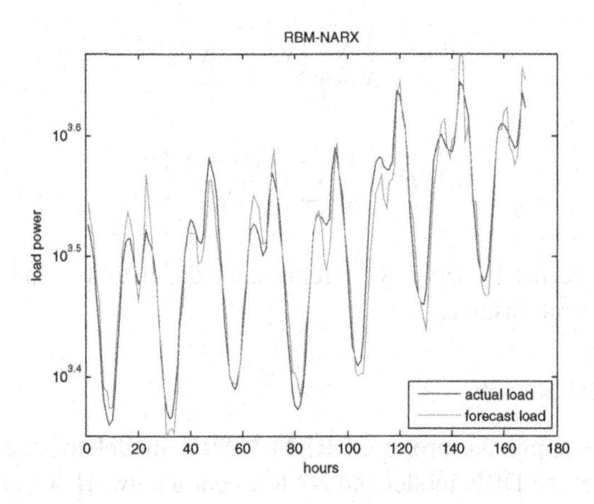

Fig. 6. The forecast results by RBM-NARX model for the last week of test data.

From Table 1, the MAPE of the RBM-NARX network is the minimum which is 0.0346. The MAPE of the DBN is 0.0381. Thus, it can be observed that our proposed RBM-NARX prediction model outperforms the other models. From the above figures, it is also evident that the predicted values of the proposed method are closer to the actual values.

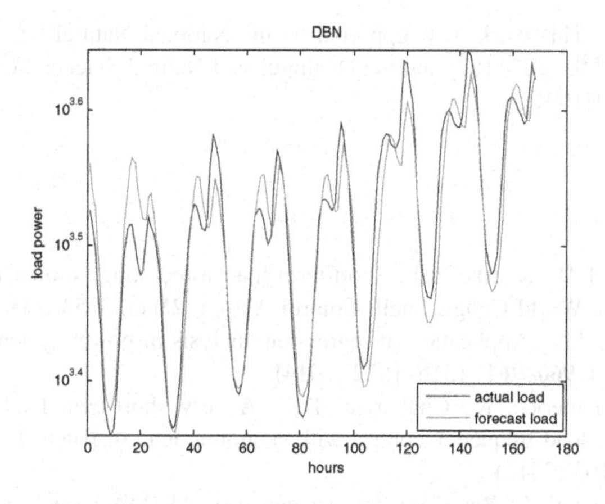

Fig. 7. The forecast results by DBN model for the last week of test data.

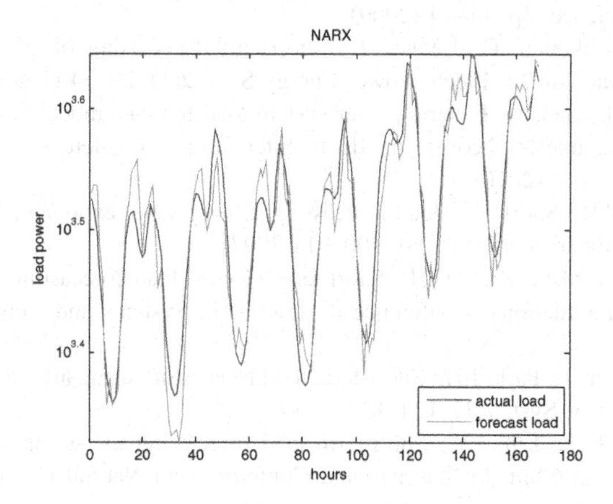

Fig. 8. The forecast results by NARX model for the last week of test data.

5 Conclusion

An accurate load forecast is required in the competitive electricity market. To overcome the shortcomings of traditional methods, this study proposes a novel method based RBM and NARX neural network for short-term load forecasting. Meanwhile, the DBN and NARX neural network comparing with the proposed model are employed to demonstrate the effectiveness of the proposed method. Experimental results show that the proposed method has outstanding performance.

Concerning future studies, first we would like to apply our method to more complex datasets. Second, the hyper-parameters of neural networks are fine-tuned by the back-propagation method which is easy to be trapped in local optima. Thus, we would like to apply advanced evolutionary algorithms [20–22] to fine-tune those hyper-parameters. Lastly, other improvements of the deep believe network for load forecasting will also be considered.

Acknowledgement. This work was supported by the National Natural Science Foundation of China (Nos. 61773390, 71571187) and the Distinguished Natural Science Foundation of Hunan Province (No. 2017JJ1001).

References

1. Chen, Y., Luh, P.B., Rourke, S.J.: Short-term load forecasting: similar day-based wavelet neural networks. World Congr. Intell. Control Autom. **25**(1), 3353–3358 (2008)
2. Chen, C., Zhou, J.N.: Application of regression analysis in power system load forecasting. Adv. Mater. Res. **960–961**, 1516–1522 (2014)
3. Shankar, R., Chatterjee, K., Chatterjee, T.K.: A very short-term load forecasting using kalman filter for load frequency control with economic load dispatch. J. Eng. Sci. Technol. Rev. **5**(1), 97–103 (2012)
4. Wei, L., Zhen-gang, Z.: Based on time sequence of ARIMA model in the application of short-term electricity load forecasting. In: International Conference on Research Challenges in Computer Science, pp. 11–14 (2009)
5. Vähäkyla, P., Hakonen, E., Léman, P.: Short-term forecasting of grid load using Box-Jenkins techniques. Int. J. Electr. Power Energy Syst. **2**(1), 29–34 (1980)
6. Li, X., Chen, H., Gao, S.: Electric power system load forecast model based on State Space time-varying parameter theory. In: IEEE International Conference on Power System Technology, pp. 1–4 (2010)
7. Christiaanse, W.R.: Short-term load forecasting using general exponential smoothing. IEEE Trans. Power Appar. Syst. **PAS-90**, 900–911(2007)
8. Koo, B.G., Kim, M.S., Kim, K.H.: Short-term electric load forecasting using data mining technique. In: International Conference on Intelligent Systems and Control, pp. 153–157. IEEE (2013)
9. Lee, K.Y., Cha, Y.T., Park, J.H.: Short-term load forecasting using artificial neural networks. IEEE Trans. Power Syst. **7**(1), 124–132 (2014)
10. Li, G., Cheng, C.T., Lin, J.Y.: Short-term load forecasting using support vector machine with SCE-UA algorithm. In: International Conference on Natural Computation, pp. 290–294. IEEE (2007)
11. Hinton, G.E., Osindero, S., Teh, Y.W.: A fast learning algorithm for deep belief nets. Neural Comput. **18**(7), 1527 (2006)
12. Sun, S., Liu, F., Liu, J.: Web classification using deep belief networks. In: IEEE International Conference on Computational Science and Engineering, pp. 768–773 (2014)
13. Arsa, D.M.S., Jati, G., Mantau, A.J.: Dimensionality reduction using deep belief network in big data case study: hyperspectral image classification. In: International Workshop on Big Data and Information Security, pp. 71–76. IEEE (2017)
14. Cheng, M.: The cross-field DBN for image recognition. In: IEEE International Conference on Progress in Informatics and Computing, pp. 83–86. IEEE (2016)
15. Zhang, X., Wang, R., Zhang, T.: Short-term load forecasting based on an improved deep belief network. In: International Conference on Smart Grid and Clean Energy Technologies, pp. 339–342 (2016)
16. Zhang, X., Wang, R., Zhang, T., Liu, Y., Zha, Y.: Effect of transfer functions in deep belief network for short-term load forecasting. In: He, C., Mo, H., Pan, L., Zhao, Y. (eds.) BIC-TA 2017. CCIS, vol. 791, pp. 511–522. Springer, Singapore (2017). https://doi.org/10.1007/978-981-10-7179-9_40

17. Takens F.: Detecting strange attractors in turbulence. In: Dynamical Systems and Turbulence, Warwick 1980, pp. 366–381. Springer, Heidelberg (1981)
18. Hegger, R., Kantz, H., Schreiber, T.: Practical implementation of nonlinear time series methods: the tisean package. Chaos **9**(2), 413–435 (1999)
19. Nonlinear Time Series Analysis (TISEAN). https://www.mpipks-dresden.mpg.de/~tisean/
20. Wang, R., Purshouse, R.C., Fleming, P.J.: Preference-inspired co-evolutionary algorithms for many-objective optimisation. IEEE Trans. Evol. Comput. **17**, 474–494 (2013)
21. Wang, R., Ishibuchi, H., Zhou, Z., Liao, T., Zhang, T.: Localized weighted sum method for many-objective optimization. IEEE Trans. Evol. Comput. **22**, 3–18 (2018)
22. Wang, R., Zhang, Q., Zhang, T.: Decomposition based algorithms using Pareto adaptive scalarizing methods. IEEE Trans. Evol. Comput. **20**, 821–837 (2016)

Time Series Prediction Using Complex-Valued Legendre Neural Network with Different Activation Functions

Bin Yang(✉), Wei Zhang, and Haifeng Wang

School of Information Science and Engineering, Zaozhuang University, Zaozhuang, China
batsi@126.com

Abstract. In order to enchance the flexibility and functionality of Legendre neural network (LNN) model, complex-valued Legendre neural network (CVLNN) is proposed to predict time series data. Bat algorithm is proposed to optimize the real-valued and complex-valued parameters of CVLNN model. We investigate performance of CVLNN for predicting small-time scale traffic measurements data by using different complex-valued activation functions like Elliot function, Gaussian function, Sigmoid function and Secant function. Results reveal that Elliot function and Sigmoid function predict more accurately and have faster convergence than Gaussian function and Secant function.

Keywords: Complex-valued · Legendre neural network · Activation function Bat algorithm

1 Introduction

Artificial neural networks (ANNs) are powerful mathematical methods that can be used to learn complex linear and non-linear continuous functions, and have been successfully applied to many areas in the past decades [1]. Due to that traditional neural network has some disadvantages such as low efficiency, long learning time and easy to fall into the local minimum solution, Legendre neural network (LNN) was proposed. LNN model has no hidden layer and could add dimensionality of the input layer with a set of nonlinear functions. Patra et al. proposed a Legendre neural network model for equalization of nonlinear communication channels with 4-QAM signal constellation [2]. Pei et al. forecasted and investigated the stock prices of the financial model by an improved Legendre neural network with random time strength function [3]. Dash et al. used a Moderate Random Search Particle Swarm Optimization Method (HMRPSO) to optimize the prameters of LNN model to predict the Bombay Stock Exchange and S&P 500 data sets [4]. Behera and Sahu proposed Pseudo inverse Legendre neural network (with *tanh* functions in the hidden layer matrix) optimized by using firefly algorithm (LNNT-FF) for nonlinear dynamic plant identification [5].

The activation function significantly affects the performance of neural networks, and the choice of proper activation functions is crucial to the neural networks. A great deal of research has focused on testing the performance of neural networks using

D.-S. Huang et al. (Eds.): ICIC 2018, LNAI 10956, pp. 204–211, 2018.
https://doi.org/10.1007/978-3-319-95957-3_22

different activation functions, and choosing the proper function structure. Wang et al. proposed a new multilayer feedforward neural networks based on different activation functions in each neuron in order to improve convergence speed [6]. Kumar and Singh investigated the wear loss prediction performance of artificial neural network using different activation functions. Results revealed that binary sigmoid function performed better for the wear loss prediction for A390 aluminum alloy [7]. Zhang proposed a new Deep Neural Network (DNN) with different activation functions in order to avoid the optimization limitation using the same activation function for all hidden neurons [8]. Malleswaran et al. compared performance of the GPS/INS data integration system by using different activation functions like Bipolar Sigmoid Function (BPSF), Binary Sigmoid Function (BISF), Hyperbolic Tangential Function (HTF) and Gaussian Function (GF) in BPN-ANN and using Gaussian function in RBF-ANN [9].

Compared with real-valued neural network, complex-valued neural network is more flexible and functional. In this paper, complex-valued Legendre neural network (CVLNN) is proposed to predict time series data. Bat algorithm is proposed to optimize the real-valued and complex-valued parameters of CVLNN model. We investigate performance of CVLNN by using different complex-valued activation functions like Elliot function, Gaussian function, Sigmoid function and Secant function.

2 Method

2.1 Complex-Valued Legendre Neural Network

Legendre neural network (LNN) was first proposed by Yang and Tseng for function approximation in 1996. LNN has less parameters and does not have hidden layer, which uses Legendre orthogonal polynomials as the output of hidden layer neurons. Due to the absence of hidden layer, LNN provides computational advantage over the MLP. In order to enchance the flexibility and functionality of LNN model, complex-valued Legendre neural network (CVLNN) is proposed in this paper. In a CVLNN model, input data and weights are complex-valued. The structure of a CVLNN is shown in Fig. 1. Suppose that complex-valued input vector $[z_1, z_2, \ldots z_m]$ and n order Legendre polynomials.

The complex-valued Legendre polynomials of each input variable ($z = x + jy$) are described as follows.

$$\begin{cases} L_0(z) = 1 \\ L_1(z) = x + yi \\ L_2(z) = 0.5[3(x^2 + 2xyi - y^2) - 1] \\ \ldots\ldots\ldots \\ L_n(z) = \frac{1}{n}[(2n-1)(x+yi)L_{n-1}(z) - nL_{n-2}(z)] \end{cases} \tag{1}$$

In order to improve the modeling ability of CVLNN, the output of f_i is defined as follows.

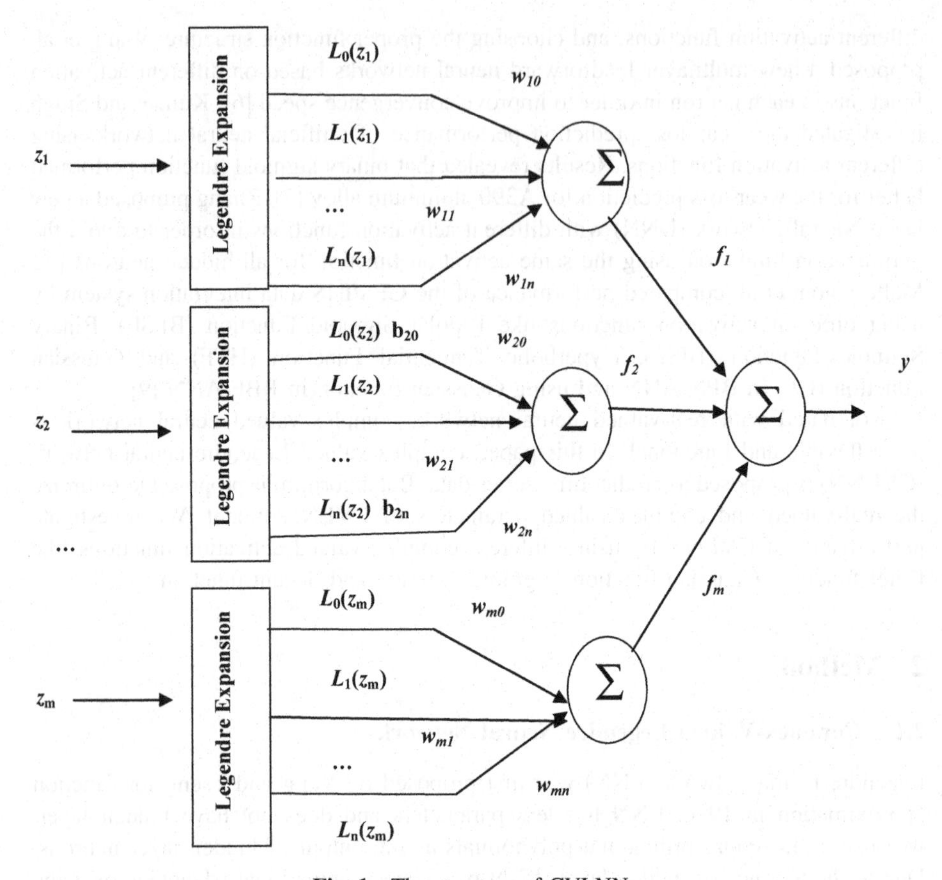

Fig. 1. The structure of CVLNN.

$$f_i = f(\sum_{k=0}^{n} L_k(z_i) \times w_{ik}). \tag{2}$$

Where $f(\cdot)$ is a complex-valued activation function (CVAF), $L_k(z_i)$ is k-th Legendre polynomial of input data z_i and weight w_{ik} is assigned.

Generally there are four kinds of CVAFs: complex-valued Elliot function, Gaussian function, Sigmoid function and Secant function, which are described as follows.

$$out_n = f(a, r, net_n) = \frac{net_n}{a + \frac{1}{r}|net_n|} \tag{3}$$

$$out_n = f(c, \sigma, net_n) = e^{-\frac{(net_n - c)^H (net_n - c)}{\sigma^2}} \tag{4}$$

$$out_n = f(net_n) = \frac{1}{1 + e^{-\mathrm{Re}(net_n)}} + j\frac{1}{1 + e^{-\mathrm{Im}(net_n)}} \quad (5)$$

$$out_n = f(net_n) = \sec(\mathrm{Re}(net_n)) + j\sec(\mathrm{Im}(net_n)) \quad (6)$$

The final output y is defined as follows.

$$y = \sum_{k=1}^{m} f_k \quad (7)$$

2.2 Bat Algorithm

Bat algorithm is a heuristic search method and could be used to search the global optimal solution, which was proposed firstly by Yang in year 2010 [10]. Because of its simple structure, few parameters, strong robustness and easy implement, bat algorithm has been applied widely in many areas, such as engineering design, image compression, data classification and prediction. In this paper, bat algorithm is proposed to optimize parameters of CVLNN model.

Bat algorithm simulates the process that bats use the sonar to detect prey and avoid obstacles. Each bat is seen as a solution and has its corresponding fitness value. Bat population follow the current optimal bat in the solution space by adjusting frequency, loudness and pulse emission rate [11]. The algorithm is described as follows in detail.

(1) Initialize randomly bat positions and create population $[x_1, x_2, \ldots x_N]$.
(2) Evaluate population and give each bat a fitness value. Search the current optimal bat position x^*.
(3) If the optimal solution condition is satisfied or the maximum iterations are achieved, then stop; otherwise go to step (4).
(4) Update the velocity and position of each bat with pulse frequency.

$$f_i = f_{\min} + (f_{\max} - f_{\min})\beta. \quad (8)$$

$$v_i^t = v_i^{t-1} + (x_i^{t-1} - x^*)f_i. \quad (9)$$

$$x_i^t = x_i^{t-1} + v_i^t. \quad (10)$$

Where f_i is pulse frequency of i-th bat, which belongs to $[f_{\min}, f_{\max}]$. β is random number from $[0,\ 1]$. v_i^t is the velocity of i-th bat at time t and x_i^t is the position of i-th bat at time t.

(5) Generate a uniformly distributed random number r_1. If $r_1 < r_i$ (r_i is pulse rate of i-th bat.), the stochastic perturbation of the current optimal solution x^* yields a new solution.
(6) Generate a uniformly distributed random number r_2. If $r_2 < A_i$ && $f(x_i) < f(x^*)$ (A_i is pulse loudness of i-th bat.), accept the new solution population from step (5).

(7) Update the pulse rate and loudness of each bat.

$$A_i^{t+1} = \alpha A_i^t. \tag{11}$$

$$r_i^{t+1} = r_i^0 [1 - e^{-\gamma t}]. \tag{12}$$

Where A_i^t is pulse loudness of i-th bat at time t, r_i^t is pulse rate of i-th bat at time t, α is loudness attenuation coefficient, rate enhancement coefficient and r_i^0 is maximum pulse rate. Go to step (3).

3 Experiments

In order to test the performance of CVLNN with different activation functions, small-time scale traffic measurements data is used. The length of this traffic data set is 36000. The front 33000 data points are used as the training set and the last 3000 data points are used as the test set. The front 10 variable is used to predict the current variable. The parameters in bat algorithm are listed in Table 1, which are empirically selected.

Table 1. Parameter setting in bat algorithm.

Parameters	Value
Initial r_i^0	[0, 1]
Initial A_i^0	[1, 2]
Bat number	50
Maximum iterations	100
$[f_{\min}, f_{\max}]$	[−1.0, 1.0]
α and γ	0.9

A comparison of actual network traffic time series data and the predicted ones by CVLNN model with four activation functions is shown in Fig. 2. In order to reveal clearly, the prediction error distribution is plotted in Fig. 3. Table 2 also summarizes the results achieved using four different activation functions. From the results, we can see that CVLNN could predict small-time scale traffic measurements data accurately, and Elliot function and Sigmoid function perform better than Gaussian function and Secant function.

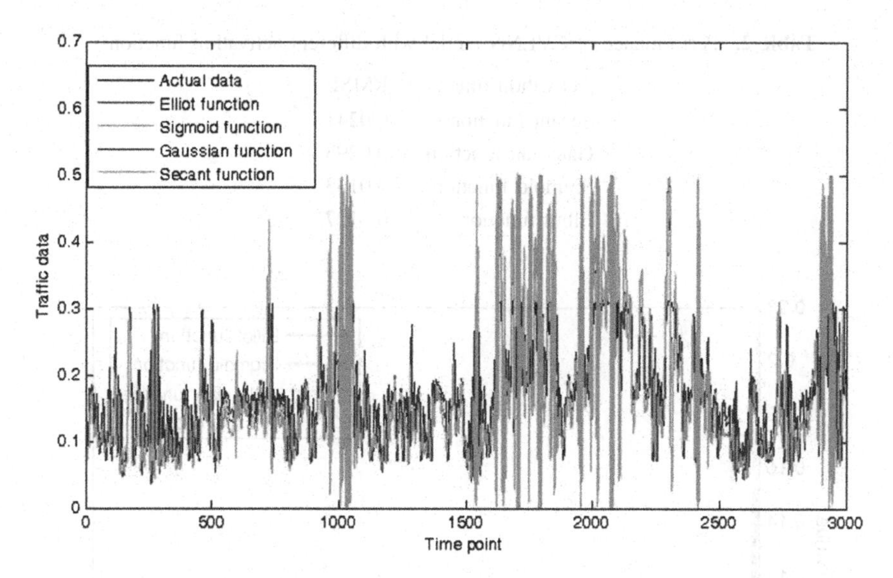

Fig. 2. A comparison of actual network traffic time series data and the predicted ones by CVLNN model with four activation functions.

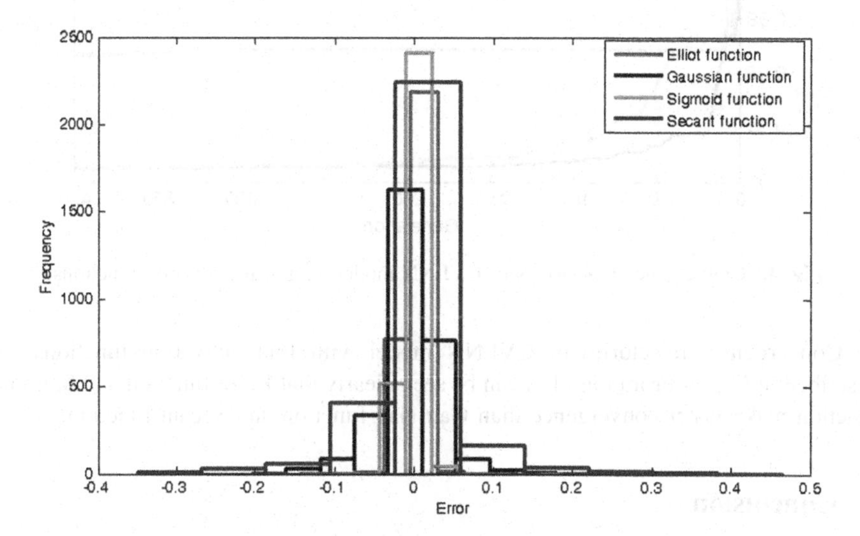

Fig. 3. The prediction error distribution.

Table 2. Performance of CVLNN model with different activation functions.

Activation functions	RMSE
Secant function	0.0241
Gaussian function	0.0293
Sigmoid function	0.0133
Elliot function	0.0127

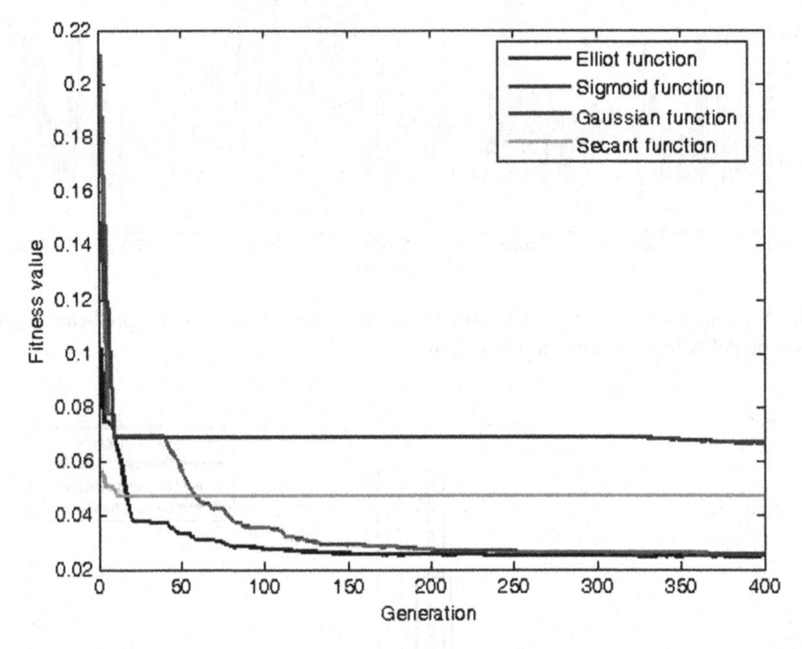

Fig. 4. Convergence trajectories of CVLNN model with four activation functions.

Convergence trajectories of CVLNN model with four activation functions are described in Fig. 4. From Fig. 4, it can be seen clearly that Elliot function and Sigmoid function make faster convergence than Gaussian function and Secant function.

4 Conclusion

As the variation of Legendre neural network, complex-valued Legendre neural network (CVLNN) is proposed to predict time series data in this paper. Bat algorithm is proposed to optimize the real-valued and complex-valued parameters of CVLNN model. We investigate performance of CVLNN by using different complex-valued activation function like Elliot function, Gaussian function, Sigmoid function and Secant function with small-time scale traffic measurements data. Results reveal that Elliot function and Sigmoid function perform better.

Acknowledgments. This work was supported by the Natural Science Foundation of China (No. 61702445), the PhD research startup foundation of Zaozhuang University (No. 2014BS13), Zaozhuang University Foundation (No. 2015YY02), and Shandong Provincial Natural Science Foundation, China (No. ZR2015PF007).

References

1. Angeline, P.J., Saunders, G.M., Pollack, J.B.: An evolutionary algorithm that constructs recurrent neural networks. IEEE Trans. Neural Networks **5**, 54–65 (1994)
2. Patra, J.C., Meher, P.K., Chakraborty, G.: Nonlinear channel equalization for wireless communication systems using Legendre neural networks. Sig. Process. **89**(11), 2251–2262 (2009)
3. Pei, A.Q., Wang, J., Fang, W.: Predicting agent-based financial time series model on lattice fractal with random Legendre neural network. Soft Comput., 1–16 (2015)
4. Dash, R., Dash, P.K.: Prediction of financial time series data using hybrid evolutionary legendre neural network. Int. J. Appl. Evol. Comput. **7**(1), 16–32 (2016)
5. Behera, S., Sahu, B.: Non linear dynamic system identification using Legendre neural network and firefly algorithm. In: 2016 International Conference on Communication and Signal Processing (ICCSP), pp. 1689–1693. IEEE, MADRAS, India (2016)
6. Wang, C., Qin, S.Y., Wan, B.W.: A novel neural network structure with fast convergence based on optimizing combination of different activation function. In: International Conference of the IEEE Engineering in Medicine and Biology Society, pp. 1399–1400. IEEE, Orlando, FL, USA (1991)
7. Zhang, L.M.: Genetic deep neural networks using different activation functions for financial data mining. In: IEEE International Conference on Big Data, pp. 2849–2851. IEEE, Santa Clara, CA, USA (2015)
8. Kumar, A.J.P., Singh, D.K.J.: A study on different activation functions for neural network-based wear loss prediction. Int. J. Appl. Eng. Res. **4**(3), 399 (2009)
9. Malleswaran, M., Dr, V.V., Angel, D.S.: Data fusion using different activation functions in artificial neural networks for vehicular navigation. Int. J. Eng. Sci. Technol. **2**(12), 7676–7690 (2010)
10. Yang, X.S., He, X.S.: Bat algorithm: literature review and applications. Int. J. Bio-Inspired Comput. **5**(3), 141–149 (2013)
11. Rahmani, M., Ghanbari, A., Ettefagh, M.M.: Robust adaptive control of a bio-inspired robot manipulator using bat algorithm. Expert Syst. Appl. **56**(C), 164–176 (2016)

Optimization Method of Residual Networks of Residual Networks for Image Classification

Long Lin[1], Hao Yuan[1], Liru Guo[2], Yingqun Kuang[3], and Ke Zhang[2(✉)]

[1] Power Systems Artificial Intelligence Joint Laboratory of SGCC, Global Energy Interconnection Research Institute Co., Ltd, Beijing 102209, China
{linlong,yuanhao}@geiri.sgcc.com.cn
[2] North China Electric Power University, Baoding 071000, Hebei, China
glr9292@126.com, zhangke41616@126.com
[3] Power Supply Service Center, State Grid Hunan Electric Power Company, Changsha 410004, China
blacktigerking@sina.com

Abstract. The activation of a Deep Convolutional Neural Network that overlooks the diversity of datasets has been restricting its development in image classification. In this paper, we propose a Residual Networks of Residual Networks (RoR) optimization method. Firstly, three activation functions (RELU, ELU and PELU) are applied to RoR and can provide more effective optimization methods for different datasets; Secondly, we added a drop-path to avoid over-fitting and widened RoR adding filters to avoid gradient vanish. Our networks achieved good classification accuracy in CIFAR-10/100 datasets, and the best test errors were 3.52% and 19.07% on CIFAR-10/100, respectively. The experiments prove that the RoR network optimization method can improve network performance, and effectively restrain the vanishing/exploding gradients.

Keywords: Image classification · RELU · Parametric exponential linear unit
Exponential linear unit · Residual networks of residual networks
Activation function

1 Introduction

In the past five years, deep learning [1] has made gratifying achievements in various computer vision tasks [2, 3]. With the rapid development of deep learning and Convolutional Neural Networks (CNNs), image classification has bidden farewell to coarse feature problems of manual extraction, and turned it into a new process. Especially, after AlexNet [4] won the champion ship of the 2012 Large Scale Visual Recognition Challenge (ILSVRC) [5], CNNs become deeper and continue to achieve better and better performance on different tasks of computer vision tasks.

To overcome degradation problems, a residual learning framework named Residual Networks (ResNets) were developed [8] to ease networks training, which achieved excellent results on the ImageNet test set. Since then, current state-of-the-art image classification systems are predominantly variants of ResNets. Residual networks of

D.-S. Huang et al. (Eds.): ICIC 2018, LNAI 10956, pp. 212–222, 2018.
https://doi.org/10.1007/978-3-319-95957-3_23

Residual networks (RoR) [13] adds level-wise shortcut connections upon original residual networks to promote the learning capability of residual networks. The rectified linear unit (RELU) [16] has been adopted by most of the convolution neural networks. RELUs are non-negative; therefore, they have a mean activation larger than zero, which would cause a bias shift for units. Furthermore, the selection of the activation function in the current DCNN model does not take into account the difference between the datasets. Different image datasets are different in variety and quality of image. The unified activation function limits the performance of image classification.

In order to effectively solve the above problem, this paper proposes an RoR network optimization method. To begin with, we analyze the characteristics of the activation function (RELU, ELU and PELU) and construct an RoR network with them. Thus, an RoR optimization based on different datasets is proposed. In addition, analysis of the characteristics of RoR networks suggest two modest mechanisms, stochastic depth and RoR-WRN, to further increase the accuracy of image classification. Finally, through massive experiments on CIFAR datasets, our optimized RoR model achieves excellent results on these datasets.

2 Related Work

Since AlexNet acquired a celebrated victory at the ImageNet competition in 2012, an increasing number of deeper and deeper Convolutional Neural Networks emerged, such as the 19-layer VGG [6] and 22-layer GoogleNet [7]. However, very deep CNNs also introduce new challenges: degradation problems, vanishing gradients in backward propagation and overfitting [15].

In order to overcome the degradation problem, a residual learning frame-work known as ResNets [8] was presented by the authors at the 2015 ILSVRC & COCO 2015 competitions and achieved excellent results in combination with the ImageNet test set. Since then, a series of optimized models based on ResNets have emerged, which became part of the Residual-Networks Family. Huang and Sun et al. [10] proposed a drop-path method, the stochastic depth residual networks (SD), which randomly drops a subset of layers and bypasses them with identity mapping for every mini-batch. To tackle the problem of diminishing feature reuse, wide residual networks (WRN) [11] was introduced by decreasing depth and increasing width of residual networks. Residual networks of Residual networks (RoR) [13] adds level-wise shortcut connections upon original residual networks to promote the learning capability of residual networks, that once achieved state-of-the-art results on CIFAR-10 and CIFAR-100 [12]. Each layer of DenseNet [14] is directly connected to every other layer in a feed-forward fashion. PyramidNet [27] gradually increases the feature map dimension at all units to involve as many locations as possible. ResNeXt [26] exposes a new dimension called cardinality (the size of the set of transformations), as a sential factor in addition to the dimensions of depth and width.

Even though non-saturated RELU has interesting properties, such as sparsity and non-contracting first-order derivative, its non-differentiability at the origin and zero gradient for negative arguments can hurt back-propagation [17]. Moreover, its non-negativity induces bias shift causing oscillations and impeded learning. Since the advent of the well-known RELU, many have tried to further improve the performance of the networks with more elaborate functions. Exponential linear unit (ELU) [17], defined as identity for positive arguments and $\exp(x) - 1$ for negative ones, deals with both increased variance and bias shift problems. Parametric ELU (PELU) [18], an adaptive activation function, defines parameters controlling different aspects of the function and proposes learning them with gradient descent during training.

3 Methodology

In this section, three activation functions (RELU, ELU and PELU) are applied to RoR and can provide more effective optimization methods for different datasets; Secondly, we added a drop-path to avoid over-fitting and widened RoR adding filters to avoid gradient vanish.

3.1 Comparative Analysis of RELU, ELU and PELU

The characteristics and performance of several commonly activation functions (RELU, ELU and PELU) are compared and analyzed as follows.

RELU is defined as:

$$f = \begin{cases} h & if \quad h \geq 0 \\ 0 & if \quad h < 0 \end{cases} \tag{1}$$

It can be seen that RELU is saturated at $h < 0$. Since the derivative of $h \geq 0$ is 1, RELU can keep the gradient from attenuation when $h > 0$, thus alleviating the problem of vanishing gradients. However, RELU outputs are non-negative, so the mean of the outputs will be greater than 0. Learning causes a bias shift for units in next layer. The more the units are correlated, the more serious the bias shift. the higher their bias shift.

ELU is defined as:

$$f = \begin{cases} h & if \quad h \geq 0 \\ \alpha\left(\exp(h) - 1\right) & if \quad h < 0 \end{cases} \tag{2}$$

The ELU incorporates Sigmoid and ReLU with left soft saturation. The ELU hyperparameter controls the value to which an ELU saturates for negative net inputs. ELUs diminish the vanishing gradients effect as RELUs do. By using a saturated negative part, the CNNs can no longer have arbitrary large negative outputs, which reduces variance. ELU outputs negative values for negative arguments, the network can push the mean activation toward zero, which reduces the bias shift.

PELU can be defined as follows:

$$f = \begin{cases} \frac{a}{b}h & if \quad h \geq 0 \\ a\left(\exp(h/b) - 1\right) & if \quad h < 0 \end{cases}, a, b > 0 \tag{3}$$

With the parameterization in the PELU function, a and b adjust the characteristics of the exponential function in the negative half axis and control the size of exponential decay and saturation point. a and b can also can adjust the slope of the linear function, to keep the differentiability. The parameters in PELU are updated at the same time as the parameters in the network weight layers during back-propagation.

3.2 RoR Networks with RELU, ELU and PELU

RoR [13] is based on a hypothesis: The residual mapping of residual mapping is easier to optimize than original residual mapping. To enhance the optimization ability of residual networks, RoR can optimize the residual mapping of residual mapping by adding shortcuts level by level based on residual networks.

Figure 2 in [13] shows the RoR (ReLU) architecture. The optimal model is 3-level RoR in [13]. Therefore we adopted 3-level RoR (RoR-3) as our basic architecture in experiments.

RoR-3 includes $3n$ final residual blocks, 3 middle-level residual blocks, and a root-level residual block, among which a middle-level residual block is composed of n final residual blocks and a middle-level shortcut, the root-level residual block is composed of 3 middle-level residual blocks and a root-level shortcut. The projection shortcut is done by 1×1 convolutions. RoR (ReLU) adopts a conv-BN-ReLU order in residual blocks.

For the saturation advantage of ELU and PELU, we designed new RoR architectures by adopting ELU and PELU, as shown in Fig. 1. The sequence of layers in residual block is Conv-PELU/ELU-Conv-BN. The batch normalization (BN) layer reduces the exploding gradient problem. We use 16, 32, and 64 convolutional filters sequentially in the convolutional layers of the three residual block groups, as shown in Fig. 1. Other architectures are the same as RoR (RELU)'s.

It can be seen from the previous section that the output of the ReLU is equal to or greater than 0, which makes the RoR network generate the bias shift when training. The bias shift directly limits the image classification performance of the RoR network. ELU and PELU outputs negative values for negative arguments, which allows the RoR network to push the mean activation toward zero. This reduces the bias shift; thus, performance of RoR is improved. Furthermore, as ELU and PELU saturate as the input gets negatively larger, the neurons can-not have arbitrary large positive and negative outputs and still keep a proper weighted sum; Thus variance is reduced. So RoR (ELU/PELU) is more robust to the inputs than RoR (ReLU).

3.3 RoR Optimization Method

PELU adopts the parameter updating mechanism to make RoR (PELU) more flexible in training. Under conditions of sufficient images in the dataset, ELU in RoR (ELU) has a constant saturation point and exponential decay for negative arguments and a constant slope for positive arguments. The deep RoR (ELU) can more easily get stuck in the saturated part during training where the gradient is almost zero–moreso than RoR (PELU) [18]. The gradient of a certain weight at the layer l in RoR (PELU) containing one neuron in each of its L layers is expressed as (4):

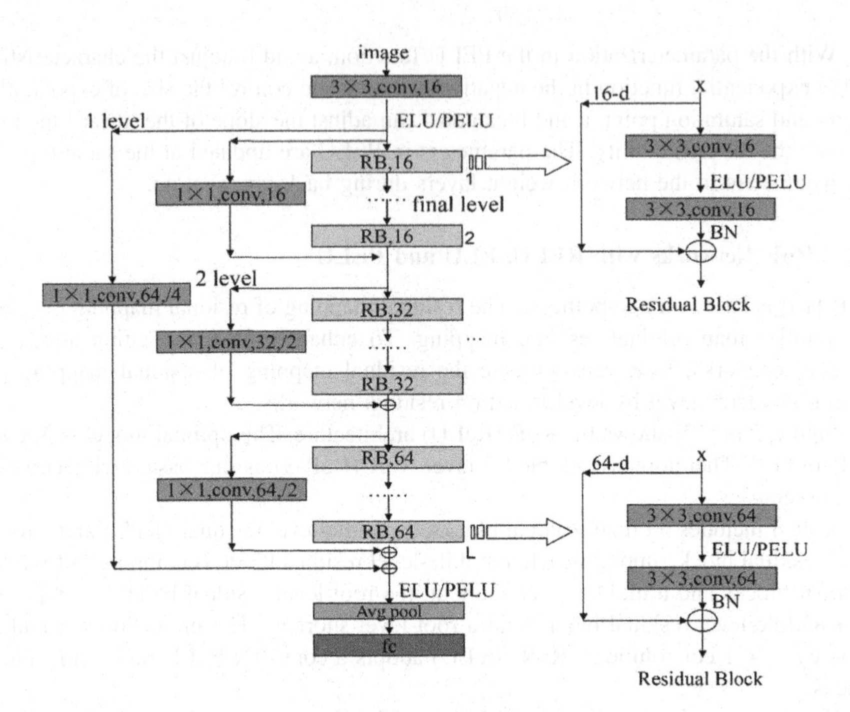

Fig. 1. RoR (ELU/PELU) architecture. The 1-level shortcut is a root-level shortcut, and the remaining three 2-level shortcuts are middle-level shortcuts. The shortcuts in residual blocks are final-level shortcuts.

$$\frac{\partial E}{\partial w_l} = h_{l-1} f'(h_l)[\prod_{j=l+1}^{L} f'(h_j) w_j] \frac{\partial E}{\partial z_L} \tag{4}$$

where h_0 is the input of the network. The output for lst layer is $z_l = f(h_l)$, where $h_l = w_l h_{l-1}$, and loss function is $E = \ell(z_L, y)$ between the network prediction z_L and label y. If $\partial E / \partial w_l$ tends to zero, it will produce vanishing gradients, making the network difficult to converge. One way to overcome this is when:

$f'(h_j)w_j \equiv f'(h_{j-1} w_j) w_j \geq 1$, which means $f'(wh)w \geq 1$. Substituted into the PELU definition, the solution lies in to get:

$$f'(wh)w = \begin{cases} w\frac{a}{b} & if \quad h \geq 0 \\ w\frac{a}{b}\exp(wh/b) & if \quad h < 0 \end{cases} a, b > 0 \tag{5}$$

Vanishing gradients problem is controlled when meeting $f'(wh)w \geq 1$. If $h \geq 0$, $w \geq b/a$ need to be met; $wb/a \cdot \exp(wh/b) \geq 1$ needs to be met, which pushes out $|h| \leq l(w) = |\log(b/aw)|(b/w)$ There is a maximum value of $l(w)$ when $w = \exp(1)b/a$; $l(w) = a\ \exp(-1)$ which means $l(w) \leq a\ \exp(-1)$.

For RoR (ReLU) and RoR (ELU), countering vanishing gradient is mostly possible with positive activations, which causes the bias shift. But for RoR (PELU), a can be adjusted to increase $a\exp(-1)$ and allow more negative activations h to counter vanishing gradients. Moreover, a and b can be adjusted to make which can $w \geq b/a$, which can make RoR (PELU) more flexible in training to eliminate vanishing gradients.

However, in RoR (PELU), additional parameters are added to the activation function compared to RoR (ReLU) and RoR (ELU). Adding the parameter update layer by layer makes the network model more complex. We believe that the parameters in the activation function have a greater impact on performance than parameters of other weight layers. Thus, when the number of images in each category is relatively small, RoR (PELU) is more likely to produce overfitting in training than RoR (ELU). Therefore, ELU will perform better under such conditions. Based on the above analysis, for different datasets of image classification, RoR (PELU) and RoR (ELU) complement each other. So, for different image datasets, we propose an optimization method for RoR based on activation functions:

For RoR, the datasets with more class images (such as CIFAR-10), they should be optimized by PELU, and the RoR (PELU) structure should be adopted. Meanwhile, the datasets with relatively fewer images in each category (such as CIFAR-100) should be optimized by ELU, and the RoR (ELU) structure should be adopted.

3.4 Stochastic Depth and Depth and Width Analysis

Overfitting and vanishing gradients are two challenging issues for RoR, which have a strongly negative impact on performance of image classification. In this paper, to alleviate overfitting, we trained RoR with the drop-path method, and obtained an apparent performance boost. We mitigated the gradient disappearing by appropriately widening the network.

RoR widens the network and adds more training parameters while adding additional shortcuts, which can lead to more serious overfitting problems. Therefore, we used the stochastic depth (SD) algorithm, which is commonly used in residual networks, to alleviate the overfitting problem. We trained our RoR networks by randomly dropping entire residual blocks during training and bypassing their transformations through shortcuts, without performing forward-backward computation or gradient updates. Let p_l mean the probability of the unblocked residual mapping branch of the l th residual block. L is the number of residual blocks, and (6) shows that p_l decreases linearly with the residual block position. p_l indicates that the last residual block is probably unblocked. SD can effectively prevent overfitting problems and reduce training time.

$$p_l = 1 - \frac{l}{L}(1 - p_L) \tag{6}$$

Under the premise network of fixed infrastructure, the main way to improve network performance is to magnify network model by deepening the network. however, increasing the depth of model blindly will lead to worse vanishing gradients. WRN [11] is used to increase width of residual networks to improve the performance, compared to blindly deepened networks (causing the vanishing gradients), in the same order of magnitude, with better performance. Based on this idea, we increased the channel of convolutional layers in the RoR residual blocks from {16, 32, 64} in the original network to $\{16 \times k, 32 \times k, 64 \times k\}$. Feature map dimension extracted from the residual blocks is increased to widen the network, keeping the network from becoming too deep, and further controlling the vanishing gradients problems. A widened RoR network is represented by RoR-WRN.

4 Experiment

In order to analyze the characteristics of three kinds of networks (RoR (ReLU), RoR (ELU), and RoR (PELU)), as well as verify the effectiveness of the optimization scheme, massive experiments were planned. The implementation and results follow.

4.1 Implementation

In this paper, we used RoR for image classification on two image datasets, CIFAR-10 and CIFAR-100. CIFAR-10 is a data set of 60,000 32 × 32 color images, with 10 classes of natural scene objects containing 6000 images each. Similar to CIFAR-10, CIFAR-100 is a data set of 60,000 32 × 32 color images, but with 100 classes of natural scene objects. This dataset is just like the CIFAR-10, except it has 100 classes containing 600 images each. The training set and test set contain 50,000 and 10,000 images, respectively. Our implementations were based on Torch 7 with a Titan X. We initialized the weights as in [19]. In both CIFAR-10 and CIFAR-100 experiments, we used SGD with a mini-batch size of 128 for 500 epochs. The learning rate started from 0.1, turned into 0.01 after epoch 250 and to 0.001 after epoch 375. For the SD drop-path method, we set p_l with the linear decay rule of $p_0 = 1$ and $p_L = 0.5$. In RoR-WRN experiments, we set the number of convolution kernels as $\{16 \times k, 32 \times k, 64 \times k\}$ instead of {16, 32, 64} in the original networks. Other architectures and parameters were the same as RoR's. As for the data size being limited in this paper, the experiment adopted two kinds of data expansion techniques: random sampling and horizontal flipping.

4.2 110-Layer RoR Experiments

Three types of 110-layer RoR were used to make up the CIFAR-10/100, with the SD algorithm and without the SD algorithm classification error rate shown in Figs. 2 and 3. It can be seen from the results that RoR+SD can obtain better results than RoR without SD, indicating that SD can effectively alleviate the overfitting problem and improve network performance. Therefore, in the subsequent experiments, we trained RoR with the drop-path method-SD. The results of RoR (ELU) and RoR (PELU) on the CIFAR-

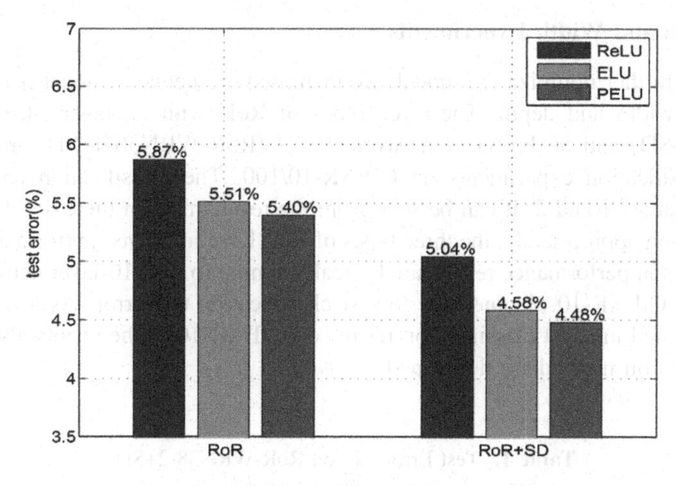

Fig. 2. Test Error (%) on 110-layer RoR

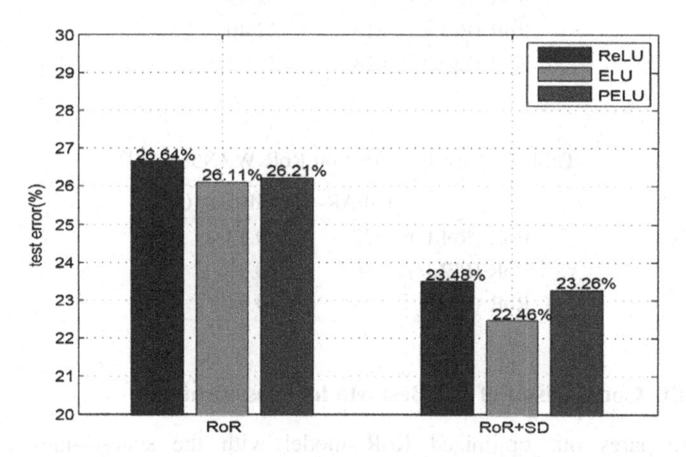

Fig. 3. Test Error (%) on 110-layer RoR

10/100 are better than that of RoR (ReLU). RoR (PELU) obtained lowest test error on CIFAR-10, RoR (ELU) obtained lowest test error on CIFAR-100.

The experimental results perfectly validate the effectiveness of the proposed optimization method. In this paper, we think that in the training, some of the input of ReLU fell into the hard saturation region, resulting in corresponding weight that cannot be updated. In addition, the ReLU output has the offset phenomenon; that is, the output mean value is greater than zero, which will affect the convergence of CNN. Using ELU and PELU, with the left side of the soft saturation, makes it more robust to the inputs, which means, we obtained better results.

4.3 Depth and Width Experiments

In order to further optimize the model, we increased the network model from the two aspects of width and depth. The three types of RoR with 38 layers, k = 2 (RoR-WRN38-2+SD) and 56 layers of network, k = 4 (RoR-WRN56-4+SD) are used for image classification experiments on CIFAR-10/100. The classification test error is shown in Tables 1 and 2. It can be seen from the results that, in the case of widening and deepening appropriately, the three types of RoR have improved performance, while the comparison performance results are basically similar to the 110-layer network. RoR (PELU) on CIFAR-10 obtained the lowest classification test error. As a result, RoR (ELU) obtained the best classification results on CIFAR-100. The results also confirm the optimization method we developed.

Table 1. Test Error (%) on RoR-WRN38-2+SD

	CIFAR-10	CIFAR-100
RoR (ReLU)	4.59	22.48
RoR (PELU)	4.19	22.40
RoR (ELU)	4.28	21.65

Table 2. Test Error (%) on RoR-WRN56-4+SD

	CIFAR-10	CIFAR-100
RoR (ReLU)	3.77	19.73
RoR (PELU)	3.56	20.39
RoR (ELU)	3.98	19.34

4.4 Results Comparison of the Best Model Classification

Table 3 compares our optimized RoR model with the state-of-the-art methods on CIFAR-10/100. It can be seen from Table 3 that the classification test error of RoR-WRN56-4+SD(PELU) and RoR-WRN56-4+SD (ELU) on CIFAR-10/100 is better than that of the original network RoR-WRN56-4+SD(ReLU), which proves the effectiveness of the proposed scheme. It can be seen from the experimental results from the optimized RoR has almost none increase in the computational cost and achieves better classification results compared with the same depth and width network. In view of the good effect of the optimization model, we attempted a deeper RoR-WRN74-4+SD model to obtain the optimal model and achieve state-of-the-art results on CIFAR-10. On the CIFAR-100, RoR-WRN74-4+SD achieves state-of-the-art results in much the same manner according to the amount of parameters as other models. Although ResNeXt and PyramidNet obtained a lower error rate on CIFAR-100, the number of model parameters was much larger than our best model.

Table 3. Test Error (%) on CIFAR-10/100 on different methods

Method(Parameters)	CIFAR-10	CIFAR-100
ELU [17]	6.55	24.28
FractalNet (30 M) [20]	4.59	22.85
ResNets-164 (2.5 M) [8]	5.93	25.16
Pre-ResNets-164 (2.5 M) [9]	5.46	24.33
Pre-ResNets-1001 (10.2 M) [9]	4.62	22.71
ELU-ResNets-110 (1.7 M) [21]	5.62	26.55
PELU-ResNets-110 (1.7 M) [18]	5.37	25.04
ResNets-110+SD (1.7 M) [10]	5.23	24.58
ResNet in ResNet (10.3 M) [24]	5.01	22.90
WResNet-d (19.3 M) [22]	4.70	–
WRN28-10 (36.5 M) [11]	4.17	20.50
CRMN-28 (more than 40 M) [23]	4.65	20.35
RoR-WRN56-4 (13.3 M) [13]	3.77	19.73
multi-resnet (145 M) [25]	3.73	19.60
DenseNet (27.2 M) [14]	3.74	19.25
PyramidNet (28.3 M) [27]	3.77	18.29
ResNeXt-29, 16 × 64*d* (68.1 M) [26]	3.58	17.31
RoR-WRN56-4+SD(PELU)(13.3 M)	**3.56**	–
RoR-WRN56-4+SD(ELU)(13.3 M)	–	**19.34**
RoR-WRN74-4+SD(PELU)(18.2 M)	**3.52**	–
RoR-WRN74-4+SD(ELU)(18.2 M)	–	**19.07**

5 Conclusion

In this paper, we put forward an optimization method of Residual Networks of Residual Networks (RoR) by analyzing performance of three activation functions. We acquired amazing image classification results on CIFAR-10 and CIFAR-100. The experiment results show optimizational RoR can give more control over bias shift and vanishing gradients and get excellent image classification performance.

Acknowledgement. This work is supported by National Power Grid Corp Headquarters Science and Technology Project under Grant No. 5455HJ170002(Video and Image Processing Based on Artificial Intelligence and its Application in Inspection), National Natural Science Foundation of China under Grants No. 61302163, No. 61302105, No. 61401154 and No. 61501185.

References

1. LeCun, Y., Bengio, Y., Hinton, G.: Deep learning. Nature **521**(7553), 436–444 (2015)
2. Hong, C., Yu, J., Wan, J., et al.: Multimodal deep autoencoder for human pose recovery. IEEE Trans. Image Process. **24**(12), 5659–5670 (2015)
3. Hong, C., Yu, J., Tao, D., et al.: Image-based three-dimensional human pose recovery by multiview locality-sensitive sparse retrieval. IEEE Trans. Industr. Electron. **62**(6), 3742–3751 (2015)

4. Krizhenvshky, A., Sutskever, I., Hinton, G.: Imagenet classification with deep convolutional networks. In: Proceedings of the Advances in Neural Information Processing Systems, pp. 1097–1105 (2012)
5. Russakovsky, O., Deng, J., Su, H., Krause, J., Satheesh, S., Ma, S., Huang, Z., Karpathy, A., Khosla, A., Bernstein, M., Berg, A.C., Fei-Fei, L.: Imagenet large scale visual recognition challenge, arXiv preprint arXiv:1409.0575 (2014)
6. Simonyan, K., Zisserman, A.: Very deep convolutional networks for large-scale image recognition, arXiv preprint arXiv:1409.1556 (2014)
7. Szegedy, C., Liu, W., Jia, Y., Sermanet, P., Reed, S., Anguelov, D., Erhan, D., Van-houcke, V., Rabinovich, A.: Going deeper with convolutions. In: Proceedings of the IEEE Conference on Computer Vision Pattern Recognition, pp. 1–9 (2015)
8. He, K., Zhang, X., Ren, S., Sun, J.: Deep residual learning for image recognition, arXiv preprint arXiv:1512.03385 (2015)
9. He, K., Zhang, X., Ren, S., Sun, J.: Identity mapping in deep residual networks, arXiv preprint arXiv:1603.05027 (2016)
10. Huang, G., Sun, Y., Liu, Z., Weinberger, K.: Deep networks with stochastic depth, arXiv preprint arXiv:1605.09382 (2016)
11. Zagoruyko, S., Komodakis, N.: Wide residual networks, arXiv preprint arXiv:1605.07146 (2016)
12. Krizhenvshky, A., Hinton, G.: Learning multiple layers of features from tiny images. M.Sc. thesis, Dept. of Comput. Sci., Univ. of Toronto, Toronto, ON, Canada (2009)
13. Zhang, K., Sun, M., Han, X., et al.: Residual networks of residual networks: multilevel residual networks. IEEE Trans. Circuits Syst. Video Technol. **PP**(99), 1 (2016)
14. Huang, G., Liu, Z., Weinberger, K.Q., et al.: Densely connected convolutional networks, arXiv preprint arXiv:1608.06993 (2016)
15. Bengio, Y., Simard, P., Frasconi, P.: Learning long-term dependencies with gradient descent is difficult. IEEE Trans. Neural Networks **5**(2), 157–166 (2014)
16. Nair, V., Hinton, G.: Rectified linear units improve restricted Boltzmann machines. In: Proceedings of the ICML, pp. 807–814 (2010)
17. Clevert, D.-A., Unterthiner, T., Hochreiter, S.: Fast and accurate deep network learning by exponential linear units (elus), arXiv preprint arXiv:1511.07289 (2015)
18. Trottier, L., Giguere, P., Chaibdraa, B.: Parametric exponential linear unit for deep convolutional neural networks, arXiv preprint arXiv:1605.09322 (2016)
19. Mishkin, D., Matas, J.: All you need is a good init, arXiv preprint arXiv:1511.06422 (2015)
20. Larsson, G., Maire, M., Shakhnarovich, G.: FractalNet: ultra-deep neural networks without residuals, arXiv preprint arXiv:1605.07648 (2016)
21. Shah, A., Shinde, S., Kadam, E., Shah, H.: Deep residual networks with exponential linear unit, arXiv preprint arXiv:1604.04112 (2016)
22. Shen, F., Zeng, G.: Weighted residuals for very deep networks, arXiv preprint arXiv:1605.08831 (2016)
23. Moniz, J., Pal, C.: Convolutional residual memory networks, arXiv preprint arXiv:1606.05262 (2016)
24. Targ, S., Almeida, D., Lyman, K.: Resnet in resnet: generalizing residual architectures, arXiv preprint arXiv:1603.08029 (2016)
25. Abdi, M., Nahavandi, S., Multi-residual networks, arXiv preprint arXiv:1609.05672 (2016)
26. Xie, S., Girshick, R., Dollr, P., et al.: Aggregated residual transformations for deep neural networks, arXiv preprint arXiv:1611.05431 (2016)
27. Han, D., Kim, J., Kim, J.: Deep pyramidal residual networks, arXiv preprint arXiv:1610.02915 (2016)

A Simple and Effective Deep Model for Person Re-identification

Si-Jia Zheng[1], Di Wu[1], Fei Cheng[2], Yang Zhao[2], Chang-An Yuan[3], Xiao Qin[3], and De-Shuang Huang[1(✉)]

[1] Institute of Machine Learning and Systems Biology, School of Electronics and Information Engineering, Tongji University, Shanghai, China
zhengsj9878@163.com, wudi_qingyuan@163.com, dshuang@tongji.edu.cn

[2] Beijing E-Hualu Info Technology Co., Ltd., Beijing, China

[3] Science Computing and Intelligent Information Processing of Guang Xi Higher Education Key Laboratory, Guangxi Teachers Education University, Nanning 530001, Guangxi, China

Abstract. Person re-identification (re-ID), which aims to re-identify a person captured by one camera from another camera at any non-overlapping location, has attracted more and more attention in recent years. So far, it has been significantly improved by deep learning technology. A variety of deep models have been proposed in person re-ID community. In order to make the deep model simple and effective, we propose an identification model that combines the softmax loss with center loss. Moreover, various data augmentation methods and re-ranking strategy are used to improve the performance of the proposed model. Experiments on CUHK03 and Market-1501 datasets demonstrate that the proposed model is effective and has good results in most cases.

Keywords: Person re-identification · Center loss · Date augmentation
Deep learning

1 Introduction

Person re-identification (re-ID) originates from the multi-camera tracking. It aims to re-identify a person that has been captured by one camera in another camera at any new location. In recent years, it has increasing attentions in the computer vision and pattern recognition research community. Despite the researches have made great efforts, person re-ID remains challenging due to the following reasons: (1) dramatic variations in visual appearance and environment caused by camera viewpoint changes, (2) background clutter and occlusions, (3) the dramatic changes of human pose due to different time and place, (4) the similar clothes and similar faces shared from different individuals.

Person re-ID is essentially to measure the similarity of the person images. Recently, convolutional neural networks [1] (CNNs) have also achieved great success in person re-ID field. The current CNN-based models can be categorized into three major types, i.e. identification model, verification model and triplet model. Generally, the

D.-S. Huang et al. (Eds.): ICIC 2018, LNAI 10956, pp. 223–228, 2018.
https://doi.org/10.1007/978-3-319-95957-3_24

organization of the input data of the triplet model is difficult. Meanwhile, there will sometime exist imbalance of the data set in verification model.

Based on the considerations above, we propose using an identification model to address the person re-ID task. Moreover, we use some tricks to make the model more effective. In a nutshell, the main contributions of this paper are summarized as follows:

(1) Before the training, we use some data augmentation methods (i.e. flipping, random erasing and style transfer) to enlarge the datasets and improve the robustness and accuracy of the model.
(2) To further minimize the intra- class variances for training, we combine the identification loss and center loss functions to get a more discriminative CNN model for person re-ID.
(3) We evaluate the proposed model on three person re-ID benchmark datasets (CUHK03 and Market-1501) and the experiment results show that the effectiveness of the proposed method.

2 Related Work

With the development of deep learning, the works on person re-ID can be divided into two aspects: hand-crafted based methods or deep learning based methods.

Hand-Crafted System for Person re-ID. In person re-ID, some works directly used low-level features in which color and texture were the most commonly used. Another works chose the attribute-based features which can be viewed as mid-level representations or semantic features as high-level representations. For distance metrics, they are categorized into supervised learning versus unsupervised learning, global learning versus local learning, etc.

Deeply-Learned System for Person re-ID. In 2014, Yi et al. [2] both used the convolution neural network (CNN) to extract the features and determine whether the input two images is the same ID or not. Since then, a large number of works based on deeply-learned were proposed for person re-ID. Some works used the identification model to output the corresponding labels of the input images. In order to boost the performance, Jin et al. [3] combined the center loss with identification loss, Xiao et al. [4] designed an Online Instance Matching (OIM) loss function to train the network. In order to push the mismatched pair farther from each other and pull the matched pair closer simultaneously, the triplet model was introduced. Although the triplet model is very effective, the identification model is simpler and easier to organize the data. In this paper, we employ the identification model and combine the center loss with softmax loss. Comparing with the network propose by Jin et al. [3],we use more robust ResNet and DenseNet as the base network. What's more, we focus on the data augmentation. Different from Xiao et al. [5] learning multiple domains features, we address the style transfer among different dataset to enlarge the train samples.

3 Proposed Method

3.1 Data Augmentation

The training of convolutional neural networks [6–9] often requires huge amounts of data, however data collection and data annotation need extraordinary efforts. In this case, data augmentation plays an important role. In order to enrich the training set, prevent the model over-fitting, in this paper, we adopt multiple methods of data augmentation, including cropping images, flipping images and other common methods. Figure 1 shows the results of image transforming.

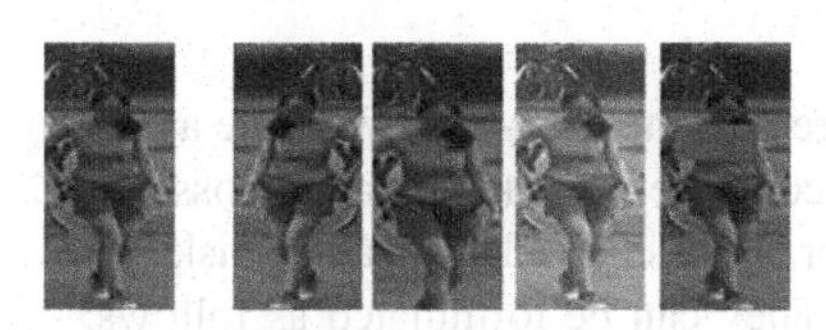

Fig. 1. Illustration of data augmentation.

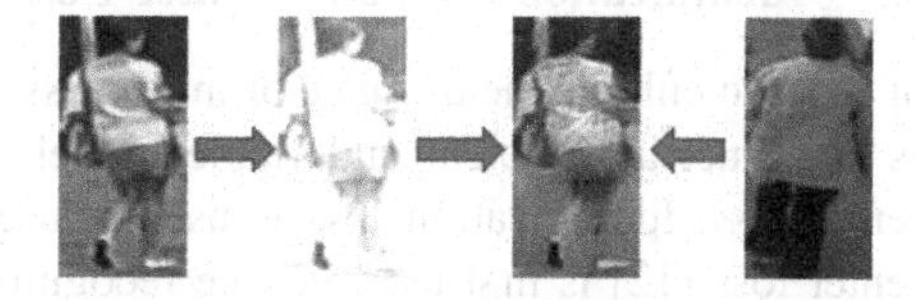

Fig. 2. Illustration of image style transfer.

In this paper, we mainly introduce the random erasing and style migration. Random erasing [10] randomly choose a rectangle region in the image and erase its pixels with random values or the ImageNet mean pixel value. This approach does not require additional parameter or memory consumption. Importantly, it improves the robustness of the model. In other word, it is a supplement of regularization.

In order to enlarge the dataset and improve the performance of the network, we use the method of image style transfer [11] on different benchmark data sets. For style transfer the output is semantically similar to the input although the background is changed a little. This method is very simple and fast. It can normalize the style of different datasets. An illustration is shown in the Fig. 2.

3.2 The Network Architecture

In this section, we present our network architecture in detail. Our model is a basically identification model with the advantages of simplicity and robustness. As shown in the Fig. 3, we first conduct the data augmentation for the training dataset. Then the feature representations are extracted from the input images by CNN model (ResNet or DenseNet). We use the pre-trained CNN model on ImageNet. ResNet50 used in this paper has been proved to be easier to optimize. Furthermore, DenseNet is further improved with the advantages of less parameters and the features reusing. Finally, we combine the identification loss with center loss. In next section, we will introduce the details of loss functions.

During the testing phase, the probe and the gallery images all pass through the trained network to get the corresponding features. After that, we use distance metric method to calculate the rank list. In order to achieve a better result, the re-ranking method is introduced to re-rank the initial rank result. The entire specific processes are shown in Fig. 4.

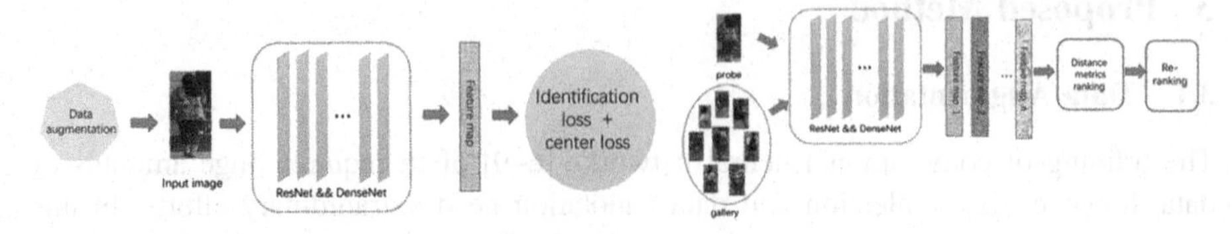

Fig. 3. The overall CNN network

Fig. 4. The overall processes of the testing

3.3 Identification Loss and Center Loss

In order to enlarge the distance of inter-class and reduce the intra-class variance as well as construct an efficient and stable model, we combine the identification loss with center loss. Identification loss is usually used for multi-class classification task. The center loss [12] is first used in face recognition. They can be formulated as follows:

$$L_i = -\frac{1}{M}\sum\nolimits_{i=1}^{M} log\frac{e^{W_{y_i}^T x_i + b_{y_i}}}{\sum\nolimits_{j=1}^{N} e^{W_j^T x_i + b_j}} \qquad L_c = \frac{1}{2}\sum\nolimits_{i=1}^{M} \left\| x_i - c_{y_i} \right\|_2^2 \tag{1}$$

On the left, M denotes the size of a mini-batch, N is the number of person identities. W_j is the j-th column of the weights of the last fully connected layer, and b is the bias term. On the right, the c_{y_i} is the y_i-th class center of deep features. In each mini-batch iteration, the centers are calculated by averaging the features of the corresponding identities and it will be updated. The gradients of L_c about x_i and the update equation of c_{y_i} are formulated as follows:

$$\frac{\partial L_c}{\partial x_i} = x_i - c_{y_i} \qquad \Delta c_j = \frac{\sum_{i=1}^{m} \delta(y_j = j) \cdot (c_j - x_i)}{1 + \sum_{i=1}^{m} \delta(y_j = j)} \tag{2}$$

There is a conditional expression δ in formula above. The center of a category only depends on the data of this class in a mini-batch. To avoid large perturbations caused by few mislabeled samples, a scalar α controls the learning rate of the centers. We combine the identification loss with center loss. The formulation is as follows:

$$\mathrm{L} = L_i + \lambda L_c = -\sum\nolimits_{i=1}^{M} log\frac{e^{W_{y_i}^T x_i + b_{y_i}}}{\sum\nolimits_{j=1}^{N} e^{W_j^T x_i + b_j}} + \frac{\lambda}{2}\sum\nolimits_{i=1}^{M} \left\| x_i - c_{y_i} \right\|_2^2 \tag{3}$$

The combination of identification loss and center loss effectively deal with the variances of inter-class and intra-class.

3.4 Re-ranking

Recently, many re-ranking methods have been obtained a good result in person re-ID. We adopt the re-ranking using k-reciprocal encoding [13] in the experiments.

4 Experiments

4.1 Datasets and Evaluation Protocol

There exist several challenging benchmark data sets for person re-identification. In this paper, we use Market-1501, CUHK03 to evaluate our network. In this paper, we use the cumulative matching characteristics (CMC) curve and the mean average precision (mAP) to evaluate our model (Table 1).

Table 1. the accuracy on the CUHK03,Maket-1501 ('-') means no available reported results).

Results on CUHK03			
Method	Rank-1	Rank-5	Rank-10
DeepLDA [15]	63.23	89.95	92.73
Gated S-CNN [16]	68.10	88.10	94.60
CNN-FRW-IC [3]	82.10	96.20	98.20
Our (ResNet50)	**84.13**	96.10	98.43
Our (DenseNet121)	**84.95**	**97.32**	98.60
Results on Maket-1501 dataset			
Method	mAP	Rank-1	Rank-5
LOMO + TMA	22.30	47.90	-
CNN Embedding [3]	59.87	79.51	90.91
IDE(R) + ML (re-ra) [13]	63.63	77.11	-
Our (ResNet50)	**82.54**	**89.91**	**94.15**
Our (DenseNet121)	**84.89**	**90.20**	**94.86**

4.2 Comparison with Other Approaches

We use the PyTorch framework to build the network and evaluate the proposed method comparing with some other state-of-the-art approaches. In our experiments, we set the training epochs to 50. The weight decay [14] and the batch size are set to 0.0005 and 32, respectively. The learning rate α of center loss is set to 0.5 and λ is set to 0.8.

5 Conclusion

In this paper, we propose an effective deep model for the person re-ID. In order to enlarge the datasets and avoid the over-fitting, we use various methods to do the data augmentation. To further minimize the intra- class variances, we combine the identification loss and center loss functions. The model can make the full use of the label information and reduce the intra-class variance at the same time. Moreover, the proposed method has good results on CUHK03 and Market-1501 datasets. In the future, we will further explore the part of style transfer between different datasets.

Acknowledgments. This work was supported by the grants of the National Science Foundation of China, Nos. 61472280, 61672203, 61472173, 61572447, 61772357, 31571364, 61520106006, 61772370, 61702371 and 61672382, China Postdoctoral Science Foundation Grant, Nos. 2016M601646 & 2017M611619, and supported by "BAGUI Scholar" Program of Guangxi Zhuang Autonomous Region of China. De-Shuang Huang is the corresponding author of this paper.

References

1. Huang, D.S.: Systematic Theory of Neural Networks for Pattern Recognition. Publishing House of Electronic Industry of China (1996). (in Chinese)
2. Yi, D., et al.: Deep metric learning for person re-identification. In: 2014 22nd International Conference on Pattern Recognition (ICPR). IEEE, pp. 34–39 (2014)
3. Jin, H., et al.: Deep Person Re-Identification with Improved Embedding and Efficient Training (2017)
4. Xiao, T., et al.: Joint Detection and Identification Feature Learning for Person Search (2017)
5. Xiao, T., et al.: Learning deep feature representations with domain guided dropout for person re-identification. In: Computer Vision and Pattern Recognition, pp. 1249–1258 (2016)
6. Huang, D.S.: Radial basis probabilistic neural networks: model and application. Int. J. Pattern Recognit. Artif. Intell. **13**(07), 1083–1101 (1999)
7. Zhao, Z.Q., Huang, D.S., et al.: Human face recognition based on multi-features using neural networks committee. Pattern Recogn. Lett. **25**(12), 1351–1358 (2004)
8. Huang, D.S., Zhao, W.B.: Determining the Centers of Radial Basis Probabilistic Neural Networks by Recursive Orthogonal Least Square Algorithms ☆. Elsevier Science Inc. (2005)
9. Huang, D.S., Du, J.-X.: A constructive hybrid structure optimization methodology for radial basis probabilistic neural networks. IEEE Trans. Neural Networks **19**(12), 2099–2115 (2008)
10. Zhong, Z., et al.: Random erasing data augmentation (2017)
11. Wei, L., et al.: Person transfer GAN to bridge domain gap for person re-identification (2017)
12. Wen, Y., Zhang, K., Li, Z., Qiao, Yu.: A discriminative feature learning approach for deep face recognition. In: Leibe, B., Matas, J., Sebe, N., Welling, M. (eds.) ECCV 2016. LNCS, vol. 9911, pp. 499–515. Springer, Cham (2016). https://doi.org/10.1007/978-3-319-46478-7_31
13. Zhong, Z., et al.: Re-ranking person re-identification with k-reciprocal encoding. In: IEEE Conference on Computer Vision and Pattern Recognition, pp. 3652–3661 (2017)
14. Huang, D.S., Jiang, W.: A general CPL-AdS methodology for fixing dynamic parameters in dual environments. IEEE Trans. Syst. Man Cybern. Part B **42**(5), 1489–1500 (2012)
15. Schumann, A., Stiefelhagen, R.: Person re-identification by deep learning attribute-complementary information. In: IEEE Conference on Computer Vision and Pattern Recognition Workshops, pp. 1435–1443 (2017)
16. Varior, R.R., Haloi, M., Wang, G.: Gated Siamese Convolutional Neural Network Architecture for Human Re-identification. In: Leibe, B., Matas, J., Sebe, N., Welling, M. (eds.) ECCV 2016. LNCS, vol. 9912, pp. 791–808. Springer, Cham (2016). https://doi.org/10.1007/978-3-319-46484-8_48

A Hybrid Deep Model for Person Re-Identification

Di Wu[1], Si-Jia Zheng[1], Fei Cheng[2], Yang Zhao[2], Chang-An Yuan[3], Xiao Qin[3], Yong-Li Jiang[4], and De-Shuang Huang[1](✉)

[1] Institute of Machine Learning and Systems Biology, School of Electronics and Information Engineering, Tongji University, Shanghai, China
wudi_qingyuan@163.com, zhengsj9878@163.com, dshuang@tongji.edu.cn

[2] Beijing E-Hualu Info Technology Co., Ltd., Beijing, China

[3] Science Computing and Intelligent Information Processing of Guang Xi Higher Education Key Laboratory, Guangxi Teachers Education University, Nanning 530001, Guangxi, China

[4] Ningbo Haisvision Intelligence System Co., Ltd., Ningbo 315000, Zhejiang, China

Abstract. In this study, we present a hybrid model that combines the advantages of the identification, verification and triplet models for person re-identification. Specifically, the proposed model simultaneously uses Online Instance Matching (OIM), verification and triplet losses to train the carefully designed network. Given a triplet images, the model can output the identities of the three input images and the similarity score as well as make the L_{-2} distance between the mismatched pair larger than the one between the matched pair. Experiments on two benchmark datasets (CUHK01 and Market-1501) show that the proposed method can achieve favorable accuracy while compared with other state of the art methods.

Keywords: Convolutional neural networks · Person re-identification
Identification model · Verification model · Triplet model

1 Introduction

As a basic task of multi-camera surveillance system, person re-identification aims to re-identify a query pedestrian observed from non-overlapping cameras or across different time with a single camera [1]. Person re-identification is an important part of many computer vision tasks, including behavioral understanding, threat detection [2–6] and video surveillance [7–9]. Recently, the task has drawn significant attention in computer vision community. Despite the researchers make great efforts on addressing this issue, it remains a challenging task due to the appearance of the same pedestrian may suffer significant changes under non-overlapping camera views.

Recently, owing to the great success of convolution neural network (CNN) in computer vision community [10–14], many CNN-based methods are introduced to address the person re-identification issue. These CNN-based methods achieve many

D.-S. Huang et al. (Eds.): ICIC 2018, LNAI 10956, pp. 229–234, 2018.
https://doi.org/10.1007/978-3-319-95957-3_25

promising performances. Unlike the hand-craft methods, CNN-based methods can learn deep features automatically by an end-to-end way [28]. These CNN-based methods for person re-identification can be roughly divided into three categories: identification models, verification models and triplet models. The three categories of models differ in input data form and loss function and have their own advantages and limitations. Our motivation is to combine the advantages of three models to learn more discriminative deep pedestrian descriptors.

The identification model treats the person re-identification issue as a task of multi-class classification. The model can make full use of annotation information of datasets [1]. However, it cannot consider the distance metric between image pairs. Verification model regards the person re-identification problem as a binary classification task, it takes paired images as input and outputs whether the paired images belong to the same person. Thus, verification model considers the relationship between the different images, but it does not make full use of the label information of datasets. As regards to the triplet model [15], it takes triplet unit (x_i, x_j, x_k) as input, where x_i and x_k are the mismatched pair and the x_i and x_j are the matched pair. Given a triplet images, the model tries to make the relative distance between x_i and x_k larger than the one between x_i and x_j. Accordingly, triplet-based model can learn a similarity measurement for the pair images, but it also cannot take advantage of the label information of the datasets and uses weak labels only [1].

We design a CNN-based architecture that combines the three types of popular deep models used for person re-identification, i.e. identification, verification and triplet models. The proposed architecture can jointly output the IDs of the input triplet images and the similarity score as well as force the L2 distance between the mismatched pair larger than the one between the matched pair, thus enhancing the discriminative ability of the learned deep features.

2 Proposed Method

As shown in Fig. 1, the proposed method is a triplet-based CNN model that combines identification, verification and triplet losses.

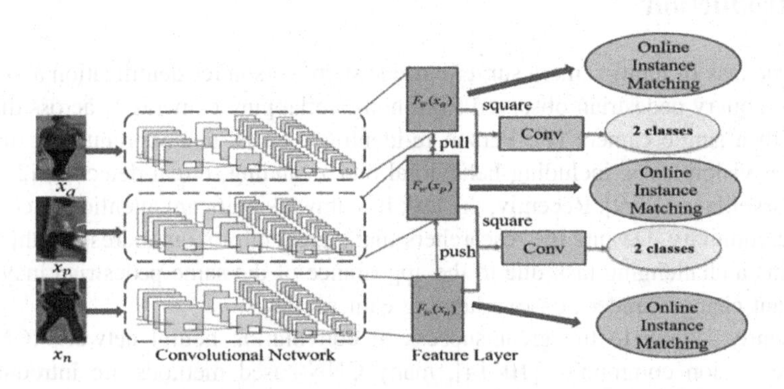

Fig. 1. The proposed model

2.1 Loss Function

Identification Loss. In this work, we utilize the Online Instance Matching [16] to instead of the cross-entropy loss for supervising the identification submodule. For one image M, the similarity score of it belong to the ID j is written as:

$$p_j = \frac{\exp(v_j^T M/\partial)}{\sum_{i=1}^{L} \exp(v_i^T M/\partial) + \sum_{k=1}^{Q} \exp(\mu_k^T M/\partial)} \tag{1}$$

where v_j^T is the transposition of the j-th column of lookup table, μ_k^T represents the transposition of the k-th ID of circular queue. ∂ is the temperature scalar.

Verification Loss. For the verification subnetwork, we treat person re-identification as a task of binary classification issue. Similar to identification model, we also adopt cross-entropy loss as identification loss function, which is:

$$\begin{aligned} q &= soft\max(\text{conv}), \\ L_2 &= \sum_{i=1}^{2} -n_i \log(q_i) \end{aligned} \tag{2}$$

in which n_i is the labels of paired image, when the pair is the same pedestrian, n_1= 1 and n_2= 0; otherwise, n_1= 0 and n_2= 1.

Triplet Loss. The triplet subnetwork is adopted to make the Euclidean distance between the positive pairs smaller than that between negative pairs. For one triplet unit R_i, the triplet loss function can be written as:

$$L_3 = [thre + d(F_w(R_i^o), F_w(R_i^+)) - d(F_w(R_i^o), F_w(R_i^-))]_+ \tag{3}$$

where *thre* is a threshold value and is a positive number, $[x]+ = \max(0, \mathrm{x})$, and *d ()* is Euclidean distance.

Hybrid Loss. The deep architecture is jointly supervised by three OIM losses, two cross-entropy losses and one triplet loss. During the training phase, the hybrid loss function can be written as:

$$L = \alpha_1 L_1 + \alpha_2 L_2 + \alpha_3 L_3 \tag{4}$$

in which α_1, α_2 and α_3 are the balance parameters.

2.2 Training Strategies

We utilize the pytorch framework to implement our network. In this work, we use the Adaptive Moment Estimation (Adam) as the optimizer of the deep model. We use two types of training strategies proposed by [17]. Specially, for large-scale dataset like Market-1501, we use the designed model directly to transfer on its training set. As for the small datasets (e.g. CUHK01), we first train the model on the large-scale person re-identification dataset (e.g. Market-1501), then fine-tune the model on the small dataset.

3 Experiments

We first resize the training images into 256*128, then the mean image is subtracted by those resized training images. For our hybrid model, it is crucial for it to organize the mini-batch that can satisfy the training purpose of both identification, verification and triplet subnetworks. In this study, we use the protocol proposed by, we sample Q identities randomly, and then all the images R in the training set are selected. Finally, we use QR images to constitute one mini-batch. Among these QR images, we choose the hardest positive and negative sample for each anchor to form the triplet units. And we randomly selected 100 paired images for verification training. As to the identification subnetwork, we use all QR images in the mini-batch for training.

(1) **Evaluation on CUHK01**
For this dataset, 485 pedestrians are randomly selected to form the training set. The remainder 486 identities are selected for test. From Table 1, we can observe that the proposed method beats the most of compared approaches, which demonstrates the effectiveness of the proposed method.

Table 1. Fig. 1. Table 1. Results (Rank1, Rank5 and Rank10 matching accuracy in %) on the CUHK01 dataset. '-' means no reported results is available.

Method	Rank-1	Rank-5	Rank-10
Deep Ranking [18]	50.41	75.93	84.07
GOG [19]	57.80	79.10	86.20
DGD [20]	66.60	-	-
Quadruplet [21]	62.55	83.44	89.71
Deep Transfer [17]	**77.00**	-	-
CNN-FRW-IC [22]	70.50	90.00	94.80
Our	76.84	**95.60**	**96.12**

(2) **Evaluation on Market-1501**
We compare the proposed model with eight state-of-the-art methods on Market-1501 dataset. We report the performances of mean average precision (mAP), Rank-1 and Rank-5. Both the results are based on single-query evaluation. From Table 2, it can be observed that the accuracies of mAP, Rank-1 and Rank-5 of the proposed model achieve 77.43%, 91.60% and 98.33%, respectively, and our method beats all the other competing methods, which further proves the effectiveness of the proposed method.

Table 2. Results (mAP, Rank-1 and Rank-5 matching accuracy in %) on the Market-1501 dataset in the single-shot. '-' means no reported results is available.

Method	mAP	Rank-1	Rank-5
Deep Transfer [17]	65.50	83.70	-
CAN [23]	35.90	60.30	-
CNN Embedding [24]	59.87	79.51	90.91
CNN + DCGAN [25]	56.23	78.06	-
IDE(R) + ML(re-ra) [26]	63.63	77.11	-
TriNet [27]	69.14	84.92	94.21
Our	**77.43**	**91.60**	**98.33**

4 Conclusions

In this paper, we design a hybrid deep model for the person re-identification. The proposed model combines the identification, verification and triplet losses to handle the intra/inter class distances. Through the hybrid strategy, the model can learn a similarity measurement and the discriminative features at the same time. The proposed model outperforms most of the state-of-the-art on the CUHK01 and Market-1501 datasets.

Acknowledgements. This work was supported by the grants of the National Science Foundation of China, Nos. 61472280, 61672203, 61472173, 61572447, 61772357, 31571364, 61520106006, 61772370, 61702371 and 61672382, China Postdoctoral Science Foundation Grant, Nos. 2016M601646 & 2017M611619, and supported by "BAGUI Scholar" Program of Guangxi Zhuang Autonomous Region of China. De-Shuang Huang is the corresponding author of this paper.

References

1. Zheng, L., et al.: Person Re-identification: Past, Present and Future. Computer Vision and Pattern Recognition (2016). arXiv: 1610.02984
2. Huang, D.S.: A constructive approach for finding arbitrary roots of polynomials by neural networks. IEEE Trans. Neural. Netw. **15**(2), 477–491 (2004)
3. Huang, D.S., et al.: A case study for constrained learning neural root finders. Appl. Math. Comput. **165**, 699–718 (2005)
4. Huang, D.S., et al.: Zeroing polynomials using modified constrained neural network approach. IEEE Trans. Neural Netw. **16**, 721–732 (2005)
5. Huang, D.S., et al.: A new partitioning neural network model for recursively finding arbitrary roots of higher order arbitrary polynomials. Appl. Math. Comput. **162**, 1183–1200 (2005)
6. Huang, D.S., et al.: Dilation method for finding close roots of polynomials based on constrained learning neural networks ☆. Phys. Lett. A **309**, 443–451 (2003)
7. Huang, D.S.: The local minima-free condition of feedforward neural networks for outer-supervised learning. IEEE Trans. Syst. Man Cybern. Part B Cybern. **28**, 477 (1998). A Publication of the IEEE Systems Man & Cybernetics Society

8. Huang, D.S.: The united adaptive learning algorithm for the link weights and shape parameter in RBFN for pattern recognition. Int. J. Pattern Recognit. Artif. Intell. **11**, 873–888 (1997)
9. Huang, D.S., Ma, S.D.: Linear and nonlinear feedforward neural network classifiers: a comprehensive understanding: journal of intelligent systems. J. Intell. Syst. **9**, 1–38 (1999)
10. Huang, D.S., Du, J.X.: A constructive hybrid structure optimization methodology for radial basis probabilistic neural networks. IEEE Trans. Neural Netw. **19**, 2099 (2008)
11. Wang, X.F., Huang, D.S.: A novel density-based clustering framework by using level set method. IEEE Trans. Knowl. Data Eng. **21**, 1515–1531 (2009)
12. Shang, L., et al.: Palmprint recognition using FastICA algorithm and radial basis probabilistic neural network. Neurocomputing **69**, 1782–1786 (2006)
13. Zhao, Z.Q., et al.: Human face recognition based on multi-features using neural networks committee. Pattern Recogn. Lett. **25**, 1351–1358 (2004)
14. Huang, D.S., et al.: A neural root finder of polynomials based on root moments. Neural Comput. **16**, 1721–1762 (2004)
15. Ding, S., et al.: Deep feature learning with relative distance comparison for person re-identification. Pattern Recogn. **48**, 2993–3003 (2015)
16. Xiao, T., et al.: Joint detection and identification feature learning for person search. In: IEEE Conference on Computer Vision and Pattern Recognition (CVPR), pp. 3376–3385. IEEE (2017)
17. Geng, M., et al.: Deep transfer learning for person re-identification (2016)
18. Chen, S.Z., et al.: Deep ranking for person re-identification via joint representation learning. IEEE Trans. Image Process. **25**, 2353–2367 (2016)
19. Matsukawa, T., et al.: Hierarchical Gaussian Descriptor for Person Re-identification. In: Proceedings of the Computer Vision and Pattern Recognition, pp. 1363–1372 (2016)
20. Xiao, T., et al.: Learning deep feature representations with domain guided dropout for person re-identification. In: IEEE Conference on Computer Vision and Pattern Recognition (CVPR), pp. 1249–1258 (2016)
21. Chen, W., et al.: Beyond triplet loss: a deep quadruplet network for person re-identification. In: Proceedings of the CVPR, pp. 1320–1329 (2017)
22. Jin, H., et al.: Deep person re-identification with improved embedding and efficient training. In IEEE International Joint Conference on Biometrics, pp. 261–267 (2017)
23. Liu, H., et al.: End-to-End comparative attention networks for person re-identification. IEEE Trans. Image Process. **26**(7), 3492–3506 (2017). A Publication of the IEEE Signal Processing Society
24. Zheng, Z., et al.: A discriminatively learned CNN embedding for person re-identification. ACM Trans. Multimedia Comput. Commun. Appl. (TOMM) **14**(1), 13 (2016)
25. Zheng, Z., et al.: Unlabeled samples generated by GAN improve the person re-identification baseline in vitro, 3774–3782 (2017). arXiv preprint arXiv:1701.07717
26. Zhong, Z., et al.: Re-ranking person re-identification with k-reciprocal encoding. In: IEEE Conference on Computer Vision and Pattern Recognition (CVPR), pp. 3652–3661 (2017)
27. Hermans, A., et al.: In Defense of the Triplet Loss for Person Re-Identification (2017). arXiv preprint, arXiv:1703.07737
28. Huang, D.S.: Systematic Theory of Neural Networks for Pattern Recognition, Publishing House of Electronic Industry of China, May 1996

A Brain Tumor Segmentation New Method Based on Statistical Thresholding and Multiscale CNN

Yun Jiang(✉), Jinquan Hou, Xiao Xiao, and Haili Deng

College of Computer Science and Engineering,
Northwest Normal University, Lanzhou 730070, China
jiangyun@nwnu.edu.cn

Abstract. Brain tumor segmentation is crucial in the diagnosis of disease and radiation therapy. However, automatic or semi-automatic segmentation of the brain tumor is still a challenging task due to the high diversities and the ambiguous boundaries in the appearance of tumor tissue. To solve this problem, we propose a brain tumor segmentation method based on Statistical thresholding and Multiscale Convolutional neural networks. Firstly, the statistical threshold segmentation method was used to roughly segment the brain tumor. Then the 2D multi-modality MRI image obtained by the rough segmentation was input into the multiscale convolution neural network (MSCNN) to obtain the tumor segmentation image. Experimental results on the MICCAI BRATS2015 [1] dataset show that the proposed method can significantly improve the segmentation accuracy.

Keywords: Brain tumor MRI image · Image segmentation
Statistical thresholding method · Multiscale convolutional neural networks
Deep learning

1 Introduction

1.1 A Subsection Sample

Among various imaging modalities, MRI shows most of the details of the brain and is one of the most common tests for the diagnosis of brain tumors [2]. MRI data for each patient included data for the four modalities T1-weighted, contrast-enhanced T1-weighted, T2-weighted and T2-FLAIR. For multi-modality MRI images of brain tumors, different modalities of the same patient emphasize different information [3]. Different patients have different tumor locations, sizes and grayscale differences [4]. Therefore, the study of automatic and semi-automatic tumor segmentation algorithm has important value in guiding clinical diagnosis and treatment, which not only can reduce the workload of artificial segmentation, but more importantly can improve the accuracy of tumor segmentation and form accurate preoperative prognosis and intra-operative monitoring, Postoperative evaluation, contribute to the development of a comprehensive surgical treatment program to improve the success rate of brain tumor surgery [5].

D.-S. Huang et al. (Eds.): ICIC 2018, LNAI 10956, pp. 235–245, 2018.
https://doi.org/10.1007/978-3-319-95957-3_26

Currently, brain tumor segmentation methods are mainly based on the atlas-based method, curve/surface evolution method, pixel-based MRI brain tumor segmentation method. Atlas-based methods rely heavily on the registration algorithm and there is no universal registration algorithm to accurately register the target image with the standard image. Therefore, such methods are generally used to provide geometry for subsequent research a priori [6]. The method based on curve/surface evolution is slow when applied to 3D image segmentation, and there are many parameters in it. There is no better way to balance these parameters for different target images [7]. Pixel-based MRI brain tumor segmentation has always been a hot research topic [3]. The main contents of the research are feature extraction, feature selection and classifier design [8]. Feature extraction methods are mainly divided into statistical method, model method and signal processing method, each with its own advantages and disadvantages [9]. The statistical method is simple and easy to implement, it has some advantages for small images, but its use of global information is not enough; the model method has great flexibility because it takes into account the local randomness and overall regularity of the texture; the deficiency is the model; the coefficient is difficult to be solved and the parameter adjustment is inconvenient; the signal processing method is good at capturing the detail information of the texture and can simultaneously present the texture features in the spatial and the frequency domain; the disadvantage is that the high frequency information is ignored and the irregular texture features are not good at extracting [10]. Different modal images of MRI can provide different texture boundary information, due to individual differences, the information displayed by the same modality of different patients is also very different [10]. From the above, no feature extraction method is suitable for all MRI brain tumor segmentation.

The convolutional neural network proposed by Yam LeCun etc. is a type of supervised depth learning method that has achieved great success in many fields such as image recognition, speech recognition, and natural language processing [11]. In the field of image recognition, it has been widely used because the network directly inputs the original image and learns to automatically obtain features that are good for classification, without considering the great differences among the various image features [12]. However, because CNN requires multiple convolutions and downsampling, the input object is usually an image with a large neighborhood value, which is not suitable for feature extraction of MRI brain tumor images with rich and varied details.

Aiming at the problem of brain tumor segmentation, this paper presents a new method of brain tumor segmentation based on statistical threshold algorithm and multiscale convolution neural network. First of all, using statistical threshold segmentation method to rough segmentation of brain tumors, and then the rough segmentation of 2D multi-modality magnetic resonance imaging (MRI) images into multi-scale convolutional neural network original features, through multi-scale input and multi-scale Under the sampling, to overcome the individual differences in brain tumors, brain tumor at the same time to adapt to the size of the differences between the image layer, weakening the edge of the tumor and normal tissue grayscale similar effects. After multilevel learning, the low-level features are transformed into high-level features. Finally, the softmax classifier at the last layer in the multi-scale convolutional neural network is used to classify each pixel in the image into two categories: tumor or normal brain tissue, according to the size of the probability value of the belonging category to obtain the binary image of the tumor and the probability image of the pixel classification. Finally, the morphological post-processing of binary images of tumor segmentation results in the final segmentation results. The test was carried out on the MICCAI BRATS2015 [1] dataset. The test results show that the proposed method can significantly improve the segmentation accuracy and separate the brain tumor accurately and effectively.

2 Related Theory

2.1 Image Preprocessing

During the MRI image acquisition, the image will contain impulse noise due to random disturbance of the instrument equipment, which needs to be filtered before the image is segmented [13]. In this paper, four modal images of brain tumor were preprocessed by median filter. Its basic idea is that a template with odd pixels roams in the figure, and the center of the template coincides with the position of a certain pixel in the picture, and reads the gray value of each corresponding pixel under the template, and changes the gray value from small to large Sort, find the median and give the corresponding template in the center of the pixel [13]. It can effectively filter out random impulse noise, while protecting the edge of the image, making the filtered image contours more clearly [13].

2.2 Rough Segmentation of Brain Tumor Images Based on Statistical Threshold

In MRI images of brain tumors, the gray values of tumor pixels are different from the gray values of other tissue pixels. Based on this characteristic of the image, we use the statistical threshold method to segment the brain tumor MRI image. The method is divided into two processes: First, a non-linear smoothing algorithm is used to smooth the MRI images of brain tumors so as to reduce the influence of noise and artifacts on the images, and then the statistical threshold is used to extract the tumor area [13].

For a given image with L gray levels, let r be a discrete random variable $p(r_i)$ representing discrete grey levels on $[0, 1, \ldots, L-1]$, indicating that the probability estimate for gray level r_i is [14]:

$$P(r_i) = \frac{n_i}{N} \quad (1)$$

Where N represents the total number of pixels in the image and n_i represents the number of gray levels r_i.

Assume that the pixels of the image are divided into two types of C_1 and C_2 by a threshold t^*, where C_1 represents a set of all pixels whose gray levels are in the $[0,\ldots,t^*]$ range, and C_2 represents a gray level greater than t^*. Then, the probability of two classes is [14]:

$$w_1 = \sum_{i=0}^{t*} p(r_i) \tag{2}$$

$$w_2 = \sum_{i=t*+1}^{L-1} p(r_i) \tag{3}$$

Among them, w_1 represents the probability of C_1 class, and w_2 represents the probability of C_2 class.

The average of two classes is:

$$E_{G1} = \sum_{i=0}^{t*} r_i p(r_i)/w_1 \tag{4}$$

$$E_{G2} = \sum_{i=t*+1}^{L-1} r_i p(r_i)/w_2 \tag{5}$$

Among them, E_{G1} represents the average gray value of the C_1 class, and E_{G2} represents the average value of the C_2 class of gray levels.

The variance of the two classes is:

$$D_{G1} = \sum_{i=0}^{t*} (r_i - E_{G1})^2 p(r_i)/w_1 \tag{6}$$

$$D_{G2} = \sum_{i=t*+1}^{L-1} (r_i - E_{G2})^2 p(r_i)/w_2 \tag{7}$$

The threshold selection criteria are:

$$D_{G(t*)} = \max(\mathrm{D}_{G1/w1}, \mathrm{D}_{G2/w2}) \tag{8}$$

The best segmentation threshold is to change $0 \sim L-1$ from t^* so that $D_{G(t*)}$ takes the minimum value. Our goal is to divide the image into two parts that are densely distributed in gray. The variance reflects the grayscale distribution of the pixels. The smaller the value, the more concentrated the grayscale distribution of pixels. Therefore, $D_{G(t*)}$ represents the maximum degree of dispersion of the grayscale distribution class after the threshold segmentation. Minimizing $D_{G(t*)}$ is to make the grayscale distribution of pixels divided into two classes as close as possible.

According to the above method, the rough segmentation of the four modal images of the patient's brain tumor MRI is completed, the results obtained by rough segmentation are selected as training samples, and the training set T is constructed as the original input of the subsequent method.

2.3 Multi-scale Convolutional Neural Network Model

The multi-scale convolutional neural network structure is shown in Fig. 1. The network structure includes a convolutional layer, a downsampling layer, a fully connected layer, and a softmax classification layer [15]. The convolutional layer obtains image features, such as the edges and texture of the tumor, through a convolution operation. For other features, the downsampling layer resamples the acquired feature image, reducing the amount of data processing, while retaining useful feature information, and the full connected layer captures the complex relationship between the output features. Softmax is a multi-class classifier for two types of classification, tumor and normal brain tissue. Its output is a conditional probability value between 0 and 1. The transfer function in the network uses the sigmoid transfer function: $y = f(x) = \frac{1}{1 + e^{-x}}$, convert the output value of the neuron node in the network from 0 to 1. Use the stochastic gradient descent method to minimize the loss function:

$$NULL = (\theta, D) = -\sum_{i=0}^{|D|} \log p(Y = y^{(i)} | x^{(i)}, \theta) \tag{9}$$

find the network parameters to get the network model [5].

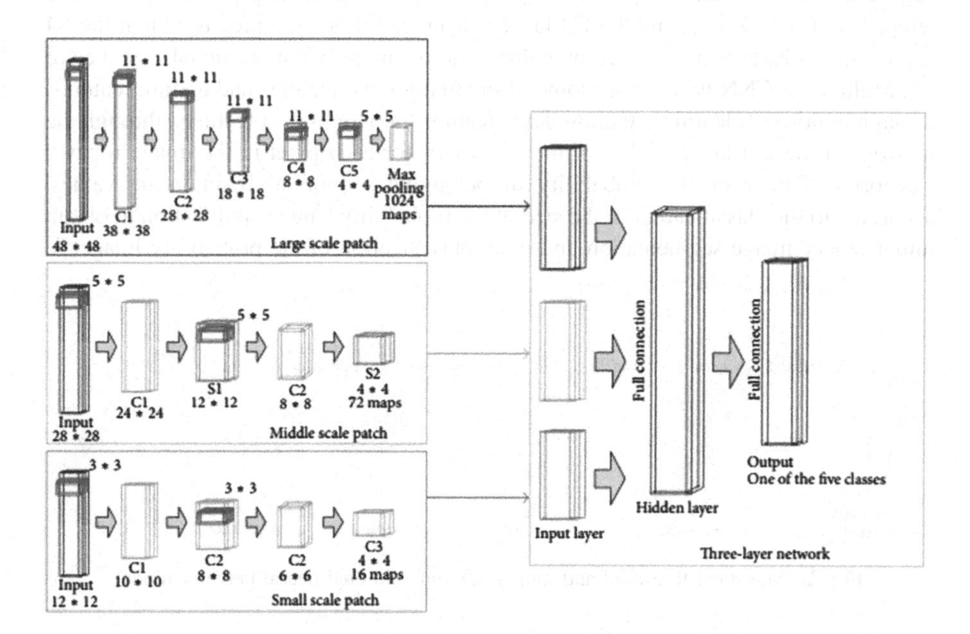

Fig. 1. Multiscale convolution neural network model

3 A Brain Tumor Segmentation New Method Based on Statistical Thresholding and Multiscale CNN

At present, in order to detect brain tumors, four brain MRI modalities are commonly used clinically, including T1-weighted, contrast-enhanced T1-weighted, T2-weighted and T2-FLAIR modalities. In this paper, a multi-scale convolutional neural network is designed, which can input four modal images of MRI in brain tumor at the same time, and select the small neighborhood of the same position of the four modalities (the specific neighborhood size is obtained based on the training data grid optimization) make up the 3D raw input layer. However, due to the changeable shape, complex structure and non-uniform grayscale of brain tumor, the boundary of brain tumor is often associated with edema, and each modality emphasizes different information. The four modal images of brain tumor are directly input into multi-scale convolution after neural network, feature extraction is not good, and it takes a long time. In order to solve this problem, this paper uses the rough segmentation method based on statistical threshold to complete the rough segmentation of four modal images of brain MRI in patients. The result of rough segmentation will be the original input of MSCNN.

As shown in Fig. 2, the small neighborhoods of the four modal image blocks obtained by the rough segmentation, such as 18×18 (the specific neighborhood size is based on the result of the rough segmentation of 2.1), constitute the original input layer of 3D ($18 \times 18 \times 4$), The original input layer is convolved with a convolution template with a size of $3 \times 3 \times 2$ shared by 6 weights to obtain six $16 \times 16 \times 3$ feature maps C1; the feature maps of the six C1 layers are respectively subjected to 2D average and the S2 layer is obtained by downsampling. After all the features of the S2 layer are summed, twelve feature maps C3 of $6 \times 6 \times 2$ are obtained after 12 convolution templates of $3 \times 3 \times 2$; and the C3 layer is averaged down-sampled to obtain the S4 layer. The S4 layer is normalized by columns to obtain a 216-dimensional eigenvector F5. Multi-scale CNN will extract some of the tumor edge features and texture features, through multi-layer learning, the low-level features into high-level features, through the network in the last layer of the softmax classifier for each pixel in the image for two categories, Tumor or the probability of belonging to normal brain tissue values, according to the classification of the size of the probability value classification to obtain tumor binary image segmentation and pixel classification of the probability image.

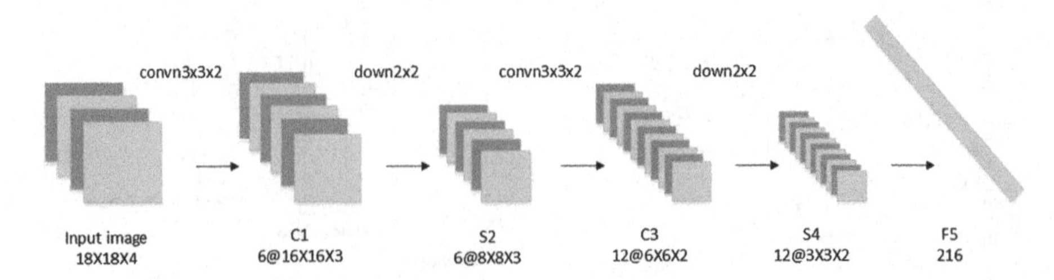

Fig. 2. Statistical threshold and multiscale convolutional neural network model

The multi-scale CNN primitive input layer is composed of four modalities of the rough segmentation in Sect. 2.1. The 3D convolution automatically extracts the differences between the modalities. The supervised learning method can be used to extract different categories for different patient differences features; downsampling makes the feature extraction contain more edge information of the structure, while eliminating redundant information and noise; multi-modal common input makes the original input requires less neighborhood information to adapt to different tumor layers of the image layer to improve brain tumor segmentation accuracy.

In summary, the algorithm of this paper is given.

Algorithm1. Brain Cancer Segmentation Algorithm Based on Statistical Threshold and MSCNN

Input: Multimodal MRI image

Output: Tumor area marker

Pretreatment: Using the method of 2.1, the four modal images of brain tumor MRI were preprocessed

Step1: Using the formula (8) to complete brain tumor MRI four modal image rough segmentation

Step2: Select the result obtained from Step1 as a training sample to construct a training set T

Step3: The original input of a multi-scale convolutional neural network is $M \times N \times 4$ ($18 \times 18 \times 4$ in this paper), where M,N is the size of the original input neighborhood

Step4: The original input layer was convolved with a $3 \times 3 \times 2$ 3D convolution kernel. The feature extraction of brain tumors was achieved through 4 levels of 3D convolution and 2D downsampling

Step5: The extracted feature map is transformed into a single neuron node, and finally it is linked to the fully connected layer F5

Step6: By using the softmax classifier to classify each pixel in the image into two categories, the probability value p that belongs to a brain tumor or a normal tissue is obtained and classified according to the size of the category-specific probability value, thereby obtaining a segmentation binary image of a brain tumor

Step7: The segmentation binary image of the brain tumor undergoes a morphological post-processing operation to obtain the final segmentation result

4 Experimental Results and Analysis

4.1 Data Sources

The data of this experiment is from the competition data set of BRATS2015. Relevant data, available for download from MICCAI 2015 [1], includes 274 patient multimodal training datasets (with standard segmentation data) of 220 high-grade tumors (HGG) and 54 low-grade tumors, and contains 110 test data sets. Each patient's data contains data for the four modalities T1-weighted, contrast-enhanced T1-weighted, T2-weighted and T2-FLAIR, and has been skull delaminated and linearly aligned with a spatial resolution of 1 mm.

When designing the experiment, 80 HGGs and 24 LGGs in the training data were used to train the segmentation model, while the remaining 140 HGGs and 30 LGGs were used to test the segmentation model. The 110 patient data contained in the test data set is used for online assessment.

4.2 Evaluation Index

Dice similarity coefficient, Predictive Positivity Value (PPV) and Sensitivity were used to evaluate the segmentation results. The Dice coefficient is the degree of similarity between the experimental segmentation result and the expert's manual segmentation result. The PPV is the ratio of the segmentation result with the correct segmentation of the tumor to the proportion of the tumor point, and the sensitivity is the ratio of the correct segmentation point to the true value of the tumor point.

4.3 Segmentation Results

Figure 3 shows the segmented data of 10 randomly selected patients from the BRATS2015 training dataset, including 8 HGGs and 2 LGGs. As can be seen from the table, high-grade tumor (HGG) segmentation results were significantly better than low-grade tumors (LGG). This is because the majority of HGG patients with advanced brain cancer patients, glioma grayscale and normal tissue was significantly different, tumor boundary is clear and easy to divide.

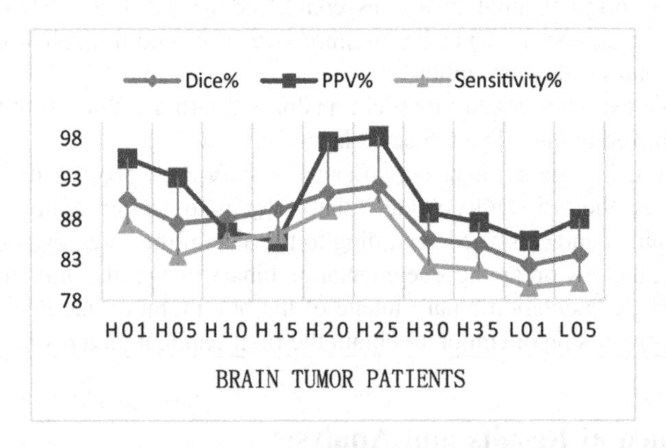

Fig. 3. Segmentation of patients with different brain tumors in training set

Figure 4 shows the segmentation results of five brain tumor patients by statistical threshold method(ST), MSCNN method and our method(ST + MSCNN), including 4 HGGs and 1 LGG. As can be seen from the data in the table, the Dice coefficient, PPV and Sensitivity of this method are higher than the statistical threshold method and the MSCNN method, and the variation range is small. This shows that the proposed method has a better segmentation result and a very stable segmentation result.

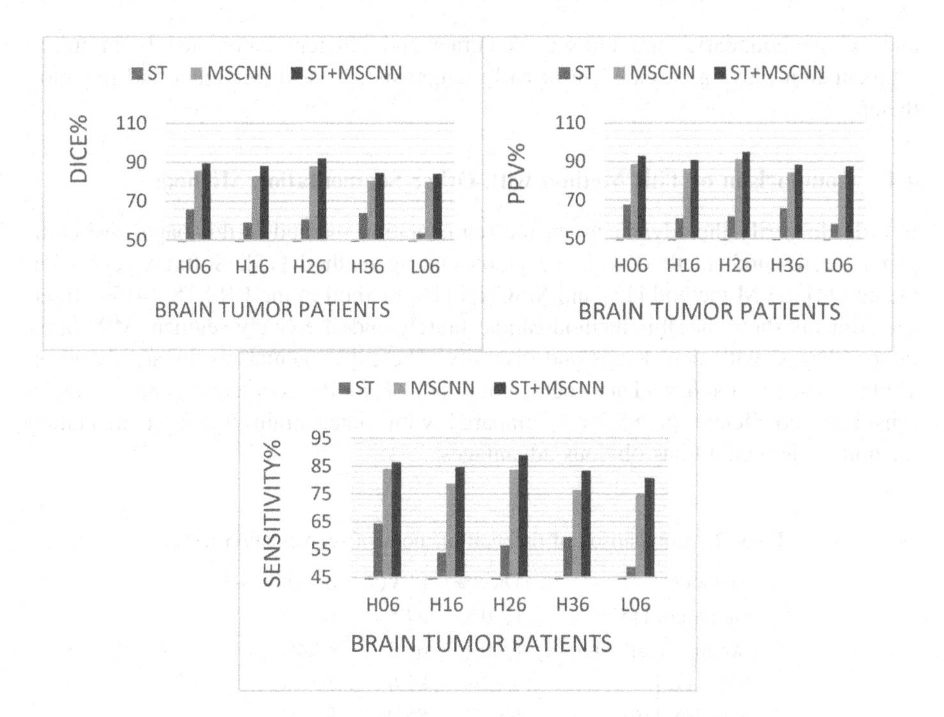

Fig. 4. Comparison of different methods of segmentation results

The results of brain tumor segmentation using this method are shown in Fig. 5 and show the brain tumor segmentation results of three patients, including two HGG and one LGG.

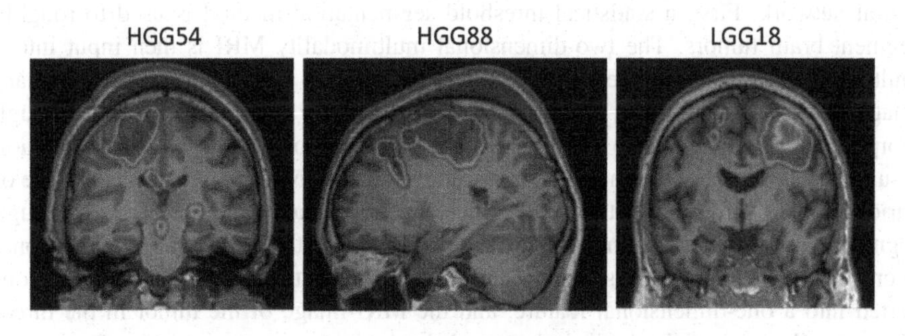

Fig. 5. Different patient segmentation diagram

Figure 5 shows the segmentation results of MRI brain tumors from HGG54, HGG88, and LGG18 patients. Among them, red is the tumor area, and the pale white color in the LGG18 tumor area is the edema area. The blue boundary line is advantageous for distinguishing the tumor area from the non-tumor area. As shown in Fig. 4, the location, size, and gray scale of tumors vary from patient to patient, and the tumor and edema boundaries are blurred. Accurate and efficient automated brain tumor segmentation plays a crucial role in early diagnosis, surgical treatment and radiation therapy.

4.4 Comparison of This Method with Other Segmentation Methods

In order to verify the advantages of the segmentation method in this paper, we compared the research method with the graph cutting method [15], Softmax regression method [4], SVM method [13] and MSCNN [16] method in the BRATS 2015 dataset. Experiments show that this method can accurately and effectively segment MRI brain tumor images, with rich details and diversity. The experimental results are shown in Table 1. The method has a Dice coefficient of 86.3%, a PPV coefficient of 88.6%, and a sensitivity coefficient of 85.2%. Compared with other brain tumor segmentation methods, this method has obvious advantages.

Table 1. Comparison of different methods of segmentation results

Methods	Dice(%)	PPV(%)	Sensitivity(%)
Graph cut [15]	62.9%	64.5%	61.4%
Softmax regression [4]	65.0%	66.8%	63.4%
SVM [13]	79.2%	81.6%	80.5%
MSCNN [16]	83.3%	85.9%	82.6%
Our Method	86.3%	88.6%	85.2%

5 Conclusion

This paper combines the statistical threshold algorithm with multi-scale convolutional neural network. First, a statistical threshold segmentation method is used to roughly segment brain tumors. The two-dimensional multimodality MRI is then input into a multi-scale convolutional neural network (MSCNN) to obtain a segmented binary image of the tumor. Finally, the tumor segmented binary image is processed through morphological operations to obtain the final segmentation result. This segmentation result makes full use of the multi-modality information of MRI, and makes better use of various modal information to achieve more accurate classification. After the rough segmentation of the brain tumor MRI image is input into the multi-scale convolutional neural network through the statistical threshold method, the 3D image is finally converted into a one-dimensional feature, and the MRI image of the tumor in the three-dimensional space of the brain is lost, and the brain tumor cannot be perfectly segmented. In the following studies, how to maintain the three-dimensional spatial characteristics of brain tumor MRI images will become the focus of attention to further improve the accuracy of brain tumor segmentation.

References

1. www.isles-challenge.org/ISLES2015/
2. Jiang, J., Wu, Y., Huang, M.: 3D brain tumor segmentation in multimodal MR images based on learning population and patient-specific feature sets. Comput. Med. Imaging Graph. **37** (7–8), 512–521 (2013)
3. Gordillo, N., Montseny, E., Sobrevilla, P.: State of the art survey on MRI brain tumor segmentation. Magn. Reson. Imaging **31**(8), 1426–1438 (2013)
4. Kamnitsas, K., Ledig, C., Newcombe, V.F.: Efficient multi-scale 3D CNN with fully connected CRF for accurate brain lesion segmentation. Med. Image Anal. **36**, 61 (2017)
5. Li, W., Jia, F., Hu, Q.: Automatic segmentation of liver tumor in CT images with deep convolutional neural networks. J. Comput. Commun. **03**(11), 146–151 (2015)
6. Conson, M., Cella, L., Pacelli, R.: Automated delineation of brain structures in patients undergoing radiotherapy for primary brain tumors: From atlas to dose-volume histograms. Radiother. Oncol. J. Eur. Soc. Ther. Radiology Oncol. **112**(3), 326–331 (2014)
7. Rajendran, A., Dhanasekaran, R.: Brain tumor segmentation on MRI brain images with fuzzy clustering and GVF snake model. Int. J. Comput. Commun. Control **7**(3), 530 (2014)
8. Schröter, D., Weber, T., Beetz, M., Radig, B.: Detection and classification of gateways for the acquisition of structured robot maps. In: Rasmussen, C.E., Bülthoff, H.H., Schölkopf, B., Giese, M.A. (eds.) DAGM 2004. LNCS, vol. 3175, pp. 553–561. Springer, Heidelberg (2004). https://doi.org/10.1007/978-3-540-28649-3_68
9. Li, L., Kuang, G.Y.: Overview of image textural feature extraction methods. J. Image Graph (2009)
10. Udayasankar, U.: Magnetic resonance imaging of the brain and spine, 5th edn. J. Magn. Reson. Imaging JMRI (2017)
11. Liu, W., Yu, Z., Lu, L.: KCRC-LCD: Discriminative kernel collaborative representation with locality constrained dictionary for visual categorization. Pattern Recogn. **48**(10), 3076–3092 (2015)
12. Ting, G.E., Ning, M.U., Li, L.I.: A brain tumor segmentation method based on softmax regression and graph cut. Acta Electronica Sinica **45**(3), 644–649 (2017)
13. Sang, L., Qiu, M., Wang, L.: Brain tumor MRI image segmentation based on statistical thresholding method. J. Biomed. Eng. Res. (2010)
14. Otsu, N.: A threshold selection method from gray-level histograms. IEEE Trans. Syst. Man Cybern. **9**(1), 62–66 (2007)
15. Aju, D., Rajkumar, R.: T1-T2 weighted MR image composition and cataloguing of brain tumor using regularized logistic regression. Jurnal Teknologi **78**(9), 149–159 (2016)
16. Zhao, L., Jia, K.: Multiscale CNNs for brain tumor segmentation and diagnosis. Comput. Math. Methods Med. **2016**(7), 1–7 (2016)

An Improved Supervised Isomap Method Using Adaptive Parameters

Zhongkai Feng[1,2(✉)] and Bo Li[1,2]

[1] College of Computer Science and Technology, Wuhan University of Sciences and Technology, Wuhan 430065, China
314788324@qq.com
[2] Hubei Province Key Laboratory of Intelligent Information Processing and Real-Time Industrial System, Wuhan University of Sciences and Technology, Wuhan 430065, China

Abstract. In this paper, to deal with the problem that nearest neighbor graph is hard to be connected in the original Isomap, a new supervised Isomap method (SS-Isomap) with adaptive parameters is proposed. This method considers the density of intra-class data points and proposes an adaptive function. The experiment results based on UCI datasets show that SS-Isomap has a better discriminant ability.

Keywords: Manifold · Density of distribution · Supervised Isomap Self-adaption · Discriminant ability

1 Introduction

In machine learning, data sparsity and difficulty of distance calculation in high dimensionality will badly influence the effect and currently of machine learning algorithms. This problem is named as "the curse of dimensionality". An effective method to solve dimension problem is dimension reduction, which is projecting the original data of high dimensionality into a lower dimensional space using some mathematical transformations [1]. After this, the density of data will be improved and distance calculation will be simplified.

Over the past decade, many classical approaches have been proposed for dimension reduction, such as PCA [2], MDS [3]. However, in reality, there are many nonlinear data which can't be projected well into lower dimensional space by PCA or MDS. Hence, some approaches based on manifold have been proposed, such as Isomap [4] and LLE [5]. For visualization, Isomap and LLE perform well. However, for classification, unsupervised algorithms Isomap and LLE don't use original class information, which result in their bad performance in classification task. To improve algorithm's classification ability, many supervised manifold methods have been proposed, such as Weightediso [6], S-Isomap [7], SLLE [8] and so on. These methods achieve dimension reduction with using class information of data. Weightediso takes class information of data into account for the first time. This algorithm alters the first step of Isomap by shortening the distances of intra-class data points, which improves Weightediso's accuracy in classification task. However, Weightediso forcibly changes original

D.-S. Huang et al. (Eds.): ICIC 2018, LNAI 10956, pp. 246–252, 2018.
https://doi.org/10.1007/978-3-319-95957-3_27

geometry of data points, that results in bad performance in visualization and weak robust. Hence, Geng [7] proposed S-Isomap that could perform well in classification and visualization tasks and have better robust. S-Isomap defines dis-similarity. This concept can keep the original geometry of data points, meanwhile, it can adjust the weight of the edges. In fact, it shortens weight of the intra-class points' edge and increases weight of the inter-class points' edge. This makes S-Isomap has stronger resistance to noise [9]. The algorithm based on Isomap is still a hot topic in recent years [11, 12].

Although S-Isomap is an effective supervised Isomap algorithm, it still has some shortcomings. The set of Isomap algorithms have a common problem that is how to choose the value of K parameter. Inappropriate value of K would result in "open circuit" of the neighborhood graph. However, S-Isomap usually improves the risk of "open circuit", because the dis-similarity defined by S-Isomap gathers intra-class points and disperses inter-class points. In fact, it should be indeed carefully to adjust the value of K to get a reasonable result when we use S-Isomap. One of the important reasons is that S-Isomap only considers the density of the whole data points, but don't consider the difference of density between the whole data points and intra-class data points. In this paper, based on the idea of S-Isomap, a new method using adaptive parameters is proposed to deal with such situation, namely SS-Isomap. SS-Isomap designs an adaptive parameter adjustment function to measure the difference of density between the whole data points and intra class points, to improve discriminant ability.

2 Related Work

2.1 Isomap

Isomap is a classical manifold algorithm for dimension reduction. It was proposed to solve the problem of nonlinear dimensional reduction. In Isomap, geodesic of two points is used to replace the straight-line distance, because geodesic can better express the true distance of two points in the same manifold. The steps of algorithm are as follow:

- Firstly, calculate Euclidean distance of any two points and use K-NN to construct the k-neighborhood graph G.
- Secondly, use Dijkstra or Floyd to measure the shortest path length of any two points as geodesic distance (namely dist) of the two points.
- Finally, use dist as element of distance matrix D and use matrix D as the input of MDS to get output.

MDS can be seen as an Eigen decomposition progress. Its input usually is a distance matrix D, which don't change after dimension reduction. We can get inner product matrix $B = Z^T Z$ of data points in lower dimensional space. Decompose the eigenvalues of the metric B, we get $B = U \Lambda U^T$. Then the data matrix is represented as $Z = \Lambda^{\frac{1}{2}} U^T$. Choose the d-th biggest eigenvalue to construct diagonal matrix Λ^*. The corresponding Eigenvector matrix is U^*. Data points in lower dimensional space is represented as a matrix $Z = (\Lambda^*)^{\frac{1}{2}} (U^*)^T$, whose one row vector represents a data point.

2.2 S-Isomap

S-Isomap uses class information of original data to improve the discriminant ability of Isomap, meanwhile, it also improves the robust. In S-Isomap, a new concept dis-similarity is defined to reconstruct distance matrix. Dis-similarity between two points is defined as follow.

$$D(x_i, x_j) = \begin{cases} \sqrt{1 - e^{\frac{-d^2(x_i, x_j)}{\beta}}} & y_i = y_j \\ \sqrt{e^{\frac{-d^2(x_i, x_j)}{\beta}}} - \alpha & y_i \neq y_j \end{cases} \quad (1)$$

In formula (1), $d^2(x_i, x_j)$ represents the square of Euclidean distance between data point x_i and x_j. The y_i and y_j represent two data points' classes, respectively. To prevent index from increasing rapidly, $d^2(x_i, x_j)$ must be divided by β whose value usually is set to mean of the whole data points' distances. The meaning of α is to ensure that dis-similarity of inter-class data points is possible less than dis-similarity of intra-class data points. S-Isomap also has three steps. Firstly, calculate dis-similarities between any two points and use these as the edges' weight to construct k-neighborhood graph G. Second step and third step are same as Isomap. Use Dijkstra to measure the shortest path length of any two points and use MDS to get the final result.

The main improvement of S-Isomap is that it considers class information and defines dis-similarity to reconstruct distance matrix. From the definition of dis-similarity, we can see that intra-class data points' dis-similarities are less than 1, while inter-class data points' dis-similarities are larger than $1 - \alpha$. Hence, the weight of intra-class data points' edges is usually less than the weight of inter-class data points' edges. This makes intra-class data points gathered and inter-class data points dispersed. The other use of α is relaxing the requirement to make construction of neighborhood graph easier. Compared to Isomap, S-Isomap has three advantages. Firstly, S-Isomap has better performance in classification task by considering class information. Secondly, S-Isomap can effectively control the influence of noise. Thirdly, S-Isomap can increase the inter-class data points' geodesic distances and shorten the intra-class data points' geodesic distances at the same time. This makes that the most data points of one region belong to the same class in neighborhood graph, which is beneficial to classification.

3 S-Isomap Using Adaptive Parameter Adjustment Function

S-Isomap has some shortcomings. Firstly, the function in formula (1) is too strict. In experiments, "open circuit" often takes place, though the parameter α can adjust the value of dis-similarity. This is because that the function is exponential. When intra-class data points are too gathered and inter-class data points are too dispersed, the weight of intra-class data points' edges and inter-class data points' edges have big gap. Secondly, the value of parameter β in S-Isomap should be decided by the density of dataset. But in [7], β is only assigned as mean of the whole data points' distances, ignoring the difference between the intra-class data and inter-class data. When the density of intra-class data is much different from the density of the whole data, β can't reflect this relationship between different class data.

Therefore, in this paper, an adaptive function is proposed. This function can adaptively adjust the value of β by considering the density of intra-class data points, to reflect the dis-similarity of intra-class data points better. This step can make it easier to construct neighborhood graph and improve the discriminant ability [10].

3.1 Adaptive Parameter Adjustment Function

In a dataset with label $X = \{x_1, x_2, \ldots, x_n\}$, $Y = \{y_1, y_2, \ldots y_n\}$, $L = \{l_1, l_2, \ldots l_c\}$, the x_i represents i-th sample and y_i represents this sample's class and L represents the set of classes. The adaptive function is as follow:

$$F(l) = \frac{\sum_{y_i \in L_l}^{n} \sum_{y_j \in L_l}^{n} d^2_{(x_i, x_j)}}{N * (N-1)} \quad (l = 1, 2, \ldots, c \text{ and } i \neq j) \tag{2}$$

In formula (2), y_j represents the j-th sample's class and L_l is the lth class. The total number of one class is N. $d^2_{(x_i, x_j)}$ represents the square of Euclidean distance between x_i and x_j.

Adaptive parameter adjustment function calculates the sum of the distance of any two data points which belong to the same class. Then the sum divides the square of the number of this class of data point. The distance between one data points to itself is zero and this distance will be not added to the sum of the distances. Because it is convenient to use the square of distance, we don't extract the root.

3.2 SS-Isomap

In order to better reflect the position relationship between the sample points and to get a distance matrix that is easy to construct the neighborhood graph, we replace the parameter β in formula (1) with the adaptive parameter adjustment function. The definition of improved dis-similarity is as follows:

$$D'(x_i, x_j) = \begin{cases} \sqrt{1 - e^{\frac{-d^2(x_i, x_j)}{F(l)}}} & y_i = y_j \text{ and } y_i, y_j \in L_l \\ \sqrt{e^{\frac{-d^2(x_i, x_j)}{\beta}}} - \alpha & y_i \neq y_j \end{cases} \tag{3}$$

In formula (3), $F(l)$ is the adaptive parameter adjustment function and β is the mean square distance of all sample points. L_l represents the l-th category. After adding the adaptive parameter adjustment function, the dis-similarity of the same class sample points can be more evenly dispersed in the range [0, 1]. While reducing the weight of edges between points of the class, it also makes the construction of the neighbor graph easier. Such improvements also make SS-Isomap obtain a better distance matrix than S-Isomap, and gain advantage in the low-dimensional coordinate representation of the data points after the Eigen decomposition.

SS-Isomap also has three steps. Firstly, calculate $D^{'}(x_i, x_j)$ for any two data points and construct K neighborhood graph using K-NN. Secondly, use Dijkstra to measure the shortest path length of any two points. Thirdly, use MDS to get the final result.

4 Experiments

This paper compares the improved method with S-Isomap, Isomap, and J4.8 on classification task. Isomap is a Global algorithm, therefore, it can't get a low dimensional projection for the new sample points. A simple way is to add these points to the original dataset and run the algorithm on the new dataset. But it is truly wasting time and effort. Hence, we adopt another method mentioned in [7], using the generalized regression neural network to get the low-dimensional projection of new data points.

The experiments have three steps. Firstly, the training samples are respectively mapped to low dimensions by using different dimension reduction methods. Secondly, construct a GRN to approximate the mapping. Thirdly, map the given query using the GRN and then predict its class label using K-NN. In the experiments, the α in SS-Isomap and S-Isomap is set in the range [0.25, 0.5], and the K in three methods is set in the range [10, 40]. For all datasets, the dimensionality of the data is reduced to half of the original.

Experiments were conducted on five datasets. Information of all the datasets is summarized in Table 1. Instances that have missing values are removed. On each dataset, ten times ten-fold cross validation is run. The experiments result is in Table 2. After getting the results, the pair wise one-tailed t-test is performed on the results of SS-Isomap paired with every other algorithm at the significance level 0.01 and the test results are in Table 3.

Table 1. Data set used in experiments.

Abbr	Data set name	Size	Classes	Attributes
liv	Liver-disorders	345	2	6
iris	Iris	150	3	4
bre	Breast-w	683	2	9
wine	Wine	178	3	13
glass	Glass	214	7	9

Table 2 reveals that SS-Isomap gives the best performance on 3 datasets, and its average performance is the best. From Table 3, compared to Isomap, SS-Isomap has better performance in all 5 datasets. Compared to S-Isomap, SS-Isomap also have improvement in most datasets. For Iris and Bre datasets, SS-Isomap are better than other methods.

From the experimental results, the classification effect of SS-Isomap on the Glass dataset is worse than S-Isomap. The reason may be that there are too much categories of Glass dataset and few samples, resulting in no obvious difference between the means of the samples of the same class and the means of the total samples. And the same class

Table 2. Experiments result.

Data set	SS-Isomap	S-Isomap	Isomap	J4.8
liv	0.6590	0.6337	0.5690	0.6656
iris	0.9733	0.9600	0.9293	0.9447
bre	0.9730	0.9620	0.9561	0.9330
wine	0.7667	0.7667	0.7556	0.7316
glass	0.6789	0.7021	0.6163	0.6796
Avg.	0.8102	0.8049	0.7653	0.7909

Table 3. Test results.

Data set	S-Isomap	Isomap	J4.8	
liv	1	1	0	0.67
iris	1	1	1	1
bre	1	1	1	1
wine	0	1	1	0.67
glass	−1	1	0	0
Avg.	0.4	1	0.6	

of sample data is too small, the use of adaptive parameters to adjust the parameters derived parameters can't fully reflect the information of density. These result in SS-Isomap disadvantage in Glass dataset.

5 Conclusion

In this paper, an improved version of S-Isomap using adaptive parameters, namely SS-Isomap, to solve the problem that S-Isomap didn't consider the data density of difference categories. To test the performance of SS-Isomap, we experiment in 5 datasets. The experiment results reveal that SS-Isomap has improvement compared with S-Isomap in classification. And SS-Isomap isn't so sensitive to the value of parameter K. However, in some unsuitable data sets, the promotion of S-Isomap isn't so large. The direction of further research will be how to modify the adaptive function to better use the information of data density and how to increase the generalization performance of the algorithm.

Acknowledgment. This work was supported by the National Natural Science Foundation of China (Grant no. 61273303 and no. 61572381). The authors would like to thank all the editors and reviewers for their valuable comments and suggestions.

References

1. Zhou, Z.-H.: Machine Learning, pp. 226–227. Tsinghua University Press, Beijing (2016). (in Chinese)
2. Comon, P.: Independent component analysis: a new concept? Signal Process. **36**(3), 287–314 (1994)
3. Cox, T.F., Cox, M.A.A.: Multidimensional Scaling. Chapman & Hall, Boca Raton (2001)
4. Tenenbaum, J.B., de Silva, V., Langford, J.C.: A global geometric framework for nonlinear dimensionality reduction. Science **290**(5500), 2319–2323 (2000)
5. Roweis, S.T., Saul, L.K.: Nonlinear dimensionality reduction by locally linear embedding. Science **290**(5500), 2323–2326 (2000)
6. Vlachos, M., Domeniconi, C., Gunopulos, D., Kollios, G., Koudas, N.: Non-linear dimensionality reduction techniques for classification and visualization. In: Proceedings of the 8th ACM SIGKDD International Conference on Knowledge Discovery and Data Mining, Edmonton Canada, pp. 645–651 (2002)
7. Geng, X., Zhan, D.-C., Zhou, Z.-H.: Supervised nonlinear dimensionality reduction for visualization and classification. IEEE Trans. Syst. Man Cybern. Part B Cybern. **35**(6), 1098–1107 (2005)
8. de Ridder, D., Kouropteva, O., Okun, O., Pietikäinen, M., Duin, R.P.W.: Supervised locally linear embedding. In: Kaynak, O., Alpaydin, E., Oja, E., Xu, L. (eds.) ICANN/ICONIP 2003. LNCS, vol. 2714, pp. 333–341. Springer, Heidelberg (2003). https://doi.org/10.1007/3-540-44989-2_40
9. Feng, H.: Application of manifold learning algorithm in face recognition. College of Optoelectronic Engineering, Chongqing University, Chongqing (2008). (in Chinese)
10. Liu, Q., Pan, C.: Face recognition based on supervised incremental isometric mapping. J. Comput. Appl. **30**(12), 3314–3316 (2010). (in Chinese)
11. Li, X., Cai, C., He, J.: Density-based multi-manifold ISOMAP for data classification. In: Asia-Pacific Signal and Information Processing Association Annual Summit and Conference (APSIPA ASC) (2017)
12. Zhou, Y., Liu, C., Li, N.: An Isomap-based kernerl-Knn classifier for hyperspectral data analysis. In: Hyperspectral Image and Signal Processing: Evolution in Remote Sensing (WHISPERS) (2015)

Robust Face Recognition Based on Supervised Sparse Representation

Jian-Xun Mi[1(✉)], Yueru Sun[1], and Jia Lu[2]

[1] College of Computer Science and Technology, Chongqing University of Posts and Telecommunications, Chongqing 400065, China
mijianxun@gmail.com

[2] College of Computer and Information Sciences, Chongqing Normal University, Chongqing 401331, China

Abstract. Sparse representation-based classification (SRC) has become a popular methodology in face recognition in recent years. One widely used manner is to enforce minimum l_1-norm on coding coefficient vector, which requires high computational cost. On the other hand, supervised sparse representation-based method (SSR) realizes sparse representation classification with higher efficiency by representing a probe using multiple phases. Nevertheless, since previous SSR methods only deal with Gaussian noise, they cannot satisfy empirical robust face recognition application. In this paper, we propose a robust supervised sparse representation (RSSR) model, which uses a two-phase scheme of robust representation to compute a sparse coding vector. To solve the model of RSSR efficiently, an algorithm based on iterative reweighting is proposed. We compare the RSSR with other state-of-the-art methods and the experimental results demonstrate that RSSR obtains competitive performance.

Keywords: Face recognition · Huber loss · Supervised sparse representation

1 Introduction

In recent years, face recognition (FR) has been studied for its broad application prospects, such as authentication and payment system. The primary task of FR consists of feature extraction and classification. Various changes including lighting, expression, pose, and occlusion can be seen in probes, which leads to that the computed feature becomes inefficient. Although facial images contain some redundant information, it is not easy to utilize the redundancy that retains discriminative information when the probe is interfered by variations. In this case, the key of a FR system is to develop a classifier making use of the most discriminative information in a probe even though containing variations and corruption.

To this end, methods based on linear regression (LR) are proposed recently, which show promising results in FR [1]. The basic assumption behind the methods is that a probe can be linearly represented by training samples from the same class of the probe. The decision of classification is in favor of the prediction which has minimum match residual. There are two typical approaches to achieve sparse representation of a probe. First, the unsupervised sparse representation(USR) incorporates collaborative representation and

D.-S. Huang et al. (Eds.): ICIC 2018, LNAI 10956, pp. 253–259, 2018.
https://doi.org/10.1007/978-3-319-95957-3_28

sparse representation together into a single optimization and the sparsity of the solution is guaranteed automatically. Inspired by theory of compressed sensing, USR formulizes sparse representation as the following optimization problem:

$$\mathbf{x}^* = \arg\min g(\mathbf{e}) + \lambda\|\mathbf{x}\|_1 \tag{1}$$

where $\mathbf{e} = \mathbf{y} - \mathbf{A}\mathbf{x} = [e_1, e_2, \cdots, e_m]$ called the residual vector, and $g(\cdot)$ is an error function which uses the squared loss $\|\cdot\|_2$, and λ is a small positive constant balancing between the loss function and the regularization term, which is called *Lasso* regression. The second approach is the SSR which uses collaborative representation itself to supervise the sparsity of representation vector with l_2-norm to regularize representation vector:

$$\mathbf{x}^* = \arg\min\|\mathbf{e}\|_2 + \lambda\|\mathbf{x}\|_2 \tag{2}$$

In this way, the solution can be computed with much lower time cost than l_1-norm based regression model, which uses a two-phase representation scheme to recognize a probe. A representative method of this kind is TPTSR [2].

In this paper, we propose a robust supervised sparse representation (RSSR) method for FR in which the Huber loss is used as fidelity term since it is superior to squared loss, when there are large outliers. To implement supervised sparse representation, the RSSR uses two-phase representation scheme. We solve the linear regression model of the RSSR by using the iteratively reweighted least squares minimization. The computational cost to recognize a probe is satisfactory.

In Sect. 2, we present the RSSR model and the classification algorithm. In Sect. 3, we validate and compare the RSSR with other related state-of-the-art methods on a public available face database.

2 Robust Coding Based Supervised Sparse Representation

2.1 Robust Regression Based on Huber Loss

In a practice, face recognition system has to deal with images contaminated with occlusion and corruption. Since a portion of pixels of the probe image is randomly distorted, the distribution of the residual $\mathbf{e}$ becomes heavy tailed. Commonly used loss functions which are less sensitive to outliers involve the absolute value function. Here we choose Huber loss which is defined as:

$$g(e) = \begin{cases} \frac{1}{2}e^2 & \text{if } |e| \le k \\ k|e| - \frac{1}{2}k^2 & \text{if } |e| > k \end{cases} \tag{3}$$

where k is a constant. As defined, Huber loss is a parabola in the vicinity of zero and increases linearly above a given level $|e| > k$. Therefore, in the paper we develop a linear representation-based FR method using Huber loss as the representation residual measure function $G(\mathbf{e}) = \sum_i g(e_i)$ where $e_i = y_i - d_i\mathbf{x}$, $y_i \in \mathbf{y}$ and d_i is the i th row of matrix $\mathbf{A}$. Here, we denote the first order derivative function of the Huber loss as

$\psi(p) = dg(p)/dp$ called the influence function which describes the extent that how error affects the cost function. Next, we use the weight function $\omega(p) = \psi(p)/p$ which assigns a specific weight to a certain pixel. The weight function of Huber loss is piecewise function that $\omega(e_i) = 1$ when $|e_i| \leq k$ and $\omega(e_i) = 1/|e_i|$ when $|e_i| > k$. It is easy to see that if a pixel has a bigger value of residual (i.e. $|e_i| > k$) it will be assigned with a smaller weight (i.e. $1/|e_i|$).

Although Huber loss and the l_1-norm fidelity employ a similar weighting function to lower the influence of outliers in a probe, their weighting functions still have large difference which is shown in Fig. 1. One can see that the assigned weight by the l_1-norm fidelity can be infinity when the residual approaches to zero, making the coding unstable. However, Huber loss views small residuals as thermal noises, and its weighting strategy is same as the l_2-norm fidelity, i.e. $\omega(e_i) = 1$.

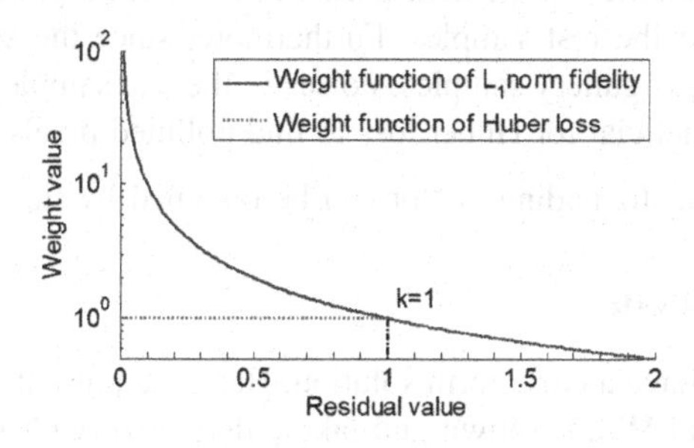

Fig. 1. Weight functions of the l_1-norm VS. Huber loss.

Huber loss is able to be transformed to the regression form of iterated reweighted-least-squares as:

$$\min \sum\nolimits_i w_{ii} e_i^2 \tag{4}$$

where w_{ii} is the weight of the i th pixel in $\bar{\mathbf{y}}$.

2.2 The Proposed Model

Our RSSR needs a two-phase representation of a probe. In the first phase, a probe is represented collaboratively and the corresponding coding vector is computed by solving the following problem:

$$\mathbf{x} = \arg\min_{\mathbf{x}} \; G(\mathbf{y} - \mathbf{A}\mathbf{x}) + \lambda \|\mathbf{x}\|_2^2 / 2 \tag{5}$$

where λ is a balance parameter. Since the Huber loss is a piecewise function, we need to transform Eq. (5) so as to facilitate the computation. First of all, we define a sign vector $\mathbf{s} = [s_1, s_2, \cdots, s_j, \cdots, s_m]^T$ where $s_j = -1$ if $e_j < -k$, $s_j = 0$ if $|e_j| \leq k$, and $s_j = -1$ if $e_j > k$ and define a diagonal matrix $\mathbf{W} \in \mathbb{R}^{m \times m}$ with diagonal elements $w_{jj} = 1 - s_j^2$. Now, by introducing $\mathbf{W}$ and $\mathbf{s}$, $G(\mathbf{e})$ can be rewritten as

$$G(\mathbf{e}, \mathbf{s}, \mathbf{W}) = \frac{1}{2}\mathbf{e}^{\mathrm{T}}\mathbf{W}\mathbf{e} + k\mathbf{s}^{\mathrm{T}}(\mathbf{e} - \frac{1}{2}k\mathbf{s}) \tag{6}$$

If all residuals are smaller than k, the second term of the right side of Eq. (6) will disappear and the fidelity term degrades to quadratic function. Actually, in Eq. (6), the entries of $\mathbf{s}$ can be considered as the indicator for large residuals, and which divide the residuals into two groups, i.e., the small residuals and the large ones. By substituting the fidelity term of Eq. (5) with Eq. (6), the optimization function now is reformed as

$$\widehat{\mathbf{x}} = \arg\min_{\mathbf{x}} \frac{1}{2}\mathbf{e}^{\mathrm{T}}\mathbf{W}\mathbf{e} + k\mathbf{s}^{\mathrm{T}}(\mathbf{e} - \frac{1}{2}k\mathbf{s}) + \frac{\lambda}{2}\|\mathbf{x}\|_2^2 \tag{7}$$

where $\widehat{\mathbf{x}}$ is estimated coding vector. By selection scheme used in TPTSR, M per cent training samples are retained in the next phase. In the second phase, the test sample is represented anew over the rest samples. Furthermore, since the second phase of representation involves less gallery samples, noises in the test sample are less likely to be compensated. It is beneficial for Huber loss to find polluted pixels. Moreover, the final Huber loss $\widehat{\mathbf{x}}' \in \mathbb{R}^M$ of the coding vector can be obtained by Eq. (7) again.

2.3 Algorithm of RSSR

The Eq. (7) does not have a close form solution. Here, we present an iterative solution. First, we assume $\mathbf{s}$ and $\mathbf{W}$ are known and take a derivative with respective to $\mathbf{x}$:

$$\mathbf{A}^T\mathbf{W}\mathbf{A}\mathbf{x} - \mathbf{A}^T\mathbf{W}\mathbf{y} + k\mathbf{A}^T\mathbf{s} + \lambda\mathbf{x} = 0 \tag{8}$$

The coding vector is obtained by:

$$\mathbf{x} = (\mathbf{A}^T\mathbf{W}\mathbf{A} + \lambda\mathbf{I})^{-1}(\mathbf{A}^T\mathbf{W}\mathbf{y} - k\mathbf{A}^T\mathbf{s}) \tag{9}$$

where $\mathbf{I}$ is an identity matrix. Then, we compute $\mathbf{s}$ and $\mathbf{W}$ respectively. And the formula of updating $\mathbf{x}$ in the t-th iteration is

$$\mathbf{x}^t = \mathbf{x}^{t-1} + u^t \mathrm{h}(\mathbf{W}^t, \mathbf{s}^t) \tag{10}$$

where $h(\mathbf{W}^t, \mathbf{s}^t) = (\mathbf{A}^T\mathbf{W}^t\mathbf{A} + \lambda\mathbf{I})^{-1}(\mathbf{A}^T\mathbf{W}^t\mathbf{y} - k\mathbf{A}^T\mathbf{s}^t) - \mathbf{x}^{t-1}$, and $0 < u^t \leq 1$ which is a step size that makes the loss value of the t-th iteration lower than that of the last. In this paper, u^t is determined via the golden section search if $t > 1$ ($u^1 = 1$). Following the representation scheme of SSR, we will use the above algorithm twice.

After converge of computing the coding vector of a testing sample, we use a typical rule of collaborative representation to make the final decision:

$$indentity(\mathbf{y}) = \arg\min_{c} d_c \tag{11}$$

where $d_c = G(\mathbf{y} - \tilde{\mathbf{A}}_c\hat{\mathbf{x}}_c, \hat{\mathbf{s}}, \hat{\mathbf{W}}) / \|\hat{\mathbf{x}}_c\|_2$ ($\tilde{\mathbf{A}}_c$ is a sub-dictionary that contains samples of class c, and $\mathbf{x}_c$ is the associated coding vector of class c).

3 Experiment Results

In this section, we compare RSSR with nine state-of-the-art methods, including ESRC [1, 3], CESR [4], RRC1, RRC2 [5], CRC [6], LRC [7], TPTSR (for TPTSR the candidate set is set to 10 percent of training set), NMR [8], and ProCRC [9]. In the following experiments, for the three parameters in RSSR, the value of k, M, and λ are set to 0.001, 10%, and 0.001 respectively.

In the experiments, we verified the robustness of RSSR against to two typical facial image contaminations, namely random pixel corruption and random block occlusion.

FR with Pixel Corruption. As in [5], we tested the performance of RSSR in FR with random pixel corruption on the Extended Yale B face database. We used only samples from subset 1 as training sample, and testing samples were from subset 3. On each testing image, a certain percentage of randomly selected pixels were replaced by corruption pixels whose values were uniformly chosen from 0 or twice as the biggest pixel value of the testing image.

In Fig. 2 the recognition rates versus different percentages of corrupted pixels are plotted. The performance of CESR under illumination conditions is the lowest when the corruption is not serious, but it drops very late, which shows its good robustness but poor accuracy. For our RSSR, the curve of recognition rate keeps straight at 100% correct rate and only starts to bend until 70% pixels are corrupted which indicates the best robustness of RSSR among all the compared methods.

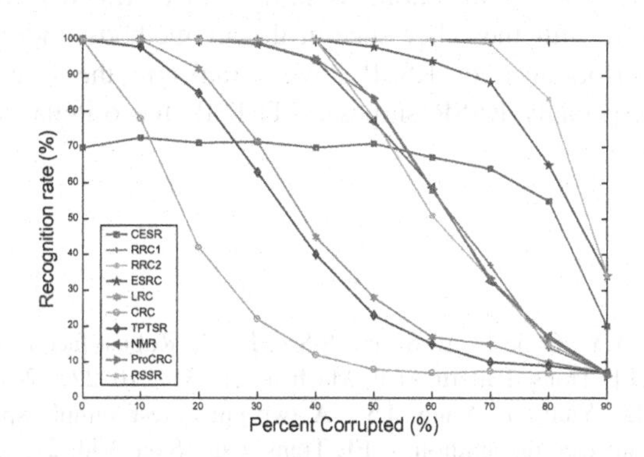

Fig. 2. The comparison of robustness of several algorithms against random pixel corruptions.

FR with Block Occlusion. To compare RSSR with other methods, we plotted, in Fig. 3, the recognition rates versus sizes of the occlusion from 0% to 50%. With increase of size of the occlusion, the recognition rates of LRC, CRC and TPTSR decreased sharply, while the recognition rates of RRC1, RRC2, ESRC, and RSSR maintained 100% up to 20% occlusion. From 30% to 50%, occlusion the recognition rates of RSSR, RRC1, NMR and RRC2 preform more stable than ESRC.

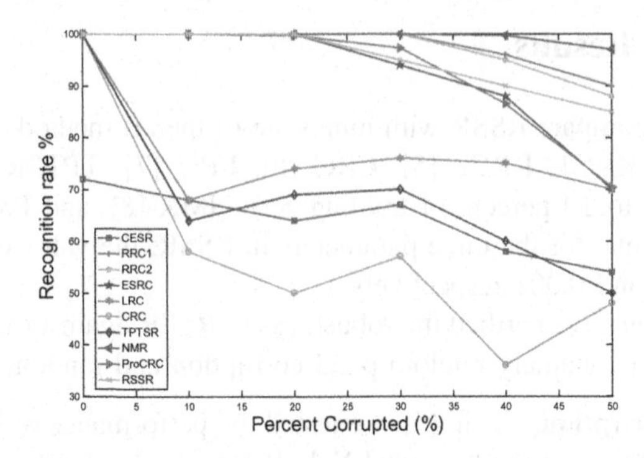

Fig. 3. Recognition rates under different sizes of block occlusion

4 Conclusions

A novel robust coding-based model, RSSR, is proposed for robust face recognition in this paper. RSSR uses a two-phase collaborative representation to implement supervised sparse representation. As to tolerate possible variations on probe images, we use Huber loss as fidelity term in cost function which reserves the more samples of correct class into the candidate set for the latter representation. Moreover, to solve the RSSR regression function we introduce two variables which can simplify the object function of RSSR in the process of optimization. As shown in experiments, we compare RSSR classification method with the other state-of-the-art methods under corruptions and occlusions. The performance of RSSR always ranks in the forefront of different comparisons and especially RSSR surpasses TPTSR, the original supervised sparse method, in all cases.

References

1. Wright, J., Yang, A.Y., Ganesh, A., Sastry, S.S., Ma, Y.: Robust face recognition via sparse representation. IEEE Trans. Pattern Anal. Mach. Intell. **31**, 210–227 (2009)
2. Xu, Y., Zhang, D., Yang, J., Yang, J.Y.: A two-phase test sample sparse representation method for use with face recognition. IEEE Trans. Circ. Syst. Vid. **21**, 1255–1262 (2011)
3. Deng, W., Hu, J., Guo, J.: Extended SRC: undersampled face recognition via intraclass variant dictionary. IEEE Trans. Pattern Anal. Mach. Intell. **34**, 1864–1870 (2012)
4. He, R., Zheng, W.S., Hu, B.G.: Maximum correntropy criterion for robust face recognition. IEEE Trans. Pattern Anal. Mach. Intell. **33**, 1561–1576 (2011)
5. Yang, M., Zhang, L., Yang, J., Zhang, D.: Regularized robust coding for face recognition. IEEE Trans. Image Process. **22**, 1753–1766 (2013)
6. Zhang, D., Meng, Y., Feng, X.: Sparse representation or collaborative representation: which helps face recognition? In: 2011 IEEE International Conference on Computer Vision (ICCV), pp. 471–478 (2011)
7. Naseem, I., Togneri, R., Bennamoun, M.: Linear regression for face recognition. IEEE Trans. Pattern Anal. Mach. Intell. **32**, 2106–2112 (2010)

8. Yang, J., Luo, L., Qian, J., Tai, Y., Zhang, F., Xu, Y.: Nuclear norm based matrix regression with applications to face recognition with occlusion and illumination changes. IEEE Trans. Pattern Anal. Mach. Intell. **39**, 156–171 (2017)
9. Cai, S., Zhang, L., Zuo, W., Feng, X.: A probabilistic collaborative representation based approach for pattern classification. In: 2016 IEEE Conference on Computer Vision and Pattern Recognition (CVPR), pp. 2950–2959 (2016)

Block-Sparse Tensor Based Spatial-Spectral Joint Compression of Hyperspectral Images

Yanwen Chong, Weiling Zheng, Hong Li, and Shaoming Pan(✉)

State Key Laboratory of Information Engineering in Surveying, Mapping and Remote Sensing, Wuhan University, Wuhan 430079, China
pansm@whu.edu.cn

Abstract. A hyperspectral image is represented as a three-dimensional tensor in this paper to realize the spatial-spectral joint compression. This avoids destroying the feature structure, as in the 2D compression model, the compression operation of the spatial and spectral information is separate. Dictionary learning algorithm is adopted to train three dictionaries on each mode and these dictionaries are applied to build the block-sparse model of hyperspectral image. Then, based on the Tucker Decomposition, the spatial and spectral information of the hyperspectral image is compressed simultaneously. Finally, the structural tensor reconstruction algorithm is utilized to recover the hyperspectral image and it significantly reduce the computational complexity in the block-sparse structure. The experimental results demonstrate that the proposed method is superior to other 3D compression models in terms of accuracy and efficiency.

Keywords: Block-sparse · Tensors · Tucker decomposition
Spatial-spectral joint compression

1 Introduction

A hyperspectral image is a three-dimensional (3D) signal, including 2D spatial information and 1D spectral information [1–3]. Due to the rapid improvement of remote sensing imaging technology, the data volume of hyperspectral image has been significantly increased, which has increased the difficulties of image transmission and storage. Therefore, effective compression and reconstruction of hyperspectral image has become an important research issue.

In the past few years, Compressed sensing (CS) [4–6] has attracted considerable attention because it provides a new mechanism for signal processing. Since the wide application of large-scale datasets, the concept of multidimensional CS is proposed [7]. Multidimensional CS is based on Tucker Decomposition (TD) [9, 10] and it is suitable for processing large-scale datasets, i.e., hyperspectral images.

Hyperspectral images as sparse signals, contain a large number of spatial and spectral information redundancy. In the past, most hyperspectral image compression approaches compress the image in the spatial and spectral domain separately. The separated operation will destroy the structure of hyperspectral image data and increase the computational and storage burden in the recovery [11–13]. Also, there are some methods that realize the spatial-spectral joint compression, but the structure features of

D.-S. Huang et al. (Eds.): ICIC 2018, LNAI 10956, pp. 260–265, 2018.
https://doi.org/10.1007/978-3-319-95957-3_29

hyperspectral image are ignored. Furthermore, non-structural reconstruction algorithms results in low efficiency. For examples, Karami et al. [14] proposed an algorithm based on Discrete Wavelet Transform and Tucker Decomposition (DWT-TD), exploiting both the spectral and the spatial information in the images. Wang et al. [15] proposed a hyperspectral image compression system based on the lapped transform and Tucker decomposition (LT-TD). Yang et al. [16] proposed a compressive hyperspectral imaging approach via sparse tensors and nonlinear CS (T-NCS).

In order to preserve the structure feature of hyperspectral image and improve efficiency, a hyperspectral image compression method based on multidimensional block-sparse representation and dictionary learning (MBSRDL) is proposed in this paper. In this scheme, the measurements can been seen as tensors and a structural linear reconstruction algorithm [17] is adopted to recovery the image. In view of that the image content has respective characteristics on each mode, the dictionary learning algorithm is used on each mode respectively for training dictionaries. These three dictionaries are applied to build the 3D block-sparse representation model of hyperspectral image. Then, TD is utilized for the spatial-spectral joint compression. Finally, the original image is reconstructed linearly by the structural reconstruction algorithm of tensor. The experimental results demonstrate the accuracy and efficiency of the proposed method is superior.

This paper is organized as follows: in Sect. 1, the block-sparse representation and TD is introduced; in Sect. 2, a hyperspectral image compression method based on multidimensional block-sparse representation and dictionary learning (MBSRDL) is depicted in detail; in Sect. 3, some experiments are performed to prove the accuracy and efficiency of the proposed method; and finally, in Sect. 4 the main conclusions are outlined.

2 Method

Given a hyperspectral image tensor $\underline{\mathbf{Y}} \in \mathbb{R}^{I_1 \times I_2 \times I_3}$, three sensing matrices $\mathbf{\Phi}_1 \in \mathbb{R}^{M_1 \times I_1}$, $\mathbf{\Phi}_2 \in \mathbb{R}^{M_2 \times I_2}$, $\mathbf{\Phi}_3 \in \mathbb{R}^{M_3 \times I_3}$ $(M_n \leq I_n, n = 1, 2, 3)$ are defined for each dimension of hyperspectral tensor. The final sensing matrix applied can be expressed as the Kronecker product: $\mathbf{\Phi} = \mathbf{\Phi}_3 \otimes \mathbf{\Phi}_2 \otimes \mathbf{\Phi}_1$. Besides sensing matrices, three dictionaries $\mathbf{D}_1 \in \mathbb{R}^{I_1 \times K_1}$, $\mathbf{D}_2 \in \mathbb{R}^{I_2 \times K_2}$, $\mathbf{D}_3 \in \mathbb{R}^{I_3 \times K_3}$ are needed for sparse representation of hyperspectral tensor and the final dictionary can be expressed as the Kronecker product: $\mathbf{D} = \mathbf{D}_3 \otimes \mathbf{D}_2 \otimes \mathbf{D}_1$. Therefore, the problem hyperspectral image compression can be represented as:

$$\underline{\mathbf{Y}} = \underline{\mathbf{X}}_{\times 1} \mathbf{\Phi}_1 \mathbf{D}_{1 \times 2} \mathbf{\Phi}_2 \mathbf{D}_{2 \times} \cdots {}_{\times N} \mathbf{\Phi}_N \mathbf{D}_N \tag{1}$$

The reconstruction of the tensor is to solve the $\underline{\mathbf{X}}$, and the typical reconstruction algorithm is Kronecker-OMP algorithm, but it has high computation complexity, and it is not suitable for the tensor with block-sparse structure. NBOMP algorithm [17] is a structural reconstruction algorithm for block- sparse tensor. NBOMP transforms complex objective functions into linear objective functions and then uses Cholesky factorization to solve the problem. Compared with Kronecker-OMP, it greatly reduced memory consumption and computation complexity.

This paper proposed a hyperspectral image compression method based on multi-dimensional block-sparse representation and dictionary learning (MBSRDL). The dictionary learning algorithm KSVD is utilized in this paper to train three dictionaries for each dimension. Then TD is adopted to compress the spatial and spectral information simultaneously as shown in Fig. 1. Finally, the original hyperspectral tensor is reconstructed by the structural reconstruction algorithm NBOMP. The details of MBSRDL can be seen in Algorithm 1.

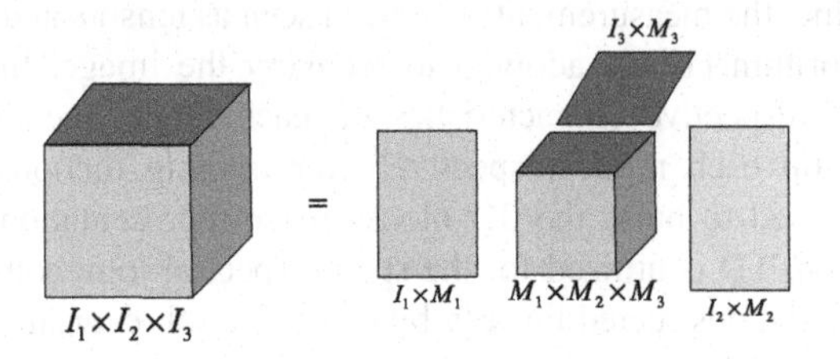

Fig. 1. Tensor decomposition

Algorithm 1: MBSRDL

Preconditioning: offline dictionaries $\mathbf{D}_1 \in \mathbb{R}^{I_1 \times K_1}, \mathbf{D}_2 \in \mathbb{R}^{I_2 \times K_2}, \mathbf{D}_3 \in \mathbb{R}^{I_3 \times K_3}$

Input: hyperspectral tensor $\underline{\mathbf{Y}} \in \mathbb{R}^{I_1 \times I_2 \times I_3}$, $\mathbf{D}_1, \mathbf{D}_2, \mathbf{D}_3$

Initialization: sampling rate of three dimensions $1/r_1, 1/r_2, 1/r_3$, sensing matrices of three dimensions $\mathbf{\Phi}_1 \in \mathbb{R}^{M_1 \times I_1}, \mathbf{\Phi}_2 \in \mathbb{R}^{M_2 \times I_2}, \mathbf{\Phi}_3 \in \mathbb{R}^{M_3 \times I_3}$ with $M_1 = I_1/r_1, M_2 = I_2/r_2, M_3 = I_3/r_3$

Procedure:

(1) compress: $\underline{\mathbf{Y}}' = \underline{\mathbf{Y}}_{\times 1}\mathbf{\Phi}_{1 \times 2}\mathbf{\Phi}_{2 \times 3}\mathbf{\Phi}_3$;

(2) code $\underline{\mathbf{Y}}'$ using NBOMP algorithm to solve $\underline{\mathbf{Y}}' = \hat{\underline{\mathbf{X}}}_{\times 1}\mathbf{\Phi}_1\mathbf{D}_{1 \times 2}\mathbf{\Phi}_2\mathbf{D}_{2 \times 3}\mathbf{\Phi}_3\mathbf{D}_3$

(3) reconstruct: $\hat{\underline{\mathbf{Y}}} = \hat{\underline{\mathbf{X}}}_{\times 1}\mathbf{D}_{1 \times 2}\mathbf{D}_{2 \times 3}\mathbf{D}_3$.

The block-sparse reconstruction algorithm significantly reduces the computation time, while greatly preserves the hyperspectral data structure of the hyperspectral image. It avoids high complexity and storage burden in non-structural reconstruction with a large number of iteration. Especially, when the sampling rate is very low, it can still present a good performance.

3 Experimental Results

Several experiments are taken in this section to investigate the performance of the proposed method. Two hyperspectral datasets (University of Pavia and Centre) are used for experiments and our software environment is: Windows 10, 3.00 GHz Intel Xeon CPU, MATLAB 8.3.

The proposed algorithm MBSRDL is compared with the DWT-TD and the non-structural reconstruction algorithm Kronecker-OMP to prove its performance in terms of accuracy and time. Tables 1, 2 and 3 show the reconstruction accuracy with the different sample rate of 0.2, 0.3 and 0.4.

Table 1. Reconstruction accuracy and time with the sampling rate of 0.2

Methods	PSNR (dB)	SSIM	Reconstruction time (s)
Kronecker-OMP	18.4919	0.3059	259.3750
DWT-TD	19.7720	0.4289	50.4327
MBSRDL	**22.8403**	**0.5938**	**34.6027**

It can be seen from Tables 1, 2 and 3 that the performance of MBSRDL is obviously better than that of DWT-TD and Kronecker-OMP. This is because that compared with DWT-TD, MBSRDL uses trained dictionaries instead of fixed dictionaries to build structural sparse representation model of hyperspectral image tensor, it is more effective to retain and restore the overall structural features of hyperspectral image at a very low sampling rate. Additionally, MBSRDL uses a structural block-sparse reconstruction algorithm, which greatly shortens the reconstruction time because of much fewer iterations. Compared with other methods, MBSRDL has the incomparable advantage of fast computation speed. Especially under the condition of scarce resources, it can meet not only the accuracy requirement, but also the efficiency requirement.

Table 2. Reconstruction accuracy and time with the sampling rate of 0.3

Methods	PSNR (dB)	SSIM	Reconstruction time (s)
Kronecker-OMP	22.6265	0.6149	386.1882
DWT-TD	24.8983	0.6678	70.0539
MBSRDL	**25.3957**	**0.7043**	**44.9183**

Table 3. Reconstruction accuracy and time with the sampling rate of 0.4

Methods	PSNR (dB)	SSIM	Reconstruction time (s)
Kronecker-OMP	26.5291	0.7079	531.6132
DWT-TD	28.0446	0.7402	100.5793
MBSRDL	**31.3536**	**0.8468**	**51.8789**

4 Conclusions

The hyperspectral image was regarded as a 3D tensor signal and a hyperspectral image compression method based on multidimensional block-sparse representation and dictionary learning (MBSRDL) was proposed in this paper. The advantages of the proposed method are that it retains the structure features of the image to a large extent because it uses trained dictionaries for each dimension. The block-sparse structure of hyperspectral tensor avoids complicated nonlinear reconstruction, which greatly reduces the iterations in the reconstruction. When the sampling rate is very low, the reconstruction performance is much better than other non-structural tensor compression models.

Acknowledgement. This work is supported by the National Natural Science Foundation of China (Nos. 61572372 & 41671382), LIESMARS Special Research Funding.

References

1. Chan, J.C., Ma, J.L., Van de Voorde, T., et al.: Preliminary results of superresolution - enhanced angular hyperspectral (CHRIS/Proba) images for land-cover classification. IEEE Geosci. Remote Sens. Lett. **8**(6), 1011–1015 (2011)
2. Toivanen, P., Kubasova, O., Mielikainen, J.: Correlation-based band-ordering heuristic for lossless compression of hyperspectral sounder data. IEEE Geosci. Remote Sens. Lett. **2**(1), 50–54 (2005)
3. Shaw, G.A., Burke, H.K.: Spectral imaging for remote sensing. Lincoln Lab. J. **14**(1), 3–28 (2003)
4. Donoho, D.L.: Compressed sensing. IEEE Inf. Theor. **52**(4), 1289–1306 (2006)
5. Candè, E.J., Wakin, M.B.: An introduction to compressive sampling. IEEE Signal Process. Mag. **25**(2), 21–30 (2008)
6. Lian, Q., Shi, B., Chen, S.: Research advances on dictionary learning models, algorithms and applications. IEEE J. Autom. Sinica **41**(2), 240–260 (2015)
7. Cesar, F.: Multidimensional compressed sensing and their applications. Wiley Interdiscip. Rev. Data Min. Knowl. Discov. **3**(6), 355–380 (2013)
8. Duarte, M.F., Baraniuk, R.G.: Kronecker compressive sensing. **21**(2), 494–504 (2012)
9. Oseledets, I.: Tensor-train decomposition. SIAM J. Sci. Comput. **33**(5), 2295–2317 (2011)
10. Fang, L., He, N., Lin, H.: CP tensor-based compression of hyperspectral images. J. Opt. Soc. Am. A Opt. Image Sci. Vis. **34**(2), 252–258 (2017)
11. Töreyin, B.U., Yilmaz, O., Mert, Y.M., et al.: Lossless hyperspectral image compression using wavelet transform based spectral decorrelation. In: Recent Advances in Space Technologies, pp. 251–254. IEEE, Istanbul (2015)
12. Lee, C., Youn, S., Jeong, T., et al.: Hybrid compression of hyperspectral images based on PCA with pre-encoding discriminant information. IEEE Geosci. Remote Sens. Lett. **12**(7), 1491–1495 (2015)
13. Zhao, D., Zhu, S., Wang, F.: Lossy hyperspectral image compression based on intra-band prediction and inter-band fractal encoding. Comput. Electr. Eng. **54**, 494–505 (2016)
14. Karami, A., Yazdi, M., Mercier, G.: Compression of hyperspectral images using discrete wavelet transform and tucker decomposition. IEEE J. Sel. Top. Appl. Earth Obs. Remote Sens. **5**(2), 444–450 (2012)

15. Wang, L., Bai, J., Wu, J., et al.: Hyperspectral image compression based on lapped transform and Tucker decomposition. Signal Process Image Commun. **36**, 63–69 (2015)
16. Yang, S., Wang, M., Li, P., et al.: Compressive hyperspectral imaging via sparse tensor and nonlinear compressed sensing. IEEE Geosci. Remote Sens. Lett. **53**(11), 5043–5957 (2015)
17. Caiafa, C.F., Cichocki, A.: Computing sparse representations of multidimensional signals using Kronecker bases. Neural Comput. **25**(1), 186–220 (2014)

DeepLayout: A Semantic Segmentation Approach to Page Layout Analysis

Yixin Li, Yajun Zou, and Jinwen Ma(✉)

Department of Information Science, School of Mathematical Sciences and LMAM, Peking University, Beijing 100871, China
{liyixin, zouyj}@pku.edu.cn, jwma@math.pku.edu.cn

Abstract. In this paper, we present DeepLayout, a new approach to page layout analysis. Previous work divides the problem into unsupervised segmentation and classification. Instead of a step-wise method, we adopt semantic segmentation which is an end-to-end trainable deep neural network. Our proposed segmentation model takes only document image as input and predicts per pixel saliency maps. For the post-processing part, we use connected component analysis to restore the bounding boxes from the prediction map. The main contribution is that we successfully bring RLSA into our post-processing procedures to specify the boundaries. The experimental results on ICDAR2017 POD competition dataset show that our proposed page layout analysis algorithm achieves good mAP score, outperforms most of other competition participants.

Keywords: Page layout analysis · Document segmentation
Document image understanding · Semantic segmentation and deep learning

1 Introduction

Page layout analysis, also known as document image understanding and document segmentation, plays an important role in massive document image analysis applications such as OCR systems. Page layout analysis algorithms take document images or PDF files as inputs, and the goal is to understand the documents by decomposing the images into several structural and logical units, for instance, text, figure, table and formula. This procedure is critical in document image processing applications, for it usually brings a better recognition results. For example, once we get the structural and semantic information of the document images, we only feed the text regions into the OCR system to recognize text while the figures are saved directly. Thus, page layout analysis has become a popular research topic in computer vision community.

Most of the conventional methods [1–5] have two steps: segmentation and classification. Firstly, the document images are divided into several regions, and then a classifier is trained to assign them to a certain logical class. The major weakness of these methods is that the unsupervised segmentation involves lots of parameters that rely on experience, and one set of parameters can hardly fit all the document layout styles.

To tackle this problem, the most straightforward way is supervised localization or segmentation. The parameters in supervised learning can be tuned automatically during

D.-S. Huang et al. (Eds.): ICIC 2018, LNAI 10956, pp. 266–277, 2018.
https://doi.org/10.1007/978-3-319-95957-3_30

the training, which avoids the large amount of human-defined rules and hand-craft parameters. On the other hand, supervised segmentation provides semantic information which means we can perform segmentation and classification simultaneously.

Since the final output of the page layout analysis is a number of bounding boxes and their corresponding labels, this problem can be framed as an object detection or localization problem. Unfortunately, the state-of-the-art object detection approaches such as Faster R-CNN [6] and Single-Shot Multibox Detector (SSD) [7] have been proven not working very well in the page layout analysis case [1]. This is because the common object detection methods are designed to localize certain objects in real life such as dogs, cars and human, and most of them have a specific boundary unlike text and formula regions and are not likely to have an extremely aspect ratio like text line in page layout analysis case. Also, the error causing by the bounding box regression is inevitable, which is usually the final stage of common object detection networks.

To address this issue, we adopt semantic segmentation approach to classify each pixel into their semantic meaning like text, formula, figure or table. The semantic segmentation model is a deep neural network trained under supervised information where parameters are learned automatically during training. Moreover, the pixel level understanding from semantic segmentation is more precise than the bounding box level understanding from the conventional object detection methods.

In this paper, we propose a page layout analysis algorithm based on semantic segmentation. The pipeline of our proposed algorithm contains two parts: the semantic segmentation stage and the post-processing stage. Our semantic segmentation model is modified on DeepLab [8] to fit our problem. As for post-processing part, we get the bounding box locations along with their confidence scores and labels by analyzing the connected components on the probability map generated from our segmentation model and adopt the run length smoothing algorithm (RLSA) locally on the original image to modify the bounding boxes. It is demonstrated in the experiments that our proposed algorithm achieves both reasonable visualization results and good quantization results on the ICDAR2017 POD competition dataset [9].

The main contribution of this paper is three fold. First, we propose a powerful and efficient approach on page layout analysis. Also, we successfully contrast a coarse-to-fine structure by combining the supervised learning algorithm (DeepLab) and unsupervised algorithm (RLSA). Finally, though the optimization targeted on POD dataset may not simply be applied to other datasets, the ideas have good extension meaning and reference value.

The rest of the paper is organized as follow: we briefly review the page layout analysis and semantic segmentation algorithms in Sect. 2. The methodology of our page layout analysis algorithm is then presented in the next section. In Sect. 4, the datasets we used and the experiments we conducted are described in detail. And discussions of the limitation and running time analysis are also given in Sect. 4. Finally, we conclude the whole paper in the last section.

2 Related Work

2.1 Page Layout Analysis

Page layout analysis has been studied for decades and a great number of algorithms have been established and developed to solve this problem. Most of the page layout analysis approaches can be roughly divided into two categories by the type of segmentation algorithms. One of them is unsupervised segmentation by hand-craft features and human-defined rules, and the other is supervised segmentation by a learning based model and supervised training data.

Most of the conventional methods [1–5] adopt a two-step strategy: an unsupervised segmentation model and a learning based classification model. An unsupervised segmentation method is either starting from the pixels then merging them into high level regions (bottom-up) [2, 3] or segmenting the page into candidate regions by projections or connected components (top-down) [4, 5]. Both need a large amount of experience-depended parameters. And as for classification step, hand-craft features are extracted to train a classifier [3, 4] or a CNN is trained to classify the segmented regions [1]. Also, some algorithms are proposed to detect the specific type of boxes or regions like equations [10, 11] and tables [12] in the PDF files or document images. As we mentioned before, the two-step algorithms mostly contain lots of human-defined rules and handcraft features which involve a large parameter set.

In recent years, supervised segmentation is introduced to solve the page layout case [13, 14]. Supervised segmentation provides semantic information which allows us to perform segmentation and classification at the same time. Oyebade et al. [13] extracts textural features from small patches and trains a neural network (fully connected) to classify them to get the document segmentation result. Due to the patch classification, there is a so-called "mosaic effect" where the segmentation boundary is rough and inaccurate. Yang et al. [14] first introduce semantic segmentation to document segmentation, and an additional tool (Adobe Acrobat) is adopted to specify the segmentation boundary. By the power of deep learning, this type of methods is normally faster and stronger than two-step ones and is easier to generalize to other type of documents.

2.2 Semantic Segmentation

Semantic segmentation is a computer vision task that aims to understand the image in pixel level by performing the pixel-wise classification. A demonstration of the semantic segmentation task is shown in Fig. 1[1]. Semantic segmentation is a deeper understanding of images than image classification. Instead of recognizing the objects in image classification task, we also have to assign each pixel to a certain object class or background to lineate the boundary of each object. In the industry, semantic segmentation is widely used in a variety of computer vision scenarios such as image matting, medical image analysis and self-driving.

[1] http://cocodataset.org/#detections-challenge2017.

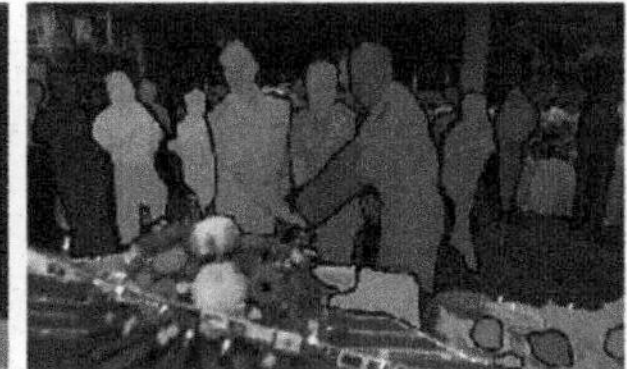

Fig. 1. Semantic segmentation task.

Before deep learning, the commonly used solutions are random forest based algorithms [15], this kind of algorithms are inaccurate and extremely slow. With CNN taking over computer vision, one of the first attempts on page layout analysis by CNN was patch classification [16] where the pixel class is assigned based on the classification result on a small image patch around it. The size of image patches need to be fixed due to the fully connected layer used in the network structure. And to keep the parameter size acceptable, the patch window which equals the receptive field needs to be small. Thus the segmentation result was still not ideal.

In 2014, Fully Convolutional Network (FCN) was proposed by Long et al. [17] which is a milestone of semantic segmentation. This model allows us to feed the images in any size to the segmentation model because no fully connected layer was used in the network structure. Also the end-to-end trainable structure makes it much faster and much stronger than the patch classification methods.

The next big problem of using CNN on segmentation is the pooling layers. Pooling layers increase the receptive field and robustness while weakening the spatial information. Spatial information is unimportant in image classification but it is essential in segmentation task. Two ways to overcome this problem are using short-cut connections to restore the spatial information [18] and using dilated (atrous) convolution layer to increase the receptive field and keep the spatial information at the same time [8]. And the segmentation model we used in our proposed page layout analysis algorithm is based on the latter solution.

3 Methodology

3.1 Overview

In this paper, we proposed a semantic segmentation based page layout analysis algorithm. The pipeline can be divided into two major parts: semantic segmentation part and post-processing part. The segmentation model we use along with some training details and the post-processing procedures are introduced in this section.

The whole pipeline of our proposed algorithm is shown in Fig. 2. First, a saliency map is generated by our semantic segmentation model. Connected component analysis is adopted to the generated saliency map to restore the bounding boxes. Then run length smoothing algorithm is applied to the local document image of detected logical boxes to specify the boundaries and get final detection results.

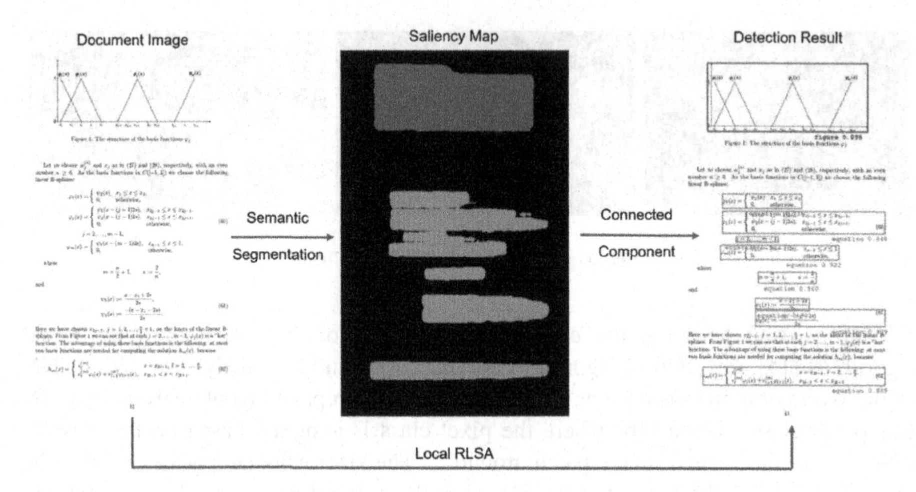

Fig. 2. Pipeline of the proposed page layout analysis algorithm.

Noticed that our proposed deep learning based page layout analysis algorithm takes only document image as input, unlike previous work taking benefits from structural information in PDF file [11] or applying additional commercial software to localize the logical units [14].

3.2 Semantic Segmentation Model

Fully Convolutional Network (FCN) [17] represents a milestone in deep learning based semantic segmentation. End-to-end convolutional structure is first introduced and deconvolutional layers are used to upsample the feature maps. However, loss of spatial information during pooling stage makes the upsampling produce coarse segmentation results which leaves a lot of room for improvement.

DeepLab [18] is proposed to overcome this problem and achieves state-of-the-art performance. So we choose DeepLab v2 structure as our segmentation model, and take ResNet-101 [19] as the backbone network. The key point in the network structure is that we use a set of dilated convolution layers to increase the receptive field without losing the spatial information or increasing the number of parameters.

DeepLab v2. As we mentioned before, pooling layers are used in deep neural networks to increase the receptive field but it causes loss of "where" information which is what we need in semantic segmentation. Dilated (also known as atrous) convolution [20] is one of the solutions to this problem. Holes in the dilated convolution kernels make them have the field of view same as a large convolution kernel and the number of parameters same as a small convolution kernel. Also, there is no decrease of spatial dimensions if the stride is set to 1 in dilated convolution.

Atrous Spatial Pyramid Pooling (ASPP) is proposed in DeepLab [18] to aggregate parallel atrous (dilated) convolutional layers. In our model, dilated convolutional layers with multiple sampling rates are designed to extract features in the multi-scale manner and are fused by ASPP layer.

Consistency Loss. The semantic segmentation model is designed to capture objects with any shapes but the logical units in document images are all rectangles. Inspired by Yang et al. [14], we implement the consistency loss to penalize the object shapes other than rectangle. The training loss is the segmentation loss combining with the consistency loss. And the consistency loss is defined as follow:

$$\mathcal{L}_{\text{con}} = \frac{1}{|gt|} \sum\nolimits_{p_i \in gt} (p_i - \bar{p})^2 \tag{1}$$

Where gt is the ground truth bounding box, $|gt|$ is number of pixels in the ground truth box, p_i is the probability given by the segmentation Softmax output, and $\bar{p}$ is the mean value of all the pixels in the bounding box.

Training Details. The segmentation model we use is based on DeepLab v2 [18]. All the layers except the last prediction layer are restored on the model pretrained on MSCOCO dataset [21]. And the last layer is random initialized to predict four classes: background, figure, formula, and table. Parameters like learning rate, weight decay and momentum are inherited from the Tensorflow implementation of DeepLab v2[2].

The ground truth mask is given by the ground truth bounding boxes, that is, the ground truth label of pixels in the bounding boxes are set to the positive class number same as the bounding box and the label of pixels that are not in any bounding boxes are set to zero which represents background.

We random scale the input document images and the corresponding ground truth masks during training to improve the robustness over multi-scale input. The model is optimized by Adaptive moment estimation (Adam) to minimize the cross-entropy loss over pixels between prediction masks and ground truth masks.

3.3 Post-processing Procedures

At inference time, a pixel-wise prediction mask is generated by the segmentation model. And the post-processing step is to restore the bounding boxes from the prediction mask and get the final layout analysis result. Our main contribution of this part is that we adopt the local Run Length Smoothing Algorithm (RLSA) on the original document image along with the connected component analysis on the prediction mask to specify the region boundary.

Conditional Random Field (CRF). CRF [22] is a standard post-processing step in deep learning based semantic segmentation model to improve the final results. CRF assumes that similar intensity tends to be the same class then construct a graphic model to smooth the boundary of each object. Usually, CRF can improve the segmentation mean IOU for 1–2%. On the contrary, CRF decreases the segmentation result in our page layout case which is shown by the experiments in Sect. 4. It is due to the differences between natural images and document images.

[2] https://github.com/DrSleep/tensorflow-deeplab-resnet.

Connected Component Analysis (CCA). To restore the bounding boxes from the saliency map predicted by our segmentation model, we extract connected components on each class then take the bounding rectangles as candidate bounding boxes. Label of each candidate bounding box is the same as the connected component and the confidence score is calculated by the average of the pixel segmentation Softmax scores.

Run Length Smoothing Algorithm (RLSA). RLSA is widely used in the document segmentation to aggregate the pixels in the same logical unit for last few decades [2, 3]. The input of RLSA is binary image or array where 0 s represent black pixels and 1 s represent white pixels. The aggregation procedure is under the rule that 1 s are changed to 0 s if the length of adjacent 1 s is less than a predefined threshold C. An example of RLSA on 1d array (RLSA threshold C is set to 3) is shown in Fig. 3(a) and an example of RLSA document segmentation procedure is shown in Fig. 3(b).

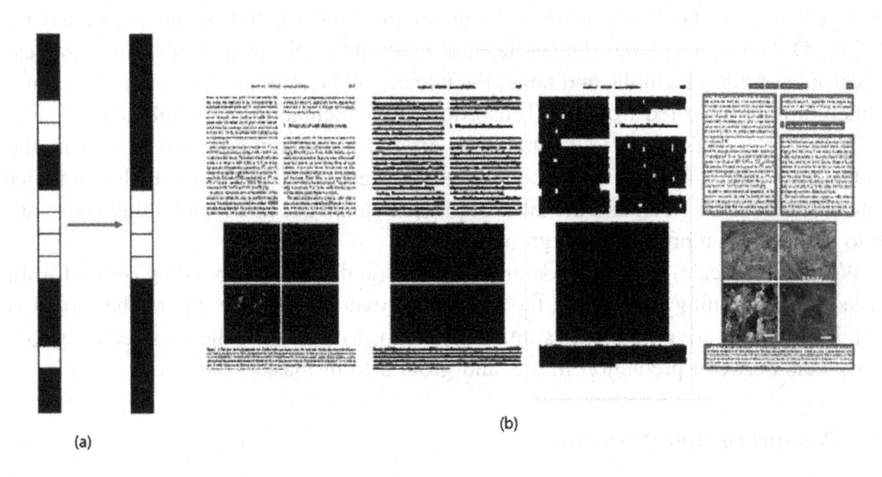

Fig. 3. (a) A 1d example of RLSA procedure. (b) A document segmentation example by RLSA.

RLSA is under the assumptions that the horizontal or vertical distance between black pixels in the same logical region is less than C while distance between different logical region is large than C. But there are several cases that do not meet these assumptions. For example, image captions are usually very close to the image, but they are different logical regions. Thus the determination of threshold C could be very hard and experience dependent.

In our case, since we have semantic meaning of each pixel by our segmentation model, we are able to apply RLSA on each connected component where pixels are in the same logical meaning. For the caption case, the figure and its caption are processed separately, they thus won't be aggregated together no matter how close they are. Semantic segmentation model gives us the logical class label of each pixel but the boundary is rough which is shown in Fig. 2, and local RLSA is adopted to gives us the exact boundary of each logical region.

Some Other Processing Steps. We investigate the ground truth of POD competition dataset [9] and design several rules to improve the mAP score. Note that this part is designed specifically for the POD competition dataset, so it may not be able to generalize to all types of document images. We briefly introduce the problems instead of solutions, for this part is not the focus of this paper.

We noticed that each subfigure is annotated separately in the POD dataset and the segmentation model tends to predict a single figure, so we set series of rules to split the figure regions into several subfigures. Tables normally have a clear boundary, so besides removing small regions, there is no additional processing step for tables.

Standard of equation annotation is unclear in POD dataset. Most of equations are annotated as "equation line" in POD dataset where multiline equation is labeled as multiple equation annotations. But some equations are annotated in "equation block". Also, the equation number is annotated in the same box with the corresponding equation. Equation number may very far away from the equation which leaves the annotated box a large blank space. Therefore, some human-defined rules are designed to split equations into single lines and aggregate equation number and equation itself. The result is still far from ideal, for the splitting procedure creates a new problem (the start and stop indexes of "Σ" and "Π") which will be discussed in Sect. 4.

4 Experiments

4.1 Datasets

Since our segmentation model is deep learning based, we need annotated data for training. We choose the open dataset for ICDAR2017 Page Object Detection Competition [9]. The competition dataset contains 2,400 annotated document images (1600 for training and 800 for testing) and the extended dataset which is not open for the competition contains about 10,000 annotated images. The document images are annotated by bounding boxes with three classes: figure, table and formula. The competition dataset can be downloaded on the official website[3].

4.2 Experimental Results

The evaluation metric we use to quantize the segmentation performance is mean IoU. Mean IoU is a standard metric in semantic segmentation [8, 17, 18] which calculates the mean intersection over union metrics on all classes. Also, mean average precision (mAP) and average F1 measure are calculated to evaluate the final page layout analysis results. Mean average precision is a rough estimation of area under precision-recall curve and is the most common used evaluation on object detection [6, 7]. F1 measure is the harmonic mean value of precision and recall. In particular, mAP and F1 are also the evaluation metrics used by POD competition [9].

The segmentation models are trained on POD dataset and POD extended dataset to show the performance gain from more training data. And post-processing methods are applied to prediction maps generated from the segmentation model. The results of different training datasets and different post-processing steps are shown in Table 1.

[3] https://www.icst.pku.edu.cn/cpdp/ICDAR2017_PODCompetition/index.html.

The results are evaluated by official evaluation tool of POD competition. And IoU threshold is set to 0.8 in mAP and average F1 calculation.

Table 1. The segmentation and final detection performance of our proposed methods.

	Mean IoU	Average F1	Mean AP
Base	0.869	0.518	0.498
Base + RLSA	0.869	0.763	0.666
Base(+) + RLSA	**0.908**	**0.801**	**0.690**
Base(+) + CRF + RLSA	0.886	0.745	0.611
Base(+) + CL + RLSA	0.897	0.776	0.662

In Table 1, "base" represents our segmentation model, "RLSA" represents our whole post-processing steps, "(+)" means the extended dataset is used to train the segmentation model, "CRF" represents segmentation model with fully connected CRF layers [22] and "CL" represents consistency loss considered during training.

From Table 1 we can see that the best result comes from segmentation model trained on extended dataset, follow by our proposed post-processing procedure including RLSA. The most significant performance gain is from our proposed RLSA post-processing which boosts the average F1 for 0.21 and mAP for 0.12. In the segmentation network, a larger training dataset gives us a 4% gain, for the deep learning structure we use heavily relies on large amount of training data. And the fully connected CRF which usually improves the segmentation results on real life objects, does not work well on page layout case. The reason is that objects in natural images have a clear boundary while logical units in document images have holes and blanks which is inadequate for CRF post-processing. Also the consistency loss is supposed to penalize the predictions with shapes other than rectangle. But in our experiments, some predictions vanished to have a zero consistency loss, thus the segmentation and final results are not what we expected.

Then we compare our best method with the top results submitted in POD competition [9]. The overall performance and performances on three specific classes (figure, table and formula) are shown in Table 2. Noticed that we come to a good place but not the best of all. All the methods in POD competition [9] have not been presented by academic papers, so the algorithms of their approaches and the authenticity of the results are unknown.

Most of the figures, equations and tables in document images can be correctly recognized and localized by our proposed page layout algorithm. Some visualization results are shown in Fig. 4. Green boxes represent equations detected by our proposed page layout analysis approach, red boxes represent figures and blue boxes represent tables. The numbers under the bounding boxes are the confidence scores.

Table 2. Results on POD competition.

Team	F1-measure				Average precision			
	Formula	Table	Figure	Mean	Formula	Table	Figure	Mean
PAL	0.902	0.951	0.898	0.917	0.816	0.911	0.805	0.844
Ours	**0.716**	**0.911**	**0.776**	**0.801**	**0.506**	**0.893**	**0.672**	**0.690**
HustVision	0.042	0.062	0.132	0.096	0.293	0.796	0.656	0.582
FastDetectors	0.639	0.896	0.616	0.717	0.427	0.884	0.365	0.559
Vislint	0.241	0.826	0.643	0.570	0.117	0.795	0.565	0.492
SOS	0.218	0.796	0.656	0.557	0.109	0.737	0.518	0.455
UTTVN	0.200	0.635	0.619	0.485	0.061	0.695	0.554	0.437
Matiai-ee	0.065	0.776	0.357	0.399	0.005	0.626	0.134	0.255

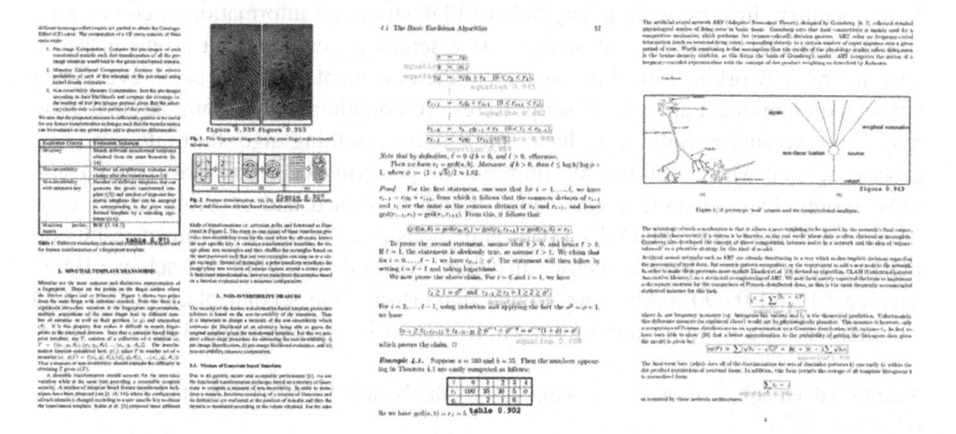

Fig. 4. Visualization results of our proposed page layout analysis algorithm. (Color figure online)

4.3 Limitations

There are still several scenarios our proposed algorithm might fail. As we can see in Table 2, the F1 measure and average precision score of equations is much lower than tables and figures. After analyzing the visualization results, we found that the evaluation scores are crippled by equations with "Σ", "Π", and equation number, for the reasons that the segmentation model tends to predict the equations with equation number into two separate equation regions, and RLSA separates the start and stop indexes of "Σ" and "Π" into three lines.

To tackle the equation number problem in the future work, one can increase the receptive field of semantic segmentation model to merge the equations and equation numbers in prediction map. As for the start and stop index problem, we trained a classification model to recognize "Σ" and "Π", and then merge the indexes. This procedure did bring us a precision gain on equations but is still not perfect. Therefore, there is still some room for the improvement of equation issue.

4.4 Running Time

Our proposed algorithm consists of two main parts: semantic segmentation and post-processing. Our segmentation is a deep neural network and a single inference takes 0.25 s on GPU (a single GTX1080). It should be at least twice faster if running on a decent GPU like Titan X. Our post-processing step can be efficiently done on CPUs (56 cores E5-2680v4) in 100 ms. In general, our whole system can process approximately 3 document images ($\sim$1300 * 1000) per second.

5 Conclusion

We have proposed a deep learning algorithm for page layout analysis, DeepLayout, which is capable of recognizing and localizing the semantic and logical regions directly from document images, without any help of PDF structural information or commercial software. We treat page layout analysis as a semantic segmentation problem, and a deep neural network is trained to understand the document image on pixel level. Then connected component analysis is adopt to restore bounding boxes from the prediction map. And we successfully bring local run length smoothing algorithm into our post-processing step which significantly improve the performance on both average F1 and mAP scores. Our semantic segmentation model is trained and experiments are conducted on ICDAR2017 POD dataset. It is demonstrated by the experiment results on POD competition evaluation metrics that our proposed algorithm can achieve 0.801 average F1 and 0.690 mAP score, which outperforms the second place of the POD competition. The running time of the whole system is approximately 3 fps.

Acknowledgement. This work was supported by the Natural Science Foundation of China for Grant 61171138.

References

1. Yi, X., Gao, L., Liao, Y., et al.: CNN based page object detection in document images. In: Proceedings of the International Conference on Document Analysis and Recognition, pp. 230–235. IEEE (2017)
2. Cesarini, F., Lastri, M., Marinai, S., et al.: Encoding of modified X-Y trees for document classification. In: Proceedings of the International Conference on Document Analysis and Recognition, pp. 1131–1136. IEEE (2001)
3. Priyadharshini, N., Vijaya, M.S.: Document segmentation and region classification using multilayer perceptron. Int. J. Comput. Sci. Issues **10**(2 part 1), 193 (2013)
4. Lin, M.W., Tapamo, J.R., Ndovie, B.: A texture-based method for document segmentation and classification. S. Afr. Comput. J. **36**, 49–56 (2006)
5. Chen, K., Yin, F., Liu, C.L.: Hybrid page segmentation with efficient whitespace rectangles extraction and grouping. In: International Conference on Document Analysis and Recognition, pp. 958–962. IEEE Computer Society (2013)
6. Ren, S., He, K., Girshick, R., et al.: Faster R-CNN: towards real-time object detection with region proposal networks. IEEE Trans. Pattern Anal. Mach. Intell. **39**(6), 1137–1149 (2017)

7. Liu, W., et al.: SSD: single shot multibox detector. In: Leibe, B., Matas, J., Sebe, N., Welling, M. (eds.) ECCV 2016. LNCS, vol. 9905, pp. 21–37. Springer, Cham (2016). https://doi.org/10.1007/978-3-319-46448-0_2
8. Chen, L.C., Papandreou, G., Kokkinos, I., et al.: DeepLab: semantic image segmentation with deep convolutional nets, atrous convolution, and fully connected CRFs. IEEE Trans. Pattern Anal. Mach. Intell. **40**(4), 834–848 (2018)
9. Gao, L., Yi, X., Jiang, Z., et al.: Competition on page object detection. In: Proceedings of the International Conference on Document Analysis and Recognition, ICDAR 2017, pp. 1417–1422. IEEE (2017)
10. Chu, W.T., Liu, F.: Mathematical formula detection in heterogeneous document images. In: Technologies and Applications of Artificial Intelligence, pp. 140–145. IEEE (2014)
11. Gao, L., Yi, X., Liao, Y., et al.: A deep learning-based formula detection method for PDF documents. In: Proceedings of the International Conference on Document Analysis and Recognition, pp. 553–558. IEEE (2017)
12. Hassan, T., Baumgartner, R.: Table recognition and understanding from PDF files. In: International Conference on Document Analysis and Recognition, pp. 1143–1147. IEEE (2007)
13. Oyedotun, O.K., Khashman, A.: Document segmentation using textural features summarization and feedforward neural network. Appl. Intell. **45**(1), 198–212 (2016)
14. Yang, X., Yumer, E., Asente, P., et al.: Learning to extract semantic structure from documents using multimodal fully convolutional neural networks. arXiv preprint arXiv:1706.02337 (2017)
15. Shotton, J., Fitzgibbon, A., Cook, M., et al.: Real-time human pose recognition in parts from single depth images. In: Computer Vision and Pattern Recognition, pp. 1297–1304. IEEE (2011)
16. Ciresan, D., Giusti, A., Gambardella, L.M., et al.: Deep neural networks segment neuronal membranes in electron microscopy images. Adv. Neural. Inf. Process. Syst. 2843–2851 (2012)
17. Shelhamer, E., Long, J., Darrell, T.: Fully convolutional networks for semantic segmentation. IEEE Trans. Pattern Anal. Mach. Intell. **39**(4), 640 (2017)
18. Ronneberger, O., Fischer, P., Brox, T.: U-Net: convolutional networks for biomedical image segmentation. In: Navab, N., Hornegger, J., Wells, W.M., Frangi, A.F. (eds.) MICCAI 2015. LNCS, vol. 9351, pp. 234–241. Springer, Cham (2015). https://doi.org/10.1007/978-3-319-24574-4_28
19. He, K., Zhang, X., Ren, S., et al.: Deep residual learning for image recognition. In: Proceedings of the IEEE Conference on Computer Vision and Pattern Recognition, pp. 770–778 (2016)
20. Yu, F., Koltun, V.: Multi-scale context aggregation by dilated convolutions. In: International Conference on Learning Representations (2016). arXiv:1511.07122
21. Lin, T.-Y., et al.: Microsoft COCO: common objects in context. In: Fleet, D., Pajdla, T., Schiele, B., Tuytelaars, T. (eds.) ECCV 2014. LNCS, vol. 8693, pp. 740–755. Springer, Cham (2014). https://doi.org/10.1007/978-3-319-10602-1_48
22. Koltun, V.: Efficient inference in fully connected CRFs with Gaussian edge potentials. In: International Conference on Neural Information Processing Systems, pp. 109–117. Curran Associates Inc. (2011)

CDC-MRF for Hyperspectral Data Classification

Yuanyuan Li[1], Jingjing Zhang[1], Chunhou Zheng[2](✉), Qing Yan[1], and Lina Xun[1]

[1] College of Electrical Engineering and Automation, Anhui University, Hefei 230601, China

[2] College of Computer Science and Technology, Anhui University, Hefei 230601, China

zhengch99@126.com

Abstract. This paper presents a new hyperspectral classification algorithm based on convolutional neural network (CNN). A CNN is first used to learn the posterior class distributions using a patch-wise training strategy to better utilize the spatial information. In order to further extract hyperspectral feature information, we propose a method of extracting features twice (CDC-MRF). In our method, we first use 2D CNN to extract spectral and spatial information of the hyperspectral data. Then, we use deconvolution layer to expand the size of sample which can make the patch contain more useful information. After that we make a second extraction of the features, use 2D CNN for secondary extraction features. After that, the spatial information is further considered by using a Markov random field prior, which encourages the neighboring pixels to have the same labels. Finally, a maximum posteriori segmentation model is efficiently computed by the α-expansion min-cut-based optimization algorithm. Experimental results show that, the proposed method achieves state-of-the-art performance on two benchmark HSI datasets.

Keywords: Convolutional neural network · Hyperspectral classification Deconvolution

1 Introduction

With the development of hyperspectral remote sensing technology, hyperspectral data is becoming more and more popular. Hyperspectral data classification [12] and target detection [1–4] have always been one of the core contents of remote sensing technology. However, high-dimensional hyperspectral data, a large number of redundant information between bands make hyperspectral data classification algorithms face severe challenges.

In recent years, with the development of deep learning, neural network have achieved excellent performance in the classification of hyperspectral image. Chen et al.

Y. Li and J. Zhang contributed equally to the paper as first authors.

D.-S. Huang et al. (Eds.): ICIC 2018, LNAI 10956, pp. 278–283, 2018.
https://doi.org/10.1007/978-3-319-95957-3_31

[10] proposed 1D CNN that extracts deep features to classify hyperspectral images. However, the input is one-dimensional input vector which loses spatial information of hyperspectral images. An improved CNN network [15] takes a three-dimensional patch as the input to remedy this shortcoming. Zhang et al. [11] proposed a novel dual-channel convolutional neural network (DC-CNN) framework. This framework included 1D CNN and 2D CNN can extract much hierarchical spectral features and hierarchical space-related features. Nevertheless the combination of 1D CNN and 2D CNN will extract a lot of redundant information.

In this paper, we propose a new framework CDC-MRF for hyperspectral classification. Inspired by [12], we use 2D CNN to extract spectral information and spatial information. To increase the extracted spectral information and spatial information, we present a double extraction method. After the first feature extraction, we use 2D deconvolution to expand the sample size, the sample can recover a large amount of spectral information and spatial information. Then we use 2D CNN to do the second feature extraction. The whole structure of the model we designed is shown in Fig. 1. Finally, we used Markov Random Field (MRF) to further optimize the classification results.

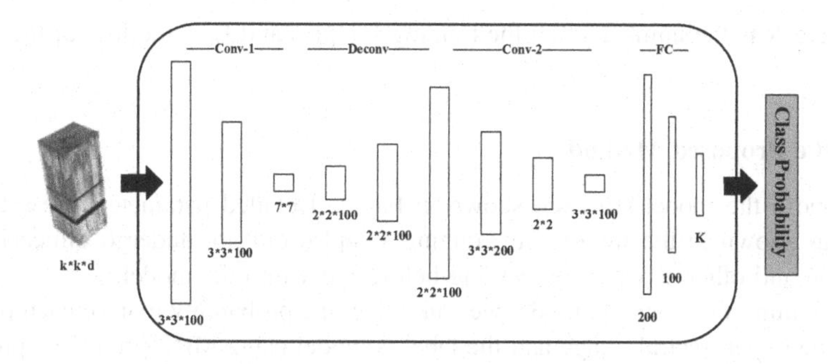

Fig. 1. The network structure of the proposed method used as our discriminative classifier

2 Materials and Methods

2.1 CNN

Our proposed model includes several convolution layers. The input sample is 3D-patch, so $x = (x_1, x_2, \ldots, x_T) \in R^{k*k*d}$ is input vector. In the first convolutional layer, a set of d filters $f_1, f_2, \ldots, f_d$, of receptive field size are applied to the input vector via convolution operation (*) to get the feature map:

$$F = (f_1, f_2, \ldots, f_T) = g(x * \{\phi_1, \phi_2, \ldots, \phi_d\}) \quad (1)$$

where $g(\cdot)$ is a nonlinear activation function.

We obtained the deep features after convolution layers and maximum pooling layers. The extracted deep features will be sent to the full connection layer after flattened into a column vector. Finally, the features output by the full connection will be sent to the classifier, where we use a softmax activation function to calculate the prediction probability for all categories. This is done as (2):

$$p(y = k|x) = \frac{\exp\left(w_k^T x + b_k\right)}{\sum_{k'=1}^{T} \exp\left(w_{k'}^T x + b_{k'}\right)} \tag{2}$$

Where w_k^s and b_k^s are the weight and bias vectors, and k is the number of categories. In the softmax classifier, the cross-entropy loss is widely used as (3):

$$L_i = -\log(p(y_i|x_i)) \tag{3}$$

The loss takes the average of the losses for every training example as (4):

$$L = \frac{1}{N} \sum_{i=1}^{N} L_i \tag{4}$$

Where N is the number of all the training samples and L_i is the loss of the sample index i.

2.2 The Proposed Method

We propose the model structure shown in Fig. 1. Detailed parameters are set in the model as shown in the figure. Our training samples did not undergo dimensionality reduction and other series of processing before input into the model.

By training the designed model we can obtain the probability map of each pixel on the whole hyperspectral image and the label is model using MRF prior. The proposed segmentation model is final given by (5)

$$\tilde{y} = \underset{y \in \kappa^n}{\operatorname{argmax}} \left\{ \sum_{n}^{i=1} \log P(y_i|x_i, W^*, b^*) + \mu \sum_{(i,j) \in C} \delta\left(y_i - y_j\right) \right\} \tag{5}$$

This objective function is actually a combinatorial optimization problem, which is difficult to solve because it contains many pairs of interacting items. In this paper, the α-expansion minimum cut method [15] is used because of its fast computational efficiency.

3 Experiment

In order to verify the validity of our proposed algorithm in different scenarios, our experiments have done on the two real-world Indian Pines and Pavia University benchmark datasets. In the experiment, we randomly chose 1% of the available labeled samples as the training set on the Indian pine dataset.

We compare our method with several advanced method HSI segmentation methods: (1) method with MRF [13–17]: MLRsub, MLM, SVM-3D and CNN; (2) method without MRF [5–9, 11]: R-ELM, Deepo, GF-FSAE, SSD-CNN, SSDL, DC-CNN. The results obtained in the experiment are the average of the data obtained by repeating the experiment 10 times. We set patch size k as 11 and smoothness parameter μ as 18.

3.1 Comparison with MRF

On the Indian pine data set, the classification results maps of all the comparison methods are shown in Fig. 2, and Table 1 shows AA, OA and Kappa. From the table, we can see that when the comparison method does not include MRF, our method is the best one of the comparison methods, and we can also find that the method with MRF can further improve the classification accuracy.

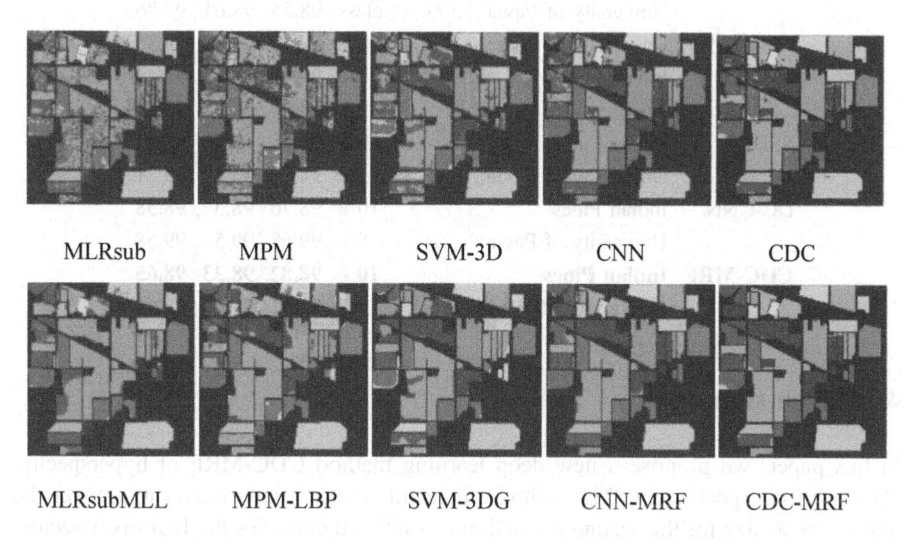

Fig. 2. Classification/segmentation maps obtained by all methods on the Indian Pines dataset

Table 1. Overall, average and individual class accuracies and kappa statistics of all methods on the Indian pines image test set.

Class	Classification algorithms					Segmentation algorithms				
	MLRsub	MLM	SVM-3D	CNN	CDC	MLRsubMLL	MLM-LBP	SVM-3DG	CNN-MRF	CDC-MRF
OA	55.84	64.94	66.76	91.42	**94.48**	65.19	70.18	71.51	95.55	**97.76**
AA	49.79	59.51	45.41	90.92	**91.52**	59.43	65.45	46.99	94.71	93.23
Kappa	60.83	54.82	61.71	90.22	**93.70**	65.54	58.07	67.20	94.93	**97.44**

3.2 Comparison with the State-of-the-Art Methods

We also compare our method with the state-of-the-art deep learning method of hyperspectral classification without MRF. Although not all methods select training set of the same size, important criteria for the development and validity of the proposed method can still be evaluated. As shown in Table 2, our method shows better performance than the other methods.

Table 2. Classification performance of different methods.

Method	Datasets	Training set	OA	AA	Kappa
R-ELM	Indian Pines	50%	95.95	95.45	95.39
	University of Pavia	50%	99.05	98.48	98.75
Deepo	Indian Pines	10%	97.45	95.91	96.36
	University of Pavia	300 per class	98.35	98.61	97.86
GF-FSAE	University of Pavia	30%	99.51	99.35	99.35
SSD-CNN	Indian Pines	10%	90.76	85.52	89.44
	University of Pavia	5%	93.34	92.2	91.95
SSDL	Indian Pines	10%	91.6	93.96	90.43
	University of Pavia	5%	94.88	93.29	93.21
DC-CNN	Indian Pines	10%	98.76	98.5	98.58
	University of Pavia	5%	99.68	99.5	99.58
CDC-MRF	**Indian Pines**	**10%**	**98.82**	**98.33**	**98.65**
	University of Pavia	**5%**	**99.71**	**99.66**	**99.61**

4 Conclusion

In this paper, we propose a new deep learning method CDC-MRF of hyperspectral classification. Specifically, this method utilizes the deconvolution layer to recover the feature patch size for the second extraction feature, and classifies the HSI pixel vectors in combination with Markov random fields to take full account of spatial and spectral information. Then, we use the optimized α-expansion graph cut algorithm to efficiently study the segmentation results. The experimental results on two benchmark HSI datasets show that our method outperforms the most advanced methods, including both deep learning and non-deep learning models.

Acknowledgments. This work is supported by Anhui Provincial Natural Science Foundation (grant number 1608085MF136), the National Science Foundation for China (Nos. 61602002 & 61572372).

References

1. Huang, C., Davis, L.S., Townshend, J.R.G.: An assessment of support vector machines for land cover classification. Int. J. Remote Sens. **23**, 725–749 (2002)
2. Li, J., Marpu, P.R., Plaza, A., Bioucas-Dias, J.M., Benediktsson, J.A.: Generalized composite kernel framework for hyperspectral image classification. IEEE Trans. Geosci. Remote Sens. **51**, 4816–4829 (2013)
3. Wu, H., Prasad, S.: Dirichlet process based active learning and discovery of unknown classes for hyperspectral image classification. IEEE Trans. Geosci. Remote Sens. **54**, 4882–4895 (2016)
4. Shaw, G., Manolakis, D.: Signal processing for hyperspectral image exploitation. Sig. Process. Mag. IEEE **19**, 12–16 (2002)
5. Liang, H., Li, Q.: Hyperspectral imagery classification using sparse representations of convolutional neural network features. Remote Sens. **8**, 99 (2016)
6. Yue, J., Zhao, W., Mao, S., Liu, H.: Spectral-spatial classification of hyperspectral images using deep convolutional neural networks. Remote Sens. Lett. **6**, 468–477 (2015)
7. Yue, J., Mao, S., Li, M.: A deep learning framework for hyperspectral image classification using spatial pyramid pooling. Remote Sens. Lett. **7**, 875–884 (2016)
8. Wang, L., Zhang, J., Liu, P., Choo, K.K.R., Huang, F.: Spectral-spatial multi-feature-based deep learning for hyperspectral remote sensing image classification. Soft. Comput. **21**, 213–221 (2017)
9. Chen, Y., Zhao, X., Jia, X.: Spectral-spatial classification of hyperspectral data based on deep belief network. IEEE J. Sel. Top. Appl. Earth Obs. Remote Sens. **8**, 2381–2392 (2015)
10. Chen, Y., Jiang, H., Li, C., Jia, X., Ghamisi, P.: Deep feature extraction and classification of hyperspectral images based on convolutional neural networks. IEEE Trans. Geosci. Remote Sens. **54**, 6232–6251 (2016)
11. Zhang, H., Li, Y., Zhang, Y., Shen, Q.: Spectral-spatial classification of hyperspectral imagery using a dual-channel convolutional neural network. Remote Sens. Lett. **8**, 438–447 (2017)
12. Shen, W., Li, D., Zhang, S., Ou, J.: Analysis of wave motion in one-dimensional structures through fast-Fourier-transform-based wavelet finite element method. J. Sound Vib. **400**, 369–386 (2017)
13. Yedidia, J.S., Freeman, W.T., Weiss, Y.: Constructing free-energy approximations and generalized belief propagation algorithms. IEEE Trans. Inf. Theory **51**, 2282–2312 (2005)
14. Cao, X., Zhou, F., Xu, L., Meng, D., Xu, Z., Paisley, J.: Hyperspectral image segmentation with markov random fields and a convolutional neural network. In: Computer Vision and Pattern Recognition (2017)
15. Cao, X., Xu, L., Meng, D., Zhao, Q., Xu, Z.: Integration of 3-dimensional discrete wavelet transform and Markov random field for hyperspectral image classification. Neurocomputing **226**, 90–100 (2017)
16. Li, J., Bioucas-Dias, J.M., Plaza, A.: Hyperspectral image segmentation using a new Bayesian approach with active learning. IEEE Trans. Geosci. Remote Sens. **49**, 3947–3960 (2011)
17. Li, J., Bioucas-Dias, J.M., Plaza, A.: Spectral-spatial classification of hyperspectral data using loopy belief propagation and active learning. IEEE Trans. Geosci. Remote Sens. **51**, 844–856 (2013)

The Application Research of AlphaGo Double Decision System in Network Bad Information Recognition

Hui Chen[1(✉)], Zeyu Zheng[2], Qiurui Chen[1], Lanjiang Yang[1], Xingnan Chen[1], and Shijue Zheng[1]

[1] College of Computer Science, Central China Normal University, Wuhan 430079, China
chmj@mails.ccnu.edu.cn
[2] Texas Department of State Health Services, Austin, TX, USA

Abstract. As the carrier of information transmission, the internet inevitably contains much bad information. In view of this phenomenon, with the purpose of identifying the bad information in the network, we combine existing Chinese text mining technology for experimental research. In combination with the idea of AlphaGo double decision system, the experiment will deal with the text information identification and classification using two system models, so that more accurate results can be obtained. In the experiment, a system does text segmentation and feature selection. Another system uses this method based on rules and statistics to compare text to determine whether or not it is bad information based on the established bad information database. And finally the two system carry out the text classification work. In the meantime, the two-system model worked together to identify and classify the bad information. AlphaGo's strategy was used to combine the former decentralized methods to make the system as a whole. This enables the system to improve the execution efficiency without reducing the recall rate, and the identification and classification accuracy.

Keywords: AlphaGo · Value network · Falling selector · Bad information Text classification

1 Introduction

Nowadays, while the Internet brings convenience to people's lives, it also brings some negative effects. Since 2014, some vulgar internet languages have become popular buzzwords and even become a mantra for some teenagers. In the meantime, the "network violence" incident that frequently broke out in recent years has polluted the language environment of our entire society. How to adopt a variety of methods and means to purify the network language environment in our country, standardize the network language has become a hot issue in the current network language work.

AlphaGo defeated the world Go champion, Li Shishi, in 2016. AlphaGo improves the play of chess by working with two different "brain" neural networks. The strategy of the two "brain" is an important treatment. This technical strategy is considered as the

D.-S. Huang et al. (Eds.): ICIC 2018, LNAI 10956, pp. 284–295, 2018.
https://doi.org/10.1007/978-3-319-95957-3_32

advent of the new artificial intelligence era and also provides new theoretical guidance for the classification of unhealthy texts in the field of Chinese information.

In view of the problem of poor text filtering, researchers at home and abroad have done many useful attempts. Su et al. proposed the construction and classification based on approximate categories [1]. Xiong et al. put forward the idea of using the concept of network technology for text representation, and then using the support vector product to classify [2]. Foltz uses latent semantic indexing techniques to represent the text as a compressed concept combination and then text-filters it [3]. The above methods for recognizing and classifying unhealthy texts are basically independent and based on the theme of text classification. These methods have some limitations.

We use microblogs and comments as experimental data to carry out the segmentation and classification of bad information. The experiment is conducted under two brains working together and coordinated to achieve better segmentation results and classification results.

2 Related Technologies

AlphaGo is playing chess through two neural network “brains”. AlphaGo’s first neural Network is the “Policy Network,” which monitors the learning of the board and attempts to find the best next step. In fact, it predicts the best probability for each step, so the first one is the one with the highest probability. This is also known as the “drop selector”. The second brain predicts the probability of each player winning a chess game in the given position. This is the “Value Network”, which can assist the “drop selector” through the overall situation. They train and adjust the parameters to make the next execution better [4].

2.1 Mature Automatic Segmentation Technology

A dictionary-based word segmentation method requires a larger dictionary. The text should be segmented according to certain scanning rules and then matched with the dictionary. It’s also called a string matching method [5].

Based on the statistical participle method, it is a statistical analysis of the combination frequency of adjacent words in the corpus. In the corpus of N words, the probability of the independent occurrence of Chinese character A can be expressed as $P_{(A)} = N_A/N$, where N_A represents the number of occurrences of Chinese character A in the corpus. Assuming that A and B are completely independent in the corpus, the number of times that AB appears at the same time is about $P_{(A)} * P_{(B)} * N = (N_A * N_B)/N$, and N_{AB} represents the number of times AB actually appears in the text. So the correlation between the two words, C_{AB} can be represented by the following formula.

$$C_{AB} = \frac{\text{The actual number of times that AB appears}}{\text{The number of random combinations of AB}} = \frac{N_{AB}}{N_A \cdot N_B / N} = \frac{N_{AB} \cdot N}{N_A \cdot N_B} \quad (1)$$

If the C_{AB} exceeds a certain threshold, the probability that AB appears in the adjacent position is greater than the probability of the occurrence of a single occurrence, then the AB can be used as a phrase [6].

2.2 Distributed Caching and Data Processing Systems

In order to achieve high-speed processing of large-scale data, it is often required that the system has a short response time and a high processing speed. During data processing, the system is still working and there is still a lot of data entered. Therefore, it is of practical significance to design a distributed data processing system to improve query processing speed and load optimization of the system. To effectively balance the trade-off between I/O performance and fault tolerance, we adopt a complete solution called "AutoReplica" – a replica manager in distributed caching and data processing systems with SSD-HDD tier storages. In detail, AutoReplica utilizes the remote SSDs (connected by high speed fibers) to replicate local SSD caches to protect data [7].

3 Information Recognition and Classification Based on AlphaGo

The identification and classification of bad information operate under two "brain". A brain first performs a crawler program to obtain text, then extracts features from the text, and then identifies the information. This is a comparison to AlphaGo's "drop selector" network. A brain uses rule-based and statistical methods to compare and classify bad information bases. This is comparable to AlphaGo's "value network".

The two "brain" work together, the first "brain" grabs the text, performs the participle, and assists in identifying the word of the network by comparing the words of the network, so that the word segmentation can be carried out faster. The second "brain" compares and categorizes the words that are identified, and then studies autonomously (Fig. 1).

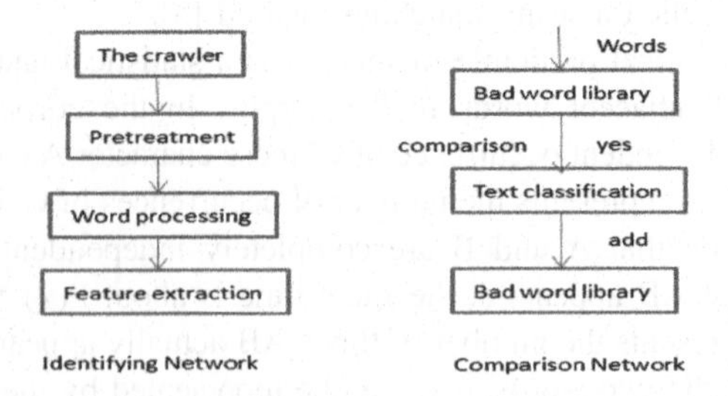

Fig. 1. Information recognition and classification based on AlphaGo

3.1 Information Recognition

AlphaGo's first neural network predicts the best probability of each next step, finding the most likely step, and making the choice. The first "brain" first needs to collect the texts of the online media, and then perform data processing and analysis to find out the words whose probability of word segmentation is greater than a certain threshold. This article uses the python language crawler framework and scrapy for data acquisition. After obtaining the text that needs analysis, this paper adopts a dictionary-based mechanical segmentation method, and selects the inverse maximum matching method with higher accuracy for word segmentation. The word segmentation dictionary uses the thesaurus provided by Sogou Lab (Fig. 2).

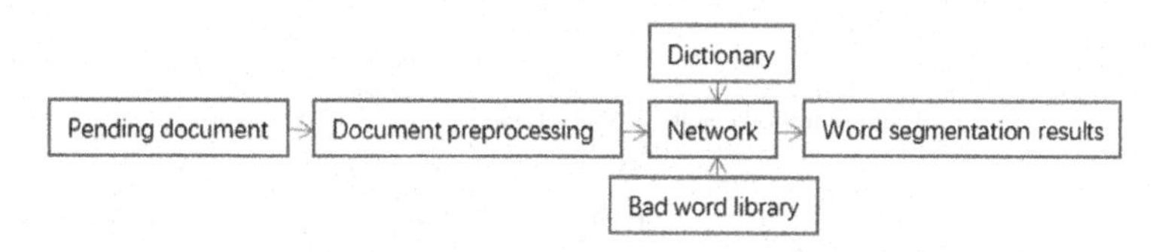

Fig. 2. Word segmentation process

(1) Segmentation processing. Documents after word segmentation should be subject to word frequency statistics. At the same time, the dependency relationship between the words in the document is obtained. It is represented by the triples $<w_i, w_j, r>$, where w_i denotes the word itself and r denotes the dependency of the word with other words.
(2) Stop-words filtering. As a document contains more words, some of the words have less influence on the classification of documents and affect the speed and accuracy of the classifier. Therefore, it is necessary to filter the documents first, such as "“的”" and "是" in the text.
(3) Getting the dependency set. Each sentence in the document is an independent semantic unit that contains multiple words. The dependency relationship between the words can represent the semantics of the sentence. And the main component of the sentence has a greater influence on the classification of the document, so the word set of the main component in the sentence is selected as the feature word.
(4) Calculation of weight values. The word frequency-inverse document frequency method is used to calculate the weight values of the words and sort them in descending order. The larger the weight value, the greater the influence of the word on the classification of the document. Therefore, setting topk = 10 filters out the words as the feature words of the document.
(5) Selection of feature items. Because the weight calculation method only considers the influence of word frequency on documents, and the criteria of dependency relationship can analyze the importance of words on semantic expression, this topic selects the selected feature words in the two feature selection methods as the final text features.

In the segmentation process, the comparison network quickly compares the entries in the thesaurus with the grabbing texts, thereby helping the segmentation to be completed quickly.

3.2 Information Identification and Classification

The second brain "value network" of Alpha Go is to predict the probability of each player winning the game given the position of the piece. The second "brain" of this paper will identify and classify the identified texts by comparing the bad information bases, predicting the uncivilized probability of each word and the classification probability of uncivilized words. The "brain" conducts autonomous deep learning and masters more uncivilized words and their classification (Fig. 3).

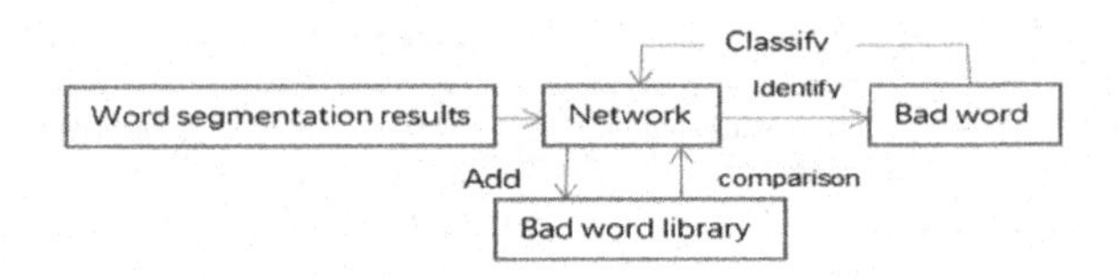

Fig. 3. Information determination and classification process

This paper uses the HowNet trend word set and the NTUSD3 merge to rebuild the uncivilized lan-guage library. This article combines the positive and negative emotional words of HowNet's emotional words (4566 positive emotional words, negative emotional words, 4370) with the Simplified Chinese Emotion Polarity Dictionary of Taiwan University's NTUSD (2810 positive emotional words, negative emotional words, 8276) to re-acquire a set of words as a basic sentiment dictionary. Then we combine the negative words in the two-dictionary resource pack to re-generate negative emotional word sets, which numbered 16,396. In this way, the basic sentiment lexicon is obtained. Then the negative emotion word set is segmented to construct the non-civilized language dictionary.

After the text has been identified by the system, some word vectors to be analyzed will flow out. These word vectors are sent into the comparison network to analyze non-civilized terms, and the cosine similarity is used to compare the word vectors with the non-civilized term lexicon. It is concluded that some words are non-civilized terms, and some are not non-civilized terms. Then the classification of non-civilized terms is processed.

The main steps of text classification are composed of two steps: training the classifier and predicting the classification effect of the classifier. The naive Bayes classifier model will be used to classify the review text. Naive Bayes classification is divided into three phases:

The first is the preparation phase, which is prepared for Naive Bayesian classification. The main tasks include data collection, invalid data elimination, data cleaning, and noise data processing. After obtaining valid data, feature attributes are selected according to the classification criteria to form a training sample set. The selection of feature attributes in this stage has a very important influence on the quality of the classification. The second stage is the classifier training phase. At this stage, a certain percentage of sample data is selected to train the classifier. Calculate the probability that the two categories appear in the sample, and the conditional probability that each feature is divided into two classes. Enter the resulting conditional probability values to train the classifier [8]. The third stage is the application stage. Classifiers are used to classify processed non-civilized terms. After the classification of feature words based on the thesaurus and dependency-based relationships, Naïve Bayes algorithm is used to classify them into three categories. One is a violent word, the other is a vulgar word, and the other is a typo.

Let $P(y_1|x)$, $P(y_2| x)$, …, $P(y_n|x)$ respectively represent the probability values of the class y_j corresponding to the sample x. The largest value is the category to which the sample belongs. It is expressed as an expression:

$$P(y_j|x) = \max\{P(y_1|x), P(y_2|x), \ldots, P(y_n|x)\} \tag{2}$$

Sample X contains m attributes, and $P(a_i|y_j)$ represents the conditional probability estimate of each feature attribute under each category. Since naive Bayes assumes that there are no dependencies between attributes, it is deduced from the Bayesian theorem as follows:

$$P(y_j|x) = \frac{P(x|y_j)P(y_j)}{P(x)} \tag{3}$$

The sample probability P(x) is the same for all classes. And it is a constant. So the denominator in the formula can be ignored, and the properties are independent of each other. The formula can be transformed into:

$$\begin{aligned} P(y_j|x) &= P(x|y_j)P(y_j) = P(a_1|y_j)P(a_2|y_j)\ldots P(a_m|y_j) \\ &= P(y_j)\prod_{i=1}^{m} P(a_i|y_j) \end{aligned} \tag{4}$$

Finally, the maximum probability value of the corresponding category is calculated, which is the class [9].

4 Simulation

Nowadays, microblog is very popular, but there are too many non-civilized expressions on it. This paper takes microblog comments as an example to carry out information identification and classification experiments.

4.1 Experimental Data Acquisition

This paper uses Python crawler technology. The language materials for crawling are mainly based on the online media materials provided by the national language resources network. The corpus has collected 48 language resources of 19 scientific research institutions. It also includes the open language resources of other universities, institutes and social institutions, as well as online news, blogs, micro-blogs and BBS.

The total number of characters is 150 billion, of which the total number of Chinese characters is 130 billion [10].

The development environment for this experiment is Scrapy 1.0 and Python 2.7 [11]. The contents of the "Chinese-Korea football wars" microblog of the National Language Council language resource network are crawled, and the acquired data is stored in a json (JavaScript object notation) format file (Table 1).

Table 1. Examples of microblog retweets and comments

id	context
1	辣鸡//@姐有点帅：韩国 19 号 记住这个垃圾
2	Mlgb 忍不了
3	！！！什么鬼！太 jian 了！
4	记住了//@real：垃圾//@专用润滑剂：辣鸡，记住辣鸡
5	转发微博
6	我艹你奶奶的，是不是有病
7	bichi
8	韩国人都是傻 X 转世的吧
9	这种垃圾在主场都这么嚣张哦。

(Note: retain the Chinese word segmentation.)

4.2 Analysis of Test Data

In the raw data collected, there are some repetitive data or blank data, which is not conducive to the quantitative and characteristic statistical process of text data. Therefore, we need to preprocess the data first. Reptile collection of original data to be saved as first .CSV format file, in which data can be expressed in the template as: comment on the user name, "text", "thumb up behavior", "comment period". Because data comments such as "/ / @ + content of the text: text" format text, the text after the "@" for microblog users, is not what we focus on the analysis of the object, the content is the content of the ":" behind. It is the key point of this topic research object. The data needs to be further processed, and the data is parsed using the column operation of the excel table. After data processing, the results are shown in the figure (Table 2).

Table 2. Processed data

id	nickname	comment
1	上古奇书	Mmp
2	未茗_Fearless	辣鸡
3	白曳	我也是
4	Jason-M	爆眼子娃娃你可能走不出长沙
5	汽车美图	身为一个企业号，我想骂人吗
6	吃糖闲步看鱼游	真垃圾
7	袅妇	这种垃圾在主场都这么嚣张哦
8	顾溪您祖宗	Mlgb 忍不了
9	浪荡女 yxyx	生气
10	Elena-婉儿	逼，不要脸
11	满江红 0000	再次看看小丑的德行
12	恭候春夏的轮替 y	转发微博
13	思源哥哥呀	垃圾
14	林夕 QAQ	很气，很想干他
15	Wi 嘉琪	草泥马

(Note: retain the Chinese word segmentation.)

In JSON format without keys to return the result of an array of information sorted according to the id, the cont, pos, the parent, relate order, respectively, the serial number of words in a sentence, participles, part of speech, parent node id number, interdependence syntactic relations (Fig. 4).

[[[[0, "这种", "r", 1, "ATT"], [1, "垃圾", "n", 6, "SBV"], [2, "在 ", "p", 6, "ADV"], [3, " 主 场 ", "n", 2, "POB"], [4, " 都 ", "d", 6, "ADV"], [5, " 这 么 ", "r", 6, "ADV"], [6, " 嚣 张 ", "a", -1, "HED"], [7, " 哦 ", "u", 6, "RAD"], [8, " 。 ", "wp", 6, "WP"]]]]

Fig. 4. JSON format data results (Note: retain the Chinese word segmentation.)

By calculating the word segmentation results in the text content, the word frequency and the frequency value of the reverse document are calculated, then the weight values are adjusted according to the result of the dependency syntax, and the feature items are obtained. Examples of feature screening results are as follows (Table 3):

Table 3. Feature weight table

Id	Cont	Relate	Parent	Tf-idf	Modify-weight
0	这种	ATT	1	0.5015	0.1223
1	垃圾	SBV	6	0.5990	0.5828
2	在	ADV	6	0.5015	0.1422
3	主场	POB	2	0.5930	0.3889
4	都	ADV	6	0.5342	0.1389
5	这么	ADV	6	0.5221	0.1439
6	嚣张	HED	-1	0.6279	0.5990
7	哦	RAD	6	0.0000	0.1211
8	。	WP	6	0.0000	0.0000

(Note: retain the Chinese word segmentation.)

4.3 Final Experimental Results

Now compare some words with the uncivilized dictionary, and the results are as follows (Table 4):

The experimental data were derived from sina microblog comments, which included three major categories of microblog, social micro-blog and economic category, as well as the data set (TAN) provided by Dr. TAN songbo of the Chinese academy of sciences. The microblog comments will be sorted, ensuring that the number of comments in each category is 20000. Cross validation was used in the experiment, and 40% of the data was collected as training set and 60% as test set. Finally, the accuracy and recall rate evaluation criteria were used for comprehensive evaluation. The classification accuracy calculation method is the ratio of the correct classification document number to the total number of documents, and the recall rate calculation method correctly classifies the number of documents and the number of documents to be classified.

Table 4. Comparison table

Id	The word vector	Bad language
1	Mmp	y
2	辣鸡	y
3	主场	y
4	垃圾	y
5	草泥马	y

(Note: retain the Chinese word segmentation.)

The following were selected different sizes of test methods to improve the test results before and after the effect of a number of experiments, the recognition accuracy of the comparison chart:

From Figs. 5, 6 and 7, we can see that the improved classification accuracy and recall rate have been improved. It can be seen that the identification and classification of non-civilized terms through the two systems designed by AlphaGo's thought are more accurate and can better reflect the category attributes of the text, thereby improving the classification performance of the classifier.

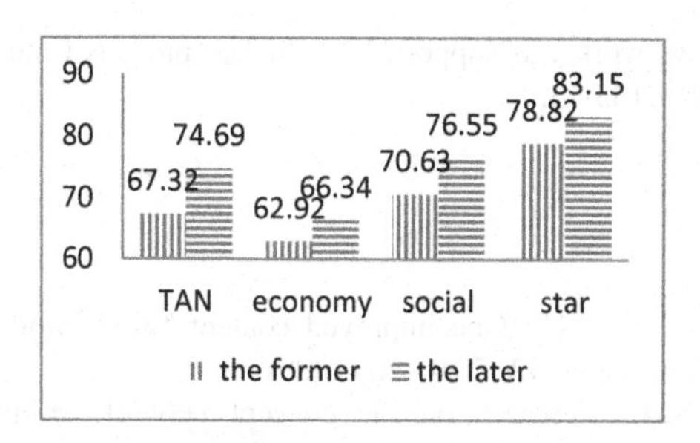

Fig. 5. Identification accuracy comparison

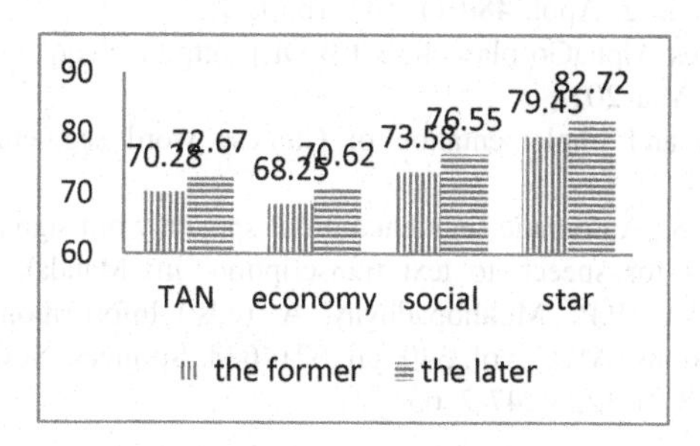

Fig. 6. Accuracy of text classification

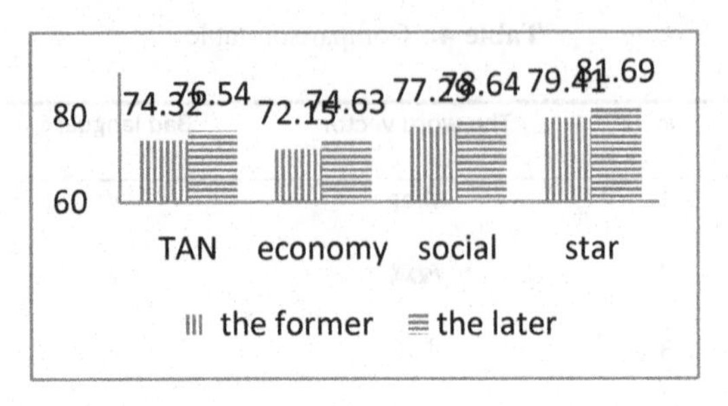

Fig. 7. Text classification recall rate

5 Conclusions

This paper integrates the research on the identification of bad information and the classification through the two "brain" strategies based on AlphaGo. Using AlphaGo's thought strategy to combine the previously dispersed methods into a whole, the system improves the efficiency of execution and the accuracy of recognition and classification without reducing the recall rate. In order to identify the indistinguishable similar text, this paper adopts the classification method of uncivilized word library and text. In the case of the lack of combination of existing word segmentation and classification, this paper proposes a method based on AlphaGo model for the identification of bad information in rules and statistical networks. Simulation results show the effectiveness of the proposed method.

Acknowledgments. This work was supported by the key projects China Language committee Research Project NO. ZDI135-13.

References

1. Su, G.Y., Ma, Y.H., Li, J.H.: One improved content based information filtering model. J. Shanghai Jiao Tong Univ. **12**, 2030–2034 (2014)
2. Xiong, J.X., Li, S.H.: Research on the concept network technique to the bad text information. Comput. Eng. Appl. **03**, 183–186 (2016)
3. Fan, X.H., Wang, P., Zhou, P.: Two-step text orientation identification based on feature extension. Comput. Eng. Appl. **48**(01), 162–165 (2012)
4. Dong, F.: How does AlphaGo play chess[EB/OL]. http://tech.qq.com/a/20160311/050990.html. Accessed 21 Mar 2018
5. Qin, Z.: Research and implementation of Chinese word segmentation algorithm. Jilin University (2016)
6. Roy, A., Phadikar, S.: Automatic segmentation of spoken word signals into letters based on amplitude variation for speech to text transcription. In: Mandal, J.K., Satapathy, S.C., Sanyal, M.K., Sarkar, P.P., Mukhopadhyay, A. (eds.) Information Systems Design and Intelligent Applications. AISC, vol. 340, pp. 621–628. Springer, New Delhi (2015). https://doi.org/10.1007/978-81-322-2247-7_63

7. Yang, Z., Wang, J., Evans, D., et al.: AutoReplica: automatic data replica manager in distributed caching and data processing systems. In: Performance Computing and Communications Conference. IEEE (2017)
8. Chu, X.L.: Machine Learning Based Patent Categorization. Shanghai Jiao Tong University (2008)
9. Veber, J.: Text classification: classifying plain source files with neural network. J. Syst. Integr. **1**(4), 39–44 (2010)
10. National Language Commission Language Resources Network. http://www.clr.org.cn/clr.jsp. Accessed 11 Apr 2018
11. Liu, T., Che, W.X., Li, Z.H.: Language technology platform. J. Chin. Inf. Process. **25**(06), 53–62 (2011)

Classification of Emotions from Video Based Cardiac Pulse Estimation

Keya Das(✉), Antony Lam, Hisato Fukuda, Yoshinori Kobayashi, and Yoshinori Kuno

Graduate School of Science and Engineering, Saitama University, Saitama, Japan
{keya0612, antonylam, fukuda, yoshinori, kuno}@cv.ics.saitama-u.ac.jp

Abstract. Recognizing emotion from video is an active research theme with many applications such as human-computer interaction and affective computing. The classification of emotions from facial expression is a common approach but it is sometimes difficult to differentiate genuine emotions from faked emotions. In this paper, we use a remote video based cardiac activity sensing technique to obtain physiological data to identify emotional states. We show that from the remotely sensed cardiac pulse patterns alone, emotional states can be differentiated. Specifically, we conducted an experimental study on recognizing the emotions of people watching video clips. We recorded 26 subjects that all watched the same comedy and horror video clips and then we estimated their cardiac pulse signals from the video footage. From the cardiac pulse signal alone, we were able to classify whether the subjects were watching the comedy or horror video clip. We also compare against classifying for the same task using facial action units and discuss how the two modalities compare.

Keywords: Video PPG · Cardiac pulse · Facial action units
Emotion recognition · Physiological signal processing

1 Introduction

Emotions play an essential role in many aspects of our daily lives, including decision making, perception, learning, and actions. Assessing emotions is a key to understanding human behavior. Thus, there has been much work on systems that identify emotional states. A popular approach is to identify emotion from facial expressions.

For example, Cohen et al. [1] proposed a method for facial expression recognition from video. They introduced a Tree-Augmented-Naive Bayes (TAN) classifier that learns the dependencies between facial features and provide an algorithm for finding the best TAN structure. In Zhang et al. [2] they propose a method of facial expression recognition based on local binary patterns (LBP) and local Fisher discriminant analysis (LFDA). However, the main limitation with facial expression recognition systems is that sometimes emotions do not have distinct facial expressions. Moreover, it has also been reported that subjects have experienced emotions, which were manifested through physiological responses without showing any visible changes in facial expressions [18]. Indeed, emotions can be defined as a mental state that occurs spontaneously

D.-S. Huang et al. (Eds.): ICIC 2018, LNAI 10956, pp. 296–305, 2018.
https://doi.org/10.1007/978-3-319-95957-3_33

without any conscious effort and is accompanied by physiological changes [18]. There are many examples of work in this area. For example, Kim et al. [3] reported an emotion recognition system with 78.4% and 6 1.8% accuracy for the recognition of 3 and 4 classes of emotions using ECG, skin temperature variation, and electrodermal activity. Zong et al. [4] used 25 features from ECG, electromyogram, skin conductivity (SC), and respiration changes by the Hilbert-Huang transform to obtain 76% accuracy for 4 classes. Picard et al. [5] used 40 features from heart rate, muscle tension, temperature, and SC to get 81% recognition accuracy on 8 classes. Guillaume et al. [6] obtained 80% recognition accuracy on 3 classes using an electroencephalogram (EEG). These researchers reinforce the finding that physiological changes primarily respond to emotion.

In our previous paper [23], we used remote video-based heart rate sensing to detect the presence of emotional arousal without relying on facial expression cues and had the benefit that physiologically based emotional arousal could be sensed without the need for attached sensors or even special equipment. We compared the heart rate data from our vision based video method to a wearable sensor (Fitbit) and found that the two methods gave us comparable heart rate estimates. Specifically, we found that the correlation between our vision based video method and the wearable sensor had a high correlation of 0.9. We also found that emotional arousal would result in a statistically significant increase in heart rate and our vision based video method had similar performance to the wearable sensor. However, our past work was only aimed at showing that emotional arousal could be detected via remote video based heart rate sensing. We did not classify the types of emotions (e.g. joy vs. fear) that the subjects experienced.

In this paper, we propose using our vision based video method to estimate cardiac pulse signals (as opposed to only the heart rate) and recognize emotion from the pulses. We show that from the cardiac pulses alone, we are able to determine the type of video (comedy vs. horror) the human subject was watching. We also compare our approach with using facial action units for emotion recognition.

2 Estimating Heart Rates from Video

The estimation of heart rate (HR) from conventional RGB video has received attention in recent years. These approaches have the benefit that a physiological response (HR) can be estimated without any special equipment and without the need for any attached sensors. Being unobtrusive, such methods have a number of applications in affective computing, human computer interaction, and can even be used to extract clinically useful information. For example, the subtle changes in the length of heart beats are associated with the health of the autonomic nervous system.

Thus, many approaches have been proposed. For example, Balakrishnan et al. [17], used subtle head oscillations that accompany the cardiac cycle to estimate HR. However, approaches considering minute changes in skin color are still the most common. This is based on the well-known photoplethysmographic (PPG) technique in which a pulse oximeter makes contact with skin, illuminates it, and changes in light absorption due to cardiac pulse is observed. The PPG approach is accurate but requires contact with the skin. Fortunately, it has been found that remote video based PPG is possible.

For example, Verkruysse et al. [11], found that conventional RGB cameras such as the Canon Powershot can capture small changes in light absorption that correspond well with the cardiac pulse. However, their tests were conducted under controlled conditions. Later on, several authors extended their work to account for more realistic settings [7, 10, 12, 13, 15]. These approaches have their strengths and weaknesses. In this paper, we choose to use the algorithm by Lam and Kuno [12] because it is effective for changing illumination. The changing lighting conditions encountered in real-world settings can be detrimental to color based HR estimation.

For completeness, we briefly summarize the method of Lam and Kuno [12]. In their work, they devised a basic model that assumes skin consists a hemoglobin component, where light absorption is influenced by cardiac activity, and pigments such as melanin that are not influenced by cardiac activity. Formally, their model expresses the pixel value for a single channel camera of a given point I on skin at time t as

$$I(t) = a_m a_l(t) \int R_m(\lambda) L(\lambda, t) C(\lambda) d\lambda + a_h(t) a_l(t) \int R_h(\lambda, t) L(\lambda, t) C_k(\lambda) d\lambda \quad (1)$$

where $R_m(\lambda)$ and $R_h(\lambda)$ are the normalized reflectance spectra of the melanin and hemoglobin pigments at wavelength λ, respectively, $L(\lambda, t)$ is the normalized light source spectrum at wavelength λ and time t. Similarly, $C(\lambda)$ is the camera's spectral response. The a terms are constants for scaling the different spectra inside the integrations.

Equation 1 indicates that if the light source spectrum (e.g. from a video display) changes colors over time, this affects the amount of light at different wavelengths. As a result, at any given time, the same light spectrum will affect the reflectance of light from the melanin and hemoglobin components differently. So the observed color changes in skin color appearance can be thought of as a mixture of two signals. It is well-known that techniques such as Independent Components Analysis (ICA) can be used to perform linear blind source separation (BSS) of signals. So these findings suggest that the hemoglobin (cardiac) signal can be estimated by taking two points on a person's face, obtaining the pixel traces of those points, treating them as signals, and applying ICA to separate the melanin and hemoglobin influenced color changes. With the hemoglobin part of the color changes determined, HR can be estimated.

However, Lam and Kuno showed that not all pairs of skin surface points could be used with ICA for linear BSS. For example, in the case where two skin points are illuminated by different colored light, the melanin at the different skin points could reflect very different colors and thus there would be three components (the two different colored light sources and the hemoglobin signal). This means that linear ICA (with only two input signals) would not be applicable to determine the two independent signals. However, if two points are illuminated by the same color spectrum (even at different brightness) and the ratios of melanin to hemoglobin between the two skin points are different, ICA can be used to effectively separate out the hemoglobin component of the signal.

Since appropriate skin point pairs would be difficult to know a priori, they decided to randomly test using ICA on different pairs of points of the face and observe the histogram of estimated HRs. Then the most common HR in the histogram was used as the final estimate of the HR. Their intuition is that randomly chosen point pairs that are good for HR estimates should consistently give the same HR estimate. But point pairs that are not suitable for HR estimates should give less consistent HR estimates. Thus, performing a majority vote on the many independent HR estimates from the random point pairs should give a robust estimate for a single final HR.

Their approach is effective for dealing with lighting changes but the basic implementation of their original algorithm did not account for pose changes of the head. In this paper, we use an improved version of their algorithm presented in Lam et al. [12]. In this version, tracked facial landmarks were used in combination with Delaunay Triangulation to determine triangular regions of interest on the face (as opposed to square patches). In addition, like Lam et al. [21] we also employed the RGB channel weighting scheme from Haan and Jeanne [22]. See Fig. 1 for the basic flow of the algorithm and a sample region pair.

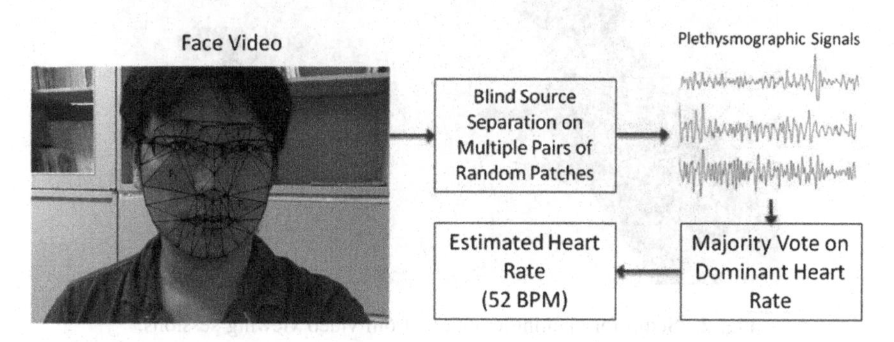

Fig. 1. Basic flow of the video PPG algorithm for estimating heart rate.

In this paper, our goal is actually to use the cardiac pulse signal (PPG signal) to perform emotion recognition. So what we actually do is that after estimating the final HR, we go back and identify all the PPG signals that contributed to the final HR estimate. We then normalized each of those PPG signals and compute their average. The final averaged PPG signal is then used as the estimated cardiac signal.

3 Dataset

In this work, our goal is to determine the feasibility of using cardiac pulse signals for recognizing different emotional states (joy vs. fear) and how this compares with the use of facial expressions. In works related to emotion recognition, acquiring emotionally stimulated physiological data is challenging because of the subjective nature of emotions and complex cognitive dependence of physiological signals.

Researchers have used different methods to elicit the target emotions such as visually showing images, using audio media such as music, and audio-visual stimuli in the form of short film video clips. In this work, emotions were induced by using short video clips. Twenty six subjects aged 20–25 years old participated in our study to rate the emotions they experienced while watching video clips. They were all of Asian descent and in sound health. We recorded videos of each subject's face for a total of 10 min while the subject watched movie video clips. In each video segment, the first 5 min consisted of the subjects watching a comedy video clip, and for the last 5 min, they watched a horror video clip. A wearable heart rate sensor (FitBit) was also used for verification of their cardiac responses. Figure 2 shows the experimental setup for building our dataset. Figure 3 shows sample screenshots from the videos used as emotional stimuli in our study.

Fig. 2. Setup for eliciting emotions from video viewing sessions.

After collecting our videos, we chose 30 s of the videos where the peak emotional responses would be expected. For example, in the case of the comedy video clip, we chose the 30 s segment of the video with the punchline for the joke. In the case of the horror video clip, we chose the 30 s segment with the scariest part of the video. We then used the OpenFace facial landmark tracker [19] to estimate facial landmarks for each subject on the 30 s segments in both the comedy and horror cases. We then used the tracked facial landmarks in conjunction with our version of Lam and Kuno's remote PPG algorithm (detailed in Sect. 2) for estimating cardiac pulses. Figure 4 shows sample screenshots from the videos of the subjects and their estimated cardiac pulses.

We also used OpenFace to obtain intensities for 17 facial action units [20] for each frame in the videos we analyzed. These facial action units were obtained to use as a means of comparison between our cardiac based approach and a facial expression based approach. For each 30 s clip, we determined the average intensity of each facial action unit resulting in 17 values associated with each 30 s video segment to be used for emotion recognition.

Fig. 3. Screenshots from the comedy (left) and horror (right) videos clips used as emotional stimuli in our study.

4 Results and Discussion

In this section, we present results on emotion classification using remotely estimated cardiac pulse signals from RGB video. In each case, the video viewing sessions lasted about 30 s and the videos were captured at about 30 FPS. This would imply that each cardiac pulse signal would result in 900 dimensions as there would be 900 frames per viewing session. Since the timings were not exact, some of the estimated cardiac pulse signals had slightly more than 900 dimensions. For example, 905 dimensions. We chose to simply discard the small number of extra dimensions so that all cardiac pulse signals would consist of exactly 900 dimensions. We also compare against using 17 facial action units to classify emotional response. Our classification task is to determine which genre of video the subject watched. In the case of the horror clip, we expect most people would experience fear and in the case of the comedy clip, we expect people would experience a combination of joy and happiness. Thus if we can classify the genre of video the human subject watched, we essentially classify their emotional reaction. In order to classify emotional responses, we took each of the 900 dimensional cardiac pulse signals and treated them as vectors. We then used a linear SVM to train a model based on ground truth labels of the video clip's genre. In order to evaluate generalization performance, we performed leave-one-out cross-validation on our dataset, which consists of 26 instances of reactions to horror and 26 reactions to comedy. Thus in each leave-one-out cross-validation test, 51 instances were used for training and 1 instance was used for testing. The average accuracy was then determined. For the case of facial action units, we also performed the same tests. In addition, we decided to test preprocessing our data using PCA (mainly because of the high dimensionality of the cardiac pulse signals). We chose to project both the cardiac pulse signals and facial action units onto their respective eigenvectors such that 90% of the variance would be retained. The results can be seen in Table 1.

In Table 1, we see from the "Video PPG" results that emotional reactions can be classified from remotely sensed cardiac activity alone. Furthermore, the cardiac pulse signals were estimated from a conventional RGB camera so no special equipment was required. From the first row's result labelled, "Linear SVM with Video PPG", we see a result of 65.4% accuracy. Since the dimensionality of the cardiac pulses was high (900 dimensions), we also decided to test using PCA to reduce the dimensionality of the

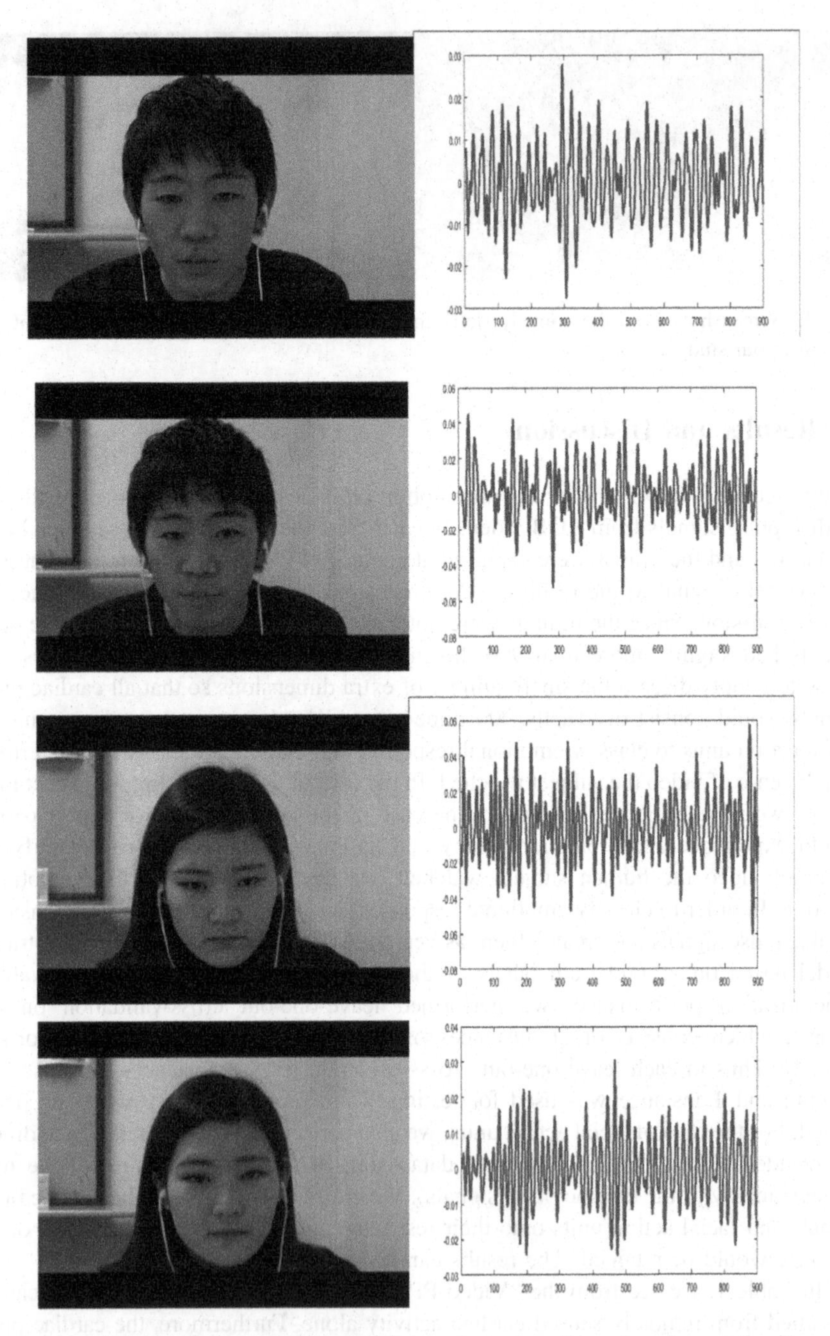

Fig. 4. Sample screenshots of subjects and their estimated cardiac pulses from video. For each subject, the first row shows them watching the horror clip. The second row for each subject shows them watching the comedy clip.

Table 1. Leave One Out Cross-Validation Results. We used linear SVMs to learn the classifiers in all cases. The term "Video PPG" denotes the cases where remotely estimated cardiac pulse signals from RGB video were used as the features for learning. Without using PCA to reduce dimensionality, we see that video PPG results are negatively impacted. With PCA, the results are not as good as in the case with facial action units but our proposed approach has the advantage that not relying on facial features, the emotional responses cannot be as easily faked.

Feature type	Leave one out cross-validation accuracy
Linear SVM with Video PPG	65.4%
Linear SVM with Video PPG (PCA)	67.3%
Linear SVM with Facial Action Units	76.9%
Linear SVM with Facial Action Units (PCA)	78.8%

cardiac pulse signals. We chose to project the cardiac pulse data onto eigenvectors such that 90% of the variance would be retained and then test again using Linear SVMs. We found that doing so did increase the accuracy to 67.3%. In the case of facial action units, we used 17 features and found that the leave one out cross-validation accuracy without PCA resulted in a 76.9% accuracy and with PCA, 78.8%. These classification results are better than our proposed cardiac-based classification approach. However, we believe this is because we did not instruct the participants to hide or fake their genuine emotions. As a result, our dataset was unfortunately not ideal for illustrating the effects of hidden or faked emotions on facial action unit based emotion recognition. Despite this setback, we did find that our "Video PPG" based approach is able to recognize emotional reactions with reasonably good performance. The major advantage of our approach is that since we do not rely on facial expressions, the emotional reactions we detect cannot be easily faked. This is because our sensing is based on physiological reactions.

5 Conclusion and Future Work

We have presented an approach to emotion recognition that is based on physiological responses rather than facial expressions. As a result, the emotions we detect cannot be easily faked. In addition, these cardiac pulse signals were entirely estimated from videos captured by a conventional RGB camera so no special equipment is required. Essentially, we have a system that operates completely using only computer vision techniques.

A drawback of the current study is that our dataset does not have human subjects that intentionally tried to hide or fake their emotions and so it was not ideal for our tests. However, we were able to show that even with a naïve approach like taking the estimated cardiac pulse signals, performing dimensionality reduction using PCA, and then leaning via linear SVM, we achieved surprisingly good accuracy. In the future, we will continue to investigate better ways to extract relevant features from the cardiac pulse signals to improve accuracy. Given the fact that our system only requires a conventional RGB camera, it would be interesting to explore various applications in affective computing in the future.

Acknowledgments. This work was supported by JSPS KAKENHI Grant Numbers JP17K12709, JP17K18850 and the Tateisi and Technology Foundation.

References

1. Cohen, I., Sebe, N., Garg, A., Chen, L.S., Huang, T.S.: Facial expression recognition from video sequences. In: Proceedings of the IEEE International Conference on Multimedia and Expo, vol. 2, pp. 121–124 (2002)
2. Zhang, S., Zhao, X., Lei, B.: Facial Expression recognition based on local binary patters and local fisher discriminant analysis. In: PMC (2011)
3. Kim, K.H., Bang, S.W., Kim, S.R.: Emotion recognition system using short-term monitoring of physiological signals. Med. Biol. Eng. Comput. **42**, 419–427 (2004)
4. Zong, C., Chetouani, M.: Hilbert-Huang transform based physiological signals analysis for emotion recognition. In: Proceedings of the IEEE International Symposium on Signal Processing and Information Technology, pp. 334–339 (2009)
5. Picard, R.W., Vyzas, E., Healey, J.: Toward machine emotional intelligence: analysis of affective physiological state. Proc. IEEE Trans. Pattern Anal. Mach. Intell. **23**(10), 1176–1189 (2001)
6. Chanel, G., Kierkels, J.J.M., Soleymani, M., Pun, T.: Short-term emotion assessment in a recall paradigm. Int. J. Hum. Comput. Stud. **67**, 607–627 (2009)
7. Li, X., Chen, J., Zhao, G., Pietkainen, M.: Remote heart rate measurement from face videos under realistic situations. In: Proceedings of the IEEE Conference on Computer Vision and Pattern Recognition, pp. 4264–4271 (2014)
8. Monkaresi, H., Sazzad, M., Calvo, R.A.: Using remote heart rate measurement for affect detection. In: The Twenty-Seventh International Flairs Conference of the Florida Artificial Intelligence Research Society Conference, pp. 119–123 (2014)
9. Wu, H.Y., Rubinstein, M., Shih, E., Guttag, J., Durand, F., Freeman, W.T.: Eulerian video magnification for revealing subtle changes in the world. ACM Trans. Graph. **31**(4), 1–8 (2012)
10. Kwon, S., Kim, H., Park, K.S.: Validation of heart rate extraction using video imaging on a built-in camera system of a smartphone. In: Proceedings of the IEEE Engineering in Medicine and Biology Society (EMBC), pp. 2174–2177, August 2012
11. Verkruysse, W., Svaasand, L.O., Nelson, J.S.: Remote plethysmographic imaging using ambient light. Opt. Express **16**, 21434–21445 (2008)
12. Lam, A., Kuno, Y.: Robust heart rate measurement from video using select random patches. In: Proceedings of the IEEE International Conference on Computer Vision, pp. 3640–3648 (2015)
13. Poh, M.Z., McDuff, D., Picard, R.: Advancements in noncontact, multiparameter physiological measurements using a webcam. Proc. IEEE Trans. Biomed. Eng. **58**(1), 7–11 (2011)
14. Tulyakov, S., Alameda-Pineda, X., Ricci, E., Yin, L., Cohn, J.F., Sebe, N.: Self-adaptive matrix completion for heart rate estimation from face videos under realistic conditions. In: Proceedings of the IEEE Conference on Computer Vision and Pattern Recognition, CVPR 2016 (2016)
15. Kaewkannate, K., Kim, S.: A comparison of wearable fitness devices (2016). https://bmcpublichealth.biomedcentral.com/articles/10.1186/s12889-016-3059-0
16. Evenson, K.R., Goto, M., Furberg, R.D.: Systematic review of the validity and reliability of consumer-wearable activity trackers. Int. J. Behav. Nutr. Phys. Act. **12**, 159 (2015)

17. Balakrishnan, G., Durand, F., Guttag, J.: Detecting pulse from head motions in video. In: Proceedings of the IEEE Computer Vision and Pattern Recognition, CVPR 2013 (2013)
18. Chakraborty, P.R., Zhang, L., Tjondronegoro, D., Chandra, V.: Using viewer's facial expression and heart rate for sports video highlights detection, pp. 371–378. ACM (2015)
19. Amos, B., Ludwiczuk, B., Satyanarayanan, M.: OpenFace: a general-purpose face recognition library with mobile applications. CMU-CS-16-118, CMU School of Computer Science, Technical report (2016)
20. Ekman, P., Rosenberg, E.L.: What the Face Reveals: Basic and Applied Studies of Spontaneous Expression Using the Facial Action Coding System (FACS). Oxford University Press, New York (1997)
21. Lam, A., Otsu, K., Das, K., Kuno, Y.: Towards taking pulses over youtube to determine interest in video content. In: Proceedings of the IEEE International Conference on Computer Vision (IW-FCV). IEEE (2018)
22. de Haan, G., Jeanne, V.: Robust pulse rate from chrominance-based rPPG. IEEE Trans. Biomed. Eng. **60**(10), 2878–2886 (2013)
23. Das, K., Ali, S., Otsu, K., Fukuda, H., Lam, A., Kobayashi, Y., Kuno, Y.: Detecting inner emotions from video based heart rate sensing. In: 13th International Conference on Intelligent Computing, ICIC 2017, pp. 48–57 (2017)

Face Recognition Using Improved Extended Sparse Representation Classifier and Feature Descriptor

Mengmeng Liao and Xiaodong Gu(✉)

Department of Electronic Engineering, Fudan University,
Shanghai 200433, China
xdgu@fudan.edu.cn

Abstract. Representing query samples, many methods based on sparse representation do not take into account the different importance of atoms. In this paper, we propose a new extended sparse weighted representation classifier (ESWRC). In ESWRC, we introduce a representativeness estimator, and use it to estimate the atom representativeness. The atom representativeness is used to construct the weights of atoms. The weighted atoms are used to represent the query samples. In addition, we propose a distinctive feature descriptor, called logarithmic weighted sum (LWS) feature descriptor, which combines the advantages of discrete orthonormal S-transform feature, Gabor feature, covariance and logarithmic operation. We combine ESWRC and LWS for face recognition and call it improved extended sparse representation classifier and feature descriptor (IESRCFD) method. Experimental results show that IESRCFD outperforms many state-of-the-art methods.

Keywords: Image recognition · LWS feature descriptor
Extended sparse weighted representation classifier

1 Introduction

Face recognition is a hot topic in the fields of computer vision and pattern recognition, and many researchers have been working on it [1]. Various techniques have been used for face recognition such as principal component analysis (PCA) [2], linear discriminant analysis (LDA) [3], local preserving projections (LPP) [4] and Eigenface [5] during the past few years. However, these early methods can only deal with simple face recognition problems well. Extreme learning machine (ELM) [6] is a single hidden layer network which can solve face recognition problems very quickly. But its recognition rate is limited. Although sparse representation classifier (SRC) [7] performs better than methods like PCA and LDA, it has no robustness for large continuous occlusion. Yang et al. [8] proposed a regularized robust coding (RRC) model to improve the robustness of SRC, which could regress a given signal with regularized regression coefficients. By assuming that the coding residual and the coding coefficient are respectively independent and identically distributed, RRC seeks for a maximum

D.-S. Huang et al. (Eds.): ICIC 2018, LNAI 10956, pp. 306–318, 2018.
https://doi.org/10.1007/978-3-319-95957-3_34

posterior solution of the coding problem. However, it can only mitigate the occlusion in face recognition to some extent.

In face recognition, the intra-class difference which is caused by variable expressions, illuminations, and disguises, can be shared across different subjects [9]. Based on this idea, Deng et al. [9] proposed the extended sparse representation-based classifier (ESRC). ESRC uses the intra-class variant dictionary and training samples to represent the test samples, which improves the recognition performance for occlusion or non-occlusion face images. In ESRC, a test sample is represented by several training samples that come from the same class. For training samples within a same class, ESRC holds that they have equal importance when representing other samples together. Because different samples even from the same class holds different amount of information or content, which suggests that they should be considered distinctively.

The statistical properties of features such as covariance matrix and distribution characteristics can also be regarded as a new feature and can be used for face recognition. Because they can reflect the characteristics of the original feature. Logarithmic image processing is a mathematical framework based on abstract linear mathematics [10]. It replaces the linear arithmetic operations with a non-linear one which more accurately characterizes the response of human eyes [11]. Texture feature is a kind of global feature. Discrete orthonormal S-transform (DOST) [12] can measure it. Hence, the coefficient of DOST can be view as a global feature. Furthermore, DOST can preserve the phase information. Gabor feature is a local feature. It has a good spatial locality and directional selectivity. Hence, DOST and Gabor feature can be used to construct the discriminative features.

In this paper, we propose an improved extended sparse representation classifier and feature descriptor (IESRCFD) method. In IESRCFD, we first define a logarithmic weighted sum (LWS) feature descriptor which combines advantages of global feature, local feature, statistical properties of features and logarithmic operation. Next, we estimate the atom representativeness using the proposed representativeness estimator. Then we propose an extended sparse weighted representation classifier (ESWRC) by using the representativeness. In ESWRC, the representativeness is used to reflect the importance of the atom when representing other samples. The atom representativeness is incorporated into the sparse representation process as a weight coefficient. Finally, IESRCFD is obtained by combining ESWRC and LWS feature descriptor.

The main contributions of this paper are as follows.

(1) Defined a logarithmic weighted sum (LWS) feature descriptor which describes images in a more accurate way. It has combined the advantages of discrete orthonormal S-transform feature, Gabor feature, covariance and logarithmic operation. Gabor feature plays the most important role in our algorithm among them.
(2) Proposed ESWRC considering the importance of each atom in representing the query samples by assigning a weight to each atom, which improves the recognition rate.
(3) Proposed to use ESWRC and LWS together as IESRCFD to achieve a very high recognition rate.

The rest of paper is organized as follows. Section 2 introduces logarithmic weighted sum feature descriptor. Section 3 introduces the proposed representativeness estimator. Section 4 is the proposed extended sparse weighted representation classifier. Section 5 is the proposed improved extended sparse representation classifier and feature descriptor (IESRCFD) method. Section 6 gives the experimental results. Section 7 is the conclusion.

2 Logarithmic Weighted Sum Feature Descriptor

In this paper, we propose a logarithmic weighted sum (LWS) feature descriptor which combines those advantages of discrete orthonormal S-transform (DOST) feature, Gabor feature, covariance and logarithmic operation. Figure 1 is the construction process of LWS feature descriptor. In Fig. 1, w_1 and w_2 are the weights.

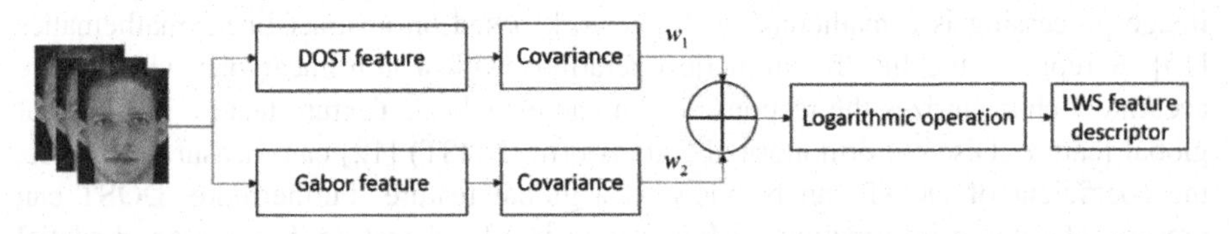

Fig. 1. The construction process of LWS feature descriptor.

2.1 Discrete Orthonormal S-Transform (DOST) Feature

Recent work shows that a discrete orthonormal S-transform (DOST) basis can be used to accelerate the calculation of S-transform (ST) [13] and eliminate the redundancy in the space-frequency domain [14]. Besides, DOST can preserve the phase information and allows for an arbitrary partitioning of the frequency domain, which achieves the zero redundancy. Those advantages make DOST has a wide range of applications, such as texture classification [12] and signal analysis [14].

DOST is a pared-down version of the fully redundant ST [14]. The main idea of DOST is to form N orthogonal unit-length basis vectors. Each base vector corresponds to a specific region in the time-frequency domain. And each specific region can be determined by the following three parameters: v, γ and τ. v is the center of each frequency domain band, γ is the width of that band, τ is the location in time domain. Thus the λ th basis vector $D[\lambda]_{[v,\gamma,\tau]}$ can be defined as

$$D[\lambda]_{[v,\gamma,\tau]} = \begin{cases} ie^{-i\pi\tau} \frac{e^{-i2\alpha(v-\gamma/2-1/2)} - e^{-i2\alpha(v+\gamma/2-1/2)}}{2\sqrt{\gamma}\sin\alpha}, & \alpha \neq 0 \\ -\sqrt{\gamma} ie^{-i\pi\tau}, & \alpha = 0 \end{cases} \tag{1}$$

where α equals to $\pi(\lambda/N - \tau/\gamma)$, and it represents the center of temporal window. $\lambda = 0, 1, \ldots, N-1$.

For any one dimensional signal $h[\lambda]$, its length is N. Its DOST coefficients $S_{[v,\gamma,\tau]}$ for the region corresponding to the choice of $[v, \gamma, \tau]$ can be obtained by formula (2).

$$S_{[v,\gamma,\tau]} = \frac{1}{\sqrt{\gamma}} \sum_{f=v-\gamma/2}^{v+\gamma/2-1} e^{-i\pi\tau} e^{i2\pi\frac{\tau}{\gamma}f} \left[\sum_{\lambda=0}^{N-1} e^{-i2\pi\frac{\lambda}{N}f} h[\lambda] \right] \quad (2)$$

where the content in the square brackets is $H[f]$.

For each input image, the DOST coefficient matrix can be obtained by using discrete orthonormal S-transform. The DOST coefficient matrix has the same size of the input image.

2.2 Gabor Wavelet Feature

Two dimensional Gabor wavelet transform is a kind of wavelet transform. It has a good time-frequency localization characteristic. Besides, the function of two dimensional Gabor wavelet is similar to enhancing bottom image features including edge and peak, as well as local features. The extracted Gabor feature not only has a good spatial locality and directional selectivity, but also is robust to illumination and pose variations. Hence, in this paper, we choose the two dimensional Gabor wavelet to extract the local feature of an image.

For each pixel in an image, a vector $F_{x,y}$ can be obtained by

$$F_{x,y} = [I(x,y), x, y, |G_{0,0}(x,y)|, \cdots, |G_{0,7}(x,y)|, |G_{1,0}(x,y)|, \cdots, |G_{4,7}(x,y)|] \quad (3)$$

where $I(x,y)$ represents the intensity value of position (x,y), $G_{\varphi,\sigma}(x,y)$ is the response of a two dimensional Gabor wavelet centered at (x,y) with orientation φ and scale σ:

$$G_{\varphi,\sigma}(x,y) = \frac{k_\sigma^2}{4\pi^2} \sum\nolimits_{t,s} e^{-\frac{k_\sigma^2}{8\pi^2}((x-s)^2 + (y-t)^2)} \left(e^{ik_\sigma((x-t)cos(\theta_\varphi) + (y-s)sin(\theta_\varphi))} - e^{-2\pi^2} \right) \quad (4)$$

where $k_\sigma = \frac{1}{\sqrt{2^{\sigma-1}}}$, $\theta_\varphi = \frac{\pi\varphi}{8}$. Let $\varphi = 8$, $\sigma = 5$.

For an image Q, we assume that the row number of Q is $\hat{A}$, the column number of Q is $\hat{B}$. Then, its Gabor wavelet feature $\mathbf{O}$ is denoted by

$$\mathbf{O} = [F_{1,1}^T, \cdots, F_{1,\hat{A}}^T, F_{2,1}^T \cdots, F_{2,\hat{A}}^T, \cdots, F_{\hat{B},1}^T, \cdots, F_{\hat{B},\hat{A}}^T]^T \quad (5)$$

2.3 LWS Feature Descriptor

For image Q, its Gabor wavelet feature is $\mathbf{O}$. The covariance matrix of $\mathbf{O}$ is denoted by C_L. Meanwhile, image $\hat{Q}$ with the size of 43×43 is obtained by down-sampling the image Q. The DOST coefficient matrix $\tilde{\mathbf{O}}$ of image $\hat{Q}$ is obtained by formula (2). The covariance matrix of $\tilde{\mathbf{O}}$ is denoted by C_G. Taking image Q as an example, we define a new logarithmic weighted sum (LWS) feature descriptor.

$$\boldsymbol{FD} = \log[\tilde{\lambda} \cdot \boldsymbol{C}_G + (1-\tilde{\lambda}) \cdot \boldsymbol{C}_L] \quad (6)$$

where $\boldsymbol{FD}$ is the LWS feature descriptor of Q, $\log[\cdot]$ represents the logarithmic operation, $\tilde{\lambda}$ is the weight coefficient, $0 \leq \tilde{\lambda} \leq 1$. The LWS feature descriptors of other images in data set are obtained in this way.

3 Representativeness Estimator

In methods based on sparse representation, a test sample is represented by several atoms. For different atoms, because they contain different amounts of information, they have different representativeness to represent other samples generally. In this paper, we propose a representativeness estimator, and use it to estimate the atom representativeness.

The main steps to estimate the representativeness by using representativeness estimator are as follows.

Firstly, for any sample (e.g., image), we remove its relativity by whitening. Then its representativeness is estimated by computing its information entropy. This process is as follows.

Given an arbitrary sample (e.g., image) $\boldsymbol{S}$, its size is $M \times N$.

3.1 Remove the Relativity of Sample

(1) Remove the mean of $\boldsymbol{S}$.

$$\boldsymbol{X} = \boldsymbol{S} - \bar{\boldsymbol{S}} \quad (7)$$

where $\bar{\boldsymbol{S}}$ is the mean of $\boldsymbol{S}$.

(2) $\boldsymbol{X}$ is arranged into a column vector.

(3) Compute the covariance matrix $\boldsymbol{\Pi}$.

$$\boldsymbol{\Pi} = \mathrm{E}[\boldsymbol{X}\boldsymbol{X}^T] \quad (8)$$

where $\mathrm{E}[\cdot]$ represents the mathematical expectation.

(4) Singular value decomposition.

$$\boldsymbol{\Pi} = \mathbf{U} * \boldsymbol{\Lambda} * \mathbf{U}^T \quad (9)$$

where $\boldsymbol{\Lambda}$ is a diagonal matrix.

(5) Compute the whitening matrix $\tilde{\boldsymbol{M}}$.

$$\tilde{\boldsymbol{M}} = \boldsymbol{\Lambda}^{-1/2}\mathbf{U}^T \quad (10)$$

(6) Compute the whitened matrix $\mathbf{Z}$.

$$\mathbf{Z} = \tilde{\mathrm{M}} * \boldsymbol{X} \quad (11)$$

3.2 Compute the Information Entropy

(1) Z is arranged into a matrix $\tilde{Z}$, and the size of $\tilde{Z}$ is $M \times N$.
(2) Compute the information entropy of $\tilde{Z}$.

$$H_{\tilde{Z}} = -\sum_{i=0}^{255} p_i \log p_i \quad (12)$$

where $H_{\tilde{Z}}$ is the information entropy of $\tilde{Z}$, p_i represents the proportion of pixel whose grayscale is i.

In this paper, $H_{\tilde{Z}}$ represents the representativeness of image S, and the greater the value of $H_{\tilde{Z}}$, the stronger its representativeness.

4 Extended Sparse Weighted Representation Classifier

In ESRC, for atoms with the same class, because they contain different amounts of information, they should have different representativeness to represent other samples. That is to say, their importance is not the same when they together represent other samples. In order to reflect their importance, we propose an extended sparse weighted representation classifier (ESWRC).

We assume that $\psi = [\psi_1, \psi_2, \psi_3, \ldots, \psi_k] \in \Re^{d \times n}$ are the training samples, where $\psi_i \in \Re^{d \times n_i}$ are the training samples with class i. Each column of ψ_i represents a training sample with class i. That is to say, ψ_i contains n_i training samples.

Let $\mathbf{B} = [\mathbf{B}_1 - c_1 e_1, \ldots, \mathbf{B}_l - c_l e_l] \in \Re^{d \times m}$ represents the intra-class variant bases, where $e_i = [1, \ldots, 1] \in \Re^{1 \times m_i}$, $c_i \in \Re^{d \times 1}$ is the class centroid of class i, $\mathbf{B}_i \in \Re^{d \times m_i}$, $i = 1, 2, \ldots, l$, $\sum_{i=1}^{l} m_i = m$. $\mathbf{B}_i$ are randomly selected from ψ_i, and they are used to obtain the intra-class variant bases. Each column of $\mathbf{B}_i$ represents a training sample with class i. That is to say, $\mathbf{B}_i$ contain m_i training samples.

The main idea of ESWRC can be illustrated by formula (13).

$$\eta = (\mathbf{W}_\psi \otimes \psi)x + (\mathbf{W}_\mathbf{B} \otimes \mathbf{B})\beta + z \quad (13)$$

where η is a test sample. $\mathbf{W}_\psi$ and $\mathbf{W}_\mathbf{B}$ are the weight coefficients. They correspond to ψ and $\mathbf{B}$ respectively. x and β are the sparse vectors. z is a noise term with bounded energy $\|z\|_2 < \varepsilon$, and $z \in \Re^d$. '$\otimes$' is the Hadamard product operator. Figure 2 is the diagram of ESWRC.

$$\mathbf{W}_\psi = [\mathbf{W}_{\psi_1}, \mathbf{W}_{\psi_2}, \mathbf{W}_{\psi_3}, \ldots, \mathbf{W}_{\psi_k}] \quad (14)$$

$$\mathbf{W}_\mathbf{B} = [\mathbf{W}_{\mathbf{B}_1}, \mathbf{W}_{\mathbf{B}_2}, \mathbf{W}_{\mathbf{B}_3}, \ldots, \mathbf{W}_{\mathbf{B}_l}] \quad (15)$$

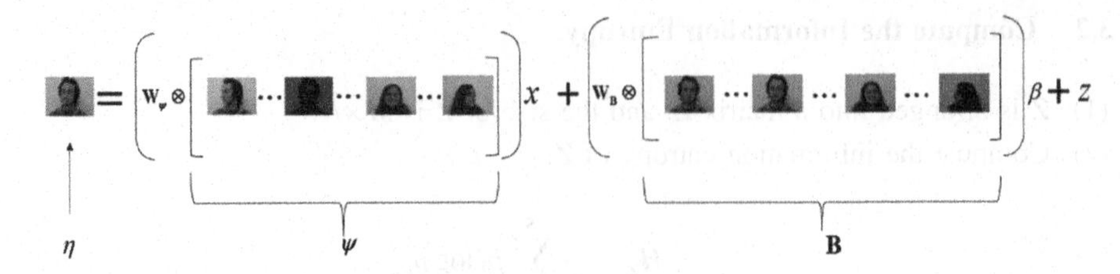

Fig. 2. The diagram of ESWRC.

where $\mathbf{W}_{\psi_i} \in \Re^{1 \times n_i}$ corresponds to ψ_i, $i = 1, 2, \ldots, k$. $\mathbf{W}_{\mathbf{B}_j} \in \Re^{1 \times m_j}$ corresponds to $\mathbf{B}_j - c_j e_j$, $j = 1, 2, \ldots, l$.

For all samples with class i, the expression of $\mathbf{W}_{\psi_i}$ is as follows.

$$\mathbf{W}_{\psi_i} = [\Gamma_{i1}, \Gamma_{i2}, \Gamma_{i3}, \ldots, \Gamma_{in_i}] \tag{16}$$

where Γ_{ij} represents the contribution factor of jth image in class i.

$$\Gamma_{ij} = \frac{H_{ij}}{\sum\limits_{j=1}^{n_i} H_{ij}} \tag{17}$$

where H_{ij} represents the information entropy of jth image in class i.

Let $(\mathbf{W}_{\psi} \otimes \psi) = \psi^{\bullet} = [\psi_1^{\bullet}, \psi_2^{\bullet}, \psi_3^{\bullet}, \ldots, \psi_k^{\bullet}] \in \Re^{d \times n}$, where $\psi_i^{\bullet}$ represent those new samples with class i. Each column of $\psi_i^{\bullet}$ represents a new sample. $(\mathbf{W}_{\mathbf{B}} \otimes \mathbf{B}) = \mathbf{B}^{\bullet} = [\mathbf{B}_1^{\bullet} - c_1^{\bullet} e_1, \mathbf{B}_2^{\bullet} - c_2^{\bullet} e_2, \ldots, \mathbf{B}_l^{\bullet} - c_l^{\bullet} e_l]$, where $e_i = [1, \ldots 1] \in \Re^{1 \times m_i}$, $c_i^{\bullet} \in \Re^{1 \times m_i}$ is the class centroid of class i, $\mathbf{B}_i^{\bullet} \in \Re^{1 \times m_i}$, $i = 1, 2, \ldots, l$, $\sum\limits_{i=1}^{l} m_i = m$. $\mathbf{B}_i^{\bullet}$ are the new training samples, and they are used to obtain the intra-class variant bases. Each column of $\mathbf{B}_i^{\bullet}$ represents a new training sample with class i. That is to say, $\mathbf{B}_i^{\bullet}$ contain m_i new training samples.

Firstly, the dimensions of $\psi^{\bullet}, \mathbf{B}^{\bullet}$ and η are reduced by PCA. Then $\tilde{\psi} = [\tilde{\psi}_1, \tilde{\psi}_2, \tilde{\psi}_3, \ldots, \tilde{\psi}_k] \in \Re^{\tilde{d} \times n}$, $\tilde{\mathbf{B}} = [\tilde{\mathbf{B}}_1 - \tilde{c}_1 e_1, \ldots, \tilde{\mathbf{B}}_l - \tilde{c}_l e_l] \in \Re^{\tilde{d} \times m}$ and $\tilde{b} \in \Re^{\tilde{d} \times 1}$ are obtained, where $\tilde{\psi}_i \in \Re^{\tilde{d} \times n_i}$, $\tilde{\mathbf{B}}_i \in \Re^{\tilde{d} \times m_i}$ and $\tilde{c}_i \in \Re^{\tilde{d} \times 1}$ correspond to ψ_i, $\mathbf{B}_i$ and c_i respectively.

Let $\tilde{\mathbf{A}} = [\tilde{\psi}, \tilde{\mathrm{B}}]$, $\tilde{x} = \begin{bmatrix} x \\ \beta \end{bmatrix}$, $G(\tilde{x}) = \|\tilde{x}\|_1$, $H(\tilde{x}) = \tilde{b} - \tilde{\mathbf{A}}\tilde{x}$.

We have the following sparse representation problem.

$$\min_{\tilde{x}} \|\tilde{x}\|_1 \quad \text{sub.to } \tilde{b} = \tilde{\boldsymbol{A}}\tilde{x} \tag{18}$$

where $\tilde{b}$ is the test sample that needs to be represented, $\tilde{\boldsymbol{A}}$ is the sparse dictionary matrix.

We use the augmented Lagrangian method to solve (18), then obtain the solution of (18), and denoted by $\tilde{x}^* = \begin{bmatrix} \breve{x} \\ \breve{\beta} \end{bmatrix}$. After that, the residual is computed by

$$R_i = (\tilde{b}) = \left\| \tilde{b} - \tilde{\boldsymbol{A}} \begin{bmatrix} \delta_i(\breve{x}) \\ \breve{\beta} \end{bmatrix} \right\|_2 \tag{19}$$

where $i = 1, 2, \ldots, k$, $\delta_i(\breve{x})$ is a new vector whose only nonzero entries are the entries in $\breve{x}$ those are associated with class i.

The category of probe sample $\tilde{b}$ can be obtained by

$$\text{Identity } (\tilde{b}) = \arg\min_i R_i(\tilde{b}) \tag{20}$$

Hence, the label of η is Identity $(\tilde{b})$.

Figure 3 is the flow chart of EWSRC.

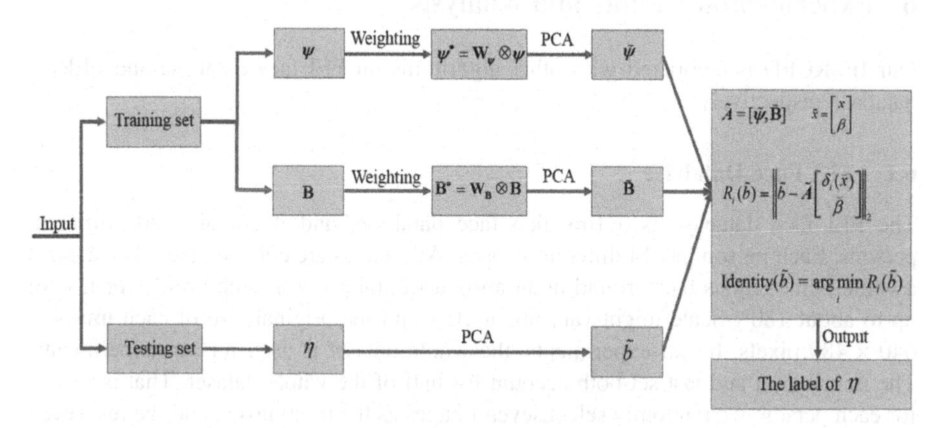

Fig. 3. The flow chart of EWSRC.

5 Improved Extended Sparse Representation Classifier and Feature Descriptor Method

On the one hand, the defined LWS feature descriptor is robust to illumination variations and pose variations. On the other hand, EWSRC has a strong robustness to occlusion and non-occlusion. In order to combine the advantages of LWS feature descriptor and EWSRC, we propose an improved extended sparse representation classifier and feature descriptor (IESRCFD) method. IESRCFD is as follows. First of all, LWS feature descriptor of each image is obtained by formula (9). Then the obtained LWS feature descriptors are used as the input of ESWRC. Finally, the labels of testing samples are obtained. The time complexity of IESRCFD is about $O(n^2)$. Figure 4 is the flow chart of IESRCFD.

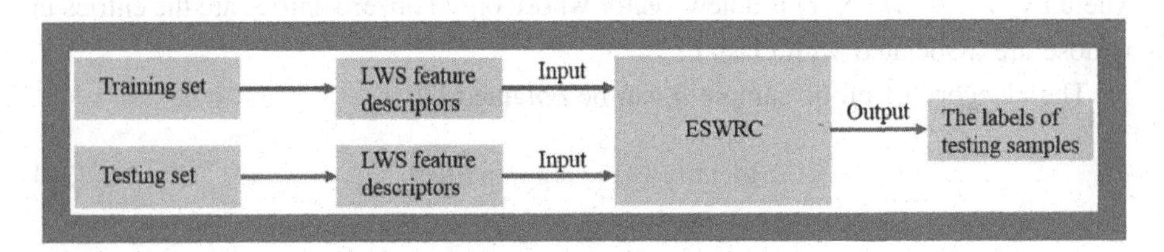

Fig. 4. The flow chart of IESRCFD.

6 Experimental Results and Analysis

Our IESRCFD is compared with other algorithms on FEI face database and FERET database respectively.

6.1 FEI Face Database

The FEI face database is a Brazilian face database, and it contains 200 different persons. Each person has 14 different images. All images are colorful and taken against a white homogenous background in an upright frontal position with profile rotation of up to about 180°. Scale might vary about 10% and the original size of each image is 640×480 pixels. In our experiments, the whole dataset is used for our experiments. The training set and test set both account for half of the whole dataset. That is to say, for each person, we randomly select seven images as the training set, and the rest seven images are used as the test set. The size of each image is down sampled to 64×64. Figure 5 shows some examples of image variations from the FEI face database.

Fig. 5. Some examples of image variations from the FEI face database.

Table 1 lists the recognition rates of different algorithms on FEI face database.

Table 1. Recognition rates of different algorithms on FEI face database ($\tilde{\lambda} = 0.1$)

Algorithm	Recognition rate
RRC [8]	56.4%
ESRC [9]	79.1%
SRC [7]	79.1%
KNN [15]	66.1%
NN [16]	66.1%
ELM [6]	61.5%
ELMSRC [17]	80.7%
RSC [18]	71.3%
KSRC [19]	77.9%
IESRCFD	**91.6%**

From Table 1 we can see that the recognition rate of our IESRCFD is 91.6%, which achieves the highest recognition rate, with about 35%, 12%, 12%, 25%, 30%, 11%, 20% and 14% improvements over RRC, ESRC, SRC, KNN, ELM, ELMSRC, RSC and KSRC respectively.

6.2 FERET Database

As in [20, 21], the "b" subset of FERET database is used to verify the recognition performance of various algorithms. The "b" subset of FERET consists of 198 subjects, each subject contains 7 different images. The training set images contain frontal face images with neutral expression "ba", smiling expression "bj", and illumination changes "bk". While the test set images involve face images of varying pose angle: "bd"- $+25°$, "be"- $+15°$, "bf"- $-15°$, and "bg"- $-25°$. As in [20], all images are down sampled to 64×64. Some samples of face images in "b" subset of FERET database are shown in Fig. 6.

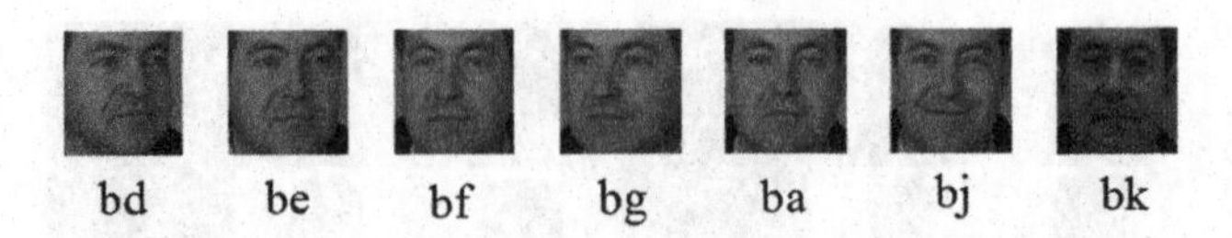

Fig. 6. Some samples of face images in "b" subset of FERET database.

Table 2 lists the recognition rates of different algorithms on the "b" subset of FERET database.

In Table 2 one see that when the test set are "bd", "be", "bf" and "bg" subsets, the recognition rates of our IESRCFD are 84.5%, 99.0%, 99.5% and 90.5% respectively, which are higher than those of other algorithms. The average recognition rate of IESRCFD is 93.4%, which not only achieves the highest recognition rate, but also shows a 7%, 30%, 37%, 42%, 33%, 39%, 31% and 35% better performance compared with that of GSRC, LogE-SR, TSC, ESRC, KNN, ELM, ELMSRC and KSRC respectively.

Table 2. Recognition rates of different algorithms on "b" subset of FERET database ($\tilde{\lambda} = 0.1$)

Algorithm\Dataset	bd	be	bf	bg	Average
RSR [20]	79.5%	96.5%	97.5%	86.0%	89.9%
GSRC [22]	77.0%	93.5%	97.0%	79.0%	86.6%
LogE-SR [23]	34.5%	81.0%	91.0%	46.5%	63.3%
TSC [21]	36.0%	73.0%	73.5%	44.5%	56.8%
ESRC [9]	38.5%	59.0%	68.0%	41.0%	51.6%
KNN [15]	56.5%	60.5%	74.0%	48.0%	59.8%
NN [16]	56.5%	60.5%	74.0%	48.0%	59.8%
KSRC [19]	50.5%	67.0%	60.0%	55.5%	58.3%
ELM [6]	45.5%	58.5%	61.0%	53.5%	54.6%
ELMSRC [17]	56.5%	63.0%	65.0%	65.5%	62.5%
IESRCFD	**84.5%**	**99.0%**	**99.5%**	**90.5%**	**93.4%**

7 Conclusion

In this paper, we propose an improved extended sparse representation classifier and feature descriptor (IESRCFD) method by using proposed LWS feature descriptor and ESWRC simultaneously. Experimental results showed that our IESRCFD performs better than many algorithms. The recognition rate of IESRCFD is about 20%, 14%, 12% and 11% higher than those of RSC, KSRC, ESRC and ELMSRC respectively on FEI face database. The recognition rate of IESRCFD is about 31%, 7% and 4% higher than those of ELMSRC, GSRC and RSR respectively on FERET face database. Meanwhile, experimental results illustrated that IESRCFD could convergence and meet the real time requirement.

Acknowledgments. The work is supported in part by National Natural Science Foundation of China under grants 61771145, 61371148.

References

1. Huang, Z.W., Shan, S.G., Wang, R.P., Zhang, H.H., Lao, S.H., Kuerban, A., Chen, X.L.: A benchmark and comparative study of video-based face recognition on COX face database. IEEE Trans. Image Process. **24**, 5967–5981 (2015)
2. Adbi, H., Williams, L.J.: Principal component analysis. Wiley Interdiscip. Rev. Comput. Stat. **2**, 433–459 (2010)
3. Etemad, K., Chellapa, R.: Discriminant analysis for recognition of human face images. J. Opt. Soc. Am. A **14**, 125–142 (1997)
4. He, X.F., Niyogi, P.: Locality preserving projections. In: Advances in Neural Information Processing Systems, Canada (2003)
5. Turk, M., Pentland, A.: Eigenfaces for recognition. J. Cognit. Neurosci. **3**, 71–86 (1991)
6. Huang, G.B., Ding, X., Zhou, H.: Optimization method based extreme learning machine for classification. Neurocomputing **74**, 155–163 (2010)
7. Wright, J., Yang, A.Y., Ganesh, A., Sastry, S.S., Ma, Y.: Robust face recognition via sparse representation. IEEE Trans. Pattern Anal. Mach. Intell. **31**, 210–227 (2009)
8. Yang, M., Zhang, L., Yang, J., Zhang, D.: Regularized robust coding for face recognition. IEEE Trans. Image Process. **22**, 1753–1766 (2013)
9. Deng, W.H., Hu, J.N., Guo, J.: Extended SRC: undersampled face recognition via intraclass variant dictionary. IEEE Trans. Pattern Anal. Mach. Intell. **34**, 1864–1870 (2012)
10. Jourlin, M., Pinoli, J.C.: Logarithmic image processing: the mathematical and physical framework for the representation and processing of transmitted images. Adv. Imaging Electron. Phys. **115**, 130–196 (2001)
11. Mandal, D.: Human visual system based object detection and recognition and introduction of logarithmic local binary patterns for face recognition. M.S. thesis, TUFTS Univ (2012)
12. Drabycz, S., Stockwell, R.G., Mitchell, J.R.: Image texture characterization using the discrete orthonormal s-transform. J. Digit. Imaging. **22**, 696–708 (2009)
13. Gibson, P.C., Lamoureux, M.P., Margrave, G.F.: Letter to the editor: stokwell and wavelet transforms. J. Fourier Anal. Appl. **12**, 713–721 (2006)
14. Stockwell, R.G.: A basis for efficient representation of the S-transform. Digit. Signal Proc. **17**, 371–393 (2007)
15. Weinberger, K.Q., Saul, L.K.: Distance metric learning for large margin nearest neighbor classification. J. Mach. Learn. Res. **10**, 207–244 (2009)
16. Boiman, O., Shechtman, Irani, M.: In defense of nearest-neighbor based image recognition. In: Proceedings of IEEE Conference Computer Vision Pattern Recognition, pp. 1–8 (2008)
17. Luo, M.X., Zhang, K.: A hybrid approach combining extreme learning machine and sparse representation for image recognition. Eng. Appl. Artif. Intell. **27**, 228–235 (2014)
18. Yang, M., Zhang, L., Yang, J., Zhang, D.: Robust sparse coding for face recognition. In: Proceedings of IEEE Conference Compute Vision Pattern Recognition, pp. 625–632 (2011)
19. Gao, S.H., Tsang, I.W.H., Chia, L.T.: Sparse representation with kernels. IEEE Trans. Image Process. **22**, 423–434 (2012)
20. Harandi, M.T., Sanderson, C., Hartley, R., Lovell, B.C.: Sparse coding and dictionary learning for symmetric positive definite matrices: a kernel approach. In: Proceedings of IEEE European Conference on Computer Vision, pp. 216–229 (2012)

21. Sivalingam, R., Boley, D., Morellas, V., Papanikolopoulos, N.: Tensor sparse coding for region covariance. In: Proceedings of IEEE European Conference on Computer Vision, pp. 722–735 (2010)
22. Yang, M., Zhang, L.: Gabor feature based sparse representation for face recognition with Gabor occlusion dictionary. In: Proceedings of IEEE European Conference on Computer Vision, pp. 448–461 (2010)
23. Guo, K., Ishwar, P., Konrad, J.: Action recognition using sparse representation on covariance manifolds of optical flow. In: Proceedings of IEEE International Conference on Advanced Video and Signal Based Surveillance, pp. 188–195 (2010)

Automatic License Plate Recognition Based on Faster R-CNN Algorithm

Zhi Yang[1], Feng-Lin Du[1], Yi Xia[1], Chun-Hou Zheng[2], and Jun Zhang[1(✉)]

[1] College of Electrical Engineering and Automation, Anhui University, Hefei 230601, Anhui, China
wwwzhangjun@163.com

[2] College of Computer Science and Technology, Anhui University, Hefei 230601, Anhui, China

Abstract. This paper proposed a method based on Faster R-CNN algorithm to locate and recognize Chinese license plate. Faster R-CNN is composed by Region Proposal Network (RPN) and fast R-CNN. To make Faster R-CNN locate and recognize license plate more effective, we optimize the training process. To validate performance of the proposed method, two datasets (standard dataset and real scene dataset) are created. Faster R-CNN with three different model are used. The experimental results show that the proposed method achieve better performance contrasting six traditional methods. In standard dataset (simple situation), three modes achieve similar recognition results. However, in real scene dataset, more deeper model achieve better recognition performance.

Keywords: Faster R-CNN · License plate recognition · Object detection

1 Introduction

With the development of intelligent transportation system, the traditional license plate recognition has been widely employed in vehicle management. Currently, the widely used traditional methods are template matching [1], SVM classifiers [2] and neural network algorithm [3]. However, the recognition rate of these methods is still affected by many factors such as the shooting angle, insufficient natural light situation in the actual condition. In these complex practical application scenarios, traditional methods cannot meet the requirements and achieve better recognition performance. Since Yan et al. [4] first applied deep convolutional neural networks in handwritten digit recognition in 1998, deep learning [5] has achieved many successful applications and employed in classification, object detection and other research fields. Usually, deep learning is a end-to-end method, it can extracted the features from the image automatically and has strong distortion tolerance ability for image application. In this paper, Faster R-CNN algorithm was used in the Chinese license plate recognition. To make the experimental results are more confident, two datasets are collected. Totally three deep models including ZF model, VGG_CNN_M1024 model, and VGG16 model are

D.-S. Huang et al. (Eds.): ICIC 2018, LNAI 10956, pp. 319–326, 2018.
https://doi.org/10.1007/978-3-319-95957-3_35

used in this research. Contrasting traditional methods, the proposed method achieve better recognition and location performance.

2 The Proposed Methods and Training Process

2.1 Faster R-CNN for Object Detection

Convolutional neural network (CNN) is a kind of deep learning network model for image classification and object detection. Based on CNN, R-CNN [6], fast R-CNN [7] and Faster R-CNN [8] are developed for object detection. Faster R-CNN can be viewed as two parts: Region Proposal Network (RPN) network and Fast R-CNN network. RPN network is used to replace the Selective Search [9] in Fast R-CNN. All the object detection modules in Faster R-CNN are placed in a unified deep convolutional neural network framework, which make the detection faster and more accurate.

2.2 Region Proposal Network

The Region Proposal Network (RPN) is a special neural network which consists of two parts: the convolution layer and a pair of fully connected layers. A sliding window is used to execute a convolution on the feature graph of the last layer to obtain some candidate frames. Each candidate frame contains two kinds of information: the probability and location of the object. A multidimensional vector which is obtained between the sliding window and the feature graph is sent to the two network layers, one is the cls layer (the classification layer) and the other is the reg layer (border regression) [12]. The structure of the RPN network is shown in Fig. 1.

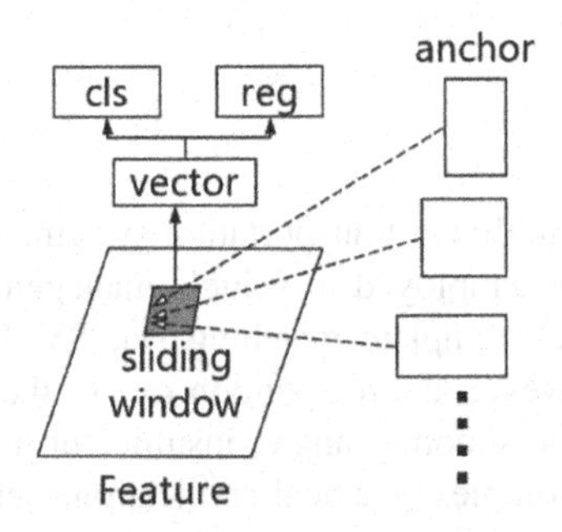

Fig. 1. RPN schematic diagram of network structure

2.3 Faster R-CNN Training Process

RPN and Fast R-CNN share the convolutional layer feature, therefore the whole training process is divided into three steps: (1) The network parameters are trained by Image Net dataset [10], the pre-trained model is used for RPN network. (2) RPN network is used to extract the object area and these object is used to train the Fast R-CNN. (3) RPN and Fast R-CNN form a unified network which is fine tuned for object detection. The training process is shown in Fig. 2.

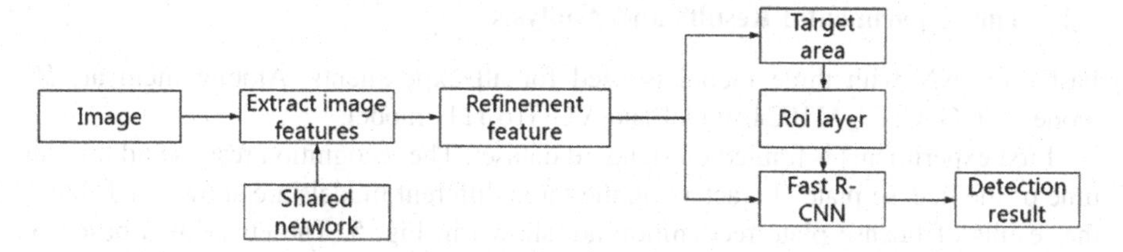

Fig. 2. Schematic diagram of convolutional layer feature sharing

3 Experimental Results

3.1 Data Set

In order to validate the recognition performance of the proposed method, two different datasets (standard dataset and real scene dataset) are used in this paper. Each data set include 27 provinces and cities of China as well as four municipalities. Both license plate characters of two data sets are divided into 66 categories, namely, the number 0–9 (10 categories), the letter A–Z (25 categories) and the provinces and cities (31 categories). The standard dataset consist of 10873 pictures and one sample picture is shown in Fig. 3. is composed of ideal picture with no any noise. Real scene data sets are collected from the internet, which contains 4711 pictures and the sample is shown in Fig. 4. The second dataset is more complicated than first one. The license plate has different colors such as blue, white, black and yellow. Besides, the tilt angle, different location and size of the license plate is very normal situation, which increases the difficulty of license plate recognition.

Fig. 3. Standard data set (Color figure online)

Fig. 4. Real scene dataset

3.2 The Experimental Results and Analysis

Faster R-CNN with three model is used for all experiments. Among them are ZF model, VGG_CNN_M1024 model and VGG16 [11] model.

First experiments is trained on standard dataset. The recognition results and training time of the license plate characters on the three different models are shown in Table 1, the results of license plate recognition are shown in Fig. 5. The curve area based on recall and accuracy is called average precision (AP), while MAP is the mean value of multiple class AP. It is an important index to measure the recognition accuracy in target detection. The recognition results of numbers, letters and Chinese characters in the license plate characters are shown in Table 2 respectively.

Table 1. The recognition rate and training time of three different network structure models

Network model	ZF	VGG_CNN_M1024	VGG16
MAP	0.9861	0.9859	0.9860
Average time (unit: s)	0.0282	0.0325	0.0565

Fig. 5. The recognition results of standard license plate data

Table 2. Recognition rate of license plate character on three different models

Model / Accuracy rate / Category	ZF	VGG_CNN_M1024	VGG16
Number	0.9453	0.9451	0.9453
Letter	0.9926	0.9922	0.9926
Chinese characters	0.9940	0.9940	0.9938

As shown in Table 1, the MAP values of the three methods can reach more than 0.98 and the difference of MAP values is very small due to the simple scene pictures Among them, the accuracy of ZF model recognition was the highest (MAP = 0.9861), 0.02% and 0.01% higher than that of VGG_CNN_M1024 model and VGG16 model respectively. This is because the convolution kernel of the ZF model is less than the convolution kernel of the VGG_CNN_M1024 model, while the convolution kernel of the VGG_CNN_M1024 model is more complex, which can extract more deeper features. So computation complexity of the model was increased and the performance was influenced. Through the analysis of the results, we found out that the accuracy of the recognition does not get higher with the increase of the number of layers. A model with low complexity and small computational complexity should be given when satisfying certain accuracy. Therefore, for those images with simple background and low noise, the shallow network model is a priority to select because it can save training resources and time.

The second experiments is carried on the real scene dataset. Because this dataset is more complicated, the proposed algorithm is used to locate the license plate position and recognize the characters of the license plate. Also, three different models (ZF model, VGG16 model and VGG_CNN_M1024 model) are used. Moreover, six methods including 6 traditional methods (SOBEL edge detection, color location, text location and combination of two methods) are selected to compare with the proposed method. Table 3 lists the rates of license plate location in the real dataset, including Faster R-CNN with three different models and six traditional identification methods.

Table 3. The experimental results on nine different types of neural network models

Method	Model	Positioning rate
6 traditional methods	SOBEL	39.66%
	COLOR	71.34%
	CMSER	77.05%
	SOBEL&COLOR	45.02%
	SOBEL&CMSER	45.16%
	COLOR&CMSER	70.26%
Faster R-CNN	ZF	88.11%
	VGG_CNN_M1024	89.26%
	VGG16	89.12%

In location experiment, it is very clear to find that Faster R-CNN achieve better performance contrasting six traditional methods. One demonstration location result is shown in Fig. 6 (Left is original picture and right is the location result). The position of the license plate is accurately located, so Faster R-CNN has a strong advantage over other traditional methods in the license plate location.

In license plate recognition experiment, 4239 pictures are used as train samples and 472 pictures as test samples. The overall experimental results are shown in Table 4, the results of license plate recognition are shown in Fig. 7. With the increase of network layer, the MAP value gradually increased. The accuracy of VGG16 model is 3.25% higher than that of ZF model and is 1.55% higher than that of VGG_CNN_M1024 model. It shows that the increase of the number of convolutional layers may extract more effective features, the recognition accuracy can be improved. Table 4 also shows detection time of each models. It takes 0.0314 s on VGG_CNN_M1024 model, which

Fig. 6. License plate location experiment

is faster than the ZF model (0.0271 s), while theVGG16 model takes almost twice time contrasting ZF model. This is because the computational complexity of the model increase with the number of network layers increasing.

Table 4. The recognition rate and training time of three different network structure models

Network model	ZF	VGG_CNN_M1024	VGG16
MAP	0.9590	0.9760	0.9915
Average time (unit: s)	0.0271 s	0.0314 s	0.0532 s

Fig. 7. Real scene license plate data recognition results

The recognition results of numbers, letters and Chinese characters in the license plate characters are shown in Table 5 respectively. It shows similar results with Table 4.

Table 5. Recognition rate of license plate character on three different models

Model / Accuracy rate / Category	ZF	VGG_CNN_M1024	VGG16
Number	0.9786	0.9800	0.9811
Letter	0.9563	0.9689	0.9955
Chinese characters	0.9548	0.9806	0.9918

4 Conclusion

Based on deep learning framework, Faster R-CNN algorithm with three different models was proposed to recognize the license plate. Two dataset (standard dataset and real scene dataset) were constructed to validate the performance of the proposed method. Contrasting traditional algorithm, the proposed method achieved better performance in two dataset. In standard dataset, three models achieved similar recognition results. It can be conclude that the deeper model cannot be fully exploited in this simple scene situation. However, in real scene dataset, with the model deeper, the recognition performance was also increased. But the detection time was also increased. In the further research, we need to take more attention in model optimization.

Acknowledgement. This work is supported by Anhui Provincial Natural Science Foundation (grant number 1608085MF136).

References

1. Ko, M.A., Kim, Y.M.: License plate surveillance system using weighted template matching. In: Applied Imagery Pattern Recognition Workshop, 2003, Proceedings (2003)
2. Jin-Jun, W.U., Shu-Xin, D.U.: Application of SVM to character recognition of vehicle license plate. J. Circuits Syst. **13**(1), 84–87 (2008)
3. Nagare, A.P.: License plate character recognition system using neural network. Int. J. Comput. Appl. **25**(10), 36–39 (2011)
4. Lécun, Y., et al.: Gradient-based learning applied to document recognition. Proc. IEEE **86** (11), 2278–2324 (1998)
5. Krizhevsky, A., Sutskever, I., Hinton, G.E.: ImageNet classification with deep convolutional neural networks. In: International Conference on Neural Information Processing Systems (2012)
6. Girshick, R., et al.: Rich feature hierarchies for accurate object detection and semantic segmentation. In: Computer Science, pp. 580–587 (2014)
7. Girshick, R.: Fast R-CNN. In: IEEE International Conference on Computer Vision (2015)

8. Ren, S., et al.: Faster R-CNN: towards real-time object detection with region proposal networks. IEEE Trans. Pattern Anal. Mach. Intell. **39**, 1137–1149 (2016)
9. Uijlings, J.R., et al.: Selective search for object recognition. Int. J. Comput. Vis. **104**(2), 154–171 (2013)
10. Russakovsky, O., et al.: ImageNet large scale visual recognition challenge. Int. J. Comput. Vis. **115**(3), 211–252 (2015)
11. Simonyan, K., Zisserman, A.: very deep convolutional networks for large-scale image recognition. Computer Science (2014)

Object Detection of NAO Robot Based on a Spectrum Model

Laixin Xie[1,2] and Chunhua Deng[1,2](✉)

[1] College of Computer Science and Technology, Wuhan University of Science and Technology, Wuhan, China
dchzx@wust.edu.cn
[2] Hubei Province Key Laboratory of Intelligent Information Processing and Real-Time Industrial System, Wuhan, China

Abstract. NAO robots often need to detect objects to accomplish its task. At present, color segmentation is popular with NAO robot vision tasks because of its lower-end specification. A spectrum segmentation algorithm is proposed to realize real time detection in this paper. Spectral model is the foundation of human visual system, which can separate objects of distinctive color characteristics from complex illumination. Compared with current methods, color threshold in our method is trained by objective color and background color, which can automatically separate foreground and background. In addition, this paper employs Support Vector Machine (SVM) to recognize segmented regions to increase detection accuracy. Experimental results demonstrate effectiveness of the proposed method.

Keywords: NAO robot · Spectrum model · Color segmentation
Object detection

1 Introduction

NAO robot is a new generation of development and research platform, assembling various sensors, such as sonar, gyroscope and two cameras with a small intersection of field of view (7.94°, calculating from the data in Fig. 1, which can not meet the basic requirements of binocular algorithm), as shown in Fig. 1. As a humanoid robot, its motion is very smooth, which shows high robustness of built-in programs and mechanical structures.

These advantages simplify many unnecessary research steps so that NAO attracts more and more researchers [1–3]. However, as mobile robot, the performance of NAO is relatively weak. NAO has a notebook-level CPU, ATOM Z530, with the frequency of 1.6 GHz, 1 GB RAM (512 M used by system) and 2 GB flash memory.

When used to solve the problem of object recognition, the monocular vision algorithm of NAO is of low complexity [4–6]. There are mainly two kinds of these algorithms, color segmentation and feature descriptor.

The former algorithm [7, 8] separates objects from background by its color feature in specific color space. The segmentation algorithm is of low computation consuming, and it highly depends on the selection of hyper-parameters [9, 10], such as thresholds

D.-S. Huang et al. (Eds.): ICIC 2018, LNAI 10956, pp. 327–338, 2018.
https://doi.org/10.1007/978-3-319-95957-3_36

of color channel and color space selection. Another kind of algorithms are based on feature descriptors [11, 12], such as HOG and SIFT etc., which need to be combined with classifiers. These algorithms are robust to illumination variation and do not require hyper-parameters. However, these algorithms are difficult to run on NAO robots in real-time. In this research, an adaptive threshold segmentation algorithm based on spectral model is proposed, aiming at finding a better way to identify objects with obvious color characteristics.

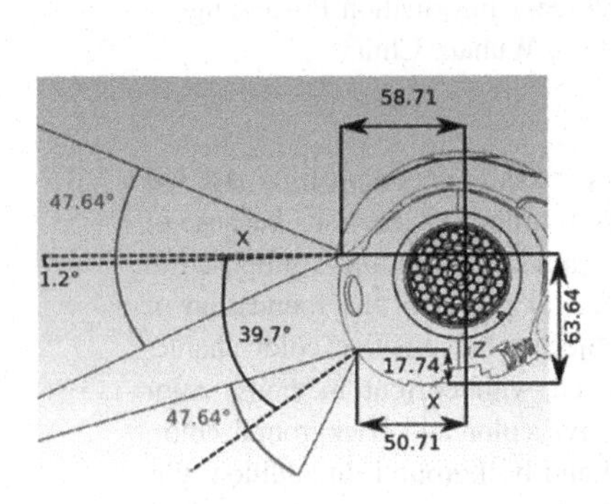

Fig. 1. Head of NAO robot

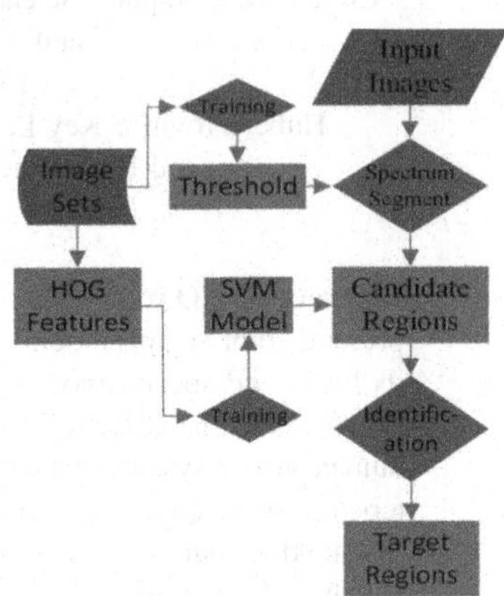

Fig. 2. Algorithm flowchart

A beam of light can be decomposed into basic rays of different wavelengths, which means each color can be decomposed into a combination of different basic colors. Inspired by the sensitivity of human vision system to basic colors, this paper proposes to classify objects based on their basic colors.

According to spectral principle, the difference between basic color and luminance value can reflect its contribution of each basic color. Based on this, a color threshold segmentation model is established. Combined with the idea of OTSU [13] and statistical histogram characteristics, a fast optimizing method is presented. In order to enhance the performance and robustness of recognition algorithm, this research introduces gradient directional histogram (HOG) [14] and SVM to further eliminate false alarm in the candidate regions of segmentation result.As NAO robots have low-end specification, segmentation algorithms are easier to achieve real-time performance than descriptor-based algorithms. The main contributions of our approach are: (i) A color threshold segmentation model based on spectral principle is proposed. (ii) We propose an adaptive threshold optimization training method. (iii) We employ SVM to recognize segmented regions to increase detection accuracy.

2 Adaptive Color Segmentation

The proposed algorithm is adaptive and robust to illumination changes, but it is disturbed by background objects. Its process is shown in Fig. 2.

2.1 Input Images

Same image of different resolutions will affect the efficiency and efficiency of algorithm. Hence we need to make trade-offs in resolution, so that a algorithm can be efficient in the case of superior effect. There are three different resolutions of NAO, which are 1280*960, 640*480 and 320*240.

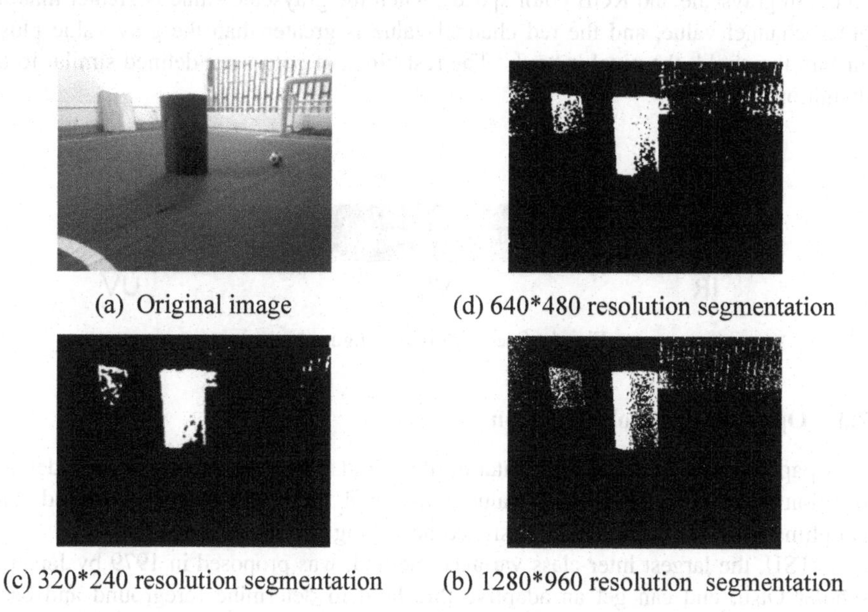

(a) Original image (d) 640*480 resolution segmentation

(c) 320*240 resolution segmentation (b) 1280*960 resolution segmentation

Fig. 3. Segmentation experiments in different resolution

Although the detail information on resolution 1280*960 image is rich, there are more noise information. Noises have little effect on overall segmentation performance, it will result in segmentation region deformation and the increasing of segmentation error rate, as shown in Fig. 3(b). When the resolution is 320*240, decreasing resolution increases the speed of calculation. It also reduces the accuracy of segmentation, especially when the color feature area is relatively small, such as Fig. 3(c). When the resolution is 640*480, objects with corresponding color characteristics can be formed after segmentation, as shown in Fig. 3(d). The precision of picture is sufficient for algorithm and choosing a smaller resolution can speed up algorithm while not affecting the robustness of the algorithm. All in all, selecting the 640*480 resolution can achieve a good balance between precision and efficiency.

2.2 Spectral Principle

The spectrum is a pattern that the monochromatic light is arranged in the order of wavelength (or frequency) after the chromatic light is divided by dispersion system (such as prism and grating), which is called optical spectrum, as shown in Fig. 4. If the color values of R, G, B color channel of a pixel are same, this pixel intuitively looks as gray in RGB color space and the grayscale value is calculated from the weighted average of the three channel values. If a channel value is larger than the gray value, then visually, the point will display the color corresponding to that channel. Because in spectrum, red is close to green, that is, in RGB space, when a pixel is red in the visual system, its green channel value is also big [15, 16]. Then we propose a definition of "red": in grayscale and RGB color spaces, when the grayscale value is greater than the green channel value, and the red channel value is greater than the gray value plus a custom threshold, the pixel is "red." The rest kinds of colors are defined similar to the definition.

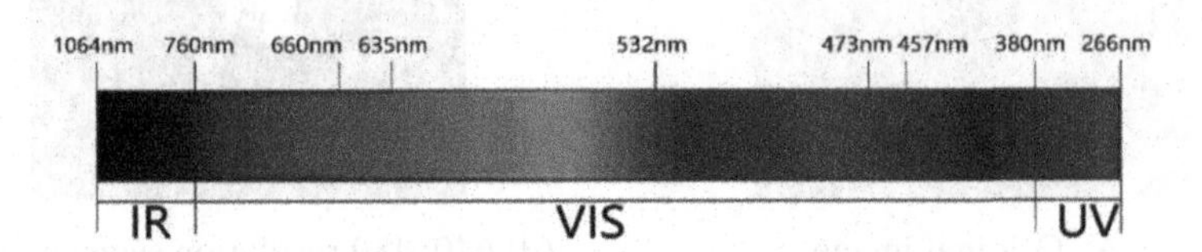

Fig. 4. Spectrum (Color figure online)

2.3 Optimal Threshold Function

This paper obtains optimal segmentation threshold by searching, we need to define a function first. When the function value is maximal, the corresponding threshold value is optimal. And this function is designed according to the idea of OTSU.

OTSU, the largest inter-class variance method, was proposed in 1979 by Japanese scholar Otsu, and can get an adaptive threshold to determine foreground and background. This method is based on global binary algorithm, which assigns the pixel value that is greater than a threshold of the image to 255 (the background), or 0 (the foreground). And finally, it transforms a grayscale image to a binary image to describe the classification result. When choosing the best threshold, the difference between the two classes, which is quantified by a formula (1), should be the largest.

$$\arg\max_{t} \sigma^2 = \omega_1(\mu_1 - \mu)^2 + \omega_2(\mu_2 - \mu)^2 \tag{1}$$

Where t is the threshold to distinguish between foreground and background, ω_1 and ω_2 are the proportions of the number of pixels in foreground and background to the total pixel, μ_1 and μ_2 are the mean values of the grayscale value in foreground and background.

Inspired by the idea of OTSU, method in this paper defines "red", definition mentioned in 2.2, as the foreground in OTSU. And the remaining pixels are defined as the background in OTSU. So hopefully, pictures are divided into two regions. In background region, the "non-red" pixels should be maximized while in foreground region, "red" pixels should be maximized. In training procedure, foreground region can be artificially selected with a bounding box, and the remaining region is background region. Finally, the function of obtaining optimal threshold sums up the proportion of the number of "red" pixels to the total pixel in foreground region, which is $\omega_1(t)$, and the proportion of the number of "non-red" pixels to the total pixel named $\omega_2(t)$ in the background region, which is shown below:

$$\omega_1(t) = \frac{\sum (red_{inside} > gray_{inside} + t\&gray_{inside} > green_{inside})}{a*b}$$
$$\omega_2(t) = \frac{\sum (red_{outside} < gray_{outside} + t|gray_{outside} < green_{outside})}{M*N-a*b} \tag{2}$$
$$\sigma(t) = \omega_1(t) + \omega_2(t)$$

Where the inside subscript expresses that pixels are in the foreground region, the outside subscript indicated that pixels are in the background region, red, green to represent pixels in red channel or green channel in RGB image, gray to represent pixels in grayscale image, *M*, *N* to represent height and width of the whole image, *a*, *b* is the height and width of the manual bounding box.

The formula here only takes "red" as the feature. When taking the "blue" and other features as characteristics, modifying definitions of the variables in the formulas. For example, the function corresponding to "yellow" feature shows below:

$$\omega_1(t) = \frac{\sum (green_{inside} > gray_{inside} + t\&\&gray_{inside} > blue_{inside})}{a*b} \tag{3}$$

Where blue, green represents the pixels in the blue channel or green channel in RGB image, the rest of the variables keep unchanged.

2.4 Searching Threshold and Optimization

In the implementation of the above model, we need to spend $O(n^2)$ to traverse image. At the same time, we need to perform this algorithm for every threshold we search. So the entire process needs to take a time complexity of $O(n^3)$ before making any optimization. Here we optimize the efficiency of the algorithm on code.

In the above algorithm, we will check whether a pixel satisfies the condition of ω_1 in Formula (2) for each threshold t. In this process, not only do we repeatedly count the pixels whose grayscale values are greater than green channel values, and the difference between red channel value and grayscale value is repeatedly calculated to compare with t. In fact, for each threshold, the number of pixels that satisfy our definition is fixed, so that we can calculate and save them before searching the threshold.

The first part in Algorithm 1 describes the above process.

This approach produces a problem. In the very first definition, we define pixel whose red channel value is greater than greyscale value plus a specific threshold as red. Hence when the sum of gray value and threshold value are greater than 255, all pixel whose red channel values of 255 should still be defined as red, because value greater than 255 in image values 255. This characteristic of pixel values is not considered in this optimization method, which causes ignorance of some pixels that conform to the "red" definition. So the second part of algorithm is to fix this.

The complete process is shown in Algorithm 1.

This time complexity of this approach is $O(n^2)$, which is needed to calculate a histogram.

After using this method, the time complexity of algorithm only promotes from $O(n^3)$ to $O(n^2)$, but the time it takes to calculate a histogram is actually very small, which is due to the optimization of matrix calculation in Numpy. However, the amount of time spent on repeated calculations is huge, which is because the nature of python. So this optimization algorithm, or, technically, a code technique, can give you a significant performance boost in actual running, which can be seen in Fig. 5.

In addition, the method of searching for the best threshold is only for a single labeled picture, and in experiment, we start by picking up photos from a couple of different environments to perform the algorithm, and selected a threshold with the best effect, relatively, from the obtained threshold to carry out the following experiment.

The complete process is in Algorithm 1.

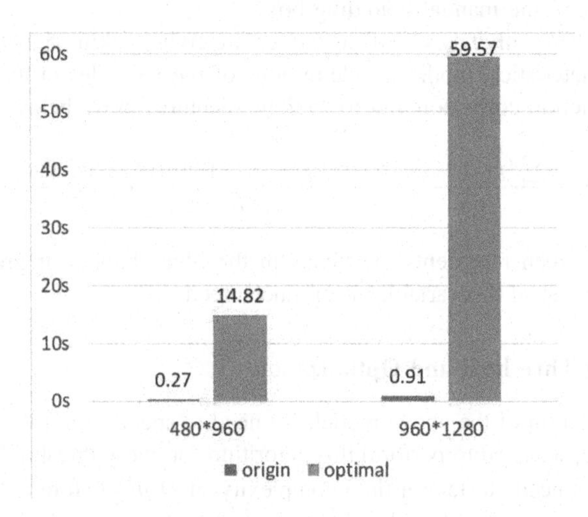

Fig. 5. Time consuming comparison. This result is performed on NAO robot and we only compare the time required to get the optimal threshold.

3 Classification of Object in Bounding Box

After obtaining candidate regions of the object through the segmentation algorithm in 1, it is necessary to classify the segmented regions by HOG and SVM because objects with the same color feature may exist in the background.

Algorithm 1 Optimization of searching threshold method

1: Calculate the *grayscale, red channel* matrix and *green channel* matrix of origin image;
2: Initialize *complement matrix* with 0;
3: **for** each position *x, y* in image **do**
4: **if** *red channel*[*x, y*] > *grayscale*[*x, y*] and *grayscale*[*x, y*] > *green channel*[*x, y*] **then**
5: *difference matrix*[*x, y*] = *red channel*[*x, y*] – *grayscale*[*x, y*];
6: **if** *red channel*[*x, y*] = 255 **then**
7: *complement matrix*[*x, y*] = *difference matrix*[*x, y*];%prepare for the second part
8: **end if**
6: **else**
7: *difference matrix*[*x, y*] = 0;
8: **end if**
9: **end for**
10: *labeled matrix* is a part of *difference matrix* in the same position as labeled region;
11: Calculate *difference histogram* with *difference matrix* and *labeled histogram* with *labeled matrix*;
13: **for** intensity *i* from 254 to 1 **do**
14: *difference histogram*[*i*] += *difference histogram*[*i+1*];
15: *labeled histogram*[*i*] += *labeled histogram*[*i+1*];
16: **end for**
17: %first part
18: *labeled complement matrix* is a part of *complement matrix* in the same position as labeled region;
19: Calculate *complement histogram* with *complement matrix*;
20: Calculate the histogram of *labeled complement matrix* and name it *labeled complement histogram*;
21: *difference histogram*[1] += *complement histogram*[1];
22: *labeled histogram*[1] += *labeled complement histogram*[1];
23: **for** intensity *i* from 2 to 255 **do**
24: *complement histogram*[*i*] += *complement histogram*[*i-1*];
25: *labeled complement histogram*[*i*] += *labeled complement histogram*[*i-1*];
26: *difference histogram*[*i*] += *complement histogram*[*i*];
27: *labeled histogram*[*i*] += *labeled complement histogram*[*i*];
28: **end if**
29: **for** threshold *t* from 0 to 254 **do**
30: ω_1 = *labeled histogram*[*t*+1] / *total pixel number of grayscale*;
31: ω_2 = (*total pixel number of grayscale* – *difference histogram*[*t*+1] – *pixel number in labeled area* 32: +*labeled histogram*[*t*+1]) / (*total pixel number of grayscale* – *pixel number in labeled area*);
33: **if** max(ω_1 + ω_2) **then**
34: *optimal threshold* = *t*;
35: **end if**
36: **end for**
37: **output** *optimal threshold*

3.1 HOG Feature Descriptor

Histogram of Oriented Gradient(HOG) [14] is a feature descriptor describing an image with local gradient histogram and local orientation histogram. Here is the process of calculating HOG feature.

The proposed region is firstly divided into 8*8 blocks, in which we define 4 cell units of 4*4. In each cell, its horizontal gradient and vertical gradient of each pixel (x, y) are computed. And then its amplitude and amplitude angle are calculated. After calculation of a cell, the amplitude histogram and the amplitude-angle histogram of the cell are computed. Since the range of amplitude angle is 0–360°, assuming that the size of each bin in histogram is 45°, there are totally 8 bins. The value of each bin equals to the number of all pixels whose amplitude angle is within the corresponding range. After the histogram of cells is computed, normalization of histogram is done to increase robustness to illumination change. Then the histogram of all cells in a block is spliced together to form a block histogram. Eventually, this approach iterates over all blocks of entire region, performing the same operation on each block to get the corresponding block histogram, which is HOG feature.

As possible regions are already proposed by the segmentation algorithm, this method merely need to calculate HOG feature for the possible region instead of entire picture, which makes time complexity decrease significantly.

3.2 SVM

Support Vector Machine [17], or SVM, is a classification model, whose basic model is a linear classifier with the largest class interval defined on the feature space.

The following is the process of SVM.

After calculating HOG feature of the segmented region, the hyperplane in the feature space is obtained by SVM classifier.

Then you can get the decision function. Each time when the object area is segmented, the size of the area is reduced by decreasing the sample rate, or the area is enlarged by interpolation, which makes the split area be the same size as the sample area. Then computed HOG feature of the partition region is substituted to the decision function. When the function evaluates to 1 and the object in the region is the target object.

3.3 Training

We use both the HOG and SVM methods in OpenCV to train. During calculation of HOG feature, we will first resize positive sample to 60*20. And block size is 10*10. Block stride is 5*5. Cell size is 5*5. Still, there are 9 bins in the histogram of HOG. While training the SVM classifier, two parameters, kernel type and svm type, are set to LINEAR and C-SVC respectively. We input positive and negative samples, the ratio of which is about 1 to 2, into SVM classifier for training, thus obtaining a classifier model. After the segmentation algorithm is completed, the classifier model will classify the candidate regions.

4 Experiments

In the experiment part, the segmentation method based on HSV space is reproduced, and the segmentation result is compared with the method presented in this paper. However, because the segmentation method based on HSV space is impossible to use same parameter, the parameters used in the reproduction are obtained by adaptive parameter acquisition method proposed in this paper.

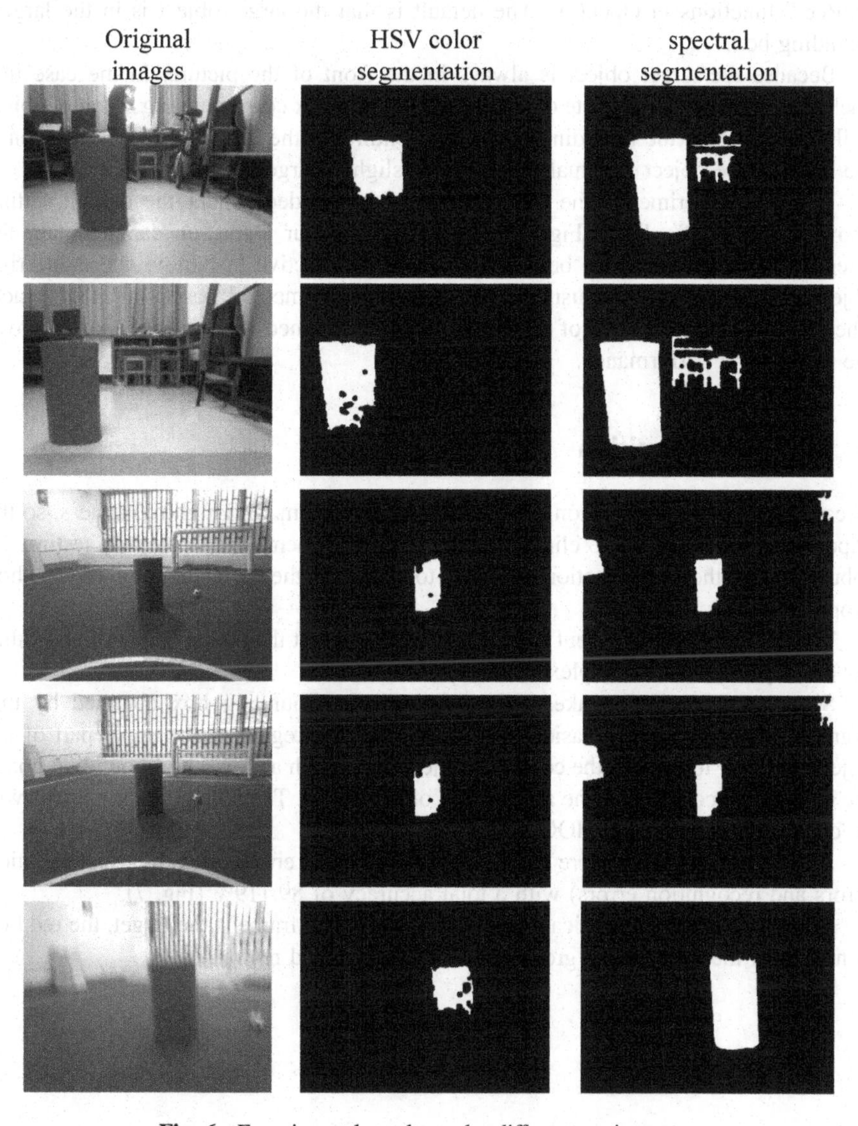

Fig. 6. Experimental results under different environment

4.1 Color Segmentation

The experiments were carried out under the same threshold conditions in light dynamic environment of simple background, dark light environment of simple background, light environment of complex background, dark light condition of complex background, and contrasted with each other.

The results of the experiment are treated with same removal of small area and morphological closed operation. The video processing is the same as the picture. And the method of generating bounding box is to use the "findContours" and "boundingRect" functions in OpenCV. The default is that the target object is in the largest bounding box.

Because the target object is always in the front of the picture. In the case that background contains complete objects, the bounding box corresponding to target object will be larger than the bounding box corresponding to the interference object, even if the interference object is actually as large or slightly larger than the target object.

In actual experiments, the accuracy rate for all videos under the default setting above is 98.8764%. From Fig. 6, we can see that our algorithm can maintain the integrity of the target shape better, but it is more effective to remove the interfering objects in the background using color segmentation method based on HSV space. Therefore, the segmentation of candidate regions combined with classifier can improve the recognition performance.

5 Object Recognition

Because HOG and SVM belongs to the mature algorithm with high robustness, so the experiment 4.1 is mainly to eliminate the error of segmentation, separately testing the robustness of the segmentation in order to illustrate the advantages of the method proposed in this paper.

This part of the experiment is primarily to prove that the effectiveness of algorithm for the object of more complex features.

Since the experiment takes red as feature, the bounding box obtained by this segmentation algorithm is basically located in the red region of the upper part of the object. In order to contain the complete object, the length and width of bounding boxes are magnified according to the aspect ratio of the object. Then the bounding boxes will be classified by SVM and HOG algorithm.

A total of 193 images were tested, of which 21 were errors (including segmentation errors and recognition errors) with a total accuracy of 89.119% (Fig. 7).

Where the green box indicates that the object in the frame is the target, the red box is not. Both the red and the green box is the segmented region.

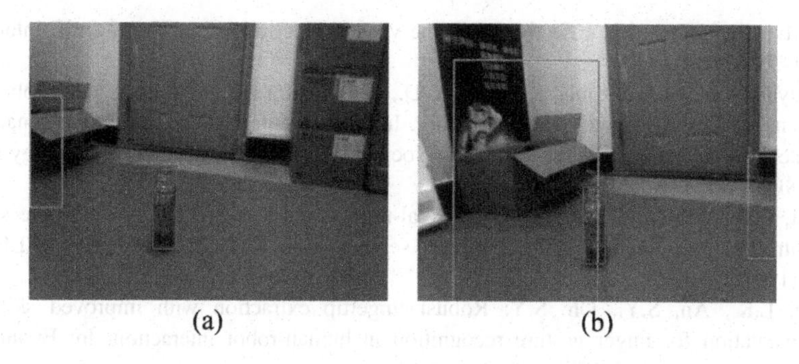

(a) (b)

Fig. 7. Some recognition results (Color figure online)

6 Conclusion

In this paper, a color segmentation algorithm based on spectral principle is proposed to improve the robustness of color segmentation by changing the way to define color. The method of adaptive training parameters can enhance the effectiveness of parameters. For complex object recognition tasks, pre-proposed segmentation region can speed up the HOG and SVM algorithm by reducing the size of the input. However, our algorithm is mainly applied to target detection with obvious color features.

Acknowledgment. This work was supported by the Hubei Province Training Programs of Innovation and Entrepreneurship for Undergraduates, 201710488036; Scientific and technological innovation fund for College Students of Wuhan University of Science and Technology, 17ZRC131; Scientific and technological innovation fund for College Students of Wuhan University of Science and Technology, 17ZRA116; Scientific and technological innovation fund for College Students of Wuhan University of Science and Technology, 17ZRA121.

References

1. Jimenez, J.J.A., Perez, D.H., Barbera, H.M.: Robust feedback control of ZMP based gait for the humanoid robot Nao. Int. J. Robot. Res. **32**(9–10), 1074–1088 (2013)
2. Catley, E., Sirlantzis, K., Howells, G., et al.: Preliminary investigations into human fall verification in static images using the NAO humanoid robot. J. Urol. **192**(4), 1298–1299 (2014)
3. Motoc, I., Sirlantzis, K., Spurgeon, S.K., et al.: A stable and robust walking algorithm for the humanoid robot nao based on the zero moment point. J. Cell Biol. **107**(5), 1911–1918 (2014)
4. Alam, M., Vidyaratne, L., Wash, T., et al.: Deep SRN for robust object recognition: a case study with NAO humanoid robot. In: Southeastcon, pp. 1–7. IEEE (2016)
5. Li, C., Wang, X.: Visual localization and object tracking for the NAO robot in dynamic environment. In: IEEE International Conference on Information and Automation. IEEE (2017)
6. Müller, J., Frese, U., Röfer, T.: Grab a mug - Object detection and grasp motion plan ning with the Nao robot. In: IEEE-Ras International Conference on Humanoid Robots, pp. 349–356. IEEE (2013)

7. Martín, F., Veloso, M.: Effective real-time visual object detection. Prog. Artif. Intell. **1**(4), 259–265 (2012)
8. Nguyen, T.L., Boukezzoula, R., Coquin, D., et al.: Color recognition for NAO robot using sugeno fuzzy system and evidence theory. In: 2015 Conference of the International Fuzzy Systems Association and the European Society for Fuzzy Logic and Technology (IFSA-EUSFLAT-15). Atlantis Press (2015)
9. Wei, G.Q., Arbter, K., Hirzinger, G.: Real-time visual servoing for laparoscopic surgery. Controlling robot motion with color image segmentation. Eng. Med. Biol. Mag. IEEE **16**(1), 40 (1997)
10. Lee, L.K., An, S.Y., Oh, S.Y: Robust fingertip extraction with improved skin color segmentation for finger gesture recognition in human-robot interaction. In: Evolutionary Computation, pp. 1–7. IEEE (2012)
11. Deng, C., Cao, Z., Xiao, Y., Chen, Y., Fang, Z., Yan, R.: Recognizing the formations of CVBG based on multiviewpoint context. IEEE Trans. Aerosp. Electron. Syst. **51**(3), 1793–1810 (2015). https://doi.org/10.1109/TAES.2015.140141
12. Deng, C., Cao, Z., Xiao, Y., Fang, Z.: Object detection based on multi-viewpoint histogram. In: 2015 Chinese Automation Congress, pp. 616–621 (2015)
13. Otsu, N.: A thresholding selection method from gray-level histogram. IEEE Trans. Syst. Man Cybern. **9**(1), 62–66 (1979)
14. Dalal, N., Triggs, B.: Histograms of oriented gradients for human detection. In: IEEE Computer Society Conference on Computer Vision & Pattern Recognition, pp. 886–893. IEEE Computer Society (2005)
15. Wang, Y., Li, J., Stoica, P.: Spectral analysis of signals: the missing data case. Synth. Lect. Signal Process. **1**(1), 90–108 (2005)
16. Snowball, P.S.: Spectral Analysis Of Signals. Leber Magen Darm **13**(2), 57–63 (2005)
17. Joachims, T.: Making large-scale SVM learning practical. Technical reports, 8(3), 499-5 (1998)

Dimensionality Deduction for Action Proposals: To Extract or to Select?

Jian Jiang[1,2], Haoyu Wang[1,2], Laixin Xie[1,2], Junwen Zhang[1,2], and Chunhua Deng[1,2](✉)

[1] College of Computer Science and Technology, Wuhan University of Science and Technology, Wuhan, China
dchzx@wust.edu.cn

[2] Hubei Province Key Laboratory of Intelligent Information Processing and Real-Time Industrial System, Wuhan, China

Abstract. Action detection is an important item in machine vision. Currently some works based on deep learning framework have achieved impressive accuracy, however they still suffer from the problem of low speed. To solve this, researchers have introduced many feasible methods and temporal action proposals method is the most effective one. Fed with features extracted from videos, these methods will propose time clips which may contain actions to reduce computational workload. It is a common way to use 3D convolutions (C3D) to extract spatio-temporal features from videos, nonetheless the dimension of these features is generally high, resulting sparse distribution in each dimension. Thus, it is necessary to apply dimension reduction method in the process of temporal proposals. In this research, we experimentally find that in action detection proposal task, reducing the dimension of features is important. Because it cannot only accelerate the process of subsequent temporal proposals but also makes its performance better. Experimental results on the THUMOS 2014 dataset demonstrate that the method of feature extraction reduction is more suitable for temporal action proposals than feature selection method.

Keywords: Action detection · 3D convolutions · Action proposals Spatio-temporal · Dimension reduction

1 Introduction

One of the crucial tasks in big data area is to process videos since massive deployment of cameras have generated huge numbers of videos every minute which contain valuable information. A large proportion of surveillance video content is about people's behaviors. If computer can interpret the surveillance videos in real-time effectively, the work intensity of the security personnel will be greatly reduced. Behavior detection based on video is an important technology for intelligent interpretation of surveillance systems, and is one of the research hotspots in the field of computer vision [15, 16].

At present, object recognition and behavior recognition methods have achieved satisfactory results. Object detection is to find some target objects in a static image and classify them. The previous popular algorithms were proposed on basis of region CNN

D.-S. Huang et al. (Eds.): ICIC 2018, LNAI 10956, pp. 339–348, 2018.
https://doi.org/10.1007/978-3-319-95957-3_37

[9, 10] framework, which used multi-scale sliding windows to search the entire image to generate region proposals then identify them in the proposed areas. In recent years, one of the best methods in object recognition area is YOLO [6, 7], which is completely different from the above algorithms. It achieves state-of-the-art results in high speed and can deal with almost 9000 classes [6]. Action recognition is designed to identify action categories in short video clips containing actions. The most representative previous study is the Dense Trajectory (DT) method introduced by Wang [3], which is based on tracing points of samples. Bilen [4] and other scholars have proposed the concept of dynamic Image (DI) which can capture changes in key movements. DT describes the features of action and has achieved state-of-the-art results on action recognition, while it often fails to capture trends of action. DI is good at catching action evolution trends, but its representation ability is slightly weak. In view of both advantages and disadvantages of the above two methods, Jian et al. [8] combine dense trajectories and dynamic features, and their method can capture action trends and predict behavior more accurately. Recently, action recognition methods based on deep learning could achieve very good performance. However, studies of action recognition generally pre-process videos by dividing them into short clips which contain a certain type of action. In real life, video records are usually long and untrimmed. For example, monitoring video duration can reach hundreds of hours, but actions may last only a few minutes or even a few seconds. In long duration video, detection methods are more suitable. At present, performance of action detection algorithm is not satisfactory. The main reason is that there are many useless background fragments in the long duration video. Besides the duration of different behaviors varies greatly. Recently, action proposal methods can quickly determine the scope of behavior without using multi-scale sliding.

If an action recommendation method is combined with action classifiers for action detection, the performance will be significantly improved [12–14]. Existing works typically generate action proposals by using sliding windows of multiple scales and retrieve action proposals. Because they slide windows over a video many times, there are redundant computations on the process. To solve this problem, Buch et al. introduce the Single-Stream Temporal Action Proposals (SST) [5], a new effective and efficient deep architecture. By using a fixed scale window, it can run continuously in a single stream over very long input video sequences, without the need to divide input into short overlapping clips or temporal windows for batch processing. And it outperforms the state-of-the-art on the task of temporal action proposals generation. In SST network, data is processed on the basis of C3D features extracted from videos. We find that the prevailing dimension of C3D features is 4096, and a dimension reduction algorithm is adopted to reduce the dimension in SST. But there is just a passing description of dimension reduction in the paper of SST. When experimentally running SST network, we found that if we did not use any dimension reduction method, the result was far from theoretical results. And after dimension reduction, It got improved on both speed and performance. Therefore in view of action proposals, to what degree we reduce the dimension can we get the best trade-off between speed and performance? And what kind of dimension reduction method is more suitable for the action proposals? In view of these two problems, this research makes comparative experiments based on SST framework.

The main contributions of our paper are: (i) We experimentally find that in action detection proposal task, reducing the dimension of video features is crucial. It not only accelerates the speed and reduces the workload in the computation process but also makes the performance better. (ii) Experiments show that feature compression algorithm is more suitable for temporal action proposal than feature selection.

2 Concrete Algorithm Process

2.1 C3D Features

In order to get feature representations of videos efficiently, we need to carry out feature extraction of videos for subsequent processing. Convolutional Neural Network (CNN) is a sort of deep learning model that can act directly on raw inputs to do feature extraction work. But CNN can only deal with 2D inputs, and the features of results lack motion modeling information which cannot be used for action recognition work. To solve this problem and capture both spatial and temporal dimensions for adjacent frames, Ji et al. [2] introduce a novel model named 3D CNN for action recognition. It is a quite effective and developed model to recognize human actions in real-world environment, and it achieves superior performance without relying on handcrafted features. The model extracts feature in spatio-temporal dimensions by performing 3D convolution, capturing motion information in multi-scale adjacent frames. The model generates multiple channels of information from the input frames, and the final feature representation is obtained by combining information from all channels. In this paper, we do the feature extraction at time resolution $\delta = 16$ and every time step we get a result, the length of which is 4096.

2.2 Dimensionality Reduction Methods Used in C3D Features

For a video set of n videos, if processed by C3D, a video V_i of length L will have $T = L/\delta$ features and end up with a matrix of $T \times 4096$, so for all videos the matrix is $n \times T \times 4096$. When n becomes very large, the processing time of subsequent work will increase sharply.

Dimensionality reduction is the process of reducing the number of random variables under consideration by obtaining a set of principal variables. It can be divided into linear and non-linear reduction according to whether the operation used in the method is linear or not. The non-linear dimension reduction methods are time-consuming. For example, t-sne [11] has an excellent reduction effect but its execution time is much longer than linear dimension reduction and its memory usage is very high. Thus, it extremely demands on hardware settings, and it cannot meet the need of daily process or real-time system. In this paper we adopt linear dimension reduction which can be divided into feature selection and feature extraction.

Feature Extraction. Feature extraction transforms the data to a space of fewer dimensions from a high-dimensional one. The new dimension is lower than original one, so the dimension decreases. In this process, the features have undergone changes and the original features have disappeared although the new features retain some properties of the original features. In this paper, we choose principal component analysis (PCA). It is the most representative method of feature extraction and this linear dimension reduction method is by far the most commonly used. This method takes the variance to measure information using unsupervised learning.

Feature Selection. Feature selection approaches try to find a subset of the original variables. It selects feature subset d from n features ($d < n$) and the other $n - d$ features are discarded. So, the new feature is just a subset of the original feature. There is no feature change in the process. This is the main difference between feature extraction and feature selection dimension reduction. And we choose Feature Generating Machine (FGM) [1] a sparse solution with respect to input features to SVM. It iteratively generates a pool of violated sparse feature subsets and then combines them via efficient Multiple Kernel Learning (MKL) algorithm. FGM shows great scalability to non-monotonic feature selection on large-scale and very high dimensional datasets.

2.3 Gated Recurrent Unit (GRU)

In order to deal with C3d sequence features, we adopt GRU network to process data before and after the combination of feature information. The GRU is simpler than the LSTM network. Each GRU unit has only two doors, reset gates R and update gates Z.

H is a hidden layer, which contains information extracted from the current GRU unit. In this paper, we use time resolution $\delta = 16$ frames $t\{t_1, t_2, \ldots\}$ to feed GRU network: the GRU unit obtains h_{t-1} from previous unit, and uses Z_t as effective degree of information at the previous t. The smaller the Z_t is, the more information is used by h_{t-1}.

2.4 Non-maximum Suppression

In terms of a same action, there may be several similar proposals. Therefore, we use non-maximal suppression to choose the best advice. It inhibits proposals of which time repetition rate is too high to get a better one. First, we use the proposal of highest confidence as a reference to delete those proposals of which the proportion of public time is greater than threshold. The rest of proposals (except for the best advice in the previous one) are recursively executed until there is no proposal left after the deletion.

3 Experiments

3.1 Framework

SST is currently the best method to process long and untrimmed video sequence in action proposal work. In our comparative experience we adopt TensorFlow version of SST to test performance of PCA and FMG.

3.2 Data Set

THUMOS Challenge 2014 (THUMOS 2014): The data set includes two tasks: action recognition and temporal action detection. Its training set is the UCF101 dataset, including 101 types of actions, with a total of 13320 segmented video clips. THUMOS2014's validation set and test set include 1010 and 1574 untrimmed video respectively. In the temporal action detection task, only 20 classes of action are labeled with temporal information. Validation set includes 200 videos (3007 pieces) and test set includes 213 videos (3358 pieces). These marked untrimmed videos can be used to test the temporal action detection model.

Because the essence of action proposal is to put forward the temporal information of actions, we choose these 413 videos as test samples. In terms of training samples, we select 2765 short videos of 20 classes in UCF101 for positive samples, and randomly selected 430 short videos of 40 kinds of actions unrelated to the 20 classes as negative samples.

3.3 Feature from Video

As mentioned in the algorithm section, all videos used in the experiment are extracted with the standard experimental parameters. And in procedures of C3D the feature presentations are extracted with 4096 features per 16 frames. The total features obtained from the 2765 training positive samples are a matrix of 21446×4096.

3.4 Training

The purpose of our training is to get dimensions which can better describe the characteristics of the 20 target classes than others from 4096 dimensions. According to the dimensions obtained from training, the corresponding dimensions of test samples are selected, and other dimensions are discarded. Then test samples are put into SST for testing. For PCA method, because it is unsupervised training and does not require a sample label, we only need to provide it with C3D features from 20 target classes. For feature selection method such as FMG, we need to provide a sample label of each C3D features, in short, we need tell the model which sample is positive and which is negative. Therefore, we provide negative samples for the training of FGM as well. According to the previous training effect of FGM, the negative samples should be twice more than positive samples to let samples become more distinctive. So, we provide 40 classes of actions which have no relation with the 20 target classes, with a total number of 59485×4096 matrix as negative samples.

PCA—We use PCA function built in matlab to deal with the feature matrix of 21446×4096, and finally get the result matrix of 4096×4096. The column of this matrix is a reordering of the original 4096 dimensions based on the ability to describe or generalize the features. And if we need to reduce the dimension to the D dimension, just pick the former D dimensions. The row of this matrix is used to do the matrix multiplication. We multiply the feature matrix of the test sample $Z \times 4096$ with our $4096 \times D$ matrix to reduce the dimension to $Z \times D$ matrix.

FGM—We also use the matlab version of FGM. The number of positive and negative samples together reach a matrix of 80904×4096. The positive and negative samples are labeled with +1 and −1 respectively. After processing, we get feature indexes of Z interested dimensions among 4096. Then we selected the dimensions in the test sample corresponding to the feature indexes while other features are discarded. So, the matrix eventually reduced the original matrix to $Z \times D$.

4 Analysis of Experimental Results

We put the test samples from different dimension reduction method into the SST framework. And we assess performance from the number of proposals, intersection over Union (IoU) and average recall.

If we use the set of results and Ground Truth to divide the intersection of them then the result is IoU, that is, the accuracy of the detection. For temporal proposal, it is the time range we predicted divided by the time range of ground truth. We evaluate recall rate of an experimental result combined with the number of proposed proposals, and it is better to obtain a higher recall rate with fewer proposals. Besides, recall rate is considered with multi-scale IoU. When IoU = 1, it means that the proposal is completely consistent with ground truth, which is obviously impossible since even human beings cannot fully confirm the start and end time of an action. So as for temporal proposal, values of IoU around 0.8 are more important, meaning that the time segments of the action can be relatively accurately identified. For much smaller IoU values, such as 0.2, that means that our proposal is far from the actual time segment of the action, and the value is of little practical significance.

In the paper of SST, the authors just make horizontal comparison to compare performance of several action temporal proposal algorithms in current mainstream. However, they do not do the work of vertical comparison to discuss concrete effect of dimensionality reduction method in the same method. And in SST, the IoU threshold values varies from 0.5–1.0 with the step of 0.05, and the recall value is calculated separately. Then it counts the average recall.

Through experiments, we find the model of SST is inclined to choose recalls in high threshold as a measurement of performance. Therefore, after the features dimension are reduced by PCA or FGM, we set the IoU threshold to 0.7 and observe the results with the effect of unreduced original features. We find that, on the interval with high confidence score, the recall of unreduced results is slightly higher than that of the reduction, see Fig. 1(b). After analysis we think that in the process of dimensionality reduction, the number of features is greatly compressed, and the useless features are heavily filtered. Therefore, it will significantly increase the weight of the proposal of low confidence score, but weight of higher ones will be influenced slightly, which we can see from Fig. 2.

In Fig. 1(b), we compare the results of 2000 dimensions' feature reduced by PCA and FGM respectively with the original one of 4096 dimensions. FGM2000 has the best effect, because C3D features are very sparse. However, there is no significant change in performance of PCA2000. In combination with Fig. 3(a) and (b), we can see that the performance of FGM100 is very bad, while PCA100 performs well. Because

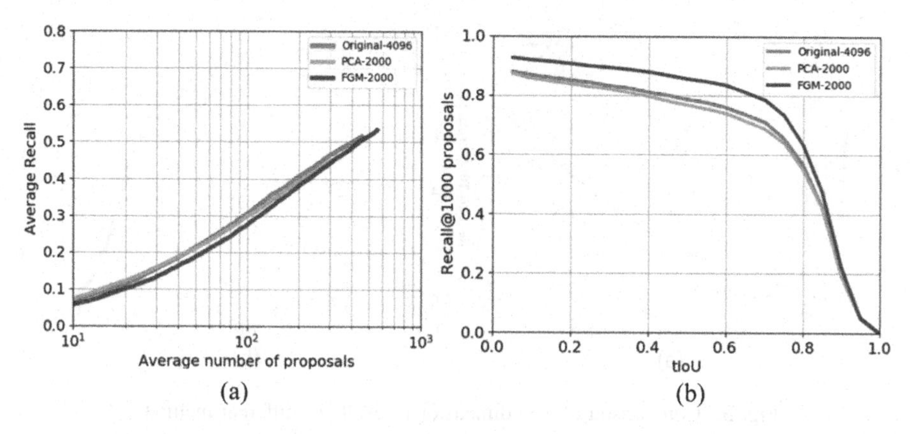

Fig. 1. Comparison of 2000 dimensional results of different methods

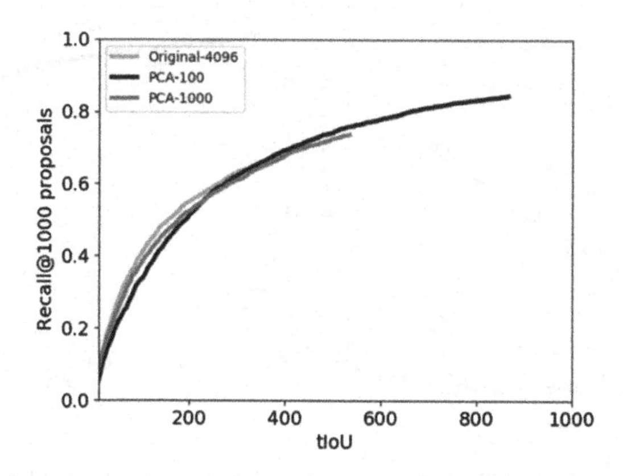

Fig. 2. Comparison of PCA dimensionality reduction results of different degrees

the reduction of PCA is based on feature extraction according to the original features, the remaining 100 dimensions are highly generalized. The FGM choose 100 dimensions from 4096 dimensions intact and lose a lot of important information, so the performance is not well. This indicates that the high intensity reduction of PCA is suitable for the action proposals.

From Figs. 4(a) and 5(a), it can be seen that the total proposal number of different dimensions is different, that is, the maximum value each method can reach on the horizontal coordinate is different. And we can see that the smaller feature dimension is, the more total proposals are. The reason is that feature dimension set restrictive condition for the confidence score of each time clips. And it is easier to accumulate confidence score in the condition of less constraint to reach the threshold. So, tests of low feature dimension are inclined to make more proposals. This explains why average proposal of high feature dimension cannot reach 1000. It is because the conditions are too harsh, and after non-maximum suppression processing, the number of proposals is greatly reduced. In Fig. 5(b), the effect of FGM1000 and FGM100 are relatively good.

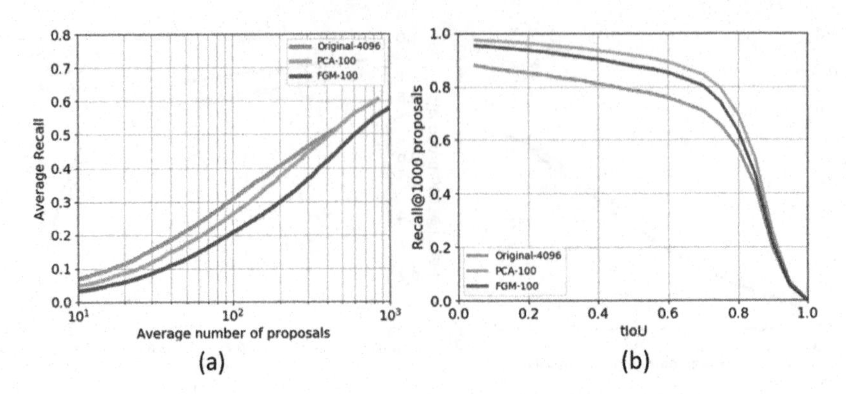

Fig. 3. Comparison of 100 dimensional results of different methods

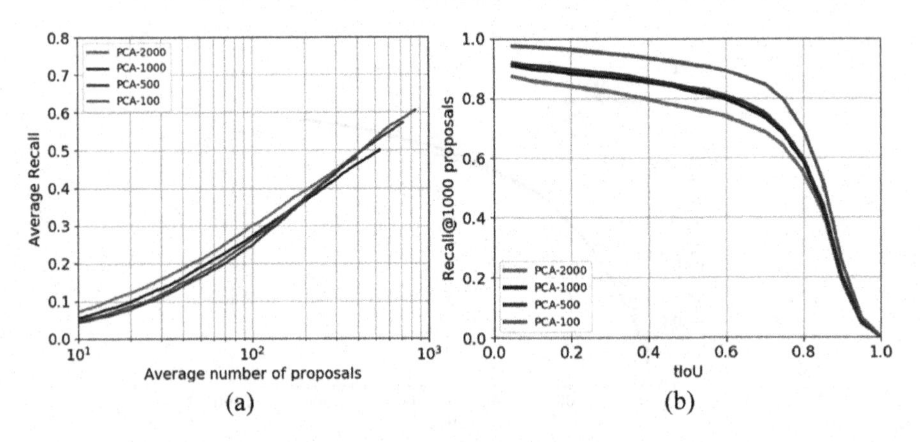

Fig. 4. Comparison of PCA dimensionality reduction results of different degrees

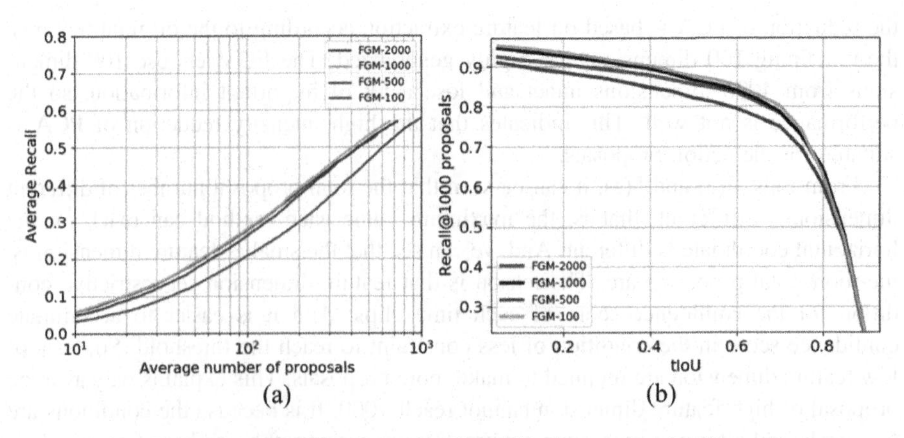

Fig. 5. Comparison of FGM dimensionality reduction results of different degrees

However, according to Fig. 5(a) and the analysis of Fig. 2, the overall recall rate of FGM100 mainly depends on the low point proposal, so the performance is still very poor. This also indicates that FGM is not suitable for high dimension reduction. Moving on to Fig. 4(a) and (b), we can see, the larger the reduction of PCA, the better performance of the recall rate it can get. Therefore we could demonstrate again that the high intensity reduction of PCA is suitable for action temporal proposal.

5 Conclusion

In this paper, by comparing results of different feature dimensions running on SST framework, we experimentally found that in action detection proposal task, reducing the dimension of feature is crucial since it cannot only accelerate the speed, alleviate the computing work in the process but also make the performance better. And on the basis of analyzing the characteristics of experimental results and dimensionality reduction methods, we argue that feature extraction reduction method is more suitable for temporal action proposals than feature selection method.

Acknowledgment. This work was supported by the Hubei Province Training Programs of Innovation and Entrepreneurship for Undergraduates, 201710488036; Scientific and technological innovation fund for College Students of Wuhan University of Science and Technology, 17ZRC131; Scientific and technological innovation fund for College Students of Wuhan University of Science and Technology, 17ZRA116; Scientific and technological innovation fund for College Students of Wuhan University of Science and Technology, 17ZRA121.

References

1. Tan, M., Wang, L., Tsang, I.W.: Learning sparse SVM for feature selection on very high dimensional datasets. In: International Conference on International Conference on Machine Learning, pp. 1047–1054. Omnipress (2010)
2. Ji, S., Xu, W., Yang, M., et al.: 3D convolutional neural networks for human action recognition. IEEE Trans. Pattern Anal. Mach. Intell. **35**(1), 221–231 (2012)
3. Wang, H., Aser, A.K., Schmid, C., et al.: Dense trajectories and motion boundary descriptors for action recognition. Int. J. Comput. Vis. **103**(1), 60–79 (2013)
4. Bilen, H., Fernando, B., Gavves, E., Vedaldi, A., et al.: Dynamic image networks for action recognition. In: The IEEE Conference on Computer Vision and Pattern Recognition (2016)
5. Buch, S., Escorcia, V., Shen, C., Ghanem, B., Niebles, J.C.: SST: Single-stream temporal action proposals. In: Computer Vision and Pattern Recognition, pp. 6373–6382. IEEE
6. Redmon, J., Farhadi, A.: Yolo9000: better, faster, stronger, pp. 6517–6525
7. Redmon, J., Divvala, S., Girshick, R., et al.: You only look once: unified, real-time object detection. In: IEEE Conference on Computer Vision and Pattern Recognition, pp. 779–788. IEEE Computer Society (2016)
8. Jiang, J., Deng, C., Cheng, X.: Action prediction based on dense trajectory and dynamic image. In: Chinese Automation Congress (CAC) (2017)
9. Girshick, R.: Fast R-CNN. In: IEEE International Conference on Computer Vision, pp. 1440–1448. IEEE Computer Society (2015)

10. Girshick, R., Donahue, J., Darrell, T., Malik, J.: Rich feature hierarchies for accurate object detection and semantic segmentation. In: CVPR (2014)
11. Laurens, V.D.M.: Accelerating t-SNE using tree-based algorithms. J. Mach. Learn. Res. **15**(1), 3221–3245 (2014)
12. Caba, F., Niebles, J.C., Ghanem, B.: Fast temporal activity proposals for efficient detection of human actions in untrimmed videos. In: CVPR (2016)
13. Escorcia, V., Caba Heilbron, F., Niebles, J.C., Ghanem, B.: DAPs: deep action proposals for action understanding. In: Leibe, B., Matas, J., Sebe, N., Welling, M. (eds.) ECCV 2016, Part III. LNCS, vol. 9907, pp. 768–784. Springer, Cham (2016). https://doi.org/10.1007/978-3-319-46487-9_47
14. Shou, Z., Wang, D., Chang, S.: Temporal action localization in untrimmed videos via multi-stage CNNs. In: CVPR (2016)
15. Karaman, S., Seidenari, L., Del Bimbo, A.: Fast saliency based pooling of fisher encoded dense trajectories. In: ECCV THUMOS Workshop, vol. 1, p. 6 (2014)
16. Ke, Y., Sukthankar, R., Hebert, M.: Event detection in crowded videos. In: 2007 IEEE 11th International Conference on Computer Vision, pp. 1–8. IEEE (2007)

Histopathological Image Recognition with Domain Knowledge Based Deep Features

Gang Zhang, Ming Xiao, and Yong-hui Huang(✉)

School of Automation, Guangdong University of Technology,
Guangzhou 510006, China
gracehuang02@163.com

Abstract. Automatic recognition of histopathological image plays an important role in building computer-aid diagnosis system. Traditionally hand-craft features are widely used for representing histopathological images when building recognition models. Currently with the development of deep learning algorithms, deep features obtained directly from pre-trained networks at less costs show that they perform better than traditional ones. However, most recent work adopts common pre-trained networks for feature extraction and train a classifier with domain knowledge which generates a gap between the common extracted features and the application domain. To fill the gap and improve the performance of the recognition model, in this paper we propose a deep model for histopathological image feature representation in a supervised-learning manner. The proposed model is constructed based on some pre-trained convolutional neural networks. After supervised learning, the feature learning network captures most domain knowledge. The proposed model is evaluated on two histopathological image datasets and the results show that the proposed model is superior to current state-of-the-art models.

Keywords: Histopathological image recognition · Deep learning
Convolutional neural network · Feature extraction · Ensemble learning

1 Introduction

Histopathological image recognition can reveal the cause and/or severity of many diseases in most medical department [1]. In some cases, histopathological image analysis becomes the golden standard for final diagnosis [2]. Generally speaking, histopathological image recognition can be divided into two categories according to their targets. The first category is annotation with a set of terms indicating several concerning histopathological characteristics within the task. When annotating a histopathological image, an annotation term may correspond to one or some local regions within the image. And on the contrary, one local region may be annotated by several terms. The second category is classification or grading a histopathological image [3, 4]. The target is to give a decision of which disease can be indicated by a given image, or how serious the disease is. For the tasks falling into this category, histopathological characteristics are not explicitly concerned and the model only outputs a classification label or a score describing the severity of a certain disease [5].

D.-S. Huang et al. (Eds.): ICIC 2018, LNAI 10956, pp. 349–359, 2018.
https://doi.org/10.1007/978-3-319-95957-3_38

Figure 1 shows an example of a histopathological image of a skin tissue, as well as the potential histopathological characteristics. We manually mark the interested local regions with black circles which may correspond to some histopathological characteristics. Both categories of histopathological image recognition pose significant challenges on building recognition models. For the first category, the annotation task is in fact a multiple-instance multiple-label (MIML) learning problem, in which the learning model has to capture the correlation between instances and labels. In such case either hand-craft or automatic-learning features of the whole image cannot be directly applied to the MIML model, which is a majority challenge. Our previous work [6] attempted to solve the first category problem within a probabilistic MIML learning framework. Our recently work [7] showed that a local region-based feature extraction with the famous convolutional neural network (CNN) can achieve better performance compared to hand-craft features under the condition of proper region partition. For the second category, the challenge is that feature extraction for the whole histopathological image cannot easily capture the essential features for classification or grading. According to medical knowledge and experience, the final decision is mainly based on the recognition of histopathological characteristics reflected by local regions.

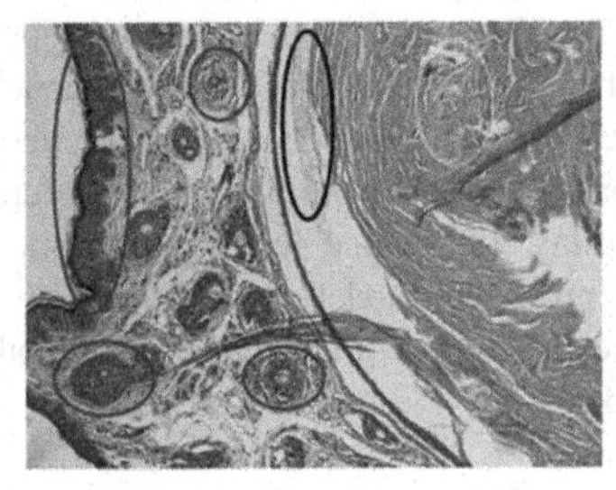

- basal cell liquefaction degeneration
- acanthosis
- hyperpigmentation of basal cell layer
- retraction space

Fig. 1. Histopathological image and characteristics.

These challenges prevent us from constructing recognition models by either simply adopting an end-to-end deep network, or a pre-trained common model [8]. In this paper, we propose a histopathological image recognition model with deep domain knowledge-based features. The model is composed of a deep feature extraction network trained in a supervised-learning manner, and a classifier. The proposed model is consistent with both categories recognition tasks mentioned above. The motivation of this work is twofold. On the one hand, it was reported that pre-trained models which perform well in common image understanding tasks have been adopted as feature extractors for histopathological image recognition. However, it does not incorporate any supervised information when adopting pre-trained models. It is believed that the pre-trained model for common image understanding effectively provides common visual features which are not specific for histopathological image recognition [9]. Hence a supervised model can make use of the domain knowledge, leading to a more effective feature representation [10]. On the other hand, a united framework that can solve both categories are preferred. It is widely accepted that the final diagnosis is derived from the recognition and analysis of local patches, as well as the relation between them. In this sense, a united framework that can recognize local characteristics and global ones reflects different depth of the model structure, which can be solved via deep learning methods.

The rest of this paper is organized as following. In Sect. 2 we briefly review some recent work related to this study. In Sect. 3 we present the histopathological image recognition model. In Sect. 4 the evaluation results of the proposed model on two benchmark datasets are reported. And the paper is concluded in Sect. 5.

2 Related Work

We briefly review some recent work related to this study. Urdal et al. [11] proposed a classification model for bladder cancer. Their work is conducted using the local binary pattern and variance operators for texture features extraction and the final classifier is RUSBoost. Their model is effective for histopathological images of sliders of bladder tissue, but it is not easy to apply to other types of tissue, since the extracted feature is designed for images of bladder tissue. Wang et al. [2] proposed a bilinear CNN (BCNN) based model for histopathological image classification. An image is decomposed into hematoxylin and eosin stain components and then BCNN is applied to each of the component for better feature representation. BCNN can be regarded as an encoder for each stain channel, which makes the fused feature more effective. However, domain knowledge is not applied in their feature extraction model and the decomposition into different stain components may not be suitable in some cases. Zheng et al. [10] proposed a probabilistic content-based histopathological image retrieval (CBHIR) model, in which a probability map regarding the malignancy of breast tumors is calculated. They construct a probability map of sub-region of whole slide image for effective image retrieval. Their work establishes the relationship between sub-region and the whole slide image. However, their work required fully annotated images as training dataset when constructing the probability map, which limited its application scope. Zhang et al. [7] proposed a deep learning based model for histopathological image feature representation. They improved the traditional CNN model by adding a sub-sampling layer in order to satisfy different magnification rates of the input images. Their work shows that CNN model can extract more effective features than hand-craft features. Spanhol et al. [12] presents a deep learning based feature extraction method for breast cancer image classification. They build the classification model by combining a CNN and a softmax layer as the classifier.

3 Recognition Model with Domain Knowledge Based Deep Features

Before going further, we give a formal definition of the problem. Let $D = \{(x_i, y_i), i = 1, 2, \ldots, m\}$ be a data set with labels. $x_i \in X$ is a histopathological image and $y_i \in Y$ is a classification label. In this study, only binary classification label is considered. The goal is to learn a function $f : X \rightarrow Y$ which achieving good performance on both the training dataset and the unseen data. Since the input data is a set of images, feature extraction is an important step when constructing the recognition model. We adopt some pre-trained model for feature extraction and refining it by incorporating the domain knowledge.

3.1 Feature Extraction

Several recent proposed deep convolutional neural networks are adopted for feature extraction. A training histopathological image is scaled to fit the dimensions of the feature extraction network. Figure 2 illustrates two convolutional layers with different channels in the ResNet model [13].

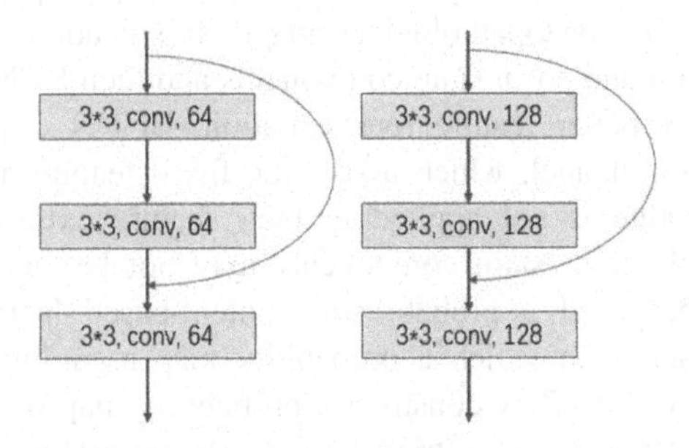

Fig. 2. Convolutional layers with different channels in the ResNet.

ResNet limits the number of the model parameters while keeping the performance by adding shortcut connections from input to output of a convolutional block. However, the theoretical explanation of the layers is not clear when examining the feature maps. Intuitively, it is widely accepted that the features extracted from the top layers representing high level concepts. We hence develop a feature selection and ensemble strategy by introducing a representation ratio r_i, as Eq. (1).

$$r_i = \frac{c_i}{rank(p_i) + rank(q_i)} \tag{1}$$

In Eq. (1), c_i stands for the number of channels in the ith layer. $rank(p_i)$ and $rank(q_i)$ stand for the rankings of accuracy and diversity of some weak learner trained with the feature map of layer i. p_i and q_i are defined as Eqs. (2) and (3).

$$p_i = 1 - \frac{L(h, D, f)}{c_i} \tag{2}$$

$$q_i = \sum_{x \in D} \Delta_f(h(f_i(x))) \tag{3}$$

In Eq. (2), the function L evaluates the loss of a model h on an evaluation dataset D with representation f. We perform the loss evaluation in a channel-wise manner, i.e. the feature maps from a certain channel are combined to form a feature representation. Note that only accuracy is necessary in this study. For multiple-classification problems, additional evaluation metrics should be adopted. Equation (3) evaluates the difference between a model h with feature representation f_i and other representations for all data samples in the evaluation dataset.

Non-Dominated Sort (NDS) [14] is applied to obtain a reasonable ranking of the performance of all the feature maps. It accepts a measurement matrix whose rows indicating performance metrics for a set of models. Its output is a ranking of the model set according to the metrics. Algorithm 1 illustrates the details of NDS algorithm.

3.2 Domain Knowledge Based Feature Learning

After NDS, a small subset of feature maps is picked out, i.e. the top $p\%$ of the sorted list. An ensemble style training method is proposed to make use of the domain knowledge. Feature maps from different convolutional layers may have different sizes and cannot be directly applied to the feature learning model. The goal of this step is to learn a feature representation by incorporating domain knowledge. Motivated by the weighted ensemble method [15], we propose a method for combining the feature maps extracted from the network model. The method is composed of two steps. In the first step, a weight vector w is initialized as $w_i = 1/|w|$. In the second step, a convex optimization problem is solved to find the best combination weights, which is shown in Eq. (4).

$$\arg\min_w Loss(w, h, X) + ||w||^2 \tag{4}$$

In Eq. (4), the target is to minimize both the ensemble model loss on the training dataset and the model complexity measured by $|w|$. Algorithm 2 summaries the main steps of the method.

Algorithm 1. Non-Dominated Sort

Require:
measurement matrix: M
model set: $H = \{h1, h2, \cdots, h_n\}$
Ensure:
the ranking of E: *ranking*

```
1: F = Ø, ranking = Ø
2: for i = 1 to n − 1 do
3:    S_i = Ø, t_i = 0
4:    for j = i to n do
5:       if h_i dominates h_j related to M then
6:          Add h_i to S_i
7:       else if h_j dominates h_i related to M then
8:          t_i = t_i + 1
9:       end if
10:   end for
11:   if t_i = 0 then
12:      add S_i to F
13:   end if
14: end for
15: F = sort(F)
16: while F ≠ Ø do
17:   Q = Ø
18:   for all s in F do
19:      for all c in s do
20:         if t_c ≠ 0 then
21:            t_c = t_c − 1
22:         end if
23:         add c to Q
24:      end for
25:   end for
26:   add sort(Q) to ranking
27: end while
28: return ranking
```

Algorithm 2. Weighted ensemble learning

Require:
 training dataset: X
 ranking of the feature maps: $ranking$
Ensure:
 the optimized weights: w^*
1: initialize w to $w_i = 1/|w|$
2: **for** $i = 1$ to $|ranking|$ **do**
3: $X_i = X(ranking_i)$
4: $h_i = train(X_i)$
5: **end for**
6: $h = \{h_1, h_2, \cdots, h_{|ranking|}\}$
7: solve the convex optimization problem defined by Eq. (4)
8: **return** w^*

In Algorithm 1, the procedure **sort** stands for a common sorting method for single metric, e.g. quick sort, heap sort. From line 2 to 14 the model subsets are generated according to the order of each metric. From line 15 to 27, the subsets are collected together to form a ranking. If two models cannot dominate each other, the ranks of all metrics are summarized and then compared to obtain a ranking.

Algorithm 2 finds an optimization weights to combine the feature maps obtained via Algorithm 1. Note that it is a supervised learning procedure and the domain knowledge implied in the training dataset can be learned and stored in the form of weights.

4 Evaluations

The proposed method is evaluated on two benchmark histopathological image datasets. The first dataset is a breast cancer histopathological dataset, BreakHis [16]. There are 7909 histopathological images of 82 patients. Table 1 shows the details of the BreakHis dataset.

Table 1. The details of BreakHis dataset.

Magnification	Benign	Malignant	Total
40X	652	1370	1995
100X	644	1437	2081
200X	623	1390	2013
400X	588	1232	1820
Total of images	2480	5429	7909

The second dataset is a tissue image dataset containing 2828 images, His2828 [3]. Table 2 shows the details of His2828. Figure 3 shows 4 sample images of His2828.

Table 2. The details of His2828 dataset.

Label	Size
connective	483
epithelial	804
muscular	515
nervous	1026

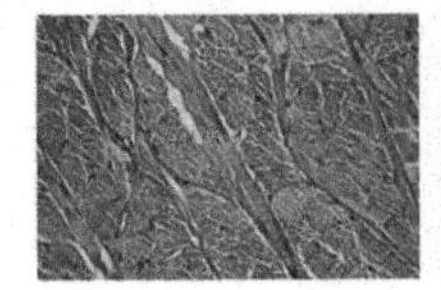 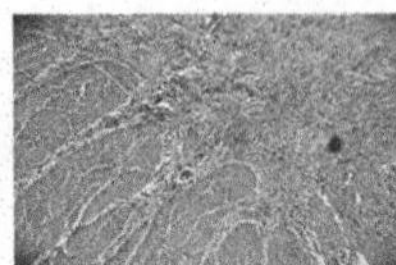 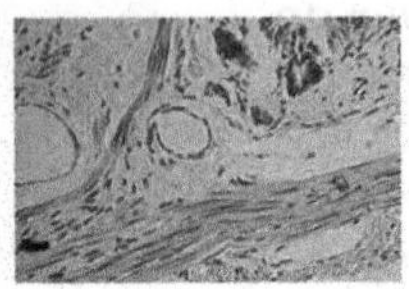 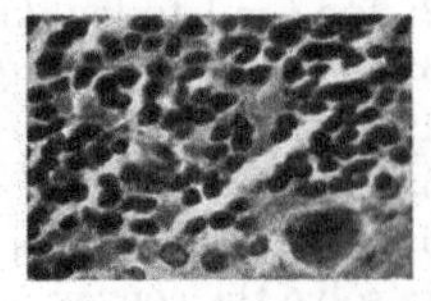

Fig. 3. Sample images of His2828

The ResNet [13] and GoogleNet [17] are adopted as feature extractor. Since the input dimension is 224 * 224 * 3, each training image is scaled before feeding to the network. To avoid overfitting, we enlarge the dataset by cropping, adding noisy points and flipping as described in [18]. For cropping, 4 corners and the center regions with 90% width and height of the original image are cropped. Each image is added 10% white noisy points to generate a noisy image. Vertical and horizontal flipping are performed. Finally, the size of the dataset is increasing by 5 + 1 + 2 = 8 times. The enlarged dataset is then divided into training set, test set and evaluation set at the ratio 8:1:1. 10-fold validation is used to make the evaluation result stable. The dataset is randomly divided into 10 parts and each time one part is used as test data.

For feature extraction, only the outputs of the convolutional layer are selected. We randomly select 50% the feature maps of the convolutional layers with the same channel size. The implementation of ResNet used in this study is **imagenet-resnet-152-dag** and that of GoogleNet is **imagenet-googlenet-dag**. Finally, there are 47 different groups of feature maps are selected. In feature map ranking stage, we set $p = 40$, i.e. the top 40% items of the ranking list generated by NDS are selected. A standard version support vector machine (SVM) is used as the final classifier, which is implemented with LibSVM [19]. Two current state-of-the-art methods are evaluated as comparison. Table 3 shows the details of these methods.

Table 3. Two methods for comparison.

No.	Name	Reference
1	Context-based CBIR (C-CBIR)	Zheng et al. [10]
2	Boost Convolutional Neural Network (BCNN)	Lee et al. [9]

The method C-CBIR is based on image retrieval which searches for regions with similar content for a region of interest (ROI) from a database consisting of historical cases. It is a region-based classification method utilizing region-of-interest (ROI) to determine the label of a test image. The method BCNN adopts Adaboost to establish an ensemble of a set of weak learners for image classification. Both methods achieve excellent performance on the evaluation datasets. Table 4 illustrates the evaluation results of the proposed method and the methods for comparison.

Table 4. Evaluation results (accuracy/variance).

Method	BreakHis	His2828
The proposed method	**81.2% ± 2.5%**	**83.6% ± 3.1%**
C-CBIR	78.6% ± 3.9%	82.0% ± 2.7%
BCNN	79.5% ± 2.2%	80.4% ± 3.6%

We highlight the best results of each column. It can be seen from Table 4 that the proposed method has the best performance among all three methods. The results indicate that the domain knowledge improves the representation ability of the extracted features of the pre-trained deep network, which illustrates the effectiveness of our method.

Finally, we show how the quantity of ensemble size affects the model performance. The parameter p is varied from 15% to 45% by step 5% and the model performance is recorded. Figure 4 illustrates the result of the proposed method on two evaluation datasets.

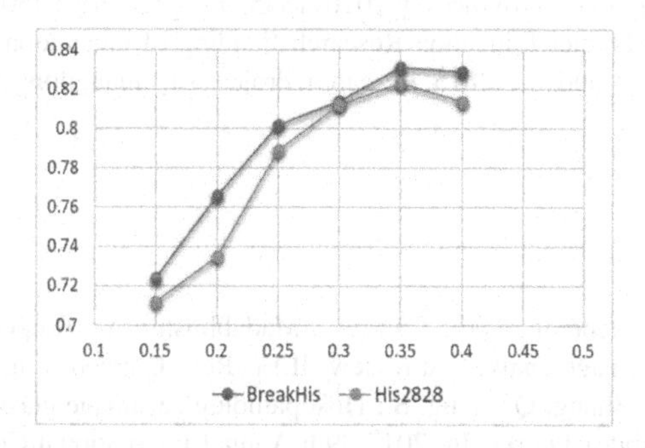

Fig. 4. The effect of ensemble size to the model performance

It can be concluded from Fig. 4 that a relative large ensemble size would lead to better model performance. We owe this to the fact that the ensemble of more feature maps reflects concepts of more levels. However, an ensemble feature maps with large size requires more computational resource. Hence, a trade-off between the complexity of feature maps and the cost should be considered when constructing the model.

5 Conclusions

In this paper we propose a deep learning based method for histopathological image recognition. We adopt a pre-trained convolutional neural network model to extract feature maps from different layers which representing different concepts. An ensemble style learning method is applied for the injection of the domain knowledge and a feature representation is generated via a supervised learning manner. The proposed method can be applied to either histopathological image classification or characteristic recognition. The proposed method is evaluated with two benchmark histopathological image datasets and the result shows that our method is superior to the state-of-the-art methods.

In the future we will focus on building the learning framework with domain knowledge for histopathological image recognition. By applying transfer learning, pre-trained models with large number of parameters can be used as an effective feature extractor for the problems of different application background. Domain knowledge can be incorporated not only in feature learning, but also in model initialization and training. Hence this work can be extended and applied in different machine learning framework, as well as other tasks related to CAD system.

Acknowledgments. This work is supported by the National Natural Science Foundation of China (No. 81373883, 81573827), the Science and Technology Planning Project of Guangdong Province (No. 2016A030310340), the Special Fund of Cultivation of Technology Innovation for University Students (No. pdjh2016b0150), the College Student Career and Innovation Training Plan Project of Guangdong Province (yj201611845593, yj201611845074, yj201611845075, yj201611845366), the Higher Education Research Funding of Guangdong University of Technology (No. 2016GJ12) and the 2015 Research Project of Guangdong Education Evaluation Association (No. G-11).

References

1. Gurcan, M.N., Boucheron, L.E., Can, A., Madabhushi, A., Rajpoot, N.M., Yener, B.: Histopathological image analysis: a review. IEEE Rev. Biomed. Eng. **2**, 147–171 (2009)
2. Wang, C., Shi, J., Zhang, Q., Ying, S.: Histopathological image classification with bilinear convolutional neural networks. In: 2017 39th Annual International Conference of the IEEE Engineering in Medicine and Biology Society (EMBC), pp. 4050–4053, July 2017
3. Caicedo, J.C., Cruz, A., Gonzalez, F.A.: Histopathology image classification using bag of features and Kernel functions. In: Combi, C., Shahar, Y., Abu-Hanna, A. (eds.) AIME 2009. LNCS (LNAI), vol. 5651, pp. 126–135. Springer, Heidelberg (2009). https://doi.org/10.1007/978-3-642-02976-9_17
4. Altunbay, D., Cigir, C., Sokmensuer, C., Gunduz-Demir, C.: Color graphs for automated cancer diagnosis and grading. IEEE Trans. Biomed. Eng. **57**(3), 665–674 (2010)
5. Bunte, K., Biehl, M., Jonkman, M.F., Petkov, N.: Learning effective color features for content based image retrieval in dermatology. Pattern Recogn. **44**(9), 1892–1902 (2011)
6. Zhang, G., Yin, J., Su, X.-y., Huang, Y.-j., Lao, Y.-r., Liang, Z.-h., Ou, S.-x., Zhang, H.-l.: Augmenting multi-instance multilabel learning with sparse Bayesian models for skin biopsy image analysis. Biomed. Res. Int. **2014**, 13 (2014). Article ID 305629

7. Zhang, G., Hsu, C.-H.R., Lai, H., Zheng, X.: Deep learning based feature representation for automated skin histopathological image annotation. Multimed. Tools Appl., 1–21 (2017)
8. Liu, X., Liu, Z., Wang, G., Cai, Z., Zhang, H.: Ensemble transfer learning algorithm. IEEE Access **6**, 2389–2396 (2018)
9. Lee, S.J., Chen, T., Yu, L., Lai, C.H.: Image classification based on the boost convolutional neural network. IEEE Access **6**, 12755–12768 (2018)
10. Zheng, Y., Jiang, Z., Zhang, H., Xie, F., Ma, Y., Shi, H., Zhao, Y.: Histopathological whole slide image analysis using context-based CBIR. IEEE Trans. Med. Imaging **PP**(99), 1 (2018)
11. Urdal, J., Engan, K., Kvikstad, V., Janssen, E.A.M.: Prognostic prediction of histopathological images by local binary patterns and rusboost. In: 2017 25th European Signal Processing Conference (EUSIPCO), pp. 2349–2353, August 2017
12. Spanhol, F.A., Oliveira, L.S., Cavalin, P.R., Petitjean, C., Heutte, L.: Deep features for breast cancer histopathological image classification. In: 2017 IEEE International Conference on Systems, Man, and Cybernetics (SMC), pp. 1868–1873, October 2017
13. He, K., Zhang, X., Ren, S., Sun, J.: Deep residual learning for image recognition. In: 2016 IEEE Conference on Computer Vision and Pattern Recognition (CVPR), pp. 770–778 (2016)
14. Li, X.: A non-dominated sorting particle swarm optimizer for multiobjective optimization. In: Cantú-Paz, Erick, et al. (eds.) GECCO 2003, Part I. LNCS, vol. 2723, pp. 37–48. Springer, Heidelberg (2003). https://doi.org/10.1007/3-540-45105-6_4
15. Zhang, C., Ma, Y.: Ensemble Machine Learning: Methods and Applications. Springer, Boston (2012). https://doi.org/10.1007/978-1-4419-9326-7
16. Spanhol, F.A., Oliveira, L.S., Petitjean, C., Heutte, L.: A dataset for breast cancer histopathological image classification. IEEE Trans. Biomed. Eng. **63**(7), 1455–1462 (2016)
17. Szegedy, C., Liu, W., Jia, Y., Sermanet, P., Reed, S., Anguelov, D., Erhan, D., Vanhoucke, V., Rabinovich, A.: Going deeper with convolutions. In: 2015 IEEE Conference on Computer Vision and Pattern Recognition (CVPR), pp. 1–9, June 2015
18. Krizhevsky, A., Sutskever, I., Hinton, G.E.: Imagenet classification with deep convolutional neural networks. In: Advances in Neural Information Processing Systems, vol. 25(2), pp. 1097–1105 (2012)
19. Chang, C.-C., Lin, C.-J.: Libsvm: a library for support vector machines. ACM Trans. Intell. Syst. Technol. **2**(3), 27:1–27:27 (2011)

Optical Character Detection and Recognition for Image-Based in Natural Scene

Bochao Wang, Xinfeng Zhang(✉), Yiheng Cai, Maoshen Jia, and Chen Zhang

Faculty of Information Technology, Beijing University of Technology, Beijing 100124, China
zxf@bjut.edu.cn

Abstract. In recent years, Optical Character Recognition(OCR) is widely used in machine vision. In this paper, we investigated the problem of optical character detection and recognition for Image-based in natural scene. The Optical Character Recognition is divided into three steps: (1) Selecting the candidate regions through image preprocessing. (2) The detection neural network is used to classify each region. The purpose is to retain text regions and remove non-text regions. (3) The recognition neural network is used to identify the characters in the text regions. We propose a novel algorithm. It integrates image preprocessing with Maximally Stable Extremal Regions(MSER), the neural network architecture of detection and the neural network architecture of recognition. Compared with previous works, the proposed algorithm has three distinctive properties: (1) We propose a new process of OCR algorithm. (2) The application scene of OCR algorithm is the images of natural scene. (3) The training data of recognition does not need artificial labels and can be generated indefinitely. Moreover, the algorithm has achieved good results in detection and recognition.

Keywords: OCR · Image detection · Image recognition · Deep learning

1 Introduction

With the strong revival of deep learning which is mainly promoted by the development of Deep Convolution Neural Network(DCNN), deep learning has made a surprising achievement in many fields. In this paper, we focus on the problem in computer vision: optical character detection and recognition. OCR is the process which determines shape of characters by detecting dark and light patterns. Then, the shape of characters are translated to computer words. In real world, OCR is widely used in many fields such as image review, image digitalization processing, annotation of article portrait and license plate recognition.

There are lots of papers about OCR. The algorithm in [1, 2] first separates individual characters. Then, it classifies and recognizes each character. The algorithm in [3] takes the problem as image classification and recognizes the words in a huge lexicon. On the other hand, some companies such as Alibaba, Baidu and Tencent have been engaged in OCR research in recently years. Google has been launched their OCR tool: Tesseract. Overview the algorithms above, most of them deal with the language of

D.-S. Huang et al. (Eds.): ICIC 2018, LNAI 10956, pp. 360–369, 2018.
https://doi.org/10.1007/978-3-319-95957-3_39

English which only has 60–70 dimensions and the scene is text images. In this paper, we mainly deal with Chinese which has 14300 dimensions. Moreover, our application scene is the images of natural scene. It is more complex and our algorithm is not restricted by the length of text.

The overall process of the algorithm in paper is shown in Fig. 1. Algorithm of OCR is divided into three steps. First, the algorithm of MSER is used to preprocess the image. It will find many candidate regions. Then, the neural network architecture of detection is used to identify the text regions and remove non-text regions. Finally, the neural network architecture of recognition is used to process text and identify the characters in the text regions. There will be more details on each step, and then it will be introduced in the paper.

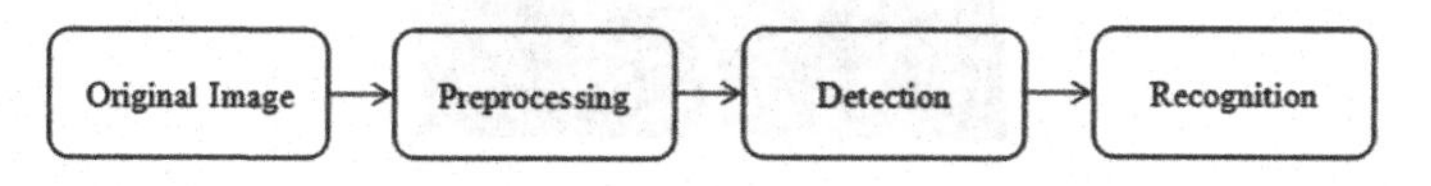

Fig. 1. The overall process of the algorithm

In Sect. 1, we introduce OCR and the superiorities of our algorithm. In Sect. 2, the preprocessing of image is introduced. In Sect. 3, the neural network architecture of detection is introduced. In Sect. 4, the neural network architecture of recognition is introduced. In Sect. 5, we make a conclusion and propose the future work.

2 Preprocessing by Maximally Stable External Regions

In this part, the algorithm of MSER is used to find candidate regions. This is the first step in the algorithm of OCR. There are many algorithms such as R-CNN, Fast R-CNN and Faster R-CNN. Candidate text regions can be detected by these algorithms. But these algorithms are complicate and have exorbitant demand of label. MSER [4] is usually used in text detection. It is the preprocess of image. The image processed by MSER has many candidate regions [5, 6]. Then, a detection network is used for classification to identify the text regions and remove non-text regions. This algorithm has good ability, pertinence and feasibility.

MSER is based on watershed algorithm. The image is binarized and the threshold ranges from 0 to 255. So, the image goes through a process from full black to full white. In this process, some connected regions change slightly with the threshold, these regions are called MSER. The formula is shown in Eq. 1.

$$v(i) = \frac{|Q_{i+\Delta} - Q_{i-\Delta}|}{|Q_i|} \tag{1}$$

Where Qi represents the regions of connected region, Δ represents a small change of threshold.

Before the MSER, the size of image is standardized, the shorter side of the image is resized to 420 pixels and the other side is scaled accordingly. It can retain useful information and improve efficiency. Then, the contrast of each channel of the image is calculated. Because MSER is also a algorithm which is based on contrast. MSER is used to process the channel with the biggest contrast to enhance the ability of MSER. MSER is used to process the image to get the result which is shown in Fig. 2. As you can see from the figure, the result of MSER contain many overlapping regions that can severely impact processing efficiency in subsequent steps. Therefore, it need to remove the overlapping regions [7, 8], the result is shown in Fig. 3. As is shown in Fig. 2, the number of overlapping regions has been greatly reduced.

Fig. 2. The result of MSER. Candidate regions: 370

Fig. 3. The result after removing the overlapping regions. Candidate regions: 136

3 The Network Architecture of Detection

In this part, the shallow CNN network is used to classify text regions and remove non-text regions [9]. This is the second step in the algorithm of OCR. The total number of dataset is 120000, including 80000 negative samples and 40000 positive samples. These samples are the color images of natural scene. The shape of images are resized to $28 \times 28 \times 3$. They have different contrast, intensity of light and background. Some samples are shown in Figs. 4 and 5.

Fig. 4. Positive samples.

Fig. 5. Negative samples.

The network architecture includes convolution layers, maxpooling layers, full connection layers and transcription layers. It is shown in Table 1.

Table 1. Network architecture of detection. The first row is the top layer. 'k', 's' and 'p' stand for kernel size, stride and padding size respectively.

Type	Configurations
Softmax	
Full Connection	#hidden units:2
Full Connection	#hidden units:512
Map-to-Sequence	
MaxPooling	Window:2 × 2, s:2
Convolution	#map:64, k:3 × 3. s:1, p:1
MaxPooling	Window:2 × 2, s:2
Convolution	#maps:32, k:3 × 3, s:1, p:1
Input	Batchsize × 28 × 28 × 3 image

In addition to this shallow network, we also tried to deepen the network, but it will reduce efficiency and there is no essential difference in accuracy. Besides, the model with the highest accuracy was not chosen because we want to reduce the false positive rate as much as possible. The false positive rate means that the model identify non-text regions into text regions. It seriously reduces the efficiency of algorithm. The accuracy of final model is 93.5%, the false positive rate is 2.5%, the recall rate is 88%. The process of training is shown in Fig. 6(a). The classifier is used to identify and remove non-text regions in Fig. 3. After processing by classifier, the regions are merged into lines and the result is shown in Fig. 6(b).

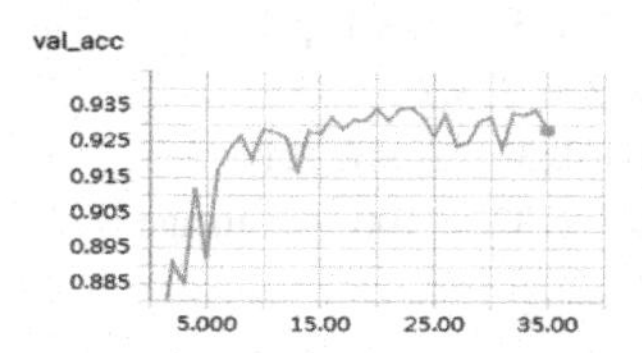

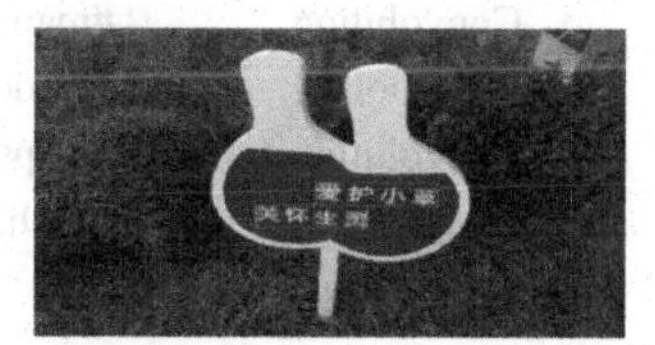

Fig. 6. (a) The accuracy of detection network. (b) The result after merging image into lines.

4 The Network Architecture of Recognition

In this part, the deep network is used to identify the characters in the text regions [10, 11]. The dataset comes from an open source project on GitHub, it is a synthetic dataset released by M. Jaderberg [12]. The recognition dataset does not need artificial labels and can be generated indefinitely. The total size of dataset is 7800000 and there are nearly 14300 dimensions. These samples are the color images of natural scene. The shape of images are resized to 32 × 100 × 3. They have different contrast, intensity of light and background. Some samples are shown in Fig. 7.

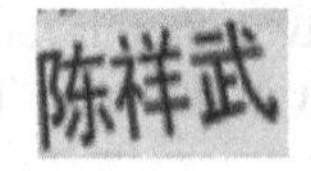

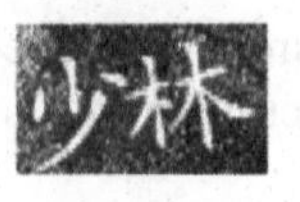

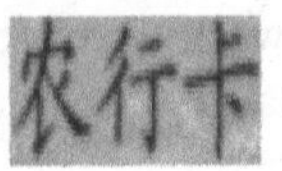

Fig. 7. Samples of recognition.

The network architecture include convolution layers, batch normalization layers, recurrent layers, full connection layers and transcription layers. It is shown in Table 2.

Table 2. Network architecture of recognition. The first row is the top layer. 'k', 's', 'x' and 'p' stand for kernel size, stride, recognition dimension and padding size.

Type	Configurations
CTC	
Full Connection	#hidden units:x
Bidirectional-LSTM	#hidden units:256
Bidirectional-LSTM	#hidden units:256
Map-to-Sequence	
Convolution	#map:512, k:2 × 2. s:1, p:0
MaxPooling	Window:1 × 2, s:2
BatchNormalization	
Convolution	#map:512, k:3 × 3. s:1, p:1
Convolution	#map:512, k:3 × 3. s:1, p:1
MaxPooling	Window:1 × 2, s:2
BatchNormalization	
Convolution	#map:256, k:3 × 3. s:1, p:1
Convolution	#map:256, k:3 × 3. s:1, p:1
MaxPooling	Window:2 × 2, s:2
Convolution	#map:128, k:3 × 3. s:1, p:1
MaxPooling	Window:2 × 2, s:2
Convolution	#maps:64, k:3 × 3, s:1, P:1
Input	Batchsize × 32 × 100 × 3 image

4.1 Batch Normalization

The training of model become more difficult because of the deeper network. The problems of vanishing gradient and slow speed of training seriously affect the training of the model. After literature research and experiment, Batch Normalization is used to overcome these problems. As is shown in Eq. 2, the distribution of data is normalized to Standard Normal Distribution by the process of whitening. After training with the same distribution of data, Batch Normalization uses transformation and reconstruction which is shown in Eq. 3 to restore the original distribution of data.

$$x'^{(k)} = \frac{x^{(k)} - E[x^{(k)}]}{\sqrt{Var[x^{(k)}]}} \quad (2)$$

Where 'E[x(k)]', 'Var[x (k)]', '$x^{(k)}$' and '$x'^{(k)}$' stand for the mean of every batch of training data, the variance of every batch of training data, input and new value of $x^{(k)}$.

$$y^{(k)} = \gamma^{(k)} x'^{(k)} + \beta^{(k)} \quad (3)$$

Where '$y^{(k)}$', '$x'^{(k)}$', '$\gamma^{(k)}$' and '$\beta^{(k)}$' stand for output, input, weight and bias.

The problem of vanishing gradient can be avoided by Batch Normalization. The problem of vanishing gradient is due to the multiplicative in backpropagation. The value of gradient is small. After a certain number of layers, the multiplication of gradient will approach to 0. As is shown in Fig. 8, the distribution of data can be normalized to Standard Normal Distribution by Batch Normalization. It means that x will fall within the range of [–1,1] with the probability of 64% and fall within the range of [–2,2] with the probability of 95%. These ranges have the largest gradient in the activation function. Therefore, it will have a large multiplicative value in backpropagation and avoid the problem of vanishing gradient.

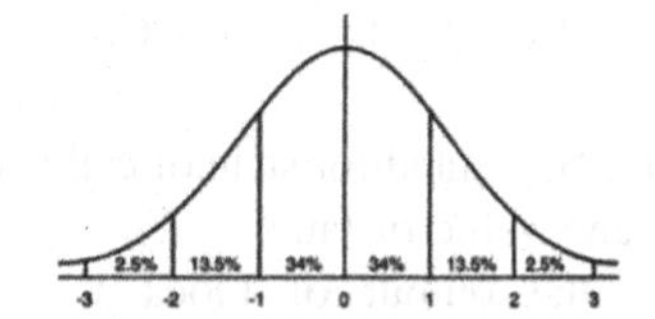

Fig. 8. Standard normal distribution

4.2 Bidirectional Recurrent Neural Network

The input has 7800000 images. The shape of each image is 32 × 100 × 3. After convolution layers and maxpooling, the data are resized to the shape of 1 × 24 × 512. OCR is not only recognize the text, but also consider the context of the text. For example, words in some phrases always appear together, these words are closely linked. RNN is a model of Natural Language Processing(NPL) [13–17]. The link between context can be captured by RNN.

A deep bidirectional Long-Short Term Memory(LSTM) is built on the top of the convolution layers. LSTM is a variant of RNN. It converts neuron into cell and effectively avoid the vanishing of gradient. It consists of a memory cell and three multiplicative gates, namely the input gate, output gate and forget gate. The overall process is shown in Fig. 9.

The first step is to determine the information which should be discarded. The function is achieved by the Equation of Sigmod which is shown in Eq. 4. Both input gate, output gate and forget gate follow this Equation.

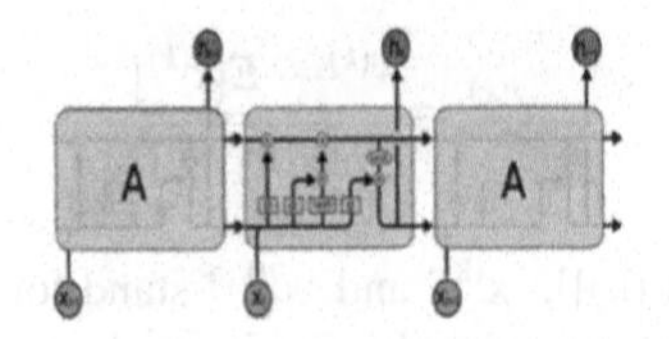

Fig. 9. overall process of LSTM

$$f_t = \sigma(W_f \cdot [h_{t-1}, x_t] + b_f) \quad (4)$$

Where 'f_t', 'W_f', 'h_{t-1}', 'x_t', 'b_f' stand for output of Sigmod, weight, output of block t–1, input of point t and bias.

As is shown in Eqs. 5 and 6, The state of current cell is determined by the second step.

$$\widetilde{C}_t = \tanh(W_C \cdot [h_{t-1}, x_t] + b_C) \quad (5)$$

Where '$\widetilde{C}_t$', 'W_C', 'h_{t-1}', 'x_t', 'b_C' stand for value of tanh, weight, output of block t–1, input of point t and bias.

$$C_t = f_t * C_{t-1} + i_t * \widetilde{C}_t \quad (6)$$

Where 'C_t', 'f_t', 'C_{t-1}', 'i_t', '$\widetilde{C}_t$' stand for state of cell t, value of forget gate, state of cell t–1, value of input gate and value of tanh.

As is shown in Eq. 7, the output of block t is determined by the third step. After LSTM, the results of LSTM are decoded. Finally, the loss is calculated by the transcription layer.

$$h_t = O_t * \tanh(C_t) \quad (7)$$

Where 'h_t', 'O_t', 'C_t' stand for output of block t, value of output gate and state of cell t.

4.3 Connectionist Temporal Classification

Connectionist Temporal Classification(CTC) is a transcription layer. It can decode the results of LSTM and calculate loss [18, 19]. Then, the SGD is used to minimize loss and find the optimum solution.

Assuming that a sequence with the length of T, which has L possibilities at each time point, then the sequence has L^T possibilities. Each possibility is called a PATH. It conditional probability is shown in Eq. 8.

$$p(\pi|x) = \prod_{t=1}^{T} y^t_{\pi(t)} \quad (8)$$

Where 'π', 'x', 't', 'T', 'y', 'p' stand for PATH, input, time point, total number of time points, probability of point t and probability of PATH π.

In CTC, the number of labels is extended from L to L + 1. It extends a label of blank character to represent the blank in the sequence. CTC separates the sequence with blanks, then removes duplicate characters and blanks in the $l^* \approx \mathrm{B}(\arg\max_\pi p(\pi|y))$ path. For example, B(a, -, a, b, -) and B(-, a, a, -, -, a, b, b) both represent aab. So, merge the PATH with the same results. The conditional probability is shown in Eq. 9.

$$P(l|y) = \sum_{\pi:\mathrm{B}(\pi)=l} p(\pi|y) \tag{9}$$

Where 'π', 'y', 'l', 'p', 'P' stand for PATH, input, the PATH after merging, probability of PATH π and probability of merging PATH l.

Because of the huge amount of computation, the efficiency of the decoding is very low. So, the highest probability Equation is used to make the prediction. The formula is shown in Eq. 10. It takes the most probable label πt at each time stamp t.

$$l^* \approx \mathrm{B}(\arg\max_\pi p(\pi|y)) \tag{10}$$

Where 'π', 'y', 'l*', 'p' stand for PATH, input, result of sequence and probability of PATH π.

After decoding, the network is trained. The objective is to minimize the negative log-likelihood of conditional probability of ground truth. The formula is shown in Eq. 11. The value of loss is calculated by Eq. 11 and the network is trained by SGD. Gradients are calculated by the back-propagation-algorithm.

$$o = -\sum_X \log p(l_i|y_i) \tag{11}$$

Where 'X', 'l', 'y', 'p' stand for training dataset, ground truth label sequence, input and probability of PATH l.

The dataset of recognition is used to train the model. The accuracy of final model is 91.97%. The process of training is shown in Fig. 10(a). The result after recognition is shown in Fig. 10(b). The recognition accuracy on the four public datasets are shown in Table 3. It includes our algorithm and some other recognition algorithms. Obviously, the situation of non-lexicon can be solved by our algorithm. It has achieved good results on the four public datasets.

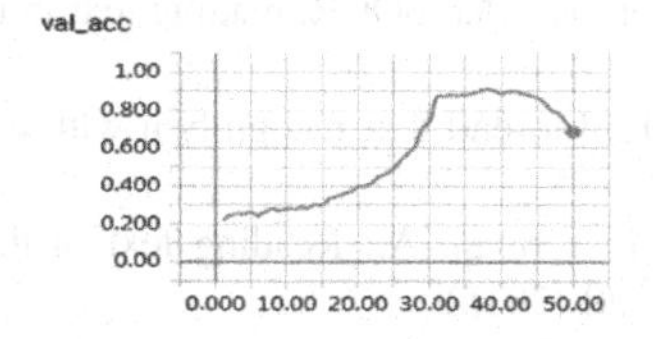

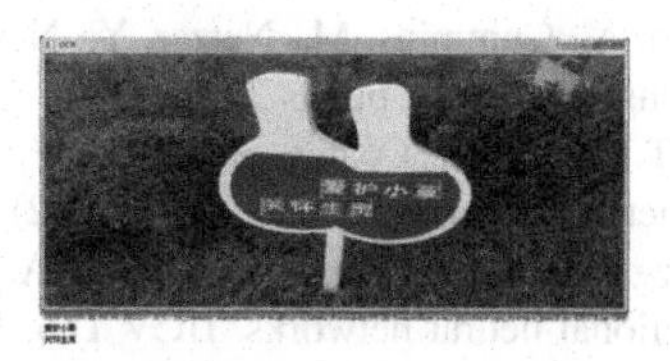

Fig. 10. (a) The accuracy of recognition network. (b) The result after recognition.

Table 3. Recognition accuracy on four datasets. "50", "1 k", "50 k" and "Full" stand for the lexicon used. "None" stand for the recognition without a lexicon.

	IIIT5 k			SVT		IC03				IC13
	50	1 K	None	50	None	50	Full	50 K	None	None
Almaz [20]	91.2	82.2	–	89.2	–	–	–	–	–	–
Yao [21]	80.2	69.3	–	75.9	–	88.5	80.3	–	–	–
Rodrguez [22]	76.1	57.4	–	70.0	–	–	–	–	–	–
Su [23]	–	–	–	86.1	–	96.2	91.5	–	–	–
Gordo [24]	93.3	86.6	–	91.8	–	–	–	–	–	–
Jaderberg [11]	95.5	85.6	–	93.2	71.7	97.8	97.0	93.4	89.6	81.8
Our paper			77.4		79.7				89.3	84.6

5 Conclusion

In this paper, we introduce a novel algorithm. It integrates preprocessing, detection and recognition. First, MSER is used to preprocess the images to gain many candidate regions. Then, the detection network is used to identify the text regions and remove the non-text regions. Finally, the recognition network is used to identify the text. The application of recognition algorithm is not limit by languages, scene and the length of text. Moreover, it does not need artificial labels and can be generated indefinitely. The algorithm is novel and perform well in both efficiency and accuracy. Besides, the first two steps of this study also can be used in detection of tongue images in Traditional Chinese Medicine. The result of tongue detection can be used in tongue segmentation and recognition in the tongue image analysis system.

In future work, we will continue to expand the dataset, enrich the quantity and languages. In addition, we will improve the algorithm to identify the vertical and oblique text regions.

Acknowledgment. Thanks to Institute of Department of information, Beijing University of Technology for supporting our work and giving us great suggestion. In addition Our work is supported by the national key research and development program (No. 2017YFC1703300) of China. At the same time, we also thank to the teachers and students who made great contribution to this study.

References

1. Bissacco, A., Cummins, M., Netzer, Y., Neven, H.: PhotoOCR: reading text in uncontrolled conditions. In: ICCV, pp. 1–7 (2013)
2. Wang, T., Wu, D.J., Coates, A., Ng, A.Y.: End-to-end text recognition with convolutional neural networks. In: ICPR, pp. 1–7 (2012)
3. Jaderberg, M., Simonyan, K., Vedaldi, A., Zisserman, A.: Reading text in the wild with convolutional neural networks. IJCV **116**, 1–20 (2015)

4. Matas, J., Chum, O., Urban, M., Pajdla, T.: Robust wide-baseline stereo from maximally stable extremal regions. Image Vis. Comput. **22**(10), 761–767 (2004)
5. Neumann, L., Matas, J.: Real-time scene text localization and recognition. In: 2012 IEEE Conference on Computer Vision and Pattern Recognition, pp. 3538–3545. IEEE (2012)
6. Neumann, L., Matas, J.: On combining multiple segmentations in scene text recognition. In: 2013 International Conference on Document Analysis and Recognition, pp. 523–527. IEEE (2013)
7. Corso, J., Hager, G.: Coherent regions for concise and stable image description. In: IEEE Computer Society Conference on Computer Vision & Pattern Recognition, vol. 2, pp. 184–190 (2005)
8. Tao, L., Jin, C., Cheng, W.: Improved maximally stable extremal region detector in color images. In: IEEE International Conference on Information & Automation, pp. 1711–1716 (2010)
9. Hu, W., Huang, Y., et al.: Deep convolutional neural networks for hyperspectral image classification. J. Sens. **2015**(2), 1–12 (2015)
10. Graves, A., Mohamed, A., Hinton, G.E.: Speech recognition with deep recurrent neural networks. In: ICASSP, pp. 1–5 (2013)
11. Jaderberg, M., Simonyan, K., Vedaldi, A., Zisserman, A.: Synthetic data and artificial neural networks for natural scene text recognition. In: NIPS Deep Learning Workshop, pp. 1–10 (2014)
12. Goel, V., Mishra, A., Alahari, K., Jawahar, C.V.: Whole is greater than sum of parts: recognizing scene text words. In: ICDAR, pp. 398–402 (2013)
13. Sutskever, I., Martens, J., Hinton, G.: Generating text with recurrent neural networks. In: ICML, vol. 336, no. 4, pp. 605–612 (2011)
14. Ji, S., Xu, W., Yang, M., Yu, K.: 3D convolutional neural networks for human action recognition. IEEE Trans. Pattern Anal. Mach. Intell. **35**(1), 221–231 (2013)
15. Socher, R., Karpathy, A., Le, Q.V., Manning, C.D., Ng, A.Y.: Grounded compositional semantics for finding and describing images with sentences. In: TACL, vol. 2, pp. 207–218 (2014)
16. Frome, A., Corrado, G.S., Shlens, J., Bengio, S., Dean, J., Mikolove, T.: DeViSE: a deep visual-semantic embedding model. In: NIPS, pp. 2121–2129 (2013)
17. Ioffe, S., Szegedy, C.: Batch normalization: accelerating deep network training by reducing internal covariate shift. In: ICML, pp. 1–11 (2015)
18. Graves, A., Fernández, S., Gomez, F.J., Schmidhuber, J.: Connectionist temporal classification: labelling unsegmented sequence data with recurrent neural networks. In: ICML, pp. 369–376 (2006)
19. Graves, A., Jaitly, N.: Towards end-to-end speech recognition with recurrent neural networks. In: Proceedings of the 31st International Conference on Machine Learning, pp. 1764–1772 (2014)
20. Almazán, J., Gordo, A., Fornès, A., Valveny, E.: Word spotting and recognition with embedded attributes. PAMI **36**(12), 2552–2566 (2014)
21. Yao, C., Bai, X., Shi, B., Liu, W.: Strokelets: a learned multi-scale representation for scene text recognition. In: CVPR, pp. 4042–4049 (2014)
22. Rodriguez-Serrano, J.A., Gordo, A., Perronnin, F.: Label embedding: a frugal baseline for text recognition IJCV **113**(3), 193–207 (2015)
23. Su, B., Lu, S.: Accurate scene text recognition based on recurrent neural network. In: Cremers, D., Reid, I., Saito, H., Yang, M.-H. (eds.) ACCV 2014. LNCS, vol. 9003, pp. 35–48. Springer, Cham (2015). https://doi.org/10.1007/978-3-319-16865-4_3
24. Gordo, A.: Supervised mid-level features for word image representation. In: CVPR, pp. 2956–2964 (2015)

Leaf Classification Utilizing Densely Connected Convolutional Networks with a Self-gated Activation Function

Dezhu Li[1(✉)], Hongwei Yang[1], Chang-An Yuan[2], and Xiao Qin[2]

[1] Institute of Machine Learning and Systems Biology, School of Electronics and Information Engineering, Tongji University, Shanghai, China
ldzlink@163.com

[2] Science Computing and Intelligent Information Processing of Guang Xi Higher Education Key Laboratory, Guangxi Teachers Education University, Nanning 530001, Guangxi, China

Abstract. Plant is of vital importance to human life, so it is necessary to research and protect them. Recently, convolutional neural network has been the most widely used technique for the task of image classification, and all kinds of convolutional neural network architecture has been proposed, including Densely connected convolutional networks (DenseNet). It has been shown that the selection of activation functions is of great importance to the training dynamics and task accuracy in deep neural networks. In this paper, we propose a new architecture that combines DenseNet with a new self-gated activation function. The experiment shows that the new architecture can get good results on the task of leaf classification.

Keywords: Convolutional Neural Network · CNN
Densely Connected Convolutional Networks · DenseNet
Swish activation function · Leaf classification

1 Introduction

Plant is of vital importance to human life, so it is necessary to research and protect them. Therefore, learning how to build an automatic system for recognizing plant is worthy of study. Image classification has been an important research area in computer vision. Among all kinds of techniques for image classification, convolutional neural network, as an efficient recognition method, has been getting development and arousing wide attention. In recent years, all kinds of convolutional neural network architecture has been proposed, and the architecture has become deeper and deeper. Densely connected convolutional networks is one of the architectures that has been proposed in recent years [1]. It is inspired by the idea that convolutional networks could be significantly deeper, easier to train and with more precision when they include shorter links between the layers near the input and the layers near the output.

The selection of activation functions is of great importance to the training dynamics and task accuracy in deep neural networks. Presently, Rectified Linear Unit (ReLU) is the most successful and popular activation function. Although all kinds of activation

D.-S. Huang et al. (Eds.): ICIC 2018, LNAI 10956, pp. 370–375, 2018.
https://doi.org/10.1007/978-3-319-95957-3_40

functions have been invented, none of them can replace ReLU due to inconsistent gains. Swish activation function is a novel activation function discovered by utilizing automatic search techniques [2]. Its effectiveness has been shown by empirical evaluation. And it is simple for practitioners to substitute Swish for ReLUs in any neural network due to its simplicity and its similarity to ReLU.

This paper describes an architecture that uses densely connected convolutional networks for leaf classification. We evaluate the effectiveness of different architectures of densely connected convolutional networks on ICL leaf database. We then replace ReLU activation function with Swish activation function, and test the effectiveness of new architecture on ICL leaf database. Finally we compare the results of different architectures on ICL leaf database.

2 Densely Connected Convolutional Networks (DenseNet)

CNN is mainly used to identify the graphics with invariance on shift, scale and other form of distortion [3, 4]. As the CNN's layers detect characteristics of the data step by step, it implicitly train itself instead of learning after direct feature extraction [7]. Due to the great development in the field of computer hardware and the invention of novel network architecture, it is now possible to efficiently train really deep networks [14, 15]. As CNNs become deeper and deeper, a new problem come into view: when information about data input and gradient go through many layers in the architecture, it can disappear by the time it reaches the layers near the input or the output [16, 18]. Different approaches have been proposed to solve this problem [19, 20]. But most of them share a common characteristic: the architectures contain short paths between layers near the input and layers near the output.

The main inspiration of DenseNet is to maximize information flow throughout the architecture. DenseNet link all the layers with the same size of feature-map directly. Every layer gets extra inputs from all preceding layers and passes its outputs to all succeeding layers. Figure 1 shows this architecture schematically.

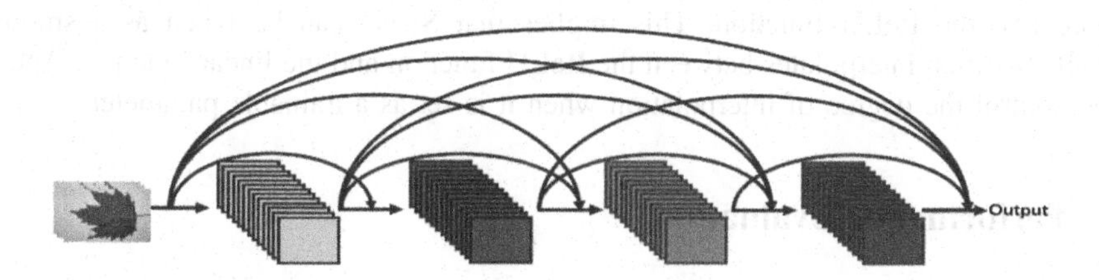

Fig. 1. Dense connectivity pattern in DenseNet

DenseNet also has the advantage that it reduces network parameters and make the network easier to train. Because of the dense connectivity pattern in the architecture, it avoids learning redundant feature-maps. Traditional network architectures pass state between layers. Each layer makes its own change to the state while forwards information that should be reserved. But DenseNet takes a different approach. It distinguishes information that is put into the architecture from information that is reserved and the number of parameters is sufficiently reduced.

The DenseNet architecture contains L layers. Each layer represents a set of composite operations H_l which include batch normalization, activation function, pooling, or convolution. As DenseNet connects every layers to all succeeding layers, the lth layer receives all previous layers' feature-maps as input. So the output of the lth layer is computed as follows:

$$x_l = H_l([x_0, x_1, \ldots, x_{l-1}]) \tag{1}$$

Where x_l denotes the output of the lth layer, $[x_0, x_1, \ldots, x_{l-1}]$ denotes the concatenation of the feature-maps from all previous layers. H_l is a composite function of three operations: batch normalization, activation function and a 3×3 convolution.

3 Swish Activation Function

The selection of activation functions in the deep networks has an important impact on the training process and task accuracy. Currently, Rectified Linear Unit (ReLU) is the most successful and popular activation function. It has been shown that ReLU can largely improve the training performance of deep networks.

Swish activation function is a novel activation function discovered by utilizing automatic search techniques, and its effectiveness has been shown by empirical evaluation. It is simple for practitioners to substitute swish for ReLU in neural network due to its simplicity and its similarity to ReLU. The swish activation function is shown in formula (2), where the β parameter can be set as a constant or make as a trainable parameter.

$$\mathrm{f(x)} = \frac{x}{1+e^{-\beta x}} \tag{2}$$

If $\beta = 0$, swish is the scaled linear function $\mathrm{f(x)} = \frac{x}{2}$. As $\beta \rightarrow \infty$, swish behaves more like the ReLU function. This implies that Swish can be taken as a smooth function which interpolates between the ReLU function and the linear function. And β can control the degree of interpolation when it is set as a trainable parameter.

4 Performance Evaluation

ICL leaf database contains more than 200 kinds of plant species, and there are 60 samples for each species. Each image of the leaves is of size 224×224, and the images all have three channels. We have randomly chosen 130 kinds of leaves for our experiment. We first do some data augmentation on ICL leaf database. The data augmentation operations include crop and flip. After data augmentation, we have 600 images per leaves. Each of the image have size of 160×160. To further reduce the numbers of the architecture parameters and to accelerate the training process, we have resize the image to size 32×32.

During training, we randomly choose 100 images for each leaf as test dataset, and the rest images are used as training dataset. The growth rate of the architecture is 24. Details of the architecture is shown in Table 1.

Table 1. Experiment network architecture

Layers	Output size	DenseNet-121	DenseNet-169
Convolution	32 × 32	7 × 7 conv, stride 2	
Dense block (1)	32 × 32	$\begin{bmatrix} 1 \times 1\,conv \\ 3 \times 3\,conv \end{bmatrix} \times 6$	$\begin{bmatrix} 1 \times 1\,conv \\ 3 \times 3\,conv \end{bmatrix} \times 6$
Transition layer (1)	32 × 32	1 × 1 conv	
	16 × 16	2 × 2 average pool, stride 2	
Dense block (2)	16 × 16	$\begin{bmatrix} 1 \times 1\,conv \\ 3 \times 3\,conv \end{bmatrix} \times 12$	$\begin{bmatrix} 1 \times 1\,conv \\ 3 \times 3\,conv \end{bmatrix} \times 12$
Transition layer (2)	16 × 16	1 × 1 conv	
	8 × 8	2 × 2 average pool, stride 2	
Dense block (3)	8 × 8	$\begin{bmatrix} 1 \times 1\,conv \\ 3 \times 3\,conv \end{bmatrix} \times 24$	$\begin{bmatrix} 1 \times 1\,conv \\ 3 \times 3\,conv \end{bmatrix} \times 32$
Transition layer (3)	8 × 8	1 × 1 conv	
	4 × 4	2 × 2 average pool, stride 2	
Dense block (4)	4 × 4	$\begin{bmatrix} 1 \times 1\,conv \\ 3 \times 3\,conv \end{bmatrix} \times 16$	$\begin{bmatrix} 1 \times 1\,conv \\ 3 \times 3\,conv \end{bmatrix} \times 32$
Classfication layer	1 × 1	4 × 4 global average pool	
		130D fully-connected, softmax	

We conduct four experiments on ICL leaf database, including DesnseNet-12+Relu, DenseNet-169+Relu, DenseNet-121+Swish, DenseNet-169+Swish. The change of the accuracy in the training process is shown in Fig. 2. And the experiments results in shown in Table 2.

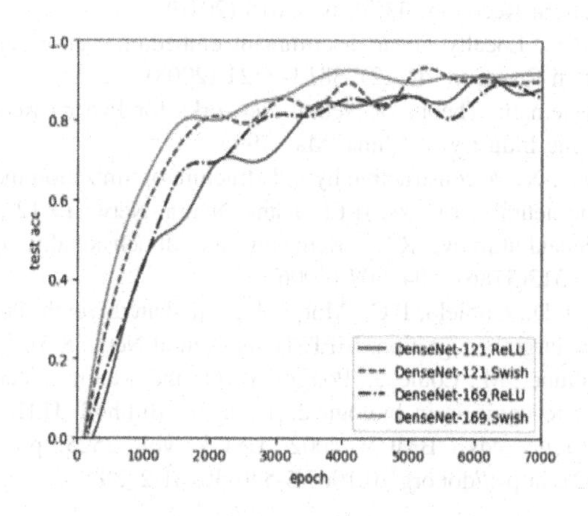

Fig. 2. Experiment results

Table 2. Experiment results

	DesnseNet-121+Relu	DesnseNet-169+Relu	DesnseNet-121+Swish	DesnseNet-169+Swish
Train precision	0.89	0.88	0.90	0.92
Test precision	0.85	0.84	0.88	0.89

5 Conclusion

This paper proposes a new architecture that combines DenseNet with a new self-gated activation function. The new self-gated activation function which is called Swish, shares some similarity to ReLU activation function. But swish does have some unique property, and the experiment shows that swish activation function has better performance than ReLU activation function on DenseNet architecture. The experiment on ICL leaf database also shows that the new architecture can get good results on the task of leaf classification.

Acknowledgements. This work was supported by the grants of the National Science Foundation of China, Nos. 61472280, 61672203, 61472173, 61572447, 61772357, 31571364, 61520106006, 61772370, 61702371 and 61672382, China Post-doctoral Science Foundation Grant, Nos. 2016M601646 & 2017M611619, and supported by "BAGUI Scholar" Program of Guangxi Zhuang Autonomous Region of China.

References

1. Huang, G., Liu, Z., et al.: Densely connected convolutional networks. In: Proceedings of the IEEE Conference on Computer Vision and Pattern Recognition, vol. 1. no. 2. 2017
2. Ramachandran, P., Zoph, B., Le, Q.V.: Swish: a self-gated activation function. arXiv preprint arXiv:1710.05941 (2017)
3. Wang, X.F., Huang, D.S., Xu, H.: An efficient local Chan-Vese model for image segmentation. Pattern Recognit. **43**(3), 603–618 (2010)
4. Li, B., Huang, D.S.: Locally linear discriminant embedding: an efficient method for face recognition. Pattern Recognit. **41**(12), 3813–3821 (2008)
5. Huang, D.S.: Systematic Theory of Neural Networks for Pattern Recognition. Publishing House of Electronic Industry of China, May 1996
6. Huang, D.S., Du, J.-X.: A constructive hybrid structure optimization methodology for radial basis probabilistic neural networks. IEEE Trans. Neural Netw. **19**(12), 2099–2115 (2008)
7. Hinton, G.E., Salakhutdinov, R.R.: Reducing the dimensionality of data with neural network. Science **313**(5786), 504–507 (2006)
8. Won, Y., Gader, P.D., Coffield, P.C.: Morphological shared-weight networks with applications to automatic target recognition. IEEE Trans. Neural Netw. **8**(5), 1195–1203 (1997)
9. Serre, T., Riesenhuber, M., Louie, J., Poggio, T.: On the role of object-specific features for real world object recognition in biological vision. In: Bülthoff, H.H., Wallraven, C., Lee, S.-W., Poggio, T.A. (eds.) BMCV 2002. LNCS, vol. 2525, pp. 387–397. Springer, Heidelberg (2002). https://doi.org/10.1007/3-540-36181-2_39

10. Huang, D.S.: Radial basis probabilistic neural networks: model and application. Int. J. Pattern Recognit. Artif. Intell. **13**(7), 1083–1101 (1999)
11. Wang, X.-F., Huang, D.S.: A novel density-based clustering framework by using level set method. IEEE Trans. Knowl. Data Eng. **21**(11), 1515–1531 (2009)
12. Shang, L., Huang, D.S., Du, J.-X., Zheng, C.-H.: Palmprint recognition using fast ICA algorithm and radial basis probabilistic neural network. Neurocomputing **69**(13-15), 1782–1786 (2006)
13. Zhao, Z.-Q., Huang, D.S., Sun, B.-Y.: Human face recognition based on multiple features using neural networks committee. Pattern Recognit. Lett. **25**(12), 1351–1358 (2004)
14. Simonyan, K., Zisserman, A.: Very deep convolutional networks for large-scale image recognition. arXiv preprint arXiv:1409.1556 (2014)
15. Srivastava, R.K., Greff, K., Schmidhuber, J.: Training very deep networks. In: Advances in Neural Information Processing Systems (2015)
16. Kaiming, H., et al.: Deep residual learning for image recognition. In: Proceedings of the IEEE Conference on Computer Vision and Pattern Recognition (2016)
17. Huang, D.S., Ip, H.H.S., Law, K.C.K., Chi, Z.: Zeroing polynomials using modified constrained neural network approach. IEEE Trans. Neural Netw. **16**(3), 721–732 (2005)
18. Huang, G., Sun, Y., Liu, Z., Sedra, D., Weinberger, K.Q.: Deep networks with stochastic depth. In: Leibe, B., Matas, J., Sebe, N., Welling, M. (eds.) ECCV 2016. LNCS, vol. 9908, pp. 646–661. Springer, Cham (2016). https://doi.org/10.1007/978-3-319-46493-0_39
19. Larsson, G., Maire, M., Shakhnarovich, G.: Fractalnet: ultra-deep neural networks without residuals. arXiv preprint arXiv:1605.07648 (2016)
20. Huang, D.S., Zhao, W.-B.: Determining the centers of radial basis probabilistic neural networks by recursive orthogonal least square algorithms. Appl. Math. Comput. **162**(1), 461–473 (2005)
21. Huang, D.S.: Application of generalized radial basis function networks to recognition of radar targets. Int. J. Pattern Recognit. Artif. Intell. **13**(6), 945–962 (1999)
22. Huang, D.S., Ma, S.D.: Linear and nonlinear feedforward neural network classifiers: a comprehensive understanding. J. Intell. Syst. **9**(1), 1–38 (1999)

Complex Identification of Plants from Leaves

Jair Cervantes[1(✉)], Farid Garcia Lamont[1], Lisbeth Rodriguez Mazahua[2], Alfonso Zarco Hidalgo[1], and José S. Ruiz Castilla[1]

[1] Posgrado e Investigación, UAEMEX (Autonomous University of Mexico State), 56259 Texcoco, Mexico
{jcervantesc, fgarcial}@uaemex.mx, jsergioruizc@gmail.com

[2] Division of Research and Postgraduate Studies, Instituto Tecnológico de Orizaba, Av. Oriente 9, 852. Col. Emiliano Zapata, 94320 Orizaba, Mexico
lisbethr08@gmail.com

Abstract. The automatic identification of plant leaves is a very important current topic of research in vision systems. Several researchers have tried to solve the problem of identification from plant leaves proposing various techniques. The proposed techniques in the literature have obtained excellent results on data sets where the leaves have dissimilar features to each other. However, in cases where the leaves are very similar to each other, the classification accuracy falls significantly. In this paper, we proposed a system to deal with the performance problem of machine learning algorithms where the leaves are very similar. The results obtained show that combination of different features and features selection process can improve the classification accuracy.

Keywords: Plant identification · Vision system · Features selection

1 Introduction

Development of vision algorithms in agronomy has been guide by several researchers to solve many real problems [1–3]. Over the last years, many algorithms to identify plants from leaves have been developed. It is a current challenge that has several applications. The plant identification from leaves is not an easy job, because it involves the solution of different problems, such as: extract the leaf features and select the best features. Moreover, there are a lot of plants on the planet, many of them possess and share one or more properties such as: shape, size, texture, color, even when they belong to different plants.

In the current literature there are many techniques for identifying plants from the leaves. However, there are no systems to automatically identify plants where the leaves are closely related or are very similar to each other. This research has been motivated of this disadvantage. In this research, were used different features selection techniques to get the most discriminative features for each subset of plant leaves.

In the results, performances of several machine learning algorithms are compared using different features to identify plants where the leaves are very similar. The extracted

D.-S. Huang et al. (Eds.): ICIC 2018, LNAI 10956, pp. 376–387, 2018.
https://doi.org/10.1007/978-3-319-95957-3_41

features allow to identify or classify the leaves. However, sometimes a high number of features introduces noise which affects the performance, i.e. the performance of the identification systems is strongly related to the features. The feature selection algorithms help to reduce the noise introduced in the classifier when the dimensionality in the data set is very large. The dimensionality of data set is reduced by eliminating features with low discriminative power. The general procedure is applied for each data set with the aim of selecting in each subset only the necessary characteristics.

The next section describes the state of the art. The third section describes the feature extraction methods used in this paper. The fourth section shows the proposed method. The fifth section shows the experiment and results. Lastly, the sixth section ends with the conclusions.

2 State of the Art

Plants identification has recently drawn attention from computer sciences. Identify a plant through leaves images is not a trivial job because it requires specialized knowledge. Current identification methods involve advanced algorithms to measure the morphological and texture features of the objects contained in the image. The best way to extract valid features is to get them from the image of the leaves. In the current literature it is shown that the external shape, chromaticity, venation and texture of the leaves give a lot of information to classify them.

Some other researches, have focused on the extraction of features from the leaf, using four important features for classification: Shape [4], texture [5], color [6], and leaf venation [7].

Leaf shape is one of the most important features of plant leaves, and the two basic approaches for these kind of analysis are the ones that are based on contour and region. One of the most employed approaches is to analyze the shape of leaves, extracting geometric characteristics such as size, elongation, ellipse, area, length, diameter, rectangularity, sphericity, eccentricity, etc. [8, 9]. Some authors have added to these basic geometric descriptors, Hu moments and Fourier moments improving performance of the classifiers [10]. The one based on region usually use moment descriptors, which includes Zernike geometric moments and Legendre moments. Some authors use basic descriptors, such as perimeter, area, circularity and elliptical, or invariant descriptors like Hu moments and Fourier descriptors for leaf contour recognition [11]. Methods based on contour, usually use methods based on the leaf curvature. Recently, some systems have been proposed to extract features describing edge variations of the leaf, using descriptors invariant to translation, rotation and size.

Texture of the leaves can be defined as the characteristics that the leaf has on its surface which is manifests as gray scale variations in the image. Texture features include local binary patterns, Gabor filters and gray level co-occurrence matrices, while the shape feature vector is modeled using Hu moment invariants and Fourier descriptors. Other researches had used a combination of geometric and textural, allowing them to use dried, wet or even misshapen leaves [12]. Some author combine both textural information and shape features to identify leaves [13, 14].

On the another hand, the first plant recognition studies used the chromaticity of the plant as an important descriptor to compare images. Very simple descriptors of chromaticity can obtain the average color in the segmented region of the leaf, average gradient in the edge or the similarity of color between two images that can be measured by comparing their color histograms. More complex color descriptors use moments of invariance which are commonly used to obtain geometrical characteristics but incorporating the information of the color variables of the leaves [15]. However, a recurring problem in the leaves of plants is that the chromaticity in the leaves is not static, it is variable with respect to time and commonly with respect to other factors. Other authors consider in addition to chromaticity and form, the texture of the leaf [16] or use combinations of descriptors to improve the classification performance [13, 14].

In other research studies the color is used as a comparison feature of images, since a simple color similarity between two images can be measured by comparing their color histograms [15]. However, a recurring problem is that the chromaticity in the leaves of plants is not static, this is variable with on the time and commonly on the other factors. Although classification approaches such as shape, texture and color are valid, it has not been documented the influence of each type of features in the performance of classification algorithms.

3 Proposed Method

In this Section, the steps of the proposed method are described in detail. After of features extraction the proposed methodology uses different techniques to select the best features of the data set. This step allows to reduce the dimensionality of the data set, reduce the training time and in some cases improve the performance of the system, this due to the elimination of features that introduce noise to the classifier (Fig. 1 and Table 5).

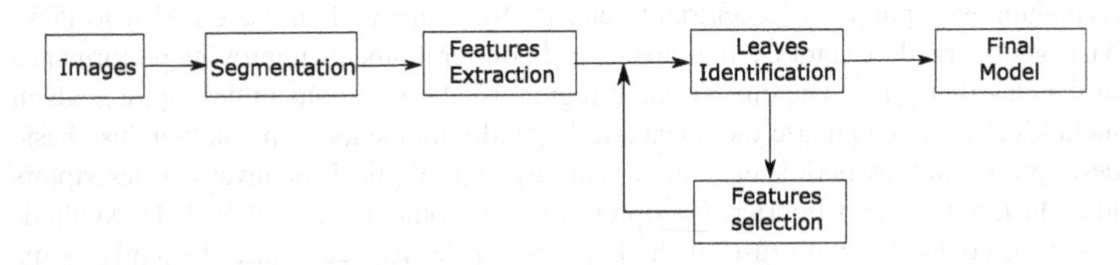

Fig. 1. Proposed methodology diagram

3.1 Segmentation Techniques

Firstly the images are preprocessed and segmented. The leaves images often are surrounded by greenery in the background. However, the images used in the experiments are images in a controlled environment (images with only leaf and white background). In all the experiments Otsu algorithm was used for segmentation. It is worth mentioning that we carry out experiments with different segmentation techniques, however the results are very similar in all cases, this is due to the images have a white background and the environment is totally controlled. Therefore, it was not necessary to use more powerful techniques to perform the segmentation of the image.

In order to obtain a good segmentation even when there are changes in global brightness conditions, the region of the leave in each image was segmented using the following steps (1) Computation of high-contrast gray scale from optimal linear combination of RGB color components; (2) Estimate optimal border using cumulative moments of zero order and first order (Otsu method). (3) Morphological operations to fill possible gaps in the segmented image. By segmenting the image, the proposed system can use only the region of the leaf, determine its edges and calculate properties by extracting features.

3.2 Features Extractors

Feature extraction is a critical process in any pattern recognition system. The feature extraction has a big influence on the final identification. Feature extraction allows us to represent the image using a set of numerical and/or categorical values. In order to improve the performance the features obtained must be invariant to scaling, rotation and translation, enabling the classifier to recognize objects despite having different size, position and orientation. All these features play an important role in the algorithm performance and allow the classifier to discriminate between different classes in an appropriate manner. In our experiments geometric, chromatic and textural features were obtained.

Geometric Features. The geometric features are one of the most important visual properties used to classify an object. The geometric features provide information on the size and shape of the previously segmented region. Elementary geometric features provide intuitive information of the basic properties of the region to be recognized, such as area of the region, roundness of the leaf, length of the edge of the leaf, elongation defined by the length and width of the leaf, the coordinates x and y of gravity center, rectangularity, projection (on the components x, y), eccentricity, center of gravity (components x, y), Danielson factor, equivalent diameter, axis length (x, y), orientation, solidity, extencion, area convex, filled area, ellipse (variance, orientation, eccentricity, area, major axis, minor axis, ellipse center x, *and*). However, an efficient classification system should be able to recognize leaves regardless of their orientation, location and size, i.e. it must be invariant to scaling, rotation and position.

Moments are commonly used in image recognition, they can recognize these images regardless of their rotation, translation or inversion. Invariant moments were initially introduced by Hu. Other used features were ellipse descriptors, region convexity, Flusser moments $(F_1, \ldots, F_4)$, R Moments $(R_1, \ldots, R_{10})$, Fourier descriptors (first 8 descriptors). 57 geometric features were extracted from each image. The geometric feature vector X_g obtained can be represented as:

$$X_g = [X_1, X_2, \ldots, X_{57}] \tag{1}$$

$$X_g = \left[X_{gb}, X_{Hu}, X_F, X_R, X_{DF}\right] \tag{2}$$

where X_{gb} represents the elemental geometric features ($X_{gb} = [X_1, \ldots, X_{28}]$), X_{Hu} represents the Hu invariant features ($X_{Hu} = [X_{29}, \ldots, X_{35}]$), X_F represents the Flusser invariant moments ($X_F = [X_{36}, \ldots, X_{39}]$), X_R represents the invariant moments to changes in illumination ($X_R = [X_{40}, \ldots, X_{49}]$), X_{DF} represents the first 8 Fourier descriptors ($X_{DF} = [X_{50}, \ldots, X_{57}]$).

Textural Features. Textural features provide information on the spatial arrangement of colors or intensities in the image. Extraction algorithms of textural features look for basic repetitive patterns with periodic or random structures in images. These structures are obtained by properties in the image such as roughness, roughness, granulation, fineness, softness, etc. Texture repeats a pattern along a surface, due to which the textures are invariant to displacements, this explains why the visual perception of a texture is independent of the position. In this paper, were used Haralick textural features and the Local Binary Patterns (LBP for its acronym in English - textit Local binary Patterns -) [17]. These features consider the distribution of intensity values in the region, by obtaining the mean and range of the following variables: mean, median, variance, smoothness, bias, Kurtosis, correlation, energy or entropy, contrast, homogeneity, and correlation. 14 textural descriptors of each image were obtained.

In total 219 textural descriptors were obtained from each image. 73 for each color channel. The textural features vector can be represented by:

$$X_t = [X_1, X_2, \ldots, X_{219}] \tag{3}$$

$$X_t = \left[X_{Rlbp}, X_{RH}, X_{Glbp}, X_{GH}, X_{Blbp}, X_{BH}\right] \tag{4}$$

where $X_{Rlbp}, X_{Glbp}, X_{Blbp}$ represents the LBP features obtained in the color channel R, G and B respectively, X_{RH},X_{GH} and X_{BH} represents the Haralick textural features in the channels R, G and B respectively. A description in detail of LBP descriptors can be found in [17].

Chromatic Features. Chromatic features provide information of the color intensity of a segmented region. These characteristics can be calculated for each intensity channel, for example, red, green, blue, grayscale, hue (Hue), Saturation (Saturation) and intensity (Value), etc. The used features were: standard intensity features, they describe the mean, standard deviation of intensity, first and second derivative in the segmented region, Hu moments with intensity information, Gabor features based on 2D Gabor functions. In experiments 122 characteristics were obtained for each channel. Since the experiments were performed in RGB only 366 chromatic features were used. Chromatic features vector X_c can be defined by:

$$X_c = [X_1, X_2, \ldots, X_{117}] \tag{5}$$

$$X_c = [X_{Re}, X_{RHu}, X_{Ge}, X_{GHu}, X_{Be}, X_{BHu}] \tag{6}$$

where X_{Re}, X_{Ge}, X_{Be} represents the elemental color features in the channels R, G and B respectively, X_{RHu}, X_{GHu} y X_{BHu} represents the Hu invariant color moments in the channels R, G and B respectively.

3.3 Feature Selection

In order to eliminate some features that do not contribute to the classifier performance, in this research several algorithms were used to extract the best combinations of features.

In the proposed algorithm, each set of features per leaf, conforms a vector defined by the number of descriptors or features. The number of features of each data set defines the size of each binary string needed to implement the genetic algorithm. The relationship between each binary string with the feature set, is that 1 is taken as a used feature and 0 as the absence of that feature. The aptitude of each individual is taken from the accuracy obtained by classifying the set corresponding to that chain.

In the proposed method, the individual with better aptitude is taken and it passes intact to the next generation, it was used two-point crosses and mutation probability of 0.08. Figure 2 shows an example of chromosomal chains used and the classifier performance when features labeled 1 are used.

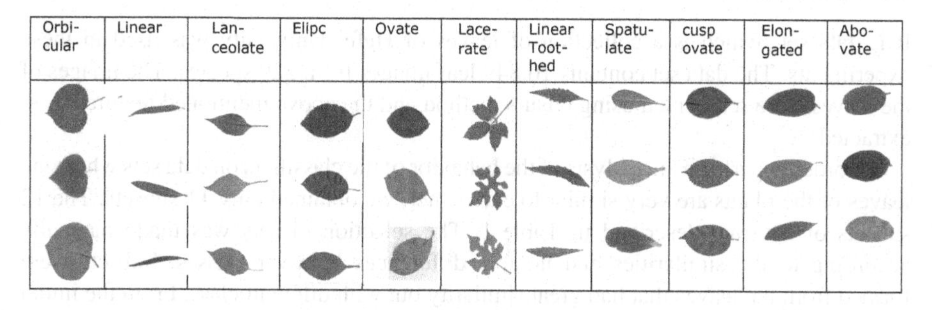

Fig. 2. Data sets

The dimensionality of data set is an important performance factor. Sometimes inappropriate attributes can affect the performance. Features selection helps to improve the performance of a classifier. This problem has been addressed by several authors, this problem is common in pattern recognition and it is commonly called course of dimensionality. An important factor when reducing characteristics, is to eliminate those that are not important to the classifier or find the combination of attributes that optimizes the performance of the classifier. Feature selection or dimensionality reduction is regularly raised as an optimization problem. In recent years, several algorithms have been used extensively to solve dimensionality problems.

Formally, given a n-dimensional data set, the features selection techniques task is to find a set of attributes in a k-dimensional space that maximizes an optimization criterion, where $k < < n$. Obtained patterns are evaluated based on two conditions, dimensionality of the data set and spacing between classes or classification accuracy.

In the experiments were used 3 techniques to reduce the dimensionality of the data sets: Correlation-based feature selection, Information Gain Based Feature Selection and a genetic algorithm. Correlation-based feature selection and Information Gain Based Feature Selection techniques are used only once and then the data sets are trained with reduced dimensionality. In the case of the genetic algorithm, this process is performed iteratively until the genetic finds the best features in the data set.

3.4 Classification Techniques

In the experiments the results were compared with some classification techniques, logistic regression, Bayesian classifier, the Backpropagation learning algorithm and support vector machine (SVM).

4 Experimental Results

In this section, parameters selection technique is shown, also data normalization and experimental results obtained with the proposed method.

4.1 Data Set

ICL data set, which is a collection of leaves of Hefei University was used in these experiments. The data set contains 16,849 leaf images from 220 species. The images of the leaves were segmented using Otsu's method and the above mentioned features were extracted.

In order to perform an analysis of the behavior of the classifiers on data sets where the leaves of the plants are very similar to each other, we obtained only 11 subsets. The 12 subsets of data are described in Table 1. The selection of sets was made manually according to the similarities and notable differences between classes. Subsets were formed from the leaves that had great similarity but with different class. From the initial 220 leaf species, 11 different subsets were formed with 169 species. The subset with fewer classes contains only 3 and the subset that has more associated classes contains 37.

Table 1. Subsets used in the experiments.

	Leaf shapes	Size	Number of clases	Labels
DS 1	Orbicular	1095	19	1–19
DS 2	Linear	348	8	20–27
DS 3	Lanceolate	962	14	28–41
DS 4	Eliptic	2481	27	42–68
DS 5	Ovate	782	12	69–80
DS 6	Lacerate	186	3	81–83
DS 7	Linear toothed	428	7	84–90
DS 8	Spatulate	832	9	91–99
DS 9	cusp ovate	1248	13	100–112
DS 10	Elongated	1632	20	113–132
DS 11	Abovate	4398	37	133–169
DS 12	Peach leaves	193	6	170–175

Figure 2 shows each subset created with some examples of leaves that were taken to form each group, in addition the reason of similarity that was taken into account to assign each leaf is described. The data set 12 contains only peach leaves. However, these are from 6 different species. As can be seen in the Fig. 3, the similarity between them is very high. The leaves are almost identical and for anyone who is not an expert, the differences could be unnoticed.

Fig. 3. Peach leaves of 6 different species

4.2 Data Normalization

The final feature vector T was stored in a m $\times$ 642 size array containing m images with 642 features. 57 geometric features, 219 textural and 366 chromatic features. All the extracted features were normalized with mean zero and standard deviation equal to 1.

4.3 Parameter Selection

In all used classifiers optimal parameters were obtained by cross-validation and grid search. Cross-validation is a model validation technique for assessing how the results of a statistical analysis will generalize to an independent data set. On the other hand, grid search exhaustively search all parameter combinations obtaining the best parameter combination.

4.4 Results

In the experiments, all data sets were normalized and cross-validation was used with k = 10. Table 2 shows the results obtained with geometric features, textural and chromaticity, as each individual features. DS_i defines the data set. For each classifier used, accuracies obtained with each individual set of characteristics are reported. The metric used to evaluate the performance of the classifier was precision and this is obtained from the classifier hits divided by the total of data set.

In the results, it is not possible to infer that the similarity between leaves significantly affect classifiers, performance of the classifiers that used very similar images of each other and dissimilar, are not contrasting. However, it is possible to appreciate that the textural features are little discriminative for most data sets, except for the set DS 6 and DS_12. One possible reason is that the size of the data set is very small.

Tables 2, 3 and 4 show the results obtained with different combination of feature techniques (chromatic, textural and geometric). In the results it is possible to see an

Table 2. Performance with different features.

	Chromatics				Textural				Geometric			
Subset	Bayes	BP	LR	SVM_{RBF}	Bayes	BP	LR	SVM_{RBF}	Bayes	BP	LR	SVM_{RBF}
DS 1	88.7	94.21	86.38	94.91	36.28	47.78	42.07	56.902	81.89	93.56	84.83	**95.408**
DS 2	81.57	84.50	82.17	88.28	38.87	68.24	70.94	73.053	85.58	90.19	87.64	**92.941**
DS 3	88.22	94.73	89.27	**95.03**	37.18	74.06	71.86	77.518	78.08	89.32	83.63	91.918
DS 4	86.25	94.56	89.61	**95.62**	36.72	67.32	65.91	75.881	74.70	90.74	85.73	93.409
DS 5	93.67	95.56	92.82	95.72	28.30	71.04	68.81	72.841	85.44	95.11	88.33	**97.289**
DS 6	99.37	**100**	98.62	**100**	75.53	91.71	76.53	93.308	93.08	99.37	98.11	99.742
DS 7	96.56	96.70	94.35	97.45	48.25	82.39	81.27	85.082	97.87	**98.75**	96.45	98.733
DS 8	95.85	97.83	96.18	98.56	56.40	84.23	83.85	87.875	95.26	98.31	96.84	**98.941**
DS 9	88.38	90.64	90.38	90.87	38.39	65.77	63.21	69.69	84.59	93.80	87.46	**94.410**
DS 10	86.31	95.85	92.48	**96.36**	39.54	78.89	75.59	82.871	80.21	93.10	86.84	94.182
DS 11	77.62	92.33	88.31	**93.91**	33.72	63.41	62.84	68.161	74.83	88.35	78.03	91.324
DS 12	54.36	62.58	52.48	**67.32**	25.49	36.23	31.43	38.76	51.21	57.28	54.83	64.69

Table 3. Performance with two combined features.

	Chromatic-textural				Chromatic-geometric				Textural-geometric			
Subset	Bayes	BP	LR	SVM_G	Bayes	BP	LR	SVM_G	Bayes	BP	LR	SVM_G
CH 1	87.35	93.42	88.26	94.26	92.37	97.24	93.73	97.59	86.47	93.80	88.15	95.27
CH 2	82.06	84.73	81.82	86.25	87.05	90.58	88.69	92.47	88.12	91.86	89.04	91.42
CH 3	90.41	95.86	90.12	96.42	92.51	96.94	94.12	97.93	85.15	93.04	83.61	94.28
CH 4	87.26	94.00	87.58	95.21	91.42	97.26	91.83	95.82	84.33	92.37	85.45	94.72
CH 5	92.24	95.95	91.34	95.30	96.33	95.11	97.43	96.01	90.21	94.60	91.46	95.08
CH 6	98.74	100	97.75	100	98.74	100	98.76	100	96.22	99.37	97.90	99.37
CH 7	96.86	95.48	96.93	97.28	95.85	96.56	95.26	97.71	96.38	96.53	95.74	98.76
CH 8	95.05	96.13	95.58	96.21	96.84	96.21	95.71	97.82	96.61	96.69	94.52	96.95
CH 9	87.77	91.48	87.04	92.10	94.41	95.07	94.91	95.23	87.55	92.13	86.35	90.17
CH 10	87.74	95.92	88.19	95.33	89.77	95.29	91.29	96.16	90.50	92.62	90.72	91.93
CH 11	80.53	92.56	82.22	93.16	83.81	93.88	85.83	94.92	84.03	92.65	87.18	92.81
CH 12	68.51	66.82	63.94	73.46	69.17	70.81	65.79	76.53	67.06	69.44	64.77	75.06

improvement in classification accuracy compared to the performances obtained with the original data set and compared with the other features selection techniques.

In all tests with different classification techniques, the results with the proposed technique improves the results obtained with other techniques. These results highlight the utility of the proposed method. The obtained results accuracy was improved in all chains using genetic, it is important to note that even though the number of features significantly decreased in all the results the combination of the three types of features is very necessary.

Table 6 shows the general results obtained with the three techniques of feature selection. In the Table ODS represents the results obtained with data set with all features, GA the results obtained with the genetic algorithm, CBFS the results obtained with the Correlation-based feature selection and IGBFS the results obtained with Information Gain Based Feature Selection algorithm.

Table 4. Performance with the three kind of features (chromatic-textural-geometric).

Subset	Bayes	BP	LR	SVM_G
DS 1	91.909	96.207	91.21	98.524
DS 2	87.739	90.038	88.41	94.941
DS 3	93.797	96.992	95.36	98.882
DS 4	90.940	96.253	94.71	97.183
DS 5	95.784	98.145	96.18	99.153
DS 6	98.742	100	99.48	100
DS 7	98.535	98.954	98.83	99.062
DS 8	96.354	98.177	97.74	99.828
DS 9	95.196	93.886	92.24	97.893
DS 10	90.027	98.292	93.54	98.531
DS 11	86.482	93.807	88.26	95.917
DS 12	71.52	72.49	65.38	79.45

Table 5. Performance with the proposed algorithm.

Classifier	DS1	DS2	DS3	DS4	DS5	DS6	DS7	DS8	DS9	DS10	DS11	DS12
Bayes	92.07	88.09	93.91	91.02	96.12	98.92	98.72	96.62	95.84	91.68	89.22	75.64
BP	96.36	91.27	97.25	96.38	98.45	100	98.95	98.22	94.68	98.6	95.87	76.02
LR	92.51	88.73	96.09	95.19	96.73	99.86	98.91	97.78	93.51	94.22	90.26	73.057
SVM_G	98.97	96.16	99.31	98.21	99.83	100	99.62	99.93	98.51	99.03	97.18	83.97

Table 6. Performance using the feature selection techniques

Classifier	ODS	GA	CBFS	IGBFS
Bayes	93.22	94.35	93.11	93.22
BP	96.43	97.54	96.34	96.43
SVM	98.17	98.86	98.17	98.17
LR	94.17	94.63	94.09	94.12

5 Conclusions

In this paper, a feature selection algorithm is proposed to improve the performance of classifiers for identifying plants that are very similar. The proposed method helps to improve the performance of the classifiers removing attributes that introduce noise. The experiments obtained show that the proposed method generates notable results by eliminating attributes that do not provide information. The main advantage of the proposed method is its ease of implementation and ease of use on small and medium size data sets. Features reduction is important to improve the response time it takes for the system to recognize a new leaf. Several issues could be considered as future works. First, the algorithm can be used for plants identification of species from leaves, however, there are leaves of different species that have a very high degree of similarity, it is necessary to add another type of characteristics such as leaf venation that were not included in this research. Second, in the results of this research only images of the front of the leaves were used. However, it is possible that the back of the leaves provides more important information than the front of the leaf. A study of this would be very important for this research.

Acknowledgments. This study was funded by the Research Secretariat of the Autonomous University of the State of Mexico with the research project 5228/2018/CI.

References

1. Huang, Y., Lan, Y., Hoffmann, W.C.: Use of airborne multi-spectral imagery for area wide pest management. Agric. Eng. Int. CIGR Ejournal Manuscr. IT **07**(010), 1–14 (2008)
2. Singh, V., Misra, A.K.: Detection of plant leaf diseases using image segmentation and soft computing techniques. Inf. Process. Agric. **4**(1), 41–49 (2017). ISSN:2214-3173
3. Ferentinos, K.P.: Deep learning models for plant disease detection and diagnosis. Comput. Electron. Agric. **145**, 311–318 (2018). ISSN:01681699
4. Du, J.X., Wang, X.F., Zhang, G.J.: Leaf shape based plant species recognition. Appl. Math. Comput. **185**(2), 883–893 (2007)
5. Sampallo, G.: Reconocimiento de tipos de hojas. Inteligencia Artificial. Rev. Iberoam. Intel. Artif. **7**(21), 55–62 (2003)
6. Cerutti, G., Tougne, L., Mille, J., Vacavant, A., Coquin, D.: Understanding leaves in natural images - a model-based approach for tree species identification. Comput. Vis. Image Underst. **117**(10), 1482–1501 (2013)
7. Larese, M.G., Namías, R., Craviotto, R.M., Arango, M.R., Gallo, C., Granitto, P.M.: Automatic classification of legumes using leaf vein image features. Pattern Recognit. **47**(1), 158–168 (2014)
8. Chaki, J., Parekh, R.: Designing an automated system for plant leaf recognition. Int. J. Adv. Eng. Technol. **2**(1), 149–158 (2012)
9. Park, J.-S., Kim, T.Y.: Shape-Based Image Retrieval Using Invariant Features. In: Aizawa, K., Nakamura, Y., Satoh, S. (eds.) PCM 2004. LNCS, vol. 3332, pp. 146–153. Springer, Heidelberg (2004). https://doi.org/10.1007/978-3-540-30542-2_19
10. Kumar N., Belhumeur P.N., Biswas A.: Leafsnap: a computer vision system for automatic plant species identification. In: Proceedings of the ECCV 2012, pp. 502–516 (2012)
11. Novotny, P., Suk, T.: Leaf recognition of woody species in Central Europe. Biosyst. Eng. **115**(4), 444–452 (2013)
12. Husin, Z., Shakaff, A.Y.M., Aziz, A.H.A., Farook, R.S.M., Jaafar, M.N., Hashim, U., Harun, A.: Embedded portable device for herb leaves recognition using image processing techniques and neural network algorithm. Comput. Electron. Agric. **89**, 18–29 (2012)
13. Liu, N., Kan, J.-m.: Improved deep belief networks and multi-feature fusion for leaf identification. Neurocomputing **216**, 460–467 (2016). ISSN:0925-2312
14. VijayaLakshmi, B., Mohan, V.: Kernel-based PSO and FRVM: an automatic plant leaf type detection using texture, shape, and color features. Comput. Electron. Agric. **125**, 99–112 (2016). ISSN:0168-1699
15. Tico, M., Haverinen, T., Kuosmanen, P.: A method of color histogram creation for image retrieval. In: Proceedings of the Nordic Signal Processing Symposium (NORSIG-2000), Kolmarden, Sweden, pp. 157–160 (2000)

16. Cope, J., Corney, D., Clark, J., Remagnino, P., Wilkin, P.: Plant species identification using digital morphometrics: a review. Expert Syst. Appl. **39**(8), 7562–7573 (2012)
17. He, D.C., Wang, L.: Texture unit, texture spectrum, and texture analysis. IEEE Trans. Geosci. Remote Sens. **28**, 509–512 (1990)

An Effective Low-Light Image Enhancement Algorithm via Fusion Model

Ya-Min Wang, Zhan-Li Sun(✉), and Fu-Qiang Han

School of Electrical Engineering and Automation,
Anhui University, Hefei, China
zhlsun2006@126.com

Abstract. In a low-light condition, the quality of a captured image may be much poorer than that obtained in a normal environment. As an effective preprocessing step, many enhancement algorithms have been proposed to improve the performance of a computer vision task. In most existing algorithms, a image is often enhanced as a whole. As a result, the image may be over-enhanced or under-enhanced due to different degree of exposure in local area. Aiming at this issue, in this paper, we propose a low-light image enhancement algorithm based on image fusion technology. In the proposed method, a fusion strategy is devised by considering the exposure extent of local area. The weight matrix for image fusion is first calculated. Then, the pixel with insufficient exposure is selected according to the adaptive threshold. Next, the multi-exposure images can be synthesized by using the estimated optimal exposure rate. Finally, we use the input image to fuse with the enhanced image for slightly under-exposed images to get the enhancement image, while the severely under-exposed images can be enhanced by fusing a reflection map based on retinex with the enhanced image. Experimental results show that our method can obtain enhancement results with less color and lightness distortion compared to several state-of-the-art methods.

Keywords: Low-light image enhancement · Image fusion
Illumination estimation

1 Introduction

In many cases, the quality of the captured images may be affected by environment, equipment and technology. Many useful information of these images may be hidden. On the one hand, images with high visibility are critical to computer vision technology including target recognition and tracking, medical research, can improve the accuracy of algorithms. On the other hand, the clear pictures have good visual effects, which are valuable images, such as space and archaeology. Therefore, the study in low-light image enhancement technology has important value and significance.

The work was supported by a grant from National Natural Science Foundation of China (No. 613 70109), a key project of support program for outstanding young talents of Anhui province university (No. gxyqZD2016013), a grant of science and technology program to strengthen police force (No. 1604d0802019), and a grant for academic and technical leaders and candidates of Anhui province (No. 2016H090).

D.-S. Huang et al. (Eds.): ICIC 2018, LNAI 10956, pp. 388–396, 2018.
https://doi.org/10.1007/978-3-319-95957-3_42

Many classic algorithms to deal with low-light images were proposed from different perspectives. Low illumination image intuitive sense is caused by the uneven distribution of pixel level. The most intuitive approach is to stretch the contrast of the image directly. However, this method of stretching contrast may lead to excessive brightness in the local area. Histogram equalization (HE) strategies [1, 2], contextual and variational contrast enhancement (CVC) [3], by seeking a layered difference representation of 2D histograms (LDR) [4], although it is possible to avoid pixel overflow and over-boost by forcing the pixel value in the range of [0,1]. However, the method only focuses on improving the contrast of the image, unable to solve the problem of color saturation. Retinex theory is a method based on human vision systems, which is widely used in image enhancement. Naturalness preserved enhancement algorithm (NPEA) was proposed in [5], the algorithm by non-uniform enlarge the contrast of brightness figure. Though solving the problem of low-light, the method may suffer from under-enhancement and color distortion. In [6], the bilateral filter can be used to generate different filter effects according to different pixel location and gray value. However, this method may appear artifacts in the edge regions of high contrast. The structure-aware smoothing model was proposed in [7], can construct the brightness map rapidly through different weighting schemes. But the algorithm will suffer color supersaturated, causing the enhanced images to be unnatural. Based on the dark channel algorithm in [8–11], the low-illumination image operates in a similar way to the fog image after inversion, and the enhancement of the images are realized by using the dark channel theory. The work proposed in [12] has applied deep learning used in image processing. Though the method has achieved good results, the global atmospheric light estimation is a challenge. The image fusion method based on camera model was proposed in [13, 14]. This algorithm can make the image enhancement more natural. However, the method is not ideal for images treatment with severe underexposure.

In this paper, we proposed different solutions to different degrees of low-light images. Experiments on a number of challenging images are conducted to reveal the advantages of our method, which can obtain results with less contrast and lightness distortion in comparison with other state-of-the-art methods.

2 Methodology

2.1 Framework Construction

In many cases, the quality of the image is not only affected by the illumination, but the exposure rate of the camera is also the cause of poor visibility. Under the same conditions, we get more information about the better exposure, as shown in Fig. 1. In [14], the mapping function of a series of different exposure rate images is obtained through the input graph, which is called the luminance transformation function (BTF).

$$P_i = g(P_0, k_i), \tag{1}$$

where g is the brightness transformation function, and k is the exposure rate. By varying k, we can get a series of images P with different exposure rates.

Fig. 1. Images with different exposure rates from the HDR dataset. The exposure ratio is −4, −2, 0, 1.7 respectively.

Many algorithms have a good effect on the low illumination image. However, they may cause over-enhancement, color distortion and unnatural to some extent. We found that images fusion technology can coordinate multiple sensor information in the same scene and help to obtain the information of target or scene more accurately and comprehensively. Therefore, in order to solve the above mentioned problems, we construct an image fusion frame.

$$E = W \circ L + (1 - W) \circ g(P,k), \tag{2}$$

where E is the output, W is the weight, the operator $\circ$ represents element-wise multiplication. In order to achieve a better enhancement, in different situations, the choice of L is different. It is specifically introduced in Sect. 2.4.

2.2 Weight Matrix and Reflection Map Estimation

In the fusion framework, the selection of weight is the key to retain the advantages of each part and avoid weaknesses. To avoid over-enhancement, the building of the weight depends on the illumination. The weighting matrix is calculated as in [13]:

$$W = T^{\mu}, \tag{3}$$

where T is the brightness map of the input image, μ is a parameter that controls the degree of enhancement, the value range of which is [0, 1]. In order to avoid over-enhancement, our algorithm selection μ = 0.5.

The Initialization of the Illumination Map

To preserve the brightness of the image, we initialize the brightness image T as:

$$\hat{T}(x) = \max_{c \in \{R,G,B\}} P^{c}(x), \tag{4}$$

for each individual pixel x. illumination map T is solved by optimization.

Optimized Illumination Map

In order to better preserve the edge information of the image, we chose the guide filter to optimize the brightness image T, which was proposed in [15].

$$e(a,b) = \sum_{i \in w_k} \left((a_k P_i + b_k - \hat{T}_i)^2 + \epsilon a_k^2 \right), \tag{5}$$

where the input image P denote the guide image, $\hat{T}$ is the rough estimate of the illumination image, ε is the regularization parameter, and w_k is the local area of the center of the pixel point k in the guide image. Parameters of a_k and b_k can be calculated by the linear ridge regression method.

Through the processing of the above, we can get further optimized illumination map T:

$$T(x) = \frac{1}{|w|} \sum_{k,i \in w_k} (a_k P_i + b_k), \tag{6}$$

where $|w|$ is the number of pixels in the local area w_k.

Reflection Estimation

Based on the Retinex theory, we can obtain the reflection map R of the scene through the illumination map T

$$R = \frac{1}{T} \circ P, \tag{7}$$

where R and T are the reflectance map and the illumination map, respectively.

2.3 Exposure Rate Estimation

Adaptive Threshold Selection

As we mentioned earlier, globally enhanced methods may lead to over-enhancement in some regions. Therefore, we proposed the adaptive threshold to filter the pixels. The pixels below the threshold were extracted, and the pixels above the threshold were ignored.

$$Q = \{P(x) | T(x) < Th\}, \tag{8}$$

where Q contains all the lower pixels. In this article, the selection of threshold Th, we use the *Otsu* threshold method, which can according to the image adaptive selection of the corresponding threshold.

The Determination of Exposure Rate

Intuitively, well-exposed images transfer to us more information than under-exposed images. Image entropy is a statistical form of feature, which reflects the average amount of information in the image. As in [13], the image entropy is calculated as follows:

$$\mathcal{H}(B) = -\sum_{i=1}^{N} p_i \cdot \log_2 p_i \tag{9}$$

where $B = \sqrt[3]{Q_r \circ Q_g \circ Q_b}$, Q_r, Q_g and Q_b are the red, green, and blue channel of Q respectively. p_i is the i-th bin of the histogram of B and N is the number of bins. The optimal k can make the maximum amount of information. As proposed in [13], we obtain the optimal exposure rate by maximizing the image entropy:

$$\hat{k} = \arg\max_k \mathcal{H}(g(B,k)), \tag{10}$$

2.4 Fusion Strategy

For Slightly Under-Exposed Images

In many cases, a poorly quality image is only partially under-exposed, and the other part is still well. In order to better use this part of the good information, we select the original map to fuse with the image that enhances the exposure rate.

$$E = W \circ P + (1 - W) \circ g(P,k), \tag{11}$$

where, P is the input image.

For Severely Under-Exposed Images

In the case of severely underexposed images, the above method is not satisfactory. However, images captured in low-light conditions are often of low visibility, which cause a lot of information to be hidden in the dark. To visualize the hidden information, we propose the following fusion method:

$$E = W \circ R + (1 - W) \circ g(P,k), \tag{12}$$

where R is the reflection of the scene.

3 Experiments

In this section, a public dataset with a standard comparison graph (Robust Patch-Based HDR Reconstruction of Synamic Scenes) is used to visualize the vividly and intuitively of our proposed algorithm. We compare our method with other seven algorithms including CVC, HE, LDR, WAHE, Dong, LIME and CAIP in our experiments.

As can be seen from the comparison of Fig. 2, our algorithm is superior to algorithms of CVC, HE, LDR, WAHE and Dong, which can enhance the picture and keep the picture color undistorted. Compared with the LIME method, our algorithm is more natural and more consistent with human visual perception. Figure 3 shows the enhanced effect of our algorithm.

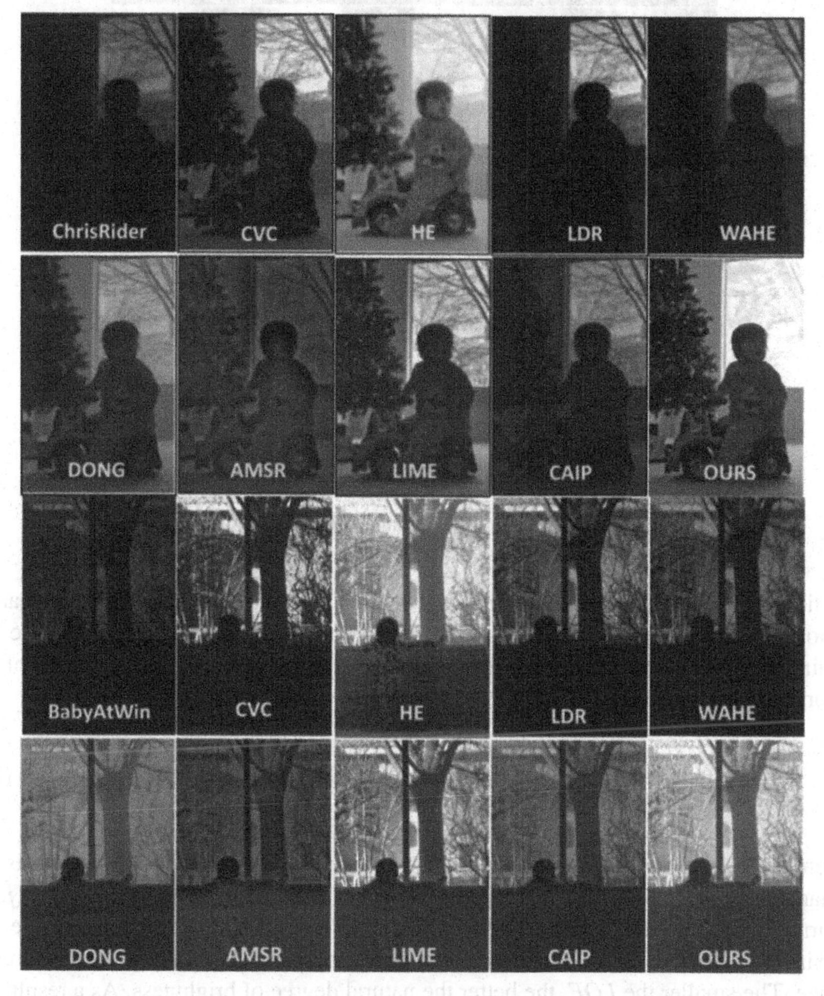

Fig. 2. Visual comparison among the competitors on the HDR dataset.

The relative order of image brightness can not only indicate the direction of light source but also the relative change of brightness. The natural degree of image enhancement is closely related to the relative order of brightness in different local regions [5, 7]. In other words, the relative order of the brightness of different regions of the picture should remain unchanged before and after the enhancement. The originally

Fig. 3. More results enhanced by our method.

relatively bright region is still brighter after enhancement, and the originally darker region still maintains a darker brightness after enhancement. Therefore, we use the luminance order error (LOE) as a measure of the brightness fidelity performance of the algorithm. The definition of LOE is as follows:

$$LOE = \frac{1}{m}\sum_{x=1}^{m}\sum_{y=1}^{m}(U(\mathbf{Q}(x), \mathbf{Q}(y)) \oplus U(\mathbf{Q}_r(x), \mathbf{Q}_r(y))), \tag{13}$$

where m is the number of pixels and $\oplus$ is the exclusive-or operator. $\mathbf{Q}$ is an enhancement image and $\mathbf{Q}_r$ is a reference image. If $p \geq q$, then the function $U(p, q)$ returns 1, otherwise it returns 0. In addition, $\mathbf{Q}(x)$ and $\mathbf{Q}_r(x)$ respectively are the maximum values of R, G and B at the x position of enhanced image and reference image. The smaller the LOE, the better the natural degree of brightness. As a result, the effect of image enhancement is better (Table 1).

Table 1. The lightness order error.

Method	HighCh	PianoM	Santas	LadyE	Feed	Christ	OnGra	AtWin
DONG	103	84	90	224	224	136	120	226
HE	132	183	175	69	109	132	173	219
CVC	121	153	185	56	97	113	139	243
LDR	128	168	192	64	81	92	128	176
WAHE	115	154	168	54	95	100	142	193
LIME	648	606	872	533	461	364	723	686
Caip	151	158	96	163	168	140	203	208
OURS	92	80	52	100	97	130	129	65

4 Conclusions

In this paper, we propose a fusion framework and adaptive enhancement algorithm. This model can better solve the problem of insufficient to enhance or increase, can better keep the naturalness of the image. The selection of weights is the key of fusion, and we use the guide filter to optimize the illumination diagram, thus obtaining more accurate weight. Moreover, different fusion schemes make our algorithm more applicable. The experimental results demonstrated that the proposed method is an effective and feasible low-light image enhancement algorithm.

References

1. Pisano, E.D., Zong, S., Hemminger, B.M., et al.: Contrast limited adaptive histogram equalization image processing to improve the detection of simulated spiculations in dense mammograms. J. Digit. Imaging **11**(4), 193 (1998)
2. Abdullah-Al-Wadud, M., Kabir, M.M., Dewan, A.A., Chae, O.: A dynamic histogram equalization for image contrast enhancement. IEEE Trans. Consum. Electron. **53**(2), 593–600 (2007)
3. Celik, T., Tjahjadi, T.: Contextual and variational contrast enhancement. IEEE Trans. Image Process. **20**(12), 3431–3441 (2011)
4. Lee, C., Lee, C., Kim, C.-S.: Contrast enhancement based on layered difference representation of 2D histograms. IEEE Trans. Image Process. **22**(12), 5372–5384 (2013)
5. Wang, S., Zheng, J., Hu, H.M., Li, B.: Naturalness preserved enhancement algorithm for non-uniform illumination images. IEEE Trans. Image Process. **22**(9), 3538–3578 (2013)
6. Elad, M.: Retinex by two bilateral filters. In: Kimmel, R., Sochen, N.A., Weickert, J. (eds.) Scale-Space 2005. LNCS, vol. 3459, pp. 217–229. Springer, Heidelberg (2005). https://doi.org/10.1007/11408031_19
7. Guo, X., Li, Y., Ling, H.: Lime: low-light image enhancement via illumination map estimation. IEEE Trans. Image Process. **26**(2), 982–993 (2017)
8. He, K., Sun, J., Tang, X.: Single image haze removal using dark channel prior. In: IEEE Computer Vision and Pattern Recognition, pp. 1956–1963 (2009)
9. Narasimhan, S.G., Nayar, S.K.: Contrast restoration of weather degraded images. IEEE Computer Society (2003)

10. Meng, G., Wang, Y., Duan, J., Xiang, S., Pan, C.: Efficient image dehazing with boundary constraint and contextual regularization. In: IEEE International Conference on Computer Vision, pp. 617–624 (2013)
11. Dong, X., Wang, G., Pang, Y., Li, W., Wen, J., Meng, W., Lu, Y.: Fast efficient algorithm for enhancement of low lighting video. In: IEEE International Conference on Multimedia and Expo, pp. 1–6 (2011)
12. Ren, W., Liu, S., Zhang, H., Pan, J., Cao, X., Yang, M.-H.: Single image dehazing via multi-scale convolutional neural networks. In: Leibe, B., Matas, J., Sebe, N., Welling, M. (eds.) ECCV 2016, Part II. LNCS, vol. 9906, pp. 154–169. Springer, Cham (2016). https://doi.org/10.1007/978-3-319-46475-6_10
13. Ying, Z., Li, G., Ren, Y., Wang, R., Wang, W.: A new image contrast enhancement algorithm using exposure fusion framework. In: Felsberg, M., Heyden, A., Krüger, N. (eds.) CAIP 2017, Part II. LNCS, vol. 10425, pp. 36–46. Springer, Cham (2017). https://doi.org/10.1007/978-3-319-64698-5_4
14. Ying, Z., Li, G., Ren, Y., Wang, R., Wang, W.: A new low-light image enhancement algorithm using camera response model. In: International Conference on Computer Vision Workshop (2017)
15. He, K., Sun, J., Tang, X.: Guided image filtering. In: Daniilidis, K., Maragos, P., Paragios, N. (eds.) ECCV 2010, Part I. LNCS, vol. 6311, pp. 1–14. Springer, Heidelberg (2010). https://doi.org/10.1007/978-3-642-15549-9_1

Pedestrian Detection Fusing HOG Based on LE and Haar-Like Feature

Jin Huang(✉) and Bo Li

Hubei Province Key Laboratory of Intelligent Information Processing and Real-Time Industrial System, College of Computer Science and Technology, Wuhan University of Sciences and Technology, Wuhan, China
179648770@qq.com

Abstract. In order to change the low detecting speed and excess redundant information of the gradient histogram (HOG), this paper proposes the pedestrian detection fusing HOG based on Laplacian Eigenmaps (LE) dimensionality reduction and Haar-Like feature. By constructing a pedestrian detection sample set mixed with HOG and Haar-Like feature set, and using LE to reduce the HOG feature, then fuse Haar-Like feature extraction with HOG-LE feature. Finally, we use Adaboost cascade strong classifier to test INRIA Person sample library, the test results are obviously better than the single feature extraction method in terms of detection rate, and the text also compares the method with some other pedestrian detection methods. The pedestrian detection fusing HOG based on LE reduced dimension and Haar-Like features proposed in this paper improves detection rate effectively in real scenarios.

Keywords: HOG · Haar-Like · LE · Adaboost · Pedestrian detection

1 Introduction

With the development of artificial intelligence, computer vision, intelligent surveillance and intelligent transportation, pedestrian detection technology is attracting more and more attention, and is widely applied in intelligent monitoring, intelligent transportation, advanced human-machine interaction and other fields. In real-life environment, pedestrian detection is faced with major difficulties due to the diversity of pedestrian posture, wearing and lighting, the different angle of observation and the change of scene. How to achieve high accuracy and real-time detection of pedestrian detection in complex background is a hot topics in this field. At present, there are two main types of pedestrian detection methods: one is pedestrian detection algorithm based on template, a pedestrian template library is set up for different posture pedestrians. After calculating the template feature of input images, we finally match them in pedestrian template library to find pedestrian template [1]. The disadvantage of pedestrian detection algorithm based on template matching is that it only applies to pedestrian detection based on pedestrian template library, and can not detect pedestrians outside of template library. Once template is missing, it will seriously reduce the detection accuracy. The other category is the pedestrian detection algorithm based on feature extraction, classifier training. In this field, the characteristics of pedestrian detection widely applied to

D.-S. Huang et al. (Eds.): ICIC 2018, LNAI 10956, pp. 397–407, 2018.
https://doi.org/10.1007/978-3-319-95957-3_43

the pedestrian detection algorithm is Haar, HOG, LBP these three kinds of mainstream. These algorithms will be used in pedestrian detection, mainly rely on the vector machine (SVM) classifier and Adaboost classifier. But these particular features in the image extraction is too single, without a complete description of the image [2]. This paper combines HOG features with Haar-Like features, uses LE to reduce the dimension of HOG features, and uses a strong Adaboost classifier to detect. The test results show that pedestrian detection fusing HOG based on LE reduced dimension and Haar-Like features improves detection rate effectively in real scenarios, and improves detection accuracy and speed.

2 Related Work

2.1 HOG Feature

HOG features mean the feature description of the object detection in the computer visual image processing. It has three main steps. The first step is to divide the image into small connected area cell, and the second step is to form histogram of orientation by cell pixel gradient in the sub area, and the last step is to combine histogram into feature descriptor [3]. HOG features are mainly divided into three steps: each step is as follows: divide the picture pixels and divide the pictures into cell small units; gather statistics of the histogram of gradient direction in each cell unit; obtain the feature vectors by means of projection.

The process of extracting and calculating the HOG features is as follows (Fig. 1):

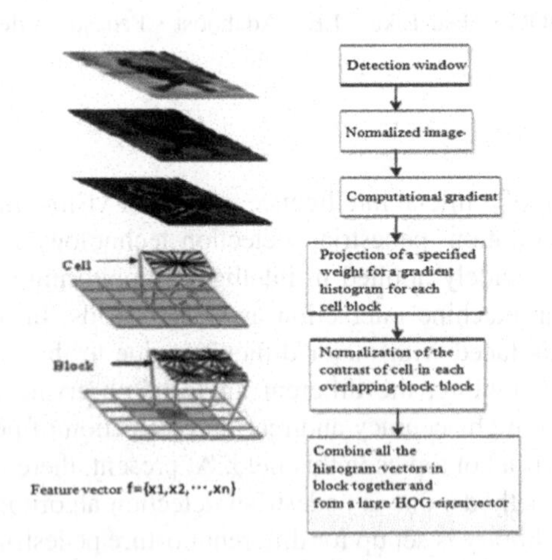

Fig. 1. HOG feature extraction algorithm

2.2 Haar-Like Feature

In the proposed method, a share nearest neighbors (SNN) graph is first constructed from the dataset. Then a hypergraph is constructed by defining the maximal cliques of the SNN graph as hyperedges. Finally, a hypergraph partition method called DSP is used to partition the hypergraph. The Haar-Like features can be called Haar feature. There are four main types of the feature template. The calculation of template eigenvalues is mainly to calculate the difference between the white area and the black area, and to describe the local gray change [4]. Figure 2 is a typical feature template map of four Haar-Like features.

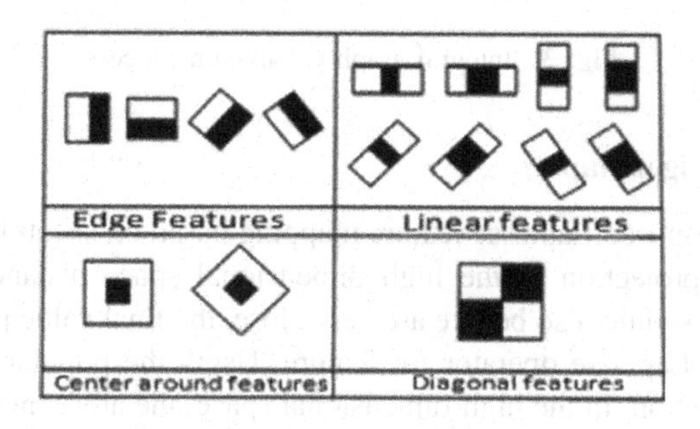

Fig. 2. Haar-Like feature template

The feature templates with different sizes and positions in Haar-Like also have different features, and with the expansion of the detection window, the numbers of features are changing. Because the Haar-Like feature dimension is large, in order to improve the efficiency of detection, we must use the algorithm that can detect a large number of window features. Its formula is like Eq. (1):

$$j(x, y) = \sum_{x' \leq x, y' \leq y} i(x', y') \tag{1}$$

Where j(x, y) is the integral map of the point value, i (x′, y′) is the pixel value of point (x′, y′).

The integral graph algorithm can be used to calculate the pixel value of any region, as shown in Fig. 3. We calculate the pixel value of the D area, assuming that j(1), j(2), j(3), j(4) are A, A+B, A+C pixels and A+B+C+D four regions sum. The sum of the pixel value of D region is:

$$sum(D) = j(4) + j(1) - j(2) - j(3) \tag{2}$$

After using the integral graph algorithm, the speed of the Haar-Like feature calculation has been greatly improved, and the repetition operation is avoided effectively, and the real-time performance of the algorithm is improved. The application of Haar-Like features in pedestrian detection can effectively identify different postures, clothing and pedestrians. The limitation is missing detection and false detection rate. Therefore, we integrate HOG features with Haar-Like features.

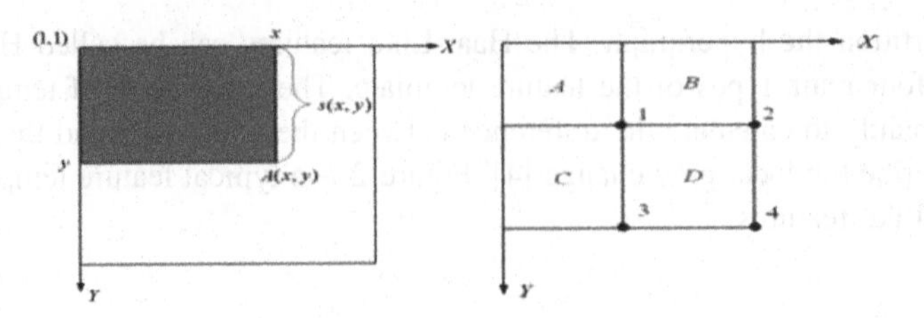

Fig. 3. Integral graph calculation process

2.3 Laplacian Eigenmap

Belkin [5] et al. Proposed Laplasse feature mapping based on spectral graph theory, the basic idea is the projection in the high dimensional space distance close to a low dimensional space should also be like are very close, the final value problem is reduced to the generalized Laplasse operator for feature. Using the popular Laplace_Beltrami operator approximation, in the high-dimensional space, the adjacent points are mapped to the low dimensional space by Laplasse mapping, and are also adjacent to the local low dimensional space. The local preserving feature of Laplasse's feature mapping algorithm makes it insensitive to outliers and noise. These methods are based on the geometric structure of the manifold, which shows the stability of the embedding.

The dimensionality reduction method has two main advantages for this experiment. After LE dimensionality reduction, the problem of feature information redundancy is solved. In some classification problems, the dimension reduction effect of LE is better than other dimension reduction methods.

3 Pedestrian Detection Fusing HOG Based on LE and Haar-like Feature

3.1 Construction of Sample Set

The positive sample size of training set in INRIA Person pedestrian library [6] created by Dalal is up to 2416, while the positive sample of test set has 1126. Its rich sample content also greatly meets the requirements of pedestrian detection. In order to ensure the accuracy of detection, before the test, you have to cut the edge of image sample, sample pretreatment is the positive samples in the original sample is 64 × 128 pictures, then select 10 resolution of 64 × 128 picture to cut, and as negative samples. Table 1 is the final composition of this experimental data set.

Table 1. Date sets in this experiment

	Positive samples numbers	Negative samples numbers
Train set	2416	12180
Test set	1126	4530

3.2 Construction of Mixed Feature Set

At first, the use of Adaboost algorithm is mainly to identify the face, and its construction of feature set is mainly designed on the basis of face recognition. Selecting the Haar-Like feature in the pedestrian detection has grayscale characteristics in object description, and has excellent resolution. This feature greatly facilitates the improvement of face recognition in the later stage, but this feature limitation is reflected in the low accuracy of pedestrian detection. Especially it can not accurately describe on the human body contour. In the detection process, affecting the accuracy of the real scene of pedestrian contour for pedestrian detection is the highest, therefore, this paper chooses HOG-LE and Haar-Like feature fusion, and fully guarantee the accuracy of image samples.

In pedestrian detection, it can not only reduce the posture of pedestrians, but also ensure the good detection and characterization. Because the Haar-Like feature requires a large number of samples to train the classifier to detect pedestrian detection in high performance, and the relationship between relevance is low, so this paper chooses the HOG characteristics of strong correlation integration, so as to reduce the number of training samples, improve the detection efficiency. It also makes up for the large amount of computation of HOG feature. Therefore, this paper will integrate HOG features and Haar-Like features through LE dimension reduction, and use Adaboost learning algorithm to learn features. Then we finally set up HOG-LE features and Haar-Like hybrid feature set.

3.3 Feature Extraction

HOG-LE Feature Dimension Reduction. Aiming at the problem of pedestrian detection, we use HOG to reduce the dimensions of the acquired pedestrians features. Due to the classification information of samples, HOG-LE feature dimensionality reduction can obtain pedestrian characteristics that are more optimized and suitable for classification and recognition, which effectively improves the pedestrian detection performance. In this paper, the dimensionality reduction of the HOG eigenvector using LE dimension reduction algorithm, HOG-LE feature extraction steps are as follows: The original input image matrix space n pedestrian image matrix; LE algorithm for feature extraction; Construct the weight matrix for the feature set sample Y; Calculate Laplace matrix; Generalize eigenvalue decomposition of Laplacian matrix and solved low dimensional embedding.

Fusion of Haar-Like Features Extraction and HOG-LE Features. Because HOG has the disadvantages of low detection speed, excessive redundant information and poor processing effect, Haar-Like feature is applied to pedestrian detection, which can effectively identify different postures and dress pedestrians. Therefore, this paper first applies LE to HOG Special dimension reduction, and then fused Haar-Like features [7] (Fig. 4).

Haar-Like feature extraction mainly in the following four steps, each step of the specific extraction process is as follows: Split the sample image that are need to detect, segmentation ratio according to the size of the image 16×16; Calculate the pixel

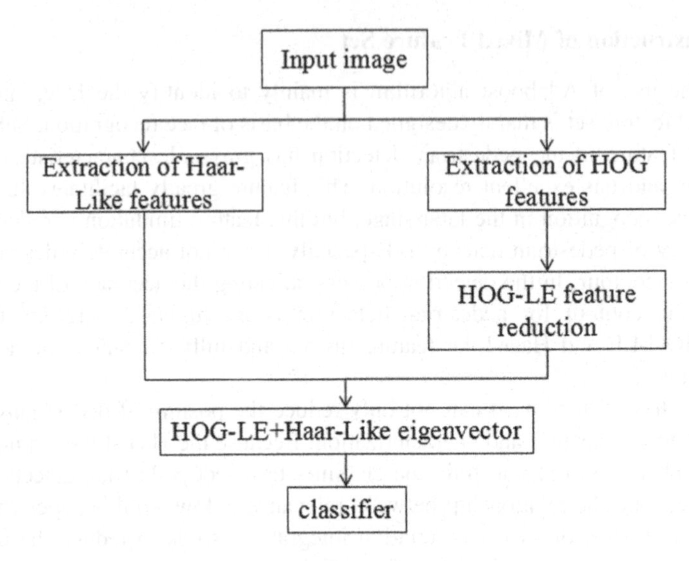

Fig. 4. Fusion process of HOG-LE features and Haar-Like features

value in Haar-Like region; Use L2-norm normalization factor to normalize the eigenvector; The image of the pedestrian detection experiment is 64 × 128 images, the detection window is 32 cells. After HOG-LE feature and Haar-Like feature are extracted, the HOG-LE + Haar-Like feature vector is concatenated.

3.4 Adaboost Strong Classifier

Adaboost strong classifier trained cascade classifier, greatly enhances the sample settings for each level of the classifier effect. At the same time, the accuracy of the image detection is improved. During the sample detection, the rate of false detection of the sample is reduced and the detection rate is increased [8]. In addition, it also reduces the false alarm rate of pedestrian detection. The strong classifier structure trained by Adaboost algorithm for pedestrian detection is as follows:

As can be seen from the Fig. 5, the composition of the cascade classifier likes a tree, its detection idea is mainly coarse to fine, for undetected windows, its specific detection steps are: first enter All Sub-windows, and then Each layer enter the classifier class, test failure will be directly judged as Reject Sub-windows, test success will enter next layer. According to the idea of this method, only the window that passes the test can be regarded as a pedestrian area. The precondition of cascade classifier detecting the image is stratified testing for each detect windows, ensure that this detection range (i.e. cascade classifier structure flow) includes pedestrians or non-pedestrians in each detection window.

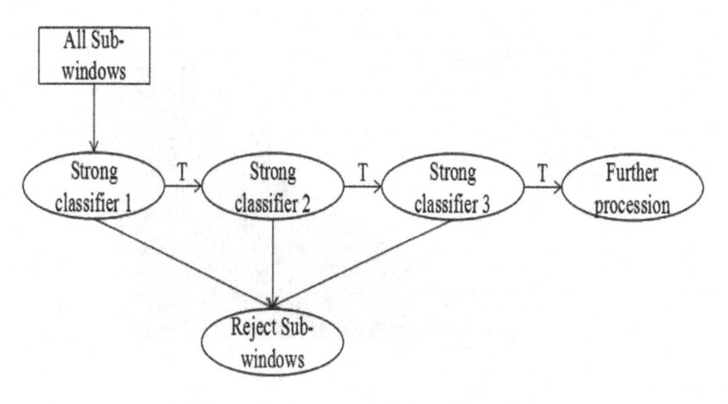

Fig. 5. Cascade classifier diagram

4 Experimental Results Analysis

The pedestrian detection based on fusion of HOG based on LE reduced dimension and Haar-Like features is mainly carried out under the Matlab (2014a) experimental platform. First, the HOG-LE + Haar-Like fusion feature library is tested, and then a single HOG, Haar-Like feature library is tested. The environment is Adaboost cascade classifier environment, and the classifier is set $d_{min} = 99.5\%$ (d_{min} refers to the minimum false detection rate of each classifier), $f_{max} = 50\%$ (f_{max} refers to the maximum false alarm rate of each classifier). Through the construction of mixed feature sets and extraction, this HOG-LE + Haar-Like fusion feature library training a 15-level class classifier, Haar-Like is 17, HOG is 13. Specific as shown in Fig. 6.

From the simulation analysis, it can be seen that the total number of weak classifiers required in the single Haar-Like or single HOG features libraries is much higher than the fusion of HOG-LE and Haar-Like feature libraries. The weak classifiers required for each feature are 1522(Haar-Like), 928(HOG), 647(HOG-LE+Haar-Like). Therefore, it can be seen from the number of required classifiers, that the fusion of HOG-LE and Haar-Like feature libraries are highly descriptive to pedestrians and pedestrian detection can be accomplished using fewer classifiers.

As can be seen from Fig. 7, the number of HOG-LE features is 400 and the number of Haar-Like features is 203. In terms of number, although the number of HOG-LE features is higher than the number of Haar-Like features, it is fully demonstrated that the features of HOG-LE are more adequate for the description of pedestrians in pedestrian detection. However, Haar-Like mainly reduces the number of weak classifiers by the merging way, which shows that this fusion of HOG-LE and Haar-Like improves the pedestrian detection performance, And reduce the detection time.

Figure 8 shows the detection results under three feature sets (fppw is false detection rate).

It can be seen from Fig. 8 that the curve of HOG-LE + Haar-Like detection is much higher than the HOG feature and the Haar-Like feature curve, and at the same time it proves that this proposed pedestrian detection based on fusion of HOG features based on LE reduced dimension and Haar-Like features is effective.

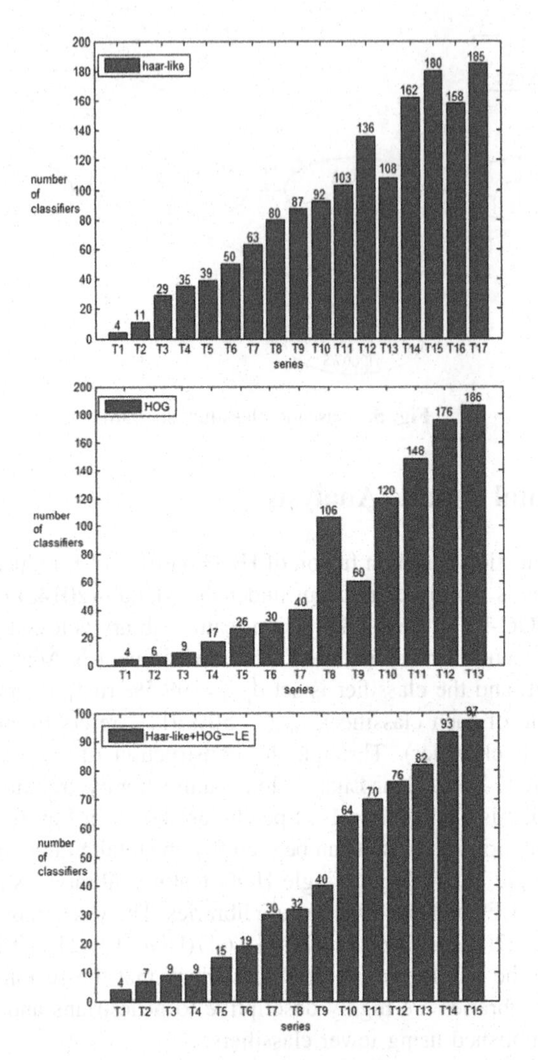

Fig. 6. The number of weak classifiers in different feature databases

In order to further verify the effectiveness of the proposed detection method, the other three advanced algorithms are compared with our method. The results are shown in Fig. 9.

It can be seen from the Fig. 9 that when the classifier threshold is set higher, the recognition rate is higher, but the missing detection increases so that the recognition rate is not high. As the threshold decreases, more false positives will occur and the recognition rate will decrease, while the false recognition rate will increase continuously. Compared with the HDMS algorithm (multifractal spectrum), the recognition rate of the ACF algorithm (Appearance Constancy and Shape Symmetry) is not greatly improved. Because the HDMS algorithm and the ACF algorithm only use the whole template detection, the recognition performance is greatly different from the

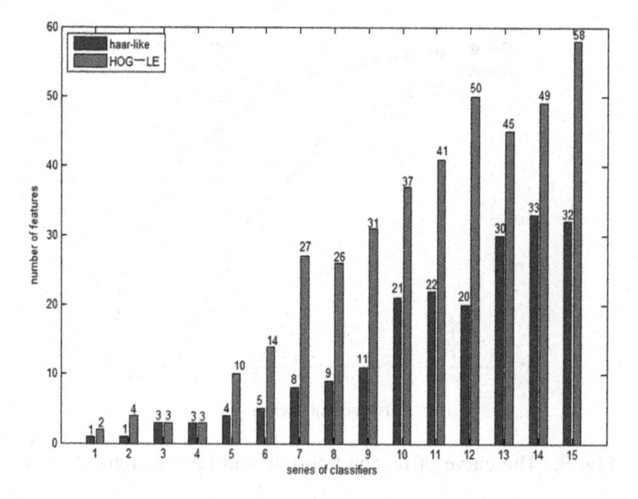

Fig. 7. The number of features that strong classifier of HOG-LE and Haar-Like fusion contains

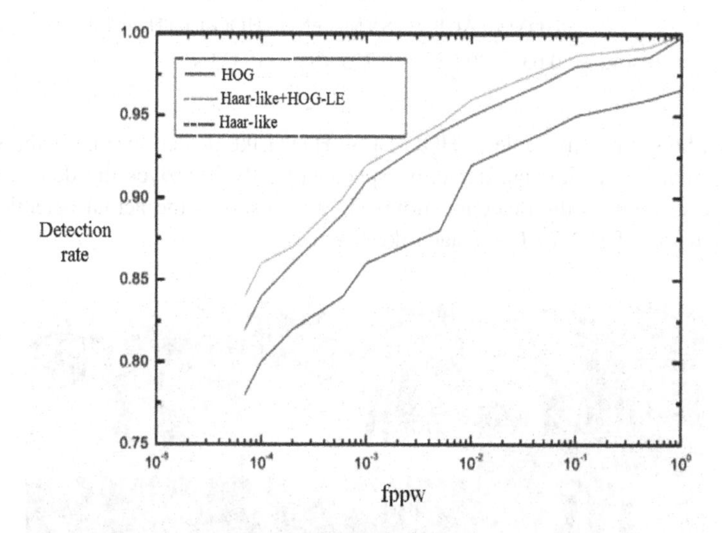

Fig. 8. Three kinds of feature sets under the test results

SVM-LeNet algorithm using multiple sets of local and global templates. However, the HOG-LE+Haar-Like algorithm adds a variety of features to optimize sorting and improve the overall detection rate, and to make up for the shortcomings of the SVM-LeNet algorithm which is slow, an increase of HOG feature initial inspection, the experiments show that the proposed HOG-LE + Haar-Like algorithm's effectiveness [9–11].

The four algorithms based on the feature extracted using the samples of the 320 × 240 images to test, then we train the cascade classifier, the four algorithms test time shown in Table 2.

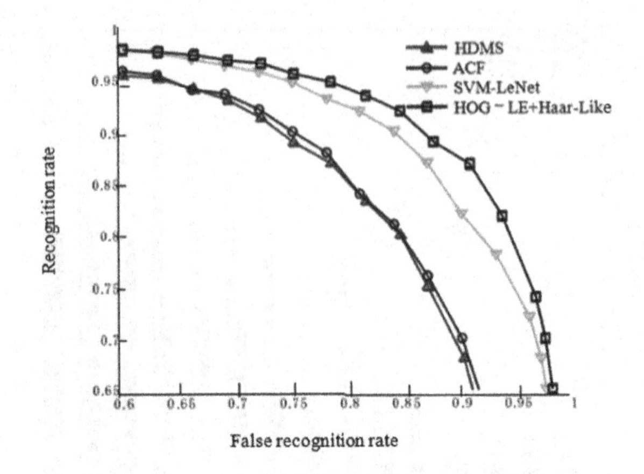

Fig. 9. The curve of recognition rate and false recognition rate

Table 2. Four algorithms test time

	HDMS	ACF	SVM-LeNet	HOG-LE+Haar-Like
Test time	0.633 s	0.556 s	0.725 s	0.323 s

As can be seen from Table 2, HOG-LE + Haar-Like detection time is the shortest, indicating that this pedestrian detection speed not only improves the detection accuracy, but also reduces the detection time. Figure 10 shows the actual detection effect based on fusion of HOG-LE + Haar-Like features:

Fig. 10. Actual test results

The experimental results show that the pedestrian detection based on fusion of HOG-LE + Haar-Like features method not only achieves the HOG feature and the Haar-Like feature complementarity, but also mix the classifier through the cascaded Adaboost algorithm. Finally, the pedestrian detection achieve high test accuracy and high efficiency. The experimental results also show that this pedestrian detection based on fusion of HOG-LE+Haar-Like features improves the performance and time of the detection has significantly decreased.

5 Conclusions

In this paper, the HOG feature is reduced by using LE, and the fusion with the Haar-Like feature is used to train and calculate the pedestrian samples. The Adaboost strong classifier is used to detect the pedestrian. The test results show that the pedestrian detection based on fusion of HOG feature based on LE and Haar-Like feature has effectively improved the detection rate in the real scene.

Acknowledgment. This work was supported by the National Natural Science Foundation of China (Grant no. 61273303, 61572381). The authors would like to thank all the editors and reviewers for their valuable comments and suggestions.

References

1. Nanda, H., Davis, L.: Probabilistic template based pedestrian detection in infrared videos. In: Intelligent Vehicle Symposium, vol. 1, pp. 15–20. IEEE (2008)
2. Ouyang, W., Wang, X.: Single-pedestrian detection aided by multi-pedestrian detection. In: IEEE Conference on Computer Vision and Pattern Recognition, pp. 3198–3205. IEEE Computer Society (2013)
3. Jiang, J., Xiong, H.: Fast pedestrian detection based on HOG-PCA and gentle AdaBoost. In: International Conference on Computer Science and Service System, pp. 1819–1822. IEEE (2012)
4. Ma, S., Bai, L.: A face detection algorithm based on Adaboost and new Haar-Like feature. In: IEEE International Conference on Software Engineering and Service Science, pp. 651–654. IEEE (2017)
5. Belkin, M., Niyogi, P.: Convergence of Laplacian eigenmaps. In: Proceedings of the Twentieth Conference on Advances in Neural Information Processing Systems, vol. 19, Vancouver, British Columbia, Canada, pp. 129–136. DBLP, December 2008
6. Taiana, M., Nascimento, J.C., Bernardino, A.: An improved labelling for the INRIA person data set for pedestrian detection. In: Sanches, J.M., Micó, L., Cardoso, J.S. (eds.) IbPRIA 2013. LNCS, vol. 7887, pp. 286–295. Springer, Heidelberg (2013). https://doi.org/10.1007/978-3-642-38628-2_34
7. Wei, Y., Tian, Q., Guo, T.: An improved pedestrian detection algorithm integrating haar-like features and HOG descriptors. In: Advances in Mechanical Engineering, p. 546206 (2013)
8. Cheng, W.C., Jhan, D.M.: A cascade classifier using Adaboost algorithm and support vector machine for pedestrian detection. In: IEEE International Conference on Systems, Man, and Cybernetics, pp. 1430–1435. IEEE (2011)
9. Cao, J., Pang, Y., Li, X.: Pedestrian detection inspired by appearance constancy and shape symmetry. IEEE Trans. Image Process. **25**(12), 5538–5551 (2016)
10. Zou, C., Cai, D., Zhao, N., et al.: A pedestrian detection algorithm based on SVM-LeNet fusion. Comput. Eng. **43**(5), 169–173 (2017). (in Chinese)
11. Luo, J., Qi, C., Zhang, J., et al.: Pedestrian detection based on multifractal spectrum. J. Tianjin Polytech. Univ. **36**(2), 59–63 (2017). (in Chinese)

Regularized Super-Resolution Reconstruction Based on Edge Prior

Zhenzhao Luo[1,2](✉), Dongfang Chen[1,2], and Xiaofeng Wang[1,2]

[1] College of Computer Science and Technology, Wuhan University of Science and Technology, Wuhan 430065, China
luozhenzhao0922@gmail.com

[2] Hubei Province Key Laboratory of Intelligent Information Processing and Real-Time Industrial, Wuhan 430065, China

Abstract. Considering that there is no edge constraint in general regularized algorithms, an improved super-resolution algorithm with additional regularization is presented. The Difference Curvature (DC) regularization which containing the image edge information is joined into the cost function, for further preserving the edge details at image reconstruction procedure. In each iteration, the DC regularization will extract the edge of high-resolution prediction frame and low-resolution observation frame. And the error between them is used to compensate the edge loss which may be smoothed by existing regularization. The reconstructed result is approximated to the original image by constraining the error between them. Then the optimum solution will be worked out by utilizing the steepest descent method. This approach is intended to constrain the edges of the image directly rather than simply avoiding the edge being smoothed. Comparing with other single regularization algorithms, experiment results indicate that the proposed algorithm can restore the edge details of reconstructed image well. And it also shows that various prior knowledge are important to image reconstruction process.

Keywords: Image reconstruction · Super-resolution · Edge Prior

1 Introduction

High-resolution pictures are widely used in remote sensing, medical diagnosis and military information gathering, etc. However, due to the high cost and physical limitations of high-precision optical systems, it is not easy to obtain desired high-resolution (HR) image. Therefore, a new approach toward enhancing the quality of image is needed to overcome these limitations. One promising approach is Super-Resolution Reconstruction (SRR). It refers to the process of constructing a HR image from one or more low-resolution (LR) images. This approach makes it possible to improve image quality without changing the hardware infrastructure.

SRR algorithm can be divided into two categories [1]: frequency domain algorithms and spatial domain algorithms. Tsai et al. [2] first proposed frequency domain based algorithm by eliminating spectral aliasing in the frequency domain to improve spatial resolution. Its advantage is the lower computational complexity makes it possible to be

D.-S. Huang et al. (Eds.): ICIC 2018, LNAI 10956, pp. 408–417, 2018.
https://doi.org/10.1007/978-3-319-95957-3_44

applied at real-time system. Reconstruction approaches based on discrete cosine transform [3] and wavelet transform [4] are similar algorithms. Among frequency domain approaches, the prior knowledge which is used to constrain or regularize the SR problem is usually difficult to express, but contrarily it can be well expressed at spatial domain. Thus, the spatial domain algorithm framework and its improved algorithm have become the research hotspots. Including the Iterative backward projection (IBP) [5], Projection on a convex sets (POCS) [6], Bayesian maximum a Posteriori (MAP) [7] and other approaches. The IBP method predicts one HR frame at each iteration and generates a corresponding LR frame from the HR prediction, then compares it with the observations to obtain an error for reconstructing the HR image. However, because of the practicality of the inverse problem, this approach has no unique solution, and it is also difficult to be combined with priori constraints [8]. The POCS based methods solve the problem of restoration and interpolation by limiting the solution to a closed convex set, but their disadvantages are non-uniqueness of the solution, slow convergence, and large amount of computation [9]. Among all the spatial domain algorithms mentioned above, the MAP method is most widely used, and the MAP solution framework is usually consist of fidelity term and regular term. Common regularized forms are: based on Markov random field (MRF) regularization [10, 11], Tikhonov regularization [12], and Bilateral total variation (BTV) regularization [13, 14].

However, in the above regularization methods, there is no direct edge constraint included in the solution framework, and the reconstruction performance is influenced by the prior model selection. So, based on the method of BTV regularization, this paper introduces a new regularization term which containing the original local structure information of the image, it re-builds the cost function of reconstruction. The new regularization term utilizes difference curvature [15] to extract the edge features in the image. Benefit from good noise resolving capability, utilizing this model can also avoid the influence caused by noise in some extent. This paper first describes the image degradation model in super-resolution reconstruction and chooses the BTV regularization method as a basic construction algorithm. Then the new regularization term which constrains edge details is added as an additional term in the cost function. Finally, the experiments verify that the proposed algorithm is better than the single regularization method, the reconstruction quality of the image has improved, and more edge detail information is retained.

2 Image Degradation Model

In general, a LR image can be regarded as a resulting image obtained from a HR image through a series of motion deformations, blurring, down-sampling, and superimposing noise. Assuming that x is the ideal HR image and Y is the LR image degraded by x, so the degraded model of the image can be expressed as:

$$Y_k = DBMx + n_k = Hx + n_k, \ k = 1, 2, 3, \ldots, p \tag{1}$$

Let us obtain p images, where H denotes a degraded matrix which is composed of motion, blurring, and down-sampling. The size of x is $R_1N_1 \times R_2N_2$, N_1 and N_2

represent the sampling factors of the rows and columns, and n_k denotes noise of the k-th frame image. Y_k is the $R_1 \times R_2$ size LR image obtained by the model in the k-th frame. SRR's task is to reconstruct the original HR image from the observed p LR image. According to the image observation model represented by Eq. (1), since H is usually an ill-conditioned matrix, the SRR process is an ill-posed inverse problem: By giving an observation frame Y, multiple result images $\boldsymbol{x}$ can be reconstructed. In order to ensure that the problem has only unique solutions, the usual approach is to add a regularization term witch can constraint the solution space and solve the Eq. (1). The formula is as follows:

$$J(x) = \sum_{k=1}^{p} \|Y_k - Hx\|_2^2 + \lambda \cdot R(x) \tag{2}$$

Where: p is the number of frames in the low-resolution observation frames, λ is the empirical parameter, and $R(\boldsymbol{x})$ is a regularization term. Accordingly, the first half part of the formula is called fidelity term. The parameter λ controls the tradeoff between the fidelity term and the regularization term. Large λ contributes much in noise suppression but produces a smoother solution. On the other hand, small λ preserve more edge information but cannot suppress noise well. The reconstruction process for solving high-resolution images converts to solve the following minimization problem:

$$\hat{x} = \arg\min_{x} J(x) \tag{3}$$

Then $\hat{x}$ will be the reconstructed high-resolution image.

3 Super-Resolution Reconstruction by Adding Regularization Term

The regularization-based algorithm present a regularization term, which can express prior knowledge of the image. In many cases, there is no unique solution to SR result image. Constraints (regularization) on the solution solve this by helping the minimization procedure discussed above to converge to a stable solution. Additional, regularization term can incorporate prior knowledge of the desirable HR image and the SR model.

The choice of regularization terms largely affects the effect of image reconstruction. The BTV regularization term contains constraints on the spatial relationship between pixels of the image, as a result of it, this reconstruction method can better avoid the edges being smoothed while suppressing the noise. The term is given as follows:

$$BTV(x) = \underbrace{\sum_{l=-p}^{p} \sum_{m=0}^{p}}_{l+m \geq 0} \alpha^{|m|+|l|} \left\| x - S_x^l S_y^m x \right\|_1 \tag{4}$$

Where α is the empirical parameter and $0 < \alpha < 1$. It indicates the contribution of pixels around in the interesting window. S_h^l, S_v^m present the matrix of translating $\boldsymbol{x}$ with l pixels in horizontal and translating with m pixels in vertical respectively. And $\left\| x - S_x^l S_y^m x \right\|$ represents the difference in pixels at different scales. To summarize Eqs. (2), (3) and (4), the final BTV regularized reconstruction cost function is:

$$J(x) = \arg\min_x [\sum_{k=1}^{p} \|Y_k - Hx\|_2^2 + \lambda \cdot \underbrace{\sum_{l=-p}^{p} \sum_{m=0}^{p}}_{l+m \geq 0} \alpha^{|m|+|l|} \left\| x - S_x^l S_y^m x \right\|_1] \tag{5}$$

The reconstruction image can be obtained by utilizing the steepest gradient method through this cost function.

4 Proposed Algorithm

4.1 Optimizing Existing Regularization Term

Some literatures show that the L_1 norm protects edge well but causes staircase effect in the result image. And using the L_2 norm can lead to blurred edges, however, it does not cause any staircase effect. So, combining two of the measures, we adjust the regularization model at sharp region and smooth region. Naturally, before proposing the additional edge constraint, it is recommended to assign the sharp region and the smooth region at predict HR frames. This can be achieved by compartmentalizing the image into $m \times n$ blocks (8×8 or 16×16 blocks are often used), then the amount of blur in each block is calculated.

The amount of blur in each block is calculated by using the gray level co-occurrence matrix (GLCM) [16]. The GLCM reflects the comprehensive information of the magnitude and direction of the image's gray level change. The texture homogeneity and texture contrast that better reflect the local texture features can be obtained from GLCM. The texture homogeneity can be presented as:

$$Homo = \sum_{i,j} \frac{P(i,j)}{1 + |i - j|} \tag{6}$$

Where *P(i, j)* is the value of GLCM result. The GLCM will generate a smaller homogeneity value if the image has more texture. So, we regard a block with small homogeneity as a sharp region, and block with large homogeneity as a smooth region. If H_q (*H* refers the homogeneity of GLCM, *q* refers the number of blocks in LR frame) is lower than a threshold, then the reconstruction process which deals with this region, is performed by using L_1 norm. On the other hand, if H_q is larger than the threshold, this region will be performed by using L_2 norm. They are expressed at Eqs. (7), (8). And the threshold is an empirical parameter.

$$REG(x) = \underbrace{\sum_{l=-p}^{p} \sum_{m=0}^{p}}_{l+m \geq 0} \alpha^{|m|+|l|} \left\| x - S_x^l S_y^m x \right\|^{p(H)} \tag{7}$$

Where:

$$p(H) = \begin{cases} 2 & H_q \geq threshold \\ 1 & H_q < threshold \end{cases} \tag{8}$$

4.2 Additional Regularization Term

For the regularization term, BTV regularization is commonly employed. During the reconstruction process, the BTV regularization can only guarantee that the edges will not be greatly smoothed, instead of preserving the edges. So, the proposed algorithm aims to find out an edge constraint which can compensate edge details directly. The algorithm extracts the edge of the predict frame at each iteration, and adds the error between the down-sampling result and the low-resolution observation frame as an edge constraint into the BTV model. There are then:

$$J(x) = \sum_{k=1}^{p} \|Y_k - Hx\|_2^2 + \lambda \cdot REG(x) + \theta \cdot \gamma(x) \tag{9}$$

$$\gamma(x) = \sum_{k=1}^{p} \|\varphi_k(Y_k) - D \cdot \varphi(x)\|_1 \tag{10}$$

Where: $\gamma(x)$ is the added regularization term, φ is an edge indicator using difference curvature as an evaluation criterion, which is more efficient to distinguishing the edge region and the smooth region. D is the down-sampling operator, and this sampling factor is same as the reconstruction sampling factor. The result of the difference curvature can be calculated from:

$$\varphi(x) = \left| |I_{\mu\mu}| - |I_{\xi\xi}| \right| \tag{11}$$

In Eq. (8), $I_{\mu\mu}$ and $I_{\xi\xi}$ represent the second derivative of the image in the tangent direction and gradient direction respectively. The formulas are given as follows:

$$I_{\mu\mu} = \frac{I_x^2 I_{xx} + 2I_x I_y I_{xy} + I_y^2 I_{yy}}{I_x^2 + I_y^2} \tag{12}$$

$$I_{\xi\xi} = \frac{I_y^2 I_{xx} - 2I_x I_y I_{xy} + I_x^2 I_{yy}}{I_x^2 + I_y^2} \tag{13}$$

At the smooth region of the image, the difference between second derivative in the tangent direction and the gradient direction is small, so the difference curvature is small. And the difference curvature of the edge region is larger than that of the smooth region. Moreover, the mentioned regularization term are mainly aimed to constrain the edges. Therefore, in order to avoid unnecessary errors caused by weak noise in the smooth region, φ can be slightly processed by experience:

$$\varphi = \begin{cases} \varphi, \varphi > T \\ 0, \varphi < T \end{cases} \tag{14}$$

To summarize Eqs. (7), (9) and (10), the cost function for reconstructing high resolution images becomes:

$$\begin{aligned} J(x) = \arg\min_x [\sum_{k=1}^{p} \|Y_k - Hx\|_2^2 + \lambda \cdot \underbrace{\sum_{l=-p}^{p} \sum_{m=0}^{p}}_{l+m \geq 0} \alpha^{|m|+|l|} \left\| x - S_x^l S_y^m x \right\|^{p(H)} \\ + \theta \cdot \sum_{k=1}^{p} \|\varphi_k(Y_k) - D \cdot \varphi(x)\|_1] \end{aligned} \tag{15}$$

Finally, using the steepest gradient method to solve Eq. (12). While *p(H)* = 1, the iterative formula is:

$$\begin{aligned} \hat{x}_{n+1} = \hat{x}_n - \beta \{ \sum_{k=1}^{p} M_k^T B_k^T D_k^T sign(DBM\hat{x}_n - Y_k) \\ + \lambda \cdot \sum_{l=-p}^{p} \sum_{m=0}^{p} \alpha^{|l|+|m|} \left[I - S_x^{-l} S_y^{-m} \right] sign\left(\hat{x}_n - S_x^l S_y^m \hat{x}_n \right) \\ - \theta \cdot \sum_{k=1}^{p} D^T \psi(x) \cdot \mathrm{sign}[\varphi(Y_k) - D \cdot \varphi(x)] \} \end{aligned} \tag{16}$$

Where $sign(\cdot)$ is the sign function. $S_x^{-l} S_y^{-m}$ is the inverse operation of $S_x^l S_y^m$. $\psi(\cdot)$ is the derivative of $\varphi(\cdot)$. The entire algorithm steps can be summarized as follows:

1. Image registration of the LR input images.
2. Set the first LR image as the initial estimated HR image and divide the initial HR image into m × n blocks, each block's homogeneity will be obtained by (6).
3. At regions with high homogeneity, L_2 norm should be used in the regularization term, and L_1 norm should be used at low homogeneity regions.
4. Update the predict HR image according to (13) until the error of two adjacent iterations meets the convergence conditions:

$$\frac{\|\hat{x}_{n+1} - \hat{x}_n\|^2}{\|\hat{x}_n\|^2} < \varepsilon \tag{17}$$

5. If the convergence condition is satisfied, the output of this round is the reconstruction result. Otherwise, the step 4 is repeated.

5 Experiments

In order to verify the effectiveness of the proposed algorithm, the BTV method and the improved method are implemented to reconstruct HR image. The above algorithms are all performed in PC (Intel I5 2.50 GHz, RAM8.0 GB), Matlab R2014a. 256 × 256 resized original images from dataset Set14 and BSD100 are selected to test two methods at the experiment. The image degradation model is applied for blurring, shifting, down-sampling and superimposing additive noise and obtain eight 128 × 128 low-resolution images. In the iteration process, the iteration step length is 0.2 and the convergence condition ε is 10^{-7}.

In this paper, the Peak Signal-to-Noise Ratio (PSNR) [17] and the Mean Structural Similarity Index (MSSIM) [18] are used to quantify the effectiveness of the experimental results. PSNR is a relatively mainstream evaluation method in evaluating the image quality. It can measure the degree of distortion of the result image relative to the original image. The larger the PSNR, the less distortion, and the better effect of reconstructing image. SSIM is a perceptual-based computing model that can take into account the tiny changes in the image's structural information over human perception. The larger the SSIM, the closer the two images are. The formula is given as follows:

$$PSNR = 10\log_{10}\frac{s^2}{MSE} \tag{18}$$

$$MSSIM = \frac{1}{M}\sum_{i=1}^{M} SSIM(i) \tag{19}$$

$$SSIM_{xy} = \frac{(2\mu_x\mu_y + c_1)(2\sigma_{xy} + c_2)}{(\mu_x^2 + \mu_y^2 + c_1)(\sigma_x^2 + \sigma_y^2 + c_2)} \tag{17}$$

Where: MSE represents the mean square error of the current image and the reference image. s is the value of the largest pixel in the image, μ and σ represent the mean and variance respectively, x and y identify two different comparison regions. The average structural similarity is the result of calculating the structural similarity by dividing the image into multiple regions.

The results are shown in Table 1, which is the evaluation data of the reconstruction implemented by Spline interpolation, BTV method, and the proposed method.

Table 1. Evaluation data of two reconstruction results.

	Spline		BTV method		Proposed method	
	PSNR(dB)	MSSIM	PSNR(dB)	MSSIM	PSNR(dB)	MSSIM
Lena	16.5076	0.6217	17.4706	0.6301	17.9793	0.6388
Butterfly	14.5942	0.5266	16.1197	0.5709	16.4963	0.5800
Pepper	18.1578	0.6004	19.9227	0.6055	20.3527	0.6089
Baby	19.0324	0.6390	20.3546	0.6386	20.6194	0.6403
Building	17.2329	0.6042	18.4534	0.6484	19.0974	0.6491
Man	18.0258	0.4936	19.6148	0.5223	20.1677	0.5436

It can be seen from Table 1 that after adding new prior knowledge, the improved reconstruction results are enhanced compared to the previous BTV reconstruction. When the regularized parameter is small, the reconstruct effect of the proposed algorithm has not be improved much. That is reasonable because what was added into the new algorithm model is more focused on retaining edge details in the procedure instead. Simultaneously, while reconstruction utilizing a small regularized parameter, the edge details will be preserved well itself. When regularization efforts (the regularized parameter) were increased, the additional prior knowledge model will show more efficiency, which is showed at table (using relatively larger parameters than usual). Large parameter makes the HR predict image over-constrained by existing regularization term, now, the additional can compensate the edge detail for the fail of the former. The details can be more preserved.

Part of the experimental reconstruction results are shown in Fig. 1. As can be seen from the figure, the reconstruction effect of proposed method is better than the basic method. After adding the prior knowledge, the reconstruction process further strengthens the reservation of the edge details.

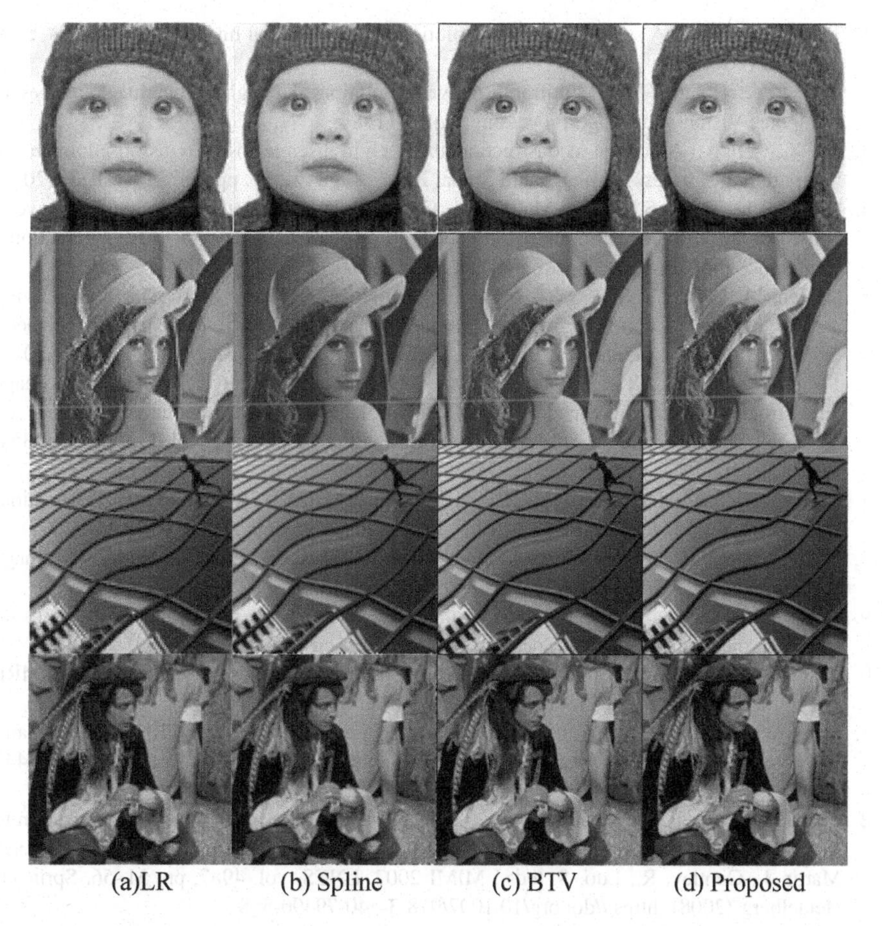

Fig. 1. LR image and results of implement by spline interpolation and BTV method and proposed method.

6 Conclusion

Considering that at the image reconstruction process, the texture information of the natural image itself should provide rich prior knowledge for the reconstruction, and the single regularization method couldn't join this prior constraint. This paper proposes an improved algorithm with an added regularization term based on the BTV regularized method. In order to better maintain the image's edge and suppress the image's distortion, the edge information of the predicted high-resolution image is extracted in each iteration, and compared with the low-resolution observation frame's edge. The error between them should be constrained. Minimize it so that the reconstructed image is closer to the original image. The experimental results show that the improved method proposed in this paper can achieve better reconstruction performance in maintaining image edge details.

References

1. Tian, J., Ma, K.K.: A survey on super-resolution imaging. Signal Image Video Process. **5**(3), 329–342 (2011)
2. Tsai, R.Y., Huang, T.S.: Multiframe image restoration and registration. In: Advances in Computer Vision and Image Processing, vol. 1 (1984)
3. Yu, X., Cen, X.: An image reconstruction approach in discrete cosine transform domain. In: International Congress on Image and Signal Processing, vol. 57, pp. 587–591. IEEE (2013)
4. Lama, R.K., Shin, S., Kang, M., Kwon, G.R., Choi, M.R.: Interpolation using wavelet transform and discrete cosine transform for high resolution display. In: IEEE International Conference on Consumer Electronics, pp. 184–186. IEEE (2016)
5. Nazren, A.R.A., Yaakob, S.N., Ngadiran, R., Hisham, M.B., Wafi, N.M.: Improving iterative back projection super resolution model via anisotropic diffusion edge enhancement. In: International Conference on Robotics, Automation and Sciences, pp. 1–4. IEEE (2017)
6. Tang, Z., Deng, M., Xiao, C., Yu, J.: Projection onto convex sets super-resolution image reconstruction based on wavelet bi-cubic interpolation. In: International Conference on Electronic and Mechanical Engineering and Information Technology, vol. 175, pp. 351–354. IEEE (2011)
7. Shen, H., Zhang, L., Huang, B., Li, P.: A map approach for joint motion estimation, segmentation, and super resolution. IEEE Trans. Image Process. **16**(2), 479–490 (2007)
8. Irani, M., Peleg, S.: Improving resolution by image registration. Cvgip Graph. Models Image Process. **53**(3), 231–239 (1991)
9. Stark, H., Oskoui, P.: High-resolution image recovery from image-plane arrays, using convex projections. J. Optical Soc. Am. Opt. Image Sci. **6**(11), 1715 (1989)
10. Shao, W.Z., Wei, Z.H.: Super-resolution reconstruction based on generalized Huber-MRF image modeling. J. Softw. **18**(10), 2434–2444 (2007)
11. Shao, W.Z., Deng, H.S., Wei, Z.H.: A posterior mean approach for MRF-based spatially adaptive multi-frame image super-resolution. Signal Image Video Process. **9**(2), 437–449 (2015)
12. Zhang, X., Lam, E.Y., Wu, E.X., Wong, K.K.Y.: Application of Tikhonov regularization to super-resolution reconstruction of brain MRI images. In: Gao, X., Müller, H., Loomes, Martin J., Comley, R., Luo, S. (eds.) MIMI 2007. LNCS, vol. 4987, pp. 51–56. Springer, Heidelberg (2008). https://doi.org/10.1007/978-3-540-79490-5_8

13. Liu, W., Chen, Z., Chen, Y., Yao, R.: An ℓ1/2-BTV regularization algorithm for super-resolution. In: International Conference on Computer Science and Network Technology, pp. 1274–1281. IEEE (2016)
14. Mofidi, M., Hajghassem, H., Afifi, A.: An adaptive parameter estimation in a btv regularized image super-resolution reconstruction. Adv. Electr. Comput. Eng. **17**(3), 3–10 (2017)
15. Chen, Q., Montesinos, P., Sun, Q.S., Peng, A.H., Xia, D.S.: Adaptive total variation denoising based on difference curvature. Image Vis. Comput. **28**(3), 298–306 (2010)
16. Bai, X.B., Wang, K.Q., Wang, H.: Research on the classification of wood texture based on Gray Level Co-occurrence Matrix. J. Harbin Inst. Technol. **12**, 021 (2005)
17. Hore, A., Ziou, D.: Image quality metrics: PSNR vs. SSIM. In: International Conference on Pattern Recognition, pp. 2366–2369. IEEE (2010)
18. Silvestre-Blanes, J.: Structural similarity image quality reliability: determining parameters and window size. Signal Process. **91**(4), 1012–1020 (2011)

An Improved Anisotropic Diffusion of Cattle Follicle Ultrasound Images De-noising Algorithm

Yong Lv[1,2](✉) and Jun Liu[1,2]

[1] College of Computer Science and Technology, Wuhan University of Science and Technology, Wuhan 430065, China
1075337670@qq.com

[2] Hubei Province Key Laboratory of Intelligent Information Processing and Real Time Industrial System, Wuhan 430065, China

Abstract. For the de-noising process of cattle follicle ultrasound images, we need to retain the edge details containing important information while removing the speckle noise. According to the traditional de-noising method, the PM anisotropic diffusion model in the selection of diffusion coefficient and diffusion threshold K properly, resulting in poor smoothing effect of ultrasound images, image detail preserving problems and other related issues, this paper proposes an improved anisotropic diffusion filtering algorithm based on several current typical anisotropic diffusion filtering. In this paper, the gradient mode of the adaptive median filter is used to replace the gradient mode of the original image, in addition, the diffusion coefficient and the selection of the diffusion threshold are also improved. PSNR, SSIM, homogeneity region contrast, and edge retention capability FOM are used to evaluate the quality of the algorithm. The experimental results show that the improved method can effectively suppress the noise of the cattle follicle ultrasonic image and better retain the edge details, providing a good basis for the subsequent processing of images.

Keywords: Cattle follicle ultrasound image · Anisotropic diffusion Adaptive median filtering · Diffusion coefficient

1 Introduction

The principle of ultrasonic imaging is to reflect, scatter and diffract the media according to the ultrasonic wave, So when the ultrasonic probe is scanning the tissue of the cattle, the tissues and organs of the cattle can interfere with the ultrasonic wave sent by the probe in the space around the follicle, a speckle field that produces random distribution of phase and amplitude, therefore, speckle noise will occur in the final imaging [1]. The noise has a great influence on the subsequent processing of the image, such as the detection and segmentation of the follicle region. So we need to deal with these noises to eliminate the interference of noise to the follow-up treatment of ultrasonic image of cattle follicle image.

In order to reduce the interference of speckle noise to the subsequent ultrasonic image processing, the scholars have put forward many feasible methods. The

D.-S. Huang et al. (Eds.): ICIC 2018, LNAI 10956, pp. 418–430, 2018.
https://doi.org/10.1007/978-3-319-95957-3_45

traditional de-noising method of ultrasonic image is mainly using the linear filtering method, for example, median filtering [2], the principle of median filtering is processed according to the local statistical features of the image. But this method has some drawbacks, it uses the same processing method for all the pixels, so it has a bad effect on protecting the edge details of the image.

Due to the shortcomings of the traditional method in the protection of the edge details of the image, and these defects also affect the subsequent detection and segmentation of the image. Therefore, in recent years, the anisotropic diffusion method based on the nonlinear partial differential equation has been proposed, and it is widely used in the field of noise reduction in ultrasonic images [3, 4]. This method overcomes the main shortcomings of the traditional methods. Perona and Malik proposed the classic PM model on this basis in 1990 [5], the method is based on the anisotropic diffusion filtering method based on the thermal diffusion equation, it is widely used because of its good protection to the edge of the image while removing noise, but it also has some defects. For example, it will fail for the strong noise while de-noising, what's more, it is difficult to control the selection of the diffusion threshold, there is an obvious "step effect" in the image after diffusion treatment. Catte and others improved the PM model in 1992 [6], then proposed the catte-PM model, the gradient modulus of the original image is replaced by the gradient mode of Gauss smoothed, and this method can solve the problem of strong noise failure of PM model to a certain extent. But because of the Gauss filter, the edge information is blurred. And the final effect depends largely on the selection of the variance of Gauss's function. You et al. In 2000 [7] proposed an anisotropic diffusion model based on the four order partial differential equation (Four-order Partial Differential Equations, F_PDE), this model eliminates the staircase effect of the PM model to a certain extent. However, because of the use of the 4 order operator, in the process of de-noising, there will be isolated impulse noise.

This paper is aimed at the characteristics of ultrasonic image of cattle follicle. Based on the analysis of the traditional de-noising algorithm, and based on the research of the PM model and the catte-PM model, in this paper, the gradient mode of the adaptive median filter is replaced by the gradient model of the Gauss smoothness in the catte-PM model, the problem of blurring the edge information after Gauss filtering is solved to a certain extent. At the same time, two order differential operators are introduced to improve the diffusion coefficient combine with the first order differential operator. The method of automatic estimation is adopted for the selection of the diffusion threshold. The experimental results show that the method proposed in this paper can effectively eliminate the speckle noise in the ultrasonic image of bovine follicular, meanwhile, it has a good effect on the preservation of the image edge details.

2 Traditional De-noising Method-BM3D

The principle of the BM3D de-noising algorithm is to divide the image into blocks of a certain size. The 2-D image blocks with similar structure are combined to form three dimensional arrays according to their similarity. Then these three dimensional arrays are processed and processed after inverse transformation, the result is returned to the original image. This is the image that we need after de-noising [8]. The algorithm is divided into two steps: the base estimate and the final estimate. The flowchart of the algorithm is shown in Fig. 1 below:

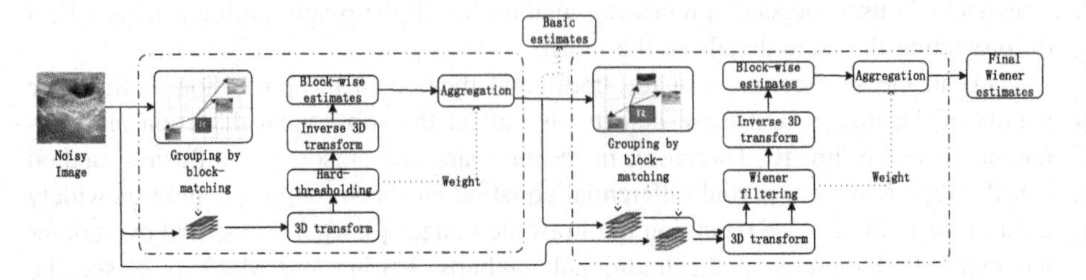

Fig. 1. BM3D algorithm flow chart

Step1: Basic Estimation
First, some K * K sized blocks are selected on the unprocessed original noise images. These blocks are reference blocks. The process of looking for similar blocks around a reference block can be shown in the following formula (1) and (2):

$$G(P) = \{Q = \mathrm{d(P,Q)} \leq \tau^{step1}\} \tag{1}$$

$$\begin{cases} Q(P) = T_{3Dhard}^{-1}(\gamma(T_{3Dhard}(Q(P)))) \\ \gamma(x) = \begin{cases} 0 & if|x| \leq \lambda_{3D}\sigma \\ x & otherwise \end{cases} \end{cases} \tag{2}$$

In formula (1), The D (P, Q) in the upper form represents the Euclidean distance between the two blocks, the final integration of three-dimensional matrix to obtain the blue R matrix in Fig. 2.1 is the lower left corner of the step1.

In formula (2), the two dimensional transformation and one dimensional transformation are represented by T_{3Dhard}. γ is a threshold, σ indicates the intensity of the noise, which is not only the standard deviation.

Step2: Final Estimate
The initial process is similar to that of step1. The difference is that this process gets two three dimensional arrays: Q^{basic} (P) and Q (P), It is the three-dimensional matrix of the noise map and the three-dimensional matrix of the basic estimation results. This process is expressed in formula (3) as follows:

$$\begin{cases} Q(P) = T_{3D\mathrm{wien}}^{-1}(w_P.T_{3D\mathrm{wien}}(Q(P))) \\ w_P(\xi) = \dfrac{\left|T_{3D}^{wien}(Q^{basic}(P))(\xi^2)\right|}{\left|T_{3D}^{wien}(Q^{basic}(P))(\xi^2)\right| + \sigma^2} \end{cases} \tag{3}$$

In formula (3), the two-dimensional and one-dimensional transformations are represented by T_{3Dwien}, W_P is the coefficient of Wiener filtering, and the σ represents the noise intensity.

Finally, with step1, all blocks are fused and placed in their original position, the difference is that at this time the weight depends on the factors that change to the Wiener filtering coefficient W_P and σ.

3 Anisotropic Diffusion Filter Based on Ultrasonic Image

3.1 Traditional PM Model

The traditional anisotropic diffusion filtering algorithm, the PM model, is a nonlinear filtering algorithm. When the noise is removed, the edge of the image can be guaranteed to be well preserved, the partial differential equation of the algorithm is shown in formula (4).

In formula (4), I_0 is the original image, ∇ is gradient, div is divergence operator, C(x) for diffusion coefficient, T is a time operator introduced, and it means De-noising process is related to the duration of diffusion.

$$\begin{cases} \frac{\partial I(x,y,t)}{\partial t} = div(c(|\nabla I|) \cdot \nabla I) \\ I(t=0) = I_0 \end{cases} \tag{4}$$

As for the selection of the diffusion coefficient, the following requirements are generally followed: The diffusion coefficient should be a monotone function, and the monotonicity is monotonically decreasing, and the diffusion coefficient and its parameters should have the following relation: when $x \to \infty$, $C(x) \to 0$.

Perona and Malik define C(x) in their paper as the following 2 functions, as shown in formula (5) and formula (6):

$$c(x) = \frac{1}{1+(x/k)^2} \tag{5}$$

$$c(x) = \exp\left[-(x/k)^2\right] \tag{6}$$

In this algorithm, in formula (5), $x = |\nabla I|$, K represents the diffusion threshold, If $x \gg K$, then the diffusion coefficient approaches 0, edge diffusion decreased, edge preserved; If $x \ll K$, then the diffusion coefficient approaches 1, non edge diffusion degree enhancement, achieve de-noising effect.

The equations in formula (4) need to be discretized, after that, it can be applied to the image de-noising process. Its discretization forms are as follows:

$$I_p^{t+1} = I_p^t + \frac{\lambda}{|\eta_p|} \sum_{q \subset \eta_p} c\left(\nabla I_{q,p}^t\right) \nabla I_{q,p}^t \tag{7}$$

In formula (7), λ is the constant to control the intensity of the diffusion, t is the number of iterations, p represents the coordinates of a pixel in a two-dimensional grid, and η_p represents the neighborhood space of pixel p.

There are some shortcomings in the PM model, first, the PM model itself is mathematically an ill conditioned equation. The existence and uniqueness of the solution cannot be guaranteed. Second, the PM model is not good for isolated noise and strong noise removal, sometimes the noise cannot be eliminated, but the noise will be enhanced.

3.2 Speckle Suppression Anisotropic Diffusion Filter

In Yu's paper, he improves the traditional anisotropic diffusion filtering algorithm, applied to ultrasonic image processing. An algorithm SARD for ultrasonic speckle suppression is proposed, the diffusion coefficient is improved accordingly. Given an intensity image $I_0(x, y)$ having finite power and no zero values over the image support Ω, the output image $I(x, y; t)$ is evolved according to the following PDE:

Its diffusion coefficient is expressed as shown in formula (9):

$$\begin{cases} \partial I(x,y;t)/\partial t = div[c(q)\nabla I(x,y;t)] \\ I(x,y,0) = I_0(x,y), (\partial I(x,y;t)/\partial \vec{n})|_{\partial\Omega} = 0 \end{cases} \tag{8}$$

$$\begin{cases} c(q) = \frac{1}{1+\left[q^2(x,y;t)-q_0^2(t)\right]/\left[q_0^2(t)(1+q_0^2(t))\right]} \\ q_0(t) = \frac{\sqrt{var[z(t)]}}{\overline{z(t)}} \end{cases} \tag{9}$$

In formula (8), where $\partial\Omega$ denotes the border of Ω, $\vec{n}$ is the outer normal to the $\partial\Omega$. $var[z(t)]$ And $\overline{z(t)}$ are the variance and average of the gray value of the homogeneous region of the image at this time, respectively. $q(x, y; t)$ is the instantaneous coefficient of change, and its definition is as shown in formula (10):

$$q(x,y;t) = \sqrt{\frac{(1/2)(|\nabla I|/I)^2 - (1/4)^2(\nabla^2 I/I)^2}{(1+(1/4)(\nabla^2 I/I))^2}} \tag{10}$$

4 Improved Anisotropic Diffusion Algorithm

4.1 Improved Diffusion Model with Adaptive Median Filter

In order to solve the defect of PM model for strong noise failure, Catte and others made corresponding improvements in 1992 based on the traditional anisotropic diffusion filtering algorithm, the PM model. We called it Catte-PM model, it replaced the gradient model of the original image with the gradient mode of Gauss smoothing, to control diffusion. The optimized model is shown as follows:

$$\begin{cases} \frac{\partial I(x,y,t)}{\partial t} = \mathrm{div}[\mathrm{c}(|\nabla G_\sigma * I|) \cdot \nabla I] \\ I(t=0) = I_0 \end{cases} \tag{11}$$

G_σ is the standard deviation for the Gauss function of sigma, $*$ represents convolution.

Because it uses Gauss filter in each iteration, so it can reduce the noise to a certain extent and protect the edges. But it is because of the use of Gauss smoothing filter, it destroys the nature of anisotropic diffusion and causes the image structure to deviate from its original position. Therefore, the blurred edge of the image is increased after Gauss filtering. In addition, the variance of Gauss's function is also an important uncertainty factor.

Median filter is widely used in image processing. It has a significant effect on the suppression of speckle noise. It can largely retain the edge information of the image and improve the edge fuzziness. The ultrasonic image used in this paper is a typical image with speckle noise, therefore, this paper is based on traditional anisotropic diffusion filtering and Catte and Yu et al., a median filter instead of Gauss filtering in Catte is proposed.

However, the effect of traditional median filtering is greatly influenced by its filter window, the details of the image cannot be effectively protected. Based on this, this paper proposed an adaptive median filter. Through scanning window, we can dynamically adjust the size of scanning window in the process of scanning images. The purpose of eliminating noise is very good. At the same time, it also enhanced the removal effect of Gauss noise and even distributed noise. The effect of image detail and edge protection is better than median filtering.

Its specific flow chart is as follows:

In this paper, the median filter initial window size Sxy is set to 3 * 3, and the median filter window Smax is set to 7 * 7.

Adaptive median filter is better than traditional median filter in image detail protection and edge blur improvement.

The improved model is shown as follows:

$$\begin{cases} \frac{\partial I(x,y,t)}{\partial t} = div[C(|\nabla(AMF * I)|) \cdot \nabla I] \\ I(x,y,0) = I_0 \end{cases} \tag{12}$$

In formula (12), AMF represents adaptive median filtering, and its algorithm flow is shown in Fig. 2. The other parameters are the same as those of the traditional anisotropic diffusion filter.

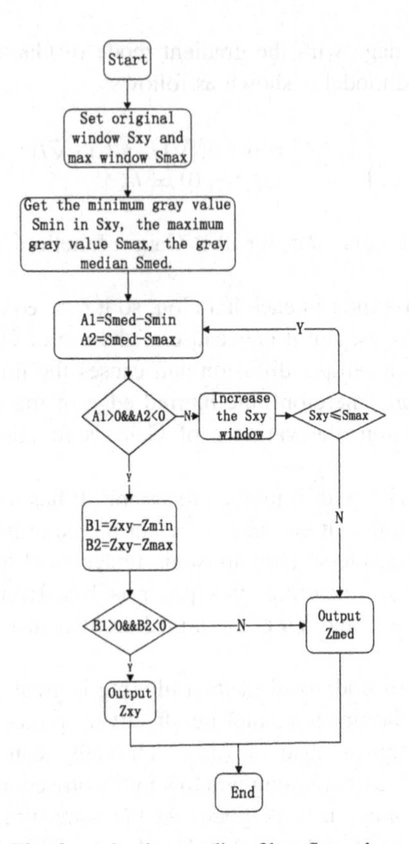

Fig. 2. Adaptive median filter flow chart

4.2 Improved Diffusion Coefficient

In addition, based on analysis, the improved diffusion coefficient is also proposed in this paper. For the improvement of the diffusion coefficient, from the above introduction, we must meet the following 3 basis points:

(1) The diffusion coefficient should be a monotone decreasing function.
(2) When the independent variable of diffusion coefficient approaches infinity, the diffusion coefficient is equal to 0.
(3) When the independent variable of diffusion coefficient approaches 0, the diffusivity equals 1.

In formula (5), (6), (10), (11), the two algorithms only use the first order partial differential operators to diffuse the images, and the details are inevitably missing. Therefore, this paper is based on the diffusion coefficient proposed by Yu, a method combining first order differential operator and two order differential operator is proposed to diffuse image. So that it can better acquire the fine features of ultrasound images during the diffusion process. The modified diffusion coefficient formula is shown as follows:

$$\begin{cases} c(q) = \dfrac{1}{1+\left[q^2(x,y;t)-q_0^2(t)\right]/\left[q_0^2(t)(1+q_0^2(t))\right]} \\ q_0(t) = \dfrac{\sqrt{\operatorname{var}[z(t)]}}{\overline{z(t)}} \\ q(x,y,t) = \sqrt{\dfrac{(1/2)(((|\nabla I|+|\Delta I|)/I)^2-(1/4)^2(((|\nabla I|+|\Delta I|)^2/I)^2}{(1+(1/4)(((|\nabla I|+|\Delta I|)^2/I)^2}} \end{cases} \tag{13}$$

In formula (13), Δ is represented by Laplace's two order differential operator. Its parameter is the same as that of speckle suppression anisotropic diffusion filtering algorithm.

5 Experimental Results and Analysis

Next, we will de-noise the ultrasonic image of the cattle follicle by the traditional BM3D algorithm, the anisotropic diffusion filter algorithm, the speckle suppression anisotropic filtering algorithm and the improved anisotropic diffusion filter algorithm. After getting out of the noisy experimental results, then compare the results of the experiment and make an evaluation.

5.1 Algorithm Evaluation Standard

For the evaluation of the de-noising algorithm, one of the most widely used and most effective objective indicators is PSNR value (Peak Signal to Noise Ratio). The meaning of this is to reach the vertex signal of the noise ratio. It is a full reference image quality evaluation index. It is defined by the mean square error between the original image and the processed image.

However, the medical ultrasound images often get contaminated by noise. We can't get the original reference image, therefore, the processed image is used as the original image, and the relative peak value SNR is taken as the evaluation standard. The calculation formula is as follows (14):

$$\begin{cases} PSNR = 10\log_{10}\left(\dfrac{(2^n-1)^2}{MSE}\right) \\ MSE = \dfrac{1}{M \times N}\sum_{i=1}^{M}\sum_{j=1}^{N}\left(X(i,j) - Y(i,j)\right)^2 \end{cases} \tag{14}$$

In formula (14), M and N represent the row and column of the original image, $2^n - 1$ represents the maximum value of the color of the image, the n represents the number of bits per sampling value, and the MSE represents the mean square error between the original and the processed images, and the unit of the PSNR is dB decibels. The higher the PSNR value is, the better the image quality is. Therefore, for a good de-noising algorithm, the PSNR value after de-noising should be higher than that before de-noising.

Because PSNR is a kind of image quality assessment based on error sensitivity. The complex visual characteristics of human eyes are not considered. Therefore, many times, its evaluation results are different from human intuition. Therefore, this paper introduces an evaluation criterion SSIM (Structural Similarity), the standard can overcome this defect to a certain extent. The basic formula is as follows:

$$\begin{cases} SSIM(X,Y) = l(X,Y).c(X,Y).s(X,Y) \\ l(X,Y) = \frac{2\mu_X\mu_Y + c_1}{\mu_X^2 + \mu_Y^2 + c_1} \\ c(X,Y) = \frac{2\sigma_X\sigma_Y + c_2}{\sigma_X^2 + \sigma_Y^2 + c_2} \\ s(X,Y) = \frac{\sigma_{XY} + c_3}{\sigma_X\sigma_Y + c_3} \end{cases} \tag{15}$$

In formula (15), μ_X and μ_Y represent the mean value of image X and Y, σ_X and σ_Y represent the variance of image X and Y, σ_{XY} represents the covariance of two images, C_1 C_2 and C_3 are constant, in order to avoid the denominator of 0, in this paper, we take $C_1 = (K_1 \times L)^2$, $C_2 = (K_2 \times L)^2$, $C_3 = C_2/2, K_1 = 0.01$, $K_2 = 0.03$, $L = 255$.

The higher the SSIM value is, the higher the image fidelity rate is, the better the de-noising effect is.

Finally, due to the characteristics of the ultrasonic image of the bovine follicle, it is a true image of speckle noise. Therefore, we should also evaluate the algorithm from the degree of image de-noising after the de-noising and the degree of image de-noising. This paper uses the following indicators to evaluate: the contrast of homogeneous area [9] and quality factor FOM [10], the formula is as follows.

$$\begin{cases} C = \frac{1}{n}\sum_{(x,y)\in\Omega} C(x,y)\log(1 + |C(x,y)|) \\ C(x,y) = 4 \times I(x,y) - \{I(x-1,y) + I(x,y-1) + I(x+1,y) + I(x,y+1)\} \end{cases} \tag{16}$$

$$FOM = \frac{1}{I_N}\sum_{i=1}^{I_A}\frac{1}{1 + \alpha \cdot d_i^2} \tag{17}$$

In formula (16), Ω is the homogenous region of the original image I, and n is the number of pixels in the region. In formula (17), $I_N = \max(I_I, I_A)$, I_I and I_A represent the number of dots on ideal and actual edge maps, respectively, α is a proportional constant, which is usually set as $\alpha = 1/9$, d is the normal distance from the actual edge point to the ideal edge line.

For a good de-noising algorithm, the value of homogeneity contrast should be much smaller than that before de-noising. For FOM, FOM is normalized, and the scope is [0, 1], for the perfect detection edge FOM = 1.

5.2 Experimental Result

The image data collected in this experiment is obtained through clinical medical equipment. The way to get it has been introduced in the last chapter. The size of the selected image is 800 * 600 pixels,

In the actual experimental operation process, it has been truncated for a certain time.

The following image is obtained after de-noising the ultrasonic images of cattle follicles by four different de-noising algorithms mentioned above.

Figure 3 shows three of the effects of image de-noising before and after de-noising. The three images of the first row represents the unprocessed original ultrasound image; the three images of the second row represents the cattle follicle ultrasound image processed by the BM3D algorithm; the three images of the third row represents the cattle follicle ultrasound image processed by the traditional anisotropic diffusion filtering algorithm; the three images of the fourth row represents the cattle follicle ultrasound image processed by the speckle suppression anisotropic diffusion filter algorithm; the three images of the fifth row represents the cattle follicle ultrasound image processed by the improved anisotropic diffusion filter algorithm.

It can be seen directly from the diagram. These four algorithms have certain effect on the noise suppression of cattle follicle ultrasound images. Next, we evaluate the four algorithms by evaluating the three algorithms mentioned in the last section. The following Table 1 is a comparison between the original ultrasound image and the PSNR value and SSIM value of the ultrasonic images after de-noising by four different algorithms. The values of 4 samples in the de-noised image are compared:

The following Tables 2 and 3 are the contrast of the homogeneity contrast and FOM between the original ultrasound image and the ultrasonic image after de-noising by four different algorithms. Similarly, the values of 4 sample images in the de-noised image are selected for comparison.

From Table 1, we can intuitively see that under the peak signal to noise ratio, compared with the budget method, the BM3D algorithm has a relatively high peak signal to noise ratio for the ultrasonic image of bovine follicle. However, we know through the previous introduction. PSNR evaluation standard is a quality evaluation standard based on error sensitivity, which does not take into account the complex visual characteristics of human eyes.

Therefore, we need to continue to introduce the remaining three indicators, SSIM value and homogeneity contrast, and FOM.

However, under the evaluation standard of SSIM similarity structure, we can get that the improved algorithm is better than the above algorithm by comparing its SSIM value.

From the data in Table 2, we can see that the improved algorithm proposed in this paper is the least in the homogeneity region, and the same is the best algorithm in the four algorithms.

As we can see from Table 3, First of all, the FOM value of the traditional BM3D algorithm is the smallest compared with the anisotropic diffusion algorithm. It also indicates that the traditional de-noising algorithm is poor for preserving the edge details of ultrasound images. Secondly, the algorithm proposed in this paper gets the highest FOM value. It also verifies the excellent effect of our algorithm in preserving edge details.

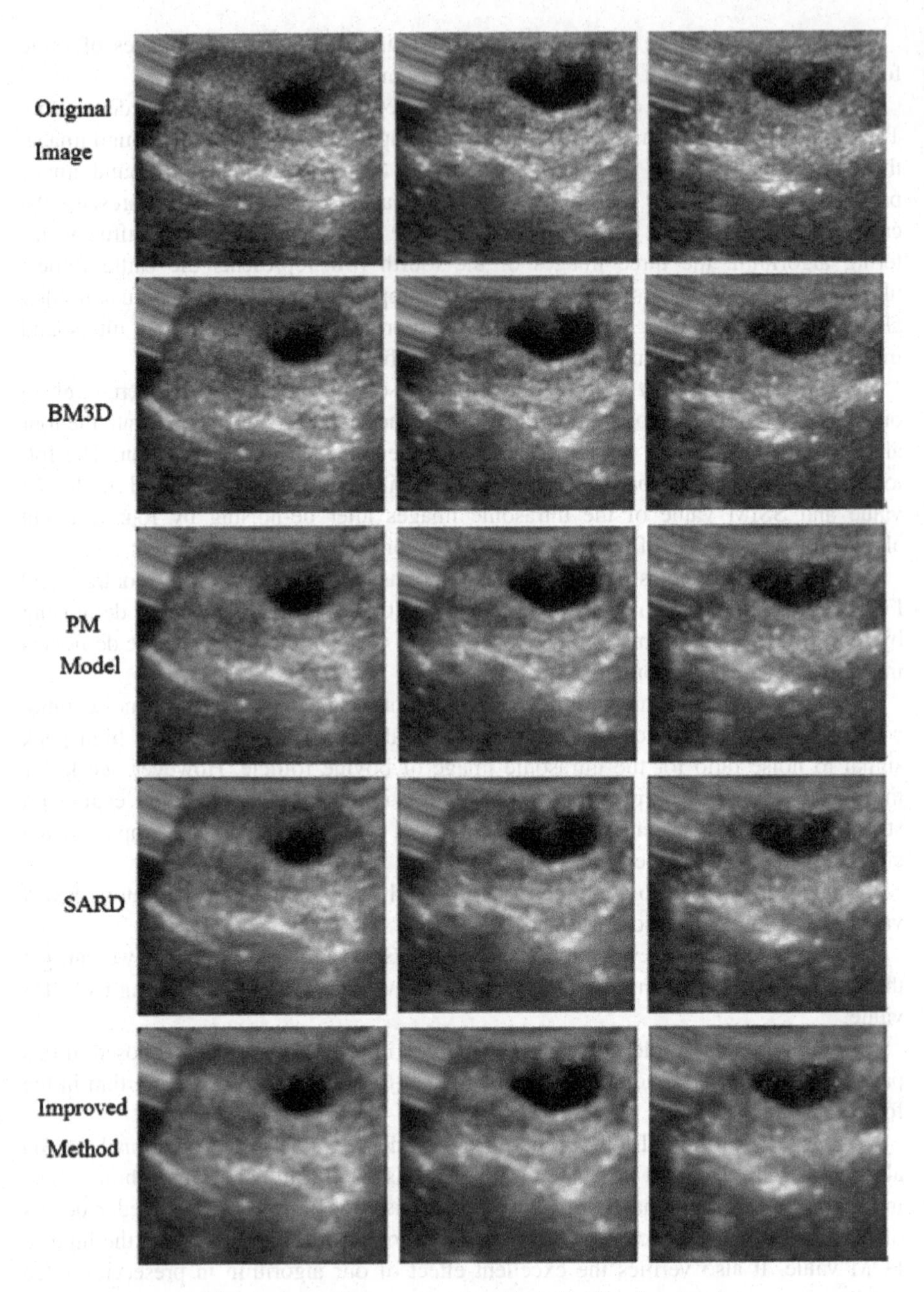

Fig. 3. De-noising results of cattle follicular ultrasound

Table 1. Three different algorithms and the PSNR and SSIM values after de-noising in this method (the upper numbers in the table represent PSNR values, and the lower decimal means the corresponding SSIM values).

PSNR SSIM	BM3D	PM model	SARD	Improved algorithm
Image 1	36.25 0.8106	35.17 0.8815	36.08 0.9014	36.18 0.9102
Image 2	38.38 0.8213	36.24 0.8934	37.29 0.9125	37.38 0.9188
Image 3	39.57 0.8824	38.35 0.9526	38.68 0.9743	38.89 0.9835
Image 4	38.03 0.9103	38.66 0.9702	39.12 0.9886	39.45 0.9899

Table 2. Three different de-noising algorithms and the improved image de-noising algorithm.

Image	Original image	BM3D	PM model	SARD	Improved algorithm
1	0.1075	0.0141	0.0115	0.0102	0.0100
2	0.1079	0.0152	0.0128	0.0118	0.0108
3	0.1095	0.0126	0.0112	0.0101	0.0092
4	0.1256	0.0236	0.0204	0.0128	0.0102

Table 3. Three different de-noising algorithms and the improved image quality factor FOM of the improved algorithm.

Image	BM3D	PM model	SARD	Improved algorithm
1	0.7836	0.7979	0.8615	0.9120
2	0.7645	0.7786	0.8546	0.9334
3	0.7756	0.7982	0.8621	0.9415
4	0.7578	0.7878	0.8301	0.9093

6 Conclusion

In this paper, at first, we introduced the traditional de-noising algorithm, the PM model for speckle noise and the improved models by others, then the defects are analyzed accordingly. The gradient module of adaptive median filter is used instead of Gauss gradient module, a new diffusion model is proposed, the diffusion coefficient is improved, and the new model is diffused.

The experimental results show that under the comprehensive evaluation of the four evaluation indexes of PSNR, SSIM, homogeneity and FOM, the improved algorithm proposed in this paper is more effective than the traditional BM3D algorithm, PM model, and SARD model to effectively eliminate the noise while preserving the edge of the image details.

References

1. Lamont, D., Parker, L., White, M., et al.: Risk of cardiovascular disease measured by carotid intima-media thickness at age 49-51: lifecourse study. BMJ Br. Med. J. **320**(7230), 273–278 (2000)
2. Huang, T., Yang, G., Tang, G.: A fast two-dimensional median filtering algorithm. IEEE Trans. Acoust. Speech Signal Process. **27**(1), 13–18 (1979)
3. Zhao, W., Zhao, J., Han, X., et al.: Infrared image enhancement based on variational partial differential equations. Chin. J. Liq. Cryst. Disp. **29**(2), 281–285 (2014)
4. Zhao, J.: Fractional differential and its application in image texture enhancement. Chin. J. Liq. Cryst. Disp. **27**(1), 121–124 (2012)
5. Perona, P., Malik, J.: Scale-space and edge detection using anisotropic diffusion. IEEE Trans. Pattern Anal. Mach. Intell. **12**(7), 629–639 (1990)
6. Catte, F., Lions, P.L., Morel, J.M., et al.: Image selective smoothing and edge detection by nonlinear diffusion. SIAM J. Numer. Anal. **29**(1), 182–193 (1992)
7. Yu, Y., Acton, S.T.: Speckle reducing anisotropic diffusion. IEEE Trans. Image Process. Publ. IEEE Signal Process. Soc. **11**(11), 1260–1270 (2002)
8. Huang, M., Huang, W., Li, J., et al.: Parameters study based on BM3D image de-noising algorithm. Ind. Control Comput. **10**, 99–101 (2014)
9. Wu, J., Wang, Y.Y., et al.: Automatic selection of based on homogeneous area of anisotropic diffusion ultrasonic image de-noising. Opt. Precis. Eng. **22**(5), 1312–1321 (2014)
10. Ranjani, J.J., Thiruvengadam, S.J.: Dual-tree complex wavelet transform based SAR despeckling using interscale dependence. IEEE Trans. Geosci. Remote Sens. **48**(6), 2723–2731 (2010)

A Fractional Total Variational CNN Approach for SAR Image Despeckling

Yu-Cai Bai, Sen Zhang, Miao Chen, Yi-Fei Pu(✉), and Ji-Liu Zhou

College of Computer Science, Sichuan University, Chengdu 610065, China
raymondbyc@gmail.com

Abstract. Synthetic aperture radar (SAR) image despeckling is an essential problem in remote sensing technology, which has a strong influence on the performance of the following processing. We propose a new despeckled algorithm combining CNN and fractional-order total variation. Through constructing a CNN model and introducing the fractional-order total variation into loss function as the regularization term, the experimental results prove that our proposed method can avoid detail ambiguity and overly smooth caused by integral-order, and preserve rich texture and details information. Therefore, the high-quality despeckled images generated by our model will significantly improve the availability of SAR images.

Keywords: Synthetic aperture radar (SAR) · Fractional differential
Despeckling · Image denoising

1 Introduction

Synthetic aperture radar (SAR) is the breakthrough of microwave remote sensing technology, with its all-weather, all-time capabilities. Meanwhile, SAR is capable of producing high-resolution images, and has a high value on the application of ground observation, especially for military use. However, because of using coherent imaging approach, SAR images are contaminated by multiplicative noise, known as the speckle, which adversely affects the performance of extraction of valuable information from SAR images. Hence, to improve the performance of following tasks, such as classification, segmentation, and recognition, it is vital to remove speckle of SAR images efficiently.

Due to the importance of denoising for SAR images, various methods have been developed in the literature in past years. Lee filter that introduces a statistical technique to define a noise model was proposed in 1981 [1]. Besides the local smoothing filters, the non-local means (NL-means) methods use all possible self-predictions to preserve texture and details [2]. PPB (probabilistic patch-based filter) uses the similarity criterion depending on the noise distribution model rather than Euclidean distance [3]. BM3D (block-matching 3D) using inter-patch correlation (NLM) and intra-patch correlation (Wavelet shrinkage) delivers excellent denoising performance [4]. However, some of these approaches are based on statistical models of signal and speckle. And the statistics depending on sensor or acquisition modality may vary in different cases.

D.-S. Huang et al. (Eds.): ICIC 2018, LNAI 10956, pp. 431–442, 2018.
https://doi.org/10.1007/978-3-319-95957-3_46

Influenced by deep learning and fractional calculus theory, we followed the architecture in [5] and proposed a SAR images despeckling model based on CNN and fractional total variation. This model consists of several convolutional layers with batch normalization and rectified linear units (ReLU), and makes use of residual learning strategy to address the issue that performance degrades with the increase of network depth. Here, CNN part is responsible for the pixel level prediction, and fractional total variation part is used to remove artifacts and preserve image texture as prior knowledge.

Meanwhile, the theory of fractional calculus has accomplished extraordinary achievements, through continuous research and development since it has been proposed. Experimental results in [5–8] have shown that, in digital image processing, the integer order is easy to produce staircase effect and lose amounts of texture details, even to lower the quality of images, though it can trade-off the denoising and smoothing somehow. In contrast, the fractional order is better than the integer order, which can nonlinearly keep the details of the image texture to the most significant extent while denoising.

This paper is organized as follows: Sect. 2 introduces the definition of G-L and other related theories. In Sect. 3, we give the details of our approach. Section 4 evaluates our approach through experiments on synthetic SAR images and real SAR images experiments. Moreover, we conclude our works in the last section.

2 Mathematical Background

The purpose of this section is to recall the necessary theoretical background about the application of fractional differential in signal processing and to make this paper self-contained for the readers not familiar with such approach.

Fractional calculus has achieved great development and progress since it has been proposed. There are three classical definition expressions, including Grumwald-Letnikov (G-L), Riemann-Liouville (R-L) and Caputo definition [9]. The G-L definition is mainly applied in signal processing, and R-L definition is mainly used to calculate analytic solutions of some relatively simple functions, while Caputo definition is mainly applicable to engineering fields. We use the G-L definition in our proposed model. The fractional-order differential expression based on G-L definition is as follows:

$$\begin{aligned} D^{v}_{G-L}s(x) &= \frac{d^{v}}{[d(x-a)]^{v}}s(x)\bigg| \\ &= \lim_{N\to\infty}\left\{\frac{\left(\frac{x-a}{N}\right)^{-v}}{\Gamma(-v)}\sum_{k=0}^{N-1}\frac{\Gamma(k-v)}{\Gamma(k+1)}s\left(x-k\left(\frac{x-a}{N}\right)\right)\right\}. \end{aligned} \quad (1)$$

where the duration of the signal $s(x)$ is $[a, x]$, and v is any real number (including a fraction). The G-L definition is directly derived from the integral calculus function, the only needed coefficient in (1) is the discrete sampling of $s(x-k((x-a)/N))$, and the derivative or integral value of the signal $s(x)$ is not required.

Actually, according to G-L definition, we get

$$\frac{d^n}{dx^n}\frac{d^v}{[d(x-a)]^v}s(x)=\frac{d^{n+v}}{[d(x-a)]^{n+v}}s(x). \tag{2}$$

where both v and n are any real numbers.

Without losing generality, suppose $a=0$ in our proposed model, one divides the duration of $s(x)$ into N equal shares. Therefore, the duration belongs to $[a,x]$. (1) can be rewritten as [12–14]

$$\begin{aligned}\left.\frac{d^v}{dx^v}s(x)\right|_{AL-1} &\cong \frac{x^{-v}N^v}{\Gamma(-v)}\sum_{k=0}^{N-1}\frac{\Gamma(k-v)}{\Gamma(k+1)}s\left(x-\frac{kx}{N}\right)\\ &=\frac{x^{-v}N^v}{\Gamma(-v)}\sum_{k=0}^{N-1}\frac{\Gamma(k-v)}{\Gamma(k+1)}s_k.\end{aligned} \tag{3}$$

3 A Fractional Total Variational CNN Approach

In this section, we present the details of our proposed model. Inspired by [5], we use the same network architecture. This network without pooling has eight convolution layers in total and applies the combination of residual learning strategy and batch normalization to guarantee a faster convergence. However, as mentioned above, it could not preserve the details very well when denoising the SAR images. To address this issue, we utilize the advantage of fractional calculus in preserving image details [10]. We add the fractional total variation loss into the loss function to abate various artifacts and retain the details as many as possible.

3.1 Network Architecture

As shown in Fig. 1, Conv, BN, and ReLU are the symbols of convolution, batch normalization and rectified linear unit, respectively. L1 and L8 denote the Conv-BN layers, and L2-L7 stands for the Conv-BN-ReLU layers. For the first layer, 64 filters of size $3 \times 3 \times 1$ are used to generate 64 feature maps and rectified linear units are then utilized for nonlinearity. For L2-L7, 64 filters of size $3 \times 3 \times 64$ are used, and batch normalization is added between convolution and ReLU. For the last layer, one filter of size $3 \times 3 \times 64$ is used to reconstruct the output. After L8, the estimated speckle component is output and then divided by the noisy image. Residual learning formulation is adopted to learn residual image, and batch normalization is incorporated to speed up training as well as boost the denoising performance. At the end of this network, a hyperbolic tangent layer is utilized to serve as a nonlinear function. Meanwhile, appropriate zero-padding is applied to accord the dimension of input with that of the output image in each layer, and the size of the filter is 3 * 3 * 64 except the last layer to get an output image.

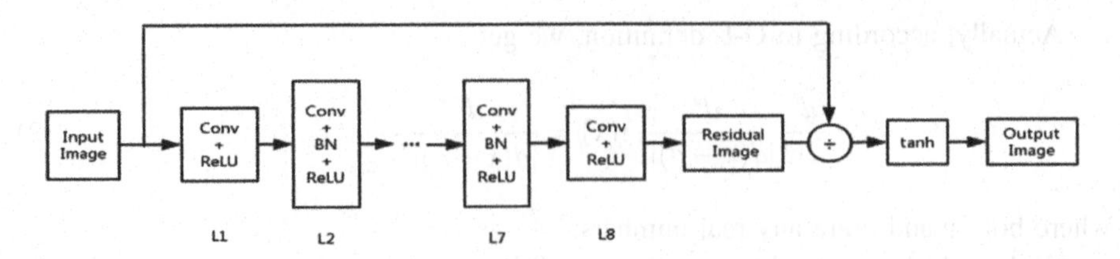

Fig. 1. Network structure of our proposed method

3.2 Fractional Total Variational Loss Function

As the object to optimize, the loss function is of vital importance in deep learning models. Most previous CNN-based image denoising models optimize the Euclidean loss, for a noisy image Y and its ground truth X, the corresponding Euclidean loss is as follows:

$$L_E(\Phi) = \frac{1}{WH}\sum_{w=1}^{W}\sum_{h=1}^{H}\left\|\Phi\left(Y^{w,h}\right) - X^{w,h}\right\|_2^2. \tag{4}$$

where Φ is the learned network after for generating the speckle output, W and H denote the width and the height of image respectively.

To reduce artifacts, an additional total variation loss can be utilized, which is defined as follows, but L1-norm usually is used to reduce the computation cost.

$$L_{TV} = \sum_{w=1}^{W}\sum_{h=1}^{H}\sqrt{\left(\hat{X}^{w+1,h} - \hat{X}^{w,h}\right)^2 + \left(\hat{X}^{w,h+1} - \hat{X}^{w,h}\right)^2}. \tag{5}$$

Where $\hat{X}$ denotes the output image after network procedure. However, experiments show that the first-order TV loss cannot preserve the texture very well. Inspired by fractional calculus theories [9], we change the integral TV loss to fractional TV loss to improve the performance of the model.

$$L_{FTV} = \sum_{w=1}^{W}\sum_{h=1}^{H}\left(\left|D_{x-}^{v}\hat{X}\right| + \left|D_{y-}^{v}\hat{X}\right| + \left|D_{x+}^{v}\hat{X}\right| + \left|D_{y+}^{v}\hat{X}\right| + \left|D_{LDD}^{v}\hat{X}\right| + \left|D_{RUD}^{v}\hat{X}\right| + \left|D_{LUD}^{v}\hat{X}\right| + \left|D_{RDD}^{v}\hat{X}\right|\right). \tag{6}$$

To control the diffusion speed of 8 directions, we apply the fractional differential mask in 8 directions. The final loss function is defined as follows:

$$\text{Loss} = L_E + \lambda_{FTV}L_{FTV}. \tag{7}$$

Where L_E is responsible for the per-pixel accuracy and L_{FTV} is related to artifacts removal and the overall image texture retention. λ_{FTV} is the weight of L_{FTV} and controls the importance of fractional TV loss in the optimization procedure, which generally greater than 0.

3.3 Numerical Implementation of Fractional Derivative

Considering neighboring image pixels have a certain relevance, we select the fractional derivative in eight directions to keep texture and details, reducing the staircase effect caused by integral-order.

The proposed method uses YiFeiPU-2 in 7 * 7 to construct the fractional differential mask respectively on the eight symmetric directions [11]. Through taking three neighboring nodes to do Lagrange 3-point interpolation and extend to a 2-D image signal, we get the numerical realization of fractional derivative in 8 directions.

The fractional-order differential mask operators in 8 directions are symmetric, correspondingly noted by W_x^-, W_y^-, W_{LDD}, W_{RUD} [12–14], shown in Fig. 2.

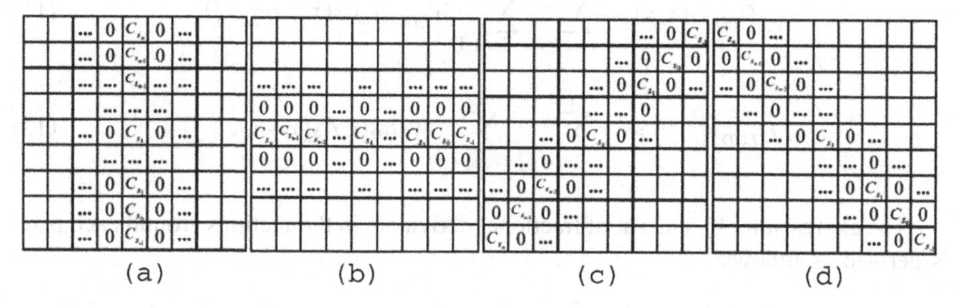

Fig. 2. Fractional differential operator on 4 directions (a) W_x^-. (b) W_y^-. (c) W_{LDD}. (f) W_{RUD}.

Take Fig. 2(a) for example, C_{s_0} is the mask coefficient on interest pixel and the fractional differential mask is $(2m+1) \times (2m+1)$, $n = 2m - 1$. The coefficient of the fractional differential mask is given by

$$\begin{cases} C_{s_{-1}} = \frac{v}{4} + \frac{v^2}{8} \\ C_{s_0} = 1 - \frac{v^2}{2} - \frac{v^3}{8} \\ \cdots \\ C_{s_k} = \frac{1}{\Gamma(-v)} \left[\frac{\Gamma(k-v+1)}{(k+1)!} \cdot \left(\frac{v}{4} + \frac{v^2}{8} \right) + \frac{\Gamma(k-v)}{k!} \cdot \left(1 - \frac{v^2}{4} \right) \right. \\ \left. + \frac{\Gamma(k-v-1)}{(k-1)!} \cdot \left(-\frac{v}{4} + \frac{v^2}{8} \right) \right] \\ \cdots \\ C_{s_{n-2}} = \frac{1}{\Gamma(-v)} \left[\frac{\Gamma(n-v+1)}{(n-1)!} \cdot \left(\frac{v}{4} + \frac{v^2}{8} \right) + \frac{\Gamma(n-v-2)}{(n-2)!} \cdot \left(1 - \frac{v^2}{4} \right) \right. \\ \left. + \frac{\Gamma(n-v-3)}{(n-1)!} \cdot \left(-\frac{v}{4} + \frac{v^2}{8} \right) \right] \\ C_{s_{n-1}} = \frac{\Gamma(n-v-1)}{(n-1)!\,\Gamma(-v)} \cdot \left(1 - \frac{v^2}{4} \right) + \frac{\Gamma(n-v-2)}{(n-2)!\,\Gamma(-v)} \cdot \left(-\frac{v}{4} + \frac{v^2}{8} \right) \\ C_{s_n} = \frac{\Gamma(n-v-1)}{(n-1)!\,\Gamma(-v)} \cdot \left(-\frac{v}{4} + \frac{v^2}{8} \right) \end{cases} \tag{8}$$

As for $nx * ny$ digital gray image $s(x, y)$, we do convolutional filter respectively in the above 8 symmetric directions. The calculation processes of 4 directions are as follows:

$$D_{x^-}^{v} s(x,y) = \sum_{i=-2m+1}^{1} \sum_{j=-m}^{m} W_x^-(i,j) s(x+i, y+j) \tag{9}$$

$$D_{y^-}^{v} s(x,y) = \sum_{i=-m}^{m} \sum_{j=-2m+1}^{1} W_y^-(i,j) s(x+i, y+j) \tag{10}$$

$$D_{LDD}^{v} s(x,y) = \sum_{i=-1}^{2m-1} \sum_{j=-2m+1}^{1} W_{LDD}(i,j) s(x+i, y+j) \tag{11}$$

$$D_{LUD}^{v} s(x,y) = \sum_{i=-2m+1}^{1} \sum_{j=-2m+1}^{1} W_{LUD}(i,j) s(x+i, y+j) \tag{12}$$

The expression of L1-norm of fractional derivative in 8 directions around each pixel is derived as follows:

$$\begin{aligned} \|D^{v} s(x,y)\|_1 = {} & \left|D_{x^-}^{v} s(x,y)\right| + \left|D_{y^-}^{v} s(x,y)\right| + \left|D_{x^+}^{v} s(x,y)\right| + \left|D_{y^+}^{v} s(x,y)\right| \\ & + \left|D_{LDD}^{v} s(x,y)\right| + \left|D_{RUD}^{v} s(x,y)\right| + \left|D_{LUD}^{v} s(x,y)\right| + \left|D_{RDD}^{v} s(x,y)\right| \end{aligned} \tag{13}$$

4 Experiment and Analysis

In this section, we will present the results of our experiments. We applied the FID-CNN algorithm to both synthetic and real SAR images. Before implementing the two experiments, we are going to perform an ablation study to compare the difference between the integral total variation and the fractional total variation. We can see the effect of the properties of fractional calculus – preserving texture details [15]. In the synthetic experiment, we will present the results of different orders and obtain the order with the best performance. In real SAR image experiment, we use the real SAR image to compare the performance of ID-CNN and FID-CNN.

The ID-CNN and FID-CNN share the same dataset – NWPU-RESISC45, scraped Google Maps images [16]. We collect more "medium_residential" type than other types in our dataset, which contain more texture details. The number of the dataset is 4294 that is approximately equal to the number of the corresponding paper. Sample images in the dataset are shown in Fig. 3.

All the images are resized to 256 * 256. We add speckle noise to these images to get synthetic images. Noise obeys the gamma distribution with unit mean and 0.125 variance.

Fig. 3. Sample images used to train our network.

We set the parameters as suggested in the corresponding paper to get the best performance in ID-CNN experiment. We adopt the network structure proposed in [15], and the structure is same as the ID-CNN. However, we change the loss function and make it better in preserving texture details. The network is trained by the adaptive moment estimation optimization method (ADAM) [16], with 16 batch-size and learning rate of 0.0002. During training, the parameter of total variation 8 is 0.0025.

4.1 Ablation Study

We demonstrate the difference between FID-CNN and ID-CNN. In ID-CNN, it uses the Euclidean loss and integral TV term in loss function. The TV term is used to remove unwanted artifacts and get better performance. However, the drawback of adding TV term is that many texture details may be lost. We proposed the FID-CNN to solve this problem. We take advantage of the property of fractional calculus to preserve the texture details. The results of ID-CNN and FID-CNN are shown in Fig. 4. This can be clearly seen by comparing the zoomed in patches shown at the top right corner of these images. The result of FID-CNN has more details. The experiments clearly show the significance of using the fractional total variation.

Fig. 4. Sample results of ID-CNN and the proposed FID-CNN (a) ID-CNN with integral TV loss and (b) FID-CNN with fractional TV loss (0.5 order).

4.2 Results on Synthetic Images

The experimental results on synthetic SAR image were presented in Fig. 5. The peak signal noise ratio (PSNR), structural similarity index (SSIM), universal quality (UQI) are used to measure the denoising performance of different methods. The corresponding results are shown in Table 1. The performance in [5] is better than existing despeckling methods, such as Lee filter [1], Kuan filter [17], probabilistic patch-based filter [3], SAR-BM3D [18], CNN [19] and SAR-CNN [20]. In our experiment, we introduced the fractional-order total variation into the loss function. In terms of long-term memory, non-locality, and weak singularity, fractional differential of an image can preserve the low-frequency contour feature in the smooth area, and non-linearly keep high-frequency edge information and texture information in those areas where graysscale changes frequently, and as well as, non-linearly enhance texture details in those areas where grayscale does not change evidently [10, 12, 21–25].

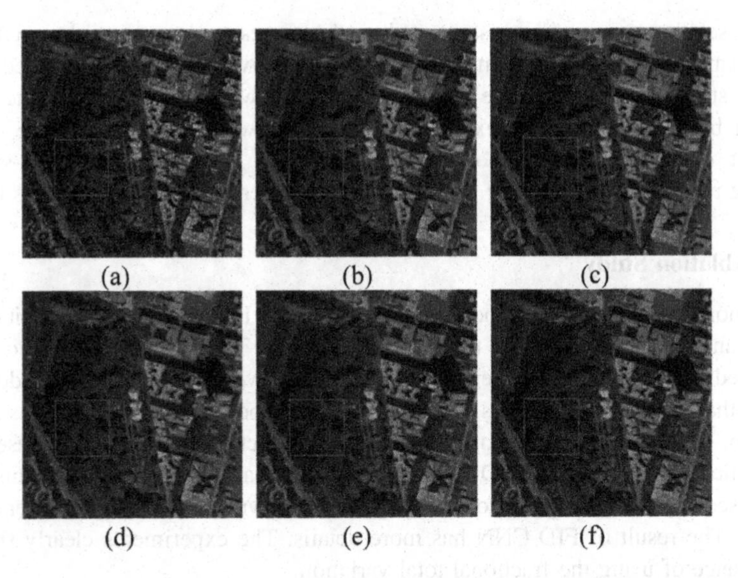

Fig. 5. (a) noisy image, (b) 0.5-order TV, (c) 0.75-order TV, (d) 1-order (integral-order) TV, (e) 1.25-order TV, (e) 1.5-order TV

Table 1. Quantitative results for various experiments on synthetic images with 0.0625 variance

	ID-CNN	v = 0.5	v = 0.75	v = 1.25	v = 1.5
PSNR	25.589	**25.754**	25.657	25.564	25.584
SSIM	0.942	**0.945**	0.943	0.942	0.942
UQI	0.941	**0.944**	0.942	0.941	0.941

From [12–14], we know that when $0 \prec \nu \prec 1$, fractional differential will enhance texture details. We perform the experiment on synthetic images, and the noise is unit mean and 0.0625 variance. From Fig. 5, the use of fractional total variation has largely enhanced the despeckling performance compared to ID-CNN. Table 1 has shown that when $v = 0.5, 0.75$, the PSNR is higher than ID-CNN, that's to say, the performance of our method is better than ID-CNN. Therefore, the combination of CNN and fractional total variation can reduce the speckle in SAR images to the largest extent meanwhile preserving the complex texture and details information and output high-quality despeckled images.

We also try the other parameter of v. When $v < 0$, the additional total variation becomes an integration instead of differential, and the performance is not satisfying. And when v is closer to 0, the total variation term is closer to 1, without obvious effect on the loss. So we choose the order beyond 0.5 as the parameter of our experiment.

The fractional differential of neighboring v is similar. Namely, it is the continuous interpolation of neighboring order fractional differential [14]. Thus, as for the SAR images, even if some textures have only a few pixels that have been known, the fractional differential can also enhance the texture details. Figures 5 and 6 are the experimental results with different fractional-order. From Fig. 5 and Table 1, it is evident that the best-despeckled performance can be gained when v = 0.5, preserving much more texture and details.

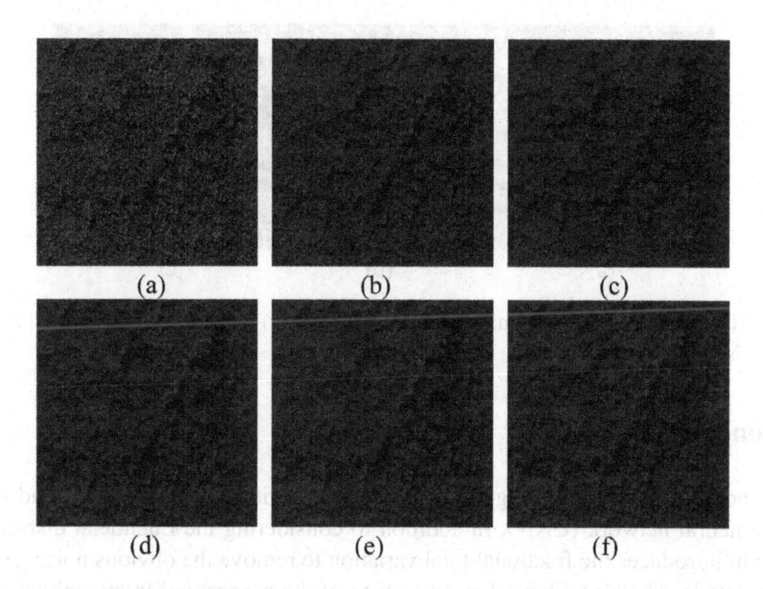

Fig. 6. (a) noisy image, (b) 0.5-order TV, (c) 0.75-order TV, (d) 1-order TV (integral-order), (e) 1.25-order TV, (f) 1.5-order TV

Furthermore, we perform comparative experiments on forest images with unit mean and 0.125 variance to measure the capacity of preserving details of FID-CNN with different level noisy. From Fig. 6 and Table 2, we can get the results that 0.5 order total variation have more details than integral-order and have better results than integral-order.

Table 2. Quantitative results for various experiments on synthetic images with 0.125 variance

	ID-CNN	v = 0.5	v = 0.75	v = 1.25	v = 1.5
PSNR	11.854	**12.143**	11.995	12.066	12.110
SSIM	0.328	**0.399**	0.361	0.351	0.354
UQI	0.318	**0.391**	0.352	0.342	0.344

4.3 Results on Real SAR Images

We use the real SAR images from the track of HJ-1C. Without the clean image, we can't know the exact difference between the target image and output image. However, we can make sure that the results of ID-CNN and FID-CNN both get good performance in denoising, with 24.549 and 24.679 of PSNR respectively. Both two algorithms remove the noise of trees and lake and make the surface of lake clean in Fig. 7. However, FID-CNN has better performance in preserving texture derails with higher PSNR, SSIM and UQI than ID-CNN. More specifically, the margin of the lake of this image is clearer. It can be clearly seen by comparing the bottom left corner of these images shown in Fig. 7.

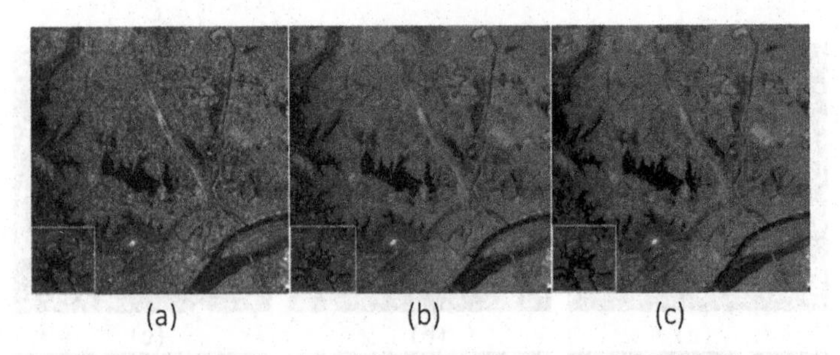

Fig. 7. The results on real SAR image (a) original image (b) the result of ID-CNN (c) the result of FID-CNN

5 Conclusion

We proposed the FID-CNN algorithm based on fractional total variation and convolutional neural network (CNN). In addition to considering the Euclidean distance, the algorithm introduces the fractional total variation to remove the obvious noise and keep texture details. The algorithm takes advantage of the property of fractional calculus to increases marginal information. Results on synthetic and real SAR images show promising qualitative and quantitative results.

References

1. Lee, J.-S.: Speckle analysis and smoothing of synthetic aperture radar images. Comput. Graph. Image Process. **17**(1), 24–32 (1981)
2. Buades, A., Coll, B., Morel, J.: A non-local algorithm for image denoising. Presented at the IEEE Computer Society Conference on Computer Vision and Pattern Recognition, vol. 2 (2005)
3. Deledalle, C.-A., Denis, L., Tupin, F.: Iterative weighted maximum likelihood denoising with probabilistic patch-based weights. IEEE Trans. Image Process. **18**(12), 2661–2672 (2009)
4. Dabov, K., Foi, A., Katkovnik, V., Egiazarian, K.: Image denoising with block-matching and 3D filtering. In: Proceedings of SPIE, vol. 6064, Art. no. 606414 (2006)
5. Wang, P., Zhang, H., Patel, V.M.: Sar image despeckling using a convolutional neural network. IEEE Signal Process. Lett. **24**(12), 1763–1767 (2017)
6. Jain, V., Seung, H.S.: Natural image denoising with convolutional networks. In: International Conference on Neural Information Processing Systems, pp. 769–776. Curran Associates Inc (2008)
7. Burger, H.C., Schuler, C.J., Harmeling, S.: Image denoising: can plain neural networks compete with BM3D? In: Computer Vision and Pattern Recognition, vol. 157, pp. 2392–2399. IEEE (2012)
8. Xie, J., Xu, L., Chen, E.: Image denoising and inpainting with deep neural networks. In: International Conference on Neural Information Processing Systems, vol. 1, pp. 341–349. Curran Associates Inc (2012)
9. Oldham, K.B., Spanier, J.: The fractional calculus: theory and applications of differentiation and integration to arbitrary order (1974)
10. Pu, Y.F., Zhou, J.L., Yuan, X.: Fractional differential mask: a fractional differential-based approach for multiscale texture enhancement. IEEE Trans. Image Process. **19**(2), 491–511 (2010)
11. Pu, Y.F.: Fractional-order Euler-Lagrange equation for fractional-order variational method: a necessary condition for fractional-order fixed boundary optimization problems in signal processing and image processing. IEEE Access **4**, 10110–10135 (2016)
12. Pu, Y.: Fractional calculus approach to texture of digital image. In: 2006 8th International Conference on Signal Processing, vol. 2. IEEE, November 2006
13. Pu, Y.F.: Fractional differential filter of digital image. Invention Patent of China, No. ZL200610021702, 3 (2006)
14. Pu, Y., Wang, W., Zhou, J., Wang, Y., Jia, H.: Fractional differential approach to detecting textural features of digital image and its fractional differential filter implementation. Sci. China Ser. F Inf. Sci. **51**(9), 1319–1339 (2008)
15. Pu, Y., Siarry, P., Zhou, J., Liu, Y., Zhang, N., Huang, G., Liu, Y.: Fractional partial differential equation denoising models for texture image. Sci. China Inf. Sci. **57**(7), 1–19 (2014)
16. Isola, P., Zhu, J.Y., Zhou, T., Efros, A.A.: Image-to-image translation with conditional adversarial networks. arXiv preprint (2017)
17. Kuan, D.T., Sawchuk, A.A., Strand, T.C., Chavel, P.: Adaptive noise smoothing filter for images with signal-dependent noise. IEEE Trans. Pattern Anal. Mach. Intell. **2**, 165–177 (1985)
18. Parrilli, S., Poderico, M., Angelino, C.V., Verdoliva, L.: A nonlocal SAR image denoising algorithm based on LLMMSE wavelet shrinkage. IEEE Trans. Geosci. Remote Sens. **50**(2), 606–616 (2012)

19. Zhang, K., Zuo, W., Chen, Y., Meng, D., Zhang, L.: Beyond a gaussian denoiser: residual learning of deep CNN for image denoising. IEEE Trans. Image Process. **26**(7), 3142–3155 (2017)
20. Chierchia, G., Cozzolino, D., Poggi, G., Verdoliva, L.: SAR image despeckling through convolutional neural networks. arXiv preprint arXiv:1704.00275 (2017)
21. Pu, Y.F.: Research on application of fractional calculus to latest signal analysis and processing. Sichuan University, pp. 4–9 (2006)
22. Pu, Y.F., Wang, W.X.: Fractional differential masks of digital image and their numerical implementation algorithms. Acta Automatica Sinica **33**(11), 1128–1135 (2007)
23. Yi-Fei, P.U.: Application of fractional differential approach to digital image processing. J. Sichuan Univ. (Eng. Sci. Ed.) **3**, 022 (2007)
24. Xu, M., Yang, J., Zhao, D., Zhao, H.: An image-enhancement method based on variable-order fractional differential operators. Bio-Med. Mater. Eng. **26**(s1), S1325–S1333 (2015)
25. Chen, D.L., Zheng, C., Xue, D.Y., Chen, Y.Q.: Non-local fractional differential-based approach for image enhancement. Res. J. Appl. Sci. Eng. Technol. **6**(17), 3244–3250 (2013)

Contrast Enhancement of RGB Color Images by Histogram Equalization of Color Vectors' Intensities

Farid García-Lamont[1(✉)], Jair Cervantes[1], Asdrúbal López-Chau[2], and Sergio Ruiz[1]

[1] Centro Universitario UAEM Texcoco, Universidad Autónoma del Estado de México, Texcoco-Estado de México, Mexico
fgarcial@uaemex.mx, chazarral7@gmail.com, jsergioruizc@gmail.com

[2] Centro Universitario UAEM Zumpango, Universidad Autónoma del Estado de México, Zumpango-Estado de México, Mexico
alchau@uaemex.mx

Abstract. The histogram equalization (HE) is a technique developed for image contrast enhancement of grayscale images. For RGB (Red, Green, Blue) color images, the HE is usually applied in the color channels separately; due to correlation between the color channels, the chromaticity of colors is modified. In order to overcome this problem, the colors of the image are mapped to different color spaces where the chromaticity and the intensity of colors are decoupled; then, the HE is applied in the intensity channel. Mapping colors between different color spaces may involve a huge computational load, because the mathematical operations are not linear. In this paper we present a proposal for contrast enhancement of RGB color images, without mapping the colors to different color spaces, where the HE is applied to the intensities of the color vectors. We show that the images obtained with our proposal are very similar to the images processed in the HSV (Hue, Saturation, Value) and L*a*b* color spaces.

Keywords: Color characterization · Histogram equalization · RGB images

1 Introduction

Color image processing has not been studied so exhaustive, relatively, because the color representation demands significant computational resources [1, 2]; but given the technological advances, the development of techniques for color image processing have been incentivized [3–7]. Several of the techniques employed for color image processing are extended versions of techniques designed for grayscale images [8–12]. But they do not always success because the nature of the chromaticity data is not considered; such techniques focus on to process, mainly, the intensity of colors. Hence, it is necessary to develop new techniques or update the current techniques to process adequately the color.

There are several models to represent colors, but the RGB space is the most employed because the image acquisition hardware uses this space to represent colors.

D.-S. Huang et al. (Eds.): ICIC 2018, LNAI 10956, pp. 443–455, 2018.
https://doi.org/10.1007/978-3-319-95957-3_47

Despite the RGB space is accepted by most of the image processing community to represent colors, such space is not suitable for color processing because the color differences cannot be computed using the Euclidean distance [13, 14].

One of the classic techniques for image processing is the HE for contrast enhancement. Usually, for RGB color images, the HE is applied in the three color channels separately [15, 16], but given the high correlation between the color channels, the chromaticity of colors is modified.

Different works overcome this problem by mapping the colors from the RGB space to a different color space where the intensity is decoupled from the chromaticity, such as the HSV and L*a*b* spaces; the HE is performed in the intensity channel and finally the colors are mapped back to the RGB space [17–19]. However, the computational load of color mapping between the RGB space and other color spaces may be high because most of the mathematical operations are not linear [1].

Therefore, the contribution of this work is a proposal of how to apply the HE for contrast enhancement of RGB color images using the colors' intensities, without mapping the colors to other color spaces and without suffering undesired chromaticity changes. The intensities of the colors are obtained by computing the magnitudes of the color vectors, creating an intensity channel; on this intensity channel the HE is applied, then, the magnitudes of the color vectors are updated. The resulting images are very similar to the images obtained if a different color space is employed to process the image.

The rest of the paper is organized as follows: in Sect. 2 the histogram equalization technique and the features of the RGB space are presented, but also, our proposal of contrast enhancement of color images is introduced. The experiments performed and the resulting images are shown in Sect. 3; but also, we compare the images obtained with our approach with respect to the images obtained using the well-known color spaces HSV and L*a*b*, where the HE is applied in their respective intensity channels. In Sect. 4 the results are discussed and their respective accumulative intensity histograms are analyzed. Conclusions in Sect. 5 close the paper.

2 Proposed Approach

The HE is an image processing technique developed for grayscale images to process the intensity of the pixels; the goal of the HE is that all the intensity levels are employed and also, all the intensity levels have the same occurrences.

For color images, the HE is applied to the intensity channel of the color space employed to represent colors. There are different color spaces where the intensity of colors is decoupled from the chromaticity. As we state before, the RGB space is often employed to model colors because the image acquisition hardware employs this space to represent colors. Thus, if the colors are planned to be processed in a different color space, then the colors must be mapped to the desired space before processing; it implies to perform, most of them, non-linear mathematical operations that may represent an important computational load. Because of space constraints within the paper, we do not show such non-linear operations, but the reader can find them in reference [1].

We claim that the contrast of RGB color images can be enhanced using the HE technique on the intensity of colors, without mapping the colors to other space; but it is important to consider the disadvantages of such space for color processing. For instance, the intensity is not decoupled from the chromaticity; that is, there is not an intensity channel. However, it is possible to build the intensity channel by computing the magnitudes of the color vectors, and then to apply the HE technique to such intensity channel, without modifying the chromaticity of the colors.

Our proposal consists on the following steps:

1. Build the intensity channel with the magnitudes of the color vectors,
2. All the color vectors are normalized,
3. The HE is applied to the intensity channel,
4. All the normalized vectors are multiplied by the corresponding scalar value obtained after HE.

Next, in this section, we explain in detail the mathematical operations of the proposed approach; but also, we give a brief explanation about the color representation in the RGB space and their features, and the HE technique.

2.1 RGB Color Space

The RGB space is based in a Cartesian coordinate system where colors are points defined by vectors that extend from the origin, where black is located in the origin and white in the opposite corner to the origin, see Fig. 1.

The color of a pixel p is written as a linear combination of the basis vectors red, green and blue, that is:

$$\phi_p = r_p\hat{i} + g_p\hat{j} + b_p\hat{k}. \tag{1}$$

Where r_p, g_p and b_p are the red, green and blue components, respectively. The colors have two features: the chromaticity or hue, and the intensity or brightness. It is important to remark that the orientation and magnitude of a color vector defines the chromaticity and the intensity of the color, respectively [1].

In order to explain the difference between chromaticity and intensity, consider the Fig. 2, we state the color of squares (a) and (b) of Fig. 2 is green because both squares have the same chromaticity although the square (a) is brighter than square (b); therefore, the color vectors of both squares have the same orientation but with different

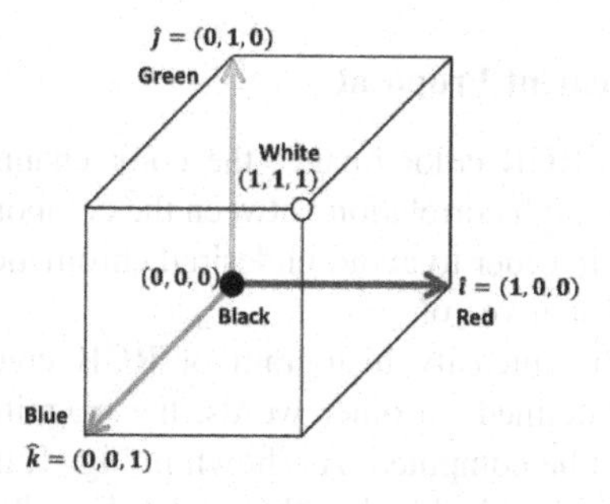

Fig. 1. RGB color space (Color figure online)

magnitude. On the other hand, we claim the colors of squares (c) and (d) are different because the chromaticity of both squares are different despite the intensities are the same; that is, the color vectors have different orientation but with the same magnitude.

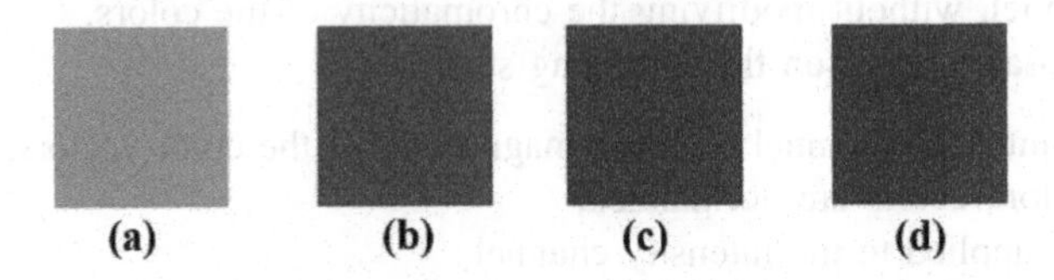

Fig. 2. Color of squares (a) and (b) with the same chromaticity but with different intensities; color of squares (c) and (d) with different chromaticity but with the same intensity (Color figure online)

2.2 Histogram Equalization Technique

The HE is an image processing technique for contrast enhancement of images. The purposes of the HE are [1]:

1. employ all the intensity levels,
2. distribute the number of intensities in all the pixels of the image, in other words, all the intensity levels have the same occurrences within the image.

The HE is computed with the following equation [1]:

$$\lambda(k) = \frac{L-1}{N}\sum\nolimits_{i=0}^{k} h(i). \tag{2}$$

Where $h(i)$ is the occurrence of intensity level i of the image to process, L is the number of intensity levels, N is the number of pixels of the image, $k = I(x,y)$ is the intensity value of the pixel located at (x,y) of the image I, and $\lambda(k)$ is the value of the equalized intensity level k.

This technique is developed for grayscale images, so, usually $L = 256$; for RGB color images, the pixel's intensity is represented by the magnitude of the vector that characterizes the pixel's color. Thus, the number of intensity levels is different; the value of L is computed as we explain in the following Sect. 2.3.

2.3 Contrast Enhancement Proposal

As mentioned before, in RGB color images the color channels cannot be processed separately; because of the high correlation between the components, the chromaticity of colors may be modified. In order to avoid undesired chromaticity changes, the color of every pixel is processed as a vector.

So as to equalize the intensity histogram of RGB color images, the range of intensity levels must be defined; in other words, the magnitudes of the smallest and largest color vectors must be computed. As shown in Fig. 1, the vector with the lowest magnitude corresponds to the black color; that is, let $\phi_b = [0,0,0]$ be the color vector of black color, therefore $||\phi_b|| = 0$. While the largest vector represents the white color, this vector extends from the origin to the opposite corner of the cube, see Fig. 1. Considering that the usual range of the red, green and blue components is $[0, 255]$, the vector representing white color is $\phi_w = [255, 255, 255]$, therefore $||\phi_w|| = 255\sqrt{3}$. So, the range of the intensity values is $\left[0, 255\sqrt{3}\right]$.

Note that L of Eq. (2) is an integer number, and the highest intensity value is $255\sqrt{3} \approx 441.673$; hence, only the integer part of this number is employed, thus $L = 441$. The magnitude of the color vectors can be modified by multiplying the respective scalar value obtained after the HE. Before this mathematical operation is performed, the color vectors must be normalized; in this way all the vectors have the same intensity without modifying their orientation that represents the chromaticity. In other words, our proposal consists on performing the following mathematical operations:

1. Let $\Phi = \{\phi_1, \ldots, \phi_N\} \subset \mathbb{R}^3$ be the set of color vectors of a given image; all the color vectors are normalized with $\hat{\phi}_i = \phi_i / ||\phi_i||$.
2. The intensity histogram $h(i)$ of the image is computed as follows

$$h(i) = \#\{k | f(||\phi_k||) = i, \forall \phi_k \in \Phi\}. \tag{3}$$

   Where # denotes the cardinality of the set, $f : \mathbb{R} \to \mathbb{Z}$ is a function that extracts the integer part of a real number.
3. Equalize the histogram using Eq. (2), where $L = 441$.
4. The normalized color vectors are multiplied by their respective scalar value computed with Eq. (2). That is:

$$\varphi_j = \lambda(k) \cdot \hat{\phi}_j, j = 1, \ldots, N. \tag{4}$$

   Where $k = f\left(\left\|\phi_j\right\|\right)$, $\lambda(k)$ is computed with Eq. (2) and φ_j is the color vector of the resulting image.
5. Replace the color vector ϕ_j by the color vector φ_j.

With this proposal the contrast of the RGB color image is enhanced without modifying the chromaticity of the colors. In Sect. 3 we present the experiments and the resulting images using our approach.

3 Experiments and Results

In this section we show the images obtained using our proposal. In previous works where the images are processed using other color spaces, the histogram of the intensity channel is equalized and then the image is mapped back to the RGB space.

We compare the images obtained with our approach and the images processed in the HSV and L*a*b* color spaces, because these color spaces are often employed for color processing [1]; but also, we show the images obtained by equalizing the histogram of each color channel of RGB color images. First column of Fig. 3 shows the input images. The images selected for experiments are extracted from the Berkeley segmentation database (BSD) of the Berkeley computer vision group[1], which is becoming the standard benchmark to test color image processing algorithms; the BSD is a database of natural images that contains 500 color images of size 481 × 321 pixels; because of space constraints within the paper, we have selected 6 images from the BSD, but we consider the resulting images show evidence our proposal is worthy.

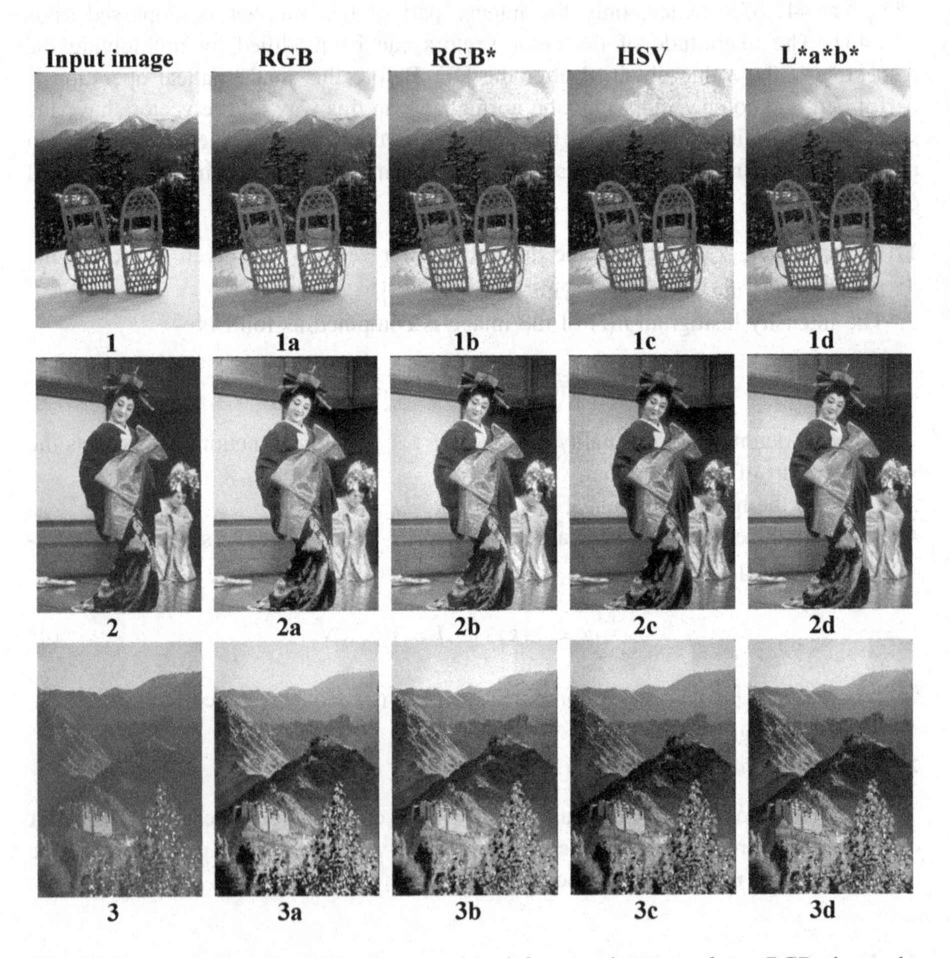

Fig. 3. First column shows the images employed for experiments; column RGB shows the images obtained by processing them with our proposal; column RGB* show the resulting images by applying the HE to each color channel; columns HSV and L*a*b* show the images obtained by using HE to the intensity channels of the respective color space. Input images and resulting images

[1] www2.eecs.berkeley.edu/Research/Projects/CS/vision/bsds/.

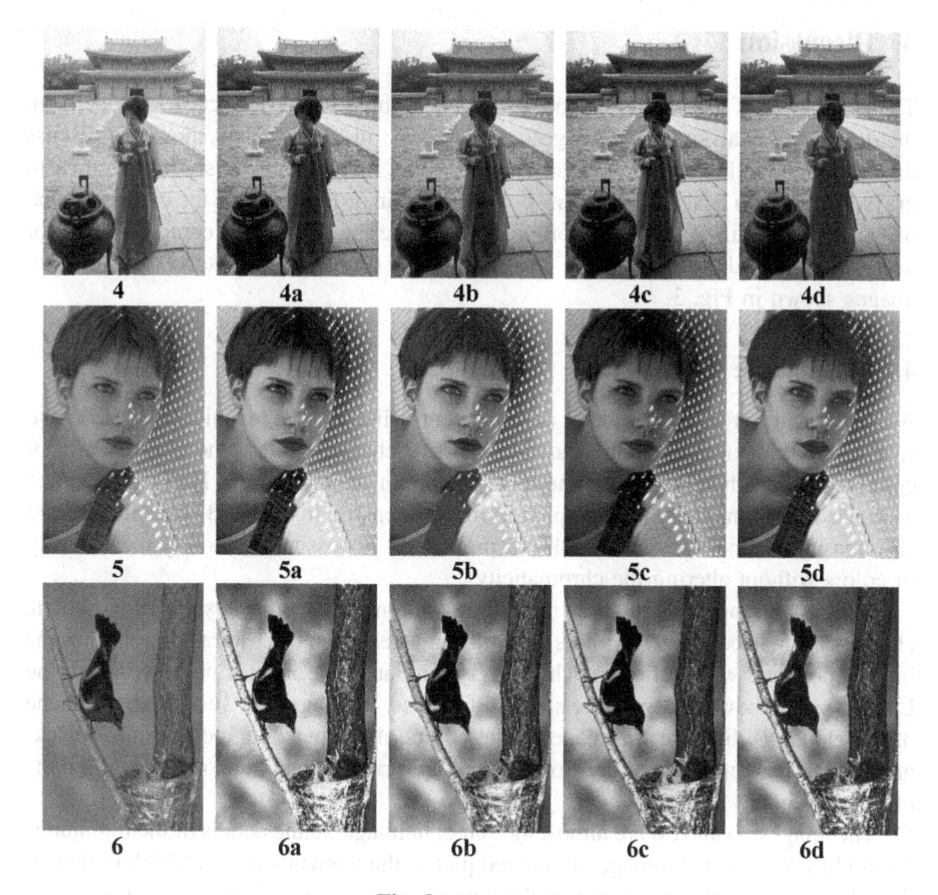

Fig. 3. *(continued)*

Columns RGB, RGB*, HSV and L*a*b* of Fig. 3 shows the images obtained by processing the input images. The column RGB shows the resulting images using our proposal; the column RGB* shows the images obtained by equalizing the histogram of each color channel separately; columns HSV and L*a*b* show the resulting images by applying HE in the intensity channels of the HSV and L*a*b* spaces, respectively.

It is easy to appreciate that the colors of the images of the column RGB* suffered chromaticity changes; perhaps, the most notable are images 2b, 3b, 5b and 6b. The images obtained with our proposal are very similar to the resulting images of the columns HSV and L*a*b*. The intensity ranges of the images processed using the HSV and L*a*b* color spaces are $[0, 255]$ and $[0, 100]$, respectively.

4 Discussion

The discussion of the results has been divided in three parts: in Sect. 4.1 we perform the qualitative analysis of the images' appearances; in Sect. 4.2 we show and discuss the accumulative histogram (AH) of each resulting images, previously transformed to grayscale, so as to analyze the contrast objectively; in order to reduce the computational of our approach, in Sect. 4.3 we propose to change the formula to compute the color vectors' magnitude, the resulting images change slightly with respect to the input images shown in Fig. 3.

4.1 Qualitative Evaluation

As mentioned before, the contrast enhancement using the HE technique for HSV and L*a*b* color images is applied to the intensity channel, where the chromaticity of colors does not change. One of the most important features of the HSV and L*a*b* spaces is the chromaticity is decoupled from the intensity. In the RGB space there is not an intensity channel; however, with our proposal it is possible to modify the intensity of colors without altering the chromaticity.

It is easy to appreciate that the images of column RGB*, Fig. 3, suffered chromatic changes in their colors. The colors of all the images are more saturated; note that the images obtained with our approach are very similar to the images processed with the L*a*b* space; however, there are small differences. In images 1a, 1c and 1d, the shape of the clouds is different. In image 2a the colors of the person's clothes of the background and person's face of the front ground are brighter than the corresponding parts of images 2c and 2d.

The images 3a and 3d are almost the same; in image 3c the details of the mountains are slightly less clear. In image 4d, the red part of the woman's dress is brighter than in the other images obtained; note that the house's roof in the background is clearer than in the original image.

The images 5a and 5c are the most similar between them; the image 5d is the brightest, it is easy to appreciate it in the hair, clothes and lips of the woman. Between the images 6a, 6c and 6d the differences are minimal; in the image 6a the branches where the bird and the nest are set, are darker than in the original images; besides, the intensity differences in the background are more notable.

4.2 Accumulative Histogram Analysis

In grayscale images, the AH of intensities is employed to analyze the contrast of the images. The shape of an image's AH with contrast enhanced is a straight line at 45°, approximately. For color images, it is not possible to compute the AH because the number of intensity levels of the color spaces is different between them. As we state before, the number of intensity levels is 441 with our proposal, while for HSV and L*a*b* spaces the number of intensity levels is 256 and 101, respectively.

However, it is possible to analyze the AH of the images if all the resulting images are converted to grayscale. From Figs. 4, 5, 6, 7, 8 and 9 the AHs of the respective images obtained in Fig. 3 are shown; the histograms are obtained by converting

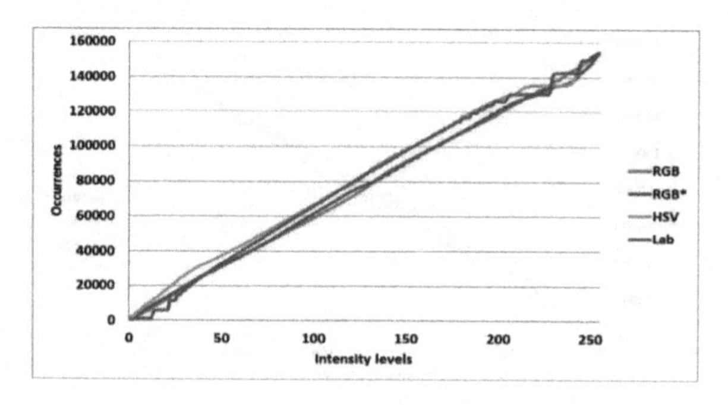

Fig. 4. Accumulative histogram of image 1

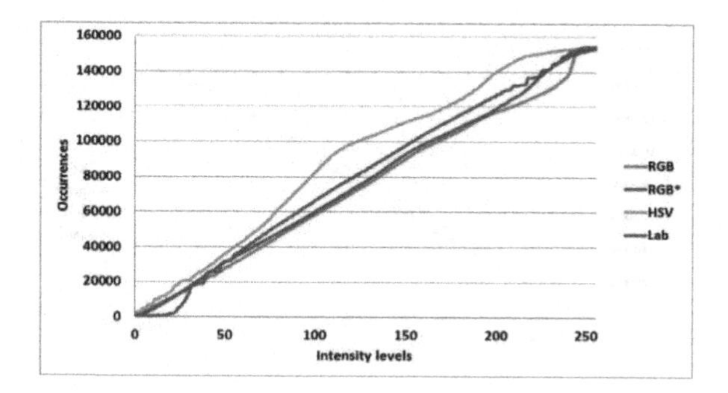

Fig. 5. Accumulative histogram of image 2

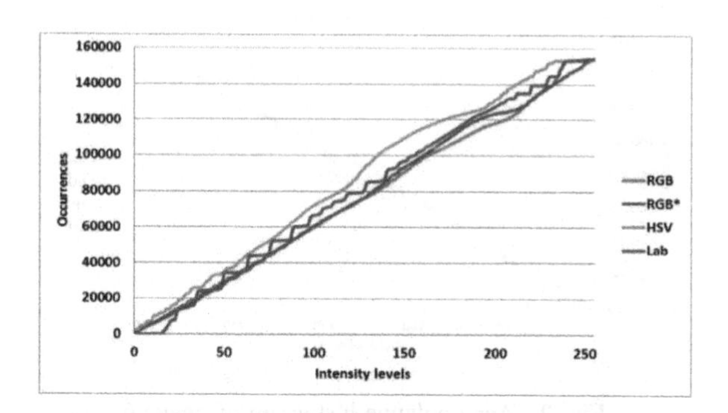

Fig. 6. Accumulative histogram of image 3

previously the images to grayscale. The blue lines represent the AHs of the images obtained with our proposal; the lines in red, green and purple represent the AHs of the images shown in the columns RGB*, HSV and L*a*b* of Fig. 3, respectively.

The AH obtained from the image processed with our proposal, shown in Fig. 4, is the straightest from the other AHs; the AHs of the other images have several peaks in the highest intensity levels.

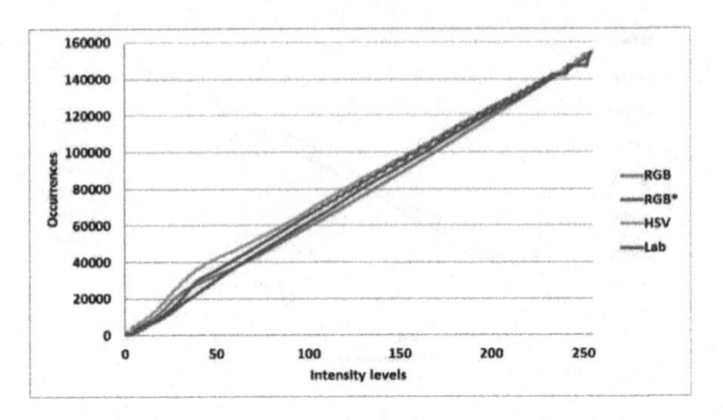

Fig. 7. Accumulative histogram of image 4

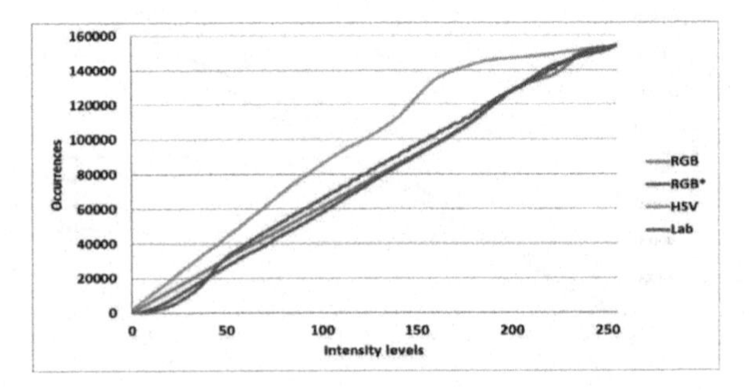

Fig. 8. Accumulative histogram of image 5

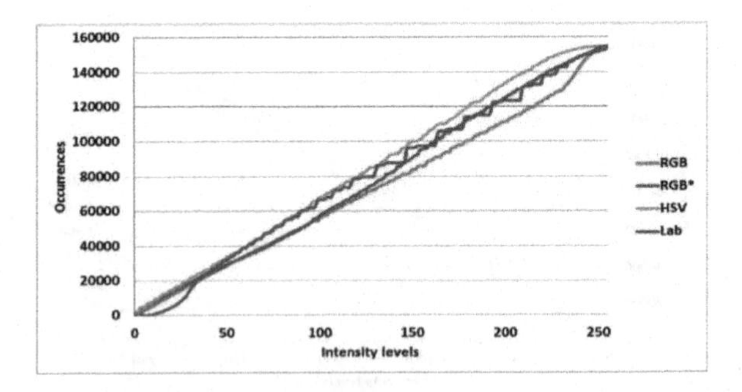

Fig. 9. Accumulative histogram of image 6

In Fig. 5 the straightest AHs are obtained with the images processed with our proposal and RGB*; it is easy to appreciate that in the other histograms there are several parts with curves. It is notorious that all the AHs shown in Fig. 6 have peaks and/or curves; however, the straightest AH is obtained with our proposal. In Fig. 7 is easy to observe that the shape of all the AH is straight, but in the lowest and highest intensity levels there are peaks and curves; nevertheless, the AH obtained with our proposal have less pronounced curves.

According to the AH shown in Fig. 8, the image processed in the HSV space has the lowest contrast because the shape of the AH is a pronounced curve. The other histograms are straighter, but the AH obtained with our proposal is the straightest. In Fig. 9 the AH obtained with our proposal is perhaps the straightest, although it has several small peaks, the AH of the L*a*b* image has the most enhanced peaks; the AH of the HSV image has a pronounced curve in the highest intensity levels, the AH of the RGB* image is perhaps the straightest histogram but it slope is higher than 45°.

4.3 Alternative Metric to Compute the Vectors' Magnitude

As we claim previously, mapping the colors between the RGB space and the HSV and L*a*b* spaces may become a huge computational load. With our proposal the mathematical operations do not represent a high computational load; due to in our proposal the Euclidean distance is employed to compute the magnitude of the color vectors, the square root operation is perhaps the mathematical operation that may demand the highest computational resources.

In order to reduce the computing resources the Euclidean distance demands, it is common to find that the magnitude of vectors is approximated by computing the average of the vector's components [1, 6]; that is, let $\phi = [r, g, b]$ be a RGB color vector:

$$\|\phi\|^{*} = \frac{r+g+b}{3} \tag{5}$$

Figure 10 shows the images obtained by processing the input images of Fig. 3 and using Eq. (5) as metric; note that the range of intensity levels is $[0, 255]$. The images shown in Fig. 10 do not suffer significant changes with respect to the images obtained

Fig. 10. Image obtained by processing the input images of Fig. 3 using as metric the Eq. (5)

with our proposal, using the Euclidean distance as metric, both in chromaticity and intensity of the colors. It is easy to appreciate that the contrast of the images does not vary significantly.

5 Conclusions and Future Work

We have introduced a proposal for contrast enhancement of RGB color images, without mapping the colors to a different color space that decouples the intensity from the chromaticity. In our proposal the magnitudes of the vectors are extracted and an intensity channel is computed, the histogram equalization is applied to this intensity channel the new intensity values are multiplied with their respective color vectors, previously normalized. The resulting images with our approach are compared with images processed in the HSV and L*a*b* spaces, where the histogram equalization is performed in their respective intensity channels. The images obtained are very similar between them, particularly the images obtained with our approach and the images processed under the L*a*b* space.

The computational load is low because it is not necessary to map the colors to a different space; the computational load may be reduced if Eq. (5) is employed as metric to compute the magnitude of vectors. The resulting images are almost the same to the images obtained using the Euclidean metric. The images obtained with our approach are evidence that the contrast of RGB color images can be enhanced.

As future work, more experiments are needed using the images of the Berkeley segmentation database; also, compare the resulting images with the images obtained with different techniques proposed in related works. Look in the state of the art for a metric, if exists, to evaluate quantitatively the contrast of the images.

References

1. Gonzalez, R.C., Woods, R.E.: Digital Image Processing, 2nd edn. Prentice Hall, Upper Saddle River (2002)
2. Jahanirad, M., Wahab, A.W.A., Anuar, N.B.: An evolution of image source camera attribution approaches. Forensic Sci. Int. **262**, 242–275 (2016)
3. Nnolim, U.A.: An adaptive RGB colour enhancement formulation for logarithmic image processing-based algorithms. Opt. Int. J. Light Electron Opt. **154**, 192–215 (2018)
4. Jun, H., Inoue, K., Hara, K., Urahama, K.: Saturation improvement in hue-preserving color image enhancement without gamut problem. ICT Express (2017). https://doi.org/10.1016/j.icte.2017.07.003
5. Qian, X., Han, L., Wang, Y., Wang, B.: Color contrast enhancement for color night vision based on color mapping. Infrared Phys. Technol. **57**, 36–41 (2013)
6. Zhang, H., Friits, J.E., Goldman, S.A.: Image segmentation evaluation: a survey of unsupervised methods. Comput. Vis. Image Underst. **110**(2), 260–280 (2008)
7. Agarwal, M., Mahajan, R.: Medical image contrast enhancement using range limited weighted histogram equalization. Procedia Comput. Sci. **125**, 149–156 (2018)
8. Rajinikanth, V., Couceiro, M.S.: RGB histogram based color image segmentation using firefly algorithm. Procedia Comput. Sci. **46**, 1449–1457 (2015)

9. Pare, S., Kumar, A., Bajaj, V., Singh, G.K.: A multilevel color image segmentation technique based on cuckoo search algorithm and energy curve. Appl. Soft Comput. **47**, 76–102 (2016)
10. Zhou, Z., Sang, N., Hu, X.: Global brightness and local contrast adaptive enhancement for low illumination color image. Opt. Int. J. Light Electron Opt. **125**(6), 1795–1799 (2014)
11. Xiao, B., Tang, H., Jiang, Y., Li, W., Wang, G.: Brightness and contrast controllable image enhancement based on histogram specification. Neurocomputing **275**, 2798–2809 (2018)
12. Tang, J.R., Isa, N.A.M.: Bi-histogram equalization using modified histogram bins. Appl. Soft Comput. **55**, 31–43 (2017)
13. Ong, S., Yeo, N., Lee, K., Venkatesh, Y., Cao, D.: Segmentation of color images using a two-stage self-organizing network. Image Vis. Comput. **20**(4), 279–289 (2002)
14. Paschos, G.: Perceptually uniform color spaces for color texture analysis: an empirical evaluation. IEEE Trans. Image Process. **10**(6), 932–937 (2001)
15. Rong, Z., Li, Z., Dong-nan, L.: Study of color heritage image enhancement algorithms based on histogram equalization. Opt. Int. J. Light Electron Opt. **126**(24), 5665–5667 (2015)
16. Li, X., Fang, M., Zhang, J.J., Wu, J.: Learning coupled classifiers with RGB images for RGB-D object recognition. Pattern Recognit. **61**, 433–446 (2017)
17. Grupt, B., Agarwal T.K.: New contrast enhancement approach for dark images with non-uniform illumination. Comput. Electr. Eng. (2017). https://doi.org/10.1016/j.compeleceng.2017.09.007
18. Ghani, A.S.A., Isa, N.A.M.: Automatic system for improving under water image contrast and color through recursive adaptive histogram modification. Comput. Electron. Agric. **141**, 181–195 (2017)
19. Gu, Z., Ju, M., Zhang, D.: A novel retinex image enhancement approach via brightness channel prior and change of detail prior. Pattern Recognit. Image Anal. **27**(2), 234–242 (2017)

An NTP-Based Test Method for Web Audio Delay

Yuxin Tang[1,3(✉)], Yaxin Li[1,3(✉)], Ruo Jia[1,3(✉)], Yonghao Wang[2,3(✉)], and Wei Hu[1,3(✉)]

[1] College of Computer Science and Technology, Wuhan University of Science and Technology, Wuhan, China
1263420797@qq.com
[2] Digital Media Technology Lab, Birmingham City University, Birmingham, UK
[3] Hubei Province Key Laboratory of Intelligent Information Processing and Real-Time Industrial System, Wuhan, China

Abstract. With the application of audio in various fields, the reliability of transmission and delay have been improved to a certain extent. However, there is no quick and effective method to test the delay of audio transmission nowadays. Therefore, it is an urgent matter to propose a reasonable method to test the audio delay. This paper takes web audio as research object, proposing a method of automatically testing web audio delay based on NTP, on the basis of which the delay of web audio is tested from two perspectives namely local delay and network delay. The experimental results show that the auto-delay testing method for web audio proposed in this paper is more flexible and efficient than other testing methods, especially for web audio testing.

Keywords: Web audio · Local delay · Network delay · Testing method

1 Introduction

With the advancement and development of technology, audio is used in different fields [1]. In the process of studying the audio transmission and modern desktop audio production applications, this paper finds that the audio triggered by events is accompanied by the influence of different factors, which causes a certain audio delay [2]. The phenomenon has influence on the expected effects to a certain extent. For web applications, the time delay between mouse and keyboard events (key down, mouse down, etc.) and a sound being heard is important. This time delay is called latency and is caused by several factors (input device latency, internal buffering latency, DSP processing latency, output device latency, distance of user's ears from speakers, etc.), and is cumulative [3–5]. The larger this latency is, the less satisfying the user's experience is going to be. In the extreme, it can make musical production or game-play impossible. At moderate levels it can affect timing and give the impression of sounds lagging behind or the game being non-responsive. For musical applications the timing problems affect rhythm [6]. For gaming, the timing problems affect precision of gameplay. For interactive applications, it generally cheapens the users experience much in the

D.-S. Huang et al. (Eds.): ICIC 2018, LNAI 10956, pp. 456–465, 2018.
https://doi.org/10.1007/978-3-319-95957-3_48

same way that very low animation frame-rates do. Depending on the application, a reasonable latency can be from as low as 3–6 ms to 25–50 ms [7–9]. Therefore, how to measure the delay of web audio effectively and reasonably is a problem that needs to be solved in this paper.

This paper proposes a test method based on webpage audio delay, so as to achieve the purpose of automatically testing audio delay by setting the frequency of audio events. At the same time, this paper compares the impact of audio delay on two different situations from local delay and web page communication. The experimental results show that the delays analyzed by different browsers and different operating systems are quite different under the study of local delays; the delay caused by network transmission is greater than the local delay.

This paper is organized as follows. Related works are described in Sect. 2, and we describe the testing method in Sect. 3. In Sect. 4, we introduce the experimental environment and design. In Sect. 5, we give the experimental results and corresponding analysis. And in Sect. 6, we draw the conclusions and give the future work.

2 Related Work

The sound signal is one of the important signals to convey information [10]. The study of the sound signal has always been valued by scholars. At present, the study of audio transmission delay spreads in all aspects. Among them, Chen Liang [11] and others in the paper of "based on auto-correlation audio transmission system delay detection method" proposed a test method of an audio transmission system delay. The test method can effectively detect the delay of more than 100 ms and can more accurately get the delay data without any professional equipment. Li Litian [12] and others proposed a fast and easy method for measuring the delay of CobraNet signal in the measurement of audio signal delay. The technology of CobraNet in the audio system is mainly used to solve the problem of audio signal transmission [13]. At present, many major projects in the country have adopted the audio system in the application of CobraNet technology such as: broadcasting system of Daya Bay Nuclear Power Station, broadcasting system of Shanghai Museum of Science and Technology, the Capital International Airport r13 Terminal broadcasting system, Shanghai F1 International Circuit audio system, etc. [14].

It can be seen that there are in-depth studies on the problem of delay existed in the process of audio transmission both at home and abroad. The essence of solving the problem is to focus on the research of wireless communication technology. However, the study of web audio events is very rare nowadays. This paper designs a automatic test method of web audio delay, which is based on the audio processing and combines the web audio application interface provided by the technology of HTML5 from the internal stress and external stress. And the external stress is mainly caused by the js load of web event. This method realizes the automatic measurement of the audio delay of the web page under different platforms. We analyzes the factors that affect the audio delay of the web page further by comparing the delay amount for different systems and different browser versions.

3 Testing Method

3.1 Web Audio Delay

Unlike the delay testing in the process of traditional audio transmission, the delay of the web audio is mainly divided into the local delay and the delay caused by the network transmission [15]. The local delay is mainly related to the operating system, the browser kernel, and so on. The generalisation of mobile devices that embed facilities for real-time audio, processing, motion sensors, and networking, has made it easy to use them for a wide range of applications. The fifth major revision of the Hypertext Markup Language (HTML5) [16] grants access to these features in a Web page from a standard Web browser, instead of developing native applications. Massively distributed applications are now within the reach of anyone, and this is of particular interest for audio applications, as one can distribute a complete audio process to a mobile device, up to an integrated speaker [17].

3.2 Web Audio Application Program Interface

Audio on the web has been fairly primitive up to this point and until very recently has had to be delivered through plugins such as Flash and QuickTime. The introduction of the audio element in HTML5 is very important, allowing for basic streaming audio playback. But, it is not powerful enough to handle more complex audio applications [18].

The Web Audio API provides a powerful and versatile system for controlling audio on the Web, allowing developers to choose audio sources, add effects to audio, create audio visualizations, apply spatial effects (such as panning) and much more. The Web Audio API involves handling audio operations inside an audio context, and has been designed to allow modular routing. Basic audio operations are performed with audio nodes, which are linked together to form an audio routing graph. Several sources—with different types of channel layout—are supported even within a single context. This modular design provides the flexibility to create complex audio functions with dynamic effects [19].

3.3 Testing Method for Web Audio Delay

Audio can introduce delays in the transmission process due to various factors which include local delays and delays of network transmission.

The test of local web audio delay can be described as follows: through the Web Audio API, the input of the web audio signal and the output time after the cache area are obtained, then the difference between the two is the local delay of the web audio. Unlike the local delay testing, the delay caused by network transmission to the web audio is mainly related to the network transmission speed, webpage loading time and other factors. Based on the time synchronization principle of NTP and the Ping-Pong model algorithm proposed by JP Lambert et al., this paper realizes the automatic testing of web page audio by calculating the error between master and slave time.

The Ping-Pong algorithm is that a client sends a time-stamp t_{ping} to a reference, that applies its own time-stamps: T_{ping} on the reception, and T_{pong} on the emission of its reply. The client then applies a final time-stamp t_{pong}. The specific process is shown in Fig. 1.

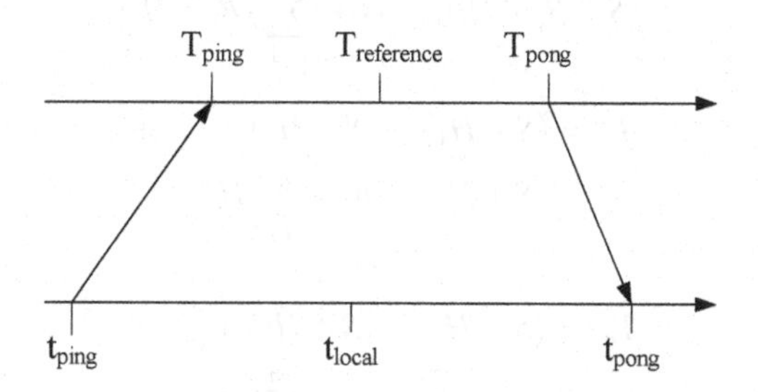

Fig. 1. Exchange of time-stamps during a ping-pong round-trip.

T is used for reference time, while t is used for local time. Given that both times are expressed with the same unit, this allows the following estimations:

$$T = T_{reference} = (T_{pong} + T_{ping})/2 \tag{1}$$

$$t = t_{local} = (t_{pong} + t_{ping})/2 \tag{2}$$

$$offset = t_{local} - T_{reference} \tag{3}$$

$$travel_{duration} = t_{pong} - t_{ping} - (T_{pong} - T_{ping}) \tag{4}$$

We assume that the reference clock T can be expressed linearly from the local clock t with an offset T_0 and a frequency ratio R, like NTP.

$$T(t) = T_0 + R \times (t - t_0) \tag{5}$$

$$R = \frac{Cov[t, T]}{Var[t]} = \frac{\overline{t \cdot T} - \overline{t} \cdot \overline{T}}{\overline{t^2} - \overline{t}^2} \tag{6}$$

In this paper, multiple sets of testing data are obtained by sampling and R can be calculated according to the formula (6), namely the rate of change of the delay amount, from which the actual offset T_0 can be obtained. The average value obtained after multiple statistics is the delay of audio signal in network transmission.

In the transmission process of webpage audio, the browser needs to establish communication with the server in above. The communication time includes: the time of connection establishment, the response time, and the processing time. Suppose the time of a connection establishment is S; a cycle C includes n Case pages; the response time of the i-th page is R_i; the processing page time of browser loading is H_i; T_m is the time spent in the m-th cycle; T is the total time tested by using HTTP/1.0 to load the page test, then the total time T can be expressed as:

$$T_1 = (S + R_1 + H_1) + (S + R_2 + H_2) + \ldots + (S + R_i + H_i) = ns + \sum_{i=1}^{n} (R_i + H_i) \tag{7}$$

$$T_2 = (S + H_1) + (S + H_2) + \ldots + (S + H_I) = ns + \sum_{i=1}^{n} H_i \tag{8}$$

$$\ldots$$

$$T_m = (S + H_1) + (S + H_2) + \ldots + (S + H_I) = ns + \sum_{i=1}^{n} H_i \tag{9}$$

$$T = (m-1)\left(ns + \sum_{i=1}^{n} H_i\right) + ns + \sum_{i=1}^{n} (R_i + H_i) = ns + (m-1)\sum_{i=1}^{n} H_i + \sum_{i=1}^{n} (R_i + H_i) \tag{10}$$

Therefore, the delay T_{delay} of the webpage audio during the network transmission is the sum of the time T when the browser establishes communication with the server and the delay T_0 of the audio signal during the network transmission.

$$T_{delay} = T + T_0 \tag{11}$$

The environment variables that are referenced between multiple pages are saved by using cookies. These environment variables store temporary testing data, which is the delay amount T_{delay}. In the testing of loading page, the start time is firstly recorded in the cookie before the first page is loaded. Cookies are stored on the local machine and the testing can continue without losing data. Cookies and Server-side components work together to summarize, process, and store the results of the tests in the database. After the testing is over, the final testing delay T_{delay} is automatically displayed in the text box of the web page, which realizes the automatic testing.

4 Experimental Environment and Design

4.1 Software and Hardware Platform

Since this experiment is focused on delay testing for web page audio, the experiment was verified to implement through programing on the Visual Studio Code. The Visual Studio Code is a lightweight and powerful code editor that supports Windows, OS X and Linux. Not only does it build in JavaScript, TypeScript, and supported by Node.js, but also it has a rich plug-in ecosystem and can support C++, C#, Python, PHP, and other languages through the installation of plug-ins.

The experiment is mainly tested on the sound delay triggered by events of different browsers under different operating systems, so the operating system is related to Windows 7, Windows 10, Linux, Mac OS; the browsers are Chrome, IE, Firefox, Safari. On this basis, we realize the purpose of automatically testing the sound delay triggered by web page events of different browsers in different operating systems to get the measurement data. The corresponding version of the browsers and kernel information in the experiment are shown in Table 1.

Table 1. Browsers and related information.

Browsers	Version	Kernel
Google Chrome	58.0.3029.81	WebKit
Internet Explore	9	Trident
Mozilla Firefox	52.0.2	Gecko
Safari	9.1.3	WebKit

The different operating systems in the experiment and the corresponding bits are shown in Table 2.

Table 2. Operating systems and corresponding bits.

Operating Systems	Bits	
Windows7	64	32
Windows7	64	32
Linux	64	32
Mac OS	64	32

4.2 Experimental Design

The audio file used in this experiment was a .WAV file, and the channel was mono. The sampling rate was 48 kHz, and the sample value was a 16-bit PCM quantization code. In this experiment, there are multiple combinations of operating systems and browsers. The tested platform and hardware are the same, but the tested operating systems and browsers are different. Thus we can analyze the audio delay of different operating systems or browsers.

This experiment is based on the delay testing method mentioned previously. The testing process is mainly divided into local delay testing and network transmission delay testing.

5 Experimental Results and Analysis

5.1 Local Delay Testing

In this part of the experiment, the audio delay triggered by the events is mainly due to the browser or the operating system, so we set the frequency of the event of different browsers to automatically test the sound delay to analyze the change of the amount of delay under different browsers and different operating systems. The average amounts of delay under different operating systems and different browsers every 50 ms are shown in Table 3.

Table 3. The average amount of delay under different situations.

Operating Systems	Google Chrome	Internet Explore	Mozilla Firefox	Safari
Windows7	18.36 ms	17.45 ms	20.11 ms	–
Windows10	20.28 ms	–	25.24 ms	–
Linux	20.23 ms	–	21.19 ms	–
Mac OS	15.16 ms	–	14.38 ms	14.29 ms

The data shown in Table 3 can be seen that the amounts of sound delay of different browsers under different operating systems are different. For the same operating system, different browsers have little impact on latency. Due to the existence of browser compatibility issues, the test data can not be displayed under IE browser of Windows 10 system. The audio delay triggered by event is greater in Windows 10 system; The audio delay triggered by event is smaller in Mac OS; As installing IE browser in the Linux system needs to install other software, the delay in this case is not considered. The experimental results indicate that the sound delay caused by the internal stress mainly related to the operating system.

5.2 Network Transmission Delay Testing

In this paper, we have tested the entire system with a 100 Mb per second local area network in the experiment of studying the effect of network transmission on the delay of Web page audio. There is one terminal on the local area network, and DHCP is enabled. The default gateway is 10.162.32.1. During the test, users can keep to browse webpages with TCP connections in the local area network. Analogous to the local delay testing, this paper also selects to test the average delay for different operating systems and browsers under the condition that the frequency of audio events is 50 ms. The statistical results are as shown in Table 4.

From the data shown in Table 4, it can be seen that the delay amount of audio corresponding to different browsers under different operating systems is also different in a local area network. For the same operating system, different browsers have little effect on delay. Compared with other systems, the audio triggered by events has a larger delay in the Win10 system; the audio triggered by events has a smaller delay in

Table 4. The average delay in different situations under local area network.

Operating Systems	Google Chrome	Internet Explore	Mozilla Firefox	Safari
Windows7	27.35 ms	33.06 ms	35.18 ms	–
Windows10	28.64 ms	–	36.36 ms	–
Linux	22.45 ms	–	23.22 ms	–
Mac OS	18.23 ms	–	16.29 ms	15.47 ms

Mac systems. Installing IE browser under Linux system requires installing other software, in which case the delay is not considered.

This article takes Google Chrome as an example to show the changes of local delay and network transmission delay under different operating systems through the line graph of Fig. 2.

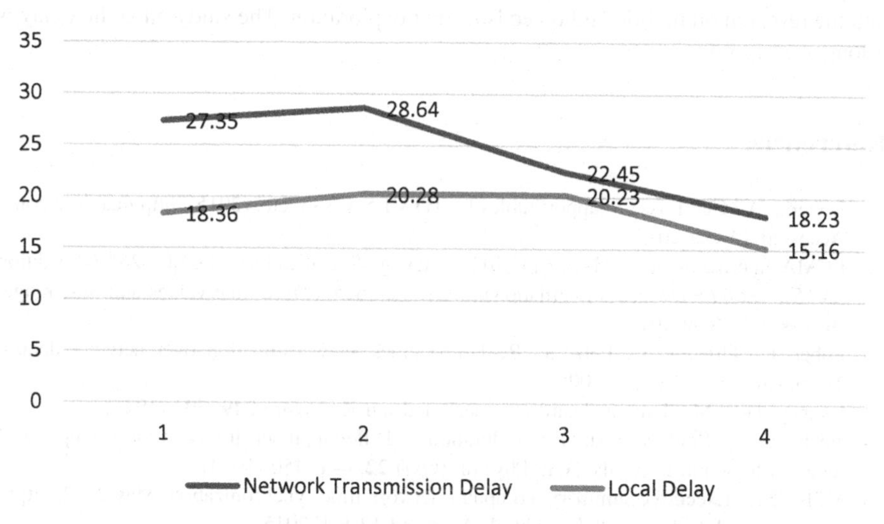

Fig. 2. Network transmission and local delay distribution.

In Fig. 2, the upper fold line represents the distribution of audio delays caused by the network transmission under four different operating systems, and the lower fold line represents the distribution of local delays under four different operating systems. The system represented by the abscissa from left to right is: Win7, Win10, Linux, and Mac OS. It can be seen from the figure that the effect of network transmission on audio delay is greater than the local delay under the same operating system, and the network transmission delay and local delay substantially vary under different operating systems, that is to say the audio delay triggered by events under the Win10 system is larger. The delay caused by network transmission to web audio is generally more than 20 ms.

6 Conclusions and Future Work

HTML5 has been widely used in the technologies related to real-time audio processing. It provides a number of application program interface related to the audio processing, which lays a foundation for the study of the audio delay triggered by events. In this paper, we proposes a method of automatic measurement of audio delay based on NTP, which aims to test the audio delay by setting the frequency of audio events. As we can see from the experimental results, the effect of network transmission on audio delay is greater than the local delay under the same operating system and the testing method for web audio proposed in this paper is more flexible and accurate than other testing methods. All the experimental results and corresponding analysis lay the foundation for the further research on the audio delay. However, there is still much work to do in the future such as how to propose a optimization method of audio delay to decrease the delay. In addition, this article is devoted to the research of audio delay for web audio, and the research on mobile audio needs further exploration. The study on audio delay is a long way to go.

References

1. Deveria, A.: Can I use… support tables for HTML5, CSS3, etc. (2015). http://caniuse.com/. Accessed 14 Oct 2015
2. ECMA International: ECMAScript 2015 Language Specification – ECMA-262 6th Edition (2015). http://www.ecma-international.org/ecma-262/6.0/#sec-time-values-and-time-range. Accessed 28 Sept 2015
3. Fober, D., Orlarey, Y., Letz, S.: Real time clock skew estimation over network delays. Technical report, Grame (2005)
4. Lunney, H.W.M.: Time as heard in speech and music. Nature **249**, 592 (1974)
5. Michon, J.A.: Studies on subjective duration: I. Differential sensitivity in the perception of repeated temporal intervals. Acta Physiol. (Oxf) **22**, 441–450 (1964)
6. Mills, D.L.: Executive summary: computer network time synchronization, May 2012. http://www.eecis.udel.edu/~mills/exec.html. Accessed 13 Feb 2015
7. Mills, D.L.: Timestamp capture principles, May 2012. http://www.eecis.udel.edu/~mills/stamp.html. Accessed 16 Feb 2015
8. Mills, D.L.: Clock filter algorithm, March 2014. https://www.eecis.udel.edu/~mills/ntp/html/filter.html. Accessed 17 Sept 2015
9. Multi-device Timing Community Group: Multi-device timing for web, community group charter (2015). https://webtiming.github.io/. Accessed 29 Sept 2015
10. Node.js Foundation: Process Node.js v4.1.1 manual & documentation, August 2015. https://nodejs.org/api/process.html#process_process_hrtime. Accessed 28 Sept 2015
11. NTP.org: Clock quality, March 2014. http://www.ntp.org/ntpfaq/NTP-s-sw-clocks-quality.htm. Accessed 13 Feb 2015
12. NTP.org: How does it work? (2014). http://www.ntp.org/ntpfaq/NTP-s-algo.htm. Accessed 06 Oct 2015
13. Georgiadis, L.: Resource allocation and cross-layer control in wireless networks. Found. Trends Netw. **1**(1), 1–149 (2006)
14. Jagannathan, K.: Delay analysis of maximum weight scheduling in wireless ad hoc networks. In: CISS 2009, Baltimore, pp. 389–394 (2009)

15. Ying, L.: On combining shortest-path and back-pressure routing over multihop wireless networks. In: INFOCOM 2009, Brazil, pp. 1674–1682 (2009)
16. Ji, B.: Delay-based back-pressure scheduling in multi-hop wireless networks. In: INFOCOM 2011, ShangHai, pp. 2579–2587 (2011)
17. Celik, G., Modiano, E.: Variable frame based max-weight algorithms for networks with switchover delay. In: ISIT 2011, Russia, pp. 2537–2541 (2011)
18. Bui, L.: Novel architectures and algorithms for delay reduction in back-pressure scheduling and routing. In: INFOCOM 2009, Brazil, pp. 2936–2940 (2009)
19. Shakkottai, S.: Scheduling for multiple flows sharing a time-varying channel: the exponential rule. Trans Am. Math. Soc. **207**, 185–202 (2002)

Inverse Halftoning Algorithm Based on SLIC Superpixels and DBSCAN Clustering

Fan Zhang(✉), Zhenzhen Li, Xingxing Qu, and Xinhong Zhang

School of Computer and Information Engineering,
Henan University, Kaifeng 475001, China
zhangfan@henu.edu.cn

Abstract. Halftone technology is widely used in the printing industry. This paper proposes an inverse halftoning algorithm based on SLIC (Simple Linear Iterative Clustering) superpixels and DBSCAN (density-based spatial clustering of applications with noise) clustering. Firstly, halftoning image is segmented by SLIC superpixels algorithm. Then the boundaries region of image is tracked by DBSCAN clustering algorithm and the boundaries of image is vectored. Secondly, the remaining part of halftoning image that boundaries have been extracted is smoothed by linear and nonlinear smoothing filters. Finally the vector boundaries and the smooth background is combined together to get the inverse halftoning image. Experimental results show that the proposed method can effectively remove halftone patterns while retains boundaries information.

Keywords: Inverse halftoning · SLIC superpixels · DBSCAN clustering

1 Introduction

Halftone is the reprographic technique that simulates continuous tone imagery through the use of dots, varying either in size or in spacing, thus generating a gradient-like effect [1]. Halftone technology is widely used in the printing industry. By the use of dots, varying either in size or in density, halftone technology simulates continuous tone image to display the gray-scale or color of image. Figure 1 shows the example of color halftoning.

Inverse halftoning algorithms try to remove halftone patterns or screen patterns caused by the process of halftoning [2]. This paper proposes an inverse halftoning algorithm based on superpixels algorithm and DBSCAN clustering. Superpixels algorithms group pixels into perceptually meaningful atomic regions which can be used to replace the rigid structure of the pixel grid [3]. Most superpixel algorithms over-segment the image. This means that most of important boundaries in the image are found.

Digital halftoning algorithms can in general be classified into two categories: ordered dither and error diffusion. Ordered dithering requires only point-wise comparisons. Error diffusion is a type of halftoning in which the quantization residual is distributed to neighboring pixels that have not yet been processed. One of the most commonly used error diffusion algorithms is the Floyd–Steinberg algorithm.

D.-S. Huang et al. (Eds.): ICIC 2018, LNAI 10956, pp. 466–471, 2018.
https://doi.org/10.1007/978-3-319-95957-3_49

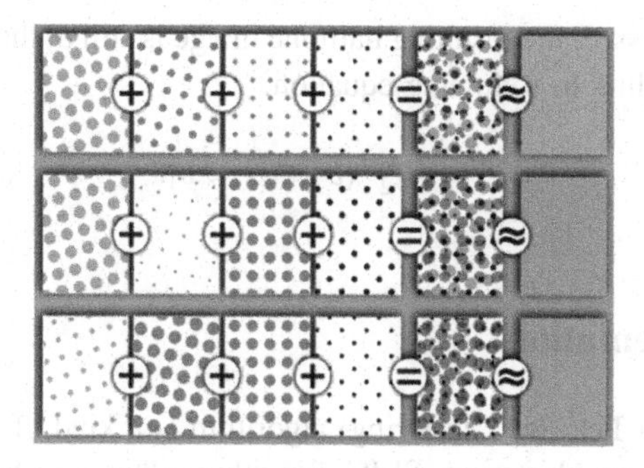

Fig. 1. An example of color halftone in CMYK separation. Varying the density of the four printing colors, cyan, magenta, yellow and black (CMYK). (Color figure online)

2 Inverse Halftoning Model

The inverse halftoning is a reconstruction technique of a gray-scale image from its halftone version. Since there can be more than one continuous tone image giving rise to a particular halftone image, there is no unique inverse halftone of a given halftoned image. A high quality of the gray-scale image obtained by the inverse halftoning is required in many applications.

Because noise is introduced in halftoning process H, noise should be removed in the process of inverse halftoning H^{-1} [4]. The simple low pass filtering can remove most of the halftoning noise, but it also removes the edge information. Let the original input is a continuous tone image x, and the output half tone image is b If the temporal dithering or the error diffusion is used and the diffusion error is θ, the halftone system model can be expressed as follows [5],

$$b = H(x, \theta), \tag{1}$$

where

$$x = \{x_{i,j} | x_{i,j} \in [0, g], 1 \leq i \leq M, 1 \leq j \leq N\},$$

$$b = \{b_{i,j} | b_{i,j} \in [0, 1], 1 \leq i \leq M, 1 \leq j \leq N\}.$$

There are two different methods to get the inverse halftoning image H^{-1}, as shown below. The first method: When the halftone image b is processed by the inverse halftone processing and get a continuous tone image $\hat{x}$. The two images are observed by human eyes in the distance range and be judged as the similar, that is, to seek a similar mapping H_1^{-1} according to following equation,

$$\hat{x} = [H_1^{-1}(b)] \cap [\hat{x} \approx b]. \tag{2}$$

The second method: When the halftone image b is processed by the inverse halftone processing and get a continuous tone image $\hat{x}$. After the reconstructed continuous tone image $\hat{x}$ is processed, the resulting halftone image is still b, that is, to seek a similar mapping H_2^{-1} according to following equation,

$$\hat{x} = [H_2^{-1}(b)] \cap [\hat{x} \approx b]. \tag{3}$$

3 Image Segmentation

SLIC (Simple Linear Iterative Clustering) algorithm is a kind of superpixels segmentation method based on clustering. SLIC algorithm adheres to boundaries as well as other segmentation algorithms, and it is faster and more memory efficient. The specific implementation process of SLIC algorithm is as follows [6]:

(1) Assuming that there are M pixels, and there are K initial clustering centers (superpixels). The size of each superpixel is M/K, then the distance between of each superpixel center is approximately $S = (M/K)^{1/2}$. In order to avoid the superpixels centering in the position of image boundaries, the clustering centers are moved to the location that corresponding to the lowest gradient in a 3×3 neighborhood.
(2) For each pixel, the similarity measure is the distance between the pixel and the nearest clustering centers.
(3) According to the similarity formula, each pixel is assigned a label of the nearest cluster center.
(4) Compute new cluster centers and compute the residual error.
(5) Repeat step 3 and step 4 until the residual error is sufficient small, and then the superpixels segmentation is done.

Figure 2 shows the experimental result of SLIC algorithm.

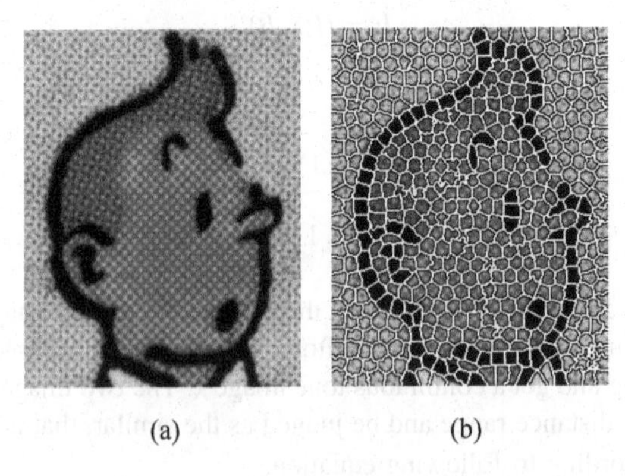

Fig. 2. (a) The original image. (b) The segmented image based on SLIC algorithm.

4 Boundary Vectorization

The purpose of this paper is to remove halftone patterns, and preserve boundaries information at the same time. Therefore, boundaries information should be extracted from halftoning image firstly.

Superpixels are an over-segmentation of an image. Many state-of-the-art superpixel segmentation algorithms rely on minimizing special energy functions or clustering pixels in the effective distance space to track the boundaries of image. Clustering is a method of grouping similar types of data. In this paper, DBSCAN clustering (density-based spatial clustering of applications with noise) algorithm is used to clustering.

DBSCAN (density-based spatial clustering of applications with noise) clustering is an important spatial clustering technique which is widely applied in many fields [7, 8]. The DBSCAN is a clustering algorithm based on the high density connected region. The region of high density will be divided into clusters, and DBSCAN algorithm can discovers arbitrary shape clusters in a spatial database with noise. Figure 3 shows a diagram for DBSCAN clustering algorithm [9–11].

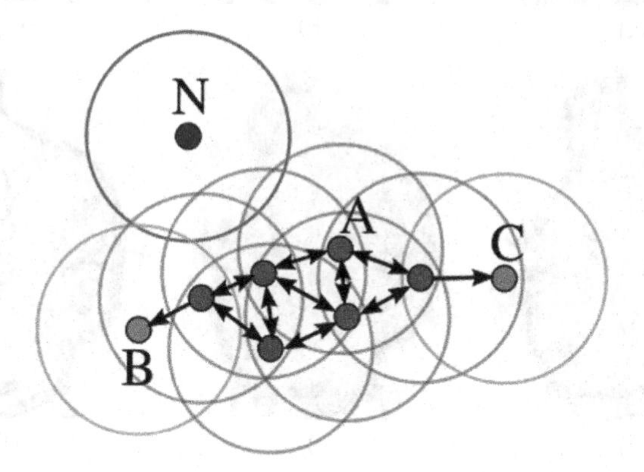

Fig. 3. Diagram for DBSCAN clustering algorithm.

5 Image Inverse Halftoning

The remaining part of halftoning image that boundaries have been extracted, is smoothed to be a continuous tone image. In this paper, we select a linear smoothing filter and a nonlinear smoothing filter to get the smooth background. The linear filter is a 9×9 Gaussian low-pass filter with the variance is 1.4. The nonlinear filter is a 3×3 median filter. Both of them are used to eliminate halftone patterns.

The inverse halftoning image is reconstructed by combining the smooth background and the vector representation of image boundaries. Halftone patterns are removed from the reconstructed image while boundaries of image are retained at the same time.

Figure 4 is the experimental comparison of proposed method with other methods. In order to quantitatively compare our method with other methods, the peak signal to noise ratio (PSNR) and the structural similarity (SSIM) of experimental images are computed. The experimental results of our algorithm are compared with mean filtering processing, Gaussian filtering processing; Wavelet based inverse halftoning processing and inverse halftoning by scanner software processing respectively. The computed results are shown in Table 1. Experimental results show that the proposed method can effectively remove the halftone patterns while retains boundaries information perfectly.

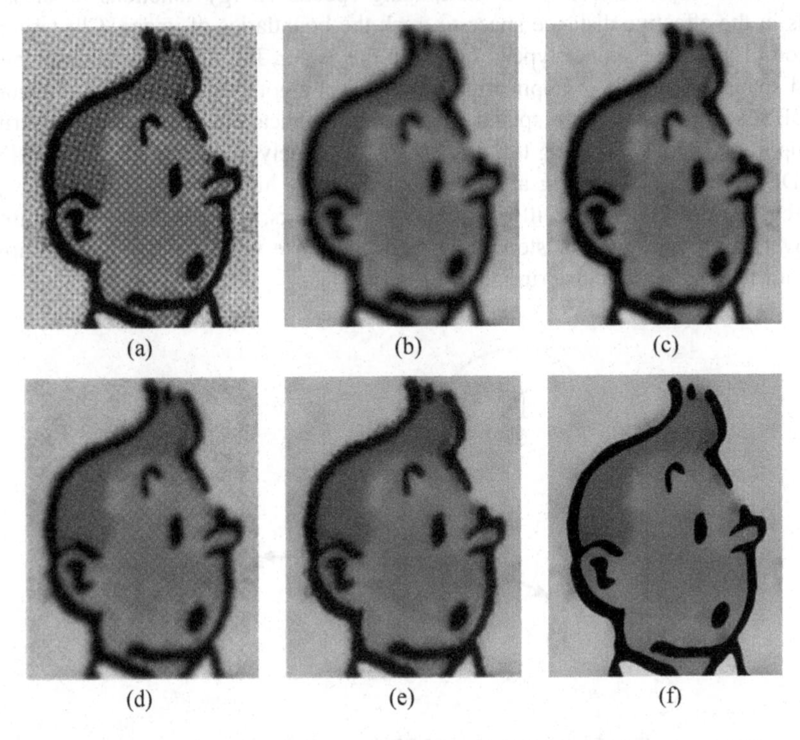

Fig. 4. The experimental results of inverse Halftoning. (a) The original image. (b) The inverse Halftoning image by mean filtering. (c) The inverse Halftoning image by Gaussian filtering. (d) The inverse Halftoning image by wavelet based processing. (e) The inverse Halftoning image by scanner software. (f) The inverse Halftoning image by algorithm of this paper.

Table 1. The comparison of experimental results.

Algorithms of inverse Halftoning	PSNR	SSIM
Mean filtering	21.6627	0.2441
Gaussian filtering	23.9786	0.2916
Wavelet based inverse halftoning	24.8281	0.2914
Inverse halftoning by scanner software	27.0168	0.3009
Our algorithm	29.7467	0.3293

6 Conclusions

This paper proposes an inverse halftoning algorithm based on SLIC superpixels and DBSCAN clustering. The reconstructed image combines the smooth background and the vector representation of image boundaries. Halftone patterns are removed while boundaries of image are retained. Experimental results show that the proposed inverse halftoning method has good performance both on PSNR and on SSIM.

Acknowledgement. This research was supported by the Natural Science Foundation of China (Grant No. 61771006, No. U1504621); Natural Science Foundation of Henan Province (Grant No. 162300410032), and International Science and Technology Cooperation Project of Henan Province (Grant No. 144300510033).

References

1. Son, C.H., Choo, H.: Local learned dictionaries optimized to edge orientation for inverse halftoning. IEEE Trans. Image Process. **23**(6), 2542–2556 (2014)
2. Son, C.H., Lee, K.W., Choo, H.: Inverse color to black-and-white halftone conversion via dictionary learning and color mapping. Inf. Sci. **299**, 1–19 (2015)
3. Achanta, R., Shaji, A., Smith, K., Lucchi, A., Fua, P., SuSstrunk, S.: Slic superpixels compared to state-of-the-art superpixel methods. IEEE Trans. Pattern Anal. Mach. Intell. **34** (11), 2274–2282 (2012)
4. Mese, M., Vaidyanathan, P.P.: Recent advances in digital halftoning and inverse halftoning methods. IEEE Trans. Circuits Syst. I Fundam. Theory Appl. **49**(6), 790–805 (2002)
5. Son, C.H.: Inverse halftoning based on sparse representation. Opt. Lett. **37**(12), 2352–2354 (2012)
6. Akyilmaz, E., Leloglu, U.M.: Segmentation of sar images using similarity ratios for generating and clustering superpixels. Electron. Lett. **52**(8), 654–656 (2016)
7. Xiang, D., Ban, Y., Wang, W., Su, Y.: Adaptive superpixel generation for polarimetric sar images with local iterative clustering and sirv model. IEEE Trans. Geosci. Remote Sens. **55** (6), 3115–3131 (2017)
8. Birant, D., Kut, A.: St-dbscan: An algorithm for clustering spatial-temporal data. Data Knowl. Eng. **60**(1), 208–221 (2007)
9. Zhong, C., Miao, D., Wang, R.: A graph-theoretical clustering method based on two rounds of minimum spanning trees. Pattern Recogn. **43**(3), 752–766 (2010)
10. Ester, M., Kriegel, H.P., Xu, X.: A density-based algorithm for discovering clusters a density-based algorithm for discovering clusters in large spatial databases with noise. In: International Conference on Knowledge Discovery and Data Mining, pp. 226–231 (1996)
11. Hou, J., Gao, H., Li, X.: DSets-DBSCAN: a parameter-free clustering algorithm. IEEE Trans. Image Process. **25**(7), 3182–3193 (2016)

JPEG Image Super-Resolution via Deep Residual Network

Fengchi Xu[1], Zifei Yan[1](✉), Gang Xiao[2], Kai Zhang[1], and Wangmeng Zuo[1]

[1] Harbin Institute of Technology, 92 West Dazhi Street, Harbin 150001, China
cszfyan@gmail.com

[2] No. 211 Hospital of PLA, 45 Xue Fu Street, Harbin 150080, China

Abstract. In many practical scenarios, the images to be super-resolved are not only of low resolution (LR) but also JPEG compressed, while most of the existing super-resolution methods assume compression free LR image inputs. As a result, the JPEG compression artifacts (e.g., blocking artifacts) are often exacerbated in the super-resolved images, leading to unpleasant visual results. In this paper, we address this problem via learning a deep residual convolutional neural network (CNN) that exploits a skips-in-skip connection. More specifically, by increasing the network depth to 31 layers with receptive field of 63 by 63, we train a single CNN model which is able to handle JPEG image super-resolution with various combinations of scale and quality factors, as well as the extreme cases, i.e., image super-resolution with multiple scale factors, and JPEG image deblocking with different quality factors. Our extensive experimental results demonstrate that the proposed deep model can not only yield high resolution (HR) images that are visually more pleasant than those state-of-the-art deblocking and super-resolution methods in a cascaded manner, but also deliver very competitive results with the state-of-the-art super-resolution methods and JPEG deblocking methods in terms of quantitative and qualitative measures.

Keywords: JPEG image Super-Resolution · Deep residual network
Skips-in-skip connection

1 Introduction

Single image super-resolution (SR), with the goal of generating a visually pleasant high-resolution (HR) image from a given low-resolution (LR) input, has been attracting intensive research attentions in the field of image modeling and computer vision. Since SR is a highly ill-posed problem as the number of pixels to be estimated in the HR image is usually much larger than that in the given LR input, the prior knowledge is widely adopted to impose constraint on the solution space. Such prior knowledge can be either predefined or learned from training data. In this sense, state-of-the-art SR methods can be generally divided into two categories. The first one comes from the maximum a posteriori estimation under the Bayesian framework where the likelihood is known while the prior is specified. Popular natural image priors include non-local self-similarity prior [1] and sparsity prior [2]. The second one concerns the discriminative

D.-S. Huang et al. (Eds.): ICIC 2018, LNAI 10956, pp. 472–483, 2018.
https://doi.org/10.1007/978-3-319-95957-3_50

learning based methods which try to learn the relationship between LR and HR space from an abundant of LR and HR exemplar pairs [3, 4].

While significant advances have been achieved with sophisticated image priors, existing SR methods generally assume the observed LR image is also clean, neglecting the potential artifacts along with the observed image. Nowadays, the down-scaled and lossy compressed images are widely transmitted over the internet due to limited network bandwidth and storage capability [5, 6]. While the downscaling scheme reduces the spatial information from the HR image, the subsequent compression scheme would lead to visually unpleasant visual artifacts, mainly known as blocking artifacts. To be clear, JPEG compression scheme divides an image into 8×8 pixel blocks and applies block discrete cosine transformation (DCT) on each block individually. Quantization which is determined by a quality factor is then applied on the DCT coefficients to reduce the storage, thus introducing the blocking artifacts. Here, the quality factor has a range from 0 to 100, and it controls the JPEG compression ratio. The larger the quality factor is, the smaller the compression ratio is and the better the compressed image (which is referred to as JPEG image in this paper) is.

Directly super-resolving the JPEG image by the plain SR methods without preprocessing would concurrently exacerbate the blocking artifacts, leading to more unpleasant visual results. The reason comes from the fact that existing SR methods only consider that an input LR image is only degraded by down-scaling or at most an additional blurring operation which is an impractical assumption under internet environment. Similar findings for noisy image super-resolution have been presented in the work by Singh et al. [7]. More formally, a general assumption for JPEG image super-resolution in the internet environment is that an input LR image is down-scaled and then compressed with a certain quality factor. Apparently, compared to the plain SR, JPEG image super-resolution is a more complex problem since it involves an additional blocking artifacts removal procedure.

An alternative solution for JPEG image super-resolution is to preprocess the image with a deblocking method, followed by applying a plain SR method. However, such an easy pipeline with existing deblocking and SR methods tend to encounter an inevitable drawback, that is, one needs to know or estimates the JPEG quality factor beforehand. Meanwhile, there are few attempts for directly super-resolving the JPEG image. Xiong et al. [5] designed a SR method which is specially designed for JPEG compressed web images. Kang et al. [6] proposed a joint SR and deblocking method for highly compressed image super-resolution via dictionary based sparse reconstruction and morphological component analysis based image decomposition. Nevertheless, those methods still suffer from the unknown JPEG image quality factor induced drawback, thereby having a limitation for practical applications. As a result, there is a demand for designing a JPEG image super-resolution method which is robust to quality factor.

Recent progresses in deep learning have shown the effectiveness and efficiency of convolutional neural networks (CNN) [8, 9] in machine vision field including high-level and low-level vision problems. For low-level vision problems, CNN has been successfully applied in several tasks such as single image deblurring [10], image super-resolution [4, 11], JPEG image deblocking [12] and image segmentation [13]. It is interesting to investigate whether we can learn a CNN model for robust JPEG image super-resolution without consideration of the quality factor. In particular, it would be very attractive if the trained model can handle the general cases with various combinations of scale and quality factors, as well as the two extreme cases, i.e., image super-resolution with multiple scale factors, and JPEG image deblocking with different quality factors. The question is whether it is possible to train such an integrated CNN model. In fact, there exist several frameworks that are capable of handling the JPEG image deblocking and single image super-resolution tasks individually. Timofte et al. [3] proposed an anchored neighborhood regression framework for single image super-resolution and then used the framework for image deblocking [14]. Kwon et al. [15] proposed a common solution to image super-resolution and compression artifact removal by using Gaussian Processes under a semi-local approximation. Dong et al. [4] proposed a CNN model for single image super-resolution and then modified the network for compression artifacts removal [12]. Chen and Pock [16] proposed a trainable nonlinear reaction diffusion framework for several image restoration problems such as single image super-resolution and JPEG image deblocking. More recently, Kim et al. [17] trained a single convolutional network for multiple scale factors super-resolution. All the above methods indicate that it is possible to learn a single CNN model for JPEG image super-resolution.

In this paper, we propose a single CNN model with a depth of up to 31 layers for JPEG image super-resolution. We refer to our proposed JPEG image super-resolution network as JSCNN. The most distinguished feature of JSCNN is that, instead of handling a certain combination of scale and quality factors by a specifically trained model, it can not only deal with the task of JPEG image super-resolution with various combinations of scale factors and quality factors, but also can suffice the extreme cases, i.e., image super-resolution with several scale factors and JPEG image deblocking with different quality factors. Specially, to enable the network for fast training, several techniques including batch normalization [18], residual learning [19], skips-in-skip connection and gradient clipping-enabled large learning rate [17] are utilized. Our extensive experiments show that our model exhibits highly competitive quantitative and visual results with state-of-the-art JPEG image super-resolution solution within a cascaded manner, while it can be efficiently implemented with GPU.

2 Proposed JSCNN Model for JPEG Image Super-Resolution

Recent years have witnessed the great success of convolutional neural networks in a wide variety of challenging applications. Such a big success can be attributed to the advances in effective and fast learning of deep CNN models. The representative achievements include the proposals of Rectified Linear Unit (ReLU) [9], batch normalization [18] and residual learning [19, 20], and viewpoint on filter size and model depth [21, 22]. Other factors such as the efficient training implementation on modern powerful GPUs and the easy access to large scale dataset are also very important. In general, training a deep CNN for a specific task involves two steps: network architecture design and network learning from training data. In the following, we will illustrate our JSCNN model from those two aspects.

2.1 The Network Architecture

Since the work by Simonyan and Zisserman [21], increasing network depth with 3 × 3 convolution filters becomes a basic rule of thumb for most of CNN models. Networks with such design possess several properties, e.g., increasing the depth would also increase the receptive field and nonlinearity capacity. Specially, the receptive field with depth of d and 3 × 3 filters would be $(2d + 1) \times (2d + 1)$ which means, for image super-resolution problem, each output pixel is connected with a patch of size $(2d + 1) \times (2d + 1)$. Thus, the increase of the depth can enable the network use more information to predict the final pixel value. We set the network depth to 31 with a receptive field of 63 × 63.

Figure 1 shows the architecture of the proposed JSCNN network. In the first layer, we use 64 filters of size 3 × 3 to generate 64 feature maps, while in the last layer, we use one filter of size 3 × 3 × 64. In the middle layers, we use 64 filters of size 3 × 3 × 64. As in [17], we pad zeros before each convolution to make sure that each feature map of the middle layers has the same size of the input image. We use one big skip connection and 14 small skip connections which in combination form a skips-in-skip connection. The big skip connection enforces the network to predict the residual image of the input while each small skip connection corresponds to a residual unit as in [20]. Besides residual learning strategy, another important technique called batch normalization can accelerate the training and boost the performance. In the rest of this subsection, we illustrate the three key elements of our JSCNN, i.e., batch normalization, residual learning and skips-in-skip connection.

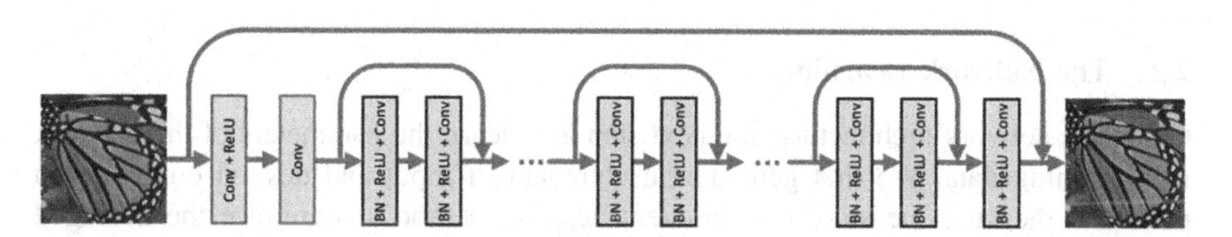

Fig. 1. The architecture of the proposed JSCNN network.

Batch Normalization. It is well-known that one key difficulty in training deep CNNs comes from the problem of vanishing/exploding gradients. This dilemma is caused by the large change in the distributions of internal nodes of a deep network during training (referred to as internal covariate shift [18]). To alleviate the internal covariate shift, the batch normalization was proposed to stabilize the distribution of nonlinearity inputs during the network training [18]. Accomplished by a normalization step that fixes the means and variances of layer inputs and a subsequent scale and shift step before the nonlinearity, batch normalization enjoys several merits. First, it enables large learning rates and thus can speed up the training. Second, it makes the model insensitive to parameter initialization. Third, it plays the role of regularizing the model, and tends to deliver a better performance.

Residual Learning. The residual learning for CNN was originally proposed to solve the performance degradation problem, i.e., performance gets saturated and then degrades rapidly along with the increase of network depth [19]. The main idea of residual learning lies in that instead of hoping each few stacked layers directly fit a desired underlying mapping, explicitly letting these layers fit a residual mapping would alleviate the degradation problem. In the more recent work by He et al. [20], it has been demonstrated that using identity mappings as the skip connections and after-addition activation can facilitate the forward and backward signals propagation from one block to any other block. As shown in Fig. 1, we follow this strategy to design the small skip connections.

Skips-in-Skip Connection. In most of single image super-resolution works, the main goal lies in predicting the high frequency information (or residual image) of the bicubically interpolated LR image. After that, the final super-resolved image can be obtained by simple addition of the interpolated LR image and the residual image. By learning the residual image rather than the clean image, Kim et al. [17] trained a deep CNN model for single image super-resolution with very promising results. Actually, this is a special case of residual learning, and it has been pointed out in [19] that if the original mapping (i.e., predicting the HR image) is more like identity mapping, then the residual learning (i.e., predicting the residual image) tends to be favorable. According to the above discussion, we can conclude that predicting the residual image can be treated as domain knowledge and it would be beneficial to CNN-based super-resolution. We implement the residual image learning by the big skip connection. As a result, the big and small connections are different levels of residual learning and they both share the merits of residual learning. More precisely, the skips-in-skip connection can result in faster training speed and better performance.

2.2 The Network Learning

Given the network architecture, the next step is to learn the parameters of the network from training data. We first generate an abundant of input and desired output patch pairs and then use the stochastic gradient descent method to minimize the averaged mean squared error loss function. To obtain a fast convergence, the most direct way is to use large learning rate. However, large learning rate is prone to cause exploding gradients problem in the beginning. Usually, there are two strategies to handle the exploding gradients problem. One is to first warm up the training by using a smaller learning rate and then use the large learning rate [20]. The other one is to limit the range of the gradients so that the effective gradient (gradient multiplied by learning rate) is small. In this paper, we adopt the following gradient clipping method proposed in [17] to enable large learning rate.

Gradient Clipping-Enabled Large Learning Rate. By using gradient clipping, Kim et al. [17] trained a CNN model with a very large learning rate of 0.1 for the first 20 epochs. The main idea is to narrow the gradients range when the learning rate is high and free the gradients range when the learning rate is low. Formally, the gradients are clipped into an adjustable interval $[-\tau/\alpha, \tau/\alpha]$ during training, where τ is a predefined threshold and α is the current learning rate. Figure 2 shows the performance of our JSCNN with respect to epochs, here the τ is set to 0.005 and the learning rate was decayed exponentially from 0.1 to 0.0001 for 30 epochs. It can be seen that our network enjoys a fast convergence. Note that even the batch normalization, fine parameter initialization and residual learning are utilized, the network collapses for convergence without above gradient clipping strategy in a few mini-batch iterations due to exploding gradients problem.

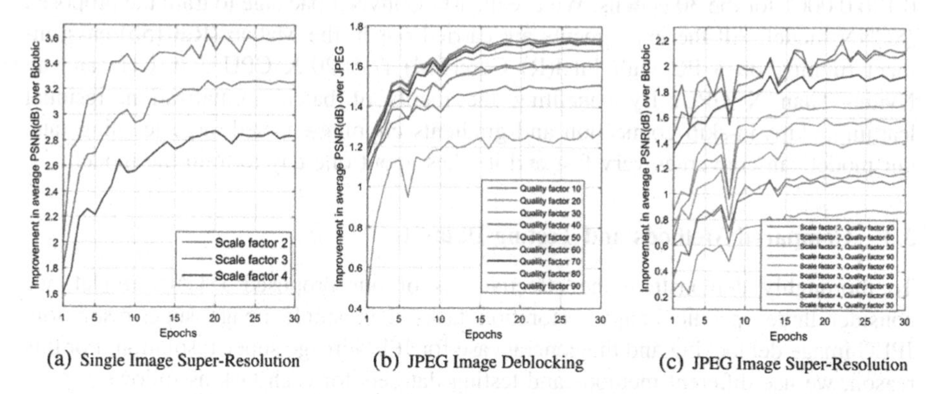

(a) Single Image Super-Resolution (b) JPEG Image Deblocking (c) JPEG Image Super-Resolution

Fig. 2. (a) Average PSNR improvement over Bicubic method for single image super-resolution with respect to epochs on Set5 datasets. (b) Average PSNR improvement over JPEG for image deblocking with respect to epochs on LIVE1 dataset. (c) Average PSNR improvement over Bicubic method for JPEG image super-resolution with respect to epochs on Set5 dataset.

3 Experimental Results

3.1 Training Data

It is widely acknowledged that convolutional neural networks generally benefit from the availability of large training data, thus, as in [17], we use a rather large dataset which consists of 91 images in [2] and 200 images from Berkeley Segmentation Dataset for training. To generate the corresponding LR JPEG image of an HR one, we first downscale the HR image with a certain scale factor to generate the clean LP image, then we compress the clean LR image into JPEG image with a certain quality factor by MATLAB JPEG encoder. Finally, we bicubically interpolated the JPEG image by the same scale factor to obtain an interpolated image such that the interpolated JPEG image and its HR image have pixel-wise correspondence. Note that the two extreme cases, i.e., single image SR and JPEG image deblocking, corresponds to the scale factor with $\times 1$ and quality factor with 100, respectively. Also note that, for color images, we convert the original images into the YCbCr color space and treat the luminance component Y as HR image. Considering the fact that the patch size should be larger than the respective field size (i.e., 63×63) in order to capture enough spatial information for better restoration, we set the training patch size to 64×64. We generate $3{,}000 \times 120$ patch pairs for training, rotation or flip based data augmentation is used during min-batch learning.

3.2 Parameter Setting

We initialize the weights by the method in [23] and use stochastic gradient descent (SGD) with weight decay of 0.0005, a momentum of 0.9 and a mini-batch size of 120. We trained 30 epochs for our model. The learning rate was decayed exponentially from 0.1 to 0.0001 for the 30 epochs. We use the MatConvNet package to train the proposed JSCNN model. All the experiments are carried out in the Matlab (R2015b) environment running on a PC with Intel(R) Core(TM) i7-5820 K CPU 3.30 GHz and an Nvidia Titan X GPU. By benefiting the merits of batch normalization, residual learning, skips-in-skip connection and gradients clipping-enabled large learning rate, our model can converges very fast and it takes about one day to train the model.

3.3 Compared Methods and Testing Datasets

To thoroughly demonstrate the effectiveness of our proposed JSCNN model, we consider three specific image restoration tasks, i.e., single image super-resolution, JPEG image deblocking and the general case for JPEG image super-resolution. For this reason, we use different methods and testing datasets for each task as follows.

For single image super-resolution, we use four state-of-the-art single image SR methods, including three external learning based methods named A+ [3], SRCNN [24] and VDSR [17], and one internal learning based method SelfEx [1]. To give a rather fair comparison, we followed the evaluation code as in [1] where the LR (color) images are provided. Thus, it is normal if the calculated metric values are slightly different from original papers. All implementation codes except [17] are downloaded from the authors' websites and we use the default parameter settings. For testing datasets, we adopt the widely used Set5 and Set14. In addition, B100 which contains 100 natural images from Berkeley Segmentation Dataset and Urban100 which is composed of 100 HR images with transformation-based repetitive structures are also used.

For JPEG image deblocking comparisons, two recently proposed JPEG image deblocking methods, i.e., AR-CNN [12] and TNRD [16], are used. The AR-CNN method trained four specific models for the JPEG quality factors Q = 10, 20, 30 and 40, while TNRD trained three models for the JPEG quality factors Q = 10, 20 and 30, respectively. Following [12], we use the Classic5 and LIVE1 as testing datasets. Classic5 contains 5 classic images and LIVE1 consists of 29 high quality images.

For the general case of JPEG image super-resolution, we use the cascaded model (denoted by AR-SRCNN) which consists of AR-CNN deblocking and a subsequent SRCNN super-resolution for main comparison. Because our model can handle the deblocking task and single image super-resolution task, similar to the above cascaded model, our model can also super-resolve the JPEG image in a cascaded manner. The cascaded model with our JSCNN is denoted as DB-SR-JSCNN. To show the effect of the plain SR method for super-resolving JPEG image, the SRCNN method is also included for comparison. We adopt the Set5 and LIVE1 as testing datasets.

3.4 Quantitative and Qualitative Evaluation

Since the three specific image restoration tasks aim to not only reconstruct the HR images but also generate visually pleasant output, we adopt two image quality metrics, i.e., PSNR and SSIM unless otherwise specified.

Single Image Super-Resolution. Table 1 shows the average PSNR and SSIM results of different single image super-resolution methods on four datasets. Since there is no available source code of VDSR, we just report the results from the original paper. As one can see, A+, SelfEx and SRCNN have very similar results. Even though our JSCNN is proposed for JPEG image super-resolution, it can achieve competing results with VDSR. Specially, our JSCNN and VDSR significantly outperform A+, SelfEx and SRCNN by a large margin. Figure 3 illustrates the super-resolved images by different methods. As can be seen, A+, SelfEx and SRCNN produce more visually pleasant results than the bicubic interpolation method; however, they tend to generate smoothed and unnatural edges. In comparison, our proposed JSCNN yields very appealing visual results as it can preserve image sharpness and naturalness.

Table 1. Average PSNR (dB)/SSIM results for *single image super-resolution* with scale factors ×2, ×3, ×4 on datasets Set3, Set14, B100 and Urban 100.

Dataset	Scale	Bicubic PSNR/SSIM	A+ PSNR/SSIM	SelfEx PSNR/SSIM	SRCNN PSNR/SSIM	VDSR PSNR/SSIM	JSCNN (Ours) PSNR/SSIM
Set5	× 2	33.64/0.9292	36.51/0.9536	36.46/0.9534	36.62/0.9534	37.53/0.9587	37.32/0.9564
	× 3	30.39/0.8678	32.58/0.9082	32.58/0.9090	32.74/0.9084	33.66/0.9213	33.71/0.9199
	× 4	28.42/0.8101	30.28/0.8599	30.32/0.8625	30.48/0.8628	31.35/0.8838	31.35/0.8822
Set14	× 2	30.22/0.8683	32.26/0.9050	32.21/0.9033	32.42/0.9061	33.03/0.9124	32.98/0.9105
	× 3	27.53/0.7737	29.12/0.8185	29.16/0.8196	29.27/0.8210	29.77/0.8314	29.76/0.8304
	× 4	25.99/0.7023	27.32/0.7492	27.39/0.7517	27.48/0.7510	28.01/0.7674	28.02/0.7666
B100	× 2	29.55/0.8425	31.21/0.8856	31.17/0.8853	31.34/0.8872	31.90/0.8960	31.86/0.8938
	× 3	27.20/0.7382	28.29/0.7831	28.29/0.7840	28.40/0.7857	28.82/0.7976	28.82/0.7960
	× 4	25.96/0.6672	26.82/0.7086	26.84/0.7086	26.90/0.7100	27.29/0.7251	27.29/0.7235
Urban100	× 2	26.66/0.8408	28.90/0.8970	29.31/0.9022	29.08/0.8966	30.76/0.9140	30.40/0.9161
	× 4	23.14/0.6573	24.33/0.7185	24.80/0.7376	24.52/0.7224	25.18/0.7524	25.27/0.7532

JPEG Image Deblocking. Table 2 shows average PSNR, SSIM and PSNR-B results for JPEG image deblocking with quality factors Q = 10, 20, 30 and 40 on datasets Classic5 and LIVE1. Note that PSNR-B is a specially designed metric for deblocked image quality assessment. It can be seen that our proposed JSCNN has an average PSNR and PSNR-B improvement larger than 0.3 dB over the state-of-the-art AR-CNN method on all the quality factors. Also, our proposed JSCNN achieves better results than TNRD. Figure 4 shows the JPEG image deblocking results of "Carnivaldolls" (LIVE1) with quality factor Q = 10. We can see that our proposed JSCNN can generate shaper edges than AR-CNN and TNRD.

Table 2. Average PSNR (dB)/SSIM/PSNR-B(dB) results for *JPEG image deblocking* with quality factors Q = 10, 20, 30 and 40 on datasets Classic5 and LIVE1.

Dataset	Quality	JPEG PSNR/SSIM/PSNR-B	AR-CNN PSNR/SSIM/PSNR-B	TNRD PSNR/SSIM/PSNR-B	JSCNN (Ours) PSNR/SSIM/PSNR-B
Classic5	10	27.82/0.7800/25.21	29.04/0.8111/28.75	29.28/0.8170/29.03	29.43/0.8202/29.04
	20	30.12/0.8541/27.50	31.16/0.8694/30.60	31.47/0.8744/31.05	31.69/0.8779/31.20
	30	31.48/0.8844/28.94	32.52/0.8967/31.99	32.78/0.8989/32.24	32.98/0.9018/32.37
	40	32.43/0.9011/29.92	33.34/0.9101/32.80	–	33.82/0.9146/33.17
LIVE1	10	27.77/0.7905/25.33	28.96/0.8217/28.68	29.15/0.8251/28.88	29.26/0.8262/28.83
	20	30.07/0.8683/27.57	31.29/0.8871/30.76	31.46/0.8906/31.03	31.66/0.8941/30.99
	30	31.41/0.9000/28.92	32.69/0.9166/32.15	32.83/0.9176/32.28	33.07/0.9206/32.26
	40	32.35/0.9173/29.96	33.63/0.9306/33.12	–	34.05/0.9346/33.22

JPEG Image Super-Resolution. Table 3 shows the average PSNR and SSIM results for JPEG image super-resolution with quality factor Q = 40, and scale factor ×2, ×3, ×4 on datasets Set5 and LIVE1. Since the bicubic interpolation method can even have better PSNR and SSIM results than SRCNN with quality factor Q = 40. Thus, we use Table 3 as a small example to show the effectiveness of our proposed JSCNN for JPEG image super-resolution. We can have the following observations. First, the plain single image super-resolution method SRCNN loses its effectiveness when the quality factor is set to 40. Second, compared to SRCNN, the cascaded model AR-SRCNN can greatly improve the results. Third, our cascaded DB-SR-JSCNN method can surpass the AR-SRCNN method, and our JSCNN can have a comparable performance with the cascaded DB-SR-JSCNN method. Figure 5 shows the visual results of different methods, we can see that SRCNN tends to exacerbate the blocking artifacts, and the super-resolved image by AR-SRCNN still contains the blocking-like artifacts. In comparison, our proposed method can generate more visually pleasant result. Figure 7 shows the average PSNR (dB) improvement of our proposed JSCNN over Bicubic method with different scale factors and JPEG quality factors on Set5 and LIVE1. It can be seen that our proposed JSCNN has a consistent improvement over Bicubic method, indicating than our JSCNN can handle the JPEG image super-resolution with various combination of scale factors and JPEG quality factors. Figure 6 gives an additional example to show the capacity of our JSCNN model. We can see that our JSCNN can produce visually pleasant output result even the input image is corrupted with different distortions in different parts. Actually, other methods such as AR-CNN or SRCNN cannot handle this case well. Thus, our JSCNN is more appealing for real applications.

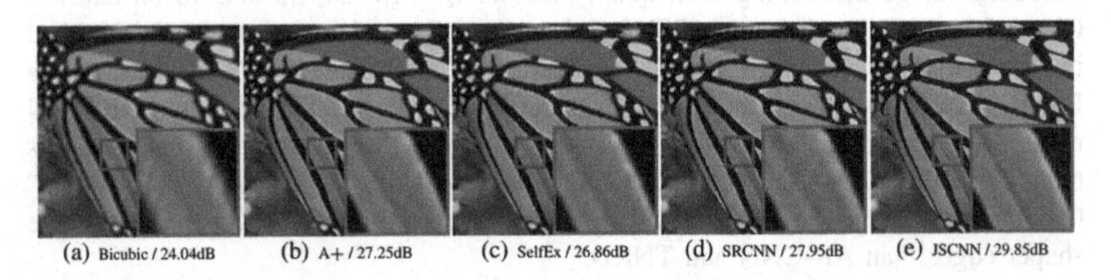
(a) Bicubic / 24.04dB (b) A+ / 27.25dB (c) SelfEx / 26.86dB (d) SRCNN / 27.95dB (e) JSCNN / 29.85dB

Fig. 3. Super-resolution results of "butterfly" (Set5) with scale factor ×3.

(a) JPEG / 28.10dB (b) AR-CNN / 28.85dB (c) TNRD / 29.54dB (d) JSCNN / 29.74dB

Fig. 4. JPEG image deblocking results of "Carnivaldolls" (LIVE1) with quality factor Q = 10.

Table 3. Average PSNR (dB)/SSIM results for *JPEG image super-resolution* with quality factors Q = 40 and scale factors ×2, ×3, ×4 on datasets Set5 and LIVE1.

Dataset	(Quality, Scale)	Bicubic PSNR/SSIM	SRCNN PSNR/SSIM	AR-SRCNN PSNR/SSIM	DB-SR-JSCNN (Ours) PSNR/SSIM	JSCNN (Ours) PSNR/SSIM
Set5	(40, ×2)	29.98/0.8559	29.57/0.8408	31.14/0.8834	31.64/0.8931	31.68/0.8941
	(40, ×3)	27.64/0.7818	27.39/0.7670	28.60/0.8196	28.95/0.8328	28.96/0.8332
	(40, ×4)	26.09/0.7137	25.87/0.7006	26.86/0.7575	27.16/0.7741	27.12/0.7700
LIVE1	(40, ×2)	27.11/0.7715	27.03/0.7675	27.78/0.7891	28.05/0.7969	28.07/0.7986
	(40, ×3)	25.39/0.6752	25.34/0.6715	25.88/0.6961	26.07/0.7039	26.05/0.7050
	(40, ×4)	24.42/0.6112	24.36/0.6073	24.81/0.6332	24.97/0.6409	24.93/0.6395

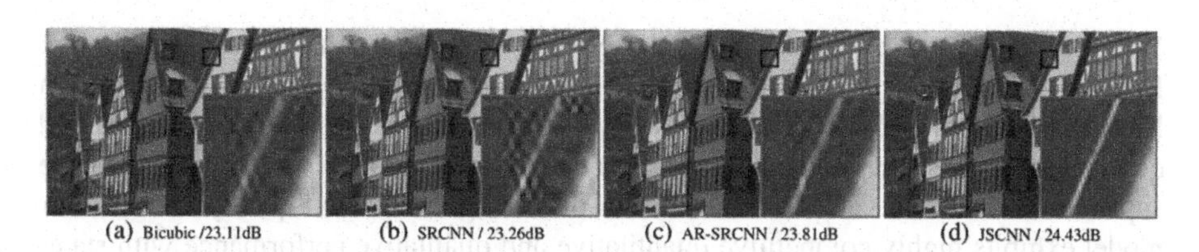

(a) Bicubic /23.11dB (b) SRCNN / 23.26dB (c) AR-SRCNN / 23.81dB (d) JSCNN / 24.43dB

Fig. 5. JPEG image super-resolution results of "Buildings" (LIVE1) with quality factor Q = 40 and scale factor ×2.

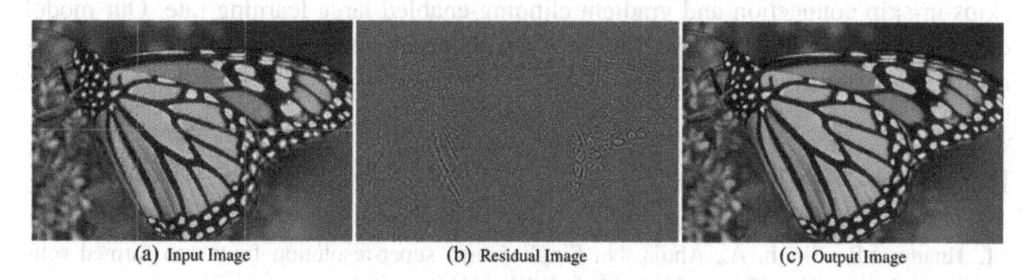

(a) Input Image (b) Residual Image (c) Output Image

Fig. 6. An example to show the capacity of our proposed JSCNN. The input image is composed by bicubically interpolated LR images with scale factor ×2 (upper left) and scale factor ×4 (lower left), JPEG images with quality factor 10 (upper middle) and quality factor 40 (lower middle), and bicubically interpolated LR JPEG image with scale factor 2, quality factor 40 (upper right) and scale factor 3, quality factor 60 (lower right). Note that the green line in the input image is just used for distinguish the six parts, and the residual image is normalized into the range of [0, 1] for visualization. Even the input image is corrupted with different distortions in different parts, the output image looks natural and does not have obvious artifacts.

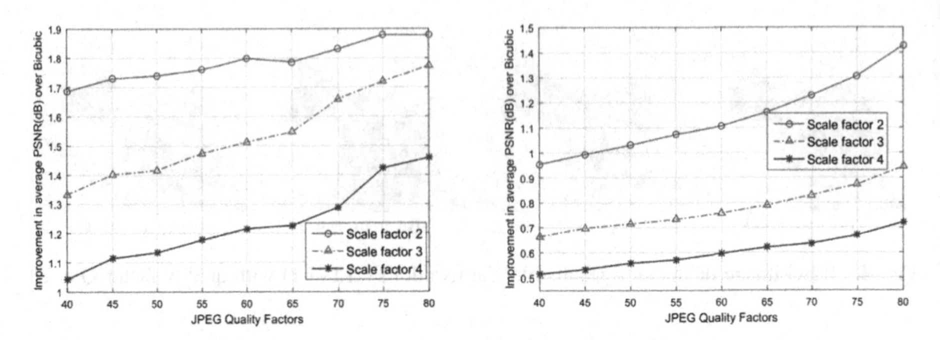

Fig. 7. The average PSNR (dB) improvement of our proposed JSCNN over Bicubic method with different scale factors and JPEG quality factors on Set5 (left) and LIVE1 (right).

Running Time. In addition to visual quality, another important aspect for practical applications is the testing speed. We use the Nvidia cuDNN-v5 deep learning library to accelerate the GPU computation of the proposed JSCNN. We do not count the memory transfer time between CPU and GPU. Our JSCNN can process an image of size 1024×1024 in 0.5 s.

4 Conclusions

We proposed a single deep residual network for JPEG image super-resolution. Different from conventional image super-resolution and deblocking methods which learn a specific model with a specific scaling factor for super-resolution or a specified quality factor for JPEG deblocking, our model can deal with the task of JPEG image super-resolution with various combinations of scaling factors and quality factors. Our model exhibits highly competitive quantitative and qualitative performance with state-of-the-art JPEG image super-resolution solutions, while it can be efficiently conducted with GPU. Furthermore, our model enjoys a fast convergence to a good solution by benefiting from various techniques including batch normalization, residual learning, skips-in-skip connection and gradient clipping-enabled large learning rate. Our model can be readily extended to other low-level vision problems such as simultaneous super-resolution, denoising and deconvolution.

References

1. Huang, J.B., Singh, A., Ahuja, N.: Single image super-resolution from transformed self-exemplars. In: CVPR, pp. 5197–5206. IEEE (2015)
2. Yang, J., Wright, J., Huang, T.S., Ma, Y.: Image super-resolution via sparse representation. IEEE TIP **19**, 2861–2873 (2010)
3. Timofte, R., De Smet, V., Van Gool, L.: A+: adjusted anchored neighborhood regression for fast super-resolution. In: Cremers, D., Reid, I., Saito, H., Yang, M.-H. (eds.) ACCV 2014. LNCS, vol. 9006, pp. 111–126. Springer, Cham (2015). https://doi.org/10.1007/978-3-319-16817-3_8

4. Dong, C., Loy, C.C., He, K., Tang, X.: Learning a deep convolutional network for image super-resolution. In: Fleet, D., Pajdla, T., Schiele, B., Tuytelaars, T. (eds.) ECCV 2014. LNCS, vol. 8692, pp. 184–199. Springer, Cham (2014). https://doi.org/10.1007/978-3-319-10593-2_13
5. Xiong, Z., Sun, X., Wu, F.: Robust web image/video super-resolution. IEEE TIP **19**, 2017–2028 (2010)
6. Kang, L.W., Hsu, C.-C., Zhuang, B., Lin, C.-W., Yeh, C.-H.: Learning-based joint super-resolution and deblocking for a highly compressed image. IEEE Trans. Multimed. **17**, 921–934 (2015)
7. Singh, A., Porikli, F., Ahuja, N.: Super-resolving noisy images. In: CVPR, pp. 2846–2853 (2014)
8. LeCun, Y., Kavukcuoglu, K., Farabet, C., et al.: Convolutional networks and applications in vision. In: ISCAS, pp. 253–256 (2010)
9. Krizhevsky, A., Sutskever, I., Hinton, G.E.: Imagenet classification with deep convolutional neural networks. In: NIPS, pp. 1097–1105 (2012)
10. Xu, L., Ren, J.S., Liu, C., Jia, J.: Deep convolutional neural network for image deconvolution. In: NIPS, pp. 1790–1798 (2014)
11. Bruna, J., Sprechmann, P., LeCun, Y.: Super-resolution with deep convolutional sufficient statistics, arXiv preprint arXiv:1511.05666 (2015)
12. Dong, C., Deng, Y., Loy, C.C., Tang, X.: Compression artifacts reduction by a deep convolutional network. In: ICCV, pp. 576–584 (2015)
13. Long, J., Shelhamer, E., Darrell, T.: Fully convolutional networks for semantic segmentation. In: CVPR, pp. 3431–3440 (2015)
14. Rothe, R., Timofte, R., Van Gool, L.: Efficient regression priors for reducing image compression artifacts. In: ICIP, pp. 1543–1547 (2015)
15. Kwon, Y., Kim, K.I., Tompkin, J., Kim, J.H., Theobalt, C.: Efficient learning of image super-resolution and compression artifact removal with semi-local gaussian processes. IEEE TPAMI **37**, 1792–1805 (2015)
16. Chen, Y., Pock, T.: Trainable nonlinear reaction diffusion: a flexible framework for fast and effective image restoration, arXiv preprint arXiv:1508.02848 (2015)
17. Kim, J., Lee, J.K., Lee, J.K.: Accurate image super-resolution using very deep convolutional networks, arXiv preprint arXiv:1511.04587 (2015)
18. Ioffe, S., Szegedy, C.: Batch normalization: accelerating deep network training by reducing internal covariate shift, arXiv preprint arXiv:1502.03167 (2015)
19. He, K., Zhang, X., Ren, S., Sun, J.: Deep residual learning for image recognition, arXiv preprint arXiv:1512.03385 (2015)
20. He, K., Zhang, X., Ren, S., Sun, J.: Identity mappings in deep residual networks, arXiv preprint arXiv:1603.05027 (2016)
21. Simonyan, K., Zisserman, A.: Very deep convolutional networks for large-scale image recognition, arXiv preprint arXiv:1409.1556 (2014)
22. Szegedy, C., Liu, W., Jia, Y., Sermanet, P., Reed, S., Anguelov, D., Erhan, D., Vanhoucke, V., Rabinovich, A.: Going deeper with convolutions. In: CVPR, pp. 1–9 (2015)
23. He, K., Zhang, X., Ren, S., Sun, J.: Delving deep into rectifiers: surpassing human-level performance on imagenet classification. In: ICCV, pp. 1026–1034 (2015)
24. Dong, C., Loy, C.C., He, K., Tang, X.: Image super-resolution using deep convolutional networks. IEEE TPAMI **38**, 295–307 (2016)

Application of an Improved Grab Cut Method in Tongue Image Segmentation

Bin Liu, Guangqin Hu(✉), Xinfeng Zhang, and Yiheng Cai

Faculty of Information Technology,
Beijing University of Technology, Beijing 100124, China
hdmh@163.com

Abstract. Grab Cut is an image segmentation method based on graph theory, and it is an improved algorithm of Graph Cut. Color images can be segmented by Grab cut. However, Grab Cut has the disadvantage of long segmentation time consuming. The application of SLIC (simple linear iterative clustering) super pixel method can reduce the time consumption. According to the particularity of the larger R value in the pixel of the tongue image, the formula of SLIC color space distance is improved, so that the super pixel produced by SLIC is more suitable for tongue image segmentation. The segmentation experiment on 300 tongue images shows that the segmentation accuracy of the improved algorithm is over 0.95, and the segmentation time is reduced greatly compared with the original Grab Cut algorithm. The algorithm can reduce the time of the tongue segmentation and improve the efficiency of the tongue segmentation, while maintaining the accuracy of the segmentation.

Keywords: Tongue image segmentation · Grab Cut · Super pixels
Improved color space distance

1 Introduction

Tongue diagnosis is an important part in the four methods of diagnosis of Traditional Chinese medicine (TCM), and it is also one of the unique characteristics of TCM diagnostic methods. Objective tongue diagnosis is an important part of the informatization of TCM, so the quality of tongue image segmentation will directly affect the subsequent operation. Many experts and scholars have studied tongue image segmentation, and proposed many algorithms. Reference [3] divides the tongue image segmentation into two categories: One is based on traditional image processing methods [4–6], because the color and brightness of the tongue body and lip are very similar, so the tongue and lip can not be well segmented by this method. The others are based on parametric active contour model (snake model) [7–9], Snake algorithm relies too much on contour initialization, so that the robustness is not strong. In addition, segment tongue images are segmented in reference [10] by support vector machine (SVM), but segmentation time is long and noise interference is serious. Thus this method leads to segmentation inefficiency and low accuracy. Because of the shortcomings of the tongue image segmentation methods above, in this paper use Grab Cut [11] to segment the tongue image. Parameters of the Gauss mixture model (GMM) is

D.-S. Huang et al. (Eds.): ICIC 2018, LNAI 10956, pp. 484–495, 2018.
https://doi.org/10.1007/978-3-319-95957-3_51

estimated in Grab cut by using each pixel in the image, and iterative method is used to further optimize the GMM. It is less dependent on the initial contour, and has higher robustness and segmentation accuracy. Due to the establishment of the GMM based on every pixel in the image, so the time consumption is long. In order to improve the segmentation speed, a lot of researches have been carried out. Reference [12] uses SLIC combined with Grab Cut to segment images, and improves the efficiency of Grab Cut segmentation. Reference [13] proves that the super pixel tightness and edge adhesion generated by SLIC algorithm are the best by experiment. Based on this, we propose a Grab Cut algorithm based on improved SLIC [14]. The features of tongue image are combined by the algorithm to modify the calculation method of spatial distance. According to the color distance between pixels, the image is divided into super pixels. The mean value of the pixels` RGB in super pixel represents the value of all pixels in the super pixel, and GMM is established with the RGB mean of each super pixel. Finally, the established GMM is used to segment the tongue image. The experimental results show that the algorithm can keep the segmentation accuracy and improve the speed of segmentation.

2 Related Research

2.1 Algorithm Introduction of Grab Cut

Graph Cut is a segmentation method based on graph theory [15]. The image is mapped into a network graph based on pixels, and each pixel is connected by edge (n-link), and each edge has a positive weight. Unlike ordinary map, two points are added in Graph Cut algorithm, respectively is the source point (S) and terminate point (T). Each of these two points is connected with image pixels by edge (t-link), and each edge has positive weights. In order to segment image, the smallest sum of two kinds of edge weights is cut by the image segmentation method. The most efficient method for solving the smallest sum of edge weight is Min-Cut/Max-Flow algorithm. The process is shown in Fig. 1.

Grab Cut algorithm is an improved algorithm for Graph Cut proposed by Rother [10] in 2004. In this algorithm, Gauss mixture model (GMM) is used to replace the gray histogram of Graph Cut, and color image segmentation is realized. Grab Cut uses Gibbs energy to represent edge weights.

The Gibbs energy of Grab Cut network graph model is defined as:

$$E(\alpha, k, \theta, z) = U(\alpha, k, \theta, z) + V(\alpha, z) \tag{1}$$

Where U represents the data item, which is the weight of the t-link, V represents the smoothing term, which is the weight of the n-link. α is the mark of foreground or background. Generally, each GMM consists of 5 Gauss components. Each pixel belongs to which Gauss component is usually determined by k-means and indicated by k. θ is the parameter of the Gauss component, z is the RGB value of each pixel.

$U(\alpha, k, \theta, z)$ is defined as:

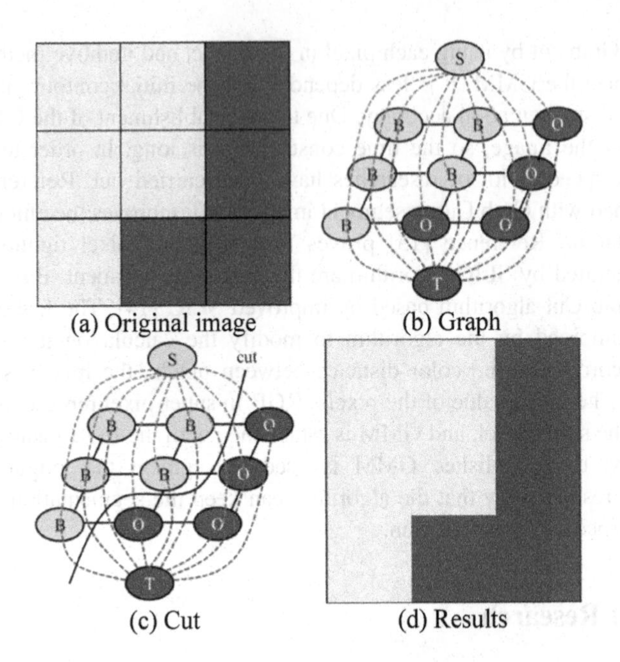

(a) Original image (b) Graph (c) Cut (d) Results

Fig. 1. The process of Graph Cut

$$U(\alpha, k, \theta, z) = \sum_n D(\alpha_n, k_n, \theta, z_n) \tag{2}$$

$$D(\alpha_n, k_n, \theta, z_n) = -\log \pi_n(\alpha_n, k_n) - \log g(x; \mu, \Sigma) \tag{3}$$

Among them, $\pi_i(\alpha_n, k_n)$ represents the mixed weight coefficient, $g(x; \mu, \Sigma)$ indicates the Gauss probability distribution.

$$g(x_i; \mu_n, \Sigma_n) = \frac{1}{\left(\sqrt{2\pi}\right)^d |\Sigma_n|^{\frac{1}{2}}} \exp\left[-\frac{1}{2}(x_i - \mu_n)^T \Sigma_n^{-1} (x_i - \mu_n)\right] \tag{4}$$

$V(\alpha, z)$ is the weighted accumulation of the color distance between all adjacent pixels which is defined as:

$$V(\alpha, z) = \gamma \sum_{(m,n) \in C} [p_n \neq p_m] \exp\left(-\beta \|z_m - z_n\|^2\right) \tag{5}$$

The pixel which belongs to the foreground or the background (α_n) will be changed when $E(\alpha, k, \theta, z)$ takes the minimum value. Calculating the updated Gibbs energy and updating α_n till α_n no longer changes. Image is segmented after the process.

Figure 2 shows the effect of tongue segmentation by Grab cut algorithm. The rectangle box shown in Fig. 2(a) is an artificial rectangle box. The pixels out of the rectangle are the background pixels and the others are the target pixels. The segmented image is shown in Fig. 2(b).

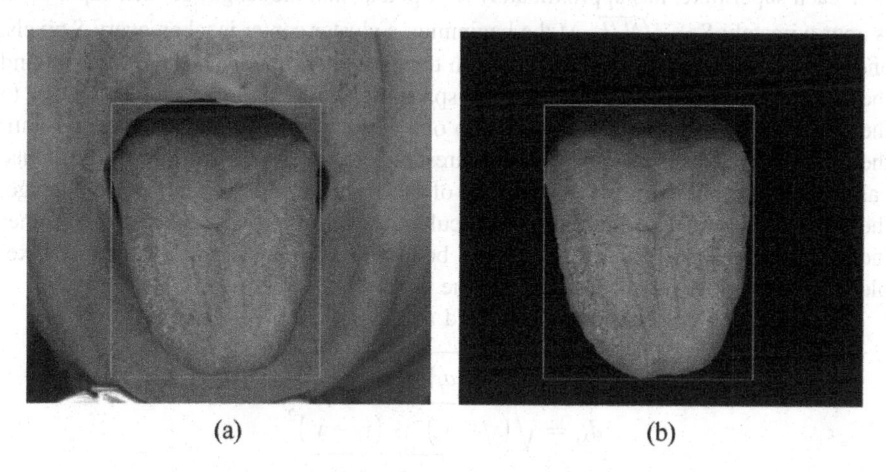

(a) (b)

Fig. 2. Grab cut algorithm for segmentation of tongue image

Table 1 is the evaluation result of Grab cut segmentation on 300 tongue images. From Table 1, we can see that Grab cut has high accuracy in tongue segmentation, and is suitable for tongue image segmentation. However each image pixel participated in the iteration when the parameters of GMM model are determined iteratively, which reduces the iterative efficiency. This is one of the drawbacks of Grab Cut. Therefore, a method is needed to reduce the amount of computation and improve the efficiency of the algorithm.

Table 1. Evaluation of Grab cut algorithm segmentation effect

Evaluation method	Grab cut
Dice-Ratio	0.9627
Recall	0.9585
Jaccard Index	0.9691
Time consuming(s)	2.6178

2.2 Introduction and Improvement of SLIC Algorithm

SLIC is a super pixel algorithm based on K-means algorithm. The reference [16] proves that SLIC is much better than Normalized Cuts, Graph-based, approach, Mean Shift and Quick Shift in generating super pixel processing speed. At the same time, SLIC generates lower block boundary error rates. The color space of RGB is converted to CIELab in SLIC algorithm. Each pixel corresponds to (L, a, b) values with (x, y) coordinates to form a 5 dimensional vectors (L, a, b, x, y).

Suppose an image is made up of N pixels. Parameter k is provided by user in the algorithm to determine that the image is divided into k super pixels. After segmentation, each super pixel has approximately N/k pixels, and the length of each super pixel is approximately $S = \sqrt{N/k}$. At the beginning, a cluster center is taken every S pixels, and move to the smallest gradient pixels in the 8 neighborhood. Then, $2S \times 2S$ around the center of the cluster is used as search space. Searching the pixels which similar to the center of the cluster and combine into one class. Updating the cluster center until the center pixel moves less than the set threshold. In this paper, an improved distance calculation formula is proposed. Because of the large R value in tongue body image, the R value is added to the distance calculation. The improved method can further increase the distance between the tongue body and lip color, so that the super pixel block can adhere to the edge of the tongue better.

The improved color distance is defined as:

$$\begin{aligned} d_{Lab} &= \sqrt{(L_i - L_j)^2 + (a_i - a_j)^2 + (b_i - b_j)^2 + \theta(R_i - R_j)^2} \\ d_{xy} &= \sqrt{(x_i - x_j)^2 + (y_i - y_j)^2} \\ d_s &= \sqrt{d_{Lab}^2 + (d_{xy}/S)^2 m^2} \end{aligned} \quad (6)$$

m is a parameter which defined by users. When the m value is large, the super pixels are more regular and compact, but they are not tightly bonded with the image boundaries. When the m value is small, the resulting super pixels are more closely bound to the image boundaries, but the pixels are not regular. Figure 3 is an effect diagram with different m values.

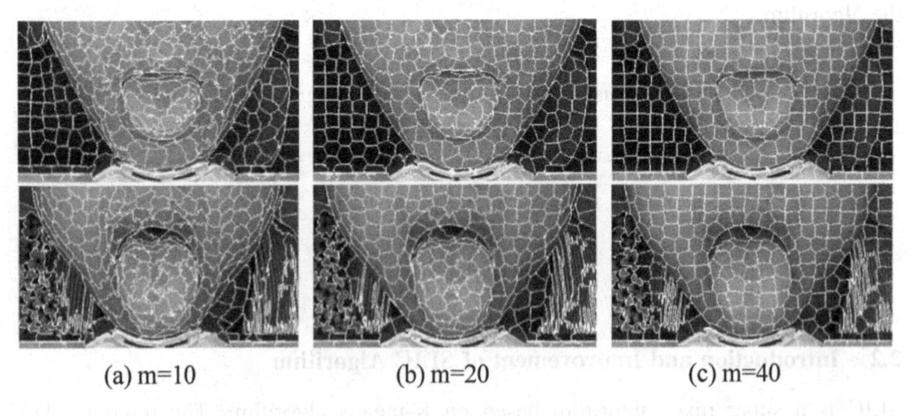

(a) m=10 (b) m=20 (c) m=40

Fig. 3. The super pixels based on k = 250 and different value of m

For SLIC algorithm, we need to choose different values based on experience. In this paper, we propose an adaptive improvement method. When the image edge is complex, a smaller m value is needed, so that the generated super pixels can better adhere to the image edge. The color image is transformed into gray image by improved algorithm, and Canny operator is used to calculate the approximate Complexity of image edge.

Figure 4(b) shows the approximate edge extracted by Canny operator. When the edge is complex, the SLIC algorithm needs smaller m values to better adhere to the edge of the image. When the edge is simple, the SLIC algorithm needs a larger m value to make the generated super pixel more regular. Suppose P is the pixel that is in the edge of Canny result image. If P has edge pixels at 225° or 315° in 8 neighborhoods, then the total number of edge pixels is increased. The representation of edge complexity is as follows:

$$\Psi = \frac{\sum_{i=1}^{I} \sum_{j=1}^{J} \delta(P_{i,j} \in E)}{I \times J} \tag{7}$$

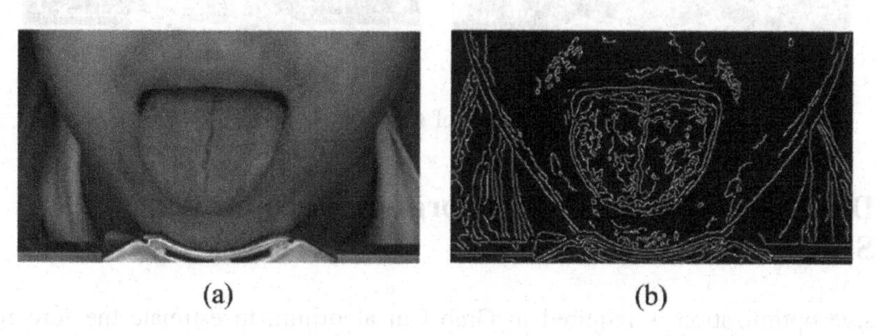

(a) (b)

Fig. 4. The result of the tongue image after using Canny operator

In which I, J represent the length and width of the image, respectively. E represents the edge points that conform to the statistical rules in the edges extracted by Canny algorithm. Formula (7) statistics the ratio of edge pixels in input image.

$$m = 40(1 - \Psi) \tag{8}$$

Formula (8) is the value of m in SLIC. The 40 in the formula is the maximum value suggested in the document [14]. It can be seen that the larger the image complexity is, the smaller the m value is, and the generated super pixels can better adhere to the image edge. The smaller the image complexity is, the bigger the m value is, and the generated super pixels can be more regular. Figure 5 is the result of super pixel segmentation for some tongue images. Where m is automatically determined and N = 400.

From Fig. 5, we can see that the improved algorithm can adaptively select the m value when super pixel segmentation is applied to tongue images. Choosing the right m value can not only better adhere to the edge of the tongue, but also ensure the regularity of the super pixels not belong to edge. The tongue image is well segmented by the improved algorithm.

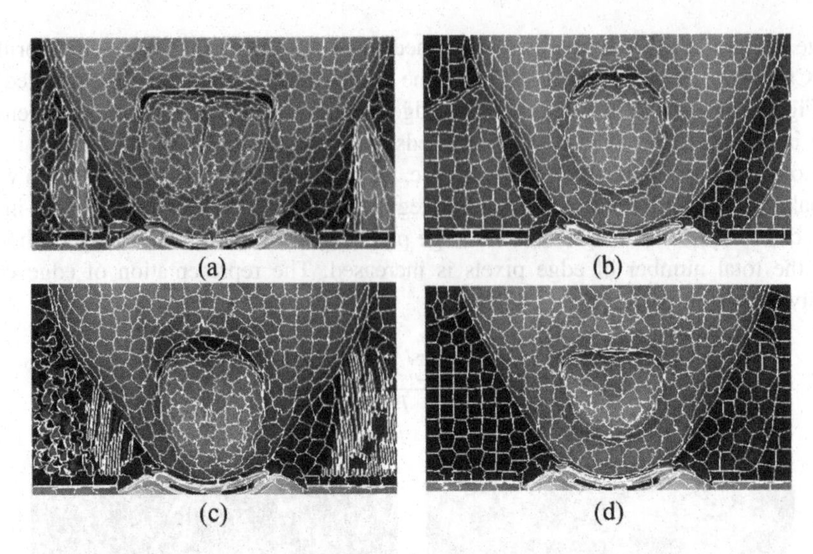

Fig. 5. The self-adaption of super pixels when N = 400

3 Description of Grab Cut Algorithm Based on Improved SLIC

Iterative optimization is required in Grab Cut algorithm to estimate the foreground GMM and background GMM parameter with each pixel in the image. The estimation efficiency is very low and takes a long time. Using the improved SLIC, the input image is converted into super pixel, and the mean value of RGB in each super pixel block is calculated. It can reduce the iteration times and time consuming when using the mean value of RGB in super pixel block to estimate the parameters in foreground GMM and background GMM. Each pixel of the original input image is segmented using two GMM models.

3.1 Super Pixel Processing of Input Image Using Improved SLIC

The input image size is 500 × 275. We calculate the image distance determined by the CIELab color space. Therefore, the input image is converted from the RGB color space to the CIELab color space. The number of pixels (k value) should be determined according to the characteristics of the input image. If the k value is too large, the time of GMM parameter estimation is increased, otherwise, the pixels cannot adhere to the tongue edge very well. Figure 6 shows the effect diagram of different k when $m = 15$.

It can be clearly seen from Fig. 6 that when the k value is between 400 and 800, the super pixel keeps good adhesion to the tongue body edge, so in this experiment the k value is determined 600.

After selecting the number of super pixels, the location of the cluster center pixel is determined according to the 1.2 section introduction method. The formula (6) is used to calculate the distance between each pixel and the center pixels of the cluster. Constantly update the pixel position of the cluster center until it is less than the set

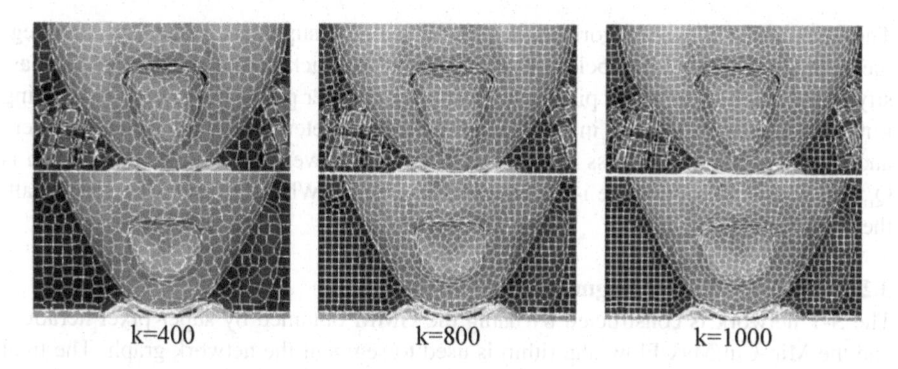

Fig. 6. The result of different k values when m = 15

threshold. The mean values of all pixels in each super pixel block are calculated, and the GMM parameters are iteratively updated by mean.

3.2 Iterative Estimation of GMM Parameters

3.2.1 Initialzing GMM Parameters

Entering the super pixel image, and manually calibrate the rectangle, as shown in Fig. 7. The foreground image is in the rectangle frame, and the background image is outside the frame. Normally, 5 Gauss components are set in GMM. The foreground images and background images are classified into 5 classes according to the RGB values by using k-means, and each class corresponds to a GMM Gauss component and initializes the Gauss component parameters. The weight of each Gauss component is the ratio of the super pixels number in Gauss component to the total super pixel.

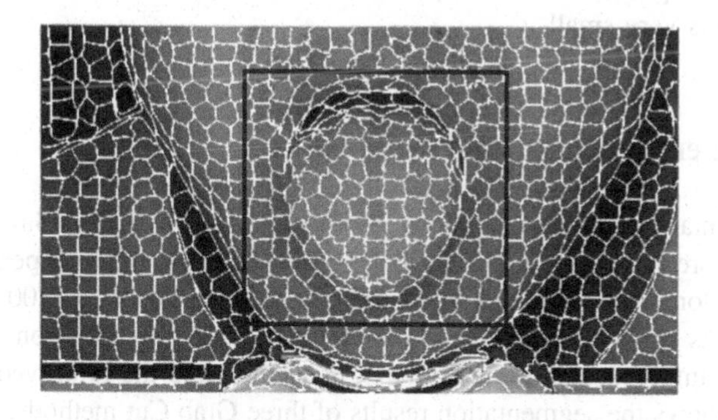

Fig. 7. Manual calibration rectangular box

3.2.2 Iterative Estimation of GMM Parameters

The S-T network graph is established according to the super pixel image, and the weight (t-link) between the super pixel and the vertex is determined by formula (3), and the weight (n-link) between the super pixels is determined by formula (5).

The Min-Cut/Max-Flow algorithm is used to segment super pixel images. After segmentation, the super pixel belongs to foreground or background is changed. Reclassifying the foreground super pixels and background super pixels into 5 classes by using k-means method mentioned in 2.1, updating the parameters in each Gauss component and the weight of each Gauss component. The sum of weight of the i segmentation is Q_i, the sum of weight of the i-1 segmentation is Q_{i-1}, When $Q_i - Q_{i-1}$ is very small, the iteration ends.

3.2.3 Using GMM to Segment Input Image

The S-T network is constructed by using the GMM obtained by super pixel iteration, and the Min-Cut/Max-Flow algorithm is used to segment the network graph. The final segmentation graph is obtained.

Algorithm steps are as following:

First step: Using ISLIC algorithm to segment the input tongue image. The number of segmented super pixels is 600.
Second step: Calculating the RGB mean of every super pixel generated by the first step of ISLIC algorithm, and replace RGB value of pixel in super pixel block by mean value.
Third step: In the second step generate super pixel picture, manually add rectangle, outside the rectangle is the background super pixel, and the target super pixel is in the rectangle.
Fourth step: Using k-means algorithm to divide the background super pixels and foreground super pixels into 5 classes. We estimate each class of Gauss model parameters and establish S-T network graph.
Fifth step: Cutting the S-T network graph into two parts by using the Min-Cut/Max-Flow algorithm.
Sixth step: Repeat the fourth step and the fifth step until the change of the super pixel label is very small.

4 Experimental Result and Analysis

The experimental platform is Windows 8, 64 bit, MATLAB R2013a, Visual Studio 2010, CPU: Core i7-5500U, RAM 8 GB. The tongue image of the experiment is taken by a standard tongue picture instrument. The size of the image is 1300×1426.

Experiments are conducted to compare the image segmentation accuracy and segmentation time with the original Grab Cut method and the improved method.

Figure 8 shows the segmentation results of three Grab Cut methods. It can be seen that the accuracy of improved SLIC Grab Cut is close to that of Grab Cut. The edge is overflowed on some tongue images by using SLIC Grab Cut. The reason is that when the SLIC is used to divide the super pixels, the super pixel block does not adhere to the edge of the tongue, which makes the estimation of the GMM parameter error. The experiment segments 300 tongue images. The effect of three Grab cut methods were compared with Dice-Ratio, Recall, Jaccard-Index and the average time of segmentation. The results show in Table 2.

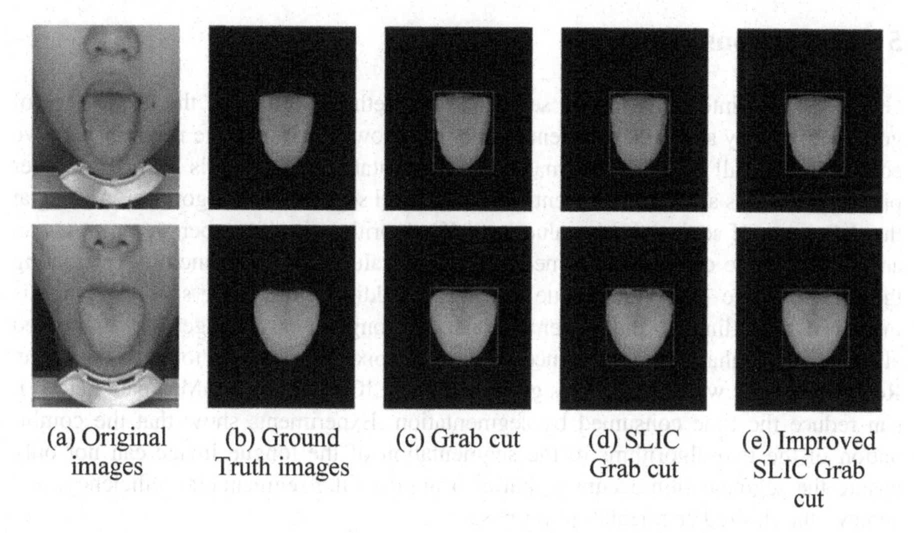

Fig. 8. Comparison of segmentation effects between three Grab Cut methods

Table 2. Comparison of mean of accuracy and time between three Grab cut methods

	Improved SLIC Grab Cut	Grab Cut	SLIC Grab Cut
Dice-Ratio	0.9604	0.9627	0.9232
Recall	0.9523	0.9585	0.9303
Jaccard-Index	0.9616	0.9691	0.9422
Time consuming(s)	1.85	2.6178	1.73

It can be seen from Table 2 that the segmentation accuracy of SLIC Grab Cut is lower than that of another two Grab cut methods. This is because the super pixels generated by the SLIC algorithm do not adhere well to the edge of the tongue body. The distance between the tongue body and lip color is increased by using the improved SLIC Grab Cut, so that the super pixel block can adhere to the edge of the tongue better. The improved algorithm can segment tongue images better. As can be seen from Table 2, the segmentation accuracy of the Improved SLIC Grab cut as well as Grab cut can be very good for the segmentation of the tongue image. It also can be seen from Table 2 that the segmentation time consuming of Improved SLIC Grab cut and SLIC Grab cut is shorter than that of Grab cut method. This is because the pixels of the input images is converted into super pixels by using SLIC, and replacing the RGB of the block pixels with the mean of RGB in the pixel block to estimate foreground GMM and background GMM iteratively. The reducing of pixel numbers and iteration times make the segmentation time of two improved Grab Cut algorithms less than the Grab Cut algorithm. Table 2 shows that the improved SLIC Grab Cut algorithm proposed in this paper can achieve good segmentation accuracy in tongue image segmentation while reducing the time of segmentation, and it can further improve the efficiency of tongue image.

5 Conclusions

Grab Cut is an interactive image segmentation method, which has the advantages of good interactivity and good segmentation effect. However, due to the need for iterative computation of all pixels of the image, the segmentation efficiency is low. SLIC super pixel algorithm is a fast and efficient boundary pixel segmentation algorithm. Aiming at the deficiency of setting the m value in SLIC algorithm, in this paper we proposes an adaptive m value determination method. The m value was determined by calculating the ratio of edge pixels of tongue images. In addition, the color space distance is improved according to the larger R value of tongue body image. The improved algorithm can enhance the adherence of the super pixel block to the tongue body edge. Replacing pixels with super pixels generated by SLIC to generate GMM parameters, it can reduce the time consumed by segmentation. Experiments show that the combination of the two algorithms to the segmentation of the tongue image can not only ensure the segmentation accuracy, but also improve the segmentation efficiency, and achieve the desired segmentation purpose.

Acknowledgment. Thanks to Institute of Department of information, Beijing University of Technology for supporting our work and giving us great suggestion. Our work is supported by the national key research and development program (No. 2017YFC1703300) of China. At the same time, we also thank to the teachers and students who made great contribution to this study.

References

1. Li, N.: Complete Diagnosis of Tongue Diagnosis in TCM. Academy Press, Beijing (1995). 1525, 12241347
2. Chiu, C.C.: A novel approach based on computerized image analysis for traditional Chinese medical diagnosis of the tongue. Comput. Methods Programs Biomed. **61**(2), 77–89 (2000)
3. Qin, W., Li, B., Yue, X.: A hybrid tongue image segmentation algorithm based on initialization of Snake contours. J. Univ. Sci. Technol. China **40**(8), 807–811 (2010)
4. Wu, W.J., Ma, L.Z., Xiao, X.Z.: Method of tongue image segmentation based on luminance and roughness information. J. Syst. Simul. (2006)
5. Li, C.H., Yuen, P.C.: Tongue image matching using color content. Pattern Recogn. **35**(2), 407–419 (2002)
6. Zhao, Z., Wang, A., Shen, L.: The color tongue image segmentation based on mathematical morphology and HIS model. J. Beijing Polytech. Univ. (1999)
7. Liu, C., Zhang, H., Yang, H.: Application of GVF Snake model based on Perona-Malik algorithm in segmentation of tongue image. Microcomput. Appl. (2017)
8. Sun, X., Pang, C.: An improved snake model method on tongue segmentation. J. Chang. Univ. Sci. Technol. **36**(5), 154–156 (2013)
9. Zhang, X., Wang, M., Cai, Y., et al.: A high robust tongue image segmentation algorithm based on an active contour model with shape priors. J. Beijing Univ. Technol. **39**(39), 1481–1487 (2013)
10. Liu, Z., Chen, J.X., Zhao, Y.M., et al.: Automatic tongue image segmentation based on visual attention and support vector machine. J. Beijing University of Traditional Chinese Medicine (2013)

11. Rother, C., Kolmogorov, V., Blake, A.: "GrabCut": interactive foreground extraction using iterated graph cuts. Trans. Graph. **23**(3), 309–314 (2004)
12. An, N.Y., Pun, C.M.: Iterated graph cut integrating texture characterization for interactive image segmentation. IEEE Comput. Graph. Imaging Vis., 79–83 (2013)
13. Song, X., Zhou, L., Li, Z., et al.: Review on superpixel methods in image segmentation. J. Image Graph. **20**(5), 0599–0608 (2015)
14. Achanta, R., Shaji, A., Smith, K., et al.: SLIC superpixels. Epfl (2010)
15. Zhou, L.: Improved image segmentation algorithm based on GrabCut. J. Comput. Appl. **33** (1), 49–52 (2013)
16. Achanta, R., Shaji, A., Smith, K., et al.: SLIC superpixels compared to state-of-the-art superpixel methods. IEEE Trans. Pattern Anal. Mach. Intell. **34**(11), 2274–2282 (2012)

Dynamic Fusion of Color and Shape for Accurate Circular Object Tracking

Thi-Trang Tran and CheolKeun Ha(✉)

School of Mechanical Engineering, University of Ulsan,
93 Daehak-ro, Namgu, Ulsan, Korea
trantrang286@ulsan.ac.kr, cheolkeun@gmail.com

Abstract. Circular shape detection and tracking in robotic systems includes a wide, promising array of research that aims to develop robotic skills for interacting with realistic environments in chasing circular objects in real-time. To this end, in this work, the authors contribute a new circular object tracking approach using dynamic feature fusion. The tracker estimates the target positions through frame sequences, and it is built on a robust fusion of possible potential target positions in different feature spaces. The proposed tracking strategy includes three main steps to perform the object chasing. First, the features are extracted from input frames. Then, in each feature space, the estimator will find the predicted position of a target. Finally, the dynamic fusing of information from different feature spaces will validate the target position. Experimental results achieved using the sequence of images obtained immediately from a pan-tilt-zoom (PTZ) camera in a real-time object detection and tracking system are provided to validate the proposed approach.

Keywords: Circle detection · Radius measurement · Object tracking
Real-time system

1 Introduction

Circle detection is fundamental to many computer vision applications such as eye detection and tracking [1–3] automatic product inspection in industrial vision applications [4], circular traffic sign recognition [5–7] and ball detection in games [8, 9]. The common main challenges of circle detection are changes in viewpoint, perspective, and noise, and time constraints in particular. This problem has received considerable attention, mainly in an attempt to improve the detection accuracy and reduce the amount of time consumed. Hough transform (HT)-based [10], approaches are commonly used. This approach is simple, yet it has massive computation and memory requirements, requires a great deal of time and is sensitive to noise. Our goal in this paper is to find a simple, fast tracker for real-time systems that runs with inexpensive cameras and does not require calibrated lenses [14].

The main contributions of this work include the proposal of a new method for robust circular object tracking based on feature fusion, which can be performed at a frame rate of 30 Hz. The proposed tracking strategy includes three main steps to perform the object chasing. First, the features are extracted from input frames. Then, in

D.-S. Huang et al. (Eds.): ICIC 2018, LNAI 10956, pp. 496–507, 2018.
https://doi.org/10.1007/978-3-319-95957-3_52

each feature space, the estimator will find the predicted position of a target. Finally, the dynamic fusing of information from different feature spaces will validate the target position.

This paper is organized as follows. The feature space and feature estimator are illustrated in Sects. 2 and 3, respectively. In Sect. 4, we present the fusion strategy. Experimental results obtained from our real-time object detection and tracking system using a pan-tilt-zoom (PTZ) camera are analyzed in Sect. 5. Finally, Sect. 6 describes the conclusions and future work.

2 Feature Space

2.1 Feature Space Selection

In order to achieve a high precise decision on target position, selected feature spaces form a robust combination. The best combination of feature spaces for object tracking needs to be adaptive to changes in illumination and perspective as the camera and/or object moves. In this work, we focus on the two most important features describing an object: color and shape. Apparently, shape information is robust under illumination changes, while color is very sensitive to these changes, as seen in Fig. 1. However, object tracking using only shape information could fail due to changes in perspective and partial occlusion, as can be seen in Fig. 2 in which the color information can help the tracker. Utilizing feature space synthesis is the key issue for a robust tracker. In the proposed method, the probability of each feature space is first calculated separately. Then, the target object position is achieved by dynamic fusion of the feature space, in which a higher priority is assigned to the feature most similar to the static model.

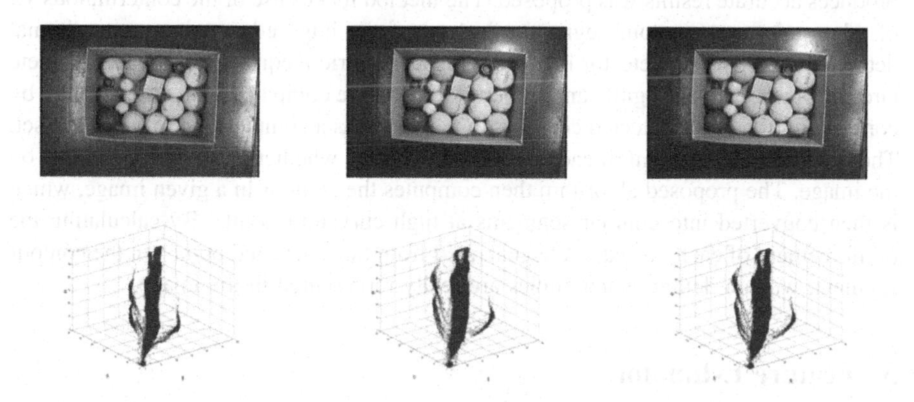

Fig. 1. Example of color illuminant changes over time

2.2 Color Space

Theo Gevers and W.M. Smeulders presented a comparison between different color models for color-based object recognition. The choice of color model depends on their

Fig. 2. Example of shape change

robustness against varying illumination and changes in object surface orientation. In this work, we need to use a color model that is robust to a viewpoint change. The HSV (Hue, Saturation, Value) color space is most appropriate because the HSV color model has two of its own strong principal advantages. Firstly, the H and S components are related to the way in which a human perceives color; therefore, the colors in this model can be clearly defined by human perception. Secondly, the H component (color information) is disassociated from the V component (the brightness of color) and S component (how concentrated the color is). Thus, the tracker can trust the H component for any purpose of object detection and recognition. In this paper, we use the hue image from the H channel in the HSV space for future histogram calculation.

2.3 Shape Feature Space

To achieve real-time performance, a fast circle detector that has high detection rates and produces accurate results was proposed. The method makes use of the conterminous set of edge points, or contour segments, instead of the huge edge points as traditional detectors do. First, the detector is based on isoperimetric inequality to extract complete circles, which shows a significant property of the circle compared with other curves, by considering the set of detected contours in the image as a complete circle candidate set. The isoperimetric quotient of each candidate indicates whether or not it is presented on the image. The proposed algorithm then computes the contour in a given image, which is then converted into contour segments at high curvature points. By calculating the reinforcement of each virtual circle generated from three selected points on the contour segment, we can define which circles are really represented in the image.

3 Feature Estimator

3.1 Color Space Estimator

The HSV color space separates the hue (color) from the saturation and brightness. In this paper, we take the 1D histogram from the H (hue) channel in the HSV space. Initially, the model H histogram, *Md*, of the target object is sampled inside the tracking window. During operation, *Md* is used as a model, or, in other words, a look-up table, to convert incoming image pixels into a corresponding probability for the model image. Let $b_{x,y}$ denote the probability of a pixel (x, y) after histogram back-projection.

For all pixels inside the tracking window

i. Compute the zero moment

$$M_{00} = \sum_x \sum_y b_{x,y} \tag{1}$$

ii. Find the first moment for x and y

$$M_{10} = \sum_x \sum_y x b_{x,y} \tag{2}$$

$$M_{01} = \sum_x \sum_y y b_{x,y} \tag{3}$$

iii. The possible object localization is determined as follows

$$x_c = \frac{M_{10}}{M_{00}} \quad ; \quad y_c = \frac{M_{01}}{M_{00}} \tag{4}$$

Record (x_c, y_c) and M_{00} for the potential tracking window in the next frame, where the window size is set to

Window width:

$$s = 2 * \sqrt{\frac{M_{00}}{256}} \tag{5}$$

Window length:

$$l = 1.2 * s \tag{6}$$

3.2 Shape Space Estimator

In circle detection, the expectation is to find triplets of (x_c, y_c, R) that completely describe a circle's center x-axis coordinate, y-axis coordinate and radius, respectively. To overcome the complex computations and time-consuming aspect of circular feature detection, a method based on the contour CCDA is proposed. The general idea is to detect the contours of an image, extract the complete circles and convert the remaining into a set of circle candidates, generate virtual circles based on the candidates, and then validate the circles that are actually present in the image. A general outline of the proposed algorithm is presented in Algorithm 1, and each step will be described in detail in the following sections.

Algorithm 1. Steps of CCDA.

1) Detect contours and extract complete circle by using isoperimetric inequality.
2) Convert the remaining contours into contour segments and make circle candidates.
3) Join the circle candidates to detect the remaining circles.
4) Output the valid circles.

Complete Circle Detection

The first step of the CCDA is to detect contours in the image as can be seen in Fig. 3. Each contour, ct_i, is stored in a contour vector, $CT = \{ct_1, ct_2, \ldots, ct_{N_1}\}$, and then complete circles are extracted from CT to generate two contour vectors, $C = \{c_1, c_2, \ldots, c_{N_{11}}\}$, which contains complete circle contours, and $T = \{t_1, t_2, \ldots, t_{N_{12}}\}$, which contains the remaining contours. N_1, N_{11} and N_{12} represent the number of original detected contours, complete circular contours, and remaining contours, respectively, where $N_1 = N_{11} + N_{12}$. To achieve this, a well-known contour detection algorithm derived from [24] is first employed. Complete circles are then detected in the contour image according to the isoperimetric inequality in Algorithm 2.

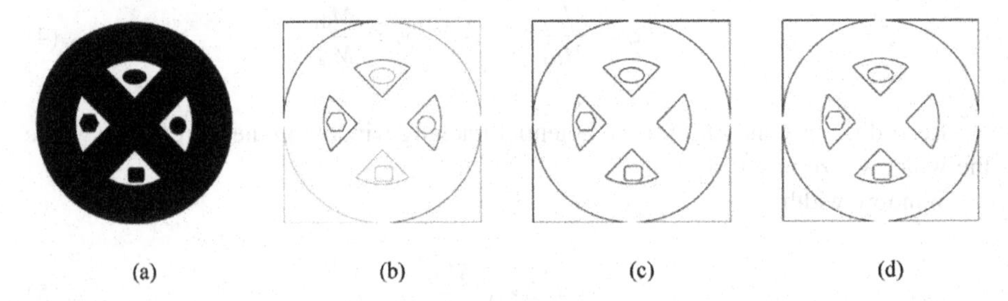

Fig. 3. (a) A sample image (906 × 918). (b) Contour detection. Each color represents a different contour. A total of 16 contours are extracted in 4 ms. (c) The remaining contours in vector T after complete circle detection. (d) A total of 60 contour segments are approximated.

Algorithm 2. Steps for complete circle detection based on isoperimetric inequality.

1) Utilize the results of a reference paper [29] to calculate the square of each contour area in the contour vector, CT.

Let n_i, S_i, and P_i denote the number of edge points in the boundary, ct_i, the square of the region inside the boundary, ct_i, and the perimeter of the boundary, ct_i, respectively, and let (x_j, y_j) denote the coordinates of point pt_j.

For all edge points, $pt_j (j \in [0, n_i - 1])$, on the boundary, ct_i:

$$S_i = \frac{1}{2}\sum_{j=1}^{n_i} x_{j-1}y_j - x_jy_{j-1} \tag{7}$$

2) Compute the perimeter of each boundary.

For all edge points, $pt_j (j \in [0, n_i - 1])$, on the boundary, ct_i:

$$P_i = \sum_{j=1}^{n_i} \sqrt{(x_j - x_{j-1})^2 + (y_j - y_{j-1})^2} \tag{8}$$

where pt_n is an additional point to close the polygon, $pt_n = pt_0$.

3) Compute the ratio between the candidate area and perimeter.

In Euclidean geometry, the isoperimetric inequality shows the privileged roles of the Euclidean circles. The isoperimetric inequality states, for the length P of a closed curve and the area S of the planar region that it encloses, that

$$4\pi S \leq P^2 \tag{9}$$

and that equality holds if and only if the curve is a circle. However, in an image plane, the isoperimetric inequality changes according to the circle definition. In this paper, according to our experimental results, we pre-set a threshold *isi* to distinguish complete circles from other closed curves. If a closed curve has $S/P^2 \geq isi$, it is determined to be a complete circle (usually the threshold is set to 0.079).

4) Find the target position.

$$x_{c_i} = \frac{1}{n_i} \sum_{j=1}^{n_i} x_j; \; y_{c_i} = \frac{1}{n_i} \sum_{j=1}^{n_i} y_j \tag{10}$$

$$R_i = \sqrt{\frac{2 \times S_i}{P_i}} \tag{11}$$

where $(x_{c_i}, y_{c_i}), R_i$ are the center and radius, respectively.

Store all complete circles in the vector, $C = \{c_1, c_2, \ldots, c_{N_{11}}\}$, and store the remaining contours in vector $T = \{t_1, t_2, \ldots, t_{N_{12}}\}$.

Contour Segments Detection

Having obtained a contour vector $T = \{t_1, t_2, \ldots, t_{N_{12}}\}$, the second step is to validate the circle candidates. Unlike traditional circle detectors which work on a set of potential edge pixels to generate candidate circles, the CCDA follows a proactive approach and works by first identifying contour corners and then removing all the detected corners to generate a new contour set, in other words, a candidate circle set. Starting with a contour, the algorithm operates by first picking two external points and connecting them with a line; it then searches to find the farthest point from the drawn line and adds this point to the approximation. The process is repeated, adding the next most distant point to the accumulated approximation until all of the points are less than the indicated distance. Satisfied point coordinates are stored in a corner vector, $V = \{v_1, v_2, \ldots, v_{N_{12}}\}$, with $v_i = \{v_{i1}, v_{i2}, \ldots, v_{in_i}\}$ being a corner vector of contour t_i and n_i being the number of corners on the contour, t_i.

The contour extraction process is implemented as follows. Given a list of corner points, $v_i = \{v_{i1}, v_{i2}, \ldots, v_{in_i}\}$, on each contour, t_i, walk over the contours, t_i, and detach the sub-contours from the contour at the corner points. Finally, the new contour segment set is generated from T and stored in a contour vector, $ST = \{st_1, st_2, \ldots, st_{N_{12}}\}$, with $st_i = \{st_{i1}, st_{i2}, \ldots, st_{in_i}\}$ being the contour segment vector of contour t_i (as shown in Fig. 1d).

Remaining Circle Detection

The CCDA detects the remaining circle based on the following ideas. A circle candidate set is first created from the contour segments, ST, and then the circles actually represented in the image are defined. For a set of contour segments ST, to be a potential candidate, it must have redundant contours; therefore, certain criteria should be released. For the purposes of our current implementation, we do not consider very short contours that are canceled as they may represent noise. Using these criteria, M satisfied contour segments are extracted to make circle candidates, and they are stored in a vector, $RC = \{r_1, r_2, \ldots, r_M\}$, with r_i and M being the ith circle candidate and the number of circle candidates of the contour, r_i, respectively, where $M \leq N_{12}$. For each extracted contour segment, i, we select three edge points, $p_{k1}(x_1, y_1)$, $p_{k2}(x_2, y_2)$, and $p_{k3}(x_3, y_3)$, and generate a virtual circle, r_i, that passes through the selected points. The virtual circle, r_i, is defined by the three parameters, x_i, y_i, and R_i, where (x_i, y_i) is the center coordinate and R_i is its radius. The virtual circle can be modeled as follows:

$$(x - x_i)^2 + (y - y_i)^2 = R_i^2 \tag{12}$$

where x_i and y_i can be calculated by the following equations

$$x_i = \frac{\det(A)}{4((x_2 - x_1)(y_3 - y_1) - (x_3 - x_1)(y_2 - y_1))} \tag{13}$$

$$y_i = \frac{\det(B)}{4((x_2 - x_1)(y_3 - y_1) - (x_3 - x_1)(y_2 - y_1))} \tag{14}$$

With

$$A = \begin{bmatrix} x_2^2 + y_2^2 - (x_1^2 + y_1^2) & 2(y_2 - y_1) \\ x_3^2 + y_3^2 - (x_1^2 + y_1^2) & 2(y_3 - y_1) \end{bmatrix} \tag{15}$$

$$B = \begin{bmatrix} 2(x_2 - x_1) & x_2^2 + y_2^2 - (x_1^2 + y_1^2) \\ 2(x_3 - x_1) & x_3^2 + y_3^2 - (x_1^2 + y_1^2) \end{bmatrix} \tag{16}$$

The radius is calculated using

$$R_i = \sqrt{(x_i - x_d)^2 + (y_i - y_d)^2} \tag{17}$$

with $p_d(x_d, y_d) \in \{p_1, p_2, p_3\}$.

After the virtual circles are generated, the next step is to validate whether the circles really exist in the image. First, the midpoint circle algorithm (MCA) is used to determine the required points for drawing circles on the image. The MCA input arguments are the center coordinate, (x_i, y_i), and the radius, R_i. It starts drawing a curve at point $(R_i, 0)$ and proceeds upwards and left by using integer additions and subtractions. It is important to ensure that it does not consider points lying outside the image plane. Assuming that the number of required points representing virtual circle r_i in the image is N_{ci}, the reinforcement, $\alpha(r_i)$, implies the matching error between the pixels on the virtual circle, r_i, and the pixels that actually exist in the image. Let $E(x_j, y_j)$ denote a function that verifies the pixel existence at (x_j, y_j).

$$E(x_j, y_j) = \begin{cases} 1 & \text{if } (x_j, y_j) \text{ is an edge point} \\ 0 & \text{otherwise} \end{cases} \quad (18)$$

Then $\alpha(r_i)$ can be calculated as follows

$$\alpha(r_i) = \frac{\sum_{j=1}^{N_{ci}} E(x_j, y_j)}{N_{ci}} \quad (19)$$

A value close to $\alpha(r_i)$ implies a better candidate. There are two ways to obtain the optimal solution: either one virtual circle (circle candidate) generates a matching reinforcement $\alpha(.)$ under the pre-established limit (typically 0.1 as has been suggested [19], or it takes the highest probability action at the end of the process.

4 Fusion Strategy

The possible target positions calculated by different estimators are combined to validate the final target position by assigning higher priority to the model's more likely features. The similarity of each statistical distribution with the model is evaluated using the Bhattacharyya coefficients. Due to the fact that the Bhattacharyya coefficients are closely related to Bayes error and their properties have been previously illustrated [2], the Bhattacharyya coefficients represent a near optimal solution. Let $q = \{q_u\}_{u=1 \ldots m}$ denote the model histogram, which is determined in the tracking initialization step. At frame k, the candidate histogram is $p(k) = \{p_u(k)\}_{u=1 \ldots m}$. The distance between two m-bin histograms is defined as

$$d(k) = d(q, p(k)) = \sqrt{1 - \rho(k)} \quad (20)$$

The Bhattacharyya coefficient $\rho(x)$ is given by

$$\rho(k) = \rho(q, p(k)) = \sum_{u=1}^{m} \sqrt{p_u(k) q_u} \quad (21)$$

The higher Bhattacharyya coefficient reflects the higher contribution to the target position decision. Let us denote possible positions found in the color space and shape feature space as (x_c, y_c) and (x_s, y_s), respectively, and the corresponding Bhattacharyya coefficients as ρ_c, ρ_s. The final target position is determined as follows:

$$x = \frac{\rho_c}{\rho_c + \rho_s} x_c + \frac{\rho_s}{\rho_c + \rho_s} x_s \tag{22}$$

$$y = \frac{\rho_c}{\rho_c + \rho_s} y_c + \frac{\rho_s}{\rho_c + \rho_s} y_s \tag{23}$$

The possible window size in any feature space that provides the largest Bhattacharyya coefficient is chosen as the adaptive window. In addition, the statistical representations of the model determined at the tracking initialization stage needs to be updated since the target and/or the camera are moving over time.

5 Experimental Results

5.1 Installation and Reinstallation After Occlusion Confirmation

The first experiment is performed while the camera is static; the frame sequence is about 1000 frames, and the target is a yellow circular object. The experiment is performed to examine the proposed system installation and reinstallation after occlusion.

First, the system is in DTS, or in other words, *whole image processing*, and when the target appears in the camera view, it is detected automatically by CCDA (as seen in Fig. 4a, b), where the detected circle is marked in the red overlay. The system decides the detected circle is the target object, learns the target histogram model, and switches to TKS mode, or in other words, *interest region processing*. When a loss is detected, the system goes back to DTS mode, as shown in Fig. 4c. The system successfully reinitializes in Fig. 4d, and DTS succeeds in finding the target after loss detection. TKS then continues to follow the target object. In the experiment, there are 4 total occlusions, 1 time fail in tracking, and 5 successful recoveries from the loss.

5.2 Performance of Object Detection

The third experiment is conducted to evaluate object detection performance of the proposed system. The test is performed in object detection and measurement of bin picking system. In this test, we compare the results between the proposed approach and approaches using color or shape information.

While the three methods have comparable performances, the proposed method using feature fusion has the best performance as can be seen in Fig. 5.

5.3 Performance w.r.t the Changing of Illumination

This experiment is performed while changing illumination through video sequence. The proposed system detects and tracks a target object.

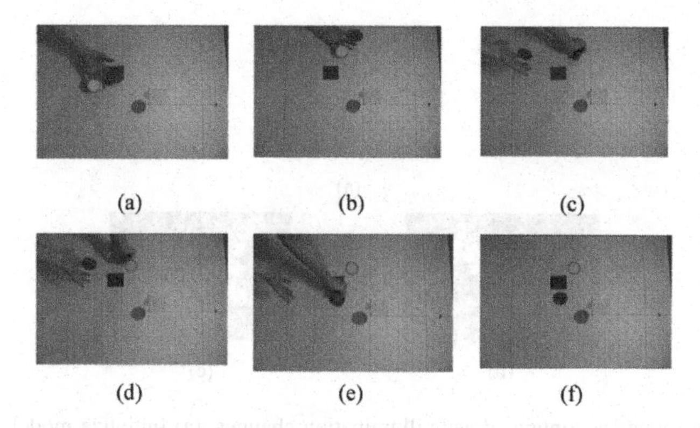

Fig. 4. Example of installation and reinstallation after occlusion.

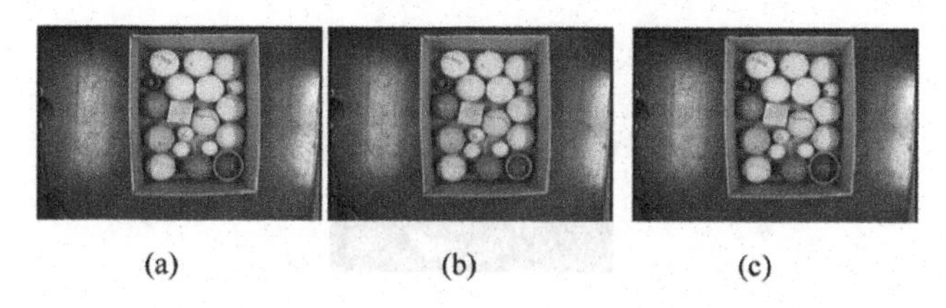

Fig. 5. Comparison between object detection and measurement approaches. (a) using proposed feature fusion; (b) using color information; (c) using shape information.

The video sequence contains the target, a yellow ball, shown in Fig. 6(a). The tracking result shows that using color cue alone is not enough to provide good result, and that the tracker fails to track the target object when the illumination changes drastically as shown in Fig. 6(c). The shape feature still preserver well the structure of the target object in this case, and the tracker that employs both shape and color features maintains good performance as shown in Fig. 6(b).

5.4 Performance w.r.t the Changing of View Point

This experiment is conducted while changing view point through video sequence. The view point changes cause flattening effects which make circular objects appear elliptic. The experiment results show that the proposed approach can detect and track a target object successfully while changing view point.

The video sequence presents challenges in viewpoint changes, shown in Fig. 7, in which the target objects are the two circles. Using shape feature only is not enough to track the target accurately as can be seen in Fig. 7(c). However, the fusion of features still tracks the target successfully as shown in Fig. 7(b).

Overall, the online fusion of color and shape feature provides a robust approach for object tracking. Color information compensate for shape changes due to viewpoint changes, partial occlusion, while shape information compensates for color variance in illumination changes condition.

Fig. 6. Video sequence contains drastic illumination changes. (a) Initialize model; (b) Tracking using proposed online feature fusion; (c) Tracking using color information fails due to illimination changes.

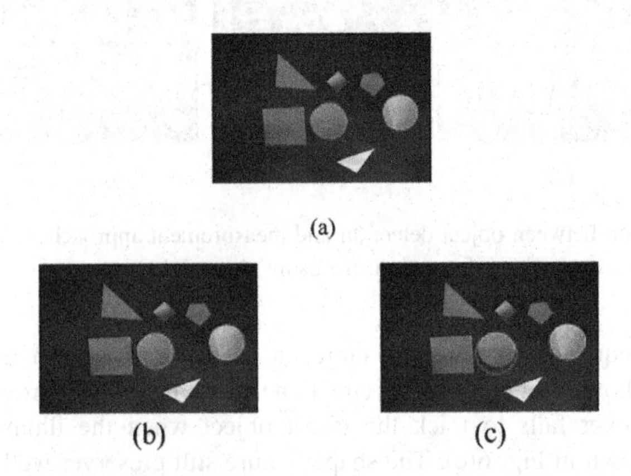

Fig. 7. Video sequence contains view point changes. (a) Initialize model; (b) Tracking using online feature fusion; (c) Tracking using shape information alone fails due to view point changes.

6 Conclusion

In this paper, we proposed a robust circular object tracking approach using image sequences at a frame rate of 30 fps. The statistical model of the object is built and updated in different feature spaces, and by fusing possible target positions in the feature spaces, the final target position is validated. The experiment results demonstrate robust tracking performance under various object tracking scenarios, including perspective changes, drastic illumination changes, and occlusions. Circular shape detection and tracking is a fundamental problem in image processing, plays an important role in sport vision-based systems, bin picking, and many other industrial applications. Thus, for future research, we plan to fuse more cues such as motion and texture to expeditiously deal with certain applications.

References

1. Wan Mohd Khairosfaizal, W.M.K., Nor'aini, A.J.: Eyes detection in facial images using circular Hough transform. In: 5th Signal Processing & its Applications (CSPA 2009), Kuala Lumpur, pp. 238–242. IEEE (2009)
2. Hassabalth, M., Ido, S.: Eye detection using intensity and appearance information. In: MVA 2009 IAPR Conference on Machine Vision Applications, Yokohama, Japan, 20–22 May 2009
3. Ito, Y., Ohyama, W., Wakabayashi, T., Kimura, F.: Detection of eyes by circular Hough transform and histogram of gradient. In: 21st International Conference on Pattern Recognition (ICPR 2012), Tsukuba, pp. 11–15, pp. 1795–1798. IEEE (2012)
4. Da Fontoura Costa, L., Marcondes Cesar Jr., R.: Shape Analysis and Classification: Theory and Practise. CRC Press, Boca Raton (2007)
5. Barnes, N., Zelinsky, A.: Real-time radial symmetry for speed sign detection. In: Intelligent Vehicles Symposium, 14–17, pp. 566–557 (2004)
6. Hoferlin, B., Zimmermann, K.: Towards reliable traffic sign recognition. In: Intelligent Vehicles Symposium, pp. 324–329 (2009)
7. Mainzer: Genetic algorithm for traffic sign detection. In: International Conference Applied Electronics 2002, Pilsen, Czech Republic, pp. 129–132 (2002)
8. Yu, X., Leong, H.W., Xu, C., Tian, Q.: Trajectory-based ball detection and tracking in broadcast soccer video. IEEE Trans. Multimed. **8**(6), 1164–1178 (2006)
9. Zhang, H.Y., Wu, Y.D., Yang, F.: Ball detection based on color information and Hough transform. In: 2009 International Conference on Artificial Intelligence and Computational Intelligence, pp. 393–397 (2009)
10. Illingworth, J., Kittler, J.: A survey of the Hough transform. Comput. Vis. Graph. Image Process. **44**, 87–116 (1988)
11. Trang, T.T., Ha, C.: Irregular moving object detecting and tracking in real-time system. In: International Conference on Management and Telecommunications (ComManTel 2013), Hochiminh city, Vietnam, pp. 415–419 (2013)
12. Tran, T.-T., Ha, C.: Slippage estimation using sensor fusion. In: Huang, D.-S., Jo, K.-H. (eds.) ICIC 2016. LNCS, vol. 9772, pp. 471–481. Springer, Cham (2016). https://doi.org/10.1007/978-3-319-42294-7_42
13. Trang, T.T., Ha, C.: An efficient approach for circular shape target recovery. In: 2015 IEEE International Conference on Advanced Intelligent Mechatronics (AIM), Busan, Korea (2015)
14. Tran, T.-T., Ha, C.: Extrinsic calibration of a camera and structured multi-line light using a rectangle. Int. J. Precis. Eng. Manuf. **19**(2), 195–202 (2018)

A Method of Ore Image Segmentation Based on Deep Learning

Li Yuan(✉) and Yingying Duan(✉)

University of Science and Technology Beijing, Beijing, China
lyuan@ustb.edu.cn, 18341256055@163.com

Abstract. In the analysis of ore particle size based on images, ore segmentation is a key link. After accurate segmentation, the geometric parameters such as the contour of these blocks, the external rectangle, the center of mass and the invariant moment can be further obtained, and the ideal ore particle size can be obtained effectively. A method of ore image segmentation based on deep learning is proposed in this paper. The method focuses on solving the problem of inaccuracy caused by mutual adhesion and shadow on the ore image. Firstly, complex environment image data set is obtained by using high resolution webcam; Next, we use the annotation data set to train HED (Holistically - Nested Edge Detection) model. This model can extract the image edge feature with strong robustness. Then, thinning edge is extracted using table lookup algorithm. The final step is labeling the connected region and getting segmented results. Our method is compared with the Watershed method based on gradient correction, and experimental results show the effectiveness and superiority of the proposed method.

Keywords: Ore segmentation · HED · Edge detection

1 Introduction

Ore particle size measurement is mainly aiming at the separation of ore images on the conveyor belt before the mine blasting and carrying into the crusher. In early days, size measurement was based on manual measurement, which not only required a large amount of manpower and material resources, but also had low accuracy and efficiency. Then researchers proposed image based method for automatic ore size measurement, which was hoped to be more accurate. Therefore, ore particle size based on images is a common and automatic means, among which the ore segmentation is a key link. The accuracy of ore segmentation can affect the subsequent production process and benefit.

The environment of ore particle size measurement based on images is shown in Fig. 1.

In the ore image data collected in outdoor environment, we can find that due to the different time period and the different intensity of light, the image will be changed in shadow and brightness. In order to meet the environmental protection department requirements, workers need to spray water on the conveyor belt to reduce the pollution caused by high dust. This causes the effect of ore moisture (dry/wet/mixture), which leads that the edge of ore in image is blurred. In addition, due to the different mine

D.-S. Huang et al. (Eds.): ICIC 2018, LNAI 10956, pp. 508–519, 2018.
https://doi.org/10.1007/978-3-319-95957-3_53

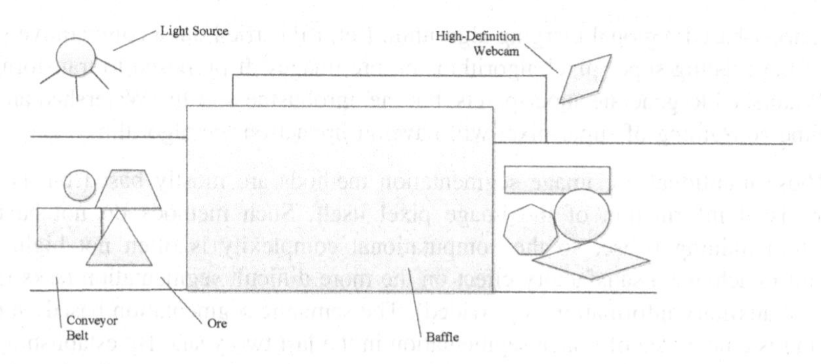

Fig. 1. The environment of ore particle size measurement based on images

break point and different amount of explosive, the ore has irregular shape, too large or too small. When it meets the colored mine, the uneven color will appear. All these phenomenons bring great difficulty for the separation of ore particle based on images. Thus, we need to solve the above problems by using a method with high segmentation accuracy and fast speed. It is instructive to improve the efficiency of mine production and save a lot of energy.

Ore image segmentation refers to dividing the image into a number of non-overlapping areas according to the characteristics of grayscale, color, texture and shape.

At present, the ore particle segmentation methods can be categorized into the following categories:

(1) The segmentation method based on threshold value. It determines the optimal gray threshold value according to a certain criterion function. Ref. [1, 2] applied double window one-dimensional information and the gray value and variance information of adjacent pixel to determine the threshold. Although some noise effects could be overcome, it was still difficult to accurately divide the ore images with high similarity and fuzzy boundary.

(2) The segmentation method based on region. It can be mainly divided into Watershed transform, region merging and so on. In Ref. [3], by using the bilateral filter to smooth image, the marker Watershed transform was implemented based on the distance transform and morphological reconstruction. This method could effectively reduce the over-segmentation rate, but most of the threshold values in the algorithm were selected by manual debugging, and the robustness was poor. In addition, the ore images on the conveyor belt have the problems such as large similarity between target and background, and the shadow region, etc. This method was not ideal for handling these problems.

(3) The segmentation method based on particular theories. Ref. [4] presented a method for the segmentation of adhesion ores with concave points. After the Harris angle points detection, it matched the concave point of the circular template, and found the best matching by the rectangular limit. This method was novel, but the limitation of the Harris algorithm limited the segmentation accuracy. Ref. [5] represented an automatic segmentation method based on multi-scale strategy to generate the pyramid and initial marker, and ore segmentation was generated by using the

marker-based regional merging algorithm. Ref. [6] carried out a comparative study of the existing super-pixel algorithms on ore images. It proposed to transform the Watershed to generate super-pixels, but the unrobustness of the Watershed and the time consuming of super-pixel will have an impact on the algorithm.

These traditional ore image segmentation methods are mostly based on the low order visual information of the image pixel itself. Such methods do not have an algorithm training phase, so the computational complexity is often not high. It is difficult to achieve a satisfactory effect on the more difficult segmentation tasks (if no artificial auxiliary information is provided). The semantic segmentation based on deep learning is a new idea of image segmentation in the last two years. By establishing the multi-level algorithm structure model through a large number of samples learning, it can obtain the deep semantic features with strong robustness [7–9]. This paper is focused on solving the difficulties mentioned above for the ore partical segmentations based on image. The appropriate convolution network HED (holier-nested Edge Detection) [10] is selected to detect ore edges. By using the multi-scale and multi-level feature learning algorithm of HED, the precise segmentation of different stones in the ore image is achieved. Experimental results on the collected images from the work site show that the proposed method is effective for the segmentation of complex ore images.

The structure of this article is as follows: Sect. 2 introduces the system structure and detailed principle of ore image segmentation based on deep learning. Section 3 details the experimental results and compares our proposed method with the Watershed method based on gradient correction. Section 4 concludes the paper.

2 Ore Image Segmentation Based on Deep Learning

2.1 System Description

In this paper, we propose a two-stage method for ore image segmentation. During the training stage, the HED network is applied to train the model of images collected in actual production process. During the ore partial segmentation stage, the edge of the test image is extracted using the trained model. Then, after thinning edge, connected region is further obtained to realize the ore segmentation. The overall system flow chart is shown in Fig. 2.

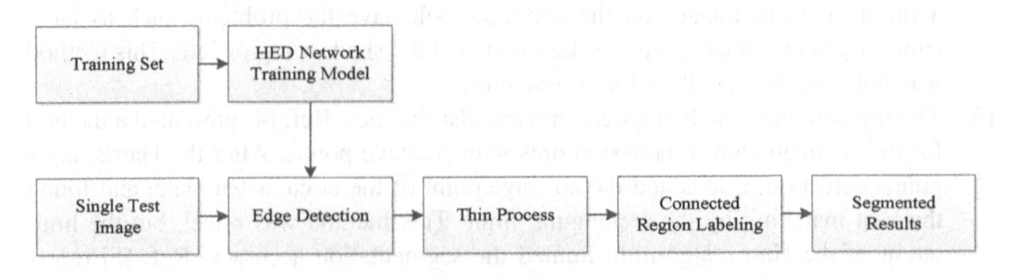

Fig. 2. System flow

The system flow describes some brief steps of the method proposed in this paper. Firstly, ore images in complex environment are collected and the edge data set is generated. Next, we use these annotation data to train the HED network, and choose the model with the best parameters. Then, we select some images that are different from the images in train set and use the trained network for edge detection. Next, for the convenience of subsequent operation the edge results are thinned. The last step is to label connected region and get segmented results.

This process eliminates the need of complicated image pre-processing, and gets rid of the disturbance of parameter adjustment, and overcomes the influence of severe noise. It not only has good detection effect in complex and changing environment, but also improves image processing speed and has high practical application value.

2.2 HED (Holistically-Nested Edge Detection) Network [10]

HED algorithm mainly solves two problems through the deep learning model, which is based on the training and prediction of the whole image, as well as the multi-scale and multi-level feature learning. Through the learning of rich classification features, HED can make a more detailed edge detection. The algorithm is improved based on the VGG network [11] and FCN network. It connects the side output layer after the last convolution layer in each stage. It also removes the last pooling of the VGG network and all the subsequent full connection layers to reduce memory resource consumption. 'Holistically' represents that the result of edge detection is based on the process of image to image, end-to-end. However, 'nested' emphasizes the process of continuously inheriting and learning from the generated output process and getting the accurate edge prediction graph. It is similar to an independent multi-network, multi-scale prediction system, while finally using the weighting strategy to fuse the multi-scale output layer into a single deep network. The network structure is shown in Fig. 3.

During the training phase, in the ground truth, the ratio of the number of non-edge pixels to the edge number is greater than 9:1. In this case, a pixel - based class-balancing weight was introduced to solve the loss balance of positive and negative samples. The formula for the loss function is:

$$l_{side}^{(m)}\left(W, w^{(m)}\right) = -\beta \sum_{j \in Y_+} \log P_r(y_j = 1|X; W, w(m)) - (1-\beta) \sum_{j \in Y_-} \log P_r(y_j = 0|X; W, w(m)) \tag{1}$$

Where $\beta = |Y_-|/|Y|$, $1-\beta = |Y_+|/|Y|$, $|Y_-|$ and $|Y_+|$ Indicates the edge and non-edge of the label set respectively. Using sigmoid function $\sigma(\cdot)$ to compute $P_r\left(y_j = 1\middle|X; W, w^{(m)}\right)$, m represents layers.

The weighted fusion layer is added to the network, and the weight value of the fusion is continuously learned during the training process, which predictions can be used directly and effectively from the side output. At this point, the fusion layer loss function is:

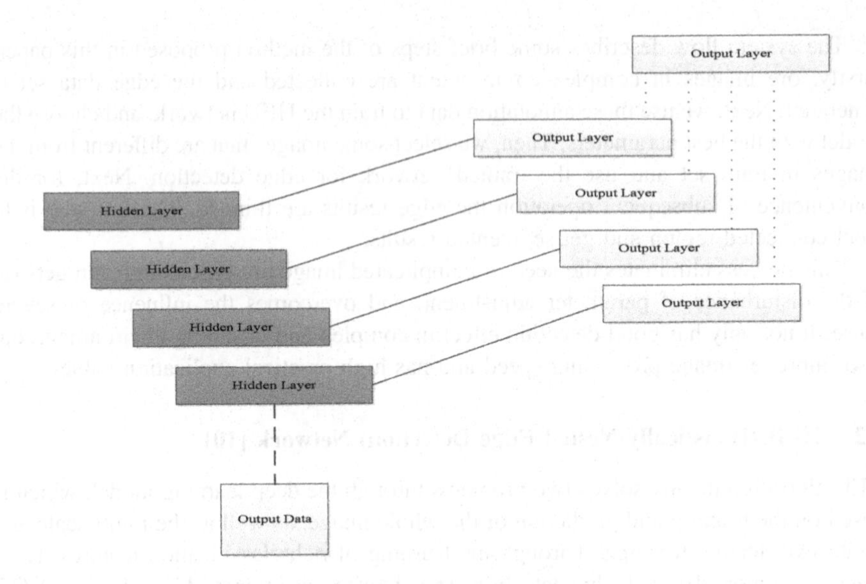

Fig. 3. Network structure of HED

$$L_{fuse}(W, w, h) = Dist\left(Y, \hat{Y}_{fuse}\right) \tag{2}$$

Among them $\hat{Y}_{fuse} \equiv \sigma\left(\sum_{m=1}^{M} h_m \hat{A}{}_{side}^{(m)}\right)$, The weight is $h = (h_1, \cdots, h_M)$, The difference between the prediction of fusion layer and the real annotation is indicated by $Dist(\bullet,\bullet)$.

By the standard stochastic gradient descent method, the minimized objective function is obtained.

$$(W, w, h)^{*} = \arg\min\left(L_{side}(W, w) + L_{fuse}(W, w, h)\right) \tag{3}$$

In the test phase, the side output layer of different scales and the weighted-fusion layer generate edge detection result.

$$\left(\hat{Y}_{fuse}, \hat{Y}{}_{side}^{(1)}, \cdots, \hat{Y}{}_{side}^{(M)}\right) = CNN(X, (W, w, h)^{*}) \tag{4}$$

$CNN(\bullet)$ is the edge detection generated by HED network.

2.3 Thinning Process

After the HED edge detection, the fuse layer result is shown in Fig. 4.

Fig. 4. Result of the fuse layer

It can be seen that the edge of the line is uneven, and most of it is very thick. When the edge is measured, its thickness is about 9 pixels, which can not be processed later, so it needs to be refined. Thinning generally refers to the operation of the skeletonization of binary images. After layers of peeling, some points are removed from the original image, but the original shape remains to be maintained until the skeleton of the image is obtained.

There are many kinds of thinning algorithms [12–14], and in this paper, the table lookup method is applied. The criterion to decide whether a point can be removed is according to the situation of eight adjacent points (eight connected). For black pixels, this method assigns different values to the eight points around it, and maps all points to the index table of 0–255 in this way. According to the situation of 8 points around, corresponding item in the index table is considered to decide whether or not to keep it. In each line of horizontal scanning, it judges the left and right "neighbors" of each point. If all are black, the point is not processed; If a black point is deleted, this method skips the right neighbor to handle the next point. Then, a vertical scan is performed. The above steps are repeated several times until the graph is no longer changed.

The thinning edge detection results are shown in Fig. 5.

Fig. 5. Thinning edge results

2.4 Connected Region Labeling

In order to further obtain the geometric parameters such as contour, external rectangle and center of mass in the ore image, we need to label the connected region on the thinning edge image.

There are many kinds of algorithms for labeling connected regions. Some algorithms only need one image traversal, and some require two or more. The method used in this paper is to establish two levels of contour. The upper layer is the four edges of the image, and the inner layer is the information of the ore connected in the image. Each disconnected region is assigned a different mark.

The core is the contour search algorithm, which marks the whole image by locating the inner and outer contour of the connected region. The specific steps of the algorithm are as follows:

Step1: make image traversal from top to bottom, from left to right;
Step 2: if A is an outer contour point and has not been marked, give A new tag. Start from point A, follow certain rules to track all the outer contour points of A, then go back to point A, and mark all points on the path as the label of A.
Step 3: if an outer contour point A' has been marked, mark points right beside it, until encountering a black pixel.
Step 4: if a marked point B is the point of the inner contours, start from point B, track the inner outline, and the points on the path are set to B's label. Because B has been marked with the same as A, the inner and outer outline will mark the same label.
Step 5: if the points on the inner outline has been walked through, mark the point on the right side with the contour mark until the black pixel is encountered.

3 Experiments

The experiment of HED edge detection is conducted under the deep learning framework Caffe. The network model is trained by GPU with the model Geforce GTX TITAN X. Edge thinning and connected region labeling are programmed by combining Opencv, Skimage, Image and other image processing libraries in Python2.7 environment. The final segmentation are compared with the results of the Watershed transform based on gradient correction method [15–17], and the validity of this method is proved.

3.1 Image Data Set Preparation

In order to solve the problem of inaccuracy caused by mutual adhesion and shadow on the ore image, we collect multi-type images in a changing environment from the production site. There are mainly two types of challenging images as shown below.

(1) Ore images taken different time periods: The ore images will show different intensity and shading situations at different time periods. As shown in Fig. 6, the left image is taken in the morning, and the right one is taken in the afternoon.
(2) Ore images taken with different degree of dryness and humidity as shown in Fig. 7. The ores are sprayed with water to reduce the dust.

The image size collected from the live webcam is 1920*1080. In actual processing, it occupies too much memory which leads to long time of the processing. Therefore, the image is interpolated and resized to 960*540. The collected different types of

Fig. 6. Different light and angle

Fig. 7. Different humidity

images are uniformly distributed in the annotation data set, and edge lines are manually depicted to form the label sets. The number of the labeled images in the annotation data set is 285.

3.2 HED Model Training and the Segmentation Process

HED network model is trained with the above mentioned annotation data set, and its training parameters are shown in Table 1. The 'base_lr' represents the basic learning rate; 'gamma' means the change index of learning rate; 'iter_size' represents the number of images per iteration; 'Ir_policy' means the learning rate attenuation strategy; 'momentum' represents the network momentum, generally using experience values; 'weight_decay' represents weight attenuation to prevent overfitting. Stochastic Gradient

Table 1. Training parameters

Parameter name	Value
base_lr	0.0000001
gamma	0.1
iter_size	10
lr_plicy	Step
stepsize	10000
momentum	0.9
weight_decay	0.0002
max_iter	30001

Fig. 8. Source images

Descent (SGD) is used in the experiment. Maximum number of iterations max_iter = 30001. Because of large size of images, we use lower 'base_lr' in order to prevent gradient explosion and Loss = Nan.

Each iteration takes approximately 9 s. When the training iteration has reached 50,000, the convergence is basically completed. Then we apply the trained model on test images, we typically selects colorful stone images collected at night and the images with shadows acquired during the day, as shown in Fig. 8. Figure 9 shows the output of the HED network fusion layer. It can be seen that the edge detection is superior, robust and granular. However, the skeleton of the edge is thick and noisy, so it needs to be thinned for subsequent processing. Figure 10 shows the effect of edge thinning by using the table lookup. Figure 11 shows the results that using the find Contours function in the Opencv library to draw the thinning edge to the original image. In this way, we can see the segmentation effect more intuitively and form the connected region. Figure 12 is the final segmentation result of the method proposed in this paper.

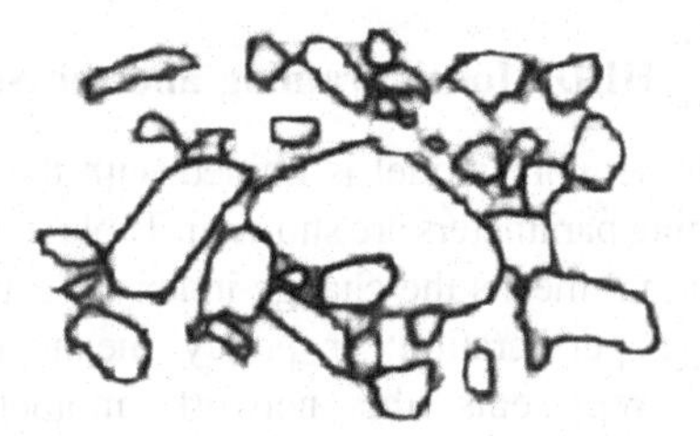

Fig. 9. The Results of fuse layer

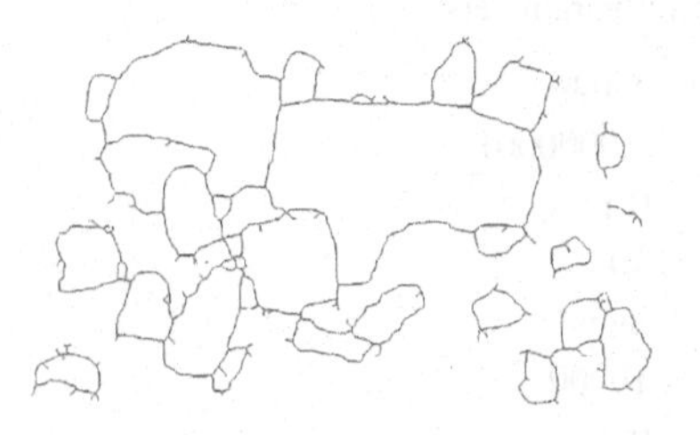
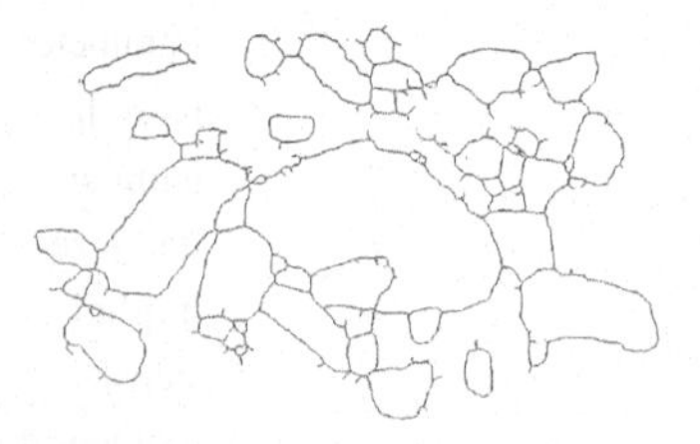

Fig. 10. Thinning the edge

Fig. 11. Connected region Labeling on the source images

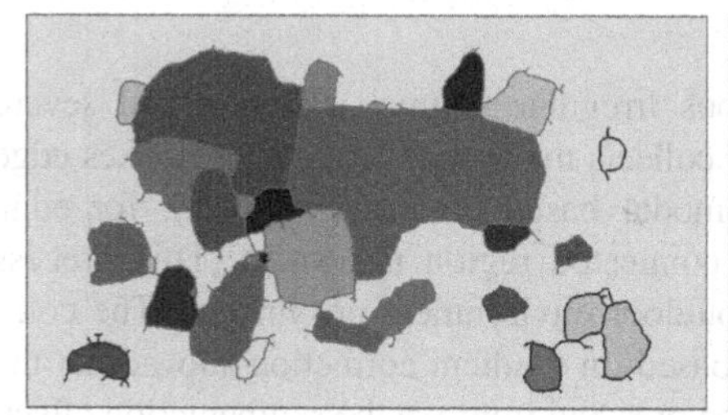
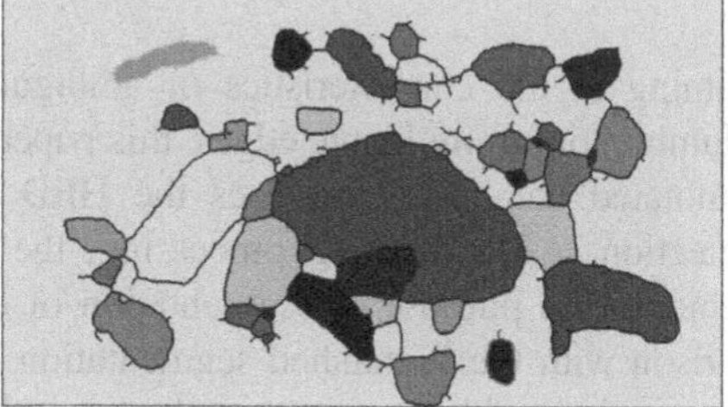

Fig. 12. Segmentation results

3.3 Comparison with the Watershed Segmentation Based on Gradient Correction

In order to demonstrate the effectiveness of the segmentation method proposed in this paper, a Watershed algorithm based on gradient correction is applied on the same ore images in Fig. 8 and the results are compared.

The basic idea of Watershed ore segmentation based on gradient correction is as follows. The Watershed algorithm has a good response to weak edges, so the noise in the ore image and the slight change of gray level on the surface of the ore will cause the over-segmentation problem. To address this problem, a series of image pre-processing is performed before the Watershed, such as adaptive histogram equalization, bilateral filter, morphological open and closed operations. Then, the foreground and background markers of the image are marked. At last, the gradient map is used to modify the Watershed algorithm. The segmentation effect is shown in Fig. 13.

From Figs. 12 and 13, we can see that our proposed method performs better on extracting the ores. In complex environment, pre-processing steps of the Watershed method based on gradient correction are complicated. The size of structural elements in open and closed reconstruction requires empirical debugging. The segmentation result of ambiguous edge is not ideal. While our method does not involve the debugging of parameters, and has better robustness, speed and segmentation effect.

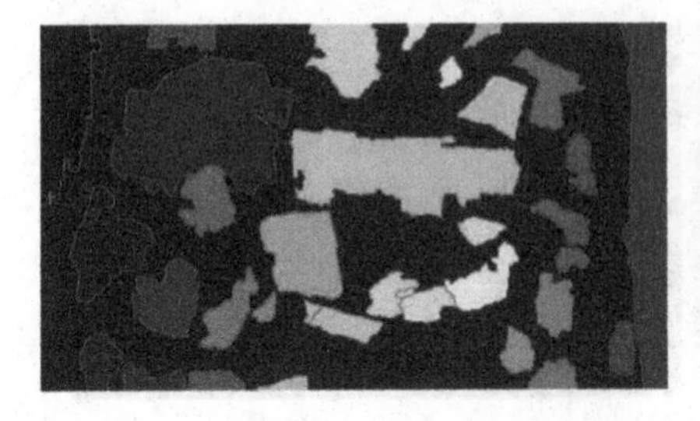
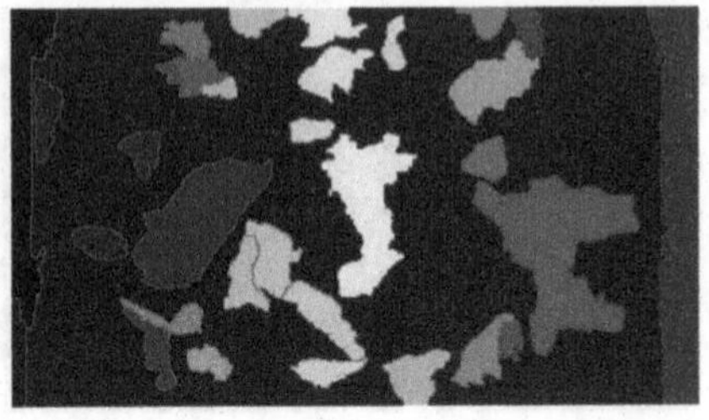

Fig. 13. Watershed segmentation result

4 Conclusion

Aiming at the characteristics of ambiguous irregular, mutual adhesion and severe illumination of ore image edges, this paper collects multi-type ore images, makes edge annotated data sets, and uses the HED model based on neural network for edge detection. After thinning processing, the connected region is formed. This process achieves the purpose of segmentation of outdoor environment ore images. The comparison with the Watershed segmentation based on gradient correction shows that the proposed algorithm can accurately detect the ores and improve the segmentation effect.

References

1. Zhang, G., Qiu, B., Liu, G.: Ore image segmentation using the one-dimensional entropy adaptive threshold based on bi-windows. In: International Conference on Information, Services and Management Engineering (2011)
2. Zhu, S., Xia, X., Zhang, Q., et al.: An image segmentation algorithm in image processing based on threshold segmentation. In: International IEEE Conference on Signal-Image Technologies and Internet-Based System, pp. 673–678. IEEE (2007)
3. Zhang, W., Jiang, D.L.: The marker-based watershed segmentation algorithm of ore image. In: International Conference on Communication Software and Networks, pp. 472–474. IEEE (2011)
4. Zhigang, N., Wenbin, S., Xiong, C.: Adhesion ore image separation method based on concave points matching. In: Balas, V.E., Jain, Lakhmi C., Zhao, X. (eds.) Information Technology and Intelligent Transportation Systems. AISC, vol. 455, pp. 153–164. Springer, Cham (2017). https://doi.org/10.1007/978-3-319-38771-0_15
5. Yang, G.Q., Wang, H.G., Wen-Li, X.U., et al.: Ore particle image region segmentation based on multilevel strategy. Chin. J. Anal. Lab. **35**(24), 202–204 (2014)
6. Malladi, S.R.S.P., Ram, S., Rodriguez, J.J.: Superpixels using morphology for rock image segmentation. In: Image Analysis and Interpretation, pp. 145–148. IEEE (2014)
7. Long, J., Shelhamer, E., Darrell, T.: Fully convolutional networks for semantic segmentation. IEEE Trans. Pattern Anal. Mach. Intell. **39**(4), 640–651 (2017)
8. Badrinarayanan, V., Kendall, A., Cipolla, R.: SegNet: a deep convolutional encoder-decoder architecture for scene segmentation. IEEE Trans. Pattern Anal. Mach. Intell. **PP**(99), 1 (2015)

9. Chen, L.C., Papandreou, G., Kokkinos, I., et al.: DeepLab: semantic image segmentation with deep convolutional nets, atrous convolution, and fully connected CRFs. IEEE Trans. Pattern Anal. Mach. Intell. **PP**(99), 1 (2017)
10. Xie, S., Tu, Z.: Holistically-nested edge detection. Int. J. Comput. Vis. **125**(1–3), 1–16 (2015)
11. Simonyan, K., Zisserman, A.: Very deep convolutional networks for large-scale image recognition. Comput. Sci. (2014)
12. Zhang, T.Y., Suen, C.Y.: A fast parallel algorithm for thinning digital patterns. Commun. ACM **27**(3), 236–239 (1984)
13. Wang, P.S.P., Zhang, Y.Y.: A fast and flexible thinning algorithm. IEEE Trans. Comput. **38** (5), 741–745 (1989)
14. Lam, L., Suen, C.Y.: An evaluation of parallel thinning algorithms for character recognition. IEEE Trans. Pattern Anal. Mach. Intell. **17**(9), 914–919 (2002)
15. Liu, Y., Zhao, Q.: An improved watershed algorithm based on multi-scale gradient and distance transformation. In: International Congress on Image and Signal Processing, pp. 3750–3754. IEEE (2010)
16. Zhang, J.M., Ju, Z., Wang, J.: Watershed segmentation algorithm based on gradient modification and region merging. J. Comput. Appl. **31**(2), 369–371 (2011)
17. Wang, X.-p., Li, J., Liu, Y.: Watershed segmentation based on gradient relief modification using variant structuring element. Optoelectron. Lett. **10**(2), 152–156 (2014)

Plant Recognition Based on Modified Maximum Margin Criterion

Xianfeng Wang, Shanwen Zhang(✉), and Zhen Wang

College of Science, XiJing University, Xi'an 710123, China
wjdw716@163.com

Abstract. The Plant recognition based on plant leaves is important for biological science, ecological science, and agricultural digitization. Because of the complexity and variation of the plant leaves, many classical plant recognition algorithms using plant leaf images are not enough for practical application. A modified maximum margin criterion (MMMC) algorithm is proposed for plant recognition by minimizing the within-class scatter, while maximizing the between-class scatter. The experimental results on the ICL leaf image database show that the proposed method is effective.

Keywords: Leaf image · Plant species recognition
Maximum Margin Criterion (MMC) · modified MMC (MMMC)

1 Introduction

Plant recognition using leaf images is a topic research [1, 2]. Feature extraction and selection from leaf images is an important steps in plant classification [3–5]. Jyotismita et al. [6, 7] proposed a plant leaf recognition method by using texture and shape features with neural classifier. Trishen et al. [8] proposed a plant leaf recognition method by using shape features and color histogram. Bama et al. [9] proposed a plant leaf recognition method by using shape, color and texture features. The facts show that most of these features are either partially or completely irrelevant or redundant to classifying plants [10]. Because of the diversity and complexity of the plant leaf shape and the large difference between within-class leaf leaves, many classical leaf based plant classification methods cannot be effective for plant classification system. In recent years, many manifold-based learning algorithms have been proposed to discover the intrinsic low-dimensional embedding of the original image data [11, 12]. Yan et al. proposed a Marginal Fisher Analysis (MFA) method for face recognition [13, 14]. Chen et al. proposed local discriminant embedding (LDE) [15] for dimensional reduction and face recognition. Li et al. [16, 17] proposed Maximum Margin Criterion (MMC) by considering the difference between-class scatter and within-class scatter as a discriminant criterion to preserve the global structures of the samples. Based on MMC, we propose a modified MMC (MMMC) algorithm for plant recognition task. The experimental results show MMMC is computationally effective by improving the classification performance.

D.-S. Huang et al. (Eds.): ICIC 2018, LNAI 10956, pp. 520–525, 2018.
https://doi.org/10.1007/978-3-319-95957-3_54

2 Maximum Margin Criterion (MMC)

Suppose a set of N multi-class data $X = [x_1, x_2, \ldots, x_n]$ belonging to C known pattern classes and $x_i \in R^D$. MMC aims to find a linear transformation matrix A to project the high-dimensionality data into a low-dimensionality subspace in which the local and non-local structure of the input data can be preserved.

Let $Y = [y_1, y_2, \ldots, y_n]$ be the projection of $X = [x_1, x_2, \ldots, x_n]$, i.e. $y_i = A^T x_i$. Similar to LDA, in MMC, the between-class scatter matrix T_b and the within-class scatter matrix T_w can be formulated as follows,

$$T_b = \frac{1}{N}\sum_{j=1}^{N}(m_j - m_0)(m_j - m_0)^T \quad (1)$$

$$T_w = \frac{1}{N}\sum_{i=1}^{C}\sum_{j=1}^{n_i}(x_i^j - m_i)(x_i^j - m_i)^T \quad (2)$$

where $m_j = \frac{1}{n_j}\sum_{i=1}^{n_j} x_j^i$ and $m_0 = \frac{1}{N}\sum_{j=1}^{C}\sum_{i=1}^{n_j} x_j^i$ denote the mean of the jth class and the mean of all samples, respectively; $x_j^i \in \mathcal{R}^D (i = 1, 2, \ldots, n_j, j = 1, 2, \ldots, C)$ denotes the ith sample in the jth class, n_j is the number of training samples in class j, then $\sum_{j=1}^{C} n_j = N$.

The objection function is

$$T(A) = tr(A^T(T_b - T_w)A) \quad (3)$$

In Eq. (3), maximizing $T(A)$ indicates that the samples are close to each other if they are from the same class, while are far from each other if they are from different classes. By eigen-value decomposition, the optimal projection axe is the eigenvector corresponding to the maximal eigen-value of Eq. (3), which can be selected as the orthonormal eigenvectors corresponding to the first d largest eigen-values, where d is the reduction dimensionality.

3 Modified Maximum Margin Criterion (MMMC)

Although the classical MMC can overcome the SSS problem occurred in LDA by PCA [18], it ignores the locality of the input samples. Thus, a modified MMC (MMMC) algorithm is proposed by an improved Fisher criterion, which is described as follows:

Suppose there are N original training samples $x_1, x_2, \ldots, x_N$ belonging to C known pattern classes $L_1, L_2, \cdots L_C$. The label of x_1 is noted as $L(x_1)$. To obtain the highest recognition rate, it is important to select the optimal projection discriminant vector. The objection function of MMMC is defined as follows:

$$J(A) = \arg\max_{A}(A^T(S_b - S_w)A) \tag{4}$$

where $A \in \mathcal{R}^{N \times d}$ is a projection matrix, S_w and S_b are within-class and between-class scatter matrices, respectively, which are calculated as follows,

$$\begin{aligned} S_w &= \frac{1}{N}\sum_{i=1}^{C}\sum_{j=1}^{n_i} H_{i,j}\left(x_i^j - m_i\right)\left(x_i^j - m_i\right)^T \\ &= \frac{1}{2}\frac{1}{N}\sum_{k=1}^{C}\frac{1}{n_k}\sum_{i=1}^{n_k}\sum_{j=1}^{n_k} H_{i,j}\left(x_k^i - x_k^j\right)\left(x_k^i - x_k^j\right)^T \end{aligned} \tag{5}$$

where $H_{ij} = \begin{cases} \exp(\frac{-\|x_i - x_j\|^2}{\eta}), \text{if } L(x_i) = L(x_j) \\ \quad \text{and } x_i \in Ner(x_j) \text{ or } x_j \in Ner(x_i) \\ 0, \quad \text{else} \end{cases}$, $Ner(x_j)$and $Ner(x_i)$ are the k nearest neighbors of x_j and x_i, respectively.

For simplicity, suppose that each class has the same sample number denoted as n, the within-class scatter matrix can be re-expressed as:

$$\begin{aligned} S_w &= \frac{1}{2}\frac{1}{N}\sum_{k=1}^{C}\frac{1}{n_k}\sum_{i=1}^{n_k}\sum_{j=1}^{n_k} H_{i,j}\left(x_k^i - x_k^j\right)\left(x_k^i - x_k^j\right)^T \\ &= \frac{1}{2}\frac{1}{N \cdot n}\sum_{k=1}^{c}\sum_{i=1}^{n}\sum_{j=1}^{n} H_{i,j}\left(x_k^i - x_k^j\right)\left(x_k^i - x_k^j\right)^T \\ &= \frac{1}{2}\frac{1}{N \cdot n}\sum_{k=1}^{C}\sum_{x_i, x_j \in L_k} H_{i,j}\left(x_i - x_j\right)\left(x_i - x_j\right)^T \\ &\propto \left(XD_wX^T - XHX^T\right) = XL_wX^T \end{aligned} \tag{6}$$

where $L_w = D_w - H$, H is a matrix constituted by H_{ij}, D_w is a diagonal matrix whose elements on diagonal are column $D_{ii,w} = \sum_{j=1}^{N} H_{ij}$.

By considering the local information, the between-class scatter matrix S_b is reformulated as follows:

$$S_b = \frac{1}{N}\sum_{j=1}^{N} W_{ij}\left(m_j - m_0\right)\left(m_j - m_0\right)^T = XL_bX^T \tag{7}$$

where $W_{ij} = \begin{cases} \exp(-\|x_i - x_j\|^2 / \eta), \text{if } L(x_i) \neq L(x_j) \\ 0, \quad \text{else} \end{cases}$, η is an adjustment parameter, $L_b = D_b - W$, W is a matrix constituted by W_{ij}, D_b is a diagonal matrix whose elements on diagonal are column $D_{ii,b} = \sum_{j=1}^{N} W_{ij}$.

Equation (4) can be solved by Lagrangian multiplier method, and the corresponding Lagrange function is as follows,

$$J(a_i, \lambda_i) = \sum_{i=1}^{d} a_i^T (S_b - S_w) a_i - \lambda_i (a_i^T a_i - 1)$$
$$\frac{\partial J(a_i, \lambda_i)}{\partial a_i} = ((S_b - S_w) - \lambda_i I) a_i = 0 \tag{8}$$

where $a_1, a_2, \ldots, a_d$ are elements of $A (A = [a_1, a_2, \ldots, a_d])$ constructing by the orthonormal eigenvectors corresponding to the first d largest eigen-values $\lambda_1, \lambda_2 \cdots, \lambda_d$ (generally $\lambda_1 \geq \lambda_2 \geq \cdots \geq \lambda_d$), derived as follows,

Then,

$$(S_b - S_w) a_i = \lambda_i a_i \tag{9}$$

After obtaining $A = [a_1, a_2, \ldots, a_d]$ from Eq. (9), any new sample x is projected to a low-dimensionality y as follows

$$x \rightarrow y = A^T x \tag{10}$$

4 Experiments and Analysis

The proposed method is tested on ICL dataset, in which, there are 17032 plant leaf images of 220 species and image number of each class is unequal. For simplicity, we select 50 species of plant leaf images with 30 samples per class, and adopt 2-fold cross validation approach to conduct plant recognition experiments. That is to say, from 1500 leaf images, half of them are selected randomly as training samples and the remaining half are used for testing samples. The input of MMMC is the gray images or binary images in which the leaf images are numerically displayed with 1 and the background is with 0. In following experiments, we select the gray leaf images as the output of MMMC. Relatively simple 1-nearest neighbor classifier is used as classifier to classify the plant leaf images. In order to show the efficiency and feasibility of MMMC, we compare MMMC with two plant recognition methods: shape features and color histogram (SFCH) [8], Fourier transform based on ordered sequence (FTOS) [10], and classical MMC [16] and probability locality preserving discriminant projections (PLPDP) [19].

We set η to be the mean distance between the k nearest neighborhoods of X_i and X_j. The k-nearest neighborhood parameter can be chosen as $k = l - 1$, where l denotes the number of training images per class [20]. The justification for this choice is that each image should be connected with the remaining $l - 1$ images in the same class, which can make the within-class images be well clustered in the observation space. The classifier is trained by the low-dimensional training vectors, and the performance of the proposed method is evaluated by the low-dimensional test vectors. In general, the performance of two methods varies with the number of dimensions. In every experiment, we record the maximum recognition rate versus the reduction dimensionality d.

In order to effectively evaluate the performance of the different algorithms in different training and testing conditions, we divide the leaf images randomly many times and repeat the experiments 100 times in the same condition, and then recode the recognition results in the form of mean recognition rate with standard deviation. The experimental results by three methods and MMMC are shown in Table 1.

Table 1. Average recognition rates and deviations (percent) by using five methods

Method	SFCH	FTOS	MMC	PLPDP	MMMC
Recognition rate	86.24 ± 2.13	87.36 ± 1.28	86.82 ± 1.57	91.24 ± 1.43	92.27 ± 1.26

From Table 1, it is seen that the proposed method outperforms other methods. The reason may be MMMC can obtain more discriminant projecting matrix for image classification. From the experimental results, we can see that our MMMC algorithm achieves the highest recognition rate. The experimental results indicate that PLPDP [20] and MMMC are effective, which obtain higher recognition rate than other three methods. Because PLPDP and MMMC are effective for the nonlinear, complicated high-dimensionality leaf images. But, PLPDP is more complex than MMMC. SFCH and SCF are two linear statistics feature extraction methods, while LAD and MMC are linear dimensionality reduction methods which ignore the local information of the leaf images. In principle, MMC can make the samples more separable than LDA by discarding the eigen-vectors whose corresponding eigen-values of $S_b - S_w$ are negative. But MMC may lose more information by discarding these eigenvectors. In the experiments, MMC is usually able to achieve similar results as LDA in a lower dimensional discriminant space.

5 Conclusions

This paper proposes a novel method for plant recognition based on MMMC, which is to seek projection matrix that preserve the local neighborhood while the projected samples of other classes are kept away. MMMC is able to get adequate information from the local neighbors for local geometry learning, and has better discriminating capability, and is directly reflected in good recognition performance, as revealed in the experimental results. The experiments are mainly designed to show the effectiveness of MMMC for leaf image recognition task. Our future work will apply MMMC algorithm to advanced visual leaf images rather than the original pixel intensity values.

References

1. Wang, X.F., Huang, D.S., Du, J.X., et al.: Classification of plant leaf images with complicated background. Appl. Math. Comput. **205**(2), 916–926 (2008)
2. Huang, D.S., Du, J.X.: A constructive hybrid structure optimization methodology for radial basis probabilistic neural networks. IEEE Trans. Neural Networks **19**(12), 2099–2115 (2008)

3. Du, J.X., Zhai, C.M., Wang, Q.P.: Recognition of plant leaf image based on fractal dimension feature. Neurocomputing **116**, 150–156 (2013)
4. Zhang, S.W., Lei, Y.K., Dong, T.B., et al.: Label propagation based supervised locality projection analysis for plant leaf classification. Pattern Recogn. **46**(7), 1891–1897 (2013)
5. Wang, X.F., Huang, D.S.: A novel density-based clustering framework by using level set method. IEEE Trans. Knowl. Data Eng. **21**(11), 1515–1531 (2009)
6. Chaki, J., Parekh, R.: Plant leaf recognition using shape based features and neural network classifiers. Int. J. Adv. Comput. Sci. Appl. **2**(10), 41–47 (2011)
7. Chaki, J., Parekh, R., Bhattacharya, S.: Plant leaf recognition using texture and shape features with neural classifiers. Pattern Recogn. Lett. **58**(1), 61–68 (2015)
8. Munisami, T., Ramsurmn, S.K., Sameerch, P., et al.: Plant leaf recognition using shape features and colour histogram with K-nearest neighbor classifiers. Procedia Comput. Sci. **58**, 740–747 (2015)
9. Bama, B.S., Valli, S.M., Raju, S., et al.: Content based leaf image retrieval using shape, color and texture features. Indian J. Comput. Sci. Eng. **2**(2), 202–211 (2011)
10. Yang, L.W., Wang, X.F.: Leaf image recognition using fourier transform based on ordered sequence. In: Huang, D.S., Jiang, C., Bevilacqua, V., Figueroa, J.C. (eds.) ICIC 2012. LNCS, vol. 7389, pp. 393–400. Springer, Heidelberg (2012). https://doi.org/10.1007/978-3-642-31588-6_51
11. Zhang, S.W., Lei, Y.K.: Modified locally linear discriminant embedding for plant leaf recognition. Neurocomputing **74**, 2284–2290 (2011)
12. Zhang, S.W., Lei, Y.K., Wu, Y.H.: Semi-supervised locally discriminant projection for recognition. Knowl. Based Syst. **24**, 341–346 (2011)
13. Xu, S.Y., Zhang, B., Zhang, H.J.: Graph embedding: a general framework for dimensionality reduction. In: Proceedings of. IEEE Conference on Computer Vision and Pattern Recognition, pp. 830–837 (2005)
14. Yan, S., Xu, D., Zhang, B., Zhang, H.J.: Graph embedding and extensions: a general framework for dimensionality reduction. IEEE Trans. Pattern Anal. Mach. Intell. **29**(1), 40–51 (2007)
15. Chen, H.T., Chang, H.W., Liu, T.L.: Local discriminant embedding and its variants. In: IEEE Computer Society Conference on Computer Vision and Pattern Recognition, vol. 2, pp. 846–853 (2005)
16. Li, H., Jiang T., Zhang K.: Efficient and robust feature extraction by maximum margin criterion. In: Advances in Neural Information Processing Systems, pp. 97–104 (2003)
17. Li, H., Jiang, T., Zhang, K.: Efficient and robust feature extraction by maximum margin criterion. IEEE Trans. Neural Networks **17**, 157–165 (2006)
18. Belhumeur, V., Hespanha, J., Kriegman, D.: Eigenfaces vs Fisher faces: recognition using class specific linear projection. IEEE Trans. Pattern Anal. Mach. Intell. **19**(7), 711–720 (1997)
19. Zhang, S.W., Wang, X.F., Wang, Z., et al.: Probability locality preserving discriminant projections for plant recognition. Trans. Chin. Soc. Agric. Eng. (Transactions of the CSAE) **31**(11), 215–220 (2015). (in Chinese with English abstract)
20. Li, B., Wang, C., Huang, D.S.: Supervised feature extraction based on orthogonal discriminant projection. Neurocomputing **73**, 191–196 (2009)

CBR Based Educational Method for the Postgraduate Course Image Processing and Machine Vision

Xin Xu[1,2](✉) and Lei Liu[1]

[1] School of Computer Science and Technology, Wuhan University of Science and Technology, Wuhan, China
xuxin0336@163.com
[2] Hubei Province Key Laboratory of Intelligent Information Processing and Real-Time Industrial System, Wuhan University of Science and Technology, Wuhan, China

Abstract. This paper discusses the urgency and importance of teaching innovation in the postgraduate course image processing and machine vision. Aiming to address the problem of deepen teaching innovation, this paper proposes a CBR based method to construct educational system. Besides, discussion on how to cultivate postgraduates' scientific literacy and application ability is also conducted to further improve the teaching quality and effect of the image processing and machine vision postgraduate courses.

Keywords: Case-Based Reasoning · Teaching innovation · Educational reform

1 Introduction

During the past 30 years of educational reforming in China, it has witnessed great achievements in the development of higher education. This has benefited from the reform of the education system. Through a series of fruitful explorations, we have the excellent situation of today's booming development of higher education. In the process of the reform of higher education in our country, we have also continuously explored reforms and innovations in teaching methods. With the advancement of China's industrialization, marketization, urbanization, information, globalization development and transformation strategy, it has injected fresh vitality and development impetus into China's higher education, and has provided more growth space and development for graduate students' training and employment opportunity. However, with the continuous enlargement of the enrollment scale of graduate students, the traditional teaching methods are facing great challenges. At present, the quality of postgraduate education has begun to show a downward trend. The reform of postgraduate teaching methods is imperative.

The teaching methods include the teaching methods employed by teachers to complete teaching tasks and the learning methods of students. Proper teaching methods can fully realize the purpose of teaching, reflect the ideological, scientific and systematic nature of teaching content and improve students' ability to acquire knowledge

D.-S. Huang et al. (Eds.): ICIC 2018, LNAI 10956, pp. 526–531, 2018.
https://doi.org/10.1007/978-3-319-95957-3_55

and master skills. Therefore, it should be emphasized that the research on teaching methods in postgraduate teaching is an important task of higher education. "Image Processing and Machine Vision" is the most basic course in the field of computer vision, and it is very important for postgraduate students in related fields to carry out research on the topic smoothly. With the development of deep learning technology in recent years, the course of "Image Processing and Machine Vision" has gradually become similar to mathematics and foreign languages. Its teaching content is no longer confined to a certain specialized field, but mainly relates to the basics of image processing in different applications, versatile concepts, technologies and applications. The technology involved in the course can provide assistance for many different disciplines such as computers, information, machinery and materials. However, due to the continuous increase in the number of graduate students, individual differences in students and other reasons may affect the effectiveness of curriculum teaching. This article will take the postgraduate course "Image Processing and Machine Vision" as an example to discuss the teaching innovation method based on CBR technology.

2 Related Works

2.1 Postgraduate Courses Image Processing and Machine Vision

Image processing and machine vision is a multidisciplinary course involving optical imaging, image processing, and artificial intelligence. This course is an important elective course for postgraduates in information related majors. The course content covers the cognitive process of studying visual information from the level of information processing, including methods, theories and descriptions in visual information processing. The course has the following two characteristics [2].

(1) Theoretical abstraction and extensive content. Computer vision involves many subject areas such as image processing, pattern recognition, machine learning, etc. The contents involved are abstract in mathematics, extensive in content and numerous in algorithms. Students find it difficult to understand;
(2) It is closely integrated with engineering applications. The construction of this course's knowledge system originates from various specialties and professional application fields. In industry, especially in various production lines, visual products are widely used. Only by allowing graduate students to experience a large number of engineering project design and implementation, can we fully appreciate the functions and functions of various computer vision algorithms and technologies.

2.2 Introduction of CBR Technology

CBR (Case-Based Reasoning) is an important content in the field of artificial intelligence. Domestic and foreign scholars have conducted relevant research on CBR cognitive models and development techniques. The main idea of CBR lies in "similar problems have similar solutions" [1]. Based on CBR technology, different applications can solve current problems by reusing programs that used similar problems in the past.

In recent years, CBR technology has been successfully used in computer-aided diagnosis [2], legal advice, e-commerce and product search [3] and so on. For example, doctors can use the CBR technology to obtain similar cases in the past, and cure the current patient only by referring to the previous treatment plan [4–6]; lawyers can also refer to the past cases provided by CBR to defend the current lawsuit.

CBR mimics this problem solving behavior of human beings. When it is difficult to formalize domain rules and cases are available, this approach should be considered. When the rules can be formalized, but more typically available input information is needed, this approach should also be considered due to incomplete problem descriptions, or when the knowledge needed to solve the problem is not available. Other uses of CBR include the fact that if too much general knowledge is not expected because of expectation or if a new plan is deduced from an old plan, it is easier to do it than to start from scratch. Much successful commercial software in these areas has demonstrated the effectiveness of this paradigm [7–9].

In order for a computer system to determine the similarity of two problems, the CBR system employs a so-called "similarity measure" which represents the mathematical form of the "similar" or "utility" of the general vocabulary. Similarity metrics usually do not describe the dependencies between the problem and the corresponding solution in detail, but only represent heuristic experience. Therefore, the choice of truly useful cases and the exactness of the output are generally not guaranteed. However, with this inaccuracy, powerful knowledge-based systems can be developed that require less effort and lower costs than traditional AI technologies that rely on complete and correct domain theory. However, CBR is not limited to empirical reuse, and CBR is also very successful in e-commerce and product search. At this point, similarity metrics are used to compare user descriptions and product descriptions, filling in the gap between customer needs and product characteristics.

In the teaching of postgraduate courses, the use of CBR technology can assist the tutor to better guide graduate students, and at the same time help tutors to improve their methods of cultivating postgraduates' innovative abilities. However, due to the high dimensionality and heterogeneity of case characteristics, the existing methods are still difficult to fully analyze the similarity between features when the existing CBR technology is applied to graduate students. In view of this, this project takes the postgraduate course of image processing and machine vision as an example, takes CBR technology as the core driving force, and studies an effective and feasible postgraduate course teaching method based on CBR technology.

3 CBR-Based Educational Methods for the Postgraduate Course Image Processing and Machine Vision

As mentioned earlier, CBR is a problem solving method and also a machine learning method. The former can often construct expert systems (knowledge-based systems), while the latter are mostly used to build data mining systems. Can CBR technology be applied to postgraduate courses of image processing and machine vision, and accordingly improve the teaching effect?

As Dr. Thomas Roth-Berghofer indicated: "similar problems have similar solutions" (CBR assumptions), CBR is just a heuristic method. Accordingly, the resulting solution is not precise, but it is precisely because of this tolerance. CBR played a role and was able to build a powerful knowledge-based system. Therefore, when CBR technology is actually applied to postgraduate courses in image processing and machine vision, case libraries need to be evaluated in advance to see if they meet the CBR hypothesis, and then the applicability of the CBR method can be judged. The specific process is shown in Fig. 1.

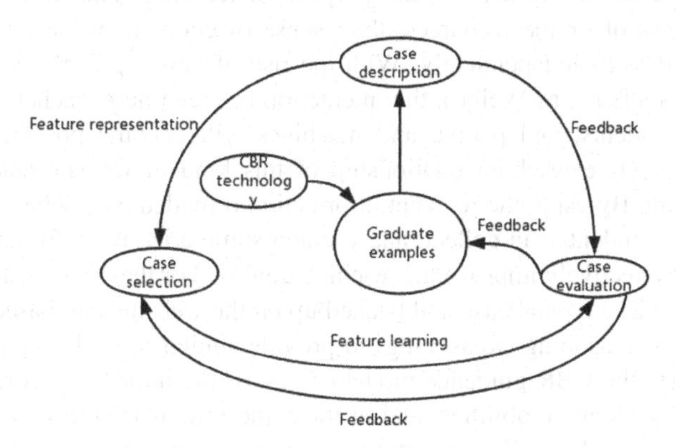

Fig. 1. Flowchart of CBR technology

First of all, according to the new forms of related postgraduate courses in universities at home and abroad, we can grasp the current status of postgraduate teaching in the field of computer vision. In-depth analysis of the field of computer vision field graduate course teaching process, rational disciplines, professional structure, individual characteristics of graduate students, with CBR technology as the core driving force, to carry out case description.

Then, taking the cultivation of postgraduates in the field of computer vision in this school as an example, we set up professional research directions for existing scientific research projects and reorganize research teams for scientific research project clusters. By tracking the frontiers of science and technology along with wide-caliber training, we form new professional growth points in combination with existing research directions, and statistically analyze the most representative examples of graduate student training and carry out case selection work.

Finally, combining with the school's characteristics of running a school, an incentive mechanism suitable for the cultivation of graduate students in the field of computer vision is constructed to promote the emergence of graduate students' innovation achievements. A practical graduate student training method that aims at computer science and promotes innovation is proposed, and case evaluation work is carried out.

In practical applications, we use the current myCBR tools at home and abroad to research and develop CBR systems (http://mycbr-project.net/index.html). In addition to traditional classroom teaching, we will strengthen the multi-directional interaction of online teaching. Through the interaction on the Internet, students can self-study through the course website, and can ask students and teachers questions that they do not understand, and teachers or classmates can answer individually. For common problems, teachers can also use the Internet radio to let the class discuss. In this way, each classmate gets exercised while also fostering the students' spirit of collaborative learning. In addition, due to the introduction of online classes, students can ask teachers questions on the course website, and students can also discuss them through the website. And because of the openness of these conversations, other students can also use these conversations to achieve the purpose of learning. This also cultivates students' awareness of online resources, their sense of cooperation, and their ability to collaborate and learn independently. With the rise of new types of network communication methods (such as Weibo), the interaction between new teachers and students, students and students, and people and machines will become possible. The multi-directional interactive teaching relationship of this kind of website teaching will be more convenient. By using the relevant information provided by teachers or classmates on the website, students can collect, analyze and summarize the information anytime, anywhere, and effectively improve the teaching quality. For these issues discussed, they can be collected as a typical case and backed up on the website, and based on students' future questions, reasoning (Reasoning) to provide similar topical review.

In summary, the CBR guidance model can assist the tutor to specifically cultivate postgraduates' innovative abilities, and at the same time help tutors to improve their graduate training methods. The specific performance is as follows:

(1) At the initial stage of training, the tutor can use the CBR technique to refer to the similar training cases of his previous students and develop a training plan for current graduate students.
(2) In the course of training, the tutor can gradually refine the individual characteristics of graduate students, compare them with previous case characteristics, and improve the previous training program based on teaching feedback.
(3) At the end of the training period, the tutor can express the postgraduate training case as a non-linear combination of high-dimensional features, which is retained as part of his experience.
(4) Mentors can also use the CBR technology to search for similar postgraduate training cases of other tutors at home and abroad, and further improve the postgraduate training program.
(5) Based on characteristic learning methods, we analyze the characteristic heterogeneity among different graduate students, and the mentors can work together to improve the postgraduate training program.

4 Conclusions

In recent years, the scale of postgraduate education in China's colleges and universities has been unprecedentedly increased and developed. Graduate education has changed from the past "elite education" to "popular education." Graduate education should establish a scientific talent cultivation system, establish a scientific talent cultivation system, determine clear goals, face the future, and meet the needs of society. Through the implementation of knowledge, ability, quality education, improve the quality of postgraduate classroom teaching, and cultivate a new generation of innovative Talent.

Higher education institutions undertake the task of cultivating high-level talents for the country. The key to improving the quality of talents is to strengthen the quality and ability. The implementation of computer vision direction of image processing and machine vision classroom teaching, the purpose of training objectives is to emphasize the cultivation of practical applications and how to combine computer vision technology and related disciplines and applied to practice. In view of the rapid update of knowledge in the field of computer vision, we must continue to improve and improve teaching methods and teaching content based on the CBR techniques in the teaching of the past. In order to adapt to the new situation, we must keep pace with the times. The new requirements for the country to cultivate qualified professionals.

Acknowledgement. This work is supported by the Educational Research Project from the Educational Commission of Hubei Province (2016234).

References

1. Aamodt, A., Plaza, E.: Case-based reasoning: foundational issues, methodological variations, and system approaches. AI Commun. **7**(1), 39–59 (1994)
2. Zhang, Y., Zhang, S., Leake, D.: Case-base maintenance: a streaming approach. In: Proceedings of the 24th International Conference on Case-Based Reasoning, pp. 222–231 (2016)
3. Sekar, A., Chakraborti, S.: Learning a region of user's preference for product recommendation. In: Proceedings of the 24th International Conference on Case-Based Reasoning, pp. 212–221 (2016)
4. Bichindaritz, I.: Data mining methods for case-based reasoning in health sciences. In: Proceedings of the 23th International Conference on Case-Based Reasoning, pp. 184–198 (2015)
5. Canensi, L., Leonardi, G., Montani, S., Terenziani, P.: A context-aware miner for medical processes. In: Proceedings of the 24th International Conference on Case-Based Reasoning, pp. 192–201 (2016)
6. Tomasic, I., Funk, P.: Potential synergies between case-based reasoning and regression analysis in assembly processes. In: Proceedings of the 22th International Conference on Case-Based Reasoning, pp. 192–201 (2014)
7. Adedoyin, A., Kapetanakis, S., Petridis, M., Panaousis, E.: Evaluating case-based reasoning knowledge discovery in fraud detection. In: Proceedings of the 24th International Conference on Case-Based Reasoning, pp. 182–191 (2016)
8. Barua, S., Begum, S., Ahmed, M.U., Funk, P.: Classification of ocular artifacts in EEG signals using hierarchical clustering and case-based reasoning. In: Proceedings of the 22th International Conference on Case-Based Reasoning, pp. 213–223 (2014)
9. Dileep, K.V.S., Chakraborti, S.: Intelligent integration of knowledge sources for TCBR. In: Proceedings of the 22th International Conference on Case-Based Reasoning, pp. 224–234 (2014)

Securing Medical Images for Mobile Health Systems Using a Combined Approach of Encryption and Steganography

Tao Jiang[1], Kai Zhang[1], and Jinshan Tang[2](✉)

[1] College of Computer Science and Technology, Wuhan University of Science and Technology, Wuhan 430065, China

[2] School of Technology, Michigan Technological University, Houghton, MI 49931, USA
jinshant@mtu.edu

Abstract. In this paper, we propose a medical image encryption scheme which can be used in mobile health systems. The proposed scheme combines RSA algorithm, logistic chaotic encryption algorithm, and steganography technique to secure medical images. In the proposed scheme, we encrypt a medical image based on chaotic sequence and encrypt the initial value of the chaotic sequence using the RSA encryption algorithm. The encrypted information by RSA is hidden in the Image. Only legitimate users can obtain the parameter information and restore the image. In the receiver side, we apply the inverse methods to get the original image after an encrypted image is arrived. We have implemented a simple application on the Android platform and have evaluated its performance. The experimental results show that the proposed image encryption scheme is practical and feasible for mobile health systems.

Keywords: Medical image encryption · RSA algorithm · Chaos
F5 steganography · Mobile phones

1 Introduction

In recent years, digital images spread mundanely around the world. It is necessary to protect the security of image information. Especially in health information systems, such as mobile health [1], telemedicine [2–6], securing the medical images is very important.

A lot of image encryption schemes have been developed in recent years and chaos-based encryption is one of the widely used methods. Chaos-based encryption was first proposed in 1989 and many chaos-based encryption algorithms have been proposed since them. Kamali [7] have presented a modification of the Advanced Encryption Standard which presents a high level of security and better image encryption. Zhang [8] have proposed an algorithm by combining chaotic encryption and DES encryption. Amitava [9] proposed a two phases encryption and decryption algorithm which was based on shuffling of image pixels using affine transform and XOR operation. Seyedzadeh [10] presented a novel chaos-based image encryption algorithm for color images using a coupled Two-dimensional Piecewise Nonlinear Chaotic Map, called

D.-S. Huang et al. (Eds.): ICIC 2018, LNAI 10956, pp. 532–543, 2018.
https://doi.org/10.1007/978-3-319-95957-3_56

CTPNCM, and a masking process. Enayatifar [11] proposed a novel image encryption algorithm based on a hybrid model of deoxyribonucleic acid (DNA) masking, a genetic algorithm (GA), and a logistic map. The proposed model used DNA and logistic map functions to create the number of initial DNA masks and applied GA to determine the best mask for encryption. A novel bit-level image encryption algorithm was proposed by Xu [12] which was based on piecewise linear chaotic maps (PWLCM). In the proposed algorithm, the plain image was first transformed into two binary sequences with the same size and then a new diffusion strategy was introduced to diffuse the two sequences mutually. After that, the binary elements of the two sequences were swapped by the control of a chaotic map. Besides the work, a new chaos-based partial image encryption scheme based on Substitution-boxes (S-box) constructed by chaotic system and Linear Fractional Transform (LFT) was proposed by Belazi [13].

Besides encryption technology, information hiding technology is another effective technology for secure transmission and storage of digital images. The technology of information hiding embeds the secret information into another carrier data so that the hidden information can't be detected. Steganography is one of the technologies to ensure the safety of image [14] and many algorithms have been proposed. Usha [15] proposed an encrypting system which combined the techniques of cryptography and steganography. In the proposed method, the cipher was hidden inside the image in the encrypted format. Thangadurai [16] presented a LSB based image steganography and its applications to various file formats. Visual cryptography and image steganography are used together [17]. A novel secure image sharing scheme with high quality stego-images was proposed by He [18] and a new technique of image steganography which using Lorenz Chaotic Encryption was proposed [19].

According to the above research status, the image encryption technology and the information hiding technology are more mature, but the research of combining them and applying it to medical image encryption are few. In view of this, a new image encryption scheme which combines RSA algorithm, chaotic encryption algorithm, and Steganography technique was proposed in this paper for securing medical images for mobile health systems. The proposed scheme encrypted medical images based on chaotic sequence and encrypt the initial value of the chaotic sequence by using the RSA public key. The encrypted information of the RSA is hidden in the image and transmitted with the image.

2 The Proposed Image Encryption Scheme

2.1 Logistic Chaotic System

Logistic mapping is derived from the dynamical system of demographics, which is a typical nonlinear chaotic equation. It has good chaotic properties such as long-term unpredictability, and non-convergence. The mapping is defined as:

$$x_{k+1} = u \cdot x_k(1 - x_k) \tag{1}$$

Where x_0 is the starting point and u is a predefined constant. When the coefficients $3.5699456 \leq u \leq 4$, $0 < x_0 < 1$, the system will enter the state called chaotic state when the iteration becomes big.

2.2 Encryption Image by Logistic Chaotic and RSA

The image encryption algorithm adopted in the proposed scheme is based on Logistic map. It uses the chaotic sequence generated by Logistic map to encrypt the image data. A plain-image image can be encrypted by use XOR operation with the integer sequence. Since the sequences generated by a one-dimensional chaotic system are easily attacked by non-linear analysis and phase reconstruction, multiple chaotic sequences will be adopted the proposed algorithm.

Let $I_{M \times N}$ be the original medical image to be encrypted, M and N are the width and height of the image. In this paper, we assume that both M and N are perfect square numbers. The specific encryption process for a medical image is as follows.

Step 1. Create T chaotic matrix sequences $r^1_{m \times n}$, $r^2_{m \times n}$, .., $r^T_{m \times n}$ using Eq. (1), which are obtained by different initial values (The number of iterations k > 1000). In this paper, the number of chaotic sequences T = 3.

Step 2.The creation of the chaotic matrix $r^t_{m \times n}(t = 1, 2, 3)$ is as follows: We first select a random x_0, then use Eq. (1) to create $x_1, x_2, .., x_{m \times n}$, where x_p is obtained by iterating Eq. (1) 1000 + p times. Then assign the created $x_1, x_2.., x_{m \times n}$ to the elements of the matrix $r^t_{m \times n}$ by

$$r^t_{m \times n}(i, j) = x_{i * n + j} \tag{2}$$

Where $0 \leq i \leq m - 1$, $0 \leq j \leq n - 1$.

Step 3. For each block. Let the gray value of the image at (i, j) is I(i, j), where $0 \leq i \leq M - 1$, $0 \leq j \leq N - 1$. Perform the following operations

$$I'(i, j) = (I(i, j) \oplus r1(i, j) \oplus r2(i, j) \oplus r3(i, j)) mod\, 256 \tag{3}$$

Where I '(i, j) indicates the gray value of I(i, j) at (i, j) after the operations. The decryption for Eq. (3) is

$$I(i, j) = (I'(i, j) \oplus r1(i, j) \oplus r2(i, j) \oplus r3(i, j)) \bmod 256 \tag{4}$$

To ensure security, we use the RSA algorithm to encrypt the different initial values. The operational process of RSA is as follows.

Step 1. Calculatersa_n = p × q where p and q are different large primes.
Step 2. Find public key e such that gcd(e,(p − 1)(q − 1)) = 1,then (e,rsa_n) is the encryption key pair.
Step 3. Find a private key d such that ed = 1 mod ((p − 1)(q − 1)), then (d,rsa_n) is the decryption key pair.

Let L consist of $l_1, l_2, \ldots, l_T$. $l_i(i = 1, 2, 3)$ is the initial values which consists of u, x_0, p for a chaotic sequence.

A message m is encrypted by calculating

$$c \equiv L^{e} (\mathrm{mod}\, rsa\_n) \tag{5}$$

The cipher text c is decrypted by

$$L \equiv c^{d} (\mathrm{mod}\, rsa\_n) \tag{6}$$

The cipher text c is embedded into image which encrypted chaotic sequence by F5 (in 2.4).

2.3 PBE

The private key of RSA is too long to remember, and it is easy to be copied when stored on the computer. Therefore, in this paper, we propose PBE (password-based password) to encrypt the local RSA private key for storage. The process is described in Fig. 1(a) and (b).

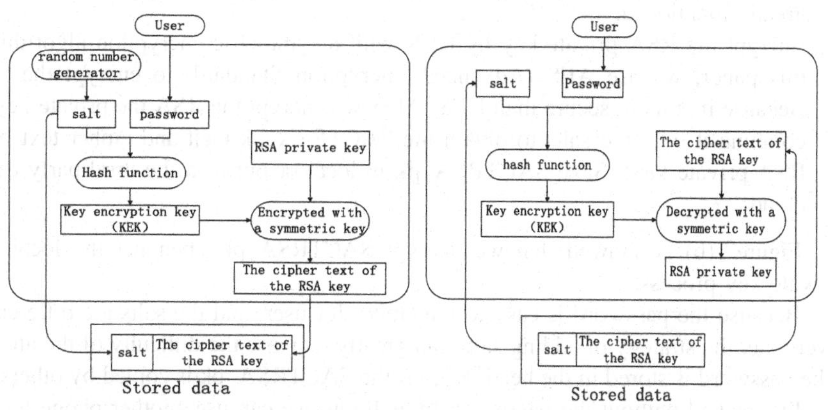

(a) Encrypt the RSA private key (b) Decrypt to get the RSA private key.

Fig. 1. The PBE flowchart to encrypt the local RSA private key

Figure 1(a) is showed that we encrypt the local RSA private key process. We used a symmetric encryption algorithm to encrypt the RSA private key. So we should find the KEK (Key Encrypting Key). The user generally used weak password, but weak passwords can easily be cracked. So we propose using a single hash function to produce the key. However, the key is hard to remember and the single hash function can be cracked by violent crack, dictionary attack and rainbow table. Therefore, we propose adding salts. It is the salts and the password that produce KEK by the single hash function. The operational process of PBE is as follows.

1. Get the Password from the user.
2. Get the salts. In order to increase the complexity of the salts. A salt consists of salt1, salt2, salt3. Salt1 is composed of case-sensitive letters, salt2 is composed of numbers, and salt3 is composed of special characters. There are three arrays. All letters are stored in Array[1], all numbers are stored in Array[2], and all special characters are stored in Array[3]. We can get the corresponding character according to its position(denoted by POS) in Arrays[i], where POS indicates the coordinate index of the array. For example, when i = 1, POS = 5, it means we get the fifth character in the array[1]. Let Array[i].length be the length of Array[i] and Random (N) be a random number from 0 to N, POS can be obtained by

$$POS = \mathrm{Random}(100000)\%\mathrm{Array}[i].\mathrm{length} \tag{7}$$

   where i = 1,2,3. In this paper, the length of salt1 is set to 30, the length of salt2 is set to 8, and the length of salt3 is set to 5.
3. Get the KEK which from the single hash function by the salt and the password. In this paper we use MD5 hash function, because it is irreversible, and cost less resources. Because of the combination of salt and password, the security of MD5 is greatly mentioned.
4. Encrypt the RSA private key by KEK with a symmetric encryption algorithm. In this paper, we use AES (Advanced Encryption Standard) to encrypt the KEK. Because it is more secure than DES. After we encrypt the RSA the private key, we combine it with the salts to obtain the SACTRSA_pk (salt and cipher text of the RSA private key) .Save SACTRSA_pk in local or put it to the third-party trusted cloud.

Figure 1(b) is showed that we decrypt SACTRSA_pk, then get the local RSA private key process.

Because the password is easy to remember for user, and the salts are different for everyone, the strategy of adding salts can greatly increase the difficulty of the attacker. The password is stored in the head. Even if the SACTRSA_pk is copied by others, it is hardly cracked without the password. In addition, we can use another phone to login after loading the SACTRSA_pk from the Could.

2.4 Embedding Secret Information by F5 Algorithm

The F5 steganography algorithm was introduced in [20]. The goal of the research was to develop concepts and a practical embedding method for JPEG images that would provide high steganography capacity without sacrificing security.

The embedding process consists of the following steps [21]:

1. Get the RGB representation of the input image.
2. Calculate the quantization table corresponding to quality factor Q and compress the image while storing the quantized DCT coefficients.
3. Initialize the pseudo-random number generator PRNG by the key.
4. Replace the DCT coefficients with PRNG.

5. The value of the parameter K is determined by the capacity of the image and the length of the secret information.
6. Calculate code word length f5_n = $2^K - 1$.
7. Embedding secret information by (1, f5_n, K) matrix coding:
   (1) Padding f5_n non-zero AC coefficients in the array;
   (2) Define a Hash function f, $\mathrm{f(a)} = \oplus_{i=1}^{f5\_n} a_i \times i$. The Hamming distance is d, producing a hash value such that x = f (a ′), and d (a, a′) $\leq$ 1;
   (3) The next k-bit information is XORed bit by bit with the hash value. Find the position of the pixel to be modified: $s = x \oplus \mathrm{f(a)}$.
   (4) If the information block is the same as f (a), there is no change in the embedding of the information block; otherwise, the change of the coefficient is decided by s, that is, the absolute value of the element is decreased by one.
   (5) Shrink test, which produces 0. If the coefficient becomes 0, subtract 1, discard from the coefficient set, the coefficient set is filled with the next non-zero coefficient, and if no contraction occurs, advance to the new coefficient. Repeat (1) step until the message is embedded.
8. Continue JPEG compression, Huffman coding.
9. To get the stego image.

2.5 Image Encryption Method Design

Figure 2 shows image encryption process in the proposed system. First, we get the Gray matrix of original image and obtain the same size chaotic sequence by the initial value, encrypt image follow the methods in 2.2. Then, we get the Cipher text which is from encrypting the initial value by RSA, to ensure security of the private key of RSA, we used the PBE to protect it in 2.3. Finally, we embed the image with the cipher text by using F5 steganography in 2.4 and sent the image which was encrypted in 2.5 to receiver. In receiver side when the message is arrived then we apply the inverse methods in reverse order to get the original image.

3 Result and Analysis

The above scheme was implemented on the phone with Android 6.0 system and applied on a medical image (size 256*256). Figure 3(a) is the original medical image. Figure 3(b) is the image after we applied Logistic Chaotic sequence to encrypt. Figure 3(c) is the image after we applied F5 steganography, which was sent to the receiver side. Figure 3(d) is the image after decryption in receiver side.

3.1 Histogram Analysis

We analyzed their corresponding histograms. These histograms are presented in Fig. 4 (a), (b), (c) and (d) is the histograms applied the encryption algorithm presented in [22]. The results show that the histograms are different. The histogram of the encrypted image is uniform, which makes it difficult to extract the pixel statistical nature of original medical image and makes it difficult to attack the encrypted image.

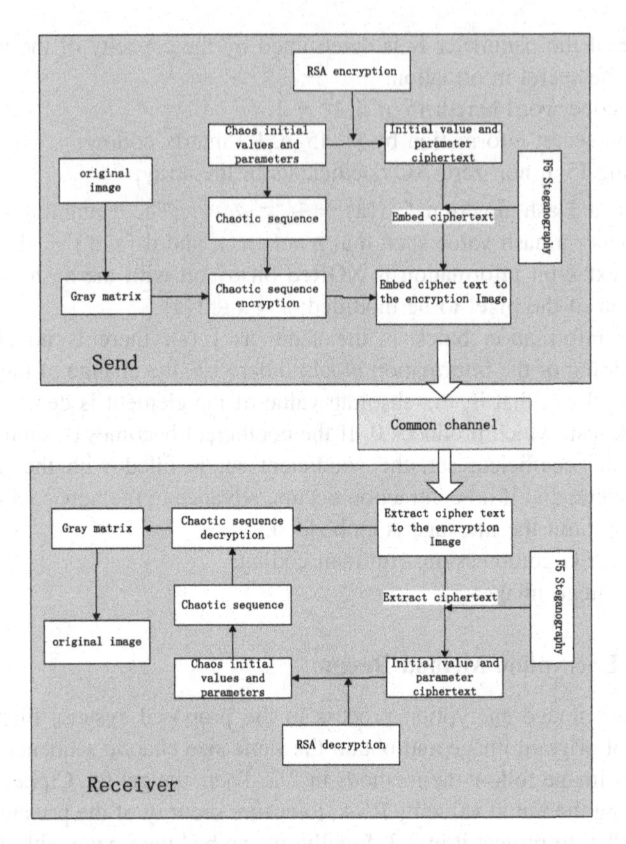

Fig. 2. Image encryption process

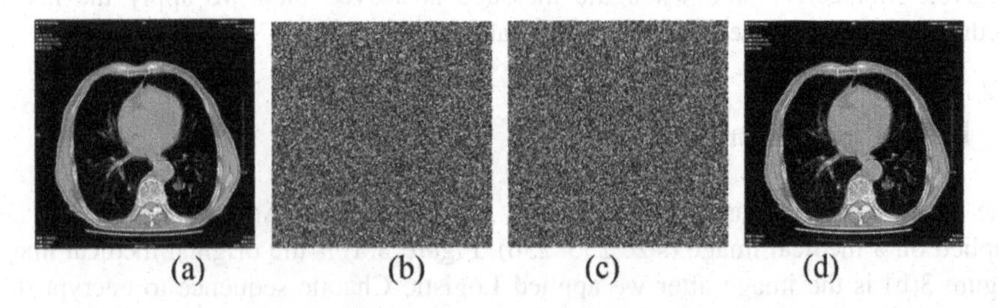

Fig. 3. (a) Original medical image; (b) Image after applying Logistic chaotic sequence to encrypt; (c) Image after F5 steganography; (d) Image after decryption.

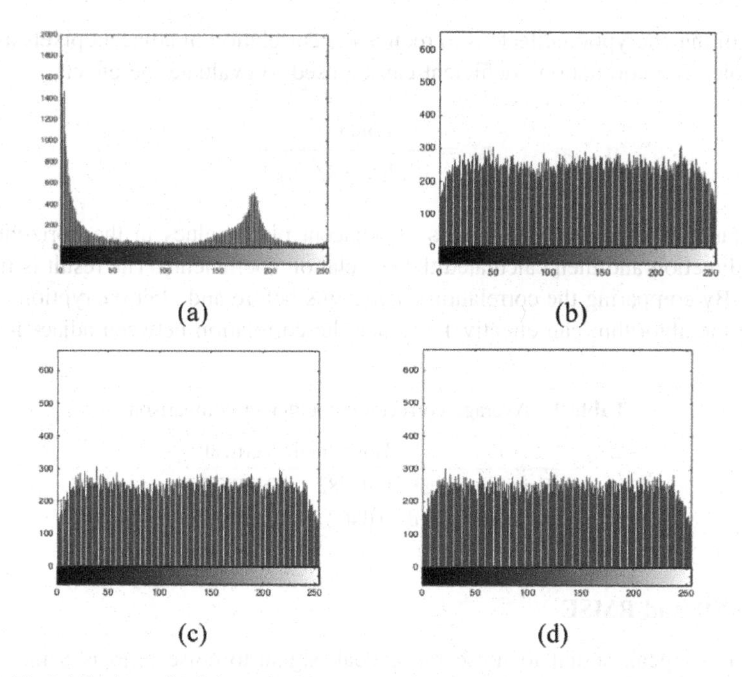

Fig. 4. (a) Original medical image; (b) Image after applying Logistic Chaotic sequence to Encrypt; (c) Image after applying F5 steganography; (d) Encrypted Image after applying the encryption algorithm presented in [22] correlation analysis

3.2 Information Entropy Analysis

Image information entropy reflects the amount of information in images. In information theory, the more ordered the system, the lower the information entropy. Therefore, the information entropy can represent the order of the system. For the calculation of the entropy $H(m)$ of the image information, the following formula can be used:

$$\mathrm{Hm} = \sum\nolimits_{0}^{255} p(i) \log p(i) \tag{8}$$

Where p(i) represents the proportion of pixels in the image where the gray value is i.

According to Fig. 4(b), the histogram of the encrypted image is uniform, it means that the gray levels have the same number of occurrence. For a grayscale image, every pixel is represented with 8 bits, the entropy of the encrypted image is closer to 8 and the better the encryption effect is. The histogram of the encrypted image using Logistic Chaotic sequence is shown in Fig. 4(b), the entropy equal to 7.9968. Figure 4(d) has an entropy equal to 7.9847(<7.9968). So the encryption algorithm in this paper is slightly better than [22], it is clear that the proposed image cryptosystem is robust to the entropy attacks.

One of the encryption effects is to reduce the correlation of adjacent pixels as much as possible. The correlation coefficient can be used to evaluate the effects.

$$r_{xy} = \frac{cov(x,y)}{\sqrt{D(x)} \cdot \sqrt{D(y)}} \tag{9}$$

We randomly selected 1000 pairs of adjacent pixel values in the horizontal and vertical direction and then calculated the correlation coefficients. The result is listed in Table 1. By comparing the correlation coefficients before and after encryption, we can find that the algorithm can effectively reduce the correlation between adjacent pixel.

Table 1. Average correlation coefficient comparison

	Horizontal	Vertical
Original image	0.95383	0.93375
After encrypted	0.0462	0.0653

3.3 PSNR and RMSE

PSNR means peak signal to noise ratio. Peak signal-to-noise ratio is a measure of image distortion. The larger the PSNR value, the less distortion. The calculation of PSNR depends on the mean square error (MSE). MSE expression is as follows:

$$\mathrm{MSE(f,G)} = \frac{\sum_{i=0}^{N-1} \sum_{j=0}^{M-1} [f(i,j) - G(i,j)]^2}{M \times N} \tag{10}$$

Where f is the input image, G is the output image, M and N are the number of rows and columns of the image.

$$\mathrm{PSNR(f,G)} = 10 \log 10 \left(\frac{(2^h - 1)^2}{MSE(f,G)} \right) \tag{11}$$

Equation (11) is used to calculate the peak signal to noise ratio PSNR. h is the number of bits per sample.

RMSE means Root Mean Squared Error. The use of RMSE is very common and it makes a general purpose error metric for numerical predictions. RMSE Reflects the degree of deviation from the true values, the smaller the value, the higher the measurement accuracy.

$$\mathrm{RMSE} = \sqrt{\frac{\sum_{i=1}^{n} \left(x_{obs,i} - x_{nodel,i}\right)^2}{n}} \tag{12}$$

We had done experiments. The PSNR and RMSE were shown in Table 2. From the table, the quality of the reconstructed image is good.

Table 2. The PSNR and RMSE

	PSNR	RMSE
Figure 3(c)	34.9291	5.4090

3.4 Time Analysis

Comparing with the algorithm in [22], the proposed algorithm was verified to be suitable for mobile devices and the encryption and decryption time was shorter than the algorithm provided in [22]. using the Logistic Chaotic algorithm and the algorithm provided in [22] on medical images with three different sizes (128 by 128, 256 by 256, 512 by 512) on a mobile phone (CPU MTK6755 2.0Ghz,8x Cortex-A53, up to 2.0 GHz) with Android system,We performed 10 experiments (The initial values of the three Logistic chaotic sequences are $L1\_x_0 = 0.25$, $L1\_u = 3.574$, $L2\_x_0 = 0.5$, $L2\_u = 3.586$, $L3\_x_0 = 0.75$, $L3\_u = 3.595$) and the average running time was taken. Table 3 lists the time. From the table, we know that the speed should meet the requirement of a mobile health system based on a phone.

Table 3. The result of the test of encryption and decryption time

Image size	This algorithm		Reference [22]	
	Encryption time (ms)	Decryption time (ms)	Encryption time (ms)	Decryption time (ms)
128 by 128	83	57	226	218
256 by 256	131	116	1443	1357
512 by 512	546	527	1693	1538

4 Conclusion

In this paper, a combined approach of encryption and steganography is proposed securing medical images in mobile health information systems. In the proposed scheme, an image is encrypted based on multiple logistic chaotic sequences and the initial value of the chaotic sequence is encrypted using the RSA encryption algorithm. Finally, we embed the image with the encrypted information of the RSA by using F5 steganography. Due to the shortage of resources on the mobile phone side, we adopt a block encryption strategy. The experimental results show that the proposed algorithm can be applied to securing medical images [23–30] in mobile health information systems and get a better encryption results measured using different measures. Compared with the image encryption algorithm in the computer, there is a certain gap in the encryption effect. In the future research, the encryption effect will be enhanced to protect the privacy of images.

Acknowledgment. This work was supported by the National Natural Science Foundation of China (Grant No. 61472293). Research Project of Hubei Provincial Department of Education (Grant No. 2016238).

References

1. Gu, Y.L., Tang, J.S.: A mobile system for skin cancer detection and monitoring. In: SPIE Mobile Multimedia/Image Processing, Security, and Applications, Baltimore, Maryland, USA, 5–9 May 2014
2. Hu, P.J., Chau, P.Y.K., Sheng, O.R.L., et al.: Examining the technology acceptance model using physician acceptance of telemedicine technology. J. Manage. Inf. Syst. **16**(2), 91–112 (1999)
3. Conde, J.G., De, S., Hall, R.W., et al.: Telehealth innovations in health education and training. Telemed. e-Health **16**(1), 103–106 (2010)
4. Weinstein, R.S., Lopez, A.M., Joseph, B.A., et al.: Telemedicine, telehealth, and mobile health applications that work: opportunities and barriers. Am. J. Med. **127**(3), 183–187 (2014)
5. Larburu, N., Bults, R.G.A., Van Sinderen, M.J., et al.: An ontology for telemedicine systems resiliency to technological context variations in pervasive healthcare. IEEE J. Transl. Eng. Health Med. **3**, 1–10 (2015)
6. Meingast, M., Roosta, T., Sastry, S.: Security and privacy issues with health care information technology. In: 2006 28th Annual International Conference of the IEEE Engineering in Medicine and Biology Society, EMBS 2006, pp. 5453–5458. IEEE (2006)
7. Kamali, S.H., Shakerian, R., Hedayati, M., et al.: A new modified version of advanced encryption standard based algorithm for image encryption. In: 2010 International Conference on Electronics and Information Engineering (ICEIE), vol. 1, pp. V1-141–V1-145. IEEE (2010)
8. Yun-Peng, Z., Wei, L., Shui-ping, C., et al.: Digital image encryption algorithm based on chaos and improved DES. In: 2009 IEEE International Conference on Systems, Man and Cybernetics, SMC 2009, pp. 474–479. IEEE (2009)
9. Nag, A., Singh, J.P., Khan, S., et al.: Image encryption using affine transform and XOR operation. In: 2011 International Conference on Signal Processing, Communication, Computing and Networking Technologies (ICSCCN), pp. 309–312. IEEE (2011)
10. Seyedzadeh, S.M., Mirzakuchaki, S.: A fast color image encryption algorithm based on coupled two-dimensional piecewise chaotic map. Sig. Process. **92**(5), 1202–1215 (2012)
11. Enayatifar, R., Abdullah, A.H., Isnin, I.F.: Chaos-based image encryption using a hybrid genetic algorithm and a DNA sequence. Opt. Lasers Eng. **56**, 83–93 (2014)
12. Xu, L., Li, Z., Li, J., et al.: A novel bit-level image encryption algorithm based on chaotic maps. Opt. Lasers Eng. **78**, 17–25 (2016)
13. Belazi, A., El-Latif, A.A.A., Diaconu, A.V., et al.: Chaos-based partial image encryption scheme based on linear fractional and lifting wavelet transforms. Opt. Lasers Eng. **88**, 37–50 (2017)
14. Cheddad, A., Condell, J., Curran, K., et al.: Digital image steganography: survey and analysis of current methods. Sig. Process. **90**(3), 727–752 (2010)
15. Usha, S., Kumar, G.A.S., Boopathybagan, K.: A secure triple level encryption method using cryptography and steganography. In: 2011 International Conference on Computer Science and Network Technology (ICCSNT), vol. 2, pp. 1017–1020. IEEE (2011)
16. Thangadurai, K., Devi, G.S.: An analysis of LSB based image steganography techniques. In: 2014 International Conference on Computer Communication and Informatics (ICCCI), pp. 1–4. IEEE (2014)

17. Seethalakshmi, K.S., Usha, B.A., Sangeetha, K.N.: Security enhancement in image steganography using neural networks and visual cryptography. In: International Conference on Computation System and Information Technology for Sustainable Solutions (CSITSS), pp. 396–403. IEEE (2016)
18. He, J., Lan, W., Tang, S.: A secure image sharing scheme with high quality stego-images based on steganography. Multimed. Tools Appl. **76**(6), 7677–7698 (2017)
19. Banik, B.G., Bandyopadhyay, S.K.: Secret sharing using 3 level DWT method of image steganography based on Lorenz chaotic encryption and visual cryptography. In: 2015 International Conference on Computational Intelligence and Communication Networks (CICN), pp. 1147–1152. IEEE (2015)
20. Westfeld, A.: F5—a steganographic algorithm. In: Moskowitz, I.S. (ed.) IH 2001. LNCS, vol. 2137, pp. 289–302. Springer, Heidelberg (2001). https://doi.org/10.1007/3-540-45496-9_21
21. Fridrich, J., Goljan, M., Hogea, D.: Steganalysis of JPEG images: breaking the F5 algorithm. In: Petitcolas, F.A.P. (ed.) IH 2002. LNCS, vol. 2578, pp. 310–323. Springer, Heidelberg (2003). https://doi.org/10.1007/3-540-36415-3_20
22. Gao, T., Chen, Z.: A new image encryption algorithm based on hyper-chaos. Phys. Lett. A **372**(4), 394–400 (2008)
23. Hu, Z., Tang, J.: Cluster driven anisotropic diffusion for speckle reduction in ultrasound images. In: Proceeding of IEEE International Conference on Image Processing, Phoenix, AZ, USA
24. Hu, Z., Tang, J., Lei, L.: Comparison of several speckle reduction techniques for 3D Ultrasound Images. In: Proceeding of IEEE International Conference on System, Man, Cybernetics, Hungary (2016)
25. Hu, Z., Tang, J.: 3D cluster-driven trilateral filter for speckle reduction in ultrasound images. In: SPIE Mobile Multimedia/Image Processing, Security, and Applications, Baltimore, Maryland, USA, 20 April 2015
26. Guo, S., Tang, J.: Content based image retrieval from chest radiography databases. In: Proceedings of the 43rd IEEE Annual Asilomar Conference on Signals, Systems, and Computers, Asilomar, Pacific Grove, California, USA (2009)
27. Guo S., Tang, J., Cuadra, E., Mason, M., Sun, Q.: Normalized wavelet diffusion for speckle reduction on 3D ultrasound images. In: Zhang, F. (ed.) Proceedings of SPIE, Medical Imaging, Parallel Processing of Images, and Optimization Techniques, MIPPR 2009. vol. 7497, p. 74971R. SPIE, Bellingham (2009)
28. Liu, X., Tang, J., Zhang, X.: A multiscale image enhancement method for calcification detection in screening mammograms. In: Proceedings of IEEE International Conference on Image Processing, Cairo, Egypt, pp. 677–680, 7–10 November 2009
29. Liu, X., Tang, J.: A multiscale contrast enhancement algorithm for breast cancer detection using laplacian pyramid. In: Proceedings of IEEE International Conference on Information and Automation (2009)
30. Tang, J., Liu, X., Xiong, S., Liu, J.: A contrast enhancement algorithm in the DCT domain with reduced artifacts for cancer detection. In: Zhang, F. (ed.) Proceedings of SPIE, Medical Imaging, Parallel Processing of Images, and Optimization Techniques, MIPPR 2009, vol. 7497, p. 749728. SPIE, Bellingham (2009)

Adaptive Image Steganography Based on Edge Detection Over Dual-Tree Complex Wavelet Transform

Inas Jawad Kadhim[1,2(✉)], Prashan Premaratne[1], and Peter James Vial[1]

[1] School of Electrical and Computer and Telecommunications Engineering, University of Wollongong, North Wollongong, NSW 2522, Australia
ijk720@uowmail.edu.au, {Prashan,Peter_Vial}@uow.edu.au

[2] Electrical Engineering Technical College, Middle Technical University, Baghdad, Iraq

Abstract. The proposed method aims for an advanced steganographic approach based on an adaptive embedding process using edge detection over Dual Tree Complex Wavelet Transform (DT-CWT). Here, subband coefficients allow for maintaining high image imperceptibility even with a dense embedding of secret data. Prior to the embedding process, the cover image is divided into multiple non-overlapping blocks and the secret data bits are indirectly concealed in the selected subbands of DT-CWT coefficients. Amount of data bits embedded on different patches depends on the high frequency elements in each patch. These high frequency regions are identified by using Canny edge detection technique. This helps to embed more bits over highly textured regions and fewer bits over smooth regions and hence significantly reduce the distortion of the stego-image. The DT-CWT provides multiple subbands along multiple orientations increasing data capacity with high cover-stego image and secret-recovered image PSNR value. The performance is evaluated on the basis of different standard benchmarks like similarity index, PSNR, payload capacity etc. to evaluate different aspects of image steganography.

Keywords: Image steganography · Adaptive embedding · Data hiding Wavelet transform · DT-CWT · Edge detection

1 Introduction

Due to the ever-increasing demand and the availability of various data formats on the internet, increased data security has become pivotal for secure communication. As a result, the use of technology for information hiding in files such as image, audio, video has also increased [1]. Image steganography is collectively known as a process of hiding information in an appropriate cover image and the processed cover image is known as stego-image [2]. The secret data embedding should be made in such a way to keep the stego image distortion as low as possible. Hence, intruders cannot even suspect the presence of hidden data inside the stego-image. The payload capacity is

D.-S. Huang et al. (Eds.): ICIC 2018, LNAI 10956, pp. 544–550, 2018.
https://doi.org/10.1007/978-3-319-95957-3_57

another criterion needs to be considered as it indicates the number of bits that can be hidden in the carrier image and is generally represented in terms of Bits Per Pixel (BPP). Robustness of the steganography method is also very important as it represents the ability of stego-image to retain the embedded data from possible attacks. The trade-off between the above said requirements is one of the main challenges in order to obtain an excellent image steganographic system.

Steganography techniques are broadly divided into two categories based on nature of embedding process [1, 2]. One method consists of embedding the secret information in the spatial domain. One of the most common approaches of spatial domain steganography was based on the method of hiding the secret message bits in the Least Significant Bit (LSB) of the image pixels [1]. Here, the information bits are directly hidden over the cover image pixel values. Another approach is to hide the information in the transform domain or frequency domain of the cover image. In the transform domain approach, the message is embedded after applying a suitable transformation of the image, allowing more bits to be embedded without altering the spatial domain pixel values of the stego-image. Even though the process is more complicated than the spatial domain method, the hidden data resides in more robust areas providing a superior resistance over statistical attacks and is unlikely to be decrypted by unintended recipients [2]. There are many transform domain techniques that can be widely used in many image processing areas [3] and data hiding [2]. Some of the most popular and widely used transforms include various forms of the Discrete Fourier Transform (DFT), the Discrete Cosine Transform (DCT) and the Discrete Wavelet Transform (DWT) [2]. DCT is a widely used in the first generation transform-based embedding systems and later was replaced by DWT due to its better embedding capacity and imperceptibility. However, DWT has some draw backs as well. It lacks directional selectivity for diagonal features and shifts invariance. Integer wavelet supports reversible data hiding [4]. The Complex Wavelet Transform (CWT) is a complex valued extension to DWT [5]. The Dual-Tree Complex Wavelet Transform (DT-CWT) is another modified version of CWT which employs dual CWT transformation using two separate sets of filter coefficients [6]. The DT-CWT has a modest amount of redundancy, but it provides shift invariance and good directional selectivity. Another advantage of DT-CWT is that the number of transform coefficients are high compared to other transforms and helps in hiding more secret bits. The researchers [5, 6] claim that due to its shift invariance and perfect reconstruction property, complex wavelet transform improves robustness against geometrical attacks.

2 Proposed Method

The flow diagram in Fig. 1 depicts the proposed embedding approach. The steps of embedding process are described below:

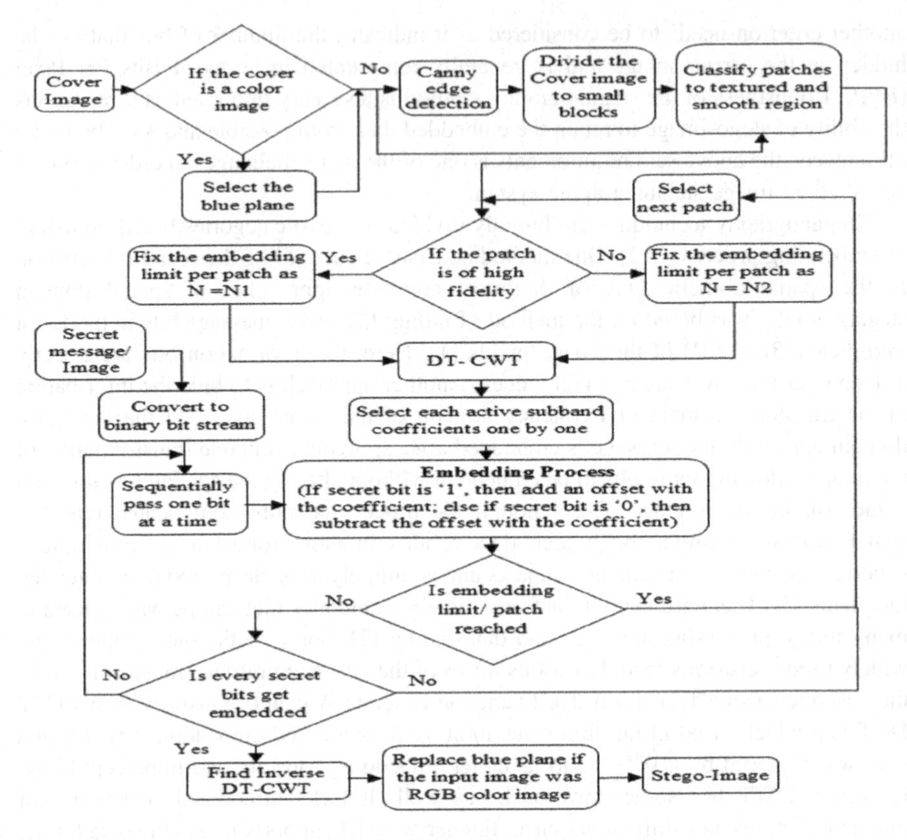

Fig. 1. Embedding scheme

2.1 Embedding Stage

1- *Selection of the carrier image and secret data:* The cover image can be either a color image or a grayscale image. The secret data can be an image, a text message or any sort of binary data.

2- *Pre-processing:* The carrier image needs to be divided into non-overlapping blocks. If the secret data is an image, or any other data form, it needs to be converted to corresponding binary format.

3- *Detect edge details of the cover image*: The edge details in the cover image are calculated using Canny edge detector with a suitable threshold [7]. Here we use a threshold of 0.5 to select moderately strong edges from the cover image.

4- *Divide the cover image into non-overlapping patches*: The cover image is divided into square blocks. On the basis of edge details present, each patch is classified into two types; patches with less edge components and patches with high edge details. If the patch has at least 10% edge pixels, it is treated as texture area and other patches are considered as smooth regions.

5- *Embedding bits on DT-CWT subbands*: The DT-CWT employs two separate discrete wavelet filter bank trees that decompose a signal into real and imaginary parts. Hence this provides more subbands for real and imaginary coefficients than that of other transforms and helps in hiding more secret bits. The DT-CWT structure is given in [8].

Here, the algorithm concentrates on embedding bits on different levels of DT-CWT subband coefficients of individual cover image patches. As the transform decomposition level increases there will be more sub-images corresponding to a patch and more bits can be embedded with. But this will in turn reduce the accuracy while recovering secret bits from the stego-image. So, in this algorithm, we limit the use of DT-CWT to 1 level for decomposition. The embedding limit is designed as *N1* and *N2* for high fidelity and smooth patches respectively. These limits are set in such a way that the maximum allowable embedding bits in a sub-image of DT-CWT coefficient of a smooth patch are lower than that of a sharp (high fidelity) patch. The nature of embedding process in a sub-image of DT-CWT coefficients are given below:

- The first task is to find the values of coefficients greater than a suitable threshold. Here, we used 0.4 based on the evaluation of results at various thresholds (The threshold which provides minimum error to payload capacity ratio).
- The active coefficients are indirectly modified according to the secret bits. An offset value will be added to such coefficients; if the bit needs to be embedded is "1" and offset will be subtracted from the active coefficient value if the secret bit is "0". The offset has much significance here and we selected "8.8" as the offset which provides best payload capacity- distortion ratio in the experimental tests.

2.2 Retrieval Stage

A detailed flow graph of the proposed retrieval scheme is shown in Fig. 2. The decoding process can be summarized into following steps:

1- Split stego-image into patches as explained in Sect. 2.1 step 1 to 4.
2- Find the DT-CWT subband coefficients of every patch as explained in step 4.
3- Find the active coefficients in the stego-image by comparing with the original cover image. The changes will be there only in those coefficient places where secret bits are embedded.
4- In such active coefficients of the stego-image subbands, if the coefficient value is greater than that of the original coefficient of the cover image, '1' is considered to be embedded and '0' for a lesser value. The process repeats until all the embedded secret bits are extracted from the embedded DT-CWT coefficients.
5- After recovering every bit, these bits are grouped into bit blocks and converted back to represent image pixels of a secret image.

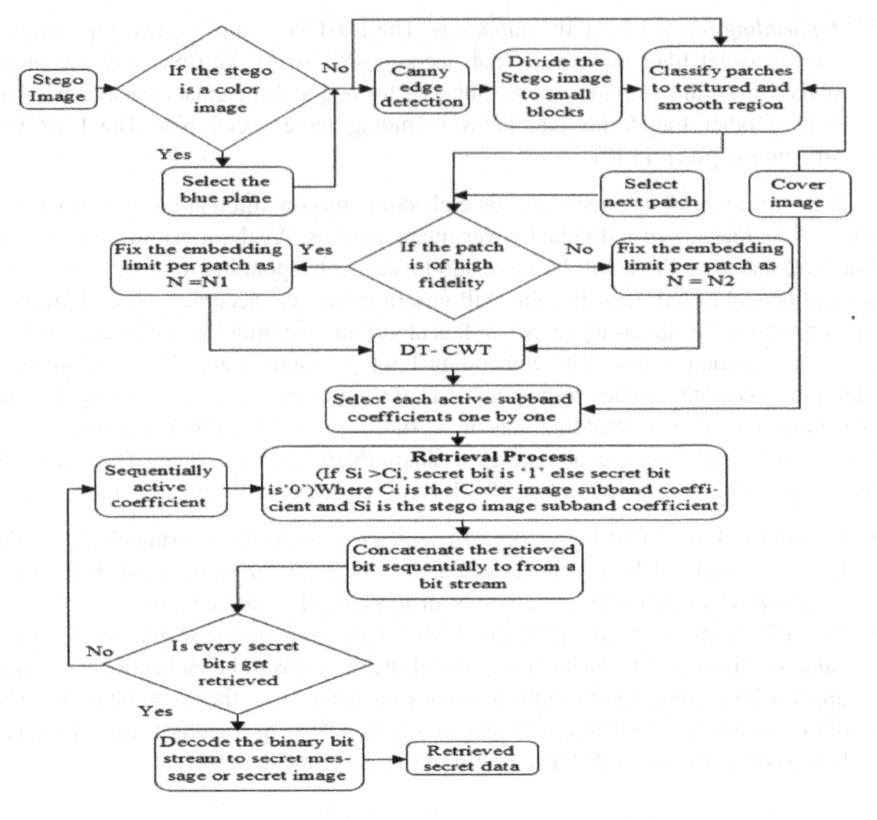

Fig. 2. Secret data retrieval scheme

3 Results and Performance Analysis

The performance is evaluated based on the Retrieval Error Rate (REB), payload capacity in terms of Bits Per Pixel (BPP), Correlation factor, Peak Signal to Noise Ratio (PSNR) and Structural Similarity Index Measure (SSIM) of both 'Cover image-Stego image' and 'Secret image– Retrieved image' [9]. The test color and grayscale images were selected from standard image processing datasets [10]. The cover and secret images are resized to multiple resolutions and are used for performance evaluation process. The results obtained from Table 1 reveals that the retrieval error rate reduces when the patch size increases. This is due to the fact that, increase in the number of patches will create more DT-CWT sub-images and more quantization errors will be there in the forward and inverse transforms. But payload capacity and PSNR degrades when patch size is high. This is because, low patch size results in higher number of individual patches in the cover image and helps to distinguish tex-ture regions and smooth regions more effectively. As a trade-off between PSNR, error and payload capacity, we chose the patch size as 64 × 64. As per the results, it can be deduced that the retrieval accuracy depends on the nature of the coefficients used for the embedding process.

Table 1. Performance statistics for various test images while using different patch sizes

Patch size	Lena			Cameraman			Mandrill		
	Max Payload (BPP)	Retrieval error	PSNR (at max payload)	Max Payload (BPP)	Retrieval error	PSNR (at max payload)	Max Payload (BPP)	Retrieval error	PSNR (at max payload)
32	5.91	0.98	47.34	5.61	1.18	48.42	5.84	0.97	47.63
64	5.65	0.81	46.21	5.45	1.01	46.28	5.1	0.53	45.49
128	5. 27	0.76	43.91	5.25	0.89	43.31	5.58	0.55	42.17

A comparison with an existing system [6] is shown in Fig. 3. The quality of the stego-image is analysed by several performance parameters and compared with average test images from similar approaches is given in Table 2. The proposed method provides better payload capacity, PSNR and another similarity index of both 'Cover image-Stego image' and 'Secret image– Retrieved image' while compared with several other approaches.

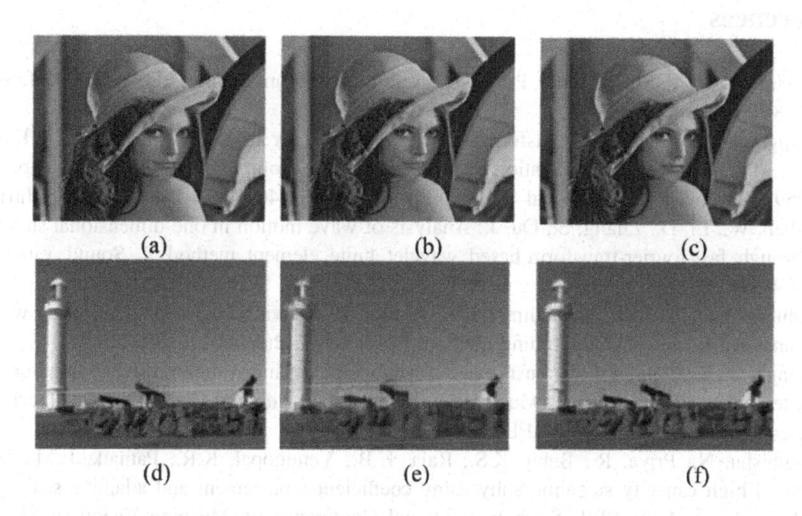

Fig. 3. (a) Cover image, (b) Stego-image [6], (c) Stego-image (Proposed method), (d) Original secret image, (e) Retrieved secret image [6], (f) Retrieved secret image (Proposed image).

Table 2. Comparison with other methods

Methods\ benchmarks	Stego-Image			Retrieved secret image			BPP
	PSNR	SSIM	Correlation factor	PSNR	SSIM	Correlation factor	
[4]	36.58	0.969	0.9994	29.96	0.829	0.948	1.18
[5]	34.83	0.941	0.9983	27.55	0.8989	0.981	2
[6]	40.57	0.945	0.9986	25.2	0.876	0.964	4.6
Proposed work	47.08	0.975	0.9998	31.08	0.9124	0.987	5.4

4 Conclusion and Future Work

The method proposed here is mainly focused on an indirect embedding system with high payload capacity without losing the quality of the cover image. The embedding is attempted on the complex coefficients obtained using DT- CWT and the embedding scheme is adaptive in nature as the embedding density is selected based on the fidelity of the cover image regions. This reduces the chance for distortion and hence more immune to reveal the presence of hidden data. Experimental results show that the stego-image achieved is having good PSNR and high similarity index compared to cover image and hence make the system more immune to intruder intervention. Even though the retrieval accuracy is good, future works are planned to reduce the error along with a few security measures to defend attacks from intruders.

Acknowledgment. The first author would like to acknowledge higher committee of education development in Iraq (HCED) for the scholarship funding.

References

1. Islam, S., Modi, M.R., Gupta, P.: Edge-based image steganography. EURASIP J. Inf. Secur. **1**, 8 (2014)
2. Singh, S., Siddiqui, T.J.: Transform domain techniques for image steganography. In: Kayem, A., Meinel, C. (eds.) Information Security in Diverse Computing Environments, pp. 245–259. Hershey, PA, IGI Global (2014). https://doi.org/10.4018/978-1-4666-6158-5.ch013
3. Shen, W., Li, D., Zhang, S., Ou, J.: Analysis of wave motion in one-dimensional structures through fast-fourier-transform-based wavelet finite element method. J. Sound Vibr. **400**, 369–386 (2017)
4. Muhammad, N., Bibi, N., Mahmood, Z., Akram, T., Naqvi, S.R.: Reversible integer wavelet transform for blind image hiding method. PLoS ONE **12**(5), e0176979 (2017)
5. Singh, S., Siddiqui, T.J.: Robust image steganography using complex wavelet transform. In: International Conference on Multimedia, Signal Processing and Communication Technologies (IMPACT), pp. 56–60. IEEE (2013)
6. Sathisha, N., Priya, R., Babu, K.S., Raja, K.B., Venugopal, K.R., Patnaik, L.M.: Dtcwt based high capacity steganography using coefficient replacement and adaptive scaling. In: Proceedings of the SPIE Sixth International Conference on Machine Vision (ICMV 13), p. 90671O, December 2013
7. Canny, J.: A computational approach to edge detection. IEEE Trans. Pattern Anal. Mach. Intell. **8**(6), 679–698 (1986)
8. Kadhim, I.J., Premaratne, P., Vial, P.J., Halloran, B.: A comparative analysis among dual tree complex wavelet and other wavelet transforms based on image compression. In: Huang, D.-S., Jo, K.-H., Figueroa-García, J.C. (eds.) ICIC 2017. LNCS, vol. 10362, pp. 569–580. Springer, Cham (2017). https://doi.org/10.1007/978-3-319-63312-1_50
9. Subhedar, M.S., Mankar, V.H.: Performance evaluation of image steganography based on cover selection and contourlet transform. In: Proceedings of the International Conference on Cloud and Ubiquitous Computing and Emerging Technologies (CUBE), Pune, pp. 172–177. IEEE, November 2013
10. http://www.ece.rice.edu/~wakin/images. Accessed 15 Mar 2017

Detecting Distributed Denial of Service Attacks in Neighbour Discovery Protocol Using Machine Learning Algorithm Based on Streams Representation

Abeer Abdullah Alsadhan[1(✉)], Abir Hussain[1(✉)], Thar Baker[1(✉)], and Omar Alfandi[2(✉)]

[1] Department of Computer Science, Liverpool John Moores University, Liverpool L33AF, UK
A.A.Alsadhan@2016.ljmu.ac.uk, {A.Hussain,T.baker}@ljmu.ac.uk

[2] College of Technological Innovation, Zayed University, Abu Dhabi, UAE
Omar.AlFandi@zu.ac.ae

Abstract. The main protocol of the Internet protocol version 6 suites is the neighbour discovery protocol, which is geared towards substitution of address resolution protocol, router discovery, and function redirection in Internet protocol version 4. Internet protocol version 6 nodes employ neighbour discovery protocol to detect linked hosts and routers in Internet protocol version 6 network without the dependence on dynamic host configuration protocol server, which has earned the neighbour discovery protocol the title of the stateless protocol. The authentication process of the neighbour discovery protocol exhibits weaknesses that make this protocol vulnerable to attacks. Denial of service attacks can be triggered by a malicious host through the introduction of spoofed addresses in neighbour discovery protocol messages. Internet version 6 protocols are not well supported by Network Intrusion Detection System as is the case with Internet Protocol version 4 protocols. Several data mining techniques have been introduced to improve the classification mechanism of Intrusion detection system. In addition, extensive researches indicated that there is no Intrusion Detection system for Internet Protocol version 6 using advanced machine-learning techniques toward distributed denial of service attacks. This paper aims to detect Distributed Denial of Service attacks of the Neighbour Discovery protocol using machine-learning techniques, due to the severity of the attacks and the importance of Neighbour Discovery protocol in Internet Protocol version 6. Decision tree algorithm and Random Forest Algorithm showed high accuracy results in comparison to the other benchmarked algorithms.

Keywords: Machine learning · Denial of service · IPV6

D.-S. Huang et al. (Eds.): ICIC 2018, LNAI 10956, pp. 551–563, 2018.
https://doi.org/10.1007/978-3-319-95957-3_58

1 Introduction

The rapid growth of the Internet usage has caused problem on Internet protocol address space. To solve the space issue of Internet Protocol version 4 IP4 addresses, Internet Protocol version 6 IP6 was created to expand the availability of address spaces. [33]. IPv6 is a new general use version of Internet Protocol; it is designed to overcome the main limitations of Internet Protocol version 4 including the lack of security and the exhaustion of Internet Protocol address space. Routers, notebooks, PCs and mobile phones are among the diverse networking devices supported by IPv6. To facilitate complete enforcement of IPv6 in the future, some devices are based on the dual stack IPv4/IPv6 mechanism. The main benefits of IP6 are its capacity for development and enhancement and its support of new Internet functionality. Given the great importance attributed to network security nowadays, effective network security is essential to take full advantage of the networks and the strengths of IPv6. Kumar et al. [24] indicated that Internet Control Message Protocol (ICMPv6) is a major protocol in IPv6 application that aids detection of neighbours and routers. However, this protocol is susceptible to attacks (e.g. ICMPv6 flood attacks) that make network services inaccessible and reduce network performance [21]. Meanwhile, unlike ICMPv4 messages, which may be blocked by some IPv4 network administrators if they pose security risks, IPv6 networks are not subject to such blanket locking as neighbour and router detection relies on ICMPv6 messages in IPv6 network operation [33].

2 Background

The most popular standard network protocol nowadays is the OSI-inspired integrated Transmission Control protocol and Internet Protocol (TCP/IP) [17]. IPv6 uses the Internet to connect users regardless of their operating system (e.g. Windows, Linux, etc.) and to represent the latest update to networking protocol version 4 [37]. The main benefits of IPv6 are its capacity for development and enhancement and its support of new Internet functionality [13]. IPv6 is a new general use version of Internet Protocol for global purposes and was built in 1998 [29]. It is designed to overcome the main limitations of IPv4 including the lack of security and the exhaustion of IP address space [15].

The continuously increasing number of utilisation of the Internet, combined with the emergence of new network services, emphasises that both of these aspects should be fully understood. Moreover, while the advantages derived from the Internet, and especially its recent growth and development, network security attack threats have become more serious. The security holes utilised by network security attacks must be accounted for to limit them in the timeliest possible manner, and this has motivated the design and implementation of Intrusion Detection Systems (IDSs) [3].

2.1 IPv6 Intrusion Detection System

Intrusion Detection Systems (IDS) are tools, which are provided to increase the security of communication and information systems [2]. With regard to performance and features, IPv6 protocols are not as well supported by Network Intrusion Detection Systems as IPv4 protocols, due to security issues [10].

Anomaly detection and misuse detection are the two major mechanisms used by IDS to detect intrusions [27]. The anomaly detection mechanism employs an established threshold to determine the distance between a suspicious activity and the norm to discover any irregularities. Meanwhile, the misuse detection mechanism uses a series of rules or signatures to identify illegitimate activities [4]. The two mechanisms are differentiated by the fact that new attacks are identified by the anomaly detection but not by the misuse detection mechanism, while the rate of false alerts is greater in anomaly detection than in misuse detection [1]. Intrusion Detection System is a critical method in information security with the growth and increasing complexity of the Internet. Since IPSec has been delegated as a built-in security measure for IPv6 with the exhaustion of IPv4 addresses and emergence of IPv6 addresses, it has become more challenging to detect intrusions [32]. Anomaly detection mechanisms can be classified into three main classes: knowledge based, statistical based, and machine learning based [25]. Most studies indicated that machine-learning approaches provide promising results in comparison to the other two techniques [18]. Machine learning can detect attacks with great accuracy. Numerous studies have provided evidence in support of the effectiveness of machine learning algorithms in IPv4, particularly in the case of intrusion detection field. Furthermore, there are two methods basis on the source of data to be analysed in Network Intrusion Detection System (NIDS): Packet-based NIDSs and flow (stream)-based NIDSs are two variants of the NIDS that have been designed to analyse varying data sources. The former examines the entire payload content in addition to headers, while the latter tends to avoid examining every packet travelling through a network link, thereby centering focus on the aggregated data of associated packets of network traffic in the form of flow. Consequently, flow-based NIDSs are not required to analyse the same large quantity of data as packet-based NIDSs [3].

2.2 Neighbour Discovery Protocol (NDP)

Neighbour Discovery Protocol (NDP) [31], one of the main protocols used by IPv6 suits, has several security concerns [14]. Moreover, the designers of IPv6 assumed that local area network (LAN) consists of trusted users. Hence, NDP trusts every device insides LAN that makes it exposed to various attacks. Therefore, in communication systems, the use of new technologies, which have security limitation due the lack of registration and authentication, makes these systems prone to attacks. In IPv6 networks, NDP allows the connected devices to generate their IP addresses and start to communicate with other devices without any registration or authentication inside the network. In addition, most of the existing operating systems are IPv6 enabled by default; thus, LANs are exposed to IPv6 attacks even if IPv6 is not the protocol that is used for communication [30].

Neighbour Discovery defines five different ICMP packet types (as mentioned in RFC 4861 [31]):

Router Solicitation RS: Hosts send Router Solicitations request to ask routers to produce Router Advertisements instantly as opposed to initially planned time.

Router Advertisement RA: In combination with a range of links and Internet parameters, routers advertise their presence in a periodic way, or they respond to the previously mentioned Router Solicitation request. Router Advertisements incorporate prefixes, which are utilised to ascertain whether an address has an identical link (on-link determination), an address configuration, or a combination of both. A suggested hop limit value is also relevant in this context, among other aspects.

Neighbour Solicitation NS: These are used for Duplicate Address Detection (DAD), to ascertain a neighbour's link-layer address, or to ensure that a neighbour can still be contacted with a cached link-layer address. To facilitate these purposes, a node transmits Neighbour Solicitation messages.

Neighbour Advertisement (NA): These represent responses to Neighbour Solicitation messages, and a node can also send unsolicited Neighbour Advertisements to broadcast a link-layer address modification.

Redirect Message (RM): These are employed to announce to hosts the issue of whether a more effective first hop towards a destination is available.

2.3 IPv6 Security Attacks

The major attacks, which could be launched against an IPv6 network, are placed into the following categories:

- Reconnaissance Attack: this attack allows an attacker to passively gather address details of available devices on a network, allowing the attacker to identify targets by starting attacks on a network. This type of attacks is likely in IPv6 because of its Multicast feature and reply from ICMPv6 messages [20].
- Man-In-The Middle Attack (MITM): an attack associated with the passive collection of confidential messages of an intention device on a network by acting as a router between the device and the communicating destination [5].
- Denial of service (Dos) attack: DoS attacks condense the target network services or resources unavailable to authorized users. DoS attack on an IPv6 network can be launched by exploiting vulnerabilities in the Multicast, Extension Headers, ICMP messages and DAD protocol [10]. Flood attack, Ping of Death, SYN attack, Teardrop attack, DDoS, and Smurf attack are the most general types of DoS attacks [23]. A DoS can be triggered by ICMPv6 in several ways, such as transmission of too many ICMPv6 packets to site destinations or broadcasting of disruptive error messages. Moreover, numerous ICMP messages are intended for transmission not just to unicast addresses but to multicast addresses as well. This makes ICMP highly significant in IPv6, but it also creates new security issues for example, communication establishment that determines session termination and invalidation of legitimate addresses or interface disabling through infiltration of fake communication establishment or maintenance messages onto a link [33, 34]. Several ICMPv6 packets with router advertisement (RA) requests are responsible for the major DoS susceptibility in an IPv6 network that is known as the Ping of Death (PoD). In August 2013, security bulletins published that a DoS might occur if a target system/server received a specially designed ICMPv6 packet from an attacker [7].

- Fragmentation Attack: In IPv6, the router load is relieved by fragmentation as this occurs at the source node [11]. The lowest Maximum Transfer Unit (MTU) of IPv6 is 1280 bytes, which means that fragments will be discarded if they are less than this value. This is intended to prevent fragment attacks, but it can also increase the likelihood of a DoS attack because the discarded fragments may be used by attackers to assess the network condition and disrupt the system by substituting the destination port, resulting in superimposed fragments on the destination host [35].

3 Data Mining Technique and IDS

As outlined by Barbara et al. [8], IDSs employ knowledge discovery processes to facilitate self-learning, and this ultimately allows them to distinguish between regular and irregular activity in the networks. Importantly, where irregular activity is detected, this is indicative of a security threat. An important distinction that should be made is between supervised and unsupervised knowledge discovery methods; where both are drawn on to facilitate consistent network activity [8]. In view of these considerations, it is evident that machine-learning algorithms constitute highly effective ways to allow IDSs to identify conventional and unconventional threats. As noted by Kruegel et al. [22], knowledge discovery primary goal is to classify the captured events as normal, malicious or as certain modes of attack.

Chan et al. [12] offered Hybrid Intrusion Detection System (HIDS) that combines the benefits of rule-based and Support Vector Machine (SVM) algorithm. The SVM is used to create rules, whereas the rule-based system is then applied to discover the DOS attacks. The rule set created by the HIDS is more accurate. The accuracy measured by comparing the HIDS with the packet sniffer tool (SNORT). Furthermore, both used to classify the DoS and non-DoS traffic in the testing dataset. The HIDS had a notable improvement by 20%.

Bahl et al. [6] used feature subset selection techniques to improve accuracy and detection rate of attacks, which had a limitation of detecting only U2R attacks efficiently while DOS and R2L attacks were not identified with the same accuracy. In addition, this method was found to be efficient only for IPv4 networks not for IPv6 networks. Salih et al. [34] suggested an adaptive methodology to detect covert channels in IPv6 using New Network Intelligent Heuristic Algorithm (NIHA) combined with Multinomial Naïve Bayes Algorithm (NBA). This approach chased only covert channels, which lead to serious security threats against legitimate targets implementing this protocol. Several data mining techniques have been introduced to improve the classification mechanism of IDS. In addition, extensive researches indicated that there are no IDS for IPv6 using advanced machine learning techniques to ward against distributed denial of service attacks.

IPv6 is vulnerable to DDoS attacks that expose the security and availability of the NDP messages. Several issues have to be addressed including:

- High error rate and low accuracy have been achieved by the existing detection systems due to their dependency on the unsuitable representation of traffics (packets representation).

- The possible existing NDP-DDoS attacks have not been completely covered by any of the existing detection systems. Therefore, if these IDSs are able to detect number of these attacks, they fail to detect other.
- The existing detection systems are built using a set of non-qualified features such as the capturing time, which leads to packets misclassification. This is because the detection model building process involves the classifier's consideration of the attack packets' time intervals as a feature that can be used to indicate an attack, which means the attacks falling outside the determined intervals are considered as secure [16].
- The absence of benchmark datasets that include all the possible NDP-DDoS attacks to be used in evaluating any proposed system.

An alternative representation can be based on stream forms similar to IPv4 dataset DARPA. The IPv4 connection is defined as a stream of TCP packets starting and ending at some well-defined time, during which data flows to and from a source IP address to a target IP address under some well-defined protocol. IPv6 traffic should be formatted using such representation by utilizing IPv4 stream (S4) definition to define a new IPv6 NDP stream (S6NDP) of its attack. S4 is defined as a number of packets shared a number of characteristics, which are IPsec, IPdst, and protocol [36]. To have streams that are created from the same NDP types packets, NDP have 5 types (RS, RA, NS, NA, and Redirect). In this research S6NDP is defined as stream of NDP packets that have the same IPv6 source address, IPv6 destination address and NDP type sent over a period of time. The use of such representation has various advantages, including: First, sending similar traffics toward the victim at a small period of time performs NDP-DDoS attacks. Using this representation for such attacks can be noticeable to any applied approach.

3.1 Features Extraction

Classification technique aims to build a model by learning the behaviours from the given training dataset then test the model based on the testing dataset. The datasets (training and testing) must be represented using a set of qualified features to allow the techniques to easily and effectively learn and model the behaviours. Qualified features should be informative, discriminative or uncertainty reducing and different among the included classes [9].

Generally, each classification problem has a set of defined features used for representing the problem's datasets. Based on these features, various researches and experiments are conducted to enhance by one of two paths; either selecting a subset of these features or improve the applied classification algorithms. For example, DARPA IPv4 dataset has defined a set of 42 different features to be used to represent the dataset for intrusion detection purpose. These features have been used in different researches of IPv4 intrusion detection [28].

However, NDP-DDoS attacks are unable to use the existing features of IPv4 attacks (such as DARPA's features) due to IPv6 and IPv4 headers differences. Moreover, the existing IDSs for IPv6 have used nonqualified features which have misclassification problem of the attacks. Therefore, there is a need to identify an initial set of features for these attacks to be the base for the attacks detection in this research and other researchers. Based on the literature to identify a new set of features, domain knowledge of the attacks are used such as in DARPA dataset and several other researchers. Domain knowledge depends on assuming a set of features that contribute to the attacks detection based on logical assumptions or by inspiring them from another similar research. To validate any set of the features, feature-ranking techniques are applied and a subset of the most related features are selected.

Based on the domain knowledge of the NDP-DDoS attacks, several features are specified to be added to the streams and representing the datasets which are:

- NDP Type feature is the first extracted features to help the classifier to differentiate between the attacks types. Most of the perfumed NDP attacks are assumed to be from one NDP type (except Replayed attacks), therefore knowing the type helps in differentiating between these attacks. Moreover, sending the same type of NDP malicious packets might perform attacks. The packets type depends on the targeted victim for example attacking router should be by RS packets.
- Packets Number feature: number of the sent packets within the stream. It also has been used in DARPA dataset. Number of packets is assumed to be small in normal streams and large in malicious stream.
- Bytes Number feature: number of bytes sent from the source to the destination within the time interval. It has been selected for the same reasons as Packets Number.
- SAT (Stream Active Time) feature: the active time of the stream between the source and destination. SAT is one of the most selected features in IPv4 datasets such as DARPA.
- Bytes Ratio feature: the ratio of the sent bytes within the SAT time. It is calculated by dividing the number of sent bytes (Bytes Number) over the active time (SAT). Ratio is assumed to be larger in case of attacks stream where huge numbers of attacking packets are sent in a small period.

Several attacking tools (such as THC-toolkit, Si6) have been proposed to tackle IPv6 network by different attacks includes NDP-DDoS attacks [10]. Attackers normally generate (send) their malicious traffic using these tools, which generate packets with almost the same header attributes. However, these header attributes are different when normal users are communicating, they send normal packets with different value of packets header attributes (such as packet length, traffic class, hop limit, etc.). Therefore, having features that model (represent) these cases will help to differentiate between NDP DDoS malicious streams and normal streams. Based on this assumption, the rest of the features are selected. The chosen features are binary features, have two values either 1 means the stream have different values of the packet header attribute or 0 means the stream have the same packet header attribute. The chosen features are as follows:

1. #Same_Length feature: 0 if all the stream's packets have the same length header attribute, 1 if at least one packet has different length (L_Diversity).
2. #Same_Traffic_Class feature: 0 if all stream's packets have the same traffic class header attribute, 1 if at least one packet has traffic class (TC_Diversity).

3. #Same_hop_limit feature: 0 if all stream's packets have the same hop limit header attribute, 1 if at least one packet has hop limit (HL_Diversity).
4. #Same_Flow_Label feature: 0 if all stream's packets have the same Flow Label header attribute, 1 if at least one packet has Flow Label (FL_Diversity).
5. #Same_Next_Header feature: 0 if all stream's packets have the same Next Header header attribute, 1 if at least one packet has Next Header (NH_Diversity).

#Same_ Checksum feature: 0 if all stream's packets have the same Checksum header attribute, 1 if at least one packet has Checksum (CS_Diversity).

Same_Payload_Length feature: 0 if all stream's packets have the same Payload Length header attribute, 1 if at least one packet has Payload Length (PL_Diversity).

To construct the streams and extract the mentioned features, several packets fields need to be extracted from the packets. These fields are the following: IPv6Src address, IPv6Dst address, NDPType, packet time, packet Length, packet Traffic_Class, packet hop_limit, packet Flow_Label, packet Next_Header, packet Checksum.

4 Experiment Design and Results

4.1 Design

The proposed framework of this research consists of five phases aim to detect the NDP-DDoS attacks. The first phase starts with capturing the traffic from the network. The second phase extracts the required attributes to build the streams and extract the features from the filtered packets. The third phase builds the streams based on the S6NDP definition as well as it extracts the features for each stream. The fourth phase aims to evaluate the selected features and excludes any nonrelated (low ranked) features using the ranking algorithm. The last phase is to apply a machine learning classifier to the built streams datasets with the selected features in order to build a detection model.

Generally, each classification problem has a set of defined features used for representing the problem datasets. Based on these features, different researches and experiments are conducted to enhance by one of two paths; either selecting a subset of these features or improve the applied classification algorithms. For example, DARPA IPv4 dataset has defined a set of 42 different features to be used to represent the dataset for intrusion detection purpose. These features have been used in different researches of IPv4 intrusion detection in the mentioned two paths [28]. However, NDP-DDoS attacks are unable to use the existing features of IPv4 attacks such as DARPA's features due to IPv6 and IPv4 header differences. Moreover, the existing IDSs for IPv6 have used nonqualified features, which have misclassification problem of the attacks. Therefore, there is a need to identify an initial set of features for these attacks to be the base for the attacks detection in this research and other researchers. Based on the literature to identify a new set of features, domain knowledge of the attacks are used such as in DARPA dataset and several other researchers. Domain knowledge depends on assuming a set of features that contribute to the attacks detection based on logical assumptions or by inspiring them from another similar research. To validate any set of the features, feature-ranking techniques are applied and a subset of the most related features are selected.

4.2 Results

In order to create our dataset, a real network has been used to capture the normal traffics using Wireshark tool. Furthermore, a virtual test bed network has been created using Graphical Network Simulator-3 GNS3 and attached to the real network [19]. Therefore, the DDoS attacks have been performed in the virtual network only that has the same setting and configuration of the real network. The malicious traffic and the normal traffic are collected to create the needed datasets.

Various software tools have been used to build the virtual testbed. GNS3 is used to build the virtual Test bed network and connect it to the real network. Oracle virtual machine is used to install different OSes. Different OSes are installed for different aims, which either generate normal traffic or performed attack. The used of the real network and the virtual network consist of different OSes and network devices in order to have as much as possible realistic level of the datasets. Having these diversity in OSes aims to make sure the created dataset includes all the possible behaviors and scenarios of the traffic, which will lead to create a solid detection models.

The created traffics have been filtered to have only NDP traffics, which are the target of this research. After that this file has been opened in Database and these traffics have been labelled into normal or attacks packets based on the pre-know information about the included traffic (normal and attack). In addition, based on the proposed methodology, the required packets attributes have been extracted from each packet in order to prepare it for the feature construction and streams building phase. These attributes are IPv6Src address, IPv6Dst address, NDPType, packet time, packet Length, packet Traffic_Class, packet hop_limit, packet Flow_Label, packet Next_Header, packet Checksum, packet Payload and Length. These header attributes are extracted from each packet before streams construction phase. The reason behind extracting them is that all of them are used to either construct the stream such as IPv6Src address, IPv6Dst address, NDPType, packet time or extracting the features such as packet Flow_Label, packet Next_Header.

The next phase has been applied to build the streams based on its definition and extract the features that are mentioned previously. This phase is applied using MYSQL queries applied on the traffic to convert the packets into streams with several features for each stream. After representing the traffic in the form of streams with 12 features for each stream, the traffic is use for the classifications.

The traffic file is loaded to train the various machine learning models using 80% of the data for training and 20% for testing. Based on the applied classifier the detection accuracy is calculated. In addition, True Positive TP and False Positive FP rate are calculated. Table 1 shows the accuracy result with the machine learning classifiers.

As shown in Table 1, Decision tree algorithm showed the highest accuracy result in comparison to the other classifiers, due to the strength of these algorithms and the strength of the selected features. In addition, decision trees are an effective supervised knowledge discovery method. One of the key points of value offered by Decision tree approach is that its classification accuracy is significant. Furthermore, decision trees have the greatest precision in the supervised subset of approaches [26].

The second dataset has been collected from Saudi Arabia. In order to create this dataset, a real network has been used to capture the normal traffics using Wireshark

Table 1. The accuracy result with different classifiers

Algorithm	Accuracy	Type	True Positive (TP) rate	False Positive (FP) rate
Decision tree algorithm	84.3	Normal	0.93	0.33
		Attack	0.66	0.06
Naive bayes algorithm	63.5	Normal	0.46	0.03
		Attack	0.96	0.53
Random forest algorithm	84.3	Normal	0.93	0.33
		Attack	0.66	0.06
MLP algorithm	75.3	Normal	0.75	0.23
		Attack	0.46	0.04

tool. Furthermore, a virtual test bed network has been created using GNS3 and attached to the real network. The DDoS attacks have been performed in the virtual network that has the same setting and configuration of the real network. The malicious traffic and the normal traffic are collected to create the needed datasets. In addition, Principal Component Analysis (PCA) has been used to rank the features. Table 2 shows the accuracy result with different classifiers.

Table 2. The accuracy result for the second dataset with different classifiers

Algorithm	Accuracy	Type	TP rate	FP rate
Decision tree algorithm	87.3	Normal	0.93	0.25
		Attack	0.74	0.05
Naive bayes algorithm	87.1	Normal	0.93	0.26
		Attack	0.73	0.06
Random forest algorithm	87.2	Normal	0.92	0.26
		Attack	0.73	0.06
MLP algorithm	82.4	Normal	0.81	0.47
		Attack	0.61	0.09

Features selection (PCA ranking) algorithms used to choose the best-related features that contribute to the attacks detection. In other words, the features selection aims to improve the evaluation metrics or at least maintain them as well as reducing the number of the used features which will improve the needed training time by the classifiers. In this Dataset PCA Algorithm has been applied and it reduced the features number to 10 instead of 12. Furthermore, Traffic Class and Flow Label features have been taken off, which result in improving the accuracy rate of DDoS detection as shown in Table 2.

5 Conclusion

In IPv6 networks, NDP allows the connected devices to generate their IP addresses and start to communicate with other devices without any registration or authentication inside the network. In addition, most of the existing operating systems are IPv6 enabled by default. Data mining techniques have been introduced to improve the classification mechanism of IDS detection. In addition, extensive researches indicated that there are no IDS for IPv6 using advanced machine learning techniques for the detection of distributed denial of service attacks. The existing NDP-DDoS attacks have not been completely covered by any of the existing detection systems.

In this research, it has been tried to include as much as possible related features based on the attacks' domain knowledge to make sure covering all the possible attacks behaviours. The features selection aims to improve the evaluation metrics or at least maintain them as well as reducing the number of the used features which defiantly will improve the needed training time by the classifiers. Combination of IGR and PCA features ranking algorithms will used to choose the most related features. The features that are highly selected by the two algorithms will be finally selected to represent the NDP attacks' dataset.

References

1. Abouabdalla, O., El-Taj, H., Manasrah, A., Ramadass, S.: False positive reduction in intrusion detection system: a survey. In: 2nd IEEE International Conference on Broadband Network & Multimedia Technology, 2009, IC-BNMT 2009, pp. 463–466. IEEE, October 2009
2. Agrawal, S., Agrawal, J.: Survey on anomaly detection using data mining techniques. Procedia Comput. Sci. **60**, 708–713 (2015)
3. Alaidaros, H., Mahmuddin, M., Al-Mazari, A.: An overview of flow-based and packet-based intrusion detection performance in high speed networks (2011)
4. Alharby, A., Imai, H.: IDS false alarm reduction using continuous and discontinuous patterns. In: Ioannidis, J., Keromytis, A., Yung, M. (eds.) ACNS 2005. LNCS, vol. 3531, pp. 192–205. Springer, Heidelberg (2005). https://doi.org/10.1007/11496137_14
5. Asokan, N., Niemi, V., Nyberg, K.: Man-in-the-middle in tunnelled authentication protocols. In: Christianson, B., Crispo, B., Malcolm, James A., Roe, M. (eds.) Security Protocols 2003. LNCS, vol. 3364, pp. 28–41. Springer, Heidelberg (2005). https://doi.org/10.1007/11542322_6
6. Bahl, S., Sharma, S.K.: Improving classification accuracy of intrusion detection system using feature subset selection. In: 2015 Fifth International Conference on Advanced Computing & Communication Technologies, pp. 431–436. IEEE, February 2015
7. Banerjee, U., Arya, K.V.: Experimental study and analysis of security threats in compromised networks. In: Sengupta, S., Das, K., Khan, G. (eds.) Emerging Trends in Computing and Communication. LNEE, vol. 298, pp. 53–60. Springer, New Delhi (2014). https://doi.org/10.1007/978-81-322-1817-3_6
8. Barbará, D.: Special issue on data mining for intrusion detection and threat analysis. ACM SIGMOD Rec. **30**(4), 4 (2001)
9. Bishop, C.M.: Pattern Recognition and Machine Learning. Springer, New York (2006)

10. Caicedo, C.E., Joshi, J.B., Tuladhar, S.R.: IPv6 security challenges. IEEE Comput. **42**(2), 36–42 (2009)
11. Campbell, P., Calvert, B., Boswell, S.: Security and Guide to Network Security Fundamentals. Thomson Course Technology, Boston (2003)
12. Chan, A.P., Ng, W.W., Yeung, D.S., Tsang, C.C.: Refinement of rule-based intrusion detection system for denial of service attacks by support vector machine. In: Proceedings of 2004 International Conference on Machine Learning and Cybernetics, 2004, vol. 7, pp. 4252–4256. IEEE, August 2004
13. Choudhary, A.R., Sekelsky, A.: Securing IPv6 network infrastructure: a new security model. In: 2010 IEEE International Conference on Technologies for Homeland Security (HST), 8–10 November 2010, USA. SEGMA Technol. Inc., Silver Spring, MD, Technologies for Homeland Security (HST), pp. 500–506. IEEE (2010)
14. Cisco: IPv6 security brief, White paper c11-678658, CISCO (2011)
15. Electronic Design: What's The Difference Between IPv4 and IPv6? p. 2 (2012). http://electronicdesign.com/embedded/whats-difference-between-ipv4-and-ipv6
16. Elejla, O.E., Belaton, B., Anbar, M., Alnajjar, A.: Intrusion detection systems of ICMPv6-based DDoS attacks. Neural Comput. Appl. **30**, 1–12 (2016)
17. Forouzan, B.: TCP/IP Protocol Suite, 3rd edn. McGraw-Hill Higher Education, New Delhi (2006)
18. Garcia-Teodoro, P., Diaz-Verdejo, J., Maciá-Fernández, G., Vázquez, E.: Anomaly-based network intrusion detection: techniques, systems and challenges. Comput. Secur. **28**(1), 18–28 (2009)
19. GNS3 (2017). https://www.gns3.com
20. Gont, F.: Implementation Advice for IPv6 Router Advertisement Guard (RA-Guard) (No. RFC 7113) (2014)
21. Hogg, S., Karpenko, J., Miller, D., Vyncke, E.: IPv6 Security: Information Assurance for the Next-generation Internet Protocol. Cisco Press, USA (2009)
22. Kruegel, C., Mutz, D., Robertson, W., Valeur, F.: Bayesian event classification for intrusion detection. In: Computer Security Applications Conference, 2003. Proceedings. 19th Annual, pp. 14–23. IEEE, December 2003
23. Kumar, A.S., Karthik, M.G., Tech, M.: An efficient detection of DDoS flooding attacks: a survey. Int. J. Sci. Eng. Technol. Res. (IJSETR) **5**(7), 2401–2405 (2016)
24. Kumar, M.A., Hemalatha, M., Nagaraj, P., Karthikeyan, S.: A new way towards security in TCP/IP protocol suite. pp. 46–50 (2010)
25. Kumar, V., Srivastava, J., Lazarevic, A. (eds.) Managing Cyber Threats: Issues, Approaches, and Challenges, vol. 5. Springer Science & Business Media (2006)
26. Laskov, P., Düssel, P., Schäfer, C., Rieck, K.: Learning intrusion detection: supervised or unsupervised? In: Roli, F., Vitulano, S. (eds.) ICIAP 2005. LNCS, vol. 3617, pp. 50–57. Springer, Heidelberg (2005). https://doi.org/10.1007/11553595_6
27. Liao, Y., Vemuri, V.R.: Use of k-nearest neighbor classifier for intrusion detection. Comput. Secur. **21**(5), 439–448 (2002)
28. McHugh, J.: Testing intrusion detection systems: a critique of the 1998 and 1999 darpa intrusion detection system evaluations as performed by lincoln laboratory. ACM Trans. Inf. Syst. Secur. (TISSEC) **3**(4), 262–294 (2000)
29. Moravejosharieh, A., Modares, H., Salleh, R.: Overview of mobile IPv6 security. In: 2012 Third International Conference on Intelligent Systems, Modelling and Simulation (ISMS), pp. 584–587. IEEE, February 2012
30. Najjar, F., Kadhum, M.M.: Reliable behavioral dataset for IPv6 neighbor discovery protocol investigation. In: 2015 5th International Conference on IT Convergence and Security (ICITCS), pp. 1–5. IEEE, August 2015

31. Narten, T., Simpson, W.A., Nordmark, E., Soliman, H.: Neighbor discovery for IP version 6 (IPv6) (2007)
32. Popoviciu, C.: Deploying IPv6 networks. Pearson Education India (2006)
33. Saad, R.M., Ramadass, S., Manickam, S.: A study on detecting ICMPv6 flooding attack based on IDS. Aust. J. Basic Appl. Sci. **7**(2), 175–181 (2013)
34. Salih, A., Ma, X., Peytchev, E.: New intelligent heuristic algorithm to mitigate security vulnerabilities in IPv6. Int. J. Inf. Secur. (IJIS), **4** (2015). https://doi.org/04.IJIS.2015.1.3
35. Satrya, G.B., Chandra, R.L., Yulianto, F.A.: The detection of DDOS flooding attack using hybrid analysis in IPv6 networks. In: 2015 3rd International Conference on Information and Communication Technology (ICoICT), pp. 240–244. IEEE, May 2015
36. Stolfo, J., Fan, W., Lee, W., Prodromidis, A., Chan, P.K.: Cost-based modeling and evaluation for data mining with application to fraud and intrusion detection. Results JAM Proj. Salvatore (2000)
37. Szigeti, S., Risztics, P.: Will IPv6 bring better security? In: Euromicro Conference, 2004, Proceedings. 30th, pp. 532–537. IEEE, September 2004

Research of Social Network Information Transmission Based on User Influence

Zhenfang Zhu[1(✉)], Peipei Wang[2], Peiyu Liu[3], and Fei Wang[3]

[1] School of Information Science and Electric Engineering, Shandong Jiaotong University, Jinan 250357, China
zhuzhfyt@163.com

[2] School of Accountancy, Shandong Management University, Jinan 250100, China

[3] School of Information Science and Engineering, Shandong Normal University, Jinan 250100, China

Abstract. Along with the rapid development of the social network, social network information transmission mode has been changed. In this case, it needs to measure the factors affecting the spread of social network in order to adapt to the changes of social network communication model, and predict the paths to social network transmission. Because of these reasons, this article proposed a social network information transmission model based on the influence of user nodes in the social networks. In this model, the function of mutual influence between the network users was defined first; Secondly, this paper proposed a model of social network information transmission based on user's relative weight; the third, the communication process and the propagation path of network is analyzed; At last, the different paths of information dissemination influence were discussed. In order to verify the validity of the model, this paper compared this model with the traditional SIR model in six kinds of social networks. From the results of contrast tests, we could see that the proposed model of social networks based on user influence could get much more excellent performance.

Keywords: Social networks · Information dissemination · User influence Propagation model

1 Introduction

As the increasing of social network sites such as Twitter, Facebook, MicroBlog, Renren community, people began to become information makers from pure information visitors, at the same time, various social networks have become an important platform for communication and information transmission. With the rapid development of social networking service, information transmission model of social networking service has changed [1].

First of all, social network information transmission has broken traditional transmission method. It takes advantage of interpersonal relationship to change relationship between people and information, which in turn affect interpersonal relationship. The mutual effect has brought about important influence on information recommendation

D.-S. Huang et al. (Eds.): ICIC 2018, LNAI 10956, pp. 564–574, 2018.
https://doi.org/10.1007/978-3-319-95957-3_59

and public opinion supervision. Next, social network information transmission is superior to the first generation of network and traditional media, especially for network emergencies, MicroBlog, Wechat and other social tools; it has a strong timeliness and liveliness, which has called for bigger challenge for disposal of internet emergencies and supervision of online public opinion. At last, as a network application, social network user could transmit and express his/her opinion freely. In addition, groundless topics, even rumors are more likely to be generated and spread, which makes it much more difficulty to control the networks.

Under the circumstance, it is required to appraise factors that influence social network transmission so as to adapt to huge change of social network transmission model, and predict transmission route of social network.

2 Relevant Researches

The research on social network started in the early 19th century. Emile [2], a French sociologist and Tonnies [3], a Germany sociologist summarized social relationship among people and put forward the concept of social network for the first time. Since then, research on social network has aroused interest of quite a many researchers. There have been a lot of research results in relevant research field for deep researches.

As online social networking services sprung up, such as Twitter, Blog and MicorBlog, researchers started to research information transmission model of online social networking service. Garg [4] and other people researched music of equal user under the circumstance of online music community, and classified network information transmission model into two kinds, such as the model influenced by the system or other user, and the model discovered by the user. Kwak [5] and other people researched topological structure of attention in Twitter, measuring quantitative distribution characteristics of "attention" number, forwarding amount, mentioning amount of the user. They discovered that the interaction among a majority of users was unidirectional. There wasn't a direct relation between attention, forwarding and mentioning. Weng [6] and others researched users with strong affect in Twitter and came to the conclusion similar to that of Kwak. Bakshy [7] and others researched whether information release by users with a large number of followers in Twitter would result in large-scale forwarding. It was indicated that forwarding had little to do with the scale of follower. In addition, they also analyzed the relationship between user link and information transmission in online social network. It was founded that strong link had a strong impact, but new information transmission was more dependent on weak link, that is to say, weak link played a stronger role in information transmission of social network than strong link.

Compared with researches abroad, China's researches on social network transmission are currently centered on social science and communication science. Researches on social network information transmission in the field of information science are mainly centered on social computing. Research institutions that research social network in China mainly include automatization institute of Chinese Academy of Sciences, Harbin Institute of Technology, Renmin University of China, Tianjin University, National University of Defense Technology, Capital University of Economics and

Business, Southeast University and Microsoft, IBM, HP social computing laboratory, etc. Xi and Hu [8] researched form and characteristics of public opinion transmission in social network and put forward opinion transmission model. In this model, the speed of information transmission is in accordance with small world theory. Some other researches also focus on social network transmission, user behavior analysis, node measurement, etc. For instance, Xie [9] and others put forward prediction algorithm for user forwarding behavior in social network by taking Sina microblog as the object and based on microblog theme as well as user characteristics. Wu [10] and others used theme model to predict directed social network link. Lu et al. [11] and others researched node influence computing mechanism on the basis of Bayes and semi-ring algebraic model in social network, There are many new researches on social network in these years.

Through the above analysis, it is discovered that information is influenced by the network status of information promulgator and his behavior during actual network transmission. It is related to promulgator influence, as well as receptivity of next user, whether the information could be transmitted from one node to the next one that will become the promulgator. However, the above factors are taken into little consideration in current research results.

3 Basic Social Network Transmission Model

3.1 SIR Virus Transmission Model

SIR model is the most classical model among epidemic models. “S” stands for the susceptible, “I” stands for the infective, R stands for the removal.

In this model, those epidemic people could be divided into three categories. S stands for the susceptible, refers to the person that is not sick but lack of immunocompetence and that is liable to get infected after contact with the infective; I stands for the infective, refers to the person that gets infected and that could transmit the disease to the person of S category; R stands for the removal, refers to the isolated person that possesses immunity after recover from an illness.

3.2 SIR Model in Social Network Service

During research of social network user, SIR model is applied widely and improved constantly. SIR in social network could be seen as a network structure. A certain user of social network is defined as a node. Friend relationship among users could be abstracted as sides between the nodes. The information is transmitted along edge between nodes.

4 Social Network Transmission Model Based on User Node Influence—URSIR Model

The example in Fig. 1 is shown to define network topology of URSIR model for the convenience of description.

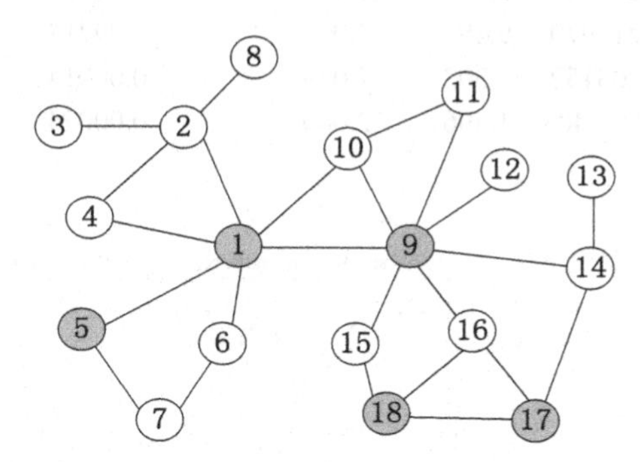

Fig. 1. Example of network topology

Pa: from authoritative node to ordinary node, such as directional route $1 \rightarrow 5$;
Pb: from ordinary node to authoritative node, such as directional route $5 \rightarrow 1$;
Pc: from authoritative node to authoritative node, such as bi-directional route $1 \leftrightarrow 9$;
Pd: from ordinary node to ordinary node, such as bi-directional route $17 \leftrightarrow 18$.

Based on the above researches, this paper puts forward information transmission model URSIR on the basis of user influence.

5 Analogue Simulations

5.1 Relevant Network Typology and Its Characteristics

We take example of popular online social networks such as Twitter [19], Sina MicroBlog [20] and Epinions [21] as three network typology of this paper. All the relevant data comes from Internet resources [22, 23]. At the same time, this paper uses related algorithm to generate three simulation networks, which are ER random network [24], NW small-world network [25] and BA scale-free network [26]. Assume the side of each network is unidirectional and unauthorized, and correlated topological characteristic parameters of each network are shown in Table 1.

5.2 URSIR Information Transmission Model

Figure 2 shows the curve that healthy node, transmission node and immune node in the above six networks change with time t, S(t), I(t), R(t).

Table 1. Correlated characteristic parameters of each network

Name	Node number	Side number	Average node	Maximum node degree	Clustering coefficient	Correlation coefficient
ER	5000	25348	5.06	14	0.00089	−0.075
NW	5000	35094	7.02	17	0.05063	0.0412
BA	5000	29970	5.99	178	0.00402	−0.0501
Epinions	22437	212970	9.49	2,031	0.09717	0.052
Twitter	145942	203152	1.392	7,079	0.00014	−0.1114
Sina blog	146091	205,408	1.406	2,000	0.00024	−0.2446

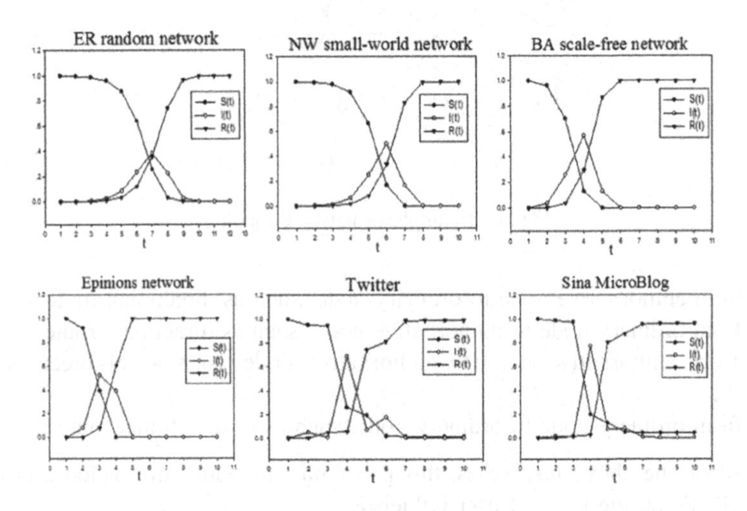

Fig. 2. Node time curve of URSIR model in each network

6 Comparison Between URSIR and SIR

In this section, the author conducts analogue simulation of SIR and URSIR models in the above six network typology. The experiment consists of two parts:

(1) Transmission test of authorized node: a node with maximum connectivity is chosen as the transmission node in each network and the rest nodes are healthy nodes.

(2) Transmission test of ordinary node: a random node with low connectivity is chosen as the transmission node and the rest are healthy nodes.

6.1 Max Infection Rate (MIR)

It is the proportion that infective nodes account for total network nodes after transmission. As all infective nodes in the model will finally become immune nodes, we regard the proportion of final immune nodes as final infection rate. MIR is used to describe the range that transmission node influences the whole network.

Transmission Test of Authorized Node. As is shown in Fig. 3, on the whole, MIR increases with λ, which means the more popular the information is, the larger the network coverage is. As for ER random network, NW small-world network, BA scale-free network, when λ is 0.5, MIR is basically 1, which means the information could cover the whole network at the time. Whereas in Epinions, Twitter and Sina microblog, network node is large and network is sparsely connected. In SIR model, MIP is between 0.4 and 0.6, therefore the information could not cover the whole network; whereas in URSIR model, MIR is close to 1, therefore the information could basically cover the whole network (Fig. 4).

Transmission Test of Ordinary Node. As is shown in Fig. 3, on the whole, MIR increases with λ like authorized node test. Except that MIR of Epinions, Twitter and Sina microblog in SIR model is between 0.4 and 0.6, MIR of other networks is close to 1, therefore the information could basically cover the whole network.

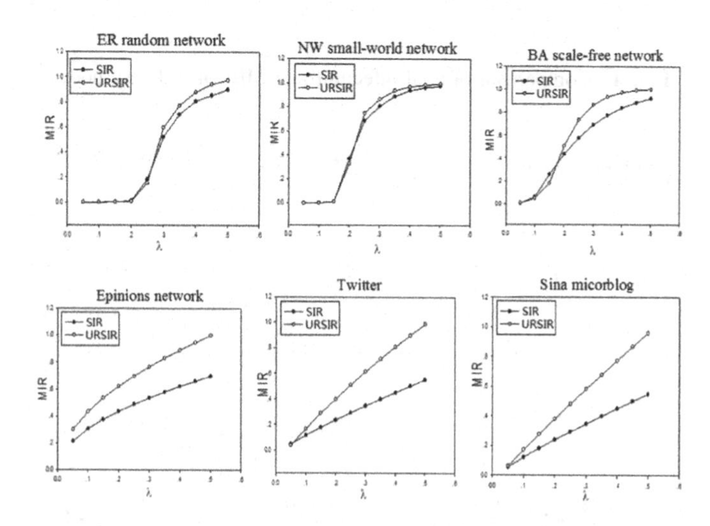

Fig. 3. Comparison of final infection rate MIR of authorized nodes

6.2 Max Infection Peak (MIP)

In the process of transmission, the percentage of total network nodes when the amount of infection nodes reaches the maximum.

(1) Transmission experiment of authorized nodes

As Fig. 5 shows, the two models in the six different network structures all show that with the increase λ, MIP increases, which means the bigger the transmission probability is, the faster the information transmission is.

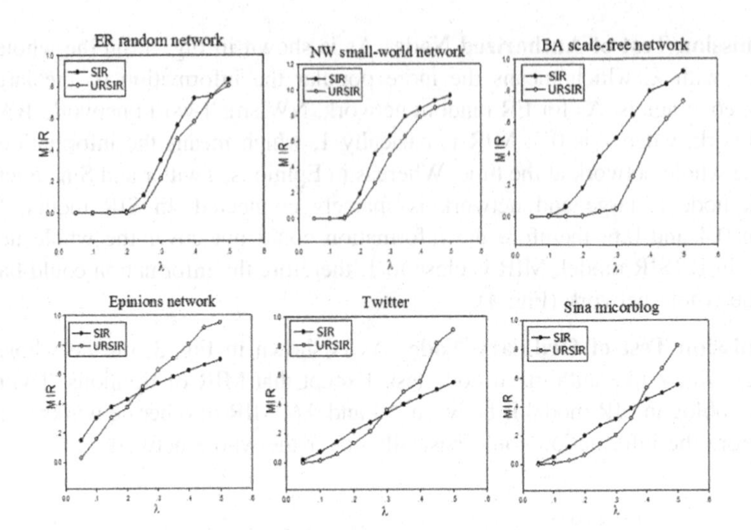

Fig. 4. Comparison of final infection rate MIR of ordinary nodes

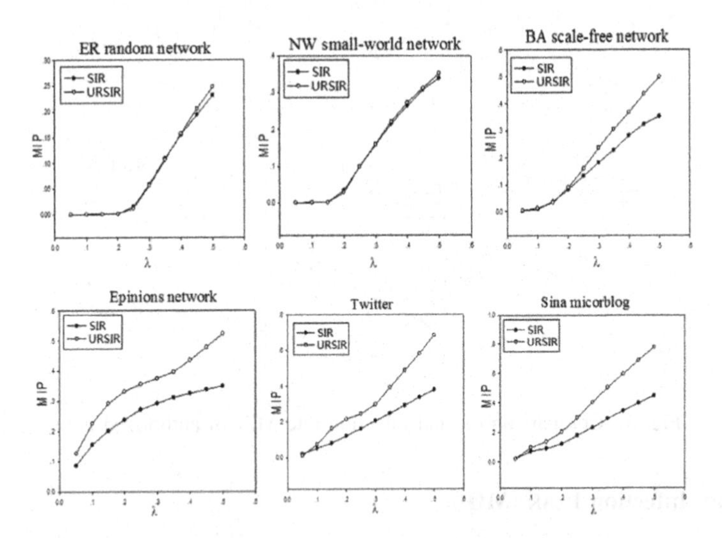

Fig. 5. Comparison of max infection peak of authorized nodes

(2) Transmission experiment of ordinary nodes

As Fig. 6 shows, MIP increases with the increase of λ. In the ER random network and NW small-world network, MIP in the URSIR is slightly lower than that in the SIR; in the BA network, MIP in the URSIR is obviously lower than that in the SIR; in the Epinions network, Twitter and Sina microblog, when λ is smaller, MIP in the URSIR is lower than that in the SIR, and opposite when λ is bigger.

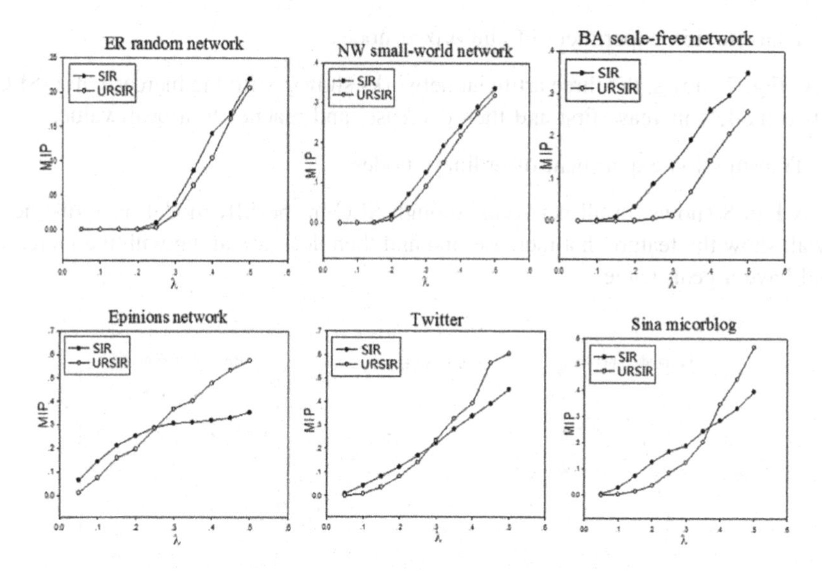

Fig. 6. Comparison of max infection peak of ordinary nodes

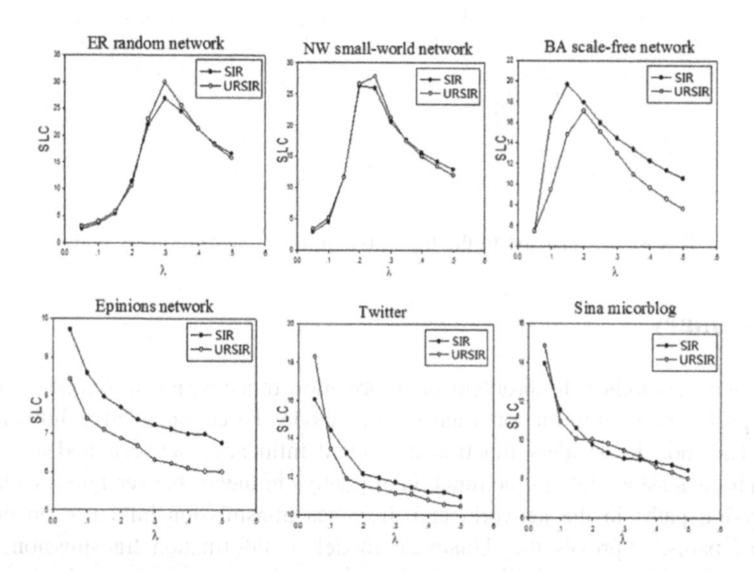

Fig. 7. Comparison to the transmission life cycle of authorized nodes

6.3 Spreading Life Cycle (SLC)

The transmission process starts from a transmission node, there is no transmission node in the whole network to the end, and the time spent in the whole process is defined as life cycle of transmission. The life cycle transmission is used to depict time spent in the overall transmission process, namely the life time of information in the Internet.

(1) Transmission experiment of authorized nodes

As Fig. 7 shows, the three artificial networks shows: with the increase of λ, SLC of the two models increase first and then decrease, and reaches to a peak value.

(2) Transmission experiment of ordinary nodes

As Fig. 8 shows, in all networks, though SLC in the SIR model have distinction, they all show the feature that increase first and then decrease along with the increase of λ and have a peak value.

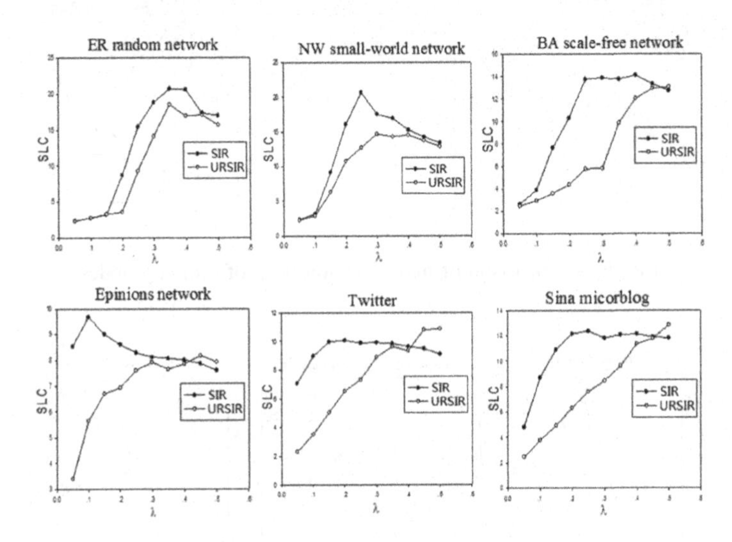

Fig. 8. Comparison to the transmission life cycle of ordinary nodes

7 Summary

This article researches the problem of information transmission in online social network, proposes an information transmission model based on relative influences of users. The model considers function of mutual influences between nodes on information transmission, defines the function of relative influence between nodes, classifies propagation paths in the network, and discusses transmission influence on different types of paths, improves the dynamical model of information transmission, which makes the model be more in line with the information transmission in the real world. The traditional SIR model and URSIR model are conducted analogue simulation under six different network topologies, and data shows there is no obvious difference between the two models in inhomogeneous networks, and obvious differences in homogeneous networks, and this could describe the information transmission of the real online social network to some extent. The article is to helpful to further understand the behavior rule of information while transmitting in the real network and effects of network topologies on the transmission of information, and lays the foundation for the research of influences of micro propagation behavior on macro network. There are disadvantages at the same time, it only considers the information transmission process in static network, and discusses too general. How to model the propagation behavior in the dynamic network and the specific research of the influence of authorized nodes and ordinary nodes on information transmission will be the research work in the future.

Acknowledgements. This research was supported by a research grant from National Natural Science Foundation (61373148), National Social Science Fund (12BXW040); Shandong Province Natural Science Foundation (ZR2012FM038, ZR2011FM030); Shandong Province Outstanding Young Scientist Award Fund (BS2013DX033), Science Foundation of Ministry of Education of China (14YJC860042), Shandong Social Science Planning Fund Program (17CHLJ30, 17CHLJ33), Shandong Provincial Education Office Science and technology project (J15WB37).

References

1. Nazari, S., Faez, K., Janahmadi, M.: A new approach to detect the coding rule of the cortical spiking model in the information transmission. Neural Netw. **99** (2018)
2. Durkheim, E.: Emile Durkheim: An Introduction to Four Major Works. Sage Publications, Beverly Hills (1986)
3. Tonnies, F., Ivars, J., Vilardebó, L.G.F., et al.: Comunitat i associació (1984)
4. Garg, R., Smith, M.D., Telang, R.: Measuring information diffusion in an online community. J. Manag. Inf. Syst. **28**(2), 11–38 (2011)
5. Kwak, H., Lee, C., Park, H., Moon, S.: What is twitter, a social network or a news media? In: Proceedings of the 19th International Conference on World Wide Web, pp. 591–600 (2010)
6. Weng, J., Lim, E.P., Jiang, J., He, Q.: Twitterrank: finding topic-sensitive influential twitters. Proceedings of the Third ACM International Conference on Web Search and Data Mining, pp. 261–270 (2010)
7. Bakshy, E., Rosenn, I., Marlow, C., Adamic, L.: The role of social networks in information diffusion. In: Proceedings of the 21st International Conference on World Wide Web, pp. 519–528 (2012)
8. Xi, X., Hu, Y.: Research on the dynamics of opinion spread based on social network services. J. Phys. **21**(16), 1505091–1505097 (2012)
9. Xie, J., Liu, G.S., Su, B.: Prediction of user's retweet behavior in social network. J. Shanghai Jiaotong Univ. **47**(4), 584–588 (2013)
10. Wu, M.D., Tang, Y.: A topic model-based method of the link prediction for directed social network. J. Southwest Univ. (Nat. Sci. Ed.) **4**(1), 1–7 (2014)
11. Lu. L.Y., Chen, D.B., Zhou, T.: New J. Phys. **13**, 123005 (2011)
12. Jiang, W., Gao, M., Wang, X., et al.: A new evaluation algorithm for the influence of user in social network. China Commun. **13**(2), 200–206 (2016)
13. Naug, D.: Structure of the social network and its influence on transmission dynamics in a honeybee colony. Behav. Ecol. Sociobiol. **62**(11), 1719–1725 (2008)
14. Kim, Y.K., Lee, D., Lee, J., et al.: Influential users in social network services: the contingent value of connecting user status and brokerage. ACM Sigmis Database Database Adv. Inf. Syst. **49**(1), 13–32 (2018)
15. Palos-Sanchez, P.R., Saura, J.R., Debasa, F.: The influence of social networks on the development of recruitment actions that favor user interface design and conversions in mobile applications powered by linked data. Mob. Inf. Syst. **2018**(1), 1–11 (2018)

16. Zhang, C., Lu, T., Chen, S., et al.: Integrating ego, homophily, and structural factors to measure user influence in online community. IEEE Trans. Prof. Commun. **PP**(99), 1–14 (2017)
17. Albladi, S.M., Weir, G.R.S.: User characteristics that influence judgment of social engineering attacks in social networks. Hum.-centric Comput. Inf. Sci. **8**, 5 (2018)
18. Wang, X., Xu, Y., Wang, L., et al.: Transmission of information about consumer product quality and safety: a social media perspective. Inf. Discov. Deliv. **45**(1), 10–20 (2017)
19. http://twitter.com/
20. http://weibo.com/
21. http://www.epinions.com/
22. http://www.datatang.com/
23. http://www.nlpir.org/
24. Erdős, P., Rényi, A.: Publ. Math. Inst. Hungarian Acad. Sci. **5**, 17 (1960)
25. Newman, M.E.J., Watts, D.J.: Phys. Lett. A **263**, 341 (1999)
26. Barrat, A.L., Albert, R.: Science **286**, 509 (1999)

Research on Cloud Storage and Security Strategy of Digital Archives

Hua-li Zhang(✉), Fan Yang, Hua-yong Yang, and Wei Jiang

City College, Wuhan University of Science and Technology,
Wuhan 430083, China
zhanghuali@wic.edu.cn

Abstract. In the process of enterprise information, more and more enterprises begin to pay attention to the construction and implementation of digital archives system. At present, the storage of the digital archives system is local storage and local area network storage, it will lead to the isolation of the archives information, the high cost of construction and management, and the difficulties in the later maintenance. This paper studies and designs the digital archives cloud storage platform, the physical architecture design of the cloud platform, the architecture and planning of the cloud storage platform, the open interface and service access design, and the cloud storage security strategy. Based on the security requirements of digital archival cloud storage, a hybrid encryption mode based on symmetric encryption and asymmetric encryption is proposed, and a digital archives encryption algorithm based on RSA and AES encryption algorithm is implemented. The construction of safe digital archives cloud storage platform can popularize the consciousness of digital archives, reduce the cost of archives management, improve the level of archives management, and ultimately promote the level of modernization management.

Keywords: Digital archives · Cloud storage · Security policy
RSA · AES

1 Introduction

With the development and popularization of Internet, cloud storage begins to be applied in all fields of information. Cloud computing is very popular because it can effectively solve the problems of inadequate capacity of computing and storage, inadequate resource utilization, big investment in IT facilities, complex system management, and so on. Gartner's data show that cloud computing has replaced virtual technology as the most technical area of global CIO [1]. With the development of the big data, people pay more and more attention to the security of big data. Big data as an important asset, they are facing serious challenge of data security. Cloud computing will become the mainstream computing mode in the future due to its high utilization of

Foundation project: Scientific planning subject of Hubei Provincial Department of Education (2016GB123).

D.-S. Huang et al. (Eds.): ICIC 2018, LNAI 10956, pp. 575–584, 2018.
https://doi.org/10.1007/978-3-319-95957-3_60

resources and cost savings. However, data security storage, including privacy protection, has become a huge obstacle to cloud computing.

According to a survey of relevant research institutions in the United States, more than sixty percent of the companies interviewed indicated that the cloud data platform was attacked in the last year, including personal privacy data [2]. The occurrence of such incidents aggravated the users' concerns about the privacy of cloud data storage technology. In addition, due to technical constraints, there is a paradox between data processing and data encryption, so that the operation of basic encrypted data can not be realized. This is because in cloud mode, once the data is encrypted, it will conflict with the retrieval system of the network, resulting in inoperable phenomena. So that cloud data security storage technology is facing the awkward situation that encryption can not be operated and data encryption can not be guaranteed.

The security of cloud computing data storage has attracted wide attention from academia and industry. Research on data security storage, security audit and ciphertext access control has begun [3]. This paper starts with data encryption and storage, and achieves cloud security storage of digital archive data. RSA algorithm and AES algorithm have been widely used in the field of data encryption. RSA algorithm uses the public key and private key to realize data encryption and decryption. It has high security and is not easy to attack and crack. However, because of the low computing efficiency, large public key size and high computing complexity, the traditional RSA encryption method can not be effectively applied to the cloud environment, and there is a big gap between the actual use [5, 6]. The design of AES encryption algorithm has the advantages of simplicity, efficiency, symmetry and modularity. However, through the research on the attack method of the algorithm, the defect of the original seed key is not high and the security is not high [7, 8].

In this paper, a hybrid data encryption method based on RSA algorithm and AES algorithm is proposed. Realize the secure encryption of the electronic data of the digital archives. Different users generate different public and private keys through the RSA algorithm. The cloud server stores the user's public key and encrypts the data with the public key. Users hold private keys to decrypt encrypted data from cloud storage. By mixing two kinds of encryption algorithms, the data can be secrecy and decrypt quickly while ensuring data security.

2 Storage Architecture

To realize the cloud storage of data, especially the cloud storage of archival data, storage platform needs to be easy to use, scalable and secure.

2.1 Physical Architecture

The digital file cloud storage platform redesigns the overall structure of the traditional single data storage file system to meet the requirements of the open service cloud storage [9]. The digital archive cloud storage physical architecture is shown in the following chart (Fig. 1).

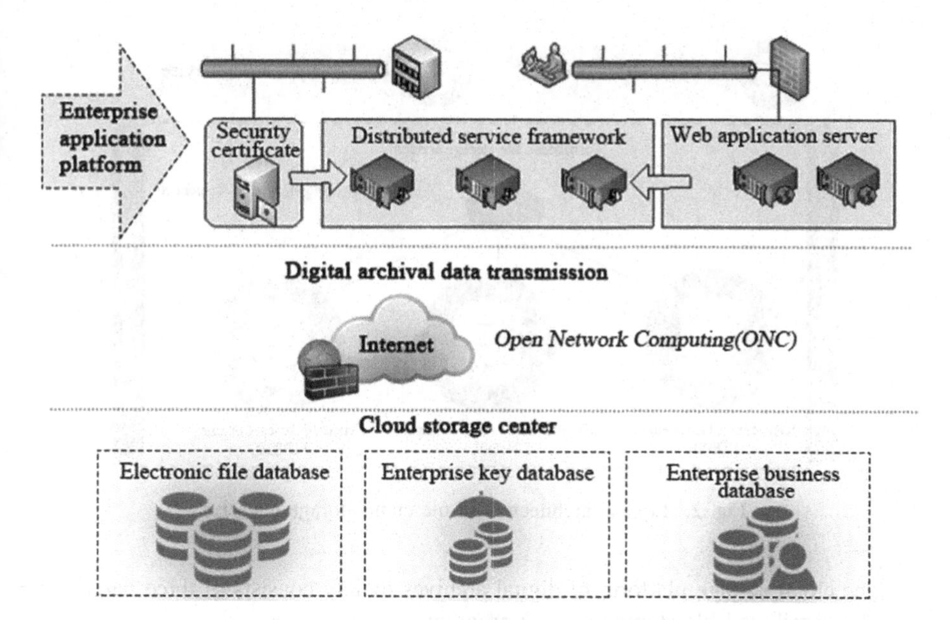

Fig. 1. Physical architecture of digital archives cloud storage

The cloud storage framework of digital archives is mainly composed of the following parts.

The cloud storage platform [10, 11].

The distributed cloud storage technology is applied to realize the extensible cloud storage platform. Cloud storage platform is used to store encrypted digital archives, electronic files, archives metadata and application service data.

The distributed service framework and application of open interface [12, 13].

The distributed service framework includes functions of distributed control, load balancing, interface permissions control, application security management and so on. The open interface provides the data cloud storage open interface to the upper layer (interface users and WEB servers).

WEB application server [14, 15].

It provides users with back office management of cloud storage and open data query service.

Security service engine [16].

In the form of AOP, it provides overall security services for the system, including data transmission encryption, data storage encryption, decryption, and security authentication.

2.2 Logical Architecture

When we design and build the file cloud storage platform, we should improve the unsuitable places in the traditional file system, so as to design a set of robust and sustainable cloud storage platform (Fig. 2).

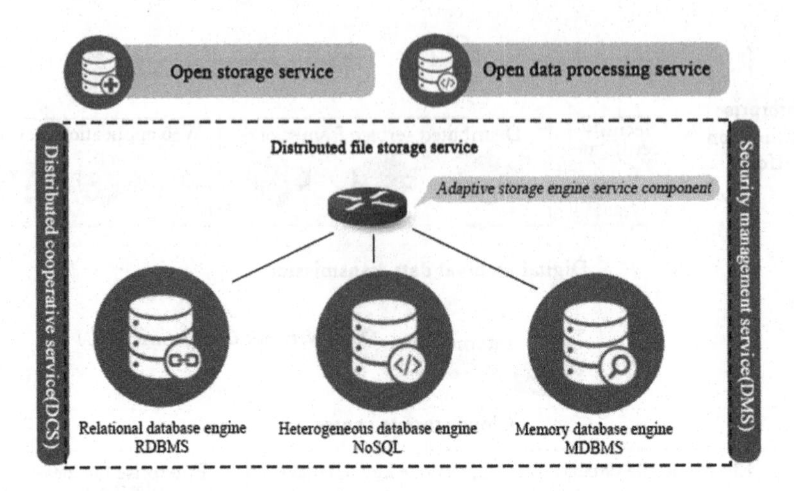

Fig. 2. Logical architecture of file cloud storage platform

The cloud storage platform of digital archives mainly consists of three parts.

The distributed file storage service program

The data storage engine is installed on different server hosts in a local area network, providing storage and query services for traditional relational data, heterogeneous metadata [19]. In order to improve system throughput and response time, a memory database engine is built to provide data cache service. At the same time, the system needs to provide an adaptive memory engine routing component, which provides a transparent storage control service to realize the adaptive storage of different data.

The data storage and query service program

Data storage and query service programs provide services for external business processes [17], including data storage service and data open service. The data storage service provides users with service directory, and allows users to store and manage metadata of system data and archives in the directory of jurisdiction. Open data processing service provides users with available data in cloud storage platform, so that users can process and use data.

System service control program

System service control service program is used to control system efficiency, security, data consistency and data redundancy function. System control service program ensures data integrity and security.

3 Security Strategy for Cloud Storage in Digital Archives

The security of cloud storage in digital archives mainly includes transmission security and storage security. In order to ensure data security, it is necessary to adopt corresponding security policies. In this paper, we use data encryption to achieve secure transmission and secure storage of digital archives.

The basic process of data encryption is to process the original document or data according to some algorithm, making it an unreadable code, usually called "ciphertext", so that it can only display the original content after the input of the corresponding key, through such a path to achieve the purpose of protecting data is not stolen and read by illegal access [18]. The reverse process is decryption, which will transform the encoded information into its original data. Encryption technology is usually divided into two categories: symmetric encryption and asymmetric encryption.

3.1 Introduction of Secret Algorithms

Symmetric Encryption Technology

Symmetric encryption is the use of the same key for encryption and decryption, usually called "Session Key", which is widely used today. For example, the DES encryption standard used by the US government is a typical "symmetric" encryption method, and its Session Key length is 56 bits [19] (Table 1).

Table 1. The main symmetric encryption algorithm

Name	Key length(bit)	Computation speed	Security	Resource consumption
DES	56	Fast	Low	Medium
3DES	112,168	Slow	Medium	High
AES	128, 192, 256.	Fast	High	Low

The Asymmetric Encryption Technology

Asymmetric encryption is not the same key used for encryption and decryption, usually with two keys, known as "public key" and "private key", and two of them must be paired, otherwise the encrypted files cannot be opened. The "public key" is publicly available, and the private key cannot be known by the holder. Its superiority is here, because the symmetric encryption method is difficult to tell the other if it is to transmit encrypted files on the network [20]. No matter what method is used, it may be heard. And the asymmetric encryption method has two keys, and the "public key" can be open, and it is not afraid of others. When the receiver declassified its own private key, it can be used so that the security of the key transmission is avoided (Table 2).

Table 2. The main asymmetric encryption algorithm

Name	Maturity	Security (depending on the key length)	Operation speed	Resource consumption
RSA	High	High	Slow	High
DSA	High	High	Slow	Only for digital signatures
ECC	Slow	High	Fast	Low

3.2 Encryption Scheme

Considering the security, operation efficiency and resource occupancy characteristics of all kinds of encryption algorithms, this paper proposes an encryption system based on RSA and AES. The enterprise service center uses RSA algorithm to generate key, including private key and public key, and public key is sent to cloud storage platform, and the enterprise maintains its private key autonomously.

Enterprise Platform Encryption Scheme Before the enterprise platform sends data, it needs to get the RSA public key from the storage platform. After the enterprise data is encrypted with the AES key, the acquired RSA public key is used to encrypt the AES key, and the encrypted data and the encrypted AES key are packaged, and then the packaged data is sent to the cloud storage platform. The encryption process is shown in Fig. 3.

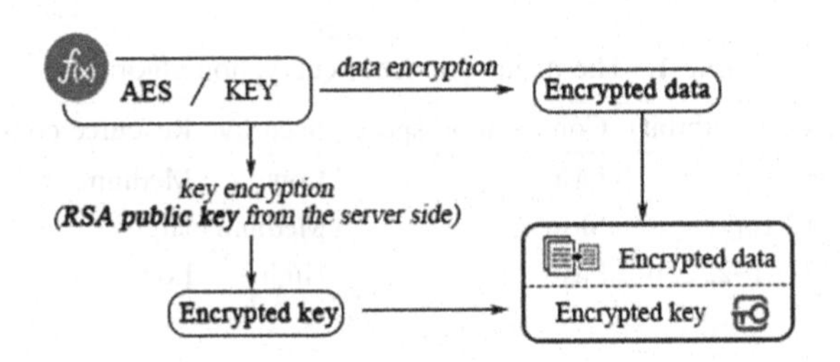

Fig. 3. Enterprise platform encryption scheme

Cloud Storage Platform Decryption Scheme
After the cloud storage platform obtains the enterprise data, it first uses the RSA private key of the enterprise to decrypt the AES key in the packet, then decrypt the encrypted enterprise data using the AES key, thus obtaining the real enterprise data. The decryption process is shown in Fig. 4.

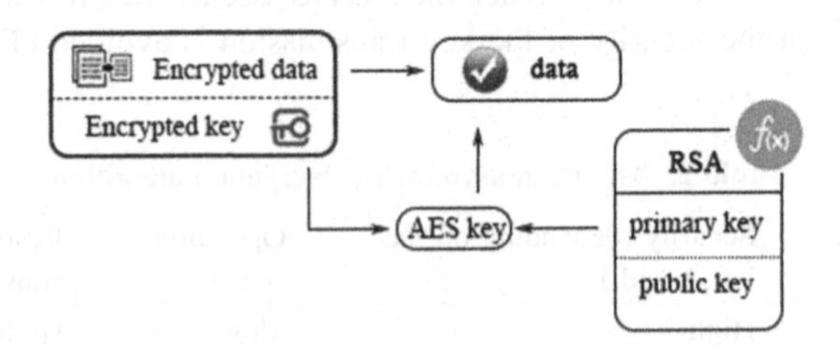

Fig. 4. Cloud storage platform decryption scheme

3.3 Application on the RSA Algorithm

The common RSA algorithm adopts two prime number algorithm. In order to improve the security of the algorithm, this paper adopts the four prime number RSA algorithm [19].

Based on the traditional double prime number RSA cipher algorithm, the number of prime numbers is 4, and the algorithm is still valid [20].

1. random selection of four different prime numbers P, Q, R and S.

$$n = pqrs, \phi(n) = (p-1)(q-1)(r-1)(s-1)$$

The secret key encryption e satisfies certain conditions, calculate the private key D, meet $de \equiv 1 \bmod \phi(n)$.

The process of encryption and decryption is still the same as that of the traditional algorithm.

Encryption algorithm: $c = E(m) \equiv m^e \bmod n$

Decryption algorithm: $m = D(c) \equiv c^d \bmod n$

Using the above signature method, if the user A wants to deny that the message M has been sent to the user B, the user B only needs to show the public key and the signature S of the A to the notary, and the correct calculation method can confirm that the user A has indeed sent the message M; if the user B forges a message, the correct signature cannot be shown to the notary because he does not know the private key of the user A. In this way, both sides of the communication must truly reflect the communication situation and effectively prevent the denial of one party from happening.

4 Encryption/Decryption Process

In the process of archival storage, it is necessary to produce a archival code for every archive data. The archival code usually includes the fund number, catalog number, file number, part number and page number. The compilation of class number is of great significance to the development of archival work, and specific rules should be followed in its compilation process. The code is unique. It can be divided into two parts: classification number and piece number [21]. In the process of file transfer and storage, the archival code is used as the unique label as the primary key of the cipher library.

4.1 Data Encryption Process

Before encrypting data, the system generates random AES encryption key. Use the AES encryption key to encrypt the data [22]. The Archival code is decomposed into the classification number and piece number [23]. Use the classification number to query the RSA public key in the database, and use the RSA public key to encrypt the AES key. Then pack encrypted data, encrypted AES key and the archival code, and the files. The data encryption process is shown in Fig. 5.

4.2 Decryption Process of Data

The encrypted data, encrypted Aes key and archival code is unpacked from the packet data. The Archival code is decomposed into the classification number and piece number, then query the RSA private key through the classification number from the

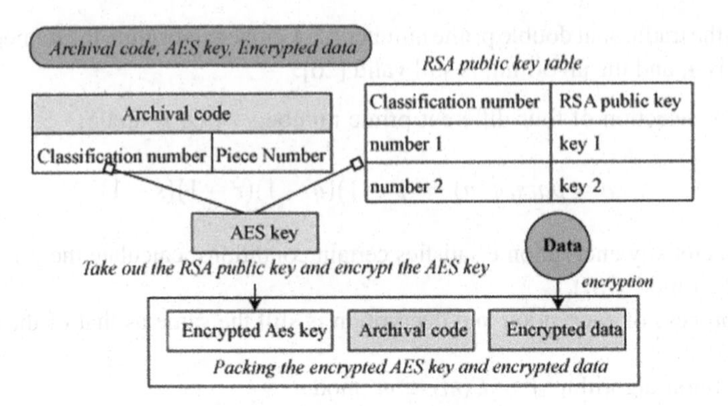

Fig. 5. Encryption process of digital archives

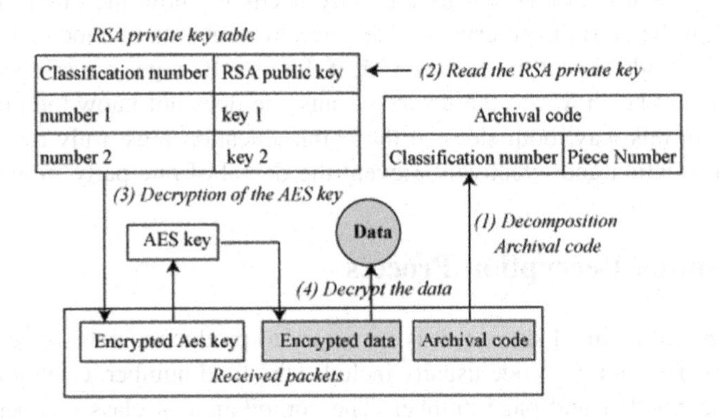

Fig. 6. Decryption process of digital archives

database. Use the RSA private key to decrypt the AES key, and then decrypt the encrypted data using the decrypted AES key. The data decryption process is shown in Fig. 6.

5 Algorithm Efficiency

In this paper, a hybrid encryption algorithm based on RSA algorithm and AES algorithm is proposed. This algorithm improves the efficiency of data encryption while ensuring encryption security. Under the pseudo model of Hadoop, different sizes of files are encrypted and decrypted, and the test results are shown in Table 3.

The encryption and decryption efficiency of the hybrid encryption algorithm is basically consistent. For the files in the 100 M range, the encryption and decryption time can be in the range of 1S, and can basically meet the encryption requirements of the archival data. The efficiency of data encryption and decryption is shown in Fig. 7.

Table 3. Data encryption/decryption efficiency

File size	Time consuming(ms)	
	Encryption	Decryption
27 KB	36	33
235 KB	252	258
8,416 KB	564	576
79,741 KB	989	876
612,363 KB	2,467	2,543

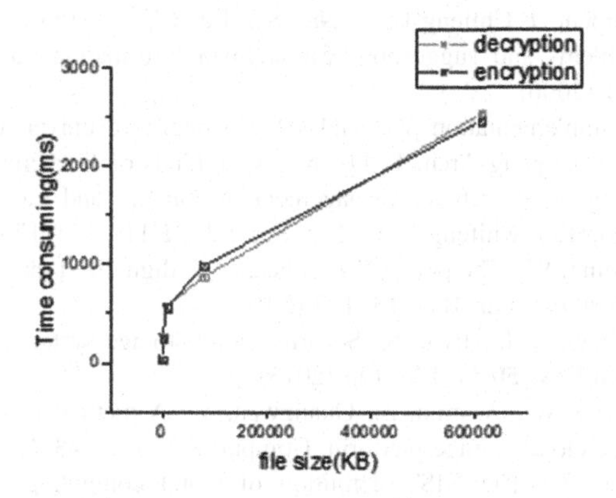

Fig. 7. Data encryption/decryption efficiency

6 Conclusion

Digital archives cloud storage can enhance the security of data storage, reduce the cost of archives management, and improve the efficiency of archives management. In this paper, a digital archives platform based on cloud is constructed. The mixed encryption algorithm of RSA algorithm and AES algorithm can enhance the security of data storage and improve the efficiency of data encryption.

References

1. Zhaosheng, F., Zhiguang, Q., Ding, Y.: Cloud data security storage technology]. J. Comput. Sci. **38**(01), 150–163 (2015)
2. Xiaojun, W.: Problems and improvement suggestions of cloud data security storage technology. Electron. Technol. Softw. Eng. **1**, 212 (2017)
3. Camastra, F., Ciaramella, A., Staiano, A.: Machine learning and soft computing for ICT security: an overview of current trends. J. Ambient Intell. Humaniz. Comput. **4**, 235–247 (2013)
4. Wenfang, Z., Dan, X., Xiaomin, W., Zhen, C., Xudong, L.: convertible. J. Comput. Sci. **40** (05), 1168–1180 (2017)

5. Quanfu, W., Wenai, S., Shunmin, Y.: Homomorphic encryption method for data security in cloud environment. Comput. Eng. Des. **38**(1), 42–46 (2017)
6. Mathur, N., Bansode, R.: AES based text encryption using 12 rounds with dynamic key selection. Procedia Comput. Sci. **79**, 1036–1043 (2016)
7. Priya, S.S.S., Karthigaikumar, P., Sivamangai, N.M., et al.: High throughput AES algorithm using parallel subbytes and mixcolumn. Wirel. Personal Commun. **95**(2), 1433–1449 (2016)
8. Yanping, L., Qiuhui's, L.: AES algorithm research and its key expansion algorithm improvement. Mod. Electron. Technol. 39 (10), 5–8+13 (2016)
9. Jie, H.: Intensive computing algorithm for mobile data terminals based on elastic cloud computing. Comput. Appl. Softw. **11**, 210–213 (2013)
10. Han, Y.J.: Research on library archives information resources sharing mechanism under the network environment. J. Chifeng Univ. Nat. Sci. Ed. **17**, 152–153 (2013)
11. Anlian, Y.: Problems and suggestions for archival information construction. China Inf. World **14**, 30–31 (2006)
12. Zuhua, H.: The implementation plan and safety strategy of university collection archives digitalization. Heilongjiang Arch. **1**, 44–45 (2014). kinds of Jincheng
13. Jingyan, X.: Strategy of archives management in colleges and universities in the era of knowledge economy. J. Chifeng Univ. Nat. Sci. Ed. **31**(11), 173–175 (2015)
14. Xiujuan, X., Yuqing, W.: The practical significance of digitalized management of university archives in the new era. Off. Bus. **13**, 132 (2016)
15. Yingxun, F., Shengmei, L., Jiwu, S.: Security cloud storage system and key technologies. Comput. Res. and Dev. **50**(1), 136–145 (2013)
16. Qing, H., Yongwei, W., Weimin, Z., Guangwen, Y.: A method to protect the privacy of users' data on the cloud storage platform. Comput. Res. Dev. **48**(7), 1146–1154 (2011)
17. Mell, P., Grance, T.: The NIST definition of cloud computing. National Institute of Standards (2011)
18. Cheng, H., Chang, A., Dengguo, F.: AB-ACCS: A cloud storage cryptographic access control method. Comput. Res. and Dev. **47**(z1), 259–365 (2010)
19. Meiyun, L., Jian, L., Chao, H.: Based on homomorphic encryption, trusted cloud storage platform. Inf. Netw. Secur. **9**, 35–40 (2012)
20. Hui, L., Wenhai, S., Fenghua, L., Boyang, W.: Summary of data security and privacy protection technology in public cloud storage service. Comput. Res. Dev. **51**(7), 1397–1409 (2014)
21. Zhi, X., Ping, W., Jiangyan, X., Weihong, C.: An attribute based enterprise cloud storage access control scheme. Comput. Appl. Res. **30**(2), 513–517 (2013)
22. Xiao, Z.-J., Hu, C., Jiang, Z.-T., Chen, H.: Optimization of AES and RSA algorithm and its mixed encryption system. Appl. Res. Comput. **31**(4), 1189–1198 (2003). Bo, Y.: Modern Cryptography. Tsinghua University press, Beijing (2003)
23. Yan, L., Li, H.: Dynamic key AES encryption algorithm based on compound chaotic sequence. Comput. Sci. 44(6), 133–138 + 160 (2017)

Natural and Fluid 3D Operations with Multiple Input Channels of a Digital Pen

Chuanyi Liu(✉), Jiali Zhang, and Kang Ma

School of Information Science and Engineering,
Lanzhou University, Lanzhou, China
liuchuanyi96@hotmail.com, jennyfocus@163.com,
mak2012@lzu.edu.cn

Abstract. We propose six 3D operation patterns with multiple input channels of a digital pen: these patterns allow users to transfer pre-existing knowledge of physical pens to digital pens on performing 3D operations simply, naturally, intuitively, and fluidly. A prototype system was designed under the patterns and implemented. An informal user study showed that eight novices grasped to perform 3D operations with the prototype system within several minutes and gained more fun than with the typical interfaces.

Keywords: 3D operation · Pen input channels · Tilt · Azimuth
Rolling angle

1 Introduction

Three-dimensional operations are widely used in many fields, e.g., CAD, education, chemistry, mechanics, and architecture. Most of the commercial 3D software is notorious for their complicated operations, typically with abstract parameters setting dialog boxes. It is very difficult to perform 3D operations even for experts. For investigating simple and intuitive 3D operations, a wide variety of research has been explored for decades. Most of those studies focused on free-form 3D modeling. Some of those studies (e.g., [13]) explored 3D modeling by suggestive sketch-based modeling systems, which sought to map rough sketches to linear geometry such as curves, planes, and polyhedrons. The other of those studies (e.g., [10]) investigated 3D modeling by literal sketch-based modeling systems, which created 3D surfaces directly from user's strokes.

Although sketch-based modeling systems have evolved for decades, current sketch-based modeling systems still have some limitations. Sketch-based operations are modeling-dependent, i.e., users have to know specific knowledge about the type of models: a given stroke or gesture can have different meanings in different modes. As an example, Schmidt et al. [15] uses gestures to initiate widgets, but also allows surficial augmentation strokes: what happens if an augmentation stroke is the same as a widget gesture? Only the user can truly know the intended meaning in this case. It is still a challenge for researchers to provide a consistent and predictable interface without modality problems.

D.-S. Huang et al. (Eds.): ICIC 2018, LNAI 10956, pp. 585–598, 2018.
https://doi.org/10.1007/978-3-319-95957-3_61

Sketch-based interfaces also suffer from the problem of self-disclosure [5]. A sketch-based modeling system typically simply provides the user with a blank window representing virtual paper, with no buttons or menus what so ever. Although it may be more usable and efficient for someone who has been given a tutorial, such an interface does not disclose any hints about how to use it. Devising elegant solutions to this problem is currently still a challenge for sketch-based modeling researchers. In some sketching-based systems, various modeling operations are performed through a sketching metaphor (e.g., [22]). Operations through a metaphor are neither natural nor intuitive enough. In fact, most sketch-based interfaces are far from natural: many require the user to draw in very specific ways to function properly, which reduces the immersion and ease of use.

Pen-based interfaces have some advantages over traditional WIMP (windows, icons, menus, and pointing devices) interfaces to be more natural and intuitive. Human beings have used various pens (e.g., Chinese writing brushes, quills) for thousands of years. Therefore, the traditional drawing or writing pens are natural tools for human beings. Digital pens, inheriting some physical properties from the traditional drawing or writing pens, permit a user convey input data not just with the overall form of drawing, but also by varying stroking pressure, pen input tilt angle, azimuth, and twist angle. All these data are utilized in calligraphy with a writing brush for centuries.

In this paper, we explore the advantages of pen-based interfaces on 3D transformations. User controlled transformations, which typically include 3D translation, rotation, and scaling, are some of the key functions of those 3D modeling or viewing systems. The transformations typically were performed discretely through click and drag with a mouse or a pen. It has been chasing by many researchers in HCI to make users closely mimic the feel of real medium with digital devices. Most 3D modeling systems aim at closely mimicking freehand drawing on paper. However, the potential of a pen to naturally and intuitively manipulate 3D objects has been rarely exploited.

We present a general operation method that is independent of any 3D model. The method includes the whole process of 3D object manipulations from 3D target selection to various 3D transformations, i.e., translation, rotation, scaling, and combined transformations. The mapping between these transformations and the pen input channels is elaborated to make all the transformations be performed with a natural, intuitive, and interactive method. A prototype system has been designed and implemented to test the mentioned 3D transformations.

2 Related Work

In this section, we discuss related work regarding sketching-based 3D operations and pen input channels.

2.1 Previous Work on Sketching-Based 3D Operations

Augmentation is a typical sketching-based modeling 3D operation, through which some new features can be added to an existing 3D model. Augmentations can be made in either a surficial (e.g., [10]) or additive manner (e.g., [6]). Besides augmentation, many sketch-based modeling systems support other sketch-based editing operations (e.g., [10]). Sketch-based 3D operations are usually modeling-dependent. The same or similar 2D sketches can be interpreted into different 3D models in different systems. On the contrary, the same or similar 3D models can also be modeled from different 2D sketches in different systems. Therefore, a sketching-based modeling system typically aims at a specific application. The applications can be roughly classified into two groups, i.e., applications in computer-aided design (CAD) [23] and applications in digital content creation. Applications in digital content creation include wide topics, e.g., plant modeling [21] and RGB-D image [8]. Some researchers (e.g., [2]) have also investigated collaborative 3D design techniques.

We are not the first to explore 3D operations with digital pens. Some researchers have explored 3D operations in pen-based interfaces, but they aim at some special 3D operation issues, not for general 3D operations. Oshita [12] presented a pen-based intuitive interface to control a virtual figure interactively. Pen input parameters, i.e., the pen positions, pressure, and tilt, were utilized to make a human figure perform various types of motions in response to the pen movements manipulated by the user. Kolhoff et al. [7] explored manipulating a virtual figure's gait with two pens "walked" on an indirect tablet. Tompkin et al. [17] proposed estimating the 3D position and 2D orientation of the pen.

2.2 Previous Work on Pen Input Channels

Besides the *X-Y* location input channel, a digital pen possesses pressure and angle input channels. And the input angles include tilt, twist, and orientation.

There are many studies concerning pressure input. These studies investigated general human ability of controlling pressure input [14], or interaction techniques utilizing pressure, e.g., 2D drawing and object manipulation [3], supplementing data for writer verification [11]. Some researchers explored general human being control ability of a certain input angle, e.g., twist [1], or the usefulness of input angles for interaction tasks, e.g., selection with the Tilt Menu [16], text entry in mobile devices [19]. The *X-Y* location input parameter was typically investigated in sketching-based interfaces, e.g., creating 3D objects from 2D sketching [20].

3 The User Interface of the Prototype System

The prototype system was implemented with JavaFX 8. The system possesses three views, i.e., full view, fixed view, and mobile view. There are menus, toolbars, fixed panels, and a mobile pie menu (MPM), in the full view (see Fig. 1). All the interfacial components in the full view remain in the fixed view except the MPM. Only the MPM exists as an interfacial widget in the mobile view. The MPM is designed based on PP-Menus [18]. PP-Menus are pressure-based pie menus, and employ pen input pressure system. values to localize traditional pie menus and own higher interactive efficiency.

The full view and the fixed view offer more interactive details and hints for novices, on the contrary, experts obtain a bonus of having more work space and higher interactive efficiency in the mobile view. A user is allowed to either click an item or cross it with the pen tip to choose a command in the MPM. Shortcut commands are available to switch between the views.

Fig. 1. The user interface of the prototype system.

4 Manipulating a 3D Virtual Object from a User's Point of View

In this section, we introduce the 3D object manipulation from a user's point of view. The operations include creating, selecting and grouping, rotating, scaling, and translating 3D objects.

4.1 Creating 3D Objects

Some kinds of simple regular 3D objects and combined 3D models are available in the prototype system. When we choose an object/model from a traditional pull-down menu or the MPM, the selected object is created in the workspace.

4.2 Selecting and Grouping 3D Objects

In the design, a pressure-piercing "pressure-based line-string" selection method [9] is employed to select 3D targets. A user slides the pen tip on the screen starting from a blank area, where there is no object's projection on XY-plane. If the pen tip pressure and the stroke length exceed the predefined thresholds, a target selection process is evoked. A blue footprint line of the pen tip is drawn on the screen and red marks are added around any selected object to give the user an intuitive feedback (see Fig. 2). We can configure the prototype system to hide the red marks to make the interface clear, but the footprint line is always displayed on the screen during the selection. All the objects crossed by the footprint line are selected when the pen pressure is kept higher than the threshold. If there is any object that the user does not want to select in the path of the selection stroke, s/he steers clear of it or reduces the pressure on the pen to a normal range without lifting the pen tip from the screen, then the object will be just crossed without being selected. The footprint line is displayed temporarily in green to show the crossing-only process. To cancel the selection of an object, the user just needs to stroke the pen back across the selected object again.

A click on a certain bezel button locks all the selected objects into one group. Objects in one group are transformed (e.g., tilted, rotated, or translated, etc.) synchronously. When we want to add or reduce any object in the group, we can click on the bezel button again and unlock the group. The pen tip does not have to be lifted from the tablet in the whole process.

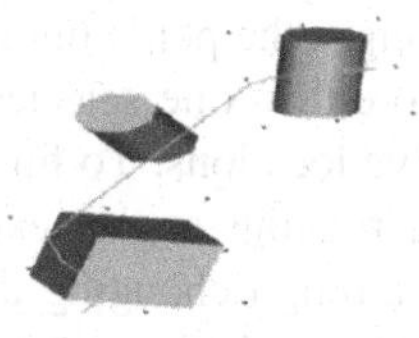

Fig. 2. Pressure-piercing "line-string" selection method (only the objects with red marks were selected).

(a) The initial state of a tellurion. (b) Rolling the tellurion to a new state by rolling the pen.

Fig. 3. Rotating objects around an axis.

4.3 Rotating 3D Objects

Rotation of a physical object can be classified into two types, i.e., rotation around a pivot and rotation around an axis. Both of the two types of rotation can be manipulated with a pen with an intuitive, natural, interactive, and continuous method.

Rotating 3D Objects around an Axis. A selected object can be rotated around an axis. The user presses the pen more heavily, when the pressure input exceeds a predefined threshold and the pen keeps stationary, a rotation axis is determined. The pen is considered to be stationary only when its locomotion and posture transformation have not exceeded the predefined thresholds. When a rotation axis is determined, it proximately coincides with the pen's longitudinal axis. After an axis has been set, the user can rotate an object by rolling the pen. When the pen's rolling speed and rolling angle exceeds the specific ranges, the object rotates around the axis concurrently with the pen's rotation. The user can "feel" the object's rotation by his/her fingertips even without seeing the screen, since the pen and the object rotate coincidentally. If there is more than one object selected, all the selected objects rotate concurrently and maintain their relative locations. Figure 3 illustrates rotating a tellurian around an axis by rolling the pen. The rotation axis can be changed interactively. Some rotation effects regarding depth can be achieved when the rotation axis is not perpendicular to the screen.

Rotating Objects around a Pivot. There are five candidates for a pivot. They are the two intersection points of a rotation axis and the object's surface, the quartile points of the line segment between the two intersection points (see Fig. 4). The pivots, from the superficial one to the deepest one, are numbered from the first to the fifth. During rotation, the pivot can be set by the pressure input values of the pen. Rotating around a pivot includes tilting and orientating around the pivot. The user tilts the pen, and then the object tilts around the pivot. Figure 5 illustrates tilting effects of an object around different pivots by tilting the pen. The user changes the pen's azimuth, and then the object's orientation is changed around the pivot. Figure 6 shows the continuous

Fig. 4. One of the five pivots is chosen to be the present rotation pivot.

orientating effects of an object around one rotation pivot. The tilting and orientating can be performed concurrently when the user concurrently changes the pen's tilt angle and azimuth. As rotating objects around an axis, if there is more than one selected object, all the objects rotate concurrently and maintain their relative locations. To filter quiver of the pen, the tilting and orientating works only when rotating angles exceed the respective thresholds. The rotation pivot can be changed through changing the corresponding rotation axis. When a rotation axis has been changed, the corresponding rotation pivot is determined by the intersection point of the object's inner surface and the axis. If there is more than one selected object and any of them is crossed by the rotation axis, the pivot is determined by the rotation axis and the inner surface of the object crossed by the axis. If there is no object crossed by the axis, the pivot is determined by the rotation axis and the last selected object's dimension.

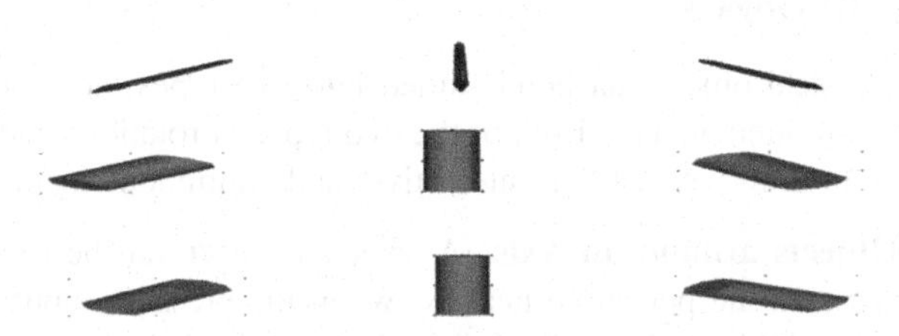

Fig. 5. Tilting an object around a pivot by tilting the pen. Tilting a pen (the top row) from its initial state (middle column) to right (left column) or left (right column) manipulates an object to tilt around different pivots (the first pivot and the fifth pivot illustrating in the middle and bottom row, respectively).

Rotating objects around a pivot and around an axis can be performed concurrently. The user rolls, tilts, and orientates the pen concurrently, then the object(s) will roll, tilt, and orientate together with the pen.

Fig. 6. Orientate an object around a pivot by orientating the pen.

4.4 Scaling 3D Objects

A user chooses the scaling menu item from a menu or the scaling button from the MPM. Then the object scaling channel is opened. The user manipulates the pen input pressure to resize all selected objects simultaneously. Pen input pressure, from the lightest to the heaviest, is divided into three layers, i.e., stretching layer, spacing layer, and shrinking layer. A colored wedge at the pen tip shows visual feedback of the input pressure to the user (see Fig. 7). Initially, the scaling operation can be performed only after the pressure goes out of the spacing layer to prevent a sharp size change of the

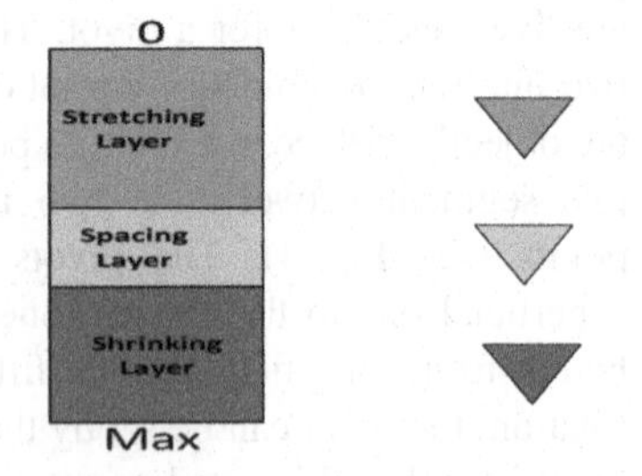

Fig. 7. Pressure layers and visual feedback for scaling.

objects; i.e., the users must adjust the pressure input into the spacing lay before scaling the objects. An object can be scaled along one of its X, Y, or Z axis at a time. The user adjusts the pen barrel and makes the angle between it and one of an object's axes less than 30°, and then s/he manipulates the input pressure to scale the object along the axis. An object is stretched by the input pressure when it is in the stretching layer; on the contrary, when the input pressure is in the shrinking layer, it shrinks the object. The user should click a bezel button to close the scaling channel to prevent a sharp size change of the selected objects before s/he lifts the pen tip from the tablet.

4.5 Translating 3D Objects

When the user wants to move the selected object(s), s/he slides the pen tip to cross the locomotion menu item. The select object(s) can be translated by the pen tip locomotion on the screen. *X-Y* locomotion of an object on the screen is similar to dragging it. The user slides the pen tip on the screen, then the objects moves together with the pen tip. The user is allowed to translate the objects in Z-direction by utilizing the pen input pressure. Like scaling, the pressure input is divided into three levels, i.e., pulling layer, spacing layer, and pushing layer, for Z-translation. Pulling layer is related to the lightest levels of pressure input, while pushing layer to the heaviest. When the pressure input is in the pulling layer, a translation towards the user is performed on the object. On the contrary, when the pressure input is maintained in the pushing layer, the object keeps moving far away. The user should press a bezel button to prevent a sharp Z-translation before s/he lifts the pen tip from the tablet.

4.6 Combined Transformations

Some of the mentioned operations can be performed concurrently (e.g., see Fig. 8). The user crosses the combined manipulation menu item to begin a combined transformation. The combined transformation sometimes produces some special operation effects, e.g., rotating an object around its center axis and translating it concurrently can simulate rolling an object on the screen. The user rolls the pen barrel and moves the pen tip on the screen simultaneously, then the object rolls on the screen. Moving and tilting a figure can make it "lurch" on the screen. Rolling and translating a whipping top can make it spin and slide on the screen. In the prototype system, all the rotating and translating related transformations are allowed to be combined together and performed concurrently.

Fig. 8. Combined transformations.

5 Function Design and Algorithm

In the design, a pen's physical feedback is exploited through keeping the coincidence of movements between the pen and a 3D target. The mapping between pen-input channels and object actions is context-aware, i.e., the same manipulation of the pen may lead to different transformations of the objects in different contexts. The operation effect is just like holding and manipulating a physical object by hand.

5.1 Selecting and Grouping 3D Objects

During the "pressure-based line-string" selection, the pressure value of the pen input is utilized as a metaphor of the depth into the screen. The pen-input pressure values are divided into to two levels, i.e., movement layer and selection layer (see Fig. 9). "Line-string" selection is like stringing an object with a needle. When the pressure of the pen tip on the screen gets heavy enough to enter the selection layer, the pen drag can be regarded as movement in the object. After an object is selected, the user can continue stroke the pen tip on the screen to select another object. All the selected objects are recorded with a list, *targetList*. Algorithm 1 shows the process of selecting and grouping objects. When the *targetList* is unlocked, the object selection process is evoked by the pen tip drag on the tablet if the pen input pressure is greater than the given threshold. Then an object can be added to or removed from the *targetList* by the pen tip drag. A click on the given bezel button switches the *targetList*'s status between locked and unlocked. When the *targetList* is locked, all recorded objects are grouped together.

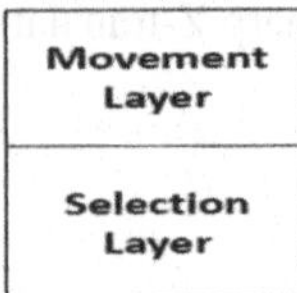

Fig. 9. Pen input pressure layers in selecting and grouping objects.

Algorithm 1. Selecting and grouping objects

```
1: scene.getRoot().getChildrenUnmodifiable()
2: .forEach(node− >{node.setOnDragEntered(value− > {
3: if targetList.locked is false then
4:     if Pen.pressure>targetList.selThreshold then
5:         if targetList.contains(node) is false then
6:             targetList.add(node);
7:         else    targetList.remove(node);
8:         end if
9:     end if
10: end if
11: });});
```

5.2 Rotating 3D Objects

In our design, the selected objects can be manipulated only when they are locked in one group. Bi et al. [1] reported that the rolling of a pen could be considered as user intentional only when the rolling speed and the rolling angle surpasses the specific ranges. We refer to their results in the rotation function design. The program monitors the change of pen input angles. The rotation will be performed when a change range of any angle exceeds a predefined threshold if the *targetList*'s rotation is permitted. Three-dimensional object rotation includes rotation around an axis and a pivot. To filter out incidental pen rotating accompanied by pen translating, we treat 3D rotation only when the tip is considered being stationary (having no *X-Y* locomotion) under 3D rotation only operation status. Algorithm 2 shows the process of rotating 3D objects around an axis. When the pen tip is determined being stationary, and the *targetList* is locked and its rotating status is true, the algorithm checks the varied amount of the pen's twist angle (*deltaTwist*) within a given time spectrum: if the *deltaTwist* is greater than the given threshold, a Rotate object is created according to *deltaTwist* and the given rotation axis (*targetList.rotateAxis*) and registered to the Transforms of each 3D object in the *targetList*. Algorithm 3 illustrates the method of tilting and orientating 3D objects around a pivot with tilt and azimuth, respectively. The algorithm calculates the varied amount of the pen's tilt (*deltaTilt*) and azimuth (*deltaOri*) within the given time spectrum when the pen tip is stationary and the *targetList* is locked and its rotating status is true. If *deltaTilt* or *deltaOri* surpasses respective predefined thresholds, a Rotate object is created from *deltaTilt* or *deltaOri* and the given rotation pivot (*targetList.rotatePoint*), and registered to each object in the *targetList*.

Algorithm 2. Twisting objects around an axis

```
1: scene.setOnTouchStationary(value− > {
2: if targetList.locked is true then
3:     if targetList.rotating is true then
4:         if Pen.deltaTwist> targetList.deltaTwistThreshold then
5:             Rotate rot=new Rotate(Pen.deltaTwist,targetList.rotateAxis);
6:             targetList. forEach(node− > node.getTransforms().add(rot););
7:         end if
8:     end if
9: end if
10: });
```

Algorithm 3. Tilting and orientating objects around a pivot

```
1: scene.setOnTouchStationary(value− > {
2: if targetList.locked is true then
3:     if targetList.rotating is true then
4:         if Pen.deltaTilt> targetList.deltaTiltThreshold then
5:             Rotate rot=new Rotate(Pen.deltaTilt, targetList.rotatePoint);
6:             targetList. forEach(node− > node.getTransforms().add(rot));
7:         end if
8:         if Pen.deltaOri> targetList.deltaOrientationThreshold then
9:             Rotate rot=new Rotate(Pen.deltaOri, targetList.rotatePoint);
10:            targetList. forEach(node− > node.getTransforms().add(rot));
11:        end if
12:    end if
13: end if
14: });
```

5.3 Scaling 3D Objects

Ramos et al. [14] explored the capability of human being in controlling pressure input with a Wacom Intuos tablet with a wireless pen, which provides 1024 levels of pressure. They reported that dividing the pressure range into 6 levels or less produced the best performance. In our design, the pressure input is divided into three layers, i.e., stretching, spacing, and shrinking layers, to perform the scaling operation. Stretching layer is related to the lightest levels of pressure input and shrinking layer to the heaviest. The stretching layer has the most pressure levels, since the pen is too sensitive to control at a low-pressure value [14]. Algorithm 4 shows the process of 3D object scaling. The algorithm judges whether the *targetList* is locked when the pen tip keeps stationary on the tablet. A scalar is calculated according to the input pressure when the scaling channel is opened. When the input pressure, P, is greater than the given shrinking threshold, P_{sk}, a scalar is calculated from the maximum pressure (P_{max}), P, and P_{sk}:

$$scalar = (P_{max} - P)/(P_{max} - P_{sk})\,. \tag{1}$$

When P is less than the given stretching threshold, P_{sh}, a scalar is calculated according to P and P_{sh}:

$$scalar = P_{sh}/P. \tag{2}$$

P_{sk} and P_{sh} are the least value of shrinking layer and the largest value of stretching layer, respectively (see Fig. 7). The algorithm sets the scalar to an object's axis L (L is one of X, Y, or Z axis of the object) only when the angle between L and the pen barrel's central axis is less than 30°.

Algorithm 4. Scaling objects

```
1: scene.setOnTouchStationary(value− > {
2: if targetList.locked is true then
3:     if targetList.resizing is true then
4:         if Pen.pressure> targetList.Psk then
5:             scalar=(Pmax-P)/(Pmax-Psk);
6:         else if Pen.pressure<tragetList.Psh then
7:             scalar=Psh/p;
8:         end if
9:         targetList.forEach(node− > {
10:        if θL < 30° then
11:            node.setScaleL(scalar);
12:        end if
13:        });
14:    end if
15: end if
16: });
```

5.4 Translating 3D Objects

The transformation regarding depth is a difficult action since the screen is actually 2D and a common mouse can only input planar points. An intuitive depth-translation is implemented in our prototype, where the pressure input of the pen is utilized as a metaphor of object depth into the screen. The change of an object's location in Z-axis is calculated:

$$deltaZ = S * P_c, \quad (3)$$

where S is a smoothing factor, which is calculated according to the pen input pressure to make a Z-translation smoother; Pc is a value calculated according to the distance from the middle point of the primitive pressure input spectrum to the current pressure input value. S is:

Algorithm 5. Translating objects

```
1: scene.setOnMouseDragged(value− > {
2: if targetList.locked is true then
3:     if targetList.translating is true then
4:         deltaZ=S*Pc;
5:         targetList.forEach(node− > {
6:         node.setTranslateX(value.deltaX);
7:         node.setTranslateY(value.deltaY);
8:         node.setTranslateZ(deltaZ); });
9:     end if
10: end if
11: });
```

- positive when the pressure is in the pushing layer,
- negative when the pressure is in the pulling layer,
- zero when the pressure is in the spacing layer.

The whole translating process is shown in Algorithm 5. To filter incidental inputs, thresholds are predefined for both *X*-*Y* and *Z*-locomotion. The translation takes effect only when a locomotion input exceeds its given threshold.

5.5 Combined Transformations

A thread scans the variations of all the input channels of the pen: when the variation of each input channel exceeds its given threshold, the thread calls the function to respond to the variation. To prevent incidental transformations, the user should press a bezel button to close the combined transformation status and determine the transformations.

6 Informal User Study

We conducted an informal user study to investigate the usability of the prototype system. Eight participants were asked to assemble a 3D chair model with a given set of components (see Fig. 10) and create any free-form 3D model.

6.1 Apparatus

The hardware used in the experiment was a WACOM Cintiq 21UX flat panel LCD graphics display tablet with a resolution of 1,600 × 1,200 pixels (1 pixel = 0.297 mm), using a wireless pen with a pressure, tilt angle, azimuth, and twist angle sensitive isometric tip (the width of the pen-tip is 1.76 mm). It reports 1024

(a) A set of components.

(b) A chair model.

Fig. 10. A chair model assembled from a set of components.

levels (ranging from 0 to 1,023, the minimum unit is 1) of pressure, 360° (ranging from 22° to 90°, the minimum unit is 1°) of tilt angle. The experimental program was implemented with JavaFX 8 running on a desktop PC with Intel CORE i5 CPU and 8 GB memory. The operating system of the PC was Windows 10 Professional.

6.2 Participants

Eight participants (2 females, ranging in age from 21 to 25 years) were all volunteers from the local university community. The participants were all right-handed. None of them had experience with a digital pen or our prototype system, but they all had 3D operation experience with some typical interfaces.

6.3 Procedure

We illustrated the functions of the prototype system within several minutes. The participants were asked to assemble a chair model with the given components quickly and precisely. Then they were each allowed to build any 3D free-form model within ten minutes. At the end of the experiment, the participants were instructed to give their subjective comments by completing a questionnaire, which consisted of four questions regarding "usability", "mental demand", "physical demand", and open comments.

6.4 Results

All the participants completed the assembly within 5 min, ranging from 2.1 min to 4.6 min. Six participants thought that the same motion direction and amplitude between the pen and the objects made the manipulation results to be predictable, and this enhanced the immersion and ease of use. All the participants felt the manipulation was interesting, and five of them hoped to have more components to create more interesting models. Figure 11 shows some samples created and manipulated by the participants.

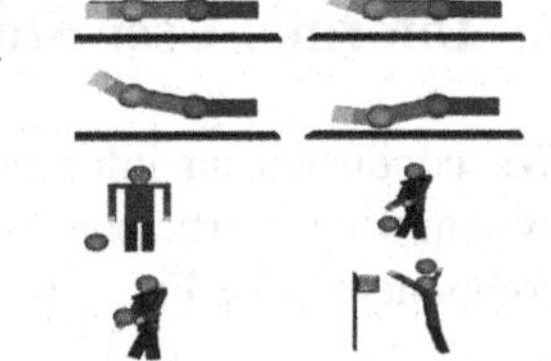

Fig. 11. Samples created by the participants.

7 Discussion

Jacob et al. [4] believed that new interaction style drew strength by building on users' pre-existing knowledge of non-digital world and reduced the gulf of execution, the gap between a user's goals for action and the means to execute those goals. Three-dimensional operations are still a challenge even for computer experts. Digital pens evolve from physical painting and drawing pens: they have the potential for natural and intuitive 3D manipulations. We proposed some operation patterns to exploit the benefits of a digital pen for 3D manipulations. We aimed at rapid, fluid, but rough 3D operations, not precise. Pen-based interfaces do have the ability for precise operations, perhaps with parameter configuration dialog boxes or precise interfacial widgets, but these interaction patterns typically prevent users from operating with the immersion, interest, fluid and easy interaction feelings.

Only some regular 3D components, e.g., boxes, cylinders, and spheres, were available in the prototype system. But the proposed interaction patterns are model-independent: they are suitable for models created by other 3D modeling systems.

8 Conclusion and Future Work

We proposed some modeling-independent 3D operation patterns, which exploited the potential of digital pens for natural, intuitive, fluid, rapid, and rough 3D transformations. These patterns helped users transfer their pre-existing knowledge of using physical painting or writing pens to 3D operations with a digital pen: this reduced the gulf of execution and made 3D operations more interesting and easier. The informal user study showed that the participants grasped the 3D operation methods easily and quickly. In our future work, we will merge free-form 3D modeling into the prototype system.

Acknowledgment. Supported by the Fundamental Research Funds for the Central Universities Grants No. lzujbky-2016-k07.

References

1. Bi, X., Moscovich, T., Ramos, G., Balakrishnan, R., Hinckley, K.: An exploration of pen rolling for pen-based interaction. In: Proceedings of UIST 2008, pp. 191–200. ACM (2008)
2. Grandi, J.G., Debarba, H.G., Nedel, L., Maciel, A.: Design and evaluation of a handheld-based 3D user interface for collaborative object manipulation. In: Proceedings of the 2017 CHI Conference on Human Factors in Computing Systems - CHI 20117, pp. 5881–5891. ACM Press (2017). https://doi.org/10.1145/3025453.3025935
3. Harada, S., Saponas, T.S., Landay, J.A.: Voicepen: augmenting pen input with simultaneous non-linguistic vocalization. In: Proceedings of ICMI 2007, pp. 178–185. ACM (2007)
4. Jacob, R.J., Girouard, A., Hirshfield, L.M., Horn, M.S., Shaer, O., Solovey, E.T., Zigelbaum, J.: Reality-based interaction: a framework for post-wimp interfaces. In: Proceeding of the Twenty-Sixth Annual CHI Conference on Human Factors in Computing Systems - CHI 2008, pp. 201–210. ACM Press (2008)
5. Joseph, J., LaViola, J.: Sketching and gestures 101. In: Proceedings of ACM SIGGRAPH 2007 courses, p. 2. ACM Press (2007)
6. Kazi, R.H., Grossman, T., Cheong, H., Hashemi, A., Fitzmaurice, G.: Dreamsketch: Early stage 3D design explorations with sketching and generative design. In: Proceedings of the 30th Annual ACM Symposium on User Interface Software and Technology - UIST 2017, pp. 401–414. ACM Press (2017)
7. Kolhoff, P., Preuß, J., Loviscach, J.: Walking with pens. In: Proceedings of EUROGRAPHICS 2005, pp. 33–36 (2005)
8. Li, Y., Luo, X., Zheng, Y., Xu, P., Fu, H.: Sweepcanvas: sketch-based 3D prototyping on an rgb-d image. In: Proceedings of the 30th Annual ACM Symposium on User Interface Software and Technology - UIST 2017, pp. 387–399. ACM Press (2017)

9. Liu, C., Ren, X.: Making pen-based operation more seamless and continuous. In: Gross, T., Gulliksen, J., Kotzé, P., Oestreicher, L., Palanque, P., Prates, R.O., Winckler, M. (eds.) INTERACT 2009. LNCS, vol. 5726, pp. 261–273. Springer, Heidelberg (2009). https://doi.org/10.1007/978-3-642-03655-2_32
10. Nealen, A., Igarashi, T., Sorkine, O., Alexa, M.: Fibermesh: designing freeform surfaces with 3D curves. In: Proceedings of ACM SIGGRAPH, pp. 41–50. ACM Press (2007)
11. Okawa, M., Yoshida, K.: Text and user generic model for writer verification using combined pen pressure information from ink intensity and indented writing on paper. IEEE Trans. Hum.-Mach. Syst. **45**(3), 339–349 (2015)
12. Oshita, M.: Pen-to-mime: a pen-based interface for interactive control of a human figure. In: Proceedings of Eurographics Workshop on Sketch-Based Interfaces and Modeling 2004, pp. 43–52. Eurographics Association (2004)
13. Pereira, J.P., Jorge, J.A., Branco, V.A., Ferreira, F.N.: Calligraphic interfaces: mixed metaphors for design. In: Proceedings of Interactive Systems: Design, Specification and Verification, DSV-IS 2003, pp. 154–170 (2003)
14. Ramos, G., Boulos, M., Balakrishnan, R.: Pressure widgets. In: Proceedings of CHI 2004, pp. 487–494. ACM (2004)
15. Schmidt, R., Singh, K., Balakrishnan, R.: Sketching and composing widgets for 3D manipulation. In: Proceedings of EUROGRAPHICS 2008, vol. 27, pp. 3–12. Blackwell Publishing, Hoboken (2008)
16. Tian, F., Xu, L., Wang, H., Zhang, X., Liu, Y., Setlur, V., Dai, G.: Tilt menu: using the 3D orientation information of pen devices to extend the selection capability of pen-based user interfaces. In: Proceedings of CHI 2008, pp. 1371–1380. ACM (2008)
17. Tompkin, J., Muff, S., McCann, J., Pfister, H., Kautz, J., Alexa, M., Matusik, W.: Joint 5D pen input for light field displays. In: Proceedings of the 28th Annual ACM Symposium on User Interface Software & Technology - UIST 2015, pp. 637– 647. ACM Press (2015)
18. Wang, X.M., Wang, P., Liu, C.Y.: PP-menus: localizing pie menus by pressure. In: Proceedings of International Conference on Control and Automation, pp. 532–540. DEStech Publications, Inc., Lancaster (2016)
19. Wigdor, D., Balakrishnan, R.: Tilttext: using tilt for text input to mobile phones. In: Proceedings of UIST 2003, pp. 81–90. ACM (2003)
20. Wu, P.C., Wang, R., Kin, K., Twigg, C., Han, S., Yang, M.H., Chien, S.Y.: Dodecapen: accurate 6dof tracking of a passive stylus. In: Proceedings of the 30th Annual ACM Symposium on User Interface Software and Technology - UIST 2017, pp. 365–374. ACM Press (2017). https://doi.org/10.1145/3126594.3126664
21. Zakaria, M.N., Shukri, S.R.M.: A sketch-and-spray interface for modeling trees. In: Butz, A., Fisher, B., Krüger, A., Olivier, P., Owada, S. (eds.) SG 2007. LNCS, vol. 4569, pp. 23–35. Springer, Heidelberg (2007). https://doi.org/10.1007/978-3-540-73214-3_3
22. Zeleznik, R.C., Herndon, K.P., Hughes, J.F.: Sketch: an interface for sketching 3D scenes. In: Proceedings of ACM SIGGRAPH 2007 Courses, pp. 19–24. ACM (2007)
23. Zhu, H., Song, Y., Nie, D., Peng, X.: Real-time 3D collaborative satellite orbit design system based on message queue and p2p structure. In: 2017 IEEE 21st International Conference on Computer Supported Cooperative Work in Design (CSCWD), pp. 503–508. IEEE, April 2017. https://doi.org/10.1109/cscwd.2017.8066745

Improved Interleaving Scheme for PAPR Reduction in MIMO-OFDM Systems

Lingyin Wang(✉) and Xiaoqing Jiang

School of Information Science and Engineering, University of Jinan, Jinan 250022, People's Republic of China
andrewandpipi@hotmail.com

Abstract. Just like orthogonal frequency division multiplexing (OFDM) systems, the large peak-to-average power ratio (PAPR) as one of the main shortcomings still exists in multi-input multi-output (MIMO) OFDM systems. Interleaving scheme is one of the attractive technologies for PAPR reduction and it can be directly used for each transmit antenna in MIMO-OFDM systems, called as ordinary interleaving scheme. In this paper, an improved interleaving scheme for PAPR reduction in MIMO-OFDM systems is proposed. Different from ordinary interleaving scheme, which transmit antenna is selected to perform the interleaving in proposed interleaving scheme is dominated by PAPR values of OFDM candidate sequences. Namely, the proposed interleaving scheme reorders the sequence from the antenna with the highest PAPR in the successive step after the PAPR of the original OFDM sequences are calculated. As a result, the proposed interleaving scheme can obtain better PAPR reduction performance compared with ordinary interleaving scheme.

Keywords: Orthogonal frequency division multiplexing
Multi-input Multi-output · Interleaving

1 Introduction

Owing to its high data rate and reliable transmission in frequency selective fading channels, multicarrier modulation, in particular orthogonal frequency division multiplexing (OFDM), has been widely adopted in wireless communications over the past several years [1]. But unluckily, an OFDM signal consists of a number of independently modulated subcarriers, which induces high peak-to-average power ratio (PAPR) when added up coherently. A large PAPR brings shortcomings like a reduced efficiency of high power amplifier and an increased complexity of the analog-to-digital (A/D) and digital-to-analog (D/A) converters, which results in system performance degradation [2]. As with OFDM, multi-input multi-output (MIMO) OFDM systems also have this disadvantage [3].

To reduce PAPR, several techniques have been proposed [4, 5], where the most promising one is the multiple signal representation schemes [6], such as interleaving [7–10], selected mapping [11–14] and partial transmit sequence [15–18]. Interleaving scheme, as one of multiple signal representation schemes, scrambles each OFDM symbol with random permutations and selects the sequence that gives the lowest

D.-S. Huang et al. (Eds.): ICIC 2018, LNAI 10956, pp. 599–607, 2018.
https://doi.org/10.1007/978-3-319-95957-3_62

PAPR. For MIMO-OFDM systems, the PAPR problem is similar to the original single-input sing-output (SISO) OFDM. Thus, interleaving scheme can also be used for optimizing PAPR performance of MIMO-OFDM systems.

Interleaving scheme is individually applied to each transmit antenna in MIMO-OFDM systems, named ordinary interleaving scheme. For the OFDM candidate sequences in each of the parallel transmit antennas, the one with the smallest PAPR is chosen. In this way, because the same number of OFDM candidate sequences can be found in each antenna, some computational complexity becomes redundant.

In this paper, an improved interleaving scheme for PAPR reduction in MIMO-OFDM systems is proposed. Different from the ordinary interleaving scheme applied in MIMO-OFDM, the proposed interleaving scheme reorders the sequence from the antenna with the highest PAPR in the successive step after the PAPR of the original OFDM sequences are calculated. As a result, the proposed interleaving scheme gives the more satisfactory complementary cumulative distribution function (CCDF) [19] curves of the PAPR compared with the ordinary one.

The rest of this paper are arranged as the follows. The basic principle of MIMO-OFDM systems and ordinary interleaving scheme are given in Sect. 2. Sections 3 and 4 introduce the proposed improved interleaving scheme and the corresponding PAPR reduction performance respectively. Finally, Sect. 5 gives a brief conclusion.

2 Background Description

2.1 MIMO-OFDM Systems

For a MIMO-OFDM system with M antennas, an OFDM signal $x_n, n = 0, 1, \cdots, N-1$ from mth antenna in the discrete time domain is described by

$$x_{m,n} = \frac{1}{\sqrt{N}} \sum_{k=0}^{N-1} X_{m,k} e^{j2\pi kn/N}, \; 1 \le m \le M, \; 0 \le n \le N-1 \tag{1}$$

where N is the number of subcarriers, $X_{m,k}$ is the symbol modulated by phase shift keying (PSK) or quadrature amplitude modulation (QAM).

The PAPR of the OFDM signal from mth antenna can be defined by the ratio of the peak power to the mean power, shown by

$$\mathrm{PAPR}(x_{m,n}) = 10 \log_{10} \frac{\max\limits_{0 \le n \le N-1} \{ |x_{m,n}|^2 \}}{E\{ |x_{m,n}|^2 \}} \mathrm{dB} \tag{2}$$

In order to estimate the PAPR reduction performance, CCDF is always employed, given by

$$CCDF = 1 - (1 - e^{-PAPR_0})^N \tag{3}$$

where PAPR_0 is a threshold of PAPR.

2.2 Ordinary Interleaving Scheme

The process of conventional interleaving scheme for OFDM systems can be performed as follows.

(1) The input symbol sequence $\boldsymbol{X}$ modulated by PSK or QAM is partitioned into several disjoint subblocks $\boldsymbol{X}_i$, $i = 1, 2, \cdots, V$.
(2) All the subblocks are permuted by employing random order to obtain different sequences. For example, after all the subblocks being reordered, the original partitioned sequence $[\boldsymbol{X}_1, \boldsymbol{X}_2, \cdots, \boldsymbol{X}_i, \cdots, \boldsymbol{X}_V]^T$ becomes $[\boldsymbol{X}_{\theta(1)}, \boldsymbol{X}_{\theta(2)}, \cdots, \boldsymbol{X}_{\theta(i)}, \cdots, \boldsymbol{X}_{\theta(V)}]^T$, where $\{i\} \rightarrow \{\theta(i)\}$ is the one-to-one mapping and $\theta(i) \in \{1, 2, \cdots, V\}$ belongs to the set of $\{1, 2, \cdots, V\}$.
(3) All the reordered sequences are transformed into time domain by inverse fast Fourier transform (IFFT) operations to generate OFDM candidate sequences. Namely, the modulated sequence $\boldsymbol{X}$ is permuted once to achieve one OFDM candidate sequence.
(4) Finally, the one OFDM candidate sequence with the smallest PAPR is selected for transmission.

The block diagram of conventional interleaving scheme for OFDM systems is shown in Fig. 1.

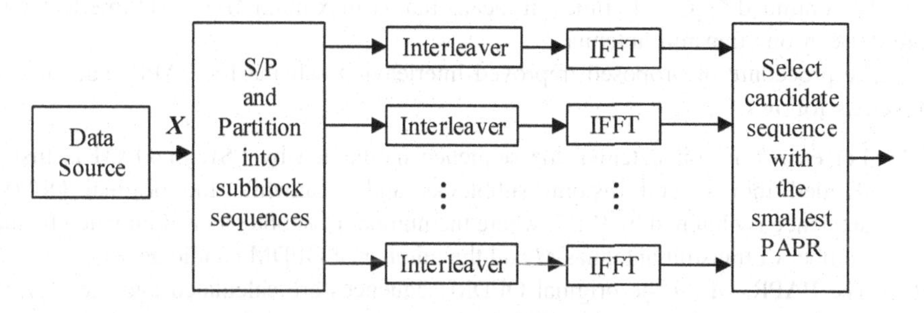

Fig. 1. Block diagram of conventional interleaving scheme for OFDM systems

In a MIMO-OFDM system with M transmit antennas, assume that the selected OFDM candidate sequence with the smallest PAPR out of the C possible is individually selected for each of the parallel transmit antennas.

For MIMO-OFDM systems, the above conventional interleaving scheme is independently applied to each transmit antenna for reducing the PAPR, called as ordinary interleaving scheme. Moreover, the side information is required for each antenna to recover the input data successfully and the number of side information bits is proportional to the number of transmit antennas M.

3 Improved Interleaving Scheme

In ordinary interleaving scheme, the transmit antennas are individually treated, which results in some unnecessary computational complexity and is not able to reduce the PAPR very well. With the number of transmit antennas increasing, more performance in PAPR reduction has been lost. For the sake of reaching better PAPR reduction performance, an improved interleaving scheme in MIMO-OFDM systems is detailedly described in this section.

The proposed interleaving scheme gets PAPR reduction by make full use of the property of MIMO communications. Assume that in ordinary interleaving scheme, C OFDM candidate sequences in each transmit antenna are generated.

The main ideas of proposed interleaving scheme is that instead of generating C OFDM candidate sequences for each of the M transmit antennas, the budget of $M(C-1)$ OFDM candidate sequences is used for improving the signal with the highest PAPR successively over the antennas. That is to say, for each transmit antenna, the original OFDM sequence is firstly obtained and its PAPR is calculated. Then, in each successive step, the OFDM candidate sequence with the highest PAPR in the previous step is considered and the interleaving is performed on the corresponding modulated sequence from the selected transmit antenna for a try of PAPR reduction. The process of finding the OFDM candidate sequence with highest PAPR and performing the interleaving on the modulated sequence in the selected transmit antenna will be continued $M(C-1)$ times. It means that at maximum $M(C-1)$ interleavings are done in one transmit antenna.

The procedure of proposed improved interleaving scheme for PAPR reduction is given as follows.

(1) For each transmit antenna, the sequence modulated by PSK or QAM is firstly divided into several disjoint subblocks and meanwhile the original OFDM sequence is obtained by IFFT, where the number of subblocks is dominated by the number of transmit antennas M and the number of OFDM candidate sequences C;
(2) The PAPRs of all the original OFDM sequences are calculated and the OFDM sequence with the highest PAPR is chosen;
(3) For the transmit antenna with the chosen OFDM sequence, the interleaving is performed for trying reducing PAPR and the new OFDM candidate sequence is obtained by IFFT;
(4) If the PAPR of the new OFDM candidate sequence is lower than that of the chosen OFDM sequence, the chosen OFDM sequence will be replaced by the new one; otherwise, the chosen OFDM sequence is retained;
(5) In the successive step, the OFDM candidate sequence with the highest PAPR is selected, the interleaving is performed on the modulated sequence from the selected transmit antenna and the new OFDM candidate sequence is achieved by IFFT operation;
(6) Compare the PAPRs of the new OFDM candidate sequence and the selected one, the one with the smallest PAPR is reserved;
(7) Repeat step (5) and step (6), the optimization process is terminated when $M(C-1)$ times interleavings are completed.

For understanding the proposed interleaving scheme clearly, the algorithm of proposed interleaving scheme have illustrated in details as shown in Fig. 2.

Given parameters: M and C;
function $[\text{PAPR}_1,\cdots,\text{PAPR}_M]$=Improved_Interleaving($[\boldsymbol{X}_1,\cdots,\boldsymbol{X}_M]$)
1: $\boldsymbol{x}_m = \text{IFFT}\{\boldsymbol{X}_m\}, m = 1,2\cdots,M$
2: calculate $\text{PAPR}_m, m = 1,2\cdots,M$
3: for $d = 1, 2, \cdots, M(C\text{-}1)$
4: $[\text{PAPR}_{\max}, m_{\max}] = \max\{\text{PAPR}_1, \text{PAPR}_2, \cdots, \text{PAPR}_M\}$
5: $\boldsymbol{x}_{\max,\text{ new}} = \text{IFFT}\{\text{Interleaving}\{\boldsymbol{X}_{m_{\max}}\}\}$
6: calculate PAPR_{new}
7: if $(\text{PAPR}_{\text{new}} < \text{PAPR}_{\max})$
8: $\boldsymbol{x}_{m_{\max}} = \boldsymbol{x}_{\max,\text{ new}}$
9: $\text{PAPR}_{m_{\max}} = \text{PAPR}_{\text{new}}$
10: end
11: end

Fig. 2. The algorithm of proposed interleaving scheme

As for computational complexity, because the same number of IFFT operations is found in both proposed interleaving scheme and the ordinary one, it leads to the same complexity.

Because the step of finding the OFDM candidate sequence with the highest PAPR is involved, the side information in proposed interleaving scheme is different from that in the ordinary one. In proposed interleaving scheme, one transmit antenna may perform at maximum $M(C - 1) + 1$ trials for achieving better PAPR reduction. Thus, $M\lceil\log_2(M(C-1)+1)\rceil$ bits are required for representing this side information to recover the input data in the receiver.

4 PAPR Reduction Performance

In this section, many simulations are done for comparing PAPR reduction performance of proposed interleaving scheme and the ordinary one, where the number of OFDM candidate sequences C in these two interleaving schemes are 2 and 8 respectively. The data symbols are transmitted by a MIMO-OFDM system with 128 subcarriers and 10^5 OFDM sequences are generated in each transmit antenna.

Here, in order to capture all the signal peaks, the oversampling factor L is adopted. It can be realized by LN-points IFFT operation of the modulated sequence with $(L-1)N$ zero-padding. It is well known that all the signal peaks could be virtually captured when the oversampling factor $L = 4$ [20].

Figure 3 gives the CCDFs of proposed interleaving scheme and the ordinary one employing QPSK with two transmit antennas.

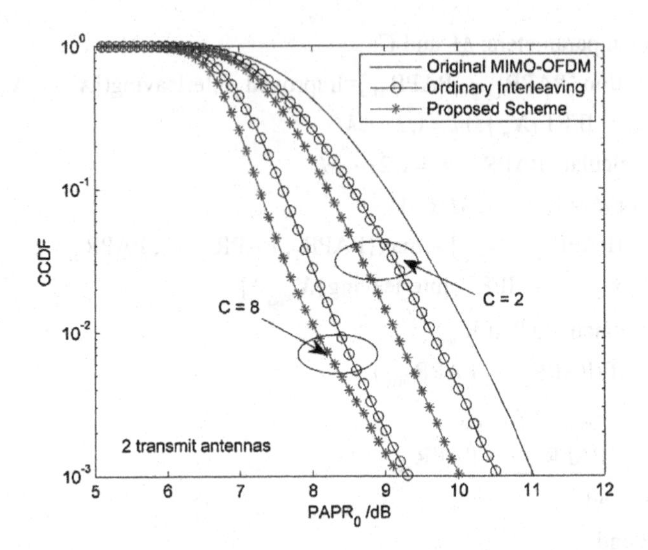

Fig. 3. The CCDFs of proposed interleaving scheme and the ordinary one employing QPSK in MIMO-OFDM with 2 transmit antennas

As we can see from Fig. 3, PAPR reduction performance of proposed interleaving scheme is much better than that of the ordinary one. For instance, at the probability of 10^{-2}, the PAPRs of proposed interleaving scheme and the ordinary one are 9.05 dB and 9.15 dB with $C = 2$, respectively; those of proposed interleaving scheme and the ordinary one are 8.05 dB and 8.38 dB with $C = 8$, respectively.

Figure 4 shows the CCDFs of proposed interleaving scheme and the ordinary one employing 16QAM with two transmit antennas.

As we can see from Fig. 4, the PAPR reduction performance of proposed interleaving scheme is much better than that of the ordinary one. For example, given CCDF = 0.01, the PAPRs of proposed interleaving scheme and the ordinary one are 9.14 dB and 9.62 dB with $C = 2$, respectively; those of proposed interleaving scheme and the ordinary one are 8.02 dB and 8.34 dB with $C = 8$, respectively.

Figures 5 and 6 show the PAPR reduction performance of proposed interleaving scheme and the ordinary one with four transmit antennas. The difference between Figs. 5 and 6 is that QPSK and 16QAM are adopted in MIMO-OFDM systems respectively.

It can be seen from Figs. 5 and 6 that, the similar results can be obtained compared with Figs. 3 and 4. In Figs. 5 and 6, the CCDF curves of proposed interleaving scheme show better PAPR reduction performance than that of the ordinary one. Thus, the proposed interleaving scheme for improving the PAPR reduction performance of MIMO-OFDM systems can keep its advantages when the number of transmit antennas is changed.

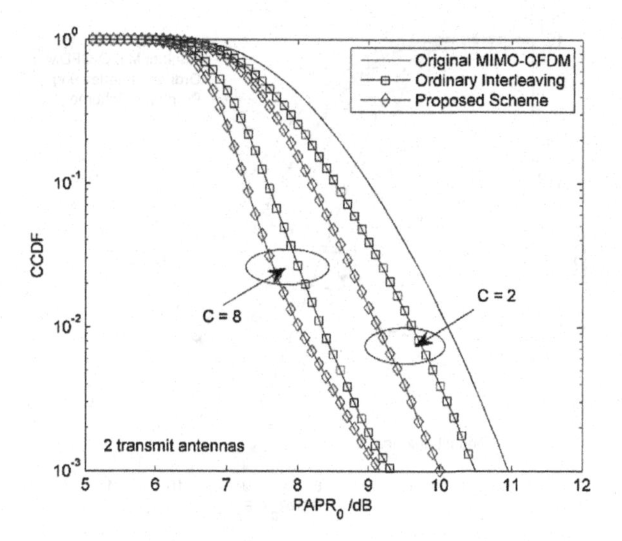

Fig. 4. The CCDFs of proposed interleaving scheme and the ordinary one employing 16QAM in MIMO-OFDM with 2 transmit antennas

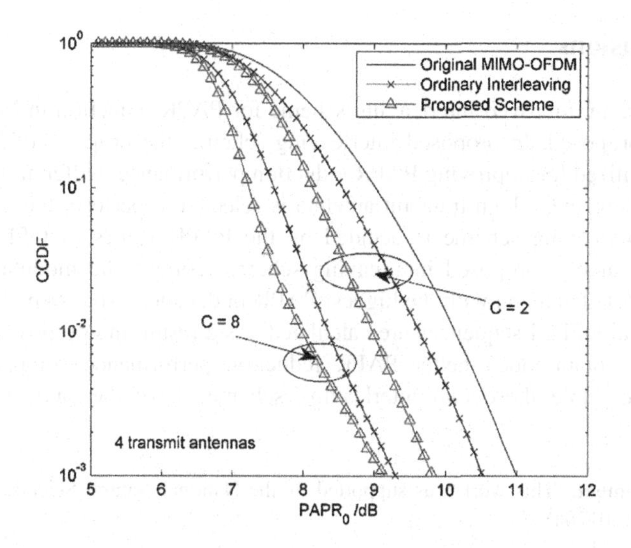

Fig. 5. The CCDFs of proposed interleaving scheme and the ordinary one employing QPSK in MIMO-OFDM with 4 transmit antennas

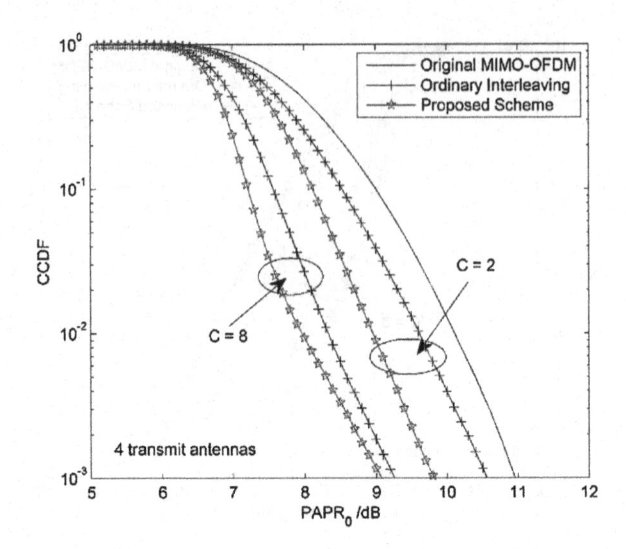

Fig. 6. The CCDFs of proposed interleaving scheme and the ordinary one employing 16QAM in MIMO-OFDM with 4 transmit antennas

5 Conclusions

In this paper, an improved interleaving scheme for PAPR reduction in MIMO-OFDM systems is proposed. In proposed interleaving scheme, the property of MIMO transmission is utilized for improving PAPR reduction performance. Different from ordinary interleaving scheme, which transmit antenna is selected to perform the interleaving in proposed interleaving scheme is decided by the PAPR values of OFDM candidate sequences. Namely, proposed interleaving scheme reorders the modulated sequence from the selected antenna with the highest PAPR in the successive step after the PAPR of the original OFDM sequences are calculated. As a result, the proposed interleaving scheme can obtain much better PAPR reduction performance compared with the ordinary one while these two interleaving schemes have the same computational complexity.

Acknowledgement. This work was supported by the National Natural Science Foundation of China (No. 61501204).

References

1. Prasad, R.: OFDM for Wireless Communications Systems. Artech House Publishers, Boston (2004)
2. Bingham, J.A.C.: Multicarrier modulation for data transmission: an idea whose time has come. IEEE Commun. Mag. **28**(5), 5–14 (1990)
3. Yang, H.: A road to future broadband wireless access: MIMO-OFDM-based air interface. IEEE Commun. Mag. **43**(1), 53–60 (2005)

4. Jiang, T., Wu, Y.: An overview: peak-to-average power ratio reduction techniques for OFDM signals. IEEE Trans. Broadcast. **54**(2), 257–268 (2008)
5. Han, S.H., Lee, J.H.: An overview of peak-to-average power ratio reduction for multicarrier transmission. IEEE Wirel. Commun. **12**(2), 56–65 (2005)
6. Jayalath, A.D.S., Athaudage, C.R.N.: On the PAR reduction of OFDM signals using multiple signal representation. IEEE Commun. Lett. **7**(8), 425–427 (2004)
7. Jayalath, A.D.S., Tellambura, C.: The use of interleaving to reduce the peak-to-average power ratio of an OFDM signal. In: IEEE Global Telecommunications Conference (GLOBECOM), vol. 1, pp. 82–86, San Francisco, USA (2000)
8. Ryu, H.-G., Kim, S.-K., Ryu, S.-B.: Interleaving method without side information for the PAPR reduction of OFDM system. In: International Symposium on Communications and Information Technologies, Sydney, Australia, vol. 1, pp. 72–76 (2007)
9. Wang, L., Yang, X.: Improved interleaving PAPR reduction scheme of OFDM signals with BPSK inputs. In: International Conference on Information Technology and Applications, Xi'an, China, vol. 1, pp. 263–268 (2014)
10. Malathi, P., Vanathi, P.T.: Improved interleaving technique for PAPR reduction in OFDM-MIMO system. In: Second Asia International Conference on Modeling and Simulation, Kuala Lumpur, Malaysia, vol. 1, pp. 253–258 (2008)
11. Bäuml, R.W., Fisher, R.F.H., Huber, J.B.: Reducing the peak-to-average power ratio of multicarrier modulation by selected mapping. IET Electron. Lett. **32**(22), 2056–2057 (1996)
12. Park, J., Hong, E., Har, D.S.: Low complexity data decoding for SLM-based OFDM systems without side information. IEEE Commun. Lett. **15**(6), 611–613 (2011)
13. Fischer, R.F.H., Hoch, M.: Directed selected mapping for peak-to-average power ratio reduction in MIMO OFDM. IET Electron. Lett. **42**(22), 1289–1290 (2006)
14. Hassan, E.S., El-Khamy, S.E., Dessouky, M.I., El-Dolil, S.A., Abd El-Samie, F.E.: Peak-to-average power ratio reduction in space-time block coded multi-input multi-output orthogonal frequency division multiplexing systems using a small overhead selective mapping scheme. IET Commun. **3**(10), 1667–1674 (2009)
15. Müller, S.H., Huber, J.B.: OFDM with reduced peak-to-average power ratio by optimum combination of partial transmit sequences. IET Electron. Lett. **33**(5), 368–369 (1997)
16. Hou, H., Ge, J., Li, J.: Peak-to-average power ratio reduction of OFDM signals using PTS scheme with low computational complexity. IEEE Trans. Broadcast. **57**(1), 143–148 (2011)
17. Lim, D.-W., Heo, S.-J., No, J.-S.: A new PTS OFDM scheme with low complexity for PAPR reduction. IEEE Trans. Broadcast. **52**(1), 77–82 (2006)
18. Yang, L., Soo, K.K., Li, S.Q., Siu, Y.M.: PAPR reduction using low complexity PTS to construct of OFDM signals without side information. IEEE Trans. Broadcast. **57**(2), 284–290 (2011)
19. Jiang, T., Guizani, M., Chen, H.-H., Xiang, W., Wu, Y.: Derivation of PAPR distribution for OFDM wireless systems based on extreme value theory. IEEE Trans. Wirel. Commun. **7**(4), 1298–1305 (2008)
20. Tellambura, C.: Computation of the continuous-time PAR of an OFDM signal with BPSK subcarriers. IEEE Commun. Lett. **5**(5), 185–187 (2001)

Failures Handling Strategies of Web Services Composition Base on Petri Nets

Guan Wang and Bin Yang(✉)

School of Information Science and Engineering,
Zaozhuang University, Zaozhuang, China
batsi@126.com

Abstract. Web services are distributed components that provide functionality applications through network. Web service-based application systems mostly adopt dynamic service composition strategies. However, the complexity of the web service composition execution engine determines the uncertainty of its application system state. It is particularly important to recover from a fault state in a timely manner after discovering a system failure. This paper, by extending the traditional petri nets, proposed the concept of dynamic petri nets, defined the concept of similar atomic services, solved dynamic replacement of transition by using similarity atomic service, and given fault treatment policy based on dynamic petri nets in atomic service and subnet level. Finally, use specific examples of Web services application system to verify the effectiveness and feasibility of the proposed method.

Keywords: Petri nets · Failures handling · Web service · Services composition

1 Introduction

Web services are now widely used as a distributed computing technology to uniformly package information, behavior and business processes, without the need to consider the application resides the environment It is also because of these excellent features, Web service technology has become the preferred way to implement SOA architecture. As Web service technology has become more sophisticated, more stable, and easier to use, Web services have been widely shared on the network. However, a single service can provide limited functionality, in order to meet the actual needs of the business, it proposed the concept of service composition. Service composition process is through mutual communication and collaboration between basic services, the relatively independent and simple service with new features combined into large-grained services to meet the service request or needs to produce value-added services [1].

Web services-based application systems mostly use the services of a dynamic portfolio strategy [2]. Dynamic service portfolio strategy is a method of service composition just in system run time. It will real-time combine some Web services based on user demand, and timely determine the quality of the combination. The system uses a combination of dynamic service strategies is that the system often provided complex and dynamic services. Such systems require real time to select some

D.-S. Huang et al. (Eds.): ICIC 2018, LNAI 10956, pp. 608–617, 2018.
https://doi.org/10.1007/978-3-319-95957-3_63

services from a number of Web services based on functional requirements and to combine them in to a complex service [3, 4].

Stability Web service composition execution engine determines the availability of their applications and the user friendliness. However, the complexity of Web services composition execution engine determines its state of uncertainty in the application system [5]. Therefore, the need for effective external monitoring mechanism to monitor the operational status of the entire system can always find the accident point lead to system failure. After the application of a system fails, the system will not produce the desired output responding in line with the outside world which is given for input, and the user's needs cannot be met. At this point, the user though knows system's failure, they face the system which still cannot be used. Therefore, after the discovery of a system's failure, it is especially important to recover from a failure state in time.

With Web service execution engine research of academia, a variety of methods have been proposed to build composite service execution engine [6]. For example, a method based on a combination of artificial intelligence planning [7], workflow-based Web service composition combination of automation technology [8] and automated service composition method based on petri nets [9]. Among them, the automatic service composition method based on petri nets due to the use of petri nets themselves with asynchronous and concurrent characteristics, gradually gained widespread attention.

Petri nets based service composition execution engine [10], introduced petri nets for describing service composition relationship, executed by the execution engine after the final planned. That simplifying the complexity of the system, making it feasible to real-time monitoring and restore the system to normal operation outside the service execution engine.

To solve the fault problem of Petri nets based application system, this paper presents the concept of dynamic petri nets by simple extended and re-defined the traditional petri nets, and base on it, this paper also presents treatment strategies base on petri nets for failure during atomic Web services composition and execution. We make transitions of petri nets and atomic Web services correspondence, and on this basis, propose the concept of similar atomic services and use this concept to do the transition replacement. Subsequently, his paper presents the way how to replace subnet of the Petri to recovery reachability of the petri nets under the circumstances replaceable atomic service cannot be found.

The remainder of this paper is organized as follows, Sect. 2 briefly introduces the basic knowledge related; Sect. 3 discusses the definition of dynamic variable petri nets, application system base on it, its application in fault recovery petri nets, and through a concrete example for the method proposed by this article is explained; Finally, in Sect. 4 of this paper, summarizes the work and noted that the work to be done next.

2 Basic Concepts

Definition 1. A net is a triple $N = (S, T; F)$

- $S \cup T \neq \emptyset$
- $S \cap T = \emptyset$
- $F \subseteq (S \times T) \cup (T \times S)$
- $dom(F) \cup cod(F) = S \cup T$

  where:
  S is a finite set of places
  T is a finite set of transitions
  F is a multiset of arcs
  The flow relation is the set of arcs:

$$dom(F) = \{x \in S \cup T \mid \exists y \in S \cup T : (x, y) \in F\} \tag{1}$$

$$cod(F) = \{x \in S \cup T \mid \exists y \in S \cup T : (y, x) \in F\} \tag{2}$$

Definition 2. A petri nets is a 4-tuple $\Sigma(S, T; F, M)$, where:

- $S, T; F$ is a Petri net graph;
- M is the initial marking, a marking of the Petri net graph.

Definition 3. Execution semantics

- firing a transition t in a marking M consumes W(s,t) tokens from each of its input places s, and produces W(t,s) tokens in each of its output places s.
- a transition is enabled (it may fire) in M if there are enough tokens in its input places for the consumptions to be possible.

$$M'(s) = \begin{cases} M(s) - 1, s \in {}^{\cdot}t - t^{\cdot} \\ M(s) + 1, s \in t^{\cdot} - {}^{\cdot}t \\ M(s), Other \end{cases} \tag{3}$$

3 Failures Handling Strategies of Web Services Composition Base on Petri-Nets

3.1 Definitions

Definition 4. (Atomic Service) On the basis of the Definition 1, transitions in the Petri net (t) and places, directly adjacent to the transition, called atomic services, denoted by H,

$$H = \{s, t, f \mid s \in S, t \in T, f \in \langle s, t \rangle\} \tag{4}$$

Definition 5. (Semantic of service) Let D as the description of domain ontology for atomic service H [12], Service parameter vector ($\vec{p}$) is consisted of input parameters and output parameters, $\vec{p} = \{i_1, i_2, \ldots, i_n, o_{n+1}, o_{n+2}, \ldots, o_m\}$. Where $i_1, i_2, \ldots, i_n$ represents the input parameters for a service, $o_{n+1}, o_{n+2}, \ldots, o_m$ represents the output parameters for a service.$func(H) = D$ and $\vec{p}$ is the semantic of atomic service H.

Definition 6. (Similar atomic services) Let A, B as atomic service defined by Definition 4. If A and B are mutually equivalent atomic services, the necessary and sufficient conditions are:

- $\forall x \in A, x \in B; \forall y \in B, y \in A$
- $func(A) = func(B)$.

Where $func(A)$ represent the semantic of service A, $func(B)$ represent the semantic of service B.

In this paper, the semantic of service is extended by the input and out parameters of an atomic service and is extended by its domain ontology.

Classical petri nets (Definition 1) is static. Application systems based on petri nets, in order to meet the individual needs of different users in real-time, always use a dynamic composition strategy. But after dynamic composition, the new petri nets is still defined by the classic definition of a petri nets, which is static. Therefore, when one transition of petri nets is abnormal excitation, in order to avoid errors, continue to spread along the petri nets, only to terminate and discard the entire petri nets, and then rebuild the petri nets.

In real applications, invoke the atomic service which is a transition of petri nets comes at a price, terminate and discard the entire petri nets often leads to unacceptable. Therefore, we must introduce a dynamic mechanism for the classical petri nets. Increase the dynamic nature of petri nets is to solve the problem of replacing equivalent atomic services, so the new definition of petri nets not only meets the classical definition, but also should meet the Definition 6 in this paper. The new dynamical petri nets are defined below (Definition 7).

Definition 7. Base on the Definition 1, $\mu PN = (S, T; F), T \subset \mu PN, \forall t \in H(T), \exists t'$ and $H(t') = H(t)$, we named μPN as dynamical petri nets. Where H is an atomic service defined in Definition 4.

3.2 Failures Handling Strategies

Application systems base on petri nets, the reasons for service failure are a variety, but the most common service failures can be attributed to failures of execution which is defined in Definition 8 below. Therefore, petri nets fault in this paper to be solved is atomic service execution failure.

Definition 8. (Failures of atomic service execution) Firability conditions of transition T are met and there are no tokens arrive in S_2 in specified time. This is called failures of atomic service execution. Where A is Atomic service, T is the transition of A of the petri nets, S_1 is the input place of the petri nets, S_2 is output place of the petri nets.

Failures Handling in Atomic Service Level. From the micro execution standpoint, base on the existing computer architecture, either concurrent formalized description on petri nets, or execution of concurrent workflow engines for Web services are ultimately used operating system thread (process) to simulate. As in the modern operating system architecture, the thread (process) is the smallest unit system scheduling. Thus, for each single thread, Web services are called serial atoms.

While allowing the presence of complex services, a combination of multiple atomic services into a single function service with a specific function, within executable petri nets, but the composite service will be divided into some atomic services after the executable petri nets transfer to executable sequence. Thus, complex services do not exist in the workflow of the execution engine practically, the fault of complex services can be described by the fault of atomic services.

The key of atomic service replacement algorithm is to find a set of services from a large number of atoms that have the same (or similar) function and satisfy the input and output of failure subnet of existing executable petri nets.

Steps of atomic service replacement algorithm base on petri nets are described in the following:

STEP 1: According to breakpoint of service execution engine, find out the corresponding Petri net transition 't'.

STEP 2: Reverse mapping transition 't' to the atomic service.

STEP 3: By Definition 5, calculated functional semantics for each atomic service, and calculate the Euclidean distance between the semantics of each service function. If the Euclidean distance is less than the threshold value θ, considered two atomic as semantically similar services.

STEP 4: According to Definition 6, combined with the semantic, lookup service functions equivalent atoms in atomic service database to obtain an equivalent set of atomic services, $A_{replace} = \{S_1, S_2, \ldots\}$.

STEP 5: Use QoS calculation method, we proposed in the literature [13], to calculate QoS of each atom service and based on it, select the best equivalent atomic service (A') from the atomic equivalent service set ($A_{replace}$).

STEP 6: The best equivalent atomic service (A') will be returned to service execution engine.

When the atomic service database of Web services is smaller or the set of similar Web services is small, equivalent atomic services set high probability is empty. At this point, failures handling need to be done in subnet level.

Failures Handling in Subnet Level. When the atomic service replacement policy failures (i.e., unable to find enough atoms these are able to satisfy both functions of input parameters and output parameters of the failure subnet of existing executable petri nets from function equivalent service set), we need to use failure subnet replacement algorithm, describe in the following, to replace the whole subsequent subnet affected by the failure subnet.

Definition 9. (Failure subnet) For petri nets *PN*, transition t_{start} is fired unexpected, and its mapped atomic service is fault which is execution fault that met the Definition 8, call

t_{start} is the start transition of fault. Assume t_{start} normal excitation, fire the following transition from t_{start} along the petri nets, all transitions in this path are called 'affected transitions'. Transitions these not have followed transitions in the set of 'affected transitions' are called 'terminal transitions', noted as T_{end}.All 'affected transitions' and there's relevant petri nets are called 'failure subnet'.

Steps of failure subnet replacement algorithm base on petri nets are described in the following:

STEP 1: Extract failure subnet (PN') which is met Definition 9 from the petri nets (PN) belong to the service execution engine. All transitions in petri nets (PN) are noted as T_{PN}.

STEP 2: Determine the function ($func_1$) of atomic service which is mapped from t_{start}, and the functions (S_{func2}) of atomic services which are mapped from T_{end}. By Dreggie service discovery method [14], search a service(e) to adapt $func_1$ of t_{start} and search services (T'_{end}) to adapt S_{func2} of T_{end} in the service database. t'_{start} and T'_{end} constitutes a new replacement set of services ($S_{replace} = \{t'_{start}, T'_{end}\}$).

STEP 3: Processing data association among all atomic services belongs to $S_{replace} \cup T_{PN}$.

STEP 4: Using the method we proposed in literature 9, fusion all sets of atomic service which are met the function of needed and met the constraint of data association with original executable petri nets (PN), to get a refreshed petri nets (PN_{new}).

STEP 5: Planning the executable path for PN_{new}, generating executable sequence.

STEP 6: Restart the service execution engine.

3.3 Examples

In the previous section, we proposed a troubleshooting algorithm which is base on dynamic petri nets. In this section, in order to verify the effectiveness and feasibility of the proposed method, we will use it on optimizing general hospital in China process.

Web services for general hospital in China process includes new patient first appointment, acquiring patient ID card, registration, patient appointment, queueing system, diagnosis, drawing medicine, medical examination and so on. All fifteen sub-services and their simple petri nets combine in to a complex petri nets (see Fig. 1).

Transitions which are described in Fig. 1 are descripted in Table 1.

When the service execution engine running the petri nets as shown in Fig. 1, if the transition $T2$ is fault, we can use the troubleshooting algorithm which is based on dynamic petri nets in atomic service level to recover the execution engine. In details, we will find $T2'$ which is described as 'appointment by human' and which meets the Definition 6. So, use $T2'$ instead of $T2$ will not affect the structure and function of the original petri nets.

In the process of actual patients, when emergency patient arrived at hospital, but hospital does not have the ability to cure, in this case, the hospital often decides to transfer the patient to a higher-level hospital in time.

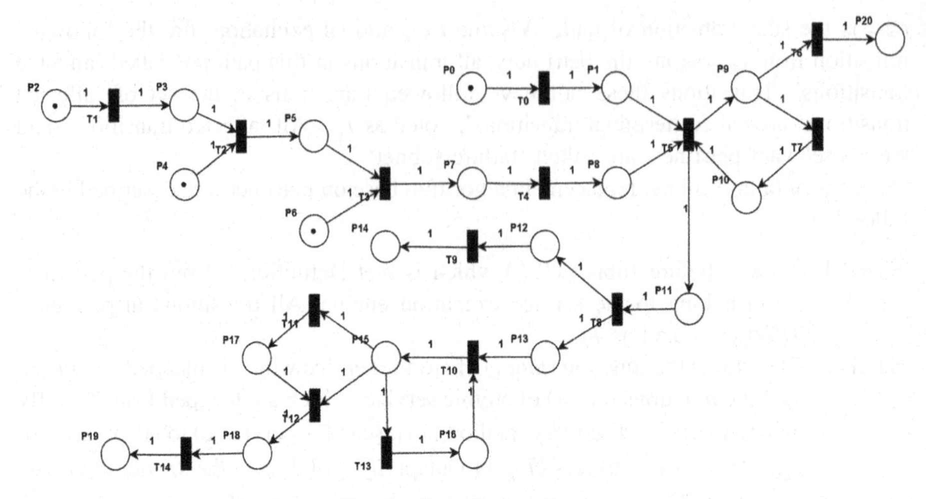

Fig. 1. Complex petri nets

Table 1. Descriptions of transitions

Transitions	Meaning
T0	Prehospital emergency care
T1	Acquiring Patient ID Card
T2	Registered
T3	Appointment
T4	Queueing
T5	Diagnosis
T6	Drawing medicine
T7	Medical examination
T8	Patient admission
T9	Medicare archives
T10	Ward Registration
T11	Inner(Outer) hospital consultation
T12	Hospitalization treatment(examination)
T13	Patient transfer
T14	Leaving hospital

This is mapped to the situation of the service execution engine running the petri nets, which is shown in Fig. 1, encounter transition T5 fault, and can not be replaced by executing the algorithm finding fault atomic service which meets the Definition 6 in the atomic service database. In this case, broader process is needed for failure recovery. We will introduce the conception of failure recovery in subnet of Petri-Net level.

Using the proposed subnet replacement algorithm which is based on dynamic petri nets to determine the fault subnet of the original petri nets, we can sure that transitions which are affected by T5 are T6 and T7. Base on that, some suitable transitions, $T_{replace} = \{T0', T1', T2'\}$, which could build the replaceable subnet, can be found in the Web service database. The replaceable subnet can be used to replace the faulty subnet of the original petri nets, and then we will get a refreshed petri nets (PN') as shown in Fig. 2. Where $T0'$, $T1'$, $T2'$ are urgent care central deployment, feedback from receiving hospital, emergency admitted.

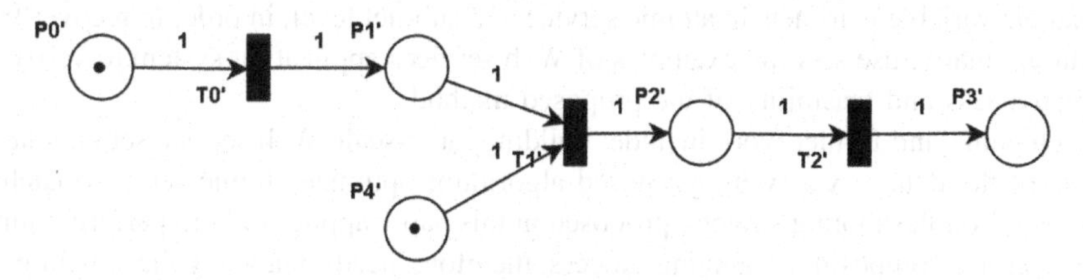

Fig. 2. Replaceable subnet $T_{replace}$

The new executable Petri-Net (PN_{new}), as shown in Fig. 3, will be made by fusion origin petri nets with the replaceable subnet (PN').

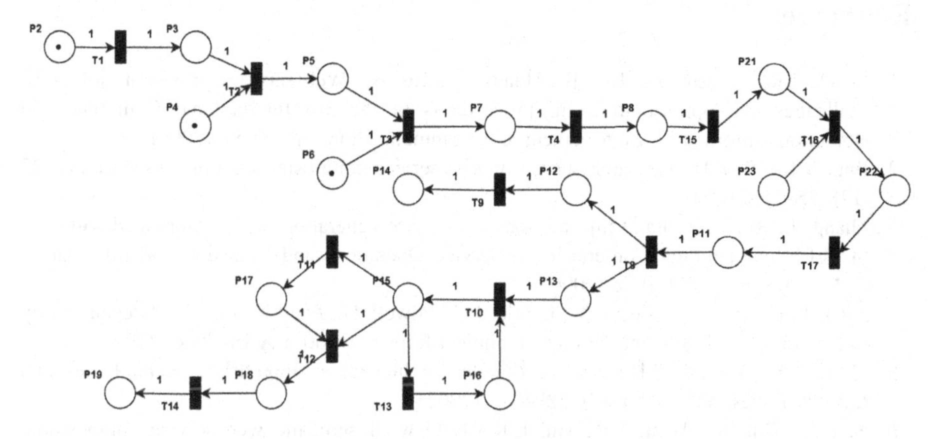

Fig. 3. The new executable petri nets, PN_{new}

The practical significance of the recovery process above is when a patient in the treatment process due to the complexity of the disease need to be referred to a higher-level hospital. Through the deployment of the emergency center, doctor will get feedback from the receiving hospital, and then determine whether or not to transfer the patient to a higher-level hospital.

Above example shows that the refreshed petri nets (PN_{new}) not only can blend with the structure of the original petri nets, and its logic function expression is also consistent with the logic functions of the original petri nets.

Through the above examples, the use of performing failure recovery algorithm proposed in Sect. 3.2 based on dynamic petri nets which is proposed in Sect. 3.1, the refreshed petri nets (PN_{new}) not only recover from the fault, make the service execution engine to run properly, and its logic functions can also be consistent with the original petri nets.

4 Conclusion and Further Work

This paper presents the concept of dynamic variable petri nets, based on this concept are given the definition of similar atomic service and the use of similar atomic service solutions in transition, belong to a petri nets, replacement. Treatment is given based on dynamic variable petri nets in atomic service and sub-net level, in order to recover from failure. Finally, use specific examples of Web services application system to verify the effectiveness and feasibility of the proposed method.

Ongoing and further work include: building large-scale Web service set on a larger scale of the data to verify the proposed algorithm; optimize atomic service matching process; troubleshooting strategy proposed in this paper applies only to perform fault in Web service composition planning process, therefore, needs analyzing the structure and dynamic nature of Web service composition based on petri nets from the perspective of the actual Web service call and service execution, study on issues from the planning process of Web service composition.

References

1. Li, Q., Liu, A., Liu, H., Lin, B., Huang, L., Gu, N.: Web services provision: solutions, challenges and opportunities. In: Proceedings of the 3rd International Conference on Ubiquitous Information Management and Communication, pp. 80–87 (2009)
2. Jing, X.L., Zi, F.H.: Research survey of web service composition. Appl. Res. Comput. **22** (12), 25–31 (2005)
3. Zhang, J., Robin Q.: Fault injection-based test case generation for SOA-oriented software. In: IEEE International Conference on Service Operations and Logistics, and Informatics, SOLI 2006, pp. 1070–1078 (2006)
4. Arlat, J., Costes, A., Crouzet, Y., Laprie, J.C., Powell, D.: Fault injection and dependability evaluation of fault-tolerant systems. Technical Report 91260, LAAS-CNRS (1991)
5. Mike, P.P., Willem, J.H.: Service oriented architectures: approaches, technologies and research issues. VLDB J **16**(3), 389–415 (2007)
6. Hua, C., Shi, Y., Wen, J.Y., Hu, L.K.: Review of semantic web service composition. Comput. Sci. **37**(5), 256–267 (2010)
7. Sheila, M., Tran, C.S.: Adapting golog for composition of semantic web services. In: Proceedings of the 8th International Conference on Knowledge Representation and Reasoning. (KR 2002), pp. 482–493 (2002)
8. Wu, G.F., Wu, N.X.: Dynamic composition of Web service workflow based on semantic description. J. Comput. Appl. **27**(11), 137–145 (2007)
9. Xiang, D.M., Ma, B.X., Zhang, Z.M.: Automatic sharing synthesis of petri nets based on semantics. J. Syst. Simul. **24**(11), 115–123 (2012)
10. Ma, B.X., Xiang, D.M., Zhang, Z.M.: Automatic generation of petri net for web services composition. J. Chin. Comput. Syst. **34**(2), 332–337 (2013)
11. Petri Net. http://en.wikipedia.org/wiki/Petri_net. Accessed 20 Jan 2018
12. Thomas, R., Gruber, A.: Translation approach to portable ontology specifications. Knowl. Acquis. **5**(2), 199–220 (1993)
13. Wang, G., Ma, B.X., Xiang, D.M.: Study on dynamic calculation method for web service QoS base on grouped function. Comput. Technol. Dev. **2013**(1), 97–105 (2013)

14. Chakraborty, D., Perich, F., Avancha, S., Joshi, A.D.: Semantic service discovery for m-commerce applications. In: Proceedings of the 20th Symposium on Reliable Distributed Systems, Workshop on Reliable and Secure Applications in Mobile Environment, pp. 25–31 (2001)
15. Du, Y., Tan, W., Zhou, M.C.: Timed compatibility analysis of web service composition: a modular approach based on petri nets. IEEE Trans. Autom. Sci. Eng. **11**(2), 594–606 (2014)
16. Cheng, J., Liu, C., Zhou, M.C.: Automatic composition of semantic web services based on fuzzy predicate petri nets. IEEE Trans. Autom. Sci. Eng. **12**(2), 680–689 (2015)
17. Du, Y.Y., Gai, J.J., Zhou, M.C.: A web service substitution method based on service cluster nets. Enterp. Inf. Syst. **11**(10), 1535–1551 (2017)
18. Chen, L., Fan, G., Zhang, H.: Petri nets-based method to model and analyze the self-healing web service composition. Int. J. High Perform. Comput. Networking **9**(1–2), 8–18 (2016)

Entity Detection for Information Retrieval in Video Streams

Sanghee Lee and Kanghyun Jo(✉)

School of Electrical Engineering, University of Ulsan, Ulsan, Korea
shlee@islab.ulsan.ac.kr, acejo@ulsan.ac.kr

Abstract. The growing amount of video data has raised the need for automatic semantic information indexing and retrieval systems. To accomplish to these needs, the text information in images and videos is proved to be an important source of high-level semantics. This paper discusses the video OCR system designed for overlay text based automatic indexing and retrieval in the video streams. The proposed framework consists of the video segmentation, the video key-frame extraction, the video text recognition, and the entity detection. The experimental results on Korean television news programs show that the proposed method efficiently realizes the automatic indexing in the video streams.

Keywords: Overlay text · Video OCR · Indexing · Named entity NLP

1 Introduction

The advances in the data capturing, storage, and communication technologies have made vast amounts of video data available to consumer and enterprise applications. It is an important task developing effective methods to manage these multimedia resources by their content. One of the goals of the multimedia indexing community is to design a system able of producing a rich semantic description in video sequences. Among the semantic features such as image, audio, and textual information, the text embedded in images is of particular interests. First, it is very useful for describing the contents of an image and video. And it enables applications such as keyword-based search, automatic video logging, and text-based indexing [1, 2].

Text in images and videos contains useful information that can help a machine to understand the content. There exist mainly two kinds of text in videos. One is the overlay text and the other is the scene text. The scene text naturally exists in the image being recorded in native environment. This text is found in the street signs, text on cars, writing on shirts in natural scenes, and so on. The appearance of the scene text is occasional, and this text usually brings less related to video information. And the difference among the different scene text is very big. On the other hand, the overlay text is called graphics text or caption in other papers. The overlay text is graphically generated and artificially overlaid on the image by human at the time of editing. This text is used to describe the content of the video or give additional information related to it. The examples of the overlay text include the subtitles in news videos, sports scores.

D.-S. Huang et al. (Eds.): ICIC 2018, LNAI 10956, pp. 618–627, 2018.
https://doi.org/10.1007/978-3-319-95957-3_64

Recognizing overlay text embedded in videos provides high level sematic clues which enhance tremendously an automatic image and video indexing. Because the overlay text contains more concise and direct description of the content of the video. Therefore, the overlay text makes it important role for the automated content analysis systems such as the scene understanding, indexing, browsing, and retrieval [3, 4].

This paper discusses the video OCR system designed for overlay text based automatic indexing and retrieval in the video streams. The currently index table generation for retrieval is done manually by person. An important human cost is induced by making the index table. And it is difficult to maintain the consistency of human subjective thoughts. Therefore, the need of automatic index table creation has raised. For this goal, this paper presents the method using the overlay text in video content. Because the overlay texts contained in video represent rich and reliable information.

The proposed framework consists of the video segmentation, the video key-frame extraction, the video text recognition, and the entity detection as shown in Fig. 1. In the Sect. 2, the proposed systems are explained in detail. Section 3 presents the experimental results, and conclusion are drawn in Sect. 4.

2 Proposed Methodology

The proposed framework consists of the video segmentation, the video key-frame extraction, the video text recognition, and the entity detection as shown in Fig. 1. The video segmentation step is used for separating a video stream into a set of the clip with the same overlay text. In the video key-frame extraction, four key frames are selected in every sub clip. And then, the video text recognition step produces text strings. The entity detection is cascaded to build automatically indexing table.

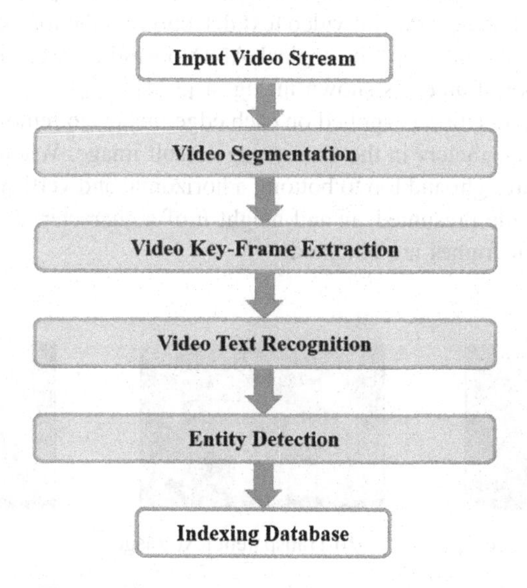

Fig. 1. The workflow of the proposed system

2.1 Video Segmentation

By observing a large quantity of the TV news programs, the appearance and disappearance of the overlay text occur suddenly or slowly in most the news video as shown in Fig. 2. By precisely locating the critical frame where each overlay text appears or disappears, the sub clips which contains the same overlay text frame are made in the input video streams.

This paper uses the identification of the overlay text beginning frame by the edge density based on Canny edge detector [3, 4]. The Fig. 3 shows the workflow for separating a video stream into sub clip which contains the same overlay text.

2.2 Video Key-Frame Extraction

Since the size and fonts of the overlay text from the same news video generally remain unchanged for a long term, this paper uses the temporal analysis of the news videos to achieve a good accuracy of video text detection. In every sub clip, the four key frames are selected, respectively [3, 4].

If the videos are played f frames per second, the overlay text stays in a fixed location for at least $2f$ consecutive frames. Let k be the nearest integer that is not less than f. This paper defines every consecutive k frames to be one round. To simplify the calculation, about only the 1st round, the four key frames are selected on frame 1, $k/3$, $2k/3$, $3k/3$. Because, the same overlay text is fixed in the same position for every consecutive k frames.

2.3 Video Text Recognition

To process accurate Optical Character Recognition (OCR), it requires a good detection of the text regions in image. To do this, this paper uses the temporal analysis of the sub clips to achieve a good accuracy of video text detection. In the four key frames of every clip selected the previous step, the logical AND operation executed on Canny edge maps of the four key frames as shown in Fig. 4 [3, 4].

The simple line deletion is applied on each edge images to remove lone lines which are unlikely to be characters in the Canny edge result image. When the edge image is scanned from left to right and top to bottom, a horizontal and vertical line is removed if its length exceeds the presumed w and height h of a character. As a result, the edge map images of four frames are obtained.

(a) Non text image

(b) Transparent text image

(c) Text image

Fig. 2. Example of overlay text appearance in the TV news video streams

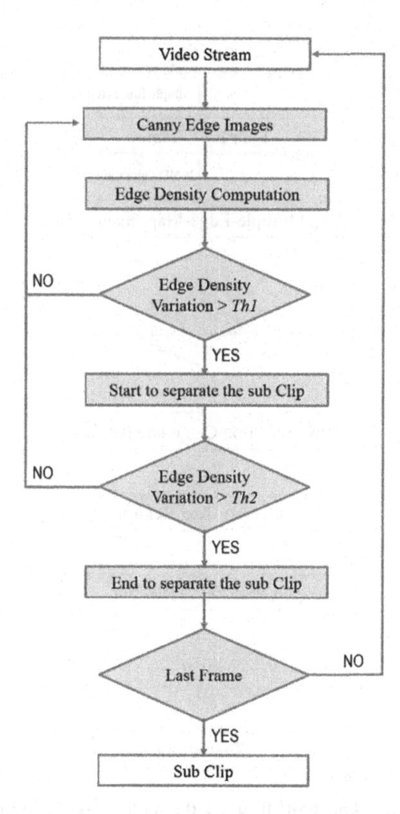

Fig. 3. The workflow of the video segmentation

Next, the logical AND operation is executed on the edge map images of the four key frames. The result image is called the Multiple-Edge-Map. Since a position (*i, j*) becomes an edge pixel if all four edge images are edge at (*i, j*), most of the background edge pixels are removed, whereas the static overlay texts are remained. The result well explains that the problem which the difficulty to distinguish whether the detected edges are really from the overlay texts is alleviated by multiple frame integration method. Because the same overlay text appears in the same location for many successive frames, while the location of background edge pixels may differ in a few pixels.

And then the overlay text candidate region is detected by utilizing the number of the black and white transition. As shown in Eq. 1, the value of N_{trans} can be obtained that a window of the presumed character size $w \times h$ slides from left to right and top to bottom on the Multiple-Edge-Map image. Where w and h are the width and height of window, and $b(*)$ is binary image. If N_{trans} is larger than threshold T_{trans} this window is masked. The union of all masked windows is the overlay text candidate region. The threshold T_{trans} depends on the character size and is obtained by $T_{trans} = \beta(w \times h)$ with β a constant which is empirically measured.

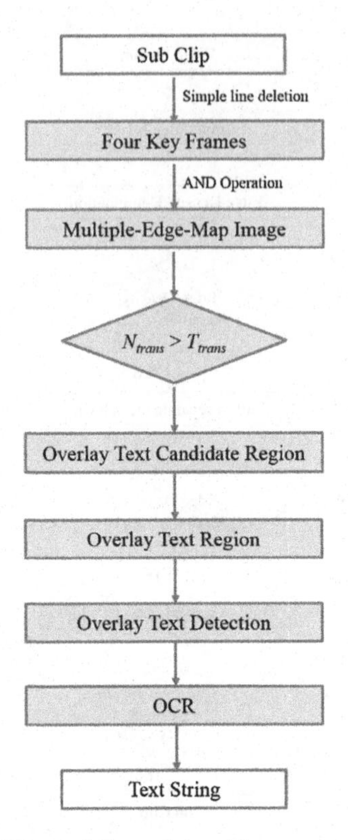

Fig. 4. The workflow of the video text recognition

$$N_{trans} = \sum_{i=0}^{h-1}\left(\sum_{j=1}^{w-1}|b(i,j) - b(i,j-1)|\right) + \sum_{j=0}^{w-1}\left(\sum_{i=1}^{h-1}|b(i,j) - b(i,-1,j)|\right) \quad (1)$$

To separate the text lines, the horizontal projection histogram must be obtained by applying on the Multiple-Edge-Map image. To scan the result image from top to bottom, and count the number of edge in a row by Eq. (2) can be gotten the horizontal projection histogram image.

$$H_{hor}(i) = \sum_{j=0}^{w-1} b(i,j), \quad i = 0, 1, \ldots, h-1 \quad (2)$$

The text line is defined as a consecutive horizontal projection histogram, if the histogram values are more than threshold. The start point and end point of ROI along the height (vertical) axis are applied to the overlay text region image. The result is the ROI text line mask image. At last, to apply on the one frame of four key frame images yields the overlay text line image.

The obtained overlay text line image becomes the text string by the optical character recognition (OCR) processing. First, the binarized image is achieved with a gray threshold value derived from Otsu's method [5]. And then this paper uses the commercial software ABBYY FineReader [6] for OCR.

2.4 Entity Detection

Named entity recognition (NER) is the information extraction task of identifying and classifying mentions of people, organizations, locations and other named entities (NES) within text. The text string obtained by the OCR has the important information such as person name, location, and organization of the video contents. This information extraction is helpful to organize the automatic indexing table. To extract the information, this paper uses the NLTK package [7] to perform the named entity recognition (NER) technology.

3 Experiments

The experimented videos were captured in TV news program in Korea. The video resolution was 720 × 480, and the frame rate was 29.97 frames per second. Each video clip may last more than 2 seconds, and the overlay text of these video clips included more than one overlay text in one frame.

The exemplary result of the proposed workflow is illustrated in Fig. 5. Figure 5(a) is the input stream. Figure 5(b) is the results of the video segmentation which have three sub clips. Figure 5(c) shows the results of the video key-frame extraction in each sub clips. By video text recognition processing proposed in this paper, Fig. 5(d) shows the text strings in each sub clips. And Fig. 5(e) is the results of the entity detection defined in this paper.

3.1 Video Segmentation

Figure 6 showed some results of the video segmentation by the overlay text beginning frame identification using Canny edge detector. The two thresholds $T_{high} = 150$ and $T_{low} = 50$ of Canny edge detector were decided on our empirical studies. Also, the reference value of abrupt difference among the frames was decided on 0.03 by the empirical studies [4].

Since the text presents many edges, the frame including the overlay text has signification changes in the edge density than the frame not containing the overlay text. As shown in Fig. 6(a) image, the period between the vertical red lines from the frame 16 to the frame 27 had sharply different from the previous frame and the frame 28 had little different of the edge density of the frame 29. The beginning frame was the frame 28. And the period between the vertical red lines from the frame 253 to the frame 261 had sharply different from the previous frame. The end frame was the frame 261. As a result, the input video stream was segmented from the frame 16 to 261, and the text period was from the frame 28 to frame 252. In case of the Fig. 6(b) image, the threshold 2 was not. As a result, beginning frame was the 23 frame, and the end frame

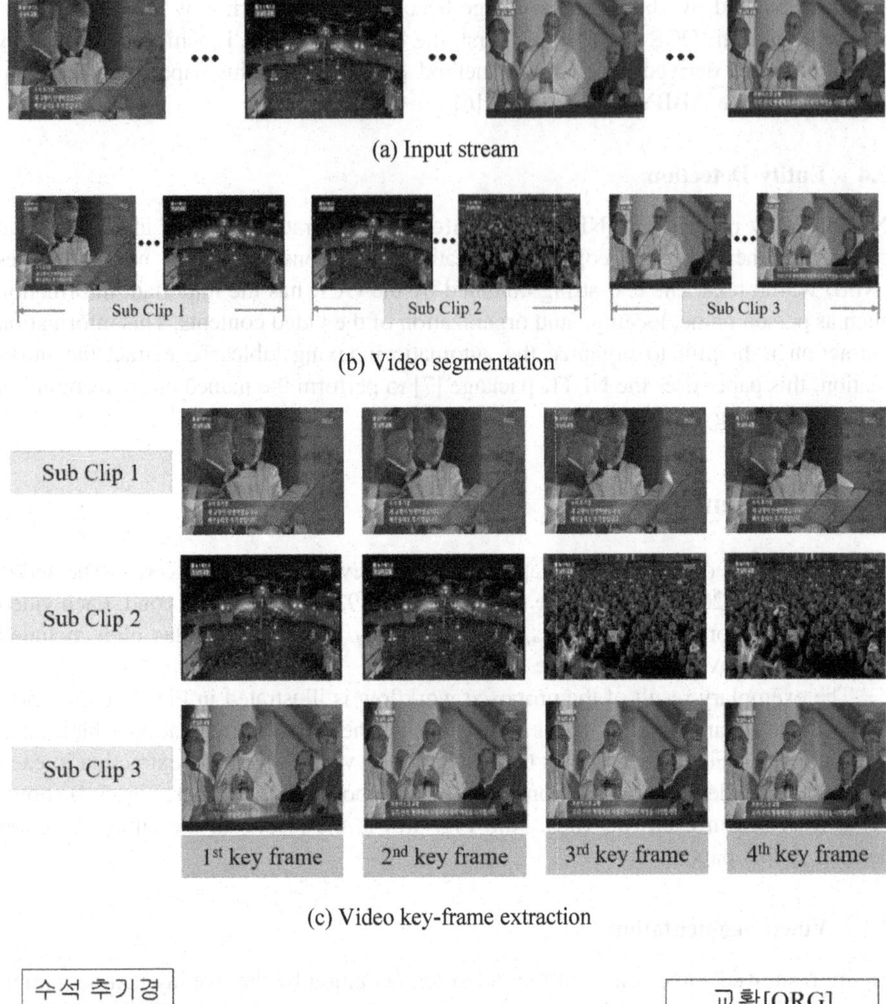

수석 추기경

새 교황이 탄생하셨습니다.

베르골리오 추기경입니다.

프란치스코 교황

우리안의 형제애와 사랑과 신뢰의 여정을 시작합시다.

(d) Video text recognition

교황[ORG]

추기경[ORG]

베르골리오[PER]

프란치스코[PER]

(e) Entity detection

Fig. 5. Exemplary results of the proposed workflow

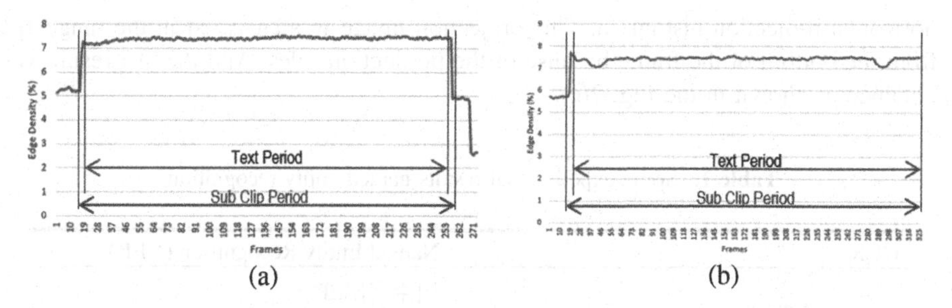

Fig. 6. The results of the video segmentation

was the 325. The input video stream was segmented from the frame 18 to frame 325, and the text period was from the frame 23 to the frame 325.

3.2 Video Text Recognition

The presumed character size $w \times h$ was 20×20 pixels. The threshold N_{trans} of the black and white transition was set to be 0.15 and the threshold of the horizontal projection histogram analysis was the presumed character width 20. Some results of the text detection showed the Fig. 7. Figure 7(a) was the beginning frame of the segmented sub clip, (b) was the Multiple-Edge-Map image, (c) was the overlay text region detected by utilizing the number of the black and white transition. (d) was the

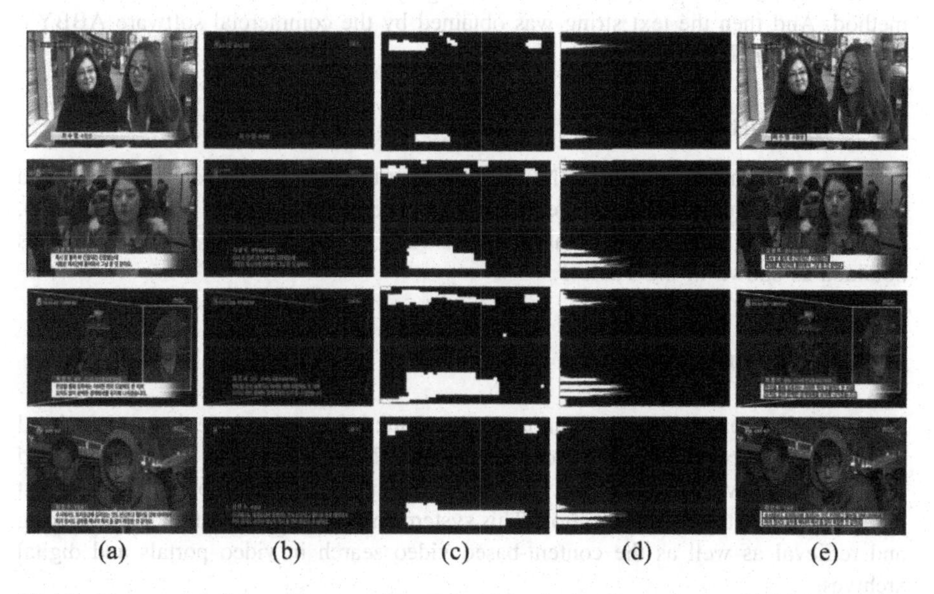

Fig. 7. The results of the video text detection. (a) beginning frame image (b) Multiple-Edge-Map image (c) overlay text region image (d) horizontal projection histogram image (e) overlay text detection image

horizontal projection histogram. This projection image was analyzed in the range 1/2 from the bottom of the frame because of the production rules. And the last result was obtained as shown in the Fig. 7(e).

Table 1. Some experimental results named entity recognition.

OCR	Named Entity Recognition (NER)
최수영 수험생	최수영[PER] 수험생[ORG]
고경주 열차 탑승 수험생	고경주[PER] 수험생[ORG]
최원석 중령 17사단 한강결사대 대대장	최원석[PER] 17사단[LOC] 대대장[ORG]
김민수 수험생	김민수[PER] 수험생[ORG]

The binarized image was achieved with a gray threshold value derived from Otsu's method. And then the text string was obtained by the commercial software ABBYY FineReader for OCR as shown in Table 1.

3.3 Entity Detection

For the named entity recognition (NER), this paper used the NLTK package to perform the NER technology. The named entity in this paper defined person name (tag: PER), organization (tag: ORG), and location (tag: LOC). The NER results of text string was obtained as shown in the Table 1.

4 Conclusion

The goal of the paper is that the automatic semantic information indexing and retrieval systems is developed based on overlay text in the TV program. The proposed method consists of the video segmentation, the video key-frame extraction, the video text recognition, and the entity detection. This system enables the automatic video indexing and retrieval as well as the content-based video search in video portals and digital archives.

In the future works, to make the more detailed indexing table, it will be necessary to improve the performance of the named entity recognition based on the convolutional neural networks. And it will be necessary to develop the method for processing the multi-language.

Acknowledgment. This research was supported by the MSIT (Ministry of Science and ICT), Korea, under the Grand Information Technology Research Center support program (IITP-2018-2016-0-00318) supervised by the IITP (Institute for Information & communications Technology Promotion).

References

1. Jung, C., Kim, J.: Player information extraction for semantic annotation in golf videos. IEEE Tran. Broadcast. **55**(1), 79–83 (2009)
2. Saidane, Z., Garcia, C.: An automatic method video character segmentation. In: International Conference Image Analysis and Recognition, pp. 557–566 (2014)
3. Lee, S., Ahn, J., Jo, K.: Comparison of text beginning frame detection methods in news video sequences. J. Broadcast. Eng. **21**(3), 307–318 (2016)
4. Lee, S., Ahn, J., Lee, Y., Jo, K.: Beginning frame and edge based name text localization in news interview videos. In: Huang, D.-S., Han, K., Hussain, A. (eds.) ICIC 2016. LNCS (LNAI), vol. 9773, pp. 583–594. Springer, Cham (2016). https://doi.org/10.1007/978-3-319-42297-8_54
5. Otus, N.: A thresholding selection method from gray level histogram. IEEE Trans. Syst. Man Cybern. **9**(1), 62–66 (1979)
6. ABBYY Cloud OCR SDK. ocrsdk.com
7. NLTK Language Toolkit. www.nltk.org

Integration of Data-Space and Statistics-Space Boundary-Based Test to Control the False Positive Rate

Jin-Xiong Lv and Shikui Tu(✉)

Department of Computer Science and Engineering, School of Electronic Information and Electrical Engineering, Shanghai Jiao Tong University, Shanghai 200240, China
{lvjinxiong, tushikui}@sjtu.edu.cn

Abstract. Many multivariate statistical methods have been applied to detect the difference between case and control population. However, it is difficult to control the false positive rate, especially under small sample size. Traditional family-wise error rate or false discovery rate adjusts the *p* values based on the distribution or ranks of *p* value in the same multiple testing. In this paper, we investigated the performance of integrating the Data-space boundary-based test (BBT) and Statistics-space BBT to control the false positive rate, under a previous proposed framework called Integrative Hypothesis Tests (IHT). The classification accuracy rate by Data-space BBT provides valuable information complementary to the *p* value from Statistics-space BBT. The simulation results demonstrated that the integration effectively controls the false positive rate even for small-sample-size cases. Experiments on the real-world dataset of bipolar disorder also validated the effectiveness of the integration.

Keywords: Integrative Hypothesis Test · Boundary-based test
False positive rate · Multivariate statistical method · Joint-SNVs analysis
Bipolar disorder

1 Introduction

Many statistical methods have been proposed in the literature to detect the difference between case and control population in the fields of social psychology, biology, and economics. One focuses on the difference between case and control population in many fields, such as social psychology, biology and economics. There are many statistical methods to detect the difference. Those methods can be divided into two groups, one for univariate and the other for multivariate methods which play a crucial role in genome-wide association study (GWAS). Recently, the multivariate methods are applied on GWAS for detection power improvement, in which single-nucleotide variants (SNVs) located in the same biological unit are collapsed into one computational unit [1–4]. However, all of them are suffering from false positive rate, especially with small sample size.

The traditional measures to control the false positive rate have two main streams. The first stream is named after family-wise error rate (FWER), which is defined as the

D.-S. Huang et al. (Eds.): ICIC 2018, LNAI 10956, pp. 628–638, 2018.
https://doi.org/10.1007/978-3-319-95957-3_65

probability of making one or more false rejections among all of the hypotheses, such as Bonferroni correction [5] and Holm correction [6]. The other stream aims to control the fraction of false rejection under the threshold α, including Benjamin-Hochberg correction [7] and Q value [8]. But they only change the scale of original p values and fix the threshold to reject less hypotheses without any other complementary information.

Recently, Integrative Hypothesis Tests (IHT) was previously proposed in [9, 10] to consider discriminating analysis and testing of case-control problems, jointly from two perspectives. One is model based such as two-sample test or model comparison to detect the difference between two populations, while the other is boundary based such as classification or model prediction about the performance of the distinguishing boundary. As preliminarily discussed in [11], the tasks of model comparison and classification were complementary to each other in nature, and it was better to jointly optimizing their performances. The advantage of IHT was demonstrated in [12] on a COPD-Lung cancer study, by combing p value (from a two-sample test) and misclassification rate in a 2D scatter plot, with a bootstrapped procedure to enhance the reliability of the ranks by the IHT. This motivated us to further investigate IHT, and following [10] Boundary-based test (BBT) was considered and empirically investigated in this paper.

Boundary-based test (BBT) is to test whether a separable plane is existed between the case population and control population [10]. It can be classified into two categories, Data-space BBT and Statistics-space BBT. Data-space BBT indicates that we seek for the separating plane in the original data space, which leads to meet traditional classification problem in the machine learning. While the Statistics-space BBT intends to ascertain the boundary between rejection region and acceptance region after calculating the statistics from the original data space. Furthermore, the Statistics-space BBT has achieved higher detection power than other multivariate methods in joint-SNVs analysis [13–15]. In order to reduce the background disturbance, the posteriori of p value (*pp* value) is introduced in the Statistics-space BBT [10].

In this paper, we investigate IHT by integrating Data-space BBT and Statistics-space BBT to control the false positive rate in the multivariate case. The effectiveness of the integration was validated on both synthetic datasets and real-world datasets. Results also empirically demonstrated that Data-space BBT and Statistics-space BBT were complementary to each other in controlling false positive rate. The *pp* value is helpful in controlling the false positive rate on the synthetic experiments.

This paper is organized in the following way. In Sect. 2 we briefly introduced IHT, as well as BBT, and then focused on the integration of Data-space BBT and Statistics-space BBT. An intuitive method for integration of them was studied. In Sect. 3, we conducted simulation experiments. In Sect. 4, we applied the intuitive integration on a SNV dataset of bipolar disorder and performed a literature search to validate the effectiveness of the integration.

2 Integrative Hypothesis Tests and Boundary-Based Test

2.1 Brief Introduction of IHT

Integrative Hypothesis Tests (IHT) was proposed in [9, 10] to consider discriminating analysis and testing of case-control problems, jointly from two perspectives, i.e., model-based perspective and boundary-based perspective, which involves four tasks as described in Table 1 of [10]. From the model-based perspective, we utilize parametric models to describe the case population and control population, and then measure the difference between two populations. From the boundary-based perspective, we detect the existence of boundary for two populations. In the following part, we focus on the task B comparison and task C classification of IHT to control the false positive rate. The task B offers *p* value to measure the difference and task C provides misclassification rate. Correspondingly, the Data-space BBT and Statistics-space BBT finish task B and task C, and both of them will be described in the following subsections.

Table 1. Annotation for top-20 genes of bipolar disorder

Class	Gene_symbol	Description
A	RYR3	Ryanodine Receptor 3
	NPAS3	Neuronal PAS Domain Protein 3
	WWOX	WW Domain Containing Oxidoreductase
	DLG2	Discs Large MAGUK Scaffold Protein 2
	DPP10	Dipeptidyl Peptidase Like 10
	CDH13	Cadherin 13
B	SHISA6	Shisa Family Member 6
	LRP1B	LDL Receptor Related Protein 1B
	ASTN2	Astrotactin 2
	PTPRD	Protein Tyrosine Phosphatase, Receptor Type D
	LRRC4C	Leucine Rich Repeat Containing 4C
	PRKCA	Protein Kinase C Alpha
	FHIT	Fragile Histidine Triad
	GALNT13	Polypeptide N-Acetylgalactosaminyltransferase 13
	PARK2	Parkin RBR E3 Ubiquitin Protein Ligase
C	THSD4	Thrombospondin Type 1 Domain Containing 4
	FAM155A	Family With Sequence Similarity 155 Member A
	ZNF664-FAM101A	Filamin-Interacting Protein FAM101A
	PRKCE	Protein Kinase C Epsilon
	USH2A	Usherin

As an early application of IHT, Jiang et al. takes the task B and task C into consideration to identify miRNAs biomarkers for the differentiation of lung cancer and Chronic Obstructive Pulmonary Disease (COPD) [12]. As illustrated in Fig. 1 in [12], a *p* value indicating difference of the two distributions and a misclassification rate indicating a separating boundary were combined in a 2D scatter plot for an IHT rank on the features. In order to improve the reliability when the sample size is small and missing value is existed, the bootstrapping method was proposed.

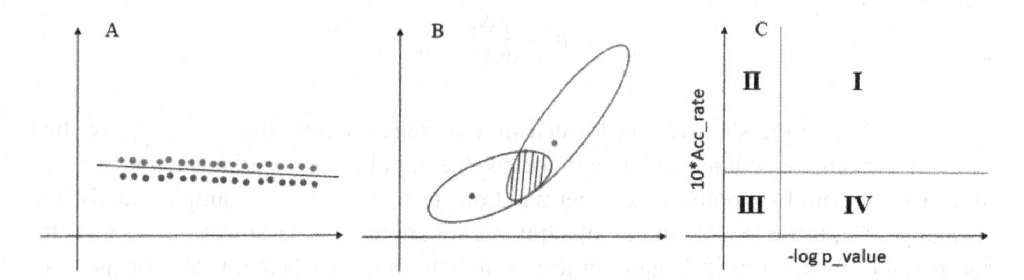

Fig. 1. Examples for explanation of integration for Data-space BBT and Statistics-space BBT.

2.2 Data-Space Boundary-Based Test

The Data-space Boundary-based test aims to seek for boundary to classify the samples into case population and control population, that is, it belongs to two-class classification problem. For the simplest case, we can defined a hyperplane,

$$g(x, \mathbf{w}) = \mathbf{w}^T x + w_0 = \mathbf{w}^T (x - \mu) \quad (1)$$

where μ is the mean of population. Then, the data can be divided into two classes by,

$$\begin{cases} \text{case,} & \text{if } g(x, \mathbf{w}) > 0 \\ \text{control,} & \text{if } g(x, \mathbf{w}) \leq 0 \end{cases} \quad (2)$$

then we constructed the statistics as misclassification rate as described in the [10],

$$s = \frac{\#X_1^{(0)} + \#X_0^{(1)}}{\#X^{(0)} + \#X^{(1)}} \quad (3)$$

where the $X^{(1)}$ indicates the case population, the $X^{(0)}$ is the control population, the $X_1^{(0)}$ means those that belong to control population but are classified into case population and $X_0^{(1)}$ indicates those that belong to case population but are classified into control population. The # means the number of candidates for one set. We can utilize support vector machine (SVM) [16] and fisher discrimination analysis (FDA) to obtain the boundary. Then we can apply the Eq. (3) to obtain the statistics. The smaller statistics are, the more separable two populations are.

2.3 Statistics-Space Boundary-Based Test

Different from the Data-space BBT, the Statistics-space BBT firstly computes statistics from the original data, such as the difference of means for two populations. We then calculate p value by permutation test. It is one-class classification problem which is to ensure whether the statistic is located in rejection region or not. Then we can describe the p value,

$$p = \frac{\#X_1^{(0)}}{\#X^{(0)}} \quad (4)$$

where the $X^{(0)}$ is the statistics set located in acceptance region, the $X_1^{(0)}$ is the set that contains wrong rejections and the # means the number of candidates for one set. Because the null hypothesis of permutation test is that the sample labels are exchangeable, those located in the rejection region are the misclassification. As a result, the p value obtained by permutation test is also the misclassification rate for the one-class classification problem.

Xu has provided four key steps of Statistics-space BBT in the Table 6 of [10]. For the multivariate case, we also summaries four main ingredients. First, we determine a rejection domain $\Gamma(\tilde{\mathbf{s}})$ based on statistic $\tilde{\mathbf{s}}$ which is obtained from case-control study,

$$\Gamma(\tilde{\mathbf{s}}) = \{\mathbf{s} : (\mathbf{s} - \tilde{\mathbf{s}})^T \mathbf{sign}(\tilde{\mathbf{s}}) > \mathbf{0}\} \quad (5)$$

where $\mathbf{sign}(\mathbf{s}) = [sign[s_1], \cdots, sign[s_m]]^T$ with $sign[u] = \pi \frac{u}{|u|}$. Second, the p value can be calculated by permutation test regardless of distribution. Third, we make full use of the principle component analysis (PCA) to remove the cross-dimensional dependence for factorization of multivariate p value (see Eq. (68) in [10]). Forth, we corrected the p value into posterior version (pp value) of it to reduce the background disturbance (see Eq. (93) in [10]). In the next section, we will also show the important role that pp value played in control of the false positive rate.

2.4 Integration for Data-Space BBT and Statistics-Space BBT via IHT

From the perspective of IHT, the Data-space BBT finishes the task C and Statistics-space BBT finishes the task B, and then we integrate the p value and misclassification rate to control the false positive rate. Their complementarity is described as follows.

The Data-space BBT is to classify the samples into two classes in the original data space, case population and control population. To some extent, it takes difference both for means and covariance into consideration. However, a separable boundary is not equivalent to existence of significant difference between two populations. As Fig. 1A described, there is a separable boundary between two populations, but there is no significant difference.

The Statistics-space BBT aims to determine whether the statistics are located in rejection region or not. The mean is often regarded as the criterion to indicate the difference for case-control study, such as Hotelling's T square test for multivariate [17]. Similarly, a significant difference of means from two populations does not indicate the existence of a separable boundary as Fig. 1B showed. In the Fig. 1B, two dots indicate means of two populations and two ellipses represent two population. Although the distance of two means is large, the shadow area is also large, which indicates that there is no separable boundary.

We have analyzed Data-space BBT and Statistics-space BBT empirically, and it is helpful to integrate both of them to obtain more accurate results, in other words, control the false positive rate. Note that the misclassification rate is replaced with accuracy rate in the 2D-scatter plot compared with the plot in the [12]. We call it p value vs. accuracy rate scatter plot.

As the Fig. 1C described, the horizontal ordinate represents negative natural logarithm of the p value and the vertical ordinate indicates accuracy rate which takes the place of misclassification rate for convenience. What's more, we multiply the accuracy rate by ten to make the scale as the same as negative natural logarithm of the p value. The scatter plot can be divided into four regions. The hypotheses located in the region I will be rejected via integration of Data-space BBT and Statistics-space BBT, while the hypotheses that belong to the rest three regions will be accepted. If we only consider the p value, the hypotheses in region I and IV are rejected. When we take the accuracy rate into consideration, those located in region I and II are rejected.

3 Simulation Framework and Corresponding Results

3.1 Simulation Framework for Type I Error

We intend to investigate the effect on the false positive rate of joint-SNVs analysis in GWAS when we integrate Data-space BBT and Statistics-space BBT via IHT. In order to mimic the pattern of the real-world data, the SNV data located in 2.5 Mb of chromosome 5 from the 1000 genomes projects was chosen and the number of the SNVs is 12,455. To remove the influence of ethnicity, we choose the Chinese Beijing (CHB) and the sample size is 97. The length of the computational unit is set to 15 kb because the length of gene is from 10 kb to 15 kb. The simulation datasets were generated by the null model,

$$y = 0.5X_1 + 0.5X_2 + \varepsilon \tag{6}$$

where y is the phenotype, X_1 is a continuous covariate generated from a standard normal distribution, X_2 is a bi-value covariate which can take 0 or 1 with a probability 0.5, and ε follows a standard normal distribution. It is of note that the generated phenotype is not associated with the genotype data when the null hypothesis holds. Then we transform the continuous phenotype into dichotomous value via the following model,

$$\operatorname{logit} P(y = 1) = \alpha_0 + 0.5X_1 + 0.5X_2 \tag{7}$$

where the logit means the logit function and α_0 is the prevalence. The α_0 is set to 0.01. The sample size is set to 100 vs. 100, 500 vs. 500 and 1000 vs. 1000. And there are 1000 replicates for them in the same parameter settings. The Data-space BBT adopted the linear SVM to seek for separable hyperplane and Statistics-space BBT followed the procedure described in the [10] to obtain joint p values.

3.2 Simulation Results

The simulation results are shown in the Fig. 2. Each row represents different sample size, 100 vs.100, 500 vs.500 and 1000 vs.1000 in order. First column described the relationship between accuracy rate and *p* value and second column is for the relationship between accuracy rate and *pp* value. The blue line indicates the type I error threshold α ($\alpha = 0.05$) and the red line means the threshold for accuracy rate (99% for 1000 vs. 1000, 95% for 500 vs. 500 and 60% for 100 vs. 100). The scatter plots are divided into four regions by the two lines as the Fig. 1C described.

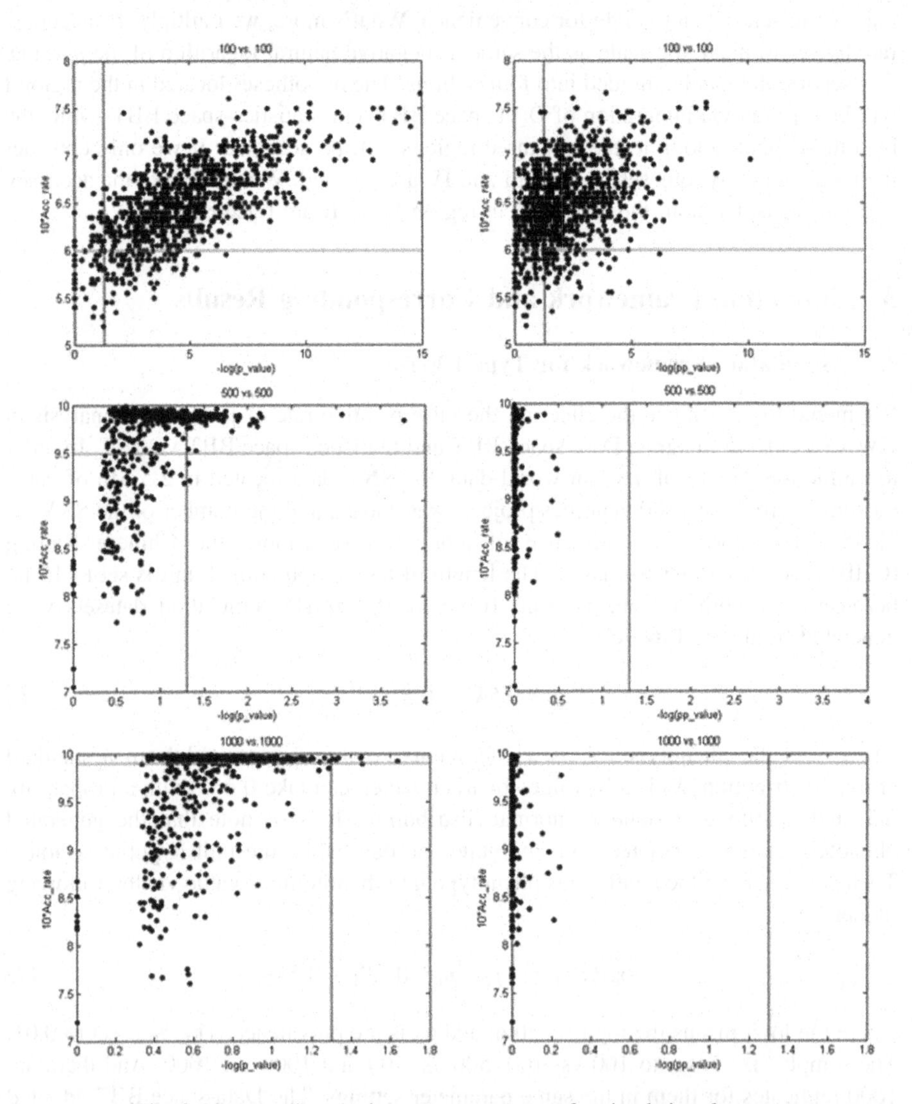

Fig. 2. Scatter plot *p* value vs. accuracy rate for simulation datasets

Altogether, there are four points to be mentioned. First, *p* value vs. accuracy rate scatter plot reduces the false positive rate under different sample size as the number of replicates decreasing in the region I. Second, for the cases of 500 vs. 500 and 1000 vs. 1000, the accuracy rate has no influence on controlling false positive rate when we use the *pp* value, while for the *p* value, the accuracy rate helps with reducing the number of replicates in the region I. Thus, the *pp* value has an effect on control of the false positive rate. Third, with the sample size increasing, false positive rate decreases. If we intend to decrease the false positive rate even further, the threshold of accuracy rate needs to be raised. Forth, if the sample size is large enough, the *pp* value can control the false positive rate efficiently.

4 Results for Bipolar Disorder SNV Dataset

4.1 Basic Information for Real-World Dataset and Data Pre-processing

Bipolar disorder is responsible for the loss of more disability-adjusted life-years and leads to high risk of suicide and self-harm [18–20]. The dataset of bipolar disorder (GSE71443) comes from GEO (Gene Expression Omnibus) database. The sample size is 65 bipolar disorder patients vs. 74 controls and the number of probes is 906600. There is no missing value to impute. We regarded the Hardy-Weinberg's equilibrium *p* value as the measure for quality control. If any of the three Hardy-Weinberg's equilibrium *p* values for case population, control population and the total population, is smaller than 1E-4, the SNVs would be filtered out. As a result, there are high-quality 722130 probes.

Then, we annotated the SNVs using the Annovar software [21]. And the unit was defined as the body of gene (exclude the upstream, downstream, intergenic and so on). Finally, we obtained 13896 genes. As the small sample size, we selected the first 20 SNVs in ascending order of their single locus *p* values to represent these genes, and then calculated the joint *p* values for them via Statistics-space BBT. All of the joint *p* values are corrected into *pp* values and the joint *p* values indicate the *pp* values. The Data-space BBT is also performed for the 20 dimensional data to calculate the misclassification rate via linear SVM.

4.2 Results for the Bipolar Disorder

After we obtained the *p* value and accuracy rate, *p* value vs. accuracy rate scatter plot of 13896 genes is shown in the Fig. 3 and the top-20 genes in the ascending order of their joint *p* values are marked in red. In addition, we also showed the relationship between two correction *p* values and original *p* values and found that both of them only change the scale of original *p* values and fix the threshold to reject less hypotheses without any other complementary information. We then conducted the literature search for them and they were described in the Table 1 (see details in the Table 5 in [15]). They are classified into three categories. Class A indicates that there is direct association with the bipolar disorder. Class B contains the genes related to the other psychiatric disorders or brain disorders, while the genes showed that there is no relationship to our best knowledge of the literature were collected in class C. Next, we will utilize the scatter plot to exclude some genes and we hope that the genes belonging to class A can be reserved as much as possible.

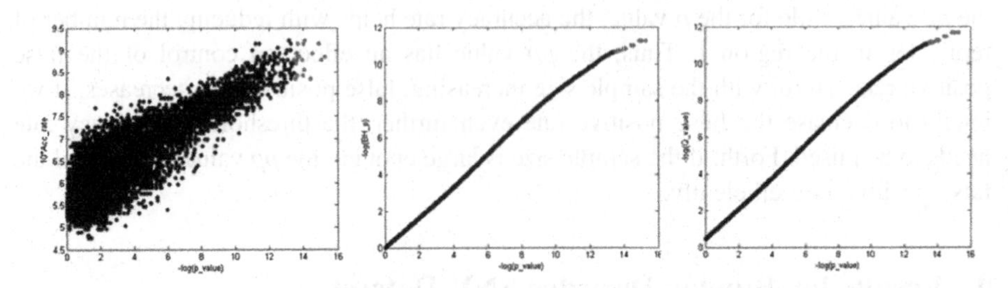

Fig. 3. Scatter plot p value vs. accuracy rate for all of 13896 genes

The top-20 genes were shown in the Fig. 4. The blue diamond represents genes in class A. the red circle means the genes in class B and the green star indicates genes in class C. The number of genes located in the region I is 8. Among them, the number of genes in class A decreased from 6 to 4, from 9 to 1 for class B and from 5 to 3 for class C. As a result, the remained genes are shown in the Table 2 and the scatter plot is able to control the false positive rate.

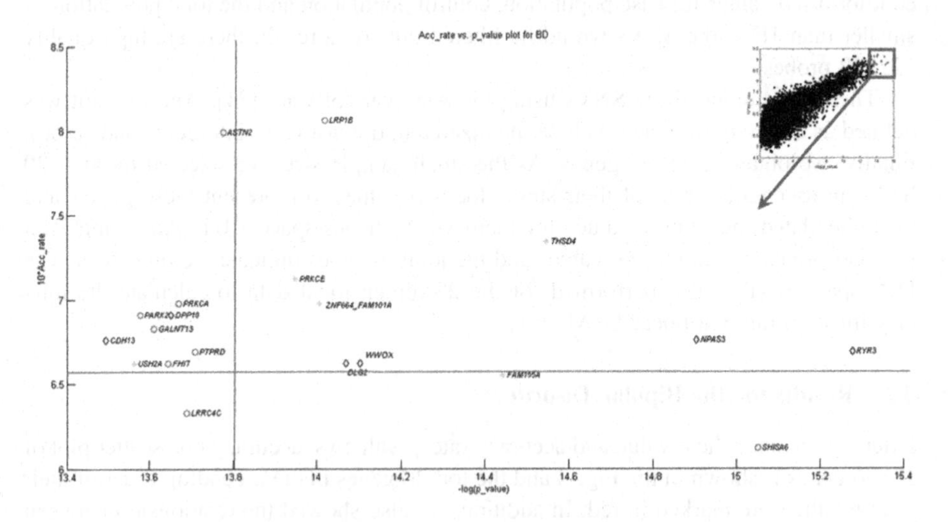

Fig. 4. Scatter plot p value vs. accuracy rate for top-20 genes

Table 2. Annotation for top-20 genes after integration

Class	Gene_symbol	Description
A	RYR3	Ryanodine Receptor 3
	NPAS3	Neuronal PAS Domain Protein 3
	WWOX	WW Domain Containing Oxidoreductase
	DLG2	Discs Large MAGUK Scaffold Protein 2
B	LRP1B	LDL Receptor Related Protein 1B
C	THSD4	Thrombospondin Type 1 Domain Containing 4
	ZNF664-FAM101A	Filamin-Interacting Protein FAM101A
	PRKCE	Protein Kinase C Epsilon

5 Discussion

We investigated the performance of integrating the Data-space BBT and Statistics-space BBT under the IHT. The simulation experiments were designed to further elucidate the effectiveness of the integration. The simulation results showed that the integration can help with controlling the false positive rate even for small sample size and the *pp* value can also control the false positive rate. While for the SNV datasets of bipolar disorder with small sample size, the integration also played a crucial role in controlling the false positive rate. In the future, there are three perspectives. First, the influence for the scatter plots under different classification methods is a worthwhile studying. Similarly, the influence for the scatter plots under different multivariate statistical methods also raises a mandatory research. Second, we will try to construct a statistic to replace the scatter plot via integrating the *p* value and accuracy rate. Third, for the scatter plot, it is essential to validate whether there is an arc line instead of the lines being parallel to the axis to control the false positive rate more efficiently. Altogether, IHT is the framework to control the false positive rate without being limited to boundary-based tests, and it is worthwhile to investigate more deeply its characteristics.

Acknowledgement. This work was supported by a grant from Shanghai Jiao Tong University, NO. WF220403029.

References

1. Han, F., Pan, W.: A data-adaptive sum test for disease association with multiple common or rare variants. Hum. Hered. **70**(1), 42–54 (2010)
2. Lee, S., Wu, M.C., Lin, X.: Optimal tests for rare variant effects in sequencing association studies. Biostatistics **13**(4), 762–775 (2012)
3. Wu, M.C., et al.: Rare-variant association testing for sequencing data with the sequence kernel association test. Am. J. Hum. Genet. **89**(1), 82–93 (2011)
4. Price, A.L., et al.: Pooled association tests for rare variants in exon-resequencing studies. Am. J. Hum. Genet. **86**(6), 832–838 (2010)
5. Dunn, O.J.: Multiple comparisons among means. J. Am. Stat. Assoc. **56**(293), 52–64 (1961)

6. Holm, S.: A simple sequentially rejective multiple test procedure. Scand. J. Stat. **6**, 65–70 (1979)
7. Benjamini, Y., Hochberg, Y.: Controlling the false discovery rate: a practical and powerful approach to multiple testing. J. Royal Stat. Soc. Series B (Methodol.) **57**(1), 289–300 (1995)
8. Storey, J.D.: The positive false discovery rate: a Bayesian interpretation and the q-value. Ann. Stat. **31**(6), 2013–2035 (2003)
9. Xu, L.: Integrative hypothesis test and A5 formulation: sample pairing delta, case control study, and boundary based statistics. In: Sun, C., Fang, F., Zhou, Z.-H., Yang, W., Liu, Z.-Y. (eds.) IScIDE 2013. LNCS, vol. 8261, pp. 887–902. Springer, Heidelberg (2013). https://doi.org/10.1007/978-3-642-42057-3_112
10. Xu, L.: Bi-linear matrix-variate analyses, integrative hypothesis tests, and case-control studies. Appl. Inform. **2**, 4 (2015). Springer, Berlin Heidelberg
11. Xu, L., Jiang, C.: Semi-blind bilinear matrix system, BYY harmony learning, and gene analysis applications. In: 2012 6th International Conference on New Trends in Information Science and Service Science and Data Mining (ISSDM). IEEE (2012)
12. Jiang, K.-M., Lu, B.-L., Xu, L.: Bootstrapped integrative hypothesis test, COPD-lung cancer differentiation, and joint miRNAs biomarkers. In: He, X., Gao, X., Zhang, Y., Zhou, Z.-H., Liu, Z.-Y., Fu, B., Hu, F., Zhang, Z. (eds.) IScIDE 2015. LNCS, vol. 9243, pp. 538–547. Springer, Cham (2015). https://doi.org/10.1007/978-3-319-23862-3_53
13. Lv, J.-X., et al.: A comparison study on multivariate methods for joint-SNVs association analysis. In: 2016 IEEE International Conference on Bioinformatics and Biomedicine (BIBM). IEEE (2016)
14. Lv, J., Tu, S., Xu, L.: A comparative study of joint-SNVs analysis methods and detection of susceptibility genes for gastric cancer in Korean population. In: Sun, Y., Lu, H., Zhang, L., Yang, J., Huang, H. (eds.) IScIDE 2017. LNCS, vol. 10559, pp. 619–630. Springer, Cham (2017). https://doi.org/10.1007/978-3-319-67777-4_56
15. Lv, J.-X., et al.: Comparative studies on multivariate tests for joint-SNVs analysis and detection for bipolar disorder susceptibility genes. Int. J. Data Min. Bioinform. **17**(4), 341–358 (2017)
16. Suykens, J.A., Vandewalle, J.: Least squares support vector machine classifiers. Neural Process. Lett. **9**(3), 293–300 (1999)
17. Hotelling, H.: The generalization of student's ratio. In: Kotz, S., Johnson, N.L. (eds.) Breakthroughs in Statistics. Springer Series in Statistics (Perspectives in Statistics), pp. 54–65. Springer, New York (1992). https://doi.org/10.1007/978-1-4612-0919-5_4
18. Anderson, I.M., Haddad, P.M., Scott, J.: Bipolar disorder. Br. Med. J. BMJ (Online) **345** (2012)
19. Pompili, M., et al.: Epidemiology of suicide in bipolar disorders: a systematic review of the literature. Bipolar Disord. **15**(5), 457–490 (2013)
20. Merikangas, K.R., et al.: Prevalence and correlates of bipolar spectrum disorder in the world mental health survey initiative. Arch. Gen. Psychiatry **68**(3), 241–251 (2011)
21. Wang, K., Li, M., Hakonarson, H.: ANNOVAR: functional annotation of genetic variants from high-throughput sequencing data. Nucleic Acids Res. **38**(16), e164–e164 (2010)

New Particle Swarm Optimization Based on Sub-groups Mutation

Zhou Fengli(✉) and Lin Xiaoli

Faculty of Information Engineering, City College Wuhan University of Science and Technology, Wuhan 430083, China
thinkview@163.com

Abstract. In order to overcome the premature convergence of particle swarm optimization (PSO) algorithm, an improved PSO algorithm based on sub-groups mutation (SsMPSO) is proposed. This algorithm has proposed the sub-groups with random directional vibrating search to mutate the global optimal position of the main swarm and changed the way of random mutation. The mutation based on sub-groups enabled the algorithm had excellent local exploit ability and circumvented the premature convergence. It used another mutation on bad particles to enhance the algorithm's global exploit ability and expand the searching space. Finally, high dimension benchmark functions have been used to test the performance of improved algorithm. The simulation results show that the proposed algorithm can effectively overcome the premature problem, the multimodal function optimization can avoid local extreme point and the convergence and convergence accuracy are greatly improved.

Keywords: Premature convergence · Random directional vibrating exploit Sub-groups · Mutation · Multimodal functions optimization

1 Introduction

Particle Swarm Optimization (PSO) algorithm was a random search optimization method based on group evolution that proposed by Kennedy et al. in 1995 [1]. It was the social sharing and co-evolution of information among individuals in population, which basic idea had originated from studying swarm behavior of the birds. PSO has achieved a better effect in many fields to solve optimization problems, such as function optimization [2, 3], system identification [4, 5], artificial neural network training [6–8] and so on. However, the PSO has some problems that is easy to be trapped into local optimization with premature convergence and converges slowly around optimal value. In order to overcome the shortcomings that PSO traps into local convergence easily and the accuracy of algorithm optimization is hard to be improved for multi-modal functions, this paper focuses on the optimization of multi-modal functions. For this purpose, the improvement research of PSO has been continued, the common improvement is dynamic modification strategy by introducing parameters [9, 10], changing evolutionary equations [11, 12], combining with other methods [13–15], and mixing these improvements with each other [15–17]. Literatures [9, 10] proposed a method of improving inertia weight based on adaptive viewpoint respectively. Literature [11] had

D.-S. Huang et al. (Eds.): ICIC 2018, LNAI 10956, pp. 639–649, 2018.
https://doi.org/10.1007/978-3-319-95957-3_66

modified the velocity equation by considering the second-order positions of particles. Literature [12] removed the particle's velocity term of equation and simplified the original second-order differential equation to first-order differential equation, so the problems of slow convergence and low accuracy caused by particle divergence can be avoided. Literature [13] reviewed some improved PSO algorithms based on algorithm integration idea. Literature [14] adopted differential algorithm and hierarchical clustering algorithm to self-adaptive and dynamically re-configure particles and populations respectively, it could guide the particles to skip the local optimum and increase the information exchange of each sub-population. Literature [15] had combined the population diversity maintenance mechanism and domain search strategy to give a good balance between local solution and global exploration of algorithms. Literature [16] proposed a hybrid improved algorithm by combining escape strategy of particle velocity and simplex method, it could equalize the global and local optimizing ability of the algorithm. Literature [17] used chaos algorithm to optimize the global learning factor of accelerated PSO algorithm which constrained only by global extremum, it could help the algorithm to skip the local optimum. In the process of PSO optimization, once the global optimum position plunges into local optima and there is no good way to get particles away from the local optimum, it will have a negative impact on population and the population evolution may stop. The improved algorithms based on mutation are completely random method [18–20], the ability of them to skip the local optimum is limited, especially when the high-dimensional multimodal problem is optimized, the precision of achieved result is low. This paper has proposed the sub-groups with random directional vibrating search to mutate the global optimal position and dynamically mutate the particles with poor fitness values. The dynamic mutation method has mutated particles with poor fitness value in each generation, that can avoid the singleness of particle search space and enhance the algorithm's global search ability. The mutation based on sub-groups introduces sub-groups with random directional vibrating search to the global optimal position of main swarm regularly, which would be mutated to provide local area deep-searching for algorithm and help it to escape from local optimum.

2 PSO Algorithm

PSO is an optimization algorithm with iterator pattern. Each particle has two properties, those are *j*-dimensional position $X_{ij}(t)$ and velocity V_{ij} of *i*-th particle. The two properties of its population are *i*-th particle's *j*-dimension previous best position $L_{ij}(t)$ and globally best position $G_j(t)$. The updating formulas of speed and location can be expressed as follows.

$$V_{ij}(t+1) = \omega V_{ij}(t) + c_1 r_1 \left(L_{ij}(t) - X_{ij}(t)\right) + c_2 r_2 \left(G_j(t) - X_{ij}(t)\right) \tag{1}$$

$$X_{ij}(t+1) = X_{ij}(t) + V_{ij}(t+1) \tag{2}$$

Where ω is inertia factor; c_1 and c_2 are learning factors; r_1 and r_2 are random numbers between 0 and 1.

3 SsMPSO Algorithm

3.1 Dynamic Mutation

The two particular cases of t-distribution have showed two kinds of boundary conditions, Gaussian Distribution has a strong local distribution, while the Cauchy Distribution has a strong global distribution. The probability density function of t-distribution is shown in Eq. (3).

$$f(x) = \frac{\Gamma\left[\frac{(n+1)}{2}\right]}{\sqrt{n\pi}\Gamma\left(\frac{n}{2}\right)}\left(1+\frac{x^2}{n}\right)^{-\frac{(n+1)}{2}} \tag{3}$$

When number n tends to be infinity, t-distribution approaches Gaussian Distribution; and when number n is equal to 1, t-distribution becomes Cauchy Distribution. The combination of these two distributions can make the global and local distribution of particles get a better integration. The dynamic mutation operator in this paper can be expressed by Eqs. (4) and (5).

$$X_{ij}(t) = X_{ij}(t) + \frac{1}{k}\rho \tag{4}$$

$$X_{ij}(t) = X_{ij}(t) + \frac{1}{k}\eta \tag{5}$$

Where $k = \frac{2}{d_j} + \frac{\sum_{i=1}^{size} D(G(t)-X_i(t))}{size}$; d_j is the j-dimensional length in optimum range; $D(G(t) - X_i(t))$ represents Euclidean Distance between the i-th particle and the global optimal position $G(t)$; $\sum_{i=1}^{size} D(G(t) - X_i(t))$ can describe the dispersal degree of population; *size* is particles' number; ρ is a random number that obeys Gaussian Distribution and η is a random number that obeys Cauchy Distribution.

To ensure that the population has a large potential search space, Eqs. (4) and (5) have been used for random mutation to each 10% of particles with poor fitness values. And the mutation is dynamic, that means each generation of evolvement will happen and it can be adjusted with the vergence degree of population. Dynamic mutation for particles with poor fitness value is not limited to individual particles. For a particle with poor fitness value at step t, it may be turn into a particle with better fitness value at step $t+1$. Relatively, a particle with a good fitness value at step t may be turn into a particle with poorer fitness value at step $t+1$. The dynamic mutation adjusts along with the population's vergence degree, especially expanding the diversity of population particles in the evolution process at later periods and enabling the population to explore potential search space at any time, so as to avoid high aggregation of particles and local optimization of population.

3.2 Sub-groups

The combination of Gaussian Distribution and Cauchy Distribution can rich the population's diversity of sub-groups. The initialization of sub-groups is generated by current global optimum position $G(t)$ of main swarm according to Eqs. (6) and (7), in which 2/3 of particles are generated by Eq. (6) and 1/3 of particles are generated by Eq. (7). Where c_3 and c_4 are particle distribution coefficients of initial sub-group, they can be set differently to satisfy the requirement of different optimization problems.

$$X_{ij}(1) = G_{ij}(t) + c_3\rho \tag{6}$$

$$X_{ij}(1) = G_{ij}(t) + c_4\eta \tag{7}$$

To ensure that the evolution of sub-groups can search potential better space, retain the optimal genetic information of main swarm, and reduce blind search, if the sub-groups' fitness value is greater than the current fitness value of main swarm, the sub-group will update the speed according to Eq. (8).

$$\begin{aligned} V_{ij}(t+1) = {} & \omega(t)V_{ij}(t) + c_1r_1L_{ij}(t) - X_{ij}(t) \\ & + [2r_3]\{c_2r_2(G_j(t) - X_{ij}(t))\} + [2r_3]\{c_2r_2(g_j(t) - X_{ij}(t))\} \end{aligned} \tag{8}$$

Where r_3 is a random number between 0 and 1; $[2r_3]$ represents integer conversion to $2r_3$; $g(t)$ is the current global optimal position of sub-groups; $G(t)$ is current global optimal position of main swarm. The evolution direction of sub-groups is randomly determined by its information and global optimal information of the main swarm according to Eq. (6), then the social cognition of sub-groups has turned to $[2r_5]\{c_2r_2(G_j(t) - X_{ij}(t))\} + [2r_5]\{c_2r_2(g_j(t) - X_{ij}(t))\}$. The social cognition consists of main swarm's optimal $G(t)$ and sub-groups' optimal $g(t)$, so the search of sub-groups will not fall into stagnation. Instead, the search will be conducted in local space with random oscillation.

3.3 Description of SsMPSO Algorithm

Evolution of Algorithm

Due to the introduction of sub-groups, the spatial distribution structure of population particles becomes two layers. One layer is the location distribution in spatial of each particle in main swarm and dynamic mutation particle, and the other layer is the location distribution in spatial of particles around the sub-groups generated by global optimum with main swarm. Dynamic mutation increases the algorithm's global search ability. Sub-groups use global search of main swarm to conduct a more comprehensive local area search. The more the sub-groups are used, the more particles are explored in local area, which increases the local search ability of this algorithm. In turn, the high-precision local search of sub-groups can lead main swarm to escape the local optimum.

Steps of Algorithm

The general steps of SsMPSO are as follows:

(a) Initializing main swarm in solution space and setting the parameters.
(b) Updating the velocity and position of particles according Eqs. (1) and (2), mutating three particles with poor fitness value according to Eqs. (4) and (5).
(c) Generating sub-groups after *M*-generation, otherwise going to step f).
(d) Generating sub-groups according to Eqs. (6) and (7), updating the velocity of particles by Eqs. (1) and (8), and updating the position of particles by Eq. (2).
(e) Comparing the global optimal position between sub-groups and main swarm, selecting the smallest fitness value as the main swarm's current global optimal position.
(f) Going to step b) if $t + 1 \geq T$, otherwise terminating the calculation and getting the optimal fitness value and global optimum position.

4 Numerical Experiment

In this paper, six typical Benchmark functions that shown in Table 1 are used to test the performance.

Table 1. Test functions

Function name	Function	Range	Optimal status and optimal value
Rosenbrock	$f_1 = \sum_{i=1}^{n-1} \left[100(x_{i+1} - x_i^2)^2 + (1 - x_i)^2\right]$	$x_i \in [-30, 30]$	$\min(f(x^*)) = f(1, \ldots, 1) = 0$
Ackley's path	$f_2 = -20\exp\left(-0.2\sqrt{\frac{1}{n}\sum_{i=1}^{n} x_i^2}\right) - \exp\left(\frac{1}{n}\sum_{i=1}^{n}\cos(2\pi x_i)\right) + 20 + e$	$x_i \in [-32.768, 32.768]$	$\min(f(x^*)) = f(1, \ldots, 1) = 0$
Griewangk	$f_3 = \frac{1}{4000}\sum_{i=1}^{n} x_i^2 - \prod_{i=1}^{n}\cos\left(\frac{x_i}{\sqrt{i}}\right) + 1$	$x_i \in [-600, 600]$	$\min(f(x^*)) = f(1, \ldots, 1) = 0$
Rastrigin	$f_4 = \sum_{i=1}^{n}\left[x_i^2 - 10\cos(2\pi x_i) + 10\right]$	$x_i \in [-5.12, 5.12]$	$\min(f(x^*)) = f(1, \ldots, 1) = 0$
Montalvo	$f_5 = \sum_{i=1}^{n-1}\left[x_i^2 - 10\cos(2\pi x_i) + 10\right]$	$x_i \in [-50, 50]$	$\min(f(x^*)) = f(1, \ldots, 1) = 0$
Levy	$f_6 = \frac{\pi}{n}\left\{10\sin^2(\pi x_1) + \sum_{i=1}^{n-1}(x_i - 1)^2\left[1 + 10\sin^2(\pi x_{i+1})\right] + (x_n - 1)^2\right\}$	$x_i \in [-50, 50]$	$\min(f(x^*)) = f(1, \ldots, 1) = 0$

$f_2 \sim f_6$ are five high-dimensional multi-modal functions and difficult to optimize. The reason for this choice is that the optimization algorithm for multi-modal functions tends to fall into local convergence easily, and the algorithm's optimization accuracy cannot be improved. However, the conventional algorithms based on completely random mutation methods are not good for escaping the local optimum, the optimization's accuracy is poor [18–20]. This paper focuses on the performance test for high-dimensional multi-modal functions. First comparing with two commonly PSO algorithms can descript that the proposed algorithm's calculation accuracy and stability are improved; then comparing with the three mutation-based PSO algorithms in references can show that this method has a better improvement than completely random mutation.

4.1 Performance Analysis for Algorithm

The dimension of test functions is set to 30 and the test parameters are set as follows: $M = 100$; main swarm's population size = 30; sub-groups' population size = 30; inertial weight $\omega(t) = 0.9 - 0.4t/T$; operation steps' number $T = 3000$; learning factors $c_1 = c_2 = 2$.

Three performance evaluation indicators are defined as follows.

(a) Mean Best Value (mean): the mean best value of this algorithm can be acquired after 20 operations that can measure the average quality of optimization.

$$\text{mean} = \frac{1}{K}\sum_{j=1}^{K} f(j)$$

Where K is the operation numbers and $K = 20; f(j)$ represents the optimal fitness value that obtained from the j-th algorithm operation.

(b) Standard deviation (Std).

$$\text{Std} = \frac{1}{K} \times \sqrt{\sum_{j=1}^{K} (f(j) - \text{mean})^2}$$

It means that the deviation value between fitness value and mean best value obtained from each iteration and can measure the optimization volatility.

(c) Worst solution (wor).

$$\text{wor} = \max(f(1) \sim f(K))$$

It can represent reliability of the algorithm.

f_1 is a unimodal function that is difficult to optimize, and the majority of existing literatures had adopted it; $f_2 \sim f_6$ are five multi-modal functions and have many local extremum points, which can easily fall into local optimum, so optimization for these complex high-dimensional functions can reflect the algorithm's coordination ability for global and local exploration.

In order to analyze the convergence process of SsMPSO, Figs. 1, 2, 3, 4, 5 and 6 show the evolution process of algorithm's fitness value. Figure 1 shows the optimization process of unimodal function f_1, LPSO and LnCPSO evolve into stagnation after 1000 operation steps and fall into local optimum, the fitness value will never change, and the algorithm cannot escape local optimum. But SsMPSO doesn't stagnate in evolution process, the fitness function value has always approached to the better and continuously skipped the local optimum, finally the optimal state and value can be approached. Figures 2, 3, 4, 5, 6 show the performance comparison of the algorithm for multi-modal functions. Here we take Fig. 2 as an example to analyze the evolution of multi-modal functions. LPSO plunges into local optimum prematurely and it has evolved towards a decreasing fitness value before 500 operation steps, then it has fallen into local optimum and could not jump out, the evolution will stagnate. LnCPSO tends to stop when it evolves to 1500 generations and falls into local optimization. SsMPSO has been evolving to global optimum all the time and a hopping of decreasing fitness value will occur every 100 generations, so this algorithm will not fall into local optimum. It can be seen from Figs. 1, 2, 3, 4, 5 and 6 that LPSO and LnCPSO will stop evolution after they fall into the local optimum, so they cannot jump out of the local optimum; while SsMPSO can continue to optimize, so that evolution will not trap in premature convergence and can converge to the global optimum.

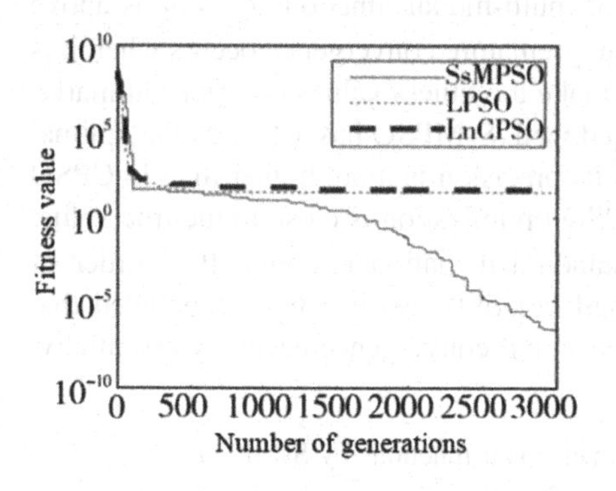

Fig. 1. Comparison to convergence of f_1

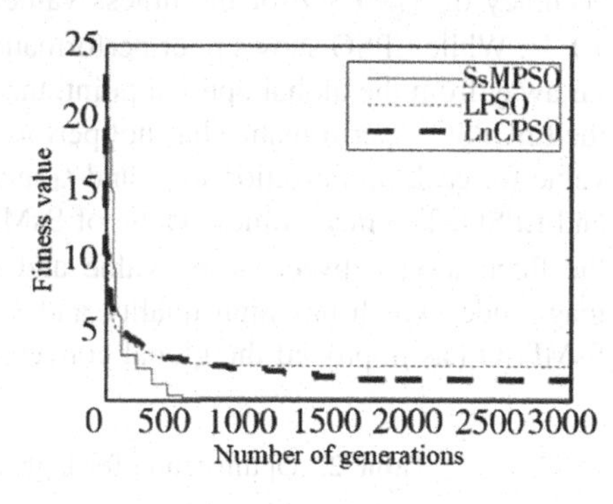

Fig. 2. Comparison to convergence of f_2

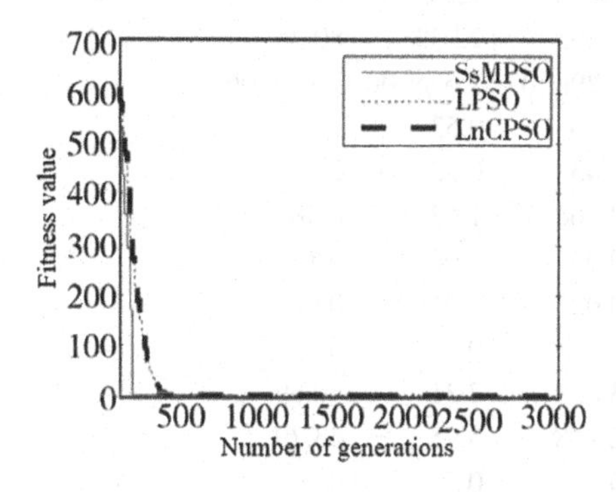

Fig. 3. Comparison to convergence of f_3

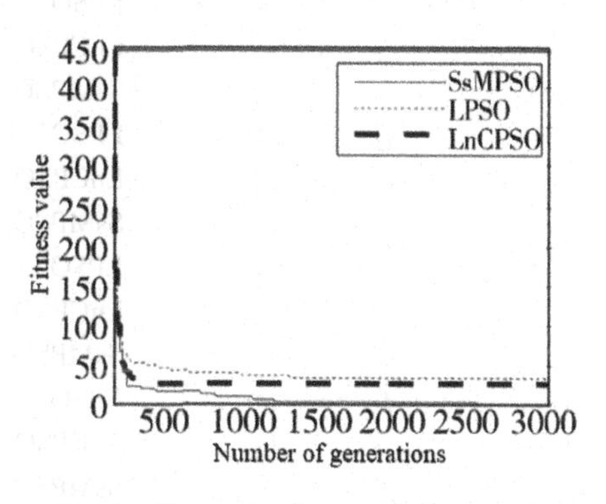

Fig. 4. Comparison to convergence of f_4

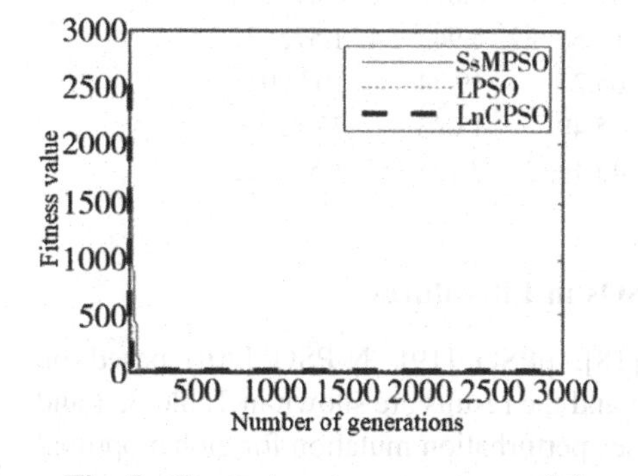

Fig. 5. Comparison to convergence of f_5

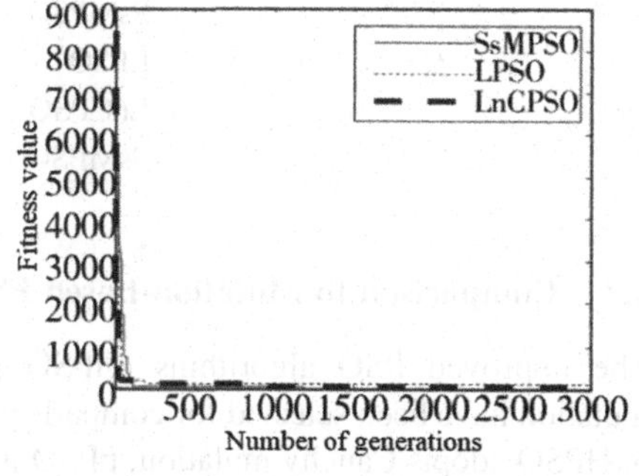

Fig. 6. Comparison to convergence of f_6

The test results for $f_2 \sim f_6$ are shown in Table 2. Compared with LPSO and LnCPSO, SsMPSO has shown a great advantage, it doesn't fall into the local optimum and premature convergence in the evolution process. SsMPSO shows good properties for unimodal function f_1 which is difficult to be optimized by many improved PSO algorithms, the precision of fitness value optimization is above 10^{-6}. Because the directed sub-groups mutation can escape the local optimal point well and the dynamic random variation expands the global performance of the algorithm, the optimization accuracy of SsMPSO for the fitness value of multi-modal functions $f_2 \sim f_6$ is above 10^{-13}. While LPSO shows poor performance, premature convergence occurs when it is far away from the global optimal point, that make the fitness values is larger. Similarly, the LnCPSO's performance has not performed well. SsMPSO has achieved the optimal value for each optimization to f_3 and f_4 and its precision is more higher than LnCPSO and LPSO. The mean fitness value of SsMPSO optimization is close to the true value, the fluctuation between worst value and standard deviation is above 10^{-6} order of magnitude, which has high quality and small volatility. So it can be concluded that SsMPSO has improved the global convergence and convergence accuracy essentially.

Table 2. Optimization for high dimensional functions by SsMPSO

Function	Optimum	Algorithm	Mean	Std	Wor
f_1	0	LPSO	30.26	12.63	83.05
		LnCPSO	38.35	22.33	80.75
		SsMPSO	5.29e−07	1.67e−06	7.29e−06
f_2	0	LPSO	2.26	0.53	3.31
		LnCPSO	1.43	0.52	2.23
		SsMPSO	2.66e−13	1.62e−13	6.38e−13
f_3	0	LPSO	0.03	0.02	0.13
		LnCPSO	0.02	0.01	0.06
		SsMPSO	0	0	0
f_4	0	LPSO	33.53	7.31	60.69
		LnCPSO	26.48	5.95	43.79
		SsMPSO	0	0	0
f_5	0	LPSO	0.03	0.02	0.11
		LnCPSO	0.01	0.01	0.02
		SsMPSO	1.36e−32	4.94e−34	1.57e−32
f_6	0	LPSO	65.24	19.74	107.90
		LnCPSO	35.49	18.33	73.72
		SsMPSO	4.61e−23	7.12e−23	1.52e−22

4.2 Comparison to Mutation-Based PSOs in Literatures

The improved PSO algorithms (HPSO [18], pPSO [19], N-PSO [20]) based on mutation have been selected for comparison and the results are shown in Table 3, 4 and 5. HPSO adopts Cauchy mutation, pPSO uses perturbation mutation for global optimal position, and N-PSO combines two types of normal variations, all of them adopt completely random mutation. The results of HPSO, pPSO and N-PSO come from literatures [18–20]. In order to analyze the improved algorithm reasonably, the test parameters are set according to literatures [18–20] respectively, and the simulation comparison is conducted again.

Table 3. Comparison with HPSO

Function	Algorithm	Mean	Std
f_2	HPSO	8.86e−06	8.58e−02
	SsMPSO	9.74e−07	1.78e−06
f_3	HPSO	3.66e−02	3.19e−02
	SsMPSO	0	0
f_4	HPSO	31.8	9.16
	SsMPSO	0	0

It can be seen from Tables 3, 4 and 5 that compared with the other three PSO algorithms, the optimization accuracy of SsMPSO has significantly improved and is more stable, which shows that the improved method based on sub-groups mutation and dynamic mutation is effective. Obviously, the SsMPSO algorithm has more improvement in computational complexity than the other three improved algorithms. As NFL [21] theorem had indicated, all performance indexes of an algorithm is impossible to be simultaneously, and the accuracy improvement of the improved algorithm is based on the increase of calculation cost.

Table 4. Comparison with pPSO

Function	Algorithm	Mean	Std
f_3	pPSO	1.21e−02	1.23e−02
	SsMPSO	0	0
f_5	pPSO	10.75	4.64
	SsMPSO	7.12e−09	1.7021e−08
f_6	pPSO	4.99e−03	8.16e−03
	SsMPSO	2.09e−15	4.67e−15

Table 5. Comparison with N-PSO

Function	Algorithm	Mean	Std
f_1	N-PSO	6.62	1.20
	SsMPSO	3.63	1.35
f_3	N-PSO	2.68e−02	1.52e−02
	SsMPSO	0	0
f_4	N-PSO	2.99e−01	9.44e−01
	SsMPSO	8.4e−03	7.60e−03

5 Conclusion

The proposed SsMPSO algorithm uses sub-groups with random directional vibrating search and dynamic mutation to improve the disadvantage that PSO is easily trapped into local optimum and premature convergence. The introduction of sub-groups in SsMPSO provides local depth search and helps the main swarm to escape the current local optimum position, and it also can provide direction for continuous evolution of main swarm and search with high-precision for main swarm. Dynamic mutation can explore potential search spaces and avoid particle aggregation that leaves the population prematurely trapped into local optima. It can be concluded from the simulation results that SsMPSO has largely overcome the shortcoming that PSO is easy to fall into local optimum and premature convergence. And SsMPSO has better stability, less volatility and high optimization accuracy, which is suitable for the optimization of high-dimensional multi-modal problems.

Acknowledgment. This work was supported in part by Hubei Province Natural Science Foundation of China (No. 2018CFB526), by National Natural Science Foundation of China (No. 61502356).

References

1. Kennedy, J., Eberhart, R.C.: Particle swarm optimization. In: Proceedings of IEEE International Conference on Neural Networks, pp. 1942–1948. IEEE Press, Piscataway (1995)
2. Ma, G., Zhou, W., Chang, X.L.: A novel particle swarm optimization algorithm based on particle migration. Appl. Math. Comput. **218**(1), 6620–6626 (2012)
3. Liang, J.J., Qin, A.K., Suganthan, P.N., et al.: Comprehensive learning particle swarm optimizer for global optimization of multimodal functions. IEEE Trans. Evol. Comput. **10**(3), 281–295 (2006)
4. Xu, X.P., Qian, F.C., Liu, D., et al.: System identification method based on PSO algorithm. J. Syst. Simul. **20**(13), 3525–3528 (2008)
5. Jin, Q.B., Cheng, Z.J., Dou, J., et al.: A novel closed loop identification method and its application of multivariable system. J. Process Control **22**(1), 132–144 (2012)
6. Chatterjee, A., Pulasinghe, K., Watanabe, K., et al.: A particle-swarm-optimized fuzzy-neural network for voice-controlled robot systems. IEEE Trans. Industr. Electron. **52**(6), 1478–1489 (2005)
7. Song, L.W., Huang, X.Y., Liu, H.S., et al.: Analog circuit diagnosis based on PSO-RBF neural network. Appl. Res. Comput. **29**(1), 72–74 (2012)
8. Li, M.S., Huang, X.Y., Liu, H.S., et al.: Dissolution model in polymer based on the gas of chaotic adaptive particle swarm and artificial neural network. Chin. J. Chem. **71**(7), 1053–1058 (2013)
9. Shen, X.J., Chi, Z.F., Yang, J.C., et al.: Particle swarm optimization with dynamic adaptive inertia weight. In: Proceedings of International Conference on Challenges in Environmental Science and Computer Engineering, pp. 287–290. IEEE Computer Society, Washington DC (2010)
10. Zhang, Z.B., Jiang, Y.Z., Zhang, S.H., et al.: An adaptive particle swarm optimization algorithm for reservoir operation optimization. Appl. Soft Comput. **18**(4), 167–177 (2014)

11. Hu, J.X., Zeng, J.C.: Second-order particle swarm optimization. Comput. Res. Develop. **44**(11), 1825–1831 (2007)
12. Hu, W., Li, Z.S.: A simpler and more efficient particle swarm optimization algorithm. Softw. J. **18**(4), 861–868 (2007)
13. Thangaraj, R., Pant, M., Abraham, A., et al.: Particle swarm optimization: hybridization perspectives and experimental illustrations. Appl. Math. Comput. **217**(12), 5208–5226 (2011)
14. Ni, H.M., Liu, Y.J., Li, P.C.: Adaptive dynamic reconfiguration multi-target particle swarm optimization algorithm. Control Decis. **30**(8), 1417–1422 (2015)
15. Wang, H., Sun, H., Li, C.L., et al.: Diversity enhanced particle swarm optimization with neighborhood search. Inf. Sci. **223**(2), 119–135 (2013)
16. Zhang, Y., Gong, D.W., Zhang, W.Q.: An improved particle swarm optimization algorithm based on simplex method and its convergence analysis. J. Autom. **35**(3), 289–298 (2009)
17. Zhao, L.P., Shu, Q.L., Wu, Y., et al.: Chaos-enhanced accelerated particle swarm optimization algorithm. Appl. Res. Comput. **31**(8), 2307–2310 (2014)
18. Wang, H., Li, C.H., Liu, Y.A.: Hybrid particle swarm algorithm with cauchy mutation. In: Proceedings of IEEE Swarm Intelligence Symposium, pp. 356–360. IEEE Computer Society, Washington DC (2007)
19. Zhao, X.C.: A perturbed particle swarm algorithm for numerical optimization. Appl. Soft Comput. **10**(1), 119–124 (2010)
20. Gao, S.G., Liu, S., Zheng, Z.T.: PSO with two types of normal variables. Control Decis. **29**(10), 1881–1884 (2014)
21. Wolpert, D.H., Macready, W.G.: No free lunch theorems for optimization. IEEE Trans. Evol. Comput. **1**(1), 67–82 (1997)

Numerical Solution of Singularly Perturbed Convection Delay Problems Using Self-adaptive Differential Evolution Algorithm

Guangqing Long, Li-Bin Liu(✉), and Zaitang Huang

School of Mathematics and Statistics,
Guangxi Teachers Education University, Nanning 530001, China
liulibin969@163.com

Abstract. In this paper, a new numerical technique is constructed to solve singularly perturbed convection delay problems. First of all, based on Taylor's series expansion, the given problem is transformed into a singularly perturbed convection-diffusion problem without delay term, which is discretized by using the rational spectral collocation method with a sinh transformation. It should be pointed out that the width of boundary layer, which is chosen as a parameter in the sinh transformation, can be determined. Then, a nonlinear unconstrained optimization problem is designed to determine the width of boundary layer. Finally, the numerical solution of the singularly perturbed problem is converted into minimizing the nonlinear unconstrained optimization problem, which is solved by using a self-adaptive differential evolution (SADE). The numerical results show that the proposed algorithm is a robust and accurate procedure for solving singularly perturbed convection delay problems. Furthermore, the obtained accuracy for the solutions using SADE is much better than results obtained using some others algorithms.

Keywords: Singularly perturbed problems
Rational spectral collocation method
Optimization problem · Self-adaptive differential evolution

1 Introduction

In this paper, we consider the following singularly perturbed differential difference equation

$$\begin{cases} L_\varepsilon u(x) := -\varepsilon u''_\varepsilon(x) - a(x)u'_\varepsilon(x-\delta) + b(x)u_\varepsilon(x) = f(x), x \in \Omega = (0,1), \\ u_\varepsilon(x) = \phi(x), \ -\delta \le x \le 0, \ u_\varepsilon(1) = \gamma, \end{cases} \tag{1}$$

where $0 < \varepsilon << 1$ is a small parameter, δ is a small delay parameter satisfying $\delta = O(\varepsilon)$. The functions $a(x)$, $b(x)$ and $f(x)$ are sufficiently smooth, and γ is a given constant. This kind of problem is sometimes addressed as two parameters problem. It is well known that the argument for small delay problems arise frequently in the mathematical modeling of various occurrences, such as signal of transmission in control theory for studying time delay problem [1], physiological control systems [2],

D.-S. Huang et al. (Eds.): ICIC 2018, LNAI 10956, pp. 650–661, 2018.
https://doi.org/10.1007/978-3-319-95957-3_67

evolutionary biology [3], first exist time problem in the modeling of the activation function of neuronal variability [4], etc.

In recent years, there has been tremendous interest in developing effective numerical methods for solving singularly perturbed differential-difference Eq. (1). Kadalbajoo and Sharma [5, 6] constructed finite difference schemes for Eq. (1). Kadalbajoo and Ramesh [7] designed a hybrid difference scheme based on central difference in the boundary layer region and midpoint upwind scheme outside the boundary layer. The authors [8] presented a fitted B-spline collocation method for singularly perturbed differential-difference equations with a small delay. Rao and Chakravarthy [9] proposed a finite difference scheme for singularly perturbed differential-difference equations with layer and oscillatory behavior. In [10, 11], the authors designed adaptive grid methods based on the upwind finite difference scheme to solve singularly perturbed differential-difference equations with a small delay. Recently, the barycentric rational spectral collocation method has been presented and applied to some differential equations [12–14]. In [15, 16], the authors used a rational spectral method in barycentric form with the following sinh transformation

$$g(x) = \lambda + \mu \sinh\left[\left(\sinh^{-1}\left(\frac{1-\lambda}{\mu}\right) + \sinh^{-1}\left(\frac{1+\lambda}{\mu}\right)\right)\frac{x-1}{2} + \sinh^{-1}\left(\frac{1-\lambda}{\mu}\right)\right], \tag{2}$$

where the parameters λ and μ represent the location and width of the boundary or interior layer, respectively. Furthermore, they used the asymptotic analysis theory to calculate the location of boundary layer for the given problem. For the width of boundary layer, they chose $\mu = \beta\varepsilon$, where β is a positive parameter. However, the parameter β represents the width of the boundary layer, which is usually not known exactly. Thus, it is necessary to design a numerical method to determine the optimal parameter β.

As far as we know, research community has implemented many evolutionary intelligence algorithms for finding accurate and reliable solution of differential equations arising in potential applications of applied science and technology. For instance, Toresch problem arising in plasma physics [17], boundary value problems of nonlinear pantograph functional differential equation [18], Jeffery Hamel flow problems in the presence of high magnetic field [19], elliptic partial differential equations [20], two-point boundary value problems [21], singularly perturbed problem [22, 23] etc. Recently, in [24], the authors first used the rational spectral collocation method to solve the singularly perturbed problems with an interior layer. Then, they presented an optimization method based on basic differential evolution (DE) algorithm [25] to determine the width of interior layer for the first time.

DE algorithm is a fast and simple yet powerful population-based stochastic search technique, which is an effective global optimizer in the continuous search domain. It is a relatively new nonlinear search and optimization method, which is suitable to solve complicate optimization problems. Recently, DE has been successfully applied in diverse fields such as neural networks [26, 27], pattern recognition [28–30]. In DE, there exist three crucial control parameters, i.e., population size *NP*, scaling factor *F* and crossover rate *CR*. The values of these parameters greatly determine the accuracy of the solution obtained and the efficiency of the search [31, 32]. For this reason, Brest et al. [33] proposed a novel approach to the self-adapting control parameter of DE.

This paper continues the general theme of the work in [24] and further develops the rational spectral collocation and DE algorithm for the singularly perturbed differential difference equations with a small delay. Generally speaking, we first obtain an approximate differential equation by using Taylor's series expansion of the convection term $u'(x-\delta)$ around x. Then, the problem (3) is discretized by utilizing the rational spectral method with the sinh transformation. Furthermore, we use the idea from [24] to construct a fitness function, which is solved by self-adapting differential evolution (SADE) algorithm presented in [33]. Finally, the optimal width of boundary layer and the corresponding numerical results for the given problem are obtained.

2 Continuous Problem

Similar to [8, 11], by using Taylor's series expansion of the convection term $u'(x-\delta)$ around x, we obtain

$$u'_\varepsilon(x-\delta) \approx u'_\varepsilon(x) - \delta u''_\varepsilon(x).$$

Thus, the above Eq. (1) can be written as follows:

$$\begin{cases} Lu(x) := -(\varepsilon - \delta a(x))u''(x) - a(x)u'(x) + b(x)u(x) = f(x), x \in \Omega = (0,1), \\ u(0) = \phi(0), u(1) = \gamma. \end{cases} \tag{3}$$

Obviously, Eq. (3) is an approximation of Eq. (1). Here, we use $u(x)$ as a different notation for $u_\varepsilon(x)$. If $a(x) \geq \alpha > 0$ and $(\varepsilon - \delta a(x)) > 0$ for $\forall x \in [0,1]$, the problem (3) has a boundary layer at $x=0$. Otherwise, the solution of problem (3) has a boundary layer at $x=1$.

3 Rational Spectral Collocation Method

3.1 The Rational Interpolation in Barycentric Form

Let $P_N(x)$ be a rational function in barycentric form which interpolates $u(x)$ at the Chebyshev-Gauss-Lobatto points $x_k = \cos(k\pi/N)$, $k = 0, 1, \cdots N$, then we have

$$p_N(x) = \sum_{k=0}^{N} L_k(x) u(x_k), \tag{4}$$

where

$$L_j(x) = \frac{\frac{\omega_j}{x-x_j}}{\sum_{k=0}^{N} \frac{\omega_k}{x-x_j}}$$

are the interpolation basis functions, $\{\omega_k\}_{k=0}^{N}$ are the nonzero numbers called barycentric weights, which are defined as follows:

$$\omega_0 = \frac{1}{2}, \omega_N = \frac{(-1)^N}{2}, \omega_k = (-1)^k, k = 1, 2, \cdots, N-1. \tag{5}$$

By simple calculation, we can obtain the n-th order derivatives of $p_N(x)$ at the point x_i as follows:

$$p_N^{(n)}(x_i) = \sum_{j=0}^{N} L_j^{(n)}(x_i)u(x_j) = \sum_{j=0}^{N} D_{ij}^{(n)}u(x_j), \quad n = 1, 2, \cdots, \tag{6}$$

where $\{D_{ij}^{n}\}_{i,j=0}^{N}$ are the n-th order differential matrices. The most commonly first and second order differential matrices are given by

$$D_{jk}^{(1)} = \begin{cases} \frac{w_k}{w_j(x_j-x_k)}, j \neq k, \\ -\sum_{i \neq j} D_{ji}^{(1)}, j = k, \end{cases} \tag{7}$$

$$D_{jk}^{(2)} = \begin{cases} 2D_{jk}^{(1)}(D_{jj}^{(1)} - \frac{1}{x_j-x_k}), j \neq k, \\ -\sum_{i \neq j} D_{ji}^{(2)}, j = k. \end{cases} \tag{8}$$

3.2 Numerical Discretization

First, by introducing the transformation $x = 0.5(y+1)$ and defining $\hat{u}(y) = u(x)$, we have $u'(x) = 2\tilde{u}'(y), u''(x) = 4\tilde{u}''(y)$. Then, the Eq. (3) can be written as follows:

$$\begin{cases} -4(\varepsilon - \delta\hat{a}(y))\hat{u}''(y) - 2\hat{a}(y)\hat{u}'(y) + \hat{b}(y)\hat{u}(y) = \hat{f}(y), y \in (-1, 1), \\ \hat{u}(-1) = \phi(0), \hat{u}(1) = \gamma. \end{cases} \tag{9}$$

Since the rational spectral collocation method applied on Chebyshev-Gauss-Lobatto points $y_k = \cos(k\pi/N)$ can't obtain satisfactory numerical results for very small ε. Therefore, we use the sinh transform given in Eq. (2) to obtain the following transformed Chebyshev points

$$\hat{y}_k = g(\cos(k\pi/N)), \tag{10}$$

where we also choose $\mu = \theta\varepsilon$, see [18, 19].

Let $\hat{P}_N(y)$ be a rational function in barycentric form which interpolates $\hat{u}(y)$ of problem (9) at the above transformed Chebyshev points $\hat{y}_k$, then we have

$$\hat{p}_N(y) = \sum_{j=0}^{N} L_j(y)\hat{u}(\hat{y}_j), \tag{11}$$

Substituting $\hat{P}_N(y)$ into the first equation of (9), yield

$$-4(\varepsilon - \delta\hat{a}(y))\sum_{j=0}^{N} \hat{U}_j L_j''(y) - 2\hat{a}(y)\sum_{j=0}^{N} \hat{U}_j L_j'(y) + \hat{b}(y)\sum_{j=0}^{N} \hat{U}_j L_j(y) = \hat{f}(y), \tag{12}$$

where $\hat{U}_j$ is the approximation solution of $\hat{u}(y)$ at point $\hat{y}_j$.

Let $y = \hat{y}_k, k = 0, 1, \cdots, N$, then the Eq. (12) can be written into the following matrix form

$$[-4ED^{(2)} - 2diag(\hat{a})D^{(1)} + diag(\hat{b})]\hat{U} = \hat{f}, \tag{13}$$

where

$$E = diag(\varepsilon - \delta\hat{a}(\hat{y}_0), \cdots, \varepsilon - \delta\hat{a}(\hat{y}_N)),$$

$$\hat{a} = (\hat{a}(\hat{y}_0), \cdots, \hat{a}(\hat{y}_N))^T, \quad \hat{b} = (\hat{b}(\hat{y}_0), \cdots, \hat{b}(\hat{y}_N))^T,$$
$$\hat{f} = (\hat{f}(\hat{y}_0), \cdots, \hat{f}(\hat{y}_N))^T, \quad \hat{U} = (\hat{U}_0, \cdots, \hat{U}_N)^T.$$

Furthermore, let $M = -4ED^{(2)} - 2diag(\hat{a})D^{(1)} + diag(\hat{b})$, then we have

$$M\hat{U} = \hat{f}. \tag{14}$$

In order to combine boundary conditions given in (3), we first use two row vectors $e_1 = (1, 0, \cdots, 0)$ and $e_{N+1} = (0, 0, \cdots, 1)$ with lengths $N+1$ to replace the first and $N+1$-th rows of matrix M, respectively. Then, we obtain a new matrix $\tilde{M}$. Similarly, by utilizing $\varphi(0)$ and γ to replace $\hat{f}(1)$ and $\hat{f}(N+1)$, respectively, we get a new right hand vector $\tilde{f}$. Finally, the numerical solution of Eq. (3) can be solved from the following linear system

$$\tilde{M}\hat{U} = \tilde{f} \tag{15}$$

4 The Parameter Optimization of Sinh Transform

4.1 Fitness Function

For a given parameter θ, let $\hat{U}^N(y_k)$ and $\hat{U}^{2N}(z_j)$ be the solutions to the discrete problem (15) computed on $\bar{\Omega}^N$ and $\bar{\Omega}^{2N}$, respectively, where

$$\bar{\Omega}^N = \{y_k\}_{k=0}^N = \{g(\cos(k\pi/N))\}_{k=0}^N,$$
$$\bar{\Omega}^{2N} = \{z_j\}_{j=0}^{2N} = \{g(\cos(j\pi/2N))\}_{j=0}^{2N}.$$

Then, we use the following formula to estimate the maximum point-wise error

$$E_\varepsilon^N = \max_{0\le j\le N} |\hat{U}^N(y_j) - \hat{U}^{2N}(z_{2j})|, \tag{16}$$

where $y_j = z_{2j}, j = 0, 1, \cdots, N$. Here, we also choose $\mu = \theta\varepsilon$, where θ is a positive constant. Thus, for each ε and N, the maximum point-wise error E_ε^N is a function about variable θ. In order to establish the optimal parameter θ, we may construct the following fitness function

$$Fitness = E_\varepsilon^N(\theta) = \max_{0\le j\le N} |\hat{U}^N(y_j) - \hat{U}^{2N}(z_{2j})|. \tag{17}$$

4.2 Original DE Algorithm Overview

Inspired by the natural evolution of species, DE algorithm was developed by Storn and Price [25]. It has gradually used in many practical cases, mainly because it has demonstrated good convergence properties and is principally easy to understand.

In the basic DE algorithm with *NP* *D*-dimensional parameter vectors, so-called individuals, each individual corresponds to a possible candidate solution, i.e. $X_{i,G} = \left\{x_{i,G}^1, \cdots, x_{i,G}^D\right\}$, where $i = 1, \cdots, NP$, G denotes one generation. DE aims to encode the candidate solutions towards the global optimum. In general, the initial population is chosen randomly with uniform distribution, and should better cover the entire search space as much as possible.

As is stated in [25], the basic DE algorithm has three operations: mutation, crossover and selection. Next, we give the basic idea of these operations as follows [34]:

A Mutation

After initialization, DE algorithm uses the mutation operation to obtain a mutant vector $V_{i,G}$ with respect to each individual $X_{i,G}$ so called target vector, in the current population. For each target vector $X_{i,G}$ so called target vector, in the current population.

For each target vector $X_{i,G}$ at the generation G, its associated mutant vector $V_{i,G} = \left\{v_{i,G}^1, v_{i,G}^2, \cdots v_{i,G}^D\right\}$ can be generated according to

$$V_{i,G+1} = X_{r_1^i,G} + F(X_{r_2^i,G} - X_{r_3^i,G}), \tag{18}$$

where the indices r_1^i, r_2^i, r_3 are mutually exclusive integers randomly generated within the range $[1, NP]$, which are also different from the index i. Note that these indices are randomly generated once for each mutant vector. F is a positive control parameter ($F \in [0, 2]$) for scaling the difference vector.

B Crossover Operation

After the mutation phase, crossover operation is used to each pair of the target vector $X_{i,G}$ and its corresponding mutant vector $V_{i,G}$ to generate a trial vector

$$U_{i,G} = (u_{i,G}^1, u_{i,G}^2, \cdots, u_{i,G}^D).$$

The basic DE algorithm employs the binomial (uniform) crossover defined as follows:

$$u_{i,G+1}^j = \begin{cases} v_{i,G+1}^j, if(rand_j \le CR)\ or\ (j = j_{rand}) \\ x_{i,G}^j,\ otherwise,\ j = 1, 2, \cdots, D, \end{cases}$$

where $CR \in [0, 1]$ is the crossover constant, which controls the fraction of parameter values copied from the mutant vector. j_{rand} is a randomly chosen index which ensures that $U_{i,G+1}$ gets at least one element from $V_{i,G+1}$. Otherwise, it is copied from the corresponding target vector $X_{i,G}$.

C Selection Operation

After crossover operation, a selection operation is carried out. Specifically, for the objective function value of each trial vector $f(U_i, G)$ and its corresponding target vector $f(X_i, G)$ in the current population, if the trial vector has less or equal objective function value than the corresponding target vector, the target vector $X_{i,G+1}$ is set to the trial vector $U_{i,G+1}$. Otherwise, the old target vector $X_{i,G}$ will retained in the population for the next generation. The selection operation can be expressed as follows:

$$X_{i,G+1} = \begin{cases} U_{i,G}, \text{if } f(U_{i,G}) \le f(X_{i,G}) \\ X_{i,G},\ otherwise. \end{cases}$$

4.3 Self-adaptive DE (SADE) Algorithm

As is stated in [25], DE algorithm is much more sensitive to the choice of F than it is to the choice of CR. We recall that D is the dimensionality of the problem, the control parameters of DE algorithm are set as follows:

$$(1)\, F \in [0.5, 1]; \quad (2)\, CR \in [0.8, 1]; \quad (3)\, NP = 10D.$$

Recently, Huang [27] used the parameters set to $F = 0.9, CR = 0.9$. Ali and Torn [35] chosen $CR = 0.5$, and F was calculated according to the following scheme:

$$F = \begin{cases} max(l_{min}, 1 - |\frac{f_{max}}{f_{min}}|), \text{if} |\frac{f_{max}}{f_{min}}| < 1, \\ max(l_{min}, 1 - |\frac{f_{min}}{f_{max}}|), otherwise, \end{cases}$$

where $F \in [l_{\min}, 1], f_{\max}$ and $f_{\min}$ are the maximum and minimum values of vectors $X_{i,G}$, respectively. $l_{\min}$ is the lower bound for F. In [33], the authors designed a self-adaptive control mechanism to change the control parameters F and CR during the run. The new control parameters or factors $F_{i,G+1}$ and $CR_{i,G+1}$ were calculated as

$$F_{i,G+1} = \begin{cases} F_l + rand_1 * F_u, \text{if } rand_2 < \tau_1, \\ F_{i,G}, \; otherwise \end{cases}, CR_{i,G+1} = \begin{cases} rand_3, & \text{if } rand_4 < \tau_2, \\ CR_{i,G}, & otherwise \end{cases},$$

and they produced factors F and CR in a new parent vector. Here, $rand_j, j \in \{1, 2, 3, 4\}$ are uniform random values. $\tau_i, i = 1, 2$ represent probabilities to adjust factors F and CR. $F_{i,G+1}$ and $CR_{i,G+1}$ are obtained before the mutation is performed.

In our paper, we also use the idea from [33] to change the control parameters F and CR. We also set $\tau_1 = \tau_2 = 0.1$ and $F_u = 0.9$. Obviously, the new F takes a value from $[0.1, 1.0]$ in a random manner, and the new CR takes a value form $[0, 1]$.

5 Numerical Experiments and Discussion

In this section, we consider the following test example to demonstrate the effectiveness of the proposed method

$$\begin{cases} L_\varepsilon u(x) := \varepsilon u_\varepsilon''(x) - (1+x)u_\varepsilon'(x-\delta) + \exp(-x)u_\varepsilon(x) = 1, \;\; x \in (0,1) \\ u_\varepsilon(x) = 1, \quad -\delta \le x \le 0, \quad u_\varepsilon(1) = 1. \end{cases}$$

We recall that U^N and U^{2N} be the optimal solutions to Eq. (3) calculated on meshes $\bar{\Omega}^N$ and $\bar{\Omega}^{2N}$ respectively. Then, we can use the following expression to estimate the maximum error

$$E_\varepsilon^N = ||\hat{U}^N - \hat{U}^{2N}||_\infty. \tag{19}$$

5.1 Experimental Environment and Parameter Settings

The SADE is validated by comparing it with six optimization algorithms, namely, differential evolution (DE) [25], Particle swarm optimization (PSO) [37], Flower pollination algorithm (FPA) [36], GA, CMAES [38] and Nelder-Mead simplex (NMS) method [39]. For the sake of fairness, we use a fixed population size for all algorithms: $n = 50$ individuals; all algorithms are executed in 30 independent runs; the number of iterations in each run for each algorithm equals 50 iterations; the range of parameter θ is $[1, 100]$. The details about other parameters settings are listed in Table 1. All algorithms are implemented in Matlab2014a to program a m-file for implementing the algorithms on a PC with a **32**-bit windows 8 operating system, a **8** GB RAM, and a 3.10 GHz-core (TM) i **5**-based processor.

Table 1. The main parameters setting of algorithms used in experimental results

Algorithms	Parameters setting
FPA	Switch probability P = 0.9, γ = 0.01, λ = 1.5
DE	Crossover factor F = 0.5, crossover probability CR = 0.1
PSO	Learning factors c1 = c2 = 2, inertia factor dropped from 0.9 to 0.4
CMAES	Parameters are fixed automatically by the algorithm
GA	Crossover probability 0.9, mutation probability 0.5, Rank-based ration 0.1

5.2 Experimental Results and Analyses

In order to confirm the algorithm's effectiveness, comparison experiments are done by using FPA, DE, PSO, GA, CMAES algorithms and SADE, respectively. At first, for $\varepsilon = 10^{-6}$, $\delta = 10^{-7}$ and $N = 32$, the statistical data about the maximum, minimum, average and variances of maximum errors E_{ε}^{N} are given in Table 2. It is shown from this table that the maximum errors E_{ε}^{N} calculated by DE and SADE algorithms are smaller than that of other algorithms. Besides, the variance of the SADE algorithm is smallest. Thus, the SADE algorithm works very well in the numerical results.

Table 2. Numerical results by using different algorithms

Method	Maximum error	Minimum error	Mean value	Variance
FPA	3.8077e−04	4.8435e−06	9.4050e−05	8.0965e−09
GA	5.6538e−04	4.0915e−06	1.4539e−04	2.8098e−08
DE	3.6189e−06	3.6188e−06	3.6189e−06	1.1732e−22
SADE	3.6189e−06	3.6188e−06	3.6189e−06	1.3805e−22
PSO	1.4749e−05	3.6212e−06	5.2385e−06	1.0846e−11
CMAES	6.1402e−03	3.6189e−06	1.2338e−03	6.2258e−06

Table 3. Numerical results by using NMS algorithm

δ	N = 32	N = 64	N = 128	N = 256
10^{-5}	3.4318e−09	8.3324e−11	7.4949e−11	9.0861e−09
10^{-6}	2.4316e−09	9.8556e−11	8.8014e−11	8.2929e−11
10^{-7}	2.5124e−09	1.0038e−10	8.9622e−11	7.4803e−11

In order to compare the SADE algorithm with the traditional optimization algorithm, we choose the NMS algorithm to solve above objective function (17). First, let θ_0 be the initial value of parameter θ. Then, when $\varepsilon = 10^{-4}$, $\theta_0 = 1$, the maximum errors calculated by using NMS method are listed in Table 3 for different values of δ and N. Meanwhile, Table 4 also give the maximum errors calculated by the SADE algorithm for the same parameters ε, δ and N. Obviously, when $N = 32$, the maximum errors obtained by using SADE algorithm are lower than that obtained by using NMS method.

Table 4. Numerical results by using SADE algorithm

δ	N = 32	N = 64	N = 128	N = 256
10^{-5}	1.5828e−10	8.3294e−11	7.4803e−11	5.9086e−09
10^{-6}	2.3565e−10	7.9966e−11	8.7934e−11	8.2772e−11
10^{-7}	2.4616e−10	1.0031e−10	8.3682e−11	7.4803e−11

6 Conclusions

In this paper, a new high accuracy numerical method based on the rational spectral collocation method and SADE algorithm has been developed for solving singularly perturbed differential difference equations. Generally speaking, we first construct a nonlinear unconstrained optimization problem based on minimum absolute error. Then, the SADE algorithm is given to solve the presented optimization problem.

Acknowledgement. This work is supported by National Science Foundation of China (11761015, 11461011, 11561009), the Natural Science Foundation of Guangxi (2017GXNSFBA198183), the key project of Guangxi Natural Science Foundation (2017GXNSFDA198014), "BAGUI Scholar" Program of Guangxi Zhuang Autonomous Region of China, and Innovation Project of Guangxi Graduate Education (JGY2017086).

References

1. Elsgolts, L.E., El-Sgol-Ts, L.E., El-Sgol-C, L.E.: Qualitative Methods in Mathematical Analysis. American Mathematical Society, Providence (1964)
2. Mackey, M.C., Glass, L.: Oscillation and chaos in physiological control systems. Science **197**, 287–289 (1977)

3. Lasota, A., Wazewska, M.: Mathematical models of the red blood cell systems. Mat. Stos **6**, 25–40 (1976)
4. Lange, C.G., Miura, R.M.: Singular perturbation analysis of boundary value problems for differential- difference equations. V. small shifts with layer behavior. SIAM J. Appl. Math. **54**, 249–272 (1994)
5. Kadalbajoo, M.K., Sharma, K.K.: Numerical analysis of singularly perturbed delay differential equations with layer behavior. Appl. Math. Comput. **157**, 11–28 (2004)
6. Kadalbajoo, M.K., Sharma, K.K.: A numerical method based on finite difference for boundary value problems for singularly perturbed delay differential equations. Appl. Math. Comput. **197**, 692–707 (2008)
7. Kadalbajoo, M.K., Ramesh, V.P.: Hybrid method for numerical solution of singularly perturbed delay differential equations. Appl. Math. Comput. **187**, 797–814 (2007)
8. Kadalbajoo, M.K., Kumar, D.: Fitted mesh B-spline collocation method for singularly perturbed differential-difference equations with small delay. Appl. Math. Comput. **204**, 90–98 (2008)
9. Rao, R.N., Chakravarthy, P.P.: A finite difference method for singularly perturbed differential-difference equations with layer and oscillatory behavior. Appl. Math. Model. **37**, 5743–5755 (2013)
10. Mohapatra, J., Natesan, S.: Uniform convergence analysis of finite difference scheme for singularly perturbed delay differential equation on an adaptively generated grid. Numer. Math. Theor. Meth. Appl. **3**, 1–22 (2010)
11. Liu, L.-B., Chen, Y.: Maximum norm a posterior error estimations for a singularly perturbed differential-difference equation with small delay. Appl. Math. Comput. **227**, 1–10 (2014)
12. Baltensperger, R.J., Berrut, P., Noël, B.: Exponential convergence of a linear rational interpolant between transformed Chebyshev points. Math. Comp **68**, 1109–1120 (1999)
13. Baltensperger, R., Berrut, J.P., Dubey, Y.: The linear rational pseudo-spectral method with preassigned poles. Numer. Alg. **33**, 53–63 (2003)
14. Luo, W.-H., Huang, T.-Z., Gu, X.-M., et al.: Barycentric rational collocation methods for a class of nonlinear parabolic partial differential equations. Appl. Math. Lett. **68**, 13–19 (2017)
15. Chen, S., Wang, Y.: A rational spectral collocation method for third-order singularly perturbed problems. J. Comput. Appl. Math. **307**, 93–105 (2016)
16. Wang, Y., Chen, S., Wu, X.: A rational spectral collocation method for solving a class of parameterized singular perturbation problems. J. Comput. Appl. Math. **233**, 2652–2660 (2010)
17. Raja, M.A.Z.: Stochastic numerical techniques for solving Troesch's problem. Inform. Sci. **279**, 860–873 (2014)
18. Raja, M.A.Z.: Numerical treatment for boundary value problems of pantograph functional differential equation using computational intelligence algorithms. Appl. Soft Comput. **24**, 806–821 (2014)
19. Raja, M.A.Z.: Numerical treatment of nonlinear MHD Jeffery-Hamel problems using Stochastic algorithms. Comput. Fluids **91**, 28–46 (2014)
20. Sobester, A., Nair, P.B., Keane, A.J.: Genetic programming approaches for solving elliptic partial differential equations. IEEE. Trans. Evolut. Comput. **12**, 469–478 (2008)
21. Fateh, M.F., Zameer, A., Mirza, N.M., et al.: Biologically inspired computing framework for solving two-point boundary value problems using differential evolution. Neural Comput. Appl. **28**, 1–15 (2016)
22. Luo, X.-Q., Liu, L.-B., Ouyang, A., et al.: B-spline collocation and self-adapting differential evolution (jDE) algorithm for a singularly perturbed convection-diffusion problem. Soft. Comput. **22**, 2683–2693 (2018)

23. Liu, L.-B., Long, G., Ouyang, A., et al.: Numerical solution of a singularly perturbed problem with Robin boundary conditions using particle swarm optimization algorithm. J. Intell. Fuzzy Syst. **33**, 1785–1795 (2017)
24. Liu, L.-B., Long, G., Huang, Z., et al.: Rational spectral collocation and differential evolution algorithms for singularly perturbed problems with an interior layer. J. Comput. Appl. Math. **335**, 312–322 (2018)
25. Storn, R., Price, K.: Differential evolution–A simple and efficient heuristic for global optimization over continuous spaces. J. Glob. Optim. **11**, 341–359 (1997)
26. Huang, D.S., Du, J.-X.: A constructive hybrid structure optimization methodology for radial basis probabilistic neural networks. IEEE Trans. Neural Netw. **19**, 2099–2115 (2008)
27. Huang, D.S.: Systematic Theory of Networks for Pattern Recognition. Publishing House of Electronic Industry of China, Beijing (1996)
28. Du, J.-X., Huang, D.S., Zhang, G.-J., et al.: A novel full structure optimization algorithm for radial basis probabilistic neural networks. Neurocomputing **70**, 592–596 (2006)
29. Du, J.-X., Huang, D.S., Wang, X.-F., et al.: Shape recognition based on neural networks trained by differential evolution algorithm. Neurocomputing **70**, 896–903 (2007)
30. Huang, D.S.: Radial basis probabilistic neural networks: model and application. Int. J. Pattern Recogn. **13**, 1083–1101 (1999)
31. Eiben, A.E., Hinterding, R., Michalewicz, Z.: Parameter control in evolutionary algorithms. IEEE. Tran. Evol. Comput. **3**, 124–141 (1999)
32. Eiben, A.E., Smith, J.E.: Introduction to Evolutionary Computing. Natural Computing Series. Springer, Berlin, Germany (2003). https://doi.org/10.1007/978-3-662-05094-1
33. Brest, J., Greiner, S., Boskovic, B., et al.: Self-adapting control parameters in differential evolution: a comparative study on numerical benchmark problems. IEEE Trans. Evol. Comput. **10**, 646–657 (2006)
34. Qin, A.K., Huang, V.L., Suganthan, P.N.: Differential evolution algorithm with strategy adaptation for global numerical optimization. IEEE Trans. Evol. Comput. **13**, 398–417 (2009)
35. Ali, M.M., Torn, A.: Population set-based global optimization algorithms: Some modifications and numerical studies. Comput. Oper. Res. **31**, 1703–1725 (2004)
36. Yang, X.-S.: Flower pollination algorithm for global optimization. In: Durand-Lose, J., Jonoska, N. (eds.) UCNC 2012. LNCS, vol. 7445, pp. 240–249. Springer, Heidelberg (2012). https://doi.org/10.1007/978-3-642-32894-7_27
37. Kennedy, J., Eberhart, R.: Particle swarm optimization. In: Proceeding of the IEEE international conference on neural networks, pp. 1942–1948. IEEE press, Piscataway, NJ, USA (1995)
38. Hansen, N., Muller, S.D., Koumoutsakos, P.: Reducing the time complexity of the derandomized evolution strategy with covariance matrix adaptation (CMA-ES). Evol. Comput. **11**, 1–18 (2014)
39. Nelder, J.A., Mead, R.: A simplex method for function minimization. Comput. J. **7**, 308–313 (1965)

PSO Application in CVAR Model

Liu Yanmin[1(✉)], Leng Rui[1], Sui Changling[2], Yuan Lian[1], and Huang Tao[1]

[1] College of Mathematics and Computer Science, Zunyi Normal University, Zunyi 563002, China
yanmin7813@163.com
[2] College of Biology and Agriculture, Zunyi Normal University, Zunyi 563002, China

Abstract. In order to solve mean variance model with the conditional value at risk (CVaR), an improvement PSO with the generalized learning and the hybrid mutation of dynamic cauchy and the normal cloud model (PSOHM) is proposed to increase the diversity of the population. In PSOHM, to enhance the ability of the population, the introduction of a generalized learning strategy is introduced to enhance flying to the optimal solution for the whole swarm, and according to swarm performance, two different mutation is stimulated to produce the new individual to guide the population flying better. In benchmark function test, the result shows that PSOHM has better performance results. In the portfolio optimization model of CVaR, PSOHM has a better results compared with other algorithms.

Keywords: Particle swarm optimizer · Cauchy mutation · Normal cloud model
Conditional Value-at-Risk

1 Introduction

Markowitz proposed the portfolio theory of mean variance in 1952, which explored the relationship between the rate of return of risk assets and the risk [1]. The theory takes variance as the risk function to solve the minimum variance portfolio in a certain income level under the minimum variance portfolio in order to provide a feasible trade-offs means. Risk value (Value-at-Risk, VaR) is a new concept proposed in 1993 by the bank of international settlements, which can not only calculate the risk in advance, but also can measure the overall portfolio. Based on the merit, VaR has gradually been a mainstream method of financial risk, and many financial institutions and business departments are considered to meet the requirements of VaR in investment choices. But in practical application, VaR has some defects. Therefore, an improved VaR model called Conditional VaR (CVaR [3]), is proposed in the literature [2]. Based on [2], the literature [3, 4] explored the optimization algorithm and the application of CVaR which was extended to the loss of general distributions, and the authors also proposed an auxiliary function method to deal with CVaR constraints in practical application. By auxiliary function, CVaR model by constructing an auxiliary function can be a portfolio optimization problem with the linear programming CVaR constraints, and the

D.-S. Huang et al. (Eds.): ICIC 2018, LNAI 10956, pp. 662–669, 2018.
https://doi.org/10.1007/978-3-319-95957-3_68

optimal solution can be obtained by solving the linear programming. However, the disadvantages of this transformation make dimension constraints become great. Therefore, seeking a way to effectively solve the portfolio optimization mode with CVaR constraints is very important for practical application. In view of this, this paper presents a generalized learning strategy and the hybrid mutation of dynamic cauchy and the normal cloud model (PSOHM) to increase the diversity of the population.

2 CVaR Model

VaR (Value at Risk) is the maximum expected losses of a financial asset or their portfolio under a certain confidence level during a certain period. In brief, in a certain period of time Δt and under a certain confidence level 1−c, a certain portfolio confronts the biggest loss. The mathematical expression is as follows,

$$p(\Delta p \leq VaR) = 1 - c \tag{1}$$

Where, Δp is a market value change of a portfolio of assets in a certain period of time Δt;$1-c$ represents that in certain holding period Δt, the portfolio loss will not exceed the maximum VaR under the confidence level of $1-c$.

However, with the development of the financial market, it is found that VaR has some defects. Therefore, Rockafellar and Uryasev [2] put forward the improvement of risk measurement in CVaR (Conditional Value-at-Risk). Its meaning can be interpreted as the conditional mean losses of VaR, which reflects the average level of excess loss.

Let $f(x, y)$ said the loss function of an investment portfolio, where y with a random variables of density $p(y)$ is the loss factor caused by the market portfolio (here, y is a continuous random variable). For any order $\alpha \in R$, let

$$\psi(x, \alpha) = \int_{f(x,y) \leq \alpha} p(y)dy \tag{2}$$

Where, $\psi(x, \alpha)$ indicates that the loss $f(x, y)$ is less than probability of α. On any $\beta \in (0, 1)$, defined,

$$F_\beta(x, \alpha) = \alpha + (1 - \beta)^{-1} \int_{y \in R^m} [f(x, y) - \alpha]^+ p(y)dy \tag{3}$$

Where, T^+ function is the maximum in {0, T}. $A_\beta(x) = \arg\min_{\alpha \in R} F_\beta(x, \alpha)$ is a non-empty, closed bounded set. Equation (4) gives value of CVaR.

$$CVaR = \min F_\beta(x, \alpha) \tag{4}$$

Here, we adopt the processing of constraints with the Eq. (5) $c_1 = c_2 = 100$)

$$\min\ F = F_{\beta}(x, \alpha) + c_1 \cdot \left|1 - \sum_{i=1}^{n} x_i\right| + c_2 \cdot \max\{\rho - x^T m, 0\} \tag{5}$$

3 PSO with the Hybrid Mutation of Dynamic Cauchy and the Normal Cloud Model (PSOHM)

3.1 Learning Strategies

The faster convergence speed in PSO can easily lead to the decline of population diversity, and premature convergence [6]. To solve this problem, scholars explore many research from different perspectives to the study of [7, 8]. In this paper, we proposed a generalized learning strategy based on the literature of [7]. The particle velocity is controlled by adjusting the particle itself in three aspects: the optimal position of other particles, the global optimal particle and particle in the neighborhood. The proposed strategies expand the particle's learning samples, and increase the diversity of population. Equations (6) and (7) give the particle evolution equation of generalized learning strategies.

$$\begin{aligned} \vec{v}_i(t+1) = \vec{v}_i(t) + \varphi_1 \cdot r_1(\vec{p}_i(t) - \vec{x}_i(t)) + \varphi_2 \cdot r_2(\vec{p}_g(t) - \vec{x}_i(t)) + \varphi_3 \cdot r_3(\overrightarrow{p_{bin(i)}}(t) - \vec{x}_i(t)) \\ \vec{x}_i(t+1) = \vec{x}_i(t) + \vec{v}_i(t+1) \end{aligned} \tag{6}$$

$$\begin{gathered} p^d_{bin(i)} = \arg\{\max[\frac{Fitness(p_j) - Fitness(p_i)}{|p_{jd} - x_{id}|}]\} \\ d \in (1, 2, \cdots, n),\ i = 1, 2, \cdots, ps \ \ j \in neighbor_i \end{gathered} \tag{7}$$

Where, $\overrightarrow{p_{bin(i)}}$ is sample of each dimension of the particle i; $neighbor_i$ is the neighbor collection of particles; $p^d_{bin(i)}$ is the d dimension of $\overrightarrow{p_{bin(i)}}$, here, $\varphi_1 = \varphi_2 = 1.49445$.

In [3], the author discusses the mutation role in promoting the performance of search algorithms, and inspired by this idea, we proposed double mutation operation, i.e., normal mutation operation based on cloud model to improve the searching ability and the dynamic Cauchy mutation operation.

3.2 Mutation Based on Normal Cloud Model

The cloud mutation sees the literature [5]. In addition, because PSO is an evolutionary algorithm based on population, it needs the algorithm as much as possible to explore the feasible region in order to increase the probability of searching the global optimal solution, however with the iteration elapse, the feasible region gradually narrowed to explore the early solution search domain. In view of this, the variation of cloud is a certain dynamic probability, i.e., with the iteration the variation probability of the global optimal particle (Gbest) and the particle itself best position (Pbest) will gradually decreases in order to extract the best solution achieved in the early search process. The expression of the dynamic mutation probability is shown in Eq. (8)

$$prob_m = t^n - 2 \times t + 1 \tag{8}$$

Where, $t = current\_\text{generation}/total\_\text{generation}$; n = 1.6.

3.3 Dynamic Cauchy Mutation

From the evolution equation of particles, it can be seen, when the global optimal particle (gbest) and its optimal particles (pbest) have no improvement for continuous several generations, the population will be trapped in local optimal solution. The literature [9] confirmed that the cauchy mutation has a positive effect.At the same time, the experiment in literature [10] can conclude that PSO in the early stages will explore feasible region as much as possible, which increase the probability of searching the global optimal solution. Therefore, a dynamic mutation probability $prob_m$ (Eq. (11)) is adopted to determine whether the *pbest* uses cauchy mutation. The principle is: when Gbest has no improvement for the consecutive 4 generation (N = 4), the cauchy mutation operation shown in Eq. (10) is conducted to generate new particles to guide population flight. Figure 1 gives the pseudo code of Cauchy mutation operation, and Fig. 2 gives the dynamic mutation probability (here n = 1.6).

```
Initialize positions and associated velocities of all particles.
While (fitcount < Max_FES) && (k<iteration)
    For each particle (i=1:ps)
        Updating velocity and position.
        Update pbest and fitness values.
        If the condition 1 is satisfied
            Cauchy mutation for each pbest.
        End if
        If the condition 2 is satisfied
            Normal cloud  mutation for each pbest.
        End if
    End for
    Update gbest
    If the condition 1 is satisfied
        Cauchy mutation for Gbest.
    End if
    If the condition 2 is satisfied
        Normal cloud  mutation for each Gbest.
    End if
    Output results
End While
```

Fig. 1. PSOHM pseudo-code

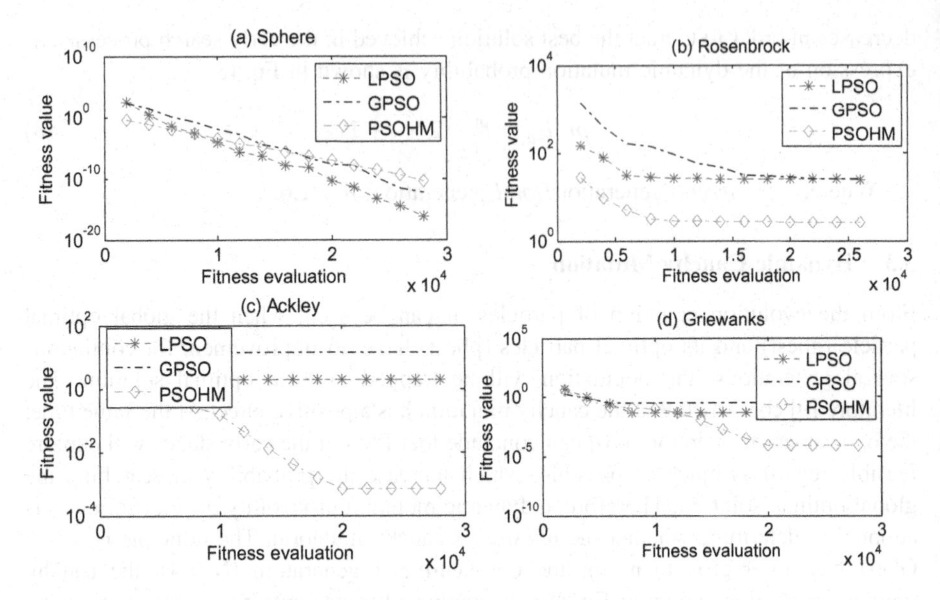

Fig. 2. Convergence characteristic curve

$$f(x) = \frac{1}{\pi} \frac{t}{t^2 + x^2} \qquad -\infty < x < \infty \tag{9}$$

$$x_i^j = x_i^j + v_i^j \cdot C(x_{\min}, x_{\max}) \tag{10}$$

Where, x_i^j is the j dimension of particle i; $C(x_{\min}, x_{\max})$ produce a random number of Cauchy distribution with Cauchy distribution function (see Eq. (10)), here parameter t = 1.

$$\begin{cases} t = \dfrac{current\_generation}{total\_generation} \\ prob_m = t^n - 2 \times t + 1 \end{cases} \tag{11}$$

In the evolution process of population, the particle's position is likely beyond the given range, thus, the Eq. (12) is used to limit the scope of each particle.

$$x_i^j = x_{\min}^j + rand \cdot (x_{\max}^j - x_{\min}^j) \tag{12}$$

Where, $x_{\max}^j, x_{\min}^j$ are the maximum and minimum values of the j dimension of particle i.

As stated earlier, mutation has active role for swarm evolution, to better use its role, so based on [11] we adopt mutation of normal cloud model and dynamic cauchy mutation in terms of swarm performance. The rule is: when pbest has no improvement, the mutation based on normal cloud model is invoked, and else the dynamic cauchy mutation is conducted. Figure 1 shows the PSOHM pseudo code.

To test the performance of PSOHM, Rosenbrock, Ackley, Griewanks and Rastrigin function is selected to test algorithm performance. The dimension of all test function is 30, and the search intervals are [−2.048, 2.048], [−32.768, 32.768], [−600, 600], [−5.12, 5.12], respectively. Additionally, we compare PSOHM with GPSO and LPSO, respectively. Each algorithm is running 30 times independently in four benchmark functions with 3×10^4 function evaluation (Fitness Evaluations). Figure 2 shows the convergence of various algorithms.

As can be seen from the Fig. 2, except for Sphere function, PSOHM have better convergence precision compared with LPSO and GPSO, which shows that the proposed PSOHM is an effective improvement of the basic PSO.

4 PSO Application in CVAR Model

4.1 Data and Parameter Settings

Data source: selecting 20 stocks on China's securities market as the research object (see Table 1) from January 8, 1999 to December 28, 2001. The data of 148 weeks are collected, and the rate of return of the corresponding yields in 147 groups of weeks. The Eq. (13) gives the formula of the rate of return with an average yield R and the lowest return rate($\rho = 0.3\%$).

$$r_t = \frac{p_t}{p_{t-1}} - 1 \tag{13}$$

Where, P_t is the stock price in the t time. To validate the effectiveness of the proposed algorithm when solving the investment model of CVaR constraint, the SwPSO in [11], genetic algorithm (GA), and the basic particle swarm optimization (PSO) are chosen. The parameter settings of other algorithm are consistent with the literature [11].

Table 1. The stock data

Stock code	Rate of return (R)	Stock code	Rate of return (R)
000002	0.41	600631	0.23
000039	0.39	600642	0.61
600058	0.68	600649	0.20
600098	0.64	600663	0.14
600100	0.11	600688	0.29
600115	0.37	600690	0.12
600183	0.39	600776	0.28
000541	0.24	600811	0.29
000581	0.57	600812	0.31
600600	0.31	600887	0.21

4.2 Experimental Results and Analysis

Table 2 gives the results for different algorithm in 50 independent running, in which the optimal value reflects the searching precision, the average value reflects the searching capability of the algorithm and the standard deviation reflects the stability of the algorithm. The time is the running time of each algorithm by using MATLAB software in (TIC, TOC) command. From Table 2 it can be seen that PSOHM has the better performance in the optimal value, average value and the standard deviation compared with other three algorithms. Additionally, From running time, PSOHM has the same order of magnitude with other algorithm and the introduction of the strategy has no incensement in computational complexity.

Table 2. Performance result for algorithm

Algorithm		Optimal value	Average value	Standard deviation	Running time (S)
$\beta = 0.99$	SwPSO	0.0691	0.0701	0.0010	4.5891
	PSO	0.0737	0.0771	0.0025	5.0314
	GA	0.0731	0.0785	0.0040	53.8974
	PSOHM	0.0612	0.0621	0.0032	7.3421
$\beta = 0.95$	SwPSO	0.0563	0.0600	0.0009	4.5773
	PSO	0.0601	0.0655	0.0026	5.0930
	GA	0.0612	0.0664	0.0039	52.2398
	PSOHM	0.0408	0.0531	0.0012	11.234
$\beta = 0.90$	SwPSO	0.0471	0.0512	0.0008	4.5882
	PSO	0.0508	0.0552	0.0018	5.0386
	GA	0.0507	0.0555	0.0026	52.5413
	PSOHM	0.0399	0.0429	0.0011	13.163

5 Conclusions

In this paper, we regarded CVaR as a tool to assess portfolio risk, and proposes an improved PSO (PSOHM) to optimize the nonlinear programming model of portfolio with CVaR constraints. In PSOHM, the generalized learning strategy is introduced to greatly enhance the ability of the population to jump out of local optimal solution, and the hybrid mutation of dynamic cauchy and the normal cloud model (PSOHM) is introduced to increase the diversity of the population. Compared with other algorithms, PSOHM has better performance when optimizing portfolio model with CVaR constraints, therefore, PSOHM can be regarded as an effective method for solving portfolio model.

Acknowledgments. This work is supported by the National Natural Science Foundation of China (Grants nos. 71461027, 71471158); Qian KeHE (NY Zi [2016]3013, LH Zi [2015]7033, J Zi LKZS[2014]06); Guizhou province natural science foundation in China (Qian Jiao He KY [2014]295); Zhunyi innovative talent team (Zunyi KH(2015)38); Science and technology talent training object of Guizhou province outstanding youth (Qian ke he ren zi [2015]06); Guizhou science and technology cooperation plan (Qian Ke He LH zi [2016]7028); Project of teaching quality and teaching reform of higher education in Guizhou province (Qian Jiao gaofa[2015]337) and 2016; 2013,2014 and 2015 Zunyi 15851 talents elite project funding; Innovative talent team in Guizhou Province (Qian Ke HE Pingtai Rencai[2016]5619); College students' innovative entrepreneurial training plan (201510664016).

References

1. Markowitz, H.: Portfolio selection. J. Finan. **7**(1), 77–91 (1952)
2. Rockafellar, R.T., Uryasev, S.: Optimization of conditional value at risk. J. Risk **2**(3), 21–41 (2000)
3. Krokhmal, P., Palmquist, J.: Portfolio optimization with conditional value-at-risk objective and constraints. J. Risk **4**(2), 43–68 (2002)
4. Alexander, G.J., Baptista, A.M.: Economic implications of using a Mean-VaR model for portfolio selection: a comparison with mean-variance analysis. J. Econ. Dyn. Control **26**(4), 1159–1193 (2006)
5. Kennedy, J., Eberhart, R.C.: Particle swarm optimization. In: Proceedings of IEEE International Conference on Neural Networks, Piscataway, USA, pp. 1942–1948 (1995)
6. Ratnaweera, A., Halgamuge, S.: Self-organizing hierarchical particle swarm optimizer with time-varying acceleration coefficients. IEEE Trans. Evol. Comput. **8**(3), 240–255 (2004)
7. Liu, Y.M., Zhao, Q.Z.: A kind of particle swarm algorithm based on dynamic neighbor and mutation factor. Control Decis. **25**(7), 968–974 (2010)
8. Mendes, R., Kennedy, J.: The fully informed particle swarm: simpler, maybe better. IEEE Trans. Evol. Comput. **8**(3), 204–210 (2004)
9. Wang, J., Liu, C.: A hybrid particle swarm algorithm with Cauchy mutation. In: IEEE Swarm Intelligence Symposium, Honolulu, USA, pp. 356–360 (2007)
10. Clerc, M., Kennedy, J.: The particle swarm explosion, stability, and convergence in a multi-dimensional complex space. IEEE Trans. Evol. Comput. **6**(1), 58–73 (2002)
11. Liu, Y.M., Zhao, Q.Z.: Improved particle swarm optimizer and its application for CvaR model. Math. Pract. Theory **41**(17), 139–147 (2011)

GreeAODV: An Energy Efficient Routing Protocol for Vehicular Ad Hoc Networks

Thar Baker[1(✉)], Jose M. García-Campos[2], Daniel Gutiérrez Reina[3], Sergio Toral[2], Hissam Tawfik[4], Dhiya Al-Jumeily[1], and Abir Hussain[1]

[1] Department of Computer Science, Liverpool John Moores University, Liverpool, UK
{t.baker,d.aljumeily,a.hussain}@ljmu.ac.uk
[2] Electronic Engineering Department, University of Seville, Seville, Spain
jmgarcam@gmail.com, storal@us.es
[3] Engineering Department, Loyola Andalucía University, Seville, Spain
d.gutierrez.reina@gmail.com
[4] School of Computing, Creative Technologies and Engineering, Leeds Beckett University, Leeds, UK
H.Tawfik@leedsbeckett.ac.uk

Abstract. VANETs allow communications among vehicles, and vehicles with the roadside infrastructure, namely Vehicle-to-Vehicle (V2V) and Vehicle-to-Infrastructure (V2I) respectively, in smart cities. Due to the number of vehicles, the infrastructure elements, the size of scenarios and mobility of nodes, the energy consumed to discover routes between source and destination nodes and to transmit applications packets can be high. In this paper, we propose a GreeDi based reactive routing protocol aimed at selecting the most efficient route in terms of energy consumption between two nodes in VANETs. The route selection is based on the power consumed by the intermediate nodes between the source and destination nodes. The proposed algorithm has been evaluated in city map-based VANET scenarios. The simulation results confirm that the proposed algorithm outperforms the original AODV in terms of power consumption. Furthermore, a computational Intelligence driven approach to address the challenge of energy efficient routing optimisation, is discussed.

Keywords: VANETs · Energy consumption · AODV · Routing protocols Computational intelligence

1 Introduction

An ad hoc network is an autonomous system of wireless mobile nodes that cooperatively form a network without a specific administration. Each node in an ad hoc network is in charge of routing information between its neighbors. We refer to nodes that are free to move randomly and organize themselves arbitrarily as Mobile Ad Hoc NETworks (MANETs) [1]. As an evolution of traditional MANETs, VANETs (Vehicular Ad hoc NETworks) [2] include communications between moving vehicles on the roads and with the roadside infrastructure. The main objective of a VANET is to

D.-S. Huang et al. (Eds.): ICIC 2018, LNAI 10956, pp. 670–681, 2018.
https://doi.org/10.1007/978-3-319-95957-3_69

provide safety and comfort to vehicle passengers. Each vehicle is integrated with wireless devices that can receive and relay messages such as traffic information, roadblock, parking, location tracking, fuel stations and weather information among others. Such massive exchange of information causes a high power consumption that leads to high fuel consumption, and consequently, CO_2 emissions.

Routing protocols for VANETs are not normally designed to take power consumption into consideration [3]. In general, they are intended for considering the high mobility of nodes and unpredictability of the wireless medium. However, saving energy is also an important matter in the design of green smart cities where VANETs are envisioned to play an important role. Therefore, it is important to propose new energy efficient routing protocols for VANETs. In this paper, we proposed GreeAODV a reactive routing protocol based on the GreeDi algorithm proposed in [4], and discussed in [5–7]. In this case, the rationale proposed in [4] is used to select energy efficient routing paths in VANETs. GreeDi [4] is based on the idea that each communication device has different power consumption when transmitting and receiving information. Using such rationale, the shortest path between two communication nodes would not always be the most efficient in terms of power consumption. In multi-hop networks like VANETs, the energy consumed by the communication routes should be calculated by using the energy information about all the participating nodes in the multi-hop communication path. Although GreeDi is designed for static networks such as the ones found in the communications between Internet nodes and data center presented in [4], we believe that its rationale for selecting routes of communications will be also valid for highly dynamic scenarios like VANETs. Consequently, in this paper we evaluate the GreeDi algorithm in dynamic networks such as VANETs.

As such, the main contributions of this paper are:

- The design of GreeAODV a novel reactive routing protocol for vehicle communications based on GreeDi [4] that selects the most efficient routes in terms of power consumption.
- The evaluation of the proposed GreeAODV in realistic urban scenarios, and its comparison with Ad Hoc On-Demand Distance Vector (AODV) routing protocol.
- A proposal of a computational Intelligence approach to the challenge under consideration.

This paper continues as follows; Sect. 2 reviews existing work on energy efficiency-based routing protocols intended for VANETs. Section 3 presents the proposed GreeAODV. Section 4 contains the simulation results obtained in realistic VANET urban scenarios. Finally, this paper ends with some conclusions in Sect. 5.

2 Related Work

Energy consumption is an important designing parameter in wireless multi-hop network routing protocols in general and in VANET scenarios in particular. Several approaches have been proposed in the last few years [8–15]. Energy consumption in multi-hop ad hoc networks gets its importance since communications are normally carried out by devices, which employ batteries and solar cells as energy sources [8, 9]. An energy efficient routing algorithm is proposed in [10], which tries to find minimum energy paths. To select them, the authors propose using two different metrics such as network lifetime and energy consumed by packets. The combination of both metrics is the criteria to select the most energy efficient path.

In [11], the authors propose ERBA, an energy efficient routing protocol for VANETs. It uses parameters such as driving patterns, vehicle category and intersection information to obtain the best routes in terms of energy. The idea behind ERBA is to select the most stable routes so as to reduce the energy consumption. Another approach based on routing stability is proposed in [12], where the authors propose Velocity Based Routing Protocol (VHRP). The authors state that route breakages impact significantly on the power consumption VHRP. Consequently, by selecting stable routes the global energy consumption of the network can be improved. VHRP considers that vehicles are grouped into different groups and also makes mobility predictions to anticipate route breakages.

In [13], the authors proposed DE-OLSR. It is a modification of the well-known proactive routing protocol OLSR. DE-OLSR improves the performance of OLSR in terms of energy consumption. For that, the network interfaces are modeled in one of the four following states: receiving, transmitting, idle or sleeping. The total energy consumption by a packet transmission is the sum of the costs incurred by the sending node and all receivers. It does not take into consideration the idle and sleeping states.

Unlike some of the previous works, which are focused on factors such as network lifetime, route stability, the way of driving, mobility predictions, and developing applications that control the traffic signals [14, 15], we focus our attention on the routing layer. We modify the well-known reactive routing protocol AODV based on the GreeDi approach proposed in [4]. The objective is to find and use the most energy efficient routes to transmit applications packets.

3 The Proposed GreeAODV

The proposed GreeAODV is based on one of the most used routing protocols for ad hoc networks like ADOV [16], and the routing selection algorithm proposed in GreeDi [4]. In its original version, AODV is hop count based. It means that AODV seeks to find the shortest route in terms of number hops between two nodes in the network. AODV [16] is a reactive protocol, so it only maintains routes if they are active. Otherwise, AODV will remove the routing information of obsoleted routes from its routing tables. AODV relies on flooding to discover routes between two nodes. Whenever AODV [16] needs a new communication route between a source and a destination node, the source floods the network with request packets (RREQ. Such request packets are forwarded once by intermediate nodes, until one of the copies reaches the destination node. Each intermediate node updates the route request packet (RREQ) packets by including the last hop followed by the packet and increasing the hop count. On receiving the request packet, the destination node replies the received request by sending back a route reply packet (RREP), which will follow the same communication path followed by the received RREQ. When the RREP reaches the source node, then, it updates its routing table and can start sending application packets to the destination using the established communication path. It is important to note that the destination node only replies once the RREQ packets. Only if the received RREQ contains a lower hop count, the destination will generate another RREP.

In the proposed GreeAODV, we modify the policy of route formation of the original AODV as proposed in GreeDi [4]. We consider the total power consumed by nodes in the found communication route as the selecting metric. The main argument is that in many occasions the shortest route is not the most energy efficient one. The primary reason is that the power consumed by each node in the network can be very different. Consequently, there can be longer routes but more efficient in terms of power consumption. The total power consumed to discover a route between source and destination nodes $P_{c(S-D)}$ is defined as:

$$P_{c(S-D)} = P_{ts} + \sum_{i=0}^{N} (P_{ri} + P_{ti}) + P_{rd} \quad (1)$$

Where P_{ts} is the power consumed by the source node to transmit a packet, P_{rd} is the power consumed by the destination node to receive a packet, and P_{ri} and P_{ti} are the powers consumed by an intermediate node receiving and transmitting an incoming packet, respectively. It is important to highlight that we only consider the power consumed by the wireless transceivers included in the nodes and used for sending and receiving packets. We do not take into account the energy consumed by the nodes for processing the received information. We make such simplification since the energy consumed by the communication operations is in general higher than the energy consumed by the nodes for processing the received information. Notice that in (1) we do not distinguish among the type of packet transmitted or received since we consider that the power consumed is independent of the type of packet (routing or application packets).

As mentioned, the proposed GreeAODV selects routes based on the total power consumed. In order to calculate the total power consumed by a route, the RREQ packets should include the cumulative power consumption from the source node until that point in the route. As in Fig. 1, the fields of the request packets are depicted. With respect to the original AODV RREQ packets, we add the power count. This field represents the energy consumed to receive and transmit the request packet and it is updated in every intermediate node. It is important to indicate that we have not modified the rest of the features of AODV [16].

Therefore, when the RREQ packet reaches the destination node, the power count field will contain the total energy consumed from the source to the destination node. For every found route the destination node will receive a different RREQ packet that contains different power consumption. The destination node will only generate a new RREP if the power count in the received RREQ is lower than the one used in current active route between the source and the destination node. Therefore, we guarantee that the most efficient route between the source and the destination nodes is used.

In order to illustrate how the retransmission of packets is carried out in GreeAODV, we depict an example in Fig. 2. In Fig. 2, we represent the discovery procedure of a

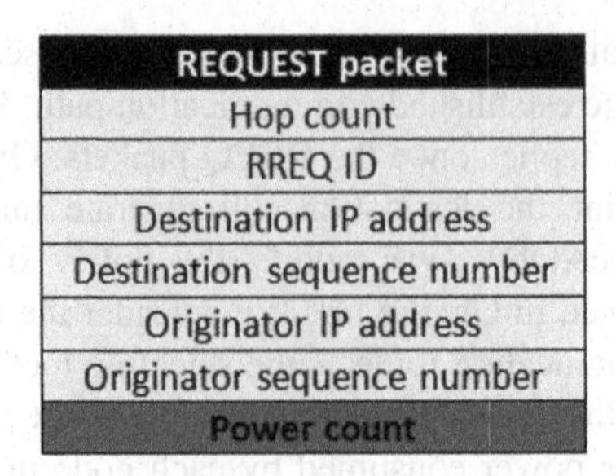

Fig. 1. RREQ packet of GreeAODV

route between a source node (S) and a destination node (D). The power consumed by the found route can be calculated using (1) as:

$$P_{c(S-D)} = P_{ts} + (P_{ra} + P_{ta}) + (P_{rb} + P_{tb}) + P_{rd} \tag{2}$$

In the example represented in Fig. 2, we show two intermediate nodes *a* and *b* that forward the RREQ packet generated by the source node *S* and targeted at the destination node *D*. In this example we only show a possible route between the source and destination nodes. However, in a real situation there can be multiple routes between the source and destination nodes. Therefore, each RREQ packet received by the destination node will contain a different power count value. Notice that the same calculation can be done for the RREP packets and also for the application packets delivered from S to D.

It is important to highlight that the proposed GreeAODV selects the most efficient route between the source and destination nodes at the moment in which the source needs a route to communicate with the destination. This route is used until a route breakage is detected. However, since VANETs are dynamic networks, another more efficient route between both nodes may appear due to the mobility of nodes. For example, if source and destination nodes get closer each other. In the current version of GreeAODV, there is not a mechanism to detect such possibility, but we will consider it as future work.

4 Simulation Results

This section includes the simulation results obtained by the proposed GreeAODV routing protocol. First, we present the simulation environment used to conduct the simulations, followed by simulation results.

4.1 Simulation Environment Settings

The main simulation parameters used are included in Table 1. We use the Network Simulator 2 (NS-2), which is one of the most popular event driven network simulators for multi-hop ad hoc networks. We implement the proposed GreeAODV using the code of the original AODV in NS-2. Regarding the mobility of vehicles, we use C4R

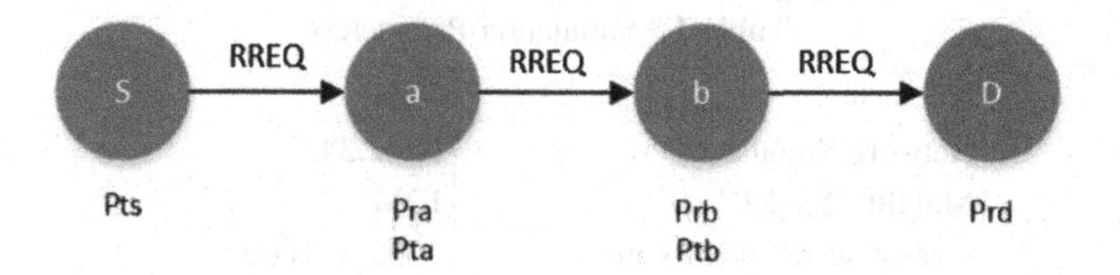

Fig. 2. Example of RREQ retransmission in GreeAODV

application [19] in order to generate the mobility of vehicles in an urban VANET scenario. It is important to highlight some specific aspects of VANET simulations such as the MAC protocol used, which is the IEEE 802.15.4, which is widely used in wireless sensor networks. Regarding the propagation model, we use the two-ray ground reflection model[1] because it gives more accurate prediction for long distances than the free space model.

Figure 3 shows the urban VANET scenario considered for the simulation study. It represents a part of the city of Seville in Spain.

The type of traffic is Constant Bit Rate (CBR), which is typically used in multi-hop scenarios with UDP transport layer. CBR traffic is suitable for real time applications. The maximum speed of nodes is 30 km/h. This maximum value is appropriate for urban scenarios, where the limited speed is about 50 km/h.

In addition, Fig. 3 also shows the movements of the vehicles in the considered scenario. Each color in Fig. 3 represents a different vehicle. To highlight the movement of the vehicles we depict the movements of some of them in terms of the begin (car picture) and the end (flag) of their trajectories.

To evaluate the performance of GreeAODV, we consider the following performance metrics:

Total Power Consumption (TPC) [mW]: It is the total power consumption consumed by the routing protocol during the simulation time. This metric considers both the routing (RREQ and RREP), and the application packets (CBR). The objective is that TPC value can be as low as possible.

$$TPC = \sum RREQ_P + \sum RREP_P + \sum APP_P \tag{3}$$

Where $RREQ_P$, $RREP_P$, APP_p represent the power consumed by the REEQ, RREP, and application packets to be sent and received by all the participating nodes from the source and destination nodes. The power consumed is calculated using (1) regardless of the type of packet.

Application Packet Power Consumption (APPC) [mW]: It is defined as the total power consumed by the application packets during the simulation time. The objective is to obtain a low value of APPC.

[1] Available at https://pdfs.semanticscholar.org/a86f/90f1238ccb90181c26335684fd762247408e.pdf.

Table 1. Simulation Parameters

Parameter	Value
Network Simulator	NS-2.34
Mobility Model	IDM
Size of scenario (m x m)	1000 × 1000
Average speed of nodes (Km/h)	30
N° Nodes	100
MAC layer	IEE 802.15.4
Routing layer	AODV, GreeAODV
Transport layer	UDP
Application layer	CBR
N° Communication flows	5, 10, 15, 20 and 25
Transmission rate (packet/s)	1
Size of application packets (bytes)	512
Transmission range (m)	250
Propagation model	Two-ray ground
Simulation time (s)	300

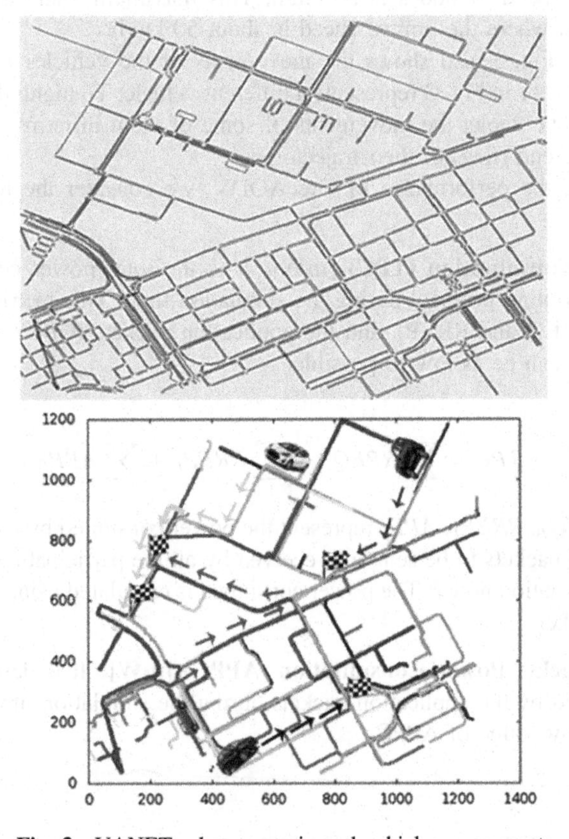

Fig. 3. VANET urban scenario and vehicles movements

$$APPC = \sum APP_P \tag{4}$$

Power Ratio (PR) [mW/pckts]: This ratio measures the power required to transmit each application packet. We seek to achieve a low PR value with GreeAODV.

$$PR = \frac{\sum APP_P}{TPC} \tag{5}$$

APP stands for the number of application packets successfully received by the destination nodes.

Request Power Consumption (RQPC) [mW]: It counts the power consumed by the RREQ packets during the discovery phase of the routing protocol. We desire a low value of RQPC.

$$RQPC = \sum RREQ_P \tag{6}$$

Reply Power Consumption (RPPC) [mW]: It measures the power consumed by RREP. We seek a low value of RPPC.

$$RPPC = \sum RREP_P \tag{7}$$

Throughput [kbps]: This metric measures the performance of the routing protocol in terms data delivery. The throughput is calculated as the number of successfully received application packets divided by the simulation time. A good routing protocol should guarantee a high throughput.

$$Throughput = \frac{\sum APP}{Simulation\ time} \tag{8}$$

The results shown in Figs. 4, 5 and 6 were obtained by averaging out the results of 50 different simulations. The source and destination nodes of the communication flows are selected randomly among all nodes in the network. The only restriction is that the source and destination nodes are not repeated in different communication flows. This condition has been demonstrated that is important to obtain reliable simulation results of routing protocols [17].

4.2 GreeAODV Parameter Settings

The following Table 2 includes the power parameters used by nodes during the simulation time. These parameters are taken from [18]. These values are based on real data information on wireless sensor network that use Xbee commercial modules. In the simulations, we use the same number of each type of nodes (power consumption). We consider that this guarantees a fair distribution.

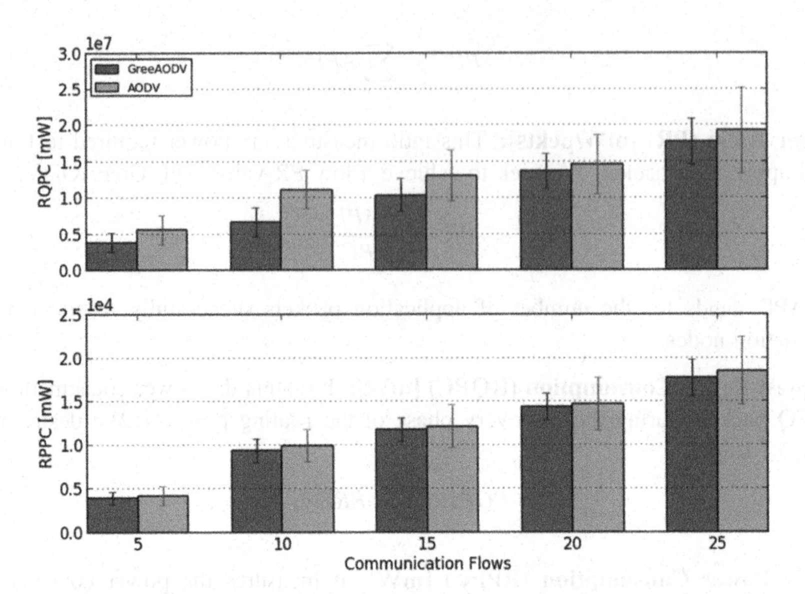

Fig. 4. RQPC and PPPC simulation results for GreeAODV and AODV

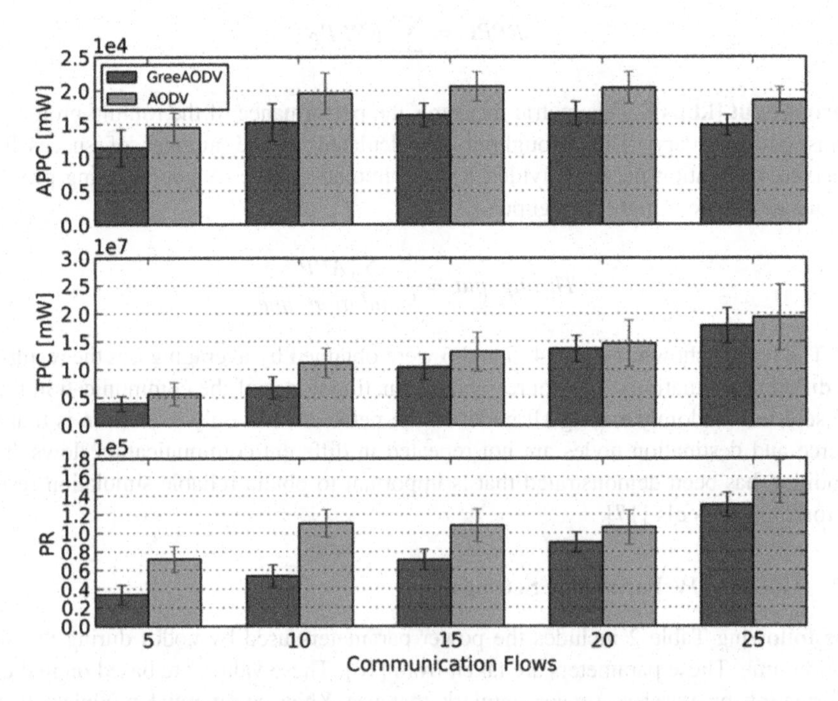

Fig. 5. APPC, TPC, and PR simulation results for GreeAODV and AODV

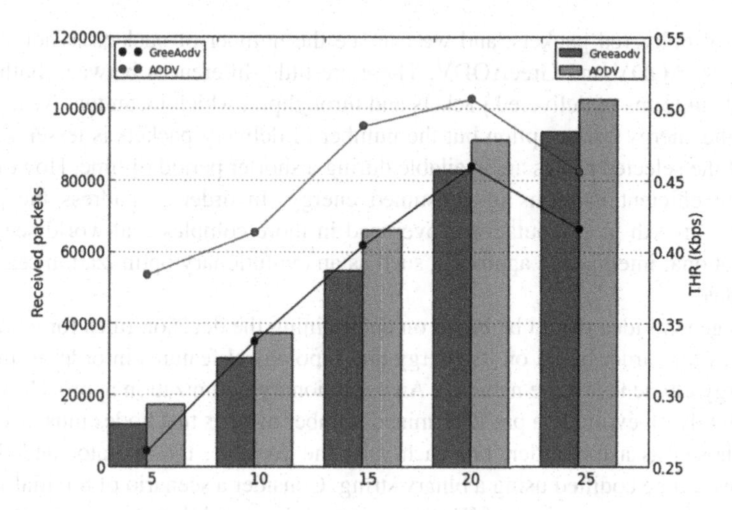

Fig. 6. Throughput simulation results for GreeAODV and AODV

Table 2. Power consumption data

Transceiver models	Power consumption [mW]
Xbee-802.15.4	1
Xbee-802.15.4Pro	100
Xbee-ZB	2
Xbee-ZBPro	50
Xbee-868	315
Xbee-900	50
Xbee-XSC	100

4.3 Simulation Results for GreeAODV

Figure 4 depicts RQPC and RPPC respectively. Related to energy consumption by the request packets (RQPC), it is lesser for GreeAODV case because the number of request packets is lesser due to the number of broken link is lesser than using AODV. As a result, it is not necessary to initiate new discovery processes. Therefore, the number of request packets is lesser and the total power consumed by the request packets will be lesser than using AODV. In addition, the power consumption by the reply packets is lesser in GreeAODV case because in this case the routes are selected based on the power consumption and the power consumed by these packets are lesser. Figure 5 depicts APPC, TPC and PR results. The application packet power consumed using GreeAODV is lesser than in the case of not using it because the routes are selected based on the power consumption. Due to the fact that the power consumed by the request packets using GreeAODV is lesser than the original AODV, the TPC is lesser because the power consumed by the Reply and application packets is also lesser. The PR is better in the GreeAODV case. It means that the power necessary to transmit an application packet is lesser than using the original AODV. In Fig. 6, we show the number of delivered packets, and we can see the number of packets is not similar in both cases, AODV and GreeAODV. There are little differences between both routing protocols in terms of delivered packets and throughput, which mean that we are able to reduce the energy consumption but the number of delivery packets is lesser due to the fact that the selected routes are available during a shorter period of time. However, they are more efficient in terms of consumed energy. In order to address the potential explosive growth in computational overhead in more complex real-world scenarios, a computational intelligence approach, such as an evolutionary optimization search, may be adopted.

The general idea would be based on determining the decision rules for a node to be chosen as forwarder based on its energy and topological features in order to maximize the energy efficiency of the network. An evolutionary optimization approach would for example rely on evolving a pre-determined number of rules that nodes must accomplish to be chosen as a forwarder. For each rule, the variable, the operator and threshold values would be codified using a binary string. Consider a scenario of 8 initial variables (features), the two operations of 'Greater or equal than' and 'Lower or equal than' and a normalized threshold value codified as an 8-bit string. This would lead to 12 bits to codify the rules and 36 bits for 3 rules. Assuming an appropriate multi-objective fitness function and that the three rules are independent and must be accomplished simultaneously and, for each individual solution a node will be chosen as a forwarder only if all the three codified rules are accomplished.

5 Conclusions and Future Work

This paper presents a novel reactive routing protocol for VANET scenarios named GreeAODV. It is aimed at reducing the power consumption of application packets in vehicular scenarios. The total power consumed by the nodes participating in the communication flows is considered as selecting parameter of route formation. GreeAODV has been implemented over AODV routing protocol, and it has been compared with the original version of AODV. The simulation results show that the proposed GreeAODV achieves better results than AODV in term of power consumption. The potential for adopting a computational intelligence paradigm to address this problem is discussed. As future work, we also seek to consider vehicle to infrastructure communications, and optimize other objectives such as latency. Furthermore, we look for combining the proposed approach with the GreeDi algorithm proposed in [4]. Therefore, the global objective will be to improve the energy consumption in the connection of vehicles to data centers.

References

1. Conti, M., Giordano, S.: Mobile Ad Hoc networking: milestones, challenges, and new research directions. IEEE Commun. Mag. **52**, 85–96 (2014)
2. Alotaibi, A., Mukherjee, B.: A survey on routing algorithms for wireless Ad-Hoc and mesh networks. Comput. Netw. **56**, 940–965 (2012)

3. Li, F., Wang, Y.: Routing in vehicular Ad Hoc networks: a survey. IEEE Veh. Technol. Mag. **2**, 12–22 (2008)
4. Baker, T., Al-Dawsari, B., Tawfik, H., Reid, D., Ngoko, Y.: GreeDi: an energy efficient routing algorithm for big data on cloud. Ad Hoc Netw. **35**, 83–96 (2015)
5. Baker, T., Asim, M., Tawfik, H., Aldawsari, B., Buyya, R.: An energy-aware service composition algorithm for multiple cloud-based IoT applications. J. Netw. Comput. Appl. **89**, 96–108 (2017)
6. Aldawsari, B., Baker, T., England, D.: Trusted energy-efficient cloud-based services brokerage platform. Int. J. Intell. Comput. Res. (IJICR) **6**(4), 630–639 (2015)
7. Baker, T., Ngoko, Y., Calasanz, R.T., Rana, O., Randles, M.: Energy efficient cloud computing environment via autonomic meta-director framework. In: 2013 Sixth International Conference on Development in eSystems Engineering (DeSE) (2013)
8. Feng, W., Alshaer, H., Elmirghani, J.: Green information and communication technology: energy efficiency in a motorway model. IET Commun. **4**, 850–860 (2010)
9. Zhang, H., Ma, Y., Yuan, D., Chen, H.-H.: Quality-of-service driven power and sub-carrier allocation policy for vehicular communication networks. IEEE J. Sel. Areas Commun. **29**, 197–206 (2011)
10. Spyropoulos, A., Raghavendra, C.: Energy efficient communications in Ad hoc networks using directional antennas. In: IEEE INFOCOM 2002, pp. 1–9 (2002)
11. Zhang, D., Yang, Z., Raychoudhury, V., Chen, Z., Lloret, J.: An energy-efficient routing protocol using movement trends in vehicular Ad hoc networks. Comput. J. Adv. **21**, 1–9 (2013)
12. Taleb, T., Ochi, M., Jamalipour, A., Kato, N., Nemoto, Y.: An efficient vehicle-heading based routing protocol for VANET networks. In: IEEE Wireless Communications and Networking Conference, pp. 2199–2204 (2006)
13. Toutouh, J., Alba, E.: An efficient routing protocol for green communications in vehicular Ad-hoc networks. In: 13th Annual Conference on Companion GECCO, pp. 719–726 (2011)
14. Dobre, C., Szekeres, A., Pop, F., Cristea, V., Xhafa, F.: Intel-ligent traffic lights to reduce vehicle emissions. In: Proceedings 26th European Conference on Modelling and Simulation, pp. 1–8 (2012)
15. Suthaputchakun, C., Sun, Z., Dianati, M.: Applications of vehicular communications for reducing fuel consumption and CO2 emission: the state of the art and research challenges. IEEE Commun. Mag. **12**, 108–115 (2012)
16. Perkins, C.E., Royer, E.M.: Ad Hoc on-demand distance vector routing. In: Proceeding of IEEE Workshop on Mobile Computing System and Applications (WMCSA), pp. 1–11 (1999)
17. García-Campos, J.M., Sánchez-García, J., Reina, D.G., Toral, S.L., Barrero, F.: An evaluation methodology for reliable simulation based studies of routing protocols in VANETs. Simul. Model. Pract. Theory **16**, 139–165 (2016)
18. Odey, A.J., Li, D.: Low power transceiver design parameters for wireless sensor networks. Wirel. Sens. Netw. **4**, 243–249 (2012)
19. C4R: CityMob for Roadmaps. http://www.grc.upv.es/Software/c4r.html

Interactive Swarm Intelligence Algorithm Based on Master-Slave Gaussian Surrogate Model

Jing Jie, Lei Zhang, Hui Zheng(✉), Le Zhou, and Shengdao Shan

Zhejiang University of Science and Technology, No. 318, Liuhe Road, Hangzhou City 310023, People's Republic of China
huizheng@zust.edu.cn

Abstract. An interactive swarm intelligence algorithm based on master-slave Gaussian surrogate model (ISIA-MSGSM) is proposed in this paper. In the algorithm, particle swarm optimization is used to act on the optimization search. During the search process, some data are sampled dynamically from the searching swarm to build the master and the slave Gaussian surrogate model, and all the particles will go through interactive evaluations based on the two kinds of surrogate models and the accurate model, which can reduce the computation cost of the objective function. At the same time, the surrogate models are managed dynamically guided by the accurate model to ensure the computational accuracy. Through the dynamical update to the master and slave model, the balance between the global exploration and the local exploitation is ensured which contributes to the efficiency of the algorithm. The experiment results on benchmark problems show this method not only can decrease the computation cost, but also has good robustness with a satisfied optimization performance.

Keywords: Intelligence computation · Swarm intelligence
Particle swarm optimization · Surrogate model

1 Introduction

Intelligent computation has been used widely to solve many complex engineering problems in real world, for examples: aerodynamic design optimization [1], integrated circuit optimization design [2], kinds of scheduling problems [3, 4], etc. However, many complex problems in fact are computationally expensive optimization problems, that means intelligent computation methods require large amounts of fitness evaluation before obtaining a feasible global optimum or near-optimal solutions, which greatly hinders the popular applications of intelligent computation in real engineering area. In order to validly solve computationally expensive optimization problems, some scholars try to introduce the surrogate models into the objective evaluation of the intelligent computation in recent years, and develop several surrogate-assisted evolutionary computations [5–9]. In the process of evaluation of intelligent computing, surrogate models can be used to replace the objective function model to calculate the fitness of the individual, which can effectively reduce the fitness evaluation cost. Generally, the surrogate model can be constructed by polynomial function (PF), artificial neural

D.-S. Huang et al. (Eds.): ICIC 2018, LNAI 10956, pp. 682–688, 2018.
https://doi.org/10.1007/978-3-319-95957-3_70

network (ANN), support vector machine (SVM), Gaussian process (GP), radial basis function (RBF), and so on.

In this paper, we try to introduce Gaussian process model into particle swarm optimization, and develop an interactive swarm intelligence algorithm based on master-slave Gaussian surrogate model (ISIA-MSGSM). The algorithm samples some data dynamically from the search swarm to build the master and the slave Gaussian surrogate model. Based on the two kinds of surrogate models and the accurate model, all the particles will be evaluated through an interactive evaluation strategy. At the same time, the master and slave model are dynamically updated under the guide of the accurate model, which can contribute the balance between the global exploration and the local exploitation, and ensure the algorithm has good robustness.

The remaining of the paper is organized as the follows. Section 2 introduces the key technologies of the proposed algorithm in details, including framework of algorithm, swarm evolution search, surrogate management strategies, and so on. Section 3 presents the experimental study, and provides some analysis about the performance of the algorithm. Finally, Sect. 4 summarizes the main work and makes a preliminary study on the future research direction.

2 Interactive Swarm Intelligence Algorithm Based on Master-Slave Gaussian Surrogate Model

2.1 Framework of Algorithm

ISIA-MSGSM consists of five components, such as a swarm, a database, an accurate function model (AFM), a master Gaussian surrogate model (MGSM) and a slave Gaussian surrogate model (SGSM), its framework is illustrated in Fig. 1.

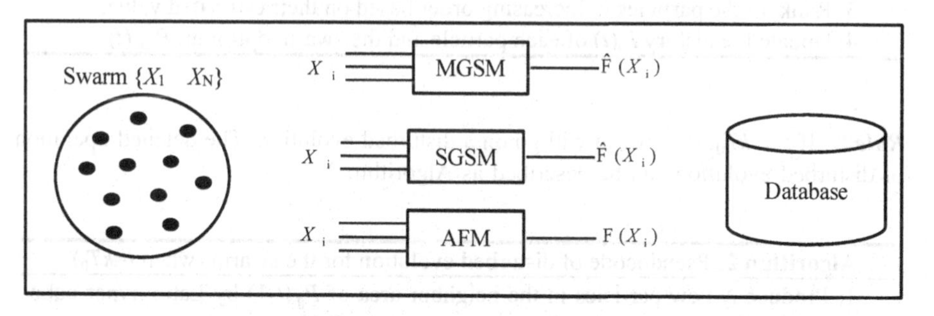

Fig. 1. Framework of ISIA-MSGSM

In ISIA-MSGSM, AFM stands for the original accurate optimization problems, MGSM is designed to approximate the AFM in a broad space, while SGMS is designed to approximate the AFM in a smaller local space around the search trajectory of the swarm, which is expected to get the better estimated value for the particles. During the optimization process, the swarm not only acts on the searching for the optimum of accurate function, but also provides the sampling data to train MGSM and SGSM.

In iterations, all the particles in the swarm will be updated through the iterative equation of PSO. After that, they will be evaluated by the AFM, MGSM or SGSM. Based on the score of evaluation, some particles will be chosen as the sampling data to record to the database. Moreover, the database also records some useful information about the swarm, for example the trajectory of its winner and the scores. Without loss of generality, the detailed technologies will be discussed based on the minimized optimization problem in the following section:

$$\min F(\mathbf{x}) \quad \mathbf{x} \in \Omega \subseteq R^D; \Omega = [a, b]^D \tag{1}$$

2.2 Swarm Evolution Search

After the initialization, the swarm will go on an evolution to search the optimum of the problem. In order to keep the trade-off between the global search and the local search, ISIA-MSGSM control the evolution of the swarm dynamically by a control parameter T_0, which can modify the evolution behaviors of the swarm periodically, the detailed rules are as the following:

***Rule* 1.** If $(k-1)T_0 < t < kT_0 (k = 1, 2, \ldots)$, the swarm will go on a convergent evolution according to the computation model of PSO. The detailed operation of convergent evolution can be described as Algorithm 1:

Algorithm 1. Pseudocode of convergent evolution for the swarm (when $t \neq kT_0$)

1: Update the location of each particle $X_i(t)$, produce the swarm $S(t)$;

2: Estimate the value of each particle in $S(t)$ by MGSM, record the estimated value as $\hat{F}_m(X_i(t))$;

3: Rank all the particles in increasing order based on their estimated value;

4: Update the history $P_i(t)$ of each particle and the swarm optimum $P_{mg}(t)$.

***Rule* 2.** If $t = kT_0$, the swarm will go on a disturbed evolution. The detailed operation of disturbed evolution can be described as Algorithm 2:

Algorithm 2. Pseudocode of disturbed evolution for the swarm (when $t = kT_0$)

1: Produce N new particles in the neighbor area of $P_g(t\text{-}1)$ by Latin hyper-cube sampling (LHS), and construct the temporary swarm $S_T(t)$;

2: Evaluate each particle in $S(t)$ and $S_T(t)$ by AFM, and record the accurate value $F(X_i(t))$;

3: Rank all the particles in $S(t)$ and $S_T(t)$ based on their accurate value in increasing order;

4: Update $S(t)$ with the first N particles;

5: Update the history $P_i(t)$ of each particle and the swarm optimum $P_g(t)$.

Obviously, the swarm distribution can be dynamically modified by the two evolution behaviors under the control parameter, which contributes to the balance of the exploitation and the exploration during the optimization search of the swarm.

2.3 Surrogate Management Strategies

In the optimization process, the surrogate model will be used to approximately evaluate the particles instead of the original AFM, and the evaluation results will directly affect the subsequent search direction. Due to the lack of sample data for the original target problem, the approximation error of the surrogate model is often large which will provide error optimization information. Therefore, to ensure optimal performance, the surrogate model is required to be monitored dynamically. An effective method is to use the AFM to monitor the surrogate model regularly during the fitness evaluation, what is called surrogate management.

Generally, surrogate management has three kinds of strategies such as individual-based, generation-based and population-based [10]. Here, ISIA-MSGSM adopts the individual-based and generation-based strategies. The real management rules are as the following:

***Rule* 3.** If $(k-1)T_0 < t < kT_0 (k = 1, 2, \ldots)$, adopt the individual-based surrogates management Strategy in each iteration;

In the individual-based surrogate management strategy, the AFM is used to evaluate some individuals during iteration and the accuracy of the surrogate is monitored by the accurate value. Considering the evaluation cost of the AFM, ISIA-MSGSM manages the MGSM and SGSM in iteration just through monitoring the two elitists found by the two surrogates respectively.

***Rule* 4.** If $t = kT_0 (k = 1, 2, \ldots)$, adopt generation-based surrogates management strategy.

Generation-based surrogate management strategy means the surrogates are used for particle evaluation in some of the generations, while the accurate model is used in the rest of the generations. In ISIA-MSGSM, the strategy mainly is used to control the update of the MGSM and SGSM. In order to synchronously manage the swarm and the surrogates well, ISIA-MSGSM also controls the update of the two surrogates under the control parameter T_0. When $t = kT_0 (k = 1, 2, \ldots)$, MGSM and SGSM will be updated based on the new collected training set under the guide of AFM.

2.4 Pseudocode of ISIA-MSGSM

In brief, the pseudocode of ISIA-MSGSM can be outlined as the following:

Algorithm 3 . The pseudocode of ISIA-MSGSM

1: let t=0, initialize the swarm , the two training sets, and the database;
2: **while** *termination criterion* is not satisfied **do**
3: **If** $t \neq kT_0$ **then**
4: Carry on convergent evolution of the swarm ;
5: manage the surrogates ;
6: **else**
7: Carry on disturbed evolution of the swarm;
8: update the surrogates;
9: Update the database;
10: t=t+1;
11: **end while**
12: Output the optimization results.

3 Experimental Study

In order to test the performance of the proposed algorithm, several experiments have been designed. The simulations are run by the tool of Matlab 2016 based on win7 system. In ISIA-MSGSM, the parameters for PSO are set as: swarm size is 30, coefficient c_1 = 1.5 and c_2 = 2.2, inertia weight w is linearly decreased from 0.95 to 0.4 over time; control parameter T_0 = 5; threshold value α = 0.01, neighbor area parameter r = 0.5, all results are from 20 independent runs based on some noted benchmark functions, such as Griewank, Ackley and Resonbrock functions. Based on the statistical results, ISIA-MSGSM is compared to two noted algorithms that are GPEME [2] and GS-SOMA [7].

In Table 1, the termination condition of algorithms is FEs = 1000, while the termination condition of algorithms in Table 2 is FEs = 8000. All the results are the mean values by 20 runs. According to the tables, it's easy to know the proposed ISIA-MSGSM outperforms GPEME and GS-SOMA greatly at all the benchmark functions. Moreover, more experiments have been designed to observe the performances of ISIA-MSGSM.

Table 1. Comparisons of ISIA-MSGSM and GPEME based on functions with 50 dimensions

AFM	Algorithm	Best value	Mean value	Worst value	Std.
Griewank	ISIA-MSGSM	**0.2829**	**1.2451**	**5.0103**	**0.9896**
	GPEME	22.5456	36.6459	64.9767	13.1755
Ackley	ISIA-MSGSM	**0.1371**	**1.7692**	**4.1413**	**1.3309**
	GPEME	9.2524	13.2327	14.9343	1.5846
Rosenbrock	ISIA-MSGSM	**45.5467**	**49.4393**	**61.6130**	**8.3051**
	GPEME	172.3547	401.4118	258.2787	80.1877

Table 2. Comparisons of ISIA-MSGSM and GPEME based on functions with 30 dimensions

AFM	Algorithm	Best value	Mean value	Worst value	Std.
Griewank	ISIA-MSGSM	**3.3e−14**	**5.5e−10**	**4.7e−9**	**1.1e−18**
	GS-SOMA	1.4e−10	0.0022	0.0154	0.0046
Ackley	ISIA-MSGSM	**1.4e−6**	**8.96e−6**	**4.34e−5**	**8.90e−11**
	GS-SOMA	2.8700	3.5800	4.2800	0.509
Rosenbrock	ISIA-MSGSM	**2.5903**	**25.3915**	**25.8689**	**49.9253**
	GS-SOMA	28.3	46.1	126	29.2

Though the algorithm performs weak on some problems sometime, it can be regarded as a robust method on the whole.

4 Concluding Remarks

The paper develops an interactive swarm intelligence algorithm based on master-slave Gaussian surrogate model (ISIA-MSGSM), also presents the algorithm framework and the main strategies in details. The proposed algorithm has been validated with selected benchmarks in comparison with noted algorithms. The experimental results and analysis show that the algorithm has good global optimization performance with a lower evaluation cost by accurate function. The future research will pay more attention to improve the estimating accuracy of the surrogate model in the optimization process and to design new valid surrogate models to solve the complex engineering problem based on the data-driven technology.

Acknowledgments. This work was supported in part by NSFC-Zhejiang Joint Fund for the Integration of Industrialization and Information (No.U1609214), National Natural Science Foundation of China (No.61203371), Zhejiang Provincial Natural Science Foundation (No. LQ16F030002), and Zhejiang Provincial New seedling talent program (No. 0201310H35).

References

1. Jin, Y., Olhofer, M., Sendhoff, B.: A framework for evolutionary optimization with approximate fitness functions. IEEE Trans. Evol. Comput. **6**(5), 481–494 (2002)
2. Liu, B., Zhang, Q., Gielen, G.G.: A Gaussian process surrogate model assisted evolutionary algorithm for medium scale expensive optimization problems. IEEE Trans. Evol. Comput. **18**(2), 180–192 (2014)
3. Gong, Y.J., Zhang, J., Chung, H.S.H., Chen, W.N., Zhan, Z.H., Li, Y., Shi, Y.H.: An efficient resource allocation scheme using particle swarm optimization. IEEE Trans. Evol. Comput. **16**(6), 801–816 (2012)
4. Yuan, H., Li, C.: Resource scheduling algorithm based on social force swarm optimization algorithm in cloud computing. Comput. Sci. **42**(4), 206–208 (2015)
5. Sun, X., Chen, S., Gong, D., Zhang, Y.: Weighted multi-output Gaussian process-based surrogate of interactive genetic algorithm with individuals interval fitness. Acta Autom. Sin. **40**(2), 172–184 (2014)

6. Wang, H., Jin, Y., Jansen, J.O.: Data-driven surrogate-assisted multiobjective evolutionary optimization of a trauma system. IEEE Trans. Evol. Comput. **20**(6), 939–952 (2016)
7. Lim, D., Jin, Y., Ong, Y.S., Sendhoff, B.: Generalizing surrogate-assisted evolutionary computation. IEEE Trans. Evol. Comput. **14**(3), 329–355 (2010)
8. Sun, C., Jin, Y., Zeng, J., Yu, Y.: A two-layer surrogate-assisted particle swarm optimization algorithm. Soft. Comput. **19**(6), 1461–1475 (2015)
9. Kong, Q., He, X., Sun, C.: A surrogate-assisted hybrid optimization algorithms for computational expensive problems. In: 12th World Congress on Intelligent Control and Automation, pp. 2126–2130. IEEE, Guilin (2016)
10. Lu, J.F.: Surrogate Assisted Evolutionary Algorithm. University of Science and Technology of China, Anhui (2013)

A Novel Image Denoising Algorithm Based on Non-subsampled Contourlet Transform and Modified NLM

Huayong Yang[1(✉)] and Xiaoli Lin[1,2]

[1] Department of Information Engineering, City College of Wuhan University of Science and Technology, Wuhan 430083, China
huayongyang1977@163.com
[2] School of Computer Science and Technology, Wuhan University of Science and Technology, Wuhan 430065, China

Abstract. A novel image denoising algorithm based on non-subsampled contourlet transform (NSCT) and modified non-local mean (NLM) is proposed. First, we utilize NSCT to decompose the images to obtain the high frequency coefficients. Second, the high frequency coefficients are used for modified NLM denoising. Finally, the NLM weight values are calculated by modified bisquare function instead of Gaussian kernel function of the traditional NLM, and each noise coefficient is corrected to get the denoised image. According to results of the simulation experiment, the denoising results of the proposed algorithm obtain higher peak signal-to-noise ratio (PSNR) and better retains structural information of image in subjective vision.

Keywords: Non-subsampled contourlet transform (NSCT)
Non-local mean (NLM) · Denoising

1 Introduction

Image denoising technology is always a core field of study in image processing and a focus issue concerned in computer vision. Generally, human eye keeps sensitive to texture and edge of image but poorly identifies the smooth part of image, which is a property of human eye taken advantage of by most algorithms to remove noise signals and retain the original information. The bearing structure of original signals usually contains a mass of such significant image contents as texture and edge [1] and to effectively maintain the bearing structure is an important goal of all kinds of denoising algorithms, which requires such denoising algorithms to enhance denoising effect while not lowering visual quality.

As denoising studies constantly go deeper, many excellent algorithms have emerged in succession and good evaluation criterion for denoising algorithms has improved continuously. Though such algorithms widely vary, they are intrinsically linked to each other in essence, which is to remove meaningless information and guarantee subjective visual effect of image on the ground of maintaining meaningful edge and texture information. So far, main domestic and foreign denoising algorithms

D.-S. Huang et al. (Eds.): ICIC 2018, LNAI 10956, pp. 689–699, 2018.
https://doi.org/10.1007/978-3-319-95957-3_71

may fall into non-local algorithms [2], random field algorithms [3], bilateral filtering algorithms [4], anisotropic diffusion algorithms [5] and statistical model algorithms [6]. Of all such denoising algorithms, non-local means (NLM) has attracted more attention mainly because of its outstanding effect in maintaining detail characteristics of image. Proposed by Buades and other scholars, NLM's central idea is: the estimated value of a target pixel is drawn from taking a mean of the target pixel and other pixels in the similar domain structure in the image weighted by their similarity. This algorithm has been noticed and constantly modified in denoising performance in recent years.

Liu et al. [7] presented an optimized NLM algorithm and skillfully applied Gaussian kernel function of weighted cosine coefficient to similarity computation, showing better denoising effect than the traditional NLM denoising algorithm; Chen et al. [2] proposed an NLM algorithm based on homogeneous similarity computation defined in the adaptive weighted neighborhood with the ability to effectively find more similar positions, particularly such points of less iteration as corner and terminal points. Zhong et al. [8] combined block-matching and 3D filtering together to divide image to image blocks and construct a 3D array according to inter-block similarity theory before processing such 3D data in the combined filtering method and drawing denoised image by inverse transform. On the ground of weight estimation of center pixel, Nguyen et al., [9] present two separate new algorithms for computation of local weight to improve denoising performance of NLM and improve the ability to maintain local structure of original image.

The above indicates that NLM filtering and denoising has received increasingly more concerns. As its modified algorithms are constantly proposed, its image denoising effect has been continuously improved. Whereas, it still has many defects such as usually too smooth denoised image. Concerning such problems, a novel denoising algorithm based on non-subsampled contourlet transform (NSCT) and modified NLM is proposed. Firstly, decomposition is made on the to-be-denoised image by NSCT to prevent the resulting image from being too smooth. Secondly, the weight value is calculated by modified bisquare function instead of the traditional NLM Gaussian kernel function and each noise coefficient is corrected to get the denoised image. According to results of the simulation experiment, the denoising result of the proposed algorithm gets higher peak signal-to-noise ratio (PSNR) and better retains structural information of image in subjective vision.

2 Non-subsampled Contourlet Transform (NSCT)

NSCT [10] mainly comprises non-subsampled pyramid filter bank (NSPFB) and non-subsampled direction filter bank (NSDFB) in cascade. Firstly, decomposition is made on the image by NSPFB and the resulting sub-bands is taken as input of NSDFB to get decomposition results of the original image in multiple dimensions and directions K-level decomposition is made on any image by NSCT to get 1 low-frequency sub-band and some high-frequency band-pass sub-bands, all of which have the same size as the original image. Both NSPFB and NSDFB are eligible for full reconstruction so NSCT is fully rebuilt as well. The steps of NSCT are shown as follows:

Firstly, image is decomposed by NSPFB in multiple dimensions to one high-frequency sub-band and one low-frequency sub-band. In multi-level decomposition, only by further iterative filtering on the low-frequency sub-band will the image be ultimately decomposed to one low-frequency sub-band and some high-frequency sub-bands. If the level of decomposition by NSPFB is *X*, the redundancy will be *X* + 1. As far as the low-pass filter and band-pass filter in the Level *X* are concerned, their ideal spans of support of frequency domain will be $\left[-\pi/2^{x-1}, \pi/2^{x-1}\right]^2$ and $\left[-\pi/2^{x-1}, \pi/2^{x-1}\right]^2 \cup \left[-\pi/2^{x}, \pi/2^{x}\right]^2$ respectively. While multi-dimension characteristics are acquired, no extra filter will be required for decomposition. Therefore, a bandpass image will generate redundancy of *X*+1 in each level, significant better than the structure of wavelet transform.

Secondly, the decomposed sub-bands by NSDFB are decomposed in multiple directions and singular points in different dimensions and directions are merged. NSDFB is also a double-channel filter bank comprising a decomposition filter(s) $U_i(z), (i = 0, 1)$ and synthesis filter(s) $V_i(z), (i = 0, 1)$ which satisfy Bézout's identity:

$$U_0(z) + V_0(z) = U_1(z) + V_1(z). \tag{1}$$

Decomposition filters $U_0(z)$ and $U_1(z)$ can decompose two channels for adopting ideal support of frequency domain. Instead of subsampling operation, upsampling operation is made on filters $U_0(z)$ and $U_1(z)$ in each level by all sampling matrices to draw direction filters in the following level. The above are major steps of decomposition of image by NSCT and schematic diagram of structure of NSCT transform is shown in Fig. 1.

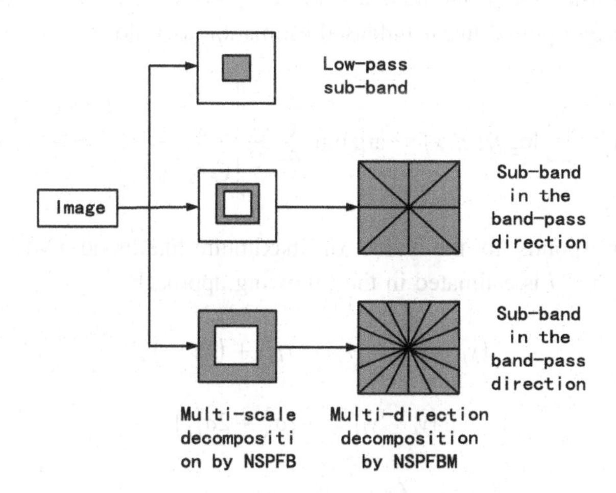

Fig. 1. Structure of decomposition by NSCT

3 Optimized NLM Denoising Algorithm

In traditional NLM, similarity computation is made by pixel values or coefficient values of transform sub-bands, coefficients of transform sub-bands are compared in fitting effect by residual, and traditional Gaussian kernel function is replaced by optimized bisquare distribution function of optimal fitting effect.

3.1 Modified Weight Kernel Function

As far as NLM denoising algorithm is concerned, it is particularly important to choose a suitable weight kernel function for computation of similarity, which will have immediate impact on the final denoising effect. A most suitable weight kernel function for computation of similarity is chosen in accordance with estimation theory [11].

In image denoising, correlation between coefficients is taken into account as well in order to more accurately describe the relationship between coefficients, as shown in Formula (2):

$$x_i = x_j + r_{i,j} \tag{2}$$

x represents NSCT coefficient in noise-free situation while r represents the residual value between coefficients. Generally assuming that the original image is polluted by white Gaussian noise only, the noise image may be represented in the following formula:

$$y_i = x_i + r_{i,j} + v_i \tag{3}$$

v refers to white Gaussian noise and meets $v_i \sim N(0, \sigma_w^2 I)$. The expected value of x is computed by optimal linear unbiased estimation as follows:

$$\hat{X}_i = \arg\min_X \sum_{j=1}^{N} \log f_y(y_j; x) = \arg\min_X \sum_{j=1}^{N} \frac{1}{2} \left\| \frac{y_j - x}{\sqrt{\sigma_w^2 + \sigma_{i,j}^2}} \right\|^2 = \frac{\sum_{j=1}^{N} \frac{1}{\sigma_w^2 + \sigma_{i,j}^2} y_j}{\sum_{j=1}^{N} \frac{1}{\sigma_w^2 + \sigma_{i,j}^2}} \tag{4}$$

where $\hat{X}_i$ corresponds to the value of maximum likelihood (ML) estimate and covariance $\sigma_{i,j}^2, i \neq j$ is estimated in the following approach:

$$(y_j - y_i) = (r_{j,j} - r_{i,j}) + (V_j - V_i),$$

$$[y_i - y_j] = \left(2\sigma_w^2 + 2\sigma_{i,j}^2\right)$$

$$\hat{\sigma}_{i,j}^2 = \max\left(0, \frac{1}{2(2K+1)} \|y_i - y_j\|^2 - \sigma_w^2\right) \tag{5}$$

Furthermore, the weight function is drawn:

$$w(i,j) = \begin{cases} \frac{4K+2}{\|y_i-y_j\|^2}, & \frac{\|y_i-y_j\|^2}{4K+2} \leq \sigma_w^2 \\ \frac{1}{\sigma_w^2}, & \frac{\|y_i-y_j\|^2}{4K+2} \leq \sigma_w^2 \end{cases} \tag{6}$$

New statistical distribution is required to get more accurate similarity weight. Therefore, robust *M*-estimator algorithm is introduced [11]. Robust M-estimator originates from ML estimate:

$$\hat{X}_i = \arg\min_X \sum_{j=1}^{N} \rho(x - y_i) \tag{7}$$

One recursion is made on the Formula (7) to draw:

$$\hat{x}_i = y_i - \lambda \sum_{j=1}^{N} \rho'(y_i - y_j) \tag{8}$$

Where $\rho'(X)$ represents gradient of $\rho(X)$:

$$\rho'(X) = xg(X) \tag{9}$$

Formula (8) may be represented as follows by robust function:

$$\begin{aligned} \hat{x}_i &= y_i - \lambda_i \sum_{j=1, j\neq i}^{N} g(y_i - y_j)(y_i - y_j) \\ &= \left(1 - \lambda_i \sum_{j=1, j\neq i}^{N} g(y_i - y_j)\right) y_i + \lambda \sum_{j=i, j\neq i}^{N} g(y_i - y_j) y_j \end{aligned} \tag{10}$$

In order to improve velocity, λ_i is worked out by Jacobi algorithm:

$$\lambda_i = \frac{1}{1 + \sum_{j=1, j\neq i}^{N} g(y_i - y_j)} \tag{11}$$

Next, Leclair robust function is taken as example, define and represented as follows:

$$\rho(r) = h^2 - h^2 \exp\left(-\frac{r^T r}{2h^2}\right) \tag{12}$$

And,

$$\rho'(r) = r \cdot \exp\left(-\frac{r^T r}{2h^2}\right),$$

$$w(i,j) = \exp\left(-\frac{||yi - yj||^2}{2h^2}\right) \tag{13}$$

By robust estimation, the following can be drawn:

$$\rho(r) = \begin{cases} \frac{1}{2} + \log\|r\| & \|r\| > h \\ \frac{\|r\|^2}{2h^2} & \|r\| \le h \end{cases} \tag{14}$$

To sum up, characteristics of robust estimate function are compared to judge characteristics of its corresponding distribution function to help select the most suitable weight function. In order to improve denoising effect, scope of application of the above formulas is expanded and noise coefficients of high-frequency sub-bands are applied to computation of similarity in the algorithm in this paper. The most suitable distribution function to statistical characteristics of residual between coefficients is obtained by experiment and regarded as the kernel function for computation of similarity. In order to get the optimal weight function, five weight functions are compared in the algorithm, which are BKF distribution, Weibull distribution, Cauchy distribution, modified bisquare function and Gaussian distribution function respectively as shown in Fig. 2.

Lena image is taken as example in Fig. 2 for decomposition by NSCT where statistical distribution of residuals between coefficients of high-frequency sub-band is shown. It is clearly seen in Fig. 2 that the modified bisquare function [11] can most accurately describe statistical distribution of residuals. Therefore, the modified bisquare function is regarded as the final kernel function for computation of similarity to work out a more accurate weight value. The modified bisquare function is defined as follows:

$$g(r_e) = \begin{cases} \left(1 - \frac{\|r_e\|^2}{h^2}\right)^8 & \|r_e\| \le h \\ 0 & \|r_e\| > h \end{cases} \tag{15}$$

where r_e represents the residual value of exponential moment while h represents the depth coefficient.

Noise pixels of high-frequency sub-band are substituted to the kernel function for computation of similarity or Formula (15), wherein, $\|r_e\|^2 = \left\|e_{y_i} - e_{y_j}\right\|^2$ represents Euclidean distance between two pixel values. In Formula (15), the similarity between pixels is drawn to work out corresponding weight value which is exactly the weight value between coefficients. The calculated weight value is used to correct each noise coefficient in the high-frequency sub-band of the noise image and secondary inverse transform is made on all the denoised high-frequency sub-bands by NSCT to draw the denoised image.

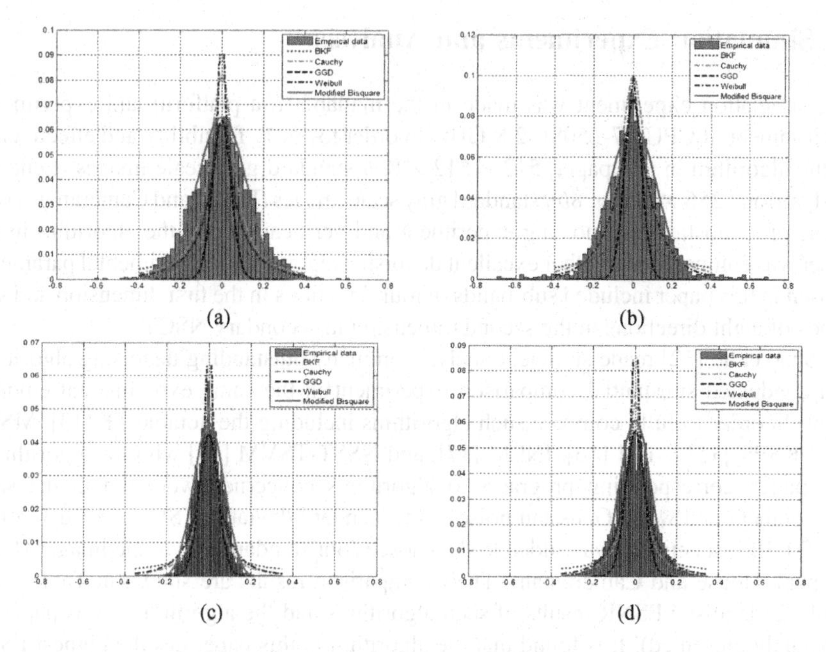

Fig. 2. Comparison of multiple weight functions: (a) Sub-band in first dimension and first direction of Barbara image; (b) Sub-band in first dimension and first direction of Lena image; (c) Sub-band in second dimension and first direction of Barbara image; (d) Sub-band in second dimension and first direction of Lena image

3.2 Algorithm

A denoising algorithm based on NSCT and modified NLM is proposed in this paper, which the major denoising steps are as follows:

Step1: Secondary decomposition is made on the noise image by NSCT to draw one low-frequency sub-band A and some high-frequency sub-bands $D_1, D_2, \cdots, D_N$.

Step2: NSCT coefficients are drawn in Step1. Centered by each coefficient in high-frequency sub-bands, NSCT coefficients are partitioned to 9×9 square blocks and each block is substituted to the kernel function for computation of similarity or Formula (15) to get the similarity between coefficients which is used to work out the corresponding weight value or exactly the weight value between coefficients.

Step3: The weight value worked out in Step2 is used to correct each noise coefficient in high-frequency sub-bands and inverse reconstruction is made on the low-frequency sub-band A and all the corrected high-frequency sub-bands $D'_1, D'_2, \cdots, D'_N$ by NSCT to ultimately draw the denoised image.

4 Simulation Experiments and Analysis

The simulation experiment was made in the matlab2015a platform and experimental environment of CPU E7-2508@2.8 GHz. In order to verify feasibility and effectiveness of the algorithm in this paper, $512 \times 512 \times 8bit$ standard gray scale images (Lena and Barbara) and $256 \times 256 \times 8bit$ standard gray scale images (House and Cameraman) were chosen for substantial denoising experiment and verification and the algorithm in this paper was compared with other excellent denoising algorithms. Experimental parameters chosen in this paper included sub-bands of four directions in the first dimension and sub-bands of eight directions in the second dimension in secondary NSCT.

Shao et al. [12] made sufficient analysis on many outstanding denoising algorithms and conducted substantial comparison experiment in the same experimental environment. In order to fully compare such algorithms including the notable TF [13], MSKR [14], KSPR [15], NLM [16], INLM [17], and NSST-TSVM [18] with the algorithm in this paper, corresponding programs to algorithms concerned were run in the same environment and white Gaussian noise in the standard deviation (SD) of 15 dB, 30 dB and 70 dB respectively was added to the chosen four standard gray scale images (Lena, Barbara, House and Cameraman). Their comparison results are shown in Table 1 and Table 2 (denoised PSNR results of such algorithms and the algorithm in this paper are separately presented). It is found that the algorithm in this paper has the highest PSNR value, indicating better image denoising ability of the proposed algorithm.

Table 1. PSNR results of multiple denoising algorithms with different image sets

σ	15	30	70	15	30	70
Methods	Lena			Barbara		
TF	28.29	23.93	10.28	28.85	24.71	11.94
MSKR	29.84	19.70	11.89	29.68	19.87	11.93
KSPR	28.35	25.94	21.97	25.64	23.98	20.97
NLM	29.48	24.19	14.92	30.12	24.44	14.95
INLM	30.25	27.39	23.96	31.34	28.35	24.36
NSST-TSVM	**30.79**	**29.04**	**24.83**	**31.68**	**28.76**	**24.93**
Algorithm in this paper	32.98	31.43	26.76	32.99	29.62	25.98

With regard to vision and visualization, Fig. 3 shows subjective denoising effect rendering of Lena in the environment where noise in the SD of 20 dB was added and comparison between the algorithm in this paper and BLS-GSM algorithm [19], Proshrink algorithm [20], SURE bivariate algorithm [21], NLM algorithm [16] and TV-model algorithm [22]. It is found that the algorithm in this paper has clearer processing results and better recovery effects. Moreover, in order to describe the ability of the algorithm in this paper to maintain the structural information of the original image, Fig. 4 shows the zoom-in rendering of local details of the denoised image. Evidently, the proposed algorithm can better maintain the structural information of the original image.

Table 2. PSNR results of multiple denoising algorithms with different image sets

σ	15	30	70	15	30	70
Methods	House			Cameraman		
TF	29.56	24.97	12.31	28.14	23.52	12.63
MSKR	32.59	20.26	18.77	28.76	20.35	13.21
KSPR	30.37	27.36	21.64	26.75	24.97	20.25
NLM	31.46	25.45	13.91	28.93	24.25	15.96
INLM	32.97	30.42	22.45	29.54	27.33	21.35
NSST-TSVM	**33.14**	**30.69**	**23.65**	**30.96**	**27.85**	**22.04**
Algorithm in this paper	34.24	32.82	23.67	31.74	28.53	23.27

Another advantage of the proposed algorithm is the reduction in the complexity of the denoising process. Table 3 shows the de-noising time of the proposed algorithm

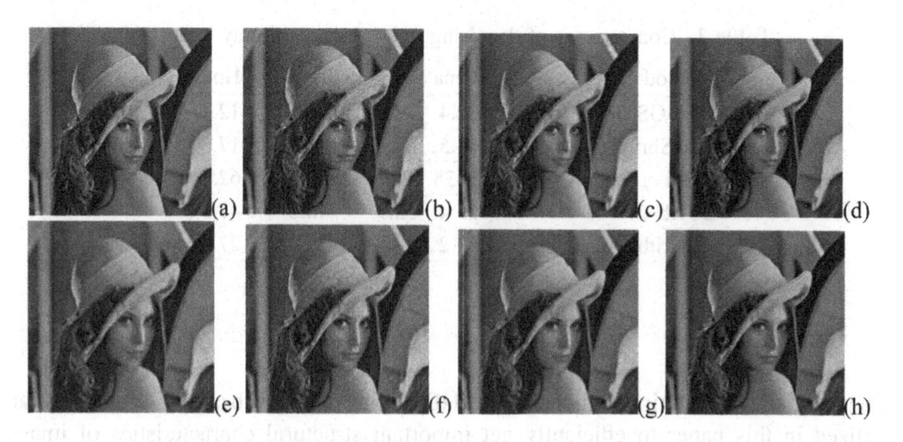

Fig. 3. Comparisons of denoising results: (a) Original image; (b) Noisy image; (c) Proshrink algorithm; (d) BLS-GSM algorithm; (e) SURE bivariate algorithm; (f) NLM algorithm; (g) TV-model algorithm; (h) Algorithm in this paper

and other 4 famous methods. It can be seen that the average time of the proposed algorithm is more efficiency compared with the other 4 algorithms. The main reason is that the images can be decomposed into low frequency and high frequency sub-bands, and only the high frequency sub-bands is processed.

By summarization of the above experimental results, the algorithm in this paper significantly improves the combined efficiency and gets good subjective and objective denoising results. This algorithm solves the problem of too smooth denoised image found in the classical NLM and replaces the original Gaussian kernel function by combining the modified Bisquare kernel function in computation to make more accurate computation of weight value to get better denoising results.

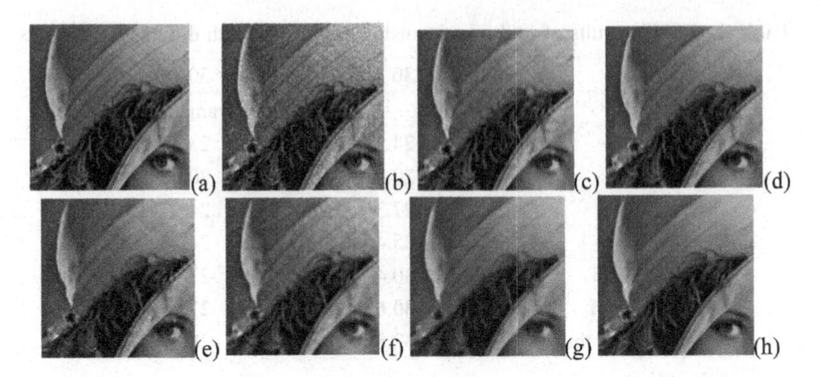

Fig. 4. Locally zoom-in rendering: (a) Part of original image; (b) Part of noisy image; (c) Proshrink algorithm; (d) BLS-GSM algorithm; (e) SURE bivariate algorithm; (f) NLM algorithm; (g) TV-model algorithm; (h) Algorithm in this paper

Table 3. Comparisons of denoising time of standard gray images (s)

Method	Lena	Barbara	Peppers	House
BLS-GSM	22.4	22.6	10.6	12.4
ProbShrink	44.5	47.2	14.1	17.6
NLM	1438	1486	594	625
SURE	55.4	53.9	46.8	47.5
Algorithm in this paper	43.2	42.7	26.9	27.4

5 Conclusions

Characteristics of multiple dimensions and directions of non-subsampled contourlet are utilized in this paper to efficiently get important structural characteristics of image, make sparse and transform the image to concentrate energy to retain more detail information in denoising. By such sparse and transform decomposition, more similarity between coefficients is captured and computation is made by bisquare kernel function in place of the original Gaussian kernel function to result in more accurate computation of weight value to get better denoising results.

Acknowledgements. This work was supported in part by National Natural Science Foundation of China (No. 61502356), by Hubei Province Natural Science Foundation of China (No. 2018CFB526).

References

1. Luo, E., Chan, S.H., Nguyen, T.Q.: Adaptive image denoising by targeted databases. IEEE Trans. Image Process. **24**(7), 2167–2181 (2015)
2. Chen, F., Zeng, X., Wang, M.: Image denoising via local and nonlocal circulant similarity. J. Vis. Commun. Image Represent. **30**(3), 117–124 (2015)

3. Schmidt, U., Gao, Q., Roth, S.: A generative perspective on MRFs in low-level vision. In: Proceedings of 2010 IEEE Conference on Computer Vision and Pattern Recognition (CVPR), , San Francisco, CA, pp. 1751–1758 (2010)
4. Zhang, W.: Image denoising algorithm of refuge chamber by combining wavelet transform and bilateral filtering. Int. J. Min. Sci. Technol. **23**(2), 221–225 (2013)
5. Cho, S.I., Kang, S.J., Kim, H.S.: Dictionary-based anisotropic diffusion for noise reduction. Pattern Recogn. Lett. **46**(3), 36–45 (2014)
6. Hill, P.R., Achim, A.M., Bull, D.R.: Dual-tree complex wavelet coefficient magnitude modelling using the bivariate Cauchy-Rayleigh distribution for image denoising. Sig. Process. **105**(12), 464–472 (2014)
7. Liu, X.M., Tian, Y., He, H.: Improved non-local means algorithm for image denoising. Comput. Eng. **38**(4), 199–207 (2012)
8. Zhong, H., Ma, K., Zhou, Y.: Modified BM3D algorithm for image denoising using nonlocal centralization prior. Sig. Process. **106**(8), 342–347 (2015)
9. Nguyen, M.P., Chun, S.Y.: Bounded self-weights estimation method for non-local means image denoising using minimax Estimators. IEEE Trans. Image Process. **26**(4), 1637–1649 (2017)
10. DaCunha, A.L., Zhou, J., Do, M.N.: The nonsubsampled contourlet transform: theory, design, and applications. IEEE Trans. Image Process. **15**(10), 3089–3101 (2006)
11. Goossens, B., Luong, Q., Pizurica, A.: An improved non-local denoising algorithm. In: International Workshop on Local and Non-local Approximation in Image Processing, Switzerland, pp. 143–156 (2008)
12. Shao, L., Yan, R., Li, X.: From heuristic optimization to dictionary learning: a review and comprehensive comparison of image denoising algorithms. IEEE Trans. Cybern. **44**(7), 1001–1013 (2014)
13. Shao, L., Zhang, H., Haan, G.: An overview and performance evaluation of classification based least squares trained filters. IEEE Trans. Image Process. **17**(10), 1772–1782 (2008)
14. Zhu, X., Milanfar, P.: Automatic parameter selection for denoising algorithms using a no-reference measure of image content. IEEE Trans. Image Process. **19**(12), 3116–3132 (2010)
15. Bouboulis, P., Slavakis, K., Theodoridis, S.: Adaptive kernel-based image denoising employing semiparametric regularization. IEEE Trans. Image Process. **19**(6), 1465–1479 (2010)
16. Buades, A., Coll, B., Morel, J.M.: A non-local algorithm for image denoising. In: IEEE Computer Society Conference on Computer Vision and Pattern Recognition (CVPR), USA, CA, San Diego, pp. 60–65 (2005)
17. Goossens, B., Luong, H., Pizurica, A., Philips, W.: An improved non-local denoising algorithm. In: International Workshop on Local Non-local Approximation Image Processing, pp. 143–156 (2008)
18. Yang, H.Y., Wang, X.Y., Niu, P.P.: Image denoising using nonsubsampled shearlet transform and twin support vector machines. Neural Netw. **57**(9), 152–165 (2014)
19. Portilla, J., Strela, V., Wainwright, M.J., Simoncelli, E.P.: Image denoising using scale mixtures of Gaussians in the wavelet domain. IEEE Trans. Image Process. **12**(11), 1338–1351 (2003)
20. Pizurica, A., Philips, W.: Estimating the probability of the presence of a signal of interest in multi-resolution single-and multi-band image denoising. IEEE Trans. Image Process. **15**(3), 654–665 (2006)
21. Luisier, F., Blu, T., Unser, M.: A new SURE approach to image denoising: interscale orthonormal wavelet thresholding. IEEE Trans. Image Process. **16**(3), 593–606 (2007)
22. Rudin, L.I., Osher, S., Fatemi, E.: Nonlinear total variation based noise removal algorithms. Phys. D **60**(1–4), 259–268 (1992)

Unambiguous Discrimination Between Mixed Quantum States Based on Programmable Quantum State Discriminators

Daowen Qiu[1,2(✉)], Hongfeng Gan[1], Guangya Cai[1], and Mateus Paulo[2]

[1] Institute of Computer Science Theory, School of Data and Computer Science, Sun Yat-Sen University, Guangzhou 510006, China
issqdw@mail.sysu.edu.cn
[2] Instituto de Telecomunicações, Departamento de Matemática, Instituto Superior Técnic, Av. Rovisco Pais, 1049-001 Lisbon, Portugal

Abstract. We discuss the problem of designing an unambiguous programmable discriminator for mixed quantum states. We prove that there does not exist such a universal unambiguous programmable discriminator for mixed quantum states that has two program registers and one data register. However, we find that we can use the idea of programmable discriminator to unambiguously discriminate mixed quantum states. The research shows that by using such an idea, when the success probability for discrimination reaches the upper bound, the success probability is better than what we do not use the idea to do, except for some special cases.

Keywords: Unambiguous discrimination · Mixed quantum state
Programmable discriminator

1 Introduction

The discrimination of quantum states is a basic task in quantum information and quantum communication [1]. A great deal of attention has been attracted into this field in recent thirty years, especially the *unambiguous discrimination* (UD) of quantum states. UD is a sort of discrimination that never gives an erroneous result, but sometimes it may fail. In the case of pure states, UD has been widely considered. In the case of two pure states, the optimum measurement for the UD of two pure states was found decades ago [2–5]. A sufficient and necessary condition for unambiguously distinguishing arbitrary pure states and upper bound on the success probability for UD of arbitrary pure states have also been given (see, for example, [6–9] and some related references therein). Indeed, a complete overview of UD of pure states can be found in two review articles [10, 11]. In the case of mixed quantum states, lots of work also have been done [12–19], which focus on the upper bound and how to get the upper bound of the success probability for discrimination. For the case of two mixed quantum states, a necessary and sufficient condition for discriminating two mixed states to reach upper bound has been derived in [16].

D.-S. Huang et al. (Eds.): ICIC 2018, LNAI 10956, pp. 700–709, 2018.
https://doi.org/10.1007/978-3-319-95957-3_72

As we know, if we want to unambiguously discriminate quantum states, we need construct some positive operator valued measurements (POVMs) according to the states. However, if the states are unknown, we cannot construct such POVMs, which means that we cannot discriminate unknown states directly. A programmable quantum state discriminator for unambiguous discrimination was first proposed by Bergous and Hillery [20] to resolve this problem. Bergous and Hillery's discriminator is a fixed measurement that has two program registers and one data register. The quantum states in the data register are what we want to identify, which is confirmed to be one of the two states in program registers. That is to say, if we want to discriminate two states $|\psi_1\rangle$ and $|\psi_2\rangle$, we assign the two sates into the two program registers, and the data register is assigned with the state which we want to identify. Here we have no idea of these two states. Now we have two input states

$$\begin{aligned} |\psi_1^{in}\rangle &= |\psi_1\rangle|\psi_2\rangle|\psi_1\rangle, \\ |\psi_2^{in}\rangle &= |\psi_1\rangle|\psi_2\rangle|\psi_2\rangle. \end{aligned} \tag{1}$$

It is easy to see that if we can discriminate $|\psi_1^{in}\rangle$ and $|\psi_2^{in}\rangle$, then we can discriminate states $|\psi_1\rangle$ and $|\psi_2\rangle$. Bergous and Hillery's discriminator makes this target successful with a fixed measurement.

Based on Bergous and Hillery's discriminator, Zhang *et al.* [21] presented an unambiguous programmable discriminator for n arbitrary quantum states in an m-dimensional Hilbert space, where $m \geq n$. If $m = n$, an optimal unambiguous programmable discriminator for n arbitrary states was given in [21]. Notably, the unambiguous programmable discriminator for two states with a certain number of copies has been discussed in [22–25]. Also, there has been several studies of the unambiguous programmable discriminator for two qudit states [26, 27]. Moreover, the unambiguous programmable discriminator for more than two quantum states with a certain number of copies has been discussed in [28].

However, all the unambiguous discriminators mentioned above concentrate on pure states. The discussion of the unambiguous programmable discriminators for mixed quantum states is still limited. In this paper, we try to deal with the problem of designing an unambiguous programmable discriminator for mixed quantum states. Our purpose is to see whether or not a programmable unambiguous discriminator for mixed quantum states can be realized.

This paper is organized as follows: In Sect. 2, we prove that there does not exist an unambiguous programmable discriminator for mixed quantum states that has two program registers and one data register. Then, however, in Sect. 3, we find that we can still use the idea of programmable quantum state discriminator to unambiguously discriminate mixed quantum states. The research shows that by using this idea, when the success probability for discrimination reaches the upper bound, the success probability is better than what we do not use the idea to do, except for some special cases. At last, we conclude the paper with a short summary.

2 Nonexistence of Programmable Discriminator for Mixed States Based on Bergous and Hillery's Model

First, we try to design an unambiguous programmable discriminator for mixed quantum states based on Bergous and Hillery's model [20]. Our purpose is to see whether or not such an unambiguous programmable discriminator for mixed quantum states can be realized. To begin with, we prove a theorem here.

Theorem 1. Two mixed quantum states ρ_1, ρ_2 can be unambiguously discriminated if and only if ρ_1^{in}, ρ_2^{in} can be unambiguously discriminated, where

$$\begin{aligned} \rho_1^{in} &= \rho_1 \otimes \rho_2 \otimes \rho_1, \\ \rho_2^{in} &= \rho_1 \otimes \rho_2 \otimes \rho_2. \end{aligned} \tag{2}$$

Proof. First let

$$\begin{aligned} \rho_1 &= \textstyle\sum_{i=1}^{n_1} \alpha_i |\varphi_i\rangle\langle\varphi_i|, \\ \rho_2 &= \textstyle\sum_{j=1}^{n_2} \beta_j |\psi_j\rangle\langle\psi_j|, \end{aligned} \tag{3}$$

be the spectral decompositions of ρ_1, ρ_2. Then

$$\begin{aligned} \rho_1^{in} &= \textstyle\sum_{i,j,k} \alpha_i \beta_j \alpha_k |\varphi_i \psi_j \varphi_k\rangle\langle\varphi_i \psi_j \varphi_k|, \\ \rho_2^{in} &= \textstyle\sum_{i,j,l} \alpha_i \beta_j \beta_l |\varphi_i \psi_j \psi_l\rangle\langle\varphi_i \psi_j \psi_l|, \end{aligned} \tag{4}$$

where $i,k = 1,\ldots,n_1$ and $j,l = 1,\ldots,n_2$. Clearly, formula (4) is also the spectral decompositions of ρ_1^{in} and ρ_2^{in}.

Suppose that ρ_1, ρ_2 can be unambiguously discriminated. Then there exist POVM elements Π_0, Π_1, Π_2 such that

$$\Pi_0 + \Pi_1 + \Pi_2 = I, \tag{5}$$

and

$$\mathrm{Tr}(\Pi_i \rho_j) = p_i \delta_{ij}, \tag{6}$$

for some $p_i > 0$, where $i,j = 1,2$. Now we construct a new set of POVM elements

$$\Pi_0^{in} = I' \otimes \Pi_0, \quad \Pi_1^{in} = I' \otimes \Pi_1, \quad \Pi_2^{in} = I' \otimes \Pi_2, \tag{7}$$

where I' denotes the identity operator on $\rho_1 \otimes \rho_2$. We can easily prove that

$$\Pi_0^{in} + \Pi_1^{in} + \Pi_2^{in} = I \tag{8}$$

and

$$\mathrm{Tr}\left(\Pi_i^{in}\rho_j^{in}\right) = p_i\delta_{ij}, \tag{9}$$

for the above $p_i > 0$, where $i, j = 1, 2$. It means that there exists a set of POVM elements which can unambiguously discriminate ρ_1^{in}, ρ_2^{in}, i.e., ρ_1^{in}, ρ_2^{in}, can be unambiguously discriminated.

On the other side, suppose that ρ_1^{in}, ρ_2^{in} can be unambiguously discriminated. Then $\mathrm{supp}\left(\rho_1^{in}\right) \neq \mathrm{supp}\left(\rho_1^{in}, \rho_2^{in}\right)$ and $\mathrm{supp}\left(\rho_2^{in}\right) \neq \mathrm{supp}\left(\rho_1^{in}, \rho_2^{in}\right)$ [14]. Here $\mathrm{supp}\left(\rho_1^{in}, \ldots, \rho_n^{in}\right)$ is defined by the Hilbert space spanned by the eigenvectors of the mixed states $\rho_1^{in}, \ldots, \rho_n^{in}$ with corresponding nonzero eigenvalues. For $\mathrm{supp}\left(\rho_2^{in}\right) \neq \mathrm{supp}\left(\rho_1^{in}, \rho_2^{in}\right)$, it means that there exist some i, j, k, where $1 \leq i, k \leq n_1$ and $1 \leq j \leq n_2$, satisfying

$$|\varphi_i\psi_j\varphi_k\rangle \neq \sum\nolimits_{i',j',l'} a_{i',j',l'}\left|\varphi_{i'}\psi_{j'}\psi_{l'}\right\rangle, \tag{10}$$

where $i', k' = 1, \ldots, n_1$ and $j', l' = 1, \ldots, n_2$. Specifically, if we choose $i' = i$, $j' = j$, then

$$|\varphi_i\psi_j\varphi_k\rangle \neq |\varphi_i\psi_j\rangle \sum\nolimits_{l'} a'_{l'}|\psi_{l'}\rangle, \tag{11}$$

and, as a result,

$$|\varphi_k\rangle \neq \sum\nolimits_{l'} a'_{l'}|\psi_{l'}\rangle. \tag{12}$$

It implies $\mathrm{supp}\left(\rho_2^{in}\right) \neq \mathrm{supp}\left(\rho_1^{in}, \rho_2^{in}\right)$. With similar discussion, we can also have $\mathrm{supp}\left(\rho_1^{in}\right) \neq \mathrm{supp}\left(\rho_1^{in}, \rho_2^{in}\right)$. Therefore, ρ_1, ρ_2 can be unambiguously discriminated. This completes the proof.

Due to Theorem 1, we further discuss whether or not there exists an unambiguous programmable discriminator for mixed quantum states based on Bergous and Hillery's model [20]. Indeed, we have the following result.

Theorem 2. There does not exist an unambiguous programmable discriminator for mixed quantum states that has two program registers and one data register.

Proof. Suppose that there exists such an unambiguous programmable discriminator for mixed quantum states. Then there also exists a fixed measurement that can unambiguously discriminate ρ_1^{in}, ρ_2^{in}, where $\rho_1^{in} = \rho_1 \otimes \rho_2 \otimes \rho_1$, $\rho_2^{in} = \rho_1 \otimes \rho_2 \otimes \rho_2$, and ρ_1, ρ_2 are guaranteed to be unambiguously discriminated. We here assume that the fixed POVM elements are Π_0, Π_1, Π_2, which satisfy

$$\begin{aligned}\Pi_1\rho_2^{in} &= 0,\\ \Pi_2\rho_1^{in} &= 0,\\ \mathrm{Tr}\left(\Pi_1\rho_1^{in}\right) &> 0,\\ \mathrm{Tr}\left(\Pi_2\rho_2^{in}\right) &> 0,\\ \Pi_0+\Pi_1+\Pi_2 &= I,\end{aligned} \tag{13}$$

for any ρ_1, ρ_2 when they can be unambiguously discriminated.
Now, we have three special mixed quantum states as follows

$$\begin{aligned}\rho_1^{'} &= a_1|\gamma_1\rangle\langle\gamma_1| + a_2|\gamma_2\rangle\langle\gamma_2|,\\ \rho_2^{'} &= b_1|\gamma_2\rangle\langle\gamma_2| + b_2|\gamma_3\rangle\langle\gamma_3|,\\ \rho_3^{'} &= c_1|\gamma_1\rangle\langle\gamma_1| + c_2|\gamma_3\rangle\langle\gamma_3|,\end{aligned} \tag{14}$$

where $\rho_1^{'}$, $\rho_2^{'}$, $\rho_2^{'}$ are mixed quantum states in m-dimension Hilbert space ($m \geq 3$), and $\{|\gamma_1\rangle, |\gamma_2\rangle, |\gamma_3\rangle\}$ consists of an orthonormal basis in this space. It is no doubt that any two of these three stats can be unambiguously discriminated. Now we use the discriminator to discriminate any two of these states.

(1) Let $\rho_1 = \rho_1^{'}$, $\rho_2 = \rho_2^{'}$. Then

$$\begin{aligned}\rho_1^{in} &= \rho_1^{'} \otimes \rho_2^{'} \otimes \rho_1^{'},\\ \rho_2^{in} &= \rho_1^{'} \otimes \rho_2^{'} \otimes \rho_2^{'}.\end{aligned} \tag{15}$$

According to (13), $\Pi_1\rho_2^{in} = 0$, $\mathrm{Tr}\left(\Pi_1\rho_1^{in}\right) > 0$, and we have

$$\begin{aligned}&\langle\gamma_1\gamma_2\gamma_2|\Pi_1|\gamma_1\gamma_2\gamma_2\rangle = 0, \quad \langle\gamma_1\gamma_2\gamma_3|\Pi_1|\gamma_1\gamma_2\gamma_3\rangle = 0,\\ &\langle\gamma_1\gamma_3\gamma_2|\Pi_1|\gamma_1\gamma_3\gamma_2\rangle = 0, \quad \langle\gamma_1\gamma_3\gamma_3|\Pi_1|\gamma_1\gamma_3\gamma_3\rangle = 0,\\ &\langle\gamma_2\gamma_2\gamma_2|\Pi_1|\gamma_2\gamma_2\gamma_2\rangle = 0, \quad \langle\gamma_2\gamma_2\gamma_3|\Pi_1|\gamma_2\gamma_2\gamma_3\rangle = 0,\\ &\langle\gamma_2\gamma_3\gamma_2|\Pi_1|\gamma_2\gamma_3\gamma_2\rangle = 0, \quad \langle\gamma_2\gamma_3\gamma_3|\Pi_1|\gamma_2\gamma_3\gamma_3\rangle = 0,\end{aligned} \tag{16}$$

and

$$\mathrm{Tr}\left(\Pi_1\rho_1^{in}\right) = \sum_{i,j,k=1}^{i,j,k=2} a_i b_j a_k \langle\gamma_i\gamma_{j+1}\gamma_k|\Pi_1|\gamma_i\gamma_{j+1}\gamma_k\rangle > 0. \tag{17}$$

(2) Let $\rho_1 = \rho_2^{'}$, $\rho_2 = \rho_1^{'}$. Then

$$\begin{aligned}\rho_1^{in} &= \rho_2^{'} \otimes \rho_1^{'} \otimes \rho_2^{'},\\ \rho_2^{in} &= \rho_2^{'} \otimes \rho_1^{'} \otimes \rho_1^{'}.\end{aligned} \tag{18}$$

According to (13), $\Pi_1\rho_2^{in} = 0$, we have

$$\begin{aligned}
&\langle\gamma_2\gamma_1\gamma_1|\Pi_1|\gamma_2\gamma_1\gamma_1\rangle = 0, \quad \langle\gamma_2\gamma_1\gamma_2|\Pi_1|\gamma_2\gamma_1\gamma_2\rangle = 0,\\
&\langle\gamma_2\gamma_2\gamma_1|\Pi_1|\gamma_2\gamma_2\gamma_1\rangle = 0, \quad \langle\gamma_2\gamma_2\gamma_2|\Pi_1|\gamma_2\gamma_2\gamma_2\rangle = 0,\\
&\langle\gamma_3\gamma_1\gamma_1|\Pi_1|\gamma_3\gamma_1\gamma_1\rangle = 0, \quad \langle\gamma_3\gamma_1\gamma_2|\Pi_1|\gamma_3\gamma_1\gamma_2\rangle = 0,\\
&\langle\gamma_3\gamma_2\gamma_1|\Pi_1|\gamma_3\gamma_2\gamma_1\rangle = 0, \quad \langle\gamma_3\gamma_2\gamma_2|\Pi_1|\gamma_3\gamma_2\gamma_2\rangle = 0.
\end{aligned} \tag{19}$$

(3) Let $\rho_1 = \rho_1^{'}$, $\rho_2 = \rho_3^{'}$. Then

$$\begin{aligned}
\rho_1^{in} &= \rho_1^{'} \otimes \rho_3^{'} \otimes \rho_1^{'},\\
\rho_2^{in} &= \rho_1^{'} \otimes \rho_3^{'} \otimes \rho_3^{'}.
\end{aligned} \tag{20}$$

According to (13), $\Pi_1\rho_2^{in} = 0$, we have

$$\begin{aligned}
&\langle\gamma_1\gamma_1\gamma_1|\Pi_1|\gamma_1\gamma_1\gamma_1\rangle = 0, \quad \langle\gamma_1\gamma_1\gamma_3|\Pi_1|\gamma_1\gamma_1\gamma_3\rangle = 0,\\
&\langle\gamma_1\gamma_3\gamma_1|\Pi_1|\gamma_1\gamma_3\gamma_1\rangle = 0, \quad \langle\gamma_1\gamma_3\gamma_3|\Pi_1|\gamma_1\gamma_3\gamma_3\rangle = 0,\\
&\langle\gamma_2\gamma_1\gamma_1|\Pi_1|\gamma_2\gamma_1\gamma_1\rangle = 0, \quad \langle\gamma_2\gamma_1\gamma_3|\Pi_1|\gamma_2\gamma_1\gamma_3\rangle = 0,\\
&\langle\gamma_2\gamma_3\gamma_1|\Pi_1|\gamma_2\gamma_3\gamma_1\rangle = 0, \quad \langle\gamma_2\gamma_2\gamma_3|\Pi_1|\gamma_2\gamma_2\gamma_3\rangle = 0.
\end{aligned} \tag{21}$$

(4) Let $\rho_1 = \rho_3^{'}$, $\rho_2 = \rho_1^{'}$. Then

$$\begin{aligned}
\rho_1^{in} &= \rho_3^{'} \otimes \rho_1^{'} \otimes \rho_3^{'},\\
\rho_2^{in} &= \rho_3^{'} \otimes \rho_1^{'} \otimes \rho_1^{'}.
\end{aligned} \tag{22}$$

According to (13), $\Pi_1\rho_2^{in} = 0$, we have

$$\begin{aligned}
&\langle\gamma_1\gamma_1\gamma_1|\Pi_1|\gamma_1\gamma_1\gamma_1\rangle = 0, \quad \langle\gamma_1\gamma_1\gamma_2|\Pi_1|\gamma_1\gamma_1\gamma_2\rangle = 0,\\
&\langle\gamma_1\gamma_2\gamma_1|\Pi_1|\gamma_1\gamma_2\gamma_1\rangle = 0, \quad \langle\gamma_1\gamma_2\gamma_2|\Pi_1|\gamma_1\gamma_2\gamma_2\rangle = 0,\\
&\langle\gamma_3\gamma_1\gamma_1|\Pi_1|\gamma_3\gamma_1\gamma_1\rangle = 0, \quad \langle\gamma_3\gamma_1\gamma_2|\Pi_1|\gamma_3\gamma_1\gamma_2\rangle = 0,\\
&\langle\gamma_3\gamma_2\gamma_1|\Pi_1|\gamma_3\gamma_2\gamma_1\rangle = 0, \quad \langle\gamma_3\gamma_2\gamma_2|\Pi_1|\gamma_3\gamma_2\gamma_2\rangle = 0.
\end{aligned} \tag{23}$$

Now using (16), (19), (21) and (23), we find that $\mathrm{Tr}(\Pi_1\rho_1^{in})$ in (17) is equal to zero, which contradicts (17) that is $\mathrm{Tr}(\Pi_1\rho_1^{in}) > 0$. It means that there does not exist such a fixed measurement. In other words, such an unambiguous programmable discriminator for mixed quantum states *does not* exist. The proof is completed.

Why does not there exist such an unambiguous programmable discriminator for mixed quantum states? The reason is not hard to find from the above proof. It is because the mixed states ρ_1^{in}, ρ_2^{in} loose the symmetry which $|\psi_1^{in}\rangle$, $|\psi_2^{in}\rangle$ have. In a way, we can say that the difference between mixed states and pure states results in Theorem 2. Also, from Theorem 2 we have seen some special features that mixed states have but pure states do not.

3 Unambiguous Discrimination Between Mixed Quantum States Based on Programmable Discriminator

It is disappointed that we do not have such an unambiguous programmable discriminator for mixed quantum states that was indicated above. We do not know whether there exists other type of discriminators for mixed quantum states, either. However, if we think about it in another way, we can find that the unambiguous programmable discriminator is a very good idea for discriminating states. We can still use the idea of unambiguous programmable discriminators here to discriminate mixed states. That is to say, if we want to discriminate two known mixed sates ρ_1, ρ_2, then we can try to discriminate two mixed states ρ_1^{in}, ρ_2^{in}. We use the idea of unambiguous programmable discriminators which have two program registers and n data registers. Specifically, if we want to discriminate two known mixed sates ρ_1, ρ_2, then we try to discriminate the following states.

$$\begin{aligned} \rho_1^{in} &= \rho_1 \otimes \rho_2 \otimes \rho_1^{\otimes n}, \\ \rho_2^{in} &= \rho_1 \otimes \rho_2 \otimes \rho_2^{\otimes n}. \end{aligned} \tag{24}$$

It is clear that if we can discriminate ρ_1^{in}, ρ_2^{in}, then we can also discriminate ρ_1, ρ_2.

First, we consider whether ρ_1^{in}, ρ_2^{in} can be unambiguously discriminated when ρ_1, ρ_2 can be unambiguously discriminated. The answer is yes. We can use the similar method in Theorem 1 to prove it. Now based on the two known states ρ_1^{in}, ρ_2^{in}, we can construct POVMs to distinguish them. Before dealing with the success probability for unambiguous discrimination between ρ_1^{in} and ρ_2^{in}, we have a simple lemma as follows.

Lemma 1. Let ρ_1, ρ_2 be two arbitrary mixed states, and let $\rho_1^{in} = \rho_1 \otimes \rho_2 \otimes \rho_1^{\otimes n}$, $\rho_2^{in} = \rho_1 \otimes \rho_2 \otimes \rho_2^{\otimes n}$. We have $F(\rho_1^{in}, \rho_2^{in}) = F(\rho_1, \rho_2)^n$, where $n \geq 1$ and $F(\cdot,\cdot)$ is the definition of fidelity in [1], i.e., $F(\rho_1, \rho_2) = \mathrm{Tr}\left(\sqrt{\sqrt{\rho_1}\rho_2\sqrt{\rho_1}}\right)$.

Remark 1. The proof of Lemma 1 follows from the simple fact as follows.

$$F(\rho_1 \otimes \rho_2, \rho_3 \otimes \rho_4) = F(\rho_1, \rho_3) \times F(\rho_2, \rho_4). \tag{25}$$

Now we discuss the failure probability of the unambiguous discrimination between ρ_1^{in} and ρ_2^{in}. According to Raynal and Lütkenhaus' work [16], if $\mathrm{supp}(\rho_1^{in}) \cap \mathrm{supp}(\rho_2^{in}) = \{0\}$ and some conditions are satisfied, the failure probability of the unambiguous discrimination between ρ_1^{in} and ρ_2^{in} can reach its low bound. Let $F(\rho_1^{in}, \rho_2^{in})$ be the fidelity of the two states ρ_1^{in}, ρ_2^{in}. Then $F(\rho_1^{in}, \rho_2^{in}) = F(\rho_1, \rho_2)^n$. We denote by P_1^{in} and P_2^{in}, the projectors onto the supports of ρ_1^{in} and ρ_2^{in}, respectively. Let P_1 and P_2 be the projectors onto the supports of ρ_1 and ρ_2, respectively. Then $P_1^{in} = P_1 \otimes P_2 \otimes P_1^{\otimes n}$ and $P_2^{in} = P_1 \otimes P_2 \otimes P_2^{\otimes n}$. We can prove $\mathrm{Tr}(P_1^{in}\rho_2^{in}) = \mathrm{Tr}(P_1\rho_2)^n$ and $\mathrm{Tr}(P_2^{in}\rho_1^{in}) = \mathrm{Tr}(P_2\rho_1)^n$ using the similar method as Lemma 1. Let η_1 and η_2 be the priori probabilities of ρ_1 and ρ_2, respectively. Now according to [16], we have

$$Q_{in}^{opt}=\begin{cases}\eta_2\frac{F(\rho_1,\rho_2)^{2n}}{\mathrm{Tr}(P_2\rho_1)^n}+\eta_1\mathrm{Tr}(P_2\rho_1)^n, & \sqrt{\frac{\eta_2}{\eta_1}}\le\frac{\mathrm{Tr}(P_2\rho_1)^n}{F(\rho_1,\rho_2)^n},\\ 2\sqrt{\eta_1\eta_2}F(\rho_1,\rho_2)^n, & \frac{\mathrm{Tr}(P_2\rho_1)^n}{F(\rho_1,\rho_2)^n}\le\sqrt{\frac{\eta_2}{\eta_1}}\le\frac{F(\rho_1,\rho_2)^n}{\mathrm{Tr}(P_1\rho_2)^n},\\ \eta_1\frac{F(\rho_1,\rho_2)^{2n}}{\mathrm{Tr}(P_1\rho_2)^n}+\eta_2\mathrm{Tr}(P_1\rho_2)^n, & \frac{F(\rho_1,\rho_2)^n}{\mathrm{Tr}(P_1\rho_2)^n}\le\sqrt{\frac{\eta_2}{\eta_1}}\end{cases} \tag{26}$$

where Q_{in}^{opt} denotes the optimal failure probability of the unambiguous discrimination between ρ_1^{in} and ρ_2^{in}. Here $\mathrm{Tr}(P_2\rho_1)\le 1$, $\mathrm{Tr}(P_1\rho_2)\le 1$, $F(\rho_1,\rho_2)^2\le\mathrm{Tr}(P_2\rho_1)$ and $F(\rho_1,\rho_2)^2\le\mathrm{Tr}(P_1\rho_2)$ (refer to [16] for more details).

The first question is whether or not $\mathrm{supp}(\rho_1^{in})\cap\mathrm{supp}(\rho_2^{in})=\{0\}$ can be satisfied? Actually, we can easily prove that if $\mathrm{supp}(\rho_1)\cap\mathrm{supp}(\rho_2)=\{0\}$, then $\mathrm{supp}(\rho_1^{in})\cap\mathrm{supp}(\rho_2^{in})=\{0\}$. It means that $\mathrm{supp}(\rho_1^{in})\cap\mathrm{supp}(\rho_2^{in})=\{0\}$ is not a stricter constraint.

Let Q_{in} denote the failure probability of the unambiguous discrimination between ρ_1^{in} and ρ_2^{in}. From [16], we know that sometimes Q_{in} here can reach Q_{in}^{opt}. When Q_{in} reaches Q_{in}^{opt}, that is, $Q_{in}=Q_{in}^{opt}$, we find that Q_{in} is smaller than Q (here Q denotes the failure probability of the unambiguous discrimination between ρ_1 and ρ_2), except for some special cases. We discuss this in what follows.

If $F(\rho_1,\rho_2)=0$, i.e., it means that the two states can be perfectly discriminated, then $Q_{in}=Q=0$. When $n=1$, we find that if Q_{in} reaches Q_{in}^{opt} then Q can also reach its optimal value, and thus $Q_{in}=Q=Q_{in}^{opt}$. Now we consider the situation where $0<F(\rho_1,\rho_2)<1$ and $n>1$:

(1) If $\frac{\mathrm{Tr}(P_2\rho_1)}{F(\rho_1,\rho_2)}\le 1$ and $\frac{F(\rho_1,\rho_2)}{\mathrm{Tr}(P_1\rho_2)}\ge 1$, then no matter which regime $\sqrt{\frac{\eta_2}{\eta_1}}$ is, we will find that if Q_{in} reaches Q_{in}^{opt}, then $Q_{in}=Q_{in}^{opt}<Q$.

(2) If $\frac{\mathrm{Tr}(P_2\rho_1)}{F(\rho_1,\rho_2)}\le\frac{F(\rho_1,\rho_2)}{\mathrm{Tr}(P_1\rho_2)}<1$, then, except for the regime $\frac{F(\rho_1,\rho_2)^n}{\mathrm{Tr}(P_1\rho_2)^n}\le\sqrt{\frac{\eta_2}{\eta_1}}\le\frac{F(\rho_1,\rho_2)}{\mathrm{Tr}(P_1\rho_2)}$ that we cannot compare, we will find that if Q_{in} reaches Q_{in}^{opt}, then $Q_{in}=Q_{in}^{opt}<Q$.

(3) If $\frac{\mathrm{Tr}(P_1\rho_2)}{F(\rho_1,\rho_2)}>1$, then, except for the regime $\frac{\mathrm{Tr}(P_2\rho_1)}{F(\rho_1,\rho_2)}\le\sqrt{\frac{\eta_2}{\eta_1}}\le\frac{\mathrm{Tr}(P_2\rho_1)^n}{F(\rho_1,\rho_2)^n}$ that we cannot compare, we will find that if Q_{in} reaches Q_{in}^{opt}, then $Q_{in}=Q_{in}^{opt}<Q$.

From the above discussion we can see that if the failure probability of the unambiguous discrimination between ρ_1^{in} and ρ_2^{in} reaches its optimization, then the failure probability of the unambiguous discrimination between ρ_1^{in} and ρ_2^{in} is better than that between ρ_1 and ρ_2 mostly. It is easy to find that the bigger n is, the smaller Q_{in}^{opt} will be. That means that if Q_{in} can reach Q_{in}^{opt} with the bigger n, then the smaller Q_{in} will be. Considering the conditions of Q_{in} being able to reach Q_{in}^{opt} in (26), we find that such conditions are not stricter when n is bigger. Especially, the conditions in the first and the third regime of (26) can be derived from $n=1$. On the other hand, even if n is small, such as $n=2$, and $F(\rho_1,\rho_2)$ is much smaller than 1, then we can also have a very small Q_{in}^{opt} here.

The rest problem is about the situation when Q_{in} does not reach its optimization. We have no solution yet. In fact, the solution of this problem depends on the solution of how to discriminate two arbitrary mixed states optimally. However, how to discriminate optimally two arbitrary mixed quantum states still seems a pending problem to be solved completely.

4 Conclusion

In this paper, we have tried to design an unambiguous programmable discriminator for mixed quantum states based on Bergous and Hillery's model [20]. We have proved that there does not exist a universal unambiguous programmable discriminator for mixed quantum states that has two program registers and one data register. However, we found that we can use the idea of programmable discriminators to unambiguously discriminate mixed quantum states. The result shows that by using such an idea, when the success probability for discrimination reaches the upper bound, the success probability is better than what we do not use the idea to do, except for some special cases. We have discussed this result in detail.

As we know, there is another important discrimination scheme, named Minimum-error discrimination [29–31]. A natural problem is whether or not we can design a minimum-error programmable discriminator for pure and mixed quantum states?

Acknowledgement. This work is supported in part by the National Natural Science Foundation (Nos. 61572532, 61272058), the Natural Science Foundation of Guangdong Province of China (No. 2017B030311011), and the Fundamental Research Funds for the Central Universities of China (Nos. 17lgjc24) and Mateus and Qiu are also funded by FCT project UID/EEA/50008/2013.

References

1. Nielsen, M.A., Chuang, I.L.: Quantum Computation and Quantum Information. Cambridge University Press, Cambridge (2000)
2. Ivanovic, D.: How to differentiate between non-orthogonal states. Phys. Lett. A **123**, 257–259 (1987)
3. Dieks, D.: Overlap and distinguishability of quantum states. Phys. Lett. A **126**, 303–306 (1988)
4. Peres, A.: How to differentiate between non-orthogonal states. Phys. Lett. A **128**, 19 (1988)
5. Jaeger, G., Shimony, A.: Optimal distinction between two non-orthogonal quantum states. Phys. Lett. A **197**, 83–87 (1995)
6. Chefles, A.: Unambiguous discrimination between linearly independent quantum states. Phys. Lett. A **239**, 339–347 (1998)
7. Chefles, A., Barnett, S.M.: Optimum unambiguous discrimination between linearly independent symmetric states. Phys. Lett. A **250**, 223–229 (1998)
8. Qiu, D.: Upper bound on the success probability for unambiguous discrimination. Phy. Lett. A **303**, 140–146 (2002)
9. Qiu, D.: Upper bound on the success probability of separation among quantum states. J. Phys. A Math. Gen. **35**, 6931 (2002)
10. Chefles, A.: Quantum state discrimination. Contemp. Phys. **41**, 401–424 (2000)

11. Bergou, J.A., Herzog, U., Hillery, M.: 11 discrimination of quantum states. In: Paris, M., Řeháček, J. (eds.) Quantum State Estimation. Lecture Notes in Physics, vol. 649, pp. 417–465. Springer, Berlin (2004). https://doi.org/10.1007/978-3-540-44481-7_11
12. Rudolph, T., Spekkens, R.W., Turner, P.S.: Unambiguous discrimination of mixed states. Phys. Rev. A **68**, 010301 (2003)
13. Fiurášek, J., Ježek, M.: Optimal discrimination of mixed quantum states involving inconclusive results. Phys. Rev. A **67**, 012321 (2003)
14. Feng, Y., Duan, R., Ji, Z.: Condition and capability of quantum state separation. Phys. Rev. A **72**, 012313 (2005)
15. Herzog, U., Bergou, J.A.: Optimum unambiguous discrimination of two mixed quantum states. Phys. Rev. A **71**, 050301 (2005)
16. Raynal, P., Lütkenhaus, N.: Optimal unambiguous state discrimination of two density matrices: lower bound and class of exact solutions. Phys. Rev. A **72**, 022342 (2005)
17. Zhou, X., Zhang, Y., Guo, G.C.: Unambiguous discrimination of mixed states: a description based on system-ancilla coupling. Phys. Rev. A **75**, 052314 (2007)
18. Herzog, U.: Optimum unambiguous discrimination of two mixed states and application to a class of similar states. Phys. Rev. A **75**, 052309 (2007)
19. Kleinmann, M., Kampermann, H., Bruß, D.: Unambiguous discrimination of mixed quantum states: optimal solution and case study. Phys. Rev. A **81**, 020304 (2010)
20. Bergou, J.A., Hillery, M.: Universal programmable quantum state discriminator that is optimal for unambiguously distinguishing between unknown states. Phys. Rev. Lett. **94**, 160501 (2005)
21. Zhang, C., Ying, M., Qiao, B.: Universal programmable devices for unambiguous discrimination. Phys. Rev. A **74**, 042308 (2006)
22. Bergou, J.A., Bužek, V., Feldman, E., Herzog, U., Hillery, M.: Programmable quantum-state discriminators with simple programs. Phys. Rev. A **73**, 062334 (2006)
23. He, B., Bergou, J.A.: Programmable unknown quantum-state discriminators with multiple copies of program and data: a Jordan-basis approach. Phys. Rev. A **75**, 032316 (2007)
24. Sentís, G., Bagan, E., Calsamiglia, J., Muñoz-Tapia, R.: Multicopy programmable discrimination of general qubit states. Phys. Rev. A **82**, 042312 (2010)
25. Zhou, T., Cui, J.X., Wu, X., Long, G.L.: Multicopy programmable discriminators between two unknown qubit states with group-theoretic approach. Quantum Inf. Comput. **12**, 1017–1033 (2012)
26. Zhou, T.: Unambiguous discrimination between two unknown qudit states. Quantum Inf. Process. **11**, 1669–1684 (2012)
27. Zhou, T.: Success probabilities for universal unambiguous discriminators between unknown pure states. Phys. Rev. A **89**, 014301 (2014)
28. Jafarizadeh, M.A., Mahmoudi, P., Akhgar, D., Faizi, E.: Designing an optimal, universal, programmable, and unambiguous discriminator for N unknown qubits. Phys. Rev. A **96**, 052111 (2017)
29. Qiu, D.: Minimum-error discrimination between mixed quantum states. Phys. Rev. A **77**, 012328 (2008)
30. Qiu, L., Li, L.: Minimum-error discrimination of quantum states: bounds and comparisons. Phys. Rev. A **81**, 042329 (2010)
31. Qiu, D., Li, L.: Relation between minimum-error discrimination and optimum unambiguous discrimination. Phys. Rev. A **82**, 032333 (2010)

A Quantum Electronic Voting Scheme with d-Level Single Particles

Yong-Zhen Xu[1], Yifan Huang[1], Wei Lu[1], and Lvzhou Li[1,2(✉)]

[1] Institute of Computer Science Theory, School of Data and Computer Science, Sun Yat-sen University, Guangzhou 510006, China
lilvzh@mail.sysu.edu.cn
[2] The Key Laboratory of Machine Intelligence and Advanced Computing, Sun Yat-sen University, Ministry of Education, Guangzhou 510006, China

Abstract. In this paper, we present a quantum anonymous voting protocol where the voting information is encoded in a traveling single d-level particle. Specifically, we take advantage of a cyclic property of a set of mutually unbiased bases and store the vote information by two types of unitary operations. Compared with existing work, our protocol is based on a single quantum d-level state instead of multi-particle entangled states, thus it is easier to implement. Furthermore, we also show that the proposed scheme satisfies most of core requirements and some additional requirements.

Keywords: Quantum anonymous voting · Mutually unbiased base
Single qudit

1 Introduction

Electronic voting systems have been one of the most remarkable and promising fields of information science and used in production and life. With the development of quantum computers, many classical voting schemes are facing the great security challenges. Some novel voting protocols are urgently needed to resist attacks by quantum computers. In this paper, we focus on quantum voting which is one of these types against quantum attacks.

In recent year, dozens of quantum voting protocols have been proposed. Let us review some representative schemes as follows. In 2007, Vaccaro et al. [1] presented quantum anonymous voting protocols for some different scenarios, in which the ballot state is an entangled state shared between at least sites. Later, Bonanome et al. [2] considered two types of quantum voting patterns: traveling voting (TV) and distributed voting (DV). TV uses entangled two-qudit state and DV uses entangled N-qudit state as the ballot state. Similarly, Wang et al. [3] make use of two kinds of entangled quantum states to design a self-tallying voting protocol, in which any voter can obtain the voting result. Recently, a quantum voting system based on quantum games was introduced by Bao and Halpern [4]. In addition, some other quantum voting schemes using different ideas have also been proposed [5–12].

However, most of the protocols mentioned have a limitation that they utilize entangled state which is hard to prepare and maintain. Based on this, we employ a

D.-S. Huang et al. (Eds.): ICIC 2018, LNAI 10956, pp. 710–715, 2018.
https://doi.org/10.1007/978-3-319-95957-3_73

single particle as the ballot state to design a novel quantum voting protocol in this paper. In the proposed protocol, the information of the vote is encoded into a travelling d-level quantum state which are chosen from a set of mutually unbiased bases (MUBs). Note that it is the first time that mutually unbiased bases are used for quantum anonymous voting, while they were ever used for quantum secret sharing [13].

This paper is structured as follows. Section 2 introduces the mutually unbiased bases which are used later. Our anonymous voting protocol is proposed in Sect. 3 and the analysis is discussed in Sect. 4. At the end, a brief conclusion is given.

2 MUBs and Its Cyclic Property

In this section, we first review the definition of MUBs [14–17]. Two groups of orthonormal bases $\{|e_1\rangle,\ldots,|e_n\rangle\}$ and $\{|f_1\rangle,\ldots,|f_n\rangle\}$ in n dimension Hilbert space are mutually unbiased bases if they satisfy

$$|\langle e_a|e_b\rangle|^2=\frac{1}{n},\forall a,b\in\{1,2,\ldots,n\}. \quad (1)$$

That is, if a particle's state belongs to one basis, then the measurement result with respect to the other basis will collapse randomly to one state of the basis with equal probabilities.

Now, we turn to a type of MUBs introduced in [13], which are used in the protocol proposed later. In a d-dimension Hilbert space, there is a set of d MUBs $\{|e_l^j\rangle\}$, where d is a prime number, j = 0, …, d – 1 is the basis, l = 0, …, d – 1 is the vector of the jth basis and

$$|e_l^j\rangle=\frac{1}{\sqrt{d}}\sum\nolimits_{k=0}^{d}\omega^{k(l+jk)}|k\rangle, \quad (2)$$

where $\omega = e^{2\pi i/d}$.

The cyclic property of this MUBs is described as follow: There is two unitary operations X_d and Y_d, for $l',j' \in \{0,\ldots,d-1\}$, we have

$$X_d^{l'}Y_d^{j'}|e_l^j\rangle=\left|e_{l+l'}^{j+j'}\right\rangle, \quad (3)$$

where $X_d=\sum\nolimits_{n=0}^{d-1}\omega^n|n\rangle\langle n|$ and $Y_d=\sum\nolimits_{n=0}^{d-1}\omega^{N^2}|n\rangle\langle n|$. The proof of this property was given in [13].

That is, the particle in the state $|e_l^j\rangle$ will move one step forward and then move one step downward with the help of X_dY_d, as shown in Fig. 1.

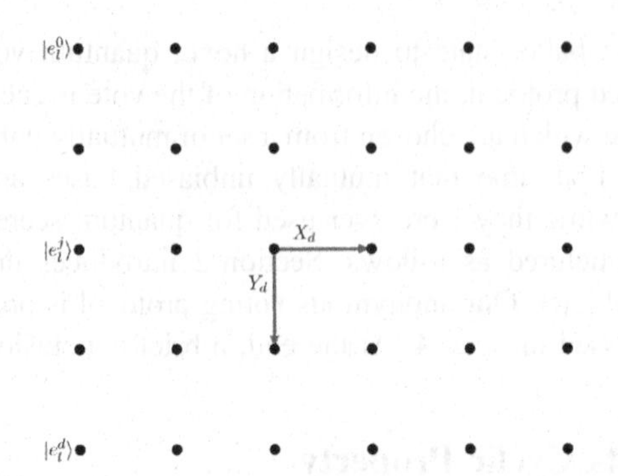

Fig. 1. The cyclic property of this type of MUBs. Each row in this lattice represents one of the complete MUBs, and each point in one row represents the individual state of this basis.

3 Quantum Voting Protocol

In this section, a two-valued voting protocol is presented, where each voter has two choices (say yes or no) for the candidate. The participants involved in our quantum voting protocol are: (1) Alice: she is a trusted third party who prepares the initial ballot state, distributes the ballot qudit to the voter, and gets the voting result from the final state in tallying phase; (2) $Bob_i, i = \{1, 2, \ldots, N\}$: he is a eligible voter. The detail process of our quantum voting scheme is as follows, as shown in Fig. 2.

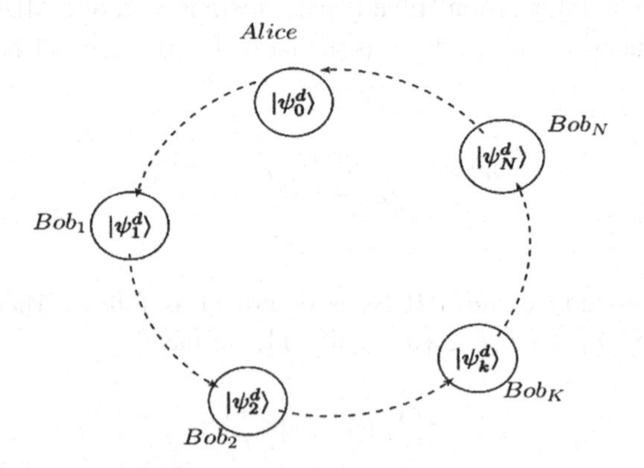

Fig. 2. A brief process of voting protocol, the empty circle represents the state of the ballot qudit after performed its corresponding operation.

(1) Alice is the distributor who prepares a single d-level particle and its state is $|e_0^0\rangle$. And she generates two random numbers x_0 and y_0 ($x_0 \in \{0, 1\}, y_0 \in \{0, 1, \ldots, d\}$). Then a unitary operation $X_d^{x_0} Y_d^{y_0}$ generated by these two random number acts on the initial state. Now the ballot qudit is denoted by $|\psi_0^d\rangle$. At last, she sends this ballot qudit to the first voter Bob_1. In addition, Alice confirms the legitimacy of all voters.

(2) For i = 1,…,N, the voter Bob_i prepares one number $x_i \in \{0(\text{no}), 1(\text{yes})\}$ depending on his choice, and generates one random number $y_i \in \{0, 1, \ldots, d\}$). After receiving the ballot state $|\psi_{i-1}^d\rangle$ sent by Bob_{i-1}, Bob_i performs the corresponding operation $X_d^{x_i} Y_d^{y_i}$ on this ballot qudit. Finally, the last voter Bob_N sends the ballot qudit $|\psi_N^d\rangle$ back to the distributor Alice. The form of $|\psi_N^d\rangle$ is

$$\begin{aligned} |\psi_N^d\rangle &= \left(\prod\nolimits_{i=0}^{N} X_d^{x_i} Y_d^{y_i}\right)|e_0^0\rangle \\ &= \frac{1}{\sqrt{d}}\left(|0\rangle + \sum\nolimits_{k=1}^{d-1} \omega^{\sum_{i=1}^{N}\left(kx_i + k^2 y_i\right)}|k\rangle\right). \end{aligned} \tag{4}$$

(3) Alice measures the qudit by a randomly selected basis $|e_l^r\rangle$ with r $\in \{0,\ldots,d-1\}$. The outcome represents by m $\in \{0,\ldots,d-1\}$.
(4) All voters randomly publish their own choice of y_i. By these values, Alice checks the validity of this round. The round is valid if

$$\sum\nolimits_{i=0}^{N} y_i = r \, mod \, d. \tag{5}$$

Otherwise the round is rejected. If the round is valid, we have

$$\sum\nolimits_{i=0}^{N} x_i = m \, mod \, d. \tag{6}$$

A limitation d > N is needed. Only in this way Alice can get the voting result v of "yes" as follows

$$\mathrm{v} = \begin{cases} m - x_0, & when \ m > x_0, \\ d - x_0 + m, & when \ m \le x_0. \end{cases} \tag{7}$$

(5) Alice performs security checks. She randomly selects subset of the rounds. And then all voters randomly publish their own choice of x_i. Alice checks the correctness of Eq. (5). If Eq. (5) holds, the round is valid, otherwise the round is rejected.

4 Analysis

An electronic voting protocol should meets all core requirements and some additional requirements as much as possible [18]. In this section, we analyze our scheme in detail from these two aspects.

4.1 Core Requirements

- Correctness. Steps (4) and (5) can ensure that all valid votes are counted correctly. And invalid votes can be detected.
- Privacy. Through security analysis in Ref. [13], an external eavesdropper and some dishonest voters cannot get voting information. And Bob_i cannot prove to anyone how he voted because no voting records exist in our scheme.
- Unreusability. If some voter casts a vote twice, Alice will find the Eq. (5) is not established and abandon this round.
- Eligibility. This requirement can be implemented by the trusted third party Alice.
- Verifiable. In our scheme, a voter cannot verify his vote is being counted or not directly. But which can be verify indirectly by the validity announced by Alice since all vote information is stored in the final state.
- Usability. Our protocol is based on a single quantum d-level state, thus it is easier to implement.

4.2 Additional Requirements

- Fairness. No one can compute the partial results ahead of Alice. Therefore, our protocol is fairness.
- Efficiency: The efficient of our protocol is only $\frac{1}{d}$. The protocol seems to be inapplicable to this situation where N is a big number. Fortunately, our protocol can be raised to $\frac{1}{2}$ through the method in Refs [19, 20].

5 Conclusions

In this paper, we proposed a quantum anonymous voting protocol using a set of complete mutually unbiased bases in d-dimensional Hilbert space. By defining two unitary operations, the transformations of different basis and vectors of the given basis in this set of complete MUBs are realized respectively. The ballot information is encoded by corresponding unitary operation on a single quantum state and only the tallyman can obtain the results by the final measurement.

Acknowledgment. This work is supported by the National Natural Science Foundation of China (Grant Nos. 61472452 and 61772565), the Natural Science Foundation of Guangdong Province of China (No. 2017A030313378), the Science and Technology Program of Guangzhou City of China (No. 201707010194), the Fundamental Research Funds for the Central Universities (No. 17lgzd29).

References

1. Vaccaro, J., Spring, J., Chefles, A.: Quantum protocols for anonymous voting and surveying. Phys. Rev. A **75**, 012333 (2007)
2. Bonanome, M., Bužek, V., Hillery, M., Ziman, M.: Toward protocols for quantum-ensured privacy and secure voting. Phys. Rev. A **84**, 022331 (2011)
3. Jiang, L., He, G., Nie, D., Xiong, J., Zeng, G.: Quantum anonymous voting for continuous variables. Phys. Rev. A **85**, 042309 (2012)
4. Wang, Q., Yu, C., Gao, F., Qi, H., Wen, Q.: Self-tallying quantum anonymous voting. Phys. Rev. A **94**, 022333 (2016)
5. Bao, N., Halpern, N.: Quantum voting and violation of arrow's impossibility theorem. Phys. Rev. A **95**, 062306 (2017)
6. Li, Y., Zeng, G.: Quantum anonymous voting systems based on entangled state. Opt. Rev. **15**, 219 (2008)
7. Horoshko, D., Kilin, S.: Quantum anonymous voting with anonymity check. Phys. Lett. A **375**, 1172 (2011)
8. Li, Y., Zeng, G.: A voting protocol based on the controlled quantum operation teleportation. Int. J. Theor. Phys. **15**(5), 2303–2310 (2016)
9. Guo, Y., Feng, Y., Zeng, G.: Quantum anonymous voting with unweighted continuous-variable graph states. Quantum Inf. Process. **15**(8), 3327–3345 (2016)
10. Cao, H., Ding, L., Yu, Y., Li, P.: A electronic voting scheme achieved by using quantum proxy signature. Int. J. Theor. Phys. **55**(9), 4081–4088 (2016)
11. Thapliyal, K., Sharma, R., Pathak, A.: Protocols for quantum binary voting. Int. J. Quantum Inf. **15**, 1750007 (2017)
12. Xue, P., Zhang, X.: A simple quantum voting scheme with multi-qubit entanglement. Sci. Rep. **7**, 7582 (2017)
13. Tavakoli, A., Herbauts, I., Zukowski, M., Bourennane, M.: Secret sharing with a single d-level quantum system. Phys. Rev. A **92**, 030302 (2015)
14. Ivonovic, I.D.: Geometrical description of quantal state determination. J. Phys. A Math. Theor. **14**(12), 3241 (1981)
15. Wootters, W.K., Fields, B.D.: Optimal state determination by mutually unbiased measurements. Ann. Phys. **191**(2), 363–381 (1989)
16. Ballester, M.A., Wehner, S.: Entropic uncertainty relations and locking: tight bounds for mutually unbiased bases. Phys. Rev. A **75**, 022319 (2007)
17. Brierley, S., Weigert, S., Bengtsson, I.: All mutually unbiased bases in dimensions two to five. Quantum Inf. Comput. **10**, 803–820 (2010)
18. Wang, K.H., Mondal, S.K., Chan, K., Xie, X.: A review of contemporary E-voting: requirements, technology, systems and usability. Data Sci. Pattern Recogn. **1**, 1 (2017)
19. Karimipour, V., Asoudeh, M.: Quantum secret sharing and random hopping: using single states instead of entanglement. Phys. Rev. A **92**, 030301 (2015)
20. Lin, S., Guo, G., Xu, Y., Sun, Y., Liu, X.: Cryptanalysis of quantum secret sharing with d-level single particles. Phys. Rev. A **93**, 062343 (2016)

Multi-template Supervised Descent Method for Face Alignment

Chao Geng[1(✉)], Zhong-Qiu Zhao[1], and Qinmu Peng[2]

[1] College of Computer and Information, Hefei University of Technology, Hefei, China
18214750850@163.com

[2] Perelman School of Medicine, University of Pennsylvania, Philadelphia, USA

Abstract. Supervised Descent Method (SDM) is a highly efficient and accurate approach for facial landmark locating and face alignment. In the training phase, it learns a sequence of descent directions to minimize the difference between the estimated shape and the ground truth in feature space. Then in the testing phase, it utilizes these descent directions to predict shape increment iteratively. However, when the face expression or direction changes too much, the general SDM cannot obtain good performance due to the large variations between the initial shape and the target shape. In this paper, we propose a multi-template SDM (MtSDM) which can maintain high accuracy on training data and meanwhile improve the accuracy on testing data. Instead of only one model is constructed in the training phase, several different models are constructed to deal with large variations on expressions or head poses. And in the testing phase, the distances between some specific landmarks are calculated to select an optimal model to update the point location. The experimental results show that our proposed method can improve the performance of traditional SDM and performs better than several existing state-of-the-art methods.

Keywords: Multi-template · Face alignment · SDM

1 Introduction

Face alignment aims at locating feature points automatically and accurately. It is essential to many facial analysis tasks such as face recognition [6], face tracking, gaze detection, face animation [11], facial attributes analysis [5] and expression analysis. The recognition performance highly depends on the accuracy of the image alignment process [6–8]. An excellent facial landmark locating approach is expected to have the follow properties: fully automatic, efficient and robust in unconstrained environment with large variations on facial expression, appearance, poses, illuminations [14], etc.

In recent years, discriminative shape regression approaches [2, 4, 15] have been proved to be one of the most popular and outstanding for accurate and robust face alignment. Among them, the SDM [16] is a representative one. It is derived from Newton Descent Method, and has high efficiency and accuracy. The SDM starts from an initial shape, e.g. the mean shape of training samples, and refines the shape using sequentially trained regressors. The shortcoming of cascaded regression approach lies

D.-S. Huang et al. (Eds.): ICIC 2018, LNAI 10956, pp. 716–721, 2018.
https://doi.org/10.1007/978-3-319-95957-3_74

in that its performance is highly depends on initialization [10, 12]. Especially when the initial shape is far from the target shape, rectifying the discrepancy completely is a difficult task to accomplish. Thereby, we propose a modified SDM, which is called multi-template SDM (MtSD-M). In our MtSDM, instead of only one model is constructed in the training phase, multiple models are constructed by different templates with various expressions or head poses. And in the testing phase, the distances between specific landmarks are calculated and utilized to selected an optimal model to update the point location.

2 Method

In this section, we briefly review the SDM approach [16], and then describe the proposed multi-template SDM in detail.

2.1 Supervised Descent Method (SDM)

For each iteration, the SDM predicts the shape increment using linear regression with the features extracted in current iteration. And the processes of training and testing start from an initial shape. So the initial shape plays an important role in producing a good prediction.

For SDM, the method just generates one sequence of $\{\mathrm{R_k}, \mathrm{b_k}\}$, and the $\{\mathrm{R_k}, \mathrm{b_k}\}$ can be learned as follows:

step 1: Minimizing Eq. 6 to get R_0 and b_0,

$$\underset{R_0,b_0}{\arg\min} \sum_{d^i} \sum_{x_0^i} \left\| \Delta x_*^i - R_0 \phi_0^i - b_0 \right\|^2 \tag{1}$$

where i denotes the ith image. Setting $k = 1$.

step 2: Updating the current estimated shape xk using Eq. 3.

step 3: Applying the update rule in Eqs. 4 and 5 to calculate the Δx_k and ϕ_k.

step 4: As in Eq. 7, learning generic descent directions R_k and bias terms b_k from a new regressor by minimizing the difference between manually labeled landmarks and the predicted shape in feature space.

$$\underset{R_k,b_k}{\arg\min} \sum_{d^i} \sum_{x_0^i} \left\| \Delta x_*^{ki} - R_k \phi_k - b_k \right\|^2 \tag{2}$$

step 5: Setting $k = k + 1$. If k < **K** (**K** represents the number of training iterations), going back to step 2.

2.2 Multi-template Supervised Descent Method (MtSDM)

Due to the large variations in facial expressions and head poses, face alignment can hardly be well accomplished starting from a single initial shape or only one model.

When the test shape has a large variation to the initial shape, the performance is usually unsatisfactory. So several models (in other words, different $\{R_k, b_k\}$) should be constructed according to different characteristics of datasets. Thereby, we propose a Multi-template Supervised Descent Method (MtSDM) in which the training and test processes can start from multiple initial shapes according to an expression or head poses predecision. For example, on some datasets with various facial expressions, such as Facewarehouse, we construct the models of mean shape, open mouth (Fig. 3), close eyes, close left eye and close right eye. In order to obtain the models with different facial expressions, these models are constructed by adjusting the training weights of the corresponding images with different expressions/head poses. For example, in order to get initial shape and regressor of mean shape model, each image is inputted for only one time. But in order to obtain those of open mouth model, the images of open mouth are inputted for several times.

In the testing phase, the mean shape model is utilized to get coarse estimation firstly, and then the euclidean distance d between specific landmarks is obtained.

$$d_{(i,j)} = \sqrt{(a_i - a_j)^2 + (b_i - b_j)^2} \tag{3}$$

where a_i and a_j represent the abscissa values of the ith and jth landmarks respectively, and b_i and b_j represent the ordinate values of the ith and jth landmarks respectively.

For mouth, the distances between the corresponding landmarks which located on the upper lip (ith, jth, kth) and the lower lip (lth, mth, nth) are calculated and added up to obtain d_{mouth}

$$d_{mouth} = d_{(i,l)} + d_{(j,m)} + d_{(k,n)} \tag{4}$$

d_{mouth} is then utilized to select the 'open mouth' model. For eyes, we use the same method to obtain the d_{left_eye} and d_{right_eye}.

$$d_{left\_eye} = d_{(e_l,g_l)} + d_{(f_l,h_l)} \tag{5}$$

$$d_{right\_eye} = d_{(e_r,g_r)} + d_{(f_r,h_r)} \tag{6}$$

The degree of opening/closing mouth/eyes is judged by comparing d_{mouth} and d_{eye} with the corresponding empirical value D_{mouth} and D_{eye}. When $d_{mouth} > D_{mouth}$, the face image is judged to be 'open mouth', and when $d_{eye} > D_{eye}$, it's judged to be 'open eye'. Thereby, the optimal model can be selected. The main process of testing phase can be seen in Fig. 1.

For head poses, the distances between the landmarks located at the end of nose and left/right ear are calculated and described as d_1 and d_2. Thereby, we obtain

$$d_{orientation} = d_1 - d_2 \tag{7}$$

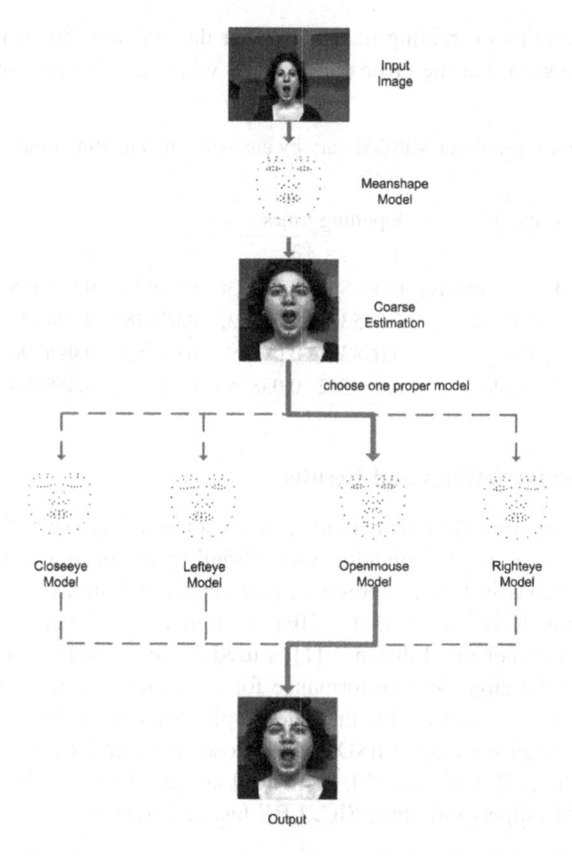

Fig. 1. Overview of the testing phase. Given a test image, the MtSDM first get coarse estimation by using mean shape. And then it selects an optimal shape model by calculating the distances between several specific landmarks and comparing them with pre-defined empirical thresholds. Finally, the MtSDM makes prediction using the proper shape model.

which is utilized to judge the head poses by comparing it with an empirical threshold Dorientation. When $d_{orientation} > D_{orientation}$, the face image is judged to be 'turn to the right'. Otherwise, the face is judged to be 'turn to the left'.

3 Experiments and Discussions

In this section, we evaluate the performance of the proposed method and compare it with several existing state-of-the-art methods on four datasets: LFPW, Helen, Face-warehouse and Multiple.

3.1 Weights of Expressions/Head Poses

Various models are repeatedly updated by adjusting the weights of training images with the corresponding expressions or head poses. The face alignment results of our MtSDM with different weights of training images on four datasets are shown in Table 1, from which, it can be seen that the error rate reduced when the weights increased.

Table 1. The error rates of our MtSDM vary by changing the inputting times of training images on four datasets.

Datasets	Inputting times			
	2	3	4	5
Facewarehouse	0.047891	0.047036	0.046843	**0.046438**
LFPW	0.075398	0.074532	**0.074189**	0.074227
Helen	0.098361	0.09689	0.09652	**0.096296**
Multiple	0.059582	0.058756	0.058352	**0.058388**

3.2 Experimental Settings and Results

In this section, we compare our method with several existing state-of-the-art methods on four challenging datasets which have been briefly introduced above. These datasets show large variations in facial expression, face shape, and number of landmarks. The training iterations **K** is set to 4, the SIFT is used as local features, and the error normalized by the inter-pupil distance [1] is used as the evaluation metric.

We evaluate the alignment performance for each sample using the standard landmarks mean error normalised by the inter-pupil distance. Table 2 shows the face alignment performance of our MtSDM and those of several existing state-of-the-art method: Fast SIC [13], LBF fast [9], SDM [16] as well. From Table 2, it can be seen that our MtSDM outperforms Fast SIC, LBF fast and SDM.

Table 2. Errors of various approaches on four datasets.

Method	Facewarehouse	LFPW	Helen	Multiple
Fast_SIC [20]	0.051691	0.089818	0.11923	0.068047
LBF fast [15]	0.049765	0.086074	0.10037	0.066365
SDM [24]	0.048372	0.078587	0.10395	0.060074
MtSDM	**0.046438**	**0.074227**	**0.09629**	**0.058352**

4 Conclusion

In this paper, we extend the traditional SDM to multiple-template SDM. In the training phase, we construct different shape models by repeatedly inputting the image with the specific expressions or head poses. In the testing, one optimal model is selected to update the point location, which provides the proper initial shape and regressor. The results demonstrate that our modification improves the accuracy of traditional SDM method and achieves better performance than other existing state-of-the-art methods.

References

1. Belhumeur, P.N., Jacobs, D.W., Kriegman, D., Kumar, N.: Localizing parts of faces using a consensus of exemplars. In: IEEE Conference on Computer Vision and Pattern Recognition, pp. 545–552 (2011)
2. Cao, X., Wei, Y., Wen, F., Sun, J.: Face alignment by explicit shape regression. Int. J. Comput. Vis. **107**(2), 177–190 (2014)
3. Chen, C., Dantcheva, A., Ross, A.: Automatic facial makeup detection with application in face recognition. In: International Conference on Biometrics, pp. 1–8 (2013)
4. Dantone, M., Gall, J., Fanelli, G., Van Gool, L.: Real-time facial feature detection using conditional regression forests, pp. 2578–2585 (2012)
5. Datta, A., Feris, R., Vaquero, D.: Hierarchical ranking of facial attributes. In: IEEE International Conference on Automatic Face & Gesture Recognition and Workshops, pp. 36–42 (2011)
6. Deng, W., Hu, J., Guo, J.: Extended SRC: undersampled face recognition via intraclass variant dictionary. IEEE Trans. Pattern Anal. Mach. Intell. **34**(9), 1864–1870 (2012)
7. Deng, W., Hu, J., Lu, J., Guo, J.: Transform-invariant PCA: a unified approach to fully automatic face alignment, representation, and recognition. IEEE Trans. Pattern Anal. Mach. Intell. **36**(6), 1275–1284 (2014)
8. Deng, W., Hu, J., Zhou, X., Guo, J.: Equidistant prototypes embedding for single sample based face recognition with generic learning and incremental learning. Pattern Recogn. **47** (12), 3738–3749 (2014)
9. Ren, S., Cao, X., Wei, Y., Sun, J.: Face alignment at 3000 fps via regressing local binary features. IEEE Trans. Image Process. **25**(3), 1685–1692 (2014)
10. Smith, B.M., Brandt, J., Lin, Z., Zhang, L.: Nonparametric context modeling of local appearance for pose- and expression-robust facial landmark localization. In: Computer Vision and Pattern Recognition, pp. 1741–1748 (2014)
11. Thies, J., Zollhöfer, M., Stamminger, M., Theobalt, C., Nießner, M.: Face2face: real-time face capture and reenactment of RGB videos. In: Proceedings of Computer Vision and Pattern Recognition (CVPR), p. 1. IEEE (2016)
12. Trigeorgis, G., Snape, P., Nicolaou, M.A., Antonakos, E., Zafeiriou, S.: Mnemonic descent method: a recurrent process applied for end-to-end face alignment. In: IEEE International Conference on Computer Vision and Pattern Recognition (2016)
13. Tzimiropoulos, G., Pantic, M.: Optimization problems for fast AAM fitting in-the-wild. In: IEEE International Conference on Computer Vision, pp. 593–600 (2013)
14. Wu, Y., Ji, Q.: Robust facial landmark detection under significant head poses and occlusion. In: IEEE International Conference on Computer Vision, pp. 3658–3666 (2015)
15. Xiong, X., De la Torre, F.: Global supervised descent method. In: Computer Vision and Pattern Recognition, pp. 2664–2673 (2015)
16. Xiong, X., Torre, F.D.L.: Supervised descent method and its applications to face alignment, vol. 9(4), pp. 532–539 (2013)

Early Software Quality Prediction Based on Software Requirements Specification Using Fuzzy Inference System

Muhammad Hammad Masood and Malik Jahan Khan(✉)

Department of Computer Science, Namal College, Mianwali, Pakistan
{mhammad,malik.jahan}@namal.edu.pk

Abstract. Software Requirements Specification (SRS) is the key fundamental document formally listing down the customer expectations from the software to be built. Any weakness or fault injected at this stage in the requirements is expected to ripple towards the following phases of software development life cycle resulting in development of a software system of poor quality. Software quality prediction promises to raise alarms about the quality of the end product at earlier stages. It becomes more challenging as we move earlier in stages because of limited information is available at earlier stages. Therefore little effort has been put in literature to predict software quality at SRS stage. This position paper presents a novel approach of prediction of software quality using SRS. SRS document is converted into a graph and different parameters including readability index, complexity, size and an estimation of coupling are extracted. These parameters are fed into a Fuzzy Inferencing System (FIS) to predict the quality of the end product. The proposed model has been evaluated on a sample of student projects and has shown reasonable performance.

Keywords: Software quality prediction · Software requirements
Fuzzy logic

1 Introduction

Requirement gathering is the foundation stone of the entire software development life cycle (SDLC). It is also the most tedious exercise, as for gathering required information from a common client having a little or no information about software engineering domain is not an easy task. Quality of a software is based on the requirements specification document which formally lays down the expectations of the client.

With the advent of time, software are getting more and more complex as our reliance on software and automation of manual process is continuously growing. To develop and maintain such a complex software, enormous amount of documentation has to be done to keep the record of software development. It leads to the problem of inconsistency and ambiguity in the documentation, which adversely affects the software development process.

The medium of interaction with the people in the world is mostly natural language which is inherently vague and ambiguous. The problem of ambiguity increases as these requirements are documented in natural language, which leads to ambiguities in the

D.-S. Huang et al. (Eds.): ICIC 2018, LNAI 10956, pp. 722–733, 2018.
https://doi.org/10.1007/978-3-319-95957-3_75

written document. Software Requirements Specification (SRS) is a document that contains all the details related to the product that is going to be developed. Foundation of a high-quality software depends upon a quality of SRS. Analysis of requirements becomes difficult with lack of consistency and clarity in SRS which adversely affects the in-time development of software product according to the expectation of the customer. Out of 100% of total flaws in a product, 48% of flaws is due to poor and insufficient requirement gathering according to Klaus (2010).

Quality is measured from different angles. According to Kitchenham and Pfleeger (1996): "The way we measure the quality depends upon the viewpoint we take". Weak quality of requirements specification at requirement gathering phase of SDLC degrades the development of an efficient and robust software. According to Software Engineering Standards Committee of the IEEE Computer Society (1998), a good SRS should be correct, consistent, complete, unambiguous, modifiable, verifiable and traceable. Errors and ambiguities induced at such an early stage produce errors in each phase of software development process, thus resulting in low quality output of the entire product. Eventually, this results in poor quality of software. It would have been quite useful if the quality of bad product could have been foreseen at an earlier stage, resulting in saving of lot of financial and human resources. Most of the existing software quality prediction methods focus on later stages of SDLC including design and implementation stages. The flaws injected at the requirements analysis stage ripple down the entire SDLC and are not addressed if the software quality is predicted based on the design or implementation metrics. This bottleneck is addressed through earlier software quality predictions based on SRS in this research work. This research work focuses on extracting useful information from the requirements and estimating quality of product based on the requirement documentation. The development of such a system that can predict the quality of software product based on the quality of requirements gathered will help the software industry manifolds. It not only reduces the cost spent on developing a faulty product but also saves the resources to be spent on producing a software on poorly written requirements.

This prediction is going to be very early prediction helping the software industry in highlighting the risks involved in development of the software based on poorly written SRS.

This research paper proposes to extract useful features from the SRS which would help in giving an insight about the quality of SRS. These features are weighed to build a bridge between requirements and their impact on quality. After quantification, these features are used as input to Fuzzy Inference System (FIS) of the proposed model. The proposed approach has been tested on different software engineering projects and shows promising performance in predicting the quality based on SRS when compared with the quality viewpoints of domain experts.

Rest of the paper is organized as: Sect. 2 presents the related work, Sect. 3 highlights the essence of research problem being studied and proposed approach to solve it, Sect. 4 explains the implementation of the proposed approach and discusses its performance and finally Sect. 5 concludes the paper highlighting the future work.

2 Related Work

In this section, various techniques of text processing and methods of estimation of software quality have been discussed.

2.1 Quality Metrics

Consistency: Consistency is an important attribute to be present in the text written in natural language. In SRS, it is given more importance so to evaluate and remove contradiction between two requirements written within the same document. Lami et al. (2004) performed consistency analysis by considering a chunk of text written together and conveying the same meaning. Quality Analyzer for Requirements Specification (QuARS) tool was used to detect the shortcoming in sentences based on the criteria explained. It also combined the parts of requirement document that contained information related to common things. QuARS determined readability metric based on the analysis of the document.

Complexity: Sharma and Kushwaha (2010) evaluated the complexity of code based on SRS. They suggest that the complexity of software can be computed using SRS. Different types of complexity metrics can be computed. Requirement-Based Complexity (RBC) can be calculated and compared with other measures (code and cognitive complexity measures). RBC's results were validated using Weyuker properties discussed by Misra (2009). Weyuker properties measure is standard complexity measurement procedure of programming languages. RBC is considered as a comprehensive technique for evaluating complexity based on SRS.

In another work, Sharma and Kushwaha (2012) focused on obtaining requirement based effort estimation by extending the previous work. Technical functional points were calculated to aid in the measurement of software development effort.

2.2 Prediction of Quality

Formal Criteria Based Prediction: Hovorushchenko and Krasiy (2015) focused on the prediction of success of the project based on the analysis of SRS. The method of success is based on prediction of software characteristics using neural networks, interpretation of characteristics obtained from the analysis of SRS and evaluation of the success of project.

Smidts et al. (1996) worked on the reliability of software during early phase of SDLC. Along with the lack in proper functional requirement, this research focused to consider faults due to the client as well as developer. A fault tree is developed to deal with the reliability, based on the series of events that cause failure. In this work, a Bayesian statistical approach is proposed to quantify the fault tree.

Christopher and Chandra (2012) worked on changes occurred during the requirement gathering process. They also tried to measure complexity based on the changes in requirements, that happened during the complete phase of requirement gathering.

Pandey and Goyal (2010) suggest that lesser number of errors and faults in a software increase the reliability of the product. They proposed a model in which faults were predicted at each stage of SDLC. The error of different phases is compared with the data obtained from the PROMISE repository.

Suanmali et al. (2009) worked on text summarization based on the sentences selected from the text. In this technique, fuzzy logic is used for text summarization.

Dargan et al. (2014) introduced a statistical model that uses requirements quality factors to predict system operational performance. Operational performance is predicted using binary regression.

Code Based Prediction: Sana et al. (2008) presented a survey of different software quality prediction techniques and compared their performances. The performance has been evaluated on the base of various evaluation metrics available in literature. Different artificial intelligence based algorithms have been compared using their performance on different publicly available datasets. Locally weighted learning has been reported as the best performer in this empirical study.

2.3 NLP Based Feature Extraction Tools

In order to extract the useful features from the text, different tools are available. Their summary is presented below:

Stanford Parser: Parser by Stanford (2015) is commonly used to work out the grammatical structure of the sentences. It groups the words which are combined to form phrases and also tag the words as subjects and objects. A lot of research has been done on Stanford parser.

Tropes: Tropes Knowledge (2014) is a well-designed tool that performs extraction of relevant information. It performs text categorization, does Part of Speech (POS) tagging, computes frequency of words and provides ontology manager. Tropes provide an easy to use graphical user interface with graphs and displays the relation between the words.

Gephi: Gephi Bastian et al. (2009) is an open source tool for network and graph analysis. It takes a number of inputs and plots them to form a graph to have a valuable insight into the data. Easy to use user interface makes it easy to work with data. Different parameters related to graphs can be computed using Gephi.

QuARS: Quality Analyzer for Requirement Specifications (QuARS) tool performs the parsing of text and helps the requirement writer to analyze the defects, inconsistencies, ambiguities etc. SRS, Quarks (2009). This tool deals with syntax-related defects and does not inspect defects related to semantics.

Text Graph: Grineva et al. (2009) work is related to the extraction of key terms from the text. For this purpose, semantics graphs of text are made. The semantically related terms are grouped together in one community, so the complete graph have different communities and each community conveys the same meaning. The higher ranked community conveys the meaning related to main term of the topic.

There is limited research work focused on developing a relationship between requirements and measures there effects on the software product. Christopher and

Chandra (2012) worked on stability of requirements. Dargan et al. (2014) focused on building a relationship between requirements and its product.

To fill the gap discovered in literature, this position paper focuses on requirements available in the form of SRS. Different parameters are extracted from the SRS text. Based on those extracted metrics, a bridge is proposed to be built between SRS and quality of product. This will help in foreseeing the effects of poor quality requirements on the software. The underlying working assumption is that poorly written SRS will result in poor quality product. Hence, quality of SRS will help foreseeing the quality of end product to be built (Fig. 1).

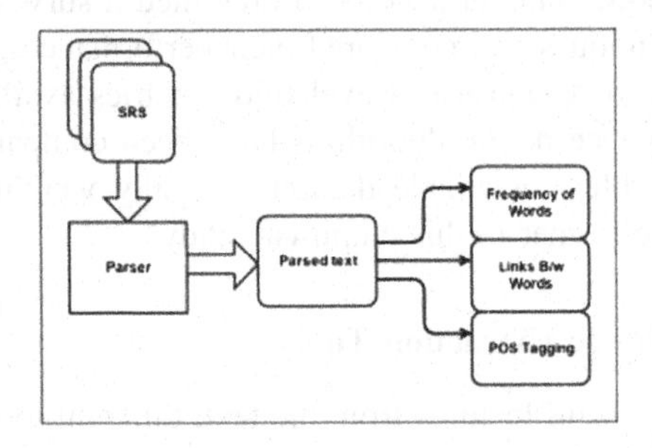

Fig. 1. Text processing.

3 Proposed Model: Early Software Quality Prediction Using SRS

In this research work, the main objective is to extract the useful attributes from the SRS. Filtered features are quantified so that they may help in predicting the quality of software under consideration. The extracted features are fed to Fuzzy Inference System (FIS) as input. FIS estimates the quality of software. The proposed approach has been explained below:

3.1 Text Processing

Different components of text summarization extracted from the SRS include word frequency, POS tagging, graphs representing different entities and their relationships and style of text.

Frequency of words is the number of times a word appears in the text. Parts of Speech (POS) tagging helps in categorizing sentence in different parts of speech. Underlying graphs have been extracted from the SRS text. Different entities are represented as nodes and links between them are represented as edges. Strongly Connected Component (SCC) of a graph is a set of nodes such that for every pair of nodes in the subset, there is a two-way connection between the nodes of the pair. If this condition holds true for the entire graph then there should be only one strongly connected component comprising of all nodes of the graph. Other extreme could be that none of the pairs of nodes qualify the condition and number of nodes are equal to number of strongly connected components. This work uses Gephi (2015) tool to generate the graphs as shown in Fig. 2. This work proposes that the ratio of number of SCCs and number of nodes of the graph helps in approximation of level of coupling and cohesion of the software to be built. Low coupling and high cohesion are desirable for good quality. As it is an early approximation, medium ratio is suggested to be the ideal ration leading to good quality. Both extremes may damage the acceptable level of coupling or cohesion hence resulting in low quality.

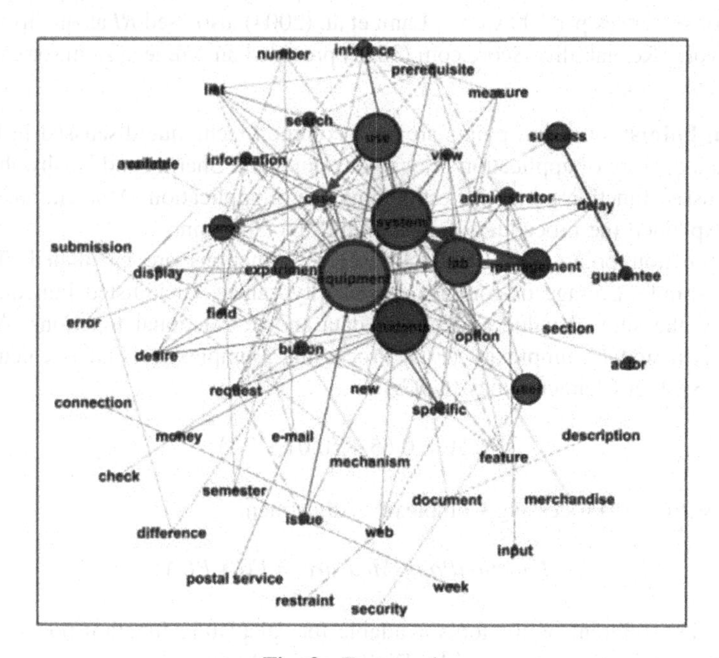

Fig. 2. Text graphs.

3.2 Quantification of Features

Following features of SRS are considered as the key features contributing towards the quality of SRS and the software to be built based on this SRS.

Complexity: According to Sharma and Kushwaha (2010), complexity is defined as degree to which a it is difficult to design and implement a system. This definition of complexity has been used to compute the complexity of projects used in this research. Complexity is calculated as Requirement based complexity (*RBC*). *RBC* is defined as:

$$RBC = (RC * IOC + SFC + DCI) * ULC \tag{1}$$

Where *RC* is Requirement Complexity, *IOC* stands for Input/output Complexity, *SFC* is System Feature Complexity, *DCI* is Design Constraint Imposed and *ULC* is User/Location Complexity.

Readability Index: Quality is also assessed based on the readability index (*RI*). *RI* helps to evaluate the ease of reading the written text. For easy to read technical document, the *RI* value is 10. Reading becomes difficult for a document with readability index greater than 15, Lami et al. (2004).

$$RI = 0.0588L - 0.296S - 15.8 \tag{2}$$

Where L is the average number of characters per 100 words and S is the average number of sentences per 100 words. Lami et al. (2004) also used *RI* as quality indicator in their work. Readablility-Score.com (2016) provided an online system for calculating the readability index.

Function Points: Function points are very common technique discussed in literature to estimate the size of application prior to development. Sharma and Kushwaha (2012) also discussed function points for estimating size of application. Albrecht and Gaffney (1983) explained the procedure of calculating function points.

For function points, data and transactional functions are calculated. They are assigned simple, average or complex complexity values. Unadjusted Function Points (*UAFP*) is the sum of value assigned to data and transactional functions. After calculating Functional Complexity (*FC*), Processing Complexity (*PC*) is calculated by "General System Characteristics (*GSCs*)".

$$PCA = 0.65 + (0.01 * PC) \tag{3}$$

Where *PCA* is Processing Complexity Adjustment.

$$FunctionPointsMeasure = FC * PCA \tag{4}$$

There are different online tools available for calculating function points. The one used in this project was provided by Divinagracia (2000).

3.3 Fuzzy Model

Fuzzy Inference System (FIS) uses fuzzy logic for implementing systems that can deal with real world decision making problems involving lack of precision and vagueness, Pandey and Goyal (2013). Fuzzy logic is capable to deal with plain linguistic input or output terms like good, bad etc. Fuzzy logic maps the numeric data onto a set of real world linguistic concepts being represented as different membership functions of respective fuzzy variables.

There are two main reasons of using FIS in the proposed model:

1. Inherent ambiguity in SRS due to natural language may be very difficult to capture otherwise
2. Early prediction of quality of the software whose requirements are vague at this stage

FIS includes the following five steps:

1. Fuzzification
2. Applying Operators
3. Implication of Rules
4. Aggregating Output
5. Defuzzification

In fuzzification, crisp values are converted to fuzzy values. Fuzzy operators are applied in the fuzzy rule-base to compute truthfulness and strength of different domain rules. Output of all rules is aggregated to confine to a solution space. Appropriate defuzzification method on the output space helps to find the output from the solution space.

3.4 Proposed Model

The proposed model is shown in Fig. 3. Steps of the underlying algorithm have been outlined in Fig. 4 and explained below:

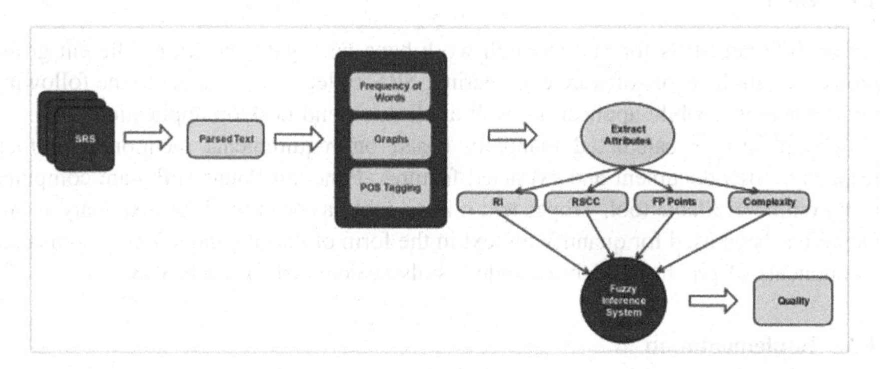

Fig. 3. Proposed model.

- Input SRS document
- Parse text of SRS
- Extract attributes of text
- Quantify the attributes
- Use FIS
- Validate quality

As explained above, SRS for the project at hand is parsed and different desired attributes are extracted including the underlying graphs. Above mentioned metrics are computed based on the extracted features including RI, RSCC, FPs and complexity. These metrics are fed into FIS which has already defined these input variables as fuzzy variables and domain rules have been fed into it. Inputs are fuzzified through their respective membership functions. FIS applies rules on the user fed input representing a

```
Input: SRS Document
Output: Predicted Software Quality
Method:
   1. RI = ExtractReadibilityIndex (SRS Document)
   2. RSCC = ComputeRSCC (SRS Document)
   3. FP = ComputeFunctionPoints (SRS Document)
   4. Complexity = ComputeComplexity (SRS Document)
   5. Fuzzy Inputs = Fuzzify (RI, RSCC, FP, Complexity)
   6. OutputRegion = Aggregate (ApplyFuzzyRules (Fuzzy Inputs))
   7. PredictedSoftwareQuality = Defuzzify (OutputRegion)
```

Fig. 4. Proposed algorithm for early software quality prediction.

particular project and using its inferencing mechanism, an aggregated output region is computed. Output region is defuzzified to output the quality approximation of the fed input.

4 Experiments and Results

4.1 Data

Set of different SRSs for this research work have been gathered from different group projects of students of software engineering. SRS collected are related to the following three domains: mobile application, web application and desktop application.

Complexity is calculated manually based on requirements mentioned in their respective SRS document and extracted features. Function Points (FPs) are computed using online available tool. Tropes text mining tool has been used for text analysis and Gephi has been used for quantifying text in the form of density and strongly connected components of graph. Readability index is also calculated for each SRS.

4.2 Implementation

In this research, Mamdani FIS has been constructed using Matlab fuzzy tool box. FIS editor provided the facility of defining number of inputs and outputs, aggregation method and defuzzification technique. Rule editor allows to add, remove and edit the rules. Different rules can be assigned different weights. Membership function editor is used to define membership functions of different fuzzy variables.

Membership Functions: All the attributes were used as input for FIS. Total there were 4 inputs and each input has 5 membership functions. FIS was used to define inputs and their membership functions. Similarly membership function for output was defined. Figure 5 shows the membership function for both inputs and an output.

Rules: Rules have been created based on data available using Matlab tool box. Sample of rules have been shown in Fig. 6. Weights have been also assigned to the rules based on their impact on quality. The generated FIS has been tested on the data through seven different projects. It is aimed to evaluate the proposed system at larger scale in future.

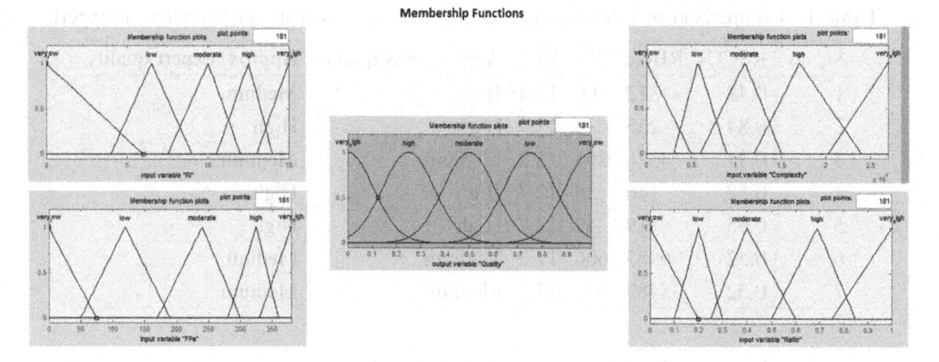

Fig. 5. Membership functions.

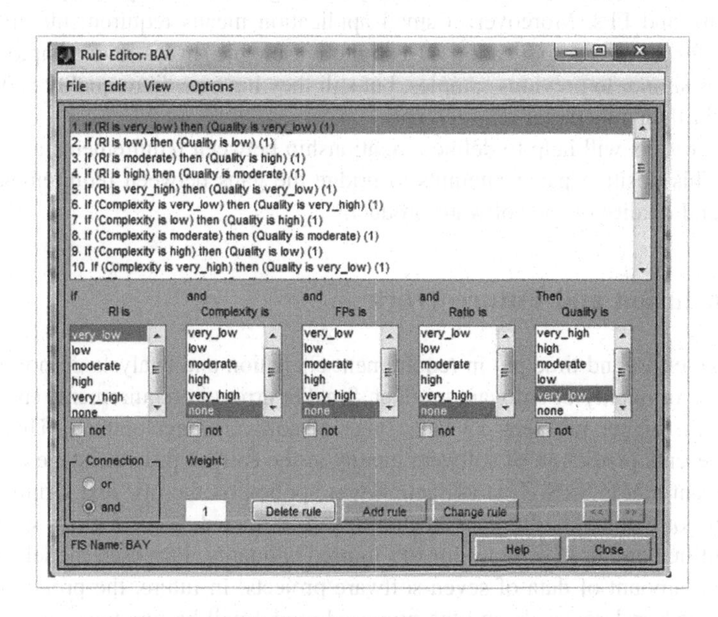

Fig. 6. FIS rules snapshot.

4.3 Results

Results were generated using the FIS which takes different metrics as input as mentioned above. Results have been summarized in Table 1. The second last column represents the fuzzy output of proposed FIS when fed with the first four columns as inputs. Last column is the view of domain expert about the same project after completion. In the initial comparison, six results produced out of seven projects are similar to the view of domain expert leading to about 85% acceptability which may not be generalized at this level but is quite promising.

The important attributes in the Table 1 are complexity and function points. It can be observed that applications with lesser complexity and fewer FPs have much better

Table 1. Comparison of FIS predicted quality with approximate quality view of expert.

Sr. no.	RSCCs	RBC	FPs	RI	Approx. FIS quality	Approx. expert quality
1	0.48	312	66	11.4	High	Medium
2	0.84	253	33	12.9	High	High
3	0.31	4376	205	13.2	Medium	Medium
4	0.58	146	84	12.7	High	High
5	0.65	119	49	12.7	High	High
6	0.52	6983	606	12.8	Medium	Medium
7	0.52	35489	641	13.1	Medium	Medium

quality. The reason is obvious that it is easy to design and develop the system with less complexity and FPs. Moreover, a small application means requirements are easy to right. As for all of the application RI is quite reasonable. In last two sample although there RI is similar to previous samples, but still they have medium quality is due to the increase in their functional requirements.

These results will help to define a relationship between requirements and software quality. This position paper attempts to bridge the gaps between requirements specification and quality of the software product.

5 Conclusion and Future Work

Significant effort and time put in requirement elicitation and analysis do not guarantee the success or quality of software product. Smaller errors in requirements specification might cause bigger problems at later stage of software development. This position papers presents prediction of software quality at the earliest possible stage of software development that is SRS. Text extraction was applied on the raw text written in SRS. As a result of extraction, different attributes were extracted. Some metrics were filtered out, quantified and used as input for FIS to predict quality. Proposed model was tested on a small amount of data of seven software projects. In future, the proposed system will be tested on large scale and the proposed model will be fine tuned in light of the rigorous testing.

References

Albrecht, A.J., Gaffney, J.E.: Software functions, source lines of codes and development effort prediction: a software science validation. IEEE Trans. Softw. Eng. **9**(11), 639–648 (1983)

Bastian, M., Heymann, S., Jacomy, M.: Gephi: an open source software for exploring and manipulating networks. In: Third International AAAI Conference on Weblogs and Social Media (2009)

Christopher, D.F.X., Chandra, E.: Prediction of software requirements stability based on complexity point measurement using multi-criteria fuzzy approach. Int. J. Softw. Eng. Appl. **3** (6), 101–115 (2012)

Dargan, J.L., Campos-Nanez, E., Fomin, P., Wasek, J.: Predicting systems performance through requirements quality attributes model. Procedia Comput. Sci. **28**, 347–353 (2014)
Divinagracia, H.R.: FP calculator (2000). http://tinyurl.com/FPC-Harvey
Gephi: The open graph viz platform (2015). https://gephi.org. Accessed 27 Jan 2016
Grineva, M., Grinev, M., Lizorkin, D.: Extracting key terms from noisy and multitheme documents. In: Proceedings of the 18th International Conference on World Wide Web, pp. 661–670 (2009)
Hovorushchenko, T., Krasiy, A.: Method of evaluating the success of software project implementation based on analysis of specification using neuronet information technologies (2015)
Kitchenham, B., Pfleeger, S.L.: Software quality: the elusive target. IEEE Softw. **13**(1), 12–21 (1996)
Klaus, P.: Requirements Engineering Fundamentals, Principles, and Techniques. Springer, Heidelberg (2010)
Semantic-Knowledge: High performance text analysis for professional users (2014). http://www.semantic-knowledge.com/tropes.htm
Lami, G., Gnesi, S., Fabbrini, F.: An automatic tool for the analysis of natural language requirements. Informe tecnico, CNR (2004)
Misra, S.: Weyuker's properties, language independency and object oriented metrics. In: Gervasi, O., et al. (eds.) ICCSA 2009. LNCS, vol. 5593, pp. 70–81. Springer, Heidelberg (2009). https://doi.org/10.1007/978-3-642-02457-3_6
Pandey, A.K., Goyal, N.K.: Fault prediction model by fuzzy profile development of reliability relevant software metrics. Int. J. Comput. Appl. **11**(6), 34–41 (2010)
Pandey, A.K., Goyal, N.K.: Early Software Reliability Prediction. Springer, India (2013). https://doi.org/10.1007/978-81-322-1176-1
QuARS: Quality analyzer for requirement specifications (2009). http://quars.isti.cnr.it/index.html
Readablility-Score.com: Measure text readability (2016). https://readability-score.com/text/
Sana, S., Hassan, A., Malik J.K., Shafay S.: Software quality prediction techniques: a comparative analysis. In: International Conference on Emerging Technologies, pp. 18–19 (2008)
Sharma, A., Kushwaha, D.: Complexity measure based on requirement engineering document and its validation. In: International Conference on Computer and Communication Technology, pp. 608–615 (2010)
Sharma, A., Kushwaha, D.S.: Applying requirement based complexity for the estimation of software development and testing effort. SIGSOFT Softw. Eng. Notes **37**(1), 1–11 (2012)
Smidts, C., Stoddard, R.W., Stutzke, M.: Software reliability models: an approach to early reliability prediction. In: Proceedings of the Seventh International Symposium on Software Reliability Engineering, pp. 132–141 (1996)
Software Engineering Standards Committee of the IEEE Computer Society: IEEE Recommended Practice for Software Requirements Specifications (1998)
Stanford: Stanford parser (2015). http://nlp.stanford.edu. Accessed 26 Jan 2016
Suanmali, L., Salim, N., Binwahlan, M.S.: Fuzzy logic based method for improving text summarization. J. Comput. Sci. **2**(1), 6 (2009)

A Robust Locally Linear Embedding Method Based on Feature Space Projection

Feng-Ming Zou[1,2], Bo Li[1,2(✉)], and Zhang-Tao Fan[1,2]

[1] School of Computer Science of Technology, Wuhan University of Science of Technology, Wuhan 430065, Hubei, China
libo@wust.edu.cn

[2] Hubei Province Key Laboratory of Intelligent Information Processing and Real-Time Industrial System, Wuhan 430065, Hubei, China

Abstract. At present, most of the manifold learning methods use the K-Nearest Neighbor (KNN) criterion to determine the adjacency relationship between data points, which is not complete in the description of the sample distribution in the original data space. When many noise data are included in the observation space, the results of constructing adjacency graph by KNN may contain too many areas outside the unsupported domain, which will produce the wrong geometric projection distance. Isometric Feature Mapping (ISOMAP) and Locally Linear Embedding (LLE) algorithm are very sensitive to noise data because of above reason. In view of the above problem, a robust locally linear embedding method based on feature space projection is proposed. Based on LLE algorithm, the feature space projection is introduced to smooth the original samples and to improve the robustness of the noise data set in the process of reducing the dimension using LLE algorithm. The experimental results prove that this method is effectiveness.

Keywords: Manifold learning · Feature space · Locally linear embedding
Robustness

1 Introduction

Manifold learning method is a hot issue in the field of machine learning. Due to its huge applications, it has attracted the attentions of many researchers. They are applied to pattern recognition [1, 2] and data visualization and other fields. Because it can well excavate the internal structure of nonlinear data, it has achieved good results in data visualization. However, when used for data classification, manifold learning methods also face some problems, such as small sample size, out-of-sample, noise sensitive, and poor data separability.

In recent years, some dimensionality reduction methods have been concentrated on. These dimensionality reduction methods are divided into the linear and the nonlinear, and the latter is further divided into kernel methods and manifold learning approaches. The methods of linear dimension reduction include Principal Component Analysis (PCA) [3], Independent Component Correlation Algorithm (ICA) [4], Linear Discriminant Analysis (LDA) [5], Local Feature Analysis (LFA) [6] et al. The nonlinear dimension reduction

D.-S. Huang et al. (Eds.): ICIC 2018, LNAI 10956, pp. 734–739, 2018.
https://doi.org/10.1007/978-3-319-95957-3_76

methods based on kernel functions include Kernel Principal Component Analysis (KPCA) [7, 8] and Kernel Independent Component Analysis (KICA) [9] et al. Manifold learning contains Isometric Feature Mapping (ISOMAP) [10, 11], Locally Linear Embedding (LLE) [12, 13], Locality Preserving Projection (LPP) [14], Local Tangent Space Alignment (LTSA) [15, 16] et al. These methods are very effective because they can explore low-dimensional nonlinear manifolds in high-dimensional data space [17, 18], and then map data points into low-dimensional embedding spaces. Although they have attractive features, these solutions are less than satisfactory for the robustness problem [19–21], which is often confronted by ISOMAP and LLE.

In this paper, based on the LLE algorithm, the projection of any point in the feature space is introduced [22], by which the robustness of LLE to noisy points can be improved. Moreover, experiments on some synthesis data are conducted to validated the performance of proposed method.

2 Method

In order to overcome the weakness of manifold learning method, i.e. sensitive to noise, a local linear embedding method based on feature space projection is proposed, which is used to improve the robustness of the method. It is the fact that the projection for any point in its feature space can be represented to the linear combination of points consisting of the feature space and the linear weights are sum-to-one. So in the proposed method, we firstly map any point into its nearest feature space and obtain the corresponding projections. Then, the original LLE is applied to these projections to achieve their low dimensional embeddings.

2.1 Calculation of Projection Points

The projection points of the sample point x_i in the neighborhood N_{i} are expressed as $\mathrm{h}^{(K)}(x_i)$, where the neighborhood N_i is determined by KNN [23, 24] as the nearest feature space. Based on the assumption that the global nonlinear local linear of the manifold is assumed, the neighborhood N_i is considered to be linear, and the method to calculate the projection point in the neighborhood is a linear representation, the projection point can be expressed linearly by the point in the neighborhood, the coefficient of the linear representation is equivalent to a weight value.

The calculation process of the projection point in the neighborhood of the sample point with $K = 2$ and $K = 3$ can be deduced from the case of $K > 3$, and the projection $\mathrm{h}^{(K)}(x_i)$ in the subspace of K point $x_1, x_2, \ldots, x_K$ can be obtained from the following equation.

$$\mathrm{h}^{(K)}(x_i) = X_{(1:K)}\left(X_{(1:K)}^T X_{(1:K)}\right)^{-1} X_{(1:K)}^T (x_i - x_1) + x_1 = \sum_{j=1}^{K} \mathrm{w}_j x_j \tag{1}$$

In the formula (2) $X_{(1:K)} = [(x_2 - x_1)(x_3 - x_1)\ldots(x_K - x_1)]$, it is a matrix with size $D \times (K-1)$, and the projection point can be linearly expressed by $x_1, x_2, \ldots, x_K$, and it also satisfies the restriction of $\sum_{j=1}^{K} \mathrm{w}_j = 1$.

2.2 The Proposed Algorithm

LLE algorithm uses the KNN criterion to determine neighborhood for any point. In order to well preserve the local structure of the manifold, in the determined neighborhood, the low dimensional embedding is obtained by minimizing the reconstruction error of the sample point. In this process, noise points often appear, resulting in the local geometry of manifold cannot be well maintained after dimensionality reduction. The local linear embedding algorithm based on the feature space projection can be divided into two steps. The first step is to calculate the projection points of all data points in the neighborhood, and the noise points are also involved in the calculation. The second step is to reduce the dimension of all projection points using the original LLE algorithm. From the following formulation, we can obtain the least reconstruction weight matrix w.

$$\min \sum_{i=1}^{N} \left\| \mathrm{h}^{(K)}(x_i) - \sum_{x_j \in N_i} w_{ij} \mathrm{h}^{(K)}(x_j) \right\|^2 \tag{2}$$

Later, with the least reconstruction weight matrix, the low dimensional embeddings can also be obtained using equation stated below:

$$\min \sum_{i=1}^{N} \left\| y_i - \sum_{x_j \in N_i} w_{ij} y_j \right\|^2 \tag{3}$$

It can be easily found that the low dimensional embeddings will be the bottom eigenvectors of $(I - w)(I - w)^T$ except that associated to eigenvalue 0.

3 Experiments

3.1 Experimental Data Set

The data sets for the experiment include Helix, Swiss roll, and S-curve. In order to verify the suppression of noise points, a number of noise points were added to the data set, which can be found from Table 1.

3.2 Experiment Results

The experimental results of the 4 data sets corresponding to the number of nearest neighbor points are shown in Fig. 1. From Fig. 1, we can see that the method proposed in this paper can gain better embedding results on three data sets with noise. The robust performance of the algorithm is better, and the original geometric structure of the manifold can be well maintained after the dimensionality reduction.

Table 1. Parameter setting of locally linear embedding method based on feature space projection

	Helix	Swiss roll	S-curve
Number of sample points	500	1500	1500
Number of noise points	75	75	150
K	11, 13	15, 17	18, 20

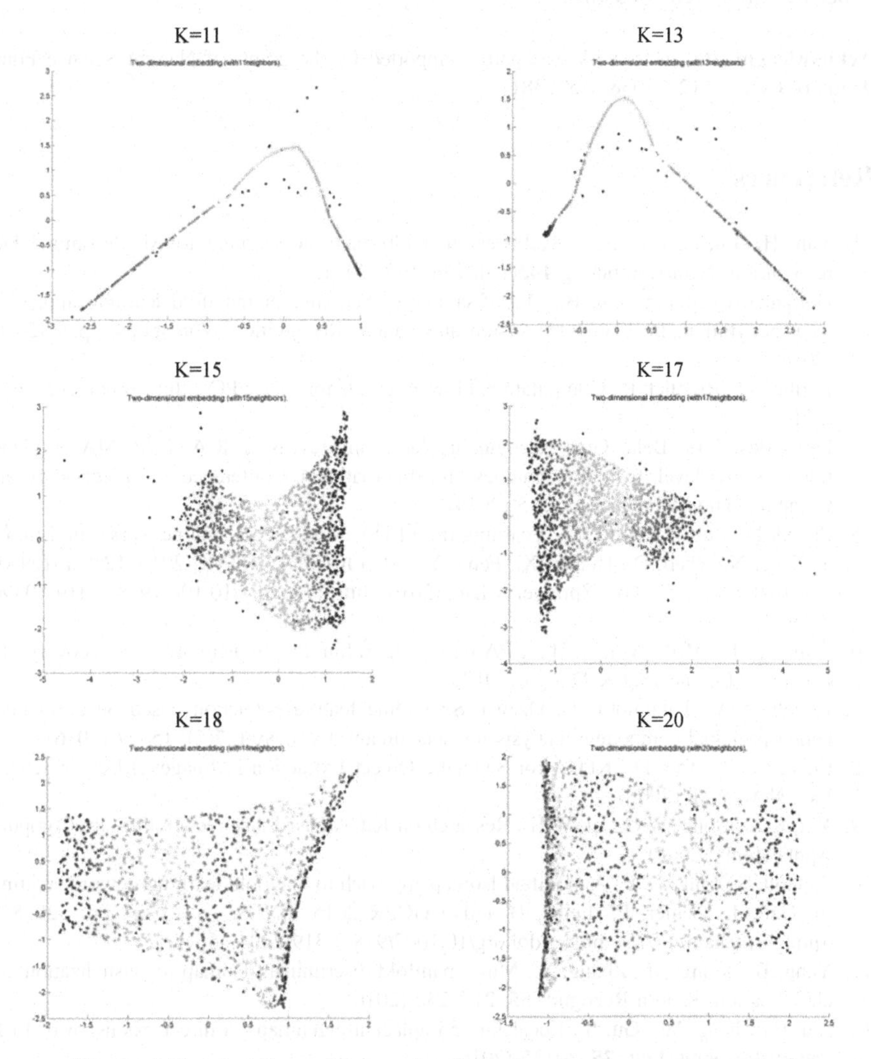

Fig. 1. 2-D embeddings of three noisy data sets using the proposed method

4 Conclusions

In this paper, a locally linear embedding algorithm based on the feature space projection is proposed. In this paper, any point in the original space will be mapped into its nearest feature space with the corresponding projection, to which the original LLE will be applied to find its low dimensional embedding. Experimental results show that the local linear embedding method based on the feature space projection proposed in this paper has improved robustness.

Acknowledgments. This work was partly supported by the grants of Natural Science Foundation of China (61273303&61572381).

References

1. Yan, H., Lu, J., Zhou, X.: Multi-feature multi-manifold learning for single-sample face recognition. Neurocomputing **143**(16), 134–143 (2014)
2. Dornaika, F., Raduncanu, B.: Out-of-sample embedding for manifold learning applied to face recognition. In: Computer Vision and Pattern Recognition Workshops, pp. 862–868 (2014)
3. Berthet, Q., Rigollet, P.: Computational lower bounds for sparse PCA. In: Computer Science (2014)
4. Fernandes, S.L., Bala, G.J.: Recognizing facial images using ICA, LPP, MACE Gabor filters, score level fusion techniques. In: International Conference on Electronics and Communication Systems, pp. 1–5 (2014)
5. Zhang, D., Luo, T., Wang, D.: Learning from LDA using deep neural networks. In: Lin, C.-Y., Xue, N., Zhao, D., Huang, X., Feng, Y. (eds.) ICCPOL/NLPCC-2016. LNCS (LNAI), vol. 10102, pp. 657–664. Springer, Cham (2016). https://doi.org/10.1007/978-3-319-50496-4_59
6. Guo, S., Li, P.Y., Wang, H.: LFA-based algorithm for IP network fast recovery. In: Computer Engineering & Design (2017)
7. Joseph, A.A., Tokumoto, T., Ozawa, S.: Online feature extraction based on accelerated kernel principal component analysis for data stream. Evol. Syst. **7**(1), 15–27 (2016)
8. Li, J., Li, X., Tao, D.: KPCA for Semantic Object Extraction in Images. Elsevier Science Inc., New York (2008)
9. Yin, K.Z., Gong, W.G., Li, W.H.: Research on KICA-based face recognition. In: Computer Applications (2005)
10. Sudholt, S., Fink, G.A.: A modified Isomap approach to manifold learning in word spotting. In: Gall, J., Gehler, P., Leibe, B. (eds.) GCPR 2015. LNCS, vol. 9358, pp. 529–539. Springer, Cham (2015). https://doi.org/10.1007/978-3-319-24947-6_44
11. Yang, B., Xiang, M., Zhang, Y.: Multi-manifold discriminant Isomap for visualization and classification. Pattern Recognit. **55**, 215–230 (2016)
12. Liu, F., Zhang, W., Gu, S.: Local linear Laplacian eigenmaps: a direct extension of LLE. Pattern Recognit. Lett. **75**, 30–35 (2016)
13. Liu, Y., Yu, Z., Zeng, M.: LLE for submersible plunger pump fault diagnosis via joint wavelet and SVD approach. Neurocomputing **185**, 202–211 (2016)
14. Wang, B., Yan, D., Chu, Y.: Face recognition based on sparse array of LPP and ELM. In: Microcomputer & Its Applications (2016)

15. Liu, Y.Q., Liang, J.G., Wang, Y.W.: Gain-improved double-slot LTSA with conformal corrugated edges. Int. J. RF Microw. Comput. Aided Eng. **1**, e21133 (2017)
16. Cui, P., Zhang, X.: Generalized improvement of LTSA algorithm based on manifold learning. In: Computer Engineering & Applications (2017)
17. Xie, Y., Chenna, P., He, J.: Visualization of big high dimensional data in a three dimensional space. In: International Conference on Big Data Computing Applications and Technologies, pp. 61–66 (2017)
18. Lo, J.T., Gui, Y., Peng, Y.: Solving the local-minimum problem in training deep learning machines. In: Liu, D., Xie, S., Li, Y., Zhao, D., El-Alfy, E.S. (eds.) ICONIP 2017. LNCS, vol. 10634, pp. 166–174. Springer, Cham (2017). https://doi.org/10.1007/978-3-319-70087-8_18
19. Bastani, O., Ioannou, Y., Lampropoulos, L.: Measuring neural net robustness with constraints (2016)
20. Carlini, N., Wagner, D.: Towards evaluating the robustness of neural networks, security and privacy (2017)
21. Chang, H., Yeung, D.Y.: Robust locally linear embedding. Pattern Recognit. **39**(6), 1053–1065 (2006)
22. Alexa, M., Adamson, A.: On normals and projection operators for surfaces defined by point sets. In: Eurographics Conference on Point-Based Graphics, pp. 149–155 (2004)
23. Zhang, S., Li, X., Zong, M.: Efficient kNN classification with different numbers of nearest neighbors. IEEE Trans. Neural Netw. Learn. Syst. **99**, 1–12 (2017)
24. Adeniyi, D.A., Wei, Z., Yong, Y.: Automated web usage data mining and recommendation system using K-Nearest Neighbor (KNN) classification method. Appl. Comput. Inform. **12** (1), 90–108 (2016)

Optimizing GPCR Two-Dimensional Topology from Contact Map

Hongjie Wu[1(✉)], Dadong Dai[1], Huaxiang Shen[1], Ru Yang[1], Weizhong Lu[1], and Qiming Fu[2(✉)]

[1] School of Electronic and Information Engineering, Suzhou University of Science and Technology, Suzhou 215009, China
Hongjie.wu@qq.com

[2] Jiangsu Province Key Laboratory of Intelligent Building Energy Efficiency, Suzhou University of Science and Technology, Suzhou 215009, China
fqm_1@126.com

Abstract. Comparative folding G-protein coupled receptors are hampered by distant homology and limited experimental templates. In order to explore the possibility of *ab initio* modeling, we proposed an optimization method of GPCR two-dimensional topologies based on the contact map that does not apply to any homologous templates or fragments, using a cost-effective 2D helix topology phase. The approach was evaluated in 12 solved GPCRs, our method has seven advantages over SWISS-MODEL, and the overall RMSD was reduced by about 20%.

Keywords: GPCR topology · Protein prediction · Optimization

1 Introduction

G-protein-coupled receptor (GPCR) is a specified transmembrane protein family. GPCRs are among the most heavily investigated drug targets of the pharmaceutical industry [1, 2], due to its critical role in various physiochemical signal-transmitting pathways. Due to the difficulty of experimentally solving the 3D structures, considerable efforts have been made to develop computational methods to model the GPCR structures [3, 4]. In this paper, we propose a new *ab initio* algorithm to assemble GPCR structures from primary sequences. In the GPCR-I-TASSER [5, 6], the process is guided by a simplistic general-purpose statistical potential that does not adapt to complex helical structures. We use sequence-based contact map predictions to guide helix beam assembly procedures.

2 Data Set

In this paper, we use 12 structured GPCRs to conduct experiments, we introduced part of the GPCRs used in the experiments (see Table 1).

D.-S. Huang et al. (Eds.): ICIC 2018, LNAI 10956, pp. 740–745, 2018.
https://doi.org/10.1007/978-3-319-95957-3_77

Table 1. Part of the GPCRs have been used in benchmark ab initio modeling methods.

#	PDB	Chain	Resolution	Name	TM-helix definition
1	2HPY	B	2.8	Rhodopsin	35-63,67-95,109-137,149-177,201-229,249-277,283-311
2	2ZIY	A	3.7	Rhodopsin	29-57,64-92,104-132,146-174,196-224,253-281,289-317
3	3EML	A	2.6	Adenosine receptor A2a	2-30,37-65,72-100,115-143,159-187,207-235,241-269
4	3RZE	A	3.1	Histamine H1 receptor	4-27,34-62,70-103,114-136,154-182,195-224,232-254

3 Methods

3.1 The Flow Chart of the Optimization for the Topologies

The flowchart of method is depicted in Fig. 1.

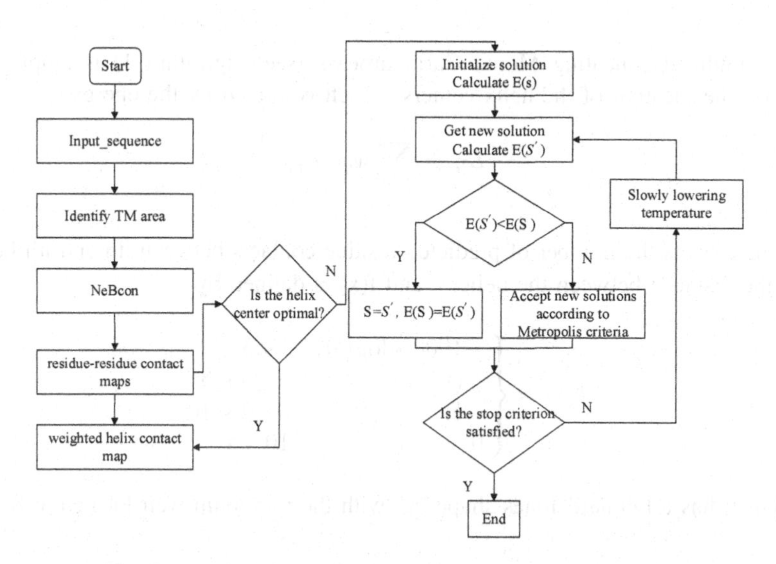

Fig. 1. The flow chat of the Optimization for the topologies

In this study, we introduced a complete *ab initio* method to simulate GPCRs without any homology template or fragment. The modeling method in this paper is mainly divided into three steps. First, we applied NeBcon to generate residue-residue contact maps. Next, we use the simulated annealing algorithm to optimize the center of the helix. Last, accord the NEBCON contact prediction to construct weighted helix contact map.

3.2 Contact-Map Guided 2D-Helix Topology Optimization

The stage aims to quickly decide the relative location of the seven helices of GPCRs using the inter-helix contact map predictions. The seven helices are projected to the membrane plane that is generally perpendicular to the helices, where the centers of the TM-helices are randomly put in the seven sectors of the membrane plane. A weighted helix contact map is then constructed based on the NeBcon contact prediction (see Fig. 2).

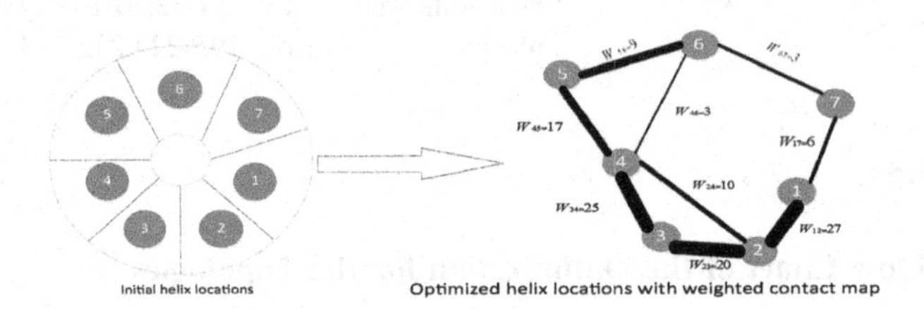

Fig. 2. Helices are reassembled by a weighted contact map potential

A simulated annealing Monte Carlo process (see Algorithm 1) is employed to optimize the location of the helix centers, which is guided by the energy

$$E_{2D} = \sum_{i \neq j} w_{ij} f\left(h_{ij}\right) \tag{1}$$

Where w_{ij} is the number of predicted residue contacts between ith and jth helices, h_{ij} is the distance between the helices, and f(x) is defined by

$$f(x) = \begin{cases} -1000 * \log\left(\frac{x}{3}\right), & x < 3 \\ -(x-3), & 3 \leq x < 8 \\ -5, & 8 \leq x < 10 \\ 0, & 10 \leq x \end{cases} \tag{2}$$

Which has a Lennard-Jones shape but with the minimum well located at 8–10 Å.

3.3 Simulated Annealing Algorithm for Helix Topology Optimization

The movements of the simulations consist of random movements of the helix points, which are accepted or rejected following standard Metropolis criterion.

Algorithm 1. Helix center optimization based on simulated annealing algorithm

```
Algorithm: Simulated annealing
Input : E(s): Function evaluation value at state S   Y(i): Indicates the current status   Y(i+1): Indicates new status   r: Used to control the speed of cooling
        T: The temperature of the system, the system should initially be in a high temperature state   T_min: The lower limit of temperatur
1.  While (T > T_min):
2.      ΔT = E(Y(i+1)) - E(Y(i+1))
3.      If (ΔT < 0):
4.          Y(i+1) = Y(i)
5.      Else:
6.          if ( exp(ΔT /T ) > random( 0 , 1 ) ):
7.                  Y(i+1) = Y(i)
8.      T = r * T
9.      i ++
10.         End if
11.     End if
12.     End while
13. Output : Optimal solutions
```

3.4 Three-Dimensional Helix Projection to the Two-Dimensional Plane

The main content of this stage is to calculate the topological structure of membrane proteins. The determination of the helix center point (COM): the whole PDB is divided into seven parts according to the transmembrane helix, each part of which is called a segment, and the first and tail two residues of each segment transmembrane helix are labeled as start and end respectively. Then the average value of the CA atomic coordinate of all residues of each segment is obtained.

$$COM = \frac{1}{N}\sum_{i}^{j} CA(x_i, y_i, z_i) \tag{3}$$

The determination of the helix center axis: the single transmembrane helix is divided into two parts, and then the helix center points of the two segments are taken respectively according to the above methods. The center point of the center axis is identified as the center point of the projection space plane. According to the 'point-Normal vector-formula', the plane equation can be obtained, and each spiral midpoint can be projected onto this plane.

4 Experiments and Discussion

In this paper, we use the helix Coordinates of the *ab initio* modeling and SWISS-MODEL methods to compare with the original coordinates and use 12 deconstructed GPCRs as subjects. A detailed list of modeling results on the 12 GPCRs is presented in Table 2.

Table 2. Comparison of the two methods

#	PDB	SWISS-MODEL/RMSD	*Ab initio* modeling/RMSD
1	2HPY	10.70	9.73
2	2LNL	1.57	7.33
3	2ZIY	12.61	11.39
4	3EML	13.34	6. 66
5	3ODU	21.68	7.29
6	3PBL	14.41	14.61
7	3RZE	4.92	11.83
8	3UON	9.80	8.19
9	3V2Y	6.92	8.75
10	3VW7	28.49	11.11
11	4AMJ	19.63	18.62
12	4DAJ	20.04	11.00
13	AVERAGE	13.68	10.90

The experimental results show that in the 12 GPCRs, our method of *ab initio* modeling has seven advantages over SWISS-MODEL, and the overall RMSD was reduced by about 20%.

5 Conclusion

This paper presents an optimization method for the topology of GPCR based on helix interaction, which is a basic problem in molecular biology calculation. This article selects two structured GPCRs as experimental subjects and can be found to provide demonstrations for further research in the field of protein three-dimensional structure prediction and other biological information.

Acknowledgements. This paper is supported by the National Natural Science Foundation of China (61772357, 61502329, 61672371), Jiangsu 333 talent project and top six talent peak project (DZXX-010), Suzhou Foresight Research Project (SYG201704, SNG201610) and the Postgraduate Research & Practice Innovation Program of Jiangsu Province (SJCX17_0680).

References

1. Fillmore, D.: It's a GPCR world. Mod. Drug Discov. **11**, 24–28 (2004)
2. Takeda, S., Kadowaki, S., Haga, T., Takaesu, H., Mitaku, S.: Identification of G protein-coupled receptor genes from the human genome sequence. FEBS Lett. **520**(1–3), 97–101 (2002)
3. Kufareva, I., Katritch, V., Participants of GD., Stevens, R.C., Abagyan, R.: Advances in GPCR modeling evaluated by the GPCR Dock 2013 assessment: meeting new challenges. Structure **22**(8), 1120–1139 (2014)
4. Kufareva, I., Rueda, M., Katritch, V., Stevens, R.C., Abagyan, R.: Status of GPCR modeling and docking as reflected by community-wide GPCR Dock 2010 assessment. Structure **19**(8), 1108–1126 (2011)
5. Zhang, J., Yang, J., Jang, R., Zhang, Y.: GPCR-I-TASSER: A hybrid approach to G protein-coupled receptor structure modeling and the application to the human genome. Structure **23** (8), 1538–1549 (2015)
6. Wu, H., et al.: Structure of a class C GPCR metabotropic glutamate receptor 1 bound to an allosteric modulator. Science **344**(6179), 58–64 (2014)

A Novel Approach Based on Bi-Random Walk to Predict Microbe-Disease Associations

Xianjun Shen[1(✉)], Huan Zhu[1], Xingpeng Jiang[1], Xiaohua Hu[1,2], and Jincai Yang[1]

[1] School of Computer, Central China Normal University, Wuhan, China
xjshen@mail.ccnu.edu.cn
[2] College of Computing and Informatics, Drexel University, Philadelphia, USA

Abstract. An increasing number of clinical observations have confirmed that the microbes inhabiting in human body have critical impacts on the progression of human disease, which provides promising insights into understanding the mechanism of diseases. However, the known microbe-disease associations remain limited. So, we proposed Bi-Random Walk based on Multiple Path (BiRWMP) to predict microbe-disease associations. Leave-one-out cross-validation (LOOCV) and 5-fold cross-validation were adopted to demonstrate the capability of proposed method. BiRWMP performed better than other methods. Finally, we listed 2 common disease and potential microbes ranked at top 10, and we demonstrated its reasonableness through looking up literatures.

Keywords: Microbe-disease associations · Bi-Random Walk
Computational prediction model

1 Introduction

The development of metagenomic and 16s RNA sequencing approach has promoted the research of microbiome. Human Microbiome Project (HMP) is to characterize microbial communities at human body sites and to look for correlations between microbiome and human disease [1–3]. Each niche of the human body represents an unique microbial community structure [4]. So, it's essential for us to decipher that some microorganisms are closely tied to a certain disease. Conventional cultivation-based experiments to find the associations between microbe and disease are not only time-consuming but also exhausting. Recently, some computational methods (Huang et al. [5]; Shen et al. [6]; Zou et al. [7]) have been implemented for identifying the microbe-disease associations. This paper is based on the hypothesis that function similar microbes share the same disease [8], and vice versa. We proposed BiRWMP which is based on that different length of path with a different parameter β dampen the contributions to microbe-disease associations.

D.-S. Huang et al. (Eds.): ICIC 2018, LNAI 10956, pp. 746–752, 2018.
https://doi.org/10.1007/978-3-319-95957-3_78

2 Methods

SM_{nm*nm} and SD_{nd*nd} represent an adjacency matrix of microbe network and disease network, respectively, where *nm* is the number of microbes and *nd* is the number of diseases. SM_{nm*nm} is normalized by Eq. (1) [9], where D_{nm*nm} is a diagonal matrix and $D(i,i) = \sum_j SM(i,j)$. The way SD_{nd*nd} is normalized is the same as SM_{nm*nm}.

$$\overline{SM} = D^{-\frac{1}{2}} * SM * D^{-\frac{1}{2}} \quad (1)$$

$$\mathrm{F}_0(m,d) = \sum_{i=1}^{nm}\sum_{j=1}^{nd} \overline{SM}(m, m(i)) * A(m(i), d(j)) * \overline{SD}(d(j), d) \quad (2)$$

$$\mathrm{F}_{t+1} = \alpha\overline{SM} * F_t * \overline{SD} + (1-\alpha)A \quad (3)$$

A_{nm*nd} represents an adjacency matrix of microbe-disease associations network, *A(m(i), d(j))* = 1 if microbe *m(i)* has known association with disease *d(j)*, otherwise *A(m(i), d (j))* = 0. Thus, all zero entries of A_{nm*nd} are unknown instead of no association between *m(i)* and *d(j)*. Random Walk on incompleteness network might be misleading. So, A_{nm*nd} as a regularization framework will be more accurate. BiRW is indicated by Eqs. (2) and (3), Eq. (2) means that initialization association probability between microbe *m* and disease *d* is defined by multiplying weight of the edge along a certain path and then summarizing of all paths, so $F_0(m,d)$ means association probabilities between *m* and *d*. We can iterate above procedure, and at each step, we add the potential microbe-disease associations into A_{nm*nd} and reconstruct microbe-disease network (see Eq. (3)). After several iterations, we can get the complete heterogeneous network, as shown in Fig. 1, the thin and thick dash lines represent the potential and known microbe-disease associations, respectively. Parameter α (0,1) not only is a decay factor, penalizing the number of steps, but also plays a role to balance the importance of potential and the known microbe-disease associations.

In each step, association probabilitiy between *m* and *d* is defined by Eq. (2), the path length *L* may be 1, or 2, or 3 from *m* to *d*. Usually, a short path tends to contribute more to the association probality than a long path. Thus, we proposed a parameter β to dampen the contribution. The formula of BiRWMP is as follows:

$$\mathrm{F}_0(m,d) = \sum_{i=1}^{nm}\sum_{j=1}^{nd} \beta * \overline{SM}(m, m(i)) * A(m(i), d(j)) * \overline{SD}(d(j), d) \quad (4)$$

$$\begin{aligned}\mathrm{F}_{t+1}(m,d) =& \alpha * \sum_{i=1}^{nm}\sum_{j=1}^{nd} \beta * \overline{SM}(m, m(i)) * F_t(m(i), d(j)) * \overline{SD}(d(j), d) \\ &+ (1-\alpha)A(m,d)\end{aligned} \quad (5)$$

When $L = 1$, $\beta = \beta_1$, and when $L = 2$, $\beta = \beta_2$, otherwise, $\beta = \beta_3$, so $1 \geq \beta_1 > \beta_2 > \beta_3 > 0$. After many iterations, association probability matrix F_{t+1} will converge.

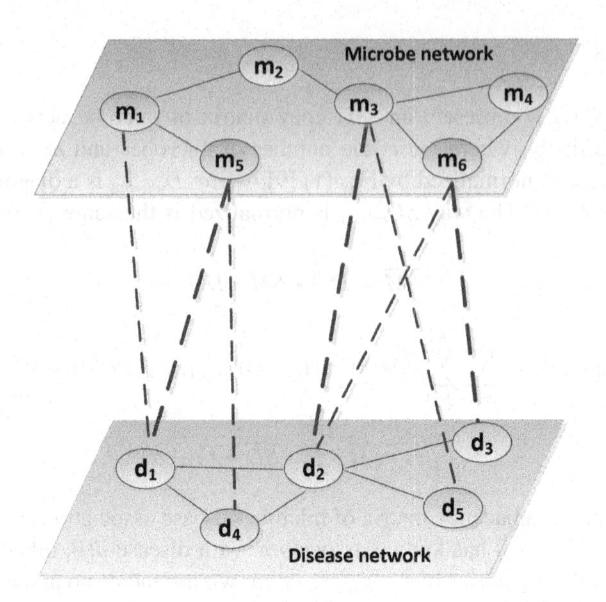

Fig. 1. Microbe-disease associations heterogeneous network.

We will end the iteration when the change between F_{t+1} and F_t measured by *L1* norm is less than 10-6.

3 Experiment and Analysis

3.1 Datasets

Ma et al. established Human Microbe-Disease Associations Database by a large-scale text mining [8, 10]. We only selected 71 microbes at the genus level. We obtained the levels information of microbe and microbial abundance data from HMP [11]. We use Mothur Output Files and V13 High Quality Trimming Pipeline, then microbial interaction network from abundance data was inferred by Spearman correlations [12]. Removed some symptoms and diseases with the same symptoms, we derived 28 diseases, and we computed disease similarity based on symptoms [13]. Finally, We obtained 127 interactions between microbe and disease.

3.2 Cross-Validation

First, we use LOOCV to evaluate the performance of BiRWMP. We removed a known microbe-disease pair in turn for testing sample, and other known microbe-disease pairs were considered as training samples, other microbe-disease pairs which without evidences would be used as candidate samples. The rank of each testing sample relative to candidate samples was further obtained. The testing samples with a higher rank than the specific threshold would be regarded as a successful prediction. So, we could obtain the corresponding true positive rates and false positive rates by varying the thresholds.

ROC curve was shown in Fig. 2. ROC curve of BiRWMP ($\alpha = 0.1$, $\beta_1 = 1$, $\beta_2 = 0.1$, $\beta_3 = 0.01$) is above on BiRW ($\alpha = 0.1$), RWRH ($\lambda = 0.9$, $\eta = 0.1$, $\gamma = 0.7$) and PBHMDA ($L = 3$, $\alpha = 2.26$), so BiRWMP showed better performance than BiRW, RWRH and PBHMDA. AUC value of BiRWMP and BiRW_avg ($\alpha = 0.4$, $l = 2$, $r = 2$) is 0.8921, 0.8811, respectively. It is considered that BiRWMP is practical for predicting new microbe-disease pairs. We also introduced 5-fold cross-validation to evaluate the effectiveness of the method. AUC value of BiRWMP and BiRW was shown in Table 1.

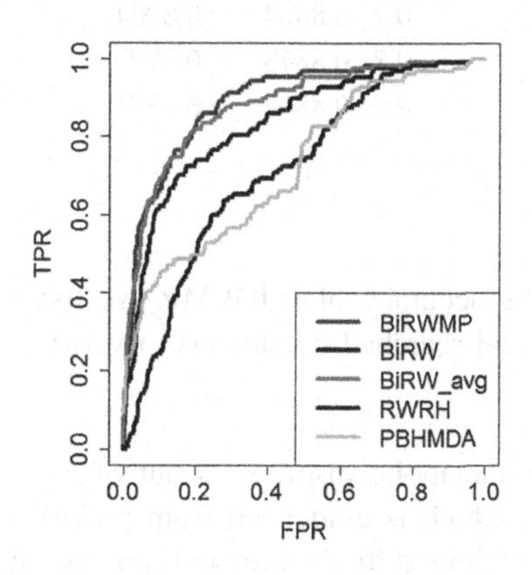

Fig. 2. ROC curve of five different methods.

Table 1. Effectiveness comparison between BiRWMP and BiRW in 5-fold cross-validation.

Method	BiRWMP	BiRW
5-fold	0.8777 ± 0.0089	0.8287 ± 0.0062

3.3 Effect of Parameters in BiRWMP

Performance of BiRWMP is dominated by α and β. We set kinds of values for them and implemented LOOCV to demonstrate the performance of method. Through constant searching, we can achieve the maxium AUC value, when α was set a fixed value and β_1, β_2, β_3 were set 1, 0.1, 0.01, respectively. When β_1, β_2, β_3 were set 1, 0.1, 0.01, respectively, we set various value for α, and we get corresponding AUC value (see Table 2). Thus, $\alpha = 0.1$, $\beta_1 = 1$, $\beta_2 = 0.1$, $\beta_3 = 0.01$ led to the best performance for predicting microbe-disease associations. α has little effect on the BiRWMP performs, however, appropriate β can enhance greatly the BiRWMP performs.

Table 2. Effectiveness of various α parameter value in the evaluation frameworks of LOOCV.

α	BiRWMP	BiRW
0.1	0.8921	0.8369
0.2	0.8916	0.8341
0.3	0.8910	0.8309
0.4	0.8904	0.8272
0.5	0.8898	0.8213
0.6	0.8892	0.8131
0.7	0.8884	0.8004
0.8	0.8873	0.7797
0.9	0.8862	0.7331

3.4 Case Analysis

For further evaluating the accuracy of BiRWMP, we listed 2 diseases and potential microbes in the top 10, and searched documents to verify.

(1) **Obesity**

Obesity is a chronic metabolic disorder. 8 out of top 10 potential microbes are associated with Obesity, which is evidenced from previous literatures (see Table 3). Obese patients with Clostridium difficile infections are considered to have a high risk of death [14]. Lactobacillus, Tannerella and Fusobacterium are enriched in obese people [15, 16]. The abundance of Prevotella in high-fat diet (HFD) mice is lower than in low-fat diet mice [17]. Body mass index provides opportunity for Streptococcus colonization [18]. Haemophilus has an effect on nutritional status and Obesity will be followed [19]. The number of Bifidobacterium spp. increases in normal weight people compared with obese people [20]. But there are no related literature to demonstrate that Vibrio and Oligella are closely related to Obesity.

Table 3. Microbes related to disease in top 10.

Rank	Microbes related to Obesity		Microbes related to Asthma	
	Microbe	Evidence (PMID)	Microbe	Evidence (PMID)
1	Clostridium	28593617	Xanthomonas	Unconfirmed
2	Lactobacillus	28555008	Burkholderia	Unconfirmed
3	Prevotella	28521862	Lactobacillus	20592920
4	Streptococcus	27997266	Lysobacter	Unconfirmed
5	Tannerella	27717180	Propionibacterium	27433177
6	Haemophilus	2788987	Veillonella	25329665
7	Fusobacterium	27717180	Streptococcus	17950502
8	Vibrio	Unconfirmed	Pseudomonas	25329665
9	Oligella	Unconfirmed	Fusobacterium	[25]
10	Bifidobacterium	28596023	Leptotrichia	28500319

(2) **Asthma**

Asthma is a common chronic inflammatory disease of the airways. 7 out of top 10 potential microbes are associated with Asthma which is confirmed from previous literatures (see Table 3). Lactobacillus plays a significant role in prevention of Asthma [21]. The level of Propionibacterium acnes could be used for judgement of Asthma VS healthy subjects [22]. Veillonella appears lower proportions in patients with Asthma [23]. Streptococcus has an effect on the progression of Asthma [24]. Pseudomonas spp. and Fusobacterium are more prevalent in Asthma than healthy controls [23, 25]. Leptotrichia species exist in people with Asthma [26]. There are no articles to confirm that Xanthomonas, Burkholderia and Lysobacter are strongly correlated with Asthma.

4 Discussion

Even if BiRWMP performed better, it still has some limitations. For instance, β has influence on experimental performance, however, we only adjust parameter according to experimental experience and can't try all value. Microbe-disease pairs is helpful in developing novel methods for curing diseases. For instance, we can alter the structure of the microbial community by transplantation microbes from healthy people to patient.

Acknowledgement. This research is supported by the National Natural Science Foundation of China (61532008), the National Key Research and Development Program of China (2017YFC0909502) and the Self-determined Research Funds of CCNU from the Colleges' Basic Research and Operation of MOE (No. CCNU17TS0003).

References

1. Peterson, J., Garges, S., Giovanni, M., et al.: The NIH human microbiome project. Genome Res. **19**(12), 2317–2323 (2009)
2. Turnbaugh, P.J., Ley, R.E., Hamady, M., et al.: The human microbiome project: exploring the microbial part of ourselves in a changing world. Nature **449**(7164), 804 (2007)
3. Turnbaugh, P.J., Gordon, J.I.: The core gut microbiome, energy balance and obesity. J. Physiol. **587**(17), 4153–4158 (2009)
4. Kinross, J.M., Darzi, A.W., Nicholson, J.K.: Gut microbiome-host interactions in health and disease. Genome Med. **3**(3), 14 (2011)
5. Huang, Z.A., Chen, X., Zhu, Z., et al.: PBHMDA: path-based human microbe-disease association prediction. Front. Microbiol. **8**(2), 233 (2017)
6. Shen, X., Chen, Y., Jiang, X., et al.: Prioritizing disease-causing microbes based on random walking on the heterogeneous network. Methods **124**, 120–125 (2017)
7. Zou, S., Zhang, J., Zhang, Z.: A novel approach for predicting microbe-disease associations by bi-random walk on the heterogeneous network. PLoS ONE **12**(9), e0184394 (2017)
8. Ma, W., Zhang, L., Zeng, P., et al.: An analysis of human microbe–disease associations. Briefings Bioinform. **18**(1), 85–97 (2016)
9. Xie, M., Hwang, T., Kuang, R.: Prioritizing disease genes by bi-random walk. In: Tan, P.-N., Chawla, S., Ho, C.K., Bailey, J. (eds.) PAKDD 2012, Part II. LNCS (LNAI), vol. 7302, pp. 292–303. Springer, Heidelberg (2012). https://doi.org/10.1007/978-3-642-30220-6_25

10. HMDAD Homepage. http://www.cuilab.cn/hmdad. Accessed 8 May 2018
11. HMP abundance data. http://hmpdacc.org/HMMCP/healthy. Accessed 8 May 2018
12. Faust, K., Raes, J.: Microbial interactions: from networks to models. Nat. Rev. Microbiol. **10** (8), 538–550 (2012)
13. Zhou, X.Z., Menche, J., Barabási, A.L., et al.: Human symptoms–disease network. Nat. Commun. **5**, 4212 (2014)
14. Nathanson, B.H., Higgins, T.L., McGee, W.T.: The dangers of extreme body mass index values in patients with Clostridium difficile. Infection **45**(6), 787–793 (2017)
15. Fernández-Navarro, T., Salazar, N., Gutiérrez-Díaz, I., et al.: Different intestinal microbial profile in over-weight and obese subjects consuming a diet with low content of fiber and antioxidants. Nutrients **9**(6), 551 (2017)
16. Maciel, S.S., Feres, M., Gonçalves, T.E.D., et al.: Does obesity influence the subgingival microbiota composition in periodontal health and disease? J. Clin. Periodontol. **43**(12), 1003–1012 (2016)
17. Kusumoto, Y., Irie, J., Iwabu, K., et al.: Bile acid binding resin prevents fat accumulation through intestinal microbiota in high-fat diet-induced obesity in mice. Metabolism **71**, 1–6 (2017)
18. Alvareza, M.D., Subramaniam, A., Tang, Y., et al.: Obesity as an independent risk factor for group B streptococcal colonization. J. Matern. Fetal Neonatal Med. **30**(23), 2876–2879 (2017)
19. Sherry, B., Weber, A., Williams-Warren, J., et al.: The impact of Haemophilus influenzae type b meningitis on nutritional status. Am. J. Clin. Nutr. **50**(3), 425–434 (1989)
20. Nicolucci, A.C., Hume, M.P., Martínez, I., et al.: Prebiotic reduces body fat and alters intestinal microbiota in children with overweight or obesity. Gastroenterology **153**(3), 711–722 (2017)
21. Yu, J., Jang, S.O., Kim, B.J., et al.: The effects of Lactobacillus rhamnosus on the prevention of asthma in a murine model. Allergy Asthma Immunol. Res. **2**(3), 199–205 (2010)
22. Jung, J.W., Choi, J.C., Shin, J.W., et al.: Lung microbiome analysis in steroid-naïve asthma patients by using whole sputum. Tuberc. Respir. Dis. **79**(3), 165–178 (2016)
23. Park, H.K., Shin, J.W., Park, S.G., et al.: Microbial communities in the upper respiratory tract of patients with asthma and chronic obstructive pulmonary disease. PLoS ONE **9**(10), e109710 (2014)
24. Preston, J.A., Essilfie, A.T., Horvat, J.C., et al.: Inhibition of allergic airways disease by immunomodulatory therapy with whole killed Streptococcus pneumonia. Vaccine **25**(48), 8154–8162 (2007)
25. Dang, H.T., Park, H.K., Shin, J.W., et al.: Analysis of oropharyngeal microbiota between the patients with bronchial asthma and the non-asthmatic persons. J. Bacteriol. Virol. **43**(4), 270–278 (2013)
26. Chiu, C.Y., Chan, Y.L., Tsai, Y.S., et al.: Airway microbial diversity is inversely associated with mite-sensitized rhinitis and asthma in early childhood. Sci. Rep. **7**(1), 1820 (2017)

Green Permutation Flowshop Scheduling: A Trade- off- Between Energy Consumption and Total Flow Time

Hande Öztop[1(✉)], M. Fatih Tasgetiren[2], Deniz Türsel Eliiyi[1], and Quan-Ke Pan[3]

[1] Department of Industrial Engineering, Yasar University, Bornova, Turkey
{hande.oztop,deniz.eliiyi}@yasar.edu.tr

[2] Department of International Logistics Management, Yasar University, Bornova, Turkey
fatih.tasgetiren@yasar.edu.tr

[3] State Key Laboratory, Huazhong University of Science and Technology, Wuhan, People's Republic of China
panquanke@hust.edu.cn

Abstract. Permutation flow shop scheduling problem (PFSP) is a well-known problem in the scheduling literature. Even though many multi-objective PFSPs are presented in the literature with the objectives related to production efficiency and customer satisfaction, studies considering energy consumption and environmental effects in scheduling is very seldom. In this paper, the trade-off between total energy consumption (TEC) and total flow time is investigated in a PFSP environment, where the machines are assumed to operate at varying speed levels. A multi-objective mixed integer linear programming model is proposed based on a speed-scaling strategy. Due to the NP-complete nature of the problem, an efficient multi-objective iterated greedy (IG_{ALL}) algorithm is also developed. The performance of IG_{ALL} is compared with model performance in terms of quality and cardinality of the solutions.

Keywords: Permutation Flowshop Scheduling · Energy efficient scheduling Multi-objective optimization · Iterated greedy algorithm · Heuristic optimization

1 Introduction

In the standard flowshop scheduling problem (FSP), there is a set of n jobs $N = \{1, \ldots, n\}$, which must be processed on m machines $M = \{1, \ldots, m\}$ following the same route, meaning that the operations of each job are always processed in the same order. It is assumed that all jobs are ready at time zero and job preemptions are not permitted. Beside these assumptions, in the PFSP, it is also assumed that the job processing order is the same for each machine. Therefore, the PFSP aims to find the job processing order (permutation) by optimizing a given performance measure.

Recently, the high-energy consumption has become one of the main issues for the industrial sector due to the increase in the fuel prices and negative ecological effects. Therefore, most of the manufacturing companies have started to search more energy

D.-S. Huang et al. (Eds.): ICIC 2018, LNAI 10956, pp. 753–759, 2018.
https://doi.org/10.1007/978-3-319-95957-3_79

efficient scheduling systems [1]. The PFSP has been extensively studied in the flowshop scheduling literature with the production efficiency measures such as makespan and total flow time. Nevertheless, objectives related to energy efficiency have been rarely considered in flowshop scheduling. A recent review on energy efficient scheduling is provided in [2]. As a well- known study, the authors proposed a turn-off strategy and concluded that energy can be saved, once the machine is turned off during idle times [3]. Even though the turn off strategy provides energy savings, it may not be applicable for some shop floors as it may shorten the service life of machines. Therefore, a speed scaling strategy was proposed for the energy-efficient FSP by treating operation speed as an independent variable that can be changed to improve energy efficiency [1]. This strategy was applied to the PFSP with the objectives of the total carbon emissions and the makespan [4]. Variable speed levels were also considered for the two machine sequence-dependent PFSP [5]. Moreover, a backtracking algorithm was developed for the energy efficient PFSP in [6].

In most of the aforementioned studies, makespan is considered with an energy-related objective. To the best of our knowledge, flow time criterion has not been studied in the literature for the energy efficient PFSP. In this study, a multi-objective IG_{ALL} algorithm and a multi-objective mixed integer programming (MIP) are presented for the energy-efficient PFSP with the total flow time criterion, employing a speed scaling strategy, where the machines can operate at varying speed levels. For small-sized instances, we obtain the Pareto-optimal solution sets by solving the MIP model on CPLEX and we test the performance of the heuristic algorithm on these instances.

2 Problem Definition

In this study, we consider a bi-objective PFSP that aims to find a feasible permutation that minimizes the two conflicting objectives of total flow time and total energy consumption (TEC). Unlike the classical PFSP, there is a finite and discrete set of l different processing speed levels $L = \{1, \ldots, l\}$ for each machine. Thus, processing time of a job can differ according to the chosen speed level. Within the job-based speed-scaling framework, it is assumed that each job is processed with the speed level at all machines.

We modified the PFSP model of [7] by introducing the speed levels and changing the objectives as TEC and total flow time. We use similar TEC calculation as proposed in two-machine PFSP model of [5]. The necessary problem notation is given in Table 1. The proposed MILP formulation is given below:

$$Minimize \sum\nolimits_{i \in N} C_{im} \tag{1}$$

$$Minimize\, TEC \tag{2}$$

$$C_{i1} \geq \sum\nolimits_{l \in L} \frac{p_{i1} y_{i1l}}{v_l} \quad \forall i \in N \tag{3}$$

Table 1. Problem notation

Parameters	
p_{ij}	Processing time of job $i \in N$ on machine $j \in M$
v_l	Speed factor of speed $l \in L$
λ_l	Processing conversion factor for speed $l \in L$
φ_j	Conversion factor for idle time on machine $j \in M$
τ_j	Power of machine $j \in M$ (kW)
D	A very large number
Decision variables	
y_{ijl}	1 if job i is processed at speed l on machine j; 0, otherwise
x_{ik}	1 if job i precedes job k; 0 otherwise ($i < k$)
C_{ij}	Completion time of job i on machine j
θ_j	Idle time on machine j
C_{max}	Maximum completion time (makespan)
TEC	Total energy consumption (kWh)

$$C_{ij} - C_{i,j-1} \geq \sum\nolimits_{l \in L} \frac{P_{ij} y_{ijl}}{v_l} \quad \forall i \in N, \ \forall j \in M : j \geq 2 \tag{4}$$

$$C_{ij} - C_{kj} + Dx_{i,k} \geq \sum\nolimits_{l \in L} \frac{P_{ij} y_{ijl}}{v_l} \quad \forall i, k \in N : k > i, \ \forall j \in M \tag{5}$$

$$C_{ij} - C_{kj} + Dx_{ik} \leq D - \sum\nolimits_{l \in L} \frac{P_{kj} y_{kjl}}{v_l} \quad \forall i, k \in N : k > i, \ \forall j \in M \tag{6}$$

$$C_{max} \geq C_{im} \quad \forall i \in N \tag{7}$$

$$\sum\nolimits_{l \in L} y_{ijl} = 1; \quad \forall i \in N, \forall j \in M \tag{8}$$

$$y_{ijl} = y_{i,j+1,l} \quad \forall i \in N, \ \forall j \in M : j < m, \quad \forall l \in L \tag{9}$$

$$\theta_j = C_{max} - \sum\nolimits_{i \in N} \sum\nolimits_{l \in L} \frac{p_{ij} y_{ijl}}{v_l} \quad \forall j \in M \tag{10}$$

$$TEC = \sum\nolimits_{i \in N} \sum\nolimits_{j \in M} \sum\nolimits_{l \in L} \frac{p_{ij} \tau_j \lambda_l}{60 v_l} y_{ijl} + \sum\nolimits_{j \in M} \frac{\varphi_j \theta_j \tau_j}{60} \tag{11}$$

$$y_{ijl} \in \{0, 1\}, C_{ij} \geq 0 \, \forall i \in N, \forall j \in M, \forall l \in L \quad x_{ik} \in \{0, 1\} \, \forall i, k \in N : k > i \tag{12}$$

The objective functions (1) and (2) minimize the total flow time and TEC, respectively. Constraint (3) states that the completion time of a job on machine 1 is greater than or equal to its processing time on that machine. Constraint (4) ensures that a job can be started on a machine only after its preceding operation on previous machine has been finished. Constraints (5) and (6) guarantee that job i either precedes job k or job k precedes job i in the sequence, but not both. Constraint (7) calculates the makespan. Constraint (8) and (9) state that exactly one speed level is selected for each job and the same speed level is used for a job at all machines. Constraint (10) calculates the idle time on each machine. Constraint (11) computes the total energy consumption as proposed by [5]. Constraint (12) defines the decision variables.

Since there are two conflicting objectives, our problem is a multi-objective problem. In multi-objective (minimization) problems, a feasible solution $\vec{r}$ dominates feasible solution $\vec{t}(\vec{r} \succ \vec{t})$ iff $\forall i; f_i(\vec{r}) \leq f_i(\vec{t})$ and $\exists i; f_i(\vec{r}) < f_i(\vec{t})$, where f_i is the ith objective function. In this study, augmented ε-constraint method is used to solve this multi-objective model, since it produces only Pareto-optimal solutions, in which any objective function cannot be improved without worsening another objective function [8].

3 Energy-Efficient IG_{ALL} Algorithm

Recently, an IG_{ALL} algorithm is proposed for the PFSP in [9]. Unlike the traditional IG algorithm, the IG_{ALL} algorithm applies an additional local search to partial solutions after destruction [9]. In this study, we propose a job-based speed-scaling strategy for the multi-objective energy-efficient IG_{ALL} algorithm. A multi-chromosome structure is proposed, where it contains a permutation of n jobs and a speed vector with three levels: fast, normal and slow speed levels. Note that the same speed vector is employed in all machines.

The initial population with size $NP = 100$ is constructed as follows. Initially, a solution is constructed by the NEH heuristic [10]. The resulting NEH solution is the initial solution for IG_{ALL} with the flow time minimization. Now, we devote 10% of the total CPU time for single-objective IG_{ALL} to obtain a good starting solution, where the destruction size is taken as $\kappa = 4$. Once the best solution π_{best} is found by IG_{ALL}, the first three solutions are obtained by assigning fast, normal and slow speed levels to each job in the best solution π_{best}. The rest of population is constructed by assigning random speed levels to each job in π_{best}. The archive set Ω is initially empty and updated with non-dominated solutions from the initial population, where the $|\Omega| = n \times 5$.

The destruction-construction procedure is a core part of IG_{ALL} algorithms. In the destruction step, κ jobs with their speeds are randomly removed from the solution. A first improvement insertion local search is applied to the partial solution by considering speed levels. Then, random speed levels are assigned to the removed jobs. These κ jobs are reinserted into the partial solution with their respective speed levels sequentially until a complete solution of n jobs is established. The dominance rule ($\succ$) in multi-objective optimization is used when two solutions are compared. Partial solutions are evaluated based on partial dominance. The complete solution is obtained by choosing the non-dominated solution among n solutions after the last removed job is inserted for n positions.

After the destruction-construction procedure, an efficient first-improvement insertion neighborhood structure is employed for each individual i in the population. Job π_{ij} and speed v_{ij} are removed from position j of solution $x_i(\pi_{ij}, v_{ij})$. A new speed level is randomly assigned to position j. Then, the local search inserts job (π^*, v^*) into all possible positions of the incumbent solution x_i. The best insertion position that dominates the incumbent solution is determined and (π^*, v^*) is inserted to that position. The archive set Ω is updated in such a way that $x^*(\pi^*, v^*)$ is replaced with any solution dominated by it. This is repeated for all the job and speed pairs. If any non-dominated solution is found, the local search is repeated again until any non-dominated solution cannot be obtained.

In addition, we propose a local search algorithm based on uniform crossover operator for speed levels only. In other words, after applying the above procedures to each individual in the population, we keep the same permutation for each individual in the population and make a uniform crossover on speed levels. For each individual x_i in the population, we select another individual from population randomly, say x_k, New offspring is generated by either taking speed levels from x_i or x_k depending on the crossover probability, which is drawn from unit normal distribution $N(0.5, 0.1)$. If x_{new} dominates x_i (i.e., $x_{new} \succ x_i$), x_i is replaced by x_{new} and the archive set Ω is updated. This is repeated for all individuals in the population. After crossover local search, we mutate the speed levels of jobs with a small mutation probability, which is drawn from unit normal distribution $N(0.05, 0.01)$.

4 Computational Results

To evaluate the performance of the algorithms, we carried out computational experiments using the well-known PFSP benchmarks of Taillard [11]. Due to the computational difficulty of this multi-objective problem, we create 30 small-sized instances: 5×5, 5×10, 5×20, by truncating the instances with 20 jobs, where the first number indicates the number of jobs and the second number specifies the number of machines. The energy related parameters used in TEC calculation are taken from [5] as follows: $L = \{1, 2, 3\}$, $v_l = \{1.2, 1, 0.8\}$, $\lambda_l = \{1.5, 1, 0.6\}$, $\varphi_j = 0.05$ and $\tau_j = 60\,\text{kW}$. We use IBM ILOG CPLEX Studio 12.8 to solve the model on a Core i7, 2.60 GHz, 8 GB RAM computer. We find very close approximations for the Pareto-optimal frontiers of instances, by determining the ε level as 10^{-2} for the augmented ε-constraint method. These finite numbers of Pareto-optimal solutions are referred as Pareto-optimal solution set (P). The energy-efficient IG_{ALL} algorithm is coded in C++ programming language on Microsoft Visual Studio 2013 and all instances are solved on a Core i5, 3.20 GHz, 8 GB RAM computer. Ten replications are made for each instance. In each replication, the algorithm is run for 25 nm ms, where n is the number of jobs and m is the number of machines.

We use following performance measures to assess the solution quality of the IG_{ALL} algorithm, where H refers to the non-dominated solution set of the IG_{ALL}: (1) Ratio of the Pareto-optimal solutions found ($R_p = |H \cap P|/|P|$), (2) Inverted generational distance [12] ($IGD = \sum_{v \in P} d(v, H)/|P|$), where $d(v, H)$ denotes the minimum Euclidean distance between v and the solution in H, and (3) Distribution spacing [13] (DS). The computational results for IG_{ALL} algorithm are reported in Table 2. As shown in the table, IG_{ALL} algorithm finds 100% of the Pareto-optimal solutions. As all Pareto-optimal solutions are found by IG_{ALL} algorithm for all instances, obviously, IG_{ALL}

algorithm has zero IGD value. In terms of distribution spacing, we can say that solutions in H are evenly distributed due to the low DS value.

Table 2. Computational results

Inst.	5 × 5			5 × 10			5 × 20		
	R_p	IGD	DS_H	R_p	IGD	DS_H	R_p	IGD	DS_H
1	1	0	0.64	1	0	0.69	1	0	0.99
2	1	0	0.66	1	0	0.80	1	0	0.80
3	1	0	0.82	1	0	0.85	1	0	0.90
4	1	0	0.58	1	0	0.68	1	0	0.59
5	1	0	1.50	1	0	0.63	1	0	0.56
6	1	0	0.79	1	0	0.53	1	0	0.68
7	1	0	0.61	1	0	0.98	1	0	0.78
8	1	0	0.62	1	0	0.53	1	0	0.81
9	1	0	0.75	1	0	0.83	1	0	0.88
10	1	0	0.51	1	0	0.82	1	0	0.86
Average							**1**	**0**	**0.76**

As mentioned above, IG_{ALL} presents an outstanding performance, as it finds 100% of the Pareto-optimal solutions for these instances. In further study, the performances of the proposed MIP model and the heuristic algorithm will be evaluated using larger instances. Furthermore, it is planned to develop different metaheuristics for the problem. Additionally, in further researches, the matrix representation can be used for the speed-scaling framework by adapting the mathematical model and IG_{ALL} algorithm.

References

1. Fang, K., Uhan, N., Zhao, F., Sutherland, J.W.: A new approach to scheduling in manufacturing for power consumption and carbon footprint reduction. J. Manuf. Syst. **30**(4), 234–240 (2011)
2. Gahm, C., Denz, F., Dirr, M., Tuma, A.: Energy-efficient scheduling in manufacturing companies: a review and research framework. Eur. J. Oper. Res. **248**, 744–757 (2016)
3. Mouzon, G., Yildirim, M.B., Twomey, J.: Operational methods for the minimization of energy consumption of manufacturing equipment. Int. J. Prod. Res. **45**(18–19), 4247–4271 (2007)
4. Ding, J.-Y., Song, S., Wu, C.: Carbon-efficient scheduling of flowshops by multi-objective optimization. Eur. J. Oper. Res. **248**, 758–771 (2016)
5. Mansouri, S.A., Aktas, E., Besikci, U.: Green scheduling of a two-machine flowshop: trade-off between makespan and energy consumption. Eur. J. Oper. Res. **248**, 772–788 (2016)
6. Lu, C., Gao, L., Li, X., Pan, Q.K., Wang, Q.: Energy-efficient permutation flow shop scheduling problem using a hybrid multi-objective backtracking search algorithm. J. Clean. Prod. **144**, 228–238 (2017)
7. Manne, A.S.: On the job-shop scheduling problem. Oper. Res. **8**, 219–223 (1960)

8. Mavrotas, G.: Effective implementation of the e-constraint method in multi-objective mathematical programming problems. Appl. Math. Comput. **213**, 455–465 (2009)
9. Dubois-Lacoste, J., Pagnozzi, F., Stützle, T.: An iterated greedy algorithm with optimization of partial solutions for the makespan permutation flowshop problem. Comput. Oper. Res. **81**, 160–166 (2017)
10. Nawaz, M., Enscore Jr., E.E., Ham, I.: A heuristic algorithm for the m-machine, n-job flowshop sequencing problem. OMEGA **11**(1), 91–95 (1983)
11. Taillard, E.: Benchmarks for basic scheduling problems. Eur. J. Oper. Res. **64**, 278–285 (1993)
12. Coello, C.A.C., Van Veldhuizen, D.A., Lamont, G.B.: Evolutionary Algorithms for Solving Multi-Objective Problems, vol. 5. Springer, New York (2007). https://doi.org/10.1007/978-0-387-36797-2
13. Tan, K.C., Goh, C.K., Yang, Y., Lee, T.H.: Evolving better population distribution and exploration in evolutionary multi-objective optimization. Eur. J. Oper. Res. **171**, 463–495 (2006)

Data Mining to Support the Discrimination of Amyotrophic Lateral Sclerosis Diseases Based on Gait Analysis

Haya Alaskar[1] and Abir Jaafar Hussain[2(✉)]

[1] Computer Science Department, Prince Sattam University, Alkharj, Saudi Arabia
h.alaskar@psau.edu.sa
[2] Department of Computer Science, Liverpool John Moores University, Liverpool L33AF, UK
A.Hussain@ljmu.ac.uk

Abstract. In many medical researches, dealing with huge dataset is crucial. However, it is difficult to use standard methodologies to analysis huge dataset. In such cases, data mining tools can support doctors for better diagnosis. Data mining tools have been utilized in clinical data analysis such as biological signals, clinical images to analysis and detect diseases. The utilization of these techniques can increase diagnostic sensitivity and specificity. It can help reduce the misdiagnosis, in addition to early prediction. This paper discusses the use of machine learning algorithms for the detection and classification of amyotrophic lateral sclerosis disease using gait data. Our analysis indicated that a number of machine learning algorithms such as the linear discriminant classifier and quadratic discriminant classifier can discriminate between normal and abnormal cases of amyotrophic lateral sclerosis disease.

Keywords: Machine learning · Neuro-degenerative disease
Amyotrophic Lateral Sclerosis

1 Introduction

Neuro-degenerative disease is described by the EU Joint Programme – Neurodegenerative Disease Research (JPND) as the progressive loss of structure or function and/or death of nerve cells. This loss raises some problems with movement (named ataxias), or mental operational (named dementias) [1]. According to National Institute of Environmental Health Sciences (NIH), millions of people are affected by neurodegenerative [2]. There are different types of neurodegenerative disease, the most popular are Alzheimer, Parkinson, Huntington, and Amyotrophic lateral sclerosis (ALS) diseases. In this study, Amyotrophic lateral sclerosis disease will be investigated. This type of disease is known as progressive brain disorder, meaning that the symptoms become worse over time. This disease has a significant adverse effect on people, including an increased risk of death and health defects. Some of the physical or mental symptoms related with this disease can be relieved by using some treatments, however, there is

D.-S. Huang et al. (Eds.): ICIC 2018, LNAI 10956, pp. 760–766, 2018.
https://doi.org/10.1007/978-3-319-95957-3_80

recently no medicine or way to reduce disease progression [2]. According to NIH symptoms in the early stages of the diseases can be similar [3].

ALS is a disorder that damages or death the motor neurons that control voluntary muscle movement [3]. Motor neurons are nerve cells that spread from the brain to the spinal cord and to muscles throughout the body. These motor neurons provide vital communication links between the brain and the voluntary muscles. This affects chewing, walking, breathing and talking. Early symptoms of ALS usually include muscle weakness affecting an arm, a leg, neck, twitching in the arm, leg, shoulder, muscle cramps, tight and stiff muscles, unclear and thick speech, problem with chewing or swallowing or stiffness. Symptoms usually appear between the ages of 55 and 75. Currently, there is no cure for ALS and no effective treatment to stop, or reduce the progression of the disease. Most people with ALS die from breathing failure, usually within 3 to 5 years from when the symptoms first appear [3]. However, about 10% of people with ALS survive for 10 or more years [3]. In 2016, in the USA, between 14,000–15,000 have ALS [3].

There are no cures for ALS, however some treatments that might help control symptoms, prevent unnecessary difficulties, and slightly make the life easier with people with ALS are available [3]. Barneoud and Curet asserted that analyzing patient data such as Gait signals could show advantageous in monitoring illness progression and in evaluating possible therapy and treatments [6].

A better understanding of Neuro-degenerative disease can help create preventative plans and thus positively mitigate, or reduce the side effects that Neuro-degenerative diseases have on individual, families, society and healthcare services. In the literature review, studies have used various biological signals for analyzing and understanding the diseases. Data mining tools and techniques have been implemented and evaluated to help understanding the hidden information in signals [7–9].

2 Gait Signal Analysis

Gait disorder is a major and common symptom of brain disease. Gait analysis can hardly be identified in the clinical setting. The analysis of gait is divided into three parts: kinematics, kinetics, and electromyography (EMG). The kinematics of the human gait studies the movements of the major joints of the lower extremity during human walk. Gait kinetics records forces and actions appearing while human move. EMG of the human gait analyzes muscle activity in human gait. Recently, development and innovation of sensors in health care has led to produce massive data values [21, 22]. Analysing these huge size of data is time consuming and in some situation, it is very difficult. In this respect, data mining tools can be used to overcome this difficult mission [7–9, 14].

Machine learning algorithms are widely applied for automatic detection of pathological gait. The most popular tool is Support Vector Machines (SVM). It has featured in several gait classification studies. Most of Studies reported that SVM achieved high rates in recognizing gait alteration [4, 16]. Zheng et al. [17] investigate the ability of using machine learning to distinguish between neurodegenerative diseases such as alzheimer based on gait analysis. They applied SVM and random forest classifiers.

Only 10 features extracted from the gait signals. Their results presented that SVM classifier with a selected four features out can obtain the best performance in the predicting of neurodegenerative diseases and healthy subjects while Bilgin et al. [10] applied Linear Discriminant Analysis and Naïve Bayesian Classifier.

3 Data Collections

In our study, gait records were collected at the Neurology Outpatient Clinic at Massachusetts General Hospital (MGH) between 1997 and 2006. These records are publicly available, via the (gaitndd) dataset, in Physionet [11]. In the gait database, there are 60 records with one record per subject. Each Subject can walk independently for 5 min. The signals are recorded using force-sensitive insoles embedded in the subject's shoes [5]. It assessed the force applied to the floor by the feet as the patient walks over them. The record has two signals one from the left and one from the right foot.

Each record has 300 samples per second, the first 20 s of each record is removed to reduce initial error [5, 15]. There are 12 subjects with ALS in which there are no specific measure of disease degree [15]. Hausdorff et al. [15] reported that ALS subjects with increased stride time showed an advanced degree of ALS.

4 Methodology

Data pre-processing phase involves a number of steps including data cleaning and filtering, standardization, transformation, feature extraction and features selection. This study has focused on the expediency of particular features in serving to separate (ALS and control) records. Since, the recording of row Gait signals is temporal, time domain features are extracted including Standard Deviation (STD), Min, Max, mean Root Mean Square (RMS), period of periodic activities, root of sum squared and Peak-magnitude-to-RMS ratio. These features can express significant information of the signals based on time domain.

However, Gait signals are transformed into other domain such as frequency domain. Features extracted from the frequency domain are peak frequency, mean frequency, Median and band power.

ALS patients do not walk at the same rate throughout the record as can be seen in Fig. 1. Hence, the distance between two segments in signals using dynamic time warping (DTW) has been extracted in these experiments. This function is able to measure similarity between two time-based sequences which may vary in time or speed. Figure 2 shows the differences between early and later segments in control and ALS groups.

Figure 3 illustrates the different between inter-stride times for 10 patients in ALS group.

The classifiers algorithms considered in this study are linear discriminant classifier (LDC), quadratic discriminant classifier (QDC), Naïve classifier and the support vector classifier (SVC). Holdout cross-validation approach is used to assess the algorithm behavior when evaluated on a new subject. It randomly separates data into a training set

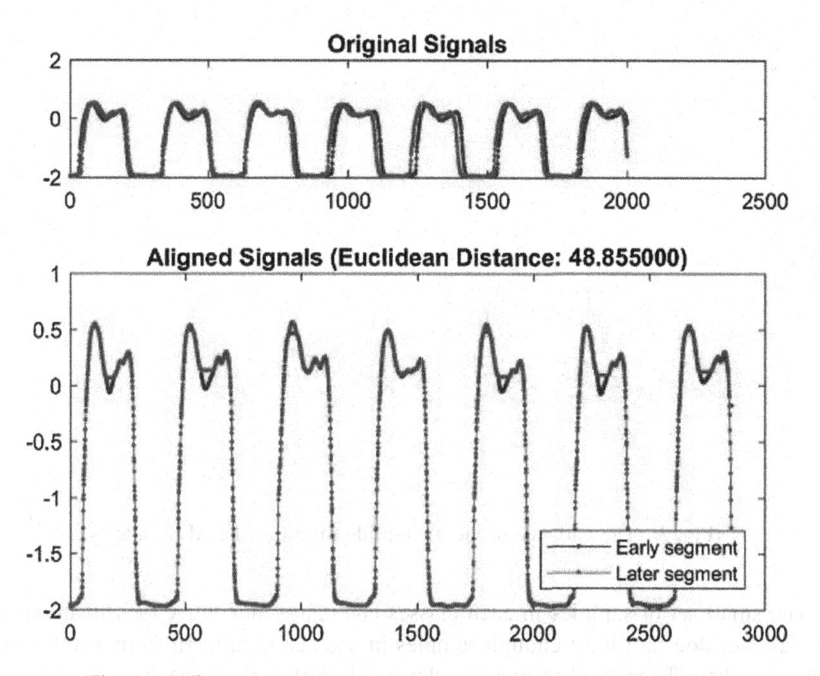

Fig. 1. Different between early and later segment in control signals

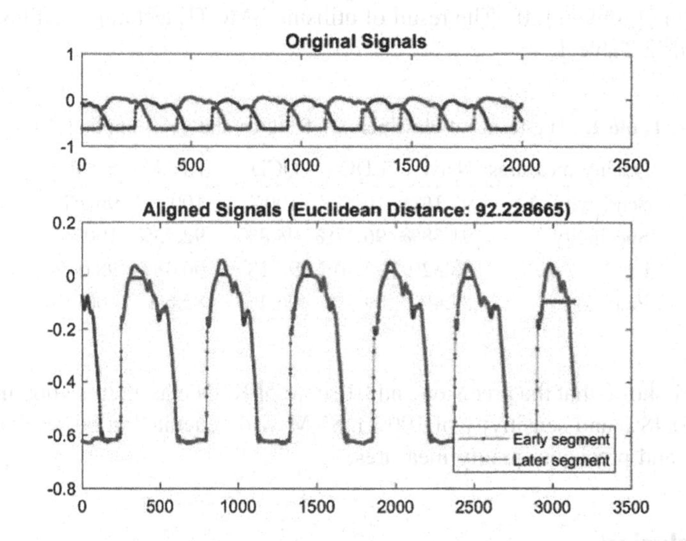

Fig. 2. Different between early and later segment in ALS signals

and a test set with stratification, that is, both training and test sets have the same class proportions. 60% of the whole dataset is selected for training and the remaining 40% for testing. To evaluate the performance of classifiers, we used accuracy, precision, F1 score, sensitivity, specify as performance metrics [9, 20].

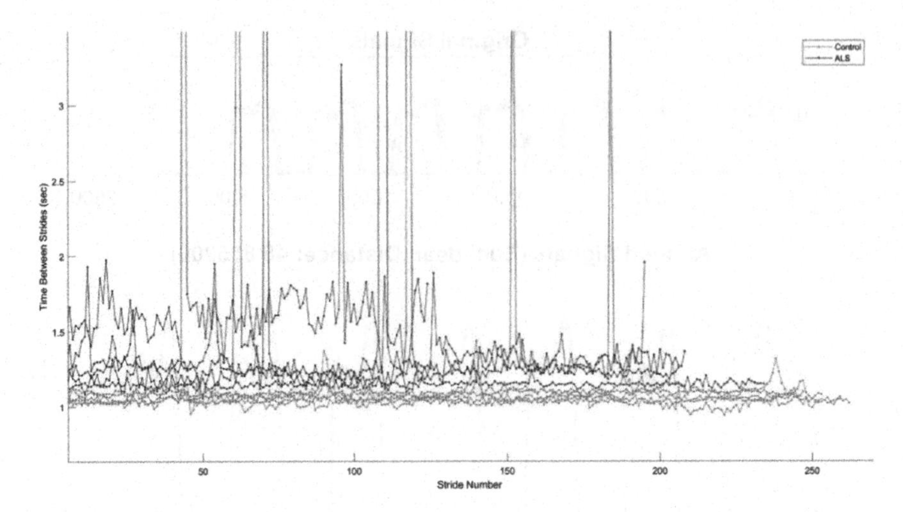

Fig. 3. The variance of the inter-stride times for the ALS patients

The small set of samples in each classes can cause ineffective classification, since the classifier does not have enough samples in each class to learn from. Oversampling techniques have been used to increase the number of each classes to 70 samples.

Numbers of studies have exposed that the SMOTE method successfully used for oversampling problem [20]. The result of utilising SMOTE technique is illustrated and represented in Table 1.

Table 1. The result of classifiers 27 features and oversampling data

Quality measures	Naive	LDC	QCD	KNN	SVM
Sensitivity	93.1%	100%	100%	100%	96.2%
Specificity	91.38%	96.47%	98.8%	92.73%	100%
F1	88.52%	97.30%	98.18%	90.48%	98.04%
Precision	84.4%	89.7%	96.4%%	82.6%	100%

Table 1 shows that the precision and F1 score of KNN classifier is good in ALS (F1 score of 90.48% and sensitivity of 100%). SVM also generated good results using the specificity and precision quality measures.

5 Conclusion

This paper evaluated a number of data mining techniques for the classification of normal and abnormality gaits in ALS disease. The methodology framework was defined through three major steps starting from preprocessing signals, then extracting features, lastly classification.

Due to small number of samples in each groups, SMOTE technique has been used to increase the numbers of samples in case and control groups. It showed good sensitivity and specificity rate. The best classification algorithms are LDC, QCD and KNN classifiers, which achieved sensitivity in ALS of 100%.

Acknowledgment. This project was funded by Prince Sattam University with grant number 2017/01/7814.

References

1. The EU Joint Programme – Neurodegenerative Disease Research (JPND): "WHAT IS NEURODEGENERATIVE DISEASE?" (2011)
2. National Institute of Environmental Health Sciences: "Neurodegenerative Diseases," National Institute of Environmental Health Services. https://www.niehs.nih.gov/research/supported/health/neurodegenerative/index.cfm. Accessed 05 Dec 2017
3. Office of Communications and Public Liaison: "Amyotrophic Lateral Sclerosis (ALS) Fact Sheet | National Institute of Neurological Disorders and Stroke." National Institute of Neurological Disorders and Stroke. https://www.ninds.nih.gov/Disorders/Patient-Caregiver-Education/Fact-Sheets/Amyotrophic-Lateral-Sclerosis-ALS-Fact-Sheet. Accessed 02 Jan 2018
4. Mannini, A., Trojaniello, D., Cereatti, A., Sabatini, A.M.: A machine learning framework for gait classification using inertial sensors: application to elderly, post-stroke and Huntington's disease patients. Sensors **16**(1), 134 (2016)
5. Hausdorff, J.M., Cudkowicz, M.E., Firtion, R., Wei, J.Y., Goldberger, A.L.: Gait variability and basal ganglia disorders: stride-to-stride variations of gait cycle timing in Parkinson's disease and Huntington's disease. Mov. Disord. **13**(3), 428–437 (1998)
6. Barnéoud, P., Curet, O.: Beneficial effects of lysine acetylsalicylate, a soluble salt of aspirin, on motor performance in a transgenic model of amyotrophic lateral sclerosis. Exp. Neurol. **155**(2), 243–251 (1999)
7. Alaskar, H.M.: Dynamic self-organised neural network inspired by the immune algorithm for financial time series prediction and medical data classification, Ph. D thesis, Liverpool John Moores University (2014)
8. Alaskar, H., et al.: Feature analysis of uterine electrohystography signal using dynamic self-organised multilayer network inspired by the immune algorithm. In: Huang, D.-S., Bevilacqua, V., Premaratne, P. (eds.) ICIC 2014. LNCS, vol. 8588, pp. 206–212. Springer, Cham (2014). https://doi.org/10.1007/978-3-319-09333-8_22
9. Alasker, H., Alharkan, S., Alharkan, W., Zaki, A., Riza, L.S.: Detection of kidney disease using various intelligent classifiers. In: 2017 3rd International Conference on Science in Information Technology (ICSITech), pp. 681–684 (2017)
10. Bilgin, S.: The impact of feature extraction for the classification of amyotrophic lateral sclerosis among neurodegenerative diseases and healthy subjects. Biomed. Signal Process. Control **31**, 288–294 (2017)
11. "PhysioNet:" https://physionet.org/. Accessed 01 Dec 2017
12. Cho, C.-W., Chao, W.-H., Lin, S.-H., Chen, Y.-Y.: A vision-based analysis system for gait recognition in patients with Parkinson's disease. Expert Syst. Appl. **36**(3), 7033–7039 (2009)
13. Chen, P.-H., Wang, R.-L., Liou, D.-J., Shaw, J.-S.: Gait disorders in Parkinson's disease: assessment and management. Int. J. Gerontol. **7**(4), 189–193 (2013)

14. Khalaf, M., et al.: Training neural networks as experimental models: classifying biomedical datasets for sickle cell disease. In: Huang, D.-S., Bevilacqua, V., Premaratne, P. (eds.) ICIC 2016. LNCS, vol. 9771, pp. 784–795. Springer, Cham (2016). https://doi.org/10.1007/978-3-319-42291-6_78
15. Hausdorff, J.M., Lertratanakul, A., Cudkowicz, M.E., Peterson, A.L., Kaliton, D., Goldberger, A.L.: Dynamic markers of altered gait rhythm in amyotrophic lateral sclerosis. J. Appl. Physiol. **88**(6), 2045–2053 (2000)
16. Shetty, S., Rao, Y.S.: SVM based machine learning approach to identify Parkinson's disease using gait analysis. In: International Conference on Inventive Computation Technologies (ICICT), vol. 2, pp. 1–5 (2016)
17. Zheng, H., Yang, M., Wang, H., McClean, S.: Machine learning and statistical approaches to support the discrimination of neuro-degenerative diseases based on gait analysis. In: McClean, S., Millard, P., El-Darzi, E., Nugent, C. (eds.) Intelligent patient management, pp. 57–70. Springer, Heidelberg (2009). https://doi.org/10.1007/978-3-642-00179-6_4
18. Lakany, H.: Extracting a diagnostic gait signature. Pattern Recognit. **41**(5), 1627–1637 (2008)
19. Bonora, G., Carpinella, I., Cattaneo, D., Chiari, L., Ferrarin, M.: A new instrumented method for the evaluation of gait initiation and step climbing based on inertial sensors: a pilot application in Parkinson's disease. J. Neuroeng. Rehabil. **12**(1), 45 (2015)
20. Fergus, P., Cheung, P., Hussain, A., Al-Jumeily, D., Iram, S., Dobbins, C.: Prediction of preterm deliveries from EHG signals using machine learning. PLoS ONE **8**(10), e77154 (2013). https://doi.org/10.1371/journal.pone.0077154
21. Khan, M.S., Muyeba, M., Coenen, F., Reid, D., Tawfik, H.: Finding associations in composite data sets. Int. J. Data Wareh. Min. **7**(3), 1–29 (2011). ISSN:1548-3924
22. Huang, R., Tawfik, H., Nagar, A.K.: Licence plate character recognition based on support vector machines with clonal selection and fish swarm algorithms. In: 11th International Conference on Computer Modelling and Simulation, pp. 101–106 (2009)

Classification of Foetal Distress and Hypoxia Using Machine Learning Approaches

Rounaq Abbas[1], Abir Jaafar Hussain[1(✉)], Dhiya Al-Jumeily[1], Thar Baker[1], and Asad Khattak[2]

[1] Department of Computer Science, Liverpool John Moores University, Liverpool L33AF, UK
R.A.Abbas@2015.ljmu.ac.uk,
{a.hussain,D.Al-Jumeily,t.baker}@ljmu.ac.uk
[2] College of Technological Innovation, Zayed University, Abu Dhabi, UAE
Asad.Khattak@zu.ac.ae

Abstract. Foetal distress and hypoxia (oxygen deprivation) is considered as a serious condition and one of the main factors for caesarean section in the obstetrics and Gynecology department. It is the third most common cause of death in new-born babies. Many foetuses that experienced some sort of hypoxic effects can develop series risks including damage to the cells of the central nervous system that may lead to life-long disability (cerebral palsy) or even death. Continuous labour monitoring is essential to observe the foetal well being. Foetal surveillance by monitoring the foetal heart rate with a cardiotocography is widely used. Despite the indication of normal results, these results are not reassuring, and a small proportion of these foetuses are actually hypoxic. In this paper, machine-learning algorithms are utilized to classify foetuses which are experiencing oxygen deprivation using PH value (a measure of hydrogen ion concentration of blood used to specify the acidity or alkalinity) and Base Deficit of extra cellular fluid level (a measure of the total concentration of blood buffer base that indicates the metabolic acidosis or compensated respiratory alkalosis) as indicators of respiratory and metabolic acidosis, respectively, using open source partum clinical data obtained from Physionet. Six well know machine learning classifier models are utilised in our experiments for the evaluation; each model was presented with a set of selected features derived from the clinical data. Classifier's evaluation is performed using the receiver operating characteristic curve analysis, area under the curve plots, as well as the confusion matrix. Our simulation results indicate that machine-learning algorithms provide viable methods that could delivery improvements over conventional analysis.

Keywords: Machine learning · Hypoxia · Foetal distress

1 Introduction

Many foetuses experience hypoxia during different stages of the pregnancy period [1]. Foetal hypoxia can be classified as acute hypoxia or chronic hypoxia according to the stage of the intrapartum foetal life [2]. The former usually occurs during the labour

D.-S. Huang et al. (Eds.): ICIC 2018, LNAI 10956, pp. 767–776, 2018.
https://doi.org/10.1007/978-3-319-95957-3_81

process and is considered more serious as its can cause significant complications, while the latter occurs during the first, second, or third trimester of the pregnancy [2]. Various methods of intrapartum foetal surveillance have been employed to detect the signs of hypoxia as early as possible and minimize the risk of life-long disability such as cerebral palsy and reduce the mortality rate among the new-borns. Intrapartum monitoring of foetuses during labour has been commonly performed by monitoring the foetal heart. Intermittent auscultation (IA) is considered the most common method of foetal surveillance in labour [3]. Many of these methods have been the subject of controversy as existing studies have found, no benefit in their use for reducing rates of cerebral palsy or peri-natal mortality [3]. Continuous labour monitoring is essential to observe the foetal well-being. There are many studies indicating foetal heart activity is the prominent source of information about foetal health and especially the detection of foetal hypoxia. However, physicians have found many challenges in identifying sufficient means of detecting foetal hypoxia through analyzing foetal heart rate [3]. From 1960, cardiotocography (continuous electronic foetal heart rate monitoring) has progressed and replaced all other traditional methods for foetal heart monitoring. Normal Cardiotocography results in nearly half of all tracings indicate that enough oxygen is delivered to the foetus [5, 6]. However; these results are not encouraging as a number of these foetuses are actually hypoxic [7]. In such cases a diagnostic test is compulsory. During the last decades, new methods have been used by antenatal care and during the labour process such as cordocentetesis (foetal blood sampling by ultrasound guided needle aspiration from the umbilical cord) [8], which can be used to detect hypoxia due to placental development abnormalities. This could cause many threats to the foetal health such as damage to the cells of the central nervous system that lead to life-long disability. Up to three quarters of infants with severe hypoxic-ischemic encephalopathy (HIE) die of multiple organ failure or lung infections caused by deregulated breathing [5]. Those who survive are commonly left with gross symptoms such as mental retardation, epilepsy, and cerebral palsy [9, 10]. James et al. [11] conducted the first research in 1958, recognising that umbilical cord blood gas analysis can give an indication of the presence of foetal hypoxic stress. Since 1962 Saling technique for detecting hypoxia has been the ideal method-using sample of blood from the foetus's scalp during labour and analyze of PH value as an indicator [8]. Westgren et al. [12] depend on this technique in detecting perinatal outcome, by using foetal scalp blood sampling to detect the PH value and lactate level, which may indicate the hypoxia and identify which one will be the accurate indicator. They selected a PH value <7.20 as cut-off value to recommend intervention, while abnormal level of lactate in the foetal scalp blood was 3.08 mmol/L. The study findings suggest the measurement of fetal scalp lactate levels as it is more often successful than PH analysis, due to the complexity of PH analysis and the demand of a relatively large amount of blood (30–50 μl), in contrast to 5 μl of blood required to detect the lactate level. Another reason for inferior results when using PH as an indicator was the sampling failure rates of 11% that have been reported.

In our research, we concentrate on PH value as indicator of respiratory acidosis, in addition to the Base Deficit Of the Extracellular fluid (BDecf) as an indicator of metabolic acidosis or compensated respiratory alkalosis [12], both derived from umbilical blood sample to predict foetal hypoxia.

Various machine learning methods have been utilised in our experimental study. Our main focus is to find machine learning techniques that have fast performance with high accuracy. We investigate classifiers with the ability to balance the error using unbalanced data set such as our data and models that can identify the most significant variable for the classification, to confirm our class division criteria. We considered several well-known algorithms that encompass a range of model architectures in order to provide suitable comparisons. Both linear and non-linear classifiers have been included, in conjunction with Principle Component Analysis (PCA) results. Our models include decision trees, represented by a Gradient Boosting model (GBM) [28] and Random Forest (RF) [31]. In addition, we consider other promising classification techniques such as Support Vector Machines (SVM), with radial basis function, kernel support [22, 32], Neural Network [18, 19], and Generalized Linear Model a Lasso and Elastic-Net Regularized (GLMNET) model [29].

Furthermore, some weaker classifiers, such as k-Nearest Neighbors (kNN) [30], have also been examined during our experimental procedure.

The reminder of this paper is organized as follows. Section 2 will discuss related research, while Sect. 3 will show the methodology. Simulation results and conclusion are shown in Sects. 4 and 5, respectively.

2 Related Research

Existing research work illustrates a range of techniques and challenges that are presented during the analysis of recorded foetal clinical data. A main focus of our research is the determination of PH and BDecf threshold values that cause adverse neurological outcomes in foetuses.

Several solutions have been proposed to diagnose foetal hypoxia depending on PH value. Malin et al. [13] report a strong association between Low arterial cord PH and clinical neonatal outcome. Following other studies where it is reported to have different predictive outcome. The mean umbilical cord artery blood PH should be 7.20 as indicated by Sykes et al. study [37]. In a similar study by Steer et al. [38], the mean PH was 7.26 and in a similar Dutch study by Berg et al. [39], the mean PH was 7.30. The American College of Obstetricians and Gynaecologists recommends that the arterial cord blood's PH should be less than 7.0 as an indicator of birth asphyxia [14, 15]. However, multiple studies confirm that birth asphyxia can cause brain damage even when umbilical cord arterial PH is greater than 7.0, as illustrated by Yeh et al. [16]. Strachan et al. [17] use PH $<=7.15$ and BDecf $= 8$ as a cut-off point for their analysis to identify which individual component of the computerised analysis of the foetal heart rate will predict acidaemia. They found that only three parameters (bradycardia, deceleration and late deceleration) could predict low PH <7.15 and BDecf > 8, indicating hypoxia as an outcome. Georgieva et al. [18] and Jeżewski et al. [19] showed that using Artificial Neural Networks (ANN) with clinical data, foetal heart rate data, PH values < 7.1 and 7.2 respectively, could be effective division criteria for classes. The performance quality represented by sensitivity and specificity showed acceptable results. Keith et al. [20] concentrated on the use of ANN to provide support to physicians and medical professionals. The objective of the study is to compare the use of intelligent computer system in labour management with the performance of experts dealing with foetal monitoring data, patient information, and foetal blood sampling. The result was indistinguishable when considering the system's performance compared with the expert decision using the 50 cases examined. In addition, Magenes et al. [21] examined three neural classifiers to discriminate normal and pathological cases, depending on the foetal heart rate. The three classifiers showed promising performance towards the prediction of foetal outcomes on the set of collected Foetal Heart Rate (FHR) signals [21]. Table 1 shows various studies using ANN and support vector machine to classify foetal behavioural states using clinical blood sampling (PH, BDecf) as division criteria for classes.

Table 1. Overview of works that presented classification results

Reference	Class	Criteria for classes	Classifier	SE %	SP %	Others
Georgieva et al. [18]	2	PH <7.1	ANN	61	68	AUC 0.64
Jeżewski et al. [19]	2	PH <7.20 or birth weight <10th perc.	ANN	67	68	ACC 67%
Warrick et al. [25]	2	BDecf <8; BDecf $\geq$ 12	SVM	70	75	
Spilka et al. [26]	2	PH <7.15	SVM	73	76	AUC 0.78
Georgoulas et al.) [23]	2	PH > 7.20; pH <7.10	SVM	70	85	AUC 0.75

3 Methodology

This research proposes the use of foetal clinical data analysis and applies machine-learning techniques for the diagnosis and detection of intra-partum foetal hypoxia. Following our analysis of related research, we selected the cutoff threshold for PH as <=7.15, and BDecf > 8, to represent all abnormal cases. Table 2 shows six well-known classifiers, which have been used to classify the data including both linear and non-linear methods. These models are the Gradient Boosting model (GBM); Generalized Linear Model Lasso, and Elastic-Net Regularized (GLMNET); K-Nearest Neighbors (KNN); Support Vector Machines (SVM) with radial basis function, kernel support; Random Forest (RF) [31, 32], and Neural Network [18, 19].

A. *Data collection*

An open source database is used for this study, provided by Physionet [27]. The data were originally collected between 27th of April 2010 and 6th of August 2012 at the obstetrics ward of the university hospital in Brno (UHB), Czech Republic. The data consisted of signal and clinical data stored in the hospital information system (AMIS) database.

Table 2. Methods of classification

Methods	Classifier	Category
GBM [28]	Stochastic Gradient Boosting	Non linear
GLMNET [29]	Lasso and elastic-net regularized generalized linear models	Linear
KNN [30]	K-Nearest Neighbor	Non linear
SVM [22]	Radial basis function Kernal support	Non linear
RF [31, 32]	Random Forest	Non linear
NN [18, 19]	Backpropagation Neural Network	Non linear

In this research, clinical data is used for the classification of hypoxia. The clinical data includes: delivery descriptions, neonatal outcome, neonatal descriptions and information about mother and possible risk factors. Table 3 includes all the parameters that are extracted from the clinical data and used for the classification. In this case, classification of the fetuses' condition, defined as either normal or pathological, depending on both PH value <=7.15 and BDecf $\geq$ 8 [12, 17]. The following parameters are taken into account for the selection of the samples.

Table 3. Patient and labour outcome statistics for the CT-UHB database

	Mean (Median)	Min	Max	Comment
Parity	0.43 (0)	0	7	
Gravidity	1.43 (1)	1	11	
Gestational age (weeks)	40	37	43	Over 42 weeks: 2
PH	7.23	6.85	7.47	
BE	−6.36	−26.8	−0.2	
BDecf (moll/l)	4.60	−3.40	26.11	
Apgar 1 min	8.26 (8)	1	10	AS1 <3: 18
Apgar 5 min	9.06 (10)	4	10	AS5 <7: 50

Abbreviations: AS1, AS5 – Apgar score at 1*st* and 5*th* minute respectively

Type of gravidity – only singleton, uncomplicated pregnancies was included.

Gestational age (weeks)- >36 weeks

Umbilical artery PH (PH) – is the most commonly used outcome measure, sign of respiratory hypoxia. Records with missing PH were excluded.

Base excess (BE) is often used in the clinical setting as a sign for metabolic hypoxia, but it often yields false positives.

Base deficit in extracellular fluid (BDecf) is a better measure of metabolic hypoxia than BE, hence in this research, records with missing BDecf values were excluded.

B. *Quality Measure Evaluations*

To evaluate the performance of the machine learning techniques for the classification of clinical hypoxia data, the receiver-operating characteristic (ROC) curve [33] and the confusion matrix are utilized to characterize the capability of the classifiers.

The ROC plot is a graphical method used to describe classifier performance by mapping the sensitivity and one minus the specificity for each value of the classification cut-off threshold. The confusion matrix metrics illustrate the true positive rate (sensitivity), true negative rate (specificity), and the accuracy of each classifier, to identify the individual and relative performance of models in the medical data analysis. Other scalar values resulting from statistical methods are considered to provide a summary of classifier performance, such as the area under the ROC curve (AUC) and Kappa values, which measure the agreement between predictions and the actual classification in the data set. Higher AUC and Kappa values indicate better classifier performance [34].

4 Results

In this section, we present classification outcomes using the proposed six machine-learning techniques as shown in Table 2. Each model was trained with a set of features derived from the foetal clinical dataset, using 10 fold-cross validations. It is significant to note that the data used are strongly imbalanced. The abnormal (pathological) class is heavily under-sampled in comparison to the normal one. Graphical mapping using PCA (linear dimensionality reduction) and tSNE (non-linear dimensionality reduction) [35] indicate that the selected features of PH values and BDecf are able to separate the data.

As discussed earlier, we used PH and base deficit values as feature indicators to detect the classes (normal, pathological) in addition to other parameters (BE, PCO2, Apgar score), reserving 70% of the data for model training and the remaining 30% as hold out for testing. The results from our experimental procedure are presented and organized in Table 4.

Table 4. Classification Performances

Model	SENS	SPEC	KAPPA	ACC	AUC
NN	0.812	0.964	0.798	0.909	0.91
SVM	0.687	0.928	0.641	0.84	0.80
RF	0.875	1	0.899	0.954	0.94
KNN	0.937	0.678	0.55	0.772	0.88
GBM	0.875	0.892	0.757	0.886	0.94
GLMNET	0.812	0.785	0.575	0.795	0.83

The experimental results from the study identify the Random forest model as the most accurate classifier considering the area under the curve quality measure (0.95). Subsequently, the second best ranking classifier was the NN model, with a predictive ability measuring (0.91) for both ACC and AUC. The simulation results indicate that GBM model showed a high AUC (0.94), similar to RF, but conversely exhibit less accuracy (0.88). Both classifiers showed an ability to detect (0.87) the true positive cases (pathological), however, the GBM has failed to recognise 11% of the true negative cases (normal), as its specificity measured 0.89. In contrast, the RF succeeded in detecting the true negative cases with specificity (1). On the other hand, classifier performances for SVM and GLMNET models show reasonable prediction ability, with AUC values of 0.80 and 0.83, respectively. Though both models yield less overall accuracy (0.84, 0.79 for SVM and GLMNET, respectively) compared to other classifiers. Although KNN is the best classifier in detecting the true positive cases, with sensitivity value of 0.93, it shows a lower overall accuracy value of 0.77 and is the weakest classifier for detection the true negative cases (specificity 0.67). The KAPPA metric was used to compare the overall accuracy to the expected random chance accuracy [37]. Kappa measures the agreement between the prediction and the actual labels of each classifier. Simulation results indicate that the RF has the best agreement as it has the higher kappa metric (0.90) compared to other models. NN with (0.80) ranked second highest in this respect. The Kappa result for GBM confirmed that although the AUC of both RF and GBM are almost the same, the RF is a superior classifier. The classifier that is showing the least agreement between the prediction and the ground truth was the KNN model, exhibiting a Kappa of (0.55). Figure 1 shows ROC plots for each classifier. As mentioned previously, we consider the use of both PH and BDecf as predictor variables, prompting a further study of the influence of these variables on the performance of our classifiers. All the models identified PH values as the most important variable for our data classification, with the exception of the SVM and KNN, while the BDecf was the most important variable detected by both SVM and KNN. BDecf was the second most important variable detected by ANN and the third most important for both RF and GBM.

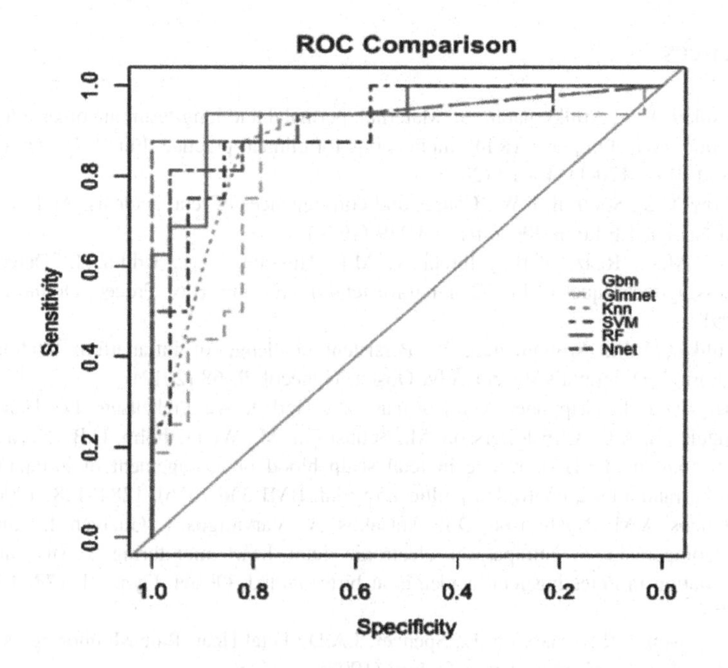

Fig. 1. ROC plot comparison

5 Conclusion

In this paper, we have presented various machine learning approaches for the classification of foetal hypoxia. We reviewed related works and problem parameters including the main characteristics of foetal distress and the essential nature of early prediction. In our study, we have investigated the application of machine learning for foetal hypoxia classification. Experimental data was obtained from open source Physionet dataset. We applied a series of test models to extracted data, whose performances were evaluated using both graphical forms of analysis, including the ROC plot, and scalar summary indices including sensitivity, specificity, Kappa, and overall accuracy. A classification scheme is proposed in which the foetal experience of hypoxia is confirmed by a blood gas and acid-base assessment with evidence of a significant respiratory (PH value) and metabolic acidosis (BDecf value). Although some studies show that resampling of imbalanced data may lower the model performance, in our study results remained promising. Statistical and graphical comparison between various classification methods has shown that RF and NN are the strongest classifiers within our trials of foetal hypoxia detection, and the most models used for such classification compared to other classifiers.

Acknowledgement. The authors would like to thanks Liverpool John Moores University for the scholarship to complete this research. In addition, this research work was partially supported by Zayed University Research Cluster Award # R18038.

References

1. Talaulikar, V.S., Arulkumaran, S.: Maternal, perinatal and long-term outcomes after assisted reproductive techniques (ART): implications for clinical practice. Eur. J. Obstet. Gynaecol. Reprod. Biol. **170**(1), 13–19 (2013)
2. Bobrow, C.S., Soothill, P.W.: Causes and consequences of fetal acidosis. Arch. Dis. Child. Fetal Neonatal Edition **80**(3), F246–F249 (1999)
3. Hasan, M.A., Reaz, M.B.I., Ibrahimy, M.I., Hussain, M.S., Uddin, J.: Detection and processing techniques of FECG signal for fetal monitoring. Biol. Proced. Online **11**(1), 263 (2009)
4. Talaulikar, V.S., Arulkumaran, S.: Persistent challenge of intrapartum fetal heart rate monitoring. Dasgupta's Recent Adv. Obstet. Gynecol. **9**, 68 (2012)
5. Wiberg-Itzel, E., Lipponer, C., Norman, M., Herbst, A., Prebensen, D., Hansson, A., Bryngelsson, A.L., Christoffersson, M., Sennström, M., Wennerholm, U.B., Nordström, L.: Determination of pH or lactate in fetal scalp blood in management of intrapartum fetal distress: randomised controlled multicentre trial. BMJ **336**(7656), 1284–1287 (2008)
6. Vintzileos, A.M., Nochimson, D.J., Antsaklis, A., Varvarigos, I., Guzman, I., Knuppel, R. A.: Comparison of intrapartum electronic fetal heart monitoring versus intermittent auscultation in detecting fetal acidemia at birth. Am. J. Obstet. Gynecol. **173**, 1021–1024 (1995)
7. Ingemarsson, I., Ingemarsson, E., Spencer, J.A.D.: Fetal Heart Rate Monitoring. A Practical Guide. Oxford University Press, Oxford (1993)

8. Bretscher, J., Saling, E.: pH values in the human fetus during labor. Am. J. Obstet. Gynecol. **97**, 906–911 (1967)
9. Tuffnell, D., Haw, W.L., Wilkinson, K.: How long does a fetal scalp blood sample take? BJOG **113**, 332–334 (2006)
10. Goldaber, K.G., Gilstrap, L.C., Leveno, K.J., Dags, J.S., McIntire, D.D.: Pathologic fetal acidemia. Obstet. Gynecol. **78**, 1103–1107 (1991)
11. James, L.S., Weisbrot, I.M., Prince, C.E., Holaday, D.A., Apgar, V.: The acid-base status of human infants in relation to birth asphyxia and the onset of respiration. J. Paediatr. **52**(4), 379–394 (1958)
12. Westgren, M., Kuger, K., Ek, S., Grunevald, C., Kublickas, M., Naka, K., et al.: Lactate compared with pH analysis at fetal scalp blood sampling: a prospective randomised study. Br. J. Obstet. Gynaecol. **105**, 29–33 (1998)
13. Malin, G.L., Morris, R.K., Khan, K.S.: Strength of association between umbilical cord pH and perinatal and long-term outcomes: systematic review and meta-analysis. BMJ **340**, c1471 (2010)
14. ACOG Committee on Obstetric Practice: ACOG Committee Opinion No. 348, November 2006: Umbilical cord blood gas and acid-base analysis. Obstetrics and gynaecology, 108(5), p. 1319 (2006)
15. https://www.abclawcenters.com/practice-areas/diagnostic-tests/hypoxic-ischemic-encephalopathy-and-umbilical-cord-blood-gases/
16. Yeh, P., Emary, K., Impey, L.: The relationship between umbilical cord arterial pH and serious adverse neonatal outcome: analysis of 51 519 consecutive validated samples. BJOG: Int. J. Obstet. Gynaecol. **119**(7), 824–831 (2012)
17. Strachan, B.K., Sahota, D.S., Wijngaarden, W.J., James, D.K., Chang, A.M.: Computerised analysis of the fetal heart rate and relation to acidaemia at delivery. BJOG Int. J. Obstet. Gynaecol. **108**(8), 848–852 (2001)
18. Georgieva, A., Payne, S.J., Moulden, M., Redman, C.W.: Artificial neural networks applied to fetal monitoring in labour. Neural Comput. Appl. **22**(1), 85–93 (2013)
19. Jeżewski, M., Czabański, R., Wróbel, J., Horoba, K.: Analysis of extracted cardiotocographic signal features to improve automated prediction of fetal outcome. Biocybern. Biomed. Eng. **30**(4), 29–47 (2010)
20. Keith, R.D., Beckley, S., Garibaldi, J.M., Westgate, J.A., Ifeachor, E.C., Greene, K.R.: A multicentre comparative study of 17 experts and an intelligent computer system for managing labour using the cardiotocogram. BJOG Int. J. Obstet. Gynaecol. **102**(9), 688–700 (1995)
21. Magenes, G., Signorini, M.G., Arduini, D.: Classification of cardiotocographic records by neural networks. In: Proceedings of the IEEE-INNS-ENNS International Joint Conference on Neural Networks, IJCNN 2000, vol. 3, pp. 637–641. IEEE (2000)
22. Abdillah, A.A.: Suwarno: diagnosis of diabetes using support vector machines with radial basis function kernels. Int. J. Technol. **7**(5), 849–858 (2016)
23. Georgoulas, G., Stylios, C., Groumpos, P.: Feature extraction and classification of foetal heart rate using wavelet analysis and support vector machines. Int. J. Artif. Intell. Tools **15** (03), 411–432 (2006)
24. Georgoulas, G., Gavrilis, D., Tsoulos, I.G., Stylios, C., Bernardes, J., Groumpos, P.P.: Novel approach for fetal heart rate classification introducing grammatical evolution. Biomed. Signal Process. Control **2**(2), 69–79 (2007)
25. Warrick, P.A., Hamilton, E.F., Kearney, R.E., Precup, D.: Classification of normal and hypoxic fetuses using system identification from intrapartum cardiotocography. IEEE Trans. Biomed. Eng. **57**, 771–779 (2010)

26. Spilka, J., Chudáček, V., Koucký, M., Lhotská, L., Huptych, M., Janků, P., Georgoulas, G., Stylios, C.: Using nonlinear features for fetal heart rate classification. Biomed. Signal Process. Control **7**(4), 350–357 (2012)
27. https://physionet.org/physiobank/database/ctu-uhb-ctgdb/
28. Friedman, J.H.: Greedy function approximation: a gradient boosting machine. Annals of statistics, pp. 1189–1232 (2001)
29. Hastie, T., Qian, J.: Glmnet Vignette (2014)
30. Altman, N.S.: An introduction to kernel and nearest-neighbour nonparametric regression. Am. Stat. **46**(3), 175–185 (1992)
31. Liaw, A., Wiener, M.: Classification and regression by random Forest. R News **2**(3), 18–22 (2002)
32. Breiman, L.: Random Forests. Statistics Department, University of California, Machine learning (2001)
33. Bradley, A.P.: The use of the area under the ROC curve in the evaluation of machine learning algorithms. Pattern Recogn. **30**(7), 1145–1159 (1997)
34. Witten, I.H., Frank, E.: Data Mining: Practical Machine Learning Tools and Techniques. Morgan Kaufmann, Burlington (2005)
35. Maaten, L.V.D., Hinton, G.: Visualizing data using t-SNE. J. Mach. Learn. Res. **9**, 2579–2605 (2008)
36. Noreen, E.W.: Computer intensive methods for hypothesis testing: An introduction (1989)
37. Sykes, G.S., Molloy, P.M., Johnson, P., Stirrat, G.M., Turnbull, A.C.: Fetal distress and the condition of newborn infants. Br. Med. J. (Clin. Res. Ed.) **287**(6397), 943–945 (1983)
38. Steer, P.J.: Fetal monitoring—Present and future. Eur. J. Obst. Gynecol. Reprod. Biol. **24**(2), 112–117 (1987)
39. Berg, P., Schmidt, S., Gesche, J., Saling, E.: Fetal distress and the condition of the newborn using cardiotocography and fetal blood analysis during labour. BJOG Int. J. Obst. Gynecol. **94**(1), 72–75 (1987)

Assessment and Rating of Movement Impairment in Parkinson's Disease Using a Low-Cost Vision-Based System

Domenico Buongiorno[1], Gianpaolo Francesco Trotta[1], Ilaria Bortone[2], Nicola Di Gioia[1], Felice Avitto[1], Giacomo Losavio[3], and Vitoantonio Bevilacqua[1(✉)]

[1] Department of Electrical and Information Engineering, Polytechnic University of Bari, Bari, Italy
vitoantonio.bevilacqua@poliba.it
[2] Institute of Clinical Physiology (IFC), National Research Council (CNR), Pisa, Italy
[3] Medica Sud S.R.L., Viale della Resistenza n.82, Bari, Italy

Abstract. Assessment and rating of Parkinson's Disease (PD) are commonly based on the medical observation of several clinical manifestations, including the analysis of motor activities. In particular, medical specialists refer to the Movement Disorder Society – sponsored revision of Unified Parkinson's Disease Rating Scale (MDS-UPDRS), the most widely used scale for rating PD. The UPDRS scale also considers the observation of some subtle motor phenomena that are either difficult to capture with human eyes or subjectively considerate abnormal. In this scenario, an automatic system able to capture the considered motor exercises and rate the PD severity could be used as a support system for the healthcare sector. In this work, we implemented a simple and low-cost clinical tool that can extract motor features of two main exercises required by the UPDRS scale (the finger tapping and the foot tapping) to classify and rate the PD severity. Sixty two participants were enrolled for the purpose of the present study: thirty three PD patients and twenty nine healthy paired subjects. Results showed that an SVM using the features extracted by both considered exercises was able to classify healthy subjects and PD patients with great performances by reaching 87.1% of accuracy. The results of the classification between mild and moderate PD patients indicated that the foot tapping features were the most representative ones to discriminate (81.0% of accuracy). We can conclude that developed tool can support medical specialists in the assessment and rating of PD patients in a real clinical scenario.

Keywords: Parkinson's Disease · Microsoft Kinect · Finger tapping
Foot tapping · UDPRS · Vision system

D.-S. Huang et al. (Eds.): ICIC 2018, LNAI 10956, pp. 777–788, 2018.
https://doi.org/10.1007/978-3-319-95957-3_82

1 Introduction

Parkinson's Disease (PD) is one of the most spread neurodegenerative disorders. The main clinical PD features related to body movement include tremor, rigidity, bradykinesia, slowness of movement, and, in some patients, tremor. The symptoms may be alleviated by treatments aiming at stimulating dopamine receptors in the brain. The dopamine precursor levodopa is the most effective drug. While in the early stage of PD advanced therapy such as levodopa-carbidopa intestinal gel infusion (LCIG) continuous subcutaneous apomorphine infusion or deep brain stimulation can be highly successful in treating motor fluctuations, in the advanced stage of PD, higher doses could reduce motor fluctuations but contribute to the development of involuntary movements. Therefore, optimization of therapy in terms of dosage and dosing frequency is a challenge when the patient has reached the state of motor fluctuations [1]. To adjust dosage and medication, physician's decisions are based on historical information from the patient, regarding motor function during activities of daily living and clinic observation, using clinical rating scale such as Unified PD Rating Scale (UPDRS) [2] and Hoehn and Yahr staging scale [3].

In particular, the UPDRS was developed as a standardized test that has proved very useful as a global measure of function. The UPDRS is straightforward to assess in the clinic, although the assessment is subjective and requires considerable experience and monitoring to minimize interrater variability. Objective assessments of motor symptoms of PD patients from clinical observation or from their home environments have been previously tested by means of several technologies. The main problem is the evaluation of the severity of specific symptoms such as freezing of gait [4–6], dysarthria [7], tremor [8–11], bradykinesia [12–14] and dyskinesia [15–20].

According to the current trends in machine learning for medicine [21–28], the purpose of the present study is to provide healthcare professionals with a mainstream tool with a two-fold objective: detect and recognize typical PD motor issues and facilitate assessment of motor alterations both at the clinic and at home. As a result, the motor outcomes of the patients can be detected and categorized, so that the patient can be monitored and supported in his therapeutic plan since the progression of PD can be assessed non-invasively.

The main novel contributions respect to the state of the art presented above are: (1) we designed, developed and evaluated with both healthy subjects and PD patients a low-cost vision-based system to capture specific movements of two of the main UPDRS scale exercises (Finger Tapping - Subsection III.4 and Foot Tapping - Subsection III.7); (2) we have developed and compared several classifiers able to assess and rate the movement impairment of PD patients using a specific set of features extracted by the recorded movements (Finger tapping and Foot tapping).

In the sections that follow, we present the experimental setups, the feature extraction process and the classification steps. We present the data acquisition protocol and the experimental setup, which is based on vision system, first. We then report the feature extraction steps, that are based on image processing techniques. After we report the comparison between several classifiers based on the SVM approach. Finally, we discuss the main findings of the study indicating the usability in a real clinical scenario of the proposed system.

2 Materials and Methods

2.1 Participants

Thirty three PD patients (mean age 71.6 years, SD 9.0, age range 54–87) and twenty nine healthy subjects (mean age 71.1 years, SD 9.2, age range 57–90) participated in the experiments after giving a written informed consent. The 33 PD patients were examined by a medical doctor and rated according to MDS-UPDRS Part IV for motor complications that considers a scoring with five levels (Normal, Slight, Mild, Moderate and Severe). In detail, fourteen (mean age 67.2 years, SD 9.8, age range 54–81) and nineteen (mean age 74.1 years, SD 7.1, age range 63–87) patients were classified as *mild* and *moderate* PD patients, respectively. None of the patients was classified as either *slight* or *severe* PD patient.

2.2 Experimental Set-Up

As written above, in this study we focused on two assessment tasks of the Section III of the MDS-UPDRS scale: Finger Tapping (Subsection III.4) and Foot Tapping (Subsection III.7). We developed two separate vision-based systems able to acquire the movement of the thumb, the index finger and the toes. Both acquisition systems are based on passive markers made of reflective material and the Microsoft Kinect One RGBD camera (Fig. 1). A detailed description of two acquisition systems follows.

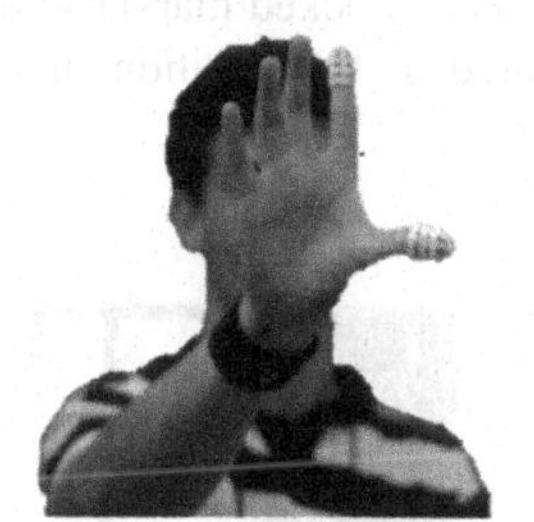
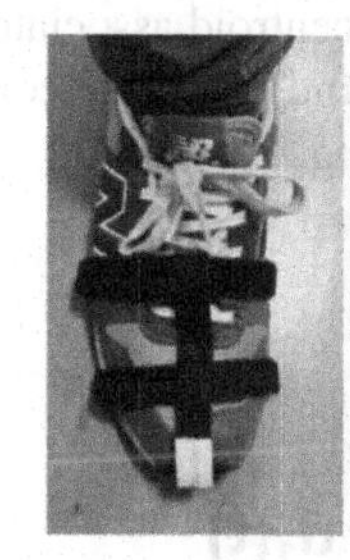
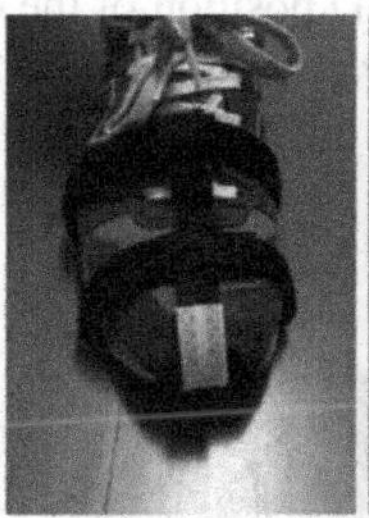

Fig. 1. Left image shows a healthy subject wearing the two passive finger markers. The images reported on the right show the foot of a subject doing the foot tapping exercise while he is wearing a passive marker on the toes.

Finger Tapping Exercise Setup. This test considers the examination of both hands separately. The tested subject is seated in front of the camera and is instructed to tap the index finger on the thumb ten times as quickly and as big as possible. During the task the subject wears two thimbles made of a reflective material on both the index finger and thumb.

Foot Tapping Exercise Setup. The feet are tested separately. The tested subject sits in a straight-backed chair in front of the camera and has both feet on the floor. He is then instructed to place the heel on the ground in a comfortable position and then tap the toes ten times as big and as fast as possible. A system of stripes with a reflective material are positioned on the toes.

2.3 Image Processing for Movement Estimation

The two vision-based acquisition systems use passive reflective markers to track the position of the thumb, the index finger and the toes. After the movement acquisition, an image processing phase is needed to recognize the marker in each acquired video frame and compute the 3D position of a centroid point associated to the specific marker. This post-processing phase has been conducted using the OpenCV library running the flowing steps on each image frame:

- Conversion to a grayscale image;
- Extraction of the pixels associated to the reflective passive markers with a thresholding operation;
- Blurring and thresholding operations in sequence;
- Eroding and dilating operations in sequence;
- Dilating and eroding operations in sequence.

After the post-processing phase, all the found blobs are extracted using an edge detection procedure. Only the blobs having sizes comparable with markers' size are kept for the next analysis. As final step, the centroid of each blob (only one blob for the foot tapping and two blobs for the finger tapping) is computed. Given the position of the centroid, its depth information and the intrinsic parameters of the used camera, we then computed the 3D position of the centroid associated to each tracked marker in the camera reference system. Such centroid has then considered as the position of the specific finger or of the foot's toes (Fig. 2).

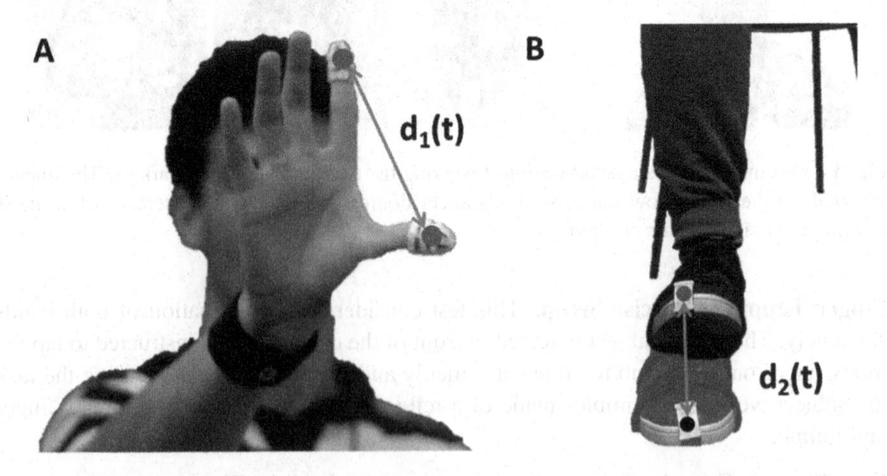

Fig. 2. (A) Finger Tapping. The signal $d_1(t)$ is the distance between the two centroids (red filled circles) of the passive finger markers. (B) Foot Tapping. The signal $d_2(t)$ is the distance between the centroid of the toes' marker and the centroid of the same marker when the toes are completely on the ground. (Color figure online)

2.4 Feature Extraction from Movements

The entire post-processing analysis described above produces the 3D positions of toes' marker (Foot Tapping) and of the two fingers' markers (Finger Tapping). Given the position of each marker we then extracted the following signals over time:

- $d_1(t)$ - the distance between the two fingers' markers over time (Finger Tapping);
- $d_2(t)$ - the distance between the position of the toes' marker over time and the position of the same marker when the toes are completely on the ground (Foot Tapping).

Both signals have been normalized to make them range in [0, 1]. Given the entire acquired signal, all the single trials (ten finger tappings and ten foot tappings) have been extracted for each side. We then extracted the same set of features for both computed signals, i.e. $d_1(t)$ and $d_2(t)$. The set of the extracted features contains features of the time domain, space domain and frequency domain. In particular, the features are:

1. meanTime: averaged execution time of the single exercise trial;
2. varTime: variance of the execution time of the single exercise trial;
3. meanAmplitude: averaged space amplitude of the single exercise trial;
4. varAmplitude: variance of the space amplitude of the single exercise trial;
5. tremors: number of peaks detected during the entire acquisition;
6. hesitations: number of amplitude peaks detected in the velocity signal during the entire acquisition;
7. periodicity: periodicity of the exercise computed as reported in [29];
8. AxF: (amplitude times frequency) the averaged value of the division between the amplitude peak reached in a single exercise trial and the time duration of the trial.

2.5 Classification

As reported above, in this study we investigated whether the set of feature extracted from the two exercises (finger tapping and foot tapping) can be used both to classify a subject either as healthy or PD affected, and to infer the rate of the disease. In particular, in this study we focused on the discrimination between mild and moderate PD patients.

Healthy Subjects vs PD Patients.
In the first part of the study we focused on the "*Healthy subjects vs PD patients*" classification. We trained three different binary support vector machine (SVM) classifiers based on following three sets of features:

- Set 1: all the finger tapping features;
- Set 2: all the foot tapping features;
- Set 3: both finger and foot tapping features;

The training process is based on 5-fold cross-validation and evaluate in sequence the following types of SVM classifiers: (1) linear SVM, (2) quadratic SVM, (3) cubic SVM, (4) gaussian SVM.

Mild PD Patients vs Moderate PD Patients.
In the second part of the study we focused on the "*Mild PD patients vs Moderate PD patients*" classification. As for the "*Healthy subjects vs PD patients*" classification, we trained three different binary support vector machine (SVM) classifiers based on following three sets of features:

- Set 1: all the finger tapping features;
- Set 2: all the foot tapping features;
- Set 3: both finger and foot tapping features;

Also in this case, the training process is based on 5-fold cross-validation and evaluate in sequence the following types of SVM classifiers: (1) linear SVM, (2) quadratic SVM, (3) cubic SVM, (4) gaussian SVM.

3 Results and Discussion

In this section we report the classification results obtained on test sets using the most accurate SVM classifier among the four types listed above discussing the main findings. In particular, we reported and analyzed the confusion matrices and the following indices:

$$Accuracy = \frac{TP + TN}{TP + TN + FP + FN}, Sensitivity = \frac{TP}{TP + FN}, \text{ and } Specificity = \frac{TN}{TN + FP},$$

where TP, TN, FP and FN stand for true positive, true negative, false positive and false negative, respectively.

Healthy Subjects vs PD Patients Classification.
In this subsection, we reported and discussed the results about the "*Healthy subjects vs PD patients*" classification. In particular we refers to the patients with *positive* and to the healthy subjects with *negative*.

Set 1: Finger tapping features.

The best classifier was the Gaussian SVM with an accuracy of 71.0%, a sensitivity of 75.7% and a specificity of 65.5% (Table 1).

Table 1. Confusion matrix. "Healthy subjects vs PD patients" classification with finger tapping features.

Predicted condition	True condition	
	PD patient	Healthy subj.
PD patient	25	10
Healthy subj.	8	19

Set 2: Foot tapping features.

The best classifier was the Gaussian SVM with an accuracy of 85.5%, a sensitivity of 91.0% and a specificity of 79.0% (Table 2).

Table 2. Confusion matrix. "Healthy subjects vs PD patients" classification with foot tapping features.

Predicted condition	True condition	
	PD patient	Healthy subj.
PD patient	30	6
Healthy subj.	3	23

Set 3: Both finger and foot tapping features.

The best classifier was the Quadratic SVM with an accuracy of 87.1%, a sensitivity of 87.8% and a specificity of 86.0% (Table 3).

Table 3. Confusion matrix. "Healthy subjects vs PD patients" classification with both finger tapping and foot tapping features.

Predicted condition	True condition	
	PD patient	Healthy subj.
PD patient	29	4
Healthy subj.	4	25

Discussion.

In the first part of the study, we investigated the ability of the extracted features to distinguish between healthy subjects and PD patients using a SVM-based classifier. As first step, we analyzed the finger tapping (FiT) and the foot tapping (FoT) features independently, we then analyzed the features extracted from both exercises movements together. The main findings of this analysis indicate that:

- the features extracted from the foot tapping exercise lead to a better classification in terms of all the three computed indices: accuracy (85.5% FoT vs 71.0% FiT), sensitivity (91.0% FoT vs 75.7% FiT) and specificity (65.5% FoT vs 79.0% FiT);
- using the features extracted from both exercises (FoT and FiT) the SVM classifier performs better than the two classifiers based either on the FiT features or the FoT features. In particular, the classifier based on both feature sets reached a better accuracy (87.1%), a better specificity (86.0%) and a slight lower sensitivity (87.8%).

This first analysis indicates that the set of features we selected from the movement acquired during both exercises (FoT and FiT) can be used to capture the abnormal motor activity of a PD patient with great results.

Mild PD Patients vs Moderate PD Patients Classification.
In this subsection, we reported and discussed the results about the "*Mild PD patients vs Moderate PD patients*" classification. In particular we refers to the moderate PD patients with *positive* and to the mild PD patients with *negative*.

Set 1: Finger tapping features.

The best classifier was the Gaussian SVM with an accuracy of 57.0%, a sensitivity of 100% and a specificity of 0% (Table 4).

Table 4. Confusion matrix. "Mild PD patients vs Moderate PD patients" classification with finger tapping features.

Predicted condition	True condition	
	Moderate	Mild
Moderate	19	14
Mild	0	0

Set 2: Foot tapping features.

The best classifier was the Gaussian SVM with an accuracy of 81.0%, a sensitivity of 84.0% and a specificity of 78.0% (Table 5).

Table 5. Confusion matrix. "Mild PD patients vs Moderate PD patients" classification with foot tapping features.

Predicted condition	True condition	
	Moderate	Mild
Moderate	16	3
Mild	3	11

Set 3: Both finger and foot tapping features.

The best classifier was the Gaussian SVM with an accuracy of 78.0%, a sensitivity of 89.0% and a specificity of 64.0% (Table 6).

Table 6. Confusion matrix. "Mild PD patients vs Moderate PD patients" classification with both finger tapping and foot tapping features.

Predicted condition	True condition	
	Moderate	Mild
Moderate	17	5
Mild	2	9

Discussion.

In the second part of the study, we investigated the ability of the extracted features to distinguish between mild PD patients and moderate PD patients using a SVM-based classifier. As done for the "*Healthy subjects vs PD patients*" classification, we first analyzed the finger tapping (FiT) and the foot tapping (FoT) features independently, we then analyzed the features extracted from both exercises movements together. The main findings of this analysis indicate that:

- the FiT extracted features are not representative of the difference between moderate and mild PD subjects (accuracy 57.0%, sensitivity 100% and specificity 0%);
- the FoT extracted features lead to best accuracy (81.0%) and specificity (78.0%);
- the SVM classifier that use both feature sets is characterized by both a slight lower accuracy (78.0%) and specificity (64.0%) than the FoT-based SMV, and the best sensitivity level that is equal to 89.0%.

This second analysis indicates that the set of features we selected from the movement acquired during both exercises (FoT and FiT) lead to a good "*Mild PD patients vs Moderate PD patients*" classification results. But, it is worth noting that the "*Mild PD patients vs Moderate PD patients*" classification scores are slight lower than the "*Healthy subjects vs PD patients*" classification ones. It is also worth noting that the FoT features are the most important ones to achieve the best accuracy and specificity levels, and that FiT features are completely not representative of the motor differences between mild and moderate PD patients. Concerning the "*Mild PD patients vs Moderate PD patients*" classification, from what emerged in this second analysis we can state that FiT features lead to lower accuracy and specificity levels. Only when the FiT are used together with the FoT features the SVM classifier present a better sensitivity level at the expense of both the accuracy and specificity.

4 Conclusion

Nowadays, the PD evaluation and rating heavily rely on human expertise. In this work we designed, implemented and tested a low-cost vision-based tool to automatically track two of the main exercises evaluated by the UPDRS scale. We then extracted eight main features from the trajectories of the acquired movements using image processing techniques. Several SVM classifiers have been trained and evaluated to investigated whether the selected sets of features can be used to detect the main differences between healthy subjects and PD patients, first, and then between mild PD patients and moderate PD patients. Results showed that an SVM using the features extracted by both considered exercises, i.e. finger tapping and foot tapping, is able to classify between healthy subjects and PD patients with great performances by reaching 87.1% of accuracy, 86.0% of specificity and 87.8% of sensitivity. The results of the classification between mild and moderate PD patients indicated that the foot tapping features are the most representative ones compared to the finger tapping features. In fact, the SVM based on the foot tapping features reached the best score in terms of accuracy (81.0%) and specificity (78.0%). From our findings, we can conclude that automatic vision system based on the Kinect v2 sensor together with the selected extracted features could represent a valid tool to support assessment of postural and spatiotemporal characteristics during walking in participants affected or not by the PD. In addition, the sensitivity of the Kinect v2 sensor could supports medical specialists in the assessment and rating of PD patients. Finally, the low-cost cost feature and the easy and fast setup phase of the designed and implemented tool support and encourage its usability in a real clinical scenario.

Acknowledgment. This work has been funded from the Italian project ROBOVIR (BRIC-INAIL-2017).

References

1. Memedi, M., Sadikov, A., Groznik, V., Žabkar, J., Možina, M., Bergquist, F., Johansson, A., Haubenberger, D., Nyholm, D.: Automatic spiral analysis for objective assessment of motor symptoms in Parkinson's disease. Sensors **15**, 23727–23744 (2015). https://doi.org/10.3390/s150923727
2. Goetz, C.G., Tilley, B.C., Shaftman, S.R., Stebbins, G.T., Fahn, S., Martinez-Martin, P., Poewe, W., Sampaio, C., Stern, M.B., Dodel, R.: Movement Disorder Society-sponsored revision of the Unified Parkinson's Disease Rating Scale (MDS-UPDRS): scale presentation and clinimetric testing results. Mov. Disord. **23**, 2129–2170 (2008)
3. Goetz, C.G., Poewe, W., Rascol, O., Sampaio, C., Stebbins, G.T., Counsell, C., Giladi, N., Holloway, R.G., Moore, C.G., Wenning, G.K., Yahr, M.D., Seidl, L.: Movement Disorder Society Task Force report on the Hoehn and Yahr staging scale: status and recommendations. Mov. Disord. **19**, 1020–1028 (2004). https://doi.org/10.1002/mds.20213
4. Djuric-Jovicic, M.D., Jovicic, N.S., Radovanovic, S.M., Stankovic, I.D., Popovic, M.B., Kostic, V.S.: Automatic identification and classification of freezing of gait episodes in Parkinson's disease patients. IEEE Trans. Neural Syst. Rehabil. Eng. **22**, 685–694 (2014). https://doi.org/10.1109/TNSRE.2013.2287241
5. Tripoliti, E.E., Tzallas, A.T., Tsipouras, M.G., Rigas, G., Bougia, P., Leontiou, M., Konitsiotis, S., Chondrogiorgi, M., Tsouli, S., Fotiadis, D.I.: Automatic detection of freezing of gait events in patients with Parkinson's disease. Comput. Methods Programs Biomed. **110**, 12–26 (2013). https://doi.org/10.1016/j.cmpb.2012.10.016
6. Bortone, I., et al.: A novel approach in combination of 3D gait analysis data for aiding clinical decision-making in patients with Parkinson's Disease. In: Huang, D.-S., Jo, K.-H., Figueroa-García, J.C. (eds.) ICIC 2017. LNCS, vol. 10362, pp. 504–514. Springer, Cham (2017). https://doi.org/10.1007/978-3-319-63312-1_44
7. Tsanas, A., Little, M.A., McSharry, P.E., Ramig, L.O.: Accurate telemonitoring of parkinsons disease progression by noninvasive speech tests. IEEE Trans. Biomed. Eng. **57**, 884–893 (2010). https://doi.org/10.1109/TBME.2009.2036000
8. Mellone, S., Palmerini, L., Cappello, A., Chiari, L.: Hilbert-Huang-based tremor removal to assess postural properties from accelerometers. IEEE Trans. Biomed. Eng. **58**, 1752–1761 (2011). https://doi.org/10.1109/TBME.2011.2116017
9. Heldman, D.A., Espay, A.J., LeWitt, P.A., Giuffrida, J.P.: Clinician versus machine: reliability and responsiveness of motor endpoints in Parkinson's disease. Park. Relat. Disord. **20**, 590–595 (2014). https://doi.org/10.1016/j.parkreldis.2014.02.022

10. Rigas, G., Tzallas, A.T., Tsipouras, M.G., Bougia, P., Tripoliti, E.E., Baga, D., Fotiadis, D. I., Tsouli, S.G., Konitsiotis, S.: Assessment of tremor activity in the parkinsons disease using a set of wearable sensors. IEEE Trans. Inf Technol. Biomed. **16**, 478–487 (2012). https://doi.org/10.1109/TITB.2011.2182616
11. Bevilacqua, V., et al.: A RGB-D sensor based tool for assessment and rating of movement disorders. In: Duffy, V., Lightner, N. (eds.) AHFE 2017. AISC, vol. 590, pp. 110–118. Springer, Cham (2018). https://doi.org/10.1007/978-3-319-60483-1_12
12. Salarian, A., Russmann, H., Wider, C., Burkhard, P.R., Vingerhoets, F.J.G., Aminian, K.: Quantification of tremor and bradykinesia in Parkinson's disease using a novel ambulatory monitoring system. IEEE Trans. Biomed. Eng. **54**, 313–322 (2007). https://doi.org/10.1109/TBME.2006.886670
13. Dai, H., Lin, H., Lueth, T.C.: Quantitative assessment of parkinsonian bradykinesia based on an inertial measurement unit. Biomed. Eng. Online **14**, 68 (2015). https://doi.org/10.1186/s12938-015-0067-8
14. Griffiths, R.I., Kotschet, K., Arfon, S., Xu, Z.M., Johnson, W., Drago, J., Evans, A., Kempster, P., Raghav, S., Horne, M.K.: Automated assessment of bradykinesia and dyskinesia in Parkinson's disease. J. Parkinsons Dis. **2**, 47–55 (2012). https://doi.org/10.3233/JPD-2012-11071
15. Keijsers, N.L.W., Horstink, M.W.I.M., Gielen, S.C.A.M.: Automatic assessment of levodopa-induced dyskinesias in daily life by neural networks. Mov. Disord. **18**, 70–80 (2003). https://doi.org/10.1002/mds.10310
16. Lopane, G., Mellone, S., Chiari, L., Cortelli, P., Calandra-Buonaura, G., Contin, M.: Dyskinesia detection and monitoring by a single sensor in patients with Parkinson's disease. Mov. Disord. **30**, 1267–1271 (2015). https://doi.org/10.1002/mds.26313
17. Saunders-Pullman, R., Derby, C., Stanley, K., Floyd, A., Bressman, S., Lipton, R.B., Deligtisch, A., Severt, L., Yu, Q., Kurtis, M., Pullman, S.L.: Validity of spiral analysis in early Parkinson's disease. Mov. Disord. **23**, 531–537 (2008)
18. Westin, J., Ghiamati, S., Memedi, M., Nyholm, D., Johansson, A., Dougherty, M., Groth, T.: A new computer method for assessing drawing impairment in Parkinson's disease. J. Neurosci. Methods **190**, 143–148 (2010). https://doi.org/10.1016/j.jneumeth.2010.04.027
19. Liu, X., Carroll, C.B., Wang, S.Y., Zajicek, J., Bain, P.G.: Quantifying drug-induced dyskinesias in the arms using digitised spiral-drawing tasks. J. Neurosci. Methods **144**, 47–52 (2005). https://doi.org/10.1016/j.jneumeth.2004.10.005
20. Loconsole, C., et al.: Computer vision and EMG-based handwriting analysis for classification in Parkinson's disease. In: Huang, D.-S., Jo, K.-H., Figueroa-García, J.C. (eds.) ICIC 2017. LNCS, vol. 10362, pp. 493–503. Springer, Cham (2017). https://doi.org/10.1007/978-3-319-63312-1_43
21. Bevilacqua, V., Salatino, A.A., Di Leo, C., Tattoli, G., Buongiorno, D., Signorile, D., Babiloni, C., Del Percio, C., Triggiani, A.I., Gesualdo, L.: Advanced classification of Alzheimer's disease and healthy subjects based on EEG markers. In: Proceedings of the International Joint Conference on Neural Networks (2015). https://doi.org/10.1109/ijcnn.2015.7280463
22. Buongiorno, D., Barsotti, M., Sotgiu, E., Loconsole, C., Solazzi, M., Bevilacqua, V., Frisoli, A.: A neuromusculoskeletal model of the human upper limb for a myoelectric exoskeleton control using a reduced number of muscles. In: 2015 IEEE World Haptics Conference (WHC), pp. 273–279 (2015). https://doi.org/10.1109/whc.2015.7177725
23. Bevilacqua, V., Mastronardi, G., Piscopo, G.: Evolutionary approach to inverse planning in coplanar radiotherapy. Image Vis. Comput. **25**, 196–203 (2007). https://doi.org/10.1016/j.imavis.2006.01.027

24. Menolascina, F., Bellomo, D., Maiwald, T., Bevilacqua, V., Ciminelli, C., Paradiso, A., Tommasi, S.: Developing optimal input design strategies in cancer systems biology with applications to microfluidic device engineering. BMC Bioinf. **10**, S4 (2009). https://doi.org/10.1186/1471-2105-10-S12-S4
25. Menolascina, F., Tommasi, S., Paradiso, A., Cortellino, M., Bevilacqua, V., Mastronardi, G.: Novel data mining techniques in aCGH based breast cancer subtypes profiling: the biological perspective. In: 2007 IEEE Symposium on Computational Intelligence and Bioinformatics and Computational Biology, CIBCB 2007, pp. 9–16 (2007). https://doi.org/10.1109/cibcb.2007.4221198
26. Bevilacqua, V., Tattoli, G., Buongiorno, D., Loconsole, C., Leonardis, D., Barsotti, M., Frisoli, A., Bergamasco, M.: A novel BCI-SSVEP based approach for control of walking in virtual environment using a convolutional neural network. In: Proceedings of the International Joint Conference on Neural Networks (2014). https://doi.org/10.1109/ijcnn.2014.6889955
27. Manghisi, V.M., Uva, A.E., Fiorentino, M., Bevilacqua, V., Trotta, G.F., Monno, G.: Real time RULA assessment using kinect v2 sensor. Appl. Ergon. **65**, 481–491 (2017). https://doi.org/10.1016/j.apergo.2017.02.015
28. Bevilacqua, V., et al.: Retinal Fundus Biometric Analysis for Personal Identifications. In: Huang, D.-S., Wunsch, D.C., Levine, D.S., Jo, K.-H. (eds.) ICIC 2008. LNCS (LNAI), vol. 5227, pp. 1229–1237. Springer, Heidelberg (2008). https://doi.org/10.1007/978-3-540-85984-0_147
29. Kanjilal, P.P., Palit, S., Saha, G.: Fetal ECG extraction from single-channel maternal ECG using singular value decomposition. IEEE Trans. Biomed. Eng. **44**, 51–59 (1997). https://doi.org/10.1109/10.553712

Genetic Learning Particle Swarm Optimization with Diverse Selection

Da Ren, Yi Cai(✉), and Han Huang

South China University of Technology, Guangzhou, China
edwardyam@outlook.com, {ycai,hhan}@scut.edu.cn

Abstract. Particle swarm optimization (PSO) is a widely used heuristic algorithm. However, canonical PSO may lead to premature convergence. To solve this problem, researchers try to hybridize PSO with genetic algorithm (GA) which facilitates global effectiveness. One of the successful algorithms is genetic learning PSO (GL-PSO). However, we find that the selection in GL-PSO reduce the diversity of particles. It may lead premature convergence in some test functions. To solve this problem, we figure out a genetic learning particle swarm optimization with diverse selection (GL-PSODS). We test our proposed algorithm in test functions of CEC2014. Our experiments show that GL-PSODS has an improvement in some test functions compared to PSO and GL-PSO.

Keywords: Particle swarm optimization · Genetic algorithm · GL-PSO

1 Introduction

Particle swarm optimization (PSO) has attracted more and more researchers' attention [3]. It is a nature-inspired optimization algorithm which is introduced by Kennedy and Eberhart [4, 9]. PSO has been widely used in many areas such as industrial electronics [1, 13], wireless sensor networks [10] and feature selection [11].

However, researchers find that canonical PSO may lead to premature convergence [5, 7]. To relieve this problem, Gong et al. propose genetic learning particle swarm optimization (GL-PSO). In GL-PSO, GA and PSO are hybridized in a cascade manner. In this way, particles in PSO are no longer simply guided by the global best-so-far solution (gbest) and personal best experience (pbests), but are guided by the exemplars constructed by GA [6].

Although GL-PSO gets a remarkable improvements in many test functions. We consider that the selection process in GL-PSO reduce diversity of particles. Therefore, we propose genetic learning particle swarm optimization with diverse selection (GL-PSODS) whose selection process can lead more diverse particles. We test our algorithm and standard GL-PSO in the test functions of CEC2014. Our experiments show that our algorithm performs better in some test functions.

This paper is structured as follows. In Sect. 2, we introduce the related works. We explain our proposed algorithm in Sect. 3. We conduct our experiments in Sect. 4 and conclude the paper in Sect. 5.

D.-S. Huang et al. (Eds.): ICIC 2018, LNAI 10956, pp. 789–794, 2018.
https://doi.org/10.1007/978-3-319-95957-3_83

2 Related Works

PSO is a widely used algorithm because it's conceptual simplicity and high efficiency [6]. However, it is easy to lead these particles to premature convergence. To solve this problem, some researchers try to hybridize PSO with GA in hybrid methods [2, 8, 12, 14, 15].

However, Gong et al. consider that GA and PSO are loosely coupled in this hybridization mechanism in previous work [6]. Therefore, the effect coming from the interaction of GA and PSO is ambiguous to recognize [6]. The experiments of Gong et al. show that GL-PSO get improvements in many test functions.

3 Proposed Algorithm

3.1 Particle Swarm Optimization

In canonical PSO, each particle learns from its own pbest and gbest. Suppose that the velocity of the i-th particle is $V_i = \left[v_{i,1}, v_{i,2}, \ldots, v_{i,D}\right]$, where D is the dimension of test functions. The position of the i-th particle is $X_i = \left[x_{i,1}, x_{i,2}, \ldots, x_{i,D}\right]$. Let $P_i = \left[p_{i,1}, p_{i,2}, \ldots, p_{i,D}\right]$ denote the pbest of the i-th particle and $G = [g_1, g_2, \ldots, g_D]$ denote the gbest of the whole swarm. The d-th dimension of particle i is update according to Eqs. 1 and 2.

$$v_{i,d} \leftarrow w \cdot v_{i,d} + c_1 \cdot r_{1,d} \cdot \left(p_{i,d} - x_{i,d}\right) + c_2 \cdot r_{2,d} \cdot \left(g_d - x_{i,d}\right) \quad (1)$$

$$x_{i,d} \leftarrow x_{i,d} + v_{i,d} \quad (2)$$

where w denotes the inertia weight, c_1 and c_2 are accelerate coefficients determining the relative importance of P_i and G, and $r_{1,d}$ and $r_{2,d}$ are random numbers uniformly selected within [0, 1].

However, a particle learns from both P_i and G may hence oscillate if two exemplars locate on opposite side of X_i. To solve this problem, researchers propose variant PSO learning schemes which is described in the following.

$$v_{i,d} \leftarrow w \cdot v_{i,d} + c \cdot r_d \cdot \left(e_{i,d} - x_{i,d}\right) \quad (3)$$

$$e_{i,d} = \left(c_1 \cdot r_{i,d} \cdot p_{i,d} + c_2 \cdot r_{2,d} \cdot g_d\right) / \left(c_1 \cdot r_{i,d} + c_2 \cdot r_{2,d}\right) \quad (4)$$

where $\mathrm{E_i} = \left[\mathrm{e_{i,d}}, \mathrm{e_{i,2}}, \ldots, \mathrm{e_{i,D}}\right]$ is a single exemplar which is built to attract particles.

3.2 Genetic Learning Particle Swarm Optimization

From Eq. 3, we can know that E_i determines the search trajectories of particles. Gong et al. propose a genetic learning scheme that applies GA to breed exemplars [6]. The detailed implementation of the genetic learning scheme is presented [6].

Crossover. For each particle i, crossover operation is first conducted on P_i and G to generate an offspring $O_i = \left[o_{i,1}, o_{i,2}, \ldots, o_{i,D}\right]$ [6].

$$o_{i,d} = \begin{cases} r_d \cdot p_{i,d} + (1 - r_d) \cdot g_d, & \textit{if } f(P_1) < f(P_{k_d}) \\ P_{k_d,d}, & \textit{otherwise} \end{cases} \tag{5}$$

where r_d is a random number uniformly distributed in [0, 1], $k_d \in 1, 2, \ldots, M$ is the index of a random particle, and f is the objective function [6].

Mutation. For each dimension $\boldsymbol{d}$, a random number $\boldsymbol{r_d} \in [\mathbf{0}, \mathbf{1}]$ is generated. If $\boldsymbol{r_d}$ is smaller than $\boldsymbol{pm}$, the corresponding dimension of $\boldsymbol{O_i}$ is reinitialized in the search place [6].

$$o_{i,d} = rand(lb_d, ub_d), \quad \textit{if } r_d < pm \tag{6}$$

where lb_d and ub_d stand for the lower and upper bounds of the d-th dimension.

Algorithm 1. Diverse Selection

1. /* Diverse Selection*/
2. if $f(O_i) < f(E_i)$ then
3. $\quad E_i = O_i$
4. else
5. $\quad$ if $r_d < p_{accept}$ then
6. $\quad\quad E_i = O_i$
7. $\quad$ end if
8. end if

Selection. Selection is an important process to determine whether the offspring or the current exemplar survives in this generation. Selection is executed by Eq. 7.

$$E_i \leftarrow \begin{cases} O_i, & \textit{if } f(O_i) < f(E_i) \\ E_i, & \textit{otherwise} \end{cases} \tag{7}$$

In Eq. 7, the exemplar remains unchanged if it is better than the new offspring [6]. Gong et al. employ the $20\%M$ – tournament selection to update the particle's exemplar if a particle ceases improving for sg generations [6].

3.3 Diverse Selection

The selection in GL-PSO only choose the prior one to survive, we consider that it may lead GL-PSO to be premature convergence. In GL-PSODS, there are probability p_{accept} in selection to select the worse offsprings. Diverse offsprings can lead particles jump out local optimum. Diverse selection is described in Algorithm 1. In Algorithm 1, $\boldsymbol{r_d}$ is a random number between $[\mathbf{0}, \mathbf{1}]$ and $\boldsymbol{p_{accept}}$ is the probability to accept worse offspring.

4 Experiment

In this section, we explain our experiment. We test our algorithm in CEC2014 which contains 30 test functions. We compare our algorithm with PSO, variant PSO, GL-PSO. The parameters can be found in Table 1. In our experiments, the iteration time is set to be 5000, the population is set to be 100. We run 30 times for each test function. The p_{accept} of GL-PSODS is set to be 0.01. The best fitness value for each test function in dimension 30 can be found in Table 2.

Table 1. Parameters of algorithm.

PSO		Variant PSO		GL-PSO		GL-PSODS	
w	0.7	w	0.7	w	0.7	w	0.7
c_1	2	c_1	2	c	1.5	c	1.5
c_2	2	c_2	2	pm	0.01	pm	0.01
		c	3	sg	7	sg	7

From Table 2, we can know that there are no an algorithm can perform better than the others in all test functions. However, we can find that GL-PSO and GL-PSODS always get a better value than PSO and variant PSO. Therefore, we can give a conclusion that global searching ability have an improvement by combining GA in exemplars learning process. Variant PSO tends to perform better than PSO since the changes of variant PSO relieve the problem of local optimum. What's more, GL-PSODS can get a better value than GL-PSO in some test functions (e.g. Func4 in dimension 30). It shows that GL-PSODS has a better ability to fly out the local optimal that GL-PSO.

Table 2. Best fitness value

	PSO	Variant PSO	GL-PSO	GL-PSODS
Func1	4332978.1	167500.8	**20374.84**	41806.02
Func2	209.0966	**200**	**200**	**200**
Func3	346.7397	**300**	300.0004	300.0021
Func4	485.7976	494.8554	476.2116	**400.1217**
Func5	520.6517	**519.9997**	519.9999	519.9999
Func6	606.2293	608.6304	605.7999	**603.6704**
Func7	700.0004	**700**	**700**	**700**
Func8	810.0166	840.7933	**800**	**800**
Func9	**922.885**	954.7227	942.7832	927.8588
Func10	1241.997	1798.642	**1000.021**	1000.042
Func11	3054.905	2760.512	**2049.009**	2678.006
Func12	1201.246	1200.144	1200.059	**1200.055**
Func13	1300.236	1300.257	1300.181	**1300.172**

(*continued*)

Table 2. (*continued*)

	PSO	Variant PSO	GL-PSO	GL-PSODS
Func14	1400.183	1400.213	1400.184	**1400.173**
Func15	1508.622	1505.584	**1502.499**	1502.929
Func16	1610.655	1609.672	**1607.35**	1608.167
Func17	62994.45	8879.679	4279.749	**3443.515**
Func18	1913.937	1919.278	1902.885	**1864.777**
Func19	1904.759	1906.9	1904.574	**1903.591**
Func20	2265.864	2085.635	2050.259	**2039.288**
Func21	22777.28	4806.601	3879.102	**2744.895**
Func22	2294.949	2341.62	**2222.164**	2243.029
Func23	2616.198	2615.432	2615.317	**2615.307**
Func24	2625.347	2624.96	**2623.763**	2623.928
Func25	2706.777	2706.882	**2705.283**	2705.833
Func26	2700.21	2700.251	2700.198	**2700.177**
Func27	3103.533	3102.056	3101.586	**3101.342**
Func28	3863.291	3838.717	3672.156	**3602.529**
Func29	**3531.896**	4026.338	3840.636	3826.614
Func30	4890.104	5757.383	**4233.964**	4487.374

5 Conclusion

In the paper, we figure out GL-PSODS which contains additional random factor in GL-PSO. GL-PSO gets a remarkable improvement. However, we find that the selection in GL-PSO reduce the diversity of particles. Therefore, we apply a diverse selection into GL-PSO to improve the ability of global search. We test our algorithm in 30 test functions of CEC2014. Our experiments show that GL-PSODS gets a better result in some test functions. However, we can find that GL-PSODS′ cannot perform much better than GL-PSO. How to combine GA and PSO more properly is our future work.

Acknowledgment. This work is supported by the Fundamental Research Funds for the Central Universities, SCUT (NO. 2017ZD048,2015ZM136), Tiptop Scientific and Technical Innovative Youth Talents of Guangdong special support program (No. 2015TQ01X633), Science and Technology Planning Project of Guangdong Province, China (No. 2016A030310423), Science and Technology Program of Guangzhou (International Science and Technology Cooperation Program No. 201704030076) and Science and Technology Planning Major Project of Guangdong Province (No. 2015A070711001).

References

1. Alrashidi, M.R., El-Hawary, M.E.: A survey of particle swarm optimization applications in electric power systems. IEEE Press (2009)
2. Arumugam, M.S., Rao, M.V.C.: On the improved performances of the particle swarm optimization algorithms with adaptive parameters, cross-over operators and root mean square (RMS) variants for computing optimal control of a class of hybrid systems. Appl. Soft Comput. **8**(1), 324–336 (2008)
3. Cai, Y., Chen, Z., Li, J., Li, Q., Min, H.: An adaptive particle swarm optimization algorithm for distributed search and collective cleanup in complex environment. Int. J. Distrib. Sens. Netw. **2013**(4), 1–9 (2013)
4. Eberhart, R., Kennedy, J.: A new optimizer using particle swarm theory. In: International Symposium on MICRO Machine and Human Science, pp. 39–43 (2002)
5. Frans, V.D.B., Engelbrecht, A.P.: A cooperative approach to particle swarm optimization. IEEE Trans. Evol. Comput. **8**(3), 225–239 (2004)
6. Gong, Y.J., Li, J.J., Zhou, Y., Li, Y., Chung, H.S., Shi, Y.H., Zhang, J.: Genetic learning particle swarm optimization. IEEE Trans. Cybern. **46**(10), 2277 (2016)
7. Janson, S., Middendorf, M.: A hierarchical particle swarm optimizer and its adaptive variant. IEEE Trans. Syst. Man Cybern. Part B Cybern. **35**(6), 1272–1282 (2005)
8. Kao, Y.T., Zahara, E.: A hybrid genetic algorithm and particle swarm optimization for multimodal functions. Appl. Soft Comput. **8**(2), 849–857 (2008)
9. Kennedy, J., Eberhart, R.: Particle swarm optimization. In: Proceedings of IEEE International Conference on Neural Networks, vol. 4, pp. 1942–1948 (2002)
10. Kulkarni, R.V., Venayagamoorthy, G.K.: Particle swarm optimization in wireless-sensor networks: a brief survey. IEEE Trans. Syst. Man Cybern. Part C **41**(2), 262–267 (2011)
11. Lane, M.C., Xue, B., Liu, I., Zhang, M.: Particle swarm optimisation and statistical clustering for feature selection. In: Cranefield, S., Nayak, A. (eds.) AI 2013. LNCS (LNAI), vol. 8272, pp. 214–220. Springer, Cham (2013). https://doi.org/10.1007/978-3-319-03680-9_23
12. Ling, S.H., Iu, H.H., Chan, K.Y., Lam, H.K., Yeung, B.C., Leung, F.H.: Hybrid particle swarm optimization with wavelet mutation and its industrial applications. IEEE Trans. Syst. Man Cybern. Part B Cybern. A Publ. IEEE Syst. Man Cybern. Soc. **38**(3), 743 (2008)
13. Ruiz-Cruz, R., Sanchez, E.N., Ornelas-Tellez, F., Loukianov, A.G., Harley, R.G.: Particle swarm optimization for discrete-time inverse optimal control of a doubly fed induction generator. IEEE Trans. Cybern. **43**(6), 1698–1709 (2013)
14. Shi, X.H., Liang, Y.C., Lee, H.P., Lu, C., Wang, L.M.: An improved GA and a novel PSO-GA-based hybrid algorithm. Inf. Process. Lett. **93**(5), 255–261 (2005)
15. Valdez, F., Melin, P., Mendoza, O.: A new evolutionary method with fuzzy logic for combining particle swarm optimization and genetic algorithms: the case of neural networks optimization, vol. 574, pp. 1536–1543 (2008)

An Effective Artificial Bee Colony for Distributed Lot-Streaming Flowshop Scheduling Problem

Jun-Hua Duan[1], Tao Meng[2], Qing-Da Chen[3], and Quan-Ke Pan[2](✉)

[1] Computer Center, Shanghai University, Shanghai 200444, People's Republic of China
[2] School of Mechatronic Engineering and Automation, Shanghai University, Shanghai 200072, People's Republic of China
Panquanke@shu.edu.cn
[3] State Key Laboratory of Synthetical Automation for Process Industries, Northeastern University, Shenyang 110819, People's Republic of China

Abstract. This paper proposes an effective discrete artificial bee colony (DABC) algorithm for solving the distributed lot-streaming flowshop scheduling problem (DLFSP) with the objective of minimizing makespan. We design a multi-list based representation to represent candidate solutions, where each list is corresponding to a factory. We present a multi-list based swap and insertion operators to generate neighboring solutions. We redesign the employ bee phase, onlooker bee phase, and scout bee phase according to the problem-specific knowledge, representation and information collected in the evolution process. The parameters for the proposed DABC algorithm are calibrated by means of a design of experiments and analysis of variance. A comprehensive computational campaign based on 810 randomly generated instances demonstrates the effectiveness of the proposed DABC algorithm for solving the DLFSP with the makespan criterion.

Keywords: Scheduling · Flowshop · Artificial bee colony · Makespan

1 Introduction

The distributed flowshop scheduling problem (DFSP), which deals with the production allocated to each factory along with the scheduling, has become a hot topic for research in recent years [1–3]. In many practical environments, a job consists of many identical items. A job is allowed to overlap its operations between successive machines by splitting it into a number of smaller sub-lots. The process of splitting jobs into sub-lots is called lot-streaming [4, 5]. This paper considers the distributed lot-streaming flowshop scheduling problem (DLFSP) that has potential applications in modern industries. The objective is to minimize makespan. The regular flowshop scheduling problem with makespan criterion is already NP-hard. With lot-streaming constraints and multi-factories involved, the DLFSP is NP-hard as well.

The first work for the DFSP is due to Naderi et al. [6], where 14 heuristics and two variable neighborhood decent method (VND) were proposed for optimizing makespan.

D.-S. Huang et al. (Eds.): ICIC 2018, LNAI 10956, pp. 795–806, 2018.
https://doi.org/10.1007/978-3-319-95957-3_84

Later, Liu and Gao [7] presented an electromagnetism metaheuristic (EM). Gao and Chen [8, 9] proposed an improved NEH heuristic and a hybrid genetic algorithm (HGA). Gao *et al.* [10] introduced a revised VND. Gao *et al.* [11] developed a tabu search (TS) method. Lin *et al.* [12] presented a modified iterated greedy (MIG) method. Wang *et al.* [2] investigated an estimation of distribution algorithm (EDA). Recently, Naderi and Ruiz [3] presented a scatter search (SS) algorithm. Fernandez-Viagas and Framinan [13] proposed a bound-search iterated greedy (BIG) algorithm. Xu *et al.* [14] introduced a hybrid immune algorithm.

For the lot streaming flowshop scheduling problem (LFSP), Yoon and Ventura [15] presented a hybrid genetic algorithm (HGA). Tseng and Liao [16] developed a discrete particle swarm optimization (DPSO) algorithm. Pan *et al.* [17] presented a discrete artificial bee colony (DABC) algorithm. Pan and Ruiz [5] presented an estimation of distribution algorithm. Pan et al. [18] introduced a harmony search (HS) algorithm. Meng and Pan [19] presented an improved migrating birds optimization method. Han et al. [20] proposed an improved NSGA-II algorithm for multi-objective lot-streaming flow shop scheduling problem.

From the above short review, we can see that the distributed lot-streaming flow shop scheduling problem (DLFSP) has not been addressed so far. It is worthwhile to develop effective algorithms for such an important problem with the objective of minimizing makespan.

The basic artificial bee colony was presented by Karaboga [21] for optimizing continuous functions. Pan *et al.* [17] introduced a discrete version of the basic artificial bee colony algorithm, DABC for short, to solve the LFSP with the total earliness and tardiness criterion. Since then, DABC has been successfully applied to many scheduling problems including the permutation flowshop, the permutation flowshop with blocking constraints, the no-wait flowshop, the hybrid flowshop problem, and the flexible job shop [22–24]. Following the successful applications of the DABC algorithm, this paper uses the DABC algorithm to solve the DLFSP with makespan criterion.

The rest of the paper is organized as follows. In Sect. 2, the DLFSP with the makespan criterion is described. The presented DABC algorithm is detailed in Sect. 3. In Sect. 4, we calibrate the presented DABC algorithm, and report the computational comparisons in Sect. 5. Finally, we give the concluding remarks and some suggestions for future work in Sect. 6.

2 Problem Description

In the considered DLPFSP, there are a set of identical factories $F = \{1, 2, \ldots, f\}$ that consist of m machines $M = \{1, 2, \ldots, m\}$. The machines are arranged into a flowshop. A set of jobs $N = \{1, 2, \ldots, n\}$ have to be assigned to the f factories. In each factory, the jobs are processed through all the m machines in the same route. For reducing leading times and accelerating the production, each job j can be split into s_j

smaller sub-lots with equal size. Once the operation of a sub-lot is finished on a machine, it can be immediately transported to the downstream machine. The processing time of a sub-lot of job j on machine i is denoted as $p_{ij} > 0$. Generally, it is assumed that all the jobs are independent and available for processing at time 0. All the machines are ready at time 0. At any time, no sub-lot can be processed on more than one machine and no machine can process more than one sub-lot. The problem is then to decide how to assign jobs to factories and how to sequence the jobs assigned in each factory so that the makespan is minimized.

3 Discrete Artificial Bee Colony

This section proposes a discrete artificial bee colony (DABC) algorithm to minimize makespan for the DLFSP. We will describe the solution representation, population initialization, and search phases in detail and then give the pseudo codes.

3.1 Solution Representation

The *f*-list representation has been widely used for the distributed flowshop scheduling problem [3]. In this representation, each list is corresponding to a factory. The list is a permutation of all the jobs assigned to the factory, which indicates the order in which the jobs are launched to the flowshop. We adopt this representation here. To be specific, a solution is represented as $\pi = \{\pi_1, \pi_2, \ldots, \pi_f\}$, where $\pi_k = (\pi_{k,1}, \pi_{k,2}, \ldots, \pi_{k,n_k})$, $k = 1, 2, \ldots, f$ is a list related to factory k. $\pi_{k,1}, \pi_{k,2}, \ldots, \pi_{k,n_k}$ represent the jobs assigned to factory k. The jobs are processed following that order.

3.2 Population Initialization

To obtain an effective DABC algorithm for solving the DLFSP with makespan criterion, we propose a simple heuristic based on the shortest processing time (SPT) rule and the earliest completion time (ECT) rule. That is, we first generate a full job permutation, consisting of all the jobs to be processed, using the SPT rule. Then, we assign the jobs one by one to the factory that can complete it at earliest time. This heuristic tries to finish the jobs as soon as possible and balance the workload among factories. It is expected to obtain a solution with a high level of quality. To get an initial population with a high level of diversity, we consider another heuristic where a full job permutation is generated randomly. We also assign the jobs according to the randomly generated permutation to the factories using ECT rule. This heuristic is quite different from the first one since the seed permutation is generated randomly. To yield an initial population of *PS* solutions, we generate a solution use the SPT-ECT heuristic and perform the random-ECT *PS*-1 times to obtain the rest solutions.

3.3 Employed Bee Phase

After the initial population is generated, the DABC algorithm performs its search process including three consecutive phase, *i.e.*, employed bees, onlooker bees, and scout bees, based on the population. The three phases are repeated until a given termination criterion is met.

In the employed bee phase, the DABC algorithm associates each solution from the population with an employed bee. In other words, each solution is regarded as an employed bee. For each current solution, a neighboring solution is generated. This process is called employed bee search in the DABC. If the neighboring solution has better makespan value than the current solution, the neighboring solution will replace the current solution and become a new member of the population. To generate a neighboring solution, we employ the swap operator. We firstly randomly choose two different factories k_1 and k_2 with uniform distribution. Then randomly select Position 1 from the list π_{k_1} of factory k_1 and Position 2 from the list π_{k_2} of factory k_2. Next, we exchange the jobs that occupy the two selected positions. The pseudo codes of the presented swap operator are given as follows (Fig. 1).

Procedure Swap(π)

$k_1 = rand[1,2,..f]$

$k_2 = rand[1,2,..f]$

$\text{Pos}_1 = rand\left[1,2,..\text{n}_{k_1}\right]$

$\text{Pos}_2 = rand\left[1,2,..\text{n}_{k_2}\right]$

Interchange job π_{k_1,pos_1} and job π_{k_2,pos_2}

return π

Fig. 1. The swap operator for DLSFP

For each solution in the population, we perform the above swap operator to generate a neighboring solution.

3.4 Onlooker Bee Phase

In the onlooker bee phase, there are a total of *PS* onlooker bees. Each onlooker bee corresponds to a solution that is selected according to the winning probability of the solutions. We consider the binary tournament selection that has been successfully utilized in the DABC algorithm [22, 23]. After the selection, the onlooker utilizes the insertion operator to produce a neighboring solution. For the insertion, we firstly select two factories k_1 and k_2 randomly with uniform distribution, then randomly remove a job from the list π_{k_1} and insert the job into a randomly chosen position at list π_{k_2}. The pseudo code is given as follows (Fig. 2):

Procedure Insertion (π)
$k_1 = rand[1,2,..f]$
$k_2 = rand[1,2,..f]$
$\mathrm{Pos}_1 = rand\left[1,2,..\mathrm{n}_{k_1}\right]$
$\mathrm{Pos}_2 = rand\left[1,2,..\mathrm{n}_{k_2}\right]$
Remove job π_{k_1,pos_1} from π_{k_1}
Insert job π_{k_1,pos_1} into Pos_2 at π_{k_2}
return π

Fig. 2. The insertion operator for DLSFP

After a neighboring solution is generated, we compare it with the worst solution in the population. If the neighboring solution has better makespan value than the worst solution and there is no other identical solution in the population, the neighboring solution will replace the worst solution and enter the population. The selection and updating process are repeated *PS* times.

3.5 Scout Bee Phase

The scout phase abandons a solution that has not been improved in a given number of consecutive iterations and inserts a new solution into the population. To generate a promising solution, we perform a number of swap operators to the abandoned solution. The generated solution enters the population.

3.6 Computational Procedure of the DABC Algorithm

With the above design, the procedure of the DABC algorithm for solving the DLFSP with makespan is outlined in Fig. 3.

```
Procedure DABC(PS, α)
  Generate an initial population {x_1,...,x_PS} using heuristics
  x* := Best solution in the population
  while (teriminiation criterion is not satisfied) do
    for i := 1 to PS do   //empolyed bees phase
      x'_i := a swap to x_i
      if TF(x'_i) < TF(x_i) do
        x_i := x'_i
      endif
    endfor
    for i := 1 to PS do   //onlooker bees phase
      x := solution selected by tourmonent selection
      x' := an insertion to x
      x^w := worst solution in the population
      if TF(x') < TF(x^w) and no solution is idenctial to x', then
        x^w := x'
      endif
    endfor
    for i=1 to PS do   //scout bees phase
      if x_i is not improved in α continuous iterations, then
        x'_i := scout solution generated
        x_i := x'_i
      endif
    endfor
    x^b := best solution among {x_1,...,x_PS}
    if TF(x^b) < TF(x*) then
      x* = x^b
    endif
  endwhile
return x*
```

Fig. 3. The presented DABC algorithm

4 Experimental Calibration

In this section, we calibrate the presented DABC algorithm by a design of experiments (DOE) and analysis of variance (ANOVA). We conduct a preliminary experiment to determine the levels for each parameter. *PS* is fix at four levels: 20, 40, 60, and 80; α is set at six levels: 100, 200, 300, 400, 500, and $+\infty$, where $\alpha = +\infty$ means that the scout bees are not used. This leads to a total of 4 * 6 = 24 configurations for the presented DABC algorithm.

We generate a total of 81 instances with $f \in \{2,4,6\}$, $m \in \{5,10,20\}$, $n \in \{100,200,300\}$ and $s_j \in \{20,30,40\}$. Each instance is corresponding to a combination of f, m, n, and s_j. The processing time for each sub-lot s_j is generated between 1 and 99 with a uniform distribution, *i.e.*, $p_{ij} \in U(1,99)$.

We code the DABC algorithms using Microsoft Visual Studio 2015. We carry out experiments based on the above 81 instances. For each configuration of the DABC algorithm, we carry out five independent replications on the Intel (R) Core (TM) i7-2600 CPU @ 3.40 GHz with 8.00 GB RAM in the Windows 7 Operation System. The termination criterion is set as the maximum CPU time of $t = 30 \cdot m \cdot n$ milliseconds. We compute the relative percentage increase (RPI) as a response variable as follows:

$$RPI = \frac{100 \times \left(C_{\max} - C^*_{\max}\right)}{C^*_{\max}} \tag{1}$$

where C_{max} is a solution generated in a replication for an instance. C^*_{max} is the best solution found in the calibration experiments. Following Ruiz and Stützle, Pan and Ruiz, and many others [25–28], all the results are analyzed by ANOVA. All the two parameters are statistically significant at a 95% confidence level. The factors are fixed as follows: $\alpha = 300$ and $PS = 40$.

5 Computational Evaluation

We compare the presented DABC algorithm with the Scatter Search (SS) algorithm [3] and the hybrid immune (HI) algorithm [14]. These two algorithms were very effective for solving the distributed permutation flowshop scheduling problem with makespan criterion. We adapt them to the DLFSP with makespan criterion and calibrate them using ANOVA.

Table 1. Average RPI values for $t = 10mn$ milliseconds

$n \times m \times s_i$	$f = 2$			$f = 4$			$f = 6$		
	DABC	HI	SS	DABC	HI	SS	DABC	HI	SS
100 × 5 × 20	1.65	2.08	2.53	1.65	1.88	2.39	1.20	1.73	1.58
100 × 5 × 30	1.72	2.33	2.23	1.69	1.70	2.19	1.56	1.77	2.02
100 × 5 × 40	1.95	2.19	2.21	1.99	1.78	2.25	1.67	1.61	1.99
100 × 10 × 20	1.75	1.98	2.30	1.84	1.94	2.32	1.67	1.57	1.97
100 × 10 × 30	1.79	2.00	2.55	1.77	2.12	2.27	1.37	1.47	1.95
100 × 10 × 40	1.71	2.13	2.37	1.85	1.82	2.21	1.37	1.90	1.77
100 × 20 × 20	1.76	2.21	2.33	1.93	2.06	1.99	1.29	1.65	1.84
100 × 20 × 30	1.96	2.23	2.38	1.55	1.74	2.13	1.36	1.87	1.94
100 × 20 × 40	2.13	1.95	2.56	1.56	1.95	2.09	1.50	1.75	1.94
200 × 5 × 20	2.13	2.09	2.26	1.92	2.06	1.96	1.28	1.58	1.95
200 × 5 × 30	2.05	2.22	2.17	1.77	1.76	2.29	1.57	1.83	1.67
200 × 5 × 40	1.69	2.03	2.16	1.53	1.92	2.10	1.46	1.69	2.00
200 × 10 × 20	1.76	2.11	2.51	1.90	1.77	2.18	1.34	1.86	1.63
200 × 10 × 30	2.10	2.29	2.33	1.70	1.83	1.98	1.66	1.72	2.02
200 × 10 × 40	1.92	2.08	2.14	1.54	2.02	2.25	1.40	1.87	2.04
200 × 20 × 20	1.96	2.04	2.55	1.95	2.15	2.19	1.30	1.66	1.81
200 × 20 × 30	1.99	1.94	2.45	1.53	1.97	2.30	1.36	1.64	1.83
200 × 20 × 40	1.80	2.21	2.12	1.90	2.00	1.91	1.32	1.76	1.64
300 × 5 × 20	2.13	2.21	2.47	1.86	2.10	2.05	1.58	1.72	1.59
300 × 5 × 30	1.84	2.00	2.48	1.50	1.87	1.95	1.49	1.61	1.83
300 × 5 × 40	2.08	2.33	2.51	1.95	1.92	2.31	1.20	1.59	1.63
300 × 10 × 20	1.82	2.36	2.11	1.83	2.11	2.05	1.50	1.63	1.94
300 × 10 × 30	2.12	2.18	2.46	1.98	2.01	2.39	1.42	1.50	1.83
300 × 10 × 40	1.68	1.99	2.43	1.61	1.70	1.96	1.57	1.58	1.98
300 × 20 × 20	1.86	2.01	2.41	1.87	1.85	1.99	1.49	1.59	1.65
300 × 20 × 30	1.87	2.15	2.38	1.75	2.08	2.18	1.57	1.55	1.58
300 × 20 × 40	1.73	2.07	2.32	1.74	1.75	2.36	1.61	1.85	1.65
Mean	1.89	2.13	2.36	1.77	1.92	2.16	1.45	1.69	1.83

We generate a total of 810 instances with $f \in \{2, 4, 6\}$, $m \in \{5, 10, 20\}$, $n \in \{100, 200, 300\}$ and $s_j \in \{20, 30, 40\}$. That is, for each combination of f, m, n, and s_j, we randomly yield 10 instances. All the algorithms are tested to solve the 810 instances. We stop the algorithms when the elapsed CPU time limit of $t = \rho \cdot m \cdot n$ milliseconds is reached, where ρ has been tested at 3 values: 10, 20, and 30. Five independent replications are executed for each algorithm. The average RPI values, grouped by instance size, *i.e.*, $n \times m \times s_i$, are reported in Tables 1, 2 and 3.

Table 2. Average RPI values for $t = 20mn$ milliseconds

$n \times m \times s_i$	$f = 2$			$f = 4$			$f = 6$		
	DABC	HI	SS	DABC	HI	SS	DABC	HI	SS
100 × 5 × 20	1.55	2.07	2.26	1.31	1.57	1.92	1.00	1.43	1.50
100 × 5 × 30	1.68	1.80	2.05	1.35	1.64	2.05	1.24	1.43	1.72
100 × 5 × 40	1.64	1.88	2.11	1.68	1.75	1.95	1.07	1.24	1.76
100 × 10 × 20	1.54	1.95	2.04	1.68	1.56	1.72	1.46	1.44	1.51
100 × 10 × 30	1.60	1.90	1.91	1.40	1.56	1.86	1.07	1.20	1.78
100 × 10 × 40	1.46	2.08	1.94	1.45	1.79	1.78	1.09	1.25	1.37
100 × 20 × 20	1.66	2.00	2.20	1.71	1.78	1.98	1.29	1.24	1.78
100 × 20 × 30	1.81	2.19	2.05	1.38	1.54	1.95	1.16	1.26	1.61
100 × 20 × 40	1.49	1.81	2.18	1.55	1.90	2.04	1.20	1.25	1.63
200 × 5 × 20	1.77	1.77	1.93	1.73	1.93	1.81	1.17	1.53	1.61
200 × 5 × 30	1.90	1.90	1.93	1.41	1.57	2.15	1.31	1.31	1.62
200 × 5 × 40	1.66	1.82	1.94	1.47	1.77	1.83	1.20	1.33	1.35
200 × 10 × 20	1.64	1.74	2.23	1.57	1.52	2.17	1.07	1.38	1.47
200 × 10 × 30	1.50	1.79	1.90	1.46	1.59	1.80	1.33	1.48	1.62
200 × 10 × 40	1.58	1.72	2.09	1.40	1.73	1.92	1.30	1.64	1.44
200 × 20 × 20	1.80	1.83	2.26	1.50	1.65	1.88	1.14	1.25	1.68
200 × 20 × 30	1.50	1.71	1.97	1.40	1.52	1.78	1.11	1.20	1.78
200 × 20 × 40	1.59	1.90	2.10	1.74	1.54	1.83	1.19	1.51	1.62
300 × 5 × 20	1.55	1.83	2.29	1.54	1.93	1.99	1.10	1.55	1.49
300 × 5 × 30	1.57	1.84	1.98	1.47	1.71	1.87	1.03	1.44	1.70
300 × 5 × 40	1.88	2.02	2.28	1.76	1.55	1.91	1.03	1.54	1.42
300 × 10 × 20	1.46	2.02	2.02	1.69	1.57	1.87	1.05	1.56	1.67
300 × 10 × 30	1.79	1.85	2.02	1.63	1.72	1.80	1.09	1.22	1.62
300 × 10 × 40	1.51	1.87	2.11	1.38	1.67	1.96	1.48	1.26	1.73
300 × 20 × 20	1.48	1.76	2.19	1.76	1.69	1.89	1.10	1.35	1.37
300 × 20 × 30	1.66	2.07	1.95	1.62	1.63	1.73	1.28	1.40	1.35
300 × 20 × 40	1.48	2.05	2.22	1.74	1.57	1.85	1.02	1.33	1.46
Mean	1.62	1.90	2.08	1.55	1.66	1.90	1.17	1.37	1.58

It can be seen from Tables 1 through 3 that the mean average RPI values generated by the presented DABC algorithm are much smaller than those of the HI and SS algorithms nevertheless different CPU times and different factories involved. For a given termination condition and factory configuration, our DABC algorithm performs much better than the HI and SS algorithms almost for all the instance sizes. With the CPU time increases, all algorithms improve their results. But, on average, our DABC algorithm still outperforms the HI and SS algorithms at a considerable margin.

Table 3. Average RPI values for $t = 30mn$ milliseconds

$n \times m \times s_i$	$f = 2$			$f = 4$			$f = 6$		
	DABC	HI	SS	DABC	HI	SS	DABC	HI	SS
100 × 5 × 20	1.54	1.64	2.01	1.20	1.43	1.61	0.93	1.23	1.47
100 × 5 × 30	1.42	1.75	1.76	1.18	1.49	1.97	0.95	1.28	1.25
100 × 5 × 40	1.60	1.63	2.01	1.32	1.38	1.62	1.00	1.13	1.49
100 × 10 × 20	1.54	1.92	2.03	1.51	1.53	1.60	1.12	1.24	1.43
100 × 10 × 30	1.58	1.80	1.78	1.25	1.40	1.69	0.98	1.08	1.31
100 × 10 × 40	1.36	1.89	1.85	1.37	1.62	1.64	1.08	1.10	1.35
100 × 20 × 20	1.40	1.78	2.07	1.46	1.54	1.67	1.03	1.12	1.66
100 × 20 × 30	1.32	1.71	1.84	1.31	1.50	1.63	1.00	1.08	1.50
100 × 20 × 40	1.46	1.70	2.08	1.34	1.60	1.82	0.85	1.24	1.38
200 × 5 × 20	1.39	1.56	1.82	1.45	1.36	1.58	1.01	1.45	1.51
200 × 5 × 30	1.37	1.68	1.90	1.31	1.49	1.94	1.03	1.16	1.23
200 × 5 × 40	1.35	1.62	1.77	1.35	1.70	1.55	0.97	1.27	1.27
200 × 10 × 20	1.30	1.58	2.10	1.49	1.39	1.58	1.00	1.16	1.46
200 × 10 × 30	1.34	1.66	1.81	1.17	1.47	1.78	0.93	1.26	1.49
200 × 10 × 40	1.50	1.65	2.05	1.23	1.65	1.89	0.95	1.25	1.42
200 × 20 × 20	1.32	1.75	2.11	1.35	1.54	1.62	1.03	1.15	1.42
200 × 20 × 30	1.34	1.70	1.95	1.20	1.37	1.58	1.01	1.08	1.39
200 × 20 × 40	1.31	1.56	2.06	1.44	1.52	1.72	0.91	1.50	1.59
300 × 5 × 20	1.39	1.68	1.84	1.19	1.41	1.79	0.86	1.14	1.48
300 × 5 × 30	1.42	1.56	1.77	1.47	1.56	1.82	0.98	1.09	1.59
300 × 5 × 40	1.45	1.92	2.19	1.41	1.53	1.63	0.89	1.45	1.22
300 × 10 × 20	1.33	1.77	1.95	1.16	1.46	1.79	0.93	1.55	1.51
300 × 10 × 30	1.44	1.82	2.00	1.53	1.72	1.79	0.98	1.07	1.42
300 ×10 × 40	1.33	1.71	1.96	1.37	1.54	1.92	1.08	1.10	1.22
300 × 20 × 20	1.38	1.61	2.10	1.48	1.45	1.64	0.95	1.11	1.20
300 × 20 × 30	1.53	1.67	1.78	1.52	1.45	1.56	1.25	1.29	1.31
300 × 20 × 40	1.32	1.95	1.90	1.19	1.42	1.74	0.97	1.18	1.22
Mean	1.41	1.71	1.94	1.34	1.50	1.71	0.99	1.21	1.40

6 Conclusions

In this paper, we dealt with a distributed lot-streaming flowshop scheduling problem with makespan criterion. We presented a discrete artificial bee colony algorithm for solving such a complex problem with potential applications. In the algorithm, we adopt the multi-list based representation to represent candidate solutions and design an insertion operator and a swap operator to generate neighboring solutions. The proposed DABC algorithm started from an initial population formed by two problem-knowledge-based heuristics, and then repeats with the employ bee phase, the onlooker bee phase, and the scout bee phase. Numerical experiments demonstrated that the presented DABC algorithm performed much better than the existing scatter search algorithm and hybrid immune algorithm for solving the distributed lot-streaming flowshop problem with the objective of minimizing makespan. In the future, we will further improve the DABC algorithm for solving scheduling problems including hot-rolling scheduling, steelmaking scheduling, flexible job shop, and etc.

Acknowledgements. This research is partially supported by the National Science Foundation of China 51575212 and 61174187, and Shanghai Key Laboratory of Power station Automation Technology.

References

1. Pan, Q.K., Ruiz, R.: Local search methods for the flowshop scheduling problem with flowtime minimization. Eur. J. Oper. Res. **222**(1), 31–41 (2012)
2. Wang, S.Y., Wang, L., Liu, M., Xu, Y.: An effective estimation of distribution algorithm for solving the distributed permutation flow-shop scheduling problem. Int. J. Prod. Econ. **145**, 387–396 (2013)
3. Naderi, B., Ruiz, R.: A scatter search algorithm for the distributed permutation flowshop scheduling problem. Eur. J. Oper. Res. **239**, 323–334 (2014)
4. Chang, J.H., Chiu, H.N.: A comprehensive review of lot streaming. Int. J. Prod. Res. **43**(8), 1515–1536 (2005)
5. Pan, Q.K., Ruiz, R.: An estimation of distribution algorithm for lot-streaming flow shop problems with setup times. Omega **40**(2), 166–180 (2012)
6. Naderi, B., Ruiz, R., Zandieh, M.: Algorithms for a realistic variant of flowshop scheduling. Comput. Oper. Res. **37**, 236–246 (2010)
7. Liu, H., Gao, L.: A discrete electromagnetism-like mechanism algorithm for solving distributed permutation flowshop scheduling problem. In: Proceedings of the 6th International Conference on Manufacturing Automation, ICMA 2010, pp. 156–163. IEEE Computer Society (2010)
8. Gao, J., Chen, R.: An NEH-based heuristic algorithm for distributed permutation flowshop scheduling problems. Technical report SRE-10-1014. College of Information Science and Technology, Dalian Maritime University, Dalian (2011)
9. Gao, J., Chen, R.: A hybrid genetic algorithm for the distributed permutation flowshop scheduling problem. Int. J. Comput. Intell. Syst. **4**, 497–508 (2011)
10. Gao, J., Chen, R., Deng, W., Liu, Y.: Solving multi-factory flowshop problems with a novel variable neighborhood descent algorithm. J. Comput. Inf. Syst. **8**, 2025–2032 (2012)
11. Gao, J., Chen, R., Deng, W.: An efficient tabu search algorithm for the distributed permutation flowshop scheduling problem. Int. J. Prod. Res. **51**, 641–651 (2013)
12. Lin, S.W., Ying, K.C., Huang, C.Y.: Minimising makespan in distributed permutation flowshops using a modified iterated greedy algorithm. Int. J. Prod. Res. **51**, 5029–5038 (2013)
13. Fernandez-Viagas, V., Framinan, J.M.: A bounded-search iterated greedy algorithm for the distributed permutation flowshop scheduling problem. Int. J. Prod. Res. **53**, 1111–1123 (2015)
14. Xu, Y., Wang, L., Wang, S.Y., Liu, M.: An effective hybrid immune algorithm for solving the distributed permutation flow-shop scheduling problem. Eng. Optim. **46**(9), 1269–1283 (2014)
15. Yoon, S.H., Ventura, J.A.: Minimizing the mean weighted absolute deviation from due dates in lot-streaming flow shop scheduling. Comput. Oper. Res. **29**(10), 1301–1315 (2002)

16. Tseng, C.T., Liao, C.J.: A discrete particle swarm optimization for lot-streaming flowshop scheduling problem. Eur. J. Oper. Res. **191**(2), 360–373 (2008)
17. Pan, Q.K., Tasgetiren, M.F., Suganthan, P.N., Chua, T.J.: A discrete artificial bee colony algorithm for the lot-streaming flow shop scheduling problem. Inf. Sci. **181**, 2455–2468 (2011)
18. Pan, Q.K., Suganthan, P.N., Liang, J.J., Tasgetiren, M.F.: A local-best harmony search algorithm with dynamic sub-harmony memories for lot-streaming flow shop scheduling problem. Expert Syst. Appl. **38**, 3252–3259 (2011)
19. Meng, T., Pan, Q.K.: An improved fruit fly optimization algorithm for solving the multidimensional knapsack problem. Appl. Soft Comput. **50**, 79–93 (2017)
20. Han, Y.Y., Gong, D.W., Sun, X.Y., Pan, Q.K.: An improved NSGA-II algorithm for multi-objective lot-streaming flow shop scheduling problem. Int. J. Prod. Res. **52**(8), 2211–2231 (2017)
21. Karaboga, D.: An idea based on honey bee swarm for numerical optimization, Technical report TR06, Erciyes University, Engineering Faculty, Computer Engineering Department (2005)
22. Pan, Q.K., Wang, L., Li, J.Q., Duan, J.H.: A novel discrete artificial bee colony algorithm for the hybrid flowshop scheduling problem with makespan minimization. Omega **45**, 42–56 (2014)
23. Pan, Q.K., Gao, L., Li, X.Y., Gao, K.Z.: Effective metaheuristics for scheduling a hybrid flowshop with sequence-dependent setup times. Appl. Math. Comput. **303**, 89–112 (2017)
24. Cui, Z., Gu, X.S.: An improved discrete artificial bee colony algorithm to minimize the makespan on hybrid flow shop problems. Neurocomputing **48**(19), 248–259 (2015)
25. Pan, Q.K.: An effective co-evolutionary artificial bee colony algorithm for steelmaking-continuous casting scheduling. Eur. J. Oper. Res. **250**, 702–714 (2016)
26. Ruiz, R., Stützle, T.: A simple and effective iterated greedy algorithm for the permutation flowshop scheduling problem. Eur. J. Oper. Res. **177**, 2033–2049 (2007)
27. Ruiz, R., Stützle, T.: An iterated greedy heuristic for the sequence dependent setup times flowshop problem with makespan and weighted tardiness objectives. Eur. J. Oper. Res. **187**, 1143–1159 (2008)
28. Ruiz, R., Maroto, C., Alcaraz, J.: Two new robust genetic algorithms for the flowshop scheduling problem. Omega **34**, 461–476 (2006)

Extremized PICEA-g for Nadir Point Estimation in Many-Objective Optimization

Rui Wang[1(✉)], Meng-jun Ming[1], Li-ning Xing[1], Wen-ying Gong[2], and Ling Wang[3]

[1] College of Systems Engineering, National University of Defense Technology, Changsha 410073, Hunan, People's Republic of China
ruiwangnudt@gmail.com

[2] School of Computer Science, China University of Geosciences, Wuhan 430074, People's Republic of China

[3] Department of Automation, Tsinghua University, Beijing 100084, People's Republic of China

Abstract. Nadir point, constructed by the worst Pareto optimal objective values, plays an important role in multi-objective optimization and decision making. For example, the nadir point is often a pre-requisite in many multi-criterion decision making approaches. Along with the ideal point, the nadir point can be applied to normalize solutions so as to facilitate a comparison and aggregation of objectives. Moreover, nadir point is useful in visualization software catered for multi-objective optimization. However, the estimation of nadir point is still a challenging problem, particularly, for optimization and/or decision-making problems with many objectives. In this paper, a modified preference-inspired coevolutionary algorithm using goal vectors (PICEA-g) called *extremized* PICEA-g is proposed to estimate the nadir point. The *extremized* PICEA-g, denoted as e-PICEA-g, is an $(N+N)$ elitist algorithm and employs a two-phase selection strategy. In the first-phase $(N+K)$ solutions are selected out from the overall $2N$ solutions based on the dominance-level and an angle based closeness indicator. In the second-phase the selected $(N+K)$ solutions are further filtered by removing K poor ones in terms of their fitness calculated by a slightly modified PICEA-g fitness scheme. By the two-phase selection strategy, the e-PICEA-g skillfully harnesses the advantages of *edge-point-to-nadir* and *extreme-to-nadir* principles. Experimental results demonstrate the efficiency and effectiveness of the e-PICEA-g on many-objective optimization benchmarks with up to 13 objectives.

Keywords: Nadir point · Evolutionary algorithm · PICEA-g · Extremized Many-objective optimization

1 Introduction

It has been widely accepted that the nadir point plays an important role in multi-objective optimization and/or multi-criteria decision making procedures. Along with the ideal point (which is constructed by the best objective vector in each objective function), the nadir point can be applied to normalize objective values. This enables

D.-S. Huang et al. (Eds.): ICIC 2018, LNAI 10956, pp. 807–814, 2018.
https://doi.org/10.1007/978-3-319-95957-3_85

evolutionary multi-objective algorithms to work more reliably and efficiently, avoiding the influence of different objective scales. Also, with the aid of ideal and nadir points, the Pareto optimal front can be properly visualized. A proper visualization is helpful in a number of ways such as the algorithm performance comparison, the selection of preferred solutions. Meanwhile, decision-makers can appropriately adjust their expectations on objectives, focusing on a desired region of the Pareto optimal front. Lastly, the nadir point is a pre-requisite in many existing multi-criteria decision making tools such as the GUESS method [1] and the NIMBUS method [2]. However, unlike the estimation of ideal point, finding the nadir point remains challenging, in particular, for problems with more than three objectives.

Evolutionary based approaches for nadir estimation can be classified into three types – *surface-to-nadir*, *edge-to-nadir*, and *extreme-to-nadir* [3]. In order to harness the advantages of both *edge-to-nadir*, and *extreme-to-nadir* principles, *extremized* PICEA-g is proposed. The algorithm, denoted as e-PICEA-g, builds on an early many-objective optimizer PICEA-g [4], and adopts an $(N+N)$ elitist framework where N is the initial population size. Moreover, a two-phase selection strategy is used in the e-PICEA-g. The first-phase selects $(N+K)$ solutions based on the solutions' dominance-level and their closeness to extreme points. In other words, $2N-(N+K)$ solutions that are neither close to extreme points nor close to edge of the PF are removed. In the second-phase, N solutions are selected based on their fitness from the $N+K$ solutions resulted from the first-phase. The first-phase is prone to extremized solutions while the second-phase, by using a modified PICEA-g fitness scheme, prefer solutions that are close to or at the edges of the Pareto optimal front. The combination of *edge-like* and *extreme-like* solutions enables e-PICEA-g to strike a balance between convergence and diversity, and thus, leading an excellent performance in terms of nadir point estimation for multi-objective benchmarks with up to 13 objectives. Experimental results demonstrates that e-PICEA-g finds more accurate nadir point than the PICEA-g and the *extremized* crowded NSGA-II [5].

The rest of this paper is organized as follows: in Sect. 2 the basic PICEA-g is introduced. Section 3 elaborates the *extremized* PICEA-g. Experimental description, experimental results and discussion are provided in Sects. 4 and 5, respectively. Finally, Sect. 6 concludes the paper and identities future studies.

2 Basic PICEA-g

The preference-inspired co-evolutionary algorithm using goal vectors (PICEA-g) is a recently proposed algorithm [4, 11] which has shown promising performance on multi/many-objective problems. It is implemented within an $(N+N)$ elitist framework as the NSGA-II [6]. Specifically, candidate solutions and goal vectors, S and G, of size, N and N_g, are co-evolved for a fixed number of generations. In each generation t, parents $S(t)$ are subjected to (representation-appropriate) genetic operators (e.g., the simulated binary crossover and polynomial mutation operators [6]) to produce N off-spring, $Sc(t)$. Simultaneously, N_g new goal vectors, $Gc(t)$, are randomly re-generated in objective space based predefined bounds. $S(t)$ and $Sc(t)$, and $G(t)$ and $Gc(t)$ are then pooled respectively and the combined population is sorted according to the fitness.

Truncation selection is applied to select the best N solutions as new parent population, $S(t+1)$ and $G(t+1)$.

3 *Extremized* PICEA-g for Nadir Point Estimation

The *extremized* PICEA-g employs the same elitist framework as PICEA-g while a different selection mechanism. Specifically, in each generation of e-PICEA-g, a two-phase selection is conducted, from the union of $2N$ parent and offspring solutions, to select N parent solutions for the next generation. The first-phase selects $N+K$ solutions, and the second-phase further removes K solutions. Next we describe how solutions are selected in the first and second phase, respectively.

- **The first-phase selection:** solutions are ranked based on their dominance-level measured by the Pareto-dominance relation and their closeness to extreme points measured by the angle between the solution and the extreme points which is calculated by Eq. (1).

$$\theta^{s} = \min_{i=1,2,\ldots,m}(\theta_i), \theta_i = \arccos(\mathbf{z}^{*}\mathbf{s} \cdot \mathbf{z}^{*}\mathbf{e}_i) \quad (1)$$

  where $\mathbf{z}^{*}$ is the estimated ideal point, i.e., $\mathbf{z}^{*} = \left(f_1^{min}, f_2^{min}, \ldots, f_m^{min}\right)$. θ_i is the angle between solution $\mathbf{s}$ and the ith extreme point $\boldsymbol{e}_i$. For each solution, θ^{s}, the minimum of θ_i, is taken to measure the closeness finally. Note that the dominance-level is the first criterion while the closeness indicator θ^{s} serves as an addition criterion. That is, a lexicographical order is built by the two criteria to select $N+K$ solutions from the whole $2N$ solutions.
- **The second-phase selection:** the remaining $N+K$ solutions are ranked based on their fitness computed using a slightly modified PICEA-g fitness scheme, see Eq. (2).

$$\mathrm{Fit}_{s} = \max_{\forall s}(rank(\mathbf{s})) - rank(\mathbf{s}) + exp^{(-1/1+F_s)} \quad (2)$$

  where $rank(\mathbf{s})$ refers to the non-dominance rank of $\mathbf{s}$; $F_{\mathbf{s}}$ denotes the original fitness calculated by the PICEA-g fitness assignment scheme [4, 10]. Note that the original PICEA-g fitness assignment is not Pareto-dominance compliant while by Eq. (2) the fitness becomes Pareto-dominance compliant.
  In addition, in PICEA-g goal vectors are randomly generated inside the hypercube bounded by the estimated ideal and nadir point. However, in the e-PICEA-g the generation of goal vectors varies along with solutions. That is, for each solution $\mathbf{s}$, two goal vectors are randomly generated inside the hypercube bounded by the $\mathbf{s}$ and $1.2 \times \mathbf{s}$ in objective space. Therefore, there are in total $N_g = 2N$ new goal vectors generated in each generation.
  It is worth mentioning that K is a parameter in the two-phase selection. In the study, K gradually increases as the search progresses, see Eq. (3). This indicates that at the beginning of the search, the first-phase selection is dominant while as the search

progresses the second-phase selection becomes dominant. Such a design aims to strike a balance between convergence and diversity.

$$K = \lfloor N \times t/(maxGen - 1) \rfloor \tag{3}$$

Overall, the pseudo-code of e-PICEA-g is show as follows.

Algorithm 1. e-PICEA-g

Input: Initial candidate solutions, S of size N, initial goal vectors, G of size N_g, maximum number of generations, $maxGen$, the number of objectives, m, offline archive, $archiveS$, archive size, $Asize$, mating probability, p, neighbourhood size, num

Output: nadir point

```
1   S = initializeS (N);
2   FS = objectiveFunction (S);
3   G = goalGenerator (N_g, [FS, 1.2 ×FS]);
4   for t =1 to maxGen do
5       S' = matingRestriction (S, p, num );
6       Sc = geneticOperation (S');
7       FSc = objectiveFunction (Sc);
8       (JointS, JointF) = multisetUnion (S, Sc, FS, FSc);
9       RankJointF = nondominatedSort (JointF);
10      ExtremeAngle = extremeAngle(extremeP, JointF);
11      ixS = phase1sel ([RankJointF, ExtremeAngle],N + K);
12      Gc = goalGenerator (N_g , [FS, 1.2 ×FS]);
13      JointG = multisetUnion(G, Gc);
14      (FitJointF, FitJointG) = fitness(JointF(ixS), JointG);
15      (S, FS, G) =  phase2sel(FitJointF, JointF, JointS, JointG, N);
16      archiveS = updateArchive(archiveS, FS, Asize);
17      nadir =  nadirByextremeP(archiveS);
18  end
```

4 Experiment Description

In order to examine the performance of e-PICEA-g, test problems 2–9 from the WFG test suite [7], invoked in 2-, 7- and 13-objective instances, are used. In each case the WFG position parameter and distance parameter are both set to 12, though other parameters can be used, too. The ideal and nadir points for these problems are [0, 0,…, 0] and [2, 4,…, 2 m], respectively.

The *extremized* crowded NSGA-II [5], denoted as e-NSGA-II, is selected as the competitor algorithm. The e-NSGA-II is built on the NSGA-II wherein a modified crowding distance is adopted. That is, solutions on a particular non-dominated front are first sorted from 1 to N_f (assuming that there are N_f solutions on this front) based on each objective. Solutions that are closer to either extreme objective values get higher ranks compared to that of intermediate solutions. Thus, the rank of the *i*th solution for the *m*th objective $R_i^m$is $\max\{R_i^m, N_f - R_i^m + 1\}$. The *extremized* crowding distance of a solution is set to the maximum value of the assigned rank amongst different objectives. Also, the original PICEA-g is considered as a reference algorithm.

For all test problems, each algorithm is run for 31 runs subjected to a statistical analysis, and each run for $maxGen = 100$ generations. The population size N is set to 100. The SBX and PM operators are chosen as genetic operators. As recommended in [8, 9], the SBX control parameters p_c and η_c are set to 1 and 30, respectively. The PM control parameters p_m and η_m are set to $1/n$ and 20 where n is the number of decision variables.

To evaluate the accuracy of the estimated nadir point, the error metric, E is used [10]. The definition of E is presented in Eq. (4). The metric describes the distance from the estimated nadir point to the true nadir point. Thus, a smaller E indicates a better estimation of nadir point.

$$E = \sqrt{\sum\nolimits_{i=1}^{m} \left(\frac{z_i - z_i^{ide}}{z_i^{nad} - z_i^{ide}} \right)^2} \quad (4)$$

where $\mathbf{z} = (z_1, z_2, \ldots, z_m)$ is the estimated nadir point, and z_i^{ide} and z_i^{nad} denote the real ideal and nadir point.

5 Results and Discussion

Comparison results of e-PICEA-g with e-NSGA-II and PICEA-g are shown in Table 1 for 2-, 7- and 13-objective problems. The non-parametric Wilcoxon-Ranksum two-sided method at the 95% confidence level is applied to test whether the results are statistically different.

From the table, we can observe that

- For 2-objective problems, e-PICEA-g performs better than e-NSGA-II for three problems. Meanwhile, for WFG4, WFG7 and WFG8, e-NSGA-II is better than e-PICEA-g. For the remaining WFG3 and WFG5, the two algorithms perform comparably. With respect to the comparison of e-PICEA-g and PICEA-g, they are comparable on five problems. PICEA-g is better than e-PICEA-g on WFG6 and WFG8. The e-PICEA-g is better only for WFG2.
- For 7- and 13-objective problems, the superiority of e-PICEA-g becomes apparent. The algorithm outperforms e-NSGA-II for all problems except for WFG2 and WFG3. Also, it beats PICEA-g for all problems.

The comparison results clearly demonstrate that the proposed e-PICEA-g is more efficient in terms of nadir point estimation, especially for problems with more than three objectives. The reason for the good performance of e-PICEA-g for high dimensional problems is as follows. The base algorithm PICEA-g is a better many-objective optimizer in terms of capturing the whole PF [4], compared to NSGA-II.

Table 1. The E metric (mean/std) comparison results. The symbol −, = or + means the considered algorithm is statistically worse than, comparable to or better than e-PICEA-g. Note that the statistical results for 7- and 13-objective problems are the same, and thus results for m = 7 are not shown here.

		e-NSGA-II	PICEA-g	e-PICEA-g
m = 2	WFG2	0.2348(0.1057)−	0.2140(0.1157)−	0.1620(0.1336)
	WFG3	0.0232(0.0049)=	0.0242(0.0064)=	0.0231(0.0072)
	WFG4	0.0141(0.0031)+	0.0142(0.0026)=	0.0166(0.0022)
	WFG5	0.0308(0.0016)=	0.0310(0.0026)=	0.0312(0.0028)
	WFG6	0.0423(0.0071)−	0.0389(0.0072)+	0.0401(0.0035)
	WFG7	0.0058(0.0017)+	0.0059(0.0016)=	0.0063(0.0009)
	WFG8	0.0891(0.0069)+	0.0827(0.0088)+	0.0927(0.0068)
	WFG9	0.0318(0.0163)−	0.0224(0.0129)=	0.0255(0.0134)
m = 13	WFG2	0.9289(0.2654)+	3.3927(0.0289)−	2.1595(0.1873)
	WFG3	0.4251(0.0981)+	3.3484(0.0338)−	1.4843(0.3128)
	WFG4	0.4677(0.0515)−	3.1340(0.0807)−	0.2677(0.1918)
	WFG5	0.6247(0.1019)−	3.0368(0.0997)−	0.2814(0.1400)
	WFG6	0.6063(0.0613)−	3.0335(0.0909)-	0.3092(0.1880)
	WFG7	0.5017(0.0295)−	3.0258(0.1731)−	0.2870(0.1620)
	WFG8	0.6493(0.0992)−	2.8728(0.1764)−	0.4241(0.0663)
	WFG9	0.5700(0.0157)−	2.8995(0.0756)−	0.3661(0.0799)

The PICEA-g does not suffer from the *dominance-resistant* issue [9]. The e-PICEA-g inherits this good feature, exhibiting better convergence performance.

Since the Pareto-dominance relation is able to differentiate solutions in 2-objective problems. The e-NSGA-II is effective on 2-objective problems. This explains the results that e-PICEA-g shows no superior performance on 2-objective problems. In addition, though the results for 4-objective problems are not presented here, the results show that e-NSGA-II still works comparably with e-PICEA-g on six problems.

It is also noted that e-PICEA-g performs worse than e-NSGA-II on 7- and 13-objective WFG2 and WFG3 problems. The reason might be that the PFs of the two problems are not *sphere-like* as the other WFG problems. This creates difficulty for e-PICEA-g. More specifically, the superiority of e-PICEA-g over e-NSGA-II is mainly attributed to the second-phase selection in which the fitness scheme of PICEA-g is used instead of the Pareto-dominance relation. Observed from Table 1, none of the methods (including e-NSGA-II) has found satisfactory nadir points for WFG2 and WFG3. This means that within 100 generations, the first-phase selection has not found enough solutions near the extreme regions of the PF. Thus, the effect of the second-phase selection cannot be activated, i.e., selecting more converged solutions amongst the solutions returned from the first-phase selection. When the maximum number of generations is increased to 250, experimental results show that e-PICEA-g becomes no worse than e-NSGA-II on both problems.

The above discussion also reveals that the linear adaptation of K as defined in Eq. (3) might not be the best. The adaptation is essentially problem dependent.

Our preliminary results show that for diversity-hard problems [11], a power value larger than 1 is helpful (e.g., $K = \lfloor N * (t/(maxGen - 1))^2 \rfloor$) since this implicitly enables the first-phase selection to play an active role, producing more solutions near the extreme regions of the PF. Likewise, for diversity-easy problems, a power value smaller than 1 is more helpful. However, this requires more systematic investigations.

6 Conclusion

Finding nadir point is an age-old problem while remains challenging for problems with three or more objectives. Nadir point is constructed by the worst objective value of the Pareto optimal solutions in each objective. Along with the ideal point, nadir point can be of great help to both multi-objective optimization and multi-criteria decision making, for example, normalizing solutions, assisting visualization, specifying decision-makers preferences. Till now, there have been a number of approaches including both non-heuristic, heuristic approaches investigating the estimation of nadir point. However, most of approaches have been shown less than satisfactory on many-objective problems. This study therefore proposes an effective evolutionary multi-objective approach, *extremized* PICEA-g to estimate the nadir point. In the e-PICEA-g, a two-phase selection mechanism is employed. The first-phase selects $N + K$ solutions from the whole population of size $2N$ based on the dominance-level and an angle indicator (measuring the closeness to extreme points). The second-phase selects N solutions from the $N + K$ solutions based on the fitness computed by the modified PICEA-g scheme. The e-PICEA-g implicitly harnesses the advantages of both *edge-to-nadir* and *extreme-to-nadir* principles, and thus, is found to be able to find more accurate nadir points than the state-of-the-art method, namely, *extremized* crowded NSGA-II on multi-objective problems with up to 13 objectives.

With respect to future studies, first, the e-PICEA-g seems to face difficulty on WFG2 problem whose PF is disconnected. This necessitates the research of e-PICEA-g on problems having complex geometries, and also, real-world problems. Second, though a linear adaptation of the parameter K provides good performance on the test problems, more effective strategies are required. Lastly, the two-phase selection mechanism might also be extended other types of MOEAs. Source code and experimental results of this study are available at http://ruiwangnudt.gotoip3.com/optimization.html.

Acknowledgment. This work was supported by the National key research and development plan (2016YFB0901900), the National Natural Science Foundation of China (Nos. 61773390), National Natural Science Fund for Distinguished Young Scholars of China (61525304) and Natural Science Fund for Distinguished Young Scholars of Hunan Province (2017JJ1001).

References

1. Buchanan, J.T.: A naive approach for solving MCDM problems: the GUESS method. J. Oper. Res. Soc. **48**(2), 202–206 (1997)
2. Miettinen, K., Mäkelä, M.M.: Synchronous approach in interactive multiobjective optimization. Eur. J. Oper. Res. **170**(3), 909–922 (2006)
3. Deb, K., Miettinen, K., Chaudhuri, S.: Toward an estimation of nadir objective vector using a hybrid of evolutionary and local search approaches. IEEE Trans. Evol. Comput. **14**(6), 821–841 (2010)
4. Wang, R., Purshouse, R.C., Fleming, P.J.: Preference-inspired co-evolutionary algorithms for many-objective optimisation. IEEE Trans. Evol. Comput. **17**(4), 474–494 (2013)
5. Deb, K., Chaudhuri, S., Miettinen, K.: Towards estimating nadir objective vector using evolutionary approaches. In: Proceedings of the Genetic and Evolutionary Computation Conference, GECCO 2006, pp. 643–650. ACM, New York (2006)
6. Deb, K., Pratap, A., Agarwal, S., Meyarivan, T.: A fast and elitist multiobjective genetic algorithm: NSGA-II. IEEE Trans. Evol. Comput. **6**(2), 182–197 (2002)
7. Huband, S., Hingston, P., Barone, L., While, L.: A review of multiobjective test problems and a scalable test problem toolkit. IEEE Trans. Evol. Comput. **10**, 477–506 (2006)
8. Wang, R., Mansor, M.M., Purshouse, R.C., Fleming, P.J.: An analysis of parameter sensitivities of preference-inspired co-evolutionary algorithms. Int. J. Syst. Sci. **46**(13), 423–441 (2015)
9. Purshouse, R.C., Fleming, P.J.: On the evolutionary optimization of many conflicting objectives. IEEE Trans. Evol. Comput. **11**, 770–784 (2007)
10. Wang, H., He, S., Yao, X.: Nadir point estimation for many-objective optimization problems based on emphasized critical regions. Soft. Comput. **21**, 1–13 (2015)
11. Wang, R., Xiong, J., Ishibuchi, H., Wu, G., Zhang, T.: On the effect of reference point in MOEA/D for multi-objective optimization. Appl. Soft Comput. **58**, 25–34 (2017)
12. Wang, R., Purshouse, R.C., Fleming, P.J.: On finding well-spread pareto optimal solutions by preference-inspired co-evolutionary algorithm. In: Proceedings of the 15th Annual Conference on Genetic and Evolutionary Computation, GECCO 2013, pp. 695–702. ACM, New York (2013)
13. Wang, R., Fleming, P.J., Purshouse, R.C.: General framework for localised multiobjective evolutionary algorithms. Inf. Sci. **258**(2), 29–53 (2014)

A New Selection Without Replacement for Non-dominated Sorting Genetic Algorithm II

Yechuang Wang[1], Zhuanghua Zhu[2], Maoqing Zhang[1], Zhihua Cui[1(✉)], and Xingjuan Cai[1]

[1] Complex System and Computational Intelligent Laboratory, Taiyuan University of Science and Technology, Taiyuan 030024, China
zhihua.cui@hotmail.com
[2] Shanxi Finance & Taxation College, Taiyuan 030024, Shanxi, China

Abstract. NSGA-II has shown good performance in solving multi-objective optimization problems, However, the tournament selection strategy in NSGA-II always generates many duplicate individuals. This phenomenon not only affects the crossover, mutation and updating operations and finally deteriorates the performance significantly. To overcome this problem, this paper introduces a new strategy, namely selection strategy without replacement, which can produces different individuals to increase the diversity. Simulation results show the proposed tournament selection without replacement achieve better performance.

Keywords: Multi-objective optimization algorithms · NSGA-II
Duplicate individuals · Selection strategy · Replacement

1 Introduction

With the development of society, many complex optimization problems [1–3] are proposed. To solve them, many researchers have proposed a number of effective algorithms [4–7] and tried to apply these algorithms to practical engineering problems [8–10]. In the 1980s, evolutionary algorithms were proposed, and quickly got spread due to its multiple features such as parallelism, intelligence, adaptability and self-organization. In 1985, Schaffer [11] tried to modify genetic algorithm for multi-objective optimization and proposed vector evaluation genetic algorithm (VEGA). In 2002, Zitzler and Thiele [12] proposed SPEA2. In the same year, Deb [13] proposed the classic algorithm of NSGA-II.

As a well-known algorithm, NSGA-II achieves good performance partly due to its fast non-dominated sorting mechanism and crowding distance assignment. Up to now, many improvements are proposed. Generally speaking, the rigorous selection is beneficial to the exploration of NSGA-II, while loose selection is beneficial to the exploration. According to this rule, Patel et al. [14] improved the convergence of the algorithm effectively by changing the focus of the selection operation. In order to solve the problem of independent task assignment, Salimi et al. [15] proposed one fuzzy adaptive mutation operator. Tran [16] propose adaptive parameter adjustment methods,

D.-S. Huang et al. (Eds.): ICIC 2018, LNAI 10956, pp. 815–820, 2018.
https://doi.org/10.1007/978-3-319-95957-3_86

so that the NSGA-II algorithm can adjust the size, crossover and mutation of the population according to the different characteristics of the corresponding problem. However, few research work consider it. Therefore, in this paper, we try to analyze the weakness of NSGA-II, and employ the selection strategy without replacement to reduce the duplicate individuals.

The rest of this paper is organized as follow: Sect. 2 the selection strategy without replacement is introduction briefly introduces key concepts and two classic strategies for the NSGA-II. After that, to investigate the influence, our methods are tested and compared with several classical algorithms in Sect. 3. Conclusion is draw in Sect. 4.

2 The NSGA-II with Selection Strategy Without Replacement

The population updating operation in NSGA-II is an important selection operation from the parent population to generate the offspring population. Generally, the tournament selection strategy is employed in NSGA-II. However, this strategy provides large chances for the good individuals to be selected into the offspring population. However, several duplicate individuals may appear in the same population. According to non-domination sorting strategy, there is no non-dominant relationship between repeated individuals. The repeated individuals are assigned the same rank, due to they are no non-dominant relationship. Regardless of whether have the duplicate individuals, the former K rank individuals will be selected into the new population if the number of individuals is less than N. When adding a K + 1 layer, the crowding distance of each individual is calculated if the population exceeds N, and the individual with larger distance is selected to the offspring population. However, the distance between the individual is not necessarily less than that of the non-repeated individual. As shown in Fig. 1, let $p_1, p_2, p_3, p_4, p_5, p_6$ belong to layer k, if p_2 and p_3 are two points with the same location, the crowding distance of $p_2(p_3)$ is obviously larger than the crowding distance of p_5, so the repeated individual $p_2(p_3)$ is preferentially selected into the new population. In summary, there are many duplicate individuals in R_l.

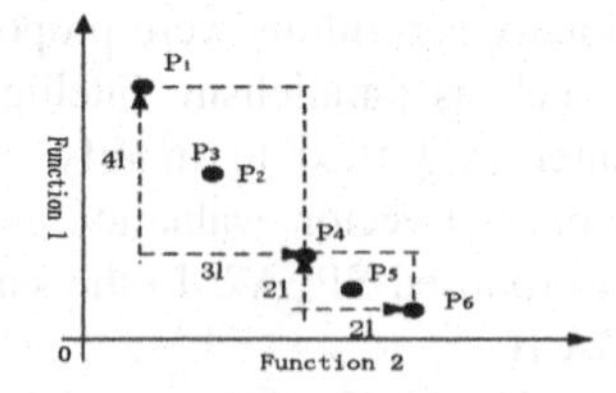

Fig. 1. Illustration of crowding distances of duplicate individuals

To verify our assumption, we did some experiments for NSGA-II, in which the experiment environment and parameter setting are the same as Sect. 4. The experiments show that for SCH, the final generation, there is only 75 different individuals, while for ZDT1, ZDT2, ZDT3, ZDT4 and ZDT6, the duplicate individuals are 23, 45, 39, 36 and 38, respectively. Of course, if we run NSGA-II with different times, this value is not the same. However, we perform NSGA-II 100 times, and the duplicate individuals are always no less than 20. This phenomenon implies that there are at least 20% individuals are duplicate for NSGA-II. Therefore, in this paper, the selection strategy without replacement is employed to avoid such case.

3 Simulation Results

3.1 Test Functions and Performance Measures

In order to test the performance of the algorithm, this paper chooses the ZDT [13] test suit and SCH [11] test suit. GD [17] and SP [17] are employed to evaluate the proposed method. The stopping criterion is a fixed number of o iteration (setting to 100), while population size n = 50 for all algorithms. The external population of size is set as 100 (Table 1).

Table 1. Means and variances of the performance for different selection strategies

Problems	Metrics	S1	S2	S3	S4	S5	S6
SCH	mean (GD)	7.63E–02	6.45E–02	7.42E–02	6.77E–02	6.89 E–02	**6.33E–02**
	std (GD)	3.93E–03	4.52E–03	3.55E–03	2.76E–03	3.55E–03	4.0E–03
	mean (SP)	2.88E–02	2.35E–02	2.57E–02	2.67E–02	2.30E–02	**2.27E–02**
	std (SP)	5.34E–03	2.51E–03	6.75E–03	2.12E–02	3.66E–03	1.33E–02
ZDT1	mean (GD)	6.88E–04	5.98E–05	6.14E–08	3.55E–04	3.47E–04	**2.83E–11**
	std (GD)	2.64E–04	2.99E–05	2.35E–08	2.16E–04	2.27E–04	**1.75E–11**
	mean (SP)	2.18E–03	3.25E–03	4.66E–03	2.05E–03	3.01E–03	4.8E–03
	std (SP)	2.15E–03	2.65E–04	2.33E–03	1.94E–03	1.82E–03	3.6E–03
ZDT2	mean (GD)	6.77E–04	5.47E–05	7.44E–11	6.02E–05	5.24E–05	**5.91E–11**
	std (GD)	2.67E–03	2.48E–05	3.55E–10	2.34 E–04	2.19E–04	**9.65E–11**
	mean (SP)	4.57E–03	4.42E–3	5.01E–03	3.18E–03	3.96E–03	4.8E–03
	std (SP)	3.99E–03	5.78E–04	5.66E–03	2.15E–03	4.32E–04	4.3E–03
ZDT3	mean (GD)	3.29E–04	3.75E–06	3.66E–10	2.75E–04	2.66E–05	**7.5E–11**
	std (GD)	2.52E–04	3.16E–06	3.57E–10	1.01E–04	1.87E–06	1.07E–10
	mean (SP)	5.97E–04	3.15E–03	3.22E–02	4.77E–03	2.57E–03	6.3E–03
	std (SP)	6.62E–04	7.52E–04	4.32E–02	5.11E–03	4.64E–04	4.5E–03
ZDT4	mean (GD)	2.13E–02	5.02E–02	3.15E–04	1.05E–02	2.55E–02	**4.28E–05**
	std (GD)	3.19E–02	3.88E–02	2.89E–03	5.30E–03	4.17E–03	**6.07E–06**
	mean (SP)	3.51E–03	3.37E–03	4.04E–03	3.09E–03	3.12E–03	1.5E–03
	std (SP)	8.21E–03	7.60E–03	6.15E–03	3.55E–02	3.72E–02	2.9E–03
ZDT6	mean (GD)	2.44 E–03	2.37E–03	2.15E–04	2.02E–03	2.05E–03	**2.89E–04**
	std (GD)	3.15E–04	3.27E–04	3.01E–05	1.88E–03	2.02E–04	1.21E–05
	mean (SP)	2.51E–03	2.47E–03	2.99E–03	2.08E–03	2.15E–03	4.4E–03
	std (SP)	7.44E–04	6.89E–04	7.84E–03	6.78E–03	5.43E–03	6.5E–03

3.2 Comparison of Different Selection Strategies

In this experiment, we mainly focus on performance of different selection strategies with/without replacements. The following several selection strategies are compared:

- Roulette wheel selection (S1, in briefly)
- Stochastic universal Sampling (S2, in briefly)
- Tournament selection (S3, in briefly)
- Roulette wheel selection without replacement (S4, in briefly)
- Stochastic universal sampling without replacement (S5, in briefly)
- Tournament selection without replacement (S6, in briefly).

In this experiment, different selection strategy is incorporated into NSGA-II. Simulation results can be found in Table 2. It is obviously that, from the GD index, the tournament selection without replacement achieves the best performance.

Table 2. Means and variances of the performance metrics

Problems	Metrics	PNIA	SPEA2	LPNSGA-II	NSGA-II	WRNSGA-II
SCH	mean (GD)	3.81E–02	4.41E–02	6.55E–02	6.17E–02	6.33E–02
	std (GD)	1.93E–03	1.14E–03	3.11E–03	2.96E–03	4.0E–03
	mean (SP)	4.96E–03	7.01E–03	1.66E–02	1.75E–02	2.27E–02
	std (SP)	3.66E–03	1.61E–03	7.80E–03	1.02E–02	1.33E–02
ZDT1	mean (GD)	6.56E–04	2.80E–05	1.54E–08	2.58E–04	2.83E–11
	std (GD)	1.64E–04	1.34E–05	6.80E–08	1.17E–04	1.75E–11
	mean (SP)	1.04E–03	1.94E–03	6.50E–03	5.09E–03	4.8E–03
	std (SP)	1.06E–03	3.38E–04	4.45E–03	3.92E–03	3.6E–03
ZDT2	mean (GD)	8.77E–04	1.27E–05	8.40E–11	8.42E–05	5.91E–11
	std (GD)	1.18E–03	1.21E–05	3.05E–10	1.32E–04	9.65E–11
	mean (SP)	2.37E–03	2.13E–11	6.27E–03	2.81E–03	4.8E–03
	std (SP)	3.17E–03	8.83E–04	7.57E–03	4.36E–03	4.3E–03
ZDT3	mean (GD)	1.89E–04	4.43E–06	2.45E–10	1.40E–04	7.5E–11
	std (GD)	1.01E–04	1.71E–06	5.58E–10	9.98E–05	1.07E–10
	mean (SP)	4.09E–04	1.89E–03	1.90E–02	5.56E–03	6.3E–03
	std (SP)	5.47E–04	5.99E–04	2.35E–02	3.81E–03	4.5E–03
ZDT4	mean (GD)	1.25E–02	6.95E–02	4.40E–04	2.68E–01	4.28E–05
	std (GD)	1.03E–02	4.19E–02	1.69E–03	1.30E–01	6.07E–06
	mean (SP)	1.46E–03	4.37E–03	6.84E–03	5.09E–03	1.5E–03
	std (SP)	3.21E–03	4.64E–03	9.81E–03	1.11E–02	2.9E–03
ZDT6	mean (GD)	1.78E–03	1.33E–03	3.91E–04	5.49E–03	2.89E–04
	std (GD)	2.31E–04	1.84E–04	2.04E–05	1.27E–03	1.21E–05
	mean (SP)	1.35E–03	1.45E–03	7.16E–03	4.43E–03	4.4E–03
	std (SP)	9.87E–04	5.74E–04	5.30E–03	2.05E–03	6.5E–03

3.3 Comparison with State-of-Art Algorithms

In this subsection, we will compare NSGA-II, SPEA2, PNIA, LPNSGA-II and NSGA-II with S6. The parameters in SPEA2, PNIA, LPNSGA-II and NSGA-II are set according to the original studies by developers. In order to demonstrate the solutions of these algorithms, typical simulation results of NSGA-II,WRNSGA-II, LPNAGA-II, PNIA and SPEA2 are presented in Table 2. It is obvious for ZDT3, that NSGA-II with S6 outperforms NSGA-II significantly in terms of convergence.

For SCH, ZDT1 and ZDT2, ZDT4 and ZDT6, NSGA-II and SPEA2 can not converge to the global optimal front effectively. However, the NSGA-II with S6 presented in this paper can converge to the ZDT4 optimal front effectively. For ZDT6, the convergence of NSGA-II with S6 performs better than other algorithm. The experiment results show that the proposed NSGA-II with S6 is promising in solving multi-objective problems.

4 Conclusions

In the past few years, few research work [18–21] focus on the duplicate individuals in NSGA-II [22–24]. In this paper, the selection strategy without replacement is employed to increase the diversity of NSGA-II. To test the performance, we compare it with six different selection strategies with/without replacement. Simulation results show the proposed tournament selection without replacement for NSGA-II achieves the better performance compared with other algorithms. Although the performance has been greatly improved, we also needs to further optimize its running time and complexity.

Acknowledgement. The paper is supported by the Natural Science Foundation of Shanxi Province under Grant No. 201601D011045, and Graduate Educational Innovation Project of Shanxi Province under Grant No. 2017SY075.

References

1. Heller, L., Sack, A.: Unexpected failure of a greedy choice algorithm proposed by Hoffman. Int. J. Math. Comput. Sci. **12**(2), 117–126 (2017)
2. Zhu, H., He, Y., Wang, X., Tsang, E.C.C.: Discrete differential evolutions for the discounted 0–1 knapsack problem. Int. J. Bio-Inspired Comput. **10**(4), 219–238 (2017)
3. Pisut, P., Voratas, K.: A two-level particle swarm optimisation algorithm for open-shop scheduling problem. Int. J. Comput. Sci. Math. **7**(6), 575–585 (2016)
4. Wang, H., Wang, W., Sun, H., Shahryar, R.: Firefly algorithm with random attraction. Int. J. Bio-Inspired Comput. **8**(1), 33–41 (2016)
5. Cai, X., Gao, X.: Improved bat algorithm with optimal forage strategy and random disturbance strategy. Int. J. Bio-Inspired Comput. **8**(4), 205–214 (2016)
6. Cui, Z., Xue, F., Cai, X., Cao, Y., Wang, G., Chen, J.: Detection of malicious code variants based on deep learning. IEEE Trans. Ind. Inform. https://doi.org/10.1109/tii.2018.2822680
7. Cui, Z., Cao, Y., Cai, X., Cai, J., Chen, J.: Optimal LEACH protocol with modified bat algorithm for big data sensing systems in Internet of Things. J. Parallel Distrib. Comput. (2017). https://doi.org/10.1016/j.jpdc.2017.12.014

8. Chen, W., Xiang, T., Xu, J.: Team evolutionary algorithm based on PSO. Pattern Recog. Artif. Intell. **28**(6), 521–527 (2015)
9. Wang, H., Ni, Z., Wu, Z.: Multi-tenant service customization algorithm based on map reduce and multi-objective ant colony optimization. Pattern Recog. Artif. Intell. **27**(12), 1105–1116 (2014)
10. Eswari, R., Nickolas, S.: Modified multi-objective firefly algorithm for task scheduling problem on heterogeneous systems. Int. J. Bio-Inspired Comput. **8**(6), 379–393 (2016)
11. Schaffer, J.D.: Multiple objective optimization with vector evaluated genetic algorithms. In: International Conference on Genetic Algorithms, pp. 93–100. Lawrence Erlbaum Associates Inc. (1985)
12. Zitzler, E., Laumanns, M., Thiele, L.: SPEA2: Improving the strength Pareto evolutionary algorithm. In: Evolutionary Methods for Design, Optimization and Control with Applications to Industrial Problems, pp. 95–100. Springer , Berlin (2002)
13. Deb, K., Pratap, A., Agarwal, S., et al.: A fast and elitist multi-objective genetic algorithm: NSGA-II. IEEE Trans. Evol. Comput. **6**(2), 182–197 (2002)
14. Patel, R., Raghuwanshi, M., Malik, L.: An improved ranking scheme for selection of parents in multi-objective genetic algorithm. In: International Conference on Communication Systems and Network Technologies, pp. 734–739. IEEE (2011)
15. Salimi, R., Motameni, H., Omranpour, H.: Task scheduling with load balancing for computational grid using NSGA-II with fuzzy mutation. In: IEEE International Conference on Parallel Distributed and Grid Computing, pp. 79–84. IEEE (2013)
16. Tran, K.D.: An improved non-dominated sorting genetic algorithm-II (ANSGA-II) with adaptable parameters. Int. J. Intell. Syst. Technol. Appl. **7**(4), 347–369 (2009)
17. Schott, J.R.: Fault tolerant design using single and multicriteria genetic algorithm optimization. Cell. Immunol. **37**(1), 1–13 (1995)
18. Philip, F.: Sums of squares of Krawtchouk polynomials, Catalan numbers, and some algebras over the Boolean lattice. Int. J. Math. Comput. Sci. **12**(1), 65–83 (2017)
19. Andreas, B., Anargyros, F.: On octonion polynomial equations. Int. J. Math. Comput. Sci. **11**(2), 59–73 (2016)
20. Lei, Y., Gong, M., Jiao, L., Shi, J.: An adaptive coevolutionary memetic algorithm for examination timetabling problems. Int. J. Bio-inspired Comput. **10**(4), 248–257 (2017)
21. Lydia, B., Ta Minh, T.: A clustering algorithm based on elitist evolutionary approach. Int. J. Bio-inspired Comput. **10**(4), 248–257 (2017)
22. Badih, G.: Half a dozen famous unsolved problems in mathematics with a dozen suggestions on how to try to solve them. Int. J. Bio-inspired Comput. **11**(2), 257–273 (2016)
23. Zhang, M., Wang, H., Cui, Z., Chen, J.: Hybrid multi-objective cuckoo search with dynamical local search. Memet. Comput. **10**(2), 199–208 (2017). https://doi.org/10.1007/s12293-017-0237-2
24. Henrik, S.: Methods for the summation of infinite series. Int. J. Math. Comput. Sci. **11**(2), 109–113 (2016)

Comparison of Two Swarm Intelligence Algorithms: From the Viewpoint of Learning

Guo-Sheng Hao[1(✉)], Ze-Hui Yi[2], Lin Wan[1], Qiu-Yi Shi[1], Ya-Li Liu[1], and Gai-Ge Wang[3]

[1] School of Computer Science and Technology, Jiangsu Normal University, Xuzhou 221116, Jiangsu, China
hgskd@jsnu.edu.cn

[2] School of Computer Science and Technology, Huaiyin Normal University, Huai'an 223300, China

[3] College of Information Science and Engineering, Ocean University of China, Qingdao 266100, China

Abstract. It is always said that learning is at the core of intelligence. How does learning work in swarm intelligence algorithms (SIAs)? This paper tries to answer this question by analyzing the learning mechanisms in two new emerged swarm intelligence algorithms: Krill Herd algorithm, cuckoo search. Each algorithm generates new solutions by learning to explore/exploit the promising subspace. For the new solutions generators in each algorithm, we study the learning mechanism from the viewpoint of learning scheme includes learning subject, learning object, learning result and learning rule. Also we analyze their ability of exploration and exploitation. The above study not only enables theory researchers to get the similarities and differences among SIAs, but also helps them understand the integration of different SIAs together.

Keywords: Swarm intelligence · Learning mechanism · Solution generators Exploitation · Exploration

1 Introduction

For many decades, researchers have been developing increasingly more sophisticated nature-inspired metaheuristic algorithms (NIOAs) [1] for solving hard optimization problems. Some new algorithms are put forward, including Krill Herd Algorithm (KHA) [2], Cuckoo Search (CS) [3, 4] and so on. Facing many algorithms, a question is often asked: what is the differences among these algorithms? For this question, we compare these new algorithms from the viewpoint of learning mechanism.

From the implementation viewpoint, the difference among NIOAs only exists in the process of generating new solutions. Although different NIOAs may have different mechanisms, almost all NIOAs correspond to the flowchart shown in Fig. 1, in which the difference among NIOAs is the generation of new solutions.

We put forward the learning scheme in NIOAs in [5] from 3 aspects: who learns, from whom to learn and how to learn, and the learning scheme is studied from 4 elements: learning subject X_S, learning object X_O, learning result X_{new} and learning rule.

D.-S. Huang et al. (Eds.): ICIC 2018, LNAI 10956, pp. 821–827, 2018.
https://doi.org/10.1007/978-3-319-95957-3_87

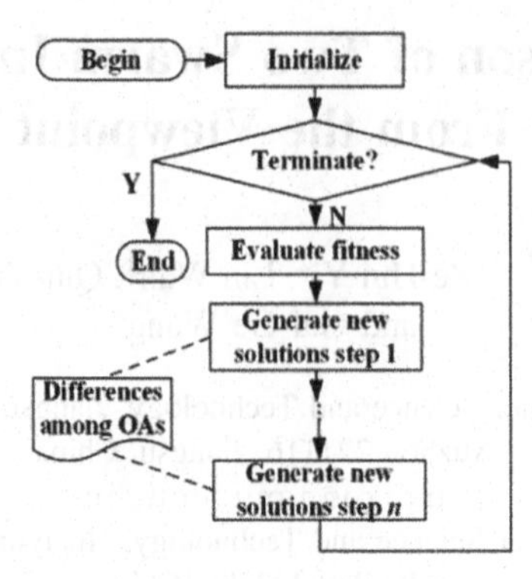

Fig. 1. General flowchart of optimization algorithms.

Based on this scheme, we analyze the learning mechanism of two kinds of NIOAs [5, 6]. This paper mainly analyzes the learning mechanism of the new emergence algorithm recently, including KHA, CS.

2 Learning Mechanism in Krill Herd Algorithm

Krill herd algorithm (KHA) is inspired by the herding of the krill swarms when searching for food in nature. Krill individuals try to maintain a high density and move based on their mutual effects. In KHA, the location of krill corresponds to points in the search space. The fitness function is decided by the distances of each krill from food and from highest density of the krill swarm [2].

Similar to PSO, KHA does not adopt an explicit selection function. The absence of a selection mechanism in KHA is compensated by the use of the global best solution to guide the search. There is no notion of offspring generation in KHA.

Generally, a solution update in KHA consists of the updating two of its properties: velocity update and location update. The partial flowchart is shown in Fig. 2 and for the rest part of the flowchart of KHA, one is referred to Fig. 1.

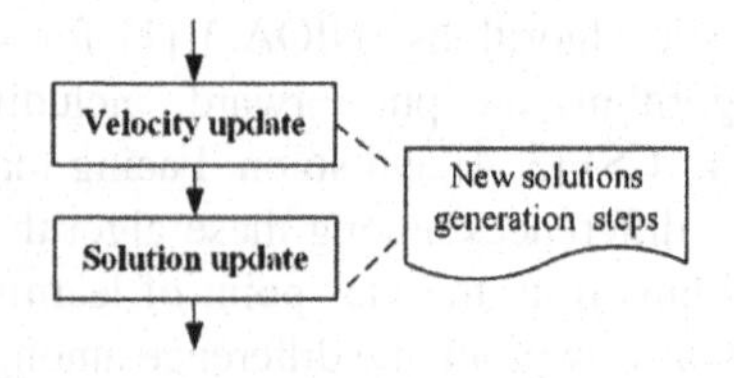

Fig. 2. Generation new solutions in KHA.

2.1 Velocity Update

The velocity update for each krill consists of three parts: (i) movement induced by other solutions; (ii) foraging motion for food, and (iii) random diffusion for being attacked. The new velocity of a krill $\mathbf{X}_i$ is denoted as $\Delta\mathbf{X}_{new\_1,i}$. The velocity update is:

$$\Delta\mathbf{X}_{new\_1,i} = \Delta\mathbf{X}_{i,1} + \Delta\mathbf{X}_{i,2} + \Delta\mathbf{X}_{i,3} \tag{1}$$

where $\Delta\mathbf{X}_{i,1}, \Delta\mathbf{X}_{i,2}$, and, $\Delta\mathbf{X}_{i,3}$ are the velocities induced by other krill individuals, foraging motion and physical diffusion respectively.

From the viewpoint gene sense unit (GSU) [7], there is:

$$\Delta x_{new\_1}^{(k)} = \Delta x_{i,new1}^{(k)} + \Delta x_{i,new2}^{(k)} + \Delta x_{i,new3}^{(k)} \tag{2}$$

where $\Delta x_{i,new1}^{(k)}, \Delta x_{i,new2}^{(k)}$ and $\Delta x_{i,new3}^{(k)}$ are the components of velocities of $\Delta\mathbf{X}_{i,1}, \Delta\mathbf{X}_{i,2}$, and, $\Delta\mathbf{X}_{i,3}$ respectively. The learning rule for each of the velocity update are shown in (3), (4) and (5) respectively. The learning rule $L(\cdot)$ is:

$$\mathbf{X}_S = \Delta\mathbf{X}_{i,1},\ S_{objective} = \left\{\mathbf{X}_i, \mathbf{X}_j, \mathbf{X}_{best}\right\}, i \neq j, \mathbf{X}_j \in P$$

$$\begin{aligned} \Delta x_{i,new1}^{(k)} &= \omega_1 \Delta x_{i,1}^{(k)} + \sum\nolimits_{j=1}^{N} \left(\frac{f(\mathbf{X}_i) - f(\mathbf{X}_j)}{f(\mathbf{X}_{best}) - f(\mathbf{X}_{worst})} \cdot \frac{\mathbf{X}_i - \mathbf{X}_j}{\|\mathbf{X}_j - \mathbf{X}_i\| + \varepsilon} \right) + \\ & 2\left(rand + \frac{t}{T}\right) \cdot \frac{f(\mathbf{X}_{best}) - f(\mathbf{X}_i)}{f(\mathbf{X}_{best}) - f(\mathbf{X}_{worst})} \cdot \frac{\mathbf{X}_{best} - \mathbf{X}_i}{\|\mathbf{X}_{best} - \mathbf{X}_i\| + \varepsilon} \end{aligned} \tag{3}$$

where ω_1 is called as the inertia weight, $\Delta x_{i,1}^{(k)}$ as the current velocity induced by other krill individuals, N as the number of neighbours of $\mathbf{X}_i$, $\mathbf{X}_{best}$ as the best solution so far, t as the current iteration or generation, and T as maximum number of iteration or generations.

Secondly, for the velocity of foraging motion, the virtual center of food concentration is labelled $\mathbf{X}_C$ and define it as: $\mathbf{X}_C = \frac{\sum_{j=1}^{N} \frac{\mathbf{X}_j}{f(\mathbf{X}_j)}}{\sum_{j=1}^{N} \frac{1}{f(\mathbf{X}_j)}}$. Then the learning mechanism for foraging motion is:

$$\mathbf{X}_S = \Delta\mathbf{X}_{i,2}$$

$$\begin{aligned} & \Delta x_{i,new2}^{(k)} = \omega_{21} \Delta x_{i,2}^{(k)} + \\ \mathbf{X}_O = \{\mathbf{X}_C, \mathbf{X}_{best}\} & \omega_{22}\left(1 - \frac{t}{T}\right) \cdot \frac{f(\mathbf{X}_C) - f(\mathbf{X}_i)}{f(\mathbf{X}_{best}) - f(\mathbf{X}_{worst})} \cdot \frac{\mathbf{X}_C - \mathbf{X}_i}{\|\mathbf{X}_C - \mathbf{X}_i\| + \varepsilon} + \\ & \omega_{23} \left(\frac{f(\mathbf{X}_{best}) - f(\mathbf{X}_i)}{f(\mathbf{X}_{best}) - f(\mathbf{X}_{worst})} \cdot \frac{\mathbf{X}_{best} - \mathbf{X}_i}{\|\mathbf{X}_{best} - \mathbf{X}_i\| + \varepsilon} \right) \end{aligned} \tag{4}$$

where ω_{21} is called as the inertia weight of the foraging motion. From Eq. (4), we can see that the exploration/exploitation effect of velocity update is similar to that of PSO.

Third, the velocity update of physical diffusion is defined as:

$$\Delta x_{i,new3}^{(k)} = \omega_3 \left(1 - \frac{t}{T}\right) rand \tag{5}$$

where ω_3 is the maximum diffusion speed and *rand* is a random number.

2.2 Solution Update

The solution update in KHA is similar to that in PSO. The learning scheme is:

$$\mathbf{X}_S = \mathbf{X} \in P, \mathbf{X}_O = \{\Delta \mathbf{X}_{new\_1}\}, x_{new\_2}^{(k)} = x^{(k)} + \omega_k \cdot rand \cdot x_{new\_1}^{(k)} \tag{6}$$

where parameter ω_k works as a factor of the speed, and it should be carefully set according to the optimization problem. A simple obtaining method for ω_k is [2]:

$$\omega_k = c \sum_{k=1}^{d} \left(x^{(k)U} - x^{(k)L}\right) \tag{7}$$

where $\Delta x^{(k)U}$ and $x^{(k)L}$ are upper and lower bounds of variable $x^{(k)}$ respectively. The constant value c is found that should be in the range of [0, 2]. The low values of c let the algorithm search the space more precisely.

Both $\Delta x_{i,new1}^{(K)}$ and $\Delta x_{i,new2}^{(K)}$ contain the knowledge from global and local optimum solution, and they work in parallel, making KHA a powerful algorithm.

Furthermore, in order to enhance the performance of the KHA, the operators, such as crossover and mutation inspired from the classic DE algorithm, are introduced into KHA, which with crossover operator is called as KHA II, and that with crossover and mutation operators is called as KH IV.

3 Cuckoo Search

Cuckoo search (CS) is developed by Yang and Deb. It is based on the brood parasitism of some cuckoo species [4]. Additionally, CS is enhanced by the so-called Lévy flights [8], rather than by simple isotropic random walks. Recent studies show that CS is potentially far more efficient than PSO and GA [9]. There are mainly two steps in CS and they are Lévy based solution location update and solution selection. The partial flowchart is shown in Fig. 3. For the rest part of the flowchart of CS, one is referred to Fig. 1.

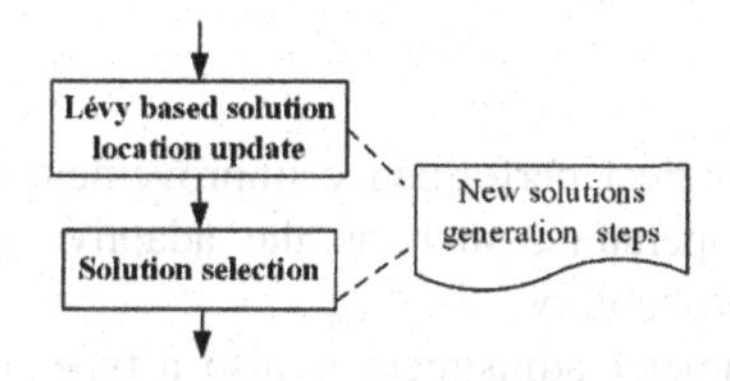

Fig. 3. Generation new solutions in CS.

3.1 Lévy Based Solution Location Update

The generation of solutions is shown in (8).

$$\mathbf{X}_S = \mathbf{X}_i \in P, \mathbf{X}_O = \emptyset, x_{new\_1}^{(k)} = x_i^{(k)} + \omega \cdot L\acute{e}vy(\lambda) \tag{8}$$

where $\omega > 0$. Since there is $\mathbf{X}_O = \emptyset$, the learning belongs to the self-taught. Here the random walk via Lévy flight is more efficient in exploring the search space, as its step length is much longer in the long run, and its random step length is drawn from a Lévy distribution as [9]:

$$L'\acute{e}vy(\lambda) \sim \mu = t^{-\lambda} \tag{9}$$

where t is the generation number. With Lévy flight, some of the new solutions should be generated around the best solution obtained so far. Moreover, their locations should be far enough from the current best solution to ensure that the system will not be trapped in a local optimum [9]. The exploration and exploitation effect of the solution location update is generally determined by the Lévy flight and the modality of the problem.

3.2 Solution Selection

The solution update rule is:

$$\mathbf{X}_{new\_2} = \begin{cases} \mathbf{X}_{new\_1}, & \mathbf{X}_{new\_1}, \succ \mathbf{X}_i \\ \mathbf{X}_{i,} & \mathbf{X}_i \succeq \mathbf{X}_{new\_1} \end{cases} \tag{10}$$

An advantage of CS is that the number of parameters is fewer than those of GA and PSO. Furthermore, it is reported that the convergence rate is not sensitive to the parameters used. This means that no fine adjustment is needed for any given problems. It is also reported that CS has a fine balance of exploration and exploitation than GA and PSO, and Lévy flight is the most suitable for the randomization on the global scale [9].

4 Discussion

There have been many perfect performance improvement methods by adjusting the parameters of learning operators, such as the adaptive parameters adjustment of crossover and mutation probability.

In fact, adaptive parameter adjustment is also a type of learning. The algorithm adapts its learning mechanism according to the environment or evolution state. With different learning mechanisms, the algorithm can achieve similar performance. In KHA, Amir proposed a method to control the crossover and mutation probability [2], in which, the best solution can be preserved from crossover and mutation. With a suitable learning mechanism, the algorithm will perform better than those without any suitable learning mechanism. Similarly an algorithm with a suitable learning mechanism may be used to solve more optimization problems, while those algorithms without suitable learning mechanisms only suitable for certain optimization problems.

An issue is the solution generation operators may favour a quick convergence towards local optima. From the above study of learning mechanism of operators, we can see that almost every operator can act as the role of exploration or exploitation, which is closely related to the coefficients or learning objects. It is rather difficult or impossible to control exploration and exploitation off-line by the proper settings of control parameters, because it is the general case that both the problem to be optimized and the evolution process are unknown in advance. Therefore, in order to get desirable results, it is a good choice to dynamically or adaptively adjust coefficients or learning objects.

5 Conclusion

We have studied the learning mechanism of NIOAs by analysing the learning scheme in the learning operators of KHA, CS. The similarities and differences are studied from the viewpoint of learning mechanism. The analysis of learning mechanisms enables the researchers have more insight of NIOAs.

Acknowledgement. Partly supported by the National Natural Science Foundation of China under Grant No. 61673196, 61503165, 61702237.

References

1. Smith-Miles, K., Lopes, L.: Measuring instance difficulty for combinatorial optimization problems. Comput. Oper. Res. **39**(5), 875–889 (2012)
2. Gandomi, A.H., Alavi, A.H.: Krill Herd: a new bio-inspired optimization algorithm. Commun. Nonlinear Sci. Numer. Simul. **17**(12), 4831–4845 (2012)
3. Gandomi, A.H., Talatahari, S., Yang, X.S., Deb, S.: Design optimization of truss structures using cuckoo search algorithm. Struct. Des. Tall Spec. Build. **22**(17), 1330–1349 (2013)
4. Yang, X.S., Deb, S.: Cuckoo search via Lévy flights. In: World Congress on Nature & Biologically Inspired Computing, NaBIC 2009, pp. 210–214 (2009)

5. Hao, G.S., Shi, Q.Y., Wang, G.G., Zhang, Z.J., Zou, D.X.: Comparison of GA-based algorithms: a viewpoint of learning scheme. In: International Conference on Communications, Information Management and Network Security (CIMNS 2017), pp. 288–293 (2017)
6. Hao, G.S., Wang, G.G., Zhang, Z.J., Zou, D.X.: Comparison of PSO and ABC: from a viewpoint of learning. In: International Conference on Artificial Intelligence: Techniques and Applications (AITA 2017), pp. 108–112 (2017)
7. Hao, G.S., Gong, D.W., Shi, Y.Q., Zhang, Y., Liu, T.H.: Relation algebra based genetic algorithm model and its applications. J. Southeast Univ. (Nat. Sci. Ed.) **34**(Suppl.), 58–62 (2004)
8. Gutowski, M.: Lévy flights as an underlying mechanism for global optimization algorithms. arXiv Mathematical Physics e-Prints, June 2001
9. Yang, X.-S.: Nature-Inspired Metaheuristic Algorithms. Luniver Press, Bristol (2010)

Using Spotted Hyena Optimizer for Training Feedforward Neural Networks

Jie Li[1], Qifang Luo[1,2(✉)], Ling Liao[1], and Yongquan Zhou[1,2]

[1] College of Information Science and Engineering,
Guangxi University for Nationalities, Nanning 530006, China
l.qf@163.com

[2] Guangxi High School Key Laboratory of Complex System and Computational Intelligence, Nanning 530006, China

Abstract. Spotted hyena optimizer (SHO) is a novel heuristic optimization algorithm based on the behavior of spotted hyena and their collaborative behavior in nature. In this paper, we design a spotted hyena optimizer for feedforward neural networks (FNNs). Training feedforward neural networks is regard as a challenging task, because it is easy to fall into local optima. Our objective is to apply heuristic optimization algorithm design to tackle these problems better than the mathematical and deterministic methods, in order to confirm that SHO algorithm training FNN is more effective. a classification datasets about Heart is applies to benchmark the performance of the proposed method. The more basic SHO is compared to other acclaimed state-of-the-art optimization algorithm, the results show that the proposed algorithm can provide better results.

Keywords: Spotted hyena optimizer · Feedforward neural networks
Classification datasets · Function-approximation

1 Introduction

One of the most important inventions in the field of soft computing is neural networks. (NN), It is inspired by the neurons of human brain. The basic concept of NN is first to establish a mathematical model by McCulloch and Pitts [1]. In feedforward neural network (FNN) [2], this is the focus of the problem study, the connection weights between different layers of neural network and the biases of neural network layers are the most important parameters.

There are more different types of heuristics that have emerged on training FNNs, such as Montana and Davis was first applies GA to improve learning of NNs [3]. The PSO algorithm was also employed as the trainer for FNN in many studies [4]. Some other heuristic-based learning algorithms in the literature are as follows: DE based trainer [5], gravitational search algorithm (GSA)-based trainer [6] and grey wolf optimizer (GWO) [7] so on. Despite the many advantages of metaheuristic algorithms, the problem of falling into the local optimum still exists.

Spotted hyena optimizer (SHO) [8] is a new metaheuristic optimization method, SHO is inspired from social hierarchy and hunting behavior of spotted hyena. Select SHO training feedforward neural network is derived from the proposed algorithm has a strong

D.-S. Huang et al. (Eds.): ICIC 2018, LNAI 10956, pp. 828–833, 2018.
https://doi.org/10.1007/978-3-319-95957-3_88

global search ability and faster convergence rate. Other contribution of this work is comparison of different heuristic algorithms on classification datasets.

2 The Feedforward Neural Networks

FNNs are those NNs that have only one-way and one-way connections between neurons, data is passed between neural networks [2]. As shown in Fig. 1, test samples are input from the input layer. Input the products of the test sample data are multiplied by the weight consider as the hidden layer input. The same form also appears between the hidden layer and the output layer. The actual output of the neuron is calculated from the transfer function.

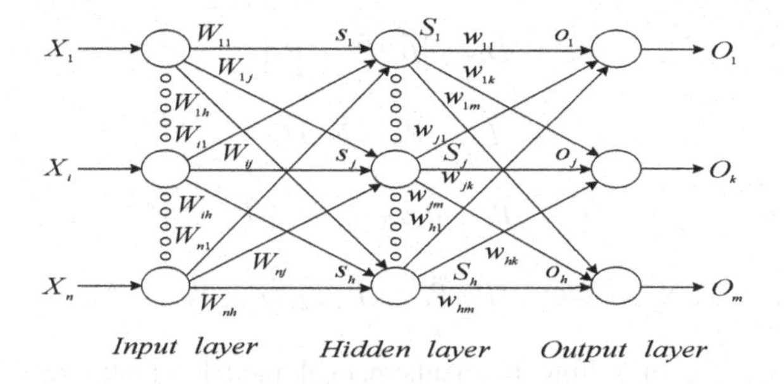

Fig. 1. Three-layer FNN

Input test sample data, weights and biases are provided for FNN, the output of the FNN is calculated as follows: [7] the weighted sums of inputs are calculated as follows:

$$s_j = \sum_{i=1}^{n} (W_{ij}.X_i) - \theta_j, \quad j = 1, 2, \ldots h \tag{1}$$

$$S_j = sigmoid(s_j) = \frac{1}{(1 + \exp(-s_j))}, \quad j = 1, 2, \ldots h \tag{2}$$

The output of the networks as follows:

$$o_k = \sum_{j=1}^{h} (W_{jk}.S_j) - \theta'_k, \quad k = 1,\ 2,\ \ldots,\ m \tag{3}$$

$$O_k = sigmoid(o_k)\frac{1}{(1 + \exp(-o_k))}, \quad k = 1,\ 2,\ \ldots,\ m \tag{4}$$

where n is the number of the input nodes, W_{ij} shows the connection weight from the ith node in the input layer to the jth node in the hidden layer, θ_j is the bias (threshold) of the jth hidden node, and X_i denotes the ith input. A sigmoid function defines the final output of each hidden node as follows:

3 Spotted Hyena Optimizer

Spotted hyenas is mainly acquired the prey through three steps to achieve the capture of process, including encircling prey, hunting prey, attacking prey. Encircling prey is set up the equations as following [8]:

$$\vec{D}_h = \left|\vec{B} \cdot \vec{P}_p(t) - \vec{P}(t)\right| \tag{5}$$

$$\vec{P}(t+1) = \vec{P}_p(t) - \vec{E} \cdot \vec{D}_h \tag{6}$$

The coefficient vector $\vec{B}$, $\vec{E}$ are calculated as follow $\vec{B} = 2 \cdot \vec{rd}_1$, $\vec{E} = 2\vec{h} \cdot \vec{rd}_2 - \vec{h}$ and $\vec{h} = 5 - (t * (5/T))\, t = 1, 2, 3, \cdots, T$. Hunting stage mathematically equations expressed as follows:

$$\vec{D}_h = \left|\vec{B} \cdot \vec{P}_h - \vec{P}_k\right| \tag{7}$$

$$\vec{P}_k = \vec{P}_h - \vec{E} \cdot \vec{D}_h \tag{8}$$

$$\vec{C}_h = \vec{P}_k + \vec{P}_{k+1} + \cdots + \vec{P}_{k+N} \tag{9}$$

$$N = count_{nos}(\vec{P}_h, \vec{P}_{h+1}, \vec{P}_{h+2}, \cdots, (\vec{P}_h + \vec{M})) \tag{10}$$

For the purpose of setting up mathematical model for attacking the prey, the mathematical formulation for attacking the prey is as follows:

$$\vec{P}(t+1) = \frac{\vec{C}_h}{N} \tag{11}$$

The pseudo code of the SHO algorithm from Dhiman and Kumar is presented as follows [8]:

Algorithm. pseudo code of SHO

1. Initialize the spotted hyena population P_i $(i = 1, 2, \cdots, n)$, h, B, E and N
2. Calculate the fitness of each search agent
3. $\vec{P}_h$ = the best search agent, $\vec{C}_h$ = the group or cluster of all far optimal solutions
4. **While** (t<Max number of iterations) do
5. **for** each search agent do
6. Update the position of the current search agent by Eq.(11)
7. **end for**
8. Update h, B, E ,N and P_h
9. Check if any search agent goes beyond the search space and revamp it
10. Calculate the fitness of each search agent
11. *t=t+1*
12. **end while**
13. Return P_h

4 SHO-Based Feedforward Neural Network

As discussed in introduction, the purpose of training feedforward neural network is to find the best weight and bias value. The most important issue is finding befitting variables of weights and biases for meta-heuristic algorithm. The vector of an FNN is shown as follows:

$$\vec{V} = \left\{\vec{W}, \vec{\theta}\right\} = \{W_{1,1}, W_{1,2}, \ldots, W_{n,n}, h, \theta_1, \theta_2, \ldots, \theta_h\} \tag{12}$$

Where n is the number of the input nodes, $W_{i,j}$ shows the connection weight from the i-th node to the j-th node, θ_h is the bias (threshold) of the j-th hidden node.

The performance metric for evaluating FNN is the Average Mean Square Error that is calculated as follows:

$$\overline{MSE} = \sum_{k=1}^{s} \frac{\sum_{i=1}^{m} (o_i^k - d_i^k)^2}{s} \tag{13}$$

where s is the number of training samples, m is the number of outputs, d_i^k is the desired output of the i-th input unit when the k-th training sample is used, and o_i^k is the actual output of the i-th input unit when the k-th training sample appears in the input. To sum up, with problem representation and the average MSE for the SHO algorithm, the problem of training FNN can be illustrated as follows:

$$Minimize: \ F(\overline{V}) = \overline{MSE} \tag{14}$$

The next section verifies the merits of the SHO algorithm for training FNN in practice.

5 Simulations and Comparisons

In this section, we plan to use the SHO algorithm based on feedforward network training to test a standard classification datasets obtained from the University of California at Irvine (UCI) Machine Learning Repository [9]: Heart.

For data verification, SHO algorithm is compared to GWO [7], HS [10], GGSA [11], and DE [12]. It is supposed that the optimization process set out generating random weights and biases in the range of [−10, 10] for all datasets. The initial parameters of the SHO algorithm compared with other algorithms are shown in Table 1.

Each algorithm in this paper was run 10 times. AVE and STD may be a good combination to evaluate the performance of the algorithm in avoiding local minima. Another comparison metric is the classification rate. The statistical results are presented in Tables 2. To fair compare SHO with other algorithms, we unify set to the population size 200 and maximum number of generations 250. Please note that we name the propose SHO algorithm and highlight best results in boldface in the table of results.

Table 1. The initial parameters of algorithms

Algorithm	Parameter values
SHO	The parameter is $\vec{h} \in [0, 5]$ over the course of iterations
GWO [7]	Components $\vec{a} \in [0, 2]$ over the course of iterations
HS	Harmony memory considering rate $HMCR = 0.75$ Pitch adjusting rate $PAR = 0.7$
GGSA	Accelerating coefficient $C_1 = (-2t^3/T^3) + 2$. Accelerating coefficient $C_2 = 2t^3/T^3$ Gravitational constant $G_0 = 1$ Coefficient of decrease $\alpha = 20$
DE	Scaling factor $F = 0.5$ Crossover Probability $CR = 0.5$

Table 2. Statistical results for Heart dataset

Algorithm	AVE	STD	Classification rate (%)
SHO-FNN	**0.052145**	**0.005304**	**91.25**
GWO-FNN [7]	0.122600	0.007700	75.00
HS-FNN	0.220286	0.012695	75.00
GGSA-FNN	0.140443	0.019178	70.00
DE-FNN	0.164107	0.007461	81.25

Heart Classification Problem

This dataset is used to diagnose cardiac Single Proton Computed Tomography (SPECT) images. All features are in binary format and the output indicates whether the patient's condition is normal or abnormal [7]. We trained FNN with the structure of 22-45-1 using 80 instances and examined them for the remaining 187 instances.

The experimental results are presented in Table 2. Heart is the most difficult dataset in terms of classification problems. The results of AVE and STD in this table show that SHO-FNN has the best performance in this dataset. The classification accuracy of the SHO-FNN is much higher. The convergence curves in Fig. 2 show that SHO-FNN has very fast convergence compared to the other algorithms. As shown in Fig. 3, the SHO-FNN, HS-FNN and GGSA-FNN Provide a similar level of stability for the Heart Dataset.

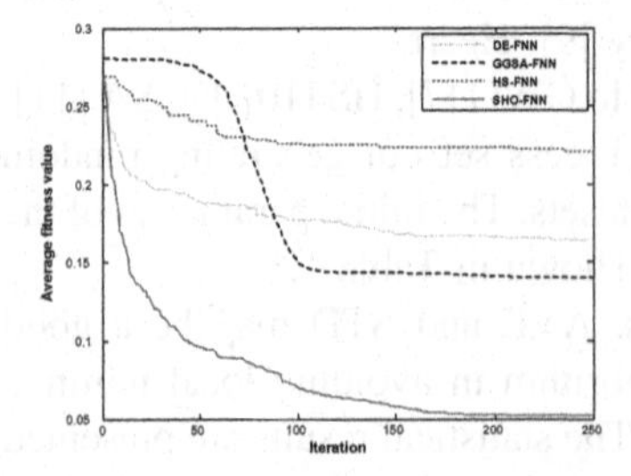

Fig. 2. Convergence curves of algorithms for Heart

Fig. 3. Anova test of algorithms for Heart

6 Conclusions

In this paper, we have proposed a new training based on the recently proposed SHO algorithm for FNNs. SHO update location corresponds to optimize the FNN weight and biases. The objection function was to minimize the average of MSE on all training samples. The performance of the proposed FNN trainer was benchmarked on standard classification problems about Heart problem. The results of the proposed SHO-FNN algorithms are compared to the other population-based optimization algorithm in the literature such as GWO, HS, GGSA and DE. The results demonstrate that the proposed method is able to be very efficient in training FNN in terms of exploration and avoidance of local minima. Moreover, the superior performance of SHO for training FNNs in term of accuracy of results and convergence rate can clearly be seen from the results. We also find that the results of SHO-FNN were better on the majority of the datasets in classification accuracy. Therefore, it can be concluded that the proposed SHO-FNN improves the problem of trapping in local minima with very good convergence rate.

Acknowledgment. This work is supported by National Science Foundation of China under Grants No. 61563008, and by Project of Guangxi Natural Science Foundation under Grant No. 2016GXNSFAA380264.

References

1. McCulloch, W.S., Pitts, W.: Bull. Math. Biophys. **5**, 115 (1943)
2. Annema, A.J.: Feed-forward neural networks. Compr. Chemom. **13**(4), 27–31 (1995)
3. Montana, D.J., Davis, L.: Training feedforward neural networks using genetic algorithms. In: International Joint Conference on Artificial Intelligence, pp. 762–767 (1989)
4. Meissner, M., Schmuker, M., Schneider, G.: Optimized Particle Swarm Optimization (OPSO) and its application to artificial neural network training. BMC Bioinform. **7**(1), 125 (2006)
5. Ilonen, J., Kamarainen, J.-K., Lampinen, J.: Differential evolution training algorithm for feed-forward neural networks. Neural Process. Lett. **17**(1), 93–105 (2003)
6. Mirjalili, S., Hashim, S.Z.M., Sardroudi, H.M.: Training feedforward neural networks using hybrid particle swarm optimization and gravitational search algorithm. Appl. Math. Comput. **218**(22), 11125–11137 (2012)
7. Mirjalili, S.: How effective is the Grey Wolf optimizer in training multi-layer perceptrons. Appl. Intell. **43**(1), 150–161 (2015)
8. Dhiman, G., Kumar, V.: Spotted hyena optimizer: a novel bio-inspired based metaheuristic technique for engineering applications. Adv. Eng. Soft. **114**, 48–70 (2017)
9. Asuncion, A., Newman, D.: UCI Machine Learning Repository (2007)
10. Zong, W.G., Kim, J.H., Loganathan, G.V.: A new heuristic optimization algorithm: harmony search. Simul. Trans. Soc. Model. Simul. Int. **76**(2), 60–68 (2016)
11. Mirjalili, S., Lewis, A.: Adaptive gbest-guided gravitational search algorithm. Neural Comput. Appl. **25**(7–8), 1569–1584 (2014)
12. Slowik, A., Bialko, M.: Training of artificial neural networks using differential evolution algorithm. In: Conference on Human System Interactions, vol. 6, no. 4, pp. 60–65 (2008)

A Complex-Valued Encoding Satin Bowerbird Optimization Algorithm for Global Optimization

Sen Zhang[1], Yongquan Zhou[1,2(✉)], Qifang Luo[1,2], and Mohamed Abdel-Baset[3]

[1] College of Information Science and Engineering, Guangxi University for Nationalities, Nanning 530006, China
yongquanzhou@126.com
[2] Key Laboratory of Guangxi High Schools Complex System and Computational Intelligence, Nanning 530006, China
[3] Faculty of Computers and Informatics, Zagazig University, El-Zera Square, Zagazig 44519, Sharqiyah, Egypt

Abstract. The real-valued satin bowerbird optimization (SBO) is a novel bio-inspired algorithm which imitates the 'male-attracts-the-female for breeding' principle of the specialized stick structure mechanism of satin birds. SBO has achieved success in congestion management, accurate software development effort estimation. In this paper, a complex-valued encoding satin bowerbird optimization algorithm (CSBO) is proposed aiming to enhance the global exploration ability. The idea of complex-valued coding and finds the optimal one by updating the real and imaginary parts value. With Complex-valued coding increase the diversity of the population, and enhance the global exploration ability of the basic SBO algorithm. The proposed CSBO optimization algorithm is compared against SBO and other state-of-art optimization algorithms using 20 benchmark functions. Simulation results show that the proposed CSBO can significantly improve the convergence accuracy and convergence speed of the original algorithm.

Keywords: Complex-valued encoding · Satin bowerbird optimization
Benchmark functions · Motor parameter identification

1 Introduction

Metaheuristics algorithm have become very popular over the past decade. This popularity is due to several key reasons: flexibility, lack of a gradient mechanism, simple structure and easy-to-understand features. The most popular of which is the genetic algorithm (GA) [1], which simulates Darwin's evolutionary theory, the principle is that optimization is a set of stochastic solutions for specific problems. In the case of the evaluation objective function, it updates the variables according to their fitness values. Similarly, there are many evolutionary algorithms, such as Differential Evolution (DE) [2]. These are evolutionary algorithms. The second type of algorithms are the

D.-S. Huang et al. (Eds.): ICIC 2018, LNAI 10956, pp. 834–839, 2018.
https://doi.org/10.1007/978-3-319-95957-3_89

ecosystem simulation algorithms, including Biogeography-Based Optimization (BBO) [3], Weed Colonization Algorithm (WCA).

While the third group are swarm intelligence algorithms, which representative are Particle Swarm Optimization (PSO) [4] algorithm, Ant Colony Optimization (ACO) [5]. There are some other swarm intelligent algorithms, such as, Artificial Bee Colony (ABC) algorithm [6], Bat Algorithm (BA) [7], Cuckoo Search (CS) algorithm [8], Satin bowerbird optimizer (SBO) [9].

The remainder of this paper is organized as follows: Sect. 2 discusses the basic principles of the complex-valued encoding satin bower bird optimizer (CSBO) Algorithm. Section 3 tests the performance of the algorithm using standard test functions and discusses the experimental results of standard test functions. Section 4 uses the CSBO algorithm to solve motor parameter identification problem and compare it with other algorithms. Finally, Sect. 4 contains some conclusions and future research directions.

2 Complex-Valued Satin Bower Bird Optimizer (CSBO)

2.1 Initialize the Complex-Valued Encoding Population

Let the range of the function argument be interval $[\text{var}_{R\min}, \text{var}_{R\max}]$. Randomly generate n complex modulus ρ_n and the amplitude θ_n, the resulting modulus vector satisfies the following relation.

$$R_n + iL_n = \rho_n(\cos\theta_n + i\sin\theta_n)$$

$$\rho_n = \left[0, \frac{\text{var}_{R\max} - \text{var}_{R\min}}{2}\right], \ \theta_n = [\text{var}_{\text{Imin}}, \text{var}_{\text{Imax}}] \tag{1}$$

In Eq. (1), $\text{var}_{R\max}$ denotes the upper bound of the real part and $\text{var}_{R\min}$ denotes the lower bound of the real part. var_{Imax} denotes the upper bound of the imaginary part, which is set as 2π, var_{Imin} denotes the lower bound of the imaginary part, and is set to -2π. The n real and imaginary parts are assigned to the real and virtual genes of the bower according to produce a bird's bower.

The Updating Method of CSBO

$$\begin{aligned} X_{Rn}^{new} + iX_{In}^{new} = X_{Rn}^{old} + iX_{In}^{old} &+ \lambda_n\left(\left(\frac{X_{Rj} + X_{Relite}}{2}\right) - X_{Rn}^{old}\right) \\ &+ i\lambda_n\left(\left(\frac{X_{Ij} + X_{Ielite}}{2}\right) - X_{In}^{old}\right) \end{aligned} \tag{2}$$

In Eq. (2), X_{Rn} is nth real parts bower or solution vector. X_{Relite} represents the real parts position of the elite. The value X_{Rj} is calculated by the roulette wheel procedure. $\text{var}_{R\max}$ and $\text{var}_{R\min}$ mean the upper and lower bounds of the real part. x_{In} is nth imaginary parts bower or solution vector. x_{Ielite} Represents the imaginary parts position of the elite. The value x_{Ij} is calculated by the roulette wheel procedure. var_{Imax} and var_{Imin} mean the upper and lower bounds of the real part.

Bower Mutation

$$X_{Rn}^{new} + iX_{In}^{new} = X_{Rn}^{old} + iX_{In}^{old} + (\sigma_R * N(0,1)) + i(\sigma_I * N(0,1)) \quad (3)$$

$$\begin{aligned} \sigma_R &= z * (\mathrm{var}_{R\max} - \mathrm{var}_{R\min}) \\ \sigma_I &= z * (\mathrm{var}_{\mathrm{Imax}} - \mathrm{var}_{\mathrm{Imin}}) \end{aligned} \quad (4)$$

In Eq. (4), the real and imaginary parts of the bird's bower position obey normal distribution N. $\mathrm{var}_{\mathrm{Imax}}$ and $\mathrm{var}_{\mathrm{Imin}}$ mean the upper and lower bounds of the real part. $\mathrm{var}_{\mathrm{Imax}}$ and $\mathrm{var}_{\mathrm{Imin}}$ mean the upper and lower bounds of the virtual part.

Fitness Calculation

In order to solve the fitness function, the complex-valued of bower must be converted into a real number, the number of modulo as the size of the real number, the symbol determined by the amplitude. The specific approach is shown in Eqs. (5) and (6)

$$\rho_n = \sqrt{X_{Rn}^2 + X_{In}^2} \quad (5)$$

$$X_n = \rho_n \mathrm{sgn}(\sin(\frac{X_{In}}{\rho_n})) + \frac{\mathrm{var}_{R\max} + \mathrm{var}_{R\min}}{2} \quad (6)$$

where ρ_n denotes the nth multidimensional modulus, X_{Rn} and X_{Ln} denote the real and imaginary parts of the complex modulus, respectively, and X_n is the transformed real number independent variable.

3 Experimental Results and Discussion

3.1 Simulation Platform

All calculations run in Matlab R2016a. The CPU is Intel i5 processor. The operating system is win10.

3.2 Compare Algorithm Parameter Settings

We have chosen five classical optimization algorithms to compare with CSBO, including the artificial bee colony optimization algorithm [6], bat algorithm [7], Cuckoo search optimizer (CS) [8] and the original version Satin Bowerbird Optimization algorithm [9].

3.3 Benchmark Test Functions

In this section, the CSBO algorithm is based on 20 Standard test functions. These 20 reference functions are classical functions used by many researchers. In spite of the simplicity, we chose these test functions to compare our results with the current heuristic results. These reference functions are listed in Table 1, where *Dim* represents the dimension of the function, *range* is the boundary of the function search space (Figs. 1 and 2).

Table 1. Benchmark functions

Benchmark test functions	Dim	Range
$F_1 = \sum_{i=1}^{D-1} [100(x_{i+1} - x_i)^2] + (x_i - 1)^2$	30	[−30, 30]
$F_2 = -20\exp\left(-0.2\sqrt{\frac{1}{n}\sum_{i=1}^{n} x_i^2} - \exp\left(\frac{1}{n}\sum_{i=1}^{n}\cos 2\pi x_i\right)\right) + 20 + e$	30	[−32, 32]

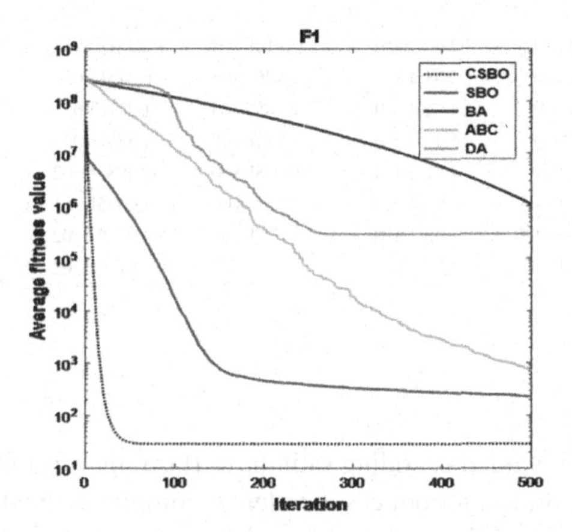

Fig. 1. Convergence curves of algorithms

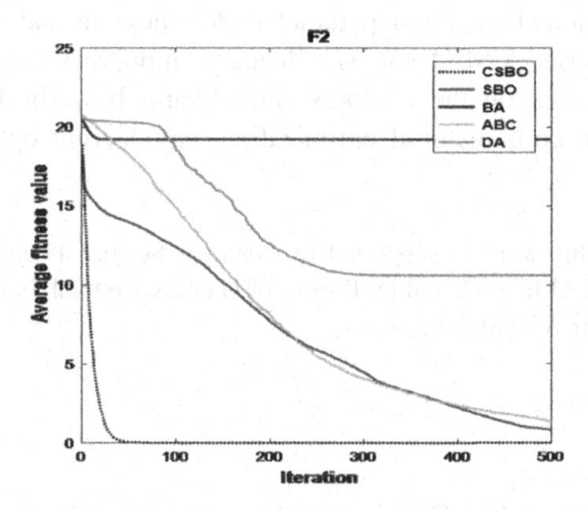

Fig. 2. Convergence curves of algorithms

The results of the standard test function are shown in Table 2. It can be observed that the accuracy of the CSBO algorithm exceeds that of other algorithms. From the convergence graph, we can see that the convergence speed of the CSBO algorithm is faster than other algorithms. The variance map shows that the CSBO algorithm results are stable. Summary, the experimental results in this paper demonstrates that CSBO shows competitive results among the three benchmark functions and outperforms other well-known algorithms and points out that CSBO has better in more other kinds of algorithms. The next section examines the CSBO solution to engineering problems.

Table 2. Benchmark functions test results

Algorithms functions		ABC	BA	DA	SBO	CSBO	RANK
F1	Best	1.114E+01	2.607E+01	7.710E+04	2.331E+01	2.876E+01	1
	Worst	8.705E+01	2.485E+02	7.876E+05	6.273E+02	2.890E+01	
	Mean	4.191E+01	4.760E+01	1.585E+05	1.311E+02	2.880E+01	
	Std.	2.206E+01	5.470E+01	1.744E+05	1.573E+02	3.5176E−02	
F2	Best	5.1376E−05	1.810E+01	6.058E+00	2.4987E−02	8.8817E−16	1
	Worst	3.1280E−03	1.903E+01	1.157E+01	1.0035E−01	7.9936E−16	
	Mean	7.8133E−04	1.922E+01	8.627E+00	4.0233E−02	3.1974E−16	
	Std.	8.0517E−04	3.3607E−01	1.335E+00	2.0166E−02	2.0859E−15	

4 Conclusions

This paper a complex-valued encoding satin bowerbird optimization algorithm (CSBO) is proposed. This algorithm introduces the idea of complex-valued coding and finds the optimal one by updating the real and imaginary parts value. The proposed CSBO optimization algorithm is compared against real-valued SBO and other bio-inspired optimization algorithms using 3 standard test functions including unimodal and multimodal functions, induction motor parameter identification and p-value test. Results show that the proposed CSBO can significantly improve the performance metrics. Future research focuses on the complex-valued satin bowerbird optimizer (CSBO) algorithm is used in combinatorial optimization, engineering optimization and other applied fields.

Acknowledgment. This work is supported by National Science Foundation of China under Grants No. 61563008, 61463007, and by Project of Guangxi Natural Science Foundation under Grant No. 2016GXNSFAA380264.

References

1. Goldberg, D.E., Holland, J.H.: Genetic algorithms and machine learning. Mach. Learn. **3**, 95–99 (1988)
2. Storn, R., Price, K.: Differential evolution–a simple and efficient heuristic for global optimization over continuous spaces. J. Glob. Optim. **11**, 341–359 (1997)

3. Simon, D.: Biogeography-based optimization. IEEE Trans. Evol. Comput. **12**, 702–713 (2008)
4. Kennedy, J.: Particle swarm optimization. In: Sammut, C., Webb, G.I. (eds.) Encyclopedia of machine learning, pp. 760–766. Springer, Boston (2011). https://doi.org/10.1007/978-0-387-30164-8_630
5. Dorigo, M., Birattari, M., Stutzle, T.: Ant colony optimization. IEEE Comput. Intell. Mag. **1**, 28–39 (2006)
6. Karaboga, D., Basturk, B.: A powerful and efficient algorithm for numerical function optimization: artificial bee colony (ABC) algorithm. J. Glob. Optim. **39**, 459–471 (2006)
7. Yang, X.-S.: A new metaheuristic bat-inspired algorithm. In: González, J.R., Pelta, D.A., Cruz, C., Terrazas, G., Krasnogor, N. (eds.) Nature Inspired Cooperative Strategies for Optimization (NICSO 2010). SCI, vol. 284, pp. 65–74. Springer, Heidelberg (2010). https://doi.org/10.1007/978-3-642-12538-6_6
8. Yang, X.-S., Deb, S.: Cuckoo search via Lévy flights. In: World Congress on Nature Biologically Inspired Computing, NaBIC 2009. IEEE (2009)
9. Moosavi, S.H.S., Bardsiri, V.K.: Satin bowerbird optimizer: a new optimization algorithm to optimize ANFIS for software development effort estimation. Eng. Appl. Artif. Intell. **60**, 1–15 (2017)

Powell-Based Bat Algorithm for Solving Nonlinear Equations

Gengyu Ge, Yuanyuan Pu, Jiyuan Zhang, and Aijia Ouyang(✉)

School of Information Engineering, Zunyi Normal University,
Zunyi 563006, Guizhou, China
ouyangaijia@163.com

Abstract. As the bat algorithm (BA) has defects such as slow convergence and poor calculation precision, it is likely to result in local extremum, and Powell algorithm (PA) is sensitive to the initial value. To resolve the above defects, advantages and disadvantages of PA and bat algorithm are combined in this paper to solve nonlinear equations. The hybrid Powell bat algorithm (PBA) not only has strong overall search ability like bat algorithm, but also has fine local search ability like Powell algorithm. Experimental results show that the hybrid algorithm can be used to calculate solutions to various nonlinear equations with high precision and fast convergence. Thus, it can be considered a positive method to solve nonlinear equations.

Keywords: Bat algorithm · Powell algorithm · Hybrid algorithm
Nonlinear equations · Optimization

1 Introduction

With the development of science and technology as well as the extensive application of electronic computers, there's an increasingly more nonlinear problems in many fields of science and engineering calculation. As one of the main research topics of mathematic science, the study on nonlinear problem is not only the need of modern scientific and technological development but also a certain result of the rapid development of modern computer technology. When solving nonlinear problem with the computer, it will ultimately be shifted to the finite dimension nonlinear problem, or the nonlinear algebraic problem. Thus, solutions to nonlinear algebraic problem have become an important research topic of modern calculation mathematics.

BA is a new meta-heuristic search algorithm proposed by Yang from University of Cambridge in 2010 by simulating bats echo location behavior. The method is characterized by easy modeling and achievable programming. It is also likely to result in local optimization, prior convergence, and slow convergence in the later stage.

The Powell BA is introduced in this paper to resolve the problem of nonlinear equations. Being characterized by simple principles, high precision of solving nonlinear equations, and high convergence rate, it is a good algorithm to solve the problem of nonlinear equations.

D.-S. Huang et al. (Eds.): ICIC 2018, LNAI 10956, pp. 840–845, 2018.
https://doi.org/10.1007/978-3-319-95957-3_90

2 The Mathematic Model

Assume that a nonlinear equations is composed by n equations, and it is related to m unknown quantities, the form of which is shown below:

$$\begin{cases} f_1(x_1, x_2, \ldots, x_m) = A_1 \\ f_2(x_1, x_2, \ldots, x_m) = A_2 \\ f_n(x_1, x_2, \ldots, x_m) = A_n \end{cases} \quad (1)$$

Where, $f_i(X) = A_i(i = 1, 2, \ldots, n)$ is the nonlinear equations, $X = [x_1, x_2, \ldots, x_m]$ is the unknown vector of the equations, and $A_i(i = 1, 2, \ldots, n)$ is the constant term. The fitness function can be constructed:

$$F(X) = \sum_{i=1}^{n} |f_i(x) - A_i| \quad (2)$$

In this way, the root problem of solutions to nonlinear Eqs. (1) has been transformed to an unconstrained optimization problem of calculating the minimum value of the fitness function $F(X)$.

3 Basic Algorithms

3.1 Bat Algorithm

BA [1] is a new swarm intelligence algorithm which simulates the echo location principle of mini bats. It is assumed the search is conducted in D dimension space, and one bat species is composed by m bats, so the current location of the i th bat can be expressed as the vector $x_i = (x_{i1}, x_{i2}, \ldots, x_{in})$, and its rate can be recorded as vector $V_i = (v_{i1}, v_{i2}, \ldots, v_{iD})$.Currently, the global optimal location searched among the whole bat group is $X^* = (x_1, x_2, \ldots, x_D)$, where the location update formula of $i = 1, 2, \ldots, m$ is shown below:

$$f_i = f_{\min} + (f_{\max} - f_{\min})\beta \quad (3)$$

$$V_i^t = V_i^{t-1} + (x_i^{t-1} - x^*)f_i \quad (4)$$

$$x_i^t = x_i^{t-1} + V_i^t \quad (5)$$

Where, t is the current iteration quantity, and β is the random variable evenly distributed at [0, 1].

$$X_{new} = X_{old} + \varepsilon A^t \quad (6)$$

$\varepsilon \in [-1, 1]$ is a random value, and $A^t = <A^t>$ is the mean loudness of all bats at some time.

$$A_i^{t+1} = \alpha A_i^t \quad (7)$$

$$r_i^{t+1} = r_i^0[1 - \exp(-\gamma t)] \quad (8)$$

Here, α and γ are constant. For any $0 < \alpha < 1$ and $\gamma > 0$, there will be $A_i^t \to 0$ and $r_i^t \to r_i^0$. When $t \to +\infty$, it is generally set in literatures that $\alpha = \gamma = 0.9$. In the iteration process, the loudness and pulse rate of each bat are set at the same value, and they change with the formulas (7) and (8). The search with BA will generally be suspended when the number of iterations reaches the maximum set value or the precision of the search value meets certain requirements.

3.2 Powell Algorithm

PA [2] is a pattern search strategy as well as an acceleration method in the conjugate direction. By making use of the randomly given initial location, extremum search of various dimensions along the positive and negative directions is carried out with the method. Only the function information is utilized without the calculation of derivatives with this method, so it is considered one of the most effective strategies among direct search methods in finding the minimum of functions.

4 Powell Bat Algorithm

As BA is similar to other bionic swarm intelligence algorithms by having poor calculation precision and being easy to trap into local optimization. However, PA can help to achieve fast convergence and strong local fine search. PA can be conducted for each bat to strengthen the local fine search capacity of the algorithm, so as to improve the global optimization competence of the algorithm. Therefore, the Powell bat algorithm (PBA) based on PA is designed. The steps of PBA algorithm are as follows:

Step 1: When $t = 0$, carry out random initialization for all bats. Their values shall be within the set range, and their location vectors are $x_{id}, i \in [1, m]$, in which, m is the number of bats, $d \in [1, D]$, and D is the particle dimension. The loudness of the initialized pulse is A_0, and the pulse emission rate is r. The maximum frequency is $Q_{\max}$ and the minimum frequency is $Q_{\min}$.
Step 2: p_{gd} is the location of the optimal particle in the species by calculation. Conduct in-depth search with PA to obtain the new optimal location p'_{gd}.
Step 3: Combine formulas (3), (4), (5) and (6), update the i-th location x_{id} and speed V_{id} of particles, and check whether the location x_{id} has crossed the line.
Step 4: Judge if *rand* is larger than r; If true, choose a solution from the optimal solutions, and form a local solution near the selected optimal solution to replace the current solution x_{id}^*. If not, skip this step directly.
Step 5: If the new fitness $f(x'_{id})$ of the bat i is superior to the fitness $f(x_{id})$ of the current extremum x_{id}, or *rand* < loudness A, replace x_{id} with the current particle location $f(x'_{id})$, otherwise, $f(x_{id})$ will not be updated.

Step 6: If the fitness of the bat i in the current iteration is superior to the fitness $f(p'_{gd})$ of the global optimal value p'_{gd}, replace the current location p'_{gd} with the new location x_{id} of the bat i, otherwise, p_{gd} will not be updated.
Step 7: Update the loudness A and the pulse emission rate by using formulas (7) and (8).
Step 8: If the operated iteration times reach the preset maximum value, terminate the search, and output the result. The global optimal solution will be p'_{gd}, and the corresponding target function $f(p'_{gd})$ will be calculated, otherwise, return to step 2 for search continuously.

5 Experiments and Analyses

Recently, by integrating the conjugate direction method (CD) and particle swarm optimization algorithm, the above problem is resolved by Mo et al. [3].

By combining chaotic search and Newton-type method, Luo et al. [4] manage to find six different solutions a to solve the above problem, too. Based on the PBA, two solutions are found. The methods for obtaining the best algorithm are listed in Table 1 and are compared with the result of Mo [3] and Luo [4] respectively.

Table 1. Comparison results of PBA with Mo et al. [3] and Luo et al. [4]

Methods	b	h	t	$f_1(x)$	$f_2(x)$	$f_3(x)$
PBA	11.42561	1.96537	15.32894	165	9369	6835
PBA	2.569164	21.31687	15.68435	165	9363	6835
Mo et al. [3]	8.943089	23.27148	12.91277	251.2378	9369	6835
Luo et al. [4]	12.5655	22.8949	2.7898	408.6488	9544.3	7213.1
Luo et al. [4]	−12.5655	−22.8949	−2.7898	408.6488	9544.3	7213.1
Luo et al. [4]	8.943089	23.27148	12.91277	251.2378	9369	6835
Luo et al. [4]	−8.94308	−23.2715	12.91277	251.2378	9369	6835
Luo et al. [4]	−2.3637	35.7564	3.0151	−334.038	9369	6835
Luo et al. [4]	2.3637	−35.7564	−3.0151	−334.038	9369	6835

Apparently, the exact solution calculated with the PBA algorithm is superior to the other two algorithms according to results of Table 1.

There are ten variables and ten equations for the system [5]. Mo et al. [3] integrate the conjugate direction method (CD) and particle swarm optimization algorithm to solve the above problem [6]. Optimal solutions obtained via PBA are listed in Table 2, and the results are compared with that published by Mo [3] previously.

Obviously, it can be seen from Table 2 that results of PBA are very close to precise solutions, and its accuracy is much superior to the method proposed by Mo [3].

Meanwhile, the following problems are solved subsequently. Optimal solutions to PBA and other algorithms are listed in Table 3.

Table 2. Comparison results of PBA Mo et al. [3]

Solutions and equations	Mo et al. [3]	PBA
x_1	0.915551	−0.39205
x_2	−0.222256	−0.34808
x_3	−0.414654	−0.45236
x_4	−0.439254	−0.43982
x_5	0.420892	−0.4459
x_6	−0.354588	−0.44284
x_7	−0.135767	−0.44487
x_8	0.427562	−0.44277
x_9	0.752203	−0.43673
x_{10}	−0.440697	−0.27523
$f_1(x)$	−3.17E−06	−0.42385
$f_2(x)$	3.52E−07	−0.44563
$f_3(x)$	−1.70E-06	9.20E-16
$f_4(x)$	1.77E−06	8.96E−16
$f_5(x)$	−1.68E+00	1.11E−16
$f_6(x)$	2.53E+00	−1.21E−16
$f_7(x)$	−8.42E−01	−1.13E−16
$f_8(x)$	−3.91E−07	−1.16E−16
$f_9(x)$	6.81E−07	9.63E−16
$f_{10}(x)$	2.34E−07	-9.89E−16

Table 3. Optimal results of case study

Case study	Methods	x_1	x_2	x_3	x_4	x_5	x_6
Case 3	Mo et al. [3]	−1	1	−1	1	−1	1
	Krzyworzcka	−1	1	−1	1	−1	1
	PBA	−1	1	−1	1	−1	1
Case 4	Mo et al. [3]	4	3	1			
	PBA	4	3	1			
Case 5	PBA	0.1689525	0.9427728	1			
	PBA	0.8275276	0.3012106	1			
Case 6	PBA	0.5	0	−0.532891			
Case 7	PBA	−0.301621	1.0132462				
		−0.801231	−0.816325				

6 Conclusions

BA is to utilize the echo location principle of mini bats. In this algorithm, bats can perceive the movement of other bats, and the optimal solution can be approached by adjusting the loudness and pulse rate, thus completing the process of making the species from being disordered to being ordered. The BA is found to have defects like slow convergence, poor calculation accuracy and high possibility of resulting in local extremum like other bionic swarm intelligence algorithms. To overcome these defects, the PBA (PBA) is put forward by this paper. The main idea of this algorithm is to implement PA for each bat, so that the local fine search ability with the algorithm is enhanced, and the global optimization capacity of the algorithm is improved. Through a series of experiments, it is verified that the algorithm proposed by this paper has advantages like high convergence speed and non-trapping in local extremum.

Acknowledgements. The research was partially funded by the science and technology project of Guizhou ([2017]1207), the training program of high level innovative talents of Guizhou ([2017]3), the Guizhou province natural science foundation in China (KY[2016]018), the Science and Technology Research Foundation of Hunan Province (13C333).

References

1. Yang, X.S., Gandomi, A.H.: Bat algorithm: a novel approach for global engineering optimization. Eng. Comput. **29**(5), 464–483 (2012)
2. Ma, W., Sun, Z.X., Li, J.L.: Cuckoo search algorithm based on powell local search method for global optimization. Appl. Res. Comput. **32**(6), 1667–1675 (2015)
3. Mo, Y.B., Liu, H.T., Wang, Q.: Conjugate direction particle swarm optimization solving systems of nonlinear equations. Comput. Math Appl. **57**(11), 1877–1882 (2009)
4. Luo, Y.Z., Tang, G.J., Zhou, L.N.: Hybrid approach for solving systems of nonlinear equations using chaos optimization and quasi-Newton method. Appl. Soft Comput. **8**(2), 1068–1073 (2008)
5. Krzyworzcka, S.: Extension of the Lanczos and CGS methods to systems of nonlinear equations. J. Comput. Appl. Math. **69**(1), 181–190 (1996)
6. Hueso, J.L., Martinez, E., Torregrosa, J.R.: Modified Newton's method for systems of nonlinear equations with singular Jacobian. J. Comput. Appl. Math. **224**(1), 77–83 (2009)

Ensemble Face Recognition System Using Dense Local Graph Structure

Dipak Kumar[1(✉)], Jogendra Garain[1], Dakshina Ranjan Kisku[1(✉)], Jamuna Kanta Sing[2], and Phalguni Gupta[3]

[1] Department of Computer Science and Engineering, National Institute of Technology Durgapur, Durgapur 713209, India
dipakcsi@gmail.com, jogs.cse@gmail.com, drkisku@cse.nitdgp.ac.in

[2] Department of Computer Science and Engineering, Jadavpur University, Kolkata 700032, India

[3] Department of Computer Science and Engineering, Indian Institute of Technology Kanpur, Kanpur 208016, India
pg@cse.iitk.ac.in

Abstract. This paper presents an ensemble face recognition system which makes use of a novel Dense Local Graph Structure (D-LGS). This descriptor uses additional graph structure along with its own local graph structure. This additional local graph structure is generated from original symmetric LGS by finding additional corner pixels through bilinear interpolation of neighbourhood pixels. These corner pixels lead to most stable features and information related to local deformation of the image. In this proposed ensemble system, three classifiers namely K-NN, Chi-square and correlation coefficient are used. Further the proposed approach fuses the decisions of individual classifiers through OR rule, majority voting and AND rule. To evaluate the performance of proposed ensemble system the experiment is conducted with two face databases viz. the AT&T (formerly the ORL database) and the UFI (Unconstrained Facial Image) database. The experimental results exhibit considerable improvement of the ensemble face recognition system on the use of novel descriptor.

Keywords: Ensemble face recognition system · Local binary pattern (LBP) Symmetric local graph structure (SLGS) · Classifiers · Fusion rules

1 Introduction

Face recognition is one of the most challenging issues in the field of computer vision, because the recognition accuracy depends on many constraints of the faces like pose, background changes, illumination variation, expression changes, age variation (aging) etc. Several researchers have proposed various face recognition systems with moderate accuracy, which use face images taken in control environment [1, 2]. However the systems do not give satisfactory accuracy when the imaging environment is changed from constrained to unconstrained environment. This issue of unconstrained environment is considered to be a challenging problem in face recognition. However many works [3, 6–8] have been done on unconstrained face recognition in the last few year

D.-S. Huang et al. (Eds.): ICIC 2018, LNAI 10956, pp. 846–852, 2018.
https://doi.org/10.1007/978-3-319-95957-3_91

showing encouraging results. In an ensemble system, input feature vector is distributed in a parallelly arranged different classifiers which separately produce the proximity of the belongingness in a class. Afterward the individual classification results are fused by one or more combination rules. One such ensemble method is proposed in [8] which explores 3D face recognition system and detects SIFT keypoints for finding similarity between two faces. The authors used two classifiers such as K-NN and K-means in their proposed ensemble system. Another work [6] discusses ensemble approach which uses many kernel random discriminant analysis (MK-RDA) classifier to discover discriminative patterns. A makeup based face recognition system is introduced in [7] that uses patch based ensemble learning method. The way the works have been presented might not be sufficient to overcome the dynamicity of unconstrained face characteristic to achieve the reliability, convenience and universality. Therefore it is important to emphasize to use feature information correctly that allow to deal with various issues of face recognition without compromising the security and performance much in order to achieve the state-off-the-art ensemble face recognition system where a multiclassifier fusion approach is being adopted. This type of ensemble system fuse the proximities which are determine form individual classifiers to combined identity evidence (discriminating information) in order to increase the system performance. Ensemble system is the multi-classifier system. Ensemble system can be divided into three categories, viz. homogeneous, heterogeneous and hybrid. In homogeneous system all classifiers are of same type. In heterogeneous systems all classifiers are of different types where as if some classifiers are different types and some are the same types then it belongs to hybrid ensemble system. In this proposed work, three different classifiers are used, therefore the proposed model is termed as heterogeneous ensemble system.

Figure 1 shows the block diagram of the proposed ensemble face recognition system which consists of five subsystems namely, data acquisition, feature extraction, template storage, multi-classifier and decision. The first subsystem does not employ data acquisition as the system uses two benchmark databases. Next subsystem is employed novel feature descriptor (D-LGS) for feature extraction. Multi-classifier subsystem uses different classifiers such as Chi-square [11], K-NN [8] and correlation coefficient [12] for producing matching proximity. Finally, three fusion rules (OR, AND and majority voting) [11, 13] are applied in the decision making subsystem.

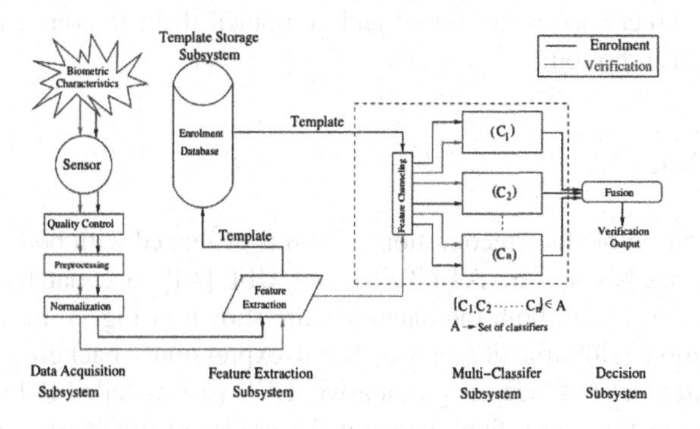

Fig. 1. Block diagram of proposed ensemble system.

2 Proposed Methodology

The main shortcoming of the state-of-the-art local feature descriptor is that they are not stable as because they do not consider the auxiliary corner pixels. For example, the given image size of 64 × 64 pixels have only 3720 SLGS codes. If the corner pixels (more neighbours) are taken into the account then additional 3599 SLGS codes along with 3720 SLGS codes can be achieved. Therefore the description with more SLGS code will be more stable and discriminative in nature. Symmetric Local Graph Structure (SLGS) [4] is a texture based descriptor which is derived from basic local graph structure [5].

The proposed ensemble face recognition system uses a novel local graph structure called Dense Local Graph Structure (D-LGS) which is derived from SLGS. Besides regular pixels in SLGS, corner pixels are also determined through bilinear interpolation and this leads to dense local graph structure because it increases the density of considered pixels. The dense local graph structure is shown in Fig. 2.

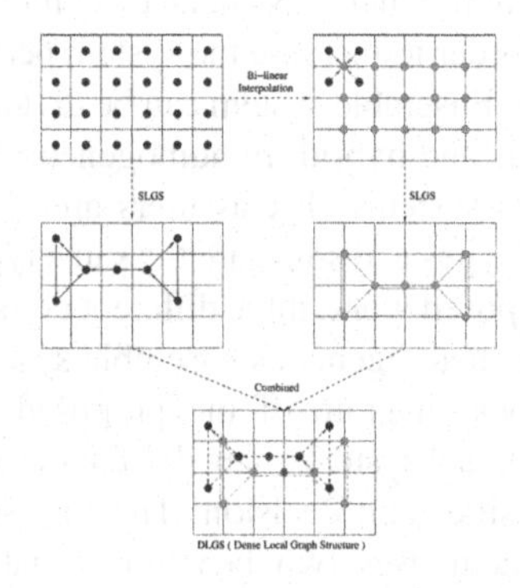

Fig. 2. Flow diagram of D-LGS (Color figure online).

In Fig. 2, blue dots in top-left diagram represent the original pixels which are used to find the corner pixels by using bilinear interpolation shown in top-right diagram (green dots). Then the symmetric local graph structures of both original pixels and newly found (corner) pixels are found and combined them to construct a dense LGS (D-LGS) shown in bottom.

3 Evaluation

The proposed ensemble face recognition system is evaluated with both constrained and unconstrained databases, viz. AT&T [9] and UFI [10] face database respectively. Sample face images from both the databases are shown in Fig. 3. In order to cope up with the variations (illumination, pose, facial expressions, background, accessories, occlusion, clutter, etc.) of face images, a novel LGS variant called D-LGS is applied to face images for feature extraction. Then the feature set obtained from each face image

Fig. 3. Sample face images from the AT&T face database [9] and UFI face database [10].

is passed through three different classifiers, viz. K-NN, Chi-square and correlation coefficient, and matching proximities are determined. Finally, matching proximities are fused together using AND rule, OR rule and majority voting to obtain consolidated decision for acceptance or rejection.

3.1 Results and Discussion

The experimental results are shown in Table 1. In this table, two accuracies corresponding to each classifiers, first accuracy determined when FAR and FRR values are closest to each other which signifies that the system is considering same cost of FAR as well as FRR. On the other hand the second value of accuracy is obtained corresponding to a threshold value for which the equal error rate (EER) is minimum i.e., the accuracy is maximum.

Table 1. The table shows accuracy, FAR (False Acceptance Rate), FRR (False Rejection Rate) determined before and after fusion on AT&T database.

AT&T database	Classifier/fusion	FAR (%)	FRR (%)	Accuracy (%)
Classifier	K-NN	3.9744	5.0000	95.5128
		1.2179	5.0000	96.8910
	Chi square	2.5000	2.5000	97.5000
		1.7308	2.5000	**97.8846**
	Correlation coefficient	76.8590	77.5000	22.8205
		0.0000	100.000	50.0000
Fusion	AND rule	97.5000	97.5000	2.5000
		0.0000	100.000	50.0000
	Majority voting	3.7821	5.0000	95.6090
		1.2179	5.0000	96.8910
	OR rule	0.0000	0.0000	100.000
		0.0000	0.0000	**100.000**

Here, FAR and FRR values may differ from each other but it always gives the highest possible accuracy like the accuracy for KNN, Chi-square and correlation coefficient are 96.8910%, 97.8846% and 50.000% respectively shown in the second row of each classifier in Table 1. The highest accuracy among these three classifiers is

97.8846% with the FAR (1.7308%) and FRR (2.5000%) which is obtained using Chi-square classifier. The lower half of the Table 1 shows the accuracy of the same classifiers but after applying three fusion rules. The accuracies are placed in the table where it is observed that the accuracy is improved extreme for OR rule. The highest accuracies for before fusion and after fusion are shown in bold face. Figure 4 shows ROC curves.

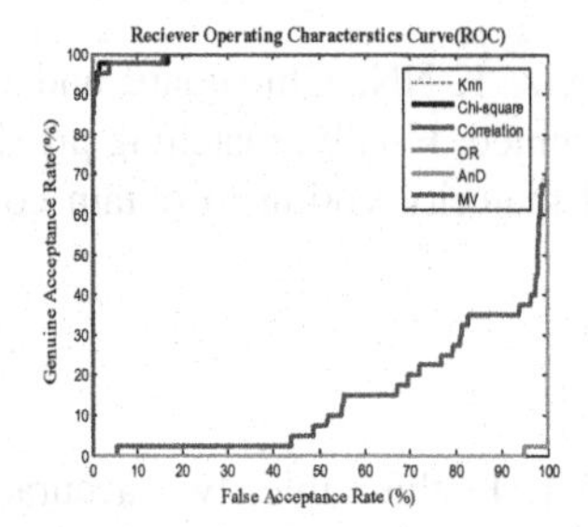

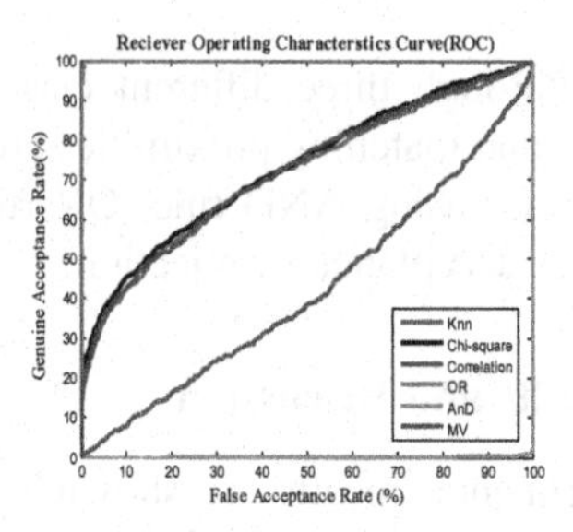

Fig. 4. Receiver operating characteristics (ROC) curves determined on AT&T (Left) and UFI (Right) databases.

Table 2 shows FAR and FRR values which may differ from each other but it always provide the highest possible accuracy like the accuracy for KNN, Chi-square and correlation coefficient are 67.3587%, 67.9949% and 50.0592% respectively shown in the second row of each classifier in Table 2. The highest accuracy among these tree classifiers is 67.9949% with the FAR (16.2416%) and FRR (47.7686%) which is obtained using Chi-square classifier. Figure 4 shows the ROC curves determined on UFI database.

Table 2. The table shows accuracy, FAR (False Acceptance Rate), FRR (False Rejection Rate) determined before and after fusion on UFI database.

UFI database	Classifier/fusion	FAR (%)	FRR (%)	Accuracy (%)
Classifier	K-NN	33.8591	34.3802	65.8804
		18.0100	47.2727	67.3587
	Chi Square	34.4357	34.0496	65.7573
		16.2416	47.7686	**67.9949**
	Correlation coefficient	56.0303	55.7025	44.1336
		0.7080	99.1736	50.0000
Fusion	AND rule	98.8422	98.8430	1.1574
		0.0000	100.000	50.0000
	Majority voting	34.0928	33.8843	66.0114
		15.1314	50.5785	67.1451
	OR rule	0.8374	0.8264	99.1681
		1.3023	0.0000	**99.3488**

4 Conclusion

This paper has presented an ensemble face recognition system based on novel descriptor D-LGS which uses bilinear interpolation to enhance the density of the pixels which are taken into consideration while forming the descriptive image from the input image. That is why it performs well on both constrained and unconstrained face images. Investigation shows that even the descriptor itself shows superior performance while the proposed ensemble system using correlation coefficient is not found impressive whereas the system gives cent percent accuracy after using OR fusion rule. So the proposed ensemble system can be more robust on proper selection of classifier and fusion rule. The proposed descriptor can also be investigated with other classifiers that have been used in state-of-the-art methods to fairly compare the robustness of the descriptor as well as the ensemble system.

References

1. Gumus, E., Kilic, N., Sertbas, A., Ucan, O.N.: Evaluation of face recognition techniques using PCA, wavelets and SVM. Expert Syst. Appls. **37**(9), 6404–6408 (2010)
2. Lin, W.H., Wang, P., Tsai, C.F.: Face recognition using support vector model classifier for user authentication. Electron. Commer. Res. Appl. **18**, 71–82 (2016)
3. Roychowdhury, S., Emmons, M.: A survey of the trends in facial and expression recognition databases and methods. CoRR abs/1511.02407 (2015)
4. Abdullah, M.F.A., Sayeed, M.S., Muthu, K.S., Bashier, H.K., Azman, A., Ibrahim, S.Z.: Face recognition with symmetric local graph structure (SLGS). Expert Syst. Appl. **41**(14), 6131–6137 (2014)
5. Abusham, E.E.A., Bashir, H.K.: Face recognition using local graph structure (LGS). In: Jacko, J.A. (ed.) HCI 2011. LNCS, vol. 6762, pp. 169–175. Springer, Heidelberg (2011). https://doi.org/10.1007/978-3-642-21605-3_19
6. Jiang, R., Al-Maadeed, S., Bouridane, A., Crookes, D., Celebi, M.E.: Face recognition in the scrambled domain via salience-aware ensembles of many kernels. IEEE Trans. Inf. Forensics Secur. **11**(8), 1807–1817 (2016)
7. Chen, C., Dantcheva, A., Ross, A.: An ensemble of patch-based subspaces for makeup-robust face recognition. Inf. Fusion **32**(Part B), 80–92 (2016)
8. Schimbinschi, F., Schomaker, L., Wiering, M.: Ensemble methods for robust 3D face recognition using commodity depth sensors. In: 2015 IEEE Symposium Series on Computational Intelligence, pp. 180–187 (2015)
9. Cambridge University Computer Laboratory AT&T Database of Faces (2005). http://www.cl.cam.ac.uk/research/dtg/attarchive/facedatabase.html
10. Lenc, L., Král, P.: Unconstrained facial images: database for face recognition under real-world conditions. In: Lagunas, O.P., Alcántara, O.H., Figueroa, G.A. (eds.) MICAI 2015. LNCS (LNAI), vol. 9414, pp. 349–361. Springer, Cham (2015). https://doi.org/10.1007/978-3-319-27101-9_26

11. Rakshit, R.D., Nath, S.C., Kisku, D.R.: An improved local pattern descriptor for biometrics face encoding: a LC-LBP approach toward face identification. Trans. Chin. Inst. Eng. Comput. Eng. **40**(1), 82–92 (2016)
12. Sharma, A.K.: Text book of Correlations and Regression, p. 212. Discovery Publishing House, New Delhi (2005)
13. Kittler, J., Hatef, M., Duin, R., Matas, J.: On combining classifiers. IEEE Trans. Pattern Anal. Mach. Intell. **20**(3), 226–239 (1998)

Improved Indel Detection Algorithm Based on Split-Read and Read-Depth

Hai Yang(✉), Daming Zhu, and Huiqiang Jia

School of Computer Science and Technology,
Shandong University, Jinan 250101, Shandong, China
yh_sdjtu@163.com

Abstract. Indel is a molecular biology term for an insertion or deletion of bases in the genome. Lots of researches have demonstrated structural variations including indel are closely related to the disease and health of human. So the detection of indel is very important in the life science. Currently, there is a certain number of structural variation detecting algorithms based on next generation sequencing technology. But some of them are not accurate or too sensitive. In this paper, an improved indel detection algorithm based on split-read and read-depth using dynamic programming has been proposed, which is named IDSD. The IDSD algorithm extracts all the discordant paired-end reads. After clustered, the discordant paired-end reads are mapped to the reference sequence using dynamic programming so as to find out the breakpoints in split reads. Then insertions and deletions could be recognized according to the information of structural variations. Furthermore the read depth is also used to detect more deletion and improve higher confidence in the recognition of deletion. The experiment was taken using IDSD and other algorithms. The results show IDSD algorithm has higher efficiency in detecting indel variations and more practical value.

Keywords: Indel · Insertion · Deletion · Split read · Read depth
Dynamic programming

1 Introduction

Genome structural variations are extensive present in the human genome. The scale of genome structural variation is between single nucleotide polymorphisms (SNPs) and chromosome mutation [1]. Its scale is usually more than hundreds of base pairs. The biggest structural variation can even reach millions of base pairs level. There is a number of researches have demonstrated structural variations are closely related to the disease and health of human. So the detection of structural variation is very important for the human's health.

There are many types of structural variations including insertion, deletion, inversion, duplication, translocation and so on. Insertion and deletion are collectively called indel (Insertion/Deletion). The number of small and medium structural variations is the second-most in the human genome, surpassed only by SNP [2]. These small and medium

D.-S. Huang et al. (Eds.): ICIC 2018, LNAI 10956, pp. 853–864, 2018.
https://doi.org/10.1007/978-3-319-95957-3_92

structural variations have a significant effect on individual phenotype and diseases. So the research of indel is getting more and more popular especially the small indel.

More recently, Zhang et al. presented the correlation between chicken performance traits and indel polymorphism which is of 31-bp in the PAX7 gene [3]. Shi et al. studied that the special indel of 17-bp in the SMAD3 gene can change the transcriptional level of cow [4].

Currently, there are many structural variation detecting algorithms based on next generation sequencing platform including read-pair method, read-depth method, split-read method, sequence assembly method and micro-array method [5]. Especially the paired-end sequencing technology has been widely applied in detecting structural variations on the whole genome level.

Although a certain amount of structural variation detection algorithms have been proposed, but some of them are not accurate or too sensitive [6]. In this paper, the indel is systematically elaborated and then an improved Indel detection algorithm based on split-read and read-depth using dynamic programming has been proposed.

2 Related Concepts and Characteristic of Indel

2.1 Insertion and Deletion

The insertion is one primary type of structural variation in genetics. It refers to the addition of one or more nucleotide base pairs into a DNA sequence. Insertions can often happen in microsatellite regions due to the DNA polymerase slipping, and also can be anywhere in size from one base pair incorrectly inserted into a DNA sequence to a section of one chromosome inserted into another. An insertion could be simply demonstrated as Fig. 1.

The deletion is another primary type of structural variation in genetics. It refers to a part of a chromosome or a sequence of DNA is lost during DNA replication. Any number of nucleotides can be deleted, from a single base to an entire piece of chromosome. The smallest single base deletion mutations are believed to occur by a single base flipping in the template DNA, followed by template DNA strand slippage, within the DNA polymerase active site. An deletion could be simply demonstrated as Fig. 2.

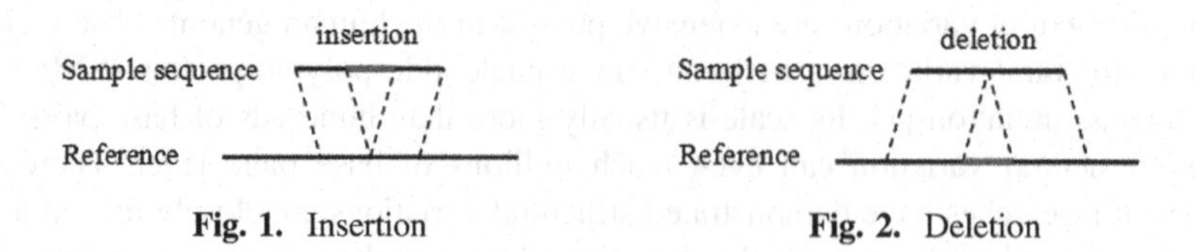

Fig. 1. Insertion **Fig. 2.** Deletion

2.2 Split Read

If there is a structural variations in the sample sequence, all the base symbols which are at the edge of structural variation couldn't be mapped onto reference sequence totally. But the prefix and postfix of these base symbols could mapped. In order to mapping onto reference sequence, the discordant read will be split into smaller reads by the breakpoints [7]. Figures 3 and 4 demonstrate the insertion variation and deletion variation.

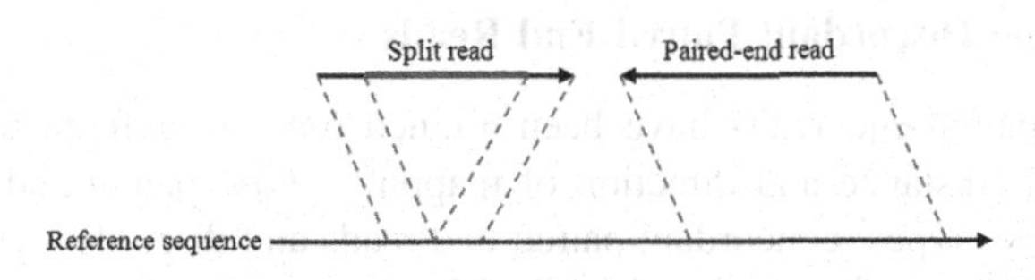

Fig. 3. Insertion structural variation

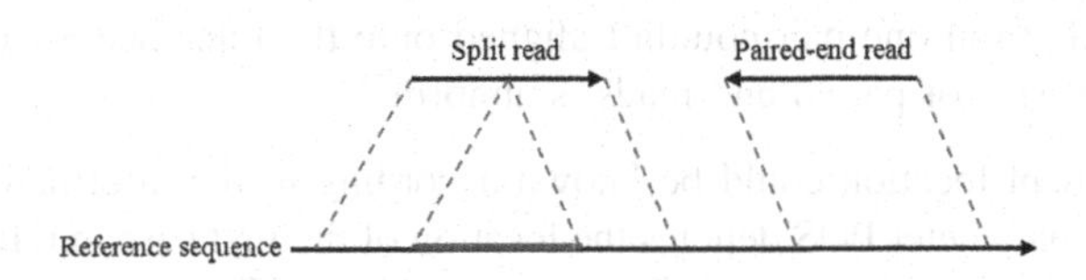

Fig. 4. Deletion structural variation

Nowadays, the reads generated by most of gene sequencing machines is size of no more than 1000 bp. Therefore the insertion variations which can be detected are comparatively small. Moreover, split-read method has a relatively high algorithm complexity. This is thought to be due to large amount of reads data and short length of reads.

2.3 Characteristic of Indel

Indel is a molecular biology term for an insertion or deletion of bases in the genome of an organism, which occurs in a large numbers in genomes [8]. Indel refers to insertions (or deletions) variation of nucleotide fragments at the same site in the genome sequence between the same or closely related species. Indel is a gap in sequence derived from alignment of the homologous sequence, which is classified among small genetic variations, measuring from 1 to 10 000 base pairs in length. In terms of distribution density, the number of indel is the second most only to SNP, and much more than SSR. There is one indel in 7.2 kp in the human genome, meanwhile there is one indel in 5 kp in the chicken and dog genomes.

Indel can be used as genetic markers in natural populations, especially in phylogenetic studies. It has been shown that genomic regions with indels can also be used for species-identification procedures.

An indel change of a single base pair in the coding part of an mRNA results in a frameshift during mRNA translation that could lead to an inappropriate (premature) stop codon in a different frame. Indels that are not multiples of 3 are particularly uncommon in coding regions but relatively common in non-coding regions. There are approximately 192–280 frameshifting indels in each person. Indels are likely to represent between 16% and 25% of all sequence polymorphisms in humans. In fact, in most known genomes, including humans, indel frequency tends to be markedly lower than that of SNP, except near highly repetitive regions, including homopolymers and microsatellites.

3 Improved Indel Detection Algorithm Based on Split-Read and Read-Depth

3.1 Extracting Discordant Paired-End Reads

Firstly, all of paired-end reads have been aligned onto the reference genome. Then according to the distance and direction of mapping, those paired-end reads could be classified into two types: concordant paired-end reads and discordant paired-end reads. The principles of classification are described below:

(1) The mapping direction of the paired-end read is opposite to its original direction.
(2) Both reads from one pair couldn't aligned onto the reference gonome.
(3) The insert size of paired-end reads is abnormal.

The alignment location could be known according to the information storaged in SAM file. The parameter POS denotes the location of the first base in reference genome where one paired-end read aligned. The parameter PNEXT denotes the location of the first base in reference genome where the other paired-end read aligned.

Let p denote parameter POS, q denote parameter PNEXT, L denote the length of paired-end read, d denote the insert size of every paired-end read, which could be calculated as follows:

$$d = q - p - L + 1 \tag{1}$$

The average insert size of all paired-end reads is μ, which could be calculated as follows:

$$\mu = \frac{\sum_{i=1}^{n} d_i}{n} = \frac{\sum_{i=1}^{n} (q_i - p_i - L + 1)}{n} \tag{2}$$

The standard deviation of insert size is σ:

$$\sigma = \sqrt{\frac{1}{n}\sum_{i=1}^{n} (d_i - \mu)^2} = \sqrt{\frac{1}{n}\sum_{i=1}^{n}\left(d_i - \frac{\sum_{i=1}^{n} (q_i - p_i - L + 1)}{n}\right)^2} \tag{3}$$

Take genetic sample TCGA-A6-2684-01A-01D-A46W-08 for example, the statistical histogram of insert size of all paired-end reads from this genetic sample is shown as Fig. 5. The horizontal coordinate axis denotes the insert size (in bp) of every pair of reads. The vertical coordinate axis denotes the frequency.

Shown in Fig. 5 most of paired-end reads have insert size greater than 250 bp and less than 500 bp. Meanwhile the number of paired-end reads have insert size greater than 500 bp or less than 250 bp is relatively few.

According to formula (2) and formula (3), the average insert size of all paired-end reads μ and the standard deviation of insert size σ can be calculated. If the insert size of

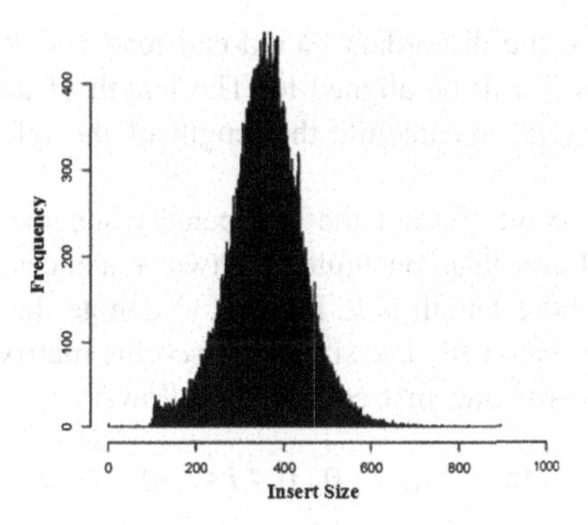

Fig. 5. The statistical histogram of insert size of all paired-end reads

one paired-end read satisfies $d > \mu + 3\sigma$ or $d < \mu - 3\sigma$, the paired-end read is abnormal which is called discordant. Extract all discordant paired-end reads as source data.

3.2 Clustering of Discordant Paired-End Reads

For each any discordant paired-end read, we let its $(POS, PNEXT)$ recorded as (p_i, q_i), so the median of mapped location of discordant read i is calculated as follows:

$$M(p_i, q_i) = \frac{1}{2}(p_i + q_i) \tag{4}$$

Next, cluster all the discordant paired-end reads according to the conditions follows:

(1) The discordant paired-end reads aligned onto the same chromosome.
(2) The discordant paired-end reads have the same mapping direction.
(3) The insert size of discordant paired-end reads in the same cluster should satisfy formula (5) and formula (6):

$$M(p_{k_{i+1}}, q_{k_{i+1}}) - M(p_{k_i}, q_{k_i}) < d_{\max} \tag{5}$$

$$\max_{ij}((q_{k_i} - p_{k_i}) - (q_{k_j} - p_{k_j})) \leq 2d_{\max} \tag{6}$$

Where $d_{\max} = \max\{d_1, d_2, \cdots, d_n\}$. So all of the discordant paired-end reads in the same cluster are near by with each other.

3.3 Finding Out the Breakpoints in Split Reads

After clustering, the breakpoints in split reads could be found out by means of mapping split reads onto reference sequence using dynamic programming.

Let $S = s_1s_2...s_s$ be the discordant paired-end read and $R = r_1r_2...r_r$ be the reference sequence which S will be aligned to. The length of the discordant paired-end read S is s, denoted by |S|, meanwhile the length of the reference sequence R is r, denoted by |R|.

Firstly, set the score matrix and the gap penalty scheme. Let $s(x, y)$ denote the similarity score of the bases that constituted the two sequences S and R. Let W_k denote the penalty of a gap whose length is k. Let $\sigma(x, y)$ denote the penalty function.

Construct a scoring matrix M. The size of the scoring matrix M is $(s+1) \times (r+1)$. Then initialize the first row and first column as follows:

$$M_{i0} = M_{0j} = 0,\ 0 \le i \le s, 0 \le j \le r$$

Fill the scoring matrix M using the equation as follows:

$$M_{ij} = \max\begin{cases} M_{i-1,j-1} + s(s_i, r_j) \\ \max_{k\ge 1}\{M_{i-k,j} - W_k\} \\ \max_{l\ge 1}\{M_{i,j-l} - W_l\} \\ 0 \end{cases} \quad (1 \le i \le s, 1 \le j \le r) \tag{7}$$

Where $M_{i-1,j-1} + s(s_i, r_j)$ is the score of aligning s_i and r_j. $M_{i-k,j} - W_k$ is the score if s_i is at the end of a gap of length k. $M_{i,j-l} - W_l$ is the score if r_j is at the end of a gap of length l. 0 means there is no similarity up to s_i and r_j.

Finally, It's the backtrace procedure. Starting at the highest score in the scoring matrix M and ending at a matrix cell that has a score of 0, trace back based on the source of each score recursively to generate the best local alignment.

Take two sequences TGTTACGG and GGTTGACTA for example according to the scheme below:

$$s(s_i, r_j) = \begin{cases} +3 & s_i = r_j \\ -3 & s_i \ne r_j \end{cases},\ \text{gap penalty } W_k = 2k,\ \sigma(s_i, r_j) = 0 \tag{8}$$

The scoring matrix will be calculated which is shown in Table 1 below:

Table 1. Scoring matrix.

		T	G	T	T	A	C	G	G
	0	0	0	0	0	0	0	0	0
G	0	0	3	1	0	0	0	3	3
G	0	0	3	1	0	0	0	3	6
T	0	3	1	6	4	2	0	1	4
T	0	3	1	4	9	7	5	3	2
G	0	1	6	4	7	6	4	8	6
A	0	0	4	3	5	10	8	6	5
C	0	0	2	1	3	8	13	11	9
T	0	3	1	5	4	6	11	10	8
A	0	1	0	3	2	7	9	8	7

As shown in Table 2, the gray cells demonstrate the highest score and the backtrace path. Note that an element can receive score from more than one element, each will form a different path if this element is traced back. In case of multiple highest scores, backtrace should be done starting with each highest score. The backtrace process is shown below on the right. The best local alignment is generated in the reverse direction.

Table 2. Backtrace procedure.

		T	G	T	T	A	C	G	G
	0	0	0	0	0	0	0	0	0
G	0	0	3	1	0	0	0	3	3
G	0	0	3	1	0	0	0	3	6
T	0	3	1	6	4	2	0	1	4
T	0	3	1	4	9	7	5	3	2
G	0	1	6	4	7	6	4	8	6
A	0	0	4	3	5	10	8	6	5
C	0	0	2	1	3	8	13	11	9
T	0	3	1	5	4	6	11	10	8
A	0	1	0	3	2	7	9	8	7

3	6	9	7	10	13
G	T	T	-	A	C
|	|	|		|	|
G	T	T	G	A	C

The alignment result of sequences TGTTACGG and GGTTGACTA is:

```
G  T  T  -  A  C
|  |  |     |  |
G  T  T  G  A  C
```

Next, the breakpoints could be determined according to the alignments of the discordant paired-end reads after the dynamic programming.

A deletion variation could lead to split reads, since those two subsequences of the split read end go away from each other on the reference genome in a longer distance than they are given. An insertion can cause those two subsequences of the split end to go overlapping on the reference genome, or make one subsequence of that split read end go in a shorter distance with the non-split end of the read.

The potential breakpoints are related to the gap loci. The length of insertion or deletion is longer than that of SNP, So one gap or few continuous gaps will be ignored. The breakpoints can be determined by filtering among all the gap loci according to the number of continuous gaps.

In the improved algorithm, a threshold value of the number of continuous gaps is given, which is denoted as t_{gap}. Usually, the number of continuous gaps nearby one breakpoint is at least three, so the threshold value t_{gap} can be set $t_{gap} = 3$. On the other hand because of the limitation of the split-read method, the breakpoints of one read are not at the edge positions. In other words, the distance from one breakpoint to one end of read couldn't be less than 10 bp. So the breakpoints can be determined according to the threshold value t_{gap} and the distance limitation mentioned above.

3.4 Indel Recognition

After finding the optimal alignment results, the information of indel structural variations could be extracted from the alignment results. The direction of the indel stucture variation could be known using the concordant paired-end reads which mapped to the reference sequence. The specific coordinates of structural variations also can be calculated using the insert size and length values of paired-end reads.

A deletion could be recognized according to two conditions below:

(1) The mapping distance of paired-end reads is greater than the upper limit of insert size according to the genome library.
(2) There is at least one breakpoint in the discordant paired-end read and then if the structural variation occurred like Fig. 6.

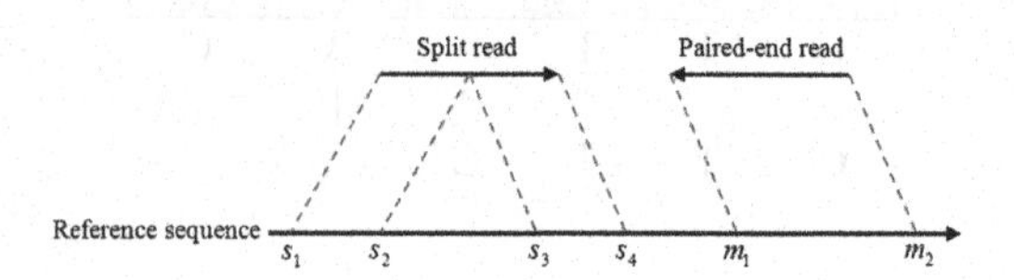

Fig. 6. Deletion recognition by split read

Assume that m_1 and m_2 denote the coordinates of the concordant paired-end read which successfully mapped to the reference sequence. The starting coordinate of the split read could be calculated as follows:

$$m_1 = m_2 - L \tag{9}$$

$$s_4 = m_1 - I \tag{10}$$

Where L is the length of paired-end read, I is the insert size.

A Insertion could be recognized if a structural variation occurred like Fig. 7. The starting coordinate of the split read could be calculated similar to deletion structural variation.

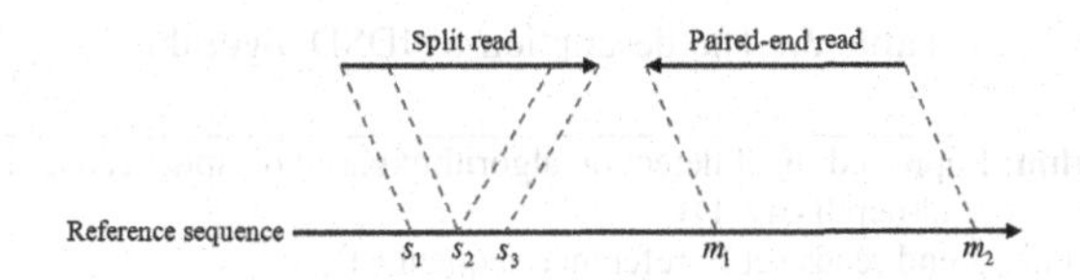

Fig. 7. Insertion recognition by split read

3.5 Confidence and Expandation of Indel with Read-Depth

When high-throughput genome sequencing using next generation sequencing, the probability distribution of sequencing loci depth usually obeys normal distribution or poisson distribution. The paired-end reads gained by sequencing basically cover every site of individual's genome sequence. After the paired-end reads mapped to the reference sequence, the average coverage depth of deletion region is far less than the normal region's coverage depth.

The coverage depth of paired-end reads could be counted easily by means of the depth command in SAMtools. Not only the average coverage depth of every base should be counted in the deletion region, but also the ones of the upstream region and downstream region nearby the deletion should be counted.

According to the statement above, two conclusions could be gained. Firstly, if the average coverage depth of any one region is noticeably smaller than others, there is a deletion in a high probability. Secondly, the smaller coverage depth is, the higher confidence in the recognition of deletion using the algorithm proposed by this paper.

3.6 The Description of IDSD Algorithm

In this paper, an improved Indel detection algorithm based on split-read and read-depth (IDSD for short) has been proposed according to the algorithmic idea mentioned above. The IDSD algorithm is described as Table 3 below:

4 Experimental Results

The experiments are taken using real data from the 1000 Genomes Project. The experimental data include 55 genetic samples of three regions. All of these genetic samples are storaged in FASTQ format which include 22 Yoruba samples, 9 Mende samples and 24 East Asian samples. The insertion and deletion variations will be detected by using Pindel algorithm, Meerkat algorithm and IDSD algorithm proposed in this paper respectively. The experimental results are shown in Table 4.

The results reflect that the number of insertion and deletion structural variation detected by IDSD algorithm is the most, which benefited from the read depth strategy brought into the algorithm. So the IDSD algorithm are the most superior among three algorithms.

Table 3. The description of IDSD algorithm

Algorithm: Improved Indel detection algorithm based on split-read and read-depth (IDSD)
Input: Paired-end reads set P, reference sequence R
Output: the number of insertion Num_ins, the number of deletion Num_del, the starting coordinates of insertion and deletion, the structural variation set SV.

```
Step 1: For each paired-end read in P
            Calculate insert size d according to formula (1)
        End For
        Calculate the average insert size of all paired-end reads μ according to
        formula (2)
        Calculate standard deviation of insert size σ according to formula (3)
Step 2: For each paired-end read in P
            Extract discordant paired-end read if d      +3σ or d      -3σ
            Put discordant paired-end read dr_i into set DR
        End For
Step 3: Cluster the discordant paired-end reads.
        For each dr_i in DR
            Calculate the median of mapped location according to formula (4)
            Cluster paired-end read dr_i according to the three clustering condi-
        tions.
        End For
Step 4: Find out the breakpoints in split reads.
          For each dr_i in DR
            Construct a scoring matrix M of s+1 columns and r+1 rows.
            s is the length of discordant paired-end read. r is the length of ref-
        erence.
            Then initialize the first row and first column as 0.
            For i=0 to r
              For j=0 to s
                Calculate the value of every cell in scoring matrix M
                Calculate the value of every cell in backtrace matrix T
              End For
            End For
            Find out the optimal backtrace path with highest score.
            Record the information, starting coordinates of the structural varia-
        tion into SV
          End For
Step 5: For each sv in SV
            Judge the structural variation type of sv.
            If sv is insertion, Num_ins= Num_ins+1
            If sv is deletion, Num_del= Num_del+1
        End For
Step 6: Calculate the coverage depth of paired-end reads after clustering.
        Calculae the average coverage depth of every base in the deletion re-
        gion.
        Calculae the average coverage depth of the upstream region and down-
        stream region nearby the deletion.
        If the average coverage depth of any one region is noticeably smaller
        than others
            Num_del=Num_del+1
            the small coverage depth results in the high confidence in the
        recognition of deletion.
        End If
```

Table 4. Results of the insertion and deletion

	Insertion	Deletion
IDSD	1126	14103
Pindel	1087	13956
Meerkat	1109	13872
DELLY	1098	14127
LUMPY	1131	14016

The running time of IDSD algorithm is shown in Fig. 8.

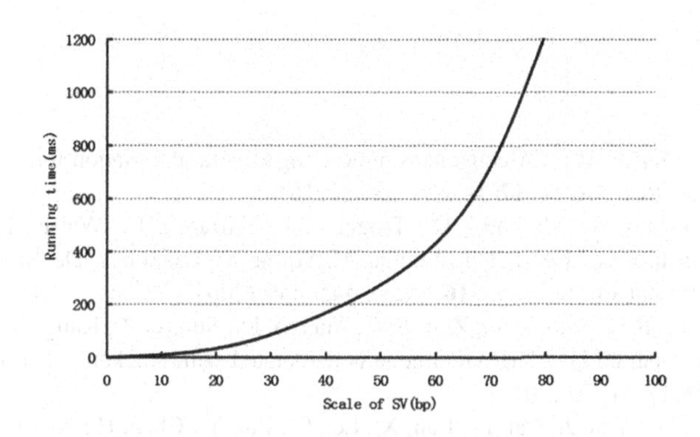

Fig. 8. The running time of IDSD algorithm

The major operations of IDSD algorithm is dynamic programming, so the execution efficiency of IDSD is subject to the time complexity of dynamic programming. When the scale of structural variation is longer than 70 bp, the running time of IDSD will become longer sharply. This is one of disadvantages need to be further improved.

The second simulation experiment is taken to verify the accuracy performance of IDSD algorithm. The simulation data is generated by GWAsimulatorv2.1 in contrast to Williams 82 Chromosome 15, which are consist of 600 insertions and 800 deletions. The vast majority of structural variations in simulation genome have the length less than 1000 bp and the insert size from 45 bp to 50 bp. The insertion and deletion variations will be detected by using IDSD algorithm, Pindel, Meerkat, DELLY and LUMPY respectively. The accuracy performances of those algorithms are shown in Table 5.

Table 5. The accuracy performance of IDSD algorithm compared with other four softwares

	IDSD		Pindel		Meerkat		DELLY		LUMPY	
	Total	Find	Total	Find	Total	Find	Total	Find	Total	Find
Insertion	600	521	600	511	600	516	600	520	600	532
Accuracy rate (%)	86.83%		85.17%		86.0%		86.67%		88.67%	
Deletion	800	719	800	702	800	703	800	738	800	717
Accuracy rate (%)	89.88%		87.75%		87.88%		92.25%		89.63%	

5 Conclusion

In this paper an improved indel detection algorithm based on split-read and read-depth was proposed, which is named IDSD. The IDSD algorithm extracts all the discordant paired-end reads mainly according to insert size. After clustering, the discordant paired-end reads are mapped to the reference sequence using dynamic programming so as to find out the breakpoints in split reads. Then insertions and deletions could be recognized according to the information of structural variations such as insert size, length, direction and coordinates etc. Furthermore the read depth is also used to detect more deletion and improve higher confidence in the recognition of deletion.

References

1. Claudia, B.C., James, R.L.: Mechanisms underlying structural variation formation in genomic disorders. Nat. Rev. Genet. **17**(4), 224–238 (2016)
2. Sebat, J., Lakshmi, B., Malhotra, D., Troge, J., Lese-Martin, C., Walsh, T., Yamrom, B., Yoon, S., Krasnitz, A., Kendall, J., Leotta, A.: Strong association of De Novo copy number mutations with autism. Science **316**(5823), 445–449 (2007)
3. Zhang, S., Han, R.L., Gao, Z.Y., Zhu, S.K., Tian, Y.D., Sun, G.R., Kang, X.T.: A novel 31-bp indel in the paired box 7 (PAX7) gene is associated with chicken performance traits. Br. Poult. Sci. **55**(1), 31–36 (2014)
4. Shi, T., Peng, W., Yan, J., Cai, H., Lan, X., Lei, C., Bai, Y., Chen, H.: A novel 17 bp indel in the SMAD3 gene alters transcription level. Arch. Anim. Breed. **59**(1), 151–157 (2016)
5. Yoon, S., Xuan, Z., Makarov, V., Ye, K., Sebat, J.: Sensitive and accurate detection of copy number variants using read depth of coverage. Genome Res. **19**(9), 1586–1592 (2009)
6. Abyzov, A., Urban, A.E., Snyder, M., Gerstein, M.: CNVnator: an approach to discover, genotype, and characterize typical and atypical CNVs from family and population genome sequencing. Genome Res. **21**(6), 974–984 (2011)
7. Rausch, T., Zichner, T., Schlattl, A., et al.: DELLY: structural variant discovery by integrated paired-end and split-read analysis. Bioinformatics **28**(18), 333–339 (2012)
8. Mills, R.E., Luttig, C.T., Larkins, C.E., et al.: An initial map of insertion and deletion (INDEL) variation in the human genome. Genome Res. **16**, 1182–1190 (2006)

Author Index